# Top 100 Flowers

Flowers are universally admired. Flowers are the leading gift expressing love and respect. They prominently decorate our homes and cities, and they accompany all the major events in human affairs, from birth through marriage and bereavement. Flowers are also of tremendous economic importance, representing billion-dollar industries in numerous countries. Hundreds of thousands of plant species produce flowers, but relatively few dominate the world of cultivated ornamental plants. *Top 100 Flowers* presents key information and illustrations of the world's most popular, beautiful, and commercially valuable ornamental flowering plants.

Toward these goals, basic information is presented on the identification, appearance, names, history, growth requirements, economic aspects, and problem issues, as well as sources of additional library and online resources. Throughout the text, extensive photos, paintings, and diagrams are provided to illustrate the beauty and applications of the plants. Although the book is encyclopedic in nature, the information given is reduced to essentials and presented in non-technical language. The world's leading flowering plants are stunningly attractive, and the thousands of illustrations presented were chosen not only for their explanatory and educational value but also to reflect the beauty and charm of the plants. The text relates many of the wonderfully entertaining stories that reflect the intimate relationships of people and our beloved flowers.

## FEATURES

- A unique compilation of authoritative, practical, and entertaining information and illustrations addressing the world's most gorgeous and popular flowers.
- Text is accessible, user-friendly, concise, and well-organized, making numerous topics comprehensible and readable, not only by students, but also by the average layperson.
- Wonderfully illustrated: approximately a billion photos, paintings, and diagrams were surveyed, and the more than 3,500 chosen for this book represent the finest floral art ever assembled.

The goal of *Top 100 Flowers: The World's Most Popular, Beautiful, and Commercially Valuable Ornamental Flowering Plants* is to provide basic knowledge of practical interest to the horticultural industry, especially students, to choose, develop, and market profitable floral crops, and the public, to choose, grow, and appreciate flowers in their gardens and homes.

# Top 100 Flowers

## The World's Most Popular, Beautiful, and Commercially Valuable Ornamental Flowering Plants

Ernest Small and Brenda Brookes

CRC Press
Taylor & Francis Group
Boca Raton London New York

CRC Press is an imprint of the
Taylor & Francis Group, an **informa** business

Designed cover image: Shutterstock

First edition published 2025
by CRC Press
2385 NW Executive Center Drive, Suite 320, Boca Raton FL 33431

and by CRC Press
4 Park Square, Milton Park, Abingdon, Oxon, OX14 4RN

*CRC Press is an imprint of Taylor & Francis Group, LLC*



ISBN: 978-1-032-90227-2 (hbk)
ISBN: 978-1-041-00284-0 (pbk)
ISBN: 978-1-003-58622-7 (ebk)

DOI: 10.1201/9781003586227

Typeset in Times
by Apex CoVantage, LLC

# Contents

# Executive Summary

This book is dedicated to the world's most popular, beautiful, and commercially valuable ornamental flowering plants. The goal is to provide key basic information of practical interest to both professional horticultural students and amateur gardeners. Although the book is encyclopedic in nature, the information given is reduced to essentials and presented in non-technical language. The world's leading flowering plants are stunningly attractive, and the thousands of illustrations presented were chosen not only for their explanatory and educational value but also to reflect the beauty of the plants.

Most of the world's leading flowers have been associated with people for thousands of years, and the text relates many of the wonderfully entertaining stories that have been recorded. The first chapter reviews the historical, botanical, psychological, and economic relationships between ornamental plants and society. This is followed by 100 chapters, each individually dedicated to an individual flower (lily, rose, tulip, etc.). A common format (expanded for several of the leading floral crops, where special information is desirable) is adopted:

- Common and Scientific Names
- Introduction
- Overview
- History
- Economic Importance
- Ornamental Uses
- Other Uses
- Problems
- Cultivation
  - *Climate and Site*
  - *Propagation and/or Purchasing*
  - *Planting*
  - *Soil and Moisture Conditions*
  - *Maintaining Plants*
- Fascinating Facts, Fables, Folklore, and Frivolity
- Key Publications
- Key Websites

# Preface

We humans are a highly visual species, and in most environments, flowers are the most universal colorful objects that attract our enthusiastic approval. There is extensive evidence of our love affair with flowers. We spontaneously admire them so much that they decorate our homes, workplaces, and even ourselves. The finest artists in history have left a legacy of stunning floral art. Every major city has established botanical parks and conservatories featuring highly artistic displays of flowering plants, both for entertainment and for education. Beautiful flowers accompany all the major events in human affairs, from birth through marriage and death, and are the leading gift expressing love and respect. Flowers present a visual feast, and indeed, one of our objectives in preparing this book has been to compile the most attractive photos and paintings we could obtain of the world's most attractive plants. This book presents more than 3,500 illustrations in color (arranged in about 900 plates) featuring the world's most popular beautiful flowering plants. Unlike with almost all "flower books" – which present endless, mostly indistinguishable close-ups of flowers – our illustrations show all essential parts of the plants as well as their use in and around buildings, in parks, and by the horticultural industry. We trust that everyone viewing these figures will appreciate their beauty.

As professional agricultural botanists, we are concerned with assisting the horticultural industry to choose, develop, and market profitable floral crops and to educate the public about their appropriate uses in gardens, inside homes, and as floristic creations. The first chapter in this book reviews key aspects of how flowers interact intimately with society, and each of the following 100 installments is dedicated to one of the world's leading flowers. Throughout, the text, paintings, and photos are designed to provide basic information on the identification, appearance, correct name(s), growth requirements, and problem issues for each of the 100 flowering plants examined. A selective list of key technical literature is also provided for each plant, where horticultural students can find specialized information, and there are also links to online cultivation information. Each chapter ends with a section called "Fascinating Facts, Fables, Folklore, and Frivolity," which provides particularly charming, humorous, or memorable information about the featured plant.

This work provides extensive information gathered from the literature, and some error, omission, or misinterpretation is inevitably incorporated into compilations of this type. Moreover, scientific knowledge concerning the material is rapidly evolving. Liability arising directly or indirectly from the use of any of the information is specifically disclaimed.

# Preface

# Acknowledgments

Some of the material for this book is adapted from several of our previous publications in the journal *Canadian Botanical Association Bulletin* (www.cba-abc.ca/resources/bulletin/). Constructive reviews of the manuscript were provided by Richard Hinchcliff, Joseph David Shorthouse, and Graham Thurston. Randy Brehm, Senior Editor, Life Sciences and Medicine/CRC Press/Taylor and Francis, very efficiently managed the production of this large project. We thank Jessica Hsiung, our staff illustrator, who prepared several outstanding plates. Illustrations purchased from the commercial provider Shutterstock are cited according to the required format. Sources of other illustrations are given in their accompanying captions. Many of the figures are reproduced consistent with the following Creative Commons Licenses: CC BY 1.0 (Attribution 1.0 Generic): https://creativecommons.org/licenses/by/1.0/; CC BY 2.0 (Attribution 2.0 Generic): http://creativecommons.org/licenses/by/2.0/; CC BY 2.5 (Attribution 2.5 Generic): https://creativecommons.org/licenses/by/2.5/deed.en; CC BY 3.0 (Attribution 3.0 Unported): http://creativecommons.org/licenses/by/3.0/. CC BY 4.0 (Creative Commons Attribution 4.0 International license): https://creativecommons.org/licenses/by/4.0/deed.en. CC BY ND 2.0 (Attribution NoDerivs 2.0 Generic): https://creativecommons.org/licenses/by-nd/2.0/; CC BY SA 1.0 (Attribution ShareAlike 1.0 Generic): https://creativecommons.org/licenses/by-sa/1.0/; CC BY SA 2.0 (Attribution ShareAlike 2.0 Generic): https://creativecommons.org/licenses/by-sa/2.0/. CC BY SA 2.1 JP (Attribution ShareAlike 2.1 Japan): https://creativecommons.org/licenses/by-sa/2.1/jp/deed.en; CC BY SA 2.5 (Attribution ShareAlike 2.5 Generic): https://creativecommons.org/licenses/by-sa/2.5/; CC BY SA 3.0 (Attribution ShareAlike 3.0 Unported): http://creativecommons.org/licenses/by-sa/3.0/; CC BY SA 4.0 (Attribution ShareAlike 4.0 International): https://creativecommons.org/licenses/by-sa/4.0/deed.en; and CC0 1.0 (Universal Public Domain Dedication): https://creativecommons.org/publicdomain/zero/1.0/deed.en.

# About the Authors

**Dr. Ernest Small** received a doctorate from the University of California at Los Angeles in 1969. He has since been employed with Agriculture and Agri-Food Canada, the country's national department of agriculture, where he presently holds the status of principal research scientist. He specializes on the evolution and classification of economically important plants, dealing particularly with food, forage, biodiversity, and medicinal species. The species *Trigonella smallii* (Small's sweetclover) was named in his honor, and he himself has named dozens of new species. He is the author of 16 previous books, 6 of which received or were nominated for major awards. He has also authored over 400 scientific publications, mostly on economically important plants. Dr. Small's career has included dozens of appearances as an expert witness in court cases, acting as an adviser to national governments, presenting numerous invited university and professional association lectures, participating in international societies and committees, journal editing, and media interviews. He has been an adjunct professor at numerous universities and continues to supervise doctoral candidates. Dr. Small has received several professional honors, including election as a fellow of the Linnean Society of London; the G. M. Cooley Prize of the American Association of Plant Taxonomists for work on the marijuana plant; the Agcellence Award for distinguished contributions to agriculture; the Queen Elizabeth Diamond Jubilee Medal for contributions to science; the George Lawson Medal, the most prestigious award of the Canadian Botanical Association, for lifetime contributions to botany; the Lane Anderson Award, a $10,000 prize for science popularization; the Industry Leadership Award of the Canadian Hemp Trade Association (subsequently renamed in his honor); the Outstanding Paper in Plant Genetic Resources Award of the Crop Science Society of America; and appointment to the Order of Canada, the nation's highest recognition of achievements.

**Brenda Brookes** received a master of science degree from Carleton University in Ottawa, Canada, in 1992. She recently retired from Agriculture and Agri-Food Canada, after more than 44 years as a botanist. She has authored three previous books and dozens of scientific publications, mostly on economically important plants. Ms. Brookes is also a professional dog breeder, specializing in keeshonds and Pomeranians, and has also bred pleasure horses. She is well-known in the canine world, her dogs having been awarded numerous championships and international awards. She has acted as a judge in many dog shows, taught aspiring dog breeders and show dog handlers, and lectured extensively on the subject. Not least among her many talents, Ms. Brookes is a professional artist, computer graphics expert, and web designer. She is an admirer of the great ecological conservationist Sir David Attenborough, with whom she has exchanged correspondence for years, and has long been a supporter of environmental protection.

# 1 Introduction

## 1.1 ORNAMENTAL PLANTS AS HIGHLY VALUED BEAUTIFUL DECORATIONS, LIVING COMPANIONS, SPIRITUAL SYMBOLS, AND EXPRESSIONS OF LOVE

Ornamental plants are employed to decorate our homes, businesses, public and recreational spaces, and even ourselves. They not only beautify day-to-day life but also are especially on display during holidays (notably Mother's Day, Saint Valentine's Day, and Christmas), personal celebrations (like birthdays and anniversaries), professional and award ceremonies, and key life events (such as graduations, weddings, and funerals). There are innumerable public celebrations and festivals dedicated to flowers, notably Bloemencorso Zundert, the largest flower parade in the world, which occurs in Zundert, Netherlands, and the Rose Bowl Parade in Pasadena, California. Ornamental plants are prominent subjects of gardening books, articles in magazines and newspapers, shows on radio and television, lectures, and tours by "master gardeners." Many societies are dedicated to particular plants, such as roses, peonies, and lilacs. Gardening is the world's leading pastime, and attractive flowers are the first choice of most people.

Ornamental plants satisfy the essential aesthetic and emotional needs of humans, much as do music, theater, art, and literature. People once lived predominantly outdoors, but as civilization increasingly dictated that we spend most of our lives inside buildings, it seems that a primitive need for the color, texture, fragrance, and indeed, essential life of plants has stimulated us to bring them inside as beloved companions. Today, about half of the world's population lives in urban settings (even more in technologically advanced countries), and ornamental plants are the chief reminders for most people that there is a natural world that is vital for human survival. Curiously, even images and imitations of ornamental plants evoke pleasant feelings, as witnessed by

FIGURE 1.1 The two flower gardens widely considered to be the world's most beautiful: (a) Keukenhof Garden, Netherlands. (b) Butchart Gardens, British Columbia.

*Source:* **(a) Photo by Shutterstock/Neirfy. (b) Photo by Master-Shake (public domain).**

FIGURE 1.2 Floral print attire, illustrating respect and love for flower images.

*Source:* **(a) Photo by Shutterstock/Svyatoslava Vladzimirska. (b) Photo by Shutterstock/Oneinchpunch. (c) Photo by Shutterstock/Nas creatives.**

FIGURE 1.3 Example of floral-themed floats in the world's most famous flower parades (cut flowers have been affixed to formed structures): (a) Bloemencorso in Zundert, Netherlands. (b) Rose Bowl Parade in Pasadena, California.

*Source:* **(a) Photo by Erwin Martens (CC BY SA 4.0). (b) Photo by Joe McGowan (CC BY 2.0).**

DOI: 10.1201/9781003586227-1

the popularity of paintings and photographs, silk flowers, plastic plants, wallpaper with floral motifs, and floral prints on dresses, lingerie, and bedsheets. Perhaps more so than any other category of plant (even food), ornamental plants reflect basic human aspirations, sentiments, and values – they are subjects of poetry, gifts between friends, expressions of sympathy for victims of illness and tragedy, and symbolic backgrounds for religious and secular ceremonies ranging from birth to death. The ornamental plant industry is dominated by plant species that produce attractive flowers. This book is dedicated to the giants of the commercial ornamental plant industry of the world, the most beautiful plants that people purchase, maintain, and value.

**FIGURE 1.4** Artistic examples of gifts of fresh cut flowers expressing love: (a) A child with a handful of wildflowers for her mother. (b) *The Love Letter*, accompanied by a bouquet, by Spanish painter Raimundo de Madrazo y Garreta (1841–1920). (c) Vintage postcard showing an angelic gift of flowers.

*Source:* **(a) Die_Gartenlaube, 1891 (journal, translated as *The Gazebo*, illustrated family bulletin), edited by Adolf Kröner, published by Ernst Keil's successor, Leipzig and Berlin (public domain). (b) Public domain. (c) Photo by Ewan Traveler (CC BY 2.0).**

**FIGURE 1.5** Flowery vintage (early 20th century) Mother's Day greeting cards.

*Source:* **Public domain.**

**FIGURE 1.6** Floral tributes to the dearly departed: (a) A mountain of flowers accumulated in Green Park, London, following the passing of Queen Elizabeth II in 2022. (b) A church graveyard with headstones decorated with flowers. (c) Flower-covered tomb of film star Charlie Chaplin (1889–1977) near Vevey, Switzerland.

*Source:* **(a) Photo by Matt Brown (CC BY 2.0). (b) Photo by 12019 from Pixabay (public domain). (c) Photo by Giramondo1 (CC BY 2.0).**

## 1.2 HOW THE CHOICE OF THE TOP 100 FLOWERS WAS MADE

Professional botanists define "flowers" as complex organs of plants that include male (stamen) and/or female (pistil) reproductive structures and associated parts (particularly sepals and petals, collectively calyx and corolla, respectively). Technically, there are about 400,000 species of plants that have flowers (ferns, mosses, and algae are examples of plants lacking true flowers). However, to most people, flowers are either attractive plants, particularly those with pretty flowers, or the beautiful flowers (blooms) per se. If you doubt this, just consider the behavior of visitors photographing plants in large public gardens: Almost all photos are close-ups of the colorful flowers. This is paralleled on the web: Photos of plants are, once again, dominated by floral close-ups. The most attractive plants produce impressively colorful floral displays. In nature, colorful (and often fragrant) flowers are designed to attract pollinators, not people. However, horticulturalists have greatly increased the floral beauty of plants to meet market demand. Wind-pollinated plants such as grasses and conifers, and most palms, have no need of colorful flowers, which is not to deny that plants without beautiful flowers can be beautiful. However, this book is dedicated to the world's most popular ornamental plants grown predominantly for their beautiful flowers.

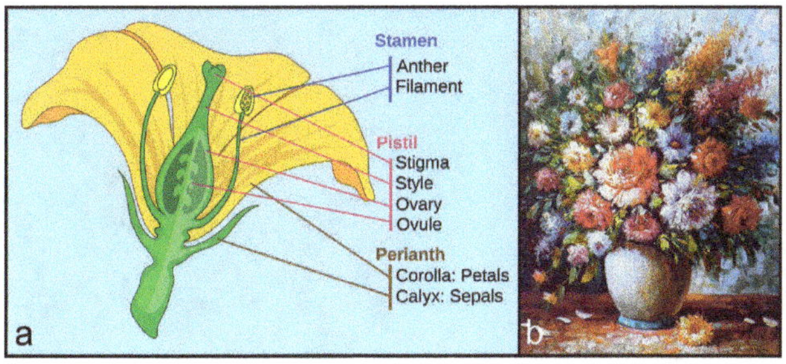


FIGURE 1.7 Scientific and artistic views of flowers: (a) Anatomy of a simple flower. (b) A floral still life by an anonymous artist, widely circulated on the web.

*Source:* **(a) Illustration by LadyofHats (public domain). (b) Photo by Counselling from Pixabay (public domain).**

FIGURE 1.8 Examples of popular ornamental plants that are attractive for their colorful foliage, not their flowers, and so are excluded from this book: (a) Ornamental cabbage (*Brassica*). (b) Coleus (*Coleus* × *hybridus*). (c) Japanese maple (*Acer palmatum*). (d) Hosta (*Hosta*) cultivar "Fire and Ice." (e) Caladium (*Caladium bicolor*). (f) Croton (*Codiaeum variegatum*, not the genus *Croton*).

*Source:* **(a) Photo by GirishkSharma31 from Pixabay (public domain). (b) Photo by Rina Pitucci (CC BY ND 2.0). (c) Photo by John Fowler (CC BY 2.0). (d) Photo by Shutterstock/Haboston. (e) Photo by Forest and Kim Starr (CC BY 2.0). (f) Photo by Shutterstock/Bryan Pollard.**

FIGURE 1.9 Colorful foliage developed in the autumn at Toji Pagoda, Kyoto, Japan. As noted in the text, plants valued for their fall leaf coloration are excluded from this book.

*Source:* **Photo by Shutterstock/Sean Pavone.**

FIGURE 1.10 Examples of popular ornamental plants that are attractive for their colorful bracts (colored modified leaves associated with the flowers that serve as the main attraction to pollinators). In all these example, small flowers are dwarfed by the colored bracts. The most attractive of such species are included in the book (although purists might exclude them since they are not attractive for their true flowers): (a) Poinsettia (*Euphorbia pulcherrima* cultivar RedFox). (b) Bougainvillea (*Bougainvillea*). (c) *Pachystachys lutea* (lollipop plant, golden candle plant, golden shrimp plant). Note the relatively inconspicuous white flowers. The plant is mostly grown for its attractive yellow bracts. (d) Bird-of-paradise (*Strelitzia*). (e) Flamingo lily (*Anthurium*). The enveloping red bract is called a "spathe." (f) *Heliconia rostrata* (lobster claw, false bird-of-paradise).

*Source:* **(a) Photo by Kenpei (CC BY SA 3.0). (b) Photo by Josch13 from Pixabay (public domain). (c) Photo by Photo by Salicyna (CC BY SA 4.0). (d) Photo by Josch13 from Pixabay (public domain). (e) Photo by Nicolas Perrault II (public domain). (f) Photo by Abhishek Jacob (CC BY SA 2.0).**

Nature has produced flowers in unlimited variety, and even flower-like displays by colorful "bracts," which are modified leaves closely associated with the flowers. Plants with colorful bracts mimic the beauty of true flowers for the purpose of attracting pollinators. Examples include bird-of-paradise, bougainvillea, bromeliads, calla lily, poinsettia, and many others. Although such species aren't valued for their true flowers (which are often tiny), the most attractive examples are included in this book because, for most people, this distinction isn't important.

Like pet dogs, cats, birds, and fish, our ancestors often adopted wild species, at least in part, for aesthetic reasons – they were pleasingly attractive – and over the centuries, by selection and breeding, increased their beauty. Beautiful blooms are especially colorful and frequently are available in many shades. They are usually either large or occur in large numbers, and often the flowering displays last for a long time. The plants may be tiny herbs or big trees, but a key criterion for inclusion in this book is that the floral display at some time of the season dominates their appearance (accordingly, many popular plants, such as anthurium, boxwood, coleus, ginger, hosta, and peace lily, have not been included because they are mostly foliage). Fragrance is often a consideration but by no means is as important to most people as appearance. Cost (reflective of the difficulty of cultivation) is also important – many consider orchids to be the most beautiful of flowers, but species that are extremely expensive are beyond the means of the majority.

Selecting the most popular flowers is not easy, because different criteria can be employed. Species such as roses and tulips are so prevalent and universally beloved that their inclusion is mandatory, but inevitably, plants vary in popularity depending on climate, geography, local availability, and taste. Moreover, with the continuing introduction of exciting new cultivars, and sometimes with the appearance of diseases that make continued cultivation difficult, the popularity of flowers changes. The internet is an excellent guide to popularity, and indeed, simply inserting the name(s) of particular plants in a search engine returns thousands to millions of files, reflective of the relative interest in given plants. The web itself presents many "top 10," "top 20," and sometimes "top 50" flowers, but very rarely "top 100," as in this book. Online lists are more or less agreed on plants that should be in the top 20 or 30, but there are exceptions, and lists that attempt to rank flowers in popularity often disagree, reflecting the preferences and knowledge of the individuals who prepared them. Important guides to popularity include trade-based lists prepared by florist companies, sales lists compiled on the basis of international trade, and production and sales statistics of various countries (dominated, as is well-known, by the United States and The Netherlands). Still another indication of floral popularity is how much relative space is dedicated to plants in major gardening encyclopedia. The number of publications (books and articles) on particular plants is also a rough sign of how much interest they have elicited. How many cultivars are available for species also provides clues to their relative importance. We have examined all these indicators, some of which are discussed in the following sections. Nevertheless, almost everyone has a personal favorite ornamental, usually considers it the most beautiful of plants, and may be offended by its not being included here.

## 1.3 THE CATEGORIES OF UNDERGROUND REPRODUCTIVE "BULBS"

Many cultivated plants have annual herbaceous aboveground parts that die at the end of a season of growth but survive a following period which is hostile, because of extreme temperature (cold or heat) or drought, by dormant underground storage organs which vegetatively reproduce the plant the next growing season. Such specialized structures are often swollen with food reserves, typically starch. These are classified as bulbs, corms, rhizomes, and tubers. In contrast to these structures, many species overwinter by roots that do not appear swollen. True bulbs (exemplified by onion, tulip, and daffodil) are spheroidal masses made up mostly of modified fleshy storage leaves ("scales") which appear as rings in a cross section.

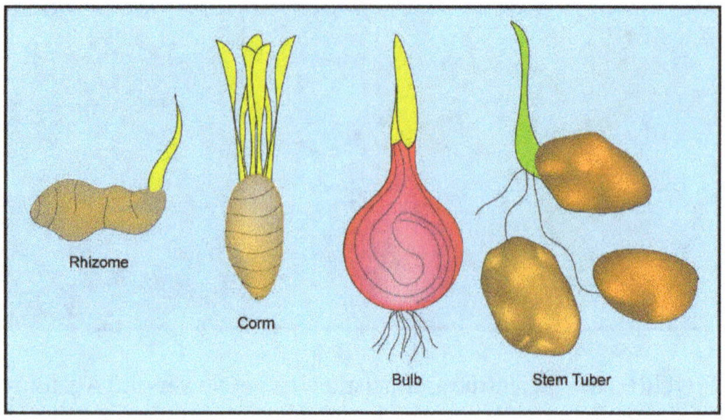


FIGURE 1.11 The four basic categories of underground reproductive "bulbs."

*Source:* **Photo by Shutterstock/KKT Madhusanka.**

Corms (e.g., crocus, banana) are similar in shape but made up mostly of a solid mass of compacted stem tissue lacking internal layers like true bulbs. Rhizomes (e.g., ginger, iris), like corms, are made of fleshy stem tissue, but unlike corms, which orientate vertically, rhizomes elongate laterally quite substantially. "Stem tubers" (such as potato and true yam – *Dioscorea* species) are essentially short (i.e., not substantially elongated), compact, swollen rhizomes. "Root tubers" or "tuberous roots" (such as sweet potato, cassava, daylily, and dahlia) are simply swollen roots and are internally composed of root tissues. (Stems ["shoots"] arise from meristems [growth points] that tend to grow upward; roots arise from meristems that tend to grow downward. There are also anatomical differences between roots and stems.) "Daughter" bulbous structures, or "offsets," are produced by many plants with bulb-like storage organs, and these serve to reproduce the plant in the spring (the original or mother bulb frequently dies). In the horticultural industry, these botanical distinctions are often not made, or the technical terms are confused (correct terminology is presented in this book; note Table 1.1). In most garden catalogues, and in most popular literature (including most information available on the internet), the term "bulb" usually includes all fleshy underground storage structures, regardless of its technical classification.

## 1.4 NAMES OF FLOWERING ORNAMENTAL PLANTS

This book presents essential information about the common and scientific names of all 100 ornamental plants discussed in detail in this book. Quattrocchi (1999) is the most comprehensive and authoritative source of information on the origin of the Latin genus names of plants (cited references are listed at the end of this chapter). The best sources of information on the origin of English names of plants are dictionaries, particularly the multivolume Oxford English Dictionary (version 3 of the 2nd edition became available as a CD-ROM in 2002). As will be noted, authorities do not always agree on the etymology of names. Although

**TABLE 1.1**
**Summary of Key Features of the 100 Flowering Ornamentals Treated in This Book**

| Flowering Ornamental | Features |
|---|---|
| African violet (*Streptocarpus*, formerly *Saintpaulia*) | Small perennial herbs; the world's most popular houseplant. |
| Allium (*Allium*) | Mostly perennial herbs grown in temperate climates, overwintering by bulbs. |
| Alstroemeria (*Alstroemeria*) | Mostly herbaceous perennials, growing from rhizomes or tuberous roots, suitable for outdoor growth in hot climates, primarily grown as a florist's cut flower. |
| Amaryllis (*Hippeastrum*) | Mostly tropical and subtropical herbs native to Central and South America, primarily grown from bulbs as a houseplant. |
| Anemone (*Anemone*) | Mostly perennial herbs overwintering by corms or rhizomes, grown especially in temperate-area gardens. |
| Angel's trumpet (*Brugmansia*) | Tropical shrubs and trees with very fragrant, beautiful, gigantic flowers, grown as an annual in temperate regions. |
| Aster (*Aster*, *Callistephus*, *Symphyotrichum*, and related genera) | Daisylike annual and perennial herbs often grown in temperate-region gardens and employed as florist's cut flowers. |
| Baby's breath (*Gypsophila paniculata* and relatives) | Annual and perennial herbs occasionally grown in temperate-region gardens but mostly cultivated for impressive sprays of tiny white flowers employed as florist's cut flowers. |
| Begonia (*Begonia*) | Mostly perennial herbs; very popularly grown as annuals and sometimes as houseplants in temperate regions, and as perennials in tropical and semi-tropical regions. |
| Bird-of-paradise (*Strelitzia*) | Beautiful perennial subtropical herbs growing from rhizomes; cultivated as an annual in temperate regions and as a perennial in hot climates. |
| Black-eyed Susan (*Rudbeckia*) | Mostly perennial daisylike herbs growing from rhizomes, adapted to temperate regions. |
| Bougainvillea (*Bougainvillea*) | "Vinous shrubs" (like common grapevine) adapted to subtropical regions; grown as an annual (usually potted) in temperate regions, very widely grown as a perennial in subtropical and tropical countries. |
| Bromeliads (Bromeliaceae family) | Mostly tropical herbaceous perennials; grown as houseplants in temperate regions, and outdoors in subtropical and tropical countries. |
| Butterfly bush (summer lilac, *Buddleja*) | Frost-sensitive shrubs; underground parts survive winter in mild regions; often grown as an annual in temperate countries and as a perennial in warm regions; the world's premier plant grown for butterflies. |
| Calceolaria (*Calceolaria*) | Perennial herbs from the Andes regions; produces dazzling pouch-shaped flowers that have been compared to orchids; grown as short-term houseplants in temperate climates. |
| California poppy (*Eschscholzia*) | A popular large-flowered herb usually grown as an annual in temperate areas. |
| Calla lily (*Zantedeschia*) | Tropical and subtropical herbs; some cultivars tolerate mild freezing; popular florist's cut flower; occasional houseplant; develops a showy leaflike, funnel-shaped bract (the spathe), inside of which there is a central rodlike columnar structure (the spadix) with many small, inconspicuous true flowers. |
| Camellia (*Camellia*) | Evergreen shrubs and small trees, flowering especially in the late winter in frost-free areas, where they are very popular. |
| Canna (canna lily, *Canna*) | Large perennial tropical herbs arising from rhizomes; intolerant of frost, so grown as annuals in temperate climates. |
| Carnation (*Dianthus*) | Popular bedding herbs, mostly grown as annuals in temperate regions, also as perennials in warmer regions. |
| Catnip (*Nepeta*) | Perennial herb adapted to temperate climates; well-known for its effects on cats, but also available in ornamental cultivars. |
| Celosia (cockscomb, woolflower, *Celosia argentea*) | Frost-sensitive herbs generally grown as annuals, but also as a short-lived perennial in subtropical areas. |
| Cherry blossom (*Prunus*) | Various trees that produce spectacular white and pink flowering displays in the spring in temperate regions. |
| Christmas cactus (also "Thanksgiving cactus" and "Easter cactus"; *Schlumbergera*) | Shrubs adapted to cool semi-tropical and tropical regions, primarily grown as long-lived houseplants; second only to poinsettia as the most popular Christmas holiday gift plant. |
| Chrysanthemum (*Chrysanthemum*) | Extremely popular herbs, known for fall flower displays; available as perennials for temperate regions and as gift potted plants for use as patio and houseplants; also a florist's cut flower. |
| Clematis (*Clematis*) | Probably the most popular of the perennial, woody flowering vines adapted to temperate climates. |
| Clivia (*Clivia*) | A popular herbaceous houseplant arising from tuberous roots; also grown as a perennial bedding plant in subtropical climates. |

(*Continued*)

**TABLE 1.1 (*Continued*)**
**Summary of Key Features of the 100 Flowering Ornamentals Treated in This Book**

| Flowering Ornamental | Features |
|---|---|
| Columbine (*Aquilegia*) | Popular, hardy perennial garden herbs growing from rhizomes; adapted to cool temperate regions. |
| Crab apple (*Malus*) | Extensively planted small trees; the leading flowering woody ornamental in cooler climates of temperate regions. |
| Crape myrtle (*Lagerstroemia*) | Shrubs or small trees adapted to tropical and subtropical climates; some cultivars grow well in the milder portions of temperate regions, where they are often extensively grown. |
| Crocus (*Crocus*) | One of the early-flowering spring perennial herbs; perhaps second in popularity only to the tulip; emerges from overwintering corms. |
| Cyclamen (*Cyclamen*) | Perennial herbs growing from tubers; adapted to mild, temperate regions; known particularly for the florist's cyclamen, a houseplant commonly acquired as a gift. |
| Daffodils (*Narcissus*) | One of the most popular of the ornamental perennial herbaceous spring-flowering bulb plants adapted to temperate regions. |
| Dahlia (*Dahlia*) | Very popular herbaceous perennials growing from tuberous roots which are too frost-sensitive to be left to overwinter in ground that freezes significantly. |
| Daylily (*Hemerocallis*) | Very popular herbaceous perennials adapted to temperate regions; regrowing annually from underground parts (mostly fibrous roots). |
| Delphinium (*Delphinium*) | Annual ("Consolida group") and perennial herbs, often tall, adapted to temperate regions. |
| Echinacea (*Echinacea*) | Hardy, tall, perennial herbs adapted to temperate regions; popular in gardens. |
| Forget-me-not (*Myosotis*) | Diminutive herbs grown as annuals and short-lived perennials; adapted to temperate regions; grown in gardens and sometimes as a houseplant and a florist's cut flower. |
| Foxglove (*Digitalis*) | Usually tall biennial herbs, adapted to temperate climates; traditional garden plants; very toxic. |
| Freesia (*Freesia*) | Frost-sensitive perennial herbs reproduced from corms; best known as "florist's freesia," marketed for cut flowers. |
| Fuchsia (*Fuchsia*) | Herbaceous perennial bedding plants or shrubs in semi-tropical climates but usually grown as an annual pot plant or in hanging baskets in temperate regions. |
| Gardenia (*Gardenia*) | Tropical and subtropical shrubs and small trees with gorgeous flowers and a heavenly fragrance; occasionally grown (with difficulty) as a houseplant in temperate climates. |
| Gazania (*Gazania*) | Low-growing, daisylike herbs; grown as annuals in temperate regions but as evergreen perennials in semi-tropical and tropical areas. |
| Geranium (*Pelargonium*) | Frost-intolerant herbs grown as perennials in subtropical climates; widely grown as annual bedding plants in temperate regions. |
| Gerbera (one of several genera called "African daisy," *Gerbera*) | Grown as perennial herbs in frost-free climates, but as annuals in temperate regions; one of the world's leading cut flowers and potted plants. |
| Gladiolus (*Gladiolus*) | Perennial herbs reproducing by corms; neither the plants nor the corms withstand freezing, so in areas with cold winters, the plants are grown as annuals; very popular cut flowers and forced temporary pot plants. |
| Gloxinia (*Gloxinia*) | A small perennial herb reminiscent of the African violet but with much larger and more colorful flowers; grown as a short-lived houseplant; too delicate to cultivate in most gardens. |
| Heath (*Erica*) and heather (*Calluna vulgaris*) | Small shrubs originating from Europe that are widely cultivated in temperate climates. |
| Hibiscus (*Hibiscus*) | Herbs, shrubs, and trees, worldwide; some with huge flowers, but *H. rosa-sinensis* is the most important and is grown as a perennial shrub or tree in tropical and subtropical climates, and *H. syriacus*, the second most important species, is grown as a winter-hardy deciduous shrub in temperate climates. |
| Hollyhock (*Alcea*) | Mostly short-lived, traditionally grown, and admired tall perennial herbs with large flowers. |
| Hyacinth (*Hyacinthus orientalis*) | A perennial herb overwintering by bulbs, adapted to temperate climates; one of the world's major "bulb class" early-flowering ornamentals, ranking behind tulips and daffodils. |
| Hydrangea (*Hydrangea*) | Very popular shrubs and small trees, many adapted to temperate climates. |
| Impatiens (*Impatiens*) | The most important garden forms are frost-intolerant perennial herbs grown as annuals in temperate regions. |
| Iris (*Iris*) | Perennial herbs with very attractive large flowers; many reproducing by rhizomes or bulbs; many adapted to temperate climates; among the most popular of cultivated ornamentals. |
| Jacaranda (*Jacaranda mimosifolia*) | A popular subtropical tree producing a spectacular bluish flowering display; widely planted in warm-temperate, subtropical, and highland tropical regions. |
| Jasmine (*Jasminum*) | Tropical and warm-temperate shrubs; the most important species is *J. officinale*, a twining vine or shrub with very fragrant flowers; hardy in subtropical areas and the warmer parts of temperate regions. |

**TABLE 1.1 (Continued)**
**Summary of Key Features of the 100 Flowering Ornamentals Treated in This Book**

| Flowering Ornamental | Features |
|---|---|
| Kalanchoe (*Kalanchoe*) | Mostly tropical herbs and shrubs; the most important species are small perennial herbs grown as potted houseplants, like African violets. |
| Lantana (*Lantana*) | Large shrubs in subtropical and tropical areas; tolerate mild temperate-region climates, but often grown as potted annuals; produce small flowers in attractive, multicolored heads. |
| Lavender (*Lavandula*) | Mostly perennial herbs adapted to temperate climates and admired for their striking purple coloration and attractive fragrance. |
| Lilac (*Syringa*) | Beloved shrubs and small trees; adapted to cool-temperate climates; admired for their spectacular, fragrant flowering displays in the spring. |
| Lilies (*Lilium*) | The most important of the many "lilies"; very popular herbaceous perennials overwintering by bulbs; adapted to temperate climates. |
| Lily of the valley (*Convallaria majalis*) | Inconspicuous, delicate herbaceous perennial, overwintering by a branched rhizome; adapted to temperate climates; beloved as a florist's cut flower in mixed floral bouquets. |
| Lisianthus (*Eustoma grandiflorum*) | Perennial herb which can be grown in gardens in the warmer parts of the temperate zone but is primarily a florist's cut flower because it looks like a rose but is cheaper to produce. |
| Lobelia (cardinal flower, *Lobelia*) | Perennial herbs grown in gardens; some suitable as perennials in temperate climates; others grown as annuals in the cooler parts of temperate areas. |
| Lupin (lupine, *Lupinus polyphyllus* hybrids) | Perennial herbs popular in temperate-region gardens; many with spectacularly colored tall flowering spikes. |
| Magnolia (*Magnolia*) | Popular ornamental trees and shrubs in milder areas of temperate climates. |
| Mandevilla (*Mandevilla*) | Tropical perennial herbs and vines; in temperate areas, primarily encountered as a gorgeous houseplant and florist's cut flower. |
| Marguerite (*Argyranthemum*) | Mostly frost-tender perennial herbs grown as garden annuals in temperate climates. |
| Marigold (*Tagetes*) | Annual and perennial herbs and shrubs native to the cool tropics of the New World; widely grown as annual herbaceous garden plants. |
| Morning glory (*Ipomoea*) | Most species are twining (or at least scrambling) perennial herbaceous vines, but some are annual and some are woody (shrubs or trees); in temperate climates, annual cultivars are grown, but in tropical areas, the plants are grown as perennials. |
| Nasturtium (*Tropaeolum*) | Frost-sensitive perennials ranging from dwarf shrubs to vines climbing to 3 m, which are cultivated as annuals in regions with cold winters. |
| Oleander (*Nerium oleander*) | A subtropical shrub or small tree, widely planted in subtropical and tropical areas, sometimes grown as a protected pot plant in temperate climates. |
| Osteospermum (one of the several genera called "African daisy"; *Osteospermum*) | Usually grown as annuals in temperate climates, but many can be grown as tender perennial herbs or subshrubs in climates that are frost-free or mostly so. |
| Pansies, violets (*Viola*) | Mostly small perennial herbs with big flowers; adapted to temperate climates; extremely popular in gardens. |
| Passionflower (*Passiflora*) | Subtropical and tropical perennial vines with interesting large flowers and edible fruit; sometimes woody; a few of the species capable of being grown in the warmer parts of temperate climates. |
| Peony (*Paeonia*) | Stout branching perennial herbs and diffuse shrubs growing from rhizomes or tuberous roots, which serve as the overwintering phase; depending on cultivar, capable of growing from subtropical to subarctic climates, but do best in milder areas of temperate regions; many consider peonies to be the most attractive of all garden ornamentals. |
| Periwinkle (*Catharanthus* and *Vinca*) | Low-growing, creeping perennial herbs or subshrubs employed as groundcover, bedding, edging, and in containers; various cultivars adapted to subtropical or temperate regions. |
| Petunia/Calibrachoa/Petchoa (*Petunia*, *Calibrachoa*, × *Petchoa*) | Petunias are extremely popular frost-sensitive perennial herbs grown as annual garden plants, notable for very impressive floral displays; *Calibrachoa* and the hybrid genus × *Petchoa* are often marketed as petunias, which they very closely resemble. |
| Phalaenopsis (*Phalaenopsis*) | Tropical epiphytic orchids (which are perennial herbs) grown as houseplants; accounts for three-quarters of all marketed orchids. |
| Phlox (*Phlox*) | The genus *Phlox* includes several popular ornamental herbs grown as annuals and perennials in gardens in temperate climates. |
| Plumeria (frangipani, *Plumeria*) | Tropical shrubs and trees which produce numerous long-lasting, gorgeous, colorful, extremely sweet-smelling flowers; occasionally, these are grown as portable potted plants that can be overwintered indoors, but these true tropical species will mostly be encountered in the hotter regions of the world. |

*(Continued)*

**TABLE 1.1 (*Continued*)**

**Summary of Key Features of the 100 Flowering Ornamentals Treated in This Book**

| Flowering Ornamental | Features |
|---|---|
| Poinsettia (*Euphorbia pulcherrima*) | A subtropical shrub that has become the world's most popular Christmas-season short-term potted houseplant. |
| Poppies (*Papaver*) | Very popular annual and perennial herbs widely grown in gardens in temperate climates. |
| Primrose (*Primula*) | Mostly small perennial herbs (including species formerly in *Dodecatheon*) grown in gardens in temperate climates and as houseplants. |
| Ranunculus (*Ranunculus asiaticus*) | Persian ranunculus, which reproduces by tuberous roots, produces gorgeous flowers that have been compared to camellia, peonies, and roses; grown as an annual herb in cool areas, but as a perennial in the warmer areas of temperate regions; employed as garden and container ornamental plant but is principally used in bouquets and flower arrangements and is one of the dominant staples of the cut flower industry. |
| Rhododendron (includes "azaleas"; *Rhododendron*) | Very popular shrubs and some trees, spectacularly beautiful in flower; grown from the tropics to the subarctic regions; also available as potted houseplants. |
| Rose (*Rosa*) | Mostly deciduous, prickly shrubs (occasionally with scrambling or trailing stems) adapted to temperate and subtropical areas; the most admired of all ornamental plants. |
| Royal poinciana (*Delonix regia*) | A medium to large tree producing a spectacular display of flowers that is so attractive the species is cultivated worldwide in tropical and subtropical countries; many consider it to be one of the most beautiful trees in the world; regrettably, its cultivation is largely restricted to the hottest countries. |
| Sacred lotus (*Nelumbo nucifera*) | A so-called "water lily"; an herbaceous perennial aquatic plant growing from a tuberous rhizome; sacred in Asian and Oriental cultures; basically subtropical, but in temperate climates, sometimes maintained in heated outdoor ponds and in indoor conservatories. |
| Salvia (*Salvia*) | Popular perennial herbs and shrubs, differing considerably in ability to overwinter in cold climates; often grown as annuals in cold climates, but as perennials in warmer-climate zones. |
| Snapdragon (*Antirrhinum majus*) | The common snapdragon is the dominant species; an herbaceous perennial usually grown as an annual; very popular garden plant, with considerable demand in the seed, nursery, and cut flower trades. |
| Stock (*Matthiola incana*) | The dominant cultivated species of *Matthiola*; a well-established garden and florist's cut flower staple, valued for its wide range of colorful, fragrant flowers displayed in impressive spires; a perennial herb, usually grown as an annual. |
| Sunflower (*Helianthus*) | Interspecific hybrids of *H. annuus* are the forms most often found as ornamentals; usually fast-growing annual herbs sometimes exceeding 3 m in height. |
| Sweet pea (*Lathyrus odoratus*) | An annual herbaceous vine; a relative of the common edible pea but is quite poisonous; everlasting pea (*L. latifolius*), a less-popular perennial, also grown in temperate regions and can tolerate moderately cold winters. |
| Trumpet creeper (*Campsis*) | Large impressive woody perennial vines with spectacular, colorful, exotic-looking flowers; cultivated from north temperate areas to the subtropics; often the chief complaint against them is that they simply grow too well; require so little attention that it has been said that anyone incapable of growing them should not attempt to grow any plant. |
| Tulip (*Tulipa*) | Spring-flowering herbaceous perennials growing from overwintering bulbs; the most widely grown ornamental in temperate climates; the most profitable of all marketed ornamental plants. |
| Wallflower (*Erysimum cheiri*) | A perennial, sometimes woody herb, grown as an annual in cold temperate climate but as a biennial in warmer regions; an old-fashioned but still popular spring-flowering ornamental widely planted in gardens for its fragrant, colorful flowers. |
| Water lily (*Nymphaea*, *Victoria*) | Aquatic herbaceous perennials that are rooted in soil which is usually covered by water; rooted base produces rhizomes, stolons, or tubers which overwinter the plants, normally protected underwater; species adapted to temperate and tropical regions; specialty garden plants. |
| Wisteria (*Wisteria*) | Woody twining vines producing extraordinarily attractive massive flower displays; most popular species are grown in temperate climates. |
| Zinnia (*Zinnia*) | Very popular annual or perennial herbs; frost-sensitive, so grown in temperate regions as annuals. |

technical, information on the conventions of naming plants is indispensable to reliably distinguish and identify them. This topic is examined in the following (also see Brickell et al. 2016; Gledhill 2008; Mahr 2007; Smith 2017; Turland et al. 2018).

### 1.4.1 Ambiguity of Common Names

There are over 400,000 species of plants that produce flowers, so to prevent confusion, it is critical to have unique names for them. Unfortunately, for important species, there are often several "colloquial," "common," or "vernacular" English names available, frequently different in different regions. Moreover, the same name is often spelled differently. In England, the white water lily (*Nymphaea alba*) has 15 common names. Often, a given common name is used for two or more species. For example, "lavender" can refer to different species and hybrids of the genus *Lavandula* grown in different countries, but also several unrelated plants. Of course, this ambiguity can make it difficult or even impossible to know exactly what is meant.

### 1.4.2 Binomial Scientific Names of Species

The scientific names of species are in the form of a pair (termed a "binomial") of italicized Latinized words, the first indicating the genus (a *genus* is a group of species, or sometimes just one species). The initial letter of a genus name is capitalized. For example, the scientific name of hyacinth is *Hyacinthus orientalis.* Once it is clear what species is referred to, the first word is abbreviated, for example, *H. orientalis*, in the manner that *Tyrannosaurus rex* is shortened to *T. rex*. For bibliographical convenience, botanical taxonomists have adopted a system of citation of the authors of names, usually abbreviated. For example, the scientific name for common sunflower is sometimes presented as *Helianthus annuus* L., where the "L." refers to Linnaeus, the author or "authority" who coined the name. For most well-known species (as in this book), it isn't necessary to include the authority.

### 1.4.3 Plant Scientific Names Are Governed by a Code Applicable to All Plants

The application of scientific names is governed by a formalized, quasi-legal system used by biologists to avoid the ambiguity of common names. Taxonomists (specialists in the naming of living things) have devised rules for coining new scientific names, and for deciding on which name must be adopted when a group has received more than one name. The International Code of Botanical Nomenclature for Algae, Fungi, and Plants (Turland et al. 2018) provides many rules for deciding which names should be employed. Only one name is the "correct" name, and other names are called "synonyms." Sometimes it is important to know which name is a synonym of what other name, because information in the literature may be recorded under a synonym. In this book, synonyms are given only when there is some probability that they will be encountered rather than the correct name.

### 1.4.4 Cultivated Plants Are Often of Hybrid Origin and May Have "×" in Their Scientific Names

Many horticultural plants, including the majority of those discussed in this book, are of hybrid origin – the progeny of two or more species. For example, the cabbage rose (*Rosa centifolia*) is a complex hybrid of *R. gallica*, *R. moschata*, *R. canin*a, and *R. damascena*. Frequently, the name of the wild parental species that is responsible for most of the genetic composition of the cultivated forms is also applied to the latter. For example, cultivated kalanchoes are mostly hybrids of *Kalanchoe blossfeldiana*, and these are identified as *K. blossfeldiana*, ignoring the contributions of other species. Often (especially when hybrids are well established in horticulture), a new scientific name is coined to designate the hybrid cultivated forms. For example, an amaryllis hybrid between *Hippeastrum vittatum* and *H. reginae* was given the name *Hippeastrum johnsonii.* When a name designates hybrids, the mathematical multiplication symbol, ×, is often incorporated into the scientific name, for example: *Hippeastrum* × *johnsonii* (frequently, no space is left between the × and the final Latin word, for example, *H.* ×*johnsonii*, but this is optional). Sometimes, a contrived illegitimate, pseudo-scientific name employed primarily by horticulturalists is invented (and should be ignored) to designate the hybrids collectively. For example, petunia hybrids are often termed *P.* × *hybrida*, an incorrect but widely used name. Sometimes, a "formula" made up of the parental names separated by × is used to label hybrids. For example, Persian lilac (*Syringa* × *persica*) can be represented as *Syringa laciniata* × *S. afghanica*. Some scientific names are based on inter-generic hybrids, not just inter-species hybrids. For example, the hybrid between *Petunia* and *Calibrachoa* is ×*Petchoa* (a contrived name combining elements of ***Pet**unia* and *Calibra**choa***). The average person is unfamiliar with the relevant technicalities, which can be avoided by just referring to the scientific name of the genus (albeit this is less precise). For example, almost all cultivated petunias are assignable to the scientific name *P.* × *atkinsiana*, but to avoid this technical nomenclature, all petunias can be designated by the genus name *Petunia*. As discussed next, where more precise identification is desirable, cultivar names can be employed.

### 1.4.5 The Names of Cultivated Plant Varieties ("Cultivars") Are Governed by a Special Code for Cultivated Plants

The word "variety" is very important for cultivated plants, and it is important to understand the different ways that it is used. The term is ambiguous when applied to living things: It may have several different meanings. First, the term *variety* may simply be used in a general sense to mean a kind or type or variation – something noticeably distinct – but without a name. However, there are two kinds of formally named varieties that are important to people interested in cultivated plants.

One formal meaning of *variety* is in relation to the codified system of entirely Latin names of plants governed by the International Code of Botanical Nomenclature for Algae, Fungi, and Plants, referred to earlier. In this case, "variety" (abbreviated "var.") is synonymous with a formal category (technically called "varietas") below a subspecies. For example, the name *Rosa rugosa* var. *rubra* (a purplish-flowered category of the rugosa rose, first recognized by the famous Harvard taxonomist Alfred Rehder in 1902) formally designates a variety in this sense. Such recognized "varieties" are usually entities that grow wild in nature, but sometimes the terminology is also applied to cultivated plants (as in the preceding example).

For most ornamental plants, the word "variety" is a synonym of the word "cultivar" as defined in the International Code of Nomenclature for Cultivated Plants (of which the latest version is Brickell et al. 2016). This codified system of plant names (which is complementary to the International Code of Botanical Nomenclature for Algae, Fungi, and Plants) combines Latin and any other language in the formation of names designating given plant varieties. In this system, the term "cultivar" is predominantly used to designate artificially selected kinds of plant that are of interest in agriculture, forestry, and horticulture. As in the first system, the name is composed partly of the scientific name of the genus or species, which is made up of Latinized words; however, rather than completing the phrase name with another Latin word, a descriptive word or words (such as Giant or Black Beauty) is used. The non-Latin part of the name is indicated in Roman characters, never in italic type, as are the Latinized words in scientific names. Cultivar names are constructed by combining two elements. The first element is any one of the following: (1) the scientific name of a plant; (2) the genus name of a plant (permissibly followed by the name of a group of cultivars); or (3) an unambiguous common name. The second element is a term or phrase in the common language (i.e., not in Latin) and enclosed in single quotation marks. The following examples illustrate these three accepted ways of constructing a complete cultivar name: (1) *Planta alba* 'Magnificent Red'; (b) *Planta* 'Magnificent Red' or *Planta* Group Northern 'Magnificent Red'; and (c) strawberry 'Magnificent Red' or 'Magnificent Red' strawberry. Many journals and book editors adhere to model (1) shown earlier. (Before the obsolete 1995 code of nomenclature for cultivated plants made the practice unacceptable, an alternative format was to use the abbreviation "cv." [for "cultivated variety"] in lieu of single quotes, in the fashion *Planta alba* cv. Magnificent Red.) Such cultivar names are long, and in practice in popular writing, such as found in magazines and newspapers (and occasionally in this book), the non-Latin portion of the name is used as the name of the cultivar when it is clear what the species or genus is.

Many horticultural companies offer trademarked plants with names reminiscent of cultivar names. Such names, of course, are not available for use without permission, unlike cultivar names, which are free for everyone to use. Many horticultural companies offer patented plants, which have legal intellectual property protection, although their names may be cited. When nurseries sell plants with attached names, the status of the names may be unclear (correct identification is an additional issue).

### 1.4.6 The Hidden Meanings of Scientific Names

The scientific names of plants often originate from Latin or Greek and, in any event, are in Latinized form. Both parts of the binomial name were adopted by the person who coined the name for some reason (which is sometimes not clear today). For example, consider the scientific name of the dog rose, *Rosa canina*. The genus name *Rosa* is based on the ancient Latin name for *roses*, *rosa*, which in turn originated from Akkadian (the ancient language of Mesopotamia) word *russu*, *red*. The "specific epithet" *canina* means "canine-like," descriptive of the stem prickles resembling a dog's formidable teeth.

More than half of the Latin genus names encountered in this book are also employed as common names for them (e.g., allium, alstroemeria, anemone, aster, begonia, bougainvillea) or are very similar to the common names (e.g., the lavender genus is *Lavandula*, the rose genus is *Rosa*, and the primrose genus is *Primula*). Many scientific names were crafted to have some meaning, although many others originate simply from old languages. Often, a knowledge of what the names mean is helpful in remembering the names, and often, there are interesting stories and facts behind the names. To assist readers in memorizing the names, background information is provided for all the genus names.

Numerous genus names and many specific epithets commemorate male Europeans (frequently royalty, nobility, and doctor-botanists, since physicians widely employed medicinal plants), reflecting their dominance in the science of botany in past centuries. Many specific epithets based on geography or other considerations are inaccurate or objectionable by modern standards, and some people commemorated by Latin names were not the nicest individuals. Biological nomenclature is based on priority, so the earliest names cannot be changed simply because their literal translation is erroneous or the people honored are undeserving (the worst example: A beetle from Slovenia named *Anophthalmus hitleri* in 1937 for the infamous Nazi). For a critique of a proposal to abandon currently accepted names that are offensive to particular groups, see Jiménez-Mejías et al. (2023).

### 1.4.7 Changing Classifications and Names

Classifications change because of advances in botanical knowledge, and the result often is that names also change. For example, the *Chrysanthemum* genus was recently split into eight different genera, so many species were transferred to different genera. While most of the scientific community rapidly adopts these changes, the horticultural industry is notoriously slow to use changed names, and many commercial catalogues and less-knowledgeable writers stick to the old names.

### 1.4.8 Pronunciation of Names

Many of the Latinized generic names and associated English common names based on European botanists are challenging to pronounce. Traditional Latin pronunciation is not of much help for most generic names, which usually originate from foreign languages other than Latin, and may be difficult for English speakers to pronounce. Advice is provided in this book on the names that are more difficult to pronounce. For example, the genus *Fuchsia*, named for German botanist Leonhart Fuchs (1501–1566), used to be pronounced with the awkward, hard German pronunciations (fuk-si-a; fook-sy-a), but this has changed to a soft English (fyu-sha; foo-shia), the same way the color fuchsia is pronounced. And *kalanchoe* is preferably pronounced with four syllables, accenting the third syllable and pronouncing the *ch* as *k*, that is, kal-un-KOE-ee. Stearn (1992a, 1992b) provides guides to modern pronunciation. Academic, classical Latin pronunciation is the ideal for international use, but simply pronouncing scientific names as if they were English words usually suffices. In any event, one finds differences in pronunciation of Latin names even among professional botanists. A good way of pronouncing scientific names is to copy the pronunciations of experts who have used the names for a long time and have become familiar with how most botanists say them.

## 1.5 SUMMARY OF KEY FEATURES OF THE 100 FLOWERING ORNAMENTALS

Table 1.1 provides a highly condensed review of key features of the 100 flowering ornamentals for which this book presents detailed information.

## 1.6 THE IMPORTANCE OF FLOWERING PLANTS TO PEOPLE AND SOCIETY

Plants are used for many indispensable purposes, notably for food, fuel, construction materials, medicine, and clothing. It may come as a surprise, therefore, that more plant species are employed for an apparently dispensable purpose – ornament – than for any other goal. Possibly as many as 100,000 species have been grown simply for display. Well over a thousand species and millions of cultivars (cultivated varieties) of ornamental plants are regularly sold to the public. A narrow-minded, indeed snobbish, perspective of ornamental plants is that they are of a much lower order of importance than plants grown for harvestable products, like food for humans and livestock, fibers for multiple purposes, lumber, industrial chemicals, and medicines. This perspective is flawed in several respects: (1) The ornamental plant industry is gigantic, employing millions and generating trillions of dollars annually on a world basis. The industry is capital-intensive, with excellent potential to become progressively profitable. (2) Ornamental plant research today is astonishingly multidisciplinary, incorporating all the fields of biology and addressing issues which are as intellectually and scientifically challenging as those examined by other sciences; the results are theories and technologies of immense value to society. The ornamental plant industry is "high-tech," requiring specialists who are as competent as those in any other field of biology. (3) Ornamental plant producers utilize resources and space more efficiently than any other area of agriculture; the industry is a model of intensive, high-yield, sustainable production. (4) Ornamental plants are so widely grown that they pose important ecological problems which need to be addressed; the responsible cultivation of ornamentals can enhance, rather than degrade, the ecosystems of the planet. (5) Ornamental plants provide key opportunities to develop favorable public relations between the general public and professional botanists. Most people are more interested in ornamental plants than any other kind of plant, and most professional botanists receive more inquiries about ornamentals than any other kind of plant. This provides a critical opportunity for scientists to educate the public about botanical and environmental issues, and to enlist support for biological research.

Ornamental plants are prominent in daily life. They are grown outdoors in private and public gardens, parks, and conservatories; indoors usually as relatively small houseplants; and often in very large planters in commercial and public buildings and shopping malls. The nursery trade specializes in potted plants. The florist trade may include potted plants but today deals primarily in cut flowers for bouquets and floral arrangements. Several horticultural fields are also included in the ornamentals industry and so are specialty areas, such as turf (sod) and Christmas trees. Ornamental plants are often the subject of craftsmanship,

FIGURE 1.12 Flower beds complementing famous public buildings: (a) The Taj Mahal in India. The building often attracts 10 million visitors annually, and in 2007, it was declared to be one of the New Seven Wonders of the World initiative. (b) Schönbrunn Palace in Vienna, the most visited tourist site in Austria. (c) Bolshoi Theatre in Moscow, Russia, renowned for its architecture and home of the Bolshoi Ballet.

*Source:* **(a) Photo by Fintan Corrigan of http://friendly-hotels.com (CC BY SA 4.0). (b) Photo by Paasikivi (CC BY SA 4.0). (c) Photo by Shutterstock/Andrew Koturanov.**

including landscaping, topiary (the shaping of shrubs and trees into figures, especially resembling animals), and bonsai (production of interestingly shaped dwarfed woody plants). Professional artistic specialization dedicated specifically to flowers (i.e., blooms) include creation of color-patterned floral beds and cut flower bouquet arrangements.

*Topiary* is the horticultural practice of clipping the foliage and twigs of living perennial plants so that they develop into clearly defined shapes. A hedge so manipulated is a simple form of topiary, but the practice has been elevated into a high art, with plants often trained to look like animals. "American portable-style topiary" uses steel wire frames within which plants are placed and forced to conform to desired predetermined shapes, rather than relying on the sculptural talents of the horticulturalist. While purists may object to the technique, it has resulted in spectacular displays at the Disney resorts and parks, as well as other large theme parks throughout the world. Much of topiary is concerned only with shaping of leafy branches and foliage, but perhaps the most spectacular concerns the presentation of exhibits covered by flowers. Such "floral sculpture" is usually achieved by attaching cut flowers to forms looking like animals, cars, and other familiar objects, mounted on stationary frameworks, but more often on mobile floats employed in parades. Of course, the blooms are destined to die in a few days, but the displays are spectacular and extremely popular.

Frequently, ornamental plants serve a useful function in addition to beautification. Groundcovers, especially on sloping lands, can serve to prevent erosion. Trees provide shade, which not only increases human comfort but, when trees are strategically planted near or even on top of buildings, can also result in considerable energy savings. Hedges can act as privacy screens, as traffic barriers, and for noise reduction. Trees and shrubs can act as wind and snow barriers, making streets safer. Ornamental plants attract birds, squirrels,

FIGURE 1.13 Artistic-patterned floral beds based on plants in bloom: (a) Garden of Versailles (Paris). (b) Royal Pavilion, Chiang Mai, Thailand. (c) Tulips at the Beijing Botanical Garden. (d) The Sunken Garden, Kensington, London.

*Source:* **(a) Photo by Daniel Stockman (CC BY SA 2.0). (b) Photo by Shutterstock/Think4photop. (c) Photo by Shutterstock/Haines. (d) Photo by Shutterstock/Geert Van Keymolen.**

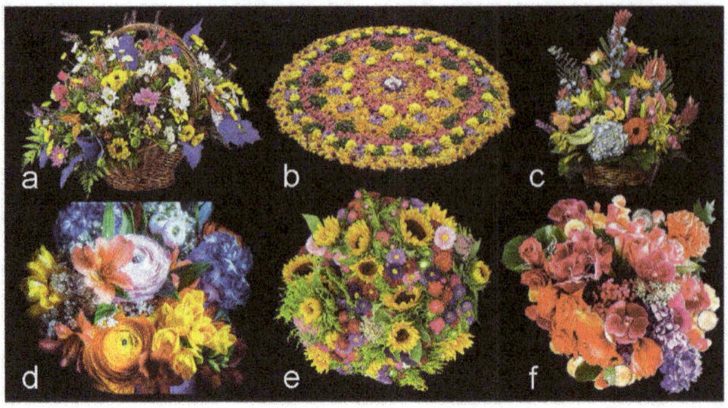

FIGURE 1.14 Artistic cut flower arrangements.

*Source:* **(a) Photo by Shutterstock/GLandStudio. (b) Photo by Prasad Pillai (CC BY 2.0). (c) Photo by Shutterstock/Nature's Charm. (d) Photo by Michael Lang (CC BY SA 3.0). (d) Photo by Garten-gg from Pixabay (public domain). (f) Photo by Allestimeni Matrimoni (public domain).**

FIGURE 1.15 Artistic bonsai (miniature woody plants). (a) Forsythia (*Forsythia*). (b) Japanese quince (*Chaenomeles japonica*). (c) Desert rose (*Adenium obesum*).

*Source:* **(a) Photo by IlonaBurschl from Pixabay (public domain). (b) Photo by Jerry Norbury (CC BY ND 2.0). (c) Photo by Shutterstock/Goldenjack.**

FIGURE 1.16 Artistic floral "sculptures." Cut flowers have been attached to the surface of a framework: (a) A snail. (b) An elephant. (c) A boy. (d) A duck. (e) A zebra. (f) A puppy sculpture by Jeff Koons, in Spain.

*Source:* **(a) Photo by Daniel Ramirez (CC BY 2.0). (b) Photo by Isobelyf from Pixabay (public domain). (c) Photo by Wildschuetz from Pixabay (public domain). (d) Photo by Shutterstock/Taljat David. (e) Photo by Yercaud Eelango (CC BY SA 4.0). (f) Photo by Jose María Ligero Loarte (CC BY SA 4.0).**

FIGURE 1.17 Artistic floral floats: (a) Giant turtle in the 2017 Bloemencorso Netherlands flower parade. (b) Lions and flamingos in the 2008 Bloemencorso Netherlands flower parade. (c) Horses in the 2012 Spalding, UK, flower parade. (d) A duck in the 2013 Bloemencorso Netherlands flower parade.

*Source:* **(a) Photo by Shutterstock/Pluut. (b) Photo by Debs-eye (CC BY 2.0). (c) Photo by Jim Linwood (CC BY 2.0). (d) Photo by Jac Janssen (CC BY 2.0).**

FIGURE 1.18 A modern interpretation of the legendary Hanging Gardens of Babylon, dated at the 6th to the 8th centuries BC. The plants (primarily fruit trees, such as dates) did not actually "hang" but were elevated on roofs and terraces of the royal palace in Babylon.

*Source:* **Photo by Shutterstock/Turbo.cgi.**

and butterflies, providing entertainment and education (and sometimes deer and rabbits, providing snacks for the others of God's creatures). "Horticultural therapy" is the cultivation of plants by people to relieve stress, furnish psychological comfort, improve memory and concentration, encourage social interaction, and provide exercise. Ornamental plants, especially the "urban forest" (city plants, especially trees, collectively), are so widely cultivated today that they serve as an important carbon sink, binding ("sequestering") carbon so that the concentration of the atmospheric gas carbon dioxide is controlled, thereby moderating the climate change problem. Ornamental plants can also be used for phytoremediation – to reclaim land polluted by industrial wastes, or to remove pollutants from sewage water. "Living walls" of plants are now used to purify water in buildings as well as to provide an attractive appearance.

## 1.7 A BRIEF HISTORY OF ORNAMENTAL FLOWER GARDENS

Anthropologists (see Ghose 2013) discovered a prehistoric grave site in Israel, 14,000 years of age, covered with wild culinary herbs, such as mint and sage, believed to be the oldest authenticated example of "grave flowers." Clearly, while we humans especially value plants for food, beautiful flowers also appeal to our aesthetic and sentimental instincts. The earliest deliberate cultivation of plants was certainly for nutrition, medicine, and other survival purposes, but nevertheless, plants were grown just for ornament in very early times.

Probably, gardens designed primarily for visual pleasure have existed for over 5,000 years. For millennia, attractive plants, many with symbolic significance, have been used for embellishing homes and places of worship. About 1570 BC, Queen Hatshepsut of Egypt sponsored plant-hunting expeditions to beautify her royal gardens. The Hanging Gardens of Babylon (in modern-day Iraq), one of the Seven Wonders of the Ancient World, are thought to have been built by King Nebuchadnezzar II about 600 BC and were destroyed by earthquakes after the 2nd century BC. While some aspects of these gardens appear to be legendary, certainly the technology of growing plants in stone vessels and on terraces was well-known at that time. In any event, the rich and politically powerful have long flouted their status with extravagant, even garish displays of living plants, both outdoors and in their homes. Perhaps the most flagrant example was set by King Louis XIV (1638–1715) of France, who, while keeping most of his subjects in abject poverty, built a fabulous 800 ha garden beside his opulent Palace of Versailles (together, a UNESCO World Heritage Site today). Common people, too, especially since the times of Louis XIV, have added flowering plants to their homes, albeit on a much more modest scale. However, reflecting the rise of a large middle class, gardens dedicated to flowers just for ornament did not become common in Western nations until the 19th century.

The first settlers to North America imported planting stock of species that were popular in western Europe, much of which originated from Eurasia, often from exotic regions of the world, but infrequently from North America. Just as almost all major food crops grown in North America today are of foreign origin, the same is true for the majority of ornamental plants grown in the New World. Nevertheless, there is now a movement to increase utilization of native plants for landscape purposes, because they are naturally adapted to local conditions and usually require less care, water, fertilizer, and pest control agents (see "Use of Native Plants as Ornamentals," presented later in this chapter). Until the 20th century, transportation from overseas ports was so slow that living plants suffered greatly when

FIGURE 1.19 A flower garden at the Palace of Versailles.

*Source:* **Photo by Constantine H. (CC BY 2.0).**

FIGURE 1.21 A contrast of the two fundamental ornamental garden planting styles: (a) A highly informal ("natural") garden, the Curé Garden in the region of Auvergne-Rhône-Alpes, France. Note that the plants are arranged in random, overlapping locations. (b) A highly formalized ("controlled") garden in Da Lat, Vietnam. Note the precise planting locations arranged in a geometrical pattern, as well as an emphasis on deliberate placement to contrast color and other features of the species.

*Source:* **(a) Photo by Zairon (CC BY SA 4.0). (b) Photo by Shutterstock/Huy Thoai.**

FIGURE 1.20 Vintage commercial flower catalogues (public domain) showing ornamental plants that were popular over a century ago: (a) Dingee & Conard 1898 "Our new guide to rose culture" catalogue. (b) Childs' spring catalogue of seeds, bulbs, and plants for 1910, featuring sweet pea. (c) Miss C. H. Lippincott 1897 catalogue, featuring hollyhocks. (d) Farquhar's 1902 seed catalogue, featuring hollyhocks. (e) Childs' combination catalogue for 1910, featuring pansies. (f) Peony (*Paeonia moutan*) in 1900 catalogue of the Yokohama Nursery Co. Ltd. Japan.

*Source:* **Photos from the US Department of Agriculture, National Agricultural Library.**

shipped. This resulted in the establishment of nurseries and greenhouses in North America to supply most of the needs of the continent. However, air freight has resulted in globalization of much of the ornamental plant trade, with some negative consequences for North America and Europe economically and environmentally.

## 1.8 MODERN ORNAMENTAL GARDENS

Gardens differ greatly in their size, style, composition, and location (e.g., such as rooftop, courtyard, and balcony gardens), often including not only plants but also complementary artistic objects, such as statues and boulders, animals (such as fishponds and butterfly enclosures), and features of relevance to people with special needs or special interests. Some gardens specialize in plants of certain habitats (e.g., alpine/rock, deserts, forests) or particular groups (e.g., cacti, orchids, roses). Garden design differs most notably in the degree to which the space occupied is formally designed or seemingly randomly occupied by plants and other constituents (like nature). "Flower gardens" are dedicated to plants with attractive flowers, including annuals, perennial herbs, and woody plants. The expression "pleasure garden" has been historically applied to large elaborate gardens established by nobles and rulers so they could be isolated from the stresses of their regular lives. "Walled gardens" are usually smaller versions with the same purpose, but even modest residential gardens can be remarkable stress relievers. Very elaborate ornamental gardens have been developed in many countries as expressions of national pride and attract huge audiences. Increasingly, plants grown in gardens intended for the entertainment of the public are those with very attractive flowering displays, including unusual (often tropical) species, since this is the desire and expectation of most visitors. "Arboreta," by definition, are tree gardens but usually include shrubs as well, and often, herbaceous plants are also exhibited, chosen to flower at times when the woody plants are not in bloom. Special-purpose flower gardens are sometimes available, such as scented flowers for the blind, pollinator gardens, and edible flowers. "Sensory gardens" appeal to specific senses: Scent gardens are the most frequent kind, sound gardens may offer artificial sounds or plants that naturally rustle in wind, while texture gardens offer artificial textures or plants that are naturally pleasant and interesting to touch. Accommodating special physical and psychological needs is important, given the therapeutic and pleasurable value of plants. However, gardens and gardening in general promote emotional well-being for everyone, and flowers in particular provide appreciation of the values of nature.

FIGURE 1.22 Sensory gardens, emphasizing fragrance, sound, and/or touch. Such gardens are especially valuable for those with sensory limitations but are also invaluable to expand everyone's perception of the subtleties of nature. (a) A sign inviting garden visitors to touch and listen. (b) A group tour in the Fragrant Garden of the Brisbane (Australia) Botanic Gardens.

*Source:* **(a) Photo by Laura Blanchard (CC BY SA 2.0). (b) Photo by John Robert McPherson (CC BY SA 4.0).**

FIGURE 1.23 Therapeutic gardening ("horticultural therapy") with flowers provides a sense of accomplishment, pride, and self-worth for those with limitations.

*Source:* **(a) Illustration by Shutterstock/brgfx. (b) Illustration by Shutterstock/Graphicyes.**

There are innumerable guides to gardening (for suggestions, see "Key Literature" at the end of this and every other chapter and "Key Cultivation Websites" at the end of the 100 chapters, each treating a particular plant or plant group). Whether or not one chooses to create and maintain a garden, visiting one is extremely rewarding. Gardens are compositions of beauty – like great art, architecture, literature, and music – and the great gardens of the world are truly stunning artistic achievements.

## 1.9 HOUSEPLANTS

Houseplants (or sometimes house plants) are mostly decorative perennial plants grown in pots indoors (aquatics and many bulbs can be grown in water; epiphytes on bark or spongy, water-retentive surfaces; and some tiny plants in terraria, the counterparts of aquaria). Houseplants are usually of tropical or semi-tropical origin and require protection from the local outdoor climate, at least during the winter. Most wild species cannot tolerate the environment found in residences, but some find it to their liking. In past times, only the wealthy could afford to grow plants indoors.

FIGURE 1.24 World-famous ornamental gardens featuring especially beautiful flowering displays: (a) Doi Tung Garden, Chiang Rai Province, northern Thailand. (b) Butchart Gardens, Victoria, British Columbia, Canada.

*Source:* **(a) Photo by Shutterstock/Ntdanai. (b) Photo by Shutterstock/2009fotofriends.**

FIGURE 1.25 "Flowers on the Windowsill," painting by Swedish artist Carl Larsson (1853–1919).

*Source:* **Public domain.**

In the early 18th century in Europe, windows began to be larger in the houses of the lower classes, expanding the growth of plants indoors. Today, ornamental plants are found in most houses and workplaces, but few people realize that the popularization of electric lighting in the early

20th century was the most important prerequisite for this. However, most plants grown within buildings (both private residences and large structures housing hundreds or even thousands of people) are cultivated primarily for their foliage. A chief limitation for plants by comparison with humans is the need for strong light for photosynthesis. Floral tissues have very limited capacity to photosynthesize compared to leaves. The result is that except on windowsills and below skylights (especially in shopping centers), most buildings don't supply enough light for plants that produce good displays of attractive flowers (the subject of this volume). Nevertheless, as detailed in this book, there are species such as African violets and orchids that are naturally tolerant of shady conditions and are capable of developing very attractive flowers under low illumination. Most flowering houseplants are naturally small or are dwarfed by being grown in small containers, and the limited size suits them to the restricted space available on windowsills. Of course, supplemental light can be supplied, and LED lamps are an efficient means of providing the higher light levels required by many plants with attractive blooms.

Flowering houseplants are especially attractive, and there is a very large industry supplying potted plants that are just beginning to flower, especially for sale at holidays like Christmas (notably poinsettia) and Easter (especially lilies). These include bulbs like tulips and daffodils that are forced to flower outside of their normal outdoor season. By comparison with bouquets of cut flowers, which rarely remain in good condition as long 2 weeks, such potted plants provide attractive floral displays sometimes lasting for months. Usually, they are discarded when the flowers fade, although the plants can often be rejuvenated.

**FIGURE 1.26** Popular windowsill flowers: (a) Blooming phalaenopsis (*Phalaenopsis*) orchids. (b) African violets (*Streptocarpus ionanthus*).

***Source:*** **(a) Photo by Shutterstock/Mariia Boiko. (b) Photo by Shutterstock/Ludmila Kapustkina.**

**FIGURE 1.27** Popular temporary (short-lived) holiday houseplants: (a) Poinsettia (*Euphorbia pulcherrima*). (b) Easter lilies (*Lilium longiflorum*) on the altar of Baltimore's Our Savior Lutheran Church during Easter 2018. (c) Daffodil (*Narcissus*), forced from bulbs.

***Source:*** **(a) Photo by Shutterstock/Asife. (b) Photo by Schetm (public domain). (c) Photo by Shutterstock/Kuttelvaserova Stuchelova.**

## 1.10 GLASSHOUSES FOR PLANTS

In northern countries, cold temperature is a principal limitation of plant growth. Huge acreages of greenhouses (also called glasshouses and hothouses) are now dedicated to production of ornamental plants. Growing plants in the cold season traces to classical Roman times, when crude greenhouses were constructed with translucent, mica-like sheets or oiled cloth. Roman emperor Tiberius (42 BC–AD 37) demanded a regular supply of cucumbers in the winter, requiring gardeners to grow the plants in wheeled carts (reputedly covered with translucent selenite sheets), taken outdoors on sunny days and protected in enclosures at night. Much later (beginning in the 14th, but especially in the 15th and 16th centuries), this simple process of harvesting a food plant – by protecting it from a period of cold in a building but moving it outdoors to get sunlight when the climate warms up – was repeated, but on a monumental scale, in Europe. The key innovations were advances in the technology of producing glass, and this resulted in buildings called orangeries (indeed, to produce oranges in cold climates), with windows concentrated on the walls receiving the most sunlight, so that fruit trees could be grown inside during the winter but moved outdoors in the summer. Throughout Europe and in parts of the New World colonized by Europeans, orangeries were established, and many still exist. Although orangeries were initially concerned with producing citrus fruit, the ability to maintain exotic ornamental plants throughout the year resulted in competition among the privileged classes to construct more specialized glasshouses for plants. Private "conservatories" (greenhouses designed specifically for growing and displaying ornamental plants throughout the year, not just overwinter) became popular, and (as with the previous craze for orangeries) their size and luxuriousness reflected the financial and aristocratic status of the owners. Rare, weird, beautiful, and/or spectacular plants became greatly valued and stimulated increasing expansion of the ornamental horticultural trade, which continues to this day.

The orangeries – which had relatively little glass in their outer walls and none in their roofs – gave way to monumental glass-clad structures in the Victorian era. In 1851, J. C. Paxton, gardener for the Duke of Devonshire, designed and built the Crystal Palace in London, England, a magnificent glass-and-iron structure that has since inspired the construction of numerous other large public greenhouses and botanical conservatories. Huge conservatories have been established in large cities and are especially welcome in winter, when outdoor gardens are dormant.

FIGURE 1.28 *Orangeries.* Precursors of greenhouses, these buildings were popular with the very rich in Europe by the 16th century, in order to grow citrus trees in winter, but mostly to flaunt the wealth of the owners. By pruning and with limited water and nutrients, sweet orange trees were made to produce fruit year-round. These opulent structures had large windows, mostly south-facing, to provide sunlight for the trees, but lacked glass on the roof, unlike modern greenhouses. Orangeries were initially established in Europe to provide oranges to the nobility and rich merchants, but by the 16th century, the fruit was commonly available, and orangeries served mainly to impress visitors. (a) The orangery beside the Versailles Palace, France, completed in 1686 for the glory of King Louis XIV. This is one of the most monumental structures ever built to maintain living plants. In the winter, the Versailles Orangery shelters indoors more than 1,000 trees in boxes, mostly orange trees, which are kept outdoors from May to October. In this photo, the trees are growing in large wheeled boxes. (b) An artistic reconstruction of the interior of the orangery at Choisy, France, showing orange trees in tubs. (The building was constructed about 1690 at the Château de Choisy, a royal castle located on the banks of the Seine River, but only remnants have survived.)

*Source:* **(a) Photo by Nono vlf (CC BY SA 4.0). (b) Illustration by Franck Devedjian (CC BY SA 4.0).**

FIGURE 1.29 The Crystal Palace in London, England, the largest all-glass-clad building constructed up to the 20th century. It was 564 m long, 39 m high, and had 92,000 m$^2$ of exhibition space.

*Source:* **Nash, J., L. Haghe, and D. Roberts. "1852" (1854). *Dickinson's Comprehensive Pictures of the Great Exhibition of 1851.* Dickinson Brothers, London.**

FIGURE 1.30 Exotic flower exhibits inside large public conservatories: (a) Artificial sloth in the Princess of Wales Conservatory, Kew Gardens, near London, UK. (b) US Botanic Garden, Washington, DC. (c) Botanic Gardens, Auckland, NZ. (d) Garfield Park Conservatory, Chicago.

*Source:* **(a) Photo by JRennocks (CC BY SA 4.0). (b) Photo by USCapitol (public domain). (c) Photo by Brendan Purdy (CC BY 2.0). (d) Photo by Mheisel (CC BY 2.0).**

## 1.11 CUT FLOWERS

### 1.11.1 Development of the Cut Flower Industry

In old times, harvested flowering plant species were placed indoors mostly as "strewing herbs" – fragrant plants especially employed at weddings, parties, and funerals to mask household odors. By the 18th century, vases designed to hold cut flowers had become widespread. Common people in past times were mostly dependent on gathering pretty wildflowers from nature rather than cultivating the plants. By contrast, wealthy and noble families once maintained "cutting gardens" of plants that were intended to provide regular displays of cut flowers for the interiors of mansions. (More practically, many houses, even humble ones, were accompanied by "kitchen gardens," which were designed to bring vegetables, fruit, and flavoring herbs into the house and not simply flowers for ornament.) In Victorian times (the ruling era of Queen Victoria, 1837–1901), commercial cutting gardens were widespread, and the trade in cut flowers was extensive.

By the beginning of the 20th century, floriculture was well established. The development of rapid air transportation in the last several decades has greatly widened the appearance of beautiful plants. The most important sector of the international ornamental plant trade is harvested flowers, because they are lighter than potted plants, and so transportation costs are comparatively low. Vast quantities

FIGURE 1.31 Paintings (public domain) of cut flowers showing their importance in Victorian times. These illustrations by French artist Louis Marie de Schryver (1862–1942) are two of many titled "The Flower Seller," showing Avenue de L'Opera in Paris in the 1890s.

FIGURE 1.32 Examples of cut flower bouquets in vases.

*Source:* **(a) Photo by Buntysmum from Pixabay (public domain). (b) Photo by MarrtDesign from Pixabay (public domain). (c) Photo by Buntysmum from Pixabay (public domain).**

FIGURE 1.33 Marketing of cut flower bouquets by florists.

*Source:* **(a) Photo by Shutterstock/CandyBox Images. (b) Photo by Shutterstock/Francesco83.**

of gorgeous bouquets of cut flowering stems (referred to as "cut flowers" in the floral trade) are now cheaply transported from tropical and semi-tropical climates to all parts of the world.

Today, stunningly attractive, relatively cheap cut flowers are available from most supermarkets as well as florists and other outlets, including internet suppliers, and are in considerable demand. More than a dozen countries have cut flower industries valued at several billion dollars annually. European consumers buy cut flowers (and also flowering potted plants) almost weekly and annually spend ten times more per person on such purchases than North Americans. In the United States, cut flowers represent only about 6% of the total annual floricultural wholesale value of about $6 billion. Cut flowers account for about 50% of worldwide trade in floriculture products, and cut foliage an additional 10%. Characteristics that make for good cut flower species include ease of cultivation and handling, color (white and pink are popular for spring weddings; orange and copper are sometimes popular in the fall), fragrance, and stem length (long stems are often preferred). The cut flower industry is worldwide, and there is considerable competition among nations. Cut flowers are harvested both from outdoor and greenhouse-grown plants. In Northern nations, the cost of maintaining greenhouses in the cold part of the year is a considerable disadvantage. On the other hand, local production is advantageous where flowers do not ship well. Information on the world cut flower industry is presented later.

### 1.11.2 Care of Cut Flowers

A gift of fresh-cut flowers expresses emotional thoughts such as "Congratulations," "Thank you," "I'm sorry," "Get well," "You're special," or "I love you." Unfortunately, cut flowers have a limited lifespan. Most freshly acquired cut flowers will last 5 to 7 days without care, but it is often possible to extend vase life, along with their message, for up to 2 weeks or more. Two factors cause cut flowers placed in water to wilt prematurely: (1) Air taken up by the stems blocks the uptake of water; (2) bacteria growing in the water clog the stems. To prevent air uptake, the cut stems should be placed in water as soon as possible, then recut underwater, at an angle (to prevent the stems from sitting flat on the bottom of the vase; an angled cut, per se, doubtfully increases water uptake), using a sharp knife, not scissors or a serrated knife (professionals will sometimes sterilize the blade with bleach or alcohol), removing about 2.5 cm from the stem (this procedure works because a film of water prevents air from entering the stem). Special tools are often used by florists for this task, and special "hydration" products are available to assist water to move up the stem. The second problem, microbial growth, is due to bacteria, yeasts, and other microorganisms growing on the sugars, amino acids, proteins, and other organic substances released by the cut stems. To control such growth, commercial preservatives with antimicrobials are added, commonly quaternary

FIGURE 1.34 Examples of extremely colorful flowers: (a) Tulips at the Keukenhof Garden in The Netherlands. (b) Flower show at the Royal Hospital, Chelsea, London.

*Source:* **(a) Photo by Krysie (CC BY 3.0). (b) Photo by Kent Wang (CC BY SA 2.0).**

ammonium, hydroxyquinine salts, aluminum sulfate, and slow-release chlorine compounds. Both of the aforementioned problems can be exacerbated by the presence of excess foliage on the cut inflorescences: Leaves above the waterline can increase water loss, stressing the stem, and leaves below the waterline will rot, increasing the bacterial problem. In addition to controlling wilting from lack of water, the life of some cut flowers (notably gladiolus) can be extended with "holding" solutions that provide carbohydrate (sugar) nutrition. Commercial solutions (often called "cut flower preservatives" and "cut flower food") that both inhibit microorganisms and provide nutrition can be purchased from florists and supermarkets, and florists sometimes provide a package of chemicals to extend vase life along with the purchase of cut flowers. Controlling acidity of the water is important (pH 3.5 is good), but some specialists have warned that aspirin and vinegar, which have sometimes been recommended, should be avoided. Two simple homemade recipes are: (1) Combine a lemon-lime soda with three parts of water (do not employ diet drinks, which lack sugar, or colas, which can be too acidic) to make 1 L, and add 1/4 tsp household bleach (like Chlorox). Or (2) add 2 tbsp of lemon juice (or the equivalent in bottled juice), 1 tbsp of sugar, and 1/4 tsp of bleach in 1 L of water (for both alternatives, add 1/4 tsp of bleach to the vase every 4 days). Other practices for extending the life of cut flowers include: Ensure the vase is clean, to eliminate bacteria and fungi. Add water daily, if necessary, to keep the waterline above the bottom of the stems. Avoid both softened water (which contains sodium) and hard water (use demineralized water if naturally soft water is unavailable). And keep the flowers cool – avoid sunny, warm, or drafty locations, especially windows and heat-generating appliances (notice that florists usually maintain their cut flowers in a cool display room). At the minimum, water should be changed if it becomes murky, and the stems should be recut underwater at that time. Not all techniques work ideally for all flowers. For some species, florists sometimes use commercial solutions that control ethylene, a gas which generally promotes ripening of fruit but causes some flowers to look wilted or to fail to open. Stems that ooze latex (poppies, poinsettia) can clog quickly when cut (one proposed solution is to remove the sealing latex by immersing the bottom 5 cm or 2 in in boiling water for 10 to 30 seconds). Recommendations on which combinations of techniques are appropriate for given species can be found in the information resources cited later, but these are of interest mainly to specialists.

## 1.12 COLOR, THE CHIEF ATTRACTION OF FLOWERS

Ornamental plants are valued for a variety of characteristics, such as trees for shade, turf grasses for lawns, and vines for hiding imperfections of buildings and fences. However, the ornamental plant industry is predominantly based on the attractiveness of flowers, and this is due mostly to displays of color. The flowers of most ornamental plants are their chief attraction to most people, so much so that the word "flower" is widely used in the vernacular to refer not just to true flowers but to herbaceous plants with attractive flowers as well. The feature of flowers of dominant interest to most people is simply their color. In nature, flowers have characteristics that are attractive to pollinators, and of these, color is particularly important because some animals prefer certain flower colors. For example, hummingbird flowers are frequently red; some butterflies prefer yellow or white, while others prefer blue (but most do not seem to have a strong color preference); and many flies prefer white or dull colors, while many social insects and birds prefer bright colors. For humans, however, flower color does not reflect a natural, symbiotic relationship. As noted previously, ornamental plants can serve several practical functions, but their main value is usually just the attractiveness of their floral color or colors to humans. Ornamental cultivars have been selected for displays that are spectacularly more flamboyant than existed in the relatively small and evanescent flowers of their wild ancestors. Today, the ornamental plant industry is most concerned with "selling color," and so some understanding of the psychology of color attractiveness is desirable.

Color preferences of humans appear to be (1) innate, selected as survival responses during evolution because they are particularly associated with either pleasant or unpleasant objects, or (2) learned. Early researchers on color preferences in humans suggested that, in general, people prefer, in decreasing order, blue, red, green, violet, orange, and yellow (a sequence that leaves out white flowers, which are widely admired). Blue has been shown to be the most pleasant color to the majority of people (perhaps explaining why baby boys are traditionally honored with blue), possibly because a blue sky indicates calm weather. By contrast, dark yellow and orange are the colors of urine, feces, vomit, and rotting food, and there is a general lack of preference for these hues in all human cultures. We are all born with an innate attraction to reddish nipples to obtain nourishment from breasts, perhaps accounting for why red is such an attention-getter and explaining the advertising slogan "red sells." There have been claims that women prefer red slightly more than men do (supposedly why little girls are dressed in pink), speculatively explained as useful in our ancestors, who relied on women gatherers to spot red berries against green foliage while the men hunted. The psychological effects of red have been extensively studied but remain controversial; certainly, red is an extraordinarily attracting stimulus and is the most popular of colors inso far as ornamental plants are concerned. Saudi Arabia once banned the selling of red flowers in the week preceding Valentine's Day on the theory that they encourage immoral relations between unmarried men and women. Purchasers of commercial cut flower bouquets have been found to prefer reddish flowers (especially when bought by men) and to avoid yellowish flowers.

Of all possible colors, black is the most unnatural for flowers. The phrase "black flowers" is used loosely to refer

FIGURE 1.35 The red rose: the most psychologically influential of all flowers.

*Source:* **(a) Photo by Susan-lu4esm from Pixabay (public domain). (b) Photo by Shutterstock/Ironika.**

to very dark flowers of some ornamental species. Usually, such flowers are a deep purple, and almost always, a close look reveals that the flowers are not truly black. Orchids, tulips, carnations, calla lilies, roses, and pansies are among the species for which black-flowered variants have been selected. Black is associated with evil and death, which limits the use of black flowers in many situations. Nevertheless, the rarity and novelty of black as a floral color have stimulated breeders to select very dark-flowered varieties, and they add considerable interest to a garden or bouquet.

On the whole, color preferences for flowers seem to be determined mostly by individual taste (although green flowers are obviously not desired because plants are already overwhelmingly green). Moreover, a variety of colors is usually considered attractive, both in bouquets and in gardens (although, as a rule, it is recommended that bedding plants of a particular color should be planted close together to accentuate the display). In foliage plants, variegated leaves have been selected in numerous cultivars, which also demonstrates a pattern preference for variation in color.

As most people realize, a given color of apparel may be appropriate for one occasion or setting and not others. Similarly, flower color is traditionally determined in part by the circumstance: white and pastel hues for a funeral, bright flowers for winter. Learning plays a very large role in color preferences, and it is important to be aware of ethnic, religious, and social traditions with respect to color. Hinduism, Buddhism, and Confucianism have special respect for yellow and orange, Muslims have considered green to be a sacred color, and Christianity has suggested special, salvation-related meaning to red, white, and black. (In addition, regardless of color, some particular species of ornamental plants often have special significance for given religions.) In modern times, social constraints limit men from displaying pink and purple, and this has extended to recommendations for giving flowers to men (although the reverse gifting is still predominant). New flower colors are continually appearing for the leading ornamental plants, but there is a tendency for traditionally available colors to remain especially popular. In the case of roses, for example, the most popular colors are red, pink, white, yellow, lavender, and purple, and for carnations the leading colors are red, pink, and white. Today, numerous ornamental plant species are available in a wide range of flower colors, and often, the chief concern of many is simply to coordinate flower color with dress or decor.

## 1.13 FRAGRANCE OF ORNAMENTAL PLANTS

In nature, flowers often attract pollinators, not just by vision, but also by odor, and frequently the smell is also attractive to people. In some cases, the chemicals responsible for pleasant smells are extracted and used in perfumes and in preparations employed in aromatherapy, most notably for rose, jasmine, lily, and violet. The floral-odor chemicals may vary from five or so to about 100 and are most frequently terpenes (not surprising, since this is the largest and most variable class of naturally occurring organic compounds). The aromatic compounds are most often synthesized in the petals in specialized cells (osmophores). Plants that release odor during the day are usually bee-pollinated. Conversely, those that open in the evening and release odors at night are typically pollinated by night-flyers, such as bats and hawk moths. Flowers that attract night-flying pollinators are often white (to reflect as much light as possible) and are also frequently quite odorous (to reinforce the weak visual display). The night-blooming tuberose (*Polianthes tuberosa*) is exceptionally aromatic and is used not just as a cut flower but also in perfumery. Other ornamental plants esteemed for their pleasant odor include daffodil, hyacinth, cyclamen, and primula. The species within genera, and indeed cultivars of a particular species, often differ greatly in intensity and pleasantness of odor. For example, "scented geraniums" refers to species of *Pelargonium* that are especially fragrant, and the Stargazer and Casa Blanca cultivars of lily are renowned for their "heavenly odor" (which can be overpowering in a confined room). The wild ancestors of some ornamental plants are often much more fragrant, because modern cultivars have often been selected for their visual appearance at the cost of decreasing their pleasant scent (some modern roses, lilacs, and carnations are notably

FIGURE 1.36 Ornamental flowering plants with pleasant scents are widely valued, as reflected in these great paintings: (a) *Sweet Flowers*, by Belgian artist Jean-François Portaels (1818–1895); (b) *Little Gardener Sniffs Flowers*, by French artist Jean-Paul Haag (1854–1906); (c) The Soul of the Rose, by English artist John William Waterhouse (1849–1917).

*Source:* **All are public domain.**

lacking in odor). Pleasant-smelling flowers are not always welcome: Worried about allergies, some health authorities have requested florists to exclude highly scented flowers, like lilac and hyacinth, in bouquets delivered to hospitals and residential facilities. Females have been conclusively demonstrated to generally have slightly greater olfactory prowess compared to men. Women have more threshold sensitivity to odors, as well as greater discrimination ability, and in some experiments have been shown to have greater preference for some odors (like rose). Blind and visually impaired people can enjoy ornamental plants that are especially fragrant.

The most surprising ornamental species from a fragrance perspective is allium. Alliums include garlics and onions, infamous for their objectionable odor. The foliage of some ornamental alliums does have a slight onion/garlic odor, but the odor of ornamental varieties seems to have been mostly suppressed by breeders.

Normally, humans are repelled by foul odors, but two of the world's stinkiest plant species attract large audiences when displayed in public greenhouses. These include the carrion or corpse flower, *Rafflesia*, and the corpse plant or titan arum, *Amorphophallus*, which attract pollinating insects with a stench like rotting flesh.

## 1.14 APPEARANCE OF ORNAMENTAL PLANTS

In the same way that people differ in their tastes for dog breeds, styles of architecture, and manner of dress, ornamental plants attract different people. There are numerous societies dedicated to particular ornamental plants (irises, roses, bromeliads, daffodils, orchids, etc.). The range of such plant "fan clubs" shows that there is considerable variation in preferences. Preference for a particular group of plants (often a genus or family of plants) is usually based on characteristics of floral structure, which in turn usually depend on plant family or genus.

This book is concerned primarily with the flowers (blooms) of ornamental plants, but plants necessarily also have stems or trunks, branches, and foliage, all of which are considered when people evaluate the attractiveness of given cultivated species. Variation of appearance of species in the natural world is stunning, and many wild plants are so interesting that they make good ornamentals without being changed by breeding. However, humans do have strong preferences, and virtually all popular ornamentals have been altered by domestication. Some animals have been selected (at least in part) for their appearance, notably dogs, cage birds, and goldfish. The selected breeds of these animals illustrate that people like novelty, beauty (in human terms), and bizarre characteristics, and the same is true for ornamental plant varieties. The flowers of several of the plants included in this book are available in forms so highly divergent that it is hard to believe they belong to the same species. Probably the most bizarre flower included in this book is cockscomb or coxscomb (*Celosia argentea* var. *cristata*), the blooms of which are highly contorted, appearing ruffled and/or wrinkly, looking like coral or brain, an appearance not to the taste of many.

### 1.14.1 Vegetative Features

Plant species differ greatly in vegetative (i.e., non-floral) characteristics. The limited humidity in many homes during the winter makes thick-skinned (hence water-retentive) foliage plants like *Sansevieria* (mother-in-law's tongue) species and rubber plants appropriate. The cacti (Cactaceae), which appear to be leafless, provide more houseplants than any other plant family, probably because most are small and easy to care for (i.e., they withstand neglect well). Some species have fernlike, highly dissected leaves; others have gigantic, elephant-ear leaves; and plants with all manner of leaves are cultivated. Not content with nature's offerings of foliage, horticulturalists have selected many leaf mutations. Cut-leaved

FIGURE 1.37 Some of the world's stinkiest plants: (a) The bloom of rafflesia (*Rafflesia arnoldii*), found in the rainforests of Borneo and Sumatra. Famous as the largest flower in the world, it emits a stench similar to that of decaying meat, to attract the flies that pollinate the plant. (b) Corpse plant or titan arum (*Amorphophallus titanum*). The huge phallic spike is made up of numerous tiny flowers, and the whole structure is commonly misinterpreted as a gigantic single flower. Like rafflesia shown in (a), it produces an odor like rotting meat to attract pollinating insects.

***Source:*** **(a) Photo by Rahmat Andriansa (CC BY SA 4.0). (b) Photo by Lothar Grünz (public domain).**

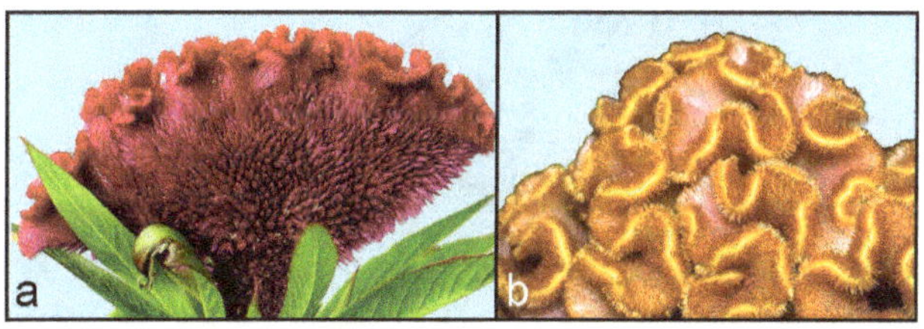

FIGURE 1.38 Examples of weird cockscomb (*Celosia argentea* var. *cristata*) "flowers."

***Source:*** **(a) Photo by Manfredrichter from Pixabay (public domain). (b) Photo by ThomasWolter from Pixabay (public domain).**

(i.e., dissected) cultivars of some woody plants (silver maple and several birch species, for example) are popular. Numerous cultivars have been selected that have the normal foliage pigments altered, so leaves are yellowish (frequently described as copper, bronze, or gold) or have anthocyanin pigments emphasized so the leaves are purplish or reddish. Cultivated plants with variegated leaves (i.e., with sectors devoid of chlorophyll) are extremely common (such mutations are almost always eliminated in nature). Stems or trunks have also been altered in many ornamentals. Several tree species (white mulberry and some beech species, for example) have forms selected for drooping branches, like weeping willow. Corkscrew (spiraling) stems have also been selected. Spiny stems protect many plants in the wild, but people do not like to be stabbed, and many spineless forms have been bred. Habit (overall appearance) of plants has frequently been altered, producing pyramidal, spreading, or other shapes. Some of the variations that have been selected by breeders are due to infection by viruses, so that sick, diseased plants that nature would normally eliminate are being maintained by humans. Indeed, virtually all characteristics selected in ornamentals would be deleterious in their wild ancestors but nevertheless are viewed by humans as attractive. Probably, the chief conclusion that can be drawn from examination of how ornamental plants have been changed by people is that humans are greatly entertained by variation, which is fortunate, because there is an enormous variety of plants to admire and protect.

### 1.14.2 Size

Size is a critical characteristic of outdoor ornamental plants. For gardens, a well-known design principle is that taller plants are placed at the back, shorter ones in front, so that all can be viewed. Among woody plants, dwarf forms have been extensively selected to meet the need for foundation perennials that complement, not overwhelm, the houses that they border. Size is also a critical characteristic of houseplants. Diminutive African violets, primroses, and kalanchoes are ideal for small rooms.

The ultimate in "garden miniaturization" is the terrarium. A *terrarium* (plural *terraria* or *terrariums*) is a glass (mainly or entirely) container in which plants are grown in soil. (The term *terrarium* is also often employed for containers for displaying reptiles and amphibians.) The simplest container is just a big bottle, and glass aquaria can be used to maintain plants rather than fish. Terraria are often sealed, at least most of the time, so that essentially the only input is sunlight. Because of the heat usually accumulated in terraria, tropical plants are especially grown.

### 1.14.3 Daisies (Asteraceae Family)

The word "daisy" in a general sense refers to a member of the daisy family (technically named the Asteraceae or Compositae), which is frequently considered to be the largest of the 400 or so recognized plant families, with about 30,000 species. The 100 chapters in this book dedicated to particular flowers include the following 11 "daisies:" aster, black-eyed Susan, chrysanthemum, dahlia, echinacea, gazania, gerbera, marguerite, marigold, sunflower, and zinnia. What is it about a daisy flower that makes it so attractive? The answer is that the so-called flower is actually a head (technically an "inflorescence") composed of tiny flowers ("florets"), and while the individual flowers are very small, the head is usually so large that it is naturally attractive. As well, just by having more species than any other plant family, on a random statistical basis, it seems that the daisy family was pre-adapted to producing more popular ornamentals than other families. In the flowers of wild daisy species, usually the outermost florets (not the central florets) are the only ones with prominent petals, but horticulturalists have selected varieties with all the florets developing petals (so-called "doubled flowers") because this makes the flower heads more attractive. Hydrangea is not a member of the daisy family, but it too often has flower heads, with the outermost florets developing especially prominent petals.

FIGURE 1.39 The importance of anticipating the ultimate size of plants in their choice and location.

*Source:* **Illustration by Shutterstock/Tim Carillet.**

FIGURE 1.40 Terraria: glass containers maintaining ornamental plants in an isolated, warm, high-humidity environment.

*Source:* **(a) Photo by Marietta555 (CC BY SA 4.0). (b) Photo by Shutterstock/Hamik.**

FIGURE 1.41 Flowers of the daisy family (Asteraceae). The top row shows flowers in which only the outermost florets have prominent petals (the stereotypical "daisy flower"). The bottom row shows "doubled flowers," in which all or almost all the florets have prominent petals (and usually are sterile, incapable of producing seeds, as a result). All plants are garden hybrids (not identifiable as a particular species): (a) Gerbera (*Gerbera*). (b) Echinacea (*Echinacea*). (c) Black-eyed Susan (*Rudbeckia*). (d) Gazania (*Gazania*). (e) Gerbera (*Gerbera*). (f) Chrysanthemum (*Chrysanthemum*). (g) Dahlia (*Dahlia*). (h) Marigold (*Tagetes*).

***Source:* (a) Photo by ThomasWolter from Pixabay (public domain). (b) Photo by MatthiasBoeckel from Pixabay (public domain). (c) Photo by Joshua Mayer (CC BY 2.0). (d) Photo by Mamjodh (CC BY 2.0). (e) Photo by ThomasWolter from Pixabay (public domain). (f) Photo by Liz West (CC BY 2.0). (g, h) Photos by Adalhelma from Pixabay (public domain).**

### 1.14.4 Size of Flower Display

Plants can attract pollinators by producing a mass of small flowers (florets) or a restricted number of large flowers. Ornamental plants, similarly, are attractive to people by using one or the other strategy. Alliums, baby's breath, crab apple, celosia, cherries, lavender, lilac, and lily of the valley exemplify the production of a very large number of very small flowers. By contrast, huge but relatively few flowers are developed by angel's trumpet, hibiscus, magnolia, sacred lotus, trumpet creeper, and water lily. Probably, most people enjoy gardens in which both kinds of flower displays are present.

### 1.14.5 Woody Plants That Produce Good Flowering Displays

Trees and shrubs represent major commitment to one's property, since they sometimes outlive their owners and can dominate a space, preventing other uses. Careful choice of species and cultivars and attention to their location can prevent damage to property. Plants that shed fruits on pathways are a particular nuisance. However, woody plants especially enhance the appearance and value of property and should be a priority. In temperate-climate areas, crab apples, lilacs, hydrangeas, and rhododendrons ("azaleas") are very dependable and, indeed, are common choices for those who appreciate flowering displays. A much greater range of choices becomes available as the growing climate becomes more hospitable southward, and indeed, one of the few benefits of climate change is that, in some regions, a greater range of plants will be growable. There is always temptation to grow warm-season trees like magnolias and shrubs like butterfly bush in locations beyond their traditional northern limit, and one can only applaud brave gardeners willing to take the risk. For most people, advice regarding cultivars that are suited to local conditions is invaluable.

FIGURE 1.42 Several of the world's most beautiful flowering trees: (a) Jacaranda (*Jacaranda mimosifolia*) in Australia (confined to tropical and semi-tropical climates). (b) Flowering cherries (*Prunus*) in Japan (suitable for temperate climates). (c) Royal poinciana (*Delonix regia*) in Australia (confined to tropical and semi-tropical climates).

***Source:* (a) Photo by Shutterstock/Chris de Blank. (b) Photo by Shutterstock/Sean Pavone. (c) Photo by John (CC BY 2.0).**

### 1.14.6 Vines That Produce Good Flowering Displays

Ten of the 100 flowering ornamentals treated in this book are climbing vines (bougainvillea, clematis, jasmine, mandevilla, morning glory, nasturtium, passionflower, sweet pea, trumpet creeper, and wisteria), and climbing/scrambling varieties of roses are available. Vines are relatively troublesome to cultivate, since supporting structures like trellises need to be provided, the stems have to be trained to grow in desirable directions, and they also need to be pruned for appearance and to prevent excessive growth. Some vines are capable of physically damaging houses and, indeed, plants on which they grow. However, small vines display well in hanging baskets, which do not require much effort. Flowering vines are among the most attractive of all plants, and some provide extraordinary beautification of gardens and buildings.

## 1.15 FLOWERS AND GENDER

### 1.15.1 Flowers and Women

Differences in preferences between male and female humans may be due to genetics or social constraints (especially learned behaviors and values). Regardless of why,

FIGURE 1.43 Spectacular mixed bougainvillea vines over a gate entrance in Canary Islands, Spain. Bougainvillea is widely considered to be the premier flowering vine, although its tropical nature limits where it can be grown as a perennial outdoors.

*Source:* **Photo by Shutterstock/Dmitry Morgan.**

FIGURE 1.44 Use of flowers by women for personal adornment: (a) Hawaiian hula dancer wearing floral leis. (b) Girls wearing floral headdresses. (c) Wrist corsages.

*Source:* **(a) Photo by Shutterstock/Milkovasa. (b) Photo by Shutterstock/Nina Buday. (c) Photo by StretchyBill (CC BY 2.0).**

FIGURE 1.45 Paintings of young women by French artist Émile Vernon (1872–1920), who characteristically showed beautiful females associated with beautiful flowers.

*Source:* **All figures are public domain.**

FIGURE 1.46 "Most beautiful women of the world," their beauty enhanced by flowers: (a) *Helen of Troy*, painting by English artist Evelyn De Morgan (1855–1919). In Greek mythology, Helen was said to be the most beautiful woman in the world. The artist has embellished her beauty with roses, claimed to be the most beautiful of flowers. (b) Tripura Sundari, a Hindu goddess whose name means "the most beautiful woman across the three realms" (heaven, earth, and hell). In this painting, she is surrounded by various exotic flowers.

*Source:* **(a) Public domain. (b) Photo by Ckvicky1992 (CC BY SA 4.0).**

women place much more value on flowers than men do, and it has been shown that this difference is already very strongly developed between young girls and young boys. Women are responsible for most purchases of ornamental plants (Russia is a notable exception, with both sexes buying flowers equally). Indeed, dramatically more than any other class of economic plants, the cut flower industry is oriented toward female consumption. Not only do women love actual flowers; they also love floral prints – particularly on clothing, but also on a huge variety of consumable products, especially those associated with femininity. Flowers are the leading present for women, and floral motifs typically decorate gifts intended for women. It is probable that flowers have become part of the social "sexual signaling" that is most evident in manner of dress. By convention (at least in Western society), conservative dress for men is dominated by black, blue, and gray, but (notwithstanding the popularity of the "little black dress") women dress far more colorfully. Women appear to be instinctively aware that their beauty and attractiveness are accentuated when associated with beautiful, colorful flowers.

It has been speculated that the human attraction to flowers traces to their value in indicating to our ancient ancestors that a habitat had good potential for resources. Wildflower species come into flower at particular times, so they are indeed indicators of when they and other species, particularly edible tubers and fruits, could be harvested. Honey produced by bees would also become available when flowers are abundant. Women, traditionally gatherers of plant resources from nature (while men were primarily hunters), may have evolved special sensitivity to flowers. Women are well-known to be more sensitive to odor than men and seem to appreciate pleasant fragrance more. In the past, when humans were primarily nomadic, finding temporary campsites where flowers indicated food was abundant would have been particularly beneficial for childbirth. This may explain the much greater appreciation of beautiful flowers by women and, consequently, the value of flowers as gifts from men to women. It has been demonstrated that the presence of attractive flowers lowers stress, and an offering of flowers by a man to a woman almost certainly has this effect. For additional discussion of why people love flowers, see Hůla and Flegr (2016), cited later.

### 1.15.2 Flowers and Masculinity

As discussed in the preceding section, women are immensely more appreciative of ornamental flowers than men. Delicate, extremely attractive flowers naturally represent characteristics considered to be feminine. By the Victorian era, violets represented the qualities of an ideal wife: humility, faithfulness, and modesty. Subsequently, as noted in the following discussion, it appears that, for men, flowers have come to represent qualities unrepresentative of masculinity, and accordingly, there is now a widespread male reluctance to associate personally with flowers.

FIGURE 1.47 Gentlemen wearing boutonnières.

*Source:* **(a) Photo by Shutterstock/Mary981. (b) Photo by Shutterstock/Tsezarina.**

FIGURE 1.48 A very rare example of men conspicuously wearing flowers: a carnival in Kranj, Slovenia, in 2009.

*Source:* **Photo by Ekke (CC BY 2.0).**

In ancient times, a flower worn on one's coat was viewed as a talisman protecting against evil spirits, odors, and disease. A descendant of this practice is the boutonnière, the only decorative floral apparel regularly worn by men and the equivalent of a corsage for women. The word is French for "buttonhole," because the flower (traditionally a white carnation for formal occasions) is affixed by sticking the stem through a designated buttonhole on the left lapel (although today usually just pinned to the lapel, because modern jackets are often made without a loop on the back of the lapel to hold the flower, and sometimes even without a buttonhole). Boutonnières have become uncommon today, mostly reserved for marriage ceremonies.

Irish literary figure Oscar Wilde (1854–1900), who was gay, habitually wore a green boutonnière, and ever since the symbol has been used to suggest homosexuality (although dyed green boutonnières are sometimes worn on St. Patrick's Day simply to reflect the association of the color with Ireland). Prejudice against male homosexuals has often been expressed by pejorative slang terms based on flowers, reflecting the perceived lack of masculinity associated with them. In particular, "pansy" refers to a weak or effeminate male, or a male homosexual, meanings which appear to have become prominent about a century ago. Other flowers, notably daisies, lavender, and violets, also have been associated with male homosexuality. The bright colors of flowers are now represented in the spectral rainbow that has become symbolic of LGBTQ+ gender rights.

Given the lack of male enthusiasm for ornamental blooms, it comes as a surprise to many that during the two world wars, hundreds of combat vessels were named after flowers. In particular, the British Royal Navy constructed about 300 so-called corvettes, many designed for minesweeping and escort operations near ports (they could navigate in waters less than 4 m deep), especially during the Second World War. These were called "flower-class" ships because all of them were assigned unique flower names like buttercup, geranium, pansy, petunia, poppy, and rosemary.

### 1.15.3 Flowers and Romance

Men widely recognize the value of flowers for women, and the unique social situation where it is considered normal (indeed, commendable) for men to purchase and carry flowers is as gifts for women.

FIGURE 1.49 Men carrying flowers, presumably as gifts for women.

*Source:* **(a) Photo by Shutterstock/PeopleImages.com-Yuri A. (b) Photo by Shutterstock/Volodymyr Tverdokhlib.**

FIGURE 1.50 Flowers play prominent roles in courtship: (a) "The Bouquet," by English illustrator Arthur R. Rackham (1867–1939), from the periodical *The Delineator* (New York) in 1924. (b) "Bouquet de primevères" (1893), by French artist Alfred Guillou (1844–1926). (c) Hindu deity Lord Krishna courting his beautiful consort Radha, goddess of love, tenderness, compassion, and devotion, by adorning her with a garland of flowers.

*Source:* **(a) public domain. (b) Photo by Grégory Lejeune (public domain). (c) Illustration from Infinite Eyes (public domain).**

Flowers are the essential organs of sexual reproduction of higher plants, designed to combine male and female elements. How appropriate, therefore, that humans have adopted flowers as the chief symbols of their own reproduction! They are a regular expression of affection during courtship (flowers say the sweetest things!). Weddings are a florist's best sales opportunity. Most marriage ceremonies feature a bridal bouquet, bridesmaid bouquets, boutonnières, corsages, flower girl flowers, large floral arrangements for the altar and reception halls, bouquets for the dining tables, and flowers for the cake.

## 1.16 FLOWERS AND THE WELFARE OF THE ENVIRONMENT

### 1.16.1 Flowers and Pollinators

Close to 90% of flowering plants are at least partly dependent on animal pollination for optimum seed set, while

FIGURE 1.51 Royal British weddings accompanied by flowers: (a) Princess Diana and Prince Charles, 1981. Her wedding bouquet included stephanotis, freesias, gardenias, lily of the valley, odontoglossum orchids, and a golden yellow rose. (b) Kate Middleton and Prince William, with bridesmaids and page boys. Her wedding bouquet featured lily of the valley, sweet William, and hyacinth.

*Source:* **(a) Photo by Joe Haupt (CC BY SA 2.0). (b) Photo by Toby Jagmohan (CC BY ND 2.0).**

FIGURE 1.52 Pollinator–flower posters: (a) Butterfly and moth pollinators. (b) A variety of different animal pollinators.

*Source:* **Public domain. (a) United States Department of Agriculture – Natural Resources Conservation Service. (b) Postermade available by the US Forest Service, US Fish and Wildlife Service, Bureau of Land Management, Natural Resources Conservation Service, US Botanical Gardens, and the NAPPC (North American Pollinator Protection Campaign). Artist: Paul Mirocha.**

almost 20% use other means, primarily wind, for pollination. Hundreds of thousands of species of animals, mostly insects, but also some mammals and birds, are pollinators. Pollination is essential for the productivity of cross-pollinated fruit and seed crops. Pollinators, especially native bees, such as bumblebees, are needed for one-third of the crop species that humans consume, although the major cereals (wheat, rice, corn, etc.) are wind-pollinated. Of course, most wild plants also cannot afford to lose their pollinators, or they would become extinct. In recent years, there has been a worldwide decline in wild pollinators, attributed to the loss of suitable habitats, introduction of harmful species, and use of pesticides. Moreover, managed bees (mostly honeybees) are also in decline because of diseases. The current catastrophic global decline in pollinators is known as the "pollinator crisis" and has generated great concern. Gardeners can help alleviate the growing scarcity of flowers by growing species that suit the needs of local pollinators and by avoiding pesticides that endanger them.

### 1.16.2 Invasive Ornamental Plants

Beautiful ornamental plants have an ugly side: Many have become invasive weeds, harming native biodiversity, increasing the costs of agricultural crop and agroforestry production, and damaging the property of people. In Australia, two-thirds of the continent's invasive plants originated from introduced ornamentals. The problem is less significant in Northern countries, such as Canada, because so many ornamentals originate from much hotter countries and cannot survive the winter. Nevertheless, more than a quarter of Canada's alien species (about one in every eight plant species growing in the country) were introduced as new ornamental plants. Quite unlike many unattractive animal invasives (such as insects, snakes, and rats), most plants do not have features considered to be ugly by people, and introduced weeds with attractive flowers often meet little resistance despite the considerable damage that they cause.

FIGURE 1.53 Beautiful garden ornamentals that have escaped from cultivation and become troublesome invasives: (a) Lupin (*Lupinus polyphyllus*) in New Zealand. (b) Purple loosestrife (*Lythrum salicaria*) in Finland.

*Source:* **(a) Photo by Shutterstock/Andrii Slonchak. (b) Photo by Shutterstock/Antero Topp.**

### 1.16.3 Chemical Pesticides and Ornamental Plants

All plants are harmed by animals, microorganisms (fungi, bacteria, viruses), and other plants (especially weeds), and the most common protection provided is the application of pesticides. Such chemicals are usually toxic not just to the target organisms but also to other species, sometimes including humans. Gardening is the world's most popular pastime, and numerous gardeners employ over-the-counter pesticides. As well, professional horticulturalists extensively employ especially toxic pesticides, and despite the necessity of professional training to apply these, enormous quantities are being applied. A very concerning environmental issue is the extensive (indeed, excessive) use of pesticides in many Third World nations now producing ornamental plants for international trade. This is not only an issue of human health and environmental health but also represents unfair competition with countries that strictly

FIGURE 1.54 Pests in the garden. Inevitably, pests and pathogens affect garden plants. Ideally, minor damage should simply be tolerated, but if control measures are adopted, they should not damage the ecology of the site or harm innocent, non-targeted species.

*Source:* **Figures by Shutterstock/White Space Illustrations.**

FIGURE 1.55 Excessive use of toxic chemical pesticides to protect ornamental plants.

*Source:* **Prepared by Jessica Hsiu.**

control pesticide use. In advanced Western countries, integrated pest management principles, including the use of biological control agents, are now widely established in the ornamental plant industry, partly because of more stringent regulations regarding pesticides, partly due to increasing pesticide resistance in some pests, and partly due to regulations regarding exports to other countries.

### 1.16.4 Use of Native Plants as Ornamentals

Environmentalists are encouraging increased utilization of native plants for landscape purposes, because they are naturally adapted to local conditions (climate, soil, predators, and pollinators) and usually require less care, water, fertilizer, and pest control agents. However, virtually all the 100 plants featured in this book are mostly or entirely grown where they are not native. The reason local plants are uncommonly grown is simple: In the opinion of most people, they are much less attractive than plants of foreign origin. Requesting people to grow less-attractive local plants is like asking them to eat less-tasty food, wear less-attractive clothes, and bathe less frequently. Such practices may be healthier, more ethical, and wiser but are unlikely to persuade the majority. Nevertheless, in the light of how much damage has been associated with the cultivation of flowering ornamental plants, it makes sense to point out the ecological advantages and disadvantages of growing particular species in certain regions, where they are potentially invasive. Also, environmental issues are destined to become increasingly acute, so it is important to understand the considerations that can determine wiser ecological choices. Generally, planting trees and long-lived shrubs is better for the environment than cultivating annuals (far less energy expenditure is required). Species and cultivars that require minimal water and fertilization are highly desirable. Plants that feed pollinators are often quite beautiful, and most gardeners welcome visits by butterflies and hummingbirds. Growing heirloom varieties, often supplied by people dedicated to maintaining rare plants, is another way of showing concern for the welfare of the world. Growing uncommon ornamentals so that a garden is quite heterogenous is a simple way of discouraging the growing disease problems associated with major ornamentals. (In agriculture, large monocultures, that is, large plantings of a uniform cultivar, are obvious targets for pests, and the same is true for ornamental plants.) As noted earlier, for residential cultivation of ornamentals, the goal can and should be to eliminate the use of synthetic pesticides, and to encourage tolerance of minor damage by pests, and native plants are usually very resistant to most local pests. Often, the most attractive wild plants are rare (this is frequently the case for wild orchids). It needs to be emphasized that such uncommon wild plants should never be transplanted to personal gardens, as this could endanger them in the wild. Nurseries are increasingly available that can supply starter plants and seeds of wild plants that are appropriate for local cultivation.

FIGURE 1.56 Use of native plants as flowering ornamentals: (a) A nursery specializing in the sale of local native plants in Texas. (b) A native plant garden in San Diego, California, with sunflower, mallow, fuchsia, and sage in flower.

***Source:*** **(a) Photo by Valley Nature Center (CC BY SA 4.0). (b) Photo by Arlene Schag (public domain).**

## 1.17 TOXICITY AND EDIBILITY OF FLOWERS

### 1.17.1 Edibility of Flowers

Probably the most consumed flower is cauliflower, although it is not usually understood to represent floral tissue (the similar broccoli includes not just undifferentiated flowers but also the stalks of the inflorescence). The flowers of some other vegetables provide excellent culinary fare, particularly those of okra, pumpkins, and squashes, and some chrysanthemums are consumed in Japan. Herbal teas are frequently prepared with flowers, most notably those of chamomile species. Rose petals are widely used in jam, tea, and numerous other food preparations. Nasturtium (*Tropaeolum*, not the genus *Nasturtium*) flowers are often consumed. Violet (or pansy) flowers have frequently been candied. Other species whose flowers have been consumed include begonia, carnation, gladiolus, lavender, lilac, and tulip (but note that certain of these species have toxic non-floral parts, and this is sufficient for some to recommend against consumption of any parts of the plants). Often, some species of a genus or some cultivars of a species are exceptionally tasty, whereas most are not. Usually, advanced cultivars with doubled flowers are much less tasty than older forms with simpler, natural flowers. The edible flowers of most species that possess them are usually not

FIGURE 1.57 Edible flowers: (a) A worker preparing and packing edible flowers for commercial sale. (b) A plate of assorted edible flowers.

***Source:*** **(a) Photo by Shutterstock/Trading Ltd. (b) Photo by Shutterstock/June Marie Sobrito.**

outstandingly pleasant in taste, and flowers are often incorporated into meals simply because of their appearance and the allure of consuming exotic fare. Some greenhouse firms specialize in the production of edible flowers, especially for high-end restaurants. It is interesting to note that, like the use of parsley as a garnishment for meals, flowers seem to be incorporated into many meals more for their ornamental than their food value. There are dangers in consuming flowers: wrong identification (common names can be very misleading), allergies (particularly to pollen), and the fact that information on the toxicity of most flowers is extremely limited. "Lily" flowers require particular mention. Daylily (*Hemerocallis*) flowers are traditional in Chinese recipes. However, lily of the valley (*Convallaria majalis*), glory lily (*Gloriosa superba*), calla or arum lily (*Zantedeschia aethiopica*), peace lily (*Spathiphyllum* spp.), and some other so-called "lilies" are quite toxic. Tiger lily (*Lilium lancifolium*) is occasionally recommended as an edible flower but is sometimes stated to be toxic. For extensive information on edible flowers, see Small (2006), cited later.

FIGURE 1.58 Castor bean (castor oil plant; *Ricinus communis*), reputedly the world's most toxic plant: (a) An ornamental plant in seed in a yard in Florida. (b) The toxic seeds. The seeds are 9–17 mm long, with a mottled coat and an "elaiosome" at the base (an edible appendage that attracts ants and other insects which distribute the seeds). The deadly toxin ricin is released if the seeds are chewed.

***Source:*** **(a) Photo by Julie Corsi (CC BY 2.0). (b) Photo by H. Zell (CC BY SA 3.0).**

### 1.17.2 Toxicity of Flowering Ornamentals

Numerous plants are poisonous, and this includes many grown as ornamentals in and around homes; note Table 1.2. While most people are wary of eating wild plants, the need for caution in a garden is less apparent. However, foraging for edible plants in a garden can be as dangerous as collecting in the wild. Numerous plants can cause dermatitis from contact, poisoning from ingestion, and allergic reactions from either route. On the whole, toxicity of ornamental plants is not a major threat to health, but it is one that bears mention because children and pets are especially at risk,

TABLE 1.2
**Examples of Toxic Ornamental Plants Discussed in This Book That Can Affect Humans**

| Plant | Part | Effects |
|---|---|---|
| Amaryllis (*Hippeastrum*; not *Amaryllis*, which is also toxic) | Bulb | Gastric irritation |
| Angel's trumpet (*Brugmansia*, but the same is true for *Datura* species) | All parts | Delirium, convulsions |
| Azalea (*Rhododendron*) | All parts | Nausea, vomiting, depression, difficulty breathing, prostration, coma; potentially fatal |
| Cyclamen (*Cyclamen persicum*) | Corm | Gastric irritation |
| Daffodil (*Narcissus*) | Bulbs | Nausea, vomiting, diarrhea; potentially fatal |
| Delphinium, larkspur (*Delphinium*) | All parts | Digestive upset, nervousness, depression; potentially fatal |
| Foxglove (*Digitalis purpurea*) | Leaves | Dangerously irregular heartbeat and pulse, digestive upset, mental confusion; potentially fatal |
| Hyacinth (*Hyacinthus orientalis*) | Bulbs | Nausea, vomiting, diarrhea; potentially fatal |
| Iris (*Iris*) | Bulbs | Severe digestive upset |
| Jasmine (*Jasminium*) | Berries | Digestive disturbance, nervous symptoms; potentially fatal |
| Lily of the valley (*Convallaria majalis*) | Leaves, flowers | Irregular heartbeat and pulse, digestive upset, mental confusion |
| Oleander (*Nerium oleander*) | Leaves, branches | Affects heart; severe indigestion; potentially fatal |
| Poinsettia (*Euphorbia pulcherrima*) | Sap, leaves, stems | Mild dermatitis in some people, slight potential for gastric irritation, but only if large amounts are consumed; reports of toxicity greatly exaggerated |
| Sweet pea (*Lathyrus odoratus*) | Seeds, seedlings, pods | Slow pulse, respiratory distress, convulsions |
| Tulip (*Tulipa*) | Bulbs | Abdominal pain, sweating, nausea, vomiting, skin irritation; considered to have limited toxicity, but potentially fatal if large amounts are ingested |
| Wisteria (*Wisteria*) | All parts, especially seeds, pods | Vomiting, diarrhea, gastric irritation |

FIGURE 1.59 Ornamental plants toxic to pets: (a) Houseplants poisonous to cats. (b) A Shetland sheepdog in front of azalea bushes, the plants that are considered most likely to poison dogs.

*Source:* **(a) Photo by Shutterstock/New Africa. (b) Photo by Shutterstock/KE Walker.**

and fatalities are possible. Location, both of houseplants inside the home and plants in the yard, is important to limiting exposure to children and pets. Most accidental human poisoning by ingestion of plants occurs in children under the age of 5, but even in this young age group, poisoning from plants is rare compared to poisoning from common substances (such as drugs and cleaning materials) in and around the home. Information on both edibility and toxicity is provided for particular plants in the following chapters. However, it is important to appreciate that people differ greatly in their susceptibility, and that repeated contact with particular species (as occurs with florists) can result in increased sensitivity.

Should flowering plants be in sick rooms? Beautiful flowers have the power to raise spirits. Plants can cleanse some pollutants out of the air and generate oxygen during the day, but it is doubtful that there are significant benefits. On the negative side, highly scented plants could cause allergies. In theory, pollen from flowers could also cause allergies, but attractive flowers, if they even have pollen, are usually not wind-pollinated, and the pollen is too heavy to be transported by wind currents in rooms. Summerbell et al. (1989) concluded that pathogenic fungi in indoor plant soils constitute a serious hazard for immunosuppressed patients. However, gifts of flowers are mostly cut flowers, which are maintained in vases of water. All things considered, for most people, plants are welcome companions during periods of illness.

Cats and dogs are the leading pets for which poisoning from plants has been reported; note Table 1.3. Dogs are much less likely than cats to be poisoned by ornamental plants. A study of poison control reports in the United States found that of about 116,000 poison occurrences in dogs, about 11% were plant-based, while of about 19,500 poison incidents involving cats, more than 25% were plant-based (see Knight 2006, cited later). Lilies (notably Easter lilies) are often considered to be the ornamental plants most frequently causing death or serious illness for cats. Cats eat a wider variety of plants than dogs, and when kept indoors most of the time, they might find houseplants attractive. Azaleas (*Rhododendron*) may be the ornamental plants most frequently causing serious illness for dogs, which, however, are most often poisoned by consuming chocolate.

**TABLE 1.3**
**Examples of Ornamental Plant Genera Discussed in This Book That Are Toxic to Dogs, Cats, and Other Pets, Showing Organ System Affected**

| Genera | Organ System Affected |
|---|---|
| *Allium* (allium) | Blood |
| *Amaryllis* (amaryllis) | Digestive |
| *Brugmansia* (angel's-trumpet) | Nervous, ocular |
| *Cyclamen* (cyclamen) | Digestive |
| *Digitalis* (foxglove) | Digestive, cardiac |
| *Gladiolus* (gladiolus) | Digestive |
| *Hyacinthus* (hyacinth) | Digestive |
| *Ipomoea* (morning glory) | Nervous |
| *Iris* (iris) | Digestive |
| *Kalanchoe* (kalanchoe) | Digestive, cardiac |
| *Lilium* (lily) | Renal |
| *Narcissus* (daffodil) | Digestive |
| *Nerium* (oleander) | Digestive, cardiac |
| *Rhododendron* (rhododendron) | Digestive, cardiac |
| *Tulipa* (tulip) | Digestive |
| *Zantedeschia* (calla lily) | Digestive |

*Source:* After Burrows and Tyrl (2001).

## 1.18 ECONOMICS OF ORNAMENTAL FLOWERING PLANTS

### 1.18.1 Floriculture in the World

Like other commodities and services, economists measure gross value in terms of annual domestic production and/or sales, as well as exports and imports. Sometimes employment statistics (e.g., number of florists) and acreages (including greenhouses) devoted to particular crops are useful indicators of value.

As with most specialized disciplines, the ornamental plant industry has its own special vocabulary, including jargon terms (see Table 1.4 for an explanation of some key words). The meaning of "ornamental plant" is straightforward: a plant or plant part grown or displayed for its attractive appearance. The ornamental plant trade includes plants grown not only for the appearance of their flowers but also for foliage (this is the case for most shrubs and trees and for most houseplants). Foliage for floral bouquets in many countries is collected mostly from nature (and is often included in "non-timber forest products," or NTFPs, and sometimes in a category termed "cut greens"). These considerations complicate evaluating plants grown mainly for their flowers (the subjects of this book), as most statistics do not distinguish ornamental plants grown for floral display from plants grown for foliage display. However, the majority of ornamental plants in trade are valued mainly for their flowers, and statistics for the cut flower trade are almost entirely based on flowers. Cut flowers are the most important sector of international trade in ornamental plants (followed by potted plants, particularly starter bedding plants). The world cut flower market was valued at about $36 billion

**TABLE 1.4**
**Key Horticultural Terminology Relevant to Ornamental Flowering Plants**

| Term | Description |
|---|---|
| **Bedding plants** | Plants that are naturally annual or treated as annuals, grown outdoors for one season. Bedding plants can be grown as individual display plants, as potted plants, and in hanging pots, but a frequent recommendation is to grow them as massed plantings (or in borders) that feature a particular color. Examples: begonia, geranium, impatiens, marigold, pansy, petunia. |
| **Cultivar** | A genetically distinctive named (in part with a nonscientific, that is, non-Latin, word or phrase) assemblage of plants, usually selected by plant breeders or horticulturists and maintained in cultivation for its usefulness to people. Names are of the nature of *Rosa canina* 'Magnificent Red', or *Rosa* 'Magnificent Red', or rose 'Magnificent Red'. Naming of cultivars is governed by the *International Code of Nomenclature for Cultivated Plants* (Brickell et al. 2016). |
| **Cut flowers** | Cut inflorescences (flowering stems) sold individually as "stems" (especially roses) or in bunches, arrangements, or corsages. |
| **Floral design (floral arts)** | The art and trade of creating flower arrangements in vases, bowls, baskets, and other containers and making bouquets, wreaths, corsages, and more complicated arrangements of ornamental plant materials. (Sometimes, *floral design* is considered synonymous with *floristry*.) Ikebana (Japanese flower arrangement) is a particularly disciplined art form. |
| **Floriculture** | A branch of horticulture concerned with the production of showy plants for sale. |
| **Floriculture crop** | Any crop used for ornamental display; includes cut flowers, bedding plants, potted flowering and foliage plants, material grown for greens, and vegetative material harvested for propagation. Woody plants (shrubs and trees grown in nurseries) are usually excluded. Floricultural crops (e.g., ferns) do not always have flowers. |
| **Floristry** | The cultivation of ornamental plants, their arrangement, and their sales. (Not to be confused with *floristics*, the geographical study of distribution and evolutionary spatial relationships of plant species.) |
| **Foliage plants** | Ornamental plants with relatively inconspicuous flowers and fruits grown for their attractive stems and leaves. The phrase includes potted ferns and fern allies (which lack flowers and fruits). |
| **Greens (cut foliage, decorative foliage)** | Ornamental stems, branches, or leaves used as filler or background in cut flower arrangements and bouquets. Greens include both cultivated and wildcrafted (wild-collected) material. |
| **Nursery** | An area where young plants are grown to establish stock for transplanting. In hot climates, floricultural nurseries can grow all classes of plants outdoors, but in cold climates, the ornamental nursery trade concentrates on woody, cold-hardy plants (bushes and trees) and sod for lawns, and the ornamental greenhouse trade on annuals and on planting stock for herbaceous gardens. |
| **Ornamental plant** | Any plant employed primarily to beautify, decorate, or enhance the appearance or functionality of the human-made environment. This includes floricultural crops (as defined earlier), but also turf grass and many shrubs and trees, whether grown for their attractive flowers or for shade, hedges, or landscaping. Vines grown for foliage are also usually included here. Christmas trees are frequently excluded but are best considered as ornamental plants because, in principle, they are like cut flowers, but on a grander scale. Plants grown for food are excluded, although there is considerable interest today in "edible landscaping." |

FIGURE 1.60 Commercial flower production greenhouses: (a) Geraniums in Slovenia. (b) Orchids in The Netherlands.

*Source:* **(a) Photo by Jernej Furman (CC BY 2.0). (b) Photo by Shutterstock/Gertjan Hooijer.**

FIGURE 1.61 Commercial retail ornamental flower sales outlets.

*Source:* **(a) Photo by Shutterstock/Kurhan. (b) Photo by Shutterstock/Monstar Studio.**

in 2022 and predicted to grow to $56 billion in 2032. By a considerable margin, roses are the dominant cut flower.

Commercial production of flowering plants for sale is strongly determined by climate and labor costs, both of which, in turn, are influenced by geography. Greenhouse cultivation of ornamentals in Northern nations during the winter is very challenging, especially in view of increasing costs for energy, especially for heat. Unfortunately, many ornamental plants are not well suited to grow in the low light levels of winter, even with the use of light supplementation, which is expensive. Until fairly recently, export trade in ornamental plants was mainly from developed nations, particularly including Europe, Japan, and North America. In recent decades, much of the trade has shifted to Third World areas with climates with very long growing seasons and cheap labor costs, notably, Ecuador, Ethiopia, Kenya, Columbia, and India, and to a lesser extent, South Africa and Malaysia. Israel also accounts for a small proportion of the floral trade. China is expected to become a major player (it is one of the world's largest producers of flowers but currently exports relatively little). Columbia (and, to a lesser extent, Ecuador) is the major exporter of cut flowers to North America. In Europe, most imported flowers sold there are from equatorial East Africa (mostly Kenya and Ethiopia).

While hundreds of ornamental flowering plants are grown commercially, a small number dominates international trade. Roses and tulips are the two giants of the world of commercial flowers. The plants listed in Table 1.5 have proven to have long-term popularity as cut flowers in international commerce. However, trade is variable from year to year, and some plants are much more popular in some countries than in others. Other plants that are very important internationally include gladiolus, anthurium, and aster.

In addition to the cut flower sector, potted plants are also exchanged internationally, albeit to a much lesser degree. Most potted plants are grown mainly for their foliage. Of the plants grown specifically for their flowers, the popular species include kalanchoe, African violet, chrysanthemum, spathiphyllum, hyacinth, rose, and primula.

The widespread popularity of a small number of plant species reflects two facts: (1) To be successful commercially, ornamental plants must be both attractive to consumers and have features that facilitate their efficient growth by producers and sale by retailers; plants that meet these criteria are likely to be successful everywhere, provided that marketplace features are comparable. (2) In fact, markets for many products have become very similar among countries purchasing ornamental plants. Most international trade is to well-off First World consumers, who live in an amazingly homogenous marketplace with the same products offered almost everywhere (for example, note the near-universal presence of the major fast-food chains).

**TABLE 1.5**

**Ornamental Plants Usually Listed as Members of the Top 10 Cut Flowers in International Trade**

| |
|---|
| Alstroemeria |
| Carnation |
| Chrysanthemum |
| Orchids (phalaenopsis, cymbidium, etc.) |
| Freesia |
| Gerbera |
| Gypsophila |
| Lily |
| Rose |
| Tulip |

## 1.18.2 Floriculture in the United States

The United States is the world's largest producer and the world's largest consumer of ornamental plants (Netherlands, Columbia, and other countries are large producers, but primarily for export). Statistics are available from the US National Agricultural Statistics Service (www.usda.gov/media/blog/tag/national-agricultural-statistics-service; note Quick Stats page: https://quickstats.nass.usda.gov/ and highlights: www.nass.usda.gov/Publications/Highlights/2022/Floriculture_Highlights_07.pdf). (Note: Unless stated otherwise, $ figures are wholesale.) In the United States, the total value of annual sales of domestically produced material across all floriculture crops is about $7 billion (see below for an analysis by categories, most of which include both plants grown for flowers and plants grown for

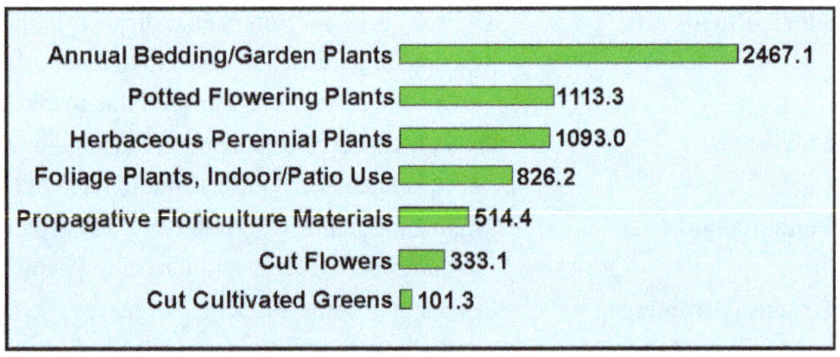


FIGURE 1.62 Wholesale value (in million dollars) of floricultural crops by category for 2022 (based on operations with $100,000 in annual sales).

***Source:*** **US Department of Agriculture. 2023, May. *National Agricultural Statistics Service 2022 NASS Highlights. 2022 Floricultural Crops.* www.nass.usda.gov/Publications/Highlights/2023/Floriculture_Highlights.pdf.**

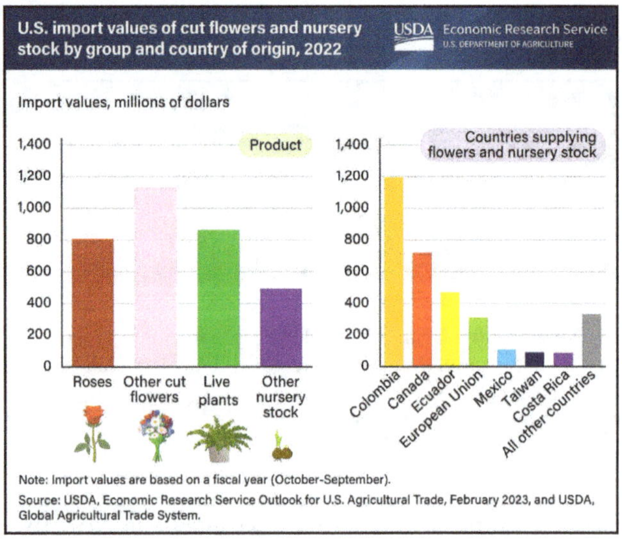


FIGURE 1.63 Value of imported cut flowers and nursery stock into the United States in 2022.

***Source:*** **US Department of Agriculture (public domain).**

foliage). The United States annually imports ornamental plants valued at over $3 billion, mostly from Columbia, Ecuador, and Canada. The majority of these are cut flowers, valued at almost $2 billion annually. Three-quarters of US consumption of cut flowers is imported. California is America's leading cut flower producer, followed by Florida. However, foreign imports dominate the US flower market. Colombia is the chief exporter, especially of roses, carnations, spray chrysanthemums, and alstroemeria. Ecuador is second, with large exports of roses, delphiniums, asters, baby's breath, and mixed bouquets. In 2022, Colombia and Ecuador accounted for about 90% of all roses, 98% of all carnations, and 95% of all chrysanthemums sold in the United States. Table 1.6 provides sales figures for cut flowers in the United States and, accordingly, is a measure of the relative popularity of cut flower species in the United States (similarly, Table 1.7 is a measure of the relative popularity of potted and garden flower species in the United States).

See Table 1.7 for the annual sales value of outdoor flowering plants, and Table 1.8 for the annual sales value of indoor flowering plants.

### 1.18.3 Floriculture in The Netherlands

The Netherlands is the world's largest exporter of ornamental plants, chiefly cut flowers and bulbs for many perennials overwintering as bulbs, most notably tulips. The country makes up only 10% of the world's total production

**TABLE 1.7**
**Retail Value in 2022 of Top-Selling Domestically Produced Flowering Bedding Plants in the United States**

| "Annual Bedding Plants" (In Flats, Pots, and Hanging Baskets) | | "Herbaceous Bedding Plants" (In Pots) | |
|---|---|---|---|
| **Species** | **Sales ($000s)** | **Species** | **Sales ($000s)** |
| Begonia | 184,018 | Chrysanthemum | 242,802 |
| Calibrachoa | 68,811 | Daylily | 26,705 |
| Geranium | 183,346 | Dianthus | 211,407 |
| Impatiens (except New Guinea hybrids) | 83,260 | Ferns | 33,578 |
| New Guinea hybrids | 30,631 | Hosta | 62,046 |
| Marigolds | 70,672 | Lavender | 26,082 |
| Pansy/violets | 155,281 | Peony | 19,007 |
| Petunia | 227,497 | Salvia | 22,270 |
| Vinca | 69,634 | Sedum | 23,051 |
| Others | 946,262 | Others | 616,003 |

*Source:* Based on USDA's National Agricultural Statistics Service, Floriculture Crops (https://quickstats.nass.usda.gov/).

**TABLE 1.6**
**Wholesale Value of Cut Flowers in the United States**

| Crop | 2021 Sales ($1,000s) | 2022 Sales ($1,000s) |
|---|---|---|
| Asters | 7,422 | 6,099 |
| Pompon chrysanthemums | 4,075 | 3,908 |
| Dahlias | 5,832 | 4,533 |
| Delphinium and larkspur | 3,828 | 2,309 |
| Gerbera daisy | 22,247 | 1,890 |
| Gladioli | 14,952 | 15,569 |
| Iris | 6,678 | 6,528 |
| Lilies, all | 58,308 | 55,875 |
| Lisianthus | 2,839 | 2,335 |
| Orchids, all | 8,251 | 8,505 |
| Peony | 10,804 | 11,861 |
| Protea | 5,044 | 1,669 |
| Roses, all | 13,713 | 12,115 |
| Snapdragon | 5,855 | 5,547 |
| Sunflower | 12,096 | 7,000 |
| Tulips | 73,834 | 63,311 |
| Wax flower | 4,779 | 4,323 |
| Other cut flowers | 99,242 | 81,104 |

*Source:* Based on USDA's National Agricultural Statistics Service, Floriculture Crops (https://quickstats.nass.usda.gov/).

**TABLE 1.8**
**Retail Sales in 2022 of Top-Selling Domestically Produced "Indoor Flowering Potted Plants" in the United States**

| Plant | Sales ($000s) |
|---|---|
| Begonia | 44,827 |
| Chrysanthemum | 41,405 |
| Hibiscus | 40,623 |
| Hydrangea | 42,751 |
| Lily (excluding Easter) | 18,344 |
| Lily (Easter) | 32,145 |
| Orchids[1] | 329,190 |
| Poinsettia | 213,711 |
| Rose | 37,480 |
| Spring-flowering bulbs | 52,759 |
| Others | 260,029 |

[1] Commercial orchid production in the United States is primarily in Hawaii and California.

*Note:* In previous reports, azaleas were included in this category, with a value of ca $40,000,000. Potted foliage plant sales are of the order of $600,000,000 annually.

*Source:* Based on USDA's National Agricultural Statistics Service, Floriculture Crops (https://quickstats.nass.usda.gov/).

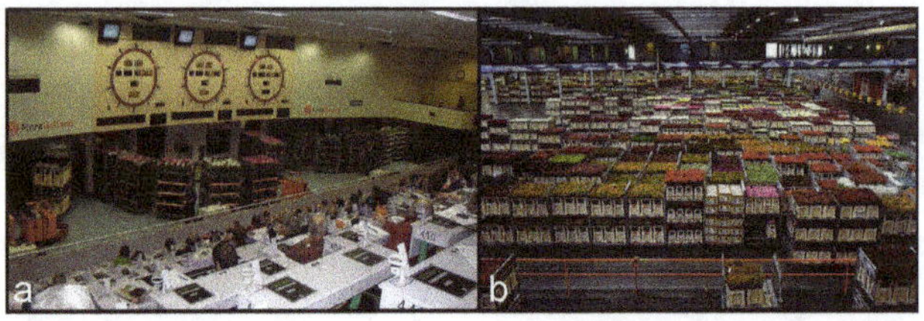

**FIGURE 1.64** Scenes showing activities of the commercial floral industry of The Netherlands: (a) A "Dutch auction" of flowers. (a) Photo by Druifkes (CC BY 3.0). (b) "Dutch flower buckets" (patented shipping crates for cut flowers) awaiting international transportation in an enormous warehouse.

*Source:* **(a) Photo by Druifkes (CC BY 3.0). (b) Photo by 12019 from Pixabay (public domain).**

**FIGURE 1.65** Scenic views of The Netherlands commercial tulip fields.

*Source:* **(a) Photo by Shutterstock/FamVeld. (b) Photo by Shutterstock/Neirfy.**

**FIGURE 1.66** Plant breeding as a source of exciting, new ornamental flowering plants.

*Source:* **Prepared by Jessica Hsiu.**

but markets over 40% of the world's cut flower exports. The Netherlands is also the leading exporter of bulbs for growing spring-flowering plants, such as tulips, which are the world's most profitable flowering plant. The Dutch floriculture market was estimated to be worth about \$6 million in 2023. The Netherlands is a central marketplace for cut flowers in Europe. The "Dutch clock system" is widely employed in the world's large auctions of ornamental plants (the phrase "Dutch auction" typically refers to this system but is also used for different types of auction). A high price is first set for an allotment of plants, and the price is gradually reduced. The first bidder is awarded the plants at the price current at the time of the bid. Bidding too soon means that the bidder overpays; however, delaying one's bid too long means that a competitor is likely to get the flowers. The system today is computerized, and bids are automatically registered to authorized participants. Sellers generally obtain more money than they would with traditional, ascending-price bidding.

## 1.19 ORNAMENTAL PLANT BREEDING

Breeding of ornamental plants is a major activity in many countries. As with plant breeding generally, hybridization, cross-pollination, back-crossing, selection, and evaluation are major steps in the creation of new cultivars. For centuries, gardeners have been selecting attractive mutations of plants, many of which can only be reproduced vegetatively. Some selections are the result of hybridization associated with loss of sexual capacity, once again requiring vegetative propagation. A large proportion of ornamentals are now reproduced by cuttings and tissue culture. Over the centuries, numerous ornamental selections have been made, some of them so attractive that they are still available as heirloom varieties. Unfortunately, many old ornamental selections are disappearing. Indeed, today, even new cultivars have a very short market existence, and companies breeding ornamental plants are now in a very competitive marketplace. Traditionally, cultivars have been selected with large or numerous flowers, doubled flowers (with extra petals), new flower colors, and for an extended flowering period. More recently, stress has been on selecting for vigor, disease resistance, and tolerance of other kinds of stress. To increase production efficiency, plant breeders are developing cultivars with a shorter flowering response time, and also compact plants, so that more can be grown in a given area. Traditional selection techniques remain the main strategy for breeding ornamentals, but modern technologies, notably tissue culture, marker gene–assisted plant breeding, and genetic engineering, are now being widely employed for the major species.

## 1.20 THE FUTURE OF ORNAMENTAL PLANTS

Several "macro trends" are expected to have major influences on the development of ornamental horticulture: (1) An aging population is expected to increase spending on garden and garden-related activities during retirement. (2) The generally poor economic outlook is expected to lead to retention of family homes and increasing investment in them, particularly with respect to ornamentals, which are well-known to increase property values. (3) The generally increasing concern for the environment and for "green" values will also lead to increasing investment in ornamental plants, not just by individuals, but by businesses and governments. (4) Competition from imported ornamental plants is likely to increase. (5) In common with agriculture generally, costs of production are increasing because of rising costs of energy and labor and decreasing availability of water and suitable land for production. (6) Organic production is likely to increase, particularly in view of increasing concern about the high use of pesticides in the ornamentals industry. (7) Marketers have become very sensitive to the importance of impulse buying and so sell plants in full floral display. (8) Increasingly, consumers are purchasing ornamental plants that are mature, to showcase homes and patios, rather than as planting stock to establish gardens or flower beds. This is consistent with the trends for "instant gratification" (compare microwave ovens, cell phones, and e-mail) and time constraints of modern life leaving little free time. (9) Also, along the lines of instant gratification, online-ordered delivery sales are progressively increasing. (10) Economies of scale and specialization are key drivers of market development generally, and these trends are strongly influencing ornamental plant development and marketing. Breeding and production are increasingly being carried out by large companies, often specializing in a limited number of ornamental crops; sales of ornamentals are increasingly being made by large chain stores. (11) In common with breeding of agricultural and horticultural crops generally, large investments are being made in breeding ornamental crops using advanced technologies (genetic engineering, tissue culture propagation, trait selection using marker genes, etc.), providing employment opportunities for biologists. (12) Plant breeder rights and patents are becoming more important as new varieties are being released; the associated legal issues, as well as the same moral and environmental issues that have become prominent for genetically engineered food and fiber plants, are likely to be raised for ornamental plants. (13) Top ornamental plant species have become staple offerings in large chain stores, which usually concentrate on sales of these high-volume commodities; however, a very large number of additional species are in demand by consumers, providing continued opportunities for smaller, specialized nurseries and greenhouses. (14) The world is an amazing source of new ornamentals, which are likely to continue to be discovered, developed, and introduced to the marketplace. This includes exotic introductions (of principal interest to greenhouse producers), and also local species that are likely to be especially suited to the region's climate and soil.

FIGURE 1.67 Winter temperature is the most important climate factor determining the ability of plants to survive outdoors as perennials.

*Source:* **(a) Illustration by Shutterstock/Dream'sDesigns. (b) Illustration by Shutterstock/Bedlovska Liana.**

## 1.21 CLIMATES WHERE ORNAMENTAL PLANTS CAN BE GROWN

The 2012 updated version of the US Department of Agriculture hardiness zone map for the United States (available online at https://planthardiness.ars.usda.gov/) is a highly respected guide to the coldest (most Northern) locations where the perennial plants discussed in this book can be grown. This is based on minimum killing winter temperatures that the plants are known to survive, and this is, indeed, the most critical factor limiting Northernmost survival. (Heat tolerance and/or a physiological need for a cold-induced period of dormancy is the most critical factor limiting Southernmost survival.) Northernmost ratings are provided for the species examined in this book. Cultivars of many of the species are hardier than others of the same species, and often, local nurseries are the best source of locally adapted plants. Of course,

**Average Annual Extreme Minimum Temperature 1976-2005**

| Temp (F) | Zone | Temp (C) |
|---|---|---|
| -60 to -50 | 1 | -51.1 to -45.6 |
| -50 to -40 | 2 | -45.6 to -40 |
| -40 to -30 | 3 | -40 to -34.4 |
| -30 to -20 | 4 | -34.4 to -28.9 |
| -20 to -10 | 5 | -28.9 to -23.3 |
| -10 to 0 | 6 | -23.3 to -17.8 |
| 0 to 10 | 7 | -17.8 to -12.2 |
| 10 to 20 | 8 | -12.2 to -6.7 |
| 20 to 30 | 9 | -6.7 to -1.1 |
| 30 to 40 | 10 | -1.1 to 4.4 |
| 40 to 50 | 11 | 4.4 to 10 |
| 50 to 60 | 12 | 10 to 15.6 |
| 60 to 70 | 13 | 15.6 to 21.1 |

FIGURE 1.68 US Department of Agriculture hardiness zone scale, 2015 version, first published in 2012.

*Source:* **Public domain.**

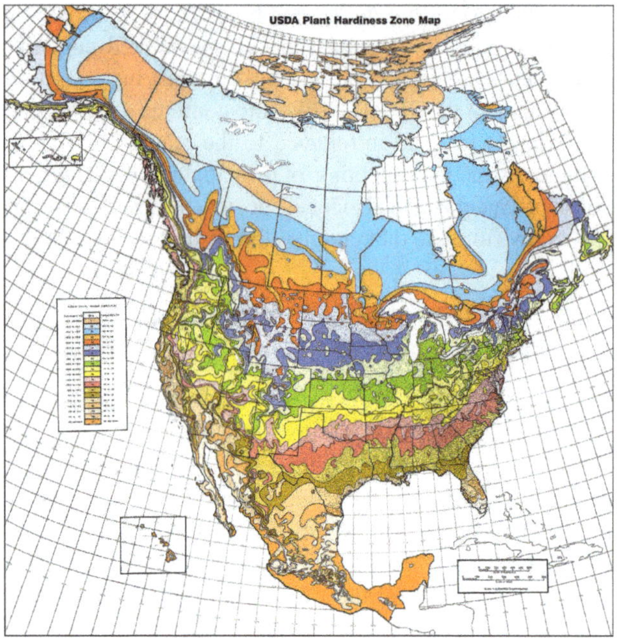


FIGURE 1.69 US Department of Agriculture hardiness zone map for North America, 2001 revision.

*Source:* **Public domain.**

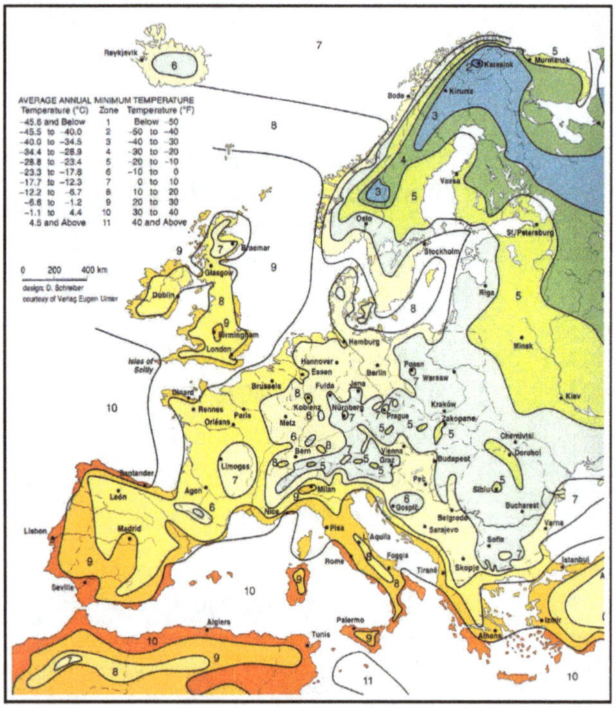


FIGURE 1.70 Hardiness zone map for Europe (employing US Department of Agriculture climate zones).

*Source:* **www.ulmer.de.**

annuals are not limited by winter temperatures (but may be constrained by summer temperatures and shortness of season). Moreover, many perennials can be usefully grown as annuals, and some perennials are maintained as houseplants. Local factors may alter the significance of minimum winter temperatures presented on climate zone maps. Most importantly, snow cover decreases the extent of frost penetration into the ground, and areas of consistent heavy snow may be "warmer" from a plant's viewpoint than indicated by the map's isotherms. Regions experiencing midwinter thaws that melt snow cover will be able to grow a smaller range of plants than places where the snow stays all winter. And climate change is rapidly altering the applicability of former guidelines. Plant hardiness zone evaluations should be thought of as estimates and should be treated as flexible guidelines.

For those living outside the United States, the USDA hardiness zones can be roughly translated by finding out how low temperatures can reach locally and observing what zone this falls into. Zone maps have been published for Europe (e.g., www.gardenia.net/guide/european-hardiness-zones), Australia (e.g., www.gardenia.net/guide/australian-hardiness-zones), Canada (www.gardenia.net/guide/canadian-hardiness-zones), and other countries.

A 12-zone hardiness zone map (https://ucanr.edu/blogs/dirt//blogfiles/37486_original.jpg) for North America, based on days of exposure to heat over 30°C, was developed by the American Horticultural Society. This is not nearly as widely employed as the USDA system, but information is often available for tolerances with respect to both systems and can be especially useful at the Southern semi-tropical or tropical limits of species.

Available plant hardiness maps have limitations. Climatic stresses include not only temperature but also humidity, wind, photoperiod (duration of light, especially important for induction of flowering), and light intensity. Some climatic factors can be controlled for outdoor plants, at least to some extent. Plants that require a long, warm growing season may often be grown in short-season areas if started indoors well in advance of being transplanted outdoors. The microclimate (the climate in the immediate vicinity) of outdoor plants can be modified to some extent. Often, it is possible to grow a plant one zone north of its stated maximum northward limit simply by varying its microclimate. Protection from cold, drying winds can result from shelter belts, hedges, or a cozy location. In the winter, aboveground parts of susceptible plants, such as evergreens, can be covered. Mulches of leaves or other materials can be used to cover some perennials that survive underground during the winter, lowering the possibility of frost injury from deep freezing of the ground. By growing plants near reflective stone walls open to the sun, the temperature near the plants can be raised. Planting close to house foundations, where some heat is lost through the walls and produces warmer soil, can make the difference between success and failure with marginally hardy plants. Southern exposure is sometimes desirable to increase temperature. On the other hand, northern exposure may be useful to prevent deadly freeze–thaw cycles in the spring. Radiation frosts are a constant danger in "frost pockets," low-lying places that are situated where they receive cold-air drainage. Many cultivated plants can be "hardened off" by appropriate cultural treatment and so will withstand colder temperatures than they would otherwise.

Withholding fertilizer in the fall, to prevent the development of tender growth, is one technique. Some plants just seem to become hardier as they mature.

## 1.22 FASCINATING FACTS, FABLES, FOLKLORE, AND FRIVOLITY

- A famous collection of over 3,000 astonishingly realistic models of painted glass flowers of 947 different plant species is housed in the Museum of Natural History at Harvard University. These were prepared as teaching models but have become a major tourist attraction for those visiting the Boston area. The glass flowers were created by Leopold and Rudolf Blaschka from 1887 to 1935 in Germany.

FIGURE 1.71 Part of the exhibit of glass flowers at Harvard's Museum of Natural History.

*Source:* **Photo by Rhododendrites (CC BY SA 4.0).**

- The "language of flowers" was a coded, Victorian-era system of communicating private emotions by mentioning plants or flowers in letters or placing them in small bouquets called tussie-mussies, which were very fashionable at that time. For example, red roses meant passionate, romantic love; white roses, virtue and chastity; yellow roses, friendship or devotion; gerbera, innocence or purity; daffodil, regard; and geranium, gentility. Most messages were positive or neutral, but some were negative; for example, lobelia meant malevolence; hydrangea, heartlessness; begonia, beware; and lime blossoms, illicit sex.
- Numerous superstitions are associated with flowers. For example, stealing flowers from graves will result in an early grave for the thieves; bouquets of red and white flowers should be avoided because they are reminiscent of blood and bandages; drooping flowers are an omen of bad luck; bouquets should have an odd number of flowers, as even numbers are reserved for funerals; flowers are unlucky on board a ship (because they could be used for a funeral); and flowers are also unlucky aboard an airplane.
- In 1910, the "Florists' Telegraph Delivery" was founded by American florists. Now known by the initials FTD, it coined the well-known phrase "Say it with flowers."
- St. Valentine's Day is the leading day for flower sales worldwide, in the United States alone amounting to about $2 billion. A Roman priest, he was executed on the 14th of February in 269, and in 496, the Roman Pope Gelasius I declared that the day should be celebrated as St. Valentine's Day.

FIGURE 1.72 An early 20th-century vintage postcard employing the phrase "Say it with flowers."

*Source:* **Public domain.**

FIGURE 1.73 Heirloom St. Valentine's Day greeting cards.

*Source:* **Public domain.**

The celebration of St. Valentine' Day may trace in part to pagan Roman traditions, and in modern times, it is non-religious, the gift-giving, romantic aspects originating largely in Victorian England.

- "Hippies" were young people in the 1960s and 1970s who advocated freedom from conventional mores, love for everyone, and the consumption of psychedelic drugs. Hippies often employed flowers to symbolize peace and love. "Flower children," a subset of hippies, especially wore flowers in their hair and painted floral motifs on their possessions, especially clothing and vehicles. The phrase "Flower Power" was coined in 1965 by American poet Allen Ginsberg (1926–1997) and employed as a mantra of the anti-war demonstrations of the hippie movement.

FIGURE 1.74 Floral art inspired by the hippie era.

*Source:* **Public domain, by artists (a) Cristian Pozzessere, (b) Cosmiiccreations from Pixabay, (c) ArtSpark from Pixabay, (d) Cosmiiccreations from Pixabay, and (e) ArtSpark from Pixabay.**

FIGURE 1.75 Flower-like fractal photographs.

*Source:* **All from public domain. (a) Prepared by Franky-Syggy from Pixabay. (b) Prepared by PublicDomainPictures. (c) Prepared by TommyVideo from Pixabay. (d) Prepared by CopperScaleDragon. (e) Prepared by Geralt from Pixabay. (f) Prepared by Krencin from Pixabay.**

- *Fractals* are complex geometric figures containing smaller parts that are more or less identical to the larger parts. The repetitive shapes of many natural objects are often fractals, for example, DNA, broccoli and cauliflower, and numerous trees and ferns. Many photographers have presented fascinating fractal figures generated by computer as art. The extraordinary beauty of many fractals is astonishingly reminiscent of flowers.
- There is no unanimity concerning the identity of "the world's most beautiful flower." However, many believe that orchids are the most beautiful of all plants, and the opinion has been expressed that the most beautiful of orchids is the curled odontoglossum (lace orchid; *Oncidium alexandrae*), which, unfortunately, is extremely difficult to grow. (*Oncidium* species are known as dancing-lady orchids and golden shower orchids.)
- Carving edible floral-shaped figures out of fruits and vegetables is a popular pastime. Preparing "radish roses" by slicing pieces from a radish is perhaps the most familiar example, but experts prepare quite realistic figures. In Thailand, the craft is exceptionally advanced.
- The "Dubai Miracle Garden" is a famous commercial flower park in Dubai, United Arab Emirates. Launched in 2013, it occupies over 72,000 $m^2$ and is claimed to be the world's largest flower garden, with over 250 million plants. The site features impressive flower-covered sculptures. The scorching climate requires closure during the hottest part of the year, and the limited supply of fresh water necessitates conservation and recycling of the limited supply.

FIGURE 1.76 *Oncidium alexandrae*, claimed to be the most beautiful of all orchids, and perhaps the most beautiful of all flowers.

*Source:* **(a) Lindenia, *Iconography of Orchids* (Brussels), vol. 7, 1894 (public domain). (b) *The Floral Magazine* (London), vol. 9, 1870 (public domain).**

FIGURE 1.77 Floral-themed "edible flowers" carved from vegetables and fruits: (a) Thai carver preparing floral sculptures from watermelons. (b) Carved radish roses. (c) "Edible flowers" carved from vegetables and fruits. (d) A flower-like bouquet composed of fruits and vegetables.

*Source:* **(a) Photo by Cotaro70s (CC BY ND 2.0). (b) Photo by Peter Alfred Hess (CC BY 2.0). (c) Photo by Shutterstock/Worradirek. (d) Photo by Fruitlife (CC BY SA3.0).**

FIGURE 1.78 Scenes in the "Dubai Miracle Garden," said to be the world's largest flower garden.

*Source:* **(a) Photo by Dennis Sylvester Hurd (public domain). (b) Photo by Shutterstock/Dmytro Nekipelov. (c) Photo by Martin (CC BY ND 2.0).**

FIGURE 1.79 A flower book.

*Source:* **Frontispiece of Pallas, P. S. 1784. *Flora Rossica*, vol. 1. Imperiali J. J. Weitbrecht, Petropoli (Saint Petersburg), Russia. Colored by B. Brookes.**

## 1.23 KEY INFORMATION

We are in an age of explosive expansion of information. Much knowledge on plants does not go out of date, or at least does so very slowly, so that major print publications retain much of their usefulness for a very long period. On the other hand, new information may be critical, such as when it is discovered that a given plant has previously unappreciated toxic or economic properties. Most of the published information about plants that are employed as ornamentals is concerned with chemistry, perfumery, cosmetics, food technology, and human and veterinary pharmacology, which are marginally relevant to cultivation and marketing ornamental aspects. The following selected publications and internet sites have been chosen to assist students of floriculture to find supplementary information on the plants discussed in this book. The selection of literature surveyed for this book is overwhelmingly English, and readers should keep this limitation in mind.

The web is an unavoidable and growing source of information. As the internet is a transient, volatile medium, URL addresses may not continue to be accurate. Internet links change, move, and often rapidly become defunct. Moreover, the quality and reliability of internet information are extremely variable. Web documents cited in this book were chosen for their value, and even if many of the URLs become obsolete, their titles furnish keywords or phrases for a search. Sites that are maintained by governments, universities, and other major research organizations tend to produce quite reliable information, although much knowledge tends to become dated. Sites providing cultivation information often are associated with commercial products, but this is the price of much internet use today.

## KEY LITERATURE

Armitage, A.M. and J.F. Laushman. 2008. *Specialty Cut Flowers: The Production of Annuals, Perennials, Bulbs and Woody Plants for Fresh and Dried Cut Flowers*, Revised edition. Timber Press, Portland, OR. 636 pp.

Biondo, R.J. 2000. *Floriculture: From Greenhouse Production to Floral Design*. Interstate Publishers, Prairie Village, KS. 662 pp.

Brickell, C.D., C. Alexander, J.J. Cubey, J.C. David, M.H.A. Hoffman, A.C. Leslie, et al. 2016. *The International Code of Nomenclature for Cultivated Plants*, 9th edition.

International Society for Horticultural Science, Leuven, Belgium. 190 pp.

Burrows, G.E. and R.J. Tyrl. 2001. *Toxic Plants of North America*, 2nd edition. Wiley-Blackwell, Hoboken, NJ. 1392 pp.

Byczynski, L. 1997. *The Flower Farmer: An Organic Grower's Guide to Raising and Selling Cut Flowers*. Chelsea Green, White River Junction, VT. 224 pp.

Creasy, R. 1982. *The Complete Book of Edible Landscaping: Home Landscaping with Food-bearing Plants and Resource-Saving Techniques*. Sierra Club Books, San Francisco, CA. 379 pp.

Dole, J. and H. Wilkins. 2005. *Floriculture: Principles and Species*, 2nd edition. Prentice Hall, Upper Saddle River, NJ. 1023 pp.

Farrar, L. 2016. *Gardens and Gardeners of the Ancient World: History, Myth and Archaeology*. Windgather Press, UK. 304 pp.

Ghose, T. 2013. Oldest grave flowers ever found unearthed in Israel. *NBC News*. www.nbcnews.com/sciencemain/oldest-grave-flowers-ever-found-unearthed-israel-6C10506405.

Gledhill, D. 2008. *The Names of Plants*, 4th edition. Cambridge University Press, Cambridge. 436 pp.

Hessayon, D.G. 1998. *The House Plant Expert*. Transworld Publishers Ltd., London, UK. 256 pp.

Hůla, M. and J. Flegr. 2016. What flowers do we like? The influence of shape and color on the rating of flower beauty. *PeerJ* 4: e2106. https://doi.org/10.7717/peerj.2106.

Jiménez-Mejías, P., S. Manzano, C. Acedo, I. Álvarez, J.V. Crisci, V. Gowda, et al. 2023. *Protecting Stable Biological Nomenclatural Systems Enables Universal Communication*. Preprint. www.researchgate.net/publication/375446195_Protecting_stable_biological_nomenclatural_systems_enables_universal_communication.

Jozwik, F.X. 1992. *The Greenhouse and Nursery Handbook: A Complete Guide to Growing and Selling Ornamental Container Plants*. Andmar Press, Casper, WY. 511 pp.

Knight, A.P. 2006. *A Guide to Poisonous House and Garden Plants*. Teton NewMedia, Jackson, WY. 324 pp.

Larson, R.A. (ed.). 1992. *Introduction to Floriculture*, 2nd edition. Academic Press, Cambridge, MA. 636 pp.

Llamas, K.A. 2003. *Tropical Flowering Plants*. Timber Press, Portland, OR. 423 pp.

Madison, M. 1998. *Growing Flowers for Market: A Practical Manual for Small-Scale Field Production and Marketing of Fresh Flowers*. Yolo Press, Davis, CA. 270 pp.

Mahr, S. 2007. *What's in a Name? Understanding Botanical or Latin Names*. https://mastergardener.extension.wisc.edu/files/2015/12/plantnames.pdf.

Maree, J. and B.-E. van Wyk. 2010. *Cut Flowers of the World: A Complete Reference for Growers and Florists*. Timber Press, Portland, OR. 400 pp.

Marimuthu, S., K. Kalaimughilan, S. Uma, and B. Suthanthiram. 2018. Critical evaluation of the benefits and risks of genetically modified horticultural crops. In: *Genetic Engineering of Horticultural Crops,* edited by G.R. Rout and K.V. Peter, 315–351. Academic Press, Cambridge, MA.

McDaniel, G.L. 1982. *Ornamental Horticulture*, 2nd edition. Reston Publishing, Reston, VA. 526 pp.

Small, E. 2006. *Culinary Herbs*, 2nd edition. NRC Press, Ottawa. 1036 pp.

Smith, J.P., Jr. 2017. *The Scientific Names of Plants*. Humboldt State University Digital Commons. https://digitalcommons.humboldt.edu/cgi/viewcontent.cgi?article=1010&context=botany_jps.

Spoerke, D.G., Jr. and S.C. Smolinske. 1990. *Toxicity of Houseplants*. CRC Press, Boca Raton, FL. 244 pp.

Stearn, W.T. 1992a. *Stearn's Dictionary of Plant Names for Gardeners*. Cassell, London. 364 pp.

Stearn, W.T. 1992b. *Botanical Latin: History, Grammar, Syntax, Terminology, and Vocabulary*, 4th edition. David & Charles, Newton Abbot, Devon, UK. 546 pp.

Stevens, A. 1998. *Field Grown Cut Flowers: A Practical Guide and Sourcebook*. Commercial Field Grown, Fresh and Dried Flower Production, Avatar's World, Edgerton, WT. 392 pp.

Still, S.M. 1994. *Manual of Herbaceous Ornamental Plants*, 4th edition. Stipes, Champaign, IL. 814 pp.

Summerbell, R.C., S. Krajden, and J. Kane. 1989. Potted plants in hospitals as reservoirs of pathogenic fungi. *Mycopathologia* 106: 13–22.

Turland, N.J., J.H. Wiersema, F.R. Barrie, W. Greuter, D.L. Hawksworth, P.S. Herendeen, et al. (eds.) 2018. *International Code of Nomenclature for Algae, Fungi, and Plants*. Koeltz Botanical Books, Glashütten, Germany. 254 pp. www.iapt-taxon.org/nomen/main.php.

United States Department of Agriculture Plant Hardiness Zone Map. https://planthardiness.ars.usda.gov/.

Wells, D. 1997. *100 Flowers and How They Got Their Names*. Algonquin Books, Chapel Hill, NC. 288 pp.

Woods, M. and A.S. Warren. 1988. *Glass Houses: A History of Greenhouses, Orangeries and Conservatories*. Rizzoli, NY. 216 pp.

## Key Plant Encyclopedia

Bailey, L.H. and the Staff of the Bailey Hortorium. 1949. *Manual of Cultivated Plants Most Commonly Grown in The Continental United States and Canada*. Macmillan, New York. 1116 pp.

Bailey, L.H. and E.Z. Bailey. 1976. *Hortus Third*. Revised by the staff of the L.H. Bailey Hortorium. Macmillan, New York. 1290 pp.

Brickell, C. (ed.) 2019. *Royal Horticultural Society Encyclopedia of Plants and Flowers*. Royal Horticultural Society, London. 744 pp.

Brickell, C. and H.M. Cathey, (eds.) 2004. *The American Horticultural Society A – Z Encyclopedia of Garden Plants*. DK Publishing, New York. 1098 pp.

Bryant, K. and T. Rodd. 2005. *The Ultimate Plant Book. Raincoast Books*. Vancouver, Canada. 1024 pp.

Dirr, M.A. 1997. *Dirr's Hardy Trees and Shrubs: An Illustrated Encyclopedia*. Timber Press, Portland, OR. 494 pp.

Dirr, M.A. 2002. *Dirr's Trees and Shrubs for Warm Climates: An Illustrated Encyclopedia*. Timber Press, Portland, OR. 448 pp.

Dirr, M.A. 2006. *The Reference Manual of Woody Plant Propagation: From Seed to Tissue Culture*. Timber Press, Portland, OR. 424 pp.

DK (corporate author). 2014. *Smithsonian Encyclopedia of Garden Plants for Every Location*. DK, London. 400 pp.

Fayaz, A. 2011. *Encyclopedia of Tropical Plants: Identification and Cultivation of Over 3000 Tropical Plants*. Firefly Books, Richmond Hill, ON, Canada. 720 pp.

Graf, A.B. 1985. *Exotica, Series 4 International: Pictorial Cyclopedia of Exotic Plants from Tropical and Near-Tropic Regions: A Treasury of Indoor Ornamentals for Home, the Office, or Greenhouse – in Warm Climates the Patio and the Garden Outdoors*, 2 Vols. Roehrs Co., East Rutherford, NJ.

Graf, A.B. 2003. *Tropica: Color Cyclopedia of Exotic Plants and Trees for Warm-region Horticulture – In Cool Climate the Summer Garden or Sheltered Indoors: 7,000 Photographs in Living Color*, 5th edition. Roehrs Co., Farmindale, NJ. 1142 pp.

Hanelt, P. (ed.). 2001. *Mansfeld's Encyclopedia of Agricultural and Horticultural Crops*, 6 Vols. Springer-Verlag, Heidelberg, Germany.

Huxley, A., M. Griffiths, and M. Levy. (eds.). 1992. *The New Royal Horticultural Society Dictionary of Gardening*, 4 Vols. Stockton Press, New York.

Quattrocchi, U. 1999. *CRC World Dictionary of Plant Names: Common Names, Scientific Names, Eponyms, Synonyms, and Etymology*, 4 Vols. CRC Press, Boca Raton, FL.

Rehder, A. 1951. *Manual of Cultivated Trees and Shrubs Hardy in North America*, 2nd edition. Macmillan, NY. 996 pp.

Shoemaker, C.A. 2001. *Encyclopedia of Gardens: History and Design*. Routledge, London. 1590 pp.

Tenenbaum, F. 2003. *Taylor's Encyclopedia of Garden Plants: The Most Authoritative Guide to the Best Flowers, Trees, and Shrubs for North American Gardens*. Houghton Mifflin Harcourt, Boston, MA. 464 pp.

Wiersema, J.H. and B. León. 2013. *World Economic Plants. A Standard Reference*, 2nd edition. CRC Press, New York. 1336 pp. [This work provides standard names for economic plants. The United States Department of Agriculture Germplasm Resources Information Network (GRIN), www.ars-grin.gov/, is based on this work but provides updated information].

# 2 African Violets (*Streptocarpus*/ *Saintpaulia* Species and Hybrids)

**Scientific Names:** The genus name *Streptocarpus* is based on the Greek *streptos*, "twisted," and *karpos*, "fruit," which is descriptive of many of the species, but not the African violet species. *Streptocarpus ionanthus* is the most widely cultivated species. The specific epithet *ionanthus* in the name is based on the Greek adjective *ion*, meaning "violet," and the noun *anthos*, "flower."

**Common Names:** The name *African violet* is based on the African origin of the plants and the superficial resemblance to true violets species of *Viola*. The old genus name *Saintpaulia* and the common name saintpaulia commemorate Baron Walter von Saint Paul-Illaire (1860–1940), district commissioner of Tanga province, Tanganyika (now Tanzania), who collected African violets in 1892, in what was then a German colony called German East Africa.

## 2.1 INTRODUCTION

In modern throwaway, planned-obsolescence times, ostentatiously flowering ornamentals like chrysanthemums, roses, and poinsettias have become the leading stars of the potted houseplant industry. However, they are intended to last for only a few weeks before being discarded. By contrast, a considerably more modest plant – the African violet – has become the world's most popular flowering potted plant maintained indoors, which is a considerable accomplishment, since it was virtually unknown to science a little over a century ago and currently is being eliminated from nature at an alarming rate. African violets are dainty, fitting on windowsills, mantels, tables, and desks, where almost all other ornamentals would take too much space. They are also unusually well adapted to indoor conditions that are too challenging for most plants. For their small size, they are remarkably appealing, developing compact rosettes of velvety leaves and colorful flowers which bloom for months. While modern cultivars do not deserve the former reputation of the plant as being difficult to grow, they nevertheless will not thrive unless their particular requirements are satisfied, as outlined in this chapter.

## 2.2 OVERVIEW

The genus *Saintpaulia* has been shown to be part of *Streptocarpus* subgenus *Streptocarpella*, and the ten or so species formerly in *Saintpaulia* have been transferred to *Streptocarpus*. The traditional recognition of the genus *Saintpaulia* is familiar to most gardeners, and the new,

FIGURE 2.1 Walter von Saint Paul-Illaire, German colonial officer stationed in eastern Africa, after whom the scientific name *Saintpaulia* and the common name saintpaulia are named.

*Source:* **Drawing (public domain) from Allers, C. W., and H. Kraemer. 1895. Union Deutsche Verlagsgesellschaft, Stuttgart.**



DOI: 10.1201/9781003586227-2

FIGURE 2.2 Cultivars of African violet (*Streptocarpus ionanthus*).

*Source:* **Photos from Shutterstock, contributed by (a, b) Galyna Syngaievska; (c) AnnHirna; and (d) Alexander52 (cultivar Arctic Frost).**

FIGURE 2.3 Close-up photos of flowers showing the range of colors of African violet (*Streptocarpus ionanthus*).

*Source:* **(a) Photo by Bff (CC BY SA 3.0). (b, d, e) Photos by Walrus 36 (CC BY 2.0). (c) Cultivar Alchemy Morning Star. Photo by HQ (CC BY SA 2.0). (f) Cultivar Snow Shadow. Photo by HQ (CC BY SA 2.0).**

correct nomenclature is scarcely known. Species traditionally placed in *Saintpaulia* are small, short-stemmed perennial herbs with succulent, dark-green, soft-hairy ("velvety") evergreen leaves usually borne in a rosette. All the wild species have blue-violet (rarely white) flowers. The tiny seeds are produced in a capsule. The number of species that were once assigned to *Saintpaulia* has been debated (from 6 to 25 have been recognized), and although they should all be called *Streptocarpus*, the number of species remains controversial. Most populations are interfertile, and hybrids contribute to the difficulty of delimiting species. Many of the "species" recognized are, in fact, assignable to *Streptocarpus ionanthus* (*Saintpaulia ionantha*).

African violets are endemic to the Eastern Arc Mountains of Eastern Africa. They range from the south of Kenya to eastern Tanzania, with the center of diversity in the Usambara Mountains of northeast Tanzania. The species are often found in small populations at altitudes

FIGURE 2.4 *Streptocarpus teitensis* (*Saintpaulia teitensis*; Taita African violet). This "critically endangered" species, with less than 2500 individuals outside of cultivation, is confined to an area less than 1 km² in southern Kenya. A shade-loving plant, it grows in sheltered spots on rocks, among tree roots, and occasionally, on the lower trunks of moss-covered trees.

*Source:* **(a) Photo of a cultivar by Daderot (public domain). (b) Wild plants sprawling over a boulder. Photo by Gerard D. Hertel (CC BY 3.0).**

from lowlands to 2000 m. They grow in a variety of forest types, on hills, rocky mountain peaks, and cliffs; in gorges; in moss and on wet rocks; and occasionally, as epiphytes (mostly on trees). Although often in cloud forests, some species are relatively drought-tolerant. Crassulacean acid metabolism, a form of photosynthesis often associated with drought-tolerant succulents, has been reported in *S. ionanthus*. Habitat degradation and climate change are having devastating effects on African violets. It is ironic that in little more than a century, horticultural selections have been spread to offices and homes around the world, while in their homeland in Kenya and Tanzania, several of the wild species, including *S. ionanthus*, are on the brink of extinction.

*Streptocarpus ionanthus* and its hybrids dominate the African violet industry. Thousands of cultivars, including some transgenic (genetically modified) forms, have been bred. The flowers of modern cultivars may be single or double and can be white or shades of pink, red, blue, purple, lavender, bicolored, or multicolored. The yellow anthers of the two stamens are responsible for the prominent yellow centers of the flowers of numerous cultivars. Some selections have ruffled, star-shaped, or fringed petals, and the flowers can even look like highly doubled rose blooms. Individual flower clusters can last 3 to 6 weeks, and plants can be maintained in a more or less continuous state of flowering. The leaves may be oval to heart-shaped, spooned, or pointed, and the margins ruffled, fringed, or scalloped. Cultivars with variegated leaves (with areas or spots that are white, cream, or pink) are available. Some forms have trailing stems. Although the plants are relatively small, there are especially small cultivars called miniatures, minis, or sometimes "teacup plants." 'Mickey Mouse' is a popular miniature. On the other hand, since the plants grow clonally, experienced growers sometimes produce impressively large specimens growing in single pots.

FIGURE 2.5 Old paintings (public domain) by professional botanical illustrators showing very early garden forms of African violet (*Streptocarpus ionanthus*).

***Source:*** **(a)** ***Curtis's Botanical Magazine*** **(London), vol. 121, 1895. (b) Regel, E. A. von. 1893.** ***Gartenflora (Erlangen, Germany)*****, vol. 42. (c)** ***Revue horticole*** **(Paris), vol. 74, 1902.**

FIGURE 2.6 (a) Advanced cultivars of African violet (*Streptocarpus ionanthus*) with bicolored (white and pink) doubled flowers (extra petals and extra stamens) and petals with scalloped margins. (b) Cultivar Ultra Eclipse. (c) Cultivar Kentucky Gooseberries.

***Source:*** **(a) Photo by Andrew Butko (CC BY SA 3.0). (b) Photo by Kor!An (CC BY SA 3.0). (c) Photo by HQ (CC BY SA 2.0).**

FIGURE 2.7 Variegated-leaf forms of African violet.

***Source:*** **(a) Photo by Mk2010 (CC BY SA 3.0). (b) Cultivar Tahitian Sunset. Photo by HQ (CC BY SA 2.0). (c) Photo by A. Benedito (CC BY SA 4.0).**

FIGURE 2.8 Postage stamps issued by Canada Post in 2010 showing Canadian-bred cultivars of African violets: (a) Cultivar Decelles Avalanche. (b) Cultivar Picasso.

FIGURE 2.9 Economic activity associated with African violets: (a) Commercial greenhouse cultivating the plants. (b) Plants on sale in a store.

***Source:*** **(a) Photo by Shutterstock/Rikard Stadler. (b) Photo by Shutterstock/Nataliia Kulikovska.**

African violets are easily confused with their very similar relatives gloxinias (see the chapter on gloxinia, the genus *Sinningia*). Similarities include a basic rosette form and softly hairy leaves. Gloxinias are usually larger, with bigger (and often fewer) flowers (up to 10 cm in diameter), a more developed corolla tube (evident only in side views), and the flowers usually displayed more evidently above the leaves. Gloxinia cultivars often have very ruffled petal edges and are sometimes doubled, but so are some African violet cultivars. African violets can grow and bloom continuously, but the florist's gloxinia grows from a tuber for a few months, blooms, then loses its leaves and becomes dormant for several weeks or months before resuming growth.

## 2.3 HISTORY

African violet species were brought to Europe in the late 19th century and were grown in subsequent decades in limited numbers. Hardier and fancier hybrids became available in the 1960s, considerably expanding the popularity of African violets as houseplants. There are over 2,000 cultivars of African violets, mainly derived from *S. ionantha*.

## 2.4 ECONOMIC IMPORTANCE

African violets are addictive for many, and hardcore hobbyists often maintain hundreds of plants, spending considerable sums on indoor gardening supplies. In the United States, the annual retail value exceeds $25,000,000.00.

## 2.5 ORNAMENTAL USES

Unlike numerous flowering ornamentals, African violets have no associated cut flower market, and at least in temperate regions, they are rarely grown in gardens, although occasionally cultivated outdoors in warm climates. African violets are mainly sold as gift or personal plants for home and office use. Although sales of throwaway flowering pot plants such as chrysanthemum and poinsettia exceed those for African violets, the latter occupies a special niche as the most popular of the flowering potted plants intended

FIGURE 2.10 Potted plants of African violet (*Streptocarpus ionanthus* hybrids).

*Source:* **(a) Cultivar Mac's Elegant Emerald. Photo by HQ (CC BY 2.0). (b) Photo by Bff (CC BY SA 3.0). (c) Photo by Ibolya (CC BY 2.0).**

FIGURE 2.11 African violets on sentimental greeting cards.

*Source:* **(a) A vintage (public domain) valentine card showing Cupid pulling a bouquet of African violets. (b) Photo by Shutterstock/Tarskaya Tatiana.**

for long-term cultivation in relatively limited personal spaces, where small, compact, long-flowering specimens that bloom at any time are ideal. As with most potted plants commonly marketed, commercial outlets prefer cultivars producing impressive flowering displays at time of sale. These pretty little plants are especially attractive to impulse buyers who see them on display in supermarkets, but they are also staples of dedicated enthusiasts.

## 2.6 OTHER USES

African violets are linked with motherhood and are a common gift on Mother's Day. They are also associated with Easter and Valentine's Day. Since the plants have been known in cultivation for not much more than a century, these symbolic associations are quite recent and have been generated mostly by the retail floral industry, which benefits from increased sales of flowers for sentimental reasons.

The flowers of true violets (*Viola* species) are often candied and consumed, but African violet flowers are not suitable for culinary purposes. However, African violets are "pet-safe" and are one of the houseplants recommended where companion animals and children might contact them. Neither the flowers nor the foliage is considered toxic to dogs, cats, birds, horses, or people, although contact dermatitis has rarely been recorded.

## 2.7 PROBLEMS

African violets are relatively easy to grow but are so widely cultivated by people with limited knowledge of their particular requirements for care that they often develop poorly and serve to discourage people from cultivating houseplants.

## 2.8 CULTIVATION

### 2.8.1 Climate and Site

African violets grow well in the average home or office, but it is advisable to be aware of conditions that are most suitable. Air currents, such as from air-conditioning vents, can be very harmful, as can contact of foliage with windows. Occasionally, African violets are grown in terraria (reminiscent of fish tanks, but for plants; see Chapter 1). A daytime temperature of 22–24°C and a night temperature of 17–18°C, not less than 18°C, are appropriate (although some cultivars will tolerate 16°C for brief periods).

The light meters that most people are likely to possess are calibrated in foot-candles or lux to indicate the density of light falling on a surface (average indoor lighting ranges from 100 to 1,000 lx, and average outdoor sunlight is about 50,000 lx). More accurate quantum PAR meters measure the micromoles of photons per square meter per second ($\mu mol/m^{-2}/s$), and still more accurate meters consider photosynthetically active radiation and integrate exposure over a day. Typically, for African violets, about 1,000 fc (5,000 lx or 8 mol $day^{-1}m^{-2}$) is adequate. Plants flower poorly or remain vegetative when maintained in a location receiving 500 fc maximum light level at noon on a sunny day (which roughly provides a daily photon flux of less than 1.5 mol $day^{-1}m^{-2}$). Light intensity above 1400 fc (280 $\mu mol\ s^{-1}m^{-2}$) causes bleaching or burning of the foliage. At least 12 hr, but preferably 14 hr, is needed (up to 18 hr has been recommended under fluorescent lights), and longer times of light availably compensate for lower light intensity. For windowsill growers, light from an east window may be needed to promote flowering, light from a north window may be inadequate, and a west window may require filtering curtains to tolerate a hot afternoon sun. Serious home growers of African violets tend to raise them in fluorescent light stands, and although standard cool whites can be satisfactory, balanced spectrum (daylight) bulbs are better, and the color of the plants appears optically to be more intense. Typically, fluorescent lights are placed 30–38 cm above the foliage, or 15–30 cm above the flowers.

### 2.8.2 Propagation and/or Purchasing

When buying plants, ensure that they are free of insects. Some pests, such as thrips, are very difficult to eliminate, and African violet specialists isolate new plants for more than a month before adding them to their collection to ensure that insect eggs are not present. African violets are usually propagated vegetatively, since most attractive cultivars are hybrids that will not reproduce true from seed. Plants with variegated foliage are chimeras and are most reliably propagated by splitting them (plantlets produced from leaves as described in the following may or may not

FIGURE 2.12 Reproducing African violet vegetatively by inserting a leaf in soil: (a) Placing the leafstalk in potting soil. (b) Plantlets growing at the base of the leafstalk.

*Source:* **(a) Photo by Shutterstock/Osobystist. (b) Photo by RobertoMM (CC BY SA 3.0).**

reproduce the variegation). Commercial propagation may be by tissue culture or micropropagation, and "artificial seeds" (encapsulated tissue cultures) are also available. For home growers, a plant may simply be divided. Mature plants occasionally produce small plantlets or shoots from the side, which can be removed and potted. A portion of leaf blade or leafstalk will produce one or more new plantlets if left in a moist chamber or on moist sand for 6 weeks or so (immersion in water is not recommended). Leaves for propagating are best chosen between the outermost and the innermost leaves of the rosette.

### 2.8.3 Planting

Plants should be repotted annually, keeping in mind that they grow best in proportionately small pots, and a larger pot is only necessary when the plant has become strongly root-bound. In theory, an African violet clone is immortal, and some hobbyists have reported potted plants that endured for more than a decade; but a given crown eventually runs down and is best perpetuated vegetatively. A rough rule of thumb is that, when repotting, the new pot should have a width of about a third of the diameter of the rosette of leaves. Experienced hobbyists strongly recommend plastic over clay pots, which require more frequent watering. If not washed away, accumulated salts on clay pots can harm contacted stems or foliage.

### 2.8.4 Soil and Moisture Conditions

Ideal soils are porous and well-drained but have a high humus content (a soilless preparation formulated for African violets is ideal). The crown of the plant should be about even with the top of the pot. To promote drainage from the crown, the soil should slope down at the edge to about 1.3 cm below the rim. Tolerated pH ranges from 5.5 to 7.0 (recommended range is 6.4–6.9). A soluble houseplant fertilizer (preferably formulated specifically for African violets) may be applied every 2 weeks. Some African violet specialists recommend liquid fertilizers high in K and P (high N encourages foliage growth at the expense of blooms); others simply prefer balanced N:P:K. Many recommend the fertilizer be diluted (e.g., quarter-strength). The soil should be maintained moist, but not wet. Soft water is best (rainwater or demineralized water may be substituted for hard or limy [calcareous] tap water). Commercial growers recommend an ambient humidity of 50 to 70%. Placing the pot on a humidifying tray filled with moist pebbles can promote humidity around the plant, and some specialists advise placing pots on moist sand. Although sometimes the foliage of African violets is misted with warm or tepid water, this practice is inadvisable because the fuzzy foliage tends to retain water, which may cause decay or discoloration. Cold water can promote foliage fungi and, in contact with leaves, often causes white spots; water at room temperature is best. Commercial growers commonly irrigate African violets by placing them on capillary mats (spongy pads saturated with water) or by drip irrigation, both of which keep the foliage dry, preventing leaf spotting. There are innumerable systems marketed in garden supply stores for watering African violets in the home. Although bottom watering is recommended, this can eventually result in harmful buildup of salts in the soil, which should be flushed thoroughly with clear water every month or two.

### 2.8.5 Maintaining Plants

Senescent leaves and flowers should be removed, and as noted earlier, regular repotting is needed.

## 2.9 FASCINATING FACTS, FABLES, FOLKLORE, AND FRIVOLITY

- "Enantiostyly" has been defined as the strong deflection of the style (receptive part of the pistil, the female part of the flower) to the left or right of the main floral axis at anthesis, with flowers of

FIGURE 2.13 Sexual organs of *Streptocarpus ionanthus*. As pointed out in the text, the style is pointed in one direction in half the flowers of a given plant, and in an opposite direction in the other half of the flowers. The label "left-facing" is from the perspective of the flower, not the viewer.

*Source:* **Photo by Calimo (CC BY SA 3.0).**

different "morphs" (types) occurring at the same frequency within a plant. All species of African violets are enantiostylous. ("Left" and "right" are arbitrarily defined for the bilaterally symmetrical flowers of *Streptocarpus* with respect to the stamens and from the view that the flowers are facing). The left- and right-hand flowers are essentially mirror images of each other. African violets are in the gesneriad family (Gesneriaceae), in which enantiostyly seems to always involve both flower types occurring in the same plant or same flowering stem. At least a dozen families of plants have independently evolved enantiostyly. This feature is often linked to "buzz" (vibrational) pollination by bees and is known to promote outcrossing.

- As pointed out earlier, misting the foliage of African violets is inadvisable because the fuzzy foliage tends to retain water, encouraging fungal infection. However, in the droughty cloud-forest habitat where many of the species occur, the ability of the velvety hairs to trap suspended water particles is advantageous, allowing the plants to obtain scarce moisture from the air.
- The popularity of the African violet is reflected in the wide usage of the phrase as a name for diverse entities, including cities, racehorses, and pets. "African violet" is also used as a color designation and is employed for paint, lipstick, markers, crayons, and eyeshadow.
- In 1984, 25,000 African violet seeds were transported into space by the NASA space shuttle and became part of the experiments in the "Long Duration Exposure Facility" testing the effects on living things of long-term cosmic radiation and lack of gravity. After orbiting Earth for 6 years, the seeds were found to have acquired mutations and became the basis of new commercial varieties sometimes called "space violets" (www.optimara.com/everfloris.html).

FIGURE 2.14 "Pane-sill murder," the death of windowsill plants that have been neglected, typically unwatered during vacations by the owners.

*Source:* Prepared by B. Brookes.

- Tactile mechanical stress ("thigmomorphogenesis") such as brushing sometimes toughens and benefits plants (for example, brushing tomatoes makes them less leggy). The velvety foliage of African violets tempts people to stroke their leaves. However, caressing African violets, especially when hand lotions have been used recently, damages them.
- "Pane-sill murder" is a colorful phrase designating the death of ornamental plants such as African violets located on windowsills that die when residents go on vacation, leaving them home alone to die from freezing or desiccation.

## KEY PUBLICATIONS


Baatvik, S.T. 1993. The genus *Saintpaulia* (Gesneriaceae) 100 years: History, taxonomy, ecology, distribution and conservation. *Fragmenta Floristica et Geobotanica Supplementum* 2(1): 97–112.

Bodner, M. and W. Larcher. 1989. Chilling susceptibility of wild *Saintpaulia* spp. of different altitudinal origin. *Angewandte Botanik* 63: 501–512.

Brotton, J.C. and J.C. Cole. 2009. Brushing using a hand coated with body lotion or in a latex glove decreases African violet plant quality and size. *HortTechnology* 19: 613–616.

Christenhusz, M.J.J. 2012. On African violets and Cape primroses – towards a monophyletic *Streptocarpus* (Gesneriaceae). *Phytotaxa* 46: 3–9.

Clarke, G.P. 1998. Plants in peril, 24: Notes on lowland African violets (*Saintpaulia*) in the wild. *Curtis's Botanical Magazine* 15: 62–67.

Daniel, M.K. 1953. Enantiostyly in *Saintpaulia*. *Proceeding of the Pennsylvania Academy of Science* 27: 61–64.

Darbyshire, I. 2006. Gesneriaceae. In: *Flora of Tropical East Africa,* edited by H.J. Beentje and S.A. Ghazanfar, 1–74. Royal Botanic Gardens, Kew.

Dimitrov, D., D. Nogués-Bravo, and N. Scharff. 2012. Why do tropical mountains support exceptionally high biodiversity? The Eastern Arc Mountains and the drivers of *Saintpaulia* diversity. *PLoS One* 7(11): e48908. https://doi.org/10.1371/journal.pone.0048908.

Eastwood, A., B. Bytebier, H. Tye, A. Tye, A. Robertson, and M. Maunder. 1998. The conservation status of *Saintpaulia*. *Curtis's Botanical Magazine* 15: 49–62.

Harrison, C.J., M. Möller, and Q.C.B. Cronk. 1999. Evolution and development of floral diversity in *Streptocarpus* and *Saintpaulia*. *Annals of Botany* 84: 49–60.

Johansson, D.R. 1978: Saintpaulias in their natural environment with notes on their present status in Tanzania and Kenya. *Biological Conservation* 14: 45–62.

Kolehmainen, J. and P. Mutikainen. 2007. Population stage structure, survival and recruitment in the endangered East African forest herb *Saintpaulia*. *Plant Ecology* 192: 85–95.

Linqvist, C. and V.A. Albert. 1999a. A high elevation ancestry for the Usambara Mountains and lowland populations of African violets (*Saintpaulia*, Gesneriaceae). *Systematics and Geography of Plants* 71: 37–44.

Linqvist, C. and V.A. Albert. 1999b. Phylogeny and conservation of African violets (*Saintpaulia*: Gesneriaceae): New findings based on nuclear ribosomal 5S non-transcribed spacer sequences. *Kew Bulletin* 54: 363–377.

Möller, M. and Q.C.B. Cronk. 1997a. Origin and relationships of *Saintpaulia* (Gesneriaceae) based on ribosomal DNA internal transcribed spacer (ITS) sequences. *American Journal of Botany* 84: 956–965.

Möller, M. and Q.C.B. Cronk. 1997b. Phylogeny and disjunct distribution: Evolution of *Saintpaulia* (Gesneriaceae). *Proceedings of the Royal Society of London, Biological Science* 264: 1827–1836.

Möller, M. and Q.C.B. Cronk. 1999. New approaches to the systematics of *Saintpaulia* and *Streptocarpus*. In: *Taxonomy of Cultivated Plants: Third Symposium,* edited by S. Andrews, A.C. Leslie, and C. Alexander, 253–264. Royal Botanic Gardens, Kew.

Moore, H.E. 1957. *African Violets, Gloxinias and Their Relatives, a Guide to the Cultivated Gesneriads*. Macmillan, New York. 323 pp.

Nishii, K., M. Hughes, M. Briggs, E. Haston, F. Christie, M.J. DeVilliers, et al. 2015. *Streptocarpus* redefined to include all AfroMalagasy Gesneriaceae: Molecular phylogenies prove congruent with geographical distribution and basic chromosome numbers and uncover remarkable morphological homoplasies. *Taxon* 64: 1243–1274.

Pilon, F. 2012. *Saintpaulia: The History and Origins of the African Violet*. Imprimerie Guavin, Gatineau, QC. 182 pp.

Robey, M.J. 1980. *African Violets: Queens of the Indoor Gardening Kingdom*. A.S. Barnes, New York. 199 pp.

Stork, K. and J. Stork. 2007. *You Can Grow African Violets*. The Official Guide Authorized by the African Violet Society of America, Inc. iUniverse Inc., New York. 216 pp.

Teixeira da Silva, J.A., Y.H. Dewir, A. Wicaksono, L. Sahijram, H. Kim, S. Zeng, et al. 2017. African violet (*Saintpaulia ionantha* H. Wendl.): Classical breeding and progress in the application of biotechnological techniques. *Folia Horticulturae* 29(2): 99–111.

## Key Cultivation Websites

Note: There are hundreds of African violet societies, many providing advice.

African Violet Production Guide. www.plantgrower.org/uploads/6/5/5/4/65545169/african_violet_production_guide.pdf.

African Violet Society of America. www.avsa.org/.

African Violet Society of Canada. www.avsc.ca/.

Chen, J. and R.J. Henry. *Cultural Guidelines for Commercial Production of African Violets (Saintpaulia ionantha)*. http://edis.ifas.ufl.edu/ep360.

La Société des Saintpaulia de Montréal. www.saintpaulia-montreal.com/.

Missouri Botanical Garden: African Violets. www.missouribotanicalgarden.org/Portals/0/Gardening/Gardening%20Help/Factsheets/African%20Violets2.pdf.

WikiHow. *Caring for African Violets*. www.wikihow.com/Grow-African-Violets#:~:text=African%20violets%20thrive%20best%20when,you%20could%20chill%20the%20roots.

WikiHow. *How to Get African Violets to Flower*. www.wikihow.com/Get-African-Violets-to-Bloom.

WikiHow. *How to Grow African Violets Indoors*. www.wikihow.life/Grow-African-Violets-Indoors.

WikiHow. *How to Repot African Violets*. www.wikihow.life/Repot-African-Violets.

# 3 Allium (*Allium* Species and Hybrids)

**Scientific Names:** The genus name *Allium* is based on the classical Latin name for garlic. Another interpretation suggests that the name traces to the ancient Greek *aleo*, "to avoid (because of offensive smell)."

**Common Names:** Ornamental *Allium* species are sometimes called "flowering onions," less frequently "flowering garlic" or "flowering leek."

## 3.1 INTRODUCTION

There are about 800 species of *Allium*, mostly native to temperate parts of the Northern Hemisphere. The most important are cultivated as vegetables and spices, notably onion, garlic, leek, chive, scallion, and shallot. Many species are cultivated as ornamentals, some little-changed from the wild species, others representing advanced cultivars and hybrids. The most popular ornamental species include *A. giganteum* and *A. hollandicum*. Ornamental alliums are usually not palatable or are simply poor in taste, but some are edible. The chief attraction of ornamental alliums is their usually spherical flower clusters, described as "colorful fireworks frozen in time."

## 3.2 OVERVIEW

*Allium* species are perennial (sometimes biennial) herbs usually overwintering by bulbs or thickened bases of leaves near the soil. A few species (like chives, *A. schoenoprasum*) grow from rhizomes. The leaves are long and narrow (often described as grass-like and strap-like). The flowers are subtended in umbels – a class of flower cluster in which the bases of the individual flower stalks all converge to a single point. The flower clusters frequently look like pompoms (globe-shaped, with dozens, or even hundreds, of tiny star-shaped blooms), but other arrangements are found. The largest clusters can be 20 cm wide. The flower clusters are displayed on top of long, slender, leafless stalks positioned over a meter above the ground in some cultivars. Some cultivars are no taller than 30 cm; others occasionally reach 2 m in height. Most ornamental allium species produce shades of purple, violet, lavender, blue, or pink or are white. A few are reddish, orange, or yellow, and some produce flower clusters that appear to be bicolored (often, younger and older flowers are differently hued). Flowering usually starts in the late spring or early summer, but some species are fall-blooming. The foliage of most early-blooming alliums tends to die back in the summer, as the plants become dormant. The foliage of ornamental alliums has a slight onion-like scent when crushed. The characteristic aroma of alliums is due to sulfur-containing volatile oils (organosulfoxides), some of which react with air to produce the irritant chemical allyl sulfate.

FIGURE 3.1 Displays of ornamental alliums: (a) Alliums at a flower show in Birmingham, England. (b) Alliums at a flower show in Chelsea, England. (c) A large bouquet of cut flowers.

*Source:* **(a) Photo by Andy Mabbett (CC BY SA 3.0). (b) Photo by Simon (CC BY 2.0). (c) Photo by Shutterstock/Look Studio.**

FIGURE 3.2 Popular ornamental *Allium* species and cultivars: (a) *Allium* 'Red Mohican'. (b) *Allium sphaerocephalon* (round-headed allium, drumstick allium). (c) *Allium moly* (yellow garlic, golden garlic, lily leek). (d) *Allium angulosum* (mouse garlic). (e) *Allium karataviense* (Turkestan onion). (f) *Allium* 'Millenium' (registered by this incorrect spelling). (g) *Allium oreophilum* (pink lily leek). (h) *Allium hollandicum* (Dutch onion, Persian onion). (i) *Allium* 'Red Mohican'.

*Source:* **(a) Photo by Mike Peel (CC BY SA 4.0). (b) Photo by Shutterstock/Alex Manders. (c) Photo by Udo Schröter (CC BY SA 2.0). (d) Photo by Shutterstock/Nomis H. (e) Photo by Shutterstock/Edita Medeina. (f) Photo by F. D. Richards (CC BY SA 2.0). (g) Photo by Agnieszka Kwiecień, Nova (CC BY SA 4.0). (h) Photo by Нащу (CC BY SA 3.0). (i) Photo by Mike Peel (CC BY SA4.0).**

## 3.3 HISTORY

*Allium* species have been cultivated for food for millennia. Onions and chives in flower are quite attractive and

DOI: 10.1201/9781003586227-3

doubtless in the past were sometimes grown simply for their appearance. Interest in the ornamental use of *Allium* was greatly stimulated in the late 1800s when Russian botanists affiliated with the Imperial Botanical Garden in St. Petersburg collected Central Asian species and introduced them to European gardeners. Since then, considerable hybridization has occurred. The parentage of many ornamental cultivars has not been determined.

## 3.4 ECONOMIC IMPORTANCE

The main value of *Allium* is as a source of human nutrition, and the "alliaceous" vegetables and spices collectively rank as one of the world's major food staples. The flowering ornamentals are relatively recent but are becoming quite popular, both in private gardens and as flowering landscapes in public places.

## 3.5 ORNAMENTAL USES

Alliums are employed in flower beds (pure or mixed), borders, and containers and as cut flowers both for fresh and dried arrangements. They are often grown on a large scale in beds for mass effect, and generally, at least a dozen should be planted at a site for visual appearance. Several of the small cultivars are suited to rock gardens. Some of the species are deliberately naturalized as groundcover, but there is some danger of the plants becoming invasive weeds in the area.

## 3.6 OTHER USES

The bulbs and leaves of alliums are familiar, indeed indispensable, foods, especially onions, garlic, chives, and leeks. Some wild alliums are widespread in temperate-climate woods (notably *A. ursinum* in Europe and *A. tricoccum* in eastern North America), and such "wild garlics" or "wild onions," as they are frequently termed, are popular with wild food collectors and, indeed, are sometimes cultivated.

FIGURE 3.3 Large-scale field cultivation of *Allium* 'Globemaster' in The Netherlands.

*Source:* **Photo by Shutterstock/O de R.**

FIGURE 3.4 Landscape uses of ornamental alliums: (a) Borders of a pathway in an elaborately designed garden. (b) Spherical flower clusters complementing spherical sculptures. (c) Huge allium beds in Boston Public Garden, Massachusetts. (d) Border beside a building.

*Source:* **(a) Photo by Shutterstock/Gardens by Design. (b) Photo by Shuttterstock/Steve Horsley. (c) Photo by Daderot (public domain). (d) Photo by Shutterstock/Debbie Smeaton.**

FIGURE 3.5 *Allium* vegetables and spices, including onion, garlic, scallion, shallot, leek, and chives.

*Source:* **Photo by Shutterstock/Madlen.**

Some alliums, especially garlic, are employed medicinally, and indeed, there is extensive research indicating that compounds in *Allium* species have therapeutic value.

## 3.7 PROBLEMS

Alliums employed as food, especially garlic, can produce gastric distress in humans, pets (especially cats and dogs),

FIGURE 3.6 *Allium ursinum*, an example of a widespread species locally harvested as food: (a) In its natural habitat in Germany. (b) Grown in pots.

*Source:* **(a) Photo by Haselburg-müller (CC BY SA 3.0). (b) Photo by Maja Dumat (CC BY 2.0).**

and livestock. More serious, several mimics of *Allium* are very toxic (notably death camas, *Toxicoscordion venenosum*, in North America) and are a potential threat to wild food collectors who misidentify them as edible species. Alliums have caused contact dermatitis in some susceptible individuals.

Some species of *Allium* are significant weeds. The most harmful is the Eurasian *A. vineale*, known as field garlic, crow garlic, and wild onion. The species is a noxious weed in lawns, field, and meadows. It is a very troublesome contaminant of cereals and other crops, and in pastureland, it can lead to off-flavors in milk and meat, and even to severe anemia in livestock animals. Nevertheless, an ornamental cultivar called "Hair" is marketed in Europe and the United States.

## 3.8 CULTIVATION

### 3.8.1 Climate and Site

Most *Allium* ornamental cultivars are suited to grow as perennials in climate zones 4 to 10 (the food species are often grown as annuals and, of course, can be grown in the summers of colder areas). The plants are best provided with full sun but can grow in part shade. Tall cultivars may need staking in windy locations. The foliage of early-blooming alliums senesces in the summer, and to hide the unsightly foliage, gardeners often place the plants behind or among other ornamental species, especially annuals or low-growing herbaceous perennials (so they will not be shaded during early growth).

### 3.8.2 Propagation and/or Purchasing

Seeds are available for some species. Chive (*A. shoenoprasum*), although primarily a food plant, is remarkably ornamental and usually established from seeds (it does not produce bulbs). However, ornamental alliums are usually established from purchased bulbs. *Allium* bulbs are sometimes unavailable from garden centers, and online sources usually offer a better selection. In many cases, cultivars will not breed true from seeds. After growing for a season or two, small offset bulbs will develop. The plants can be excavated in the fall, and the offsets replanted, but they are likely to require two seasons to produce flowers. Rhizome-forming alliums can be lifted and divided at any time of the season, once the clump appears to be crowded.

### 3.8.3 Planting

Allium dormant bulbs should be planted in the fall; container-grown starter plants are best transplanted in the spring. Spacing of 15 to 30 cm is recommended, depending on cultivar. A rule of thumb is that depth of planting should be two to three times (or even four times) the diameter of the bulb. The rhizomes of rhizomatous species can be planted (just below the soil surface) anytime. Allium bulbs are frequently interplanted with the bulbs of other species.

### 3.8.4 Soil and Moisture Conditions

Alliums can grow in various soils but do best in sandy to gritty soil enriched with organic matter. The soil should be well-draining, with a pH of 5.5 to 6.5. Top dressing with rich but well-rotted material is often recommended. Alliums are considered to be drought-tolerant, but the soil should be maintained at "medium moisture." When the plants have become dormant, damp soil may induce the bulbs to rot. The addition of a balanced fertilizer to poor soil is advisable, but otherwise, fertilization may not be required.

### 3.8.5 Maintaining Plants

The mature flower heads of alliums are attractive, and some gardeners prefer to maintain the seed heads on the plants, even overwinter. However, to promote bulb development after flowering, the flower stalks can be pruned off (to direct energy to the bulbs, not fruit development).

## 3.9 FASCINATING FACTS, FABLES, FOLKLORE, AND FRIVOLITY

- Of all the folklore associated with allium species, the most memorable is the legend that garlic served as a defense against vampires, particularly Dracula. The tale is based on the Romanian tyrant Vlad the Impaler (ca. 1431–1476), known for his bloody cruelty, if not for his vampirism. Vlad is thought to have inspired the English writer Bram Stoker's

FIGURE 3.7 The legend of garlic protecting against the vampire Dracula.

*Source:* **(a) Photo by Shutterstock/Anton Brand. (b) Photo by Shutterstock/Scc.comics.**

19th-century *Dracula*, a novel that was the basis for numerous vampire films. Vlad the Impaler's father was called Vlad Dracul; *Dracul* means "the devil" in Romanian. It has been suggested that Count Dracula himself was responsible for spreading the reputation of garlic as a defense against vampires. Since it is clear that garlic is a blood thinner, lowering the clotting action of blood, Count Dracula and other vampires would have feasted more easily on those who consumed garlic.

- The city of Chicago derives its name from the Menomini Indian word *shika'ko*, meaning "skunk place," a reference to the odor of the wild *Allium* that was prevalent at the location.
- Many people confuse true shallots, green onions, and leeks. All but the leek are members of the onion species *A. cepa*. Green onions have a definite bulb with the typical concentric arrangement of scale leaves that the dry onion has, although they are mostly grown for their edible leaves. Leeks (*A. ampeloprasum* var. *porrum*) lack a bulb but do develop a swollen stem base just above the roots ("shallot" is occasionally used to designate leeks – a confusing and undesirable usage). The true shallot is distinguished by its distinctive bulbs, which are made up of a clump of cloves (or small bulbs), reminiscent of garlic, but unlike garlic, the individual bulbs are not closely compacted together by a common membrane.
- "Alliaceous" (pronounces ALL-ley-AY-shus) means pertaining to garlic, onions, and relatives, especially their smell and taste. The word is particularly used for food and breath.
- Major Stephen Long (1784–1864) of the US army led a pioneering scientific expedition into the Great Plains. While exploring Nebraska and Colorado in the 1820s, about 100 men in his party died of scurvy. Local Indians provided a wild allium to the survivors, who did not also perish due to the high vitamin C content of the herb. A decade later, Prince Maximilian of Wied (1782–1867), a German naturalist and explorer, was conducting a 2-year expedition of the Missouri River Valley and also fell ill with scurvy. A cook who had been with Major Long remembered the cure, acquired wild allium from Native Americans, and healed the prince.
- Many *Allium* species are infamous for their distinct, objectionable odor, often causing tears. The volatile oils that help give species of *Allium* their unique flavors contain a class of organic molecules known as amino acid sulfoxides. Peeling, cutting, or crushing *Allium* tissue releases enzymes called allinases, which convert these molecules to sulfenic acids. The sulfenic acids, in turn, spontaneously rearrange to form a volatile gas (syn-propanethial-S-oxide), the chemical that triggers tears. The gas mixes with water in the eyes, forming a weak acid that causes the tear ducts to flood the eye to get rid of the irritant. Rubbing the eyes is a natural reaction, but if juices from the plant are present on the hands, it will probably worsen the irritation. Some solutions: Cut the bulbs under water or run the tap over them as they are sliced. Chill them in the refrigerator prior to cutting (this somewhat reduces the release of the gas). Wear goggles.
- *Jainism* is a religious sect which originated in India in the 6th century BC. There are more than 2 million Jains, mostly in India. The vegetarian dietary code is demanding – no eating at night (animals must be respected, and tiny flying creatures could be inadvertently eaten), no consumption of parts of plants that are essential to their lives (such as roots), and no eating of any member of the onion family.
- "Edible landscaping" refers to plants that serve simultaneously as ornamentals and food. The concept is both ecologically and financially responsible. Of the food species of *Allium*, perhaps the most beautiful is chive (*A. schoenoprasum*), which produces the most impressive flowering display.
- An interesting allium mimic grown as an ornamental: the blood lily (*Scadoxus multiflorus*) of

FIGURE 3.8 Chives (*Allium schoenoprasum*), a food plant that can serve as edible landscaping: (a) A beautiful cultivar in flower. (b) A flowering ornamental border.

*Source:* **Photos from Shutterstock, contributed by (a) Radka Palenikova and (b) Peter Turner Photography.**

FIGURE 3.9 The blood lily (*Scadoxus multiflorus*), a near-doppelganger of the alliums: (a) Wild plants in Zimbabwe. (b) Cultivated plants in a pot.

*Source:* **(a) Photo by BetterWORLDphoto (CC BY SA 3.0). (b) Photo by Shutterstock/Elroma15.**

tropical Africa and Asia. It produces its flowers in umbels, reproduces by bulbs, like alliums, and is in the same family (Amaryllidaceae). However, it is extremely toxic and has been employed as an arrow poison.

## KEY PUBLICATIONS

Block, E. 2010. *Garlic and Other Alliums: The Lore and the Science*. Royal Society of Chemistry, Cambridge, UK. 480 pp.

Davies, D. 1992. *Alliums: The Ornamental Onions*. Timber Press, Portland, OR. 168 pp.

De Hertogh, A.A. and K. Zimmer. 1989. Allium – ornamental species. In: *CRC Handbook of Flowering*, Vol. 6, edited by A.H. Halevy, 22–33. CRC Press, Boca Raton, FL.

Friesen, N., R. Fritsch, and K. Bachmann. 1997. Hybrid origin of some ornamentals of *Allium* subgenus *Melanocrommyum* verified with GISH and RAPD. *Theoretical and Applied Genetics* 95: 1229–1238.

Fritsch, R. 2015. *Checklist of Ornamental* Allium *Species and Cultivars Currently Offered in the Trade*. E-Book. 64 pp. www.researchgate.net/publication/312948238_Checklist_of_ornamental_Allium_species_and_cultivars_currently_offered_in_the_trade.

Khan, S.A., M. Jameel, S. Kanwal, and S. Shahid. 2017. Medicinal importance of *Allium* species: A current review. *International Journal of Pharmaceutical Science and Research* 2: 29–39.

Nomura, Y., M. Saito, and H. Komori. 2005. Breeding of autumn-flowering interspecific hybrids of *Allium*. *Acta Horticulturae* 673: 141–147.

### Key Cultivation Websites

Better Homes and Gardens. *16 Pretty Allium Varieties to Add an Explosion of Color to Your Garden*. www.bhg.com/gardening/flowers/bulbs/alliums-for-your-garden/#slide_555605ba-932e-42fc-a9d8-6cae51804764.

Bulb Blog. *Complete Guide to Growing Allium*. www.bulbblog.com/complete-guide-to-growing-allium/.

Dutch Grown. *How to Grow Alliums*. www.dutchgrown.com/blogs/the-dutchgrown-blog/how-to-grow-alliums.

Horticulture Magazine. *How to Grow Alliums 'Ornamental Onions'*. https://horticulture.co.uk/allium/.

The Spruce. *How to Grow & Care for Allium (Ornamental Onion)*. www.thespruce.com/how-to-grow-alliums-ornamental-onions-1402878.

WikiHow. *How to Plant Allium Bulbs*. www.wikihow.life/Plant-Allium-Bulbs.

# 4 Alstroemeria (*Alstroemeria* Species and Hybrids)

**Scientific Names:** The genus *Alstroemeria* is based on the Swedish naturalist Baron Claus von Alstroemer (1736–1794). Also named for him is the cultivar *Alstroemeria* 'Red Baron'.

**Common Names:** Alstroemeria (sometimes misspelled *alstromeria* because *Alstroemer* can be spelled with a diaresis: *Alströmer*), Peruvian lily (applied to all genera of the Alstroemeriaceae), Inca lily, lily of the Incas. Some species are called parrot lily and princess lily. The names with "lily" reflect the lily-like appearance of the flowers. The odd name "Ulster Mary" is a deliberate mispronunciation of alstroemeria.

## 4.1 INTRODUCTION

Because of their beauty and ability to survive after harvest, cut flowers of alstroemeria (species of the genus *Alstroemeria*) have become one of the florist's favorite flowers. This is remarkable, since alstroemeria is the plant most feared by florists for its ability to produce severe dermatitis. In mild climates, cultivars of alstroemeria are also grown as garden ornamentals and in public landscape beds.

FIGURE 4.1 Baron Claus von Alstroemer (at age 35), Swedish naturalist commemorated by the genus name *Alstroemeria*, coined by his teacher and friend Carl Linnaeus, the father of biological nomenclature. Alstroemer collected plants in southern Europe for Linnaeus.

*Source:* **Public domain painting.**

## 4.2 OVERVIEW

*Alstroemeria* is a genus of about 80 species of South America, mostly southern South America and Eastern Brazil, with centers of diversity in eastern Brazil and Central Chile. The species are found in diverse habitats, ranging from the Andes Mountains (up to 4500 m in altitude) and highland forests to deserts and ocean coastlines. Many are

FIGURE 4.2 Close-up photos of alstroemeria flowers. Their notably lily-like appearance is responsible for their being called "lilies" (true lilies are species of *Lilium*).

*Source:* **(a) Photo by Crabchick (CC BY 2.0). (b) Photo by HQ (CC BY SA 2.0). (c) Photo by Shutterstock/Lepas.**

FIGURE 4.3 Flowering displays of alstroemeria.

*Source:* **(a) Photo by Shutterstock/Prykhodkoira. (b) Photo by Ruth Hartnup (CC BY 2.0). (c) *Alstroemeria aurantiaca*. Photo by HQ (CC BY SA 2.0). (d) Photo by Jacques Gaimard from Pixabay (public domain). (e) Cultivar Indian Summer. Photo by Russell Harris (public domain). (f) Photo by Halley Pacheco de Oliveira (CC BY SA 3.0).**



DOI: 10.1201/9781003586227-4

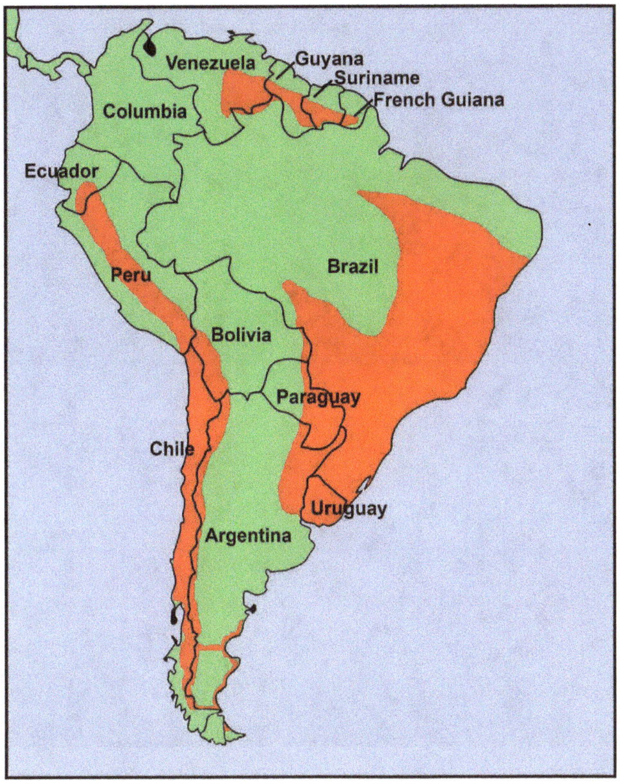


FIGURE 4.4 Native geographical distribution of *Alstroemeria*, in orange.

*Source:* **After Sanso et al. (2005).**

very restricted geographically, and some are threatened by habitat degradation, especially from agricultural development. The genus was once included in the Amaryllidaceae or the Liliaceae but is now placed in the Alstroemeriaceae, a small family of four genera. The northernmost Indigenous range includes southern Venezuela and northwest Brazil. Several species have become naturalized outside of their native South America, and some have become invasive.

The plants are mostly herbaceous perennials, growing both vegetative and reproductive shoots from creeping rhizomes or clusters of tuberous roots (true roots thickened to become storage and perennating organs, like those of dahlias and sweet potatoes; by contrast, stem tubers like those of the common potato develop from thickened rhizomes). The "tubers" can accumulate considerable starch. The flower clusters (with few to several dozen blooms) tend to be on the ends of branches. Sexual reproduction is by many small globose seeds produced in explosive capsules. The leaves are more or less elongated, often without an evident petiole, and as they unfold, they become *resupinate* (they turn upside down because of twisting at the base so that the lower and upper surfaces are reversed). The flowers are lily-like (trumpet form), with six stamens and six petaloid tepals. The outer three tepals are uniform, while one of inner set of three is different from its companions. The flowers of the wild species vary considerably in color, including shades of white, yellow, pink, red, purple, and violet. The wild species are of quite limited commercial value because their flowers are often small, cut flowering stems have limited shelf life, and the plants often become dormant for part of the year.

Some cultivars grow as tall as 2.5 m, but dwarf forms (about 30 cm in height) are also available (the "Princess" series is popular). Hundreds of hybrids are available, some with very large solitary flowers, others with clusters of smaller blooms. Flower colors of cultivars may be lavender, orange, peach, pink, purple, red, white, or yellow. The inner set of tepals has colored lines, which are nectar guides for pollinators. Cut flowering stems bearing flowers with solid colors are often presented in uniform bouquets or, alternatively, in mixtures of different shades. Many recent cultivars bear flowers with striking bicolor and multicolor patterns of streaks, freckles, and spots. For growth of plants in pots or in warm climates, cultivars have often been selected that have evergreen foliage and bloom continuously, often for months. Most species and cultivars are not noticeably fragrant, but recent scented lines have been produced.

Many alstroemeria cultivars are divided into two groups based on flowering habit. The "butterfly type" flowers for 9–12 months annually, depending on cultivar and environment. These have shorter growth habits and larger, more open flowers and are better suited to be potted plants. Most butterfly-type alstroemerias are polyploid (with an extra set of chromosomes) and can produce seeds. The "orchid type" flowers mainly for a period of 3–5 months in the spring, with limited or no flowering at other times, remaining vegetative until the spring. These cultivars are tall and produce a large number of flowers in a short period and so are employed mostly as cut flowers. Most orchid-type alstroemerias are diploid (with just one set of chromosomes) and sterile and so are propagated vegetatively.

FIGURE 4.5 *Alstroemeria* paintings by Pierre Joseph Redouté (1759–1840). Redouté, a Flemish (Belgian) painter specializing on flowering plants, is regarded as, arguably, the finest botanical illustrator of all time. (a) *Alstroemeria pelegrina* and (b) *A. ligtu* from Les liliacées (1805–1816; Paris), (c) *Alstroemeria pelegrina* from Choix des plus belles fleurs et des plus beaux fruits (1833; Paris).

*Source:* **Public domain.**

**FIGURE 4.6** Early European botanical paintings (public domain) showing old garden alstroemeria cultivars: (a) *Alstroemeria aurea*. (b) Mixed cultivars. (c) Mixed cultivars.

*Source:* **(a) Van Eeden, A. C. 1872–1881. *Album van Eeden, Haarlem's flora, afbeeldingen in kleurendruk van verschillende bol- en knolgewassen*. (b) *Revue horticole* (Paris), vol. 56, 1884. (c) Van Houtte, L. (ed.). 1845. *Flore des serres et des jardins de l'Europe (Gand)*, vol. 1.**

**FIGURE 4.7** Commercial production and marketing of alstroemeria: (a) Greenhouse production. (b) Bouquets on sale in a flower shop.

*Source:* **(a) Photo by Shutterstock/Lidian Neeleman. (b) Photo by Shutterstock/Medvedeva Oxana.**

**FIGURE 4.8** Bouquets of alstroemeria cut flowers in vases.

*Source:* **(a) Photo by Shutterstock/Pitroviz. (b) Photo by Kowloonese (CC BY SA 3.0). (c) Photo by ProFlowers (CC BY 2.0).**

**FIGURE 4.9** Alstroemerias used as hair flowers.

*Source:* **(a, b) Shutterstock/Freya-photographer. (c) Shutterstock/Africa Studio.**

## 4.3 HISTORY

Collections were brought back to Europe in the 16th century by explorers. Early breeding, especially by hybridization, involved *A. aurea* (*A. aurantiaca*), *A. ligtu*, *A. pelegrina* (all from Chile), and *A. psittacina* (from Brazil). In more recent times, breeding has been conducted in several countries, notably The Netherlands, England, the United States, and Japan, and at least a dozen more species have been employed as parents in the production of new hybrids. The requirement for a period of dormancy was a problem in some early cultivars, but this has been overcome.

## 4.4 ECONOMIC IMPORTANCE

Alstroemeria is primarily of value as a cut flower used for bouquets and flower arrangements. It is among the world's ten leading commercial cut flowers. In Europe and North America, alstroemeria's popularity expanded considerably in the 1970s. Alstroemerias are rather similar to lilies of the genus *Lilium*, and in some countries, cut flower alstroemeria production is similar in scale to cut flower lily production.

## 4.5 ORNAMENTAL USES

Aside from their primary use as commercial cut flowers, alstroemerias are grown outdoors as bedding plants in warm climates.

## 4.6 OTHER USES

Alstroemeria is one of the largest, most colorful of cut flowers supplied by florists and so is pre-adapted to use as a personal floral decoration, especially as a hair flower.

## 4.7 PROBLEMS

All parts of alstroemeria plants can cause significant contact dermatitis, especially for florists and growers (it has been claimed that the plant is the most common cause of allergic contact dermatitis among florists). The flowers contain the most concentrated amounts of allergens. The main allergens are 6-tulipalin A and tulipalin A. Tulipalins are also the cause of dermatitis due to tulips. Gardeners are advised to wear impermeable plastic gloves when working with the plants. The allergens can pass through vinyl, such as polyvinyl chloride gloves; nitrile gloves have been recommended, and for those who are concerned about the toxic potential of bouquets, the same advice is warranted. Although the American Society for the Prevention of Cruelty to Animals (ASPCA) website states that alstroemeria is non-toxic to dogs, cats, and horses, this is contradicted by many other online sources, which indicate that pets ingesting the plant may experience mild to moderate gastric toxicity. Cutting or damaging the plant can even release allergens into the air, and although the risk of airborne toxicity is limited, there is documentation showing that this has caused pulmonary difficulties in some people. Negative responses may take a day or more to develop, and sensitization (additional contact eliciting more severe reactions) may occur. While alstroemeria has toxic potential, except for people who regularly handle it as part of their occupation, reports of adverse reactions are uncommon and have not prevented the plant from becoming one of the world's top 10 cut flowers. The dangers posed by alstroemeria are comparable to those posed by tulips (but tulip bulbs are more toxic than alstroemeria tubers). Alstroemerias are so widely handled as bouquets – even used as boutonnières and corsages – that contact with skin is inevitable. Nevertheless, the plant has rarely been a concern.

## 4.8 CULTIVATION

### 4.8.1 Climate and Site

Alstroemerias are mainly grown for the cut flower trade but are also useful as outdoor container and garden plants, especially in warm climates. Recently, cultivars have been bred that are relatively cold-hardy and capable of growing as outdoor perennials in colder regions than possible in the past. In colder climates, alstroemeria beds can be established in the spring for growth as annuals. Some garden catalogues offer cold-hardy cultivars claimed to be able to overwinter as far north as climate zone 5. However, most cultivars are cold-sensitive, and rhizomes or tubers will only tolerate a few degrees of freezing. Overwintering outdoors in colder areas may be possible if planted against a warm foundation and a mulch is provided for insulation. For some species, excessive heat results in much better tuber development at the expense of sparse flowering. Because of the limited hardiness of the hardiest cultivars, gardeners may prefer to grow the much more cold-tolerant Asiatic lily (*Lilium auratum*). Alstroemerias grow well in pots, but as they do best in sunny locations, they are not ideal houseplants. For outdoor growth, plants should be in locations that provide full sunlight or partial shade (filtered shade is recommended for hot climates) and protection from wind. The plants do well in raised beds. When grown indoors, they are best maintained in a cool area. Potted plants moved to gardens may bloom as early as late spring.

FIGURE 4.10 Examples of landscape uses of alstroemeria: (a) Flower bed with mixed cultivars. (b) A large garden bed. (c) A border beside a house. (d) Patterned garden beds.

***Source:* (a) Photo by Dhruv 1103 (CC BY SA 4.0). (b) Photo by Shutterstock/J. Hamilton. (c) Photo by Anne and David (public domain). (d) Photo by Jean Jones (CC BY 2.0).**

### 4.8.2 Propagation and/or Purchasing

In cold climates, it is not feasible to grow plants outdoors from seeds, so plants that have been established indoors are transplanted, or tubers or rhizomes are planted. Seeds are often not available for hybrid cultivars (and, of course, they probably will not reproduce beautiful flowers identical to their parents'), and seeds for some species are difficult to germinate (for advice, see the Pacific Bulb Society website, cited later). Commercially, several modern cultivars are frequently started from tissue culture (i.e., by micropropagation). Usually, the simplest way to establish plants is to purchase potted plants or tubers (tuberous roots). Tubers for planting outdoors are best obtained from specialty "bulb" suppliers, who may be able to provide the hardiest cultivars. In warm climates, underground parts can be divided and planted where desired. It is possible to dig up the tubers or rhizomes in the fall and store them overwinter in moist potting mix, at 2–5°C, in the dark. However, in warm areas, where long-term beds are to be established, digging up the tubers is inadvisable for 2 to 3 years following transplantation, so that the underground parts will develop well.

### 4.8.3 Planting

The tubers or rhizomes are often brittle and need to be handled gently. If a planting depth is not specified by the supplier, they may be placed 10–20 cm below the soil surface. Deep cultivation of the soil (to 35 cm or more) to improve drainage will reduce possible infection by root rot, to which alstroemeria is susceptible. If spacing is not specified by the supplier, the tubers may be positioned 30 cm apart. Tall varieties may require staking or trellising.

### 4.8.4 Soil and Moisture Conditions

Cultivated alstroemerias grow best in moist, well-drained, rich organic soils. A pH between 5.5 and 6.5 is recommended.

Well-rotted manure or compost is best. Alternatively, a balanced fertilizer may be applied occasionally (some say once per season, others every 2–3 weeks). Some authorities recommend application of a half-strength, soluble bloom-type fertilizer once flower buds are initiated.

### 4.8.5 Maintaining Plants

Most people encounter alstroemerias in bouquets. The cut flowers can have a very long post-harvest vase life, lasting 2 weeks or longer. Indeed, alstroemerias are likely to be the last flowers enduring in vase bouquets. The following recommendations have been made to prolong vase life: Ensure that the vase is clean. Fill it with water at room temperature. Add cut flower food. Diagonally trim away 3–5 cm from the bases of stems using a sharp, clean knife or secateurs. Ensure leaves do not hang in water. Place vase away from drafts, in full sun, and away from fruit (ethylene emissions hasten flower senescence, and alstroemeria is particularly sensitive). Regularly top up water in vase. Remove wilted flowers.

With reference to garden plants, it is widely thought that *cutting* alstroemeria flowers will slow the growth and flower production of the plant. Accordingly, to harvest the flowering stems for display in a vase, it has been strongly recommended that the stalks be grasped close to the ground and *pulled* out of the rhizome or tuber with a rapid upward-twisting, picking movement.

## 4.9 FASCINATING FACTS, FABLES, FOLKLORE, AND FRIVOLITY

- Although alstroemeria is not considered to be an edible plant, starch employed as a medicinal food was obtained from the thicker roots of *A. ligtu* by Indigenous peoples in South America. Native people have frequently learned how to process different starch-containing toxic plants, as also exemplified by their use of some extremely toxic cycads (tropical allies of coniferous trees that look like palm trees).
- Numerous ornamental plants have naturally acquired symbolic meanings over the centuries, both positive and negative. However, old scary superstitions can frighten people from bringing particular plants inside their home, so the florist trade not only avoids mentioning darker stories but has also often invented inspirational symbolic associations to stimulate sales. It has been alleged that alstroemerias symbolize friendship, and even that the six petals stand for *understanding*, *humor*, *patience*, *empathy*, *commitment*, and *respect*.

FIGURE 4.11 A large bouquet of cut alstroemeria flowers on a windowsill.

*Source:* **Photo by Shutterstock/Pitroviz.**

## KEY PUBLICATIONS


Aagesen, L. and M. Sanso. 2003. The phylogeny of the Alstroemeriaceae, based on morphology, rps16 Intron, and rbcL sequence data. *Systematic Botany* 28: 47–69.

Adams, R.M., A.D. Daily, R.R. Brancaccio, I.P. Dhillon, and E.C. Gendler. 1990. Alstroemeria. A new and potent allergen for florists. *Dermatologic Clinics* 8: 73–76.

Aker, S. and W. Healy. 1990. The phytogeography of the genus *Alstroemeria*. *Herbertia* 46: 76–87.

Aros, D., M. Suazo, C. Rivas, P. Zapata, C. Úbeda, and M. Bridgen. 2019. Molecular and morphological characterization of new interspecific hybrids of alstroemeria originated from *A. caryophlleae* scented lines. *Euphytica* 215: 93. https://doi.org/10.1007/s10681-019-2415-4.

Babarabie, M., H. Zarei, and F. Varasteh. 2016. Potential of increasing the vase life and improvement of some physiological characteristics of *Alstroemeria* cut flowers by using non-harmful compounds environmentally. *Journal of Chemical Health Risks* 6: 1–8.

Bridgen, M.P. 2018. Alstroemeria. In: *Ornamental Crops, Handbook of Plant Breeding 11*, edited by J. Van Huylenbroeck, 231–236. Springer International.

Bridgen, M.P., E. Kollman, and C. Lu. 2009. Interspecific hybridization of *Alstroemeria* for the development of new ornamental plants. *Acta Horticulturae* 836: 73–78.

Chacón, J., M. Camargo de Assis, A.W. Meerow, and S. Renner. 2012. From East Gondwana to Central America: Historical biogeography of the Alstroemeriaceae. *Journal of Biogeography* 39: 1806–1818.

Finot, V.L., C.M. Baeza, E. Ruiz, O. Toro, and P. Carrasco. 2017. Towards an integrative taxonomy of the genus *Alstroemeria* (Alstroemeriaceae) in Chile: A comprehensive review. In: *Selected Studies in Biodiversity*, edited by B. Şen and O. Grillo. IntechOpen, https://doi.org/10.5772/intechopen.71823. www.intechopen.com/books/selected-studies-in-biodiversity/towards-an-integrative-taxonomy-of-the-genus-alstroemeria-alstroemeriaceae-in-chile-a-comprehensive-.

Hofreiter, A. and E. Rodríguez. 2006. The Alstroemeriaceae in Peru and neighbouring areas. *Revista Peruana de Biología* 13: 5–69.

Hoshino, Y. 2008. Advances in alstroemeria biotechnology. In: *Floriculture, Ornamental and Plant Biotechnology. Advances and Topical Issues*, Vol. 5, edited by J. da Silva, 540–547. Global Science Books, Ikenobe, Japan.

Kim, J.S. and J.H. Kim. 2013. Comparative genome analysis and phylogenetic relationship of order Liliales: Insight from the complete plastid genome sequences of two lilies (*Lilium longiflorum* and *Alstroemeria aurea*). *PLoS One* 8(6): e68180. https://doi.org/10.1371/journal.pone.0068180.

Kristiansen, K. and L.P. Christensen. 1998. Allergen contents in *Alstroemeria* can be reduced by breeding. *Euphytica* 101: 367–375.

Lim, S.-S., S.-I. Lee, S.-C. Kang, and J.-B. Kim. 2012. *Alstroemeria* plants and its biotechnological applications. *Journal of Plant Biotechnology* 39: 219–224.

Marks, J.G. 1988. Allergic contact dermatitis to *Alstroemeria. Archives of Dermatology* 124: 914–916.

McCluskey, J., M. Bourgeois, and R. Harbison. 2014. Tulipalin A induced phytotoxicity. *International Journal of Critical Illness and Injury Science* 4: 181–183.

Muñoz, M. and A. Moreira. 2003. *Alstroemerias de Chile: Diversidad, Distribución y Conservación*. Taller La Era, Santiago, Chile. 139 pp.

Nasri, F., M.K. Saba, N. Ghaderi, A.A. Mozafari, and T. Javadi, T. 2014. Improving germination and dormancy breaking in *Alstroemeria ligtu* hybrid seeds. *Trakia Journal of Science* 12: 38–46.

Park, T.-H., I.-S. Han, and J.-B. Kim. 2010. Review on the development of virus resistant plants in *Alstroemeria. Journal of Plant Biotechnology* 37: 370–378.

Reid, A. 2006. *Alstroemeria. Bulletin 4688*. Department of Agriculture and Food, Western Australia, Perth. 8 pp.

Sanso, A.M. 1996. El género *Alstroemeria* (alstroemeriaceae) en Argentina. *Darwiniana* 34: 349–382. (In Spanish, English Abstract.)

Sanso, A.M., M. Camargo de Assis, and C.C. Xifreda. 2005. *Alstroemeria*: A charming genus. *Acta Horticulturae* 683: 63–77.

Sanso, A.M. and C. Xifreda. 2001. Generic delimitation between *Alstroemeria* and *Bomarea* (Astroemeriaceae). *Annals of Botany* 88: 1057–1069.

Sarwar, A.K.M.G., H. Hoshino, and H. Araki. 2010. Pollen morphology and infrageneric classification of *Alstroemeria* L. (Alstroemeriaeae). *Grana* 49: 227–242.

Shahriari, A.G., A. Bagheri, and Z. Shahriari. 2013. In vitro regeneration of *Alstroemeria*: A review. In: *Bulbous Plants: Biotechnology*, edited by K.G. Ramawat and J.M. Mérillon, 270–283. CRC Press: Boca Raton.

Uphof, J.C.T. 1952. A review of the genus *Alstroemeria. Plant Life (Herbertia Edition)* 8: 37–53.

Wagstaff, C., U. Chanasut, F.J. Harren, L.J. Laarhoven, B. Thomas, H.J. Rogers, et al. 2005. Ethylene and flower longevity in *Alstroemeria*: Relationship between tepal senescence, abscission and ethylene biosynthesis. *Journal of Experimental Botany* 56(413): 1007–1016. https://doi.org/10.1093/jxb/eri094.

Wilkins, H.F. 1986. The history of alstroemeria production in North America. *Herbertia* 41: 15–19.

## Key Cultivation Websites

Healy, W.E. and H.F. Wilkins. *Alstroemeria Culture. https://hortscans.ces.ncsu.edu/file/2018/07/alstroem_537ba198e59b3.pdf.*

National Cut Flower Centre. *Alstroemerias (Alstroemeria Hybrids) as a Tunnel-Grown Cut Flower Crop*. https://projectblue.blob.core.windows.net/media/Default/Imported%20Publication%20Docs/Alstroemerias%20as%20a%20tunnel%20grown%20cut%20flower%20crop.pdf.

Pacific Bulb Society. *Alstroemeria*. www.pacificbulbsociety.org/pbswiki/index.php/Alstroemeria.

The Spruce. *How to Grow and Care for Peruvian Lily (Alstroemeria)*. www.thespruce.com/peruvian-lily-alstroemeria-plant-profile-4690186.

WikiHow. *How to Grow Alstroemeria*. www.wikihow.com/Grow-Alstroemeria#:~:text=Plant%20when%20temperatures%20do%20not,10%C2%B0%20C)%20at%20night.

# 5 Amaryllis (*Hippeastrum* Species and Hybrids)

**Scientific Names:** This chapter examines cultivated forms of the large South American genus (over 90 species) *Hippeastrum*, usually termed "amaryllis." This has led to confusion with the related genus *Amaryllis*, which includes only two South African species, *A. belladonna* and A. *paradisicola*. The use of the name amaryllis for two different genera traces to past failure to recognize their separateness. The name amaryllis is thought to trace to a classical Greek tale for a shepherdess who shed her blood (to impress the man she loved), which turned into the flower. The name *Hippeastrum* was coined by English botanist William Herbert from the Greek *hippeus*, "knight," and *astron*, "star." One explanation of the name's origin is that the flower supposedly resembles a "knight's star," a spiky ball on the end of a long handle employed as a medieval weapon. The name *H. hybridum* is often used as a collective phrase for the cultivars.

**Common Names:** "Amaryllis" and "hippeastrum" are both employed as English names for *Hippeastrum* species, but the former is much more frequent. Grammatical websites usually give the plural of "amaryllis" as "amaryllises," but in practice, "amaryllis" is frequently treated as both singular and plural (so is the genus name *Amaryllis*). Similarly, "hippeastrum" is often treated as both singular and plural (so is the genus name *Hippeastrum*), although "hippeastrums" is also common as the plural. Some of the species are termed "lilies," such as Barbados lily and St. Joseph's lily. "Cluster amaryllis" refers to species of *Lycoris*; 'blue amaryllis" is *Worsleya*; "orchid amaryllis" is *Sprekelia* (all these are also in the Amaryllidaceae family).

FIGURE 5.1 Amaryllis the shepherdess, after whom the plant name allegedly originated.

*Source:* **Painting (public domain) by English artist William Holman Hunt (1827–1910).**

## 5.1 INTRODUCTION

Ornamental plants are grown mostly for their displays of colorful flowers – frequently developing either a large number of small flowers or a small number of large flowers. Amaryllis exemplifies the latter pattern, and indeed, few herbaceous plants produce equally ostentatious, indeed stunning, attention-getting giant blooms. Except in very warm climates, amaryllis is mostly cultivated indoors as a houseplant, and this "queen of winter bulbs" provides wonderful displays during often bleak winters.

FIGURE 5.2 An attractive domestic scene of a little girl contemplating an amaryllis bulb and the large plant grown from a bulb.

*Source:* **Photo by Shutterstock/Hanna Zavorotna.**



DOI: 10.1201/9781003586227-5

## 5.2 OVERVIEW

The genus *Hippeastrum* is native to Central and South America and is mostly tropical and subtropical. The species occur especially in eastern Brazil and the central southern Andes of Bolivia, Peru, and Argentina. Lara Rico et al. (2021) provide the number of species for these countries: Brazil, 38; Bolivia, 34; Peru, 21; and Argentina, 10. Several species extend north to Mexico and the West Indies. Unfortunately, many of the species are threatened by urban development. The plants are adapted to a long moist growing season, followed by a shorter dry season. At the start of the rainy season, the storage bulb produces its foliage and flowers. Flowering lasts 2 to 3 weeks, but the foliage grows throughout the moist season, allowing new flower buds to form within the bulb. The old bulb is perennial but annually buds off new bulblets. Although the foliage is shed annually in the wild, it is termed "evergreen" and, in moist environments, can persist for more than a year. At the close of the rainy season, the plant becomes dormant.

Most cultivated amaryllises are started from dormant storage bulbs, which are usually 5–12 cm in diameter. The bulbs are tunicate (with a tunic, that is, a protective dry outer layer, and fleshy concentric inner scales or leaf bases), like onions, tulips, and daffodils. The leaves are typically about 60 cm long (occasionally up to 1 m), usually strap-like (typically about 5 cm wide, sometimes up to 10 cm), evergreen, and produced from the base. When grown from bulbs, the foliage may be produced simultaneously with the flowering stalk or, quite frequently, afterward. The flowers are borne on a leafless stalk, typically 50–90 cm tall, which in most cultivars bears one or several flowers (often 2–5). The bulbs can simultaneously or sequentially produce more than one flower stalk. Cultivars typically have large (often up to 25 cm in diameter) trumpet-shaped flowers available in shades of white, pink, purple, red, orange, and salmon (rarely, some cultivars have yellow or purple flowers, although these colors are characteristic of some of the wild species). The flowers of some cultivars are bicolored, spotted, streaked, or variegated, and doubled flowers (i.e., with extra petals) are also available.

Spider amaryllis (*Hippeastrum cybister*) is an unusual species with elongated floral parts reminiscent of spider legs. 'Chico' is a popular cultivar. Cluster amaryllis (*Lycoris* species) also have similar-looking flowers, and so do a variety of unrelated plants called "spider lilies."

FIGURE 5.3 Amaryllis (*Hippeastrum*) bulbs: (a) Dormant bulb. (b) Bulb with three floral stalks in very early development. (c) Bulb with two developing floral stalks, one much more mature than the other.

***Source:*** **(a) Photo by Simon Lee (CC BY SA 2.0). (b) Photo by Teebeutel (CC BY SA 3.0). (c) Photo by Böhringer Friedrich (CC BY SA 3.0).**

FIGURE 5.5 Illustrations of old cultivars with bicolored flowers.

***Source:*** **(a) *Flore des serres et des jardins de l'Europe* (Belgium), vol. 16, 1865–1867. (b) John Lewis Childs garden catalogue, Floral Park, NY, 1892. (c) ArtsyBee from Pixabay (public domain).**

FIGURE 5.4 Pots of amaryllis exhibited at a flower show.

***Source:*** **Photo by Laura Blanchard (CC BY SA 2.0).**

FIGURE 5.6 Flowers of spider amaryllis (*Hippeastrum cybister*).

***Source:*** **Photo by SAplants (CC BY SA 4.0).**

## 5.3 HISTORY

Early domestication of *Hippeastrum* involved interspecific hybridization using a very limited number of species. *Hippeastrum* × *johnsonii*, a hybrid of *H. vittatum* and *H. reginae* made in England in 1799, is thought to have been the first hybrid. Early hybrids were produced mainly from *H. aulicum*, *H. psittacinum*, *H. puniceum*, *H. reginae*, *H. reticulatum*, *H. striatum*, and *H. vittatum*. Hybridization has been ongoing for over 200 years, particularly selecting mostly for large flower size and color, but there are also recently created multi-flowered and double-flowered varieties. Most wild species and most cultivars do not have fragrant flowers, but there have been recent efforts to breed for pleasant scent. Most wild species are diploid (with 22 chromosomes), but almost all hybrid cultivars are tetraploids (with 44 chromosomes; doubling the chromosome number tends to produce larger flowers, which is probably why the condition is common among cultivars). The number of cultivars is difficult to determine – perhaps over 1,000. Breeding programs have been conducted primarily by private companies in The Netherlands and South Africa, with interest developing more recently in the United States.

FIGURE 5.7 Paintings (public domain) of early cultivars of amaryllis (*Hippeastrum*): (a) *Hippeastrum* × *johnsonii*. (b) *Hippeastrum striatum*. Yellow flowers are rare in the genus. (a, b) Prepared by English botanist Priscilla Susan Bury (1799–1872). (c) Prepared by Beatrice Parsons.

*Source:* **Public domain. Wright, H. J., and W. P. Wright. 1922. *Beautiful Flowers and How to Grow Them*. T. C. & E. C. Jack, Edinburgh.**

FIGURE 5.8 Flowers of modern amaryllis (*Hippeastrum*) cultivars with especially large flowers: (a) 'Barbados Novelty'. (b) 'Minerva Mammoth'.

*Source:* **Photos by B. Brookes.**

## 5.4 ECONOMIC IMPORTANCE

Amaryllis is considered to be among the world's top 20 commercial flowering ornamentals. Principal exporting countries include The Netherlands, the United States, Brazil, Japan, Israel, and South Africa. The bulbs are produced commercially in field culture, and sometimes also in greenhouses. In various countries, harvested bulbs are grown by greenhouse producers for potted plants marketed for major holidays, notably Thanksgiving, Christmas, and Mother's Day. There is also a substantial market for "cut flowers" (i.e., cut flowering stems), and in The Netherlands, tens of millions are sold annually. Amaryllises are also often sold to consumers as dry bulbs for planting and growth in homes, marketed both by retail outlets and online suppliers.

## 5.5 ORNAMENTAL USES

In temperate regions, amaryllis is mostly grown indoors, often as a short-term plant acquired as a gift. Amaryllis is widely grown in containers, both indoors and outside. In warm regions, it is also grown as a garden perennial in beds and borders, and sometimes the plant is allowed to naturalize in an area.

## 5.6 OTHER USES

The Amaryllidaceae contain unique alkaloids, most commonly lycorine, which has been found in almost all species

FIGURE 5.9 Sales of amaryllis (*Hippeastrum*) in stores in Amsterdam, Netherlands: (a) Potted plants. Bulb, and seeds. (b) Bulbs.

*Source:* **(a) Photo by Thierrytutin (CC BY 2.0). (b) Photo by Michael Coghlan (CC BY SA 2.0).**

FIGURE 5.10 Amaryllises (*Hippeastrum*) grown indoors in containers.

*Source:* **(a) Cultivar Gervase. Photo by Alexey Yakovlev (CC BY SA 2.0). (b) Photo by Laredawg (public domain). (c) Cultivar Ruby Star. Photo by Barb How (CC BY 2.0).**

FIGURE 5.11 Amaryllises grown outdoors in pots in Vietnam.

*Source:* **Photos by Lienyuan Lee (CC BY 3.0).**

FIGURE 5.12 Amaryllises grown outdoors in ornamental borders and beds: (a) Amaryllis in Thailand. (b) Amaryllis in Australia. (c) *Hippeastrum correiense* in Texas. (d) Amaryllis in Hawaii.

*Source:* **(a) Photo by Shutterstock/Praneem79. (b) Photo by Shutterstock/Sandra Lass. (c) Photo by Surely Shirly (public domain). (d) Photo by Forest and Kim Starr (CC BY 3.0).**

of *Hippeastrum*. Alkaloids are responsible for the toxicity of the genus. Although the folk uses of various species are not accepted in modern medicine, the alkaloids are under investigation for a wide variety of potential therapeutic applications.

## 5.7 PROBLEMS

All parts of amaryllises are toxic and can cause nausea, vomiting, and diarrhea if ingested. The plants should be kept out of reach of children and pets. They are safe to handle for most people, but the sap may cause skin irritation in sensitive individuals.

## 5.8 CULTIVATION

### 5.8.1 Climate and Site

In cold temperate climates, amaryllises are grown primarily as houseplants or as potted plants protected indoors during the winter. By contrast, they are grown as in-ground perennial garden plants in North America, in regions rated at no colder in winter than −12°C (climate zone 7b). In cold climates, they can be established in the spring in soil

FIGURE 5.13 Pets associating with amaryllises, which could seriously poison them: (a) Shih tzu dog. (b) Tabby cat.

*Source:* **(a) Photo by Shutterstock/Suponev Vladimir. (b) Photo by Shutterstock/Crazy Nook.**

outdoors but will need to be lifted in the fall. Occasionally, they are also grown as outdoor annuals. Amaryllis can grow outdoors in both full sun and moderate shade but usually grows best in semi-shade.

### 5.8.2 Propagation and/or Purchasing

Cultivated amaryllises are sometimes propagated by seeds, but some are infertile or partly so, and seeds may be unavailable or of poor quality. Usually, named varieties are reproduced vegetatively, since the seeds will not produce plants true to type. Commercial reproduction is by vegetative techniques (offsets, scaling, and tissue culture) employing bulb tissues. Amaryllises are available from nurseries and retail stores as potted plants and as dormant bulbs intended for forcing. Bulbs are especially sold in the fall for early-winter bloom. Kits with dry bulbs already potted along with directions included are also available. Specialty online distributors usually provide a much wider choice of bulbs. Larger bulbs usually provide bigger plants, more flowering stalks, and/or bigger flowers and sometimes produce new flowering stalks after the flowers of the early-flowering stalks fade. Generally, bulbs larger than 7.5 cm in diameter are recommended (although some cultivars have naturally smaller bulbs).

### 5.8.3 Planting

**Containers:** When grown in pots, purchased dry bulbs should be established 6–8 weeks before the desired time of flowering (individual flowers last 1–2 weeks). Suitable pots for single bulbs can be small (the plants grow best when slightly potbound), allowing 5 cm of soil around the sides of the bulb. For more impressive displays, several bulbs can be planted close together (even touching). The flowering stalks can become top-heavy when in bloom, so the container should be heavy or gravel should be added to the bottom to add weight to prevent tipping. One-third to two-thirds of the

FIGURE 5.14 Bulbs planted as a group for a mass potted display: (a) Dormant bulbs. (b) Bulbs in early production of flowering stems.

*Source:* **Photos by Laura Blanchard (CC BY SA 2.0).**

pointed end of the bulbs should remain above the soil. Place the container in a cool room. When the stalk(s) begins to grow, the container should be in a location where it will receive several hours of direct sunlight and temperatures close to 24°C (reflecting its origin from warm climates).

**Outdoors:** In gardens, plants can be spaced 30 to 45 cm apart, but otherwise, the plants should be established as for containers.

### 5.8.4 Soil and Moisture Conditions

Containers employed should have drainage holes. Whether in pots or in the ground, the soil should be porous (well-draining), for example, prepared with equal parts garden loam, peat moss, and Perlite or sand. Circumneutral (pH 6–8) soil is recommended. Water when the soil becomes dry to the touch. Waterlogged soil induces fungal diseases. Fertilization is unnecessary if the plant is to be discarded after flowering.

Although amaryllises are described as "evergreen," some so-called deciduous species must be allowed to go into a dry winter dormancy, in which case the foliage senesces and dies. If these are forced to continue growing by watering, they may refuse to flower or the bulbs may rot.

### 5.8.5 Maintaining Potted Plants Indoors

During the growth of the stalk, while the plant is on a windowsill, turn the pot daily to promote even upward growth. If the stalk becomes very tall, it may require stalking.

### 5.8.6 Maintaining Potted Plants after Flowering

Rejuvenating the spent plants requires effort, but for those who are willing, the following steps are suggested. After flowering, the stalk should be cut away 5–8 cm above the bulb. The foliage will continue to grow. Watering of the vegetative plant should continue as needed, taking care not to overwater. A balanced fertilizer is recommended. After several months of indoor growth (under the best light that can be provided), the plant can be induced to become dormant by restricting water. The senescent leaves can be cut within 5 cm of the bulb and discarded. A minimum of 2–3 months of dormancy is recommended (stored dry in a cool place). The plant can be induced to flower again by rewatering. Alternatively, assuming the plant flowered indoors during the winter, the potted vegetative plant can be moved outdoors for the summer (it can be plunged into the ground) and will enter dormancy (the leaves senesce) naturally in late summer or early fall. The dormant bulb can be stored dry, in its pot, at 5–7°C, and (as noted earlier) induced to flower again by rewatering. Repotting is usually not needed for 3 or 4 years and, indeed, should be avoided earlier.

## 5.9 FASCINATING FACTS, FABLES, FOLKLORE, AND FRIVOLITY

- The Château de Beloeil (a refurbished ultra-luxurious castle) in Belgium is a major tourist attraction. It is the site of an annual competition lasting more than a week among professional florists with the goal of preparing the best amaryllis bouquet (www.chateaudebeloeil.com/en/event/amaryllis-competition-2018/).
- "Waxed amaryllis bulbs" are, true to the phrase, bulbs that have been covered in wax (usually brightly colored and often molded into attractive shapes), except for the uppermost part, to allow the flower stalks to develop. They are self-standing, and no potting is required. The wax cover minimizes water loss. The refrigerated bulbs are more or less dormant when sold or shipped, so they merely need to be brought into a warm location in advance of the desired flowering time. It is claimed that the plants don't require light, but at least medium light should be provided to prevent legginess. After flowering, they are discarded (like Christmas trees after the holiday, often brought into the house at the same time). Waxed bulbs are expensive (usually at least $20.00 for a single bulb,

FIGURE 5.15 Interior entrance of the Château de Beloeil in Belgium, decorated with amaryllis plants during the annual amaryllis competition.

*Source:* **Paul M. R. Maeyaert, photographer (CC BY SA 3.0).**

FIGURE 5.16 Waxed amaryllis bulbs (bulbs in ornamental wax containers, for use as care-free, temporary flowering houseplants). The bulbs in these containers are in the early phase of developing flowering stalks.

*Source:* **Photo by Angelbattle Bros (CC BY ND 2.0).**

often considerably more), which is why cheaper bulb species are rarely available in this form.

- Scientific (Latin) names for animals sometimes are the same as for plants. The genus name *Amaryllis* refers not only to plants but also to a genus of crustaceans.
- Amaryllis is employed internationally as a symbol of the fight against Huntington disease, a genetic neurodegenerative brain condition.
- Floriography, the use of flower arrangements to communicate secret, often sentimental messages, traces to ancient times. In Britain, a form of floriography called the Victorian "language of flowers" became particularly popular during the Victorian age (roughly 1820–1880). Depending on authority, the flowers (and different colors) may have different meanings, and for the more popular flowers, several interpretations became accepted. Today, the gift flower industry employs the Victorian symbolism to sell its products (as one would expect, only nice meanings are listed). Amaryllis is claimed to have the meanings "pride," "determination," and "radiant beauty."
- One of the only two species of the genus *Amaryllis*, *A. belladonna* (belladonna lily, March lily, naked lady, Madonna lily, Jersey lily) is widely cultivated in warm climates as a garden ornamental. It is frequently

FIGURE 5.17 *Amaryllis belladonna* – often confused with the *Hippeastrum* amaryllises: (a) Flowering tops. (b) Flowering plants. Note that flowers have developed on the stalks well before the basal foliage has appeared.

*Source:* **(a) Photo by Gailhampshire (CC BY 2.0). (b) Photo by Stan Shebs (CC BY SA 3.0).**

confused with the amaryllis discussed in this chapter. Its provocative name "naked lady" is based on the plant producing flowers before it is clothed with leaves (sometimes, hippeastrums are also called naked ladies for the same reason). The "belladonna" (meaning "beautiful lady") in the scientific name is reminiscent of the quite toxic *Atropa belladonna*, unadvisedly used by women in Renaissance Italy to enlarge their pupils and so appear more alluring (but also legitimately used medicinally).

## KEY PUBLICATIONS


Azimi, M.H. and M.K. Alavijeh. 2020. Morphological traits and genetic parameters of *Hippeastrum hybridum*. *Ornamental Horticulture* 26: 579–590.

Barnhoorn, F. 1991. Cultivars of *Hippeastrum*: Their evolution from the past and their development for the future. *Herbertia* 47: 76–79.

Bell, W.D. 1973. New potentials in amaryllis breeding. *Proceedings of the Florida State Horticultural Society* 86: 462–466.

Bell, W.D. 1977. More potentials in amaryllis breeding. *Plant Life* 33: 65–69.

Holmes, W.C. *Hippeastrum*. Flora North America. www.efloras.org/florataxon.aspx?flora_id=1&taxon_id=115497.

Jirakiattikul, Y. 1999. *Study of Intergeneric Hybridization in Hippeastrum*. Ph.D. thesis. University of Tasmania, Hobart. 203 pp.

Khaleel, T.F., S. Haven, and T. Gilg. 1991. Karyomorphology of amaryllis hybrids. *Cytologia* 56: 31–41.

Lara Rico, R.F., R. Vásquez Chávez, and M. Atahuachi Burgos. 2021. *The Genus* Hippeastrum *(Amaryllidaceae) in Bolivia*. Pacific Bulb Society, Leonia, NJ. 137 pp.

Meerow, A.W. 2000. Breeding amaryllis. In: *Breeding Ornamental Plants*, edited by D. Callaway and P. Callaway, 174–195. Timber Press, Portland.

Meerow, A.W. 2009. Tilting at windmills: 20 years of *Hippeastrum* breeding. *Israel Journal of Plant Sciences* 57: 303–313.

Meerow, A.W. 2014. The Florida series of hybrid amaryllis: Five new *Hippeastrum* cultivars. *Hortscience* 49: 1102–1107.

Meerow, A.W., T.K. Broschat, and M.E. Kane. 1992. Breeding of new *Hippeastrum* cultivars using diploid species. *Acta Horticulturae* 325: 583–590.

Narain, P. and T.H. Khoshoo. 1977. Origin and evolution of garden amaryllis. *Indian Journal of Horticulture* 34: 80–85.

Ockenga, S. 2002. *Amaryllis*. Random House, New York. 96 pp.

Okubo, H. 1993. *Hippeastrum* (Amaryllis). In: *The Physiology of Flower Bulbs*, edited by A.A. DeHertogh and M. LeNard, 321–334. Elsevier, Amsterdam.

Read, V.M. 1999. *Hippeastrum* hybridization developments. *Herbertia* 54: 84–109.

Read, V.M. 2004. *Hippeastrum, the Gardener's Amaryllis*. Timber Press, Portland. 344 pp.

Sandler-Ziv, D., A. Cohen, A. Ion, H. Efron, and D. Amit. 1997. A two-year production cycle of Israeli-grown hippeastrum bulbs from bulb chipping to Christmas flowering. *Acta Horticulturae* 430: 361–368.

Shields, J.E. 1979. The ancestors of the amaryllis. *Amaryllis Bulletin* 1: 2–6.

Steinegger, D. and J.E. Watkins. 1990. *Amaryllis Culture*. University of Nebraska-Lincoln Extension Document. G74–G188 pp.

Tallini, L.R., R.B. Giordani, J.P. de Andrade, J. Bastida, and J.A.S. Zuanazzi. 2021. Structural diversity and biological potential of alkaloids from the genus *Hippeastrum*, Amaryllidaceae: An update. *Revista Brasileira de Farmacognosia* 31: 648–657.

Traub, H.P. and A. Rosse. 1958. *The Amaryllis Manual*. Macmillan, New York. 388 pp.

Vijverberg, A.J. 1981. *Growing Amaryllis*. Grower Books, London. 57 pp.

Wang, Y., D. Chen, X. He, J. Shen, M. Xiong, X. Wang, et al. 2018. Revealing the complex genetic structure of cultivated amaryllis (*Hippeastrum hybridum*) using transcriptome-derived microsatellite markers. *Scientific Reports* 8: 1–12.

WikiHow. *How to Store Amaryllis Bulbs*. www.wikihow.life/Store-Amaryllis-Bulbs.

Xiong, M., Y. Wang, D. Chen, X. Wang, D. Zhou, and Z. Wei. 2020. Assessment of genetic diversity and identification of core germplasm in single-flowered amaryllis (*Hippeastrum hybridum*) using SRAP markers. *Biotechnology & Biotechnological Equipment* 34: 966–974.

## Key Cultivation Websites

Krumm, S.J. and Taylor, J.L. *Forcing Amaryllis*. Michigan State University Extension Document. E1848 pp. www.plantgrower.org/uploads/6/5/5/4/65545169/e1848_forcing_amaryllis.pdf.

Pacific Bulb Society. *Hippeastrum*. www.pacificbulbsociety.org/pbswiki/index.php/hippeastrum.

The Spruce. *How to Care for Amaryllis Flowers Year-Round*. www.thespruce.com/growing-amaryllis-1402227.

The Spruce. *What to Do with Amaryllis After It Blooms*. www.thespruce.com/amaryllis-care-after-blooming-5211111.

Stewart, M. *How to Plant and Care for Amaryllis Flowers*. www.marthastewart.com/1534797/plant-and-care-amaryllis.

Sullivan, S. 2019. *Make Your Own Waxed Amaryllis Bulbs*. www.greenhomediy.co/make-waxed-amaryllis-bulbs/.

WikiHow. *How to Care for Amaryllis Flowers*. www.wikihow.com/Care-for-Amaryllis-Flowers#:~:text=Move%20the%20pot%20to%20indirect,flowers%20before%20the%20buds%20open.

WikiHow. *How to Get Amaryllis to Rebloom*. www.wikihow.com/Get-Amaryllis-to-Rebloom.

WikiHow. *How to Plant Amaryllis*. www.wikihow.com/Plant-Amaryllis.

# 6 Anemone (*Anemone* Species and Hybrids)

**Scientific Names:** The genus name *Anemone* originates from Greek *anemone*, meaning "daughter of the wind" (hence the common name *windflower*). *Anemone* is a genus that has been split into different genera and combined with other genera, in light of recent studies. Different authorities classify *Anemone* differently. Two popular cultivated ornamental groups are now sometimes included in *Anemone*: the pasqueflowers (*Pulsatilla* species) and the liverleafs (*Hepatica* species).

**Common Names:** *Anemone* species are typically called anemone. They are also often termed *windflowers*, a name related to the derivation of the genus name, as explained earlier. Other occasional names include lily of the field (not to be confused with *lily of the valley*, treated in this book) and thimbleflower. *Anemone coronaria* is sometimes called Spanish gold. "Anemone dahlias" are dahlias with flowers featuring a pincushion-like center of tiny florets, surrounded by an outer ring of flat petals. "Dahlia anemone" is an animal – the sea anemone *Urticina felina* – and indeed many people understand the word *anemone* to mean the sea anemone.

## 6.1 INTRODUCTION

The anemones discussed in this chapter are cultivated ornamental herbaceous perennials (the fall-flowering species can become woody at their base) belonging to the buttercup family (Ranunculaceae). They have been popular garden ornamentals for centuries and are available in a wide range of attractive cultivars.

FIGURE 6.1 Some species that many botanists consider deserve to be reclassified from other genera to *Anemone*: (a) European pasqueflower (*Pulsatilla vulgaris*, *Anemone vulgaris*), the most popular cultivated ornamental species usually recognized as *Pulsatilla*. (b) Liverleaf (*Hepatica nobilis*, *Anemone hepatica*), the most popular cultivated ornamental species usually recognized as *Hepatica*.

*Source:* **(a) Photo by Laura Otýpková from Pixabay (public domain). (b) Photo by Per Vetlesen (CC BY SA 3.0).**

## 6.2 OVERVIEW

Many of the anemone species that have become popular in cultivation produce spectacular flowering displays carpeting large areas. The most attractive of these is *Anemone coronaria*, called poppy anemone and florist's anemone. This native of southern Europe, the Middle East, and the

FIGURE 6.2 A spectacular display of wild-flowering anemones in Israel.

*Source:* **Photo by Itamar Grinberg for the Israeli Ministry of Tourism (CC BY ND 2.0).**

FIGURE 6.3 *Anemone coronaria*: (a) Cut flowers on sale in a shop. (b) Plants with doubled flowers growing beside a house. (c) A small bouquet of flowers of different colors. (d) Flowers of the popular red form.

*Source:* **(a) Photo by Shutterstock/Bellena. (b) Photo by Shutterstock/Alybaba. (c) Photo by Alfred Grupstra from Pixabay (public domain). (d) Photo by Beverly Buckley from Pixabay (public domain).**

DOI: 10.1201/9781003586227-6

Mediterranean area, including North Africa, is the most popular of the cultivated anemones. The species is tuberous-rooted (the knobby storage organ is usually called a tuberous rhizome and a bulb), with divided leaves and large poppy-like flowers produced singly on stalks 15–20 cm (occasionally up to 50 cm) long. A single plant can produce over a dozen flowers. The flowers do not have different sepals and petals (best collectively termed *tepals* when the two kinds are not evident). The large flowers (sometimes over 10 cm in diameter) are bluish-purple, crimson, reddish, scarlet, or white, and there are also cultivars with doubled flowers and bicolored flowers. Reddish shades are most popular. The centers of single (i.e., non-doubled) flowers are usually black.

Some Asian species formerly included within the genus *Anemone* were transferred to the genus *Eriocapitella* (meaning "growing in a small woolly head," a reference to the hairy floral ovary and fruit of some of the species). These include the "Japanese anemones," which are not native to Japan, although they have been cultivated there for many centuries. They are also known as fall-blooming anemones (although they also flower during the summer). Aside from the numerous hybrids (many under the name *E.* × *hybrida*), the six species include *E. hupehensis* (the most commonly grown species), *E. japonica*, *E. rivularis*, *E. rupicola*, *E tomentosa*, and *E. vitifolia*. These are often identified under their old *Anemone* names. The plants have fibrous roots. The popular cultivars have strong branching stems, usually 60 to 120 cm in height at maturity. The cultivars have single, semi-double, or double flowers up to 8 cm in diameter that are white, pink, or purple. The hybrids (often collectively assigned to *E.* × *hybrida*) usually have white or pink blossoms. The plants often display both opened flowers and seed heads for several weeks in late summer and fall.

FIGURE 6.4 Japanese (fall-flowering) anemones often reclassified from *Anemone* to *Eriocapitella*. These are identified as *Eriocapitella hupehensis/Anemone hupehensis* or *E.* × *hybrida*/"*A.* × *hybrida*."

**Source: (a) Photo by Shutterstock/Joe Kuis. (b) Photo by Leonora (Ellie) Enking (CC BY SA 2.0). (c) Photo by Ernesto Velázquez from Pixabay (public domain). (d) Photo by Acabashi (CC BY SA 4.0).**

Some north-temperate species often considered to be anemones are placed by some authorities in the genus *Anemonoides* (the name means "anemone-like"). These include a large group of small (typically 30 cm or shorter) woodland species native to North America and Eurasia. The most frequently cultivated are European wood anemone (*Anemone nemorosa*, *Anemonoides nemorosa*) and Grecian windflower (*Anemone blanda*, *Anemonoides blanda*).

## 6.3 HISTORY

*Anemone coronaria* was introduced into England in the 16th century and subsequently became very popular in the

FIGURE 6.5 European (spring-flowering) anemones often reclassified from *Anemone* to *Anemonoides*: (a, b) European wood anemone (*Anemone nemorosa/Anemonoides nemorosa*); (c, d) Grecian windflower (*Anemone blanda/Anemonoides blanda*).

**Source: (a) Photo by Hans from Pixabay (public domain). (b) Photo by Matthias Kaiser (CC BY SA 4.0). (c) Cultivar Blue Shades. Photo by Acabashi (CC BY SA 4.0). (d) Cultivar White Splendour. Photo by Acabashi (CC BY SA 4.0).**

FIGURE 6.6 Nineteenth-century paintings (public domain) of bouquets of *Anemone coronaria* by European botanical illustrators.

**Source: (a) Step, E., D. Bois, B. Herincq, and W. Watson. 1896. *Favourite Flowers of Garden and Greenhouse*, vol. 1. Frederick Warne, New York. (b) Painting by French artist Pierre-Joseph Redouté (1759–1840). (c) Witte, H., and A. J. Wendel. 1868. *Dutch Language Book: Flora: afbeeldingen en beschrijvingen van boomen, heesters, éénjarige planten, enz. voorkomende in de Nederlandsche Tuinen*. Wolters, Groningen.**

Western world. The Japanese (fall-blooming, *Eriocapitella*) anemones may have been cultivated as far back as the 7th century in China. However, they were imported into Japan, where very considerable breeding occurred, resulting in many of the cultivars currently available (hence the name "Japanese anemones" despite the plants not being native to the country). Hybrids were also produced in England in the 19th century and soon became very popular. The European (*Anemonoides*) species have never acquired the popularity of the previously mentioned species, but cultivars of several of them have been selected.

## 6.4 ECONOMIC IMPORTANCE

Anemones are well-established garden ornamental staples, of moderate importance to the horticultural industries of numerous countries. The florist's anemone is the most widely known form of anemone. It is indeed beautiful but competes for attention with the rather spectacular, roselike flowers of its relative, ranunculus (also treated in this book).

## 6.5 ORNAMENTAL USES

Anemones are employed in massed garden beds, borders, rock gardens, and containers. They are also good sources of cut flowers and are popular for the purpose in florist shops.

## 6.6 OTHER USES

Dozens of species currently or at least recently placed in the genus *Anemone* have been extensively employed in folk medicine throughout China, North America, and Europe to treat a variety of medical conditions. Anemone is employed by herbal medicine professionals and enthusiasts, but it needs to be appreciated that the plants are toxic. Like virtually all other species in the large buttercup family, the plants are not considered edible.

FIGURE 6.7 A commercial field of anemones being grown for garden centers.

*Source:* **Photo by Shutterstock/Dirk M. de Boer.**

FIGURE 6.8 Anemone (*Anemone coronaria*) gardens.

*Source:* **(a) Photo by Shutterstock/Hama-Kid. (b) Photo by 663highland (CC BY SA 2.5). (c) Photo by Ernesto Velázquez from Pixabay (public domain).**

FIGURE 6.9 Cut flower bouquets of anemone.

*Source:* **(a) Photo by Shutterstock/Fusionstudio. (b) Photo by Shutterstock/AKaltykova.**

FIGURE 6.10 Mixed flower arrangements featuring anemones.

*Source:* **Prepared by Beverly Buckley from Pixabay (public domain).**

## 6.7 PROBLEMS

Ingestion of anemones may cause mild stomach upset, and contact may irritate skin. Gloves and other protective clothing are advisable when handling. The plants can be toxic to pets (dogs and cats), and anemone is among the common causes of cases of poisoning handled by veterinarians.

Anemones can be quite weedy, and if they become problematic in a garden, they may be difficult to eradicate. (In some areas, gardeners are advised to grow anemone in containers to prevent their possible spread.) Japanese anemones are considered invasive in Hawaii.

## 6.8 CULTIVATION

### 6.8.1 Climate and Site

*Anemone coronaria* is rated for growth as a perennial in hardiness zones 7–11. In cooler climates, the species is grown as an annual, the tubers planted in the autumn or in the fall in very cold areas. In marginal areas (zone 6), an anemone bed can be covered with mulch overwinter, or the rhizomes can be excavated and overwintered indoors (a process that is often unreliable). Potted plants also can be maintained indoors overwinter in colder areas. The fall-flowering anemones (*Eriocapitella hupehensis* and *E.* × *hybrida*) are rated to grow in climate zones 4 to 9. *Eriocapitella tomentosa*, the hardiest of the fall-flowering anemones, is rated to grow as a perennial as far north as zone 4. The European (*Anemonoides*) anemones can be extremely tolerant of cold (https://hardiness.zone/plant/?sle=Anemonoides&art=nemorosa). *Anemonoides nemorosa* can survive easily in zone 2.

*Anemone coronaria* grows well in full sun and part shade. The fall-blooming (*Eriocapitella*) cultivars thrive in light to partial shade but can tolerate full sun. *Anemonoides nemorosa* and *Anemonoides blanda* are adapted to deciduous woodlands, so dappled sun is ideal.

Despite often being called windflowers, the taller cultivars of anemones benefit from some protection from wind and may require staking (although not usually necessary).

### 6.8.2 Propagation and/or Purchasing

Anemones can be started from seeds, either indoors (preferably) or outside in suitable climates. However, cultivars usually will not reproduce their flower characteristics well. Most gardeners establish anemones from corms or rhizomes or from starter plants purchased from garden outlets. *Anemone coronaria* tubers produce flowers quickly in the current season (when they are grown from seeds, flowering is delayed until midsummer). Large corms of *Anemonoides nemorosa* and *A. blanda* can flower in the first year.

Once established in climates in which they can survive, the plants can perpetuate for years. *Anemone coronaria* can self-seed. *Anemonoides nemorosa* and *A. blanda* also self-seed and can spread and multiply by their rhizomes and corms. Fall-flowering anemones also spread vigorously

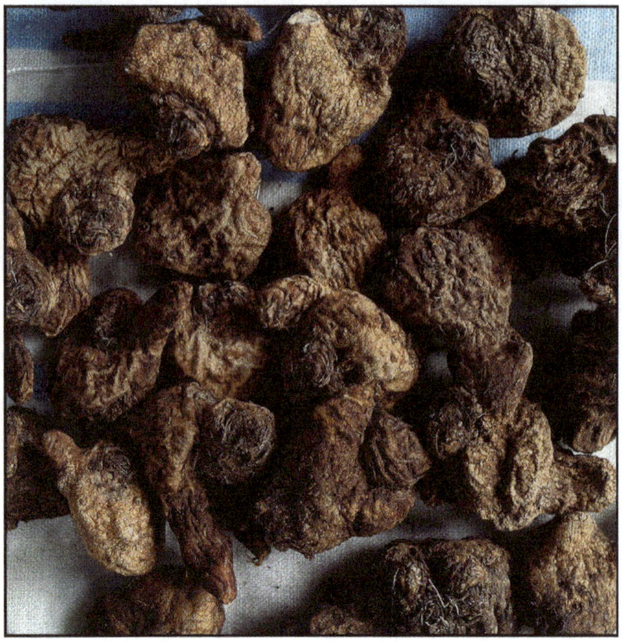

FIGURE 6.11 Dormant tubers of *Anemone coronaria*.

*Source:* Photo by Carlfbagge (CC BY 2.0).

by rhizomes, and they can be propagated by division and root cuttings taken in early spring or early fall. Established plants (at 4 or 5 years of age) benefit from the clumps being divided and replanted (a process termed "crown division"). This can be done when plants become large, about every 3 years for plants grown in full sun (this promotes rapid growth), or much less often for plants grown in shade.

### 6.8.3 Planting

The tuberous rhizomes ("bulbs") of *Anemone coronaria* are planted in gardens in the fall, about 5 to 8 cm deep and 15 to 20 cm apart, or closer according to some authorities. The fall-flowering species should be planted 30 to 60 cm apart. The rhizomes of *Anemonoides* species can be planted about 20 cm apart. The part of the tubers of *A. coronaria* studded with buds should be uppermost (if not clear, the recommendation is to plant sideways). The smaller rhizomes of fall-flowering and *Anemonoides* species can be planted about 5 cm deep and 7 to 10 cm apart and will develop well regardless of how their buds are oriented during planting. (As a general rule, the depth of planting should be about three times the size of the bulb or rhizome.)

### 6.8.4 Soil and Moisture Conditions

Most "anemones" grow well in evenly moist but well-drained, humus-enriched, fertile soil. Sandy soils are good for all species (especially the Japanese anemones), but loam soils also are suitable. A pH of 6 to 8 is appropriate for *A. coronaria* and the *Anemonoides* species, but 4.5 to 7.0 is recommended for the fall-flowering species. Anemones do not require much fertilizer but benefit from an application of an organic compost or a balanced fertilizer in early spring. Liquid fertilizer is often recommended, especially for potted plants.

### 6.8.5 Maintaining Plants

Anemones are considered to be low-maintenance. Deadheading isn't particularly necessary, although it may be desirable to prevent seeds from being dispersed in areas where they may become invasive. The dead foliage of *A. coronaria* overwintered outdoors can be cut away, but no pruning is necessary. The fall-flowering species produce interesting seed heads that attract birds. Cutting the fall-flowering plants back to the ground at the end of the season is likely to kill them (faded flower stems can be cut back in late autumn, and old dead foliage removed in early spring). *Anemone/Anemonoides nemorosa* and *A./A. blanda* will go dormant in summer, and the aboveground parts will disappear entirely.

## 6.9 FASCINATING FACTS, FABLES, FOLKLORE, AND FRIVOLITY

- In Greek mythology, *Anemone coronaria* was created by the goddess Aphrodite (equivalent to Venus of the Romans), who mixed her tears with the blood of her dying lover (wounded by a boar), the mortal Adonis. The flowers sprang up from where the blood and tears fell. *Anemone coronaria* flowers are sometimes known as Adonis flowers and symbolize the briefness of life and love.
- In Luke 12:27–28 (and, similarly, in Matthew 6:28–29), the following passage is attributed to Jesus: "Consider the lilies of the field, how they grow; they toil not, neither do they spin: And yet I say unto you, That even Solomon in all his glory was not arrayed like one of these." Identification of many of the plants recorded in the Bible is often very challenging, but most authorities regard *Anemone coronaria* as the "lily of the fields." The species is widespread in Israel.
- In Israel, anemones flower during the rainy season (in the winter in Mediterranean climates), blossoming until March or April and carpeting large areas. Hundreds of thousands of people are attracted to view the blooms in the northern Negev area. *Anemone coronaria* was chosen by the public as the national flower of Israel in 2013.
- In Christian tradition, red anemones have often been associated with the blood of Christ. Anemones are pictured on old European paintings of the Crucifixion, and of Mary mourning the event.
- It was commonly believed in Europe that anemones were symbols of death and bad luck. Many held their breath as they walked through fields of anemone. In English and Irish folktales, fairies are often said to sleep within the closed flowers.
- The flowers of some species of anemone close up at night (this frequent phenomenon among plant species often protects the flowers from water damage to the pollen).
- Sea anemones are a group of about 1,000 predatory marine invertebrates, many of which are reminiscent of anemone flowers, after which they are

FIGURE 6.12 A painting by English artist John William Waterhouse (1849–1917) showing the goddess Aphrodite, her sweetheart Adonis, and the anemones generated by their love.

*Source:* **Public domain.**

FIGURE 6.13 A carpet of red *Anemone coronaria* flowers in Shokeda forest, Israel.

*Source:* **Photo by Zachi Evenor (CC BY SA 4.0).**

FIGURE 6.14 Sea anemones (public domain print). Artist: Giacomo Merculiano (1859–1935).

*Source:* **Lydekker, R. 1893. *The Royal Natural History.* F. Warne, London.**

named. They are related to corals and jellyfish. Sea anemones are typically sessile, attached at their base to a surface such as a rock. Most are 1 to 5 cm in diameter and 1 to 10 cm long, but they can inflate their soft bodies, and some are a meter wide or over 1 m in length. A central mouth at the top is lined by one or more rings of tentacles, looking somewhat like petals. Every tentacle is covered with thousands of very small capsules which can launch tiny harpoons that inject toxins for defense or to capture prey. Subdued prey are drawn to the mouth and swallowed whole. Most sea anemones are harmless to humans, but a few are potentially lethal.

## KEY PUBLICATIONS


Bowles, E.A. and W.T. Stearn. 1947a. The history of *Anemone japonica*, part I. *Journal of the Royal Horticultural Society* 72: 261–268.

Bowles, E.A. and W.T. Stearn. 1947b. The history of *Anemone japonica*, part II. *Journal of the Royal Horticultural Society* 72: 297–308.

Ehrendorfer, F., S.N. Ziman, C. König, C.S. Keener, B.E. Dutton, and O.N. Tsarenko. 2009. Taxonomic revision, phylogenetics and transcontinental distribution of *Anemone* section *Anemone* (Ranunculaceae). *Botanical Journal of the Linnean Society* 160: 312–354.

Hao, D.C., X. Gu, and P. Xiao. 2017. Anemone medicinal plants: Ethnopharmacology, phytochemistry and biology. *Acta pharmaceutica Sinica* B7: 146–158.

Hoot, S.B., K.M. Meyer, and J.C. Manning. 2012. Phylogeny and reclassification of *Anemone* (Ranunculaceae), with an emphasis on Austral species. *Systematic Botany* 37: 139–152.

Laura, M. and A. Allavena. 2007. *Anemone coronaria* breeding: Current status and perspectives. *European Journal of Horticultural Science* 72: 241–247.

Rauter, S., M. Stock, B. Black, D. Drost, X. Dai, and R. Ward. 2022. Anemone cut flower timing, yield, and quality in a high-elevation field and high tunnel. *Horticulturae* 9(1): 2. www.mdpi.com/2311-7524/9/1/2.

Shirreffs, D.A. 1985. *Anemone nemorosa* L. *Journal of Ecology* 73: 1005–1020.

Wang, W., A.-M. Lu, Y. Ren, M.E. Endress, and Z.-D. Chen. 2009. Phylogeny and classification of Ranunculales: Evidence from four molecular loci and morphological data. *Perspectives in Plant Ecology, Evolution and Systematics* 11: 81–110.

Ziman, S.N., E.V. Bulakh, Y. Kadota, and C.S. Keener. 20008. Modern view on the taxonomy of the genus *Anemone* L. sensu stricto (Ranunculaceae). *Journal of Japanese Botany* 83: 127–155.


### Key Cultivation Websites


Gardener's Path (Blog). *How to Grow Fall-Blooming Japanese Anemone Flowers*. https://gardenerspath.com/plants/flowers/grow-japanese-anemones/.

Rauter, S. and M. Stock. 2023. *Anemone Cut Flower Production in Utah*. Utah State University Extension. https://digitalcommons.usu.edu/cgi/viewcontent.cgi?article=3353&context=extension_curall.

The Spruce. *14 Beautiful Types of Anemone Flowers*. www.thespruce.com/best-anemone-flower-varieties-4147178.

The Spruce. *How to Grow and Care for Anemone (Windflower)*. www.thespruce.com/growing-anemone-flowers-1316064.

University of Maryland, Extension. *Producing Anemone as a Cut Flower*. https://cybercemetery.unt.edu/archive/allcollections/20090117171520/http://extension.umd.edu/publications/PDFs/FS753.pdf.

Utah State University, Extension. *Anemone Cut Flower Production in Utah*. https://digitalcommons.usu.edu/cgi/viewcontent.cgi?article=3353&context=extension_curall.

WikiHow. *How to Plant Anemone Corms (Anemone Bulbs)*. www.wikihow.life/Plant-Anemone-Corms#:~:text=Anemones%20do%20best%20when%20they,area%20gets%20the%20most%20sunlight.&text=Without%20enough%20sunlight%2C%20your%20flowers,not%20achieve%20their%20bright%20colors.

# 7 Angel's Trumpet (*Brugmansia* Species and Hybrids)

**Scientific Names:** The genus name *Brugmansia* is named for Dutch botanist and physician Sebald Justinus Brugmans (1763–1819), who was a professor of natural history at Leiden in Holland.

**Common Names:** The common name *angel's trumpet* is based on the large, very attractive, trumpet-shaped flowers of *Brugmansia* species. This name is used for some species of the closely related genus *Datura*, but the flowers of the former are pendulous (hang down), while those of the latter are usually oriented upward or horizontally. In allusion to the toxic nature of the plant, some of the species are occasionally termed *trumpet of death*, *devil's breath*, and *devil's trumpet* (all names also used for *Datura*). They are also sometimes called *tree daturas* (true *Datura* species do not reach tree size). In addition to the name *angel's trumpet*, the genus name is also employed as a common name: "brugmansia."

## 7.1 INTRODUCTION

*Brugmansia* species are tropical shrubs and trees producing stunning displays of very fragrant, beautiful, gigantic flowers that demand to be photographed. The plants are deadly poisonous, perhaps making them even more intriguing.

## 7.2 OVERVIEW

*Brugmansia* species are large perennial shrubs or small trees widely distributed in the world as ornamental plants, especially in the tropical and subtropical to temperate regions, and occasionally occurring as garden escapes. However, all the species seem to be extinct in their native tropical regions of South America. About six species are recognized (most are illustrated in this chapter), as well as several hybrids (*B.* × *candida* = *B. aurea* × *B. versicolor*;

FIGURE 7.1 Photographers attracted by gorgeous *Brugmansia* flowers.

**Source:** **Figures adapted by B. Brookes. (a) Photo by Shutterstock/Viktoriia Belova. (b) Photo by Shutterstock/Lebedeus.**

FIGURE 7.2 Species of *Brugmansia* in flower: (a) *Brugmansia sanguinea*, a red-flowered species commonly pollinated by long-billed hummingbirds; the plant is native to Colombia and Chile. (b) A sword-billed hummingbird (*Ensifera ensifera*) visiting a flower of *B. sanguinea* in Ecuador. (c) *Brugmansia arborea*, from the Andes from Ecuador to northern Chile; the flowers are usually relatively short and white. (d) *Brugmansia aurea*, with yellow blooms, considered native to Ecuador and Venezuela. (e) *Brugmansia versicolor*, a native of Ecuador, often with pale apricot flowers; this has especially large blooms. (f) *Brugmansia suaveolens*, the most popular of the brugmansias. Considered to be native to Brazil, this species usually has cream-colored flowers, but other colors have been selected, as shown here.

**Source:** **(a) Photo by Janeyhenning (CC BY 2.0). (b) Photo by Andy Morffew (CC BY 2.0). (c) Photo by Manuel m.v. (CC BY 2.0). (d) Photo by Zeynel Cebeci (CC BY SA 4.0). (e) Photo by Cultivar 413 (CC BY 2.0). (f) Photo by Asit K. Ghosh (CC BY SA 3.0).**

FIGURE 7.3 Hybrids of *Brugmansia* in flower: (a) *Brugmansia* × *candida*. (b) *Brugmansia* × *insignis*. (c) *Brugmansia* × *cubensis* cultivar Charles Grimalidi.

**Source:** **(a) Photo by CEphoto, Uwe Aranas (CC BY SA 3.0). (b) Photo by Shutterstock/Sunshower Shots. (c) Photo by Cultivar413 (CC BY 2.0).**

DOI: 10.1201/9781003586227-7

*B.* × *flava* = *B. arborea* × *B. sanguinea*; *B.* × *cubensis* = *B. suaveolens* × *B. versicolor* × *B. aurea*; *B.* × *insignis* = *B. suaveolens* × *B. versicolor*). Some species reach a height of 11 m, and in hospitable climates, the plants are long-lived and long-flowering during much of the year. The foliage is evergreen in suitable climates but will drop off in response to frost. The flowers are up to 60 cm long, the floral tube at the opening as wide as 40 cm. The blooms are shades of green, orange, pink, red, white, or yellow, sometimes bicolored, sometimes doubled, and fragrant in most species (not *B. sanguinea*, which is hummingbird-pollinated). Most of the species are more fragrant in the evenings to attract pollinating moths (hence the occasional name "moonflower"). In many of the species, seeds are set infrequently.

## 7.3 HISTORY

The medicinal, ritual, and psychoactive use of *Brugmansia* plants in South America likely traces to prehistorical times. It is believed that all the species are extinct in their natural wild habitats, having been preserved initially by Indigenous people for use as medicine and spiritual intoxicants ("entheogens"), and in more recent centuries primarily as ornamentals.

## 7.4 ECONOMIC IMPORTANCE

*Brugmansia* is frequently cultivated in tropical and semitropical climates, where it is one of many attractive ornamental shrubs and small trees. It is uncommonly grown in temperate climates, where it has minor significance in horticulture.

## 7.5 ORNAMENTAL USES

The plants are suitable as specimen shrubs and small trees for various landscape uses – parks, street sides, beside large buildings, in larger residential gardens, and in containers for patios, balconies, and decks.

## 7.6 OTHER USES

Of the dozens of alkaloids present in *Brugmansia* species, scopolamine, atropine, and hyoscyamine have significant applications in medicine, but only in very limited doses. These alkaloids cause delirium and death in high doses. The plant and its preparations are employed, almost universally legally, as a recreational intoxicant, but the drugs are not popular because they are known to be very dangerous and frequently produces extremely uncomfortable symptoms. Scopolamine is useful in preventing nausea and vomiting and is commonly used after surgery. Atropine is prescribed to decrease saliva and increase heart rate, but mainly to prevent certain kinds of poisoning. Hyoscyamine is employed to treat gastrointestinal tract disorders.

FIGURE 7.4 *Brugmansia* in containers. (a) In a public park. (b) *Brugmansia versicolor* cultivar Peaches and Cream.

***Source:*** **(a) Photo by Shutterstock/Pr2is. (b) Inside a building. Photo by Nadya J, anestwithayard.com (CC BY 2.0).**

## 7.7 PROBLEMS

*Brugmansia* species are among the most toxic of ornamental plants. All parts are poisonous to humans, pets, and livestock. Just 50 mg of scopolamine, its principal psychoactive ingredient, is sufficient to kill an adult. It produces a state of intoxication, leaving people open to suggestions and unable to resist (it is one of the forms of "truth serum"), and is believed to be employed by criminals to disable victims, including use as a date rape drug. The sap can be irritating, especially if it contacts mucus membranes in the eyes and nose. There is at least one case of a gardener handling a plant, then her eye, resulting in extreme eye problems. Some authorities recommend always wearing gloves when handling or pruning *Brugmansia* (although the plants are not considered to be a significant source of dermatitis).

The plants have become naturalized in Australia, New Zealand, the Caribbean, and the Pacific Islands, where they pose minor problems as invasive weeds.

## 7.8 CULTIVATION

### 7.8.1 Climate and Site

*Brugmansia* species are adapted to frost-free climates, in sun (preferably) or part shade (especially in hot climates). In areas with mild winter frost, the roots may survive outdoors, especially if given protection. In climate zone 8, temperatures may go down as low as −12°C, and deep mulching and shelter might keep the plants from dying. The plants are rated to survive in zones 9 to 11. Some of the species originating from higher-elevation areas (of the Andes) are adapted to cooler summers and may not flower in very hot regions. The plants can be grown as annuals or as potted plants overwintered indoors in a dormant condition (in the dark, not below 7°C) or maintained as a houseplant (which appreciates a relative humidity of at least 50%). Because of the potential for poisoning children and pets, consideration should be given to planting outdoors in locations that cannot be reached and, similarly, taking care that indoors it is not accessible.

### 7.8.2 Propagation and/or Purchasing

Although they can be grown from seeds, most *Brugmansia* are propagated from root cuttings. Cultivars often will not breed true from seeds. Most people purchase starter plants in pots.

### 7.8.3 Planting

*Brugmansia* species need lots of space for roots to spread, and sometimes also for branching (if not regularly pruned). Even in a container, the plants can spread over 2 m, and much more in the ground. When grown as a perennial in the ground, the plants should be separated by a minimum of 2 m.

### 7.8.4 Soil and Moisture Conditions

*Brugmansia* species grow best in moist, fertile, well-drained, slightly acidic soil (ideally, pH 6.5 to 7). Adequate irrigation is important for good growth. The plants can grow in poor soil, but if so, fertilization is highly recommended. Soil texture (sand, loam, clay) is not important, but drainage is. The plants are considered to be heavy feeders, and flower-promoting fertilizers, such as 15–30–15 or 10–50–10, are best. In containers, a good potting soil is advisable.

### 7.8.5 Maintaining Plants

Deadheading of flowers can be done but is usually not considered essential. Some authorities strongly recommend pruning to promote open growth that makes for healthy plants. In temperate regions, the plants will die back to the ground (if not entirely), and some gardeners also cut plants in containers to near the base, although they may be stored indoors. Growing in containers is a good way to limit the size of the plants. The plants should be placed in larger pots every spring as necessary.

## 7.9 FASCINATING FACTS, FABLES, FOLKLORE, AND FRIVOLITY

- The phrase "angel's trumpet" traces to the biblical account in the Book of Revelation of seven angels each sequentially sounding a trumpet to forewarn of apocalyptic events.
- The Chibcha Indians of Colombia have been reported to have administered intoxicant preparation of *Brugmansia* to wives and slaves of a departed husband or leader and then buried them alive with the deceased.
- John Gould (1804–1881) was a British ornithologist, artist, and businessman. He published several outstanding monographs on birds, illustrated by himself; his wife, Elizabeth Gould; and several other artists and prepared by highly skilled lithographers. His *Monograph of the Trochilidae* is a comprehensive guide to hummingbirds, the five volumes containing 360 hand-colored lithographic illustrations. Gould's original works bring vast sums at auction houses around the world. Although he was fascinated by hummingbirds, Gould did not see a living specimen until he was 54 years old, when he observed a ruby-throated hummingbird in Philadelphia. Several of Gould's plates show hummingbirds obtaining nectar from *Brugmansia* species.

FIGURE 7.5 "Angel's trumpets" in art (public domain figures): (a) *The Seven Trumpets* from a biblical painting prepared about AD 1000, published in the Bamberg Apocalypse Folio 19 v7. (b) *The Last Judgement*, a fresco painted on the altar wall of the Sistine Chapel in Vatican City by the Italian artist Michelangelo from 1536 to 1641. (c) *La Renommée à cheval sur Pégase*, a statue at the Louvre in Paris constructed by anonymous sculptures from 1698 to 1792.

*Source:* **Photo by Vicki Hamilton from Pixabay.**

FIGURE 7.6 Illustrations of hummingbirds visiting flowers of angel's trumpets (*Brugmansia* species): (a) Purple-throated sunangels (*Heliangelus viola*) visiting *B. arborea*. (b) Mulsant's wood star (*Calothorax mulsanti*) visiting *B. arborea*. (c) Sword-billed hummingbirds (*Ensifera ensifera*) visiting *B. sanguinea*. The flowers of *B. sanguinea* are not fragrant and so of little interest to insect pollinators. Note how the very long flowers are adapted to the very long bills of the birds (the only ones in the world whose bills are longer than their bodies).

*Source:* **From John Gould's Monograph of the Trochilidae or family of humming-birds.**

## KEY PUBLICATIONS

Abril, R., T. Ruiz, J. Alonso, and G. Cabrera. 2017. Propagation and initial growth of *Brugmansia suaveolens*, species with ethnoveterinary uses in the Amazon region of Ecuador. *Revista Cubana de Ciencia Agrícola* 51: 359–369.

Algradi, A.M., Y. Liu, B.Y. Yang, and H.X. Kuang. 2021. Review on the genus *Brugmansia*: Traditional usage, phytochemistry, pharmacology, and toxicity. *Journal of Ethnopharmacology* 279: 113910.

Bauml, J.A., 2006. Bibliography – *Brugmansia* sp. *Cultura y Droga* 11(13): 339–350. https://revistasojs.ucaldas.edu.co/index.php/culturaydroga/article/view/6005.

Bristol, M.L. 1966. Tree datura drugs of the Colombian Sibundoy. *Botanical Museum Leaflets* 22(5): 165–227.

De Feo, V. 2004. The ritual use of *Brugmansia* species in traditional Andean medicine in Northern Peru. *Economic Botany* 58(Suppl. 1): S221–S229.

Hay, A., M. Gottschalk, and A. Holguín. 2012. *Huanduj: Brugmansia*. Kew Royal Botanic Gardens, Kew, UK. 424 pp.

Kerchner, A. and Á. Farkas. 2020. Worldwide poisoning potential of *Brugmansia* and *Datura*. *Forensic Toxicology* 38: 30–41.

Lockwood, T.E. 1973. Generic recognition of *Brugmansia*. *Botanical Museum Leaflets* 23: 273–283.

Loockwood, T.E. 1979. The ethnobotany of *Brugmansia*. *Journal of Ethnopharmacology* 1: 147–164.

Pasiecznik, N. 2014. *Brugmansia suaveolens (white angel's trumpet)*. CABI Compendium https://doi.org/10.1079/cabicompendium.10790.

Petricevich, V.L., D.O. Salinas-Sánchez, D. Avilés-Montes, C. Sotelo-Leyva, and R. Abarca-Vargas. 2020. Chemical compounds, pharmacological and toxicological activity of *Brugmansia suaveolens*: A review. *Plants* 9(9): 1161.

Preissel, U. and H.G. Preissel. 2002. *Brugmansia* and *Datura: Angel's Trumpets and Thorn Apples*. Firefly Books, Buffalo, NY. 144 pp.

Pundir, S., M.K. Shukla, A. Singh, R. Chauhana, U.R. Lala, A. Alic, et al. 2022. A comprehensive review on angel's trumpet (*Brugmansia suaveolens*). *South African Journal of Botany*. https://doi.org/10.1016/j.sajb.2022.02.023.

Rojas-Sandoval, J. and P. Acevedo-Rodríguez. 2014. *Brugmansia candida (Angel's Trumpet)*. CABI Compendium. https://doi.org/10.1079/cabicompendium.1013.

Shaw, J.M.H. 1999. Nomenclature notes on *Brugmansia*. *The New Plantsman, Royal Horticultural Society* 6: 148–151.

## Key Cultivation Websites

Curtis, J.G. 2023. *Brugmansia: A Stunning Flower with a Sinister Reputation*. Tripsitter (Blog). https://tripsitter.com/brugmansia/.

Rare Plant Fair. *Brugmansia: Growing in a Frost-Prone UK*. www.rareplantfair.co.uk/news-and-articles/brugmansia-growing-in-a-frost-prone-uk/.

The Spruce. *How to Grow and Care for Brugmansia (Angel's Trumpet)*. www.thespruce.com/angels-trumpet-brugmansia-spp-3269243#citation-5.

University of Wisconsin-Madison. *Angel's Trumpet, Brugmansia*. https://hort.extension.wisc.edu/articles/angels-trumpet-brugmansia/.

WikiHow. *How to Plant Angel Trumpet Seeds*. www.wikihow.com/Plant-Angel-Trumpet-Seeds.

WikiHow. *How to Prune Brugmansia*. www.wikihow.com/Prune-Brugmansia.

# 8 Aster (*Aster, Callistephus, Symphyotrichum*, and Related Genera + Hybrids)

**Scientific Names:** The principal "aster" ornamentals originate from the genera *Aster*, *Callistephus*, and *Symphyotrichum*. The genus name *Aster* originates from the Greek *aster*, "star," referring to the star-like appearance of the flower heads. *Symphyotrichum* is based on the Greek *sýmphsis*, "coming together" or "coalescing," and *trich*, for "hair," the combination meaning "united hairs," referring to the pappus (the hairs that attach to the seeds and function as sails or parachutes, as in dandelions). *Callistephus* is based on the Greek *kallos*, "beauty," and *stephos*, "crown," a reference to the appearance of the "seeds" (tiny fruits), which have pointy appendages, like a crown.

**Common Names:** Some botanists employ the word *aster* to designate all species (perhaps over 32,000) in the Asteraceae (sunflower family, also called the aster family and the daisy family). Much more frequently, the common name "aster" (and, less frequently, the names *Michaelmas daisy* and *starflower*) has been applied to species in what are now known to be in a dozen or more distinct genera within the tribe Astereae of the sunflower family. Although the scientific name *Aster* is now relatively restricted, the common name "aster" continues to be applied to many species, as in the past. The name *Michaelmas daisy* honors the Archangel Michael, the greatest of the angels in Christian theology, whose feast day is in the autumn, at which time only a few plants (including asters) are in flower in the north temperate world. China aster (also annual aster) is *Callistephus chinensis*, discussed in detail here. Mexican aster is *Cosmos bipinnatus*, an ornamental plant. Sea aster is *Tripolium pannonicum*, an Old World seashore plant that is harvested and eaten. Similarly, there are other plant names incorporating the word "aster" that are not related to the ornamentals discussed here.

## 8.1 INTRODUCTION

The word "aster" designates more than a dozen genera, all in the sunflower family, all with daisylike flowers. The three genera which contribute most of the garden varieties are discussed in this chapter.

China aster (*Callistephus chinensis*), usually an annual, is the best-known garden aster. More than a century ago, cultivars with gorgeously colored, doubled flowers dominated asters illustrated in seed catalogues, and they remain extremely popular. However, additional genera, noted in the following, have since also become popular, particularly grown as perennials.

In the past, the genus *Aster* included as many as 600 species. In the 1990s, it was reclassified to include about 200 species of Eurasia and just the arctic-alpine species *A. alpinus* (unusual in being spring-flowering) in North America. The remaining plants have been assigned to more than a dozen other genera. Most asters, in the comprehensive sense, are natives of northern Eurasia and North America, while a few are in South America. Most native North American "asters" are now in *Symphyotrichum*, the remainder in *Eurybia* and other genera. Because the new Latin nomenclature is recent, the scientific names for asters on the web and names used in the horticultural trade are often the old names that need to be abandoned. Frequently, information provided on the internet and at sales centers is insufficient to accurately identify the plants.

*Symphyotrichum* asters and *Aster* asters are mostly grown as perennials, valued for their late-summer and autumn displays of daisylike flowers, a time when gardens tend to be discouragingly uncolorful. China aster similarly characteristically flowers well into autumn. All these asters compete with chrysanthemums as "stars of the autumn garden." *Symphyotrichum* asters dominate the perennial cultivars. New England aster (*Symphyotrichum novae-angliae*,

FIGURE 8.1 Cover pages of old garden catalogues featuring China aster (*Callistephus chinensis*): (a) Maule's 1934 seed book, by H. Maule, Philadelphia, PA. (b) Ferry & Co. 1889 seed catalogue, by D. M. Ferry, Detroit, MI. (c) Elliott & Sons 1889 seed catalogue, by Wm. Elliott & Sons, New York, NY.

*Source:* **Public domain.**

DOI: 10.1201/9781003586227-8

**FIGURE 8.2** Cultivars of New England aster (*Symphyotricum novae-angliae*). (a) Unidentified cultivar. (b) Cultivar James Ritchie. (c) Unidentified cultivar. (d) Cutlivar Harrington's Pink.

*Source:* **(a) Photo by Shutterstock/SnnAy. (b) Photo by Shutterstock/Alex Manders. (c) Photo by Shutterstock/APugach. (d). Photo by Krzysztof Ziarnek, Kenraiz (CC BY SA 4.0).**

**FIGURE 8.3** Cultivars of New York aster (*Symphyotrichum novi-belgii*). (a) and (c): unidentified varieties. (b) Cultivar Royal Blue.

*Source:* **(a) Photo by Niepokój Zbigniew (CC BY SA 4.0). (b) Photo by CT Johansson (CC BY SA 3.0). (c) Photos by HQ (CC BY SA 2.0).**

formerly *Aster novae-angliae*) and New York aster (*S. novi-belgii*, *A. novi-belgii*) are especially popular. In North America, there are efforts to employ the many wild native North American species as ornamentals, in preference to the cultivars, especially local species, which usually are more sustainably grown, with less potential ecological damage.

## 8.2 OVERVIEW

*Callistephus chinensis* (the only species in *Callistephus*), called China aster and annual aster, is native to China and Korea. It is an old favorite, available in numerous very attractive cultivars cultivated worldwide. Although called

**FIGURE 8.4** Cultivars of bushy aster (*Symphyotrichum dumosum*): (a) Cultivar Diana. (b) Cultivar Kassel. (c) Cultivar Lady in Blue.

*Source:* **Photos by Agnieszka Kwiecień, Nova (CC BY SA 4.0).**

**FIGURE 8.5** Cultivars of (a) European Michaelmas daisy (European daisy, (b) Italian daisy, *Aster lamellus*).

*Source:* **(a) Photo by Maja Dumat (CC BY 2.0). (b) Photo by Shutterstock/Starover Sibiriak.**

**FIGURE 8.6** China aster (*Callistephus chinensis*).

*Source:* **(a) Photo by Shutterstock/Nick Pecker. (b) Photo by Shutterstock/Ole Schoener.**

annual aster, it is also sometimes a biennial. The plants are 20 to 100 cm in height. The flowering heads, 7 to 15 cm in diameter, develop at the top of the stem and on branches. The flowers are available in many colors and in doubled forms (with extra petals), and there are cultivars that produce plants with some flower heads of one color, other flower heads of different colors.

Both *Symphyotrichum* and *Aster* species are usually herbaceous, becoming woody at the base. Heights range from 20 cm to 2 m, or taller. The low-growing species are suitable for rock gardens; the tall, clump-forming forms are good for the back of beds and borders. Most perennial asters usually flower in the late summer and autumn. The flowers (i.e., flower heads) are mostly 3 to 6 cm in diameter, and the flowering displays in cultivars are attractive usually because they occur in masses rather than as large individual flower clusters. Flower colors include blue, pink, purple, red, and white. There are many cultivars with semi-doubled flowers.

## 8.3 HISTORY

China aster, from Asia, has long been recognized as distinctive from all other plants called aster and has been popular for centuries, especially for cultivars with doubled flowers grown as annuals. Some "aster" species from Eurasia and dozens of species from the Americas (mostly North American *Symphyotrichum*) have been the source of many cultivars of perennials, often with semi-doubled flowers, occasionally with doubled flowers.

## 8.4 ECONOMIC IMPORTANCE

Asters are grown commercially in greenhouses and in fields to supply potted plants for gardens and cut flowers. They are among the popular daisylike ornamentals that are staples of the world's horticultural industries.

## 8.5 ORNAMENTAL USES

Asters are suitable for beds, borders, and rock gardens. They are also useful for pollinator gardens, attracting bees and butterflies. The seed heads also attract a variety of birds.

FIGURE 8.7 Commercial production and marketing of asters: (a) A garden center display of potted asters for sale. (b) A field of asters being grown for seed (small farmers in southern Asia commonly cultivate China aster).

*Source:* **(a) Photo by Shutterstock/Africa Studio. (b) Photo by Dinesh Valke (CC BY SA 2.0).**

FIGURE 8.8 Principal ornamental uses of asters: (a) As landscape residential plantings. This photo shows a house in Northern California with a hedge of purple New York asters. (b) In containers. This photo shows a floral memorial to fallen soldiers of World Wars I and II. The red, white, and blue flower colors are official for France, the Czech Republic, and Serbia.

*Source:* **(a) Photo by Shutterstock/Emily C. McCormick. (b) Photo by Shutterstock/Beekeepx.**

FIGURE 8.9 Pollinators attracted to aster flowers: (a) Peacock butterfly on an aster. (b) Common eastern bumblebee (*Bombus impatiens*) pollinating frost aster (*Symphyotrichum pilosum*) in Wisconsin. (c) Monarch butterfly on New England aster in New Hampshire. (d) Honeybees on aster flowers.

*Source:* **(a) Photo by Shutterstock/Martina Unbehauen. (b) Photo by Joshua Mayer (CC BY SA 2.0). (c) Photo by Greg Thompson/US Fish and Wildlife Service (public domain). (d) Photo by Shutterstock/Klaus Brauner.**

## 8.6 OTHER USES

There are many so-called aster species in numerous countries that have been employed in traditional herbal medicine. A legacy of this extensive usage is that preparations incorporating asters are marketed on the web, most with debatable claims of medical effectiveness.

## 8.7 PROBLEMS

Several asters are minor invasive weeds in some countries. Like many garden plants, asters are subject to several fungal diseases, especially if overwatered or grown in poorly draining soil.

## 8.8 CULTIVATION

### 8.8.1 Climate and Site

China aster is normally grown as an annual, so winter conditions are not relevant. Smaller cultivars are suited to the front of beds and containers. The plants should be in full sun or at least in light shade. They grow better when sheltered from wind.

Depending on cultivar, perennial asters are suited to climate zones 3 to 10 (or more conservatively, 4 to 8). Most asters are adapted to cool climates and, in hot climates, need protection from the midday sun. They grow well in both full and partial sun. Asters can be planted in containers and even grown as houseplants, but this is infrequent. Many species can spread aggressively, so sites that constrain spread are preferable.

### 8.8.2 Propagation and/or Purchasing

The annual China aster is grown from seeds. Many perennial cultivars and all wild species also can be grown from seeds. However, propagation is usually by division. Most cultivars will not reproduce true to type from seeds. Gardeners usually establish plants using potted plants obtained from commercial sources. Plants can/should be divided every few years.

### 8.8.3 Planting

China aster is best spaced 30 to 45 cm apart. Spacing of perennial asters should be 30 to 100 cm apart, depending on the eventual sizes of the plants.

### 8.8.4 Soil and Moisture Conditions

Asters can be planted in a range of soils, but most prefer moist but well-drained, organically rich loams. Soil pH can range from 5.5 to 7.5 (*Aster amellus* grows well in alkaline soil).

### 8.8.5 Maintaining Plants

Flowering of the annual China aster is stimulated by cutting (so the species is a principal cut flower). For perennials, pinching the tips of branches and cutting back plants encourages bushiness. Tall varieties may need to be staked. The perennials should be divided every 3 years or so.

FIGURE 8.10 Asters as cut flowers: (a) Flowers being collected from a garden. (b) A vase of flowers collected from a garden.

***Source:*** **(a) Photo by Shutterstock/Ben Schonewille. (b) Photo by Shutterstock/Ben Schonewille.**

## 8.9 FASCINATING FACTS, FABLES, FOLKLORE, AND FRIVOLITY

- Identifying the plants that were important in ancient times is often guesswork, but nevertheless, asters are often associated with the Greek goddess Astraea. She is credited with beautifying the world by shedding tears, which transformed into aster flowers wherever they landed.
- The Cherokee have a different story explaining the origin of North American asters. Two sisters fleeing from warring tribes encounter a medicine woman who turns them into flowers, one an aster, the other a goldenrod.
- Dating back to ancient Greece, in old Europe, asters were grown in gardens or burned in the belief that either practice would ward off serpents. Possibly, this explains why snakebite remedies in medieval Europe sometimes added asters. Curiously, it appears that some North American Indigenous people independently employed asters to treat snakebite.
- The Hungarian revolution of October 31, 1918, has been nicknamed the aster revolution (and the chrysanthemum revolution) because protesters in Budapest wore the flower. Following the revolution, Hungary declared its independence and Austria-Hungary dissolved.
- The word *aster* has a quite different meaning for animals than for plants. In non-sexual (mitotic) cell division in animals, star-shaped structures termed *asters* appear and participate in cell replication (technical definition of *aster*: "a system of gelated cytoplasmic rays (aster rays) typically arranged radially about a centrosome at either end of the mitotic spindle and sometimes persisting between mitoses"). Such asters are absent from plants.

## KEY PUBLICATIONS

Burgess, E.S. 1902. Studies in the history and variations of asters – part I: History of pre-Clusian botany in its relation to aster. *Memoirs of the Torrey Botanical Club* 10: 1–5, 7, 9–93, 95–447. www.jstor.org/stable/pdf/43392214.pdf.

Dorn, R, 2004. *Asters Retreat to Eurasia*. Summer, Kelseya. www.mtnativeplants.org/wp-content/uploads/2018/07/KR-Asters-Retreat-to-Eurasia-Dorn.pdf.

Eddy, R. 1994. Production and marketing of garden asters. *Floriculture Indiana* 8: 10–13. https://hortscans.ces.ncsu.edu/uploads/p/r/producti_53bc07c6edc1a.pdf.

Gray, W.R. 2018. Where have all our asters gone? *Sego Lily* 41(2): 1–8. www.unps.org/segolily/Sego2018Spring.pdf.

McAvoy, R.J. 1993. Growing garden asters. *Connecticut Greenhouse Newsletter* 174: 13–16. https://hortscans.ces.ncsu.edu/uploads/g/r/growing__53babdf697684.pdf.

Nesom, G.L. 2020. The genus *Aster* (Asteraceae) in the strictest sense. *Phytoneuron* 56: 1–22.

Payne, R.N. and W.R. Kays 1954. *Spring Production of China Asters in the Greenhouse in Oklahoma*. Agricultural Experiment Station, Division of Agriculture, Oklahoma A. & M. College, Stillwater. 15 pp. https://shareok.org/bitstream/handle/11244/332160/oksa_B-439_1954-11.pdf?sequence=1.

Picton, P. 1999. *The Gardener's Guide to Growing Asters*. Timber Press, Portland, OR. 160 pp.

Picton, P. and H. Picton. 2015. *The Plant-Lovers Guide to Asters*. Timber Press, Portland, OR. 248 pp.

Travlos, I. 2016. *Symphyotrichum novi-belgii (New York Aster)*. CABI Compendium. https://doi.org/10.1079/cabicompendium.7557.

## Key Cultivation Websites

Hawke, R.G. 2013. A comparative study of cultivated asters. *Plant Evaluation Notes* 36: 1–11. www.chicagobotanic.org/downloads/planteval_notes/no36_asters.pdf.

Martin, S. *Asters – Color and Life in the Fall Garden*. https://piedmontmastergardeners.org/article/asters-color-and-life-in-the-fall-garden/.

Royal Horticultural Society. *How to Grow Asters*. www.rhs.org.uk/plants/aster/growing-guide.

Semple, J.C. *Astereae Lab. An Overview of "Asters" and the Tribe Astereae*. https://uwaterloo.ca/astereae-lab/research/asters.

The Spruce. *How to Grow and Care for China Aster*. www.thespruce.com/china-aster-care-guide-7556737#:~:text=China%20asters%20can%20be%20grown,regions%2C%20midday%20shade%20is%20beneficial.

WikiHow. *How to Grow Asters*. www.wikihow.com/Grow-Asters.

# 9 Baby's Breath (*Gypsophila paniculata* and Relatives)

**Scientific Names:** The genus name *Gypsophila* is based on the Greek *gypsos*, "gypsum," and *philios*, "loving," for the soils rich in gypsum (calcium sulfate), in which some of the species grow. (Calcium sulfate dihydrate is widely mined for use as a fertilizer and as the main constituent in plaster and drywall.) The plants are adapted to basic soils (pH >7.0), which may also be lime-based (with substantial content of calcium carbonate). *Gypsophila* is also a genus of six Asian birds (scientific nomenclature for plants is independent of the parallel Latin system of names for animals, and there are other such examples of the same genus name designating both plants and animals).

**Common Names:** The common name *baby's breath* (also *baby's-breath*, *babe's breath*) was first recorded in dictionaries in the late 19th century (the first use of the phrase has been claimed to trace to 1866). It has been suggested that the name arose as an indication of youthful innocence, or that the association with babies traces to its use at baby showers. However, there is little reliable indication of why the name was coined. The breath of babies usually does not have significant odor (the lack of teeth is helpful in this respect), although sometimes it has a milky fragrance, and sometimes bad smells develop because of gastrointestinal or other physiological problems. In fact, as will be noted, the odor of the flower is bad. Other occasional names include *gyp*, *maiden's breath*, *soapwort*, *soap flower* (explained later), and *bridal wreath*. *Gypsophila paniculata*, the most popular species, is usually called baby's breath but also is referred to as common baby's breath.

## 9.1 INTRODUCTION

Baby's breath varieties are grown as garden plants, but their predominant usage is as cut flowers by florists. The small size of individual flowers and the predominant white color of baby's breath flowers seem to suit their usual role of supporting or accenting larger, more colorful blooms, especially red roses, in bouquets and floral arrangements. The appearance of the flowering clusters of baby's breath – variously described as airy, cloudy, fluffy, lacy, or misty – is quite appealing to most people. The extensive slim branching on which the diminutive blooms are situated is responsible for the fluffy appearance of the flowering displays. The dainty, delicate white flowers have come to be very strongly associated with bridal bouquets, indeed with gift flowers to women on such occasions as Valentine's Day, Mother's Day, and birthdays. The lack of ostentation also seems to make the flowers appropriate for more somber occasions, particularly funerals.

## 9.2 OVERVIEW

*Gypsophila* is a genus of about 150 herbaceous species, mostly of the temperate part of the Old World. Only a few species are employed as ornamental plants. The most important is *G. paniculata*, grown on a large scale in various countries for the production of cut flowers employed primarily by florists. *Gypsophila paniculata*, originating from Europe and North Asia, is a perennial but is sometimes grown commercially as an annual crop. It has a taproot capable of growing to a depth of several meters. Some cultivars can exceed 1 m in height, but most are smaller, and some sprawl on the ground and are less than 20 cm tall. The species normally flowers from late spring to autumn and can overwinter in regions with mild freezing winters by persistent roots (while the upper shoot is frost-killed). The small flowers (3–5 mm in diameter) are most often naturally white, but sometimes pink. The blooms are doubled (with extra petals) in the most popular cultivars. The annual *G. elegans* (called annual baby's breath and showy baby's breath), also native to Eurasia, is additionally grown for cut flowers, but to a lesser degree. *Gypsophila elegans* is shorter (usually 30–40 cm) but has larger flowers.

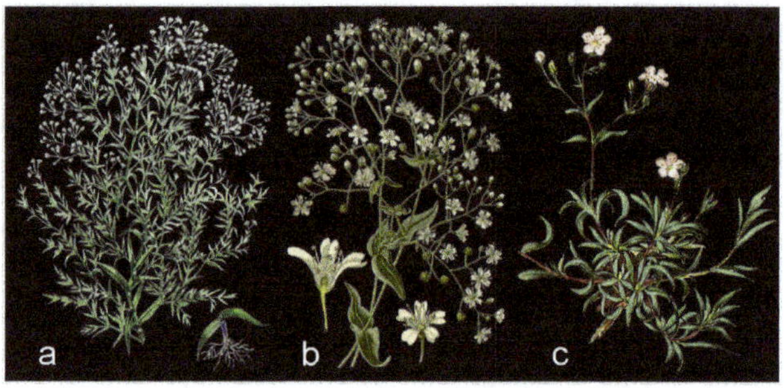


FIGURE 9.1 Paintings by botanical illustrators of the most commonly grown baby's breath (*Gypsophila*) species (public domain figures): (a) Common baby's breath (*G. paniculata*). (b) Annual baby's breath (*G. elegans*). (c) Creeping baby's breath (*G. repens*).

***Source:*** **(a) Kops, J., and F. W. van Eeden. 1898. *Flora Batava*, vol. 20. Published in Haarlem, Netherlands (www.BioLib). (b) Step, E., D. Bois, B. Herincq, and W. Watson. 1896. *Favourite Flowers of Garden and Greenhouse*, vol. 1. Frederick Warne, New York. (c) Hartinger, A. 1882. *Atlas der Alenflora. Eigenthum und Verlag des Deutschen und Oesterr.* Alpenvereins, Wien, Austria.**


DOI: 10.1201/9781003586227-9

FIGURE 9.2 Flowering displays of baby's breath (*Gypsophila*) species: (a) A cluster of flowers of G. *paniculata*. (b) Mature flowering plant of *G. repens* 'Rosea'. (c) Mature flowering plant of *G. muralis*. (d) Mature plant of *G. repens* in a rock garden. (e) Single flowers (i.e., with a single row of petals) of *G. repens* 'Rosea'. (f) Doubled flowers (with numerous petals) of *G. repens*.

***Source:*** **(a) Photo by Stefan Lefnaer (CC BY SA 4.0). (b) Photo by HQ (CC BY SA 2.0). (c) Photo by Chris Gladis (CC BY ND 2.0). (d) Photo by Udo Schmidt (CC BY SA 2.0). (e) Photo by HQ (CC BY SA 2.0). (f) Photo by André Karwath (CC BY SA 2.5).**

Several *Gypsophila* species, including *G. paniculata* and *G. elegans*, are cultivated to a relatively minor extent as garden ornamentals. Creeping baby's breath (*G. repens*) is relatively frequently encountered as a garden plant. *Gypsophila muralis* (*Psammophiliella muralis*), called annual gypsophila (it is indeed an annual), cushion baby's breath, and low baby's breath, a native of Eurasia, is sometimes grown as a garden plant. It has become naturalized in much of the world, including North America.

## 9.3 HISTORY

Queen Victoria (1819–1901) of England was a fan of baby's breath, used the flowers in her own wedding floral arrangements, and was an early influencer in establishing the popularity of the flower. As has happened frequently, once Old World plants became popular in Europe, they soon also became cultivated in North America and other countries of colonial origin.

## 9.4 ECONOMIC IMPORTANCE

Baby's breath is an established staple of the florist industry worldwide, generating large revenues for growers and sellers. In some locations, baby's breath is grown commercially outdoors, while in others it is cultivated in heated or unheated greenhouses. In Florida, for example, to obtain flowering stems for the winter–spring market, root clumps, and crowns that have been cold-stored during the summer are planted from September to early February. The species is of minor importance as a garden bedding and container plant and is mostly grown for harvest as a cut flower.

FIGURE 9.3 Commercial production of baby's breath (*Gypsophila paniculata*): (a) Mass production in a "high-tunnel" greenhouse. (b) Production of individual potted plants.

***Source:*** **(a) Photo by Shutterstock/Liu Yu Shan. (b) Photo by Shutterstock/Sharohyip.**

FIGURE 9.4 Use of baby's breath as "filler" in bouquets. Note how the sprays of tiny white flowers complement the much larger, more intensely colored blooms.

***Source:*** **(a) Photo by Pdbreen (CC BY 2.0). (b) Photo by Annca from Pixabay (public domain). (c) Photo by Positive Images from Pixabay (public domain). (d) Photo by Myriams-Fotos from Pixabay (public domain).**

## 9.5 ORNAMENTAL USES

Baby's breath is a staple cut flower in florist shops, employed (often as "filler") in bouquets for various gatherings, especially weddings (notably in bridal bouquets). It is sometimes employed as the only flower in an arrangement but usually is present in mixed preparations. Fresh flowers are widely used in corsages, gift bouquets, table centerpieces, and sometimes boutonnieres. Several species are grown as garden plants. They are particularly useful in cutting gardens (i.e., for harvesting the stem for bouquets, the flowers often lasting more than a week in vases). Baby's breath displays well in rock gardens and also grows well in containers.

The dried flowers are useful in dried preparations. Indeed, often dried preparations are so lifelike that it isn't clear whether a container of flowers is alive or dried. The flowers are extensively dyed by the floral industry. (Even worse, for those who insist on authenticity, artificial gypsophila flowers are frequently marketed.)

FIGURE 9.5 Use of baby's breath at weddings.

*Source:* **(a) Photo by Shutterstock/J. Robert Williams. (b) Photo by Sady Muñoz from Pixabay (public domain). (c) Photo by Frauke Riether from Pixabay (public domain). (d) Photo by Ulrike Mai from Pixabay (public domain).**

FIGURE 9.6 Dried, cut flowers of baby's breath. These are often placed in containers to give the appearance that the plants are in pots of soil or water: (a) Displays along the main aisle of a church in Mexico. (b, c, d) Bouquets.

*Source:* **(a) Photo by Cultivar 413 (CC BY 2.0). (b) Photo by Mahazyn kvitiv (CC BY SA 4.0). (c) Photo by Tuân Nguyen Anh from Pixabay (public domain). (d) Photo by Magoer from Pixabay (public domain).**

## 9.6 OTHER USES

Saponins – a class of chemicals widely distributed in plants – are employed commercially for many purposes, especially for their ability to produce foam. Saponins are harvested from the underground parts of *Gypsophila* species and have been used in cosmetics, detergents, medicines, and industrial preparations, such as fire extinguisher chemicals. Extracts from *Gypsophila* species have been added to liquors, flavored cheese, the traditional Middle Eastern dessert halva (halvah), and other sweets, such as Turkish delight.

FIGURE 9.7 Commercially dyed baby's breath flowers.

*Source:* **(a) Photo by Shutterstock/Liudmyla Kulias. (b) Photo by Shutterstock/Skywing.**

FIGURE 9.8 Baby's breath (*Gypsophila paniculata*) growing as an invasive weed in the United States. (a) In Montana. (b) In Utah.

*Source:* **(a) Photo by Matt Lavin (CC BY SA 2.0). (b) Photo by Steve Dewey, Utah State University, Bugwood.org (CC BY 3.0).**

## 9.7 PROBLEMS

In recent years, baby's breath has become controversial, with both supporters and detractors. The principal difficulties with the plant are outlined here.

Baby's breath is considered old-fashioned, and there has been some decrease in popularity. However, it still remains very widely employed by florists, although some look-alike plants are being substituted (for example, lily of the valley, which has equally small but much more perfumed flowers).

The cultivated species of *Gypsophila* have become significant invasive weeds in some parts of the world, including parts of North America. Bouquets that have been prepared with branches possessing seeds, when discarded, can contribute to spreading the plant. Large plants of *G. paniculata* have been found to produce about 13,000 seeds annually.

The sickly-sweet odor common in baby's breath is objectionable to many, sometimes compared to cat urine. The stench is due to the emission of isovaleric acid (3-methylbutyric acid). The odor of the wild plants is usually more unpleasant that that of garden forms. Some cultivars have been selected with little or no odor. Some florists spray scent on baby's breath to mask the odor.

Baby's breath can cause dermatitis, asthma, and eye irritation, especially as an occupational hazard for florists.

Millions have handled and smelled bouquets with the flowers without problems, but individual reaction are always possible. When ingested, the saponins can produce minor gastrointestinal upset in people and other animals, sometimes causing internal irritation, vomiting, and diarrhea. Nevertheless, baby's breath is considered to have only mild toxic potential for pets. Twigs of baby's breath are sometimes artfully placed on wedding cakes, which is generally harmless, but the mild toxicity of the plant should be borne in mind.

## 9.8 CULTIVATION

In regions where baby's breath is classified as an invasive species, it should not be planted.

### 9.8.1 Climate and Site

The cultivated perennial species of *Gypsophila* (including *G. paniculata*) can overwinter in climate zones 3 to 9. All the major garden species can simply be grown as annuals in zones 2 to 10, and *G. elegans* is naturally an annual. All garden forms do best in full sun but can grow in part shade.

### 9.8.2 Propagation and/or Purchasing

Garden plants can be propagated both by seeds and cuttings. The doubled flowers of commercial florist baby's breath (*G. paniculata*) are sterile and do not produce seeds, and so other methods of propagation (including tissue culture) are employed.

### 9.8.3 Planting

If planted directly in the ground, the seeds should be spaced 20–25 cm apart and thinned to 30–40 cm between the plants. Starter plants should also be established at this spacing. Most perennial species will not produce flowers the first year when planted outdoors from seed.

### 9.8.4 Soil and Moisture Conditions

Baby's breath is naturally adapted to dry soils, so the soil should be well-draining (clay should be avoided or amended). The plant is also naturally adapted to alkaline soils. Preparing the soil by adding compost is recommended, but afterward fertilization should be light.

### 9.8.5 Maintaining Plants

Plants grown in gardens may flop over, and if this is objectionable, stakes should be provided. When cutting flowers for display in vases, half the flower buds should not yet have opened.

## 9.9 FASCINATING FACTS, FABLES, FOLKLORE, AND FRIVOLITY

- A centuries-old, persistent, and widespread urban legend contends that cats steal a baby's breath in its sleep. Supposedly, cats are attracted to the milk scent on a baby's breath or are jealous that the owners are giving more attention to the baby. A coroner's report in 1791 concluded that this was in fact the cause of a particular baby's death. The concept is, of course, ridiculous but particularly alarms parents of babies and very young children (see Crespy1 2019).
- Sometimes, a good (or bad) legend gives rise to another legend. According to newspaper reports in 1987 in North Carolina and Virginia, a baby had been fed milk and left sleeping beside a strawberry field while its mother picked strawberries. A snake, attracted by the infant's milky breath, slithered down its throat and strangled the baby (see Brunvand, J. H. 2001. *Encyclopedia of Urban Legends*. ABC-CIO, Santa Barbara, CA).
- Baby's Breath is a Canadian national organization founded in 1973. It is focused on SIDS (sudden infant death syndrome) and, more recently, on reducing all sudden and unexpected infant deaths, including stillbirths.
- *Gypsophila* species have been called soaproot in Europe because the roots contain saponins. These

FIGURE 9.9 A cat, baby, and baby's breath: the subjects of a disturbing urban legend.

*Source:* **Photo by Shutterstock/Vasilyev Alexandr.**

FIGURE 9.10 Root of *Gypsophila paniculata*, employed as a source of commercial extracts and traditionally used in preparing soap.

*Source:* **Photo by Shutterstock/Shansh23.**

chemicals foam and act as a gentle but effective cleaner when mixed with water. Many plants contain saponins, which are toxic (probably protecting the plants from injury from various animals and microorganisms) but usually occur in low concentrations. A traditional source of saponins to make soap is *Saponaria officinalis* (*Gypsophila* and *Saponaria* are related genera; *Gypsophila* is often called *white saponaria* in Asia). Use of high-saponin plants as soap persists in various countries.

## KEY PUBLICATIONS

Barkoudah, Y.I. 1962. A revision of *Gypsophila*, *Bolanthus*, *Ankyropetalum* and *Phryna*. *Wentia* 9: 1–203.

Crespy1. 2019. *Can a Cat Really Steal a Baby's Breath, or is that an Urban Legend? Ohio State University: The Psychology of Extraordinary Beliefs*. Student (Blog). https://u.osu.edu/vanzandt/2019/04/11/can-a-cat-really-steal-a-babys-breath-or-is-that-an-urban-legend/.

Darwent, A.L. 1975. The biology of Canadian weeds: 14. *Gypsophila paniculata* L. *Canadian Journal of Plant Science* 55: 1049–1058.

Darwent, A.L. and R.T. Coupland. 1966. Life history of *Gypsophila paniculata*. *Weeds* 14: 313–318.

Doi, M., H. Nimitkeatkai, K. Inamoto, and Y. Ueda. 2008. Suppression of unpleasant odor emission from cut *Gypsophila paniculata* L. Inflorescences by isovalerate ester formation. *Acta Horticulturae* 766: 493–500.

Emery, S.M. and P.J. Doran. 2013. Presence and management of the invasive plant *Gypsophila paniculata* (baby's breath) on sand dunes alters arthropod abundance and community structure. *Biological Conservation* 161: 174–181.

Hicklenton, P.R. 1987. Flowering of *Gypsophila paniculata* cv. Bristol Fairy in relation to irradiance. *Acta Horticulturae* 205: 103–112.

Lamar, S.K. and C.G. Partridge. 2021. Combining herbarium databases and genetic methods to evaluate the invasion of a popular horticultural species, baby's breath (*Gypsophila paniculata*), in the United States. *Biological Invasions* 23: 37–52.

Marousky, F.J. and J. Nanney. 1972. Influence of storage temperatures, handling and floral preservatives on post-harvest quality of gypsophila. *Proceedings of the Florida State Horticultural Society* 85: 419–422.

Raulston, J.C., S.L. Poe, and F.J. Marousky. 1972. Cultural concepts of *Gypsophila paniculata* L. production in Florida. *Proceedings of the Florida State Horticultural Society* 85: 423–428.

Rice, E.K., H. Leimbach-Maus, C. Partridge, and J.N. McNair. 2020. Assessment of invasive *Gypsophila paniculata* control methods in the northwest Michigan dunes. *Invasive Plant Science and Management* 13: 94–101.

Shillo, R. 1985. *Gypsophila paniculata*. In: *Handbook of Flowering*, edited by A.H. Halevy, 83–87. CRC Press, Boca Raton, FL.

## Key Cultivation Websites

BBC Gardeners' World Magazine. *How to Grow Gypsophila (Baby's Breath)*. www.gardenersworld.com/how-to/grow-plants/how-to-grow-gypsophila-babys-breath/.

Gardening Channel. *How to Grow Baby's Breath*. www.gardeningchannel.com/how-to-grow-babys-breath/.

Gardening Know How. *Baby's Breath Propagation*. www.gardeningknowhow.com/ornamental/flowers/babys-breath/babys-breath-propagation.htm.

Horticulture Magazine. *Gypsophila 'Baby's Beath'*. https://horticulture.co.uk/gypsophila/.

The Spruce. *How to Grow and Care for Baby's Breath*. www.thespruce.com/babys-breath-garden-favorite-4082480.

WikiHow. *How to Grow Baby's Breath*. www.wikihow.com/Grow-Baby%27s-Breath.

# 10 Begonia (*Begonia* Species and Hybrids)

**Scientific Names:** The genus name *Begonia* commemorates Michel Bégon (1638–1710), a French administrator (the intendant of San Domingo, the western part of the island of Haiti, and the governor of French Canada). He was an amateur botanist known for collecting interesting plants.

**Common Names:** The common name "begonia," of course, just repeats the genus name. Some plants with "begonia" in their names are superficially similar to some begonias but are quite unrelated. These include strawberry begonia (*Saxifraga stolonifera*), Swedish begonia (*Plectranthus*), and wild begonia (*Rumex venosus*).

## 10.1 INTRODUCTION

Begonia lovers are sometimes affectionately termed "begoniacs," an indication that the flowers are attractive to the point of addiction. Along with petunias, impatiens, marigolds, and geraniums, begonias rank among the top five bedding plants grown for floral display in the Northern Hemisphere. Begonias also are considered to be among the most popular houseplants. The 15,000 or so begonia hybrid cultivars are astonishingly diverse, because hybridization is easy and there are over 1,500 wild species, providing breeding material. The floral cultivars are dazzlingly beautiful – the same is true of the foliage cultivars – and the colorful displays last for months. Moreover, most varieties are adapted to shade, where only a minority of ornamentals thrive.

The Begoniaceae family is made up mostly of the genus *Begonia*, but it also includes the genus *Hillebrandia*, which has only one species, *H. sandwicensis*, named for the Sandwich Islands (now known as the Hawaiian Islands). Older literature, however, recognizes several other genera in the family. *Begonia* is considered to be the fifth or sixth largest genus of flowering plants, with estimates ranging from 1,500 to 2,000 species. The primary center of diversity of this pantropical genus is the neotropics (Mexico to Argentina), where about 700 species occur. A second center of diversity is in south-central and southeastern Asia, with about 650 species. About 160 species are native to the southern, southeastern, and western parts of Africa. The Asian and American species are thought to have arisen from the African species. *Begonia* species are frequent in tropical and subtropical regions, usually in shady locations in forested areas, from sea level to 3,000 m. Most occur in moist, humid microhabitats, although some are adapted to quite dry situations.

## 10.2 OVERVIEW

Most begonias are perennials, but some are annuals, and in temperate regions, the perennials are normally grown as annuals. Most are small or medium-sized, ranging from under 5 cm to over 2 m. Most are herbaceous, with succulent, juicy stems, but some develop soft-wooded stems. A few are vine-like (described as "climbing," "trailing," or "hanging"), sometimes over 5 m in length. Some are epiphytes that live entirely on trees, the roots never reaching the ground. Most are rooted in the ground, and the underground part is usually a thick rhizome or, less often, a tuber, from which fibrous roots arise. Although virtually all begonias have fibrous roots, so-called "fibrous-rooted begonias" refer to plants lacking well-developed rhizomes or tubers. The base of the leaves is characteristically asymmetrical, and leaves may be less than 2 cm to over 50 cm in length. *Begonia thiemei* of Mexico and Honduras has petioles (leafstalks) as long as 1 m. Wild begonias produce unisexual male and female flowers that are predominantly white but may be pink, red, orange, or occasionally yellow. Most plants bear both male and female flowers, but some species produce plants with male flowers, others with female flowers. Usually male, but not female, flowers produce nectar, the female flowers deceitfully attracting pollinators by virtue of their resemblance to the male flowers. Pollination has not been extensively studied in wild plants, and the extent to which pollinators and wind are significant pollen vectors is unclear. Domesticated species are self-pollinating. The fruit is usually a dry capsule but is occasionally fleshy and berry-like. There may be hundreds of tiny dust-like seeds in the capsules. Seeds are distributed by animals, wind, gravity, and raindrops (by the "splash cup syndrome").

The international horticultural trade does not utilize the scientific classifications of *Begonia* that have been proposed but instead employs an artificial classification popularized by the American Begonia Society. The following simplified presentation summarizes the most important of these arbitrary cultivar groups. Uncommonly cultivated form classes, not discussed, include shrub-like, thick-stemmed, cane-forming, and trailing or climbing begonias. As is the case for many ornamental species, the artificial horticultural classes of cultivars that are recognized are sometimes difficult to recognize because intergroup hybrid cultivars have been generated.

### 10.2.1 Wax Begonias

Wax begonias (*Begonia* × *semperflorens-cultorum*) are the best-known kind of begonias. The common name is due to the foliage often having a waxy look. These plants are widely grown by home gardeners and are often exhibited as mass displays in public gardens and parks. They are so popular as bedding plants that wax begonias are sometimes called "bedding begonias." Wax begonias dominate so-called "fibrous-rooted begonias," and popular articles on begonias sometimes employ the phrase "fibrous-rooted begonias" to

DOI: 10.1201/9781003586227-10

FIGURE 10.1 Mixed cultivars of wax begonias.

*Source:* **Photo by Shutterstock/Cristina Ionescu.**

refer specifically to wax begonias. In Europe, wax begonias are called "Semperflorens gracilis" (a pseudo-scientific name, cultivar group name, or cultivar name). In the United States, annual sales of wax begonias are estimated to exceed $60,000,000. This kind of begonia is believed to have arisen from the crossing in 1878 of two Brazilian species, *B. cucullata* var. *cucullata* (*B. semperflorens*) and *B. schmidtiana.* Wax begonias have somewhat succulent stems and are bushy and compact, usually less than 30 cm in height. The leaves are glossy, rather rounded or spoon-shaped, sometimes bronzed or brownish ("mahogany"), with some recent cultivars producing variegated leaves and even new leaves that are white. The plants bloom continuously (the term *semperflorens* mentioned earlier means "ever-flowering"), producing numerous small flowers that are red, pink, or white, depending on the cultivar. Cultivars with semi-double or doubled flowers are available. The plants are perennials in their native habitats but are widely grown as annuals.

### 10.2.2 Tuberous Begonias

Next to wax begonias, tuberous begonias are the most popular kind. Hybrid tuberous begonias (*B.* × *tuberhybrida*) are a very popular group of hybrids first developed in the middle and latter 1800s from tuberous-rooted *Begonia* species of the South American Andes (notably *B. boliviensis*, *B. clarkei*, *B. davisii*, *B. pearcei*, and *B. veitchii*). Most cultivars are grown for their blooms. The flowers are generally large (occasionally the size of dinner plates) and colorful, sometimes produced singly (in which case they are large) or in clusters (of smaller blooms), sometimes with frilled tepal (petal) margins, in solid or variegated colors. Many cultivars produce doubled male flowers (with stamens transformed to petals), and some specialists remove the less-attractive female (single, that is, non-doubled) flowers. Doubled flowers often resemble roses and carnations. Some cultivars have trailing growth habits; others are bushy or compact. Tuberous begonias are employed as bedding plants, potted plants, and hanging baskets.

FIGURE 10.2 Mixed cultivars of tuberous begonias.

*Source:* **(a) Photo by Shutterstock/Watcharee Suphaluxana. (b) Photo by Shutterstock/Wuttisit Somtui.**

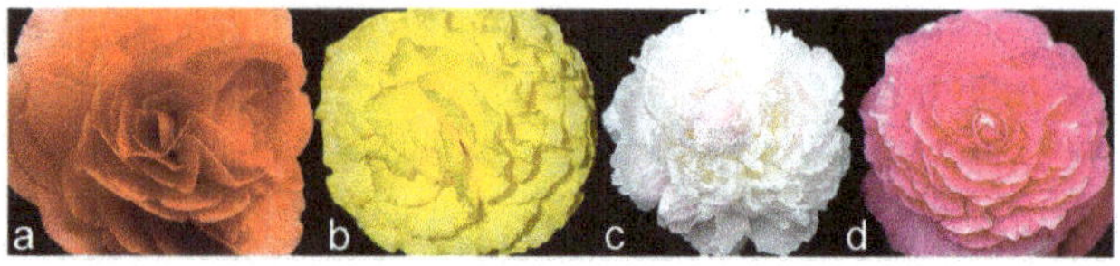

FIGURE 10.3 Roselike appearance of doubled flowers of tuberous begonias.

*Source:* **Pixabay (public domain) by photographers (a) Rebecca Matthews, (b) Frauke Feind, (c) No-longer-here, and (d) Henryk Niestrój.**

FIGURE 10.4 Old paintings (public domain) of tuberous begonias by 19th-century European botanical artists: (a) Lorraine begonia (*B.* × *cheimantha* 'Gloire de Lorraine'). (b) *Begonia veitchii*, one of the ancestors of modern hybrid tuberous begonias. (c) *Begonia* group elatior.

*Source:* **(a) *L'Illustration Horticole* (Gand, Belgium), vol. 43, 1896. (b) *La Belgique Horticole* (Liège, Belgium), vol. 29, 1879. (c) *Die Gartenwelt*, G. Schmidt, Berlin, 1897.**

Hiemalis begonias (*B.* × *hiemalis*, *B.* × *elatior*) are another kind of tuberous begonias. They arose from crosses of *B. socotrana* (named for Socotra, an island south of the Arabian Peninsula) and hybrid tuberous begonias from Peru and Bolivia. Cultivars are often encountered under the names Rieger begonia and elatior begonia and are also referred to as autumn- and winter-flowering begonias. There is a very wide range of cultivars of this group, which are among the most important begonias grown as commercial potted flowering plants. In Europe, the international begonia trade is dominated by this class of begonia. Although the flowers are smaller than those of most hybrid tuberous begonias, they are produced profusely, usually doubled, and available in many colors. After flowering, the stems tend to die back, and because the plants are difficult to rejuvenate, they are usually discarded. Cultivars are vegetatively propagated

FIGURE 10.5 Hiemalis begonias (*Begonia* × *hiemalis*): (a) A pot of mixed cultivars. (b) Begonia cultivar Karen.

*Source:* **(a) Photo by Tequila (CC BY SA 3.0). (b) Photo by Serres Fortier (CC BY 2.0).**

commercially by leaf or stem cuttings, as well as by tissue culture.

Lorraine or Christmas begonias (*B.* × *cheimantha*) are still another class of tuberous begonia. This group originated from crosses between *B. socotrana* and B. *dregei* in 1880. The first cultivar, Gloire de Loraine, is responsible for the common name Lorraine begonia. Cultivars are commonly called Scandinavian, Norwegian, and Christmas begonias (they do flower in early winter). Like the much more popular hiemalis begonias noted previously, after these plants flower, they are difficult to rejuvenate. Also, like the hiemalis begonias, they are vegetatively propagated.

### 10.2.3 Ornamental Foliage Begonias

While they do produce flowers, the popular rex begonias are mainly grown for their often variegated or multicolored leaves, which are available in various sizes and shapes. The foliage comes in shades of green, purple, red, brown, silver, and pink, often with metallic highlights. The "rex begonia" (*B. rex*) was introduced to Europe from the Himalayas in 1856. It was crossed with Asian species (*B. annulata, B. decora, B. diadema, B. robusta, B. tenuifolia*) to produce various hybrid cultivars, under the name *B.* × *rex-cultorum*. Subsequently, additional species from Mexico have been used to produce hybrids.

In addition to rex begonia cultivars, some other species are also employed as foliage ornamentals, notably *B. masoniana* (Iron Cross begonia).

## 10.3 HISTORY

*Begonia* appears to have been first cultivated at the Royal Botanic Gardens at Kew in 1777. By the mid-19th century, species from the West Indies, Mexico, and South America were being grown in Europe, and hybrids were being created. Hybridization has continued to this day, although the parentage of most cultivars has been poorly documented.

## 10.4 ECONOMIC IMPORTANCE

Begonia is one of the world's most popular garden plants, responsible for extremely large horticultural production and sales in many countries.

FIGURE 10.6 Rex begonia hybrids (*Begonia* × *rex-cultorum*).

*Source:* **(a) Photo by GCornelis (CC BY SA 3.0). (b) Photo by Nemrace (CC BY SA 3.0).**

FIGURE 10.7 Winglike leaves often formed by begonias. Also note the asymmetrical leaf bases: (a) Begonia 'Angel Wing' (*B. aconitifolia* × *B. coccinea*). (b) Begonia 'Dragon Wing Red' (*B.* × *hybrida*).

*Source:* **(a) Photo by Yercaud-elango (CC BY SA 4.0). (b) Photo by Kenpei (CC BY SA 2.1 JP).**

FIGURE 10.8 Photos showing commercial activities related to begonias: (a) A commercial greenhouse with tuberous begonias. (b) A commercial hydroponic greenhouse with wax begonias. (c) A florist with 'Angel Wing' begonias. (d) A florist with a gift pot of begonia.

*Source:* **(a) Photo by T. Kiya (CC BY SA 2.0). (b) Photo by Shutterstock/Vladimir Mulder. (c) Photo by Shutterstock/BearFotos. (d) Photo by Shutterstock/Shebeko.**

FIGURE 10.9 Wax begonia cultivars arranged in colorful patterns in the landscape.

*Source:* **(a) Photo by Shutterstock/Dirk M. de Boer. (b) Photo by Zefram (CC BY SA 3.0).**

FIGURE 10.10 Begonias in containers.

*Source:* **(a) Photo by Peter H. from Pixabay (public domain). (b) Photo by Shutterstock/Summer 1810. (c) Photo by Shutterstock/Olga Vasilek.**

## 10.5 ORNAMENTAL USES

A very wide range of begonia cultivars is available, many especially suitable for beds, borders, and various kinds of containers, including window boxes and hanging baskets.

## 10.6 OTHER USES

In the areas that begonias grow as native plants (Africa, Asia, and South America), the plants (especially the leaves) have been employed as a potherb and as a flavorant herb. In Yunnan Province of southern China, the foliage of *B. fimbristipula* is an important herbal tea. Both foliage and underground parts of begonias have been rather widely employed in folk medicine, especially in medicinal teas in areas where the plants are indigenous. Examples include *B. mooreana* herbal tea to treat stomach ailments in New Guinea, and herbal tea using *B. humilis* and B. *rotundifolia* to treat colds and fevers in South America. Despite its name, *B. deliciosa* from Borneo, which is cultivated as an ornamental, has bitter foliage (some have claimed the flowers are sweet). In Gabon, chimpanzees and gorillas commonly consume *B. mannii*.

## 10.7 PROBLEMS

Some species contain calcium oxalate crystals in the leaves, and when consumed, they need to be deactivated by first boiling them. Oxalic acid in the plants produces a tart pleasant taste, but as with spinach and rhubarb, large quantities can be toxic, especially for sensitive individuals. Begonia roots and tubers have been employed as emetics and purgatives, suggesting that their edible use is unwise. As with most plants, some sensitive individuals can experience dermatitis on contact. Dog and cat websites often list ornamental begonias as (mildly) toxic.

## 10.8 CULTIVATION

### 10.8.1 Climate and Site

The different groups of begonia have particular cultural requirements, but since most species come from tropical and semi-tropical regions, they need warm temperatures and usually can only be grown outdoors year-round in subtropical and tropical climates. In temperate climates, begonias are grown outdoors as annuals, or indoors as houseplants (in which case they should be provided with fairly bright window light). The hardy begonia (*B. grandis*) is rated to grow as a perennial in climate zones 6 to 9, but most cultivars are adapted to survive overwinter in zones 9 to 11. Although begonias are typically adapted to shade, moderate light is usually necessary to prevent legginess. In temperate regions, wax begonias grow well in both full sun and partial shade (but deep shade should be avoided). Tuberous begonias usually grow best in partial shade, but a few recent cultivars also do well in full sun. Cultivars with bronze foliage tend to do well in full sun. Wax begonias are better adapted to surviving hot, humid conditions than tuberous begonias. Many begonias have fleshy, brittle, easily broken stems and should be located away from wind and paths where they may be bumped.

### 10.8.2 Propagation and/or Purchasing

Wax begonias are often commercially propagated from seeds (especially single-flowered forms). Begonia seeds are very small, often with about 35,000 seeds per gram (or 1,000,000 seeds per ounce). The seeds are sometimes pelleted to facilitate handling. The tiny seedlings produced are initially quite fragile, and growth is slow, so home gardeners usually buy plug trays of young plants.

Tuberous begonias become dormant during the short days of fall and winter (reflecting their Andean history), and the tubers can be dug up before frost and stored overwinter in a dry, cool area, in vermiculite, peat, sawdust, or paper. Tubers sprout toward the end of winter, at which time they should be potted indoors and watered, and replanted outdoors when the weather is suitable. Some cultivars are grown from seeds, but more often, commercial growers propagate the plants by stem cuttings. Gardeners can purchase dormant tubers (specialist mail-order firms will have the best selections) or young potted plants, and indeed, almost all home gardeners start begonias employing starter plants bought from local sources.

### 10.8.3 Planting

After danger of frost, begonias can be placed outdoors. In beds, planting distance depends on ultimate size of the plants, but typically, wax begonias can be planted 15 to 20 cm apart, and tuberous begonias 25 to 30 cm apart. In pots and hanging baskets, the plants can be closer.

### 10.8.4 Soil and Moisture Conditions

Most begonias require a well-drained soil that is maintained semi-moist (this applies both to bedding plants and indoor potted plants). When watering, trying to keep the foliage dry is recommended to avoid foliage diseases. Probably, most begonias die an unnatural death from overwatering. Soils should be rich in organic matter, with a pH of 5.5 to 6.5. Supplemental inorganic fertilization should not be excessive. Phosphorus-rich fertilizer encourages flowering, while nitrogen-rich fertilizer encourages foliage development (a good idea for cultivars grown for their ornamental leaves).

### 10.8.5 Maintaining Plants

Begonias are somewhat self-cleaning (they tend to drop old blossoms), but deadheading is often advised to promote reflowering, and pruning stems that have become leggy will improve appearance. Pinching stems back encourages bushiness.

### 10.8.6 Rex Begonias

Rex begonias are usually grown as houseplants and tend to suffer from overwatering, too much heat, and too little atmospheric humidity. Moistening the leaves should be avoided, and so should use of saucers under the pots, which tend to accumulate water that soaks the soil (putting pots on a layer of stones in which water is placed to increase atmospheric humidity is recommended). Sometimes the plants become semi-dormant in winter, losing many or most of their leaves, in which case watering should be reduced until new growth resumes. The plants do well in moderate heat, and temperatures lower than 21°C may induce dormancy.

FIGURE 10.11 Old paintings (public domain) of rex begonias by 19th-century botanical artists. (a) shows a chromolithograph on woven paper prepared by German artist and lithographer Adam Gatternicht (1819–1899).

*Source:* (a) Photo from National Gallery of Canada. (b) *L'Illustration Horticole* (Ghent, Belgium), vol. 26 1871. (c) Step, E., D. Bois, B. Herincq, and W. Watson. 1896. *Favourite Flowers of Garden and Greenhouse*, vol. 1. Frederick Warne, New York.

## 10.9 FASCINATING FACTS, FABLES, FOLKLORE, AND FRIVOLITY

- Belgium is the world's largest producer of tuberous begonias, annually exporting over 60 million tubers. To promote the industry, a breathtaking "flower carpet" made up of cut tuberous begonia flowering stems has been assembled every 2 years since 1971 in Brussels, the capital. Craftsmen arrange the carpet by hand, inserting flowering stems directly into the cobblestones of the public square. The carpet is typically based on medieval carpet designs. The display lasts 3 or 4 days, the flowers kept fresh by sprinklers concealed within the pattern of the carpet.
- Kimjongilia, a tuberous begonia cultivar, was named in honor of North Korea's past president, the "Dear Leader" Kim Jong-il. The plant has been claimed to bloom annually on his birthday (February 16). When he died in 2011, his publicly displayed corpse was adorned with Kimjongilia flowers. The cultivar is highlighted in a North Korean song:

  The red flowers that are blossoming over our land
  Are like hearts: full of love for the leader
  Our hearts follow the young buds of Kimjongilia
  Oh! The flower of our loyalty!
- How popular is your name? Estimations of the popularity of names in the United States can be found at www.mynamestats.com/. "Begonia" is ranked as the 31,899th most popular given name, with an estimated population of only 135.
- Blue iridescent leaf coloration occurs in some plants growing in very shady conditions, and this is the case for some *Begonia* species. The bluish

FIGURE 10.12 "Flower carpet" made up of cut tuberous begonias assembled in Grand-Place, Brussels, Belgium, a UNESCO World Heritage Site with spectacular baroque, Gothic, neo-Gothic, and neo-classical architecture. The carpet is 75 m long by 24 m wide, with around 300 cut flowers per $m^2$. Up to 1,00,000 flowers are required and are assembled by more than 100 volunteers.

*Source:* Public domain photo from Pixabay.

FIGURE 10.13 *Begonia pavonina*. As noted in the text, the blue iridescent sheen of the foliage is related to the presence of special epidermal chloroplasts that capture light relatively efficiently in extremely shady habitats.

*Source:* **Photo by Clivid (CC BY ND 2.0).**

sheen in *B. pavonina* is due to reflectance of the blue wavelengths in modified upper (adaxial) epidermal chloroplasts ("iridoplasts"). Chloroplasts contain thylakoids, membrane-bound compartments in which the light reactions of photosynthesis occur. In iridoplasts, the thylakoids are stacked in a much more regular arrangement than in normal chloroplasts, giving them lens-like, light-concentrating properties. These special chloroplasts are adaptive in two ways: First, they are very efficient at capturing the predominantly green wavelengths available in deep shade (sunlight reaching the floor of forests occupied by *Begonia* species can be reduced by one millionth to one ten-millionth). Second, they concentrate the more abundant green and red wavelengths of light on the leaves' energy-absorbing apparatus, producing a 5% to 10% increase in energy capture.

## KEY PUBLICATIONS


Basurto-Peña, F., D. Castro-Lara, and M.A. Martínez-Alfaro. 2003. Edible begonias from the north of Puebla, Mexico. *Economic Botany* 57: 48–53.

Bowes, B.G. and E.W. Curtis. 1991. Conservation of the British National Begonia Collection by micropropagation. *New Phytologist* 119: 169–181.

Brilmayer, B. 1960. *All About Begonias*. Doubleday, Garden City, NY. 223 pp.

Catterall, E. 1991. *Begonias: The Complete Guide*. Crowood Press, Ramsbury, Marlborough, Wiltshire, UK. 199 pp.

Clement, W.L., M.C. Tebbitt, L.L. Forrest, J.E. Blair, L. Brouillet, T. Eriksson, et al. 2004. Phylogenetic position and biogeography of *Hillebrandia sandwicensis* (Begoniaceae): A rare Hawaiian relict. *American Journal of Botany* 91: 905–917.

De Wilde, J.J.F.E. 2011. Begoniaceae. In: *Families and Genera of Vascular Plants*, Vol. *10: Flowering Plants, Eudicots*, edited by K. Kubitzki, 56–71. Springer, Berlin.

Dewitte, A., A.D. Twyford, D.C. Thomas, C.A. Kidner, and J. van Huylenbroeck. 2011. *The Origin of Diversity in Begonia: Genome Dynamism, Population Processes and Phylogenetic Patterns*. Edited by O. Grillo, 27–52. Intech, Rijeka, Croatia.

Doorenbos, J., M. Sosef, and J. de Wilde. 1998. *The Sections of Begonia. Studies in Begoniaceae VI*. Backhuys, Leiden, The Netherlands. 266 pp.

Forrest, L.L. and P.M. Hollingsworth. 2003. A recircumscription of *Begonia* based on nuclear ribosomal sequences. *Plant Systematics and Evolution* 241: 193–211.

Forrest, L.L., M. Hughes, and P.M. Hollingsworth. 2005. A phylogeny from *Begonia* using nuclear ribosomal sequence data and morphological characters. *Systematic Botany* 30: 671–682.

Goodall-Copestake, W., D.J. Harris, and P.M. Hollingsworth. 2009. The origin of a mega- diverse genus: Dating *Begonia* (Begoniaceae) using alternative datasets, calibrations and relaxed clock methods. *Botanical Journal of the Linnean Society* 159: 363–380.

Goodall-Copestake, W., S. Pérez-Espona, D.J. Harris, and P.M. Hollingsworth. 2010. The early evolution of the mega-diverse genus *Begonia* (Begoniaceae) inferred from organelle DNA phylogenies. *Biological Journal of the Linnean Society* 101: 243–250.

Haegeman, J. 1979. *Tuberous Begonias. Origin and Development*. J. Cramer, Vaduz, Liechtenstein. 268 pp.

Harrison, N., R.J. Harrison, and C.A. Kidner. 2016. Comparative analysis of *Begonia* plastid genomes and their utility for species-level phylogenetics. *PLoS One* 11(4): e0153248. https://doi.org/10.1371/journal.pone.0153248.

Hughes, M. and P.M. Hollingsworth. 2008. Population genetic divergence corresponds with species level biodiversity patterns in the large genus *Begonia*. *Molecular Ecology* 17: 2643–2651.

Hvoslef-Eide, A.K., T. Fjeld, and J.W. Einset. 1995. Breeding Christmas begonia (*Begonia* × *cheimantha* Everett) for increased keeping quality by traditional and biotechnological methods. *Acta Horticulturae* 405: 197–204.

Hvoslef-Eide, A.K. and C. Munster. 2007. *Begonia*. History and breeding. In: *Flower Breeding and Genetics*, edited by N.O. Anderson, 241–275. Springer, Dordrecht, The Netherlands.

Jacobs, M., M. Lopez-Garcia, O.-P. Phrathep, T. Lawson, R. Oulton, and H.M. Whitney. 2016. Photonic multilayer structure of *Begonia* chloroplasts enhances photosynthetic efficiency. *Nature Plants* 2, Article 16162. https://doi.org/10.1038/nplants.2016.162.

Karlsson, M.G. and R.D. Heins. 1992. Begonias. In: *Introduction to Floriculture*, 2nd edition, edited by R.A. Larson, 409–427. Academic Press, New York.

Kramer, J. 1967. *Begonias: Indoors and Out*. E.P. Dutton, New York. 128 pp.

Krempin, J.L. 1993. *Know Your Begonias*. Krempin Books, Broadbeach Waters, Queensland, Australia. 119 pp.

Langdon, K.B. 1969. *The Tuberous Begonia*. Cassell, London, U.K. 98 pp.

Misono, I. 1974. *Begonias*. Translated 1978 by A.M. DeCola and H. Arakawa. American Begonia Society, Los Angeles, CA. 77 pp.

Moonlight, P.W., J.E. Richardson, M.C. Tebbitt, D.C. Thomas, R. Hollands, C.-I. Peng, et al. 2015. Continental-scale diversification patterns in a megadiverse genus: The biogeography of Neotropical *Begonia*. *Journal of Biogeography* 42: 1137–1149.

Neale, S., W. Goodall-Copestake and C.A. Kidner. 2006. The evolution of diversity in *Begonia*. In: *Floriculture, Ornamental and Plant Biotechnology*, edited by J.A. Teixeira da Silva, 606–611. Global Science Books, Middlesex, UK.

Rajbhandary, S., M. Hughes, T. Phutthai, D.C. Thomas, and K.K. Shrestha. 2011. Asian *Begonia*: Out of Africa via the Himalayas? *Gardens' Bulletin Singapore* 63: 277–286.

Stevens, M. 2002. *Begonias*. Firefly Books, Buffalo, NY. 96 pp.

Stults, D.Z. and B.J. Axsmith. 2011. First macrofossil record of *Begonia* (Begoniaceae). *American Journal of Botany* 98: 150–153.

Tebbitt, M.C. 2005. *Begonias: Cultivation, Identification and Natural History*. Timber Press, Portland, OR. 272 pp. + 64 plates.

Thompson, B. and W. Corby. 2019. *Brad's Begonia World*, 3rd edition. American Begonia Society, Sacramento, CA. 233 pp.

Thompson, M.L. and E.J. Thompson. 1981. *Begonias: The Complete Reference Guide*. Times Books, New York, NY. 356 pp. + 32 plates.

Wyatt, G.E. and M. Sazima, M. 2011. Pollination and reproductive biology of thirteen species of *Begonia* in the Serra do Mar State Park, São Paulo, Brazil. *Journal of Pollination Ecology* 6: 95–107.

Yu, N.A. and T.D. Fershalova. 2015. The creation of protected ecosystems as the conservation-friendly way to save genus *Begonia* from extinction. *Journal of Tropical Resources and Sustainable Science* 3: 164–168.

## Key Cultivation Websites

American Begonia Society. www.begonias.org/.

Montreal Botanical Garden. *Begoniaceae*. http://espacepourlavie.ca/en/botanical-garden-collections/begoniaceae#.

The Spruce. *10 Varieties of Begonias for Gardens and Containers*. www.thespruce.com/begonias-for-containers-4121751.

The Spruce. *How to Grow and Care for Begonias*. www.thespruce.com/begonias-care-guide-7108561.

WikiHow. *How to Care for a Begonia*. www.wikihow.com/Care-for-a-Begonia.

WikiHow. *How to Grow Begonias*. www.wikihow.com/Grow-Begonias.

WikiHow. *How to Make Begonias Bloom*. www.wikihow.com/Make-Begonias-Bloom.

WikiHow. *How to Plant Begonia Bulbs*. www.wikihow.com/Plant-Begonia-Bulbs.

WikiHow. *How to Plant Begonias in a Pot*. www.wikihow.com/Plant-Begonias-in-a-Pot.

WikiHow. *How to Propagate a Rex Begonia Leaf*. www.wikihow.com/Propagate-a-Rex-Begonia-Leaf.

WikiHow. *How to Prune Begonias*. www.wikihow.com/Prune-Begonias.

# 11 Bird-of-Paradise (*Strelitzia* Species and Hybrids)

**Scientific Names:** The genus name *Strelitzia* honors Queen (Sophia) Charlotte (1744–1818) of the United Kingdom, who originated from the house of Mecklenburg-Strelitz, a royal family, and its dukedom in Germany. Additionally, *reginae* in the name S. *reginae* means "queen," another reference to her. Many people honored by botanical names were prominent Europeans who had nothing to do with plants, but Charlotte made several significant contributions. She was an amateur botanist and contributed to the development of Kew Gardens in South London, which was once a royal palace and was transformed into the world's premier plant collection center. She introduced the Christmas tree to Britain, after decorating one for a Christmas party in 1800.

**Common Names:** The common name bird-of-paradise (bird of paradise) reflects similarity to birds-of-paradise. The plant is also called crane flower (crane-flower). "False bird-of-paradise" refers to species of *Heliconia*, some of which look like *Strelitzia*.

FIGURE 11.1 The roles of Queen Charlotte and Prince Albert in introducing Christmas trees to Great Britain: (a) Portrait of Queen Charlotte in 1781 by Thomas Gainsborough (1727–1788), one of the greatest of English artists. The genus *Strelitzia* is named for her (based on her German family, Mecklenburg-Strelitz). She made important contributions to botany, including adding the Christmas tree to Christmas celebrations in England. (b) December 1848 illustration of Queen Victoria, Prince Albert, and their children admiring the Christmas tree in Windsor Castle. Although Prince Albert is usually given credit for introducing the Christmas tree to England, he, in fact, simply popularized its use, which was originated by Queen Charlotte, consort of King George III.

*Source:* **(a) Public domain photo. (b) Anonymous engraver in 1848 (public domain).**

## 11.1 INTRODUCTION

*Strelitzia reginae* is one of the world's most popular flowering plants, impressive for its long-lasting, eye-catching,

FIGURE 11.2 Examples of avian "birds of paradise." The plant bird-of-paradise was so-named for the resemblance of the flower to a bird, and the exotic appearance of the plant suggested the analogy to the well-known, exotic group of birds. However, the plant bird-of-paradise is indigenous to Africa, while the animal bird of paradise is native to Asia and Australia: (a) Greater bird-of-paradise (*Paradisaea apoda*). (b) Lawes bird-of-paradise (*Parotia lawesi*). (c) Goldie's bird-of-paradise (*Paradisaea decora*).

*Source:* **Gould, J., and W. M. Hart. 1875–1888. *The Birds of New Guinea and the Adjacent Papuan Islands*. Henry Southeran, London.**

FIGURE 11.3 Views of *Strelitzia reginae*, the predominant cultivated species of bird-of-paradise.

*Source:* **(a) Photo by Peggy und Marco Lachmann-Anke from Pixabay (public domain). (b) Photo by Gerhard from Pixabay (public domain). (c) Photo by Forest and Kim Starr (CC BY 2.0). (d) Photo by Rinina25 (CC BY SA 2.5).**


DOI: 10.1201/9781003586227-11

usually orange and bright purple/blue flowers and tropical-looking foliage. Other species of *Strelitzia*, especially the giant forms, are attractive specimens often found gracing parks in near-tropical countries.

## 11.2 OVERVIEW

*Strelitzia* is a genus of five perennial, subtropical species native to Southern Africa. *Strelitzia reginae* (bird-of-paradise, crane lily, strelitzia) is the most frequently cultivated species. *Strelitzia nicolai* (blue-and-white strelitzia, giant bird-of-paradise, white bird-of-paradise, wild banana), the tallest species (growing up to 10 m in height), is also often grown. *Strelitzia juncea* is occasionally cultivated (distinguished in part by its narrow, grass-like leaves; this is closely related to *S. reginae* and interbreeds with it). The plants grow from compact rhizomes (underground stems), tending to form clumps. The evergreen leaves are large, often reminiscent of banana leaves, but with prominent leafstalks (petioles), and arranged in two ranks (in each rank, one leaf is above another). On top of the floral stalk, a hard, beak-like sheath (the spathe) is developed, from which the flowers emerge, one at a time. The spathe grows at right angles to the stem, looking like a bird's head. Nectar in the flower attracts birds and other pollinators in the African home of the plant, which consequently transfer pollen to other flowers.

FIGURE 11.4 Giant bird-of-paradise (*Strelitzia nicolai*): (a) In a park in Madeira, Portugal. (b) Beside a building in Hong Kong.

*Source:* **(a) Photo by Mike Finn (CC BY 2.0). (b) Photo by Deilanboo (CC BY SA 4.0).**

FIGURE 11.5 Rush-leaved bird-of-paradise (*Strelitzia juncea*).

*Source:* **Photos by Brent Newhall (CC BY 2.0).**

FIGURE 11.6 *Strelitzia* flowers and their pollinators: (a) Flowering stem of *Strelitzia reginae*. There are several flowers (opening sequentially) arising from the spathe (the boat-shaped, enveloping expanded green structure at the top of the flowering stalk, below the flowers). The individual flower has three orange sepals and three smaller blue petals – two of which are fused around the (male) anthers into an arrowhead shape, and a third petal transformed into a very short nectar-secreting zone at the base of the flower. The pistil, the female part of the plant, extends upward in the center of the flower. This elaborate structure (admittedly much more of interest to botanists than to the average person) is designed to attract birds to pollinate the plant. The birds push the sexual column (the fused petals, etc.) down to reach the nectar and, in doing so, pick up pollen from the flower and transfer pollen previously picked up to the flower. (b) A female malachite sunbird (*Nectarinia famosa*), a native of South Africa, perched on the spathe of strelitzia, about to feed. Sunbirds, which are common visitors to the flowers, are part of a family of about 150 Old World species. Most feed on the nectar from flowers and have downwardly curved bills. Other birds, especially outside of Africa, and insects are also attracted to the flowers.

*Source:* **Photo by Whisker Flowers from Pixabay (public domain).**

## 11.3 HISTORY

Bird-of-paradise was probably grown in South Africa before being imported to the royal garden at Kew in 1773, but evidence is scant. It was subsequently popularized as a houseplant in Europe and exported to other countries around the world. Of the species of *Strelitzia*, S. *reginae* has been most extensively bred into cultivars. However, compared to most other equally popular garden plants, a relatively small number of modern cultivars are available.

## 11.4 ECONOMIC IMPORTANCE

Bird-of-paradise is a popular nursery offering in much of the world, albeit its chief value to the horticultural industry is in semi-tropical countries. It is cultivated on a commercial scale in many subtropical parts of the world, and also to a minor extent in heated greenhouses in temperate regions. It is also a significant source of cut flowers in the international trade, employed in bouquets and flower arrangements.

## 11.5 ORNAMENTAL USES

In subtropical areas, strelitzia is suited to permanent outdoor beds (even larger "mass plantings") and borders, including street and park exhibits, and it is extensively used

FIGURE 11.7 Commercial production and sale of bird-of-paradise: (a) Potted plants of *Strelitzia reginae* for sale in a garden center. (b) *Strelitzia nicolai* being grown for sale.

***Source:* (a) Photo by Shutterstock/SariMe. (b) Photo by Shutterstock/Aappp.**

FIGURE 11.8 Cut flowers of *Strelitzia reginae*: (a) Cut flowering branches being carried to market in Harare, Zimbabwe. (b) Flower shop in Madeira, Portugal. (c) Flowers and a leaf artistically arranged in a vase.

***Source:* (a) Photo by DocTirivashe (CC BY SA 4.0). (b) Photo by Allie Caulfield (CC BY 2.0). (c) Photo by Bdf (public domain).**

for these purposes in warm climates. In cold climates, for outdoor displays, it is preferably grown in containers that can be overwintered in sheltered, heated facilities. Bird-of-paradise can be grown as a houseplant, but its requirements for large containers and strong light make it more suited as a container plant for sunrooms, greenhouses, and the atriums of large buildings. The plant is very extensively used as a cut flower in the florist trade. Cut flowers may last as long as 2 weeks. The leaves are so large that the plants (especially of *S. nicolai*) are often grown as much for their foliage as for their flowers.

## 11.6 OTHER USES

Like most plants, bird-of-paradise has been employed in its native area in herbal medicine. It has been used in traditional South African medicine to treat various illnesses. Modern research is attempting to find modern pharmacological applications.

## 11.7 PROBLEMS

Bird-of-paradise is mildly toxic, the fruit and seed, if consumed, capable of inducing drowsiness and vomiting.

## 11.8 CULTIVATION

### 11.8.1 Climate and Site

Bird-of-paradise plants are semi-tropical and tolerate only a slight amount of frost. They have been rated as appropriate for permanent outdoor plantings in zones 10 to 12, although sometimes they are planted as far north as zone 8. They do best in full sun or in partial shade in hot climates. Plants grown indoors for the winter need to be slowly acclimated to full sun when moved outdoors in the spring.

*Strelitzia reginae* and *S. juncea* can be cultivated as interior plants. When grown as a houseplant, bird-of-paradise requires a sunny location indoors during the winter and benefits from being moved outside for the summer.

### 11.8.2 Propagation and/or Purchasing

Bird-of-paradise is propagated by seeds and division. Seeds germinate slowly (up to 2 months) and may need 3 to 7 years to grow large enough to flower. By contrast, plants grown from divisions often flower in 1 to 2 years. Mature plants can be divided, or young suckers can be removed from the parent and used as planting stock.

### 11.8.3 Planting

When planted permanently in the ground (in warm climates), the plants can be grown close together as a green fence, but for good flowering, it is recommended that clumps be 2 m apart. *Strelitzia nicolai* can become as large as a small tree and may be too big for small landscape spaces. (*Strelitzia alba* and *S. caudata*, which are not grown very often, can also become treelike).

### 11.8.4 Soil and Moisture Conditions

The soil should be loamy (but both sand and clay are tolerated), well-drained, and circumneutral (slightly acidic is recommended). The soil needs to be kept moist, but not waterlogged. Indoors, potted plants should be allowed to

FIGURE 11.9 Groundskeepers transferring a large bird-of-paradise to a planter in the West Virginia State Capitol in Charleston.

*Source:* **Photo by Governor Jim Justice (public domain).**

FIGURE 11.10 Illustration of the yoga pose "bird-of-paradise" (often termed by the Sanskrit *svarga dvidasana*).

*Source:* **Photo by Shutterstock/Baleika Tamara.**

dry out moderately between waterings and maintained on the drier side in the winter. Bird-of-paradise is a heavy feeder, and plants in the soil outdoors are best treated with a water-soluble fertilizer every 2 weeks. Potted plants may be fertilized once a month.

### 11.8.5 Maintaining Plants

Dead flowers and leaves may remain on the plant indefinitely, so for appearance, spent flowers and tattered foliage should be removed. Potted strelitzias are thought to flower better when their roots are congested, as occurs when they are repotted infrequently and become pot-bound. However, until the plants are about 5 years old, pots that are too small will significantly stunt and distort growth.

## 11.9 FASCINATING FACTS, FABLES, FOLKLORE, AND FRIVOLITY

- Bird-of-paradise is the floral emblem of Los Angeles. It was chosen in 1952 to celebrate the city's 171st birthday. Officials also considered geranium, calla lily, and bougainvillea, but strelitzia won out.
- Georgia Totto O'Keeffe (1887–1986) was an American modernist artist famous for her paintings of enlarged flowers. This included a 1940 painting of white bird-of-paradise (*Strelizia nicolai*, viewable at www.nytimes.com/2018/05/24/arts/okeeffes-hawaii-paintings.html) commissioned by the Dole tropical fruit company. In 2014, the Georgia O'Keefe Museum sold a floral painting for $44 million dollars, setting a record for artwork sold by a female artist. Her work was accused of being deliberately erotic, but she rejected this contention.
- Hatha yoga involves assuming and maintaining physical postures. One of the most difficult is the "bird-of-paradise," so-named for the resemblance of the pose to the orientation of the flower. While one stands on one leg (analogized to the stalk of

the flower), the other leg is lifted upward (claimed to be like the extended flowers), while the spathe is said to be like the upper body.

- Many plants develop an edible attachment termed an "aril" on their seeds in order to attract animals for the purpose of distribution. Strelitzia seeds are dark (black or brown) and have a bright yellow or orange aril (in the form of a tuft of hairs) that is easy for animals to distinguish. The seeds are hard and can pass unharmed through the digestive systems of some large animals. Extracts from the arils are employed in a variety of preparations alleged to have medicinal value.

FIGURE 11.11 Seeds (black) of bird-of-paradise, with arils (tufts of orange hairs). The seeds (exclusive of the arils) are about 1.3 cm long.

*Source:* **Photo by Shutterstock/Kazakov Maksim.**

FIGURE 11.12 A form of *Strelitzia reginae* with yellowish-orange flowers.

*Source:* **Photo by Shutterstock/Pixel-Shot.**

FIGURE 11.13 The cultivar Mandela's Gold of *Strelitzia reginae.*

*Source:* **(a) Photo by Meshugas (CC BY 2.0). (b) Photo by Scott Bauer, US Department of Agriculture (public domain).**

- Although the flowers of the popular *Strelitzia reginae* are usually orange, they are sometimes notably yellowish. However, an extremely lemony form was discovered in South Africa and bred into the new cultivar 'Mandela's Gold'. This was named in honor of the legendary Nelson Mandela (1918–2013), anti-apartheid activist who served as the first president of South Africa from 1994 to 1999.

## KEY PUBLICATIONS


Baijnath, H. and P.A. McCracken. 2018. *Strelitzias of the World: A Historical & Contemporary Exploration.* Durban Botanic Gardens Trust, South Africa. 300 pp.

Coombs, G., S. Mitchell, and C. Peter. 2007. Pollen as a reward for birds. The unique case of weaver bird pollination of *Strelitzia reginae. South African Journal of Botany* 73: 283–283.

Coombs, G. and C.I. Peter. 2009. Do floral traits of *Strelitzia reginae* limit nectar theft by sunbirds? *South African Journal of Botany* 75: 751–756.

Criley, R.A. 1988. Propagation of tropical cut flowers: *Strelitzia, Alpinia,* and *Heliconia. Acta Horticulturae* 226: 509–518.

Criley, R.A. and A.H. Halevy. 1985. *Strelitzia reginae.* In: *CRC Handbook of Flowering,* edited by A.H. Halevy, 396–402. CRC Press, Boca Raton, FL.

Da Silva Vieira, M.R., G.P.P. Lima, D.C. de Medeiros, Â.V. de Souza, and E.C.A. de Oliveira. 2012. Genus: *Strelitzia. Journal of Horticulture and Forestry* 4: 178–180.

Dwarka, D., I. Mackraj, A. Bissessur, P. Ramharack, and H. Baijnath. 2021. Pharmacotherapeutic properties of *Strelitzia nicolai* aril extract containing bilirubin. *South African Journal of Botany* 137: 377–385.

Frost, S.K. and P.G.H. Frost. 1981. Sunbird pollination of *Strelitzia nicolai. Oecologia* 49: 379–384.

Gurung, A. and R. Kumar. 2020. Bird of paradise (*Strelitzia reginae*): A low maintenance, high potential ornamental plant. *Indian Farmer* 7: 37–43.

Jaroenkit, T. and R.E. Paull. 2003. Postharvest handling of heliconia, red ginger, and bird-of-paradise. *HortTechnology* 13: 259–266.

Sane, A., S. Sujatha, K.N. Shilpa, R.H. Laxman, and K.S. Shivashankara. 2020. Growth, yield, physiological and biochemical traits of different accessions of bird of paradise (*Strelitzia reginae* L.). *Industrial Crops and Products* 151: 112477.

Van de Pol, P.A. and T.F. van Hell. 1988. Vegetative propagation of *Strelitzia reginae. Acta Horticulturae* 226: 581–586.

## Key Cultivation Websites

College of Tropical Agriculture and Human Resources, University of Hawaii at Manoa. *Bird-of-Paradise*. www.ctahr.hawaii.edu/oc/freepubs/pdf/OF-27.pdf.

Plantura Magazine. *Bird of Paradise Plant Care: Repotting, Pruning & More*. https://plantura.garden/uk/houseplants/bird-of-paradise-plant/bird-of-paradise-plant-care.

Plantura Magazine. *Bird of Paradise Plant: Flower, Cultivation & Propagation*. https://plantura.garden/uk/houseplants/bird-of-paradise-plant/bird-of-paradise-plant-overview.

The Spruce. *5 Types of Bird of Paradise Plants*. www.thespruce.com/bird-of-paradise-plants-2132859.

The Spruce. *Bird of Paradise: Plant Care & Growing Guide*. www.thespruce.com/how-to-grow-strelitzia-1902742.

WikiHow. *How to Divide a Bird of Paradise*. www.wikihow.com/Divide-a-Bird-of-Paradise-Plant.

WikiHow. *How to Get a Bird of Paradise to Bloom*. www.wikihow.com/Get-a-Bird-of-Paradise-to-Bloom.

WikiHow. *How to Grow Bird of Paradise*. www.wikihow.com/Grow-Bird-of-Paradise.

WikiHow. *How to Prune a Bird of Paradise Plant*. www.wikihow.com/Prune-a-Bird-of-Paradise-Plant.

Wisconsin Horticulture, Extension. *Bird of Paradise, Strelitzia reginae*. https://hort.extension.wisc.edu/articles/bird-of-paradise-strelitzia-reginae/.

# 12 Black-Eyed Susan (*Rudbeckia* Species and Hybrids)

**Scientific Names:** The genus name *Rudbeckia* commemorates Olaus Olai Rudbeck (1660–1740), a Swedish physician and botanist who was the professor, mentor, employer, and patron of Carl Linnaeus, who coined the name. Nevertheless, many believe that Linnaeus was also honoring Rudbeck's father, Olof (1630–1702), who taught botany in Uppsala, Sweden, before his son did the same. A modern descendant of the Rudbecks is Alfred Nobel, originator of the Nobel Prizes.

**Common Names:** *Rudbeckia* species are often called by the equivalent vernacular name, rudbeckia. More frequent is the common name black-eyed Susan, an allusion to the dark (black or brown) centers of the "flowers." Like with other members of the daisy family (Asteraceae or Compositae), the apparent flowers are made up of tiny florets. Less frequently, the plants are called brown-eyed Susan (a name often applied specifically to *R. triloba*). Occasional names include brown betty, gloriosa daisy, golden Jerusalem, English bull's-eye, poorland daisy, yellow daisy, and yellow ox-eye daisy. *Rudbeckia* species are also called coneflowers, in reference to the conical shape of the flower heads. However, other species in the daisy family, notably in the genera *Echinacea*, *Dracopsis*, and *Ratibidia*, are also called coneflower because the blooms have the same cone-like appearance. *Rudbeckia* species are often misidentified because similar daisylike ornamental species are widespread. Black-eyed Susan vine is *Thunbergia alata*, a small tropical vine grown as an ornamental.

The name "black-eyed Susan" may trace to a poem titled "Black-Eyed Susan" by English dramatist John Gay (1685–1732). Susan is in love with a sailor, Sweet William, coincidentally the name of another well-known flower, *Dianthus barbatus*. (In English literature, Sweet William is a favorite name for lovelorn young men.) William is departing on a long sea voyage. As she bids goodbye to him, she cries uncontrollably, producing black circles around her eyes. In folklore, black-eyed Susan and sweet William are said to flower at the same time to celebrate their undying love. The story is commemorated in many paintings.

FIGURE 12.1 Paintings (public domain) of the literary figure Black-Eyed Susan and her departing lover, Sweet William: (a) Illustration by artist Walter Crane (1845–1915). (b) Illustration (1786) by artist Charles Knight (1746–c. 1826).

*Source:* **(a)** ***Pan-Pipes, A Book of Old Songs, Newly Arranged and with Accompaniments by Theo.*** **Marzials, 1884. Routledge, London. (b) Yale Center for British Art.**

## 12.1 INTRODUCTION

Black-eyed Susans have become a garden staple. In many respects, they are ideal flowering ornamentals: hardy, reliable, low-maintenance, easy to grow, extremely long-blooming, and available in a wide range of sizes and flower colors.

## 12.2 OVERVIEW

Twenty-three species of *Rudbeckia* are recognized. These are annual, biennial, or perennial herbs native to damp woodlands, meadows, and dry prairies in North America.

FIGURE 12.2 Some of the cultivars of *Rudbeckia hirta*, the most popular species of black-eyed Susan: (a) Cultivar Autumn Colours. (b) Cultivar Enchanted Sun. (c) Cultivar Toto Rustic. (d) Cultivar Cherry Brandy. (e) Cultivar Cappuccino. (f) Cultivar Toto Gold.

*Source:* **(a) Photo by Shutterstock/Alex Manders. (b) Photo by Shutterstock/Alex Manders. (c) Photo by Shutterstock/Ernie Janes. (d) Photo by Hardyplants (public domain). (e) Photo by Dwight Siple (CC BY 2.0). (f) Photo by Acabashi (CC BY SA 4.0).**


DOI: 10.1201/9781003586227-12

Most species are concentrated east of the Rocky Mountains. They also often occur as introductions in fields and along roadsides. Rudbeckias now grow outside of cultivation in all 48 contiguous US states, in all ten provinces of Canada, and in other continents. Most are perennials, although the popular *R. hirta* is mostly grown as an annual and biennial. Some grow up to 3 m in height; most are much shorter. The stems and leaves are often covered with hairs, producing a rough texture (the specific epithet of the most popular species, *hirta*, is Latin for *hairy*). The plants form clumps and produce colonies by spreading rhizomes (underground stems that grow laterally). The flower head has a prominent black or dark-brown (occasionally greenish) central cone which is surrounded by (usually) yellow petals. Cultivars may have flowers that are brown, cream, orange, peach, red, or white, and some have bicolored or doubled flowers. The plant may flower from spring to a hard freeze in the autumn and often can continue to bloom all year in warmer areas.

*Rudbeckia hirta* (sometimes called hairy black-eyed Susan) is the most popular species. It grows from 30 to 100 cm in height and has the showiest flowers in the genus. It has far more cultivars (often hybrids with other species) than any other black-eyed Susans. This short-lived perennial is often grown as an annual in the warmer part of its range, and as a biennial in the cooler parts (as a biennial, it forms a non-flowering rosette the first year). Once planted, it frequently self-seeds, producing a flowering display every year, giving the impression that it is a genuine perennial. The gloriosa variation of the species produces flowers, often doubled (with extra petals), up to 20 cm in diameter. This variant was created as a tetraploid (i.e., with double the usual number of chromosomes), which sometimes induces such large size of a plant's organs.

FIGURE 12.3 Some of the less-popular cultivated ornamental species of *Rudbeckia*: (a) *Rudbeckia subtomentosa* cultivar Henry Eilers. (b) *Rudbeckia fulgida* cultivar Goldsturm. (c) *Rudbeckia triloba*. (d) *Rudbeckia nitida* cultivar Herbstsonne.

***Source:*** **(a) Photo by Krzysztof Ziarnek, Kenraiz (CC BY SA 4.0). (b) Photo by Krzysztof Ziarnek, Kenraiz (CC BY SA 4.0). (c) Photo by Saharadesertfox (CC BY SA 3.0). (d) Photo by Staudengärtnerei Forssman (CC BY SA 2.0).**

*Rudbeckia fulgida* (with the curious name orange black-eyed Susan) is the second most popular species. Its flowers are usually smaller but more numerous than those of *R. hirta*. Unlike *R. hirta*, it is a true perennial and, once planted, continues to reproduce indefinitely.

Less commonly, planted species include *R. laciniata* (tall coneflower), *R. maxima* (large coneflower), *R. nitida* (shiny coneflower), *R. subtomentosa* (sweet coneflower), and *R. triloba* (brown-eyed Susan or three-lobed coneflower).

## 12.3 HISTORY

*Rudbeckia* species were used by many Native American tribes to treat a variety of diseases and illnesses. Philadelphia-born John Bartram (1699–1777), often called the father of American botany, sent seeds of *Rudbeckia* to Europe, and they soon acquired popularity as garden plants in the Old World.

## 12.4 ECONOMIC IMPORTANCE

Black-eyed Susans are favorite garden plants, for which there is a reliable demand. The chief value to horticultural industries is to supply seeds and starter plants.

## 12.5 ORNAMENTAL USES

Black-eyed Susans are frequently mass-planted in parks and wildlife gardens. They are also useful in mixed plantings and in borders. Occasionally, the plants are grown in containers, but they are not suitable as houseplants. The flowering stalks are useful for cut flowers for vases and dried arrangements.

## 12.6 OTHER USES

Black-eyed Susans are ideal plants for butterfly, pollinator, and wildlife gardens. The showy flowers attract bees and butterflies, and the seeds are consumed by birds (the

FIGURE 12.4 A commercial field of black-eyed Susan near Silverton, Oregon.

***Source:*** **Photo by Shutterstock/Bob Pool.**

FIGURE 12.5 Examples of landscape uses of black-eyed Susans: (a) *Rudbeckia fulgida* cultivar Oldsturm as part of the public display at the Chicago Botanic Garden. (b) In a public park bordering the US Capitol in Washington. (c) In a border by a residential fence. (d) A raccoon at a birdbath surrounded by plants.

*Source:* **(a) Photo by Cultivar 413 (CC BY 2.0). (b) Photo by Lorie Shaull (CC BY SA 2.0). (c) Photo by Patrick Standish (CC BY 2.0). (d) Photo by Shutterstock/Mike Truchon.**

FIGURE 12.6 Animals attracted to black-eyed Susans: (a) A monarch butterfly visiting a flower. (b) A female American goldfinch (*Carduelis tristis*) eating seeds of *Rudbeckia maxima*. (c) A swallowtail butterfly visiting flowers. (d) A honeybee (*Apis mellifera*) visiting a blossom of *Rudbeckia fulgida*.

*Source:* **(a) Photo by Steven Rumbalski from Pixabay (public domain). (b) Photo by Shutterstock/Danita Delimont. (c) Photo by Shutterstock/Cathy Keifer. (d) Photo by Shutterstock/Dirk Daniel Mann.**

senesced plants and their seed heads can be allowed to persist overwinter). In their native areas in the United States, the species are incorporated in seed mixes for prairie restoration, erosion control, and highway beautification.

## 12.7 PROBLEMS

Black-eyed Susans can become weeds within a garden, spreading by their underground stems and by seeds. There is limited evidence of poisoning of horses, sheep, and swine, but animals generally refuse to eat the unpalatable and rough-textured foliage. The plants are resistant to deer but will be nibbled. Skin contact with black-eyed Susans can be irritating to sensitive people. However, the plants are not considered to be a significant threat of toxicity.

## 12.8 CULTIVATION

### 12.8.1 Climate and Site

Most cultivated *Rudbeckia* are hardy in climate zones 4 to 9, and various species can grow as far north as zone 2 or as far south as zone 11. The plants prefer full sun but will tolerate partial shade.

### 12.8.2 Propagation and/or Purchasing

Most cultivars will not reproduce true from seeds, and most gardeners grow named cultivars from starter plants and by dividing them. *Rudbeckia* can be propagated by seed, and the species often self-sow in a location (the resulting plants will likely differ from the parents). Often, mature plants will produce seedlings underneath that can be transplanted.

### 12.8.3 Planting

Plants can be established 50 to 60 cm apart (closer for dwarf varieties – 15 to 30 cm). Left alone, *Rudbeckia* can become

FIGURE 12.7 *Rudbeckia nitida* covered by snow in November. Leaving such mature, seed-bearing black-eyed Susan plants overwinter attracts and feeds birds.

*Source:* **Photo by Shutterstock/Carmen Hauser.**

quite dense. Wider spacing between plants increases air circulation and helps prevent the spread of diseases. The plants can be divided every 3 to 5 years in early spring or fall to prevent overcrowding.

### 12.8.4 Soil and Moisture Conditions

*Rudbeckia* develop best in evenly moist, well-drained soils, but they are drought-tolerant once established. They grow well in average soils, whether based on sand, loam, or clay. They survive in a wide range of pH, but circumneutral or slightly acidic is ideal. Fertilizer is usually not necessary, but the addition of rich organic matter or compost worked into the soil around the plant base is beneficial.

### 12.8.5 Maintaining Plants

Once established, dead foliage and stems are best removed at the end or beginning of the season (basically to improve appearance, although removing diseased material is important). Deadheading faded old flowers encourages reblooming.

## 12.9 FASCINATING FACTS, FABLES, FOLKLORE, AND FRIVOLITY

- American artist and designer Louis Comfort Tiffany (1848–1933) is famous for preparing lamps with shades made of cut stained glass joined by soldered copper strips. Antique Tiffany lamps sell for thousands of dollars, sometimes upward of $1 million. Lamps bearing floral motifs are among the most in-demand. Blooms chosen for illustration needed to be familiar and beloved and include lilies, peonies, irises, and roses. Black-eyed Susan is also a member of this distinguished floral group.
- In the United States, the three thoroughbred horse races that make up the Triple Crown are the Kentucky Derby (2.0 km, Churchill Downs in Louisville, Kentucky), Preakness Stakes (1.9 km, Pimlico Race Course in Baltimore, Maryland), and Belmont Stakes (2.4 km, Belmont Park in Elmont, New York). The Kentucky Derby is called the "Run for the Roses," and the winning horse receives a blanket made of roses. The Belmont Stakes is known as "The Race to the Carnations," because the winning horse is covered with a cloak of white carnations after the race. The Preakness is nicknamed the "Run for the Black-Eyed Susans," and the winning horse receives a blanket of black-eyed Susans (Maryland's state flower). Another horse race termed the "Black-Eyed Susan Stakes" (for fillies) is additionally held at Pimlico Race Course. Because black-eyed Susans are not yet in flower when the Preakness is held in May, other flowers, such as carnations, are usually painted to look like Susans. The Preakness Stakes was named for Preakness (1867–1881), an American thoroughbred racehorse. After his retirement from racing, Preakness's owner, the Duke of Hamilton, shot him for misbehaving. The resulting public outcry led to new laws protecting animals.
- The mint julep (a cocktail made mostly with bourbon, sugar, water, ice, and fresh mint) is associated with the Kentucky Derby. A less-famous cocktail (less tasty, some would say), the black-eyed Susan, has been associated with the Preakness since 1973. Originally, the attempt was made to mimic the

FIGURE 12.8 Tiffany Studios black-eyed Susan table lamp, ca. 1900–1906.

*Source:* Photo by the Adverse Possessors (CC BY 2.5).

FIGURE 12.9 Winners of the 2013 Preakness horse race. The horses have been covered with the traditional blanket made of flowers that mimic black-eyed Susans: (a) Tom Nappi on American Pharoah, the 2015 winner. (b) Gary Stevens on Oxbow, the 2013 winner.

*Source:* Photos by Maryland GovPics (CC BY 2.0).

FIGURE 12.10 An idealized black-eyed Susan cocktail, associated with the Preakness horse race.

*Source:* **Prepared by Steve A. Johnson (CC BY 2.0).**

FIGURE 12.11 Tall "outhouse flowers" (*Rudbeckia laciniata* 'Hortensia') concealing a facility in Russia.

*Source:* **Photo by Carlfbagge (CC BY 2.0).**

yellow color typical of black-eyed Susan, employing juices such as pineapple, grapefruit, and peach, but the color of current versions is no longer distinctive, and a variety of liqueurs (bourbon, rum, triple sec, vodka, whiskey) have been tried.

- People try to make outhouses as inconspicuous as possible, and this is often accomplished by placing tall flowering plants around them. Such species are affectionately known as "outhouse flowers." One of the most frequently employed is *Rudbeckia laciniata*, which can grow to a height of 3 m.

## KEY PUBLICATIONS


Burlec, A.F., Ł. Pecio, C. Mircea, C. Tuchiluş, A. Corciovă, C. Danciu. et al. 2023. Preliminary phytochemical and biological evaluation of *Rudbeckia hirta* flowers. *Plants* 12(15): 2871. https://doi.org/10.3390/plants12152871.

Campbell, J.J.N. and W.R. Seymour, Jr. 2013. Towards a revision of the *Rudbeckia fulgida* complex (Asteraceae), with description of a new species from the blacklands of southern USA. *Phytoneuron* 90: 1–27.

Foote, K. 2002. Black-eyed Susan (*Rudbeckia hirta* L.). *New York Flora Association Newsletter* 13(1): 1–5. www.poison.org/-/media/files/pdf-for-article-dowloads-and-refs/foote-black-eyed-susan.pdf.

Harkess, R.L. and R.E. Lyons. 1994 *Rudbeckia hirta* L.: A versatile North American wildflower. *HortScience* 29: 134, 227.

Kochankov, V.G. and M.K. Chailakhyan. 1986. Rudbeckia. In: *Handbook of Flowering*, Vol. 5, edited by A. Halevy, 295–320. CRC Press, Boca Raton, FL.

Urbatsch, L.E., B.G. Baldwin, and M.J. Donoghue. 2000. Phylogeny of the coneflowers and relatives (Heliantheae: Asteraceae) based on nuclear rDNA internal transcribed spacer (ITS) sequences and chloroplast DNA restriction site data. *Systematic Botany* 25: 539–563.

Urbatsch, L.E. and P.B. Cox. 2006. *Rudbeckia*. In: *Flora of North America North of Mexico*, Vol. 21, edited by Flora of North America Editorial Committee, 44–63. Oxford University Press, New York.

Urbatsch, L.E. and P.B. Cox. *Rudbeckia*. Flora of North America North of Mexico (Online). www.efloras.org/florataxon.aspx?flora_id=1&taxon_id=128840.

Vardeman, E., J.T. Lyles, and CL. Quave. 2022. The genus *Rudbeckia*: A critical review of its traditional medicinal uses, phytochemistry, and pharmacology. *Journal of Herbal Medicine* 31: 100530.

Walsh, R.A. 1994. *Rudbeckia hirta*. In: *Fire Effects Information System [Online]*. U.S. Department of Agriculture, Forest Service, Rocky Mountain Research Station, Fire Sciences Laboratory. www.fs.usda.gov/database/feis/plants/forb/rudhir/all.html.


## KEY CULTIVATION WEBSITES


Blooming Backyard. *How to Divide and Transplant Black-Eyed Susan (Rudbeckia)*. www.bloomingbackyard.com/divide-black-eyed-susan/.

Clemson University, Extension. *Rudbeckia*. https://hgic.clemson.edu/factsheet/rudbeckia/.

Flower Patch Farmhouse. *How to Grow Black Eyed Susans Easily*. www.flowerpatchfarmhouse.com/how-to-grow-black-eyed-susans/.

Horticulture Magazine. *Rudbeckia, Black Eyed Susan*. https://horticulture.co.uk/rudbeckia/.

Pennsylvania State University, Extension. *Black-Eyed Susan: Beautiful and Beneficial*. https://extension.psu.edu/black-eyed-susan-beautiful-and-beneficial.

Royal Horticultural Society. *How to Grow Rudbeckia*. www.rhs.org.uk/plants/rudbeckia/growing-guide.

The Spruce. *How to Grow and Care for Black-Eyed Susan*. www.thespruce.com/choosing-and-growing-black-eyed-susan-1402860.

University of Missouri, Integrated Pest Management. *Rudbeckia*. https://ipm.missouri.edu/meg/2008/3/Rudbeckia/.

WikiHow. *How to Grow Black-Eyed Susan Flowers*. www.wikihow.life/Grow-Black-Eyed-Susan-Flowers.

# 13 Bougainvillea (*Bougainvillea* Species and Hybrids)

**Scientific Names:** The genus name *Bougainvillea* honors Louis Antoine de Bougainville (1729–1811), a French navigator and admiral. Many sources mistakenly state that he was the first to record or collect the plant, but he simply commanded the voyage in the 1760s during which the plant was discovered in Rio de Janeiro, Brazil, by the French botanist Philibert Commerson (1727–1773). Bougainville is known for Bougainville Island, part of Papua New Guinea.

**Common Names:** The common name "bougainvillea," unlike many other names for plants, is quite unambiguous, although difficult to spell. "Bougainville" was one of the words that had to be spelled correctly for the 2019 Scripps National Spelling Bee contest. "Paperflower" (based on the papery bracts of *B. glabra*) is sometimes used as a synonym.

## 13.1 INTRODUCTION

Most ornamental bougainvilleas are vining, sprawling, "climbing shrubs" reminiscent of commercial grapevines. The growing plants need to be trained and pruned, or they will produce quite irregular specimens spreading and scrambling over their surroundings. They produce rather spectacular, long-lasting flowering displays in a wide array of attractive colors. Many consider bougainvillea to be the premier flowering vine, but its tropical nature limits where it can be grown as a perennial outdoors.

FIGURE 13.1 Louis-Antoine de Bougainville, French explorer commemorated by the plant bougainvillea (public domain illustrations): (a) Portrait by French artist Joseph Ducreux (1735–1802). (b) An English satirical cartoon of Bougainville hoisting French Colors on a small rock in Magellan Straits, South America. (c) Painting of Tahitians presenting fruits to Bougainville, attended by his officers, by an unknown artist, about 1768.

*Source:* **(a) Photo from Château de Versailles. (b) Copper plate engraving published for Moore's Voyages & Travels around 1780. (c) Public domain.**

FIGURE 13.2 A large bougainvillea in Crete, Greece.

*Source:* **Photo by Shutterstock/KatarzynaZakowska.**

## 13.2 OVERVIEW

The genus *Bougainvillea* has about 18 species (the number differing according to authority), including shrubs, vines, or small trees native to South America. Several are cultivated as ornamentals, mostly as outdoor perennials in warm climates, but also as annuals and houseplants. The true flowers are inconspicuous but are surrounded by attractively colored, long-lasting, papery bracts (small modified leaves). Most people interpret the bracts as petals and view them and their associated small flowers as a single flower (a situation highly reminiscent of poinsettia). The flowers occur in clusters of three, surrounded by three to six brightly colored bracts which, depending on cultivar, may be orange, pink, purple, red, white, or yellow. Bougainvillea foliage is typically evergreen, but the plants often lose many leaves after a flush of flowering, followed by the development of new leaves. The plants are deciduous when grown in areas with a long dry season. A few cultivars have variegated leaves. The ornamental species are often multi-trunked, developing slender, climbing, or sprawling arching canes, with thorns up to 10 cm long originating in the leaf axils. With age, the basal stem can develop into a substantial trunk. Branches of one variety can easily be grafted onto another variety, and often, many branches, each representing a cultivar with different flower color, are all grafted onto one plant, producing an extraordinarily colorful specimen.

DOI: 10.1201/9781003586227-13

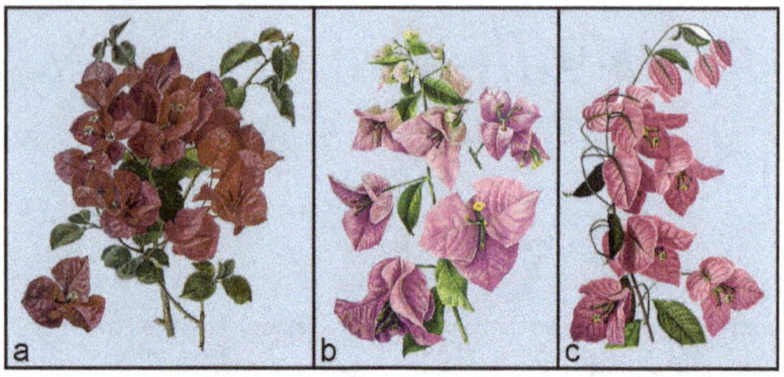


FIGURE 13.3 Technical botanical illustrations of flowering branches of *Bougainvillea* (public domain): (a) *Bougainvillea spectabilis*. (b) *Bougainvillea glabra*. (c) *Bougainvillea glabra*.

*Source:* (a) ***Revue horticole***, **sér. 4 (1852–1974). (b)** ***Revue horticole***, **sér. 4, vol. 81, 1909. (c)** ***Revue horticole***, **1889.**

FIGURE 13.4 Bougainvillea cultivars with differently colored floral bracts.

*Source:* **(a) Photo by Shutterstock/Amawasri Pakdara. (b) Photo by Shutterstock/Watcharit Jiraweeranuwat.**

FIGURE 13.5 Bougainvillea plants to which numerous different branches from differently colored cultivars have been grafted, producing a plant bearing flowers of multiple colors: (a) At a plant nursery in India. (b) On a sidewalk in Singapore. (c) In a garden.

*Source:* **(a) Photo by Anish Nellickal (CC BY SA 4.0). (b) Photo by Aruna (CC BY SA 2.0). (c) Photo by Shutterstock/Pangcom.**

## 13.3 HISTORY

Bouagainvilleas were imported to Europe from South America in the 18th century and subsequently expanded hugely in popularity. Three species are the basis of most cultivars. *Bougainvillea glabra* (paperflower, lesser bougainvillea) has pentagonal (five-angled) floral tubes and short floral bracts, and the leaves are mostly hairless, while *B. spectabilis* (great bougainvillea) has rounded floral tubes, longer floral bracts, and densely hairy leaves, but the two Brazilian species are quite similar and have been extensively hybridized. *Bougainvillea peruviana*, native from Colombia to Peru, has been hybridized with *B. glabra* (the hybrid is often termed *B.* × *buttiana*) and is the basis of many cultivars. Hybridization is so extensive that it is now difficult or impossible to identify the origin of many varieties. Horticultural selection has produced plants with extra bracts (the flowers normally have just three), bracts of different hues, bicolored bracts, and dwarf forms. Half of modern varieties have been produced in India.

## 13.4 ECONOMIC IMPORTANCE

Bougainvillea is one of the major ornamental plants in commercial horticulture in semi-tropical and tropical regions. It is a minor but significant potted plant (usually in hanging baskets) in temperate areas.

## 13.5 ORNAMENTAL USES

In subtropical and tropical areas, bougainvillea is a popular ornamental in both home and public gardens, grown as a bush-like specimen, climber, hedge, pot plant, bonsai, and groundcover. Bougainvilleas respond well to pruning and so are frequently utilized as hedges in tropical areas. Their sprawling habit also makes them appropriate as groundcover in hot, semi-arid regions. In temperate regions, the plants are often offered as potted annuals.

FIGURE 13.6 A commercial bougainvillea flower plantation in Vietnam.

*Source:* **Photo by Shutterstock/Phuong D. Nguyen.**

FIGURE 13.7 Bougainvillea hedges: (a) A hedge made up of differently colored cultivars in St. Lucia overlooking the Caribbean Sea. (b) In Australia. (c) A hedge bordering a sidewalk.

*Source:* **(a) Photo by Terri Needham (CC BY 2.0). (b) Photo by Sheba (CC BY SA 2.0). (c) Photo by Shutterstock/Noraset.Me.**

FIGURE 13.8 Bougainvillea grown as bonsai.

*Source:* **(a) Photo by Shutterstock/Walter Pall. (b) Photo by Ewen Roberts (CC BY 2.0).**

FIGURE 13.9 Landscape beautification employing bougainvillea: (a) A colorful bougainvillea walkway arch. (b) A horse-drawn carriage station in Malta framed by a bougainvillea background.

*Source:* **(a) Photo by Shutterstock/Thaisign. (b) Photo by Benreis (CC BY SA 3.0).**

## 13.6 OTHER USES

Bougainvillea has minor usage in folk medicine, primarily in its indigenous South American homeland. The floral bracts are described as edible at many websites, which often present recipes, but consumption is inadvisable.

## 13.7 PROBLEMS

The sap of bougainvillea is mildly toxic and could result in dermatitis for sensitive individuals, and in digestive upset if material is ingested in substantial quantity. Nevertheless, the plant is generally considered to be safe for people and pets. The thorns are more hazardous. Many cultivars have sharp thorns, so when handling or pruning, it is advisable to wear heavy gloves. Bougainvilleas have become naturalized in some tropical countries, but the invasive potential is relatively limited.

## 13.8 CULTIVATION

### 13.8.1 Climate and Site

Bougainvilleas are suited to climate zones 9 to 11. They are tropical and semi-tropical and do not tolerate frost or freezing temperatures. They thrive in full sun but tolerate light shade.

### 13.8.2 Propagation and/or Purchasing

Ornamental bougainvilleas can be grown from seeds (although many hybrid cultivars are sterile and do not develop seeds or produce very few). However, cultivars usually will not reproduce true from seeds. The simplest starting material is small potted plants acquired from nurseries. Many of the cultivars are easily reproduced from small cuttings.

### 13.8.3 Planting

Bougainvilleas are remarkably adaptable to ornamental usages. They can be pruned to grow into shrubs positioned as single or grouped specimens; they are often grown as hedges; they can be trained to grow upward as vines upon fences and trellises; they can be placed in high positions so the vines cascade over walls, roofs, and rocks; they can be grown in large containers positioned so the plants grown into a shape and place where desired; they are often grown in small containers as hanging baskets and as bonsai; and sometimes they are grown as groundcover.

### 13.8.4 Soil and Moisture Conditions

Bougainvilleas can grow in a variety of soils, provided they are well-drained. Acidic to neutral soils are best. A pH of 5.5–6.0 is recommended, and alkaline soils can result in iron deficiency. The plants are native to semi-arid regions and are drought-tolerant. They benefit when the soil is left to become a little dry.

### 13.8.5 Maintaining Plants

Bougainvilleas adhere to objects by their curved thorns but have minds of their own. They need to be tied to structures such as fences, walls, and pergolas or they will sprawl uncontrollably.

FIGURE 13.10 Jeanne Baret (1740–1807) disguised as a male sailor (since women were not permitted on ships), holding a plant collection. A member of Louis Antoine de Bougainville's expedition to South America, assigned as an assistant to the ship's botanist, she is recognized as the first woman to have completed a voyage of circumnavigation of the globe and speculated to be the true European discoverer of bougainvillea.

***Source:*** **Portrait by Cristoforo Dall'Acqua (1734–1787) in *Navigazioni di Cook del grande oceano e intorno al globo*, vol. 2, 1816.**

## 13.9 FASCINATING FACTS, FABLES, FOLKLORE, AND FRIVOLITY

- In parts of Africa, introduced bougainvillea flowers are collected, and the papery bracts are shredded and employed as confetti.
- Bougainvillea is Guam's territorial flower, the national flower of Grenada, and the official flower of several counties of Taiwan and several cities in semi-tropical and tropical locations.
- *Bougainvillea* is a constituent of the Nyctaginaceae, the four o'clock family. This odd name originated from another member of the family, the four o'clock (*Mirabilis jalapa*), because its flowers do not open until late in the day.
- As noted in the beginning of this chapter, the French naturalist Philibert Commerson is credited with discovering bougainvillea. However, it is believed (see http://khkeeler.blogspot.com/2016/01/plant-story-bougainvillea-from-brazil.html) that Commerson's housekeeper and lover, Jeanne Baret (Baré), disguised herself as a man and went along with him as an assistant and was the actual discoverer of the plant. Assuming that she really did accompany the Bougainville expedition, she is the first woman known to have circumnavigated the globe.

## KEY PUBLICATIONS


Abarca-Vargas R, and V.L. Petricevich. 2018. *Bougainvillea* genus: A review on phytochemistry, pharmacology, and toxicology. *Evidence-Based Complementary and Alternative Medicine* 2018: 9070927. www.ncbi.nlm.nih.gov/pmc/articles/PMC6035817/.

Bautista, M.A., Y. Zheng, D.E. Boufford, Z. Hu, Y. Deng, and T. Chen. 2022. Phylogeny and taxonomic synopsis of the genus *Bougainvillea* (Nyctaginaceae). *Plants* 11(13): 1700; https://doi.org/10.3390/plants11131700.

Chen, T. 2008. *Bougainvillea*. China Agriculture Press, Beijing, China. 118 pp.

Datta, S.K. 2021. Genetic diversity and improvement of bougainvillea. *LS – An International Journal of Life Sciences* 10: 61–79.

Datta, S.K. 2022. Breeding of bougainvillea: Past, present, and future. *The Nucleus* 65: 239–254.

Datta, S.K., R. Jayanthi, and T. Janakiram. 2017. *Bougainvillea*. New India Publishing Agency, New Delhi. 318 pp.

Datta, S.K., R. Jayanthi, and T. Janakiram. 2021. Bougainvillea. In: *Floriculture and Ornamental Plants*, edited by S.K. Datta and Y.C. Gupta, 1–34. Springer, Singapore.

Gillis, W.T. 1976. Bougainvilleas of cultivation (Nyctaginaceae). *Baileya* 20: 34–41.

Holttum, R.E. 1955a. The cultivated bougainvilleas. *Bougainvillea* × *buttiana*, its variety and hybrids. *Malayan Agri-Horticultural Association Magazine* 12: 2–11.

Holttum, R.E., 1955b. The cultivated bougainvilleas. The varieties of *B. glabra*. *Malayan Agriculture Magazine and Journal* 12: 25–36.

Iredell, J. 1994. *Growing Bougainvilleas*. Simon & Schuster, East Roseville, Australia. 96 pp.

Lack, H. 2012. The discovery, naming and typification of *Bougainvillea spectabilis* (Nyctaginaceae). *Willdenowia* 42: 117–126.

Lim, T.K. 2014. *Bougainvillea glabra*. In: *Edible Medicinal and Non Medicinal Plants*, edited by T.K. Lim, 483–488. Springer, Dordrecht.

Ohri, D. 2013. Cytogenetics of domestication and improvement of garden *Gladiolus* and *Bougainvillea*. *The Nucleus* 56: 149–153.

Salam, P., V. Bhargav, Y.C. Gupta, and P.K. Nimbolkar. 2017. Evolution in bougainvillea (*Bougainvillea* Commers.) – A review. *Journal of Applied and Natural Science* 9: 1489–1494.

Saleem, H., A. Usman, M.F. Mahomoodally, and N. Ahemad. 2021. *Bougainvillea glabra* (Choisy): A comprehensive review on botany, traditional uses, phytochemistry, pharmacology and toxicity. *Journal of Ethnopharmacology* 266: 113356.

Singh, B., R.S. Panwar, S.R. Voleti, V.K. Sharma, and S. Thakur. 1999. *The New International Bougainvillea Check List*. Division of Floriculture and Landscaping, Indian Agriculture Research Institute, New Delhi, India. 76 pp.

Singh, K.K. 2018. A review: Multiplication of bougainvillea species through cutting. *International Journal of Chemical Studies* 6(2): 1961–1965.

Srinivas, M. 1995. Bougainvillea. In: *Advances in Horticulture*, Vol. *12: Ornamental Plants*, edited by K.L. Chadha and S.K. Bhattacharjee, 140–156. Malhotra House, New Delhi.

## Key Cultivation Websites

Hackett, W.P., R.M. Sachs, and J. Debie. 1972, August 12–13. *Growing Bougainvillea as a Flowering Pot Plant*. California Agriculture. https://calag.ucanr.edu/archive/?type=pdf&article=ca.v026n08p12.

Kobayashi, K.D., J. McConnell, and J. Griffis. 2007. *Bougainvillea*. University of Hawaii at Manoa, Cooperative Extension Service. www.ctahr.hawaii.edu/oc/freepubs/pdf/of-38.pdf.

Royal Horticultural Society. *Bougainvillea*. www.rhs.org.uk/plants/bougainvillea/growing-guide.

Schoellhorn, R. and E. Alvarez. 2002. *Warm Climate Production Guidelines for Bougainvillea*. Cooperative Extension Service, University of Florida, ENH. 874 pp. www.plantgrower.org/uploads/6/5/5/4/65545169/bougainvillea.pdf.

The Spruce. *How to Grow Bougainvillea Indoors*. www.thespruce.com/grow-bougainvillea-inside-1902743.

WikiHow. *Easy Steps to Plant, Grow, and Care for Bougainvilleas*. www.wikihow.com/Grow-Bougainvillea.

WikiHow. *How to Propagate Bougainvillea*. www.wikihow.com/Propagate-Bougainvillea.

WikiHow. *How to Prune Bougainvillea*. www.wikihow.com/Prune-Bougainvillea.

# 14 Bromeliads (Bromeliaceae Species and Hybrids)

**Scientific Names:** The family Bromeliaceae (and the associated genus *Bromelia*) is named for the Swedish medical doctor and botanist Olof Bromelius (1639–1705). Although sometimes called the bromeliad family, the Bromeliaceae is more often termed the pineapple family.

**Common Names:** "Bromeliads" is a general term referring to species of the Bromeliaceae. The pineapple is the best-known bromeliad and is sometimes grown as an ornamental, occasionally called "pineapple flower," a phrase also applied to other plants. "Pineapple flowers" or "pineapple lilies" are species of *Eucomis*, which are ornamental flowering plants that resemble but are quite unrelated to pineapple plants. Other, unrelated plants are also termed *pineapple flower*. Most bromeliads do not have widely employed common names, but their genus names (such as *Aechmea*, *Bilbergia*, *Guzmania*, *Tillandsia*, and *Vrisesea*) effectively serve as common names.

## 14.1 INTRODUCTION

Botanists are rather precise in their definition of the parts of plants called "flowers," but most people are content to identify flowers as the sections of plants with especially colorful "petals." However, some widely admired plants produce tiny, inconspicuous flowers surrounded by extremely colorful "bracts" (which are essentially just leaves, usually small), and these are often interpreted by the public as the petals of flowers. Practically, such floral mimics, of which the bromeliads are examples, are employed in the same ways as more conventional ornamentals, with the advantage that bracts are much more long-lived than petals. Poinsettias and calla lilies, also discussed in this book, are similar examples, and indeed, there are also many cultivated species with quite colorful foliage. This chapter concentrates on bromeliads with the most attractive bracts.

## 14.2 OVERVIEW

The Bromeliaceae include about 80 genera and 4000 species, all in the Americas, except for one in tropical West Africa. A few are subtropical; the majority are tropical. Bromeliads are usually herbaceous perennials. Most are "monocarpic" (meaning, they produce fruit and seeds only once), the mother plant subsequently dying, but also developing offsets ("pups") which continue the growth of the original plant. Many are so-called "tank bromeliads," their weak root systems compensated by developing the basal foliage in a rosette (a circle of leaves around the stem) that forms a watertight vase or cup (technically a tank) serving to collect and store water. Some writers (as in this chapter) define *tank bromeliads* as those with a central water-impounding tank; others also include those that trap and store water in their leaf axils (the crotch between the leafstalk or base and the stem), even if there is no central tank. By contrast with bromeliads that retain water in either leaf axils or tanks, there are "atmospheric bromeliads," which are usually covered in special hairs (plant hairs are termed "trichomes") that, remarkably, absorb water from humid air.

About half of bromeliad species are "epiphytes," also termed "air plants." These are defined as plants that grow for support upon other plants, especially trees; are not rooted in the ground; and are not nutritional parasites (they do not penetrate and extract nutrients from their hosts). Epiphytes can also grow on inanimate objects, such as rocks (such species are termed "saxicolous" or "lithophytic" – some of the many technical labels that pertain to the complex

FIGURE 14.1 Spanish moss (*Tillandsia usneoides*), which is neither Spanish nor a moss. This is the best-known epiphytic (i.e., growing on trees) bromeliad, although it does not look typical and grows further north than other bromeliads. It often hangs down picturesquely from large trees in tropical and subtropical climates, from the Southern United States to South America. It is not parasitic (no bromeliad is) but absorbs nutrients from rain falling on its pale, threadlike leaves: (a) Threadlike stems and foliage hanging down. (b) Trees covered by Spanish moss in Washington, North Carolina, shown in a 1938 postcard. (c) A 1907 postcard showing a tree covered by Spanish moss in Charleston, South Carolina.

***Source:*** **(a) Photo by Raffi Kojian (CC BY SA 3.0). (b) Durwood Barbour Collection of North Carolina Postcards, University of North Carolina Library (public domain). (c) Published by the Detroit Publishing Company (public domain).**


DOI: 10.1201/9781003586227-14

lifestyles of bromeliads). Sometimes the term "terrestrial" is applied to bromeliads that root in the ground (but, virtually, all bromeliads can anchor in soil, such as in pots, if given the opportunity). Epiphytes often grow on trees in forests, where they have the advantage of gaining access to light high up on branches. They obtain water and nutrients from rain and from falling debris that accumulates on the trees, and because these resources are limited, epiphytes tend to grow quite slowly and to remain small. Epiphytes rarely cause significant harm to the plants they grow on (by contrast, vines often are so heavy that they topple trees, so dense they shade and even kill trees, and sometimes they physically strangle or actually parasitize other plants). Many bromeliads grow both as epiphytes on trees and also on the ground, and in cultivation, it is usually much easier to grow them as conventional garden or container plants. Only the orchid family (Orchidaceae) has developed more epiphytic species. Epiphytes of both orchids and bromeliads are more often than not cultivated as potted plants rather than grown upon other plants or artificial supports.

Like cacti (species of the Cactaceae family), there are hundreds of species of bromeliads that are grown (especially as houseplants) primarily for their interesting foliage

FIGURE 14.2 Colorful bromeliad rosettes: (a) A collection of species at a commercial nursery. (b) Close-up of a rosette. Note the small flowers inside the central tank of water.

*Source:* **(a) Photo by Bertknot (CC BY SA 2.0). (b) Photo by Shutterstock/Stockimo.**

FIGURE 14.3 Some of the many bromeliads with feathery flower clusters: (a) Exhibit of various bromeliad species by Bird Rock Tropicals Nursery in 2018 at the San Diego Bromeliad Society Show. (b) *Aechmea blanchetiana*. There are about 125 species of *Aechmea*, many of them cultivated. The green form shown here is typical of plants grown in the shade or over-fertilized. A golden-orange form develops in full sunlight and is a popular landscape plant in subtropical areas, including climate zones 10 and 11 (see *c*). (c) Golden-orange form of *Aechmea blanchetiana*.

*Source:* **(a) Photo by Cultivar 413 (CC BY 2.0). (b) Photo by Cliff from Arlington, Virginia (CC BY 2.0). (c) Photo by Forest and Kim Starr (CC BY 2.0).**

FIGURE 14.4 Guzmania (*Guzmania* species and hybrids) in flower. The genus is named for Anastasio Guzman (died in 1807), a Spanish pharmacist and naturalist. Although there are about 200 species, most cultivars belong to *G. lingulata*, the most popular bromeliad for indoor display.

*Source:* **(a) Photo by Cultivar 413 (CC BY 2.0). (b) Photo by Leonora (Ellie) Enking (CC BY SA 2.0). (c) Photo by Virginie from Pixabay (public domain).**

FIGURE 14.5 Vriesea (*Vriesea*) cultivars in flower. The genus, with about 200 species, is named for Willem Hendrik de Vriese (1806–1862), a Dutch botanist and physician.

*Source:* **Photo by Shutterstock/BearFotos.**

and stems, but not for their usually relatively inconspicuous flowers. The rosettes of many species are often made up of strap-shaped or sword-shaped leaves in wonderful patterns of red, pink, yellow, and green. Often, small, inconspicuous flowers develop in the centers of such rosettes, growing in the water of the central tank.

Many bromeliads produce quite attractive, feathery, flower-like structures. However, the most popular bromeliads grown for their "floral" (bract) displays are hybrid cultivars of *Guzmania* and *Vriesea*, and these are highlighted in the remainder of this chapter.

## 14.3 HISTORY

Ancient civilizations in the Americas, including the Incas, Aztecs, and Maya, employed bromeliads for food, fiber, and medicine. Europeans were introduced to bromeliads on Columbus's second trip to the New World in 1493, during which pineapples were observed being cultivated by the Carib tribe in the West Indies. It is said that when Columbus's armada brought a pineapple back to King Ferdinand of Spain, he declared it to be his favorite fruit. The pineapple became the prized fruit for royalty and aristocracy throughout Europe, often raised in heated glasshouses for the rich. Eventually, pineapple became a leading fruit throughout the world.

FIGURE 14.6 Nineteenth-century paintings (public domain) of cultivated bromeliads (Bromeliaceae) by European botanical illustrators. During the period, the wealthy classes collected and displayed bromeliads as status symbols: (a) *Vriesea carinata.* (b) *Vriesea × cardinalis.* (c) *Neoreglia carolinae.* (d) *Nidularium scheremetiewii.* (e) *Billbergia pyramidalis.* (f) *Nidularium innocentii.* (g) *Quesnelia quesneliana.* (h) *Pitcairnia staminea.*

***Source:*** **(a) *The Floral World and Garden Guide*, vol. 17, 1874. (b) *L'Illustration horticole* (Gand, Belgium), vol. 38, 1891. (c) *Gartenflora* (Berlin), bd. 6, 1857. (d) Step, E., D. Bois, B. Herincq, and W. Watson. 1896. *Favourite Flowers of Garden and Greenhouse*, vol. 4. Frederick Warne, New York. (e) Annales de la Société royale d'agriculture et de botanique de Gand (Belgium), Tome 3, 1847. (f) *Gartenflora* (Berlin), Jahrg. 44, 1895. (g) *La Belgique horticole*, vol. 32, 1882. (h) *L'Illustration horticole* (Gand, Belgium), vol. 21–22, 1874–1875.**

*Guzmania lingulata* was imported into Europe in 1776, *Aechmea fasciata* in 1828, and *Vriesea splendens* in 1840, all becoming popular exotic ornamental plants grown in protected places. From the 19th century onward, dozens of species and hundreds of cultivars were created. Many cultivars were grown in public and private conservatories in Europe, when exotic species were frequently great attractions. Often, the older varieties have been lost. However, today hundreds of species and thousands of cultivars of bromeliads are grown as ornamental plants.

## 14.4 ECONOMIC IMPORTANCE

Bromeliads constitute an important source of houseplants in international and domestic horticultural markets. The cost of the more attractive species can be high because they grow slowly, but because the plants are greatly admired, they can sell for premium prices. In subtropical and tropical regions, they are also part of the very large array of ornamental plants that are cultivated outdoors.

## 14.5 ORNAMENTAL USES

In subtropical areas, bromeliads are widely grown in garden beds and borders. In temperate climates, they are mostly grown in containers, usually as potted houseplants.

Pineapple (*Ananas comosus*) and the other (much less-edible) six *Ananas* species are sometimes employed as ornamental garden curiosities.

## 14.6 OTHER USES

Pineapple is one of the world's most important fruit crops. Many familiar fleshy fruits are formed from the ovary of a flower and contain seeds, like an apple, but some represent a cluster of fused ovaries, like a blackberry, and this is the case with pineapple. Each more or less hexagonal section represents the place where a flower was once attached. A pineapple is the result of 50 to 200 flowers coalescing. The core of a pineapple, albeit soft, is the stem around which the flowers were located. Commercial pineapples do

FIGURE 14.7 Commercial production and sale of bromeliads: (a) A large Dutch greenhouse growing bromeliads. (b) A woman purchasing a potted plant in a garden store.

***Source:*** **(a) Photo by Shutterstock/Ruud Morijn. (b) Photo by Shutterstock/Zhuravlev Andrey.**

FIGURE 14.8 Garden beds of ornamental bromeliads: (a) A mixed bed of guzmania and vriesea. (b) A mixed border outside a house in San Diego, California, with *Guzmania lingulata* and other flowering plants. (c) Guzmania beds. (d) Guzmania.

*Source:* **(a) Photo by Shutterstock/Aunyaluck. (b) Photo by Cultivar413 (CC BY 2.0). (c) Photo by Shutterstock/Ivan Smuk. (d) Photo by Leonora (Ellie) Enking (CC BY SA 2.0).**

FIGURE 14.9 Bromeliads in containers: (a) Mixed guzmania and vriesea. (b) Vriesea.

*Source:* **(a) Photo by Shutterstock/Oliver Hoffmann. (b) Photo by Shutterstock/Akarawut.**

FIGURE 14.10 Ornamental pineapple plants.

*Source:* **(a) Photo by Giorgiomonteforti (public domain). (b) Photo by Tekana (CC BY SA 3.0). (c) Photo by MarvinBikolano (CC BY SA 4.0). (d) Photo by Louise Wolff (CC BY SA 3.0). (e) Photo by Sowmi (CC BY SA 3.0). (f) Photo by Almighty Franklinstein (CC BY SA 3.0).**

FIGURE 14.11 Pineapple production, sales, and culinary preparations: (a) A pineapple field. (b) A pineapple greenhouse. (c) A supermarket display of pineapples. (d) Whole and sliced pineapples. (e) Pineapple upside-down cake. (f) Pina colada (Spanish for "strained pineapple"), a cocktail made with pineapple juice and traditionally garnished with a pineapple wedge.

*Source:* **(a) Photo by Waleed Khalid from Pixabay (public domain). (b) Photo by Ajay Suresh (CC BY 2.0). (c) Photo by Fiona Smallwood on Unsplash (public domain). (d) Photo by Joseph Mucira from Pixabay (public domain). (e) Photo by Kimberly Vardeman (CC BY 2.0). (f) Photo by Miansari66 (public domain).**

not develop seeds and are propagated vegetatively, from the top of the pineapple or from side shoots.

A few other bromeliad species are grown locally in the neotropics to provide food, drink, and fiber. Bromelain, a proteolytic enzyme extracted from pineapples, is used as a meat tenderizer and has pharmaceutical properties.

## 14.7 PROBLEMS

The flowering parts of bromeliads are very widely harvested from the wild in neotropical countries as decoration for churches on religious holidays. This threatens the survival of some of the species.

The water in tank bromeliads can be a perfect environment for mosquitoes to reproduce and subsequently bite people, spreading serious diseases, such as yellow fever, dengue fever, chikungunya, and Zika.

For those who grow their own pineapples, edible varieties or not, the following warning should be heeded. Ornamental bromeliads are not considered to pose risks of toxicity (although many have formidable spines). However, unripe pineapple is not only inedible but can also be poisonous, irritating the throat and acting as a drastic purgative. The flesh of very young fruits is toxic and has been deliberately ingested to produce abortion.

## 14.8 CULTIVATION

### 14.8.1 Climate and Site

No bromeliad will survive freezing. Bromeliads are mostly tropical plants that appreciate warm temperatures (one of the reasons they make good houseplants). Most grow best in diffused or indirect light, but some (notably pineapple) grow well in full sun.

### 14.8.2 Propagation and/or Purchasing

Bromeliads can be grown from seed, but many of the most attractive cultivars are mutations that need to be maintained vegetatively. Individual bromeliads normally die after flowering (this can be quite slow, taking 1 to 2 years, and sometimes, part of the still-living portion can be resurrected), but the offsets ("pups") that are consequently stimulated to develop can be replanted. In general, the offsets should be allowed to grow to one-third the size of their parent before being excised along with their roots. The pups can be planted in organic material (such as humus, sphagnum moss, bark, or chopped coconut fiber), with grit added to improve drainage. A pineapple can be propagated from the crown (topknot and top of the fruit).

### 14.8.3 Planting

Most frequently, pups are cut off where they attach to the mother plant and placed in potting soil, buried about 2.5 cm below the surface. They often develop at different rates and are best replanted individually when they reach sufficient size. Epiphytic species can be tied to the bark of small logs, with their roots placed in sphagnum moss. However, wiring bromeliads to objects is more appropriate for a humid greenhouse than for houses that become very dry in winter (perhaps requiring daily misting). In frost-free climates, many ornamental bromeliads can be treated as conventional garden subjects, but they often grow especially well in containers that can be easily monitored for water content.

### 14.8.4 Soil and Moisture Conditions

Bromeliads cannot stand having their roots soaked, so good drainage is essential. The central reservoir of tank bromeliads should be maintained full (but emptying and refilling it every 1 to 2 months is recommended to clean out debris and prevent buildup of mineral salts). Hard water is hard on some bromeliads, and in hard water areas, rainwater, distilled, or de-ionized water should be used. Bromeliads prefer acidic water (pH 4.0–7.0). Saline and softened water (which is also high in sodium) should be avoided. The compost that is used in pots or for the roots of air plants should be kept moist, but not soggy. Bromeliads prefer a relative humidity of 50 to 75%. Misting the foliage when the humidity is low is recommended, and a dilute fertilizer can be added occasionally to the spray. Adding fertilizer to the cup should be avoided (according to some authorities) because this encourages salts to accumulate. Only light fertilization, at about monthly intervals, when the plants are actively growing, is appropriate for bromeliad houseplants.

### 14.8.5 Maintaining Plants

As houseplants, bromeliads are not demanding. However, they are slow-growing, and patience is necessary as years may elapse before the plants flower. Adding to the frustration, they bloom just once, and the parental plants die.

Nurseries often force bromeliads into bloom by exposing them to chemicals such as ethephon which stimulate the release of ethylene, a plant growth hormone which promotes maturation. Ethylene is the only naturally occurring plant growth hormone that exists as a gas. Apples naturally release ethylene gas as they ripen and can be used to speed up the flowering of a bromeliad. A mature, healthy plant is simply sealed in a plastic bag with a large ripe apple for 7–10 days, and depending on species, flowers may develop within 6 weeks, or not for several months.

## 14.9 FASCINATING FACTS, FABLES, FOLKLORE, AND FRIVOLITY

- Carib Indians placed pineapples or pineapple crowns outside the entrances to their dwellings as symbols of friendship and hospitality. Pineapples were very expensive in past centuries in Europe, and their presence became associated with good living. The practice of placing pineapple carvings over doorways became common in Spain, in England, and later, in New England (where pineapples are also often engraved on door knockers). Pineapple symbols are still used in hotels and restaurants to signal the presence of hospitality.

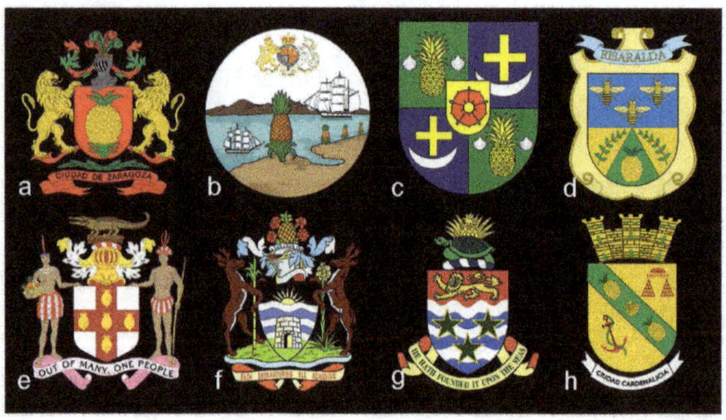


FIGURE 14.12 Examples of coats of arms decorated with pineapples: (a) Zaragoza, Colombia. (b) British Leeward Islands between 1871 and 1956. (c) Shield of coat of arms of Felix Maria Davídek, bishop of the Czech "underground" Roman Catholic Church, secretly consecrated in 1967. (d) Department of Risaralda, Colombia. (e) Jamaica (unofficial). (f) Antigua-Barbuda. (g) Cayman Islands. (h) Lajas, Puerto Rico.

*Source:* **(a) Photo by SajoR (CC BY SA 2.5). (b) Photo by LadyofHats (public domain). (c) Photo by GiMa38 (CC BY SA 4.0). (d) Photo by Milenioscuro (CC BY SA 3.0). (e) Created by BaronJaguar (CC BY SA 4.0). (f) Photo by Clker-Free-Vector-Images from Pixabay (public domain). (g) Photo by Petr Sládek (CC BY SA 3.0). (h) Photo by the Eloquent Peasant (CC BY SA 4.0).**

- The first pineapple grown in France was offered to King Louis XIV (1638–1715). The characteristically greedy and impatient monarch bit into the fruit before anyone had time to peel it, severely cutting his royal lips. Pineapple growing in France ceased until the king's death.
- Reflecting the great esteem with which pineapples have been held historically, they are remarkably widespread on coats of arms.
- *Puya raimondii* (Queen of the Andes), a native of the High Andes at altitudes of 3,000 to 4,000 m in Bolivia and Peru, is the world's largest bromeliad. Its flowering stems grow as tall as 15 m. Like most bromeliads, the mother plant dies after it first flowers and sets seeds, which may take as long as 80 years. The plant is considered to be endangered in the wild.
- Several species in the bromeliad family are carnivorous (at least insectivorous), including *Brocchinia reducta*, *B. tatei*, and *Catopsis berteroniana*. These lure insects by their bright yellow coloration, supplemented by ultraviolet reflections which insects can perceive. Probably, fragrance also helps attract insects. The plants act like pitcher plants, the waxy leaf surfaces causing the visiting insects to slip into the tank of water in the center of the plant. The insects drown and decay, providing nutrients for the plants. At least in the case of *B. reducta*, the reservoir also acts like a stomach, secreting enzymes which help digest the prey.

FIGURE 14.13 The world's largest bromeliad, *Puya raimondii*, near Huaraz, Peru.

*Source:* **Photo by Shutterstock/Christian Vinces.**

FIGURE 14.14 Carnivorous plants that kill insects by drowning them in a foliar tank of water: (a) The bromeliad *Brocchinia reducta*. (b) A leaf of a purple pitcher plant (*Sarracenia purpurea*).

*Source:* **(a) Photo by BotBln (CC BY SA 3.0). (b) Photo by Aaron Carlson (CC BY SA 2.0).**

## KEY PUBLICATIONS

Aranda-Peres, A.N. and A.P.M. Rodriguez. 2006. Bromeliads. In: *Floriculture, Ornamental and Plant Biotechnology*, edited by A. Teixeira da Silva, 645–656. Global Science Books, Middlesex, UK.

Baensch, U. and U. Baensch. 1994. *Blooming Bromeliads*. Tropical Beauty Publishers, Nassau, Bahamas. 271 pp.

Coelho de Paula, C. and H.M. Peregrino da Silva. 2022. *Bromeliads – Botany and Cultivation*. Editora UFV, Brazil. 139 pp.

Crayn, D.M., K. Winter, K. Schulte, and J.A.C. Smith. 2015. Photosynthetic pathways in Bromeliaceae: Phylogenetic and ecological significance of CAM and C3 based on carbon isotope ratios for 1893 species. *Botanical Journal of the Linnean Society* 178: 169–221.

Hornung-Leoni, C.T. 2011. Bromeliads: Traditional plant food in Latin America since prehispanic times. *Polibotánica* 32: 219–229.

Kramer, J. 2011. *Bromeliads for Home and Garden*. University Press of Florida, Gainesville. 176 pp.

Ladino, G., F. Ospina-Bautista, J. Estévez Varón, L. Jerabkova, and P. Kratina. 2019. Ecosystem services provided by bromeliad plants: A systematic review. *Ecology and Evolution* 9: 7360–7372.

McPherson, S. 2007. The carnivorous bromeliads. *Journal of the Bromeliad Society* 57(2): 80–88.

Negrelle, R.R.B., D. Mitchell, and A. Anacleto. 2012. Bromeliad ornamental species: Conservation issues and challenges related to commercialization. *Acta Scientiarum (Biological Sciences)* 34: 91–100.

Oliva-Esteve, F. 2000. *Bromeliads*. Armitano, Caracas, Venezuela. 464 pp.

Wilke, A.B., C. Vasquez, P.J. Mauriello, and J.C. Beier. 2018. Ornamental bromeliads of Miami-Dade County, Florida are important breeding sites for *Aedes aegypti* (Diptera: Culicidae). *Parasites & Vectors* 11(1): 1–7.

Zimmer, K. 1985. Bromeliaceae (Ornamentals). In: *Handbook of Flowering*, Vol. 2, edited by A.H. Halevy, 78–81. CRC Press, Boca Raton, FL.

### Key Cultivation Websites

Bromeliad Society International. www.bsi.org/new/.

Bromeliad Society of Queensland (Australia). https://bromsqueensland.com.au/resources/.

Bromeliads.info. *Top Books and Resources for Bromeliad Enthusiasts*. www.bromeliads.info/top-bromeliad-books-and-resources/.

Clemson College of Agriculture, Extension. *Bromeliads*. https://hgic.clemson.edu/factsheet/bromeliads/.

Naples Botanical Garden. *How to Divide Your Bromeliads and Expand Your Collection*. www.naplesgarden.org/how-to-divide-your-bromeliads-and-expand-your-collection/.

San Diego Bromeliad Society. www.sandiegobromeliadsociety.org/.

The Spruce. *Bromeliad: Plant Care & Growing Guide*. www.thespruce.com/grow-bromeliads-indoors-1902667.

The Spruce. *Growing Bromeliads at Home*. www.thespruce.com/growing-bromeliads-at-home-1403144.

University of Wisconsin-Madison, Extension. *Bromeliads*. https://hort.extension.wisc.edu/articles/bromeliads/.

WikiHow. *How to Care for a Bromeliad Indoors*. www.wikihow.com/Care-for-a-Bromeliad-Indoors.

WikiHow. *How to Water a Bromeliad Plant*. www.wikihow.com/Water-a-Bromeliad-Plant.

# 15 Butterfly Bush (*Buddleja* Species and Hybrids)

**Scientific Names:** The genus name *Buddleja* commemorates Reverend Adam Buddle (1662–1715), an English cleric and botanist. The name is frequently spelled *Buddleia*, but when the biologist Carl Linnaeus coined the genus, it was pronounced with a silent *j*, and *Buddleja* is the correct name by priority according to the International Code of Botanical Nomenclature. The specific epithet in the name *B. davidii* recognizes Father Armand David (1826–1900), a French missionary and explorer in China, who was the first European to recognize the uniqueness of the species (he was also the first to bring the attention of the Western world to the giant panda).

**Common Names:** Butterfly bush (butterfly-bush, butterflybush), orange eye butterfly bush (for the orange-yellow nectar guide in the throat of the flowers), orange eye, summer lilac (for its resemblance to lilac [*Syringa*], most cultivars of which are spring-flowering). *Rotheca myricoides* (*Clerodendrum myricoides*), an unrelated evergreen species native to Africa but widely cultivated in hot climates, is also known as butterfly bush.

FIGURE 15.2 *Liberta* (Italian for *freedom*), a painting illustrating the similarities of butterflies and flowers by contemporary Italian artist Carlo Scomparin.

*Source:* **Carlo Scomparin (CC BY SA 4.0).**

## 15.1 INTRODUCTION

Most people love ornamental plants for their colorful flowers (poetically described as stationary butterflies). The insects that most people greatly admire are butterflies (sometimes called flying flowers), of which there are about 18,000 species globally. "Butterfly gardens" combine the attractiveness of both. Butterfly bush, the subject of this chapter, is the

FIGURE 15.1 (a) Close-up of flowers of butterfly bush (*Buddleja davidii*) with orange-yellow nectar guides in their throats. (b, c) A visiting hoverfly (*Volucella zonaria*).

*Source:* **(a) Photo by Fritz Geller-Grimm (CC BY SA 2.5). (b, c) Photos by Martin Cooper (CC BY 2.0).**

FIGURE 15.3 Butterfly bush (*Buddleja davidii*) with many visiting butterflies sipping nectar from the flowers.

*Source:* **Photo by Shutterstock/Lantapix.**

DOI: 10.1201/9781003586227-15

world's most widely employed garden plant grown specifically to attract butterflies. This champion "butterfly magnet" is pretty, but sometimes it becomes a pretty awful pest.

## 15.2 OVERVIEW

*Buddleja* is a genus of about 100 species native to tropical and subtropical Asia, Africa, and North and South America. The plants include evergreen, semi-evergreen, and deciduous shrubs, trees, vines, and herbs. They tend to occupy riversides as well as rocky and scrub areas.

FIGURE 15.4 Some *Buddleja davidii* cultivars: (a) 'Border Beauty'. (b) 'Buzz Ivory'.

*Source:* Photos by Ptelea (CC BY SA 3.0).

FIGURE 15.5 Illustration of a butterfly bush (*Buddleja davidii*) emphasizing branching pattern. Note how most of the conical flower clusters ("inflorescences") terminate arching stems ("canes").

*Source:* Photo by Shutterstock/Vac1.

*Buddleja davidii*, the principal economic species, is native to Japan and central and southwestern China, from lowlands to elevations up to 3,500 m, where it produces thickets on rocky mountain slopes, along stream banks, and in forest clearings. It has become naturalized in many parts of the world, particularly Europe. It is a vigorous, deciduous ("semideciduous" or even evergreen in very warm climates – south of climate zone 8), multi-stemmed shrub growing as tall as 5 m but typically is 2 to 4 m in height. Some cultivars are dwarfs, no higher than 1.5 m. Although primarily grown for flowering displays, some cultivars have been selected for variegated foliage. There are also cultivated hybrids with other species, particularly *B. globosa* and *B. fallowiana*. Butterfly bush produces characteristically arching stems topped by leaning flower clusters. The stems arising from a particular plant may spread over 5 m, but much less in areas where the aboveground parts of the plant are winter-killed. Prominent masses of flowers, quite reminiscent of lilacs, are produced and bloom from early summer to autumn, sometimes to frost. The individual flowers are small, 9–15 mm long, and mildly fragrant (said to be honey-scented). Wild plants have purple-lilac flowers; cultivars additionally often have white, pink, or reddish flowers, and occasionally yellow blooms. The flowers produce considerable sugar-rich nectar and are very attractive to insects and hummingbirds. The very showy, cone-shaped flower clusters are 15 to 50 cm long, occurring mostly on the ends of branches. Fertilized flowers transform into two-valved seed capsules which split open when ripe to release up to 100 seeds. The seeds are tiny and elongated at both ends into pointed wings. Large shrubs can produce millions of seeds. The light seeds are dispersed by wind and water and can remain dormant in the ground for years.

*Buddleja globosa* (orange ball tree) is a shrub growing up to 4 m in height. It is a native of South America and is widely grown as an ornamental in mild climates (zones 7 to 9) for its globose heads of golden-yellow flowers, which attract pollinators.

FIGURE 15.6 (a) *Buddleja globosa* (orange ball tree, orange butterfly bush), a shrub native to the Andes of Chile, Peru, and Argentina which is cultivated outdoors in warm climates. (b) Bumblebee on a flower cluster.

*Source:* (a) Photo by Leonora (Ellie) Enking (CC BY SA 2.0). (b) Photo by S. Rae (CC BY 2.0).

FIGURE 15.7 *Buddleja* × *weyeriana*: (a) Cultivar Golden Glow. (b) Cultivar Sungold.

**Source: (a) Photo by Daniel Villafruela (CC BY SA 4.0). (b) Photo by Ptelea (CC BY SA 3.0).**

FIGURE 15.8 *Buddleja alternifolia.*

**Source: (a) Photo by Cillas (CC BY SA 3.0). (b) Photo by Wouter Hagens (public domain).**

FIGURE 15.9 Examples of ornamental uses of *Buddleja*: (a) A butterfly bush hedge. (b) A dwarfed (bonsai) butterfly bush. (c) A collage of cut flowers from *Buddleja* cultivars.

**Source: (a) Photo by Cattan2011 (CC BY 2.0). (b) Photo by Jerry Norbury (CC BY ND 2.0). (c) Photo by Indiana Ivy Nature Photographer (CC BY 2.0).**

*Buddleja* × *weyeriana* (the first hybrid of an Asiatic species, *B. davidii*, and a South American species, *B. globosa*) is a popular butterfly bush available in several cultivars. The flowers occur in rounded clusters and are usually yellow or purple. The plants are grown in climate zones 5 to 9 but may die back to the ground in zone 5 winters.

*Buddleja alternifolia* (alternate-leaved butterfly bush), a native of China, is another tall shrub (up to 5 m in height) which is also popular in mild climates (it is suited to zones 7 to 9). The species produces graceful hanging elongated clusters of scented, lilac-color flowers.

## 15.3 HISTORY

*Buddleja davidii* was introduced to the UK from China in the late 1800s and subsequently entered world commerce. To date, about 300 cultivars have been recorded.

## 15.4 ECONOMIC IMPORTANCE

Butterfly bushes are relatively minor but well-known garden plants. The commercial value of common butterfly bush is related to its popularity in private and commercial butterfly gardens, which generate appreciable horticultural activity.

## 15.5 ORNAMENTAL USES

Butterfly bushes are considered to be especially useful in gardens to provide flowering displays during the latter part of summer and in the fall, when relatively few other shrubs are in bloom. Indeed, the flowering period is longer than virtually for any other perennials commonly employed in butterfly gardens. Like lilac, they are popular plants for providing fresh-cut flowers. It has been claimed (but not rigorously tested) that butterflies generally prefer cultivars with lavender or pink (mauve) flowers to those with white or dark purple blooms. Among their horticultural virtues, butterfly bushes are considered resistant to mammalian herbivores, notably rabbits and deer.

## 15.6 OTHER USES

Butterfly bush (*B. davidii*) is the best-known ornamental flowering plant cultivated specifically to attract and feed butterflies and other pollinators in both outdoor and indoor settings.

### 15.6.1 Wildlife Gardens

The phrase “wildlife garden” is employed in two basic ways: First, it can designate natural areas reserved for local wildlife (both plants and animals) and protected against human activities, except for the purposes of promoting education and conservation. More commonly, *wildlife gardens* refer to gardens made up of plants that have been deliberately established to attract and sustain certain local wild animals because they are entertaining or endangered. Most animals are unwelcome in gardens since they often injure plants. The principal desired visitors are butterflies, bees, and birds (primarily hummingbirds). Other pollinators may be welcome, although bats (where they occur) are often feared. For the most part, so-called wildlife gardens focus on attracting adult butterflies to plants with flowers that produce abundant nectar. The species most widely grown for this purpose are not usually best for providing food for the caterpillar stage of butterflies (for example, in North America, there are almost no native lepidopterans that employ butterfly bush as a source of food for the larval stage). Many ecologists recommend employing locally adapted plant species to feed both the larval and adult stages of butterflies, but while this is ecologically responsible, it does compromise

FIGURE 15.10 Butterflies on butterfly bush (*Buddleja davidii*): (a) Old World swallowtail (*Papilio machaon*), found in the northern world. (b) Eastern tiger swallowtail (*Papilio glaucus*) of eastern North America. (c) Pipevine swallowtail (*Battus philenor*) of North and Central America. (d) Peacock butterfly (*Aglais io*) of Eurasia.

***Source:* (a) Photo by Fischer.H (CC BY SA 4.0). (b) Photo by Audrey (CC BY 2.0). (c) Photo by Richard Chambers (CC BY SA 3.0). (d) Photo (public domain) by Nowaja from Pixabay.**

FIGURE 15.11 Pollinators feeding on nectar from butterfly bush (*Buddleja davidii*) while hovering: (a) Hummingbird in Oregon. (b) Hummingbird hawk moth (*Macroglossum stellatarum*) of Eurasia, which uses its long proboscis to feed.

***Source:* (a) Photo by Jeri Leandera (CC BY 2.0). (b) Photo by Ander Morea (CC BY SA 3.0).**

FIGURE 15.12 Butterflies in butterfly houses: (a) Butterfly house in Vienna, Austria. (b) Butterflies feeding on fruit in a butterfly house in Mindo, Ecuador.

***Source:* (a) Photo by Shutterstock/Lals Stock. (b) Photo by Shutterstock/Matyas Rehak.**

the ornamental value of the garden, which is the principal objective of most gardeners. Butterfly bush is unequalled in most locations for attracting adult butterflies, although in some locations it is ecologically damaging.

### 15.6.2 Insectaria and Butterfly Houses

The word "insectarium" (plural *insectariums, insectaria*) is employed primarily to mean an institution exhibiting insects and other small, primarily terrestrial invertebrates (especially other arthropods) for public education and entertainment. These may be museums displaying mounted insects or zoos with living insects. Some retail stores selling preserved butterflies and similar curios also sometimes advertise themselves as insectaria. By far, the most popular kinds of insectaria are "butterfly houses" (also known as lepidopteraria) specializing in captive live butterflies and moths. "Butterfly gardens," by contrast, are usually open facilities that attract free-living indigenous butterflies (although the phrase "butterfly garden" is sometimes applied to indoor facilities). Insectaria are present in numerous cities and are superb places for the public to become more sympathetic to insects in particular and nature in general. While keeping large animals in zoos is controversial, insectaria are wonderful centers for persuading the public that "bugs" are essential components of the biota and merit conservation. Most of the large, attractive butterflies that are released in butterfly houses are raised for the purpose and so do not endanger those in the wild. Adult butterflies can be maintained on fruit and prepared solutions, but butterfly houses are always stocked with beautiful nectar-providing plants, and butterfly bush is one of these.

## 15.7 PROBLEMS

In some warm climates, butterfly bush is invasive, establishing initially from seeds dispersed from cultivated plants. It grows well in a variety of natural and disturbed landscapes. As a pioneer species, it is able to colonize barren, nutrient-poor sites, such as gravel banks and denuded areas. It has been classified as a "noxious weed" in some countries (notably New Zealand) and so is subject to regulations, which may include prohibition of sales.

FIGURE 15.13 Invasive *Buddleja davidii*: (a) Beside railway tracks in Germany. (b) Beside the River Thames in England.

***Source:*** **(a) Photo by Noebse (CC BY SA 4.0). (b) Photo by Timo Newton-Syms (CC BY SA 2.0).**

FIGURE 15.14 Fruit of *Buddleja davidii*. The mature fruit may distribute seeds, and in areas where the plant may become invasive, it is often recommended that the fruits not be allowed to develop seeds (this is required in many US states): (a) Branches with unopened young fruit. (b) Opened capsules.

***Source:*** **(a) Photo by AnRo 0002 (public domain). (b) Photo by Gmihail (CC BY SA 3.0).**

Some cultivars are sterile or can produce very little seed and are recommended for areas where there is concern about the invasive potential of the plant. (Note that sterile ornamental cultivars sometimes tend to revert to fertility.) Deadheading spent flowers before they can produce seed is often recommended but probably just postpones seed escape in some regions.

*Buddleja* plants are sometimes grown close to residences in cold climates to promote their survival overwinter. However, butterfly bush roots are notorious for growing through cracks in lime mortar construction and damaging foundations and walls.

## 15.8 CULTIVATION

### 15.8.1 Climate and Site

The most cold-hardy cultivars of *Buddleja* are claimed to survive winter temperatures down to –31°C. Most gardening sites list climate zone 4 as the northern limit of cold tolerance of the hardiest varieties of *B. davidii*. In regions with cold winters, the plants will often die to the ground overwinter, and only the belowground portion survives. A 10 to 15 cm thick layer of bark can be added in the fall to help insulate the base against the cold. In very frigid climates, butterfly bush can be grown close to the foundations of houses, where the warmer conditions may promote survival overwinter. Smaller cultivars can be grown in containers, which can be moved to protected places during the winter. Yellow (or orange) flowers of some hybrids are very attractive to many butterflies, but these varieties are not as tolerant to cold climates as many of the purple-flowered cultivars of *B. davidii*.

Butterfly bush requires full sun to perform reasonably. Moreover, butterfly gardens are ideally established in very sunny sites, preferably south-facing (butterflies are cold-blooded and need the sun's warmth to warm up for flight, so they tend to avoid shady, cooler sites). Additionally, butterflies rarely fly in windy conditions and have trouble feeding on flowers exposed to prevailing winds, so a protected situation is desirable. Most pollinators, including hummingbirds, are alarmed by nearby human activities and ideally are observed from windows and balconies in homes, so plantings are ideally near residences.

### 15.8.2 Propagation and/or Purchasing

*Buddleja davidii* can be propagated from softwood cuttings (young stems) and by seed – both slow procedures that most gardeners avoid – so buying potted planting stock saves considerable work. Established plants can be divided and transplanted (preferably in the spring or fall). Horticultural varieties are usually sold as vegetatively propagated plants (clones probably won't breed true to type from seeds).

### 15.8.3 Planting

Standard recommendations for trees and shrubs for excavating and providing soil for a planting hole should be followed. Spacing varies from about 1.5 m to 3 m, depending on cultivar. Some varieties tend to spread vigorously and can overwhelm small areas. Early spring or late autumn are usually recommended planting times.

### 15.8.4 Soil and Moisture Conditions

Butterfly bush grows well in soils of medium fertility but requires good drainage. In nature, the species tends to occur on basic soils but tolerates a range of pH from 5.5 to 8.9. An annual addition of a layer of well-rotted compost around the base of the plant is beneficial. Butterfly gardens often include mud puddles to provide the mineral salts that butterflies need (salts concentrate and accumulate when puddles dry up), and rocks to sunbathe on to warm up in the morning.

### 15.8.5 Maintaining Plants

Even when the old stems survive winter, it is often recommended that they be cut close to the ground, because the plants usually regrow and flower more vigorously from the base. Flowers are developed on new wood, so the loss of old stems is of limited significance. Plants growing in shady locations tend to become leggy, and support poles may be required for some of the branches. Needless to say, pesticides and herbicides should not be employed in wildlife gardens.

Removal of spent floral spikes is recommended to encourage continuing bloom until frost. In areas where the plants may have potential to produce seeds, cutting away the faded flower clusters will discourage naturalization.

FIGURE 15.15 Special ground features for butterfly gardens: (a) Dried-up mud puddle attracting a variety of butterflies to acquire concentrated salts. (b) Peacock butterflies (*Aglais io*) sunbathing on stone tiles to warm up.

*Source:* **(a) Photo by Donvikro/Vikramjit Kakati, India (CC BY SA 3.0). (b) Photo by Shutterstock/Christine Bird.**

FIGURE 15.16 Renewing butterfly bush at the end of the growing season by cutting away the stems close to the ground.

*Source:* **Photo by Shutterstock/Thijs de Graaf.**

## 15.9 FASCINATING FACTS, FABLES, FOLKLORE, AND FRIVOLITY

- The classification of the insect order Lepidoptera, which includes butterflies and moths, is unsettled, and these names have been applied inconsistently to both related and quite unrelated species. "Moths," which make up over 90% of the Lepidoptera, are usually distinguished from butterflies primarily as being drab and nocturnal and having antennae that are leaf- or feather-shaped, while butterfly antennae tend to be club-shaped, with a bulb at the end. Butterflies usually fold their wings back at rest, while moths flatten their wings against their bodies. However, there are exceptions to these and other distinctions.
- The French monk Père Armand David, after whom *B. davidii* is named, led an adventurous life looking for plants in wild regions. His diary records that he once shared his tent with a donkey to protect it from ravenous wolves at night.
- In post–World War II England, butterfly bush came to be known as "the bombsite bush" because it proved to thrive in the rubble of bombed areas.
- *Buddleja* species are remarkably tough plants, sometimes establishing in niches on stone walls and on the exterior walls and roofs of buildings. In the United Kingdom, large shrubs growing on derelict buildings have become symbolic of poor urban planning.

FIGURE 15.17 Butterfly bush (*Buddleja davidii*) growing on derelict buildings in the United Kingdom: (a) Huge shrub growing out of a wall. (b) Plants growing on the roof of the former Palace Theatre in Swansea (built in 1888).

*Source:* **(a) Photo (in 2021) by 70023venus2009 (CC BY ND 2.0). (b) Photo (in 2013) by Jaggery (CC BY SA 2.0).**

## KEY PUBLICATIONS


Anisko, T. and U. Im. 2001. Beware of butterfly bush. *American Nurseryman* 194(2): 46–49.

Bruner, L.L., D.J. Eakes, G.J. Keever, J.W. Baier, C.S. Whitman, P.R. Knight, et al. 2006. Butterfly feeding preferences for *Buddleja* selections in the landscape. *Journal of Environmental Horticulture* 24: 39–44.

CABI. 2021. *Buddleja davidii (Butterfly Bush)*. Invasive Species Compendium. www.cabi.org/isc/datasheet/10314#tosummaryOfInvasiveness.

Carlsson, M.K. 2017. Blitzweed: The rise and fall of *Buddleia davidii* in England (1896–2008). *Studies in the History of Gardens & Designed Landscapes* 37: 87–96.

Chau, J.H. 2017. *Systematics of Buddleja (Scrophulariaceae): Phylogenetic Relationships, Historical Biogeography, and Phylogenomics*. Ph.D. abstract. https://digital.lib.washington.edu/researchworks/handle/1773/39958.

Culin, J.D. 1997. Relationship of butterfly visitation with nectar qualities and flower color in butterfly bush, *Buddleia davidii*. *News of the Lepidopterists' Society* 39: 35–38.

Kriticos, D.J., M.S. Watt, K.J.B. Potter, L.K. Manning, N.S. Alexander, and N. Tallent-Halsell. 2011. Managing invasive weeds under climate change: Considering the current and potential future distribution of *Buddleja davidii*. *Weed Research* 51: 85–96.

Lehner, S., S. Schulz, and S. Dötterl. 2022. The mystery of the butterfly bush *Buddleja davidii*: How are the butterflies attracted? *Frontiers in Plant Science* 3315. https://doi.org/10.3389/fpls.2022.994851.

Marazzi, B. and A. De Micheli. 2019. Are sterile *Buddleja* cultivars really sterile and "environmentally safe"? *Bollettino della Società Ticinese di Scienze Naturali* 107: 55–60.

Marquand, C.V.B. 1930. Revision of the Old World species of *Buddleja*. *Bulletin of Miscellaneous Information (Royal Botanic Gardens, Kew)* 5: 177–208.

Meyer, E. 1962. *The Genus Buddleia in North America*. Ph.D. thesis. Cornell University. www.proquest.com/openview/0798bbadc24a6434d833e72eb66c6220/1.pdf?pq-origsite=gscholar&cbl=18750&diss=y.

Meyer Norman, E. 2020. *Buddleja*. Flora North America (Online). http://floranorthamerica.org/Buddleja.

Moore, R.J. 1961. Polyploidy, phylogeny, and photoperiodism in Old World *Buddleia*. *Evolution* 15: 272–280.

Starr, F., K. Starr, and L. Loope. 2003. *Buddleia davidii*. United States Geological Survey – Biological Resources Division Haleakala Field Station, Maui, Hawai'i. http://hear.its.hawaii.edu/Pier/pdf/pohreports/buddleia_davidii.pdf.

Stuart, D.D. 2006. *Buddlejas*. Timber Press, Portland, OR. 192 pp.

Tallent-Halsell, N.G. and M.S. Watt. 2009. The invasive *Buddleja davidii* (butterfly bush). *Botanical Review* 75: 292–325.

Turnbull, C. 2004. Pruning the common butterfly bush. *Tree Care Industry Magazine* 15: 26–30.


### Key Cultivation Websites


Clemson University, Extension. *Butterfly Bush*. https://hgic.clemson.edu/factsheet/butterfly-bush/.

Garden for Wildlife. *The Truth about the Butterfly Bush (and What to Plant Instead!)*. https://gardenforwildlife.com/blogs/learning-center/why-to-avoid-butterfly-bush.

Garden Rant. *The Great Butterfly Bush Debate*. https://gardenrant.com/2015/08/the-great-butterfly-bush-debate.html.

GardenersWorld.com. *How to Grow Buddleia (Butterfly Bush)*. www.gardenersworld.com/how-to/grow-plants/how-to-grow-buddlejas/.

North American Butterfly Association. https://naba.org/about/.

Plant Delights Nursery Inc. *The Butterfly Bush – Buddleia davidii*. www.plantdelights.com/blogs/articles/butterfly-bush-buddleia-davidii-plant-buddleja.

Royal Horticultural Society. *How to Grow Buddlejas*. www.rhs.org.uk/plants/buddleja/growing-guide.

The Spruce. *How to Grow and Care for Butterfly Bush*. www.thespruce.com/types-of-butterfly-bush-2132429.

WikiHow. *How to Prune a Butterfly Bush*. www.wikihow.com/Prune-a-Butterfly-Bush.

WikiHow. *When to Prune Buddleia (Butterfly Bush) & How to Do It*. www.wikihow.com/Prune-Buddleia.

# 16 Calceolaria (*Calceolaria* Species and Hybrids)

**Scientific Names:** The genus name *Calceolaria* is based on the Latin *calceolus*, "slipper," a reference to the appearance of the flower. (It has been suggested that Linnaeus, the father of biological nomenclature, who coined the name, had a secondary motive: to honor F. Calceolari, a 16th-century Italian botanist.)

**Common Names:** The principal English name for this group is "calceolaria." Other names include *pocketbook flower*, *lady's purse*, *poor man's orchid*, *pouch flower*, *slipper flower*, and *slipperwort*. The name *poor man's orchid* reflects (superficially) the orchid-like appearance of the flowers. The remaining names are based on the puffy flowers that look like pouches, slippers, and old-fashioned pocketbooks. "Florists' slipperwort" (less frequently, "florist's slipperwort") refers to *C.* × *herbeohybrida*, a group of hybrids known only in cultivation, which includes the main cultivars employed as ornamentals. The "slipper" in the common names of calceolaria sometimes leads to confusion with lady's slipper orchids (*Cypripedium* species).

## 16.1 INTRODUCTION

Calceolaria is one of the world's most beautiful flowering houseplants, its dazzling blooms competing in attractiveness with the finest orchids.

## 16.2 OVERVIEW

There are about 400 species of *Calceolaria*, natives of Central and South America, especially found in the Andean region. Several of the species are shrubs or semi-shrubs. Various parts of some of the species are covered with tiny hairs that secrete adhesive substances, and such plants can be sticky to the touch. The cultivars (mostly in the herbeohybrida group) are usually small herbaceous plants with pouch-shaped blooms growing in clusters on stems that reach above the foliage at the base of the plant. The plants are usually no taller than 25 cm but sometimes reach 75 cm (which may require staking). The leaves are heart-shaped, soft, and up to 15 cm across but are often obscured by the flowers above them. The blooms in garden varieties are typically 2 to 5 cm in diameter. The flowers have two-lipped corollas (the petals are fused together), with the much larger lower lip inflated into a pouch. The flowers are usually solid-colored, bicolored, or spotted, in shades of yellow, orange, red, and purple. Given proper care, calceolarias often bloom for 2 to 4 weeks, some varieties for longer. They are usually grown as short-term annuals, although in their native regions, they are perennials.

In addition to the herbeohybrida group, a few other species have some popularity. Bush slipperwort (*C. integrifolia*)

FIGURE 16.1 Flowering calceolaria hybrids (*Calceolaria* × *herbeohybrida*): (a) Potted plants pressed together. (b) Potted plant. (c) Isolated flower.

*Source:* **(a) Photo by Shutterstock/Maria Dryfhout. (b) Photo by Shutterstock/StockPictureGarden. (c) Photo by Virginie from Pixabay (public domain).**

FIGURE 16.2 Bush slipperwort (*Calceolaria integrifolia*): (a) Wild plant growing on the coast of Central Chile. (b) A garden hybrid. (c) Cultivar Kentish Hero.

*Source:* **(a) Photo by Dick Culbert (CC BY 2.0). (b) Photo by BotBin (CC BY SA 3.0). (c) Photo by Patrick G (CC BY SA 2.0).**


DOI: 10.1201/9781003586227-16

is a woody-based plant growing to a height of 90 cm. Several hybrids of this and other species are treated as garden perennials in warmer climates. Several alpine species are also grown in rock gardens.

## 16.3 HISTORY

Several species of *Calceolaria* were introduced to English gardens in the late 18th century. Hybridization was initiated about 1820. *Calceolaria crenatiflora* is thought to have been the most important parental plant. Most modern cultivars are in the herbeohybrida group, which includes hybrids of several species, especially *C. crenatiflora*, *C. corymbosa*, and *C. cana*.

## 16.4 ECONOMIC IMPORTANCE

Calceolarias are mostly purchased, like cut flower bouquets, as gifts or for personal use as household decorations. Their diminutive size is advantageous both to the producers, who can grow large numbers in smaller greenhouses, to retailers, who need to use shelf space efficiently, and to consumers, who want their houseplants to be easily handled and located. Calceolarias are considered to be a specialty potted plant by the greenhouse industry and represent a niche market.

FIGURE 16.3 Old paintings (public domain) of calceolaria hybrids (*Calceolaria* × *herbeohybrida*) by botanical illustrators.

*Source:* (a) Lamaire, C. A. 1884. *L'Illustration horticole*, vol. 31. (b) Weise, G. 1858. *Deutsches Magazin für Garten- und Blumenkunde*. Stuggart. (c) Step, E., D. Bois, B. Herincq, and W. Watson. 1896. *Favourite Flowers of Garden and Greenhouse*, vol. 3. Frederick Warne, New York. (d) *L'Illustration horticole* (Gand, Belgium), vol. 31, 1884. (e) Von Regel, A. (ed.). 1852. *Gartenflora, Band 1*. (f) *The Gardeners' Chronicle* (London), vol. 13, 1880. (g) *The Florist's Journal and Gardener's Record* (London), vol. 1, 1846. (h) *The Floral Magazine* (London), vol. 3, 1874.

## 16.5 ORNAMENTAL USES

Plants of the herbeohybrida group are mostly treated as short-term flowering houseplants. Potted plants tend to be marketed in late winter and early spring. Because of the environmental conditions required to rebloom, potted plants are normally discarded when flowering is completed. These short (usually no higher than 25 cm) plants can be placed in beds in front of taller plants but deteriorate rapidly with summer heat. However, some of the uncommonly grown species are maintained long-term, both in greenhouses and outdoors, by dedicated gardeners.

## 16.6 OTHER USES

Given that there are hundreds of *Calceolaria* species in Central and South America, it would be surprising if there was not Indigenous usage, and indeed there is. *Calceolaria thyrsiflora*, known as "sweet stick" due to the intense sweetness of its leaves, is employed to flavor beverages. It is also popularly

FIGURE 16.4 Commercial planting of a cultivar of calceolaria (*Calceolaria* × *herbeohybrida*).

*Source:* Photo by Shutterstock/Gina Hsu.

FIGURE 16.5 Potted flowering calceolaria hybrids (*Calceolaria* × *herbeohybrida*).

*Source:* (a) Yewchan (CC BY SA 2.0). (b) Fanghong (CC BYSA 3.0). (c) HQ (CC BY 2.0). (d) Andrey Zharkikh (CC BY 2.0). (e) Bernard Spragg (public domain). (f) Manu (CC BY SA 2.0).

used to relieve sore throats, tonsil inflammation, and lesions of the mouth, as well as for the treatment of urinary incontinence and other renal and bladder conditions. Chemical studies have shown that the plant has antimicrobial, antioxidant, and insecticidal properties. Other species of *Calceolaria* are also employed in herbal medicine. To date, there are no reliable studies indicating that the cultivated ornamental calceolaria species have toxic properties of concern in regard to accidental consumption by pets and children.

## 16.7 PROBLEMS

Unfortunately, calceolarias are in flower for only a few weeks and are too demanding in their requirements for soil, water, light, and temperature for most people to maintain after the blooming period.

## 16.8 CULTIVATION

The rather strict environmental conditions, mostly for herbeohybrida-type calceolarias, are presented in the following. An additional consideration is the set of requirements to induce flowering. Under moderate or warm temperatures, calceolaria is usually a long-day plant (continuous light needs to be provided for 14–15 hr daily, for weeks), but the plants can be day-neutral (indifferent to hours of light provided daily) under cool temperatures. However, some cultivars are indifferent to day length regardless of conditions. When plants are purchased from garden centers, this information is usually unavailable, so trying to get a plant to reflower without providing the appropriate conditions may be futile.

### 16.8.1 Climate and Site

Calceolaria can be grown outdoors as a perennial in climate zones 10 and 11. Houseplant calceolarias should be placed near a bright window, receiving indirect or diffused sunlight. The desirable temperature is 15 to 18°C. The plants can be temporarily summered outdoors in a shady location. Calceolarias are useful for decorating tabletops, decks, patios, and balconies. They should be protected from drafts indoors and from wind if outside.

### 16.8.2 Propagation and/or Purchasing

Most cultivated calceolarias are herbaceous herbeohybrida group plants purchased from florists and garden centers. When bought, they should have mostly unopened buds, as the blooming period is limited. Plants can be grown from seeds, but the most attractive cultivars will not reproduce true to type. However, dedicated amateurs and professionals often grow many of the species from their dust-like seeds. Professional growers also reproduce advanced cultivars from cuttings. Shrubby calceolarias (more popular in Europe than in North America), particularly the yellow-flowered *C. integrifolia*, can be raised from seeds (which won't produce flowers for several years in some species) but are generally propagated by cuttings, purchased as starter plants from garden outlets.

### 16.8.3 Planting

When the plants are grown from seed, the seeds are scattered on the surface of the soil (usually in pots). These may grow and flower in 4 to 5 months. Herbeohybrida group plants can be transferred outdoors to beds, planted about 30 cm apart, but as noted earlier, they do not withstand stresses well and tend to deteriorate. In suitable climates, some of the hardier species can be grown as conventional garden plants.

### 16.8.4 Soil and Moisture Conditions

Calceolaria is adapted to moist, but not waterlogged, conditions, and like most houseplants, they frequently die from overwatering. The soil should be maintained moist or at least not allowed to become significantly dry, watering (preferably with soft water) into the soil, not on the plant (which can promote fungal growth). Placing the container in a basin of water and allowing the moisture to soak up from below ("subirrigation"), followed by a period of draining, is recommended. To increase local humidity when it is low (as in residences in winter), placing the pots on trays of stones in shallow water is advisable (misting the plant's surfaces is not). Professional growers control the soil pH between 5.5 and 6.2; higher values sometimes induce leaf yellowing (a significant problem for the herbeohybrida group). Potted plants from retail outlets will normally be provided with suitable soil, but when growing new plants, they should be placed in soil that is rich in compost and well-draining, and an all-purpose, slow-release fertilizer can be added at planting time.

### 16.8.5 Maintaining Plants

Potted plants with heavy flower heads should be supported by thin bamboo or wire stakes. Senescent leaves and flowers should be trimmed off (for appearance and also to extend the life of the plant). Since the plants are usually discarded, repotting is unnecessary.

## 16.9 FASCINATING FACTS, FABLES, FOLKLORE, AND FRIVOLITY

- Most pollinators are insects, but birds are also important, particularly hummingbirds, which visit flowers to collect nectar. However, most *Calceolaria* species are pollinated by female "oil bees," which collect oil from special oil-secreting hairs on an appendage of the lower lip of the flower. The females have special oil-collecting structures on their front legs. An exception to this pattern of mutualism (cooperation between two species that benefits both) is Darwin's slipper (*C. uniflora*), a native of Tierra del Fuego and the southernmost

FIGURE 16.6 Darwin's slipper (*Calceolaria uniflora*) and its pollinator, the least seed snipe (*Thinocorus rumicivorus*), which is attracted to and consumes the white portions of the petals: (a) Darwin's slipper in flower. (b) Close-up of the flower. (c) The least seed snipe. (d) Portrait of Charles Darwin (1809–1882), the Father of Evolution, in 1840, painted by George Richmond (1809–1896).

*Source:* **(a) Photo by R. Escudero Frank (CC BY SA 4.0). (b) Photo by Michael Lejeune (CC BY SA 3.0). (c) Photo by Matias Cabezas (CC BY SA 4.0). (d) Public domain.**

continental part of South America. It was discovered by Charles Darwin, who sailed around the world from 1831 to 1836 aboard the HMS *Beagle*, which spent most of the voyage around South America. The plant is pollinated by the least seed snipe, a bird which feeds on fleshy white floral bodies on the flowers. The bird normally forages for seeds and fruits but happily eats the white floral portions (and unintentionally transfers pollen between plants). A second species of *Calceolaria*, native to the Falkland Islands, also seems to have evolved a similar relationship with birds.

## KEY PUBLICATIONS


Aimone, T. 1986. Culture notes: Calceolaria. *Grower Talks* 50(4): 18–20.

Andersson, S. 2006. On the phylogeny of the genus *Calceolaria* (Calceolariaceae) as inferred from ITS and plastid matK sequences. *Taxon* 55: 125–137.

Céspedes, C.L., J.R. Salazar, and J. Alarcon. 2013. Chemistry and biological activities of *Calceolaria* spp. (Calceolariaceae: Scrophulariaceae). *Phytochemistry Reviews* 12: 733–749.

Cosacov, A., A.N. Sérsic, V. Sosa, J.A. De-Nova, S. Nylinder, and A.A. Cocucci. 2009. New insights into the phylogenetic relationships, character evolution, and phytogeographic patterns of *Calceolaria* (Calceolariaceae). *American Journal of Botany* 96: 2240–2255.

Erwin, J. 1994. Calceolaria production. *Minnesota Commercial Flower Growers Association Bulletin* 43(5): 1–6. www.researchgate.net/publication/328759117_Calceolaria_production/link/5be113cca6fdcc3a8dc17854/download.

Frankel, L., M. Murúa, and A. Espíndola. 2022. Biogeography and ecological drivers of evolution in the Andes: Resolving the phylogenetic backbone for *Calceolaria* (Calceolariaceae). *Botanical Journal of the Linnean Society* 199: 76–92.

Owen, W.G. 2019. Response of *Calceolaria* × *herbeohybrida* cultivars to substrate pH and corresponding leaf tissue nutrient concentration effects. *HortScience* 54: 2163–2168.

Pennell, F.W. 1951. The genus *Calceolaria* in Ecuador, Colombia, and Venezuela. *Proceedings of the Academy of Natural Sciences of Philadelphia* 103: 85–196.

Rünger, W. 1975. Flower formation in *Calceolaria* × *herbeohybrida* Voss. *Scientia Horticulturae* 3: 45–64.

Sérsic, A.N. 2004. *Pollination Biology in the Genus* Calceolaria *L. (Calceolariaceae)*. Biology Centre of the Upper Austrian Museums, Linz, Austria. 121 pp. www.zobodat.at/pdf/STAPFIA_0082_0001-0121.pdf.

Sérsic, A.N. and A.A. Cocucci. 1996. A remarkable case of ornithophily in *Calceolaria*: Food bodies as rewards for a non-nectarivorous bird. *Botanica Acta* 109: 172–176.

Strømme, E., H.F. Wilkins, and W. Rünger. 1985. Calceolaria. In: *Handbook of Flowering*, Vol. 2, edited by A.H. Halevy, 108–111. CRC Press, Boca Raton, FL.

Valdés, E., C. González, K. Díaz, Y. Vásquez-Martínez, C. Mascayano, C. Torrent, et al. 2020. Biological properties and absolute configuration of flavanones from *Calceolaria thyrsiflora* Graham. *Frontiers in Pharmacology* 11: 1125. https://doi.org/10.3389/fphar.2020.01125.

White, J.W. 1975. Calceolaria, a year-round crop. *Pennsylvania Flower Growers Bulletin* 383(1): 6–9.


## Key Cultivation Websites

E-Grow Nutritional Monitoring (Blog). *Nutritional Monitoring Series: Calceolaria (Calceolaria × herbeohybrida).* https://fertdirtsquirt.org/pdf/calceolaria.pdf.

Greenhouse Management. *Cool Calceolaria.* www.greenhousemag.com/article/cool-calceolaria/.

North American Rock Garden Society. *The Genus Calceolaria.* www.nargs.org/article/genus-calceolaria.

The Spruce. *Calceolaria Plant Profile: Showstopping but Short-Lived Indoor Plants.* www.thespruce.com/calceolaria-plant-profile-5073219.

What Flower? *Calceolaria.* https://whatflower.net/houseplant/calceolaria/.

# 17 California Poppy (*Eschscholzia californica* and Relatives)

**Scientific Names:** The genus name *Eschscholzia* was named for Johann Friedrich Gustav von Eschscholtz (1793–1831; "Baltic German," that is, from an area now identified as Estonia and Latvia). He was a surgeon and naturalist who accompanied Russian expeditions to the American Pacific Coast in 1816 and 1824. The species was discovered in California by Aldebert von Chamisso, a German colleague and admirer of Eschscholtz, whose name was spelled in different ways. The "t" was left out of the original genus name (which is nevertheless sometimes incorrectly spelled *Eschscholtzia*).

**Common Names:** *California poppy* is the dominant English name. Other names include *California goldenpoppy*, *California sunlight*, *gold(en) poppy*, *flame poppy*, and *cup of gold*.

## 17.1 INTRODUCTION

California poppy is a rather charismatic ornamental plant, reflecting the glamorous nature of its native state. In its homeland, it produces spectacular, vast, golden-flowering

FIGURE 17.1 Wild California poppy (*Eschscholzia californica*) in California.

*Source:* **(a) Photo by Cultivar 413 (CC BY 2.0). (b) Photo by John Rusk (CC BY 2.0).**

FIGURE 17.2 California poppies in bloom in the Antelope Valley California Poppy Reserve, a 700 ha park of hills blanketed with poppies in April. This state-protected conserved land is located in northern Los Angeles County in Southern California and is part of the western Mojave Desert.

*Source:* **(a) Photo of entrance sign by Vahe Martirosyan (CC BY SA 2.0). (b) Photo by Rennett Stowe (CC BY 2.0). (c) Photo by Shutterstock/Saraporn. (d) Photo by Doncram (CC BY SA 4.0).**

carpets over the landscape, reflecting the "golden state" nickname. Nevertheless, the flowers are rather delicate, not overly ostentatious, but producing a rather mesmerizing, colorful display in home gardens.

## 17.2 OVERVIEW

California poppy is a native of western North America, like the other dozen or so species of the genus *Eschscholzia*. It is indigenous from Baja California to Washington State but is naturalized in much of North America. It is an herbaceous, wild plant growing either as an annual or as a short-lived perennial (which can also flower in their first year). The plants develop a deep taproot, and new branches arise from the center of the plant each year (the old foliage senesces). In nature, California poppies are adapted to surviving and thriving after fires and frequently occur on dry slopes. Cultivated plants are usually grown as annuals. The wild plants are 15 to 60 cm in height, either spreading or erect. The foliage is finely divided and usually bluish-green ("glaucous"). The cup-shaped flowers are oriented upward (facing the sky) and are produced individually or



DOI: 10.1201/9781003586227-17

FIGURE 17.3 Floral variation among cultivars of California poppy (*Eschscholzia californica*).

*Source:* (a) Photo by Shutterstock/Nick Pecker. (b) Photo by Dagmar Collins (CC BY ND 2.0). (c) Cultivar California Sunlight. Photo by Dguendel (CC BY 3.0). (d) Photo by Franklin Hunting (CC BY ND 2.0). (e) Photo by Joanna Boisse (CC BY SA 4.0). (f) Photo by Pointnshoot (CC BY 2.0).

FIGURE 17.4 Open and closed flowers of California poppy (*Eschscholzia californica*): (a) Open flowers. (b) Closed flowers.

*Source:* (a) Photo by Chris Gladis (CC BY ND 2.0). (b) Photo by Shutterstock/Mariola Anna S.

in groups, on the ends of slim upright stems. The petals are orange, reddish, or yellowish in wild forms, also white, pink, purple, and other shades in cultivars. The center of the floral cup is often darker in color. Wild plants have four-petalled flowers, but cultivars with doubled flowers (with extra petals) have been selected. Depending on cultivar and climate, plants generally flower during the summer for several weeks. The seed capsules are cylindrical and slim, containing up to 100 seeds. The capsules split longitudinally, often in an explosive fashion, at maturity, ejecting the small seeds up to 2 m from the parent plant.

The flowers of California poppy close at night and when light levels are low, such as on cloudy days or in cold, windy, or rainy weather.

## 17.3 HISTORY

California poppy was brought to England in the early 1800s. It became popular in Victorian-era gardens and has become a favorite garden plant worldwide.

FIGURE 17.5 Old paintings (public domain) of California poppy (*Eschscholzia californica*) by botanical illustrators.

*Source:* (a, b) Prepared by American botanical artist Mary Vaux Walcott (1860–1940). (c) Prepared by Mary Elizabeth Duffield, published in *The Garden: An illustrated weekly journal of gardening in all its branches*, vol. 12, 1877. (d) Paxton, J. 1837. *Magazine of Botany and Register of Flowering Plants*, vol. 3. (e) Speer, A. E. 1911. *Annual and Biennial Garden Plants, Their Value and Uses*. J. Murray, London. (f) Lounsberry, A. 1899. *A Guide to the Wild Flowers*. Stokes, New York.

FIGURE 17.6 Landscape uses of California poppy (*Eschscholzia californica*): (a, b) Rock gardens. (c, d, e) Garden beds.

*Source:* (a) Photo by Karelz (public domain). (b–e) Shutterstock: (b) Joe Kuis, (c) Nick Pecker, (d) LianeM, (e) Venygret.

## 17.4 ECONOMIC IMPORTANCE

California poppy is moderately important as a garden plant, of primary horticultural significance as a source of seeds for planting, both in ornamental gardens and in habitat restoration.

## 17.5 ORNAMENTAL USES

California poppies are used in a wide variety of ornamental landscape applications, including beds, borders (including paths and walkways), rock gardens, and containers. They are also employed in roadside beautification. The petals rather quickly fall away when the plants are handled, making the flowers unsuitable for use as cut flowers in vases or bouquets.

## 17.6 OTHER USES

American Indians used the seeds, leaves, and roots as food and medicinally, although the plants are toxic. As with numerous other plants, extracts are added to a variety of questionable herbal preparations marketed on the web and claimed to be useful for various ailments. In the western United States, California poppy is used in erosion control mixtures. California poppy has been significantly reduced in parts of its natural range, and it is the natural species to employ in rehabilitating the ecology of such regions.

## 17.7 PROBLEMS

California poppy is considered to have potential for invasiveness, although in the United States, the issue has been minor to date. It is of concern as a potential weed in Mediterranean climates (its native Californian home has the same climate) and, indeed, is considered to be an invasive weed in parts of Chile and Australia which have similar climates. California poppy is also of concern for poisoning livestock, although other plants are much more toxic.

## 17.8 CULTIVATION

### 17.8.1 Climate and Site

Although the plants are quite heat-tolerant, hot summers can stimulate the onset of dormancy (if followed by cool conditions, they may subsequently bloom or rebloom). As an annual, California poppy is able to grow in most climate zones. It has been rated as capable of perennial growth in climate zones 6 to 10. The plants are adapted to full sun. Since they are thought to be significantly toxic, consideration should be given to locating them in sites that are not easily accessible to children and pets.

### 17.8.2 Propagation and/or Purchasing

California poppy is normally grown from seeds, although some cultivars will not breed true from seed. In some climates, the plants will self-seed where they are grown. Frequently, seed packages contain a mixture of cultivars, or at least of seeds that will produce a mixture of colors.

### 17.8.3 Planting

It is frequently recommended that seeds be direct-seeded where they will grow, as the roots do not react well to transplanting, but in short-season locations, they can be started indoors in fiber pots (or purchased as young plants from a garden center). Seedlings should be thinned to 15 to 20 cm apart. Seed dormancy is widespread in wild California poppies but has been considerably reduced in most cultivars. Seed packets should provide instructions if and how much preliminary cold treatment ("stratification") is required.

### 17.8.4 Soil and Moisture Conditions

California poppies are drought-tolerant, and most cultivars prefer infrequent watering and low to moderate humidity. California poppies can thrive in poor soils and a lack of fertilization (but soil that is moderately fertile or has humus added will promote flowering). Sandy and rocky soils suffice, but heavy clay does not provide the necessary drainage. Soil pH does not seem to be critical (6.5 to 7.5 has been recommended).

### 17.8.5 Maintaining Plants

California poppy is considered to need virtually no care, except for weeding. Nevertheless, providing at least some watering and removing spent flowers will prolong flowering. Some gardeners make successive plantings to have flowering plants available over a longer season.

## 17.9 FASCINATING FACTS, FABLES, FOLKLORE, AND FRIVOLITY

- Skilled illustrators have often employed ornamental flowering plants in their cartoons, and this is the case for California poppy.

FIGURE 17.7 Cartoons (public domain) featuring California poppy.

*Source:* **(a) Part of an old advertisement. California Historical Society. (b) Gordon, M. T. 1910. *Flower Children, the Little Cousins of the Field and Garden.* Volland, Chicago. (c) A 1909 postcard.**

FIGURE 17.8 Coat of arms of Meghan, Duchess of Sussex (born 1981), the wife of Prince Harry, Duke of Sussex (born 1984). Note the four California poppies at the base. Although they are no longer working members of the British royal family, Prince Harry and Duchess Markle still hold their royal titles, acquired when they married in 2018.

*Source:* Photo by Sodacan (CC BY SA 4.0).

- Although California poppy has not been shown to be a narcotic of concern like opium poppy, among Native peoples of the western United States, it has a reputation as a sedative. The roots have been employed in Indigenous medicine as sedatives and analgesics, and plants have been placed under the bed of children to promote sleep. Experimental studies with mice have validated the sedative effect of California poppy.
- In the western United States, the pollen of California poppy has been employed by Indigenous people as a yellow dye, for body paint, and as eye shadow.
- The California poppy was recognized as the official state flower of California in 1903. It beat out two competitors for the honor: the matilija poppy (*Romneya coulteri*) and the mariposa lily (*Calochortus*). California Poppy Day, occurring on April 6, was established in 2010. Lancaster, California, hosts an annual California Poppy Festival in April. There is also a poppy week, on May 13–18, but this is not specifically for California poppy.
- Among the other official California State living (or once-living) species are amphibian (California red-legged frog), animal (California grizzly bear), bird (California quail), dinosaur (*Augustynolophus morrisi*), fish (California golden trout), fossil (saber-toothed cat), grass (purple needlegrass), insect (California dogface butterfly), lichen (lace lichen), marine reptile (Pacific leatherback sea turtle), pet (a shelter pet), reptile (desert tortoise), and tree (California redwood).
- A coat of arms created for Meghan Markle, the Duchess of Sussex, decorated with California poppies, reflecting her Californian background, was unveiled in 1918. Coats of arms date back to the 12th century. While they are honorific today, they were traditionally worn over armor in tournaments so participants could identify their opponents.

## KEY PUBLICATIONS


Clark, C. 1978. Systematic studies of *Eschscholzia* (Papaveraceae). I. The origin and affinities of *E. mexicana*. *Systematic Botany* 3: 374–385.

Clark, C. *Eschscholzia*. In: *Flora North America*. Online Version. www.efloras.org/florataxon.aspx?flora_id=1&taxon_id=112192.

Gould, R. 1976. The California poppy – garden hybrids. *Pacific Horticulture* 37(1): 6–8.

Leger, E.A. and K.J. Rice. 2003. Invasive California poppy (*Eschscholzia californica* Cham.) grow larger than native individuals under reduced competition. *Ecological Letters* 6: 257–264.

Leger, E.A. and K.J. Rice. 2007. Assessing the speed and predictability of local adaptation in invasive California poppies (*Eschscholzia californica*). *Journal of Evolutionary Biology* 20: 1090–1103.

Lim, T.K. 2014. *Eschscholzia californica*. In: *Edible Medicinal and Non Medicinal Plants*, edited by T.K. Lim, 622–632. Springer, Dordrecht.

Pearson, K.D., N.L.R. Love, T. Ramirez-Parada, S.J. Mazer, and J.M. Yost. 2021. Phenological trends in the California poppy (*Eschscholzia californica*): Digitized herbarium specimens reveal intraspecific variation in the sensitivity of flowering date to climate change. *Madroño* 68: 343–359.

Rolland, A., J. Fleurentin, M.C. Lanhers, C. Younos, R. Misslin, F. Mortier, et al. 1991. Behavioural effects of the American traditional plant *Eschscholzia californica*: Sedative and anxiolytic properties. Planta Medica 57: 212–216.

Ryan, E.M. and E.E. Cleland. 2021. Clinal variation in phenological traits and fitness responses to drought across the native range of California poppy. *Climate Change Ecology* 2: 100021. https://doi.org/10.1016/j.ecochg.2021.100021.

Smither-Kopperl, M.L. 2018. *Plant Guide for California Poppy (Eschscholzia californica)*. USDA-Natural Resources Conservation Service, CA. https://adminplants.sc.egov.usda.gov/plantguide/pdf/pg_esca2.pdf.

Still, S.M. 2021. *A Profusion of Poppy Confusion*. University of California Davis Arboretum and Public Garden Blog. https://arboretum.ucdavis.edu/news/profusion-of-poppy-confusion.

Still, S.M. and D. Potter. 2013. California poppy conundrums: Insights into relationships within tribe Eschscholtzieae (Papaveraceae). *Systematic Botany* 38: 104–117.

Wilts, B.D., P.J. Rudall, E. Moyroud, T. Gregory, Y. Ogawa, S. Vignolini, et al. 2018. Ultrastructure and optics of the prism-like petal epidermal cells of *Eschscholzia californica* (California poppy). *New Phytologist* 219: 1124–1133.


## KEY CULTIVATION WEBSITES


BlueHouse. *Easy to Grow California Poppy Guide*. www.finehomedecorideas.com/easy-to-grow-california-poppy-guide/.

California.com. *How to Grow California Poppies*. www.california.com/how-grow-california-poppies/.

GardenBeast. *California Poppy Guide: How to Grow & Care for Eschscholzia californica.* https://gardenbeast.com/california-poppy-guide/.

The Spruce. *How to Grow and Care for California Poppies.* www.thespruce.com/how-to-grow-and-care-for-california-poppies-4686987.

University of Wisconsin, Madison. *California Poppy, Eschscholzia californica.* https://mastergardener.extension.wisc.edu/files/2016/04/Eschscholzia_californica.pdf.

WikiHow. *How to Grow California Poppies.* www.wikihow.life/Grow-California-Poppies.

# 18 Calla Lilies (*Zantedeschia* Species)

**Scientific Names:** The genus name *Zantedeschia* commemorates Giovanni Zantedeschi (1773–1846), an Italian botanist and physician.

**Common Names:** The name "lily" is applied to numerous, unrelated species, but especially to the genus *Lilium* (the so-called "true lilies"). The word "calla" is applied not only to *Zantedeschia* but also to *Calla palustris*, which is a bog plant (also known as arum) in the same family (the Araceae). The word "arum lily" (or *white arum lily*) is often applied to *Zantedeschia aethiopica*, but this is confusing, because the word "arum" is usually applied to species of the genus *Arum*, which is also in the Araceae and is often grown for its attractive foliage.

FIGURE 18.1 Massed calla lilies of different colors.

*Source:* **(a) Photo by Shutterstock/Lilia Solonari. (b) Photo by Shutterstock/Zoom Zone.**

## 18.1 INTRODUCTION

*Zantedeschia* includes eight species of perennial herbs native to Central and Southern Africa, although the popular *Z. aethiopica* is not from Ethiopia, as its scientific name implies. The cultivars are among the most striking, spectacularly colored ornamental flowering plants.

The Araceae is a family of flowering plants with many species (often called "aroids") that are cultivated. What appears to be a single enormous solitary flower in the family is technically a *pseudanthium* (Greek for *false flower*; a compact inflorescence of numerous tiny flowers, the entire structure appearing to be a single flower). In *Zantedeschia*, the upper flowers tend to be male, the lower ones female. The pseudanthium consists of two basic parts, a large showy, specialized, enveloping leaflike bract (the *spathe*) shaped like a funnel, inside of which there is a central rodlike columnar structure (the *spadix*) which holds many small, inconspicuous true flowers. The spathe is initially green, but as it matures, it becomes colored. The *Zantedeschia* species discussed here are often considered to produce the most beautiful blooms in the family (although *Anthurium*, the flamingo lily, is also quite attractive). Many other aroids are well-known foliage houseplants, including monstera, peace lily, and philodendron. Although calla lilies are grown primarily for their flowering displays, some of the species have attractive foliage, which is usually deciduous, although in some climates the leaves may remain evergreen.

## 18.2 OVERVIEW

Two classes of calla lilies are recognized for commercial purposes. The first class is represented by *Z. aethiopica*, often termed *white calla*. It is the most widespread (both in gardens and in nature), hardy, and large species, often growing to over 1 m in height. It produces rhizomes (spreading underground stems), while some of the other species produce tubers (underground swollen stem-based organs, like potatoes). *Zantedeschia aethiopica* is from relatively tropical valley conditions of the Western Cape in South Africa. It prefers a very moist to semi-aquatic location and is often marketed as a bog plant or water garden plant.

Several of the remaining species have been hybridized to produce cultivars with highly colorful flowers in shades of apricot, burgundy, mauve, orange, pink, purple, red, and yellow. The ancestors of these plants originated in higher mountain regions with cool temperatures but where the ground never freezes, and the resulting cultivars are quite frost-sensitive and need warm conditions. They are also adapted to high light levels and soils that drain quickly.

## 18.3 HISTORY

Calla lilies were imported to Western countries from Africa centuries ago and have become familiar ornamental plants.

DOI: 10.1201/9781003586227-18

FIGURE 18.2 Calla lilies in art: (a) Late 19th-century chromolithograph (public domain) titled *Calla Lilies in Vase*. Painted by R. Hill, published by L. Prang & Co. (b) Vintage trade card (public domain) published by the Milwaukee Lithograph & Engraving Company. (c) A 1907 chromolithograph titled *Seated woman gazing at calla lily*.

*Source:* **(a) Photo from Boston Public Library. (b) Photo from Plantillustrations.org. (c) Photo from Boston Public Library (public domain).**

FIGURE 18.3 Commercial production of calla lilies: (a) In a field in Italy. (b) In a greenhouse in The Netherlands.

*Source:* **(a) Photo by Shutterstock/Underworld. (b) Photo by Shutterstock/Gertjan Hooijer.**

## 18.4 ECONOMIC IMPORTANCE

Extensive commercial production of *Zantedeschia* for cut flowers, houseplants, and garden plants occurs in warm countries. New Zealand, The Netherlands, and the United States are major centers of research and development.

## 18.5 ORNAMENTAL USES

There is a large worldwide market for cut flowers produced from calla lilies, often raised in semi-tropical countries and exported. Potted plants, often grown for the grow-and-throw-away market, are also popular. The relatively cold-hardy *Z. aethiopica* is grown as a garden plant in areas with mild but freezing winters, and the many popular spectacularly colored hybrids of the other species are also standard garden plants in areas that are not exposed to winter freezing. Calla lilies are also used as houseplants but are relatively less-popular.

## 18.6 OTHER USES

Despite significant toxicity, the tubers and foliage of *Zantedeschia* are sometimes cooked and eaten in some Native African cultures. In the past, the tubers of *Z. aethiopica* were boiled and fed to pigs, resulting in the occasional name "pig lily" (in Afrikaans, the plant is called *Varkoor*, which means "pig's ear"). *Zantedeschia* has also been used to a quite limited extent in traditional African medicinal practices. Warmed leaves of *Z. aethiopica* have been applied to sores, boils, and insect bites and used to treat gout and rheumatism. A decoction of *Z. albomaculata* was employed by Zulu women to prevent miscarriage and the birth of weak babies.

## 18.7 PROBLEMS

Calla lilies, especially when white, are a conventional choice for funeral arrangements and associated expressions of sympathy. Curiously, they are also a frequent choice for weddings, the whiteness indicating the bride's purity, while the phallic spadix suggests fertility. Nevertheless, the association with death has been repugnant to many, dissuading them from choosing the plant.

FIGURE 18.4 Calla lilies as sexual symbols: (a) A bride carrying a bouquet of white calla lilies. Cover of the June 1932 *Ladies' Home Journal*. (b) Calla lilies as indicators of the chastity of a young woman.

*Source:* **(a) Photo by Genibee (CC BY SA 2.0). (b) Photo by Shutterstock/Tayka Ya.**

FIGURE 18.5 Calla lilies as funeral flowers: (a) The grim reaper, accompanied by black calla lilies. (b) Funerary sympathy card.

*Source:* **(a) Photo by Shutterstock/Ottoflick. (b) Photo by Shutterstock/Vector Tradition.**

FIGURE 18.6 *Zantedeschia aethiopica* growing in wetlands.

*Source:* **(a) Photo by Shutterstock/Zoom Zone. (b) Photo by Shutterstock/Tatyana Zabara.**

*Zantedeschia aethiopica*, which grows naturally in marshy areas, has become a widely naturalized invasive weed in many parts of the world. In some areas, its cultivation is prohibited or controlled.

Most aroids tend to cause irritation or dermatitis from contact, and toxicity (sometimes vomiting and diarrhea) when consumed, because of the presence of calcium oxalate crystals in the form of tiny needles ("raphides"). They can poison domesticated pets, including dogs, cats, and horses.

## 18.8 CULTIVATION

### 18.8.1 Climate and Site

Calla lilies are relatively tender plants with respect to temperature. Most are quite frost-sensitive and are recommended for climate zones 8 to 11. *Zantedeschia* cultivars most resistant to cold climates have parentage of *Z. aethiopica and Z. pentlandii* and can be grown outdoors, sometimes as far north as zone 6. In marginally suitable areas, the plants can be mulched to provide protection overwinter. In areas too cold to maintain plants permanently outside, they can be grown in containers that can be moved indoors or excavated and protected indoors during the winter. The plants generally do well in both full sun and shade. Calla lilies are often displayed as houseplants, sometimes discarded after flowering (they are not the easiest of houseplants to revive).

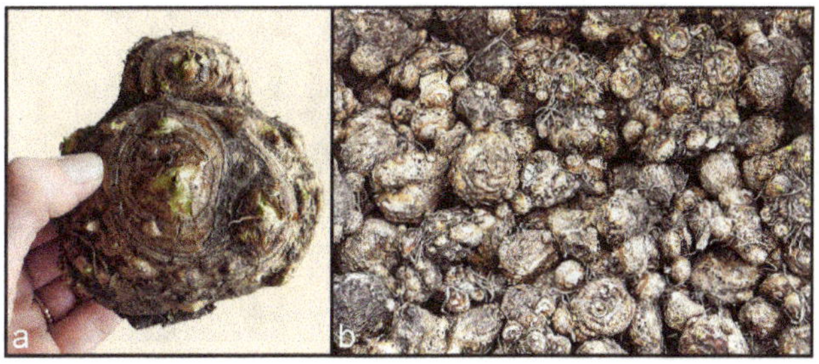

FIGURE 18.7 Tubers ("bulbs") of calla lily: (a) A single tuber (note developing daughter tubers on the periphery). (b) A mixture of harvested tubers.

*Source:* **(a) Photo by Shutterstock/Liga Cerina. (b) Photo by Shutterstock/PosiNote.**

### 18.8.2 Propagation and/or Purchasing

The simplest way of obtaining plants is to purchase specimens established in containers. Dormant rhizomes or tubers can also be bought. Plants can be grown from seed, but flowering may not occur for several years, and the resulting plants may not reproduce the appearance of cultivars maintained only as clones. Established outdoor clumps can be divided. Some cultivars can be propagated by digging them up and replanting portions of rhizomes or tubers (preferably dormant). Potted specimens grown as houseplants may eventually produce basal offsets, which can be separated and repotted.

### 18.8.3 Planting

In warm climates, as noted in the following, calla lilies can be treated as conventional garden plants. In cold climates, the plants cannot be left outdoors overwinter, so excavating them in the autumn and maintaining them indoors is necessary. Growing the plants in containers can facilitate the seasonal transfer.

### 18.8.4 Soil and Moisture Conditions

Outdoors, most species prefer moist, fertile soils rich in organic matter. Recommended soil pH is 5.5–6.5. *Zantedeschia aethiopica* can be planted in shallow water.

### 18.8.5 Maintaining Plants

Tender cultivars grown outdoors in the ground during the summer can be maintained overwinter in a dormant state. After the foliage dies down in the autumn, the underground rhizomes or tubers (sometimes called "bulbs" but more accurately termed *tuber-like rhizomes*) should be unearthed and moved to a dry, frost-free location, preferably in the dark, until spring. Watering should be minimal.

Plants maintained as houseplants also require a period of dormancy. After flowering, the foliage yellows and dies and should be trimmed away above the soil (after the leaves are entirely senescent) using clean scissors or secateurs. The pot should be kept in a dark, dry location, with temperatures about 10°C, for 2 or 3 months. When leaves start

FIGURE 18.8 Skunk cabbage (*Symplocarpus foetidus*): (a) A plant growing through snow in Japan in the early spring. (b) A clump of plants in New York State.

*Source:* **(a) Photo by Sakaori (CC BY SA 3.0). (b) House, H. D., W. B. Starr, and H. H. Snyder. 1918. *Wild Flowers of New York, Part 1* (public domain).**

to re-emerge, fertilize weakly. Relocate to a bright location with limited direct sunlight. Keep the soil moist, but allow some drying between watering.

## 18.9 FASCINATING FACTS, FABLES, FOLKLORE, AND FRIVOLITY

- According to a colorful legend, calla lily grew wherever Eve's tears of sorrow and repentance fell as she and Adam were departing from the Garden of Eden.
- Since the time of the ancient Greeks, the phallic appearance of the spadix and the vaginal appearance of the spathe have been noticed. Psychoanalyst Sigmund Freud (1856–1939), famous for saying "Sometimes a cigar is just a cigar," nevertheless was obsessed with assigning sexual symbolism to common objects. At the start of the 1900s, Freud emphasized the sexual appearance of calla lilies, setting off a search for how artists employed the plants in paintings and literature. Georgia Totto O'Keeffe (1887–1986), an American modernist artist known for her paintings of enlarged flowers, was accused of using flowers to depict sex, although she challenged Freud's view. Six of her calla lily paintings sold for $25,000 in 1928, the largest amount ever paid at that time for a group of paintings by a living American artist.
- During the Victorian period in England, gifts of specific flowers were assigned hidden messages in the "language of flowers." The Victorians interpreted a gift of calla lilies as an expression of sexual interest.
- Several of the more than 4,000 species in the Araceae are extremely malodorous, producing bad smells reminiscent of rotting flesh to attract pollinating insects. One of these is the corpse plant, or titan arum (*Amorphophallus titanum*; Figure 1.37). This native of western Sumatra is often interpreted as the world's largest flower: The spike can be longer than 3 m. Although the odor is repulsive, when specimens in botanical gardens come into flower (a rare event), people attend in great numbers to view and smell the plant. A much more common and widespread plant that is also very bad-smelling is the eastern skunk cabbage (*Symplocarpus foetidus*), which, true to its name, produces a skunky odor. It is a native of North America and, in the opinion of some botanists, also Asia (some botanists consider the Asian species to be different, albeit closely related). Remarkably, it is one of the few plants that produces heat, warming itself to about 15°C in the spring, allowing it to burn its way through snow cover and emerge much earlier than surrounding plants.

## KEY PUBLICATIONS

Kumar, A. and I. Dogra. 2020. In vitro micropropagation of calla lily: An overview. *Indian Journal of Pure & Applied Biosciences* 8: 144–153.

Letty, C 1973. The genus *Zantedeschia*. *Bothalia* 11: 5–26.

Maroyi, A. 2019. *Zantedeschia aethiopica* (L.) Spreng.: A review of its medicinal uses, phytochemistry, and biological activities. *Asian Journal of Pharmaceutical and Clinical Research* 12(8): 30–36.

Singh, Y., A.E. Van Wyk, and H. Baijnath. 1996a. Floral biology of *Zantedeschia aethiopica* (L.) Spreng. (Araceae). *South African Journal of Botany* 62: 146–150.

Singh, Y., A.E. Van Wyk, and H. Baijnath. 1996b. Taxonomic notes on the genus *Zantedeschia* Spreng. (Araceae) in southern Africa. *South African Journal of Botany* 62: 321–324.

Szuman, K. and N. Lall. 2020. *Zantedeschia aethiopica*. In: *Underexplored Medicinal Plants from Sub-Saharan Africa*, edited by N. Lall, 303–308. Academic Press, Cambridge, MA.

Tjia, B.O. 1989. *Zantedeschia*. In: *Handbook of Flowering*, Vol. 6, edited by H.F. Wilkins, 697–702. CRC Press, Boca Raton, FL.

Wei, Z., H. Zhang, Y. Wang, Y. Li, M. Xiong, X. Wang, et al. 2017. Assessing genetic diversity and population differentiation of colored calla lily (*Zantedeschia* hybrid) for an efficient breeding program. *Genes* 8(6): 168. https://doi.org/10.3390/genes8060168. www.ncbi.nlm.nih.gov/pmc/articles/PMC5485532/.

### Key Cultivation Websites

Rowland, J. and A. Bobinger. 2022. *How to Grow a Calla Lily Indoors*. www.wikihow.com/Grow-a-Calla-Lily-Indoors.

Schoellhorn, R. *Warm Climate Production Guidelines for Zantedeschia (Calla Lily) Hybrids*. University of Florida Commercial Floricultural Update. https://hort.ifas.ufl.edu/floriculture/pdfs/crop_production/Callas_ENHFL04-001.pdf.

The Spruce. *How to Grow and Care for Calla Lily*. www.thespruce.com/grow-calla-lily-indoors-1902501.

WikiHow. *How to Get Calla Lilies to Flower*. www.wikihow.life/Get-Calla-Lilies-to-Flower.

WikiHow. *How to Grow Calla Lilies*. www.wikihow.com/Grow-Calla-Lilies.

WikiHow. *How to Grow a Calla Lily Indoors*. www.wikihow.com/Grow-a-Calla-Lily-Indoors.

WikiHow. *How to Keep Cut Calla Lilies Fresh*. www.wikihow.com/Keep-Cut-Calla-Lilies-Fresh.

WikiHow. *How to Winterize Calla Lily Bulbs*. www.wikihow.com/Winterize-Calla-Lily-Bulbs.

# 19 Camellia (*Camellia* Species and Hybrids)

**Scientific Names:** The genus *Camellia* was named for Georg Joseph Kamel (1661–1706), a Moravian Jesuit missionary, pharmacist, and naturalist who described the Philippine flora and fauna.

**Common Names:** The name "camellia" mostly refers to *Camellia* species, but some ornamental species of *Stewartia* (which is also in the tea family, Theaceae) are also referred to as camellia.

## 19.1 INTRODUCTION

The genus *Camellia* has over 200 species of evergreen shrubs and small trees native to eastern and southern Asia. Several dozen species have been recorded in cultivation as ornamentals, but *C. japonica* (common camellia, Japanese camellia) and *C. sasanqua* (sasanqua camellia) are especially widely grown. Sometimes called "Queen of the Winter Flowers" because the plants bloom in mild Southern winters, camellias are among the leading flowering shrubs and trees in warm-climate areas.

FIGURE 19.1 Common camellia (*Camellia japonica*) trees.

*Source:* (a) Photo by Shutterstock/LukeLuke. (b) Photo by Shutterstock/Jun Ahn.

FIGURE 19.2 Sasquana camellia (*Camellia sasquana*).

*Source:* (a) Photo by Antakaran (CC BY SA 3.0). (b) Photo by Peter Stevens (CC BY 2.0).

FIGURE 19.3 Varied camellia flowers.

*Source:* Photo by Shutterstock/Donggraphic.

## 19.2 OVERVIEW

Camellias are evergreen shrubs and small trees grown primarily for their long-lasting flowering period. They bloom especially in the late winter in frost-free areas and in the spring. *Camellia* flowers tend to be large (up to 12 cm in diameter), usually white or pink to red, but also lavender and yellow, and even bicolored or multicolored in some cultivars.

## 19.3 HISTORY

The tea plant (*Camellia sinensis*) has perhaps been grown in China for over 5,000 years, but ornamental camellias were likely first cultivated about a thousand years ago in the gardens of China, later in Japan and Korea. The first importation of the ornamental forms into Europe was in 16th-century Portugal. Camellias were first brought to England in the mid-18th century, where serious interest in Western nations started. Very extensive selection and hybridization have since occurred, and there are now thousands of cultivars (perhaps over 30,000). *Camellia japonica* is the exclusive or main source of most ornamental cultivars, followed in importance by *C. sasanqua*. *Camellia reticulata* and *C. saluenensis* are also significantly grown. Today, camellias are popular ornamentals, with flower shows, festivals, and organizations dedicated to it.

## 19.4 ECONOMIC IMPORTANCE

The genus *Camellia* is valued commercially mainly for Chinese tea. The ornamental flowering shrubs are greatly

DOI: 10.1201/9781003586227-19

FIGURE 19.4 Classic art featuring camellias: (a) A folding screen decorated with a camellia illustration. Painted in 1929 by Japanese artist Hayami Gyoshû (1894–1935). (b) Painting titled *Choosing* by British artist George Frederic Watts (1817–1904), showing a woman deciding her preference between a large camellia flower in her right hand (most cultivars have very limited fragrance) and small (but very attractively scented) violets in her left hand.

***Source:*** **(a) Public domain. (b) National Portrait Gallery, London (public domain).**

FIGURE 19.5 Camellia flower shows. Because the plants are so large, cut flowers are exhibited: (a) A camellia flower show in Spain. (b) A camellia flower show at the Cape Fear Botanical Garden in Fayetteville, North Carolina.

***Source:*** **(a) Photo by Dorna (CC BY SA 3.0). (b) Photo by Selena N. B. H. (CC BY 2.0).**

admired (indeed, considered charismatic) and widely planted in areas that do not experience significant winter freezing, but so are many other plants.

## 19.5 ORNAMENTAL USES

Ornamental camellias are used primarily as residential garden and public landscape shrubs and trees. Some of the smaller cultivars make good container plants, and to a limited extent, they can be grown as houseplants. Camellias are sometimes employed commercially as cut flowers but are not long-lasting (vase life is 5 to 7 days).

## 19.6 OTHER USES

*Camellia sinensis*, the source of beverage tea, is not employed as an ornamental but is one of the world's most important economic plants. Oil from the seeds of oilseed camellia (*C. oleifera*) is known as tea oil, as camellia oil, and sometimes as "eastern olive oil," because it is similar in having high amounts of oleic acid and linoleic acid and low saturated fats. ("Tea tree" is an ambiguous term, and it is often not clear what "tea tree oil" means.) This edible oil has been employed for over a thousand years in China

FIGURE 19.6 Camellia plants grown in containers.

***Source:*** **(a) Photo by Shutterstock/Yevgeniy11. (b) Photo by Shutterstock/MaryAbramkina.**

FIGURE 19.7 Camellia grown as bonsai.

***Source:*** **(a) Photo by Sage Ross (CC BY SA 2.5). (b) Photo by Shutterstock/Frances A. Miller.**

and is now the main cooking oil in the country's southern provinces. More than 90% of global camellia oil production is from China. The flowers of the ornamental camellias are edible (the petals, at least, are tasty), and sometimes the flowers are pickled.

## 19.7 PROBLEMS

The frost susceptibility of camellias is the chief impediment of growing them in temperate climates. Unfortunately, camellias have acquired a reputation for being difficult and unreliable.

## 19.8 CULTIVATION

### 19.8.1 Climate and Site

Most camellias are adapted to climate zones 7 to 10, although a few hybrid cultivars will survive in zone 6.

FIGURE 19.8 A buttercream layer wedding cake decorated with edible camellia flowers.

*Source:* **Photo by Shutterstock/Elisabeth Coelfen.**

FIGURE 19.9 Camellia covered by snow. Spring freezing temperatures resulting in destruction of flower buds are the chief problem encountered by camellias in temperate climates.

*Source:* **Photo by Shutterstock/Ian Grainger.**

At their northernmost area of cold tolerance, planting in north-facing sites is advisable because unusual warm days in the spring may cause the plants to break dormancy too early and lose their flowers to frost. Planting near a building, fence, or hedge can provide protection from wind. In cold climates, camellias in containers can be moved to indoor sites. Most cultivars do well in partial shade, but some are adapted to more open sites. They do not compete well with tree roots so should not be planted near trees.

### 19.8.2 Propagation and/or Purchasing

Camellias can be propagated by seeds, but producing a mature plant is very slow. Moreover, many cultivars are hybrids, and seeds will not reproduce the plant's appearance. Most plants are established from potted plants obtained from nurseries.

### 19.8.3 Planting

Some camellias are best planted relatively shallowly, the root ball that they grew in located at or slightly above soil level (with soil piled up against the sides of the exposed root ball). To accommodate the growing shallow roots, the hole should be dug at the same depth as the root ball, but at least twice (some say three times) the diameter of the root ball. Large cultivars should be at least 2 m apart. In northern areas, spring rather than fall planting is recommended to give the plants time to establish before freeze-up.

### 19.8.4 Soil and Moisture Conditions

Camellias are adapted to organically rich, well-drained, but moderately and consistently moist soil (the plants have limited drought tolerance). Soils should be acidic (pH preferably 5.5 to 6.5), although neutral soils suffice. The plants do well in sandy and loamy soils but prefer the addition of compost or peat moss mixed into the soil and a mulch on top. Limited (infrequent) fertilization, using a fertilizer for acid-loving plants like rhododendrons, or a nitrogen-rich slow-release fertilizer, is recommended.

### 19.8.5 Maintaining Plants

Camellias tend to be slow-growing but long-lived, often requiring little care except for pruning of dead branches.

## 19.9 FASCINATING FACTS, FABLES, FOLKLORE, AND FRIVOLITY

- In Japanese folklore, a malevolent spirit resembling a beautiful woman resides inside camellia trees. If she seduces a victim, tragedy may result. Because of the perceived threat, camellia flowers are often prohibited as gifts to patients.
- Some camellias around the emperor's palace in Japan are over 500 years of age.
- The popularity of flowers often increases when they are featured in artistic creations. The camellia gained fame in 1848 when Alexandre Dumas published a novel titled *Lady of the Camellias*. Camellias also became more widespread after 1853, when Dumas's story was set to music in Guiseppe Verdi's opera *La Traviata*.

FIGURE 19.10 Paintings (public domain) illustrating Alexandre Dumas's story of *The Lady of the Camellias*.

*Source:* **(a) By French Art Deco artist George Barbier (1882–1932). (b) By Czech graphic artist Alphonse Mucha (1860–1939), renowned for his posters of actress Sarah Bernhardt.**

FIGURE 19.11 US stamps celebrating Alabama's state flower, camellia (*Camellia japonica*), and state bird, the yellowhammer (*Emberiza citronella*). *Left*, issued 1982. *Right*, issued 1969.

*Source:* **Public domain.**

- The camellia is the state flower of Alabama. The Camellia Bowl is a college football game played in Montgomery, Alabama.
- White camellias were a symbol of the women's right-to-vote suffrage movement in New Zealand. In 1893, New Zealand became the first country in the world in which all women over the age of 21 years could vote.
- Camellias were also a symbol of the slavery abolitionist movement in Brazil. Contrarily, the Knights of the White Camelia was a southern US racist organization, like the Ku Klux Klan.
- "Gibbing" is the horticultural practice of applying the plant hormone gibberellic acid to young flower buds to induce them to grow into especially large blooms. This is done mostly for flower shows, and especially for ornamentals like camellia known to produce very large flowers.

## KEY PUBLICATIONS


Bartholomew, B. 1986.The Chinese species of *Camellia* in cultivation. *Arnoldia* 46: 2–15.

Chidamian, C. 1959. *Camellias for Everyone*. Doubleday & Co., NY. 191 pp.

Edgar, L.A. 1991. *Camellias, the Complete Guide*. Crowood Press, London. 224 pp.

Ellison, D. 1997. *Camellias, a Photo Dictionary*. Flora Publications International, Brisbane, Australia. 160 pp.

Galle, F.C. 1979. *Azaleas and Camellias*. A.B. Morse, Burlington, IL. 48 pp.

Gonos, A.A. and S. Bracci. 1999. *Camellia Nomenclature*. Southern California Camellia Society. 184 pp.

Green, B. 2015. You say you want a revolution: Reinventing the garden camellia. *Acta Horticulturae* 1085: 405–408.

Griffin, F., Sr. 1964. *Camellia, a Compilation of Authoritative Information on Camellia Culture*. Vogue Press, Columbia, SC. 245 pp.

Lim, T.K. 2014. *Camellia japonica*. In: *Edible Medicinal and Non Medicinal Plants*, edited by T.K. Lim, 764–776. Springer, Dordrecht.

Macoboy, S. 1998. *The Illustrated Encyclopedia of Camellias*. Timber Press, Portland, OR. 304 pp.

Midcap, J.T., H. Clay, B.S. Brewer, and E.H. Moody. 2000. *Camellia Culture for Home Gardeners*. University of Georgia College of Agricultural and Environmental Sciences, Cooperative Extension Service, Bulletin 813. 12 pp.

Napier, J.M. 1953. *Growing Camellias as a Hobby*. Clemson Agricultural Extension Service Circular 377, Clemson, SC. 32 pp.

Parks, C.R. 2002. Basics of camellia breeding. In: *The American Camellia Yearbook 2002*, 30–40. American Camellia Society, Fort Valley, GA.

Sealy, J.R. 1958. *Review of the Genus Camellia*. Royal Horticultural Society, London. 239 pp.

Trehane, J. 1998. *Camellias, the Complete Guide to Their Cultivation and Use*. Timber Press, Portland, OR. 176 pp.

Wang, Y., H. Zhuang, Y. Shen, Y. Wang, and Z. Wang. 2021. The dataset of *Camellia* cultivars names in the world. *Biodiversity Data Journal* 9: e61646. www.ncbi.nlm.nih.gov/pmc/articles/PMC7838149/.

## Key Cultivation Websites

Horticulture Magazine. *Camellia*. https://horticulture.co.uk/camellia/.

Kessler, J.R. 2013. *The Culture of Camellias, the State Flower of Alabama*. Alabama A&M & Auburn Universities Extension ANR-0202. www.aces.edu/wp-content/uploads/2019/02/ANR-0202L.pdf.

The Spruce. *Camellia japonica: How to Grow and Care for Japanese Camellia*. www.thespruce.com/japanese-camellia-growing-guide-5210436.

The Spruce. *How to Grow and Care for Camellia*. www.thespruce.com/camellia-flowers-1316021.

Website of the American Camellia Society. www.americancamellias.com/.

Website of the International Camellia Society. https://internationalcamellia.org/en-us/.

WikiHow. *How to Care for Camellias*. www.wikihow.com/Care-for-Camellias.

WikiHow. *How to Prune Camellia*. www.wikihow.com/Prune-Camellia.

# 20 Canna (*Canna* Species and Hybrids)

**Scientific Names:** The genus name *Canna* is based on *kanna*, a Latin word tracing to Greek for a *cane* or *reed*, a reference to the tall, upright, reed-like stems of the plants. The name *C.* × *generalis* (the × indicating hybridity) is sometimes employed to designate garden cannas, since most are of hybrid origin. The species name *Canna indica* (often employed for both the pure species and also most garden hybrids) is strongly reminiscent of the name *Cannabis indica* employed for marijuana, especially when abbreviated as *C. indica*, sometimes leading to confusion.

**Common Names:** Canna plants are also known as canna lilies (sometimes resulting in confusion with calla lilies), although they are not true lilies (genus *Lilium*). Cannas are known by a variety of other names, depending on country, including *achira*, *African arrowroot*, and *Queensland arrowroot*. The plural of *canna* is either *canna* or *cannas*.

## 20.1 INTRODUCTION

Cannas are spectacular, indeed show-stopping, plants that do not need to be grown in large numbers to be impressive. In temperate regions, their very large leaves provide a distinctly exotic tropical appearance. The very unusually shaped, large, very colorful flowers also contribute to their pronounced attractiveness.

## 20.2 OVERVIEW

The genus *Canna* has ten species, native to the New World tropics ("neotropics"), from Virginia in the United States to the West Indies and northern Argentina. These are herbaceous plants ranging up to 6 m in height. The plants have rhizomes (subterranean stems), which in most of the species are short, tuberous, and starch-rich.

In *C. indica*, the rhizome may be as long as 60 cm. The stems (as tall as 2.5 m) produce up to ten leaves, which are notably large (up to 120 cm long and as wide as 45 cm) and are sometimes said to be reminiscent of banana leaves. Some

FIGURE 20.1 Attractive floral displays of garden canna.

*Source:* **(a) Photo by KimonBerlin (CC BY SA 2.0). (b) Photo by Shutterstock/Suthat Chaithaweesap.**

FIGURE 20.2 Plantings of garden canna emphasizing foliage coloration: (a) Canna cultivar Tropicanna. (b) Cultivars with different leaf colors.

*Source:* **(a) Photo by Krzysztof Ziarnek (CC BY SA 4.0). (b) Photo by Shutterstock/Aqua 214.**

FIGURE 20.3 Photos of garden canna flowers.

*Source:* **(a) Photo by Siegfried Poepperl from Pixabay (public domain). (b) Photo by In hiatus (CC BY 2.0). (c) Photo by Shutterstock/Svetlana Foote.**


DOI: 10.1201/9781003586227-20

varieties have been selected for their variegated purplish, or brown-purple, foliage. The aboveground part of the plant lasts only one season. The inflorescence (flower-bearing branch system) occurs on the top of the main stem and may be as long as 1 m. The flowers are up to 8 cm long and 15 cm in diameter and are rather complex in structure. They may be red, purple, pink, orange, yellow, or white and may have combinations of colors or colored spots and streaks. The throat and edges of flowers may also be differently colored.

## 20.3 HISTORY

For thousands of years, *C. indica* has been cultivated for its edible rhizomes in Central and South America, and there is evidence that larger rhizomes were selected for this purpose. Since the plants were first imported into Europe, canna has been selected primarily for ornamental appearance, and at least a thousand cultivars have been created. Modern breeders have concentrated on producing shorter plants with larger flowers that are more prominently displayed above the leaves, as well as longer-lasting floral displays, improved cold-hardiness, and self-cleaning plants whose spent flowers fall off automatically and do not need to be removed.

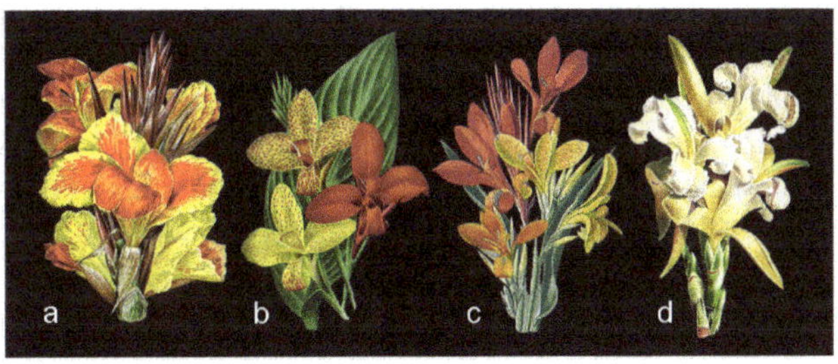

FIGURE 20.4 Illustrations of garden canna from 19th-century European illustrated botanical journals.

*Source:* **(a) *Revue horticole*. 1896. Librairie agricole de la maison rustique, Paris. (b) L' Illustration horticole (Gand, Belgium), vol. 42, 1895. (c) Bois, D. 1891–1896. *Atlas des plantes de jardins et d'appartements*, vol. 2. (d) *Revue horticole*. 1884. Librairie agricole de la maison rustique, Paris.**

FIGURE 20.5 Illustrations of garden cannas from early American garden catalogues.

*Source:* **(a) Cannas. Dreer's Garden calendar, Philadelphia, 1893. (b) Maule's seed catalogue, Philadelphia, 1898. (c) Miss Ella V. Baines, the woman florist, Springfield, Ohio, 1915.**

## 20.4 ECONOMIC IMPORTANCE

Cannas are popular garden plants, contributing to local nursery trades and the general horticultural industry. In most countries, the flowers are too expensive to produce for use as cut flowers (the vase life is as short as 2 days).

## 20.5 ORNAMENTAL USES

As ornamentals, cannas are employed primarily as outdoor plants grown mainly in beds, borders, and containers, often as featured plants, either solitary or in small numbers. While they were exhibited more for their foliage in the past, today their main attraction is their flowering display.

## 20.6 OTHER USES

The rhizomes of *C. indica* have been cultivated for their edible starch, initially in tropical America, but after European colonization of the Americas, all over the world. Like potatoes, the rhizomes are typically boiled. The flour is sometimes employed in medicinal soup. The large leaves have been used to wrap parcels and tamales. The foliage is edible, fed to livestock, and sometimes eaten by poor people. The seeds have had various minor uses, such as for

FIGURE 20.6 Commercial garden canna farms: (a) In Thailand. Corrugated aluminum-colored sheets have been laid between rows of plants to suppress weeds. (b) In North Carolina.

*Source:* **(a) Photo by Shutterstock/Sutipond Somnam. (b) Photo from University of North Carolina Libraries.**

FIGURE 20.7 Garden canna plantings: (a) Beds in City Park, Mannheim, Germany. (b) Border with mixed flower colors. (c) Bed with mixed flower colors. (d) Old car employed as a planter. (e) A large bed with rows of different flower colors.

*Source:* **(a) Photo by Shutterstock/AlinaMD. (b) Photo by Shutterstock/Guentermanaus. (c) Photo by Russell Flood from Pixabay (public domain). (d) Photo by Patrick Emerson (CC BY ND 2.0). (e) Photo by Jack Holloway (CC BY 2.0).**

FIGURE 20.8 Rhizomes of canna. These are employed to grow new plants. They are edible and often used as a source of starch.

*Source:* **(a) Photo by Maja Dumat (CC BY SA 2.0). (b) Photo by Graibeard (CC BY SA 2.0).**

making rosaries (by both Christians and Muslims), in musical rattles, and to make necklaces and bracelets. All these usages are relatively minor today.

## 20.7 PROBLEMS

*Canna indica* has been cultivated all over the tropics and the subtropics and has often escaped and become naturalized. It is an invasive weed in some countries.

## 20.8 CULTIVATION

### 20.8.1 Climate and Site

*Canna indica* is frost-sensitive, too tender to be left outdoors north of climate zone 8. In most temperate regions, the plants are grown as annuals, the perennial rhizomes often overwintered indoors. Cannas grow best in full sun but tolerate some shade. When grown in containers, they can be moved to focal locations as desired. The large leaves can be torn by high winds, and the plants are preferably placed in sheltered locations.

### 20.8.2 Propagation and/or Purchasing

Canna cultivars are mostly propagated by division of the rhizomes (sometimes termed "bulbs" in the horticultural industry). Major suppliers also employ micropropagation (using portions of tissue cultures to grow new plants).

### 20.8.3 Planting

Like most annual flowering ornamentals, cannas can be placed in herbaceous beds and borders. Rhizomes are typically planted with the tops about 3 to 6 cm below the soil surface, horizontally with the "eyes" (buds) pointed up (larger commercially sold rhizomes may have several eyes). The smaller cultivars are suitable for containers, for use on decks, patios, and window boxes. In temperate areas, cannas purchased in pots are relatively expensive and so are often featured in prominent locations or as the central attraction of gardens.

### 20.8.4 Soil and Moisture Conditions

Plants of common canna grow best in well-drained but moderately moist, rich, loamy, or sandy soil, ideally slightly acidic (pH 6.0 to 6.5). Soggy soil is tolerated relatively well (at least by some cultivars), and some gardeners put potted cannas in water gardens, the bottom of the pot only slightly contacting the water. Golden canna (*C. flaccida*) is decidedly tolerant of waterlogged conditions, and cultivars are available for planting in water gardens.

### 20.8.5 Maintaining Plants

Once established in containers or the soil, cannas require little attention other than keeping the soil moist and deadheading spent flowers. The rhizomes can be excavated in the autumn and overwintered in a dormant condition.

## 20.9 FASCINATING FACTS, FABLES, FOLKLORE, AND FRIVOLITY

- Canna seeds are small, globular, black, and very hard, resembling oversized BBs and buckshot pellets, giving rise to the common name Indian shot.

FIGURE 20.9 Seeds of garden canna: (a) Mature seed capsule. (b) Split-open seed capsule showing seeds.

*Source:* **Photos by Berthold Werner (CC BY SA 3.0).**

The seeds are actually used as shot in children's weapons. In 1857, a major uprising occurred in India ("The Indian Rebellion") against the rule of the British East India Company, the agent of the British Crown. The British are said to have used canna seeds in their guns when they ran out of bullets. Similar claims have been made that the seeds were employed in flintlock muskets and blunderbusses, but definitive proof is lacking. The seeds have actually been shot experimentally from shotguns, with considerable impact.

- Stemming from the use of canna seeds in rosaries, devout Christians have nicknamed canna as the rosary bead and often planted it in religious gardens beside statues of the Virgin Mary.

## KEY PUBLICATIONS


Al-Snafi, A.E. 2015. Bioactive components and pharmacological effects of Canna indica – an overview. *International Journal of Pharmacology and Toxicology* 5: 71–75.

Cooke, I. 2001. *The Gardener's Guide to Growing Canna*. Timber Press, Portland. 160 pp.

Darsini, I.P., S. Shamshad, and M. Paul. 2015. Canna indica (L.): A plant with potential healing powers: A review. *International Journal of Pharma and Bio Sciences* 6(2): B1–B8.

Donahue, J.W. 1965. History, breeding and cultivation of the canna. *American Horticultural Magazine* 44: 84–91.

Gade, D.W. 1966. Achira, the edible canna, its cultivation and use in the Peruvian Andes. *Economic Botany* 20: 409–413.

Gilman, E.F., C.J. Della Torre III, and L.A. Gettys. 2015. *Golden Canna:* Canna flaccida. University of Florida Extension FPS102. 4 pp.

Khoshoo, T.N. and I. Guha. 1976. Origin and evolution of cultivated cannas. In: *Glimpses in Plant Research*, edited by P.K.K. Nair, 1–81. Vikas Publishing House, New Delhi.

Lim, T.K. 2016. Canna indica. In: *Edible Medicinal and Non-Medicinal Plants*, edited by T.K. Lim, 69–81. Springer, Dordrecht.

Maas-van der Kamer, H. and P.J.M. Maas. 2008. The Cannaceae of the world. *Blumea* 53: 247–318.

Pasiecznik, N. 2008. Canna indica *(Canna Lily)*. CABI Compendium. https://doi.org/10.1079/cabicompendium.14575.

Segeren, W. and P.J.M. Maas. 1971. The genus *Canna* in northern South America. *Acta Botanica Neerlandica* 20: 663–680.

Srivastava, R. and P. Punetha. 2021. Canna. In: *Floriculture and Ornamental Plants*, edited by S.K. Datta and Y.C. Gupta, 1–22. Springer, Singapore.

Tanaka, N. 2001. Taxonomic revision of the family Cannaceae in the New World and Asia. *Makinoa, Series* 2(1): 34–43.

Tanaka, N. 2004. The utilization of edible *Canna* plants in southeastern Asia and southern China. *Economic Botany* 58: 112–114.

Tanaka, N., N. Inouch, and T. Koyama. 2006. Edible canna and its starch: An under-exploited starch-producing plant resource. *Food & Food Ingredients Journal of Japan* 211: 319–325.


## Key Cultivation Websites


Almanac. *Canna Lily Growing Guide: How to Plant Canna Bulbs*. www.almanac.com/plant/cannas.

Australia Broadcasting Corporation. *How to Lift, Dry, Store and Replant Canna Bulbs (Rhizomes)*. www.abc.net.au/gardening/how-to/dividing-cannas/9427988.

*Cannas: Planting, Growing & Caring for Magnificent Canna Plants*. https://hi-spring.com/post.php?id=24.

Garden Beast. *Canna Guide: How to Grow & Care for "Canna Lily"*. https://gardenbeast.com/canna-guide/.

Have Garden. *Winterize Canna Lilies? (Here's When & How to Do It!)* https://havegarden.com/winterize-canna-lilies/.

Iowa State University Extension. *Growing Cannas in the Home Garden*. https://hortnews.extension.iastate.edu/2015/01-16/cannas.html.

The Spruce. *How to Grow and Care for Canna Lily (Canna spp.)*. www.thespruce.com/growing-canna-lilies-in-containers-847913.

WikiHow. 2022. *How to Grow Cannas*. www.wikihow.com/Grow-Cannas.

# 21 Carnation (*Dianthus* Species and Hybrids)

**Scientific Names:** The genus name *Dianthus* is based on the Greek *dios*, "divine," referring to the god Zeus, and *anthos*, "flower," an allusion to beauty and fragrance so heavenly it is fit for the gods.

**Common Names:** The word "carnation" is of obscure origin, often applied to many of the species of *Dianthus*, but most particularly to *D. caryophyllus* and its horticultural hybrids. The word *caryophyllus* in *D. caryophyllus* (the most cultivated species) is based on the Greek *karyon*, a nut, and *phyllon*, a leaf, literally "nut-leaved," an allusion to an old name of the Indian clove tree (*Eugenia caryophyllata*) applied because the flower is often strongly scented of cloves.

## 21.1 INTRODUCTION

The carnation and its relatives are simply gorgeous. Known as the "divine flower," the carnation has been admired for over 2,000 years and has become a staple of the flower trade.

## 21.2 OVERVIEW

The genus *Dianthus* is estimated to have over 300 species. Most are native to Eurasia, many in the Mediterranean region and the mountains of southern Europe. A few are in the African mountains, and *D. repens* of arctic North America is the only one native to the Americas. The species often hybridize readily in nature. Although some are introduced in various countries, on the whole their invasiveness is limited, and the cultivated species are not notably weedy. Wild *Dianthus* are usually found in dry, sunny locations, usually on soil influenced by limestone. Most grow in rocky sites or open meadows, but a few are found in open woodlands. Most species of *Dianthus* are perennial, tufted evergreen herbs, many with somewhat woody bases, and a few are shrubs. Height ranges from less than 15 cm to more than 1 m. The leaves are usually long and slim, like grass foliage. Wild species have five petals, while numerous cultivars have doubled flowers with dozens of petals. The flowers of cultivars are available in solid colors, bicolors, "picotees" (solid colors with paler or white-edge petals), streaks, flecks, multicolored, "eyed" (with dark centers), and variegated (sectors of the petals with different hues). Carnations are also very popular flowers to dye (a practice that is offensive to right-thinking people; "just because you can doesn't mean you should"). The corolla of carnations often exhibits a very attractive ruffled crowding of petals that is complemented by scalloping of the margins. The most frequently cultivated species are *D. caryophyllus*, *D. barbatus*, *D. plumarius*, and *D. chinensis*. Information on these is presented in the following.

FIGURE 21.1 An 1895 nursery catalogue showing a broad range of ornamental carnations.

*Source:* **Public domain.**

### 21.2.1 *DIANTHUS CARYOPHYLLUS*

*Dianthus caryophyllus* is most commonly called "carnation" but is also known as clove pink, cottage pink, gillyflower (a name of obscure origin; it has been suggested "gilly" is based on "July"), and picotee (usually referring to forms with petal edges differently colored from most of the blade). This native of southern Europe and the Mediterranean islands is the most important ancestor of modern cultivars of the genus. It was collected in the Pyrenees in the 15th century, becoming popular in Europe



DOI: 10.1201/9781003586227-21

**FIGURE 21.2** Old paintings (public domain) of carnation (*Dianthus caryophyllus*) cultivars (mostly double-flowered) popular more than a century ago. Note that (a) has single-flowered forms.

*Source:* **(a) *Revue horticole* (Paris), vol. 94, 1922. (b) *Gartenflora* (Berlin), vol. 39, 1890. (c) *L'Illustration horticole* (Gand, Belgium), vol. 31, 1884. (d) B. K. Bliss and Sons, New York, illustrated handbook for the farm and garden for 1883. (e) Henry A. Dreer, Philadelphia, PA, 1910 wholesale catalogue. (f) Dingee & Conard, West Grove, PA, 1898 catalogue.**

by the 16th century. Because the species has been widely introduced, its original natural distribution is uncertain. It grows wild in some Mediterranean locations, notably Greece, Italy, Sicily, and Sardinia. Carnation flowers are easily recognized by their characteristic fused sepals that form a tube. Both single- and double-flowered forms (with up to 80 petals in a flower) and a very wide variety of color forms have been bred. "Standards" have a single large flower per stem (up to 10 cm in diameter), whereas "sprays" have many small flowers.

### 21.2.2 *Dianthus barbatus*

Known as sweet William ("sweet" for the fragrance, "William" based on a mispronunciation of the French *oeillet*), *D. barbatus* is probably the most common species of *Dianthus* encountered in gardens. It is the major species of the genus used as a summer-bedding annual. It originated in Europe and has been grown in Great Britain since the 16th century. Unlike most *Dianthus* species, sweet Williams have dense heads of many flowers.

### 21.2.3 *Dianthus chinensis*

*Dianthus chinensis* is known as China pink (it grows wild in the hills of eastern Asia and the coast of Manchuria, and cultivated forms originated in China), Indian pink (it occurs

**FIGURE 21.3** (a, b, c) Photos of carnations (*Dianthus caryophyllus*) with doubled flowers. (d) Side views of carnation flowers showing their prominent fused sepal tubes. Note the epicalyx, a group of sepals clasping the base of the main calyx.

*Source:* **(a) Photo by DaoUiDai HaKungMi 88 (CC BY SA 4.0). (b) Photo by Forest and Kim Starr (CC BY 2.0). (c) Photo by Rusty Clark (CC BY 2.0). (d) Photo of left flower by Max Pixel (public domain); right by Mike Beales (CC BY ND 2.0).**

**FIGURE 21.4** Old paintings (public domain) of *Dianthus barbatus* (sweet William) cultivars popular more than a century ago.

*Source:* **(a) Step, E. and Bois, D. 1896–1897. *Favourite Flowers of Garden and Greenhouse*, vol. 1. Frederick Warne, New York. (b) Curtis, W. (ed.). 1792–1793. *The Botanical Magazine (London)*, vol. 5–6. (c) Orbigny, C.d', F. Arago, and J.V. Audouin. 1849. *Dictionnaire Universel d'histoire Naturelle*, vol. 3. Paris.**

**FIGURE 21.5** Photos of *Dianthus barbatus* (sweet William) cultivars.

*Source:* **(a) Photo by Zirguezi (public domain). (b) Photo by Céréales Killer (CC BY SA 3.0).**

**FIGURE 21.6** Old paintings (public domain) of *Dianthus chinensis* (China pink): (a) Single-flowered forms. (b) Double-flowered forms. (c) Double-flowered form.

*Source:* **(a)** ***Gartenflora*** **(Berlin), vol. 8, 1859. (b)** ***Flore des serres et des jardins de l'Europe*** **(Gand, Belgium), vol. 13, 1845. (c) Source: Blanco, F.M. 1880–1883.** ***Flora de Filipinas.*** **Establecimiento tipografico de Plana, Manila.**

FIGURE 21.7 Photos of *Dianthus chinensis* (China pink). (a) Cultivar Telstar Picotee. (b) Cultivar Telstar Crimson. (c) Unidentified cultivar.

*Source:* **(a) Photo by David J. Stang (CC BY SA 4.0). (b) Photo by David J. Stang (CC BY SA 4.0). (c) Photo by Raul 654 (C BY SA 3.0).**

in northern India), and rainbow pink (for the variegated colors of some forms). It is too frost-tender to grow as a perennial in cold climates and is commonly grown as an annual. Most cultivars are of hybrid origin, with *D. chinensis* as one of the parents.

### 21.2.4 *Dianthus plumarius*

This "species" includes a group of hybrids usually called pink, garden pink, and common pink. The cultivars are based on *D. plumarius* as the main parent and have been grown since the 16th century, having originated in southern Russia. The wild species is native to southern Europe east to the Caucasus.

FIGURE 21.8 *Dianthus plumarius.*

*Source:* **(a) Photo by Krzysztof Ziarnek, Kenraiz (CC BY SA 4.0). (b) Step, E., and D. Bois. 1896–1897.** ***Favourite Flowers of Garden and Greenhouse,*** **vol. 1. Frederick Warne, New York.**

## 21.3 HORTICULTURAL CLASSES

In the horticultural trade, a number of categories are recognized, which may include more than one species, as well as hybrids. These are presented in the following subsections.

### 21.3.1 Pinks

The name "pink" is often applied to garden species and is ambiguous. In practice, "pinks" are low-growing, clump-forming garden *Dianthus* plants, often with dainty flowers, sometimes with attractive foliage, and often pleasantly scented. Pinks come in many colors, but rarely blue or yellow. The word "pinks" without qualification may refer to a class of domesticated *Dianthus* species, particularly those mentioned next, except for *D. caryophyllus*. The name *pink* by itself most often refers to *D. plumarius* (also known as garden pink and common pink). *Dianthus caryophyllus*, usually called carnation, is also known as clove pink and cottage pink, and *D. chinensis* is known as China pink, Indian pink, and rainbow pink. Many interpret the name "pink" as reflecting this common flower color in the genus (conversely, the color pink has been said to be based on the floral hue). More commonly, the word has been claimed to be based on the frilly margins of the petals of many of the species, seemingly trimmed with pinking shears (the Old English word *pynken*, on which "pinking shears" is based, refers to "notching" or "piercing").

### 21.3.2 Fancy Carnations

The vague phrase "fancy carnations" refers to forms with particularly interesting multicoloration within flowers, and perhaps also attractive shape and/or size.

### 21.3.3 Annual Cultivars

So-called "annual" cultivars, frequently grown from seed, include *D. barbatus* and *D. chinensis*, both biennials or short-lived perennials (which mostly lack sufficient hardiness to overwinter in cold climates). Annual cultivars flower prolifically, and despite the smaller flowers than those of border carnations mentioned next, they are very popular.

### 21.3.4 Border Carnations

The expression "border carnations" refers to garden forms (basically of *D. caryophyllus*) that are generally tall, with large attractive flowers, grown from cuttings and intended (at least in warm-temperate climates) to be grown as perennials. Many of these cultivars require staking, and they have become less-popular.

### 21.3.5 Perpetual Flowering Carnations

"Perpetual flowering carnations" (sometimes called tree carnations, American carnations, and greenhouse carnations) are forms that have long flowering periods and are designed for the cut flower trade. Derived largely from *D. caryophyllus* and *D. chinensis*, they are grown mainly in greenhouses. They usually have flowers about 7 cm across, sometimes up to 10 cm wide.

### 21.3.6 Cut Flower Spray Carnations

Cut flower "spray carnations" have smaller flowers than most perpetual flowering carnations, with many flowers on a stem. Although mainly grown by commercial flower producers, perpetual flowering carnations are popularly grown by dedicated amateurs to exhibit in competitive flower exhibitions.

### 21.3.7 Mini Carnations

"Mini carnations" are small-flowered forms which are popular because they make superb bouquets.

### 21.3.8 Rockery Species

"Rockery species" are dwarfs of alpine origin that make ideal rock garden plants. The main species are *D. alpinus* (from the Austrian Alps, it was introduced to Britain about 1759) and *D. gratianopolitanus* (*D. caesius*), from southwestern France (introduced to Great Britain in 1792).

FIGURE 21.9 "Mini carnations."

*Source:* **Photos by Kaz Andrew (CC BY SA 2.0).**

FIGURE 21.10 Species of *Dianthus* employed in rock gardens: (a) *Dianthus alpinus*. (b) *Dianthus gratianopolitanus*.

*Source:* **(a) *Gartenflora* (Berlin), vol. 4, 1855. (b) Photo by Babij (CC BY SA 2.0).**

## 21.4 HISTORY

The classical Greeks and Romans are believed to have cultivated *Dianthus* species, although most selections and hybrids have been generated within the last several centuries. There are over 30,000 cultivar names for *Dianthus* in *The International Dianthus Register* (including supplements), which is maintained by the Royal Horticultural Society in London.

William Sim, A Scots émigré to the United States, developed large-flowered, constantly flowering ("perpetual") cultivars, and more or less scentless carnations ("Sim forms"), which became the basis of the international cut flower carnation industry. These are the kinds most often found in bouquets, wreaths, and corsages and marketed by florists and supermarkets.

## 21.5 ECONOMIC IMPORTANCE

*Dianthus* species are widely grown and marketed as cut flowers, potted flowering plants, and bedding plants. Colombia has become the world's largest producer of cut flower carnations (the capital, Bogota, has been said to be the capital of carnations). Israel, Japan, Kenya, Spain, and the American states of California and Colorado also produce significant crops.

Many species of *Dianthus* are quite fragrant, some with a spicy odor. The petals of carnations contain small amounts of essential oil, which is extracted for use in perfumes (5 kg of flowers are required to produce 1 g of oil). Some well-known perfumes which contain carnation oil include Elizabeth Arden, Gucci No. 1, Lauren by Ralph Lauren, Red Door by Elizabeth Arden, and Yves Saint Laurcnt Opium.

## 21.6 ORNAMENTAL USES

Because of their attractiveness, cheapness, and long vase life, carnations are a leading cut flower and among the most popular inclusion in floral arrangements. The carnation is often used as one of the cut flowers employed to adorn parade floats. Transgenic carnations are available with an antisense gene-lowering ethylene production

**FIGURE 21.11** Commercial production and sale of carnations and sweet Williams: (a) A field of sweet Williams in Provence, France. (b) A field of carnations in France. (c) A greenhouse of carnations in Japan. (d) Bouquets of red carnations on sale by a florist.

*Source:* **(a) Photo by Shutterstock/Marina VN. (b) Photo by Caroline Ingram (CC BY ND 2.0). (c) Photo by Shutterstock/Sky Stock. (d) Photo by Nia Ran01 (CC BY SA 4.0).**

**FIGURE 21.12** Landscape uses of *Dianthus*: (a) A large bed of *Dianthus chinensis* 'Raspberry Parfait'. (b) Carnations in a planter.

*Source:* **(a) Photo by David J. Stang (CC BY SA 4.0). (b) Photo by Shutterstock/Martina Osmy.**

(which shortens flower life), extending flower longevity well beyond the normal week or so.

Most garden *Dianthus* species are also excellent subjects for gardens and borders and are not difficult to cultivate. "Pot carnations" are typically dwarf forms with numerous small flowers. Their soil should be kept moist, but not overwatered. Although more commonly grown outdoors, they can be kept indoors. Carnations are not usually grown as houseplants, although they can be.

## 21.7 OTHER USES

A boutonnière (based on the French for *buttonhole*) is a floral decoration, usually a showy flower, worn on a coat or jacket by men for formal occasions, especially proms and weddings. Although occasionally worn by women (corsages are preferable), the boutonnière is the only masculine concession to wearing flowers. Traditionally, the stem of the flower is pushed through the lapel buttonhole on the left side, to be worn over the heart. Today, almost no jacket is made that has a suitable buttonhole, and boutonnières are usually simply pinned to the jacket. A carnation is the most traditional species, white for very formal occasions,

**FIGURE 21.13** Harvested (cut flower) carnations: (a) A basket full of flowers. (b) A girl with long-stemmed flowers. (c) A flower wheel decorated with cut carnations shown at the Melbourne International Flower and Garden Show in 2010. (d) A floral "Viking float" decorated with carnations.

*Source:* **(a) Photo by Nanimo (CC BY 2.0). (b) Photo by Shutterstock/YuliaPodlesnova. (c) Photo by Rexness (CC BY SA 2.0). (d) Photo by Man Vyi (public domain).**

**FIGURE 21.14** A carnation boutonnière as worn during the early 20th century. Note that these are always placed on the heart side. This public domain figure shows an 1895 poster of Charles Armand Ménard, a French comic singer and actor.

or alternatively red. Carnations with insignificant fragrance are often used for boutonnières for men. The popularity of boutonnières has waned in the 21st century.

Carnation flowers are considered edible. Reflecting its alternative name "clove pink," the flowers and foliage can have a clove-like aroma. However, only the petals should be consumed (other floral parts are bitter), and before adding these to a food preparation, it is wise to test their taste. Carnations from florists and nurseries may have traces of toxic chemicals and should not be eaten. Flowers should also be examined for content of insects. The petals can be candied ("crystallized") or pickled, but this defeats the goal of featuring their unique taste properties. They are employed as an edible garnish in salads and to flavor butters, lemonade, vinegars, conserves, and syrups. The classical Romans spiced wine with wild carnation flowers, and carnations are one of the ingredients employed to flavor the French liqueur Green Chartreuse since the 17th century.

## 21.8 PROBLEMS

In recent decades, cut flower carnations in North America acquired an undesirable reputation as a budget (i.e., cheap), tacky flower that reflects badly on both gifters and giftees ("a flower for mothers-in-law"), reaching a social status not much higher than dandelions. Some florists have simply refused to supply carnations for weddings and other events. The problem has been simply that carnations became so cheap and ubiquitous that they no longer reflected value. For corsages, carnations came to be viewed as much less desirable than orchids, gardenias, or roses. Today, carnations are retrieving their popularity. Professional florists who prepare displays strongly recommend that to avoid the cheapness factor, carnation blooms need to be massed very closely together, hiding the stems. Another technique is to surround the lower stems with other ornamentals bearing small or delicate leaves and/or flowers. It's also best to use recently bred, spectacular carnation cultivars.

Carnations were employed in traditional folk medicine in Europe for coronary, gastrointestinal, and nervous disorders, but this is not a guarantee that consumption is safe, since many rather toxic plants have been used in herbal treatments. Workers who handle large amounts of *Dianthus* species sometimes develop occupational allergic contact dermatitis and/or pulmonary conditions. Veterinary websites list *Dianthus* species as toxic to dogs, cats, and horses. Generally, the genus is not considered to be significantly dangerous.

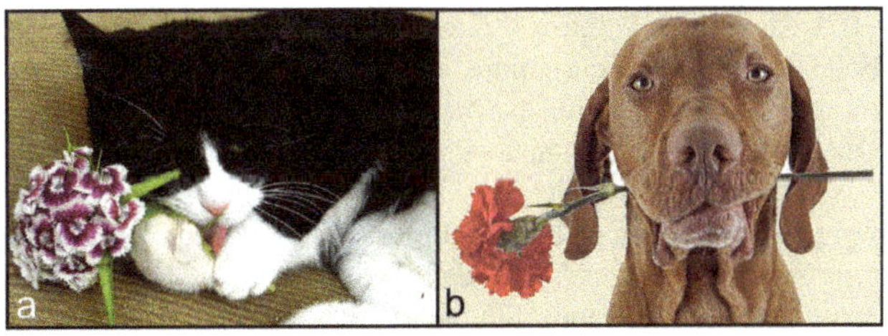


FIGURE 21.15 Pets exposed to potential toxicity of *Dianthus*: (a) Cat chewing sweet William (reportedly significantly toxic to felines). (b) A vizsla dog holding a carnation.

***Source:* (a) Photo by Tommy Hemmert Olesen (CC BY ND 2.0). (b) Photo by Shutterstock/Barna Tanko.**

## 21.9 CULTIVATION

### 21.9.1 Climate and Site

*Dianthus* species are available that can grow in climate zones between 3 and 10, or sometimes 11, but cold tolerance varies greatly among the thousands of cultivars. Plants grown as annuals of course need not consider winter conditions. Biennial or "perpetual" (perennial) forms vary greatly in adaptation to climate zones. The species mentioned in this chapter have been given the following climate zone tolerances to grow as biennials or perennials: *D. alpinus*, 4–8; *D. barbatus*, 3–9; *D. caryophyllus*, 7–10; *D. chinensis*, 7–10; *D. gratianopolitanus*, 4–8; D. *plumarius*, 4–9. Full sun and wind-protected sites are recommended. Most ornamental *Dianthus* species are short and need to be situated where they will not be obscured by taller objects. Growing plants in containers provides site flexibility.

### 21.9.2 Propagation and/or Purchasing

The most important carnation cultivars are hybrids that are perpetuated vegetatively, and planting stock is usually available in nurseries. Many *Dianthus* species grow well from seed, but plants may not flower until the second year. If a large bed is desired, seeding is the economical choice, but it is advisable in short-season climates, where plants are grown as annuals to start seeds indoors. Carnations grow easily from cuttings from the suckers formed around the base of the stem (or, less preferably, from the side shoots of the flowering stem or the main shoots, before they have developed flower buds). Well-developed plants can also be divided. Plants allowed to go to seed may result in self-seeding at a site, and if hybrids are cultivated, they likely will not reproduce true to type.

### 21.9.3 Planting

Planting distance depends on size and spread of particular cultivars, and pinching the growing tips to encourage branching can considerably increase the area covered. A general rule of thumb is that plants should be about 30 cm apart.

### 21.9.4 Soil and Moisture Conditions

The major *Dianthus* species discussed in this chapter do best in well-drained, neutral to slightly alkaline loam soil. Poor drainage is said to be the chief cause of death of *Dianthus*. Nutrient-rich soils with humus are suitable. Clays and soggy soils should be avoided. Nitrogen-rich fertilizers tend to favor vegetative growth at the expense of flowers.

### 21.9.5 Maintaining Plants

Deadheading encourages continued flowering (alternatively, flowering stems may simply be brought indoors as a cut flower). After several years, dividing and replanting may be necessary to maintain vigor. Gardeners sometimes have difficulty distinguishing the linear leaves of *Dianthus* species from those of weedy grasses, and removal of the latter needs to be done with care. A variety of pests (including rabbits) and diseases affect garden *Dianthus* species, but no more than other common ornamentals.

A bouquet of carnations may have closed buds, which will open over the next few days (florists acquire bouquets at the unopened bud stage to maximize shelf life). Stems should be trimmed at a 45° angle using a sharp knife (as explained in Chapter 1, to prevent the bottoms of the cut stems from sitting flat on the bottom of the vase, not for better water absorption) just above a node. Also, excess leaves should be removed before placement in a clean vase of water. Cut flower food should be added. Every 2 or 3 days, stems should be recut, water changed, and more food added. The vase should be kept away from drafts, direct sun, excessive heat, and ethylene (from wilting plants and ripening fruits and vegetables, which can prematurely age the flowers). Cut carnations have been known to last up to 3 weeks.

## 21.10 FASCINATING FACTS, FABLES, FOLKLORE, AND FRIVOLITY

- Sweet Williams and carnations have been familiar, respected, popular ornamental flowers in Europe for centuries, as reflected by their frequent appearance in old European paintings.

FIGURE 21.16 *Dianthus* in paintings by European artists: (a) *The Carnations* by Polish American artist Karl Witkowski (1860–1910). (b) *A Street Urchin Selling Flowers* by British-born American artist John George Brown (1831–1913). (c) *The beautiful flower girl* by Austrian artist Hans Hamza (1879–1945).

***Source:* (a) Photo from Bonhams, New York (Birmingham Museum of Art (public domain)). (b) Photo by Sean Pathasema (public domain). (c) Public domain.**

- Carnations have acquired exceptional symbolic values. They are the national flowers of Monaco, Slovenia, and Spain; the provincial flowers of the autonomous community of the Balearic Islands; and the state flower of Ohio. Carnations have represented people's movements, such as labor initiatives and socialist causes, and in Portugal, red carnations came to represent a coup d'état in 1974. In Christianity, carnations became associated with holiness, especially Mary, in her role as mother of Christ, a relationship which subsequently led to carnations becoming symbolic of feminine purity and Mother's Day. Carnations are often worn on other holidays, such as Parents Day (in Korea), St. Patrick's Day, and May Day, and are employed for sentimental occasions, such as weddings and

FIGURE 21.17 Paintings (public domain) illustrating the symbolic association of carnations with feminine purity and aristocracy: (a) Painting by French artist Jean Ranc (1674–1735) showing Queen Bárbara de Braganza of Spain holding carnations. Numerous such portraits of royal and noble women holding carnations were painted in past centuries. (b) A woman dressed in a costume representing carnations. (c) *Madonna of the Carnation* by Italian artist Leonardo da Vinci (1452–1519). (d) Portrait of Maria Luisa of Parma by German artist Anton Raphael Mengs (1728–1779).

***Source:* (a) Public domain. (b) Grandville, J. J. 1867. *Les fleurs animées*, vol. 1. Grandville (1803–1847), a French artist and caricaturist, produced a huge number of intriguing and often controversial illustrations. (c) Photo from Bavarian State Painting Collection. (d) Public domain.**

funerals. Carnations are a common funeral flower (symbolically expressing sympathy) in some countries and, accordingly, are sometimes considered unlucky in France.

- Crayons produced by the Crayola Company come in up to 200 variations. One of these, "Carnation Pink," was introduced in 1949 and is still one of the inclusions in the company's popular 16- and 24-crayon packs. New introductions generally are accompanied by retirements of older hues.
- Carnation Evaporated Milk traces to a company founded in 1899 in the state of Washington, the Pacific Coast Condensed Milk Company, which provided a sterile dairy product before refrigeration was commonly available. In 1901, the company founder noticed a display of cigars bearing the name *carnation* and transferred the word to his milk product, eventually adding carnation flowers to his cans' labels. Carnations are edible, as noted in the text, but the only association between them and milk is this curious history of clever advertising.
- In 2016, 17-year-old Hayden Godfrey of Sky View High School in Utah purchased 900 carnations for $450.00 and provided one for every girl in the school on Valentine's Day.
- The blue pigment delphinidin, which determines blue and violet flower coloration in numerous species, is absent in *Dianthus*, and breeding a truly blue or purple carnation has not been possible by conventional means. In the early 1990s, the company Florigen employed genetic engineering to transfer genes from petunia and snapdragon flowers to produce blue-mauve carnations, first commercialized as the cultivar Moondust.

FIGURE 21.18 Scenes commemorating the 1974 "Carnation Revolution" in Portugal: (a) Street art in Lisbon. Note the red carnation in the soldier's gun barrel. (b) Massed carnations displayed at the anniversary celebration in 2013 at the site of the revolution in Lisbon, Portugal.

*Source:* **(a) Photo by Jeanne Menjoulet (CC BY 2.0). (b) Photo by BKP (CC BY SA 3.0).**

FIGURE 21.19 The blue-mauve carnation cultivar Moondust, produced by genetic engineering.

*Source:* **Photo by Pagemoral (CC BY SA 3.0).**

## KEY PUBLICATIONS

Note: Cultivated species of *Dianthus* are extremely popular in Great Britain, where dozens of books have been published on the subject. Only a selection of books is listed here.

Allwood, M.C. 1954. *Carnations, Pinks and all Dianthus*, 4th edition. Allwood Bros., Hayward's Heath, Sussex, UK. 353 pp.

Bailey, S. 1982. *Carnations: Perpetual-flowering Carnations, Borders, and Pinks*. Blandford Press, Poole, Dorset, UK. 215 pp.

Bird, R. 1994. *Border Pinks*. Timber Press, Portland, OR. 174 pp.

Ebrahimzadeh, A., S. Jiménez, J.A. Teixeira da Silva, S. Satoh, and M.T. Lao. 2008. Postharvest physiology of cut carnation flowers. *Fresh Produce* 2(2): 56–71.

Fassou, G., N. Korotkova, A. Nersesyan, M.A. Koch, P. Dimopoulos, and T. Borsch. 2022. Taxonomy of *Dianthus* (Caryophyllaceae) – overall phylogenetic relationships and assessment of species diversity based on a first comprehensive checklist of the genus. *PhytoKeys* 196: 91–214.

Galbally, J. and E. Galbally. 1997. *Carnations and Pinks for Garden and Greenhouse: Their True History and Complete Cultivation*. Timber Press, Portland, OR. 310 pp.

Great Britain Ministry of Agriculture, Fisheries and Food. 1967. *A Manual of Carnation Production*, 3rd edition. H.M.S.O., London. 158 pp.

Holley, W.D. and R. Baker. 1963. *Carnation Production: Including the History, Breeding, Culture and Marketing of Carnations*. W.C. Brown, Dubuque, IA. 142 pp.

Holley, W.D. and R. Baker. 1991. *Carnation Production II*. Kendall Hunt Pub. Co., Dubuque, IA. 156 pp.

Hughes, S. 1993. *Carnations and Pinks*. Crowood Press, Marlborough, UK. 221 pp.

Iwazaki, Y., Y. Kosugi, K. Waki, T. Yoshioka, and S. Satoh. 2004. Generation and ethylene production of transgenic carnations harboring ACC synthase cDNA in sense or antisense orientation. *Journal of Applied Horticulture* 6: 67–71.

McGeorge, P. and K. Hammett. 2002. *Carnations and Pinks*. Firefly Books, Willowdale, ON. 96 pp.

Onozaki, T. 2018. *Dianthus*. In: *Ornamental Crops. Handbook of Plant Breeding*, vol. 11, edited by J. van Huylenbroeck, 349–381. Springer, Cham.

Sánchez-Fernández, C., M.L. González-Gutiérrez, M.I. Esteban-López, A. Martínez, and M. Lombardero. 2004. Occupational asthma caused by carnation (*Dianthus caryophyllus*) with simultaneous IgE-mediated sensitization to *Tetranychus urticae*. *Allergy* 59: 114–115.

Segers, A. 1987. The development of interspecific carnation hybrids. *Acta Horticulturae* 216: 373–375.

Shigeta, D.T. 1963. *Carnation Culture in Hawaii*. University of Hawaii Cooperative Extension Service Circular 395. 25 pp. https://scholarspace.manoa.hawaii.edu/server/api/core/bitstreams/4254bb79-93ef-4f5d-a055-13d9669facf2/content.

Sitch, P. 1975. *Carnations for Garden and Greenhouse*. John Gifford, London, UK. 136 pp.

Smith, F.C. 1994. *Growing Carnations and Pinks*. Ward Lock, New York. 128 pp.

Thakur, A., B.S. Dilta, D.K. Mehta, B.P. Sharma, and R.K. Gupta. 2018. Effect of plant spacing and GA3 on growth and flowering of sweet William (*Dianthus barbatus* L.). *International Journal of Farm Sciences* 8: 122–125.

Way, T. 2016. *Carnation*. Reaktion Books, London, UK. 221 pp.

Zucker, A., E. Shklarman, G. Scovel, H. Ben-Meir, M. Ovadis, and I. Neta-Sharir. 2001. Genetic engineering of agronomic traits in carnation. *Acta Horticulturae* 560: 91–94.

## Key Cultivation Websites

GardenersWorldCom. *How to Grow Pinks (Dianthus)*. www.gardenersworld.com/how-to/grow-plants/dianthus-grow-guide/.

GardenersWorldCom. *How to Take Carnation Cuttings*. www.gardenersworld.com/how-to/grow-plants/how-to-take-carnation-cuttings/.

The Spruce. *How to Grow Perennial Dianthus*. www.thespruce.com/perennial-dianthus-flower-1316045.

WikiHow. *How to Grow Sweet William Plants*. www.wikihow.com/Grow-Sweet-William-Plants.

WikiHow. *How to Help Carnations Make a Comeback*. www.wikihow.com/Help-Carnations-Make-a-Comeback.

WikiHow. *How to Pinch Carnations*. www.wikihow.com/Pinch-Carnations.

WikiHow. *How to Plant Carnations*. www.wikihow.com/Plant-Carnations.

WikiHow. *How to Water Carnations*. www.wikihow.life/Water-Carnations.

# 22 Catnip (*Nepeta cataria* and Relatives)

**Scientific Names:** The genus name *Nepeta* is said to have been derived from the town Nepete (called Nepi today), north of Rome in Italy, where catnip was once cultivated.

**Common Names:** "Nip" means a small quantity of liquor, so "catnip" is descriptive of material that intoxicates cats. Catnip is also known as *catmint*, *catnep*, *catrup*, *catwort*, *English catnip*, *field balm*, *nep*, and *nip*. Many of the ornamental species of *Nepeta* are called catmint, particularly cultivars of the sterile hybrid *N.* × *faasseni* (*N. racemosa* × *N. nepetella*, named for Dutch nurseryman J. H. Faassen, in whose nursery this hybrid first appeared). This is also known as "*N. mussinii*," an uninterpretable name applied to various species. *Nepeta* × *subsessilis*, which bears the largest flower clusters of any catmint, is called Japanese catnip or catmint (unfortunately, *Schizonepeta tenuifolia*, a medicinal herb widely used in China, Korea, and Japan, is also called Japanese catnip).

FIGURE 22.2 The euphoric effect of catnip on cats.

*Source:* **(a) Photo by Shutterstock/Anna Hoychuk. (b) Photo by Denise Mattox (CC BY ND 2.0). (c) Photo by Teresa (CC BY 2.0). (d) Photo by Bgwashburn (CC BY 2.0).**

## 22.1 INTRODUCTION

The genus *Nepeta* has close to 300 species of perennial, sometimes annual, herbaceous plants native to dry habitats in temperate Europe, Asia, northern Africa, and the mountains of tropical Africa. Several species are grown as ornamentals and as groundcovers. Catnip (*N. cataria*) is the most important species. Its popularity with people is largely determined by the popularity of cats, since people enjoy seeing the euphoric effect of catnip on cats. However, catnip is also a significant ornamental plant. Other species of the genus *Nepeta* produce relatively limited nepetalactone, the chief feline chemical attractant of catnip, but are more attractive than most cultivars of catnip.

FIGURE 22.1 Catnip (*Nepeta cataria*): (a) A display of plants in a public garden. (b) A flowering branch.

*Source:* **(a) Photo by Andre Carrotflower (CC BY SA 4.0). (b) Photo by Shutterstock/Davide Bonora.**

## 22.2 OVERVIEW

*Nepeta cataria* is an erect perennial 0.3–1.6 m high which produces small whitish or pinkish (occasionally blue or lilac) purple- or red-dotted flowers. The plant is strongly scented with a mintlike odor which many find somewhat disagreeable. Catnip is native from the eastern Mediterranean region to the western Himalayas, Central Asia, southern Siberia, and China. It has been introduced in Japan, North America, South Africa, and Java. Ornamental *Nepeta* species frequently produce sprawling flowering branches, but some cultivars are compact dwarfs and others are erect, reaching over 1 m in height. The flowers can be blue, lavender, pink, purple, or white. Yellow catmint (*N. govaniana*), a rarely cultivated native of the Himalayas, has yellow flowers. Most often, the flowering display of nepetas is blue-purple, and the appearance can resemble ornamental lavenders (see the chapter on lavender). Flowering can occur from late spring to autumn.

## 22.3 HISTORY

Catnip is thought to have been cultivated for cats by the classical Greeks and Romans. By 1265, it was a familiar herb of kitchen gardens in England. During the early medieval period, the leaves and young shoots were used as a seasoning in the kitchen. In 15th-century England, catnip leaves were rubbed on meats before cooking and also sprinkled in mixed green salads. Before modern Chinese tea became

DOI: 10.1201/9781003586227-22

FIGURE 22.3 Cultivated species of catmint (*Nepeta*): (a) *Nepeta ×faassenii* (Faassen's catmint). (b) *Nepeta discolor* (pale catmint). (c) *Nepeta grandiflora* (Caucasus catmint). (d) *Nepeta subsessilis* 'Cool Cat' (Japanese catmint). (e) *Nepeta racemosa* 'Superba' (Persian or dwarf catmint). (f) Siberian catmint (*Nepeta sibirica* 'Souvenir D'Andre Chaudron').

*Source:* **(a) Photo by A. Barra (CC BY 3.0). (b) Photo by Shutterstock/Pavaphon Supanantananont. (c) Photo by Christian Hummert (CC BY SA 3.0). (d) Photo by Cultivar 413 (CC BY 2.0). (e) Photo by Leonora (Ellie) Enking (CC BY SA 2.0). (f) Photo by Andrey Zharkikh (CC BY 2.0).**

FIGURE 22.4 Paintings of catnip (*Nepeta cataria*).

*Source:* **(a) Sturm, J. G. 1796. *Deutschlands Flora in Abbildungen, Germany.* (b) Shutterstock/EvaMira.**

FIGURE 22.5 Examples of commercial products of catnip (*Nepeta cataria*): (a) Young potted plants of catnip (*Nepeta cataria*) on sale in a store. (b) Euphoric cat playing with a catnip-laced toy banana.

*Source:* **(a) Photo by Forest and Kim Starr (CC BY 3.0). (b) Photo by Lisa Zins (CC BY 2.0).**

widely available, catnip tea was frequently consumed in England. In recent centuries, the culinary usage of the plant decreased, while the ornamental value increased.

## 22.4 ECONOMIC IMPORTANCE

*Nepeta* cultivars (especially of species other than *N. cataria*) are popular garden plants. Catnip is also cultivated as a crop for harvest of herbal material and essential oil (which is mostly nepetalactone), primarily as a stimulant for cats, but to a lesser extent for culinary and medicinal usages. Catnip and its relatives constitute a minor but significant economic commodity.

## 22.5 ORNAMENTAL USES

*Nepeta* is employed in beds, borders, groundcover, and containers. Smaller cultivars are suited to rock gardens. Catnip is sometimes cultivated as a houseplant for the benefit of cats.

## 22.6 OTHER USES

The principal modern culinary use of catnip is as a tea, which is reputedly sedative and soporific (sleep-inducing). Fresh or dried leaves and young shoots are sometimes used for flavoring sauces, soups, and cooked foods. Dried leaves are employed in herb mixtures for soups, stews, and condiment sauces. Catnip has a strongly minty, warm, pungent, bitterish, camphoraceous taste. The cultivar Citriodora contains a relatively high proportion of lemon oil and has a mild lemon aroma that is more appealing to most people for culinary purposes than regular catnip but less appealing to

FIGURE 22.6 Landscape usage of catmint (*Nepeta* species): (a) A large ornamental field of catnip (*Nepeta cataria*) in Hinode Floral Park, Hokkaido, Japan. This much-photographed tourist display is often misinterpreted as lavender. (b) A planter on a street in the town of Olsztynek in northern Poland. (c) Plants bordering a garden path. (d) A large bed in the Sensory Garden at Lake Merritt, Oakland, California. The planter wall is capped with smooth-edged bricks, facilitating movement for the visually handicapped.

*Source:* **(a) Photo by MaedaAkihiko (CC BY SA 3.0). (b) Photo by Dawid Galus (CC BY SA 3.0). (c) Photo by Shutterstock/Yolanta. (d) Photo by Stephanie Falzone (CC BY 2.0).**

FIGURE 22.7 Cats and catnip tea.

*Source:* **(a) A 1940s trade card, photo by Joe Haupt (CC BY 2.0). (b) Photo by Tambako the Jaguar (CC BY ND 2.0). (c) Photo by Tracie Hall (CC BY SA 2.0).**

cats. This lemony catnip makes a tasty tea, and the leaves can be rubbed on meat for flavoring or candied with egg white and sugar to serve as after-dinner mints. Excessive consumption of catnip tea induces vomiting in some people.

Catnip is extensively grown for extraction of its oil. Oil from catnip is used in perfumes, candies, and pharmaceuticals. Nepetalactone is employed to prepare feline attractants, sometimes used on toys for cats. The oil from catnip was used in the past as an attractant to bait wild animal traps but has been replaced by less-expensive synthetic products.

Catnip was once employed extensively in medicine, but it is now best known for inducing a euphoric response in cats. Cats purr contentedly, tear with delight, and roll in ecstasy on the crushed leaves of catnip. Catnip loses potency over time and is best stored long term in an airtight container in a freezer.

Catnip also attracts bees, but not for the same reason that it attracts cats. It is an excellent honey plant, providing nectar for honeybees and other bees. Hummingbirds, butterflies, and other insects also visit the flowers. Goldfinches are drawn to the dry seed heads in winter.

FIGURE 22.8 Wildlife attracted to catnip (*Nepeta cataria*): (a) Swallowtail butterfly. (b) Brown-banded bumblebee (*Bombus pascuorum*). (c) American goldfinch eating catnip seeds. (d) Ants on flowers in Hawaii.

*Source:* **(a) Photo by Shutterstock/Pipapur. (b) Photo by Shutterstock/HWall. (c) Photo by Indiana Ivy Nature Photographer (CC BY 2.0). (d) Photo by Forest and Kim Starr (CC BY 3.0).**

FIGURE 22.9 Catnip (*Nepeta cataria*) growing as a weed in Montana.

*Source:* **Photos by Matt Lavin (CC BY SA 3.0).**

## 22.7 PROBLEMS

Many species of *Nepeta* can become weedy, and to avoid this possibility, sterile cultivars are sometimes grown. *Nepeta cataria* is a widespread, albeit minor, weed, particularly in North America. As a weed, it is found in hedges, fencerows, roadsides, stream banks, and waste places.

## 22.8 CULTIVATION

### 22.8.1 Climate and Site

Catnip (*N. cataria*) is a hardy plant that will grow well in climate zones 3 to 9. It is best cultivated in full sun but grows well in partial shade. The other species of *Nepeta* and their cultivars have somewhat different tolerances, and this will usually be provided on labels accompanying starter plants or seed packets.

### 22.8.2 Propagation and/or Purchasing

Catnip is easily propagated by seeds, and once in place, the plant self-sows. Cultivars often will not breed true and are usually established by transplants. Propagation can also be by root division and stem tip or softwood cuttings.

### 22.8.3 Planting

*Nepeta* species tends to grow wider than tall, so they should be spaced according to the anticipated size of the cultivars. Generally, 45 cm apart suits most cultivars.

Catnip can be grown as a houseplant, and this is often done for the sake of house cats. It should be provided with moist, not soggy, soil and at least 5 hr of direct sunlight daily. The plants will do best at temperatures of 12–16°C and benefit from a commercial potting mixture supplemented with lime. Indoor catnip tends to become straggly and should be pruned to promote a bushy shape.

### 22.8.4 Soil and Moisture Conditions

*Nepeta* species are very tolerant of different soil conditions. Dry, sandy soil is recommended, but the plants also grow well (and bushier) on well-drained, moderately rich garden loam. Nepetas tolerate soil pH from 5.0 to 7.5. Fertilization is considered unnecessary. Catmints are drought-tolerant once the root system is established, but regular watering the first year is recommended.

### 22.8.5 Maintaining Plants

*Nepeta* plants need little attention, except weeding, although they themselves are often weeds. Pinching the plant at the appearance of the first flower buds produces a bushy plant. The plants can become quite unkempt after flowering, and many gardeners choose to prune them for neatness of appearance (or sometimes in the hope of stimulating a second round of blooming). The plants can be sheared back by a third or more after their first period of blooming. Every 3 or 4 years, a catmint patch should be divided (spring or early fall are best) to keep the plants vigorous. Some gardeners concerned about attracting potentially damaging cats place chicken wire over newly established or transplanted material.

## 22.9 Fascinating Facts, Fables, Folklore, and Frivolity

- The response of domesticated cats to catnip has been extensively studied. Any branches of catnip that have been bruised or broken will emit the catnip chemicals (unless so damaged, fresh plants are relatively safe from cats). The cats sniff, then lick and chew, while shaking their heads, followed by chin and cheek rubbing, a head-over roll, and body rubbing. They may become quite vocal. Because the reaction is similar to estrous rolling patterns and sexual stimulation is apparent (even in neutered cats), some have interpreted catnip as an aphrodisiac. Cats take great pleasure in rolling in and eating the foliage, returning daily to repeat the experience. Catnip does not appear to harm cats.

  While most domesticated and wild cats are affected by catnip, not all are "nipaholics." A dominant gene has been shown to be responsible for inheriting the response. Kittens less than 2 months old do not react or react only partially, and the full behavioral pattern may not develop until cats are 3 months old.

  What benefit could catnip plants derive from stimulating cats? Apparently none. The chemicals in catnip that stimulate cats do have an adaptive function, however. They have been shown to repel many insects and so protect the plants. The same or similar chemicals that protect catnip are synthesized by other plants, many of which have also been shown to attract cats. Moreover, the same or similar chemicals that protect the plants have been found in some insect species and have been shown to protect these species by repelling other insect species. Perhaps most astonishingly, cats have been shown to be strongly attracted to some ants that possess these catnip-like chemicals.
- Catnip is by no means the only plant species known to elicit the catnip response in cats. The following have also been reported to affect cats: *Actinidia kolomikta* (miyama-mata-tabi), *A. polygama* (silver vine), *Boschniakia rossica* (a plant parasite of the roots of alders, found in Northern areas), *Lippia javanica*, *Menyanthes trifoliata* (buckbean), *Nemophila menziesii* (baby blue eyes), *Origanum dictamnus* (dittany of Crete), *Teucrium marum* (marum, cat thyme), *Valeriana celtica* (a wild native of the Alps), *Valeriana officinalis* (common valerian), *Valeriana phu*, *Viburnum opulus* (cranberry bush).
- In the belief that catnip roots made even the kindest person mean, early American hangmen consumed them before executions to harden themselves for their work.
- Old European folklore held that eating jelly conserve made with catnip would keep nightmares away.
- A mixture of wine and catnip was once applied to bruises to make them heal faster.
- Laboratory experiments at Iowa State University showed that nepetalactone could be ten times as effective in driving away mosquitoes as DEET, the chemical in many commercial insect repellents. In 2001, Iowa State University submitted a patent application for the use of catnip compounds as insect repellents.

FIGURE 22.10 Catnip garden party.

*Source:* **Prepared by Jessica Hsiung.**

FIGURE 22.11 Cat grass.

*Source:* **(a) Photo by RadioMaxx/Pixabay (public domain). (b) Photo by Shutterstock/Irzhanova Asel.**

- Cat grass is a grass or a combination of grass species cultivated for cats. It is usually provided indoors, growing in small containers. Cats are motivated to chew fresh grass, and there are various hypotheses advanced to explain the behavior. It is believed to facilitate passing of hairballs or provide fiber that can act as a mild laxative or trigger vomiting to rid stomachs of non-digestible parts of ingested prey. Common forage or pasture grasses, such as oatgrass (*Arrhenatherum elatius*), intermediate wheatgrass (*Thinopyrum intermedium*), and ryegrass (*Lolium perenne*), and common cereals, such as barley (*Hordeum vulgare*) and wheat (*Triticum aestivum*), are usually grown. Cat grass, unlike catnip, is not euphoric for felines.

## KEY PUBLICATIONS


Bol, S. 2017. Responsiveness of cats (Felidae) to silver vine (*Actinidia polygama*), Tatarian honeysuckle (*Lonicera tatarica*), valerian (*Valeriana officinalis*) and catnip (*Nepeta cataria*). *BMC Veterinary Research* 13(1): 70. https://doi.org/10.1186/s12917-017-0987-6.

Eisner, T. 1964. Catnip: Its raison d'être. *Science* 146: 1318–1320.

Ferguson, J.M., W.W. Weeks, and W.T. Fike. 1988. Catnip production in North Carolina. *The Herb, Spice and Medicinal Plant Digest* 6(4): 1–4.

Gomes, E.N. et al. 2020. Catnip (*Nepeta cataria* L.): Recent advances in botany, horticulture and production. In: *Medicinal and Aromatic Plants of North America. Medicinal and Aromatic Plants of the World*, Vol. 6, edited by Á Máthé, 247–284. Springer, Cham.

Grognet, J. 1990. Catnip: Its uses and effects, past and present. *Canadian Veterinary Journal* 31: S.455–S.456.

Hawke, R.G. undated. A comparative study of cultivated catmints. *Chicago Botanic Gardens Plant Evaluation Notes*. www.chicagobotanic.org/downloads/planteval_notes/no29_catmint.pdf.

Kaiser, C. and M. Ernst. 2022. *Catnip*. Center for Crop Diversification, University of Kentucky College of Agriculture, Food and Environment. www.uky.edu/ccd/sites/www.uky.edu.ccd/files/catnip.pdf.

Lichman, B.R., G.T. Godden, J.P. Hamilton, L. Palmer, M.O. Kamileen, D. Zhao, et al. 2020. The evolutionary origins of the cat attractant nepetalactone in catnip. *Science Advances* 6: 20. www.science.org/doi/10.1126/sciadv.aba0721.

Pargaien, A.V., N. Bisht, H. Joshi, S. Pargaien, and M. Adhikari. 2020. Analysis of ethanomedicinally [sic] potential extract of *Nepeta cataria. Ecology, Environment and Conservation* 26: S90–S95. www.envirobiotechjournals.com/EEC/Vol26OctSuppl20/EEC-15.pdf.

Salehi, B., M. Valussi, A.K. Jugran, M. Martorell, K. Ramírez-Alarcón, Z.Z. Stojanović-Radić, et al. 2018. *Nepeta* species: From farm to food applications and phytotherapy. *Trends in Food Science & Technology* 80: 104–122.

Setzer, W.N. 2016. Catnip essential oil: There is more to it than making your cat go crazy. *American Journal of Essential Oils and Natural Products* 4(4): 12–15.

Small, E. 2012. Catnip – safer pesticide potential. *Biodiversity* 13: 118–126.

Tucker, A.O. and S.S. Tucker. 1988. Catnip and the catnip response. *Economic Botany* 42: 214–231.


### KEY CULTIVATION WEBSITES


A-Z Animals. *How to Grow Catnip: Your Complete Guide*. https://a-z-animals.com/blog/how-to-grow-catnip-your-complete-guide/.

Piedmontmastergardeners.org. *Catmint – A "Must-Have" Perennial*. https://piedmontmastergardeners.org/article/catmint-a-must-have-perennial/.

The Spruce. *How to Grow and Care for Catmint*. www.thespruce.com/catmints-nepeta-1402851.

WikiHow. *How to Grow Catmint from Cuttings*. www.wikihow.com/Grow-Catmint-from-Cuttings.

WikiHow. *How to Grow Catnip*. www.wikihow.com/Grow-Catnip.

WikiHow. *How to Harvest Catnip*. www.wikihow.com/Harvest-Catnip.

# 23 Celosia (*Celosia argentea*)

**Scientific Names:** The genus name *Celosia* is based on the ancient Greek *keleos*, "burning," a reference to the colorful flower heads which, in many varieties, resemble candle flames.

**Common Names:** The common name *woolflower* is employed for ornamental varieties with diffuse ("wooly") flower clusters. Edible forms are often called *Lagos spinach*. Cockscombs include cultivars with contorted, massed clusters of tissue that replace the flowers and are reminiscent of a cock's comb.

## 23.1 INTRODUCTION

Ornamental forms of celosia are very popular, easily cultivated plants with spectacularly colored flowers available in charming shapes.

## 23.2 OVERVIEW

*Celosia argentea* is an annual herb (potentially a short-term perennial in semi-tropical climates) thought to be native to the Mediterranean region or East Africa. The plants range in height from dwarfs 15 cm or so to plants 1 m tall. The flowers are tiny, hundreds densely packed in each flower cluster, blooming from the bottom up. The coloration of the flower clusters does not come from the petals of the many tiny flowers but, rather, from bracts (tiny leaves under the flowers) and the sepals of the flowers. Celosia flowers are available in many shades of gold, maroon, orange, pink, purple, red, and yellow, including multicolored flower heads. Flowering occurs from midsummer to early autumn. Some cultivars are available with foliage that is bluish, purple, reddish, or yellowish.

The species *Celosia argentea* has been classified in various ways. In the following discussion, "var." = "varietas," a taxonomic category more inclusive than the terms "cultivar" or "variety" in the common sense.

FIGURE 23.1 Weedy, wild silver cockscomb (*Celosia argentea* var. *argentea*) in Japan.

*Source:* **Photo by Kenpei (CC BY SA 3.0).**

### 23.2.1 *Celosia argentea* var. *argentea*

Cultivars domesticated for food can be placed in *C. argentea* var. *argentea*. However, this taxonomic group also includes the wild celosia or silver cockscomb, which is a weed in most tropical areas and is rather different in appearance from cultivated kinds.

### 23.2.2 *Celosia argentea* var. *plumosa*

Woolflowers include ornamental cultivars with plumed flower clusters. This is the most common kind of ornamental celosia. These are placed in *Celosia argentea* var. *plumosa* (*C. plumosa*) and have also been named plume flower, plume of feathers, and Prince of Wales's feathers.

### 23.2.3 *Celosia argentea* var. *cristata*

Cockscombs (coxcombs) are ornamental cultivars with bizarre "crested" flower clusters. These are placed in *C. argentea* var. *cristata* (*C. cristata*). "Cresting" refers to "fasciation," a normally undesirable, tumor-like growth (as in broccoli and cauliflower, the flowers do not develop as such, but the floral tissues proliferate in an undifferentiated, convoluted fashion). In the case of celosia, the flower clusters become compressed, contorted, ruffled, and/or wrinkly, looking like coral, brain, or fans, the appearance considered attractive by many. This unique class tends to produce much fewer, albeit more spectacular, floral-like displays.



DOI: 10.1201/9781003586227-23

FIGURE 23.2 Woolflowers (ornamental forms with feathery flower clusters, placed in *Celosia argentea* var. *plumosa*).

**Source: (a) Photo by Shutterstock/Tom Black Dragon. (b) Photo by Carl Lewis (CC BY 2.0). (c) Photo by Mokkie (CC BY SA 3.0). (d) Photo by Pauk (CC BY SA 3.0).**

FIGURE 23.3 Cockscombs (ornamental forms with oddly contorted flower clusters, placed in *Celosia argentea* var. *cristata*).

**Source: (a) Photo by Shutterstock/Totokzww. (b) Photo by Mike Locke (CC BY ND 2.0). (c) Photo by Cultivar 413 (CC BY 2.0). (d) Photo by Tiểu Bảo Trương from Pixabay (public domain).**

FIGURE 23.4 Wheat-type ornamental cultivars (forms with spike-like flower clusters placed in *Celosia argentea* var. *spicata*).

**Source: (a) Photo by Nur Kayat from Pixabay (public domain). (b) Cultivar Intenz. Photo by Cultivar 413 (CC BY 2.0).**

FIGURE 23.5 Commercial production of celosia: (a) Greenhouse cultivation. (b) Field cultivation.

**Source: (a) Photo by ProFlowers (CC BY 2.0). (b) Photo by Pakku (CC BY 3.0).**

### 23.2.4 *Celosia argentea* var. *spicata*

Wheat-type ornamental cultivars (which are relatively uncommon) have spiky (elongated) flower clusters resembling a wheat head and much less feathery than the plume class. These have been called *C. argentea* var. *spicata* (*C. spicata*, sometimes assigned to *Deeringia spicata*). They sometimes produce a single, rather than several, flowering stalk.

## 23.3 HISTORY

Crested (fasciated) plants, when they arise in nature, are normally eliminated because they are detrimental to normal growth. Clearly, cultivated forms of *Celosia* arose by human selection. Some authorities have suggested that this occurred in early times in India, and that the very unusual plants acquired religious significance in India, Burma, and China, where people planted (and still plant) cockscomb around their temples and in their gardens. Both crested and feathery celosia cultivars were grown in Europe dating back to the 16th century. Ornamental celosia has been cultivated in North America since the 18th century.

## 23.4 ECONOMIC IMPORTANCE

The ornamental cultivars of celosia are well-established and important in the horticultural trade of numerous countries. The edible cultivars are a minor vegetable in some tropical countries, particularly in Africa and Asia.

## 23.5 ORNAMENTAL USES

Ornamental celosias are grown in beds and borders. They are also grown in containers but are often considered relatively difficult to maintain in pots. Celosia is an excellent cut flower for vases and bouquets. Celosia flowers (both the plumed and crested forms) are also frequently dried and employed in flower arrangements. Drying usually is done by removing the foliage from the stems and hanging them upside down in a cool, dry, dark room with good ventilation. Drying is best done rapidly, but in humid areas, this may take weeks. The dried flowers are known to retain their color relatively well for months.

FIGURE 23.6 An elaborate ornamental garden showing celosia beds and borders.

*Source:* **Photo by Shutterstock/Moolkum.**

FIGURE 23.7 Ornamental celosia grown in containers: (a) Woolflower (var. *plumosa*) type, with feathery flower clusters. (b) Cultivar Merida Firossa, a wheat (var. *spicata*) type, with spike-like flower clusters.

*Source:* **(a) Photo by Shutterstock/Yuttana Joe. (b) Photo by Shutterstock/InfoFlowersPlants.**

FIGURE 23.8 Cockscomb celosia (var. *cristata* type, with contorted flower clusters) employed as a cut flower.

*Source:* **(a) Photo by Shutterstock/Shandor. (b) Photo by Shutterstock/Natalie Board.**

FIGURE 23.9 An experimental cultivar of Lagos spinach (culinary form of *Celosia argentea*) at AVRDC (The World Vegetable Center), Taiwan.

*Source:* **Photo by Michael Hermann (CC BY SA 3.0).**

## 23.6 OTHER USES

Edible cultivars of *C. argentea* var. *argentea*, called Lagos spinach (the foliage has a mild taste reminiscent of spinach), are employed as a leafy vegetable in numerous semi-tropical and tropical countries. Ornamental cultivars may be tasty when young but often become bitter after blooming. The young stems and flower clusters are also cooked and eaten. The seeds are also consumed as a grain.

## 23.7 PROBLEMS

The stems of celosia tend to be fragile, especially in tall cultivars. There has been a tendency in recent times to breed dwarf varieties, which are resistant to bending over or breaking. Celosia tends to self-seed, and if new plants are not wanted in a particular area, the flowers should be removed when they wilt, to prevent seed formation. The odd, tumor-like growth of the cristata-type cultivars is distasteful to some people.

## 23.8 CULTIVATION

### 23.8.1 Climate and Site

Because it is frost-sensitive, celosia is generally grown as an annual, so winter temperatures (and climate zones) are not relevant. However, it can be a short-lived perennial in zones 9 to 11. Full sunlight is best; partial shade is tolerated. Windy sites are undesirable as the stems are fragile.

### 23.8.2 Propagation and/or Purchasing

Celosia is usually planted from seed, but many ornamental cultivars will not reproduce true to type and are often available only as cuttings, typically sold in cell packs. Ornamental forms usually are started in peat pots (not flats, as the roots are easily disturbed) that are transplanted to their growth sites.

### 23.8.3 Planting

Plants should be spaced 20 to 30 cm apart, depending on cultivar. The plants may self-seed in locations where they are grown.

### 23.8.4 Soil and Moisture Conditions

Celosia thrives in well-drained, rich, deep, compost-enriched, loamy soil with a pH of 6.0 to 6.5 (or up to 7.0). Although the plants are drought-resistant, they grow best in moist soil. Application of a balanced fertilizer is recommended (professional growers often provide high-nitrogen fertilizers when the plants are young, low-nitrogen at the flowering stage).

### 23.8.5 Maintaining Plants

Cockscomb-type celosia blooms can make the plants top-heavy, requiring stakes for support. The stems are somewhat succulent and subject to breakage if roughly handled. The feather and fasciated classes are available in cultivars that tend to produce a single large flowering stalk bearing a very large flower cluster, or in branching forms with several flower clusters. Regardless, by pruning away branches, a single flowering stalk can be produced. Conversely, pinching (destroying) the tip of the main stem (leader) encourages side branches to develop, each with a flower cluster. These tactics work better with some cultivars than others.

## 23.9 FASCINATING FACTS, FABLES, FOLKLORE, AND FRIVOLITY

- The uncontrolled, proliferative growth of cockscomb flowers gave rise to the symbolic association of the plant with "foppery" (silliness).
- In the 1920s and 1930s, competitions were held in US state fairs for the celosia with the largest crest. One award-winning specimen had a crest wider than 50 cm (normally, the diameter of crests is 10 to 25 cm).
- Mexicans employ ornamental celosia to decorate church altars, shrines, and graves, especially during the Día de los Muertos, or Day of the Dead, celebrations.

FIGURE 23.10 Day of the Dead holiday in Mexico, featuring purple cristata-type ornamental celosia and yellow marigolds (*Tagetes erecta*).

*Source:* **(a) Photo by Marianafloresocampo (CC BY SA 4.0). (b) Photo by J Mndz (CC BY SA 2.0).**

## KEY PUBLICATIONS

Divya, B.J., M.J. Sravani, J.H. Chandana, T. Sumana, and K. Thyagaraju. 2019. Phytochemical and phytotherapeutic activities of *Celosia argentea*: A review. *World Journal of Pharmacy and Pharmaceutical Sciences* 8: 488–505.

Gill, S. and E. Dutky. 2022. *Production of Celosia as a Cut Flower*. University of Maryland, Extension. https://extension.umd.edu/resource/production-celosia-cut-flower/.

Grant, W.F. 1954. A cytological study of *Celosia argentea*, *C. argentea* var. *cristata*, and their hybrids. *Botanical Gazette* 115: 323–336.

Grant, W.F. 1961. Speciation and basic chromosome number in the genus *Celosia*. *Canadian Journal of Botany* 39: 45–50.

Grant, W.F. 1962a. Malayan fantasy – a new garden cockscomb. *The American Horticultural Magazine* 41: 201–203.

Grant, W.F. 1962b. Speciation and nomenclature in the genus *Celosia*. *Canadian Journal of Botany* 40: 1355–1363.

Iamonico, D. 2013. About the circumscription of *Celosia argentea* (Amaranthaceae) and the related Linnaean taxa. *Phytotaxa* 90: 61–64.

Kanu, C.L., O. Owoeye, I.O. Imosemi, and A.O. Malomo. 2017. A review of the multifaceted usefulness of *Celosia argentea* Linn. *European Journal of Pharmaceutical and Medical Research* 4: 72–79.

Khoshoo, T.N. and M. Pal. 1973. The probable origin and relationships of the garden cockscomb. *Botanical Journal of the Linnean Society* 66: 127–141.

Porat, R., E. Shlomo, and A.H. Halevy. 1995. Horticultural techniques to improve *Celosia plumosa* growth for cut flowers. *Scientia Horticulturae* 63: 209–214.

Ravindra, G.M., S. Sridhara, G.K. Girijesh, and H.V. Nanjappa. 2008. Weed biology and growth analysis of *Celosia argentea* L., a weed associated with groundnut and finger millet crops in southern India. *Communications in Biometry and Crop Science* 3: 80–87.

Tang, Y., H.-L. Xin, and M.-L. Guo. 2016. Review on research of the phytochemistry and pharmacological activities of *Celosia argentea*. *Revista Brasileira de Farmacognosia* 26: 787–796.

United States Department of Agriculture. 1979. *Vegetables for the Hot, Humid Tropics. Part 6. Amaranth and Celosia, Amaranthus and Celosia*. Agricultural Research (Southern Region), Science and Education Administration, U.S. Department of Agriculture, New Orleans, LA. 21 pp.

### Key Cultivation Websites

Garden Gate. *How to Grow Celosia*. www.gardengatemagazine.com/articles/flowers-plants/plant-guide/how-to-grow-celosia/.

New York Botanical Garden. *Celosia*. https://libguides.nybg.org/celosia.

The Spruce. *How to Grow and Care for Celosia.* www.thespruce.com/celosia-flowers-1315717.

UK House Plants. *How to Grow Healthy Indoor Celosia (Cockscomb).* www.ukhouseplants.com/plants/celosia-cockscomb.

University of Maryland, College of Agriculture & Agricultural Sciences. *Production of Celosia as Cut Flowers.* https://web.archive.org/web/20060627183920/www.agnr.umd.edu/MCE/Publications/Publication.cfm?ID=21.

Utah State University Extension. *Celosia Cut Flower Production in Utah.* https://digitalcommons.usu.edu/cgi/viewcontent.cgi?article=3271&context=extension_curall.

WikiHow. *How to Care for a Celosia.* www.wikihow.com/Care-for-a-Celosia.

# 24 Cherry Blossom (Flowering Cherries, *Prunus* Species, and Hybrids)

**Scientific Names:** The genus name *Prunus* is identical to the classical Latin name of the plum tree and originated from the old Greek *prunos*.

**Common Names:** The name "cherry" traces to ancient languages: the Latin *cerasus*, "cherry tree," which is from the Greek word for the cherry tree, *kerasos*, derived ultimately from *karsu* from the Akkadian language of Mesopotamia.

Cherry is part of the name of various plant species that are unrelated to cherry species. For example, "cherry currant" refers to a variety of the red currant, "cherry pepper" refers to various chili pepper varieties, "cherry tomato" refers to small tomatoes, and "cherry orange" is the kumquat.

## 24.1 INTRODUCTION

"Flowering cherries" or "cherry blossom" trees (sometimes shrubs) are cultivars of species of the genus *Prunus* with attractive, very conspicuous flowers. *Prunus* has (depending on authority) 250 to more than 400 species of woody plants distributed across the Northern Hemisphere and into the subtropics and the tropics. In addition to cherries, the genus includes other economically significant species, notably peaches, plums, apricots, almonds, and a wide variety of ornamentals and timber species. The plums, cherries, apricots, peaches, nectarines, and almonds are known as "drupe fruits." A *drupe* is a fleshy or pulpy fruit, with the inner portion of the pericarp (fruit wall) being hard and stony and normally enclosing one (sometimes more than one) seed. The inner hard portion is commonly the "pit" or "stone." Most *Prunus* species produce quite attractive flowers, but the cultivars selected as ornamentals are usually grown for their especially notable floral displays, not their edible fruits. Most species called cherries are native to eastern Asia, whereas only a little more than two dozen species are indigenous to North America. The "true cherry" species have been traditionally classified in one of the subgenera of *Prunus*: subgenus *Cerasus* (sometimes recognized as a separate genus). The most widely cultivated cherry species are sweet cherry (*Prunus avium*) and sour cherry (*P. cerasus*), which originated in the Old World and are among the world's most important fruits (they are not principally grown for flowering displays). However, not all species called cherries belong to group *Cerasus*, nor are all species called flowering cherries. Indeed, some flowering *Prunus* cultivars are sometimes referred to as "flowering plums" or "flowering almonds." Most ornamental cherries need cool temperatures to thrive and so are native to and cultivated primarily in temperate countries, although they are sometimes grown in subtropical regions.

FIGURE 24.1 Scenes of cherry blossom trees in Japan: (a) Chureito Pagoda in a sea of cherry blossoms, with Mount Fuji in the background. (b) A child in Tokyo, Japan, rejoicing in cherry blossoms. (c) *Prunus × yedoensis* 'Somei-yoshino' (Yoshino cherry), the most popular flowering cherry tree cultivar in Japan and the Asian region. (d) Fallen cherry blossom flowers carpeting the ground in Shinjuku Gyoen National Garden, Japan.

*Source:* **(a) Photo by Shutterstock/Mapman. (b) Photo (public domain) by Fg2. (c) Photo by Akiyoshi Matsuoka (CC BY SA 4.0). (d) Photo by Kakidai (CC BY SA 4.0).**

## 24.2 OVERVIEW

Cherry trees employed as ornamentals generally flower for about 2 weeks in the spring, while blooming of groups of trees typically extends over a month. Some cultivars flower later than others. The somewhat-fragrant flowers are usually shades of white or pink, some with yellow-green or white-green blooms that turn pink as they mature. Some

FIGURE 24.2 Cherry blossom trees in Vancouver, British Columbia, Canada: (a) At the University of British Columbia. (b) In Stanley Park.

*Source:* **(a) Photo by Xicotencatl (CC BY SA 4.0). (b) Photo by GoToVan (CC BY 2.0).**

DOI: 10.1201/9781003586227-24

FIGURE 24.3 Purple-leaf sand cherry (dwarf red-leaf plum, *Prunus × cisterna*). This is a hybrid between the Eurasian *P. cerasifera* and the North American *P. pumila*. It is a bush or shrubby tree, typically reaching a height of 1.5–2.5 m, and is hardy up to climate zone 3: (a) In flower. (b) In leaf.

***Source:*** **Photos by F. D. Richards (CC BY SA 2.0).**

FIGURE 24.4 The two basic colors of cherry blossoms: (a) Pink. (b) White.

***Source:*** **(a) Photo by Denis Iosub from Pixabay (public domain). (b) Photo by YD PARK from Pixabay (public domain).**

FIGURE 24.5 Doubled cherry blossom flowers.

***Source:*** **(a) Cultivar Kanzan. Photo by Famartin (CC BY SA 3.0). (b) Photo by ws0922 from Pixabay (public domain).**

cultivars have doubled flowers (i.e., with extra petals). Some species (especially shrubs) have purple foliage; others have leaves that turn purple, red, or orange in the autumn. Some varieties feature colorful bark. The trees are usually laid out in the landscape in one floral color to produce a massed display in a given location. The fruits are small, and although

FIGURE 24.6 Gigantic old weeping cherry blossom trees: (a) In Branch Brook Park, Newark, NJ. (b) In Japan.

***Source:*** **(a) Photo by Cjbvii (CC BY 3.0). (b) Photo by Camerakid (CC BY SA 3.0).**

birds and other animals will consume them, they are usually too sour for most people. Although there are centuries-old specimens (some even over 1,000 years of age), most flowering cherry trees are relatively short-lived, lasting 25–50 years. Dwarf forms and weeping kinds are available.

## 24.3 HISTORY

Respect for blossoming *Prunus* trees is ancient in Asia, most particularly in Japan, where hundreds of cultivated varieties have been selected. Today, annual spring "cherry blossom" festivals are very widespread in Asia but also occur in many countries throughout the world. Trees are widely planted in parks, on street sides, and in front of buildings. Cherry blossoms have considerable symbolic value in Japan, the annual regrowth representing renewal and hope, and the briefness of flowering representing the short transience of physical beauty and mortality. Cherry blossom trees are known as *sakura* in Japan, and the tradition of picnicking (and often drinking) beneath the trees is called *hanami*.

In 1912, Japan provided a gift of ornamental cherry trees to surround the Jefferson Memorial in Washington. Today, over 4,000 cherry trees of several varieties grow around the Tidal Basin, at East Potomac Park, and on the Washington Monument Grounds. Most trees are the Yoshino variety, Japan's favorite cultivated cherry tree that was developed about 1870. Ironically, the symbolic act of friendship between the Japanese and American nations was to be followed in two decades by a bitter war. Still more ironically, in the early 1950s, America sent cuttings of the cherry trees to restore the original collection in Japan that had originally furnished many of the trees but was damaged or deteriorated as a result of the war. In Japan, ornamental cherry trees have become symbols of remembrance of the many thousands who died from the atomic bombs dropped on Hiroshima and Nagasaki in 1945, and the anniversaries of these bombings, August 6 and 9, are occasionally marked by planting cherry trees or having ceremonies around cherry trees.

## 24.4 ECONOMIC IMPORTANCE

Flowering species of *Prunus* are an important product in the nursery trade. However, the commercial and tourist

FIGURE 24.7 Japanese paintings featuring cherry blossoms: (a) Painting of cherry blossoms, ca. 1940, by anonymous artist celebrating the Miyako Odori (spring festival) in Kyoto, Japan. (b) *Ladies in the Edo palace enjoying cherry blossoms*, a painting (public domain) by Japanese artist Chiyoda Ooku Ohanami (1838–1912) showing geishas under cherry blossom trees.

*Source:* **(a) Photo by Joi Ito (CC BY 2.0). (b) Photo credit: National Diet Library, Benichan.**

FIGURE 24.8 Cherry blossom trees in Washington, DC: (a) Washington Monument framed by cherry blossom trees, viewed from the Tidal Basin. (b) Crowd at the National Cherry Blossom Festival.

*Source:* **(a) Photo by Carol M. Highsmith (public domain). (b) Photo by Eric T. Gunther (CC BY SA 3.0).**

activities associated with cherry blossom festivals greatly outweigh the monetary values of the plants.

## 24.5 ORNAMENTAL USES

Flowering cherries are primarily employed as landscape trees and sometimes shrubs, extremely widely planted in public areas as well as on larger residential properties, primarily in temperate regions of the world.

## 24.6 OTHER USES

Cherry blossoms are edible, and a variety of animals feed on them. The flowers are sometimes pickled in salt and used in ethnic dishes, including ume vinegar (based on plums and brine), wagashi (traditional Japanese confections, usually plant-based), and anpan (Japanese sweet rolls, especially filled with red bean paste). Similarly, the leaves are also occasionally employed as food, although as noted next, they can be toxic.

FIGURE 24.9 Huge crowds at cherry blossom festivals generate extensive economic activities: (a) In Seoul, South Korea. (b) In Himeji Castle, Japan. (c) In Brooklyn, USA. (d) In Bonn, Germany.

*Source:* **(a) Photo by Korean Culture and Information Service (CC BY 2.0). (b) Photo by 663 highland (CC BY SA 3.0). (c) Photo by Ryan Somma (CC BY 2.0). (d) Photo by Spielvogel (public domain).**

FIGURE 24.10 Animals feeding on cherry blossoms: (a) Hummingbird. (b) Bee on flowers of *Prunus campanulata*. (c) Deer feeding on fallen cherry blossoms at Nara Park, Japan.

*Source:* **(a) Photo by Shutterstock/Mythaiphotography. (b) Photo by Mauro Halpern (CC BY 2.0). (c) Photo by Shutterstock/Bule Sky Studio.**

FIGURE 24.11 Traditional Japanese food preparations made from cherry blossom trees: (a) Pickled (salted) cherry blossoms. (b) Sakuramochi, a dessert commonly found in street food festivals consisting of sweet, pink-colored rice cake with a red bean paste and wrapped in a pickled cherry blossom leaf.

*Source:* **(a) Photo by Shutterstock/Mssy. (b) Photo by Shutterstock/The Wander Eddie.**

## 24.7 PROBLEMS

All members of the genus *Prunus* contain deadly cyanin in the form of prussic acid (hydrocyanic acid) or hydrogen cyanide (HCN). Generally, among plants, cyanide is found in the form of glycosides, that is, in combination with sugars. Amygdalin in bitter almonds is an example of such a glycoside; it yields prussic acid upon hydrolysis (which may occur in the digestive system). Prussic acid and the cyanides are highly toxic, inhibiting cellular oxidative processes by combining with the red blood cells, preventing the cells from carrying oxygen to the tissues. HCN is present in all parts of plants of *Prunus*, except the fleshy portion of the fruit. HCN is often concentrated in the seeds. Livestock are frequently poisoned by consuming the foliage of wild species of *Prunus*. Some cherry species, particularly choke cherry and black cherry, are noted for poisoning livestock that consume the foliage. Ingesting large quantities of chokecherry and black cherry fruits without removing the seeds has caused illness and death in children. Children have even been poisoned by chewing the twigs.

It has been claimed that, along with the ornamental cherry trees that were introduced from Japan to beautify the Washington, DC, area in 1912, came the oriental fruit moth (*Grapholita molesta*), which loves peaches and has been a major pest of peaches in North America ever since. British Columbia remains the only commercial fruit-growing area of North America that is free of this pest. Similarly, ornamental cherries can harbor various other pests that can infect fruit species of *Prunus*, but the latter are also capable of vectoring pests that can harm the former.

FIGURE 24.12 Ceremonial planting of a large ornamental cherry tree at the 2008 Cherry Blossom Festival of Greater Philadelphia.

*Source:* **Photos by Chris Yeager, Japan America Society of Greater Philadelphia (CC BY SA 3.0).**

## 24.8 CULTIVATION

### 24.8.1 Climate and Site

Cherry blossom trees may be adapted to climate zones 4 to 9, but the particular tolerance of a given variety should be checked before planting. Flowering trees are typically 5 to 10 m high, with a canopy as wide, so locations should be able to accommodate the mature plants. Trees should not be too near buildings, although shrubs can be much closer. Single specimens can be featured in lawns and courtyards, while paired trees may be used to accent the sides of an entry or gateway. Large trees can be employed for shade. In public areas, ornamental cherries are unsurpassed as rows along driveways and streets. Although many wild *Prunus* trees (and even cultivated varieties) can grow in shade, for best development, ornamental cherries should be planted in open areas, providing at least 6 hr of sunlight daily. Wind will often blow away flowers, so if possible, very windy areas should be avoided. Southwest injury or sunscald is particularly damaging to cherry trees. Bright sunshine during cold periods in midwinter warms the bark on the trunks on the southwest side during the day, and this may be followed by rapid chilling when the sun drops toward the horizon. Bark temperatures may fall from 20° to −18°C. This injures the bark and adjacent areas, which frequently die, leaving bare wood areas. Painting with white latex-based paint and shading the trunks can prevent this kind of injury.

### 24.8.2 Propagation and/or Purchasing

Many flowering cherry tree cultivars are sterile, unable to produce fruit and seeds, so propagating by seed is not an option. The best cultivars are maintained as clones, reproduced by cuttings, often grafted to hardy rootstocks. Reputable nurseries are good sources of cultivars suitable for local conditions. Numerous websites advertise cultivars, and those at least in the same country or region are more likely to provide plants that are adapted to the local climate.

### 24.8.3 Planting

Standard recommendations for planting trees include preparing a hole larger than the root ball; adding organic matter, such as compost, to the bottom of the hole; adjusting the height of the plant in the hole so that the tree is planted at

FIGURE 24.13 Flowering cherry bonsai plants.

*Source:* (a) Photo by Rumpleteaser (CC BY 2.0). (b) Photo by Shutterstock/Walter Pall.

the same depth as when originally growing; staking the tree with a rigid support and ties; watering the newly planted tree thoroughly to settle the soil; and mulching the base with bark, garden compost, or wood chips. Some compact cultivars are suitable for planting in containers or even for growth as bonsai.

### 24.8.4 Soil and Moisture Conditions

Soils for flowering trees should be fertile and well-drained, but sandy, loamy, and clay soils are all usable. The plants are tolerant of a range of soil pH, but slightly acidic soil is preferred. Most gardeners tend to overwater, and irrigation is necessary only during significant droughts. Cherry trees do well with limited fertilization, which is usually desirable only once or so annually.

### 24.8.5 Maintaining Plants

Dead or diseased wood should be cut away, but cutting for shape should be minimized (some authorities recommend removing branches that interfere with light penetration to the center; others prefer allowing the development of a haphazard habit as part of the charm of the trees). Pruning is best carried out while the plants are dormant, never before flowering.

## 24.9 FASCINATING FACTS, FABLES, FOLKLORE, AND FRIVOLITY

- "Cherry" is part of the names of various, rather subjectively defined colors, including cherry (a moderate red), cherry blossom (also a moderate red, but paler), cherry pink (pink like cherry blossoms), cherry rose (a deep pink), and cherry wine (a deep red).
- The Victorian "language of flowers" was a secret coded language in Victorian times, with flowers and plants symbolic of certain messages, so when the flower or plant was mentioned in a letter, those who knew the code could understand the hidden information. "Cherry tree" meant "good education," while "white cherry tree" stood for "deception."
- A Chinese legend speaks of the goddess Hsi Wang Mu, who had a cherry garden which ripened its fruit every 3,000 years, providing the food that the gods needed to remain immortal.
- In Japan, cherry trees have long been revered ornamentals, and indeed, an old name of the country, still often used poetically, was "Land of the Cherry Blossom."
- A Japanese story relates that cherries blossom because they are occupied by the spirit of a dog, and every spring his master commands him to bring the plants into bloom. In another story, a wet nurse prayed that the disease killing the baby she was caring for would be transferred to her. When the nurse acquired the disease and died, her spirit was transferred into the cherry trees, which bloom on the anniversary of her death. Still another tale is about conflict among plant spirits. One day, a grass prince fell in love with a beautiful cherry tree, and the two were married. Overcome with jealousy, a plum tree assembled an army of trees and killed the grass prince.
- According to a Japanese legend, there was a brave but old warrior whose fondest memory was of playing beneath a particular cherry tree during his youth. When the tree died, the man took it as a sign that it was also his time to die. The old man asked the dead tree to produce flowers one more time, vowing that if his request was granted, he would give up his long life. The tree flowered, and he committed hara-kiri beneath it. Legend holds that the tree blooms every winter on the anniversary of the warrior's death, although all other cherry trees remain dormant.
- The Japanese once made printing blocks from cherry wood, and the wood engravers customarily held memorial services for the spirits of the trees from which the wood was obtained.
- In Europe, Barbara of Nicomedia, a legendary holy figure said to have been martyred in 306, is venerated as St. Barbara. (However, in 1969, the Roman Catholic Church declared her to be a non-historical person and took her feast day [December 4] off the church calendar.) According to legend, Barbara lived in Asia Minor (Turkey). Her father, the pagan emperor Dioscorus, persecuted Christians, and when she became a Christian, he ordered that she be tortured and beheaded. While in her cell, Barbara found a dried-up cherry branch, which blossomed profusely before her execution. Immediately following her execution, Dioscorus was struck dead by lightning, explaining why St. Barbara is often called on for protection during thunderstorms. Traditionally, in German-speaking countries, particularly in Austria and the Catholic regions

of Germany, a small cherry branch is cut off and placed in water on December 4, a custom known as Barbarazweig, or Barbara branch. If the branch blooms precisely on December 25, it is regarded as a particularly good sign for the future. According to an old custom, if a single woman hangs the names of the men who are courting her on several Barbara branches, she will know who she will marry by which branch is in full bloom on Christmas Eve. It has been said that Barbara branches were the forerunners of the Christmas tree.

- A traditional carol of England relates how Joseph, much older than his wife, Mary, learns that she is pregnant. While walking through a cherry orchard, Mary asks Joseph to pick a cherry for her, but he (uncharacteristically) replies, "Let he pluck thee a cherry that brought thee with child." Christ, speaking from his mother's womb, commands the cherry trees to lower their branches so that Mary can gather their fruit. When the trees obey him, Joseph repents. Cherry blossoms came to represent the purity of Mary.
- An old rural American belief was that a slip of cherry wood in the pocket would protect against poison ivy.
- The story of first American president George Washington (1732–1799) zealously felling a cherry tree at the tender age of 6 is American history. What is generally not known is that this touching portrait of American honesty is a fabrication of clergyman-author Mason Locke Weems (1759–1825), in his *Life of George Washington*, a bestseller first published in 1800 that was so popular 59 editions were published by 1850. Parson Weem's story follows:

> When George was about 6 years old, he was made the wealthy owner of a hatchet, of which, like most little boys, he was immoderately fond; and was constantly going about chopping every thing which came in his way. One day, in the garden, where he often amused himself by hacking his mother's pea-sticks, he unluckily tried the edge of his hatchet on the body of a beautiful young English cherry tree, which he barked so terribly, that I don't believe the tree ever got the better of it. The next morning, the old gentleman, finding out what had befallen his tree, which, by the by, was a great favourite, came into the house; and with much warmth asked for the mischievous author, declaring at the same time, that he would not have taken 5 guineas for his tree. Nobody would tell him anything about it. Presently George and his hatchet made their appearance. "George," said his father, "do you know who killed that beautiful little cherry tree yonder in the garden?" This was a tough question; and George staggered under it for a moment; but quickly recovered himself; and looking at his father, with the sweet face of youth brightened with the inexpressible charm of all-conquering truth, he bravely cried out, "I can't tell a lie, Pa; you know I can't tell a lie. I did cut it with my hatchet." – "Run to my arms, you dearest boy," cried his father in transports, "run to my arms; glad am I, George, that you killed my tree, for you have paid me for it a thousand fold. Such an act of heroism in my son is worth more than a thousand trees, though blossomed with silver and their fruits of purest gold."

- Candymaker James Smith, along with his two sons, William and Andrew, originally from Scotland, moved from St. Armand, Quebec, to Poughkeepsie, New York, in 1847 and started the Smith Brothers firm. Drawings of the two bearded brothers came to grace the packages of cough drops, and by chance, the word "Trade" appeared under the picture of William, and the word "Mark" under that of Andrew, and the Smith Brothers became known to generations as Trade and Mark. Andrew died in 1895, and William continued as president of Smith Brothers almost up to his death in 1913. He was succeeded by his son, Arthur, who added the famous Smith Brothers wild cherry flavor in 1948. The ingredients that produced the "cherry" flavor are unclear, but it appears that bark extracts from wild cherry trees were added. In fact, boiled preparations of bark from various native and introduced species of cherry have long been used by several North American Indian groups as a remedy for coughs and colds.
- According to the 2002 *Guinness Book of World Records*, Al Gliniecki of the United States established a world record by knotting 911 cherry stems with his tongue in 1 hr.
- Small nectar-producing glands ("extrafloral nectaries") occur on the leaf margins of some cherry

FIGURE 24.14 Painting by American artist Grant Wood (1891–1942) showing Parson Weems and his famous fable of George Washington confessing that he chopped down a cherry tree.

***Source:*** **Photo by Amon Carter Museum of American Art (public domain).**

species. It has been found that the predaceous red-headed ant (*Formica obscuripes*) feeds on the nectaries on the leaves of black cherry. The ant seems to repay the cherry for the nectar by consuming some insects that feed on the plant.

- Overripe cherries left on a tree may ferment under hot conditions and have been reported to produce intoxication in birds that consume them in large quantities.
- In 1997, the cherry was declared to be the official state fruit of Utah.
- In 2007, the chokecherry became the official fruit of the state of North Dakota. A petition started by grade 6 schoolchildren was the key impetus that led to the legislation.
- Species of *Prunus*, in common with many tree species, often have lighter-colored bark in warmer and dryer climates. The bark color often becomes lighter toward the south in North America.

## KEY PUBLICATIONS


Aiello, A.S. 2012. Japanese flowering cherries – a 100-year-long love affair. *Arnoldia* 69(4). https://arboretum.harvard.edu/stories/japanese-flowering-cherries-a-100-year-long-love-affair/.

Chadbund, G. 1972. *Flowering Cherries*. Collins, London. 160 pp.

Cho, M.-S. and S.-C. Kim. 2019. Multiple lines of evidence for independent origin of wild and cultivated flowering cherry (*Prunus yedoensis*). *Frontiers in Plant Science* 10: 1555. https://doi.org/10.3389/fpls.2019.01555.

Jefferson, R.E. and A.E. Fusonie. 1977. *The Japanese Flowering Cherry Trees of Washington, DC; a Living Symbol of Friendship*. National Arboretum Contribution No. 4, U.S. Department of Agricalture, Washington, DC. 66 pp.

Kato, S., A. Matsumoto, K. Yoshimura, T. Katsuki, K. Iwamoto, T. Kawahara, et al. 2014. Origins of Japanese flowering cherry (*Prunus* subgenus *Cerasus*) cultivars revealed using nuclear SSR markers. *Tree Genetics & Genomes* 10: 477–487.

Kuitert, W. and A.H. Peterse. 1999. *Japanese Flowering Cherries*. Timber Press, Portland, OR. 395 pp. www.researchgate.net/profile/Kuitert-Wybe/publication/323846622_Japanese_Flowering_Cherries/links/5aaf439c458515ecebe97741/Japanese-Flowering-Cherries.pdf.

McClellan, A. 2005. *The Cherry Blossom Festival: Sakura Celebration*. Bunker Hill Publishing, Chicago. 96 pp.

Nakamura, I., H. Takahashi, and Y.I. Sato. 2014. Diversity and breeding of flowering cherry in Japan. *Advances in Horticultural Science* 28: 236–243.

Nakamura, M. and K. Blood. 2020. *Cherry Blossoms: Sakura Collections from the Library of Congress*. Smithsonian Books & Library of Congress, Washington, DC. 96 pp.

Primack, R. and H. Higuchi. 2007. Climate change and cherry tree blossom festivals in Japan. *Arnoldia* 65(2): 14–22.

Salazar, G.M. 2022. The cherry blossom and its influence on Japanese culture. *Japanese Society and Culture* 4: 155–165.


### KEY CULTIVATION WEBSITES


Chuk-ar Cherries. *Cherry Fruit Trees vs. Cherry Blossom Trees*. www.chukar.com/blog/cherry-fruit-trees-vs-cherry-blossom-trees.

Clemson University Cooperative Extension. *Ornamental Cherry, Plum, Apricot & Almond*. https://hgic.clemson.edu/factsheet/ornamental-cherry-plum-apricot-almond/.

Diversivore (Blog). *Preserved Cherry Blossoms*. www.diversivore.com/preserved-cherry-blossoms/.

Funk, F. 2009. *Flowering Cherry Trees for Your Own Garden*. Brooklyn Botanical Garden. www.bbg.org/gardening/article/pretty_in_pink.

Southern Living. *How to Grow and Care for Cherry Blossom Trees*. www.southernliving.com/garden/trees/cherry-blossom-tree.

The Spruce. *How to Grow and Care for Japanese Flowering Cherry*. www.thespruce.com/kwanzan-cherry-trees-2132134.

The Spruce. *How to Grow and Care for Purple Leaf Sand Cherry*. www.thespruce.com/purple-leaf-sand-cherry-growing-profile-3269234.

WikiHow. *How to Grow a Cherry Blossom Tree*. www.wikihow.com/Grow-a-Cherry-Blossom-Tree.

# 25 Christmas and Other Holiday Cacti (*Schlumbergera* Species and Hybrids)

**Scientific Names:** The genus name *Schlumbergera* commemorates Frédéric Schlumberger (1823–1893), a collector of cacti at his chateau near Rouen, France, and a patron of horticultural botany.

**Common Names:** "Christmas cactus" is the most widely encountered name applied to cultivars of *Schlumbergera*, but the names "Thanksgiving cactus" and "Easter cactus" are also employed for fall-flowering and spring-flowering plants, respectively. Additional information on the names, which are quite ambiguous, is presented in the following discussions. The plural of *cactus* is *cactuses* or *cacti* (the latter is more popular).

## 25.1 INTRODUCTION

Christmas cactus and its close relatives are easy to grow, long-lived, attractive flowering houseplants said to thrive on benign neglect, making them ideal companions for perhaps the majority of people. The Christmas cactus is second only to poinsettia as the most popular Christmas holiday gift plant.

## 25.2 OVERVIEW

Cacti are native to North and South America and the West Indies (some species of the primitive genus *Rhipsalis* are apparently native to Madagascar and a few other parts of the Old World) and characteristically grow in hot, dry, and hostile desert areas. The family has over 100 genera and about 1600 species, nearly all of which have succulent, spiny stems. In most cacti, there are no true leaves, but the stems and branches are often green because of the presence of chlorophyll, which transforms the energy of the sun into carbohydrates. Frequently, the branches (often the stems as well) are made up of flattened, attached segments (technically phylloclades, also called cladodes, joints, pads, and paddles). The cladodes have the function of leaves, collecting sunlight. Cacti are favorite ornamentals, and many people specialize in collecting and cultivating these unusual plants. It has been claimed that there are more species of cacti growing in homes (especially in cold regions) than species of any other plant family (about 300

FIGURE 25.1 The original interpretations of holiday cacti, as illustrated in early botanical illustrations (public domain): (a) "Thanksgiving cactus" (*Schlumbergera truncata*). (b) "Christmas cactus" (*S.* × *buckleyi* = *S. truncata* × *S. russelliana*). (c) "Easter cactus" (*S. gaertneri*).

***Source:*** **(a, b) Schumann, K. M., M. Gürke, and F. Vaupel. 1904. *Blühende Kakteen – Iconographia Cactacearum*. Photo by Peter A. Mansfeld. (c) Step, E., and W. Watson. 1896–1897. *Favourite Flowers of Garden and Greenhouse*. Frederick Warne, London.**

FIGURE 25.2 *Schlumbergera* × *buckleyi* (= *S. truncata* × *S. russelliana*). The original "Christmas cactus," but now only a small proportion of holiday cacti.

***Source:*** **(a) Pixabay (public domain). (b) Photo by Fanghong (CC BY SA 3.0). (c) Photo by BotBln (CC BY SA 3.0).**

FIGURE 25.3 *Schlumbergera truncata*. Thanksgiving cactus + most Christmas cactus.

***Source:*** **(a) Photo by Shutterstock/Steamroller Blues. (b) Photo by Shutterstock/Nadezhda Nesterova. (c) Photo by Maja Dumat (CC BY 2.0). (d) Photo by Maja Dumat (CC BY 2.0).**


DOI: 10.1201/9781003586227-25

FIGURE 25.4 Easter cactus (*Schlumbergera gaertneri*).

*Source:* (a) Photo by Shutterstock/KarenHBlack. (b) Photo by HQ (CC BY SA 2.0). (c) Photo by Marco Verch (CC BY 2.0). (d) Photo by Shutterstock/Ddok.

FIGURE 25.5 Flower variants of Easter cactus hybrids (*Schlumbergera* × *graeseri* = *S. rosea* × *S. gaertneri*).

*Source:* (a, b) Photos by Dominicus Johannes Bergsma (CC BY SA 3.0). (c) Photo by FoeNyx (CC BY SA 3.0).

species of cacti are available as ornamentals). Unfortunately, many wild species have been badly overcollected from nature. There is legislation that now protects cacti. CITES (Convention on International Trade in Endangered Species of Wild Fauna and Flora) requires appropriate permits and documentation for moving plants across international boundaries, and this applies to all cacti, except certain species used in commerce. Arizona law requires that native cacti be tagged and a fee paid before they can be moved. California also has strict requirements that deal with the movement of cacti.

The genus *Schlumbergera* includes less than ten species, all native to shady habitats in cool, humid tropical and semi-tropical coastal mountainous forests of southeastern Brazil. The wild plants grow on rocks and trees, often developing into small woody shrubs rarely exceeding 1 m in height. Several of the species are favorite houseplants. Depending on when they flower, they are commonly called Thanksgiving cactus, Easter cactus, or Christmas cactus, and collectively they are known as holiday cactuses.

Flowers develop from the edges of the stem segments, often from notches on the sides or on the tips of the segments. The flowers of *Schlumbergera* species are either erect or hang down and are adapted to being pollinated by hummingbirds. They have a tubular shape, forcing the birds to probe for nectar at the base. The wild species tend to have reddish flowers, but cultivars have been selected in which the flowers may be orange, pink, purple, red, white, or yellow. In many cultivars, flower color is modified when the plants are grown in cold temperatures (below 14°C), for example, altering normally white and yellow cultivars to pink. Large old specimens can bear hundreds of flowers.

A simplified interpretation of the three basic categories of holiday cactus is given in Table 25.1. Unfortunately, the

TABLE 25.1
**Simplified Descriptions of the Three Principal Groups of Cultivars of *Schlumbergera* ("Holiday Cacti")**

| | | | |
|---|---|---|---|
| **Common Names** | Thanksgiving cactus, false Christmas cactus, claw cactus, crab cactus, zygocactus (most plants sold as Christmas cactus are likely this species) | Christmas cactus | Easter cactus,[1] Whitsun cactus (Whitsun is derived from Pentecost, a day for baptisms) |
| **Flowering Time (Northern Hemisphere)** | Late fall | Winter | Late winter to spring |
| **Scientific Names** | *S. truncata* | *S.* × *buckleyi* (*S. truncata* × *S. russelliana*), *S. bridgesii* | *S. gaertneri*, *Hatiora gaertneri*, *Rhipsalidopsis gaertneri* |
| **Frequency** | Dominates the holiday cactus market (along with hybrids) | Infrequent (despite the name "Christmas cactus," most so-called Christmas cacti not this plant) | Rare |
| **Cladodes (Stem Segments)** | A few teeth on the margin, upwardly pointed (analogized to claws, hence the name claw cactus); often two large teeth at the top end of the segment | Rounded bulges (not pointed teeth) on margin | Obscure, rounded bulges on margin; small golden bristles on the top ends (evident when flowers not also on the ends of the segments) |
| **Flowers** | Bilaterally symmetrical lengthwise; upper half differing from the lower half | Bilaterally symmetrical | Radially symmetrical (daisylike arrangement of petals) |

[1] *Schlumbergera* × *graeseri* = *S. rosea* × *S. gaertneri* is also marketed as Easter cactus.

phrases "Christmas cactus" and "Easter cactus" are applied to different species and hybrids by different writers. Most discussions of holiday cacti (both online and in print) lack insight into the ambiguity of the names. Regardless, as with many ornamental plants, extensive hybridization has obscured the differences that exist among the wild species, so that identifying the correct scientific names for a given cultivar can be difficult or impossible. For practical purposes, the phrase "Christmas cactus" can be applied to all cultivars flowering before midwinter (all these have at least some parentage of *S. truncata*, and the flowers are quite bilaterally asymmetrical, that is, some petals unlike others), and the phrase "Easter cactus" can be applied to all cultivars flowering after midwinter (all these have at least some parentage of *S. gaertneri*, and the flowers are symmetrical, that is, like a daisy, with all the petals appearing similar).

## 25.3 HISTORY

The ancestors of modern holiday cacti were discovered in southeast Brazil in the early 19th century. Subsequent hybridization, first in Europe, and later in other countries, has produced hundreds of cultivars. The tradition of giving the plants as holiday gifts first became popular during the 19th century in Victorian England.

## 25.4 ECONOMIC IMPORTANCE

Holiday cacti are extremely popular houseplants, mostly purchased for personal use or as gifts. Annual world commercial production amounts to hundreds of millions of plants. With the exception of the cactus pear (*Opuntia ficus-indica*), grown for its edible fruits and stem segments, the Christmas cactus is probably the most widely grown of all cacti.

FIGURE 25.6 Use of holiday cacti as houseplants (all *Schlumbergera truncata*, except *a*, which is *S. gaertneri*): (a) Easter cactus in a hanging basket. (b) Christmas cactus on a windowsill. (c) Christmas cactus over a fireplace (undesirable if heated). (d) Christmas cactus grown as bonsai.

*Source:* **(a) Photo by Johan Fredriksson (CC BY SA 3.0). (b) Photo by Empereur Day (CC BY SA 3.0). (c) Photo by Jinx McCombs (CC BY ND 2.0). (d) Photo by Emmanuelm (CC BY 3.0).**

## 25.5 ORNAMENTAL USES

Holiday cacti are predominantly employed as potted plants, especially as houseplants, typically grown on windowsills. They can be regarded also as patio plants (temporarily in cold climates) and even as border plants for shady, cool locations in suitable, semi-tropical climates.

## 25.6 OTHER USES

Aside from its use as a flowering ornamental, *Schlumbergera* species do not appear to have acquired other significant uses. Unlike most other major ornamentals, they were discovered quite recently and occupied a very limited natural distribution area, so people have not had much opportunity to find other uses.

## 25.7 PROBLEMS

Occasionally, some people sensitive to holiday cacti experience contact dermatitis, but the plants are considered to be non-toxic to people and pets.

## 25.8 CULTIVATION

### 25.8.1 Climate and Site

The recommended indoor temperature range of holiday cactuses is 10 to 21°C. To provide the needed humidity in dry

FIGURE 25.7 Christmas cactus (*Schlumbergera truncata*) in a bathroom. Kitchens and bathrooms are good locations to maintain the plants (if sufficient light is available) because of the high humidity.

*Source:* **Photo by chrisbb@prodigy.ne (CC BY 2.0).**

houses, the plants can be misted, and the pot can be placed on a saucer of pebbles filled with water. Keep the plants away from heat sources and cold drafts. Indoors, given the limited light of Northern winters, it is often desirable to maintain the plants in a sunny location. In the summer, the pots can be moved outdoors until the autumn, if they are placed in a shady or semi-shady location. Excessive light will turn the foliage pink or even burn the plants.

### 25.8.2 Propagation and/or Purchasing

Holiday cactuses are widely marketed in the weeks preceding holidays and have been conditioned to flower at the appropriate time. Because the plants are so popular as gifts, most plants are obtained this way.

Plants can be produced from seeds but are unlikely to flower in less than 3 or 4 years, and the popular hybrids won't reproduce true to type. The stem segments are usually employed to grow new plants. Segments can simply be twisted off (this is recommended in contrast to cutting), the damaged ends allowed to dry and callus for a few days and inserted about 2.5 cm deep in moist potting mix. The container can be covered with plastic to provide a humid environment. Segments placed partly in water will also root, but significantly less well. It may require 3 to 12 weeks for plants to root, at which stage they should be potted individually.

### 25.8.3 Planting

As noted earlier, stem sections (preferably from the vigorously growing ends of branches) are employed as starting material. Holiday cactuses do well in smaller pots with limited root space (simulating their natural occupation of sites among rocks or in the crooks of trees). Once plants become very root-bound, they should be transferred to larger pots.

FIGURE 25.8 Christmas cactus plants (induced to flower) in small pots, on sale in a California store.

*Source:* **Photo by the Greenery Nursery and Garden Shop (CC BY 2.0).**

### 25.8.4 Soil and Moisture Conditions

The potting mix should be loose, fast-draining, and somewhat acidic (pH 5.0 to 6.0). The soil should be allowed to dry out partly between waterings. A balanced fertilizer can be applied according to directions, about every 2 to 4 weeks, or a granular slow-release formulation can be provided.

### 25.8.5 Maintaining Plants

Following flowering (which may last 4 to 6 weeks), the plants will enter a new growth cycle. Pruning to keep the shape of the plant attractive is best done after blooming. *Schlumberga* can be very long-lived, sometimes lasting 50 years, occasionally as long as a century.

To get plants to reflower, a combination of cool temperature (ideally 10 to 13°C) and short days (fewer hours of daylight; 12 to 16 hr a day of total darkness) beginning 6 to 8 weeks for Thanksgiving and Christmas cacti ahead of the desired flowering period is recommended. This is because the plants are "short-day" plants, induced to flower in the fall by shortening days. Admittedly, these conditions are not easy to provide. By contrast, Easter cactus (technically a long-day plant) is induced to flower by long days in the spring. Avoid fertilizing holiday cacti while the plants are in bloom.

## 25.9 FASCINATING FACTS, FABLES, FOLKLORE, AND FRIVOLITY

- Belief in fairies is ancient, but especially in 19th- and early 20th-century Europe, several illustrators specialized on showing fairies in costumes representing well-known flowers. Because the holiday cacti did not become well-known until the 20th century, they are rarely shown in these illustrations. The best-known Christmas cactus fairy figure was prepared by the skilled French artist J. J. Grandville (1803–1847).
- The holiday cacti (note that "holiday" is based on "holy day") are associated with Christian holidays: Thanksgiving, Easter, Christmas – a result of Christian Europeans discovering, developing, and popularizing the plants. Virtually all the established religions are ancient, dating back long before the 19th-century discovery of the holiday cacti, and the plants that are associated with the Abrahamic religions (Judaism, Christianity, Islam) are almost all indigenous to the Old World. The holiday cacti are exceptional in acquiring religious association in very recent times.
- Brazil, the native homeland of Christmas cactus, is of course in the Southern Hemisphere, where, contrary to the Northern Hemisphere, the days are long at Christmas. Because Christmas cactus is induced to flower by short days, in the Southern Hemisphere, it blooms in May, not December. Instead of calling the plant Christmas cactus, Brazilians call it *flor*

FIGURE 25.9 Christmas cactus fairy, exemplifying the popular 19th-century European practice of personifying flowers as fairies.

*Source:* **Grandville, J. J. 1846. *Fleurs Animées*. Printed by Dubuisson, Paris.**

FIGURE 25.10 Misidentification of Christmas and Easter cacti: (a) "Christmas cactus." (b) "Easter cactus."

*Source:* **(a) Illustration by Shutterstock/Larry Rains. (b) Illustration by Shutterstock/Ikinugrihi.**

*de maio* (May flower), a phrase widely used in the Northern Hemisphere for other species.

- Most cacti are conspicuously succulent and spiny, so not surprisingly, many have difficulty in recognizing Christmas and Easter cacti as genuine cactuses. Numerous succulent ornamentals are often misidentified as cacti, most notably poinsettia.

## KEY PUBLICATIONS

Boyle, T.H. 1997. Holiday and Easter cactus. In: *Tips on Growing Specialty Potted Crops*, edited by M.L. Gaston, S.A. Carver, C.A. Irwin, and R.A. Larson, 82–88. Ohio Florists' Association, Columbus, OH.

Boyle, T.H. 2007. Cacti: *Schlumbergera truncata*, *S.* × *buckleyi*, *Hatiora gaertneri*. In: *Flower Breeding and Genetics*, edited by N.O. Anderson, 360–388. Springer, Dordrecht.

Calvente, A., D.C. Zappi, F. Forest, and L.G. Lohmann. 2011. Molecular phylogeny of tribe Rhipsalideae (Cactaceae) and taxonomic implications for *Schlumbergera* and *Hatiora*. *Molecular Phylogenetics and Evolution* 58: 456–468.

McMillan, A.J.S. and J.F. Horobin. 1995. *Christmas Cacti: The Genus Schlumbergera and Its Hybrids*. David Hunt, Sherbourne, Dorset, UK. 160 pp.

Mildred, E. Mathias Botanical Garden, UCLA. 2019. *Cladode or Phylloclade*. www.botgard.ucla.edu/cladode-or-phylloclade/.

Nobel, P.S. (ed.). 2002. *Cacti. Biology and Uses*. University of California Press, Los Angeles. 280 pp.

Rünger, W. and R.T. Poole. 1985. Schlumbergera. In: *CRC Handbook of Flowering*, edited by A.H. Halevy, 277–282. CRC Press, Boca Raton, FL.

Sriskandarajah, S., E.A.D. Al-Ramamneh, and M. Serek. 2007. Biotechnology of *Schlumbergera* and *Rhipsalidopsis*. *Floriculture and Ornamental Biotechnology* 1: 14–19.

## Key Cultivation Websites

The Spruce. *How to Care for Christmas Cactus*. www.thespruce.com/christmas-cactus-4176983.

The Spruce. *How to Grow and Care for the Thanksgiving Cactus*. www.thespruce.com/thanksgiving-cactus-care-guide-5199419.

University of Massachusetts Extension Greenhouse Crops and Floriculture Program. *Commercial Production of Holiday Cacti*. https://ag.umass.edu/greenhouse-floriculture/fact-sheets/commercial-producton-of-holiday-cacti.

WikiHow. *How to Care for a Christmas Cactus*. www.wikihow.com/Care-for-a-Christmas-Cactus.

WikiHow. *How to Get a Christmas Cactus to Bloom*. www.wikihow.life/Get-a-Christmas-Cactus-to-Bloom.

WikiHow. *How to Plant a Christmas Cactus*. www.wikihow.life/Plant-a-Christmas-Cactus.

WikiHow. *How to Propagate Christmas Cactus*. www.wikihow.com/Propagate-Christmas-Cactus.

WikiHow. *How to Prune a Christmas Cactus*. www.wikihow.com/Prune-a-Christmas-Cactus.

Wisconsin Horticulture, Extension. *Holiday Cactus*. https://hort.extension.wisc.edu/articles/holiday-cactus/.

# 26 Chrysanthemum (*Chrysanthemum* Species and Hybrids)

**Scientific Names:** The name *Chrysanthemum* is based on the Greek *chrysos*, "gold," and *anthemon*, "flower." In the late 20th century, because of technical nomenclatural considerations, chrysanthemums were re-assigned to the genus *Dendranthema*. Because chrysanthemums were well-known as the genus *Chrysanthemum*, in 1995, a formal proposal was made to the International Botanical Congress to conserve ("resurrect") the genus name *Chrysanthemum*. The proposal was approved at the 1999 meeting of the International Botanical Congress and published in the resulting 2000 "St. Louis Code of Botanical Nomenclature." Most ornamental chrysanthemums are hybrids identifiable as *Chrysanthemum* × *morifolium* (synonyms: *C.* × *grandiflorum*, *Dendranthema* × *grandiflorum*, *D.* × *morifolium*).

**Common Names:** Chrysanthemum, mum, chrysanth. "Garden chrysanthemums," "hardy chrysanthemums," and "fall mums" are phrases applied to cultivars hardy enough to overwinter outdoors. "Florist's chrysanthemums" and "florist's daisy," by contrast, produce showier flowers, but the plants are rarely winter-hardy.

FIGURE 26.1 Examples of ornamental chrysanthemums (*Chrysanthemum* × *morifolium*): (a) Chrysanthemum flowers of various colors. (b) An exhibition of cultivars at a chrysanthemum festival in Japan. (c) A well-developed shrub growing in the ground. (d) Mixed cultivars growing in a pot.

***Source:* (a) Photo by Jebulon (public domain). (b) Photo by Marufish (CC BY SA 2.0). (c) Photo by KrakowiakKW from Pixabay (public domain). (d) Photo by Shutterstock/Grafvision.**

## 26.1 INTRODUCTION

The chrysanthemum is one of the world's leading ornamental plants. It is widely known as the "queen of autumn" and "queen of fall flowers," reflecting its domination of the late season for flowering plants. The hardy cultivars suited for home gardens and public flower beds produce spectacular domes of yellow, burgundy, plum, white, pink, red, orange, and lavender blossoms that can endure early frosts. In addition, an astonishing array of non-hardy floral forms are raised in greenhouses for the cut flower and indoor potted plant industries.

## 26.2 OVERVIEW

The genus *Chrysanthemum* consists of about 40 species (or fewer according to some sources) distributed mainly in temperate East Asia. About half occur in China, and some are indigenous to northeastern Europe. In the past, *Chrysanthemum* included more species than it does at present. Many have been transferred to *Argyranthemum*, *Glebionis*, *Leucanthemopsis*, *Leucanthemum*, *Rhodanthemum*, and *Tanacetum*. *Chrysanthemum* species are herbaceous perennials or subshrubs with characteristically palmately divided leaves, often with marginally toothed leaflets. The wild species have white, pink, or (especially) yellow petals. Many can be crossed with each other, and there are numerous hybrids in nature as well as in cultivation. Moreover, some intergeneric crosses are known between species of *Chrysanthemum* and species in allied genera,

FIGURE 26.2 Examples of beautiful flowers of garden chrysanthemums (*Chrysanthemum* × *morifolium*).

***Source:* (a) Photo by Nia Ran01 (CC BY SA 4.0). (b) Photo by MrNgochu from Pixabay (public domain). (c) Photo by Forest and Kim Starr (CC BY 3.0). (d) Photo by Babij (CC BY SA 2.0). (e) Photo by Alicja from Pixabay (public domain). (f) Photo by Niki Emmert from Pixabay (public domain).**

DOI: 10.1201/9781003586227-26

such as *Ajania*. Even more distantly related genera (notably *Artemisia*) can be crossed with *Chrysanthemum*, but embryo rescue techniques are necessary to obtain a viable hybrid.

The true flowers of the Asteraceae or daisy family are very small ("florets") and aggregated into heads ("capitula") that most people interpret as conventional flowers. As with the majority of the species of the family, in *Chrysanthemum*, the outermost flowers are "ray flowers," with a prominently elongated corolla (the petals, collectively) that is interpreted by most people as a single petal, and the central flowers in the head are "disk flowers," lacking a prominent corolla. In *Chrysanthemum*, both types of floret are fertile (capable of producing seeds), but the ray flowers are entirely male, while the disk flowers are bisexual, bearing both a stigma and anthers. Compared to the wild species, cultivated forms have showier flower heads, often bred to have several rows of ray flowers, rather than one row as in the wild species. As noted in the following, a spectacular variety of flowers and flower heads have been created by human selection. Petals may be strap-like, narrow and lacy, spoon-shaped, feathery, or quill-shaped. The flowers may be so numerous that the plant takes on the appearance of a mound covered with blooms. Almost all colors of the spectrum are encountered (blue is noticeably absent), and flower hues may be solid, variegated, or multicolored.

### 26.2.1 Floral Classes

Several classes of chrysanthemum cultivars are recognized on the basis of flower characteristics, although there are intermediate kinds that make identification challenging. Given that chrysanthemum is often considered to be the world's second-most important ornamental flower (after rose) and garden sources often specify these types, a summary is presented here (Table 26.1). The information presented in the preceding paragraph will assist in interpreting the floral classes. However, only the most frequently encountered kinds are illustrated here.

**TABLE 26.1**

**Floral Classes of Chrysanthemum Cultivars Recognized Initially by the United Kingdom's National Chrysanthemum Society and Accepted by Most Chrysanthemum Societies throughout the Western World**

| Class | Description |
|---|---|
| Class I – irregular incurve | These are giant blooms. The ray petals loosely incurve (i.e., the petals curve upward toward the top center of the flower head) and make fully closed centers. The lower ray petals present an irregular appearance and may give a skirted effect. |
| Class 2 – reflex | The ray petals in this class curve downward (the tips curving toward the bottom center of the flower head, exactly opposite to classes 1 and 3) and overlap, reminiscent of bird plumage. The tops of these blooms are full, but somewhat flattened. |
| Class 3 – regular incurve | A true globular bloom equal in breadth and depth. The ray petals smoothly incurve outward and upward to form a regular sphere of dense petals. |
| Class 4 – decorative | A flattened bloom with short petals. As in the previous classes, the central disk should not be visible. The upper ray petals tend to incurve, but the lower petals generally reflex. |
| Class 5 – intermediate | This bloom class is smaller than the irregular incurve, with shorter ray petals, only partially incurving with full centers, but giving a more open appearance. Many of the popular commercial incurving types are in this intermediate class. |
| Class 6 – pompon | A small globular bloom, somewhat flat when young but fully rounded when mature. Size ranges from small button types to large disbudded blooms almost 10 cm in diameter, but low bushes are frequent. The ray petals incurve or reflex in a regular manner and fully conceal the center. |
| Class 7 – single and demi-double | A daisylike flower with a central disk and one or more rows of ray petals. These cultivars tend to be winter-hardy. |
| Class 8 – anemone | These blooms are similar to the semi-doubles but have a raised cushion-like center which can extend up to half the diameter of the flower head. |
| Class 9 – spoon | Essentially the same as the semi-double, except the ray flowers are like spoons at the tips. The central disk is round and visible. |
| Class 10 – quill | The ray flowers in this type are straight and tubular, with open tips. The bloom is fully double, with no open center. |
| Class 11 – spider | Spiders have long tubular ray flowers which may coil or hook at the ends. The florets may be very fine to coarse and drape gracefully at full bloom. The center is not obvious. |
| Class 12 – brush or thistle | Fine tubular ray flowers which grow parallel to the stem and resemble thistles or an artist's paintbrushes. |
| Class 13 – unclassified | Those blooms which fit in none of the other classes. They are often exotic, with twisted ray flowers. They may also be a combination of two classes. |

FIGURE 26.3 Examples of floral classes mentioned in Table 26.1.

*Source:* **(a) Photo by +- (CC BY SA 3.0). (b) Photo by Vargazs from Pixabay (public domain). (c) Photo by Anrita1705 from Pixabay (public domain). (d) Photo by Elena Chochkova (CC BY 3.0). (e) Photo by Juliana Coutinho (CC BY 2.0). (f) Photo by Kristine Paulus (CC BY 3.0). (g) Photo by Anonymous (CC BY 2.0). (h) Photo by Tata Aka T (CC BY 2.0). (i) Photo by Jam 343 (CC BY 2.0).**

### 26.2.2 Hardiness Classes

"Florist's chrysanthemums" are usually raised in greenhouses and sold for use as houseplants or as annual patio plants. They may have been treated with dwarfing agents to produce a compact appearance. They are relatively attractive, often with large flower heads (over 7 cm in diameter), but frequently they are not winter-hardy and so usually should not be transplanted outdoors in cold climates. By contrast, "garden chrysanthemums" or "hardy chrysanthemums" are temperature-resistant enough to survive overwinter in a particular climatic zone and will regrow in the spring from underground suckers. Many garden chrysanthemums are "cushion chrysanthemums" (sometimes called "azalea mums"). These are low (typically no higher than 30 cm), bushy plants with the appearance of a mound so fully covered with flowers that the foliage is obscured. Frequently, single, pompon, and anemone floral types – all usually small-flowered – are winter hardy. Hardy chrysanthemums have been selected to bloom before a killing frost, but they can usually survive mild frosts and still produce blooms (the flowers of florist's chrysanthemums may be killed under the same frosty conditions, or the plants may simply fail to come into flower before a killing frost). Moreover, florist's chrysanthemums will usually not regrow after a cold winter. Some gardeners in cold climates treat chrysanthemums as annuals, allowing more attractive, albeit less-hardy, cultivars to be grown.

FIGURE 26.4 (a) Hardy "cushion chrysanthemums," illustrating the typical low, bushy, mound form covered with relatively small flower heads. (b) Chrysanthemum 'Jewel Box'.

*Source:* **(a) Photo by JebulonOliBac (public domain). (b) Photo by F. D. Richards (CC BY SA 2.0).**

### 26.2.3 Greenhouse Trade Form Classes

"Standards" are plants with one large terminal flower head, typically used for the cut flower trade. These are produced by removing all branches but one and removing all lateral flower buds from that branch.

"Disbuds" are like standards, but with more than one main branch. These are multi-stem plants produced by pinching the growing points of rooted cuttings, and (like standards) all lateral flower buds are removed to develop one large terminal flower head on each branch. These are mostly used for production of potted plants.

"Sprays" are multi-stemmed plants with the terminal flower bud removed to stimulate all lateral flower buds to develop into flowering heads. These are also used for production of potted plants.

### 26.2.4 Photoperiodic Flowering

Flowering in chrysanthemums is induced by photoperiod, and most cultivars are short-day plants, requiring short days to flower. (The expression "short-day" is now established, but in fact, length of the dark period is the determining factor, and plants that flower in response to shortening days should have been called "long-night plants.") Plants flower when the day length is shorter than the critical day length, but grow vegetatively otherwise. It has been discovered that there are actually two critical photoperiods in chrysanthemums – one for floral initiation, and a shorter day length for flower development. Moreover, the critical photoperiod varies with temperature. Florist's chrysanthemums typically require 8 to 14 weeks of short days to flower, and the day length period may be shorter than 11 hr or longer than 16 hr, depending on cultivar. In practice, the greenhouse trade uses black cloth curtains to shade plants for 12–15 hr daily to induce flower initiation. Temperatures above 20°C and below 22°C will delay flowering. Chrysanthemums are known as fall-flowering plants, but garden cultivars are available that will flower earlier and so are more suited to cold climates. Garden cultivars typically require 5 to 7 weeks of short days (a shorter period than needed by florist's chrysanthemums). Streetlights or other sources of light at night can inhibit the formation of flowers.

## 26.3 HISTORY

Dozens of *Chrysanthemum* species are grown, but almost all cultivated forms are hybrids. The main parent of modern hybrid chrysanthemums is *C. indicum* (*Dendranthema indicum*). It is widely dispersed in China, northern Russia, Southeast Asia, and Europe, occurring in diverse environments, including mountain slopes, grasslands, shrublands, wetlands, roadsides, and coastal saline areas. The aerial portions are employed in traditional Chinese medicine, and overharvesting and heavy grazing in recent decades have decreased its occurrence. Other species that have been considered to have contributed to the ancestry of hybrid chrysanthemums include *C. chanetii*, *C. erubescens*, *C. japonense*, *C. lavandulifolium*, *C. makinoi*, *C. ornatum*, *C. sinense*, *C. vestitum*, and *C. zawadskii*.

Chrysanthemums have been cultivated in China for at least 1,600 years, and soon after that in Korea and Japan.

FIGURE 26.5 Nineteenth-century paintings (public domain) of old cultivars of chrysanthemum by European botanical illustrators.

***Source:*** **(a) Reproductions of flowers from old flower books by Rawpixel (CC BY 4.0). (b) Flore *des serres et des jardins de l'Europe* (Ghent, Belgium), vol. 14, 1861. (c) *Annales de la Société Royale d'Agriculture et de Botanique de Gand*, vol. 5, 1849.**

FIGURE 26.6 Commercial production and marketing of chrysanthemums: (a) A chrysanthemum farm in Blue Anchor, New Jersey. (b) Potted chrysanthemums on sale at a plant nursery. (c) A florist preparing a chrysanthemum bouquet.

***Source:*** **(a) Photo by Shutterstock/TommyB Photo. (b) Photo by Wuestenigel (CC BY 2.0). (c) Photo by Shutterstock/Svyatoslav Balan.**

By the 11th century, there were about 3 dozen cultivars in China. By 1990, more than 3,000 cultivars from China were recorded. However, cultivars were not available in Europe and North America until the 1800s. Cultivars of garden chrysanthemum arose as hybrids of older cultivars, as hybrids of cultivars with wild species, and as bud mutations from older cultivars.

## 26.4 ECONOMIC IMPORTANCE

Chrysanthemums are thought to be second only to roses in the world floral market (others claim tulips are the world's most profitable crop). Chrysanthemum is the most popular flowering plant of Japan and China. It is one of the world's leading floral crops grown specifically for cut flowers (for floral bouquets and flower arrangements) and as potted plants. Chrysanthemum is one of the top three best-selling cut flowers in almost all major flower-consuming countries. Leading producers include China, Japan, India, The Netherlands, South Korea, Malaysia, Kenya, Columbia, and Brazil. The success of chrysanthemum as a commercial greenhouse product is due to diversity of floral color and form, lasting quality of the flowers, ease of cultivation, and ability to control and schedule flowering.

## 26.5 ORNAMENTAL USES

Chrysanthemums are commercially produced in huge amounts for sales as cut flowers and short-term potted plants, but there is also a substantial market for starter plants and seeds for gardens.

Chrysanthemums are typically planted outdoors in massed landscape beds or at least in small groups. They are also suitable for borders, especially the cushion mum forms, which show best as edging plants or in the foreground of a border. In addition, chrysanthemums are particularly suitable for container gardening. They are sometimes grown as bonsai, and occasionally, grafting is employed to produce extraordinary combinations of form and color.

## 26.6 OTHER USES

Although mainly employed as an ornamental, some "chrysanthemums" have culinary uses. Chop suey green (also

FIGURE 26.7 In-ground plantings of chrysanthemum: (a) An exhibition of chrysanthemums in Shilin Residence Park, Taiwan. (b) A beautiful bush.

***Source:*** **(a) Photo by Lienyuan Lee (CC BY SA 3.0). (b) Photo by Shutterstock/APugach.**

FIGURE 26.8 Chrysanthemums in containers: (a) A variety of cultivars. (b) A residential garden in Japan.

*Source:* **(a) Photo by Shutterstock/Flower Studio. (b) Photo by Caoha from Pixabay (public domain).**

FIGURE 26.9 Chrysanthemums grown as bonsai.

*Source:* **(a) Photo by Mwollohalkka (CC BY SA 4.0). (b) Photo by Gryffindor (CC BY SA 4.0).**

known as Japanese green, garland chrysanthemum, and crown daisy), *Glebionis coronaria* (*Chrysanthemum coronarium*), is an edible species popular in Chinese, Vietnamese, and Japanese cuisine. The vegetable form of the species has fleshier leaves than the kinds cultivated as ornamentals. Chrysanthemum petals are also sometimes eaten and are considered non-toxic. In a Chinese restaurant, "chrysanthemum tea" is likely to be a blend of chrysanthemum blossoms and conventional tea. The flowers of ornamental chrysanthemums (*C.* × *morifolium*) are also employed as an herbal tea in traditional Asian (especially Chinese) medicine. Sometimes, extracts are also used medicinally.

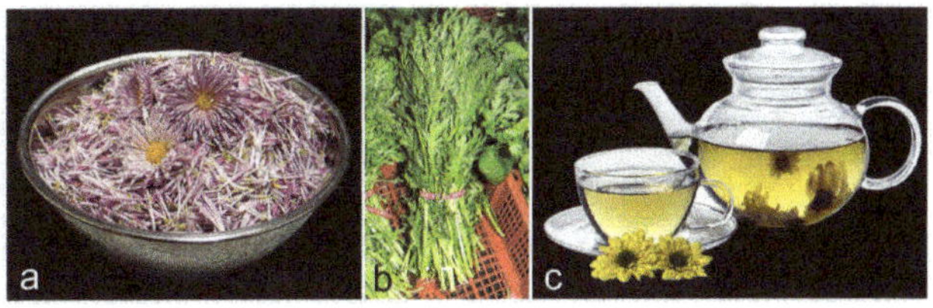


FIGURE 26.10 Food uses of chrysanthemums: (a) Edible flowers. (b) Chop suey green (*Glebionis coronaria*), an edible-leaved relative of chrysanthemum. (c) A glass teapot with Chinese chrysanthemum tea.

*Source:* **(a) Photo by T. Kiya (CC BY SA 2.0). (b) Photo by ChildofMidnight (CC BY 3.0). (c) Photo by Shutterstock/Picture Partners.**

## 26.7 PROBLEMS

Florists, nursery workers, and gardeners occasionally develop contact dermatitis after prolonged exposure to garden chrysanthemums. Arteglasin A, a sesquiterpene lactone, has been identified as a cause. Other chemicals of this class are also present in chrysanthemums and, indeed, in many other species of the Asteraceae family. All parts of chrysanthemum plants have been claimed to be potentially toxic to dogs, cats, horses, and other mammals, ingestion causing vomiting, diarrhea, salivation, or a lack of coordination. Although not considered particularly toxic, the recommendation has been made that pets and children not be exposed to chrysanthemums. As noted earlier, the flowers of chrysanthemums (*C.* × *morifolium*) are consumed by humans to a minor extent, and so is the foliage of "edible chrysanthemum" (*Glebionis coronaria*). Some people have experienced mild stomach upset from ingestion or a skin reaction from skin contact with *G. coronaria.*

*Tanacetum cinerariaefolium* (*Chrysanthemum cinerariaefolium*) is the source of pyrethrins, very widely used as a natural insecticide and insect repellent. Pyrethrins are relatively non-toxic to humans and warm-blooded animals, but cats and fish are notably susceptible. Many information sources (especially on the web) consider *T. cinerariaefolium* to be a "chrysanthemum" and confound toxicity information for it with ornamental chrysanthemum (*C.* × *morifolium*). Indeed, it is often difficult to know just what species are being referred to when the toxicity of "chrysanthemums" is being discussed.

## 26.8 CULTIVATION

### 26.8.1 Climate and Site

Chrysanthemums are adapted to a temperate climate, optimally in the range 18–21°C. Most garden chrysanthemums are hardy in zones 5 to 9. They need full sun (at least for 6 continuous hours daily). As noted earlier, chrysanthemums require a period of uninterrupted darkness for weeks in order to develop flowers in the autumn. If they are planted near streetlights or strong night lights, the artificial lighting could prevent the growth of flowers.

FIGURE 26.11 Painting of a chrysanthemum garden by American artist Daniel Ridgway Knight (1839–1924).

*Source:* **Public domain.**

### 26.8.2 Propagation and/or Purchasing

Ornamental garden chrysanthemums are mostly grown from vegetative stem cuttings or suckers (not seeds) and sold as young plants in pots intended for transplanting both in the spring and in the autumn. Plants should be selected that have unopened buds or are just beginning to produce open flowers. Plants with fading flowers, damaged stems, or discolored foliage should be avoided. Plants available in the spring may have been forced into flower (after the initial bloom has faded, these can be pruned to one-third or one-half their height, and a second blooming period will probably result in the fall). Plants acquired in the autumn are less likely to harden off sufficiently to survive winter. Alternative to planting these, they might last through the winter in a cool, dry location that does not freeze.

Chrysanthemums can become crowded as they grow and should be divided or reproduced from cuttings or suckers (preferably between 2 and 4 years of age). To divide, dig the entire clump up in the spring when new shoots appear; slice it into wedges (like a pie), discard the central (older) portion, and replant the outer portions. It is also possible to propagate the plants using cuttings about 10 cm long from the tops of young stems 15–30 cm long (dip the cut ends in rooting hormone, stick the cuttings in a container of sterile potting soil or a mix of sand and peat moss, maintain in bright but not direct light, and after roots form in 2–3 weeks, acclimate to brighter light and transplant).

### 26.8.3 Planting

Chrysanthemums can be planted in the spring (preferably) or in the fall. Plants purchased in the fall should be established in the garden at least 6 weeks before a killing frost. Garden chrysanthemums should be planted at least 0.5 m apart and up to 1 m for vigorous cultivars.

### 26.8.4 Soil and Moisture Conditions

Chrysanthemums prefer fertile, neutral or slightly acidic (ideally, pH 5.7 to 6.2), well-drained but moist soils. Light, sandy soils are not suitable. The addition of abundant organic matter in the form of peat, compost, or well-rotted manure is recommended.

### 26.8.5 Maintaining Plants

Potted chrysanthemums are routinely shaped by commercial growers. Similarly, "pinching" is recommended for many cultivars grown in the home garden, to prevent the plants from becoming tall and leggy and producing a poor floral display. "Pinching" refers to removing about 1.5 cm of the growing tips to induce branching, reduce plant height, and delay flowering. The result can be a compact bushy plant with more blooms. Cultivars have recently become available that naturally form low shrubs without pinching, but even these can benefit from some pruning. Home gardeners should pinch off the growing tips of small-flowered cultivars when they are about 18 cm high, or the plants could produce tall, weak stems with only a few flowers. (Young potted plants sold for transplanting are commonly already pruned to produce a well-branched plant.) All shoots of early varieties should be pinched every 2 weeks until early summer, and until midsummer for late-flowering varieties. Cultivars intended for the production of large, solitary flowers require removing most branches and flower buds, allowing only a few branches to survive and only an uppermost flower on a given branch to mature. Suckers are developed from the base of growing plants, and these can be removed to promote growth of the main plant.

Mulching garden plants in cold climates is recommended (alternate freezing–thawing is the chief overwintering challenge of chrysanthemums). A layer of 10 to 15 cm of pine needles or other material that will not pack solid when wet (broad leaves are best avoided) should be applied after the soil surface freezes. It has been recommended that old, dead aboveground shoots be removed in the fall to avoid disease, but leaving this task until the spring may promote overwintering.

### 26.8.6 Advice Regarding Cut Chrysanthemums

On average, cut chrysanthemum flowering stems last 1 to 2 weeks. Purchased material should have two-thirds to three-quarters of the flowers open, and the foliage should not be yellowed. Florist shops may keep bulk material in near-freezing storage, and in the home, cool locations are preferable (although troublesome, keeping a vase of stems in a refrigerator overnight will extend the life of the bouquet). The flowers should be located away from direct light and either hot or cold drafts. Stems should be recut diagonally after purchase (using a sharp knife or floral scissors intended for the purpose), discarding the bottom 1.3 cm, and the portion placed in water should be free of leaves. Re-cut the stems daily and change the water. The foliage often deteriorates faster than the flowers, and leaves in poor shape should be removed. The cut flower industry employs rehydrating solutions, "flower foods" (including sugar), and germicides that inhibit bacteria, to extend the life of cut flowers, and often flower shops provide packages of these along with purchased bouquets.

### 26.8.7 Advice Regarding Indoor Potted Chrysanthemums

Houseplant chrysanthemums are relatively easy to maintain. A cool room (e.g., 10–15°C) prolongs flowering, but the plants will tolerate warmer temperatures. Indirect but bright light is recommended, such as near a window. The soil should be kept moist at all times, but avoid waterlogging. Fertilization is unnecessary for the first few weeks, then add a light feeding (e.g., with a diluted tomato fertilizer) weekly. Deadhead old flowers, and remove yellowed foliage. The plants may remain attractive for 6 to 8 weeks, and since they can be purchased cheaply, most people will discard them rather than attempt to rejuvenate the plants (best done by cuttings).

## 26.9 FASCINATING FACTS, FABLES, FOLKLORE, AND FRIVOLITY

- In Asia, chrysanthemum has long been associated with notions of long life, because it blooms in the autumn, when most other flowers have faded. In times past, people would wipe chrysanthemum dew on their skin on Chrysanthemum Day, the ninth day of the ninth month of the year, in the hope of maintaining youthfulness.
- Chrysanthemum has considerable symbolic value in Japan. It has been considered to be the national flower since AD 910 and is used as the crest and official seal of the emperor. A chrysanthemum with 16 "petals" is the traditional emblem of the Mikado (monarch), and the monarchy is called "the chrysanthemum throne." Japan's "Order of the Chrysanthemum" is the country's highest recognition of chivalry and is awarded by the emperor.
- In Japan, chrysanthemum-based celebrations are accompanied by the preparation of "dolls" – figurines, often life-sized, elaborately and very artistically adorned with chrysanthemum flowers.

FIGURE 26.12 "Japanese dolls" decorated with chrysanthemums, prepared for a festival.

*Source:* Photo by Marufish (CC BY SA 2.0).

- The chrysanthemum is the official flower of Chicago (designated in 1961), Beijing (China, 1987), and many other cities.
- In several European countries, incurve-type chrysanthemums are emblematic of death and are displayed in funerals and on graves. Similarly, in China, Japan, and Korea, white chrysanthemums symbolize grief. Flower shops in Western nations sometimes advise that a gift of white-flowered chrysanthemums may be inappropriate because of this symbolism.
- Words are often coined by analogy with familiar objects, and in Asia, where chrysanthemums have a very long history, a canine came to be called the "chrysanthemum dog." The dog in question is the shih tzu, one of the most distinctive of the toy breeds. The name originated because its facial hair grows in all directions, resembling the "petals" of a chrysanthemum. In a similar vein, English humorist P. G. Wodehouse (1881–1976) remarked, "Why don't you get a haircut? You look like a chrysanthemum."
- A "chrysanthemum cut" is an ornamental method of preparing a root vegetable, traditionally carried out by holding a portion of vegetable upright with chopsticks held parallel to each other, slicing vertically between the chopsticks, not quite to the base, while very slowly rotating the material through 180°, producing fine strands that radiate outward from the center, reminiscent of a daisy flower head.
- In the cocktail world, a "chrysanthemum" is a combination of dry vermouth, Bénédictine, and absinthe liqueurs and a garnish of orange peel twist.
- "Football chrysanthemums" are large (hence the analogy with a football) fluffy flowers of the incurve class, traditionally presented in the United

States as a corsage by gentlemen to their dates, especially at football games (a custom that has become antiquated).

- The "chrysanthemum revolution" (also known as the aster revolution) was a revolt in 1918 in Hungary against the Austro-Hungarian Empire. Socialist count Mihály Károlyi, soldiers from the Hungarian Army, and protestors wearing daisy-like flowers seized public buildings throughout Budapest. By the end of the year, Károlyi became the prime minister of Hungary. However, in 1919, the revolution was overturned.
- In 1989, scientists at NASA (US National Aeronautics and Space Administration) published a study on the effects 19 popular houseplants had on common indoor air pollutants, toward finding ways of purifying the air in future space stations. Certain plants proved to be more efficient at removing specific pollutants than others. Chrysanthemum was the best absorber of benzene, which is present in many household items, irritates skin and eyes, and is suspected of having a variety of toxic effects. Potted chrysanthemums have been especially recommended in homes with fresh paint or new plastics, which tend to release benzene.
- "The Chrysanthemums" is a poignant short story by American writer John Steinbeck (1902–1968), winner of the Nobel Prize for Literature in 1962, best known for his novel *The Grapes of Wrath*. It has been extensively analyzed for its significance as a critique of a society that has no place for intelligent women, but this brief masterpiece is also well worth reading simply as an example of great literature.
- Like nicknames employed as substitutes for the long or difficult names of people, "mums" is a popular substitute for "chrysanthemums." American actress Goldie Hawn remarked, "A chrysanthemum by any other name would be easier to pronounce."

## KEY PUBLICATIONS

Note: Several of the most authoritative books on chrysanthemums are cited in the following, despite their age, as they contain considerable information. However, the cultivars mentioned are mostly dated (new selections are rapidly being produced), and commercial growing techniques have also changed notably since these publications appeared.


Anderson, N.O. 2007. Chrysanthemum (*Dendranthema × grandiflora* Tzvelv.). In: *Flower Breeding and Genetics*, edited by N.O. Anderson, 389–437. Springer, Berlin, Germany.

Bhattacharya, A. and J.A. Teixerra da Silva. 2006. Molecular systematics in *Chrysanthemum × grandiflorum* (Ramat.) Kitamura. *Scientia Horticulturae (Amsterdam)* 109: 379–384.

Boase, M.R., R. Miller, and S.C. Deroles. 1997. Chrysanthemum systematics, genetics, and breeding. *Plant Breeding Reviews* 14: 321–361.

Cumming, R.W. 1964. *The Chrysanthemum Book*. D. Van Nostrand, New York. 301 pp.

Fang, H.L., Q.S. Guo, H.-L. Shen, and Q.-S. Shao. 2010. Phylogeography of *Chrysanthemum indicum* L. (Compositae) in China based on trnL-F sequences. *Biochemical Systematics and Ecology* 38: 1204–1211.

Gollnow, B. 1980. A history of the cultivated chrysanthemum. *Journal of Australian Institute of Horticulture* 3: 9–14.

Gosling, S.G. 1988. *Chrysanthemum Manual*. National Chrysanthemum Society, UK. 324 pp. [Eight editions published between 1961 and 1976, as well as reprints in various years].

Hao, D.C., Y. Song, P. Xiao, Y. Zhong, P. Wu, and L. Xu. 2022. The genus *Chrysanthemum*: Phylogeny, biodiversity, phytometabolites, and chemodiversity. *Frontiers in Plant Science* 13: 973197. https://doi.org/10.3389/fpls.2022.973197.

Lazar, V., M. Cantor, R. Sima, T. Gocan, and G. Rad. 2011. Researches concerning the behavior of some *Chrysanthemum* cultivars as cut flowers in nutritive solutions. *Bulletin of University of Agricultural Sciences and Veterinary Medicine Cluj-Napoca. Horticulture* 68: 386–391.

Liu, P.-L., Q. Wan, Y.-P. Guo, J. Yang, and G.-Y. Rao. 2012. Phylogeny of the genus *Chrysanthemum* L.: Evidence from single-copy nuclear gene and chloroplast DNA sequences. *PLoS One* 7(11): e48970.

Locke, B. 1990. *Chrysanthemums: The Complete Guide*. Crowood Press, Ramsbury, UK. 191 pp.

Machin, B. 1997. *Pot Chrysanthemum Production*. Nexus Media, Kent, UK. 86 pp.

Machin, M. and N. Scopes. 1978. *Chrysanthemums: Year-Round Growing*. Blanderford Press, Poole, Dorset, UK. 233 pp.

Masuda, Y., T. Yukawa, and K. Kondo. 2009. Molecular phylogenetic analysis of members of *Chrysanthemum* and its related genera in the tribe Anthemideae, the Asteraceae in East Asia on the basis of the internal transcribed spacer (ITS) region and the external transcribed spacer (ETS) region of nrDNA. *Chromosome Botany* 4: 25–36.

Shi, Z., C.J. Humphries, and M.G. Gilbert. 2011. *Chrysanthemum*. In: *Flora of China*, Vol. 20–21, edited by Flora of China Editorial Committee, W. Zhengyi, P.H. Raven, and H. Deyuan, 669–676. Science Press, Beijing, China and Missouri Botanical Garden Press, St. Louis, MO. www.efloras.org/florataxon.aspx?flora_id=2&taxon_id=106957.

Shinoyama, H., R. Aida, H. Ichikawa, Y. Nomura, and A. Mochizuki. 2012. Genetic engineering of chrysanthemum (*Chrysanthemum morifolium*): Current progress and perspectives. *Plant Biotechnology* 29: 323–337.

Smith, J.F. 1975. *Chrysanthemums*. B.T. Batsford, London, UK. 207 pp.

Teixera da Silva, J.A., H. Shinoyama, R. Aida, Y. Matsushita, S.K. Raj, and F. Chen. 2013. Chrysanthemum biotechnology: Quo vadis? *Critical Reviews in Plant Sciences* 32: 21–52.

Trehane, P. 1995. Proposal to conserve *Chrysanthemum* L. with a conserved type (Compositae). *Taxon* 44: 439–441.

U.S. Department of Agriculture. 1976. *Growing Chrysanthemums in the Home Garden. Home and Garden Bulletin 65*. Agricultural Research Service, Washington, DC. 8 pp.

Yang, W., B.J. Glover, G.-Y. Rao, and J. Yang. 2006. Molecular evidence for multiple polyploidization and lineage recombination in the *Chrysanthemum indicum* polyploid complex (Asteraceae). *New Phytologist* 171: 875–886.

Zhao, H.B., F.D. Chen, S.M. Chen, G.S. Wu, and W.M. Guo. 2010. Molecular phylogeny of *Chrysanthemum*, *Ajania* and its allies (Anthemideae, Asteraceae) as inferred from nuclear ribosomal ITS and chloroplast trnL-F IGS sequences. *Plant Systematics and Evolution* 284: 153–169.

Zhao, H.E., Z.H. Liu, X. Hu, J.L. Yin, W. Li, G.Y. Rao, et al. 2009. *Chrysanthemum* genetic resources and related genera of *Chrysanthemum* collected in China. *Genetic Resources and Crop Evolution* 56: 937–946.

## Key Cultivation Websites

Canadian Chrysanthemum and Dahlia Society. www.mumsanddahlias.com/.

Commercial Greenhouse Production. www.ag.auburn.edu/hort/landscape/Potmum.htm.

National Chrysanthemum Society, USA. www.mums.org/.

The Spruce. *How to Grow and Care for Chrysanthemums*. www.thespruce.com/fall-garden-mums-hardy-or-no-1402718.

WikiHow. *How to Care for Mums*. www.wikihow.com/Care-for-Mums.

WikiHow. *How to Grow Chrysanthemums*. www.wikihow.com/Grow-Chrysanthemums#:~:text=Chrysanthemums%20like%20the%20morning%20sun,earlier%20hours%20of%20the%20day.&text=Ensure%20the%20soil%20is%20well,there%27s%20plenty%20of%20air%20circulation.

WikiHow. *How to Plant Mums*. www.wikihow.com/Plant-Mums.

WikiHow. *How to Root Chrysanthemum Cuttings*. www.wikihow.com/Root-Chrysanthemum-Cuttings.

# 27 Clematis (*Clematis* Species and Hybrids)

**Scientific Names:** The genus name *Clematis* is based on the Greek *clema*, a vine-twig, in reference to the vine nature of the plant.

**Common Names:** The common name clematis is usually applied to the species of *Clematis*, but other names encountered include *virgin's bower*, *leather flower*, and *vase vine*.

## 27.1 INTRODUCTION

Clematis, sometimes termed the "queen of vines," is probably the most popular of the perennial flowering vines grown in temperate regions, notable for its very impressive flowering displays, adaptability to a wide range of climates, and range of cultivars.

## 27.2 OVERVIEW

The genus *Clematis* contains about 300 species, mostly native to the temperate Northern Hemisphere. Dozens of these are cultivated. The species include perennial herbs (most gardeners are unfamiliar with these non-climbing kinds) and small-flowered vines that have not been greatly changed by breeding and are employed like ivies to cover structures. However, the most attractive ornamentals are large-flowered vines which have been extensively bred and hybridized. These are mostly woody vines that climb by long petioles (leafstalks) which spiral around, thereby attaching to objects (clematis does not have conventional tendrils; in some species, the stems also twine to a lesser extent). Some varieties can grow to 10 m in length, but the more popular cultivars are 2 to 4 m long (or high). The aboveground parts of species exposed to cold winters are herbaceous and die back at the end of the growing season, while many of the hot-climate species are evergreen. Spring-flowering species produce blooms on branches of the previous year's stems, while late-flowering species (blooming from midsummer and later) produce buds at the ends of new branches. The latter includes *C.* × *jackmannii*, a hybrid that has been very

FIGURE 27.1 Colorful flowering clematis displays.

*Source:* (a) Photo by Bill Smith (CC BY 2.0). (b) Cultivar Niobe. Photo by F. D. Richards (CC BY SA 3.0). (c) Photo by Liz West (CC BY 2.0).

FIGURE 27.2 Examples of clematis cultivars with doubled flowers.

*Source:* (a) Photo by Mike Finn (CC BY 2.0). (b) Cultivar Piilu. Photo by Shutterstock/Alesikka. (c) Photo by Shutterstock/Flower Studio.

FIGURE 27.3 Vintage (public domain) technical botanical illustrations of clematis species and cultivars: (a) Mixed cultivars. (b) Cultivar John Gould Veitch. (c) Pitcher's leather flower (*Clematis pitcheri*). (d) A mix of clematis cultivars. (e) *Clematis* × *jackmannii*. (f) Cultivar Star of India.

*Source:* (a) Nairn and Sons' 1906 horticultural catalogue (Christchurch, New Zealand). (b) *The Floral Magazine*, vol. 7, 1868. (c) *Revue horticole*, serié 4, vol. 50, 1878. (d) Lunzer, A, (artist); L. Prang & Co. (publisher), 1885. Photo from Boston Public Library. (e) Van Houtte, L. 1865–1867. *Flore des serres et des jardins de l'Europe*, vol. 16. (f) *The Floral World and Garden Guide*, vol. 14, 1871.



DOI: 10.1201/9781003586227-27

popular since the 19th century and is still the most popular clematis grown. Hybrid cultivars are available, with flowers that are solid blue, pink, purple, red, violet, or yellow, or bicolored, and some varieties have semi-doubled or doubled flowers. The flowers, which can be quite different in form (lily-shaped, bell-like, pitcher-shaped, urn-shaped, star-like, flattish, etc.) depending on species, can be as large as 25 cm in diameter. The "petals" (tepals) can be thin or wide, pointed or rounded, crinkled, twisted, or even crimped. The flowers may also be erect or nodding. Small-flowered plants can produce hundreds of flowers, while large-flowered plants may develop over 100 flowers on a single plant. Many of the species have fragrant flowers, but the hybrids tend not to be scented. Blooming time ranges from early spring to late fall, depending on cultivar. The fruits are typically feathery and ball-shaped.

## 27.3 HISTORY

Local species of *Clematis* were cultivated in China and Japan since ancient times, and in Europe at least as far back as the 16th century. In the 19th century, the Asian species were imported into Europe, and considerable hybridization and breeding resulted in the production of many new cultivars. Thousands of cultivars are now available.

## 27.4 ECONOMIC IMPORTANCE

Clematis is primarily a standard garden ornamental, and the horticultural industry in most countries is mostly concerned with supplying planting materials. There is also a minor market for cut flowers.

## 27.5 ORNAMENTAL USES

Large-flowered clematis vines (which include most garden cultivars) require structures like posts, arches, fences, rock walls, pergolas, or trellises on which to grow. Some compact cultivars can be grown on lampposts, deck rails, and balconies. The vines can be placed near trees and shrubs to grow on them, but this is often discouraged because it obscures and can harm the host plants. Some cultivars grow well in containers. Some of the small-flowered species are useful as groundcover and as less-colorful (i.e., green) cover for fences and walls. Pollinators, including bees, butterflies, and hummingbirds, are attracted to the flowers of clematis, which are

FIGURE 27.5 Residential usages of clematis vines: (a) Against a house. (b) Over an arch leading to a garden. (c) On a porch railing.

***Source:* (a) Photo by Andrew Fogg (CC BY 2.0). (b) Photo by A. Davey (CC BY 2.0). (c) Photo by Shutterstock/Anna Kraynova.**

FIGURE 27.4 Vintage trade illustrations of clematis: (a) A woman with a background of clematis. (b) Old French postcard (ca. 1900). (c) A woman beautified with clematis.

***Source:* (a) Appletons Booklovers Magazine, 1904. (b) Public domain. (c) Trade card published in 1892 by American Tobacco Company.**

FIGURE 27.6 Ornamental fruit (dry "seeds") with elongated hairs (styles): (a) Fruits in the wind, showing why the plants are often called "old man's beard." (b) *Clematis gentianoides* in fruit, illustrating how attractive the plants are at this stage.

***Source:* (a) Photo by Shutterstock/KW-photographic. (b) Photo by Peganum (CC BY SA 2.0).**

ideal in wildlife gardens. The cut flowers have excellent vase life (4 to 12 days, depending on cultivar), as reflected by the name "vase flower." The odd, feathery seedpods are used in dried flower arrangements.

## 27.6 OTHER USES

Clematis was employed in folk medicine for a variety of uses. It is still used as an herbal remedy by some, although its safety is questionable.

## 27.7 PROBLEMS

Clematis are particularly subject to several fungal diseases, sometimes requiring the application of fungicides.

Clematis is toxic when eaten, for people, dogs, cats, and horses. In humans, it can result in dizziness, confusion, fainting, and convulsions and (rarely) death. The taste is intensely bitter, discouraging consumption. Sensitive individuals may suffer contact dermatitis, so gardening gloves are recommended.

Some cultivated clematis species have escaped from gardens and become invasive.

## 27.8 CULTIVATION

### 27.8.1 Climate and Site

Most hybrid cultivars are suited to climate zones 4 to 8. Some cultivars can survive as far north as zone 3. Clematis species are usually adapted to full sun or partial shade (in very sunny regions, semi-shade is claimed to protect blossoms from fading). Protection from strong winds is desirable, but air movement around the plants is beneficial because of susceptibility to fungi. Trellises flush against walls may not provide sufficient air circulation. It is said that the head of clematis should be in the sun but its feet in the shade, because the roots do best in a cool, moist environment. This is often promoted by a 5 cm layer of mulch or, in some cases, by flat rocks placed over the root zone. Some species are quite drought-tolerant, but most garden cultivars should be grown in moist (but well-drained) soils, and it is often necessary to water the plants. Whenever a perennial is added to a garden, its potential aesthetic appearance and suitability need to be considered for a particular location, but this is particularly important for clematis vines, since cultivars vary greatly in size and appearance.

FIGURE 27.7 Pets beside clematis vines, which can seriously poison them: (a) Shetland sheepdog. (b) Cat.

***Source:*** **(a) Photo by Shutterstock/James Hime. (b) Photo by Shutterstock/Loriginals.**

### 27.8.2 Propagation and/or Purchasing

Superior cultivars (often hybrids) often do not produce seeds that breed true to type and are propagated vegetatively (by cuttings or divisions). Nurseries offer young potted starter plants, sometimes grafted onto hardy rootstocks. Online catalogues providing seeds and starter plants by mail order are good sources of the most recent and interesting cultivars.

### 27.8.3 Planting

Because clematis can live for decades and become quite large, the initial excavated hole should be large and prepared with high-quality soil. When transplanting a pot of large-flowered clematis (not the shrub and evergreen species), it is recommended that the crown be positioned at least 5 cm below the soil surface, to minimize damage by mowers or other forces (the shoots can regrow so long as the crown is intact). Support for the developing vine is necessary. The structure needs to have portions slim enough to allow the twining leafstalks to encircle them, but strong enough to support the plant, especially if it is a vigorous cultivar.

### 27.8.4 Soil and Moisture Conditions

Clematis grows best in moist, well-drained, fertile (organically rich), loamy, circumneutral (ideally 6.5–7.0) soil. However, the plants can grow in average soils and soils of various pH levels. A low-nitrogen, slow-release fertilizer is recommended to promote flowering (but once plants are established, some authorities suggest minimal fertilization, if any, is all that is necessary).

### 27.8.5 Maintaining Plants

Clematis are not considered to be maintenance-free plants, primarily because of the need to prune. The pruning requirements differ according to cultivar. In all climates, most large-flowered vine cultivars benefit from annual thinning and pruning, to promote vigorous flowering, and to control rampant growth. Those that flower on old wood (i.e., early in the season) should not be pruned vigorously before they bloom (because this will remove the flowers) but subsequently. The late-flowering vines which produce flowers on the new (current season's) growth can be pruned to less than 1 m above the ground (down to the ground is often recommended, even if it seems severe) for the dormant season (these are usually the plants that are suited to cold Northern climates).

All newly planted clematis vines should be cut back the first season to encourage basal branching (newly purchased plants frequently have only a single stem). The stem is fragile at the juncture of the roots and stem in very young plants and needs to be handled with care. Clematis do not have true tendrils, aerial rootlets that penetrate and cling to tiny crevices, or suction cups, and young plants may benefit from being initially tied to the support structures provided. Economical supports can be fashioned from chain-link fencing, heavy-gauge chicken wire, thin wood lattice, and snow fencing. A barrier can be installed around the base of the plant to protect it from mowers and trimmers. Many clematis cultivars are very slow to establish (especially for the first several years), and patience is necessary (once established, the plants may last for generations). Many clematis books originate from England, and the advice provided may not be best for other climates.

## 27.9 FASCINATING FACTS, FABLES, FOLKLORE, AND FRIVOLITY

- Clematis lovers sometimes call themselves "clematarians" ("climatarians" are people who advocate diets and food production systems that promote the health of the planet).
- Since ancient times, beggars have employed vesicants – substances that cause serious tissue damage – to produce blisters on their faces and elsewhere in order to evoke sympathy. In classical Rome, clematis (which can produce significant rashes in some people) was one of the materials employed.
- Pioneers of the American West called clematis "pepper vines" because they used the peppery stems and leaves (which are toxic) as a substitute for pepper, which was very rare and expensive at that time.
- The development of doubled flowers (i.e., with extra petals) in some varieties of clematis is rather unstable, the plants occasionally producing single flowers in one year and double the next, sometimes just single flowers when stressed, and sometimes producing both kinds at the same time.
- Cows (and other animals) that consume clematis may produce bitter milk (ranunculin from the plant is converted to the bad-tasting chemical protoanemonin).

## KEY PUBLICATIONS


Brown, M. and M. Toomey. 2013. Herbaceous clematis. *Plantsman* 12: 106–111.

Buetler, L. 2004. *Gardening with Clematis: Design and Cultivation*. Timber Press, Portland, OR. 300 pp.

Evison, R.J. 1998. *The Gardener's Guide to Growing Clematis*. Timber Press, Portland, OR. 160 pp.

Evison, R.J. 2007. *Clematis for Small Places: 150 High Performance Plants for Decks, Patios, Balconies and Boarders*. Timber Press, Portland. 272 pp.

Fair, K. and C. Fair. 1997 *Clematis for Colour and Versatility*. Crowood Press, Wiltshire, UK. 128 pp.

Fretwell, B. 1999. *A Comprehensive Guide to Clematis*. HarperCollins, UK. 160 pp.

Gardener, D. 2007. *Clematis and Climbers*. D.K. Publishing, New York. 160 pp.

Gooch, R. and J. Gooch. 2011. *Clematis: An Essential Guide*. Crowood Press, Wiltshire, UK. 256 pp.

Grey-Wilson, C. 2000. *Clematis: The Genus: A Comprehensive Guide for Gardeners, Horticulturists and Botanists*. Timber Press, Portland, OR. 219 pp.

Howells, J. 2003a. *Choosing Your Clematis*. Acc Art Books, Suffolk, UK. 144 pp.

Howells, J. 2003b. *Trouble-free Clematis: The Viticellas*. Acc Art Books, Suffolk, UK. 191 pp.

Lindgren, D.T. 2007. Clematis. In: *Flower Breeding and Genetics*, edited by N.O. Anderson, 781–799. Springer, Dordrecht.

Samarakoon, U.C. and J.E. Faust. 2020. Shortening the production cycle of clematis. *HortScience* 55(12), https://doi.org/10.21273/HORTSCI15384-20.

Toomey, M. and E. Leeds. 2010. *An Illustrated Encyclopedia of Clematis*. Timber Press, Portland, OR. 428 pp.


### Key Cultivation Websites


British Clematis Society. www.britishclematis.org.uk./.

Clematis on the Web. https://clematisontheweb.org/.

Clemson College of Agriculture, Forestry and Life Sciences. *Clematis*. https://hgic.clemson.edu/factsheet/clematis/.

Horticulture Magazine. *How to Train Clematis Against a Fence or Trellis*. https://horticulture.co.uk/clematis/training/.

International Clematis Society. www.clematisinternational.com/.

Kansas State Research and Development. *Clematis Pruning Made Simple*. www.johnson.k-state.edu/lawn-garden/agent-articles/vines/clematis-pruning-made-simple.html.

WikiHow. *How to Grow Clematis in Pots*. www.wikihow.com/Grow-Clematis-in-Pots.

WikiHow. *How to Plant Clematis*. www.wikihow.com/Plant-Clematis#:~:text=Large%2Dflowered%20clematis%20prefer%20cool,a%20shrub%20or%20small%20tree.

WikiHow. *How to Prevent Clematis Wilt*. www.wikihow.life/Prevent-Clematis-Wilt.

WikiHow. *How to Propagate Clematis*. www.wikihow.com/Propagate-Clematis.

WikiHow. *How to Train Clematis*. www.wikihow.com/Train-Clematis.

# 28 Clivia (*Clivia miniata*)

**Scientific Names:** The genus *Clivia* was named for Lady Charlotte Florentia Clive (1787–1866), later Duchess Percy of Northumberland, noted for being the governess of the future Queen Victoria. She was the first in Great Britain to grow clivia.

**Common Names:** Other names for clivia include *bush lily*, *clivia lily*, *fire lily* (especially applied to orange-flowered cultivars), *kaffir lily* (*kaffir-lily*, *kafir lily*), and *Natal lily*. The chief alternative name is *kaffir lily*, which can also refer to *Hesperantha coccinea* (*Schizostylis coccinea*). Unfortunately, the word *kaffir* is pejorative. It is considered extremely racially offensive in South Africa. *Kafir* (which can simply be a misspelling of *kaffir* and is its original language root) is an Arabic and Islamic term (sometimes also pejorative), often translated as *infidel* or *pagan* but also used in derived senses.

## 28.1 INTRODUCTION

*Clivia miniata* is one of the world's most extensively planted houseplants, valued for its attractive, trumpet-like, usually orange or red blooms. The flowers resemble those of amaryllis but are smaller and more numerous. The foliage is also quite attractive (more so than in amaryllis), and the plants are relatively easy to grow.

## 28.2 OVERVIEW

There are about six species and some hybrids of *Clivia*, including the popular *C.* × *nimbicola* (= *C. caulescens* × *C. miniata*). The wild species are native to the subtropical forests of South Africa and Eswatini (formerly known as Swaziland). *Clivia miniata* has been reported to have escaped in Australia, China, Japan, Mexico, New Zealand, and the United States. The plants mostly grow in low light under trees. Clivias are herbaceous perennials, with evergreen, strap-like leaves in two opposite fans. The flowers occur in umbels (clusters with the flower stalks all attached at one point on top of a stem). Information on the web variously refers to the underground parts of clivia as "bulbs" and "rhizomes." The aerial part of the plant sits on stem tissue (which is true for most plants). In related genera, such as *Amaryllis*, such stem tissue forms a "bulb" (which is a storage organ), but this is not the case for clivia. Fleshy "tuberous roots" (not rhizomes, which are usually interpreted as laterally spreading stems) grow from the basal stem tissue in clivia. Effectively, the storage function that enables the plant to survive in a dormant condition during an inhospitable season has been taken over by the fleshy roots, in comparison to related plants. True rhizomes (which are slim, about 2 cm in diameter) develop from the base of the plant after several years, spreading outward, and at their tip they may produce an offset (daughter plantlet). Several stems may arise (eventually) from a rhizome.

FIGURE 28.1 *Clivia miniata* (including hybrids) in flower.

*Source:* **(a) Photo by Shutterstock/LifeCollectionPhotography. (b) Photo by Shutterstock/Creative Tatsiana. (c) Photo by Bernard Dupont (CC BY SA 2.0). (d) Photo by HQ (CC BY 2.0).**

FIGURE 28.2 Flowering heads of some of the less-popular species of *Clivia* (*C. miniata* is the most popular species). Note how the flowers hang down. (a) *Clivia robusta*. (b) *Clivia nobilis*. (c) *Clivia* × *nimbicola*.

*Source:* **All photos by SAplants (CC BY SA 4.0).**

*Clivia miniata* is the most widely cultivated species. Depending on cultivar, age, and how well the plant has been grown, it may be 30 cm to 1 m in height and about as wide. The dark-green leaves are up to 60 cm long. Its flowers are produced in dense clusters of 12 to 20 flowers (sometimes more on a single flowering stem), which individually may last for only a few days, but a succession of opening flowers maintains the floral display for weeks. The flowers are sometimes weakly but sweetly fragrant. Over time, a potted plant produces offsets (daughter plants), and these can grow into multiple clumps within a single pot, magnifying the floral display. *Clivia miniata* has semi-erect, trumpet-shaped flowers, while all the other species have tubular



DOI: 10.1201/9781003586227-28

FIGURE 28.3 Close-up photos of flowers of *Clivia miniata*.

*Source:* **(a) Francesco Giordano (CC BY SA 2.0). (b) Lynn Greyling from Pixabay (public domain). (c) P. J. de Lange, Auckland Museum (CC BY 4.0). (d) Sandid from Pixabay (public domain). (e) Raul Antonio Alvarez (CC BY SA 3.0). (f) Algirdas (CC BY SA 3.0).**

flowers usually hanging down in a cluster. Flower colors are most often shades of orange and yellow but also include near-white, cream, peach, pink, and red. Bicolored forms are frequent.

## 28.3 HISTORY

In Europe, many hybrids and cultivars of *C. miniata* were produced after its discovery in the early 1850s. The most familiar hybrid is *C.* × *cyrtanthiflora*, alleged to be a cross of *C. miniata* and *C. nobilis*, and perhaps other species, was produced in the late 1850s. Clivia breeding today has become extensive in Japan, China, and Korea.

## 28.4 ECONOMIC IMPORTANCE

Clivias are popular houseplants, and there are reliable international and domestic markets. They are slow-growing, and because of the time and effort required to grow large potted plants (it may take 10 years to develop to full size), they frequently sell for premium prices.

## 28.5 ORNAMENTAL USES

In temperate climates, most clivias are grown as potted houseplants, sometimes summered outdoors. In subtropical regions, huge clumps of clivia are often observed in parks, gardens, and residential gardens. Clivias are suitable for beds, borders, and containers in semi-tropical areas.

## 28.6 OTHER USES

In South Africa, clivias are employed in a wide range of herbal medicines. Clivia has been used in Native medicine to speed up childbirth, because the lycorine content has this effect. Various components of the plant are under investigation for possible modern medical applications.

FIGURE 28.4 Nineteenth-century paintings (public domain) of cultivated clivias by European botanical illustrators: (a) *Clivia miniata*. (b) *Clivia miniata*. (c) *Clivia miniata*. (d) *Clivia* × *cryptanthiflorum*. (e) *Clivia gardenia*. (f) *Clivia nobilis*.

*Source:* **(a) *Flore des serres et des jardins de l'Europe* (Gand, Belgium), vol. 23, 1880. (b) *Revue horticole* (Paris), vol. 54, 1882. (c) *L'Illustration horticole* (Gand, Belgium), vol. 26, 1879. (d) Flore des serres et des jardins de l'Europe (Gand, Belgium), vol. 18, 1869–1870. (e) Curtis, W. (ed.). 1856. *Botanical Magazine* (London), vol. 82. (f) Edwards, S. T. 1828. *Botanical Register* (London), vol. 14.**

FIGURE 28.5 Potted clivias on sale in a garden center.

*Source:* **Photo by Ilee Wu (CC BY ND 2.0).**

FIGURE 28.6 *Clivia miniata* in ornamental plant borders.

*Source:* (a) Photo by Cultivar 413 (CC BY 2.0). (b) Photo by Bennilover (CC BY ND 2.0).

## 28.7 PROBLEMS

*Clivia miniata* and some other clivias produce alkaloids, particularly lycorine, which is toxic and presents a poison hazard for small children and pets. The concentration is small, and usually, a large amount needs to be consumed by adults in order to become seriously ill. Sap from the plant may produce dermatitis in sensitive individuals.

## 28.8 CULTIVATION

### 28.8.1 Climate and Site

Indoors, clivias should be provided with temperatures of 18–24°C and not less than 10°C at night. During the recommended 6- to 8-week fall/early winter resting period, temperatures of 5–13°C are suitable (at least at night). Some gardeners simply keep their plant on a patio in the fall for at least a month, and this suffices to induce blooming when brought inside. Pots should be away from radiators and other sources of heat. Houseplant clivias require bright, indirect light during the growing season. Clivia grows well with a humidity of about 50%, which can be difficult to achieve in winter, so use of a moist pebble tray or a room humidifier is advisable. However, misting (spraying with water) is inadvisable because it encourages disease.

Clivias are frost-sensitive plants naturally adapted to subtropical climates, and they can be grown permanently outdoors in climate zones 9 to 11. Exposure to cold winds can be harmful, and a protected site is preferable. Outdoor plants usually grow best in the shade.

FIGURE 28.7 *Clivia miniata* in pots.

*Source:* (a) Photo by Guérin Nicolas (CC BY SA 3.0). (b) Photo by Maja Dumat (CC BY 2.0). (c) Photo by Blumenbiene (CC BY 2.0).

FIGURE 28.8 Clivia – a poison hazard for children and pets: (a) Child with clivia. (b) Jack Russell terrier among clivias. (c) Cat nestled against a potted clivia.

*Source:* (a) Photo by Shutterstock/Leah-Anne Thompson. (b) Photo by Shutterstock/Dja Photography. (c) Photo by Per Ola Wiberg (CC BY 2.0).

### 28.8.2 Propagation and/or Purchasing

Most people do not grow clivias from seed, because it generally takes 3 to 5 years to reach flowering (sometimes longer), and cultivars will not usually breed true. There are rules of thumb alleging that at least 8, 10, or 12 leaves are required before a plant will flower. *Clivia miniata* plants are largely self-sterile, that is, a given plant will usually not be able to produce seeds unless cross-pollinated by a genetically different plant (one not from

FIGURE 28.9 Berries and seeds of *Clivia miniata*. The bright-red, cherry-sized spherical berries take a year or so to mature. They often contain several seeds, which should be planted when mature and still moist: (a) Berries. (b) Seeds.

*Source:* **(a) Photo by SAplants (CC BY SA 4.0). (b) Photo by Paul Venter (CC BY SA 3.0).**

the same clone). Professionals divide plants into portions that have rhizome(s), and the portions can flower in 1 or 2 years. Mature plants 3 to 4 years of age begin producing one or more offsets annually around the base of the mother plant. These can be separated and replanted after they develop at least three or four leaves and have some thick roots attached to their base. The thick, fleshy roots are somewhat fragile and make handling challenging. Clivias will not grow from roots alone. Most people simply maintain plants and buy new ones, preferably already mature. A much better choice of cultivars is available from mail-order sources compared to garden centers, but the plants are unlikely to be large.

### 28.8.3 Planting

Offsets should be detached where they meet the parent plant, using a sharp knife or razor. They should have at least three leaves 20–25 cm long. They can be planted in an 8–12 cm (3–5 in) pot with equal parts of peat moss and coarse sand or perlite. These can be kept in diffused light, watered sparingly, and as the plants develop, they should be moved to larger pots containing a potting mix. Flowering may occur in a year or so.

Old plants can be divided into separate portions using a stout knife, taking care to minimally damage the fleshy roots. The segments can be potted in 10–12 cm (4–5 in) pots.

### 28.8.4 Soil and Moisture Conditions

For houseplants, rich, well-draining, soilless potting mix is recommended. The plants should be kept somewhat moist but allowed to dry slightly between deep watering. Fertilizing twice a month with half-strength solution is recommended. Some professional growers employ a high-phosphorus fertilizer, others a balanced fertilizer. During the recommended 2-month fall/winter resting period, watering should be minimal, and fertilization discontinued. Without this rest period, the plants will continue to develop foliage rather than flowers. Following a rest period in the fall, clivias usually begin blooming in late winter.

FIGURE 28.10 Clivia and the fairy world.

*Source:* **(a) Photo by George from Pixabay (public domain). (b) Photo by Shutterstock/Kate Krasun.**

In subtropical areas, plants in the ground are quite drought-tolerant, thanks to their thick, fleshy roots that store water. The soil should be free-draining, organically rich, with a recommended pH of 5.0 to 6.0.

### 28.8.5 Maintaining Plants

In pots, the expanding roots can force the plant and potting medium up and out of the pot. Leaving the plant to become root-bound is recommended before transplanting to a larger container, because it will grow and flower better when the root system is constricted. Some authorities recommend not repotting for 3 to 5 years. The plants tend to become top-heavy as they grow, and heavy pots are advisable to prevent them from toppling. Accumulated dust on the leaves can be wiped off with a damp cloth. The flowers often develop into fruits, and unless seeds are desired, the fruits should be removed with a razor blade to prevent them from developing and draining the plant's energy.

## 28.9 FASCINATING FACTS, FABLES, FOLKLORE, AND FRIVOLITY

- In South Africa, parts of clivia are used as protective charms against evil spirits, and as protection from storms.
- For a period in the 19th century, yellow clivias were extremely rare and were greatly valued by the privileged classes. Seeds and offsets of yellow clivias sometimes sold for thousands of dollars.
- For almost 2 centuries, artists have illustrated children's stories with imaginative art featuring fairies and the most familiar, beautiful flowers. Clivia flowers have been illustrated in this tradition.

## KEY PUBLICATIONS

Conrad, F. and D. Snijman. 2011. Systematics and phylogeography of *Clivia*. *Clivia* 13: 9–25.

Crouch, N.R., D.A. Mulholland, T.L. Pohl, E. Ndlovu, and B.E. van Wyk. 2002. The ethnobotany and chemistry of the genus

*Clivia* (Amaryllidaceae). *South African Journal of Botany* 69: 144–147.

Dixon, R. 2011. Relationships in *Clivia*. *Clivia* 13: 63–81.

Duncan, G. 2008. *Grow Clivias: A Guide to the Species, Cultivation and Propagation of the Genus Clivia*, 2nd edition. National Botanical Institute, Kirstenbosch, Claremont, South Africa. 194 pp.

Duncan, G. 2016. A historical synopsis of the genus *Clivia* Lindl. *Curtis's Botanical Magazine* 33: 169–189.

Kiepiel, I. and S.D. Johnson. 2014a. Breeding systems in *Clivia* (Amaryllidaceae): Late-acting self-incompatibility and its functional consequences. *Botanical Journal of the Linnean Society* 175: 155–168.

Kiepiel, I. and S.D. Johnson. 2014b. Shift from bird to butterfly pollination in *Clivia* (Amaryllidaceae). *American Journal of Botany* 101: 190–200.

Koopowitz, H. 2002. *Clivias*. Timber Press, Portland, OR. 384 pp.

Musara, C., E.B. Aladejana, and A.E. Aladejana. 2021. *Clivia miniata* (Lindl.) Bosse (Amaryllidaceae): Botany, medicinal uses, phytochemistry and pharmacological properties. *Journal of Applied Pharmaceutical Science* 11: 012–018.

Petravich, A. 2015. Clivia breeding at Longwood Gardens. *Acta Horticulturae* 1085: 207–209.

Spies, P., J.P. Grobler, and J.J. Spies. 2011. A review of phylogenetic relationships in the genus *Clivia*. *Philosophical Transactions in Genetics* 1: 168–207.

Swanevelder, D. and R. Fisher. 2009. *Clivia Nature and Nurture*. Briza Publications, Pretoria, South Africa. 220 pp.

## Key Cultivation Websites

The Clivia Society. https://cliviasociety.com/.

Gardener's Path. *How to Grow and Care for Clivia Houseplants*. https://gardenerspath.com/plants/flowers/grow-clivia/.

GardenersWorld.com. *How to Grow and Care for Clivia miniata (Natal Lily)*. www.gardenersworld.com/how-to/grow-plants/how-to-grow-and-care-for-clivia/.

North American Clivia Society. www.northamericancliviasociety.org/index.php.

The Spruce. *How to Grow and Care for Fire Lily*. www.thespruce.com/fire-lily-plant-profile-4768477.

WikiHow. *How to Plant Bush Lily*. www.wikihow.life/Plant-Bush-Lily.

Wisconsin Horticulture, Extension. *Clivia*. https://hort.extension.wisc.edu/articles/clivia/.

# 29 Columbine (*Aquilegia* Species and Hybrids)

**Scientific Names:** The genus name *Aquilegia* is based on the Latin *aquila*, "eagle," plus *lego*, "to gather," a reference to the spurs (hollow extensions) of the five outermost "tepals" (petal-like floral segments). The spurs are said to resemble the talons of an eagle.

**Common Names:** The common name columbine is based on the Latin *columba*, "dove" (see Hill 1902, cited later, for an extensive analysis). The name *aquilegia* is also frequently employed, and the name *granny's bonnet* is occasionally encountered. *Thalictrum aquilegifolium*, another species in the buttercup family (Ranunculaceae), is called *columbine meadow-rue* and *Siberian columbine meadow-rue*.

FIGURE 29.1 Displays of flowering columbines (*Aquilegia* species): (a) Mixed cultivars at the Chelsea flower show in England. (b) A hedge of hybrid columbine (*A.* × *hybrida*). (c) A cultivar with semi-doubled flowers. (d) Flowering columbines.

*Source:* **(a) Photo by Shutterstock/David Young. (b) Photo by Shutterstock/Coulanges. (c) Photo by Indiana Ivy Nature Photographer (CC BY 2.0). (d) Photo by Ed Fladung (CC BY ND 2.0).**

## 29.1 INTRODUCTION

Columbines are hardy, easily grown, familiar garden plants with unique, beautiful flowers that make the plants extremely attractive to gardeners in cooler climates.

## 29.2 OVERVIEW

*Aquilegia* has about 80 species of perennial herbaceous plants native to the circumpolar Northern Hemisphere, many in alpine and woodland habitats. The species interbreed readily, and numerous hybrids and cultivars have been created. The plants arise from rhizomes (underground stems), which reproduce the plants after the aboveground parts die back to the ground at the end of the season. The base of the main stem sometimes becomes woody. Most species and cultivars grow between 15 and 90 cm in height. The flowers contain many tepals (a term denoting both *sepals* and *petals*, especially when they intergrade and are not clearly distinguishable). The outermost ring of five are colored like petals, and each tepal has a long spur, within which nectar is secreted, and attracts pollinators, mainly long-billed birds, such as hummingbirds. (The flowers of some hybrids with *A. vulgaris* are short-spurred or lack spurs.) A second ring of colored tepals (without spurs) is present within the first ring. Most columbines have little or no scent (*A. caerulea* is pleasantly perfumed). The flowers are available in blue, green, lavender, orange, pink, purple, red, yellow, and white; they are often bicolored and are doubled in some cultivars. Blooming occurs in late spring to early summer.

Particularly popular species include the following (note that common names of columbines, such as "mountain columbine," are often unreliable): *A. caerulea* (Rocky Mountain columbine), *A. canadensis* (Canadian columbine), *A. chrysatha* (golden columbine, yellow columbine), *A. flabellata* (fan columbine), *A. formosa* (Western columbine; "formosa" is Latin for *beautiful*), and *A. vulgaris* (European columbine). Many cultivars are hybrids of these and other species. Cultivars with hooked spurs are likely to

FIGURE 29.2 Comparison of a natural and a doubled flower: (a) Normal flower. (b) Double-flowered columbine 'Blue Barlow'.

*Source:* **(a) Photo by Albrecht Fietz from Pixabay (public domain). (b) Photo by Swallowtail Garden Seeds (CC BY 2.0).**

DOI: 10.1201/9781003586227-29

be derived from *A. vulgaris*, and those with long straight spurs are from *A. caerulea* and *A. chrysantha*.

## 29.3 HISTORY

European gardeners have cultivated columbines for at least 400 years. In North America, Native Americans employed preparations of the wild plants for a variety of diseases, ranging from heart problems to fever, and even to relieve the pain of poison ivy.

## 29.4 ECONOMIC IMPORTANCE

Columbines are traditional garden flowers that fill a niche requirement for plants that bloom in late spring after most of the bulb ornamentals. They are quite hardy in regions with cold winters, which suit them to the climate of much of the cool temperate world. The flowers are extremely attractive. These advantages make columbines a staple of the gardening industry of much of the world.

## 29.5 ORNAMENTAL USES

Columbines are primarily grown in ornamental garden beds, borders, rock gardens, and containers. They can be employed as cut flowers.

## 29.6 OTHER USES

The flowers of columbines are very attractive to hummingbirds, butterflies, hawk moths, and bees, and the seeds are consumed by birds, so the plants are frequently included in pollinator and wildlife gardens.

FIGURE 29.3 Old paintings (public domain) of columbines (*Aquilegia* species) by professional botanical illustrators: (a) *Aquilegia canadensis* (Canadian columbine, red columbine, wild columbine). (b) *Aquilegia caerulea* (Rocky Mountain columbine, blue columbine). (c) *Aquilegia flavescens* (yellow columbine, yellow mountain columbine, lemon columbine). (d) *Aquilegia flabellata* (fan columbine, fan-leaved columbine, dwarf columbine). (e) *Aquilegia formosa* (western columbine, crimson columbine, red columbine). (f) *Aquilegia pyrenaica* (Pyrenean columbine). (g) *Aquilegia skinneri* hybrid (Mexican columbine). (h) *Aquilegia vulgaris* (European columbine, granny's bonnet). (i) *Aquilegia chrysantha* (golden columbine, yellow columbine). (j) *Aquilegia alpina* (alpine columbine).

***Source:*** **(a) House, H. D., W. B. Starr, and H. H. Snyder. 1918. *Wild Flowers of New York Part 1*, University of the State of New York, State Museum, Albany. (b, c) Paintings by American artist Mary Vaux Walcott (1860–1940). Courtesy of Smithsonian American Art Museum. (d) *L'Illustration horticole* (Gand, Belgium), vol. 11, 1864. (e) Harriman, E. H., and M. C. Hart. 1901. *Alaska Expedition*. Doubleday, Page & Company, New York. (f) *The Floral Magazine* (London), vol. 6, 1867. (g) *Gartenflora* (Erlangen, Germany), vol. 14, 1865. (h) Painting by British botanical artist Lydia Penrose (1787–1842), from Royal Horticultural Society Lindley Library. (i) *The Floral Magazine* (London), vol. 2. 1873. (j) Thomé, O. W. 1885. *Flora von Deutschland, Österreich und der Schweiz*. Gera, Germany.**

FIGURE 29.4 Some attractive ornamental columbine cultivars: (a) *Aquilegia canadensis* 'Nana'. (b) *Aquilegia vulgaris* 'Clementine Rose'. (c) *Aquilegia canadensis* 'Little Lanterns'.

***Source:*** **(a) Photo by Ghislain 118 (CC BY SA 3.0). (b) Photo by Daderot (public domain). (c) Photo by David Stang (CC BY SA 4.0).**

FIGURE 29.5 Animals attracted to flowers of columbines (*Aquilegia* species): (a) White-lined sphinx (*Hyles lineata*) on *A.* × *hybrida*. (b) Ruby-throated hummingbird (*Archilochus colubris*) male on *A.* × *hybrida*. (c) Butterfly on columbine. (d) Carder bee (*Bombus pascuorum*) on *A. vulgaris*. (e) Golden-mantled ground squirrel (*Callospermophilus lateralis*) eating a columbine flower.

***Source:*** **(a) Photo by Scottk (public domain). (b) Photo by Shutterstock/Danita Delimont. (c) Photo by Buntysmum from Pixabay (public domain). (d) Photo by Shutterstock/Keith Hider. (e) Photo by Shutterstock/BEWalter.**

## 29.7 PROBLEMS

*Aquilegia* plants are toxic (so are most other plants in the Ranunculaceae, the buttercup family), especially the seeds and roots, which can cause severe gastroenteritis and heart palpitations. Contact with the sap can irritate the skin of sensitive individuals. Columbines are considered to pose a minor threat to people, pets, and livestock. Nevertheless, the flowers and foliage of some species were reputedly sometimes eaten by Native Americans, but usually in small quantities.

## 29.8 CULTIVATION

### 29.8.1 Climate and Site

Most *Aquilegia* species are hardy in climate zones 3 to 9; some are also rated for zone 10. Most are adapted to cool seasons and do not thrive in hot or very humid climates. Some species of columbines grow best in full sun (especially in cooler climates), others in part shade. The alpine species are particularly suitable for rock gardens.

### 29.8.2 Propagation and/or Purchasing

Columbines are frequently purchased as starter plants from nursery-grown stock, and many cultivars are hybrids that will not breed true if propagated from seeds. Once established, columbines can be grown from rhizomes (the entire plants, including the rhizomes, can be divided). Columbines are frequently grown from seeds (hybrid seed is available, and often, rather spectacular mixtures of flower colors are offered in a single seed packet), but the plants may not flower the first year (plant breeders are addressing this problem). The seeds of alpine species sometimes take 2 years to germinate. Columbines often self-seed in a location, but the flowers of the new plants often are not as attractively colored as those of their parents.

### 29.8.3 Planting

For most cultivars, the plants can be established at a spacing of 60 cm. Small cultivars may be planted as closely as 30 cm.

### 29.8.4 Soil and Moisture Conditions

Most columbines grow best in well-drained, "medium-moisture," sandy or loamy, humus-rich, circumneutral or slightly acidic soil (pH 6.0–7.2 is acceptable) and are considered to be relatively drought tolerant. A few species native to moist meadows are tolerant of wetter soils. A water-soluble fertilizer may be applied once a month during active growth. Fertilizer high in phosphorus (e.g., 5–10–5) is recommended.

### 29.8.5 Maintaining Plants

Division of plants every 2 or 3 years is recommended (most clumps survive naturally only for 3 or 4 years). The divided portions are often slow to recover. Deadheading of flowers often prolongs blooming and is a way of preventing self-seeding. In cold climates, mulching over the plants before freeze-up will assist their survival.

## 29.9 FASCINATING FACTS, FABLES, FOLKLORE, AND FRIVOLITY

- Petal spurs such as occur in *Aquilegia* are obviously designed to attract specialized pollinator species, particularly hummingbirds, bees, hawk moths, and butterflies. Since the nectar is usually produced at the base of the spur, spur length is critical to limiting access to pollinators with a long proboscis, tongue, or beak. The longest spurs in *Aquilegia* – up to 16 cm – occur in *A. longissima* of northern Mexico, Arizona, and Texas.
- Native Americans, including Meskawki, Ponca, Pawnee, and Omaha, rubbed ground columbine seeds on their palms in the belief that this would act as a love potion when they held the hand of their beloved.
- In Shakespeare's *Hamlet*, Ophelia made a bouquet of flowers, stating, "There's fennel for you and columbines." The symbolism of the columbines has been extensively analyzed but is not necessarily clear. At that time in England, the columbine flower was said to resemble the hat of a court jester, so the flower has been alleged to have represented folly. Another interpretation is that columbines were believed to represent ingratitude and self-occupation (often understood by 16th- and 17th-century playwrights), and this was consistent with Ophelia's character. Still another claim is that the columbines represented the adultery of the king and queen, since they received the bouquet.

FIGURE 29.6 Flower of *Aquilegia longissima*, which develops the longest spurs in *Aquilegia*.

*Source:* **Photo by Shutterstock/Vladimira Pufflerova.**

- The first US presidential aircraft officially designated as "Air Force One" was a 16-seat, piston-engine Lockheed VC-121 manufactured in 1948, named *Columbine II*. It transported President Dwight D. Eisenhower starting in 1953. The name *Columbine* was adopted because it was the official flower of Colorado, the adopted home state of Eisenhower's wife, Mamie. The *Columbine I* had been General Eisenhower's personal transport when he was in the army. *Columbine II* was replaced in 1954 by *Columbine III*. As of 2023, the phrase "Air Force One" refers to either of two highly customized Boeing 747-200B series aircraft, originally rated at a capacity of 524 passengers, but after modification just 70.
- "Harlequinade" is a British comic theatrical presentation that achieved great popularity between the 17th and the end of the 19th century. It involved pantomime and dancing (and sometimes no dialogue) that lasted sometimes for over 4 hr. Of the five main characters, Columbine was the highly intelligent, manipulative, perky young woman. There are many European paintings of her by famous artists.
- Columbines were once used on Memorial Day to honor veterans and decorate their graves in New England.
- Columbine was the symbol of the former city of Scarborough in the province of Ontario, Canada.
- Well-established and admired flowers often have colors named for them. "Columbine blue" is defined as "a moderate purplish-blue to violet" (*Webster's Third New International Dictionary*).
- The columbine, identified as *Aquilegia caerulea*, was declared the official state flower of Colorado in 1899. Although the choice was made after the plant won a vote of Colorado's schoolchildren, a colorful explanation has been advanced to justify the pick: The blue color of the flower is a symbol of the sky, the white represents snow, and the yellow center symbolizes Colorado's gold mining history. In 1925, a law was enacted prohibiting digging up the plant on public lands and limiting the gathering of blossoms to 25 in one day. "Where the Columbines Grow" was adopted as the official state song of Colorado in 1915.
- On April 20, 1999, a school mass shooting occurred at Columbine High School in the city of Columbine, a suburb of Denver, Colorado. Twelve students and a teacher were killed, and 21 people were injured. At that time, it was the deadliest mass shooting at a high school in US history and subsequently inspired more similar incidents. Columbine was founded in 1973, named after the surrounding community of Columbine, which, in turn, was named after the state flower of Colorado.

FIGURE 29.7 Artistic portrayals of Columbine, one of the basic characters of Harlequinade theater of past centuries: (a) Painting by Russian artist Konstantin Somov (1869–1939) for a Columbine costume design for Russian prima ballerina Anna Pavlova. (b) Early 19th-century print *Miss Barnett as Columbine* circulated by art publisher J. L. Marks. (c) Porcelain statue (prepared in 1738) of Columbine, the original by German sculptor Johann Joachim Kändler (1706–1775).

*Source:* **(a) Public domain. (b) Photo from Museum of London. (c) Photo by Sailko (CC BY SA 3.0).**

FIGURE 29.8 *Aquilegia caerulea*, the state flower of Colorado: (a) Photo of a wild-growing plant in south-central Colorado. (b) Photo of wild plants in Medicine Bow Routt National Forest, Colorado, at an elevation of 3,075 m. (c) Garden cultivar.

*Source:* **(a) Photo by Dave Powell, US Department of Agriculture Forest Service (CC BY 3.0). (b) Photo by Jim Morefield (CC BY SA 2.0). (c) Photo by Alice Gomez (CC BY SA 4.0).**

FIGURE 29.9 Columbines employed as a memorial tribute at the Columbine Memorial, Littleton, Colorado, to those murdered in the 1999 Columbine High School mass shooting.

*Source:* **Photo by Jeffrey Beall (CC BY ND 2.0).**

## KEY PUBLICATIONS

Bastida, J.M., J.M. Alcantara, P.J. Rey, P.Vargas, and C.M. Herrera. 2010. Extended phylogeny of *Aquilegia*: The biogeographical and ecological patterns of two simultaneous but contrasting radiations. *Plant Systematics and Evolution* 284: 171–185.

Hill, E.J. 1902. The etymology of columbine. *The Plant World* 5: 175–177. www.jstor.org/stable/pdf/43477174.pdf.

Hodges, S.A. and M.L. Arnold. 1994. Columbines: A geographically widespread species flock. *Proceedings of the National Academy of Sciences, USA* 91: 5129–5132.

Kandeler, R. and W.R. Ullrich. 2009. Symbolism of plants: Examples from European-Mediterranean culture presented with biology and history of art: MAY: Columbine. *Journal of Experimental Botany* 60: 1535–1536.

Masvidal, L.I. 1993. Development of new pot-plants from wild species of *Aquilegia*. *Acta Horticulturae* 337: 131–138.

Munz, P.A. 1946. The cultivated and wild columbines. *Gentes Herbarum* 7: 1–150.

Nold, R. 2003. *Columbines: Aquilegia, Paraquilegia, and Semiaquilegia*. Timber Press, Portland, OR. 194 pp.

Puzey, J.R., S.J. Gerbode, S.A. Hodges, E.M. Kramer, and L. Mahadevan. 2012. Evolution of spur-length diversity in *Aquilegia* petals is achieved solely through cell-shape anisotropy. *Proceedings Biological Sciences* 279: 1640–1645.

Weiler, T.C. and K.G. Shedron. 1986. *Aquilegia × hybrida*. In: *Handbook of Flowering*, Vol. 5, edited by A.H. Halevy, 18–21. CRC Press, Boca Raton, FL.

Whittemore, A.T. *Aquilegia*. Flora of North America [Online]. www.efloras.org/florataxon.aspx?flora_id=1&taxon_id=102388.

## KEY CULTIVATION WEBSITES

Better Homes & Gardens. *How to Plant and Grow Columbine*. www.bhg.com/gardening/plant-dictionary/perennial/columbine/.

The Spruce. *How to Grow and Care for Columbine*. www.thespruce.com/botany-of-columbine-flowers-2132551.

United States Department of Agriculture/Forest Service. *Aquilegia Express: Columbine Natural History*. www.fs.usda.gov/wildflowers/beauty/columbines/naturalhistory.shtml.

University of Washington Botanic Garden. *On the Edibility of Columbine Flowers*. https://depts.washington.edu/hortlib/pal/on-the-edibility-of-columbine-flowers/.

WikiHow. *How to Plant Columbine*. www.wikihow.life/Plant-Columbine#:~:text=Columbine%20does%20best%20when%20it,such%20as%20near%20a%20tree.&text=Amend%20the%20soil%20with%20compost,8%20inches%20(20%20cm).

# 30 Crab Apple (*Malus* Species and Hybrids)

**Scientific Names:** Crab apples are members of *Malus*. The genus name *Malus* is based on the Greek *melon*, an ancient name that was applied to the apple as well as to other tree fruits with a fleshy exterior. The word also traces to an older language, Akkadian, where *malum* meant "fullness" or "to be full." Malic acid, which produces the sour taste of crab apples, has the same language root.

**Common Names:** The word "apple" is perhaps derived from the Old High German *apful* or the Old Slavic *ablŭko*. In old Europe, it was common practice to coin names for new edible plants that were introduced using familiar names, and this was particularly true for the well-known *apple*. Thus, depending on language, lemons were called "Persian apples," dates were "finger apples," potatoes were "apples of the earth," eggplants were "mad apples," pomegranates were "apples of Carthage," and tomatoes were "love apples" and "gold apples." Numerous plants unrelated to the apple have "apple" in their name, for example, Adam's apple (*Tabernaemontana divaricata*), bake apple (*Rubus chamaemorus*), custard apple (*Annona* species), and sugar apple (*Annona squamosa*).

The catchall term "crab apple" is ambiguous. It is often applied to specific wild species, but also to any small sour apple, sometimes including varieties of the common domesticated apple and to apple trees that originated from the seeds of tossed apple cores. In commerce, crab apples are frequently defined as less than 5 cm in diameter. The "crab" in *crab apple* (also termed crabapple, crab-apple, crab) may be an allusion to the crab, from the French *crabbe* for this crustacean, with the suggestion that crab apples are fit only for crabs.

FIGURE 30.1 A "crab apple."

*Source:* Photo by Helen (CC BY 2.0).

## 30.1 INTRODUCTION

Crab apple trees have been described as "the jewels of the landscape." In the spring, they are the leading flowering woody ornamental in cooler climates of temperate regions, extensively planted along public boulevards and on private lawns. Varieties are available that produce dazzling floral displays, delightfully tart miniature apples with various culinary usages, and decorative fruits that last well into winter, attracting birds and other wildlife.

## 30.2 OVERVIEW

*Malus* species are deciduous shrubs or small to medium-sized trees growing as tall as 15 m. The several dozen species

FIGURE 30.2 Ornamental crab apple trees in flower.

*Source:* Photos by Ted (CC BY SA 2.0).



DOI: 10.1201/9781003586227-30

FIGURE 30.3 Big and little apples: (a) Fruit variation among varieties of the common apple (*Malus domestica*). (b) Apple, crab apples, cheese, and mouse.

*Source:* **(a) Photo by Keith Weller, US Department of Agriculture, Agricultural Research Service (CC BY 3.0). (b) Photo by Böhringer Friedrich (CC BY SA 3.0).**

FIGURE 30.4 Flowers of a crab apple in Wisconsin.

*Source:* **Photos by PaulaN (CC BY SA 4.0).**

are native to the temperate Northern Hemisphere, the majority in Eurasia, particularly in China. The plants can live for more than a century, some surviving for more than two centuries. *Malus* species are most valued for their commercial edible fruit. Most modern fruit varieties are placed in the hybrid *M.* × *domestica*. Although there are thousands of edible apple varieties, only a few dozen have commercial importance. Apple cultivars are generally mutants which have been selected from great numbers of seedlings and propagated and held true to type by budding or grafting. Trees raised from the seeds of apples rarely possess as desirable qualities as the maternal trees from which they were obtained. Depending on variety, apples range in size from a little larger than a cherry to as big as a grapefruit. The skin of an apple may be red, green, yellow, orange, or brown, and the flesh white, cream, pinkish, or greenish. Since some crab apples are little different from old conventional fruit varieties, much information applies to both.

FIGURE 30.5 Old crab apple varieties with doubled flowers (public domain illustrations): (a) Cultivar Bechtel Double Flowering Rose. (b) *Malus ioensis*.

*Source:* **(a) Stark Bro's Nurseries & Orchards Co., Louisiana, MO, 1896 catalogue. (b) *Addisonia: colored illustrations and popular descriptions of plants*, vol. 6, 1913.**

FIGURE 30.6 Common colors of crab apples.

*Source:* **(a) Photo by Shutterstock/Traveller70. (b) Photo by Shutterstock/Tatyana Mi. (c) Photo by Shutterstock/Amelia Martin.**

The flowers and fruits of ornamental crab apples essentially look like those of conventional apples, but the blooms of most are especially dense and colorful, and the small fruits are usually in clusters. Crab apples are widely grown for their attractive growth habit, spring flower display, and decorative fruits. Some dwarf ornamental forms have been selected. Frequently, the flower buds are red, but when they open, they are white or pink, the predominant flower colors (some cultivars have red flowers). Fruits are usually yellow, orange, purple, or red. Some varieties have attractive fall foliage; others have fruits that persist overwinter. A few varieties have semi-doubled or doubled flowers, and several have especially fragrant blooms. Weeping forms are also available.

FIGURE 30.7 Crab apples persisting on a tree overwinter.

*Source:* **Photo by Shutterstock/Sarah Marchant.**

FIGURE 30.8 Paintings of Japanese flowering crab apple (*Malus* × *floribunda*). Hybrids with this species provide many of the showiest ornamental cultivars.

*Source:* **Public domain. Van Houtte, L. (ed.). 1862–1865. *Flore des serres et des jardins de l'Europe*, vol. XV.**

## 30.3 HISTORY

As noted earlier, some crab apples are derived from varieties of edible apples. Several European species are the chief ancestors of modern edible apple cultivars, notably *M. pumila* (paradise apple), *M. sylvestris* (European crab apple), and *M. baccata* (Siberian crab apple). The apple has been grown by human beings at least as far back as recorded history in both Europe and western Asia and may have been the earliest of all fruits to be cultivated. Kazakhstan, dating back some 8,500 years, is often said to be the birthplace of the modern edible apple. Lake dwellers of Central Europe grew apples in neolithic times, as reflected by the discovery of stored apple seeds. The Greeks and Romans were familiar with about two dozen varieties, while more modern people are believed to have selected as many as 10,000 – indeed some claim 15,000 (almost half of them American, as reflected in the expression "American as apple pie").

Most ornamental crab apples originate from Asian species, especially Chinese flowering crab (*M. spectabilis*), Siberian crab apple (*M. baccata*), Toringo crab apple (*M. sieboldii*), and Japanese flowering crab apple (*M. floribunda*). The last mentioned is often considered to be the most beautiful of the ornamental crab apples. American species that are parents of crab apple cultivars include garland or sweet crab (*M. coronaria*), Oregon or Pacific crab apple (*M. fusca*), prairie crab apple (*M. ioensis*), and Southern crab apple (*M. angustifolia*). More than 1,000 crab apple cultivars have been recorded.

## 30.4 ECONOMIC IMPORTANCE

The apple is the most widespread of all fruits, accounting for about half of the world's output of deciduous tree fruit. Apples may be kept for 8 months in an atmosphere of reduced oxygen, regulated by carbon dioxide and low temperature, and such controlled atmosphere storage prevents continued ripening and makes tree-fresh apples available through the winter months. However, crab apples make up a very minor portion of the edible fruit market.

FIGURE 30.9 Paintings (public domain) of some of the Asian species extensively employed to breed ornamental crab apples: (a) *Malus spectabilis* (Chinese flowering crab apple). (b) *Malus sieboldii* (Toringo crab apple). (c) *Malus baccata* (Siberian crab apple).

*Source:* **(a) Illustration by Abraham Jacobus Wendel, from the Dutch language book by Witte, H. 1868. *Flora: afbeeldingen en beschrijvingen van boomen, heesters, éénjarige planten, enz. voorkomende in de Nederlandsche Tuinen*. Wolters, Groningen. (b) *Revue horticole* (Paris), 1870, sér. 4, vol. 42. (c) Van Houtte, L. B. (ed.). 1873. *Flore des serres et des jardin de l'Europe* (Gand, Belgium), vol. 19.**

Ornamental crab apples are the most commonly planted ornamental trees in the temperate world and accordingly represent a very important source of revenue for the horticultural industries of many countries.

## 30.5 ORNAMENTAL USES

Flowering crab apples are frequently planted as street trees, bordering major roadways and boulevards, and as compact lawn specimens.

FIGURE 30.10 Native American crab apples (public domain illustrations): (a) *Malus coronaria*. (b) *Malus ioensis*. (c) *Malus angustifolia*.

*Source:* **(a)** ***Revue horticole*** **(Paris), 1877. (b)** ***Curtis's Botanical Magazine*** **(London), vol. 139, 1913. (c)** ***Revue horticole*** **(Paris), 1910.**

FIGURE 30.11 Agricultural production related to crab apples: (a) Rows of young blooming columnar crab apple trees ("whips") in an orchard in the Niagara region of Ontario, Canada. (b) A worker holding handfuls of crab apples being harvested from an orchard.

*Source:* **(a) Photo by Shutterstock/Joanne Dale. (b) Photo by Shutterstock/Nalidsa.**

FIGURE 30.12 Flowering crab apple trees in public landscapes: (a) Trees lining a path in a park. (b) Trees lining a boulevard in Chelyabinsk, Russia. (c) Trees in a park.

*Source:* **(a) Photo by Mona El Falaky from Pixabay (public domain). (b) Photo by Shèyǐng Shī (CC BY SA 4.0). (c) Photo by Ted (CC BY SA 2.0).**

Crab apples are popular bonsai plants, and the smallest cultivars can be used for container gardening.

## 30.6 OTHER USES

Although some crab apples are quite edible, the small fruits are usually so tart (astringent) that they are considered inedible as fresh fruit. It has been claimed that smaller crab apples are usually less tasty than larger ones. Crab apples are suitable for jams, jellies, preserves, and cider. They are an excellent source of pectin, contributing to its usage for jellies. Crab apples can be incorporated in baked goods and chutneys, but admittedly, coring them is laborsome. Apple wood is fragrant and is burned to provide a pleasant scent and flavor to smoked or barbecued foods.

Some crab apples are used as rootstocks for domestic apples, for example, to provide cold hardiness or to dwarf the trees. They are also used as sources of pollen in apple orchards. Wildlife makes considerable use of the flowers and the fruit of crab apple trees, whether cultivated or growing in the wild.

## 30.7 PROBLEMS

The seeds of all apples, including crab apples, have trace amounts of amygdalin, which breaks down into cyanide. However, large numbers of seeds must be chewed and swallowed to produce serious toxicity (the seeds need to be

FIGURE 30.13 Bonsai crab apple trees.

*Source:* **Photos by Jerry Norbury (CC BY ND 2.0).**

FIGURE 30.14 Crab apple culinary preparations: (a) Candied crab apples on sticks for sale in Japan. (b) Crab apple preserves. (c) Crab apple jelly.

*Source:* **(a) Photo by Tuchodi (CC BY 2.0). (b) Photo by Shutterstock/MaxCab. (c) Photo (public domain) by Nutrition, Food Safety, and Health.**

FIGURE 30.15 Wildlife consuming crab apples: (a) Yellow-shafted flicker (*Colaptes auratus*). (b) Virginia opossum (*Didelphis virginiana*). (c) White-tailed deer (*Odocoileus virginianus*).

***Source:*** **(a) Photo by Shutterstock/Glenn Krahulic. (b) Photo by Normanack (CC BY 2.0). (c) Photo by Shutterstock/Breck P. Kent.**

pulverized through chewing in order to release the toxins). Wildlife frequently eat crab apples and appear resistant.

FIGURE 30.16 Crab apples persisting on trees in winter: (a) Crab apples covered in snow. (b, c) Crab apples covered by ice.

***Source:*** **(a) Photo by Shutterstock/Kobeprior99. (b) Photo by Vvenka1 (CC BY SA 2.5). (c) Photo by Lida (CC BY ND 2.0).**

## 30.8 CULTIVATION

### 30.8.1 Climate and Site

Crab apple trees are suited for regions with cold winters and warm summers. They do best in climate zones 4 to 8, but some cultivars are adapted to zone 3. Although semi-shade is tolerated, full sun is best (at least 6 hr daily). The ultimate size of the tree should be considered: not too close to a foundation, and distant from other trees as they enlarge in the future. Crab apples are superb small-specimen trees for small lots, but the largest cultivars are also useful as shade trees, and the dwarf forms are suitable for some borders. Crab apple trees often drop their fruit in the fall and early winter, and the decaying apples may stain driveways, sidewalks, and patios if located too close to them (this is much less of a problem for cultivars with persistent fruit).

### 30.8.2 Propagation and/or Purchasing

Most crab apple cultivars are propagated vegetatively, and the majority of people buy starter plants rather than attempt to grow the plant by making their own cuttings, or by employing seeds. The most attractive ornamental crab apple varieties are often grafts, composed of a cloned cultivar on the rootstock of a hardier species of *Malus*. Grafting is not practical for most amateur gardeners.

### 30.8.3 Planting

Crab apple trees are normally planted from potted nursery specimens in the fall but can be spring-planted. Most commercially sold starting stock will require several years to flower, but very large material that requires professional transplanting may already be in flower. Tree planting rules should be followed: Dig the hole twice as large as the root ball; loosen the root ball; add soil amendments, such as peat moss; and water frequently and deeply during the first few weeks.

### 30.8.4 Soil and Moisture Conditions

Crab apple trees grow well in many soil types. However, ideal soil for crab apples is rich, loamy, and well-drained, with a pH of 5.5 to 6.5. A mulch at the base of the tree (wood chips and pine bark are suitable) is recommended (turf grasses are best not grown against the trunk). Established trees usually need watering only in prolonged droughts. Most crab apples do well with minimal or no fertilization.

### 30.8.5 Maintaining Plants

Regular pruning is recommended; hitting the trunk with lawnmowers is not. Wrapping the trunk of young trees will shield them from rabbits, deer, and winter sunburn.

## 30.9 FASCINATING FACTS, FABLES, FOLKLORE, AND FRIVOLITY

- Apples are represented as symbolic of forbidden fruit and sin in several religions. In Christian tradition, the story of the Garden of Eden has left a

FIGURE 30.17 The symbolic role of the apple tree in the Garden of Eden: (a) Painting (public domain) showing Eve receiving a forbidden apple from the snake and giving one to Adam. (b) A charming, if much simpler, illustration of the story.

*Source:* **(a) Prepared (ca. 1615) by Jan Brueghel de Oude and Peter Paul Rubens, housed in the Mauritshuis Museum, Netherlands. (b) Photo by Shutterstock/Ottoflick.**

FIGURE 30.18 Espalier – training plants to grow on frames: (a) A diagram for growing fruit trees on a frame. (b) An espaliered crab apple tree.

*Source:* **(a) Image (public domain) from Edinburgh Postmaster General. 1846. Post Office Edinburgh and Leith directory. Colored by B. Brookes. (b) Photo by Shutterstock/Joanne Dale.**

stain on the reputations of all the participants – not just Adam and Eve, but also the snake and the apple tree. More favorably, the biblical interpretation of the apple as the "tree of knowledge" led to the associated tradition of giving an apple to the teacher, a symbolic way of indicating respect for the world of learning.

- The first colonists in New England naively transplanted American wild crab apple trees to their gardens, hoping that cultivation would immediately turn them into large trees producing succulent fruit, as in Europe. Of course, the wild trees merely produced crab apples.
- At the transition of the 19th and 20th centuries, "apple green" was a popular color because it was thought to be calming. Asylum walls were painted apple green to calm the patients, and husbands were advised to paint their kitchens apple green so their wives "would not insist on getting the vote."
- Twenty-five percent of the volume of an apple is air, which is why it floats.
- Why does the white flesh of apples and potatoes turn brown when exposed to air for a few hours? For the same reason that iron rusts – oxidation (chemical combining with oxygen) occurs. When oxygen becomes available due to removing the skin of the apple, the enzyme polyphenol oxidase (also known as tyrosinase) oxidizes compounds (phenolics) in the fruit, turning them brown. Browning of a peeled apple can be prevented simply by putting it in water, so that the enzyme does not have access to oxygen.
- An old (unkind) adage regarding apple tree pruning: You should be able to throw a cat through the branches without hitting any of them.
- Espalier is the horticultural practice of training the branches of fruit trees, such as apples, to grow on a frame so that the fruit can easily be harvested. By growing the plant against a sun-facing stone wall, the heat accumulated by the wall stimulates growth. As well as being practical, espalier is decorative and artistic.

## KEY PUBLICATIONS


Craig, J. and H.H. Harold. 1899. Native crab apples and their cultivated varieties. *Proceedings of the Iowa Academy of Science* 7: 123–141.

den Boer, A.F. 1959. *Ornamental Crab Apples*. American Association of Nurserymen, Washington, DC. 226 pp.

Dickson, E.E. 2021. *Malus*. Flora North America (Online). http://floranorthamerica.org/Malus.

Fiala, J.L. 1994. *Flowering Crabapples: The Genus Malus*. Timber Press, Portland, OR. 340 pp.

Harris, R.E. 1970. Crabapple 'Arctic Red'. *Canadian Journal of Plant Science* 50: 511–511.

Huckins, C.A. 1967. *Flower and Fruit Keys to the Cultivated Ornamental Crabapples*. Cornell University, NY. 190 pp.

Iles, J. 2009. Crabapples . . . with no apologies. *Arnoldia* 67(2): 2–13. http://arnoldia.arboretum.harvard.edu/pdf/issues/2009-67-2-Arnoldia.pdf#page=4.

Jefferson, R.M. 1970. History, progeny, and locations of crabapples of documented authentic origin. *National Arboretum Contribution* 2: 1–107.

Tardío, J., A. Arnal, and A. Lázaro. 2021. Ethnobotany of the crab apple tree (*Malus sylvestris* (L.) Mill., Rosaceae) in Spain. *Genetic Resources and Crop Evolution* 68: 795–808.

United States Science and Education Administration. 1978. *Growing Flowering Crabapples*, Revised edition. Dept. of Agriculture, Science and Education Administration, Washington, DC. 11 pp.

Zeng, X., H. Li, W. Jiang, Q. Li, Y. Xi, X. Wang, et al. 2022. Phytochemical compositions, health-promoting properties and food applications of crabapples: A review. *Food Chemistry* 386: 132789.


### KEY CULTIVATION WEBSITES

Colorado State University Extension. *Flowering Crabapple Trees*. https://extension.colostate.edu/docs/pubs/garden/07424.pdf.

Horticulture Magazine. *Crab Apple Tree Care & Growing Tips*. https://horticulture.co.uk/crab-apple/.

The Spruce. *How to Grow and Care for Flowering Crabapple*. www.thespruce.com/crabapple-tree-plant-profile-4801469.

University of Kentucky, Extension. *The Flowering Crabapple*. https://www2.ca.uky.edu/agcomm/pubs/id/id68/id68.pdf.

University of Maine, Extension. *Flowering Crabapples for Maine*. https://extension.umaine.edu/publications/2058e/.

Utah State University, Extension. *Crabapples in the Landscape*. https://digitalcommons.usu.edu/cgi/viewcontent.cgi?article=2987&context=extension_curall.

Virginia State University, Extension. *Flowering Crabapple, Malus spp*. https://vtechworks.lib.vt.edu/server/api/core/bitstreams/a3c8490f-0bb8-4163-af2a-5f96d733f276/content.

WikiHow. *How to Grow Crabapple Trees*. www.wikihow.life/Grow-Crabapple-Trees.

WikiHow. *How to Prune a Crabapple Tree*. www.wikihow.com/Prune-a-Crabapple-Tree.

# 31 Crape Myrtle (*Lagerstroemia indica* and Hybrids)

**Scientific Names:** The genus name *Lagerstroemia* honors Magnus von Lagerstroem (1696–1759), a Swedish merchant who was the director of the Swedish East Indies Company. Von Lagerstroem used his position to send collections of *Lagerstroemia* to his friend Carl Linnaeus, the father of biological nomenclature, who named more species than anyone else in history (over 12,000 plants and animals), including *L. indica*.

**Common Names:** The name crape myrtle (crape-myrtle, crapemyrtle, crepe myrtle) is based on the petals, variously described as crimped, crinkled, ridged, and wrinkled, which resemble crepe paper. In English, *crape* and *crepe* are synonyms. The spelling "crepe myrtle" is common in the Southern United States. The spelling "crêpe myrtle" represents confusion with the French term *crêpe*, referring to a thin pancake (although the English words *crepe* and *crape* are both derived from the French *crêpe*). The word *myrtle* is based on the myrtle family (Myrtaceae), especially the common myrtle (*Myrtus communis*). Crape myrtle is not related, but the term *myrtle* is presumed to have been applied because the leaves are reminiscent.

## 31.1 INTRODUCTION

Crape myrtle is nicknamed the "lilac of the South," a reference to its extraordinary popularity in the Southern United States. The plant has become a mainstay in mild-climate landscapes, valued for its stunning, long-lasting floral displays, as well as its ease of production and cultivation and wide range of forms that are useful for a diversity of landscape uses.

FIGURE 31.1 Wrinkled (crape-like) appearance of the petals of crape myrtle (*Lagerstroemia indica*).

*Source:* **(a) Photo by Juan Carlos Fonseca Mata (CC BY SA 4.0). (b) Photo by BFS Man (CC BY 2.0).**

*Lagerstroemia* is a genus of about 60 tropical and subtropical Old World species. "Crape myrtle" traditionally refers to *L. indica* but has also come to include a widely grown, disease-resistant hybrid (*L. indica* × *L. faurei*) with the Japanese *L. faurei*. Other species of *Lagerstroemia*, such as *L. speciosa* (Queen's crape myrtle, pride of India), are also sometimes cultivated.

## 31.2 OVERVIEW

Crape myrtle is a deciduous shrub (especially in colder areas, where the aboveground parts are winter-killed) or small tree 4 to 8 m high, often with multiple trunks if basal suckers are not removed. Some recent cultivars are dwarfs,

FIGURE 31.2 Crape myrtle (*Lagerstroemia indica*) cultivars with differently colored flowers.

*Source:* **Photo collage by Shutterstock/Noel V. Baebler.**

FIGURE 31.3 Attractive vegetative (non-reproductive) features of crape myrtle (*Lagerstroemia indica*): (a) Peeling bark. (b) A collection of colorful autumn leaves.

*Source:* **(a) Photo by Shutterstock/Kathleen Stirling. (b) Photo by Jan Haerer from Pixabay (public domain).**

DOI: 10.1201/9781003586227-31

about 1 m in height. Crape myrtle is a native of the Indian subcontinent, Southeast Asia (especially China and Korea), northern Australia, and parts of Oceania. Strips of bark peel and flake off in early summer, revealing mottled new bark that may be brown, pale cream, cinnamon, or orange. The bark color gradually fades over winter, peeling again the next summer. The young foliage of several cultivars is bronze, red, or burgundy. In the autumn, the leaves turn red, yellow, or orange. The branches droop as the tree grows. The flowers are 5–8 cm wide, characteristically crinkled, and occur in large clusters on the ends of branches. Flowering begins in spring in some cultivars, early to midsummer in most, and lasts into autumn. Flower colors may be lavender, pink, purple, red, or white. Some cultivars have bicolor flowers, and some even have flowering clusters that produce a mix of flower colors. The flowers transform into brown capsule-like fruit, attractive to birds.

## 31.3 HISTORY

*Lagerstroemia* has been in cultivation for centuries in its native Southeastern Asia. Several species, particularly *L. indica*, were introduced to Europe in the late 18th century, and subsequently to other parts of the world. In the last several decades, breeding programs in the United States and Europe have concentrated on producing improved cultivars of *L. indica* and *L. fauriei*.

## 31.4 ECONOMIC IMPORTANCE

Crape myrtle is an important, popular ornamental tree in semi-tropical regions, contributing to local horticultural industries. In the United States alone, about a million crape myrtle plants are sold by nurseries every year.

## 31.5 ORNAMENTAL USES

Some cultivars are employed as small foundation shrubs, others as specimen or shade trees for lawns, gardens, street sides, parks, and the fronts of buildings. In the United States,

FIGURE 31.4 A nursery orchard of crape myrtle.

*Source:* Photo by Shutterstock/Natalia Golubnycha.

FIGURE 31.5 Lawn plantings of crape myrtle.

*Source:* (a) Photo by Bishnu Sarangi from Pixabay (public domain). (b) Photo by Captain-tucker (CC BY SA 3.0). (c) Photo by Bill McChesney (CC BY 2.0). (d) Photo by Vhines200 (CC BY ND 2.0).

FIGURE 31.6 Street plantings of crape myrtle.

*Source:* (a) Photo by Shutterstock/Noel V. Baebler. (b) Photo by Tony Glover, Alabama Extension (public domain). (c) Photo by Shiva Shenoy (CC BY 2.0).

FIGURE 31.7 Crape myrtle grown as bonsai.

*Source:* (a) Photo by Shutterstock/Khairil Azhar Junos. (b) Photo by Deb Nystrom (CC BY 2.0). (c) Photo by Allison Meier (CC BYSA 2.0).

many states plant crape myrtle extensively along interstate highways and other right-of-ways. Many towns in the southeastern United States also use crape myrtle in landscaping.

## 31.6 OTHER USES

*Lagerstroemia* wood is strong, and the timber from several species (under the commercial names pyinma, Asian satinwood, and Cambodian satinwood) is employed to construct bridges, furniture, and railroad ties. *Lagerstroemia* species have been employed in herbal medicine in their Indigenous Old World areas, and their chemical constituents are under investigation for potential medical applications.

## 31.7 PROBLEMS

The cold tolerance of crape myrtle limits its usage in climate zone 6 and northward, although attempts at breeding cold-resistant varieties are continuing. In the United States, there are perhaps ten million plants of crape myrtle, a non-native species, and this has considerably changed the natural diets of numerous birds (consuming the seeds) and bees (eating the pollen). *Lagerstroemia indica* is significantly invasive in semi-tropical and tropical regions of the world.

## 31.8 CULTIVATION

### 31.8.1 Climate and Site

Crape myrtle grows in hardiness zones 6 to 9. It is sometimes grown in colder areas, with mixed success (climate change is extending the Northern limit where it can be grown). It is quite tolerant of hot climates and should be provided with full sun, or at least partial sun in very hot areas. *Lagerstroemia speciosa* (Queen's crape myrtle) is adapted to hardiness zones hotter than zone 9.

### 31.8.2 Propagation and/or Purchasing

Crape myrtle cultivars are reproduced commercially by stem cuttings. In some countries (including the United States), it may be illegal to do this for patented varieties.

FIGURE 31.8 Animals feeding on crape myrtle: (a) American goldfinch perched on fruit capsules full of seeds. (b) Honeybee collecting pollen.

*Source:* **(a) Photo by Shutterstock/Jake Daugherty. (b) Photo by Shutterstock/Muratart.**

FIGURE 31.9 Snow-covered crape myrtle in Virginia.

*Source:* **Photo by Christopher Connell (CC BY ND 2.0).**

### 31.8.3 Planting

The size of particular cultivars dictates how much space they require. Often, crape myrtle is positioned as an isolated specimen, but it naturally suckers and spreads. When it is planted as a single-row hedge, a spacing of 1 to 1.3 m is recommended. Dwarf crape myrtles can be grown in individual containers, or a group can be placed in a large box planter.

### 31.8.4 Soil and Moisture Conditions

Crape myrtle needs to be well-watered when young, but once it develops its root system, it is tolerant of drought. It prefers moist, well-drained, loamy-clay soils but can grow in a wide range of soil textures. The recommended pH is slightly alkaline to acidic (tolerating pH as low as 5.0, but the recommended range is 6.0 to 7.3). Heavy fertilization increases foliage production at the expense of flowers.

### 31.8.5 Maintaining Plants

New buds produced early in the season can be pinched to increase flower number and branching. Often, the plants are trained to grow as trees. The lower branches are often removed to highlight the trunk. In cooler areas, pruning should be done early in the season, as late-season removal of branches can stimulate new growth that is not cold-hardy. Some horticulturalists strongly discourage extensive pruning (calling the practice "butchering" and "crape murder"), arguing that it encourages disease, suckering, soft new wood that attracts insects, and discourages flowering.

## 31.9 FASCINATING FACTS, FABLES, FOLKLORE, AND FRIVOLITY

- Various charming stories about crape myrtle trace to ancient times in Asia. According to a Chinese folktale, there was a vicious beast which attacked many people and animals, so an ambassador from heaven went down to earth and locked the monster into a mountain, allowing it to come out once, during the new year. To supervise the beast during the new year, the ambassador remained on earth as a crape myrtle.

- The flowers of most species of *Lagerstroemia*, including *L. indica*, produce two kinds of stamens: six peripheral ones with long, curved filaments and large anthers producing fertile blue-green pollen, and a central tuft of several dozen stamens with smaller anthers that produce yellow infertile pollen that serves to attract and feed pollinators, mainly bees.
- Texas designated the crape myrtle as the "state shrub" in 1997 (Texas also recognizes an official "native shrub").

## KEY PUBLICATIONS


Al-Snafi, A.E. 2019a. Medicinal value of *Lagerstroemia speciosa*: An updated review. *International Journal of Current Pharmaceutical Research* 11: 18–26.

Al-Snafi, A.E. 2019b. A review on *Lagerstroemia indica*: A potential medicinal plant. *IOSR Journal of Pharmacy* 9: 36–42.

CAB International. 2014. *Lagerstroemia indica* (L.), Indian crape myrtle. *Invasive Species Compendium*. www.cabdirect.org/cabdirect/abstract/20107601612.

Cabrera, R.I. 2004. Evaluating and promoting the cosmopolitan and multipurpose *Lagerstroemia*. *Acta Horticulturae* 630: 177–184.

Cabrera, R.I. and R.D. Lineberger. 2003. New on-line educational and research resource for crape myrtles (*Lagerstroemia* spp.). *SNA Research Conference* 48: 259–262. https://irp.cdn-website.com/5d757b5b/files/uploaded/section05.pdf.

De Wilde, W.J. and B.E. Duyfjes. 2014. Miscellaneous information on *Lagerstroemia* L. (Lythraceae). *Thai Forest Bulletin (Botany)* 41: 90–101.

Dubois, J.J.B. and F.A. Blazich. 2008. *Lagerstroemia* L. In: *The Woody Plant Seed Manual, Agriculture Handbook 727*, edited by F.T. Bonner, R.P. Karrfalt, and R.G. Nisley, 634–636. U.S. Department of Agriculture, Forest Service. www.fs.usda.gov/nsl/Wpsm/Lagerstromia.pdf.

Egolf, D.R and A.O. Andrick. 1978. *The Lagerstroemia Handbook/Checklist*. American Society of Botanical Gardens and Arboreta, Inc., Wilmington, DE. 72 pp.

Furtado, C.X. and M. Srisuko. 1969. A revision of *Lagerstroemia* L. (Lythraceae). *The Gardens Bulletin (Singapore)* 24: 185–335.

Nepi, M., M. Guarnieri, and E. Pacini. 2003. "Real" and feed pollen of *Lagerstroemia indica*: Ecophysiological differences. *Plant Biology* 5: 311–314.

Pooler, M. 2007. Crapemyrtle. In: *Flower Breeding and Genetics*, edited by N.O. Anderson, 439–457. Springer, Dordrecht.

Rawson, J.M. and R.L. Harkess. 1997. Pinching of *Lagerstroemia* for use as a potted florist crop. *HortScience* 32: 470–470.

Rojas-Sandoval, J. and P. Acevedo-Rodríguez. 2014. *Lagerstroemia indica* (Indian crape myrtle). *CABI Compendium*. https://doi.org/10.1079/cabicompendium.296.

Stimart, D.P. 1986. *Lagerstroemia*. In: *Handbook of Flowering*, Vol. 5, edited by A.H. Halevy, 187–190. CRC Press, Boca Raton, FL.

Wang, J., W. He, X. Liao, J, Ma, W. Gao, H. Wang, et al. 2023. Phylogeny, molecular evolution, and dating of divergences in *Lagerstroemia* using plastome sequences. *Horticultural Plant Journal* 9: 345–355.


## KEY CULTIVATION WEBSITES


Clemson Cooperative Extension. *Crape Myrtle Pruning*. https://dc.statelibrary.sc.gov/bitstream/handle/10827/41664/CU_ES_HGIC_1009_2006-11.pdf?sequence=1.

Crape Myrtle Trails of McKinney. *Crape Myrtle Varieties: The Complete List*. https://crapemyrtletrails.org/varieties/complete-list-crape-myrtle/.

Louisiana State University, College of Agriculture. *The Great Crape Myrtle Controversies*. www.lsuagcenter.com/profiles/rbogren/articles/page1608734072627.

Piedmont Master Gardeners. *Crape Myrtle – A Southern Icon Gets Even Better*. https://piedmontmastergardeners.org/article/crape-myrtle-a-southern-icon-gets-even-better/.

The Spruce. *How to Grow and Care for Acoma Crape Myrtle*. www.thespruce.com/growing-acoma-crape-myrtle-5076300.

The Spruce. *How to Grow and Care for Crepe Myrtle (Lagerstroemia indica)*. www.thespruce.com/crepe-myrtle-growing-guide-5210870.

United States Department of Agriculture Natural Resources Conservation Centre. *Crape Myrtle, Lagerstroemia indica*. https://plants.usda.gov/DocumentLibrary/plantguide/pdf/pg_lain.pdf.

University of Georgia Extension. *Crape Myrtle Culture*. https://extension.uga.edu/publications/detail.html?number=C944.

WikiHow. *How to Grow Crepe Myrtle*. www.wikihow.com/Grow-Crepe-Myrtle

WikiHow. *How to Kill [Remove] Crepe Myrtle*. www.wikihow.com/Kill-Crepe-Myrtle

WikiHow. *How to Prune a Crepe Myrtle*. www.wikihow.com/Prune-a-Crepe-Myrtle

# 32 Crocus (*Crocus* Species and Hybrids)

**Scientific Names:** The genus name *Crocus* traces to the Greek *Crocus* or *Krokus*. Various explanations have been given of why the word was adopted. *Krokos* in Greek means "wire," "filament," or "hair," and it has been proposed that the word originally referred to the filamentous stigmas of the saffron crocus, employed as a spice. Alternatively, it has been claimed that the modern word *crocus* traces to a Greek myth in which a youth, Crocus, frustrated in his love life, was transformed into a flower bearing his name.

**Common Names:** Crocus; plural: crocuses, croci. The name "Dutch crocus" reflects the widespread supply from The Netherlands. "Snow crocus" is based on the frequent early development of the flowers through a covering layer of snow. *Crocus* is a member of the Iridaceae. Some species in other families also have crocus in their common names. "Autumn crocus" usually refers to the genus *Colchicum* (of the Colchicaceae), which has about 160 species, notable for the toxic alkaloid colchicine used to treat gout and in doubling the chromosome number of plants for breeding. Sometimes the phrase *autumn crocus* is employed to designate crocuses that flower in late summer and autumn, often before the foliage appears.

## 32.1 INTRODUCTION

Crocus is one of the early-flowering spring perennials emerging from "bulbs" (strictly, corms), perhaps second in popularity only to the tulip. Many of the species are exquisitely beautiful, and extensive breeding has made a very wide range of cultivars available. The exceptional tolerance of cold of most of the plants makes them especially suited to climates with frigid winters.

## 32.2 OVERVIEW

Most authorities accept between 100 and 160 species of *Crocus*, but a recent monograph (Rukšāns 2017) recognized 235. Crocuses are native to woodland, scrub, and meadows from sea level to alpine tundra (some species occur at altitudes over 3,000 m), from western Europe and northwestern Africa to western China. The center of species diversity includes the Balkan Peninsula and Turkey. Many

FIGURE 32.1 Crocuses emerging from snow in the early spring.

*Source:* (a) Photo by M. W. from Pixabay (public domain). (b) Photo by Peter Stenzel (CC BY ND 2.0).

FIGURE 32.2 *Crocus* in the Alps.

*Source:* (a) Photo by Shutterstock/FamVeld. (b) Photo by Micpicee from Pixabay (public domain). (c) Photo by Claude Perso (CC BY SA 2.0).

FIGURE 32.3 Attractive beds of mixed crocuses.

*Source:* (a) Photo by G. Gerhard from Pixabay (public domain). (b) Photo by Tony Hisgett (CC BY 2.0).

DOI: 10.1201/9781003586227-32

of the species are under threat of genetic erosion because of urbanization and global warming, and a few are considered to be endangered.

Crocuses are charming dwarf plants, the flower stalks rarely exceeding 15 cm, although the foliage may be much longer. The leaves are semi-erect, long, narrow, and lance-shaped, like blades of grass, with a white or silvery stripe (due to cells lacking chlorophyll) running along the center. The stripe differs in width depending on species but is characteristic of the genus *Crocus*. Some species or cultivars will produce just one flower; many produce several. Crocus flowers have six petals in two whorls of three, the outer set often somewhat larger and attractively marked. The petals are arranged into a cup (much more obvious in the wide-petalled species), which closes at night and in cloudy weather. Floral colors vary extensively, often shades of white, yellow, or purple, sometimes in combinations. There are three stamens and a pistil with a style divided into three stigma-tipped branches. Pollination is by bees, moths, and beetles, and many of the species are scented. Most ornamental crocuses are spring-flowering, although some are autumn-flowering. Spring-flowering plants have a long floral tube, the ovary (where seeds are produced) situated belowground, where it is protected from adverse weather and herbivores. Flowering of an individual plant may extend from 1 to 3 weeks. As the seeds ripen, the subterranean ovary elevates, pushed up by lengthening of the stem beneath it, so that the ripe seeds can be dispersed aboveground.

Crocuses are perennials, at least with respect to the underground parts, which survive an inclement season by dormant underground "corms." These produce new foliage and flowers when environmental conditions become suitable. In horticulture, so-called "bulb crops" include a variety of plants with underground storage structures, from which the plants regenerate after a period of dormancy. The storage structures may be modified roots, stems, or a combination of stem and leaf tissues, associated with bud meristems that can produce new roots and stems. Many herbaceous plants whose tops are killed annually survive a dry or cold season simply by conventional underground stems and/or roots, but many also produce specialized structures that are strongly swollen to store food reserves (commonly starch). These are distinguished technically as bulbs, corms, rhizomes, and tubers. True bulbs (exemplified by onion, tulip, and daffodil) are spheroidal masses made up mostly of modified fleshy storage leaves ("scales") which appear as rings in a cross section. Corms (e.g., crocus, banana) are similar in shape but made up mostly of a solid mass of compacted stem tissue lacking internal layers like true bulbs. Rhizomes (e.g., ginger, iris), like corms, are made of fleshy stem tissue, but unlike corms, which orientate vertically, rhizomes elongate laterally quite substantially. "Stem tubers" (such as potato and true yam – *Dioscorea* species) are essentially short (i.e., not substantially elongated), compact, swollen rhizomes. "Root tubers" or "tuberous roots" (such as sweet potato, cassava, daylily, and dahlia) are simply swollen roots and are internally composed of root tissues. All these storage structures produce both shoot and root meristems from the outer part of the storage organ (e.g., eyes of a potato). In the horticultural industry, the term "bulb" usually includes all fleshy underground storage structures. Crocus corms have a covering formed from expanded leaf bases, termed a *tunic*, which may be fibrous, papery, or eggshell-like. The roots emerge from a smooth area at the base known as the basal plate. A new corm often develops above the old crocus corm (once a given corm produces a shoot or shoots, it has used up its energy reserves and dies). Because new corms are produced annually above the old corms, the plants would eventually pop out of the ground. To prevent this, crocuses and many other bulb-like plant develop "contractile roots." These are modified adventitious roots growing from the base of the stem, and they shorten to pull the corm or bulb down to a suitable level. Depending on species, corms may generate small offshoots known as daughter corms or cormels, which, like the corms, reproduce the plant vegetatively.

FIGURE 32.4 Corms of *Crocus*: (a) *Crocus sieberi*. (b) *Crocus chrysanthus* 'Dorothy'.

***Source:*** **Photos by Mariluna (CC BY SA 3.0).**

About 3 dozen species of *Crocus* are cultivated significantly, and there are cultivars of many of these. Cultivars identified as given species, in fact, are often hybrids with other species. The alpine species *C. vernus* ("Dutch crocus") is the chief ancestor of commonly planted garden crocuses, but cultivars so identified are likely to be hybrids with other species. Many hybrid cultivars develop notably large flowers – up to 8 cm in width (hence the name "giant crocus," reminiscent of "jumbo shrimp," since the plants are dwarves by comparison with most plants). Dutch yellow crocus (*C. flavus*), from stony slopes in Southeastern Europe, is another popular spring-flowering species. Other important species include the golden or snow crocus (*C. chrysanthus*), the Tommie crocus (*C. tommasianus*), the tricolor crocus (*C. sieberi* 'Tricolor'), and the silvery crocus (*C. biflorus*). As with most ornamentals, breeding has been particularly concerned with flower size, color, and longevity, and modern cultivars have been changed considerably by comparison with their ancestors.

Saffron crocus (*C. sativus*), the source of the spice saffron, is the most important species in the genus *Crocus*. Its stigmas were employed as a spice by the classical Egyptians, Jews, Greeks, Romans, Hindus, and Muslims, and the plant continues to be much-cultivated.

**FIGURE 32.5** Examples of container-grown ornamental crocuses: (a) *Crocus tommasinianus* 'Whitewell Purple'. (b) *Crocus sieberi* 'Firefly'. (c) *Crocus chrysanthus* 'Cream Beauty'. (d) *Crocus olieveri*.

*Source:* (a) Photo by Ghislain 118 (CC BY SA 3.0). (b) Photo by Averater (CC BY 4.0). (c) Photo by JOHN19701970 (public domain). (d) Photo by Averater (CC BY 4.0).

**FIGURE 32.6** Saffron crocus (*Crocus sativus*) and its product, saffron (stigmas). (a) Illustration. (b) A worker with a container of flowers on a saffron farm in Iran, the source of most of the world's saffron. About 160 flowers are required to produce 1 gram of dry saffron, which typically sells for about $10. (c) The spice saffron (stigmas from the flowers).

*Source:* (a) Redouté, P. J. 1805. *Les liliacées*. Digitally enhanced by Rawpixel (CC BY 4.0). (b) Photo by Safa Daneshvar (CC BY SA 3.0). (c) Photo by Salonik Saffron (CC BY SA 4.0).

## 32.3 HISTORY

Serious cultivation of *Crocus* species as ornamentals appears to trace to 16th-century Netherlands, from which the garden use spread across Europe.

## 32.4 ECONOMIC IMPORTANCE

Tulips and daffodils dominate spring bulb sales in most countries, but crocus is earlier-flowering and is the most popular of the very early-flowering bulbs. In The Netherlands, large quantities of corms of the ornamental crocus species

**FIGURE 32.7** Nineteenth-century paintings (public domain) of cultivated crocuses (mostly *Crocus vernus*) by European botanical illustrators.

*Source:* (a) Thomé, O. W. 1885. *Flora von Deutschland, Österreich und der Schweiz*. Gera, Germany. (b) *Flore des serres et des jardins de l'Europe (Gand, Belgium)*, vol. 13, 1858. (c) Redouté, P. J. 1805. *Les liliacées*. Paris. (d) Robertson, C. J. 1830. *Crocus spp.*, vol. 7. Transactions of the Royal Horticultural Society of London.

**FIGURE 32.8** Commercial crocus fields in The Netherlands producing corms for the international trade.

*Source:* Photos (public domain) by Jan Nijman, Pixabay.com.

are generated for international trade, and corms are marketed throughout the temperate world.

## 32.5 ORNAMENTAL USES

Most ornamental crocuses are spring-flowering, although there are species and cultivars that bloom in late fall and early winter in mild climates. Crocuses are mostly planted outdoors for their colorful flowers and are rarely grown commercially as cut flowers. Spring-flowering crocuses are very widely employed as short-season ephemerals in lawns and under deciduous trees and shrubs. Some cultivars show up through melting snow to take advantage of the lack of competition with other plants. Because they are very short, crocuses are inappropriate for mixed planting with most other ornamentals, although they do blend well with other species of spring-flowering bulbs and are superb in rock (or alpine) gardens. They are very attractive in dedicated beds or borders that can be left fallow for most of the year or replanted with other ornamentals. Crocuses are especially suited to containers, since their limited height is not a handicap. Potted crocuses can be employed as temporary flowering houseplants, and the corms can be forced for out-of-season flowers, and indeed, they are marketed for this purpose. The period of flowering is relatively short, but mixing cultivars can extend the period of blooming.

FIGURE 32.9 Landscape uses of crocus: (a) Mixed cultivars in a garden. (b) Crocuses emerging in the spring under deciduous trees in a park in England. (c) Crocuses planted along a roadway in Vancouver, Canada. (d) A trail of crocuses established in a large public lawn.

*Source:* **(a) Photo by Shutterstock/Emilia Kun. (b) Photo by Julian Smith (CC BY 2.0). (c) Photo by Heather Harvey (CC BY SA 2.0). (d) Photo by Ellen26 from Pixabay (public domain).**

FIGURE 32.10 A dachshund puppy beside crocus plants. Ingestion of *Crocus vernus* (common spring crocus) can cause general gastrointestinal upset in cats and dogs, including drooling, vomiting, and diarrhea. (*Colchicum autumnale*, one of the various plants called autumn crocus, can cause death.)

*Source:* **Photo by Shutterstock/Tori555.**

## 32.6 OTHER USES

The use of saffron (*C. sativus*) as a spice, as mentioned earlier, is responsible for most of the economic value of the genus *Crocus*. Numerous studies have been conducted of potential medical applications of saffron, particularly in Iran, the center of world production, where it is relatively available. There are many claims of medicinal values, and saffron has been employed in folk medicine for millennia. However, because it is so expensive, its pharmaceutical usage is negligible.

## 32.7 PROBLEMS

Crocuses are significantly toxic. Saffron (from *C. sativus*) is potentially quite poisonous, and when employed as a spice, it should be used sparingly. However, little information is available on the relative safety of most *Crocus* species. The corms of true *Crocus* species are often consumed by wildlife and have occasionally been employed as food by Indigenous people but are not recommended and may cause gastrointestinal upset. The shoots (stems, foliage, flowers) of true *Crocus* species are also sometimes consumed by wildlife but are not suitable for livestock. There is danger of people confusing the relatively harmless fall-flowering species of *Crocus* (often called "autumn crocuses") with the extremely toxic *Colchicium* species (also called autumn crocuses).

## 32.8 CULTIVATION

### 32.8.1 Climate and Site

Most crocus cultivars prefer cool climates and are not heat-tolerant. The freeze tolerance of the ornamental species and cultivars varies considerably, so it is important to obtain planting stock from sources that reliably identify the ability of the plants to overwinter in the local climate. The corms usually require a period of 12 to 15 weeks of cold (2–7 °C) to set blooms. The flowers fade in heat. Covering the plants overwinter with mulch may be desirable to protect less-hardy cultivars.

Crocuses do best in full sun. They tolerate partial shade but should not be located on the north side of a building or in dense shade. They are often planted under deciduous trees because their early growth in the spring is not highly shaded. In protected situations, they may bloom weeks earlier than in open, exposed areas. Because they go dormant, a bed of crocuses can leave the area they occupy looking rather desolate. Gardeners concerned about this will interplant other low-growing flowering plants, restrict the size of the bed, or simply scatter the plants in lawns.

### 32.8.2 Propagation and/or Purchasing

Crocus seeds are slow to germinate, and the seedlings also grow slowly, so practically it is preferable to plant corms. The more attractive cultivars are hybrids that will not breed true to type from seeds; these must be reproduced vegetatively. Crocuses are relatively short-lived in many areas and need to be replanted (preferably not in exactly the same location, to prevent disease accumulation). Natural vegetative reproduction often occurs by offsets (baby corms on the sides of the mother corm). These can be separated and replanted after the foliage deteriorates. In some locations, self-sowing of seeds may occur (but the resulting plants may not be true to type).

Crocus corms are available in most garden centers in the fall for autumn planting to produce spring flowers. Rarer varieties are available from bulb catalogs on the internet.

Larger corms should be sought, and bargain-basement offerings are unlikely to be desirable. Since crocuses are grown as perennials, varieties should be chosen that are suited to one's climate.

### 32.8.3 Planting

Spring-blooming cultivars are planted in early autumn. In beds, the corms of most varieties are planted 10–13 cm deep, 5–10 cm apart, pointed (upper) end up. The pointed end can be difficult to identify, but if not planted correctly, the shoot will simply turn and grow upward. For a concentrated display in a given location, corms may be planted very closely – a dozen or so in a large hole. They can be scattered randomly in turf or a natural area that is open in the spring, but this will necessitate delaying mowing of lawns in the spring until the plants senesce. In very large lawns of public or private estates, trails of crocuses are sometimes established. Planting in pots provides the flexibility of locating them in different places and, indeed, facilitates keeping the dormant plants out of sight after they flower. Crocuses display well in alpine and rock gardens and in containers.

### 32.8.4 Soil and Moisture Conditions

Moist soil promotes growth during the spring and summer. However, a well-draining soil is important, as crocus poorly tolerates wet soils, especially after the plants go dormant. Clay should be avoided. Drier soils discourage rot when the corms are dormant, but in locations where snow cover is not present in winter, some watering may be required. Well-grown corms have sufficient energy to produce flowering plants, so fertilization is not essential. However, in poor soils, a light application of a bulb fertilizer may be added at planting time. Soils with pH 6.0–7.0 are recommended.

### 32.8.5 Maintaining Plants

The leaves will deteriorate after flowering but should not be cut back until they yellow naturally. When planted in lawns, mowing should be delayed until the flowers have faded and the leaves are yellowing and withering, in order to allow the corms to store up energy for the next spring growing season (some authorities recommend not mowing until 6 weeks after flowering). In some locations, squirrels, chipmunks, voles, and other rodents will feed on the plants, which may require protection in cages, by screening or by a scent deterrent. *Crocus tommasinianus* ("Tommies"), unlike most crocuses, are claimed to contain a bitter-tasting alkaloid that repels squirrels.

### 32.8.6 Care of Indoor Plants

Potted flowering crocuses are often available cheaply in stores for use as short-term houseplants. When purchasing potted plants, they should be free of viruses (a particular problem for crocuses), as evidenced by distortions, streaking, or buds that won't open.

FIGURE 32.11 Crocus corms are suitable for forcing in containers of water, using rocks and pebbles as a stabilizing substrate.

*Source:* **(a) Public domain diagram from Macrovector/Freepik. (b) Public domain photo from Pixabay.com.**

Corms can be forced to flower in the winter (providing that they have received the necessary period of chilling). These could be harvested from one's garden during the summer and stored, or simply purchased. Large corms should be employed, and they need to be vernalized (8–15 weeks of chilling are recommended for crocus). The corms may be potted about 3 cm deep in a pot of well-watered soil in the autumn, and chilled in a dark location for 2 to 3 months at 3.3–7.2°C. Moderate watering should be applied during this period. More simply, chilling of bulbs can be done by putting them in damp sphagnum moss in a paper bag, maintained in a refrigerator. Subsequently transferred to room temperature, the crocuses should bloom in about a month. Alternative to soil, chilled crocuses (and other bulb-like plants) can be forced by maintaining the bottom 7 to 14 mm of the corm in water. A substrate of pebbles can be employed to "root" the plants, or specialty clear hourglass-shaped vases can be used to stabilize the corms so that the plants don't flop over, while also allowing the roots to be viewed. Roots will grow into the water, and the flowers will develop from the upper portion. Forced plants have little energy left and are not ideal for transplanting outdoors.

## 32.9 Fascinating Facts, Fables, Folklore, and Frivolity

- The maritime city of Husum in Germany, near Denmark, is known as "the flower wonder of the north" for its Husum crocuses (identified as "*C. napolitanus*," but likely *C. vernus*). In the spring, around 4 million purple-flowered plants bloom on the grounds of Husum Castle, on an area of around 5 ha, transforming the park into a sea of purple.

FIGURE 32.12 Parts of the enormous display of Husum crocuses (termed "*Crocus napolitanus*," but likely *C. vernus*) in Husum Park, Germany.

***Source:* (a) Photo by Shutterstock/Mischroeder. (b) Photo by Genet at de.wikipedia (CC BY SA 3.0).**

- Crocus enthusiasts are known as "croconuts."
- The financial community sometimes calls companies or economic sectors that recover early after an economic downturn as "crocuses," in reference to the plant's ability to thrive in the early season.
- "Nyctinasty" is the movement of plant parts – usually in response to environmental conditions, such as temperature, but sometimes controlled by internal rhythms. One example of this is the "sleeping" of flowers, commonly at night. *Crocus* flowers open in sunny conditions and close when the sky is overcast. In *Crocus*, there is evidence that the phenomenon protects the flower's pollen.
- In Great Britain and Ireland, it was customary to paint the little finger of children with a purple dye to indicate that they had received the lifesaving polio vaccine. In the latter 20th century, the Rotary organization noted a resemblance of the spreading dye on fingers to the petals to an opening purple crocus and consequently adopted the crocus as a symbol for the campaign to eradicate polio. On World Polio Day (October 24), millions of purple crocus bulbs were frequently planted throughout Britain.
- Prairie crocus (*Anemone patens*) of the Ranunculaceae is the floral emblem of the Canadian province of Manitoba. "The Crocus Trail," a Manitoban section of the Trans-Canada Trail (aka the Great Trail), which connects the Atlantic and Pacific Oceans, is based on prairie crocus. Similarly, Manitoba's honorary "Order of the Crocus" is named for this species.

FIGURE 32.13 Pollinators visiting *Crocus vernus*: (a) A honeybee. (b) A European peacock butterfly.

***Source:* (a) Photo by httpwiresstock.io. (b) Photo by Paasikivi (CC BY SA 4.0).**

- Early-flowering crocuses are greatly appreciated by bees and other pollinators, since other floral sources are usually very limited in the early spring.

## KEY PUBLICATIONS

Abu-Izneid, T., A. Rauf, A.A. Khalil, A. Olatunde, A. Khalid, F.A. Alhumaydhi, et al. 2020. Nutritional and health beneficial properties of saffron (*Crocus sativus* L.): A comprehensive review. *Critical Reviews in Food Science and Nutrition*. https://doi.org/10.1080/10408398.2020.1857682.

Bowles, E.A. 1952. *Handbook of Crocus and Colchicum for Gardeners*. Bodley Head, London. 222 pp.

Erol, O., H.B. Kaya, L. Şik, M. Tuna, L. Can, and M.B. Tanyolaç. 2014. The genus *Crocus*, series *Crocus* (Iridaceae) in Turkey and 2 East Aegean islands: A genetic approach. *Turkish Journal of Biology* 38: 48–62.

Harpke, D., S. Meng, H. Kerndorff, T. Rutten, and F.R. Blattner. 2013. Phylogeny of *Crocus* (Iridaceae) based on one chloroplast and two nuclear loci: Ancient hybridization and chromosome number evolution. *Molecular Phylogenetics and Evolution* 66: 617–627.

Kerndorff, H., E.H. Pasche, and D. Harpke. 2015. The genus *Crocus* (Liliiflorae, Iridaceae): Lifecycle, morphology, phenotypic characteristics, and taxonomical relevant parameters. *Stapfia* 103: 27–65.

Kerndorff, H., E.H. Pasche, and D. Harpke. 2016. The genus *Crocus* (Liliiflorae, Iridaceae): Taxonomical problems and how to determine a species nowadays? *Stapfia* 105: 42–50.

Khansarinejad, B., M.R. Hesandokht, V. Nazeri, and A. Sorani. 2015. Evaluation of genetic diversity of some Iranian edible *Crocus* species using morphological and ISSR marker. *Iranian Journal of Horticultural Science* 46: 51–61. (In Persian, English abstract.)

Kothari, D., R. Thakur, and R. Kumar. 2021. Saffron (*Crocus sativus* L.): Gold of the spices – a comprehensive review. *Horticulture, Environment, and Biotechnology* 119. https://doi.org/10.1007/s13580-021-00349-8.

Mathew, B.F. 1982. *Crocus: A Revision of the Genus Crocus*. Timber Press, Portland. 224 pp. (= Mathew, B.F. 1982. *The Crocus. A Revision of the Genus Crocus (Iridacae)*. B.T. Batsford, London. 127 pp. + 96 plates.)

Mathew, B.F., G. Petersen, and O. Seberg. 2009. A reassessment of *Crocus* based on molecular analysis. *The Plantsman (N.S.)* 8: 50–57.

Mykhailenko, O., O. Kovalyov, O. Goryacha, L. Ivanauskas, and V. Georgiyants. 2019. Biologically active compounds and pharmacological activities of species of the genus *Crocus*: A review. *Phytochemistry* 162: 56–89.

Nematia, Z., D. Harpke, A. Gemicioglu, H. Kerndorff, and F.R. Blattner. 2019. Saffron (*Crocus sativus*) is an autotriploid that evolved in Attica (Greece) from wild *Crocus cartwrightianus*. *Molecular Phylogenetics and Evolution* 136: 14–20.

Petersen, G., O. Seberg, S. Thorsøe, T. Jørgensen, and B. Mathew. 2008. A phylogeny of the genus *Crocus* (Iridaceae) based on sequence data from five plastid regions. *Taxon* 57: 487–499.

Pieroni, A., H. Zahir, H.I.M. Amin, and R. Sõukand. 2019. Where tulips and crocuses are popular food snacks: Kurdish traditional foraging reveals traces of mobile pastoralism in Southern Iraqi Kurdistan. *Journal of Ethnobiology and Ethnomedicine* 15: 59. https://doi.org/10.1186/s13002-019-0341-0.

Prokop, P., J. Jersáková, J. Fančovičová, and M. Pipíška. 2019. Flower closure enhances pollen viability in *Crocus discolor* G. Reuss. *Flora* 250: 68–71.

Rukšāns, J. 2017. *The World of Crocuses*. Latvian Academy of Sciences, Riga. 568 pp.

Weryszko-Chmielewska, E. and M. Chwil. 2011. Structure of the floral parts of *Crocus vernus* (L.) Hill. *Acta Agrobotanica* 64: 35–46.

## Key Cultivation Websites

Alpine Garden Society, Crocus Pages by Tony Goode. *The Largest Online Resource for the Genus Crocus*. http://archive.alpinegardensociety.net/plants/Crocus/.

Challenger, C. 1996. The pleasures of crocus. *New Zealand Garden Journal* 1: 12–17. www.rnzih.org.nz/pages/crocus.htm.

Edgewood Gardens Website. www.edgewoodgardens.net/Plants_album/The%20Plants%20-%20%20Complete%20Collection/Iridaceae/Crocus/Subgen%20Nudiscapus%20Ser%20Biflori/index.html.

Guide to . . . Plant Societies Around the World that Focus Completely or Partially on Bulbs. www.pacificbulbsociety.org/pbswiki/index.php/Societies.

Pacific Bulb Society. *Crocus*. www.pacificbulbsociety.org/pbswiki/index.php/Crocus.

The Spruce. *Crocus Flowers: How to Plant, Grow, and Care for Spring Crocus*. www.thespruce.com/planting-and-caring-for-crocus-1402246#:~:text=Crocuses%20are%20generally%20low%2Dmaintenance,summer%20and%20prefer%20drier%20soil.

WikiHow. *How to Grow Crocus*. www.wikihow.life/Grow-Crocus.

WikiHow. *How to Grow Saffron*. www.wikihow.com/Grow-Saffron.

World Saffron & Crocus Collection. http://crocusbank.uclm.es/Information%20System.html.

# 33 Cyclamen (*Cyclamen* Species and Hybrids)

**Scientific Names:** The genus *Cyclamen* is based on the Greek word *kylos*, meaning "circle." This has been interpreted as referring either to the typically rounded tubers or the spiralling stem holding the fruit of some species. *Cyclamen persicum* was inappropriately named, since it does not grow naturally in "Persia."

**Common Names:** "Cyclamen" is the usual name. Occasional names include *Persian violet* (better reserved for the ornamental houseplant *Exacum affine*), *alpine violet*, *sowbread* (synonymous with the French *pain de porceau*, based on the allegations that the tuberous roots are eaten by wild boar and were historically fed to swine, which is unlikely, as they are toxic). The principal species, *C. persicum*, and its horticultural hybrids are best known as florists' cyclamen. The name is also spelled *florist's cyclamen* and *florists cyclamen*. The ambiguous scientific name *C. europaeum* has been applied particularly to *C. purpurascens*, but also to *C. hederifolium*, *C. repandum*, and *C. persicum*. The phrase "cyclamen europaeum" has frequently been the subject of traditional herbal medicines, but the name is also employed in more modern publications concerned with the medicinal properties of extracts of *Cyclamen*, and unfortunately, the identity of the species cannot usually be reliably determined.

FIGURE 33.1 Attractive potted displays of cyclamens.

*Source:* (a) Photo by Shutterstock/Sergiy Palamarchuk. (b) Photo by Laura Blanchard (CC BY SA 2.0). (c) Photo by Buntysmum from Pixabay (public domain).

## 33.1 INTRODUCTION

The florists' cyclamen (*Cyclamen persicum*) is among the most popular pot plants of the world, notable for its compact size, attractively variegated foliage, and brilliantly colored flowers in a variety of sizes, shapes, and forms. Cyclamens are petite ornamentals, easily recognized by their flowering stems, bare of leaves, each topped by a solitary arched bloom with swept-back petals. Even more striking is a bizarre corkscrewing of the flowering stem, which positions the fruit at ground level, where the sweet seeds can be picked up by ants. Several species of the genus are hardy enough to be grown as garden perennials in the warmer parts of temperate climates, but it is in the form of a houseplant and gift plant that cyclamens are best known.

FIGURE 33.2 Potted plants of florist's cyclamen (*Cyclamen persicum*).

*Source:* (a) Photo by Fanghong (CC BY SA 3.0). (b) Photo by Buntysmum from Pixabay (public domain). (c) Photo by Darkone (CC BY SA 2.0).

## 33.2 OVERVIEW

*Cyclamen* is a genus of about two dozen perennial species indigenous to Europe and the Mediterranean Basin east to Iran, mostly distributed around the Mediterranean Sea and the Black Sea. One species (*C. purpurascens*) occurs in Central Europe, and one (*C. somalense*) is isolated in the mountains of northeastern Somalia. Most species are adapted to a semi-arid Mediterranean climate, the aboveground stems arising in autumn, growing during cool-moist winters, senescing in the spring, and remaining dormant


DOI: 10.1201/9781003586227-33

FIGURE 33.3 Several ornamental species of *Cyclamen* grown as much for their ornamental foliage as for their flowers: (a) *Cyclamen coum* (eastern sowbread). (b) *Cyclamen hederifolium* (ivy-leaved cyclamen). (c) *Cyclamen africanum* (North African cyclamen). (d) *Cyclamen purpurascens* (alpine, European, and purple cyclamen).

*Source:* **(a) Photo by Ian Kirk (CC BY 2.0). (b) Photo by Vulkano (CC BY 3.0). (c) Photo by Cultivar413 (CC BY 2.0). (d) Photo by C. T. Johansson (CC BY SA 3.0).**

as a storage underground stem over the dry summer. (This is an adaptive ecological pattern exhibited by numerous other Mediterranean "geophytes" to avoid summer drought.) However, the tubers are not summer-dormant in *C. purpurascens* and *C. colchicum*, which grow outside of the Mediterranean region. Most of the species occur in evergreen or deciduous woodland and shrubland. The plants inhabit rocky terrain, often under trees or shrubs. Overcollecting for the horticultural trade has endangered several species, and all the wild species are afforded some protection under international regulations (CITES).

The leaves often have long stalks (petioles), and the blades are broad, triangular or rounded, and often variegated or mottled in shades of green and silver-gray. The solitary flowers are individually subtended by a slim long stem. This is characteristically recurved back 150–180° below the flower (the pendent flowers are said to be "nodding"). While solitary flowers are the rule in *Cyclamen*, *C. coum* is known to often produce two flowers on the flowering stem. The five petals are joined to form a short tube at the base and become curved back (reflexed) 90–180° and twisted. In the Primulaceae family, of which *Cyclamen* is a member, only *Dodecatheon* and a few *Primula* species share the characteristic of the corolla bending backward, making it appear as if the flower has been turned inside out. A few cyclamen cultivars have been bred which lack petal reflexing. Auricles (ear-like projections) may be present on the top of the floral tube, where the petals become free and reflexed. A calyx with five sepals is present. Flowers of the wild species are usually white, pink, or purple and are sometimes fragrant but lack nectaries to attract pollinators. After fertilization, in most of the species, the pedicels (stems of the flowers) curl or bend downward to place the capsules at soil level, where they mature, making the seeds available for ant distribution. However, in wild *C. persicum*, the fruit stalks haphazardly bend and twist instead of coiling back to the crown. Modern cultivars often develop non-coiling pedicels. The round fruits open at maturity, exposing numerous brown seeds, known to attract ants, which consume their sticky covering and disperse the seeds.

*Cyclamen* plants develop from subterranean tubers, which are usually spherical with a flattened or concave upper surface. Depending on species, the roots arise from the top, bottom, or other locations, while leaves and flowering stems originate from the center. "Tubers" are enlarged storage structures of plants, but so are tuberous roots, corms, bulbs, and rhizomes. These terms are separated in part by whether the new tissues originate from root or stem tissue. Some sources define *tubers* as organs derived from stems; others apply the word to structures derived from stems *or* roots. However, it appears that *Cyclamen* tubers don't fall exactly into any of these concepts. The structures of cyclamen are sometimes described as "corm-like tubers."

Cultivars may have any of numerous shades of white, pink, scarlet, red, lavender, or purple flowers (true yellow flowers are not available, although apparently, yellow-flowered

FIGURE 33.4 Coiling flower stems (pedicels) of the developing fruits of *Cyclamen*: (a) *Cyclamen africanum*. (b) *Cyclamen coum*.

*Source:* **(a) Photo by Cultivar413 (CC BY 2.0). (b) Photo by Dominicus Johannes Bergsma (CC BY SA 3.0).**

FIGURE 33.5 Corm-like tubers of *Cyclamen*: (a) A bushel of tubers of *C. hederifolium* on sale in a store. (b) The two halves of a longitudinally sectioned tuber of *C. purpurascens*.

*Source:* **(a) Photo by Frank Vincentz (CC BY SA 3.0). (b) Photo by Shutterstock/Shansh23.**

plants are offered by some websites). Some selections have doubled flowers. Some "picotee" forms have a thin white edge on the petals. Other floral forms have stripes or ruffled edges. Most cultivars of *C. persicum* have rounded or heart-shaped leaves, which are either dark green or marbled with white or silver, while the leaf margins of some cultivars are slightly toothed or lobed, and the lower leaf surfaces may be purplish. Some miniature cultivars have standard-size leaves but tiny flowers. Some kinds are lightly scented. Other than *C. persicum*, domesticated species tend to have been selected for interesting leaf coloration and patterning rather than for floral features.

Traditionally, *C. persicum* and its hybrids have been sold in winter, as the plants are naturally summer-dormant, but today production can be year-round, depending on market demand, because many of the commercial cultivars do not require a resting period or at least can be forced to grow continuously by providing favorable conditions. Commercial cultivars have often been selected for very fast growth and minimal needs for removing older yellowed leaves.

## 33.3 HISTORY

*Cyclamen persicum* appears to have been known by the classical Greeks and Romans (it was referred to as *cyclaminus* after its disk-shaped tuber). Plants were recorded in gardens in western Europe as early as the 16th century and, at that time, were illustrated in guides to medicinal plants. By 1850, selection by gardeners and early breeders had produced several ornamental forms. *Cyclamen persicum*, the florists' cyclamen, is the principal species encountered in the ornamental trade. In the latter part of the 19th century, numerous cultivars were selected, especially in Europe (particularly England and The Netherlands), with stress on increased flower size, increased range of colors, and different petal forms. Improving *C. persicum* by hybridization with the wild species of the genus is difficult because of hybrid inviability and sterility, but in the late 20th century, it was shown that these problems can be overcome by embryo rescue techniques or by doubling of chromosome number. Modern cultivars are often the result of hybridization of *C. persicum* with *C. africanum*, *C. graecum*, *C. rohlfsianum*, and other species.

FIGURE 33.6 Paintings (public domain) by European botanical illustrators showing cyclamens cultivated in the 19th century: (a) *Cyclamen coum*. (b) *Cyclamen persicum*. (c) *Cyclamen purpurascens*. (d) *Cyclamen hederifolium*. (e) A garden hybrid.

**Source:** (a) ***Flore des serres et des jardins de l'Europe*** **(Ghent, Belgium), vol. 23, 1880. (b) Thornton, R. J. (editor) and Pether, A. (artist). 1804.** ***The Temple of Flora*****. T. Bensley, London. (c) Thomé, O. W. 1885.** ***Flora von Deutschland, Österreich und der Schweiz*****, vol. 4. Gera, Germany. (d) Baxter, W. 1843.** ***British Phaenogamous Botany*****, vol. 6. J. H. Parker, Oxford. (e)** ***L'Illustration horticole*** **(Gand, Belgium), vol. 43, 1896.**

## 33.4 ECONOMIC IMPORTANCE

Cyclamens are popular pot plants of particular value to the florist trade and their suppliers. In warm climates, the species are also significant as garden bedding plants.

## 33.5 ORNAMENTAL USES

Cyclamens, especially *C. persicum* cultivars, are grown for use primarily as potted plants. Much less frequently, the flowering stems are marketed as cut flowers. Some species, such as *C. hederifolium*, are grown as outdoor perennials in the milder parts of temperate climates, and several of the species are grown as bedding plants in warmer climates. Many of the species are ideal in rock gardens, where their small size is not obscured by taller garden ornamentals. However, all species except *C. purpurascens* (one of the few woodland species) are summer-dormant, limiting attractiveness during the summer.

FIGURE 33.7 Commercial production and marketing of cyclamen: (a) Cyclamen in an advanced greenhouse. (b) Cyclamen in a simple ("low tunnel") greenhouse in Israel. (c) A florist holding a pot of cyclamen.

**Source: (a) Photo by Jonathan Billinger (CC BY SA 2.0). (b) Photo by Arie Zadok (public domain). (c) Shutterstock/Africa Studio.**

FIGURE 33.8 Large landscape plantings of cyclamen.

*Source:* **(a) Photo by Shutterstock/Apiguide. (b) Photo by Falco from Pixabay (public domain). (c) Photo by KRiemer from Pixabay (public domain).**

## 33.6 OTHER USES

Cyclamens have limited use in herbal medicine and are even sometimes eaten, but such usages are hazardous, as discussed next.

## 33.7 PROBLEMS

*Cyclamen* species, especially their tubers, contain toxic saponins (it is uncertain if the foliage and flowers are significantly toxic). According to the ASPCA, cyclamen tubers and roots are poisonous to both cats and dogs. However, despite the widespread occurrence in Europe of woodland plants, as well as their common use as houseplants, there has been very little concern about the toxic potential of cyclamen. The fresh tuber of *C. hederifolium* is a drastic purgative and has (unadvisedly) been employed in traditional herbal medicine for the purpose. There is interest in researching the toxic compounds for possible medicinal applications.

There are some rare occurrences of the (very inadvisable) culinary usage of cyclamen tubers. In Lebanese cooking, cyclamen leaves are valued (even more than grapevine and Swiss chard leaves) for stuffing with vegetables and meat dishes, but the supply of cyclamen foliage is quite limited, and they might be toxic.

## 33.8 CULTIVATION

### 33.8.1 Climate and Site

The common florists' cyclamen is too frost-intolerant to grow as a garden plant in temperate countries. While potted

FIGURE 33.9 Potential toxicity of cyclamens to pets: (a) Early 20th-century art print (public domain) showing cats beside cyclamen, a curious combination, since the plants are believed to be poisonous to felines. Increasing the hazard, cats are attracted to windowsills, where plants are typically located. The French artist Maurice Boulanger (active in the early 20th century) was known for cute and funny anthropomorphic cats. (b) Cyclamen and a canine.

*Source:* **(a) Public domain. (b) Prepared by B. Brookes.**

plants can be kept outdoors during warm weather, they require indoor protection at other times, and cyclamen is grown primarily as a houseplant. Cyclamens do best in cool temperatures. *Cyclamen hederifolium* appears to be the hardiest species (capable of surviving −18°C and occasionally claimed to tolerate even lower winter temperatures). It is rated for growth as a perennial in climate zones 5 to 9. *Cyclamen coum*, *C. parviflorum*, and *C. purpurascens* are also relatively cold-hardy. In warm climates, where cyclamens can be treated as garden perennials, the plants are best placed where these low-growing species are not blocked from view by taller plants. In their natural habitats, cyclamens grow in semi-shaded areas, and so in gardens they should be provided with semi-shade. The plants can be placed under trees and shrubs.

### 33.8.2 Propagation and/or Purchasing

*Cyclamen persicum* cultivars can be propagated by seed and are to a considerable extent in the horticultural trade. However, it requires a year to produce plants of a reasonable size. Vegetative division is also possible, notable by dividing tubers, but this is also an activity mainly conducted by professional horticulturalists. For most people, it is advisable to simply purchase a potted plant of the florists' cyclamen. When buying, select a plant with many buds that are just starting to open. The foliage should be sturdy and succulent. The species of *Cyclamen* suitable for outdoor growth in the winter are relatively difficult to obtain in cool countries, but tubers are available in warmer climates.

### 33.8.3 Planting

For outdoor planting in warm climates, "corms," or young plants, should be set out in early fall, since they bloom naturally in the winter. A planting distance of 15 cm apart is recommended. When planted, a small portion of the corm should be aboveground.

### 33.8.4 Soil and Moisture Conditions

Cyclamens should be in organically rich, well-draining soil of pH 6.0 to 7.0.

### 33.8.5 Maintaining Plants

Plants growing in warm regions may nevertheless be exposed to occasional freezing, and during such periods, they should be covered. After outdoor winter-grown plants bloom, they will go dormant and should not be watered.

### 33.8.6 Care of Potted Houseplants

Houseplant cyclamens prefer bright indirect light. The plants should be kept away from warm drafts from heat vents or cold drafts from open windows. Flowering plants are best maintained at about 20°C, with the night temperatures preferably between 6.5 and 15°C. Temperatures above 20°C may induce dormancy. However, many modern hybrids can be maintained over a wider temperature range. Potted cyclamens should be watered well when the soil is quite dry. Both excess and insufficient moisture may cause the foliage to yellow, and hot, dry conditions may induce bud drop. Some plant guides warn against pouring water over the corm in the center of the pot to avoid its rotting. Another irrigation recommendation is that it is best to water a cyclamen plant from the bottom rather than from the top (by placing the pot in a saucer of water, long enough for the soil to become moist). The low humidity in houses during winter is undesirable, and to increase humidity, the pot may be set on a tray of wet gravel. A low-nitrogen fertilizer or houseplant food for blooming plants, applied every 2 weeks, is recommended when the plants are actively growing, but excess fertilization may promote lush growth susceptible to disease.

The florists' cyclamen is typically marketed as a winter-flowering houseplant, and most cultivars go dormant during the heat of the summer (or when exposed indoors to excessive heat or other stresses). Attempting to keep this houseplant continuously in flower is misguided, since cyclamens are naturally adapted to require a period of dormancy. The plant is considered to be fairly easy to grow and often flowers for as long as 3 months. Cooler temperatures and removing spent flowers prolong flowering. However, it is difficult to provide appropriate conditions for reblooming, and consequently, most plants are discarded after they flower.

FIGURE 33.10 Houseplant cyclamens on a windowsill.

*Source:* **Photo by Shutterstock/Edward O'Neil.**

In late spring, overwintered cyclamen houseplants stop blooming, the leaves fade and drop off, and most plants go into dormancy for 2 to 3 months. For those who wish to attempt to resurrect their plants, the following steps have been recommended: Reduce watering and fertilization, and keep the dormant plant outdoors for the summer, preventing the pot from being more than slightly moist. Alternatively, gradually withhold water, and when the leaves are withered, remove the tuber from the soil and store it in dry vermiculite at about 10ºC for 6–12 weeks. Then repot the tuber in well-draining soil mix, placing the tuber so the upper half is above the surface. If a larger pot is needed, this is best done during the dormant period. Water when the tuber starts producing new leaves, and fertilize until flower buds develop. Unless strong light is provided during growth, the plants are unlikely to regain their former beauty.

## 33.9 FASCINATING FACTS, FABLES, FOLKLORE, AND FRIVOLITY

- The Greek philosopher Theophrastus (ca. 371–287 BCE) recommended cyclamen tubers as an aphrodisiac, and its reputation for this purpose persists to this day.
- The classical Greeks and Romans recorded that cyclamen tubers were put into streams to stun fish so that they could be easily collected, presumably a result of the toxic saponins poisoning the fish.
- An old superstition held that a pregnant woman should not step over a cyclamen, for fear of terminating her pregnancy. This may trace to knowledge that ingesting the toxic tubers could lead to an abortion. In medieval herb books, it was believed that simply wearing the tuber around the neck could trigger a miscarriage.
- In mediaeval Europe, cyclamen, referred to as the "bleeding nun" (since early times, it was consecrated to the Virgin Mary), was often used as a charm against bad weather.
- In China, cyclamen has been interpreted as a floral symbol of hospitality, the closed bud turning downward as if in a welcoming bow.
- During and in the decades following the First World War, there was a trend to name new allied warships after plants (e.g., amaryllis, buttercup, crocus, gladiolus, hydrangea, lupin, pansy, poppy). At least two British ships were named HMS *Cyclamen* (these were non-contemporary), as well as a French warship. Italy and Britain were allies during the First World War, but in 1917, the HMS

FIGURE 33.11 Cyclamen fairies.

*Source:* **(a) A 1905 postcard in public domain. (b) Photo by Shutterstock/Algol.**

*Cyclamen* accidentally sank an Italian submarine after mistaking it for a German U-boat.

- Cyclamens are official emblems of several countries. *Cyclamen cyprium* is the national flower of Cyprus. Cyclamen is the national flower of San Marino. Four plant species are said to represent Greece, including cyclamen. Until 2013, *C. persicum* was the national flower of Israel, at which time it was replaced by the poppy anemone, *Anemone coronaria.*
- Only the most popular flowers have colors named after them (e.g., lilac, lavender, fuchsia, rose, violet, wisteria). The color "cyclamen" has been described as "a very dark reddish purple" (*Webster's Third New International Dictionary*).
- A "fairy garden" is a miniature garden with small plants, such as cyclamens, and miniature gardening furnishings, typically created in a container or very small segregated outdoor area. These became very popular in Victorian England, and fairy garden kits remain popular. "Flower fairies" for most familiar garden plants were illustrated in the 19th and early 20th centuries, and to this day, porcelain figurines and prints of them have been a mainstay of the collector's trade. Especially popular are "cyclamen fairies," profusely decorated with the flowers.

## KEY PUBLICATIONS


Anderberg, A.A. 1994. Phylogeny and subgeneric classification of *Cyclamen* L. (Primulaceae). *Kew Bulletin* 3: 455–467.

Blasdale, W.C. 1953. *Cyclamen persicum. Its Natural and Cultivated Forms.* Stanford University Press, Stanford, CA. 49 pp.

Clennett, J.C.B. 2002. An analysis and revision of *Cyclamen* L. with emphasis on subgenus *Gyrophoebe* O. Schwarz. *Botanical Journal of the Linnean Society* 138: 473–481.

Compton, J.A., J. Chris, B. Clennett, and A. Culham. 2004. Nomenclature in the dock. Overclassification leads to instability: A case study in the horticulturally important genus *Cyclamen* (Myrsinaceae). *Botanical Journal of the Linnean Society* 146: 339–349.

Debussche, M. 2003. Frost and cyclamen under mediterranean climatic conditions. *Journal of the Cyclamen Society* 27: 14–18.

Debussche, M., E. Garnier, and J.D. Thompson. 2004. Exploring the causes of variation in phenology and morphology in mediterranean geophytes: A genus-wide study of *Cyclamen. Botanical Journal of the Linnean Society* 145: 469–484.

Debussche, M. and J.D. Thompson. 2002. Morphological differentiation among closely related species with disjunct distributions: A case study of mediterranean *Cyclamen* L. subgen. *Psilanthum* Schwarz (Primulaceae). *Botanical Journal of the Linnean Society* 139: 133–144.

Doorenbos, J. 1950a. The history of the "Persian" cyclamen. *Mededelingen Landbouwhogeschool, Wageningen* 50(3): 1–59 + figures.

Doorenbos, J. 1950b. Taxonomy and nomenclature of *Cyclamen. Mededelingen van de Landbouwhogeschool, Wageningen* 50(2): 1–13 + figures.

Ewald, A. 1996. Interspecific hybridization between *Cyclamen persicum* Mill. and *C. purpurascens* Mill. *Plant Breeding* 115: 162–166.

Grant, M. 2012. Is it a bulb, corm or tuber? *Journal of the Royal Horticultural Society* 137: 17.

Greilhuber, J. 1989. Karyotype structure and evolution in *Cyclamen* L. subgen. *Psilanthum* Schwz. (Primulaceae). *Flora* 183: 103–113.

Grey-Wilson, C. 2002. *Cyclamen: A Guide for Gardeners, Horticulturists, and Botanists*, Revised edition. Timber Press, Portland. 224 pp.

Ishizaka, H. 2008. Interspecific hybridization by embryo rescue in the genus *Cyclamen. Plant Biotechnology* 25: 511–519.

Lyons, R.E. and R.E. Widmer. 1980. Origin and historical aspects of *Cyclamen persicum* Mill. *HortScience* 15: 132–135.

Matthew, B. (ed.). 2013. *Genus Cyclamen in Science, Cultivation, Art and Culture.* Kew Publications, Royal Botanic Gardens, Kew. 574 pp.

Mazouz, W. and S. Djeddi. 2013. A biological overview on the genus *Cyclamen. European Journal of Scientific Research* 110: 7–22.

Mohammed, G.J., I.H. Hameed, and S.A. Kamal. 2018. Anti-inflammatory effects and other uses of *Cyclamen* species: A review. *Indian Journal of Public Health Research and Development* 9: 206–211.

Nightingale, G. 1982 (reprinted 1984, 1987). *Growing Cyclamen.* Timber Press, Portland. 148 pp.

Saunders, D.E. and C. Grey-Wilson. 2003. *Cyclamen: A Gardener's Guide to the Genus*, Revised edition. The Alpine Garden Society, London. 64 pp.

Schwartz-Tzachor, R., A. Dafni, S.G. Potts, and D. Eisikowitch, D. 2006. An ancient pollinator of a contemporary plant (*Cyclamen persicum*): When pollination syndromes break down. *Flora* 201: 370–373.

Spoerke, D.G., S.E. Spoerke, A. Hall, and B.H. Rumack. 1987. Toxicity of *Cyclamen persicum* (Mill). *Veterinary and Human Toxicology* 29: 250–251.

Takamura, T. 2007. *Cyclamen.* In: *Flower Breeding and Genetics Issues, Challenges and Opportunities for the 21st Century*, edited by N.O. Anderson, 459–478. Springer, Dordrecht.

Thulin, M. and A.M. Warfa. 1989. *Cyclamen* (Primulaceae) in Tropical Africa. *Plant Systematics and Evolution* 166: 249–252.

Yesson, C. and A. Culham. 2006. A phyloclimatic study of *Cyclamen. BMC Evolutionary Biology* 6(72). https://doi.org/10.1186/1471-2148-6-72.

Yesson, C. and A. Culham. 2011. Biogeography of *Cyclamen*: An application of phyloclimatic modeling. In: *Climate Change,*

*Ecology and Systematics*, edited by T.R. Hodkinson, M.B. Jones, S. Waldren, and JA.N. Parnell, 265–279. Cambridge University Press, Cambridge.

Yesson, C., N.H. Toomey, and A. Culham. 2008. *Cyclamen*: Time, sea and speciation biogeography using a temporally calibrated phylogeny. *Journal of Biogeography* 36: 1234–1252.

## Key Cultivation Websites

The Cyclamen Society. www.cyclamen.org/.

GardenersWorld.com. *How to Grow Indoor Cyclamen*. www.gardenersworld.com/how-to/grow-plants/how-to-grow-indoor-cyclamen/.

Lonsdale, J. and E. Gardens. 2020. *The Magic of Cyclamen*. https://edgewoodgardens.net/the-magic-of-cyclamen/.

The Spruce. *Cyclamen: Plant Care & Growing Guide*. www.thespruce.com/how-to-care-for-cyclamen-1402644.

The Spruce. *How to Grow and Care for Cyclamen persicum (Mini Cyclamen)*. www.thespruce.com/mini-cyclamen-cyclamen-persicum-guide-5199380.

Stefen, R. 2008, October. *Cyclamen for Winter and Spring*. Pacific Horticulture Society. www.pacifichorticulture.org/articles/cyclamen-for-winter-and-spring/.

WikiHow. *How to Care for Cyclamen Plants*. www.wikihow.com/Care-for-Cyclamen-Plants#:~:text=Cyclamen%20need%20water%20during%20their,soil%20moist%20at%20all%20times.

# 34 Daffodil (*Narcissus* Species and Hybrids)

**Scientific Names:** The genus name *Narcissus* is based on an ancient Greek name, possibly *narke*, referring to *numbness* (daffodils have a reputation for narcotic properties).

**Common Names:** Daffodil, narcissus. The plural of "narcissus" has been debated. *Webster's Third New International Dictionary* gives plurals in the order "narcissus," "narcissuses," and "narcissi." "Jonquil" is sometimes used to indicate all species of *Narcissus* but should be reserved for *N. jonquilla* and closely related hybrids.

## 34.1 INTRODUCTION

Daffodils are one of the most popular of the ornamental spring-flowering bulb plants, available in thousands of varieties. They can produce very large, beautiful flowering displays with limited cost and care and have become one of the staples of the horticultural industries of the temperate world.

## 34.2 OVERVIEW

There is disagreement about the number of species in *Narcissus*, estimates varying from less than 20 to well over 100. The taxonomic differences of opinion have been attributed to extensive hybridization, both in the wild and in cultivation, and naturalization of many hybrids, making interpretation of species boundaries problematic. Several species are endangered, and some may have become extinct in recent times.

*Narcissus* species occur mostly in southwestern Europe, especially in Spain and Portugal (the Iberian Peninsula), and in North Africa. Most species have quite restricted ranges. Several extend into France and Italy, and a few grow east to the Balkans (*N. poeticus*, *N. serotinus*, and *N. tazetta*) and the eastern Mediterranean (*N. serotinus*), including Israel (*N. tazetta*). In ancient times, some species were transported eastward, establishing a naturalized distribution in a narrow band across Eurasia, all the way to Japan. Six species and several hybrids have also been naturalized in North America.

The wild species occur in a variety of habitats, including grassland, riverbanks, rocky crevices, scrub, woods, and mountains. Most flower in the spring (a few in the autumn), usually producing a flowering stem with a single flower, but occasionally with as many as 20 flowers on the end of the stem. Height varies from 5 to 50 cm, depending on variety. In most species and cultivars, a trumpet-shaped or cup-shaped structure called a corona arises near the base of the flower and, in many plants, is differently colored than the sepals and petals at the bottom of the flower. Some cultivars have doubled flowers (with either or both the petal segments and the corona doubled), and in such flowers, sometimes

FIGURE 34.1 Daffodil (*Narcissus*) cultivars: (a) A bed of daffodils. (b) A pot of daffodils. (c) A display at the famous Chelsea Flower Show in England. (d) A basket of daffodils.

*Source:* **(a) Photo by Les Haines (CC BY 2.0). (b) Photo by Brenjbeecrafts from Pixabay (public domain). (c) Photo by Shutterstock/Andrew Fletcher. (d) Photo by Buntysmum from Pixabay (public domain).**

FIGURE 34.2 (a) Narcissus flowers with a corona (colloquially termed *floral tube*, *trumpet*, and *cup*) that is differently colored than the six white "tepals" (often considered to be three sepals and three petals). (b) Narcissus 'Spring Glorie'.

*Source:* **(a) Photo by Buntysmum from Pixabay (public domain). (b) Public domain. Van Olphen. 1910. Bol bloemen catalogus. Photo by Nikontino (CC BY 2.0).**

DOI: 10.1201/9781003586227-34

FIGURE 34.3 Daffodils with doubled flowers: (a) Narcissus 'Dick Wilden'. (b) Unidentified cultivar. (c) Narcissus 'Sir Winston Churchill'.

*Source:* **(a) Photo by Cillas (CC BY SA 4.0). (b) Photo by Christine Matthews (CC BY SA 2.0). (c) Photo by Cillas (CC BY SA 4.0).**

FIGURE 34.4 Bulbs of narcissus 'Dutch Master'.

*Source:* **Photo by Famartin (CC BY SA 4.0).**

the corona is not evident. The flowers produce nectar and are insect-pollinated (by bees, butterflies, flies, and hawk moths). The shape and size of the corona limit access to the nectary to appropriate pollinators. Most daffodils have little or only a mild scent (some do, and most jonquils and paperwhites are quite perfumed). The foliage consists of several strap- or awl-shaped leaves, which die after flowering, the plants surviving much of the summer and autumn by underground bulbs. The bulbs are onion-like, with a bud in the center surrounded by white, fleshy scales which store energy for the bulb to flower. A thin outer skin ("tunic") covers the bulb. A "basal plate" at the bottom of the bulb produces roots. A cold (vernalization) period is required for most species (not for *N. tazetta*) for growth to resume in the spring (professional growers often force bulbs to flower by providing a chilling period at 9°C, for 2–10 weeks, depending on cultivar). Toxic alkaloids protect the plants from predation.

The Royal Horticultural Society of Great Britain is the international registration organization for daffodil cultivars and has recognized over 30,000. Daffodils have been arbitrarily assigned horticulturally to 1 of 13 standard categories based mostly on flower and plant form or species ancestry, although hybridization has obscured the differences. Daffodils with trumpet-shaped or cup-shaped coronas are the most popular. "Miniature species" (as distinguished from "dwarf species") have been defined as having flowers no more than 5 cm in diameter.

Several white-flowered, usually fragrant species of *Narcissus* are called "paperwhites," and the bulbs are commonly forced to flower indoors (these are usually too tender to grow outdoors as perennials in cold climates). Some cultivars of *N. tazetta* are particularly used for the purpose. Unlike most other *Narcissus* species, they do not require chilling to flower. These produce tight clusters of four to eight or more small, usually intensely fragrant flowers.

## 34.3 HISTORY

Most garden daffodils are of hybrid origin. Commercial cultivars are usually larger and more robust than their wild parents. The corona's size, shape, and color have been principal characteristics selected by breeders. Cultivars with flowers developing trumpet-shaped coronas usually originated at least in part from *N. pseudonarcissus* (trumpet narcissus, Lent lily). Crosses between *N. pseudonarcissus* and *N. poeticus* (poet's narcissus, pheasants-eye narcissus), have produced a variety of cultivars with cup-like flowers. Many of the multi-headed cultivars arose as hybrids of

FIGURE 34.5 The most important parental species of modern garden daffodils: (a) *Narcissus pseudonarcissus.* (b) *Narcissus jonquilla.* (c) *Narcissus poeticus.* (d) *Narcissus tazetta.*

*Source:* **(a) Photo by Amanda Slater (CC BY SA 2.0). (b) Photo by Mark Pellegrini (CC BY SA 2.5). (c) Photo by Katso Käyttöehdot (public domain). (d) Photo by Zachi Evenor (CC BY 3.0).**

*N. poeticus* and *N. tazetta* (bunch-flowered narcissus, polyanthus narcissus). Other species important in breeding of cultivars include *N. bulbocodium* (hoop petticoat daffodil, petticoat daffodil), *N. cyclamineus*, *N. jonquilla* (jonquil), *N. serotinus*, *N. tazetta*, and *N. triandrus* (angel's tears). Flowers are predominantly white or yellow (green in the night-flowering *N. viridiflorus*), sometimes (especially in cultivation) partly pink or salmon, and many cultivars are "bicolors" (the perianth differing in color from the corona).

## 34.4 ECONOMIC IMPORTANCE

Daffodils have been grown commercially in Europe since the 16th century and have become one of the major ornamental bulb crops grown in temperate regions. The plants are produced as field crops, providing both bulbs and flowers for the cut flower trade. Forcing of bulbs in glasshouses furnishes flowers and pot plants throughout the year. The Netherlands is the largest producer. Other areas of Europe (notably the UK), Australia, Canada, the United States, and additional countries produce significant crops.

FIGURE 34.6 Paintings (public domain) by European botanical illustrators showing daffodils cultivated in the 19th century. (a, b) Unidentified garden hybrids. (c) *Narcissus bulbocodium*.

**Source:** **(a, b)** ***Album van Eeden. Harlem's Flora,*** **door A. C. Van Eeden & Co. 1872. (c) Curtis's botanical magazine (London), vol. 26, 1870.**

FIGURE 34.7 Commercial production and sales of daffodils: (a) A daffodil farm in The Netherlands. (b) Potted daffodils on sale. (c) A florist holding cut flowers of daffodils.

*Source:* **(a) Photo by PNWsights from Pixabay (public domain). (b) Photo by Zákupák (public domain). (c) Photo by Shutterstock/Olena Antonenko.**

## 34.5 ORNAMENTAL USES

Daffodils are extremely widely cultivated ornamental plants in gardens and parks and are extensively grown commercially for the cut flower industry. They display well in borders, containers, mixed beds, and "naturalized" (scattered) in lawns.

FIGURE 34.8 Landscape uses of daffodils: (a) A large bed. (b) Borders of a walkway. (c) Plants scattered in a park.

*Source:* **(a) Photo by Clemens v. Vogelsang (CC BY 2.0). (b) Photo by Frank Vincentz (CC BY SA 3.0). (c) Photo by Shutterstock/BM-W.**

FIGURE 34.9 Cut daffodil stems in the home.

*Source:* **(a) Photo by Shutterstock/OlesyaPogosskaya. (b) Photo by Kathryn Yengel (CC BY ND 2.0). (c) Photo by Shutterstock/Richard Griffin.**

## 34.6 OTHER USES

Daffodils were harvested from the wild or cultivation in ancient Europe and Asia for at least 2,000 years, mostly as medicinal herbs. Most of the old applications are obsolete and, indeed, rather dangerous, given the content of toxic alkaloids present. However, one of the alkaloids, galanthamine, which is reminiscent of morphine, is under examination as a possible therapeutic agent in Alzheimer's disease. It is found in various species of Amaryllidaceae, particularly in *Narcissus*, as well as in some closely related families (Agavaceae, Haemodoraceae, and Hypoxidaceae). Daffodils are also employed in perfumes, for their essential oils, not their alkaloids.

## 34.7 PROBLEMS

Occasionally, daffodil bulbs are mistaken for small onions and ingested (raw or cooked) by people, usually resulting in acute vomiting (and sometimes death) from the toxic alkaloids. Livestock (cattle, goats, and pigs) have also been poisoned, sometimes fatally. Food shortages in The Netherlands during the Second World War led to attempts to feed the bulbs to cattle, which sometimes died. Domestic pets – notably cats, but also dogs, horses, and rabbits – have been poisoned, and so have children who consumed leaves or sucked on the stalks. Wild animals (except insects) usually avoid daffodils, which are quite resistant to deer, rabbit, squirrel, and other rodent damage (although the bulbs may be dug up).

"*Narcissus* species are thought to be the most common cause of irritant contact dermatitis among florists," according to Lee and Elston (2023). "Daffodil itch" or "lily rash" is the dermatitis associated with picking or handling daffodils. Because of this, workers handling daffodils minimize contact between daffodils and their skin. Gardeners concerned about the possibility of an allergic reaction from handling the bulbs should wear gloves.

The presence of daffodils in a vase with other cut flowers (especially tulips) has a deleterious effect on them. The cause is uncertain but does not seem attributable to the alkaloids of the daffodils.

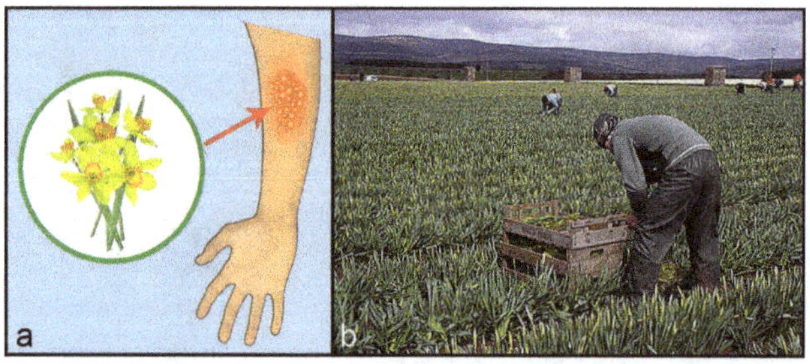


FIGURE 34.10 Daffodil dermatitis. As noted in the text, daffodils can produce dermatitis: (a) Allergic reaction. (b) Daffodil picking in Scotland. Note that the harvesters are protectively clothed to prevent dermatitis from developing.

***Source:* (a) Prepared by B. Brookes. (b) Photo by Anne Burgess (CC BY SA 2.0).**

## 34.8 CULTIVATION

### 34.8.1 Climate and Site

Although most *Narcissus* species originate from quite warm climates, most daffodil cultivars need a period of cold weather each winter to stimulate flower bud formation for the following spring. Such cultivars grow best in cooler areas but are rated for growth in climate zones 4 to 9 (many survive in zone 3). Varieties that grow well in warmer areas (zone 8 and above) should be marked as such in bulb catalogs. Jonquils (*N. jonquilla*) are rated for climate zones 4–9, ideally in climate zone 8. Paperwhites are rated for climate zones 8–11.

Daffodils are well-suited to many situations, particularly in formal landscape beds. They require a sunny location but will do well in light shade (or "dappled light"). They look attractive near deciduous shrubs and trees

FIGURE 34.11 Daffodils in snow.

***Source:* Photo by J. Blueberry from Pixabay (public domain).**

FIGURE 34.12 Daffodils growing under a deciduous tree. As noted in the text, although daffodils are often planted beside trees, extra water and fertilizer are often required to compensate for competition from tree roots.

***Source:* Photo by Evelyn Simak (CC BY SA 2.0).**

(flowering occurring in the spring, before the woody plants produce significant shade). They are also suitable in rock and perennial gardens and beside buildings (but not too close to foundations or under extended eaves, which can be excessively dry). Daffodils are also frequently deliberately planted in lawns and groundcovers. They are often placed in woodlands and meadows (contributing to their naturalization in foreign locations). Miniature or dwarf varieties are particularly appropriate for containers and rock gardens.

### 34.8.2 Propagation and/or Purchasing

Wild species and non-hybrid cultivars can be grown from seed, but it usually requires up to 5 years to produce flowering plants of reasonable size, so most gardeners buy bulbs from commercial sources (especially from catalogue sales) or acquire them from other gardeners or divide their own plants (i.e., use the daughter bulbs or offsets produced by their plants as new planting stock). If bulbs are not planted immediately on receipt, they should be stored in a cool, airy place (not a refrigerator).

Commercial growers employ a technique called "twin scaling," in which bulbs are sliced vertically into numerous segments, each with a pair of adjoining bulb scales and accompanying tissue, culturing these on growth media to produce new plants. Some ambitious gardeners sometimes slice bulbs to produce many new plants, although this is labor-intensive and slow to produce flowering plants (up to 4 years). True micropropagation (use of tissue culture for reproduction), as currently utilized for many crops, has not been commercially successful yet for daffodils.

### 34.8.3 Planting

Bulbs are ideally planted in early fall, allowing time for the plants to establish before freeze-up (although bulbs planted late nevertheless usually survive). The bulbs should be planted with the flattened base pointed downward, packing soil firmly around the bulb. In general, the tops (pointed ends) of the bulbs should be planted at a depth about twice their largest diameter in medium to heavy soils, and about three times their depth in sandy soils, if instructions on planting depth are not provided. (Another rule of thumb is that the bulbs should be planted at a depth about three times their height.) The soil should be worked deeply (25–30 cm) to facilitate root development. Daffodils have contractile roots, which shorten with age, pulling the bulb deeper into the soil. Small bulbs should be spaced about 7.5 cm apart, the largest bulbs about 15 cm apart. Closer planting can speed up the development of a good display of flowers (excessive closeness suppresses good plant development) but requires replanting of the bulbs sooner. Wider spacing (up to 30 cm) eventually produces dense stands and has the advantage of not requiring replanting for up to 10 years. Bulbs of the same type should be planted in groups (at least three to 12) for best effect.

### 34.8.4 Soil and Moisture Conditions

Daffodils are tolerant of many substrate types and conditions but grow best in well-drained, deep, fertile, slightly acidic soil (pH 6.0 to 6.5, although a pH as high as 7.5 has been considered acceptable). Daffodils have been grown in soils with considerable organic matter, although commercial daffodils are mostly raised in silty or loamy-sandy soils. A low-nitrogen commercial fertilizer with an N:P:K ratio of 1:3:3 is recommended, and it should be worked well into the soil before planting. Excessive nitrogen promotes foliage rather than floral growth and can stimulate production of too many small bulbs. Bonemeal (which is very low in nitrogen) was once recommended, but many daffodil specialists today dislike its use.

### 34.8.5 Maintaining Plants

After bulbs are planted, a 5 cm protective winter mulch is helpful, especially for bulbs planted late in the season or in a southern exposure, where early spring soil warming could result in damage. A light fertilization each spring as the leaves are developing is helpful. Daffodils planted near trees or shrubs, or in lawns, may need additional water and fertilizer to keep them healthy. Faded flowers should be removed to prevent seed and fruit development, which lowers bulb production. Foliage should be allowed to die completely before being removed. After several years, when the floral display has lessened because of crowding, the bulbs may be dug up after the tops have started to die back, and replanted immediately, or stored in a cool, shady, airy, dry location. Bulbs are typically stored commercially at 17–18°C, at a relative humidity below 75%. In cold climates (hardiness zones 3–7) plants in containers can be overwintered indoors in a protected situation, such as an unheated basement or garage.

### 34.8.6 Forcing Paperwhite Daffodils

The bulbs are typically potted close together, in a well-drained but quite moist soil. Alternatively, the bulbs are often simply placed in gravel or stones, or just in water (keeping the water level somewhat above the bases of the

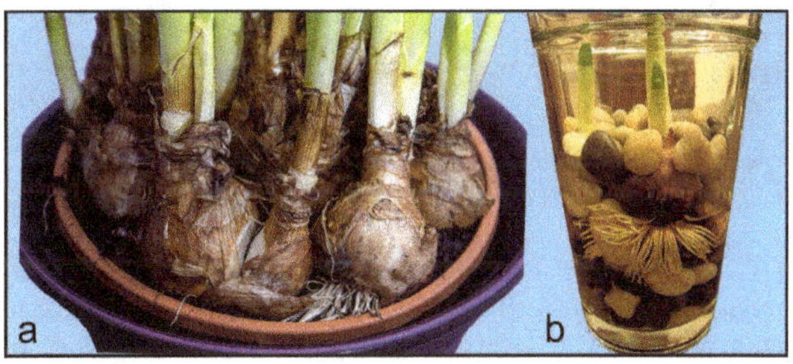


FIGURE 34.13 Planted paperwhite bulbs: (a) Bulbs planted in soil. (b) Bulbs planted in a glass of water.

*Source:* **(a) Photo by van Radic (CC BY 2.0). (b) Photo by Terren in Virginia (CC BY 2.0).**

bulbs to allow roots to develop in the water). The plants are best kept in a cool room with diffused light. However, if insufficient light is available, the stems sometimes grow tall and flop over unless staked.

## 34.9 FASCINATING FACTS, FABLES, FOLKLORE, AND FRIVOLITY

- According to Greek mythology, a young man named Narcissus fell in love with his image reflected in a pool of water. For rejecting the nymph Echo, Narcissus was punished by the goddess Nemesis, who turned him into a flower. Narcissus's fascination with himself is debatably the basis of the term *narcissism*, descriptive of people with an exaggerated, grandiose view of their own greatness (as noted earlier, the genus name *Narcissus* has been claimed to trace to the Greek *narke*, meaning "narcotic" or "capable of producing stupor").
- Another ancient Greek tale concerns the goddess Persephone, who was kidnapped by the god Hades (god of the underworld) while she was distracted by the beauty of the flowers. Accordingly, the ancient Greeks considered daffodils to be the flower of death (since it was the last flower that Persephone picked). To this day, some people consider daffodils to be unlucky and will not allow the flower to be brought into their houses. In some areas of Great Britain, it was thought that if daffodils were brought into a farmer's residence, his chickens would not lay eggs.
- The Prince of Wales is traditionally owed a symbolic rent of one daffodil for untenanted lands of the British Isles of Scilly and for a property in Cornwall. Such nominal compensation to British royalty is more common in the form of "peppercorn rent." In English law, as well as in the legal systems of some other countries, a legal contract is not binding unless each party receives something of value, even if one party receives something of much greater value than the other.
- Only well-known flowering ornamentals with characteristic flower colors have contributed their names to colors. Examples include buttercup, carnation, fuchsia, hyacinth, lavender, lilac, marigold, rose, and violet. Daffodil is a member of this elite club. "Daffodil yellow" is a bright yellow, and "paperwhite" is an off-white, resembling paper.
- In the early 1990s, the Rockefeller Foundation sponsored research to genetically engineer increased provitamin A (beta-carotene) production into rice, in order to improve the diet of many in the Third World who relied excessively on the crop. The pathway for such biosynthesis is present

FIGURE 34.15 The mythical ancient Greek goddess Persephone picking wild daffodils.

*Source:* **Public domain photo of painting by John William Waterhouse (1849–1917).**

FIGURE 34.14 Painting of Narcissus (*Narcisse changé en fleurcolas*) by Bernard Lépicié (1735–1784). This depiction is based on a version of the legend which holds that Narcissus, fond of looking at his own mirrored reflection, attempted to kiss his image on the surface of a pool, fell in, and drowned. The gods then transformed him into a scented flower which blossoms in the mountains in spring.

*Source:* **Public domain photo. Le Château de Versailles (CC BY SA 3.0).**

in green parts of rice, but not in the endosperm (the starchy part of the grain). Genes ("transgenes," notably one from daffodil) were introduced into rice, which resulted in a form with relatively high beta-carotene. High–provitamin A rice became known as "golden rice" because of the yellowish color visible after milling and polishing, due to the presence of beta-carotene in the endosperm.

- Daffodils have been recognized as official flowers of various regions. The daffodil and the leek are national symbols of Wales. The daffodil is associated with St. David (the patron of Wales), as the flower is said to be the first to bloom on his day (March 1). The daffodil has been interpreted as a national flower of Iran, and in Kurdish culture, it is also a national flower symbolizing the New Year. In 2007, New York recognized the daffodil as the city's official flower, after a tradition developed of planting more than a million bulbs producing yellow flowers (the color of remembrance) to commemorate the September 11 attack.
- Yellow daffodils have been widely adopted throughout the world as symbols of organized efforts to defeat cancer. While many national cancer societies employ the daffodil as their official symbol, the Canadian Cancer Society has become the world's largest purchaser of daffodils (often grown in British Columbia), which are marketed for fundraising. April is the Canadian Cancer Society's Daffodil Month. The 2015 annual "Daffodil Ball," described as "Canada's most exclusive gala," was held in Montreal's historic Windsor Station, featured 30,000 daffodils, and a blue-chip guest list that raised 2 million dollars for the Canadian Cancer Society.
- Spring daffodil festivals are widespread in areas where the flowers are prevalent, especially in mountainous areas of Europe.
- A popular theme for artists has been the depiction of young ladies collecting daffodils, symbolic of youthfulness, reproduction, and renewal.

FIGURE 34.16 Floats decorated with daffodils at daffodil festivals in Gérardmer, northeastern France.

*Source:* **(a) Photo by Ji-she (CC BY SA 3.0). (b, c) Photos by Superjuju10 (CC BY SA 3.0).**

FIGURE 34.17 (a) Anonymous artwork. (b) An 1893 chromolithograph trade card published by L. Prang & Co. (c) A 1923 poster advertising the Montreux (Switzerland) Festival of the Daffodils. Large annual daffodil celebrations are held on the slopes around Montreux, Switzerland and its associated riviera.

*Source:* **(a) Illustration by ChaminaGallery from Pixabay. (b) Photo from Boston Public Library. Printed by Säuberlin & Pfeiffer S. A. Vevey.**

## KEY PUBLICATIONS


Barnes, D. 1987. *Daffodils for Home, Garden and Show*. Timber Press, Portland, OR. 176 pp.

Barrett, S.C.H. and L.D. Harder. 2005. The evolution of polymorphic sexual systems in daffodils (*Narcissus*). *New Phytologist* 165: 45–53.

Bradbury, M. 2017. *Daffodil, Snowdrop and Tulip Yearbook 2017*. Royal Horticultural Society, London. 128 pp.

Caldwell, J. and T.J. Wallace. 1955. *Narcissus pseudonarcissus* L. *Journal of Ecology* 43: 331–341.

David, J.C. 2017. Evolution of the species of *Narcissus*: A review. *The Daffodil Journal* 53(3): 4–25.

Fernandes, A. 1968a. Improvements in the classification of the genus *Narcissus* L. *Plant Life* 24: 51–57.

Fernandes, A. 1968b. Keys to the identification of native and naturalized taxa of the genus *Narcissus* L. *Daffodil Tulip Year Book* 1968: 37–66.

Fernandes, A. 1975. L'évolution chez le genre *Narcissus* L. *Anales del Instituto Botánico A. J. Cavanilles* 32: 843–887.

Hanks, G.R. (ed.). 2002. *Narcissus and Daffodils*. Taylor & Francis, New York. 428 pp.

Hanks, G.R. and A.R. Rees. 1979. Twin-scale propagation of narcissus: A review. *Scientia Horticultarae* 10: 1–14.

Heath, B. and B. Heath. 2001. *Daffodils for North American Gardens*. Bright Sky Press, Albany, TX. 144 pp.

Jefferson-Brown, M.J. 1969. *Daffodils and Narcissi: A Complete Guide to the Narcissus Family*. Faber, London. 224 pp. + 49 plates.

Kamenetsky, R. and H. Hiroshi Okubo. (eds.). 2012. *Ornamental Geophytes: From Basic Science to Sustainable Production*. CRC Press, Boca Raton, FL. 553 pp.

Kingsbury, N. 2013. *Daffodil: The Remarkable Story of the World's Most Popular Spring Flower*. Timber Press, Portland, OR. 220 pp.

Lee, K.P. and D.M. Elston. 2023. Botanical briefs: Daffodils (*Narcissus* species). *Cutis* 111: 41–42.

Mather, J.C. 1961. *Commercial Production of Tulips and Daffodils*. W.H. & L. Collingridge, London. 212 pp.

Quinn, C.E. 1959. *Daffodils, Outdoors and In*. Hearthside Press, New York. 204 pp.

Rivera Nuñez, D., C. Obón De Castro, S. Ríos Ruiz, and F. Alcaraz Ariza. 2003. The origin of cultivation and wild ancestors of daffodils (*Narcissus* subgenus *Ajax*)

(Amaryllidaceae) from an analysis of early illustrations. *Scientia Horticulturae* 98: 307–330.

Salmon, M. 2017. *A Survey of the Genus Narcissus*. Published by the Author, Somerton, UK. 345 pp.

Shepherd, F.W. 1987. *Daffodils*. Cassell and the Royal Horticultural Society, London. 64 pp.

Waters, M.T., A.M.M. Tiley, E.M. Kramer, A.W. Meerow, J.A. Langdale, and R. Scotland. 2013. The corona of the daffodil *Narcissus bulbocodium* shares stamen-like identity and is distinct from the orthodox floral whorls. *Plant Journal* 74: 615–625.

Wells, J.S. 1989. *Modern Miniature Daffodils: Species and Hybrids*. Timber Press, Portland, OR. 170 pp.

Willis, D. 2012. *'Yellow Fever': A Prospect of the History and Culture of Daffodils*. Published by the Author, Sherriff Hutton, York, UK. 426 pp.

## Key Cultivation Websites

American Daffodil Society. http://daffodilusa.org/.

Great Britain Daffodil Society. http://thedaffodilsociety.com/.

Narcissus Manual. https://projectblue.blob.core.windows.net/media/Default/Imported%20Publication%20Docs/Narcissus%20Guide%20final%20low%20res%20version%20(1).pdf.

Royal Horticultural Society Daffodil Cultivar Registration. www.rhs.org.uk/plants/plantsmanship/plant-registration/daffodil-cultivar-registration.

Royal Horticultural Society Guide to Botanical Names in the Genus *Narcissus*. www.rhs.org.uk/plants/pdfs/plant-registration-forms/daffalpha.

The Spruce. *How to Grow and Care for Daffodils*. www.thespruce.com/planting-and-growing-daffodils-1402136.

WikiHow. *How to Fertilize Daffodils*. www.wikihow.com/Fertilize-Daffodils.

WikiHow. *How to Force Daffodils Indoors*. www.wikihow.life/Force-Daffodils-Indoors.

WikiHow. *How to Grow Daffodils Indoors*. www.wikihow.life/Grow-Daffodils-Indoors.

WikiHow. *How to Plant Daffodils*. www.wikihow.com/Plant-Daffodils.

WikiHow. *How to Transplant Daffodils*. www.wikihow.com/Transplant-Daffodils.

# 35 Dahlia (*Dahlia* Species and Hybrids)

**Scientific Names:** The genus name *Dahlia* honors Swedish botanist Andreas Dahl (1751–1759). *Dalea* is an independent, unrelated genus of plants.

**Common Names:** The common name "dahlia" normally refers to *Dahlia* species, but "sea dahlia" is *Coreopsis maritima* and "climbing dahlia" is *Hidalgoa wercklei*. Dahlia is one of the non-phonetic words that the Scripps National Spelling Bee (an annual contest held in the United States) has often employed.

FIGURE 35.1 Classical botanical illustrations of flowering dahlia (public domain). Note the frequency of single-flowered cultivars, as well as double-flowered forms, in past centuries.

*Source:* **(a) Annales du Muséum d'histoire naturelle, 1802, G. Dufour, et Ed. d'Ocagne, Paris. (b) Orbigny, C. d'. *Dictionnaire Universel d'Histoire Naturelle*, 1892, A. Pilon et cie, Paris. (c) Baines, E.V. 1913. *Miss Ella V. Baines the Woman Florist*. Self-Published, Springfield, OH. (d) Curtis, S. 1820. *The Beauties of Flora*. Curtis, Gamston, UK. (e) *Revue horticole*, Sér. 4, vol. 82, 1910. (f) Joseph Breck and Sons, 1886, trade catalogue, Boston.**

## 35.1 INTRODUCTION

Dahlias are an ornamental garden staple known for providing a huge range of flower sizes, shapes, and colors. Estimates of the number of cultivars range up to 50,000. The internet presents uncountable numbers of pictures of flowers, reflecting the popularity of garden species, and dahlias are among the favorite flowers that are photographed. In recognition of its late-season flowering period, the dahlia has been called "queen of the autumn garden" (chrysanthemum is similarly known as "queen of autumn" and "queen of fall flowers").

## 35.2 OVERVIEW

The genus *Dahlia* contains about 40 species of herbaceous, perennial bushes native to Mexico and Central America. The plants produce tuberous roots, which are enlarged (usually swollen) lateral roots functioning as storage organs providing energy reserves for the regrowth of the plant in the next season. These are conventionally called tubers, although true tubers (which have the same function and often look the same) are derived from stem tissue. The flower colors are nearly unlimited, ranging through shades of white, pink, red, yellow, orange, and purple, and various combinations of these (almost everything except true blue and true black). The flowers range in shape from single types (with daisylike flowers) to the more popular doubled varieties, which range in diameter from 5 to over 30 cm (the largest often compared to dinner plates). The doubled flowers often mimic carnations, peonies, roses, and zinnias. Flowering usually occurs in late summer. The single flowers (which most people identify as "daisies") include some cultivars that look like sunflowers.

FIGURE 35.2 Excavated dahlia tuberous roots.

*Source:* **Photo by Shutterstock/Aquatarkus.**

## 35.3 HISTORY

Long before Spanish colonists brought dahlia back to Europe in the 18th century, it was cultivated by the Aztecs. Allegedly, it was the favorite flower of Emperor Moctezuma II (about 1466–1520). The Aztecs employed the plants for several uses. The tuberous roots were employed as animal fodder, and apparently also as human food; the plants were used for medicine, and they were grown as ornamentals, including double-flowered forms. The hollow stems of the

DOI: 10.1201/9781003586227-35

FIGURE 35.3 Especially pretty pictures of dahlia.

*Source:* **All photos are public domain, from Pixabay: Photographers: (a) Bishnu Sarangi, (b) Günther Schneider, (c) Matthias Böckel, (d) Beverly Buckley, and (e) Ann-Marie.**

FIGURE 35.4 Tree dahlia (*Dahlia imperialis*): (a) Flowers of tree dahlia planted in Australia. (b) Non-flowering plants in a garden. (c) Tree dahlia planted in Japan, Mount Fuji in the background.

*Source:* **(a) Photo by Aussigall (CC BY 2.0). (b) Photo by Denis Barthel (CC BY SA 3.0). (c) Photo by Shutterstock/San Hoyano.**

treelike *D. imperialis* (up to 10 m long) were employed as pipes to transport water. Modern dahlias are mostly the result of hybridization. In mid-19th-century Europe, the passion for dahlias was so extreme it is said to have matched the craze for tulips, known as "tulipmania," in early 17th-century Netherlands. Estimates of the number of cultivars that have been recognized range from 20,000 to 50,000.

Ornamental dahlias are believed to have been derived mainly from four species: *D. coccinea*, *D. pinnata*, *D. merckii*, and *D. imperialis*. *Dahlia pinnata*, of particular importance, appears to be derived from a fertile hybrid between *D. coccinea* and *D. sorensenii*, produced in Mexico about 500 years ago.

## 35.4 ECONOMIC IMPORTANCE

Dahlias are grown commercially on a large scale in various warm-temperate and subtropical countries and are also a significant commodity in the international floral trade.

## 35.5 ORNAMENTAL USES

Garden dahlias come in a variety of sizes, from giants over 2 m tall to dwarfs, and (depending on cultivar) can be grown in numerous climates. They are planted in borders, formal gardens, and jumbled-jungle-style gardens. They also grow well in small and large containers. Dahlias are employed as

FIGURE 35.5 Commercial farm production of dahlia: (a) Dahlia farm in Oregon. (b) Workers with buckets of freshly picked dahlias. (c) On a flower farm near Saint Paul, Minnesota.

*Source:* **(a) Photo by Sheila Sund (CC BY 2.0). (b) Photo by Shutterstock/RossHelen. (c) Photo by Media Mike Hazard (CC BY 2.0).**

FIGURE 35.6 Ornamental dahlia gardens.

*Source:* **(a) Photo by Bilderfee from Pixabay (public domain). (b) Photo by Art Poskanzer (CC BY 2.0). (c) Akita Kokusai Dahlia Garden in Japan photo by Ja-N (CC BY SA 4.0).**

FIGURE 35.7 Bride with a bouquet of dahlias.

*Source:* **Photo by Shutterstock/Ilze Lucero.**

commercial cut flowers to a relatively minor extent and are popular wedding flowers.

## 35.6 OTHER USES

Dahlia is a minor source of inulin. Inulins are polysaccharide carbohydrates produced in storage organs by many plants, often instead of starch. Inulin, which is not digestible by humans (and so does not contribute calories), is often employed as a dietary fiber, thought to have various healthful properties, including use as a prebiotic (promoting "good bacteria" in the digestive system), while not aggravating diabetes.

The tubers were employed in Aztec times as food, and today they are occasionally consumed as a culinary curiosity. Tubers sold in stores to grow plants might be treated with fungicides and pesticides and so should not be eaten. The flowers (at least the petals) are also consumed, although like most edible flowers, they contribute mostly to the appearance rather than the taste of food. Some websites allege the plants are (mildly) toxic to pets (dogs, cats, horses).

## 35.7 PROBLEMS

Aside from requiring warm climates to overwinter outdoors and flowering quite late in the season, dahlias are considered to be superb, easy-to-grow ornamentals.

## 35.8 CULTIVATION

### 35.8.1 Climate and Site

Dahlias are frost-sensitive, and the tubers can be left in the ground only in areas that do not freeze significantly (climate zone 8 and warmer). In cold climates, the tubers are normally excavated and maintained in a dormant condition overwinter, in a cool, dry place. In cold climates, the tubers can be started in pots indoors in early spring. Dahlias should be grown in full sun. Taller varieties benefit from wind-sheltered sites.

### 35.8.2 Propagation and/or Purchasing

Seeds are available for some dahlias (mostly small-flowered forms), but the majority of garden varieties (which are hybrids) are reproduced vegetatively. Professionals often establish dahlias from root and stem cuttings, but most gardeners grow them from purchased "tubers" (actually tuber-like roots), which should be planted after danger of frost. As with many of the plants discussed in this book, starting material is available from nursery centers (and is often adapted to the local climate), but a greater range of varieties is offered online and in garden catalogues.

### 35.8.3 Planting

Depending on cultivar, the tubers are usually planted 5 to 10 cm deep, about 60 cm apart (if specific recommendations are not available). The eye (bud or growing point) of the tuber should face upward when planted (this upper side is usually evident from appearance). Most garden plants are tall, and these require supports, usually provided as stakes. Some authorities recommend allowing only one shoot to develop from a given tuber in order to produce impressive plants.

### 35.8.4 Soil and Moisture Conditions

Soils should be rich (preferably with incorporated compost), well-drained, and slightly acidic. The soil needs to be maintained moderately moist. If the soil contains rotted manure, little, if any, fertilization may be necessary; otherwise, fertilization with a balanced fertilizer (e.g., 10–10–10) or one that is lower in nitrogen (e.g., 5–10–10) is desirable, since excess N tends to favor vegetative growth rather than flower production.

### 35.8.5 Maintaining Plants

Tying growing plants to stakes (if they aren't grown in cages), keeping the soil from drying, and deadheading spent flowers are the main tasks. To encourage branching and bushiness, the growing points at the end of stems may be pinched off (a recommended practice to encourage flowering side shoots). When plants have grown to about 20 cm in height, pinch or cut off the main growing tip (a process called "disbudding"). To encourage the development of large flowers, all but the terminal flower bud on a branch can also be pinched off. When excavating the tubers in the fall, it must be appreciated that the growing point ("eye" or bud) develops only at the top of the tuber, where it contacts the base of the stem, so if the tuber is simply pulled off, the bud is destroyed and the tuber is worthless.

FIGURE 35.8 Floats made entirely of dahlia flowers in the annual Bloemencorso Zundert Netherlands dahlia parade.

*Source:* **(a) Photo by Foto van Hassel BV (CC BY SA 3.0). (b) Photo by Niels Braspenning (CC BY 3.0).**

## 35.9 FASCINATING FACTS, FABLES, FOLKLORE, AND FRIVOLITY

- *The Blue Dahlia* is a 1946 American crime film noir, so-named primarily because events occurred in a nightclub named the Blue Dahlia, although briefly a blue dahlia floral figurine is shown. "The Black Dahlia" refers to Elizabeth Short (1924–1947), an American woman whose gruesome murder became highly publicized. Her nickname is thought to have been based partly on the 1946 film released the year before her death and her habit of wearing black garments. As noted in the text, dahlia flowers are available in virtually all colors, except true blue and true black, so that both names are imaginative.
- Dahlia is the national flower of Mexico and the official flower of San Francisco and Seattle.
- In zoology, *Dahlia* is a genus of moths. However, animal and plant nomenclature are independent, although there are efforts to create a universal naming system applicable to all living things.
- Scientific (Latin) names are notoriously difficult to pronounce. The genus should be pronounced day-LEE-a if conventional botanical pronunciation of Latinized proper names is followed, but daw-LEE-a or DAW-lee-ah are also correct (like to-MAY-toe and to-MAH-toe).
- The *Bloemencorso Zundert* of Zundert, Netherlands, is the world's largest flower parade. It is dedicated exclusively to dahlia, employing millions of them to prepare very elaborate floats.

## KEY PUBLICATIONS


Aguirre, R., J. Legaria, M. Colinas, M. Juárez-Hernández, J. Mejía-Muñoz, and K. Hammett. 2023. Interspecific hybridization of *Dahlia* (Asteraceae) for the development of cultivars with ornamental value. *Acta Horticulturae* 1383: 73–80.

Albrecht, K. 2020. *Dahlia Breeding for the Farmer-Florist and the Home Gardener*. Published by the author. 73 pp.

Albrecht, K. and B. Sprinsock. 2023. *Dahlias: Seed to Bloom: The Dahlia Grower's Companion*. Independently Published. 278 pp.

Benzakein, E., J. Chai, and J. Jorgensen. 2021. *Floret Farm's Discovering Dahlias: A Guide to Growing and Arranging Magnificent Blooms Hardcover*. Chronicle Books, San Francisco. 224 pp.

Dastvan, B., J. Pate, J. Dole, T. Precht, and P. Hall. 2022. Growing dahlias, *Dahlia coccinea* Cav., for commercial cut flower production in aquaponics and autopots. https://projects.sare.org/wp-content/uploads/Journal-of-Floriculture-and-Landscaping-Report-SARE-1-29-22.pdf.

Mariña, L.J. 2015. Cultivation of the dahlia. *Cultivos Tropicales* 36: 103–110. www.haifa-group.com/sites/default/files/article/Cultivation%20of%20the%20Dahlia.pdf.

Mejía-Muñoz, J.M., I. De Luna-García, E.F. Jiménez-Ruiz, E. Sosa-Montes, C. Flores-Espinosa, G. Treviño-De Castro, et al. 2020. Research on dahlia, the national flower of Mexico. *Acta Horticulturae* 1288: 103–108.

Oliver, E., M. Stock, M. Lewis, A. Collins, A. Pratt, M. Brenneman, et al. 2022. *Dahlia Cut Flower Production in Utah*. Utah State University Extension Publication CutFlower/01. https://projects.sare.org/wp-content/uploads/Dahlia-production-fact-sheet-formatted-in-progress.pdf.

Schwab, J., S. Burnett, and B.J. Peterson. 2023. *Dahlia Cut Flower Production in the Northeastern United States* [abstract]. American Society for Horticultural Science Annual Conference. https://ashs.confex.com/ashs/2023/meetingapp.cgi/Paper/39719.

Şekerci, A.D. and O. Gülşen. 2016. Overview of dahlia breeding. *Scientific Papers-Series B Horticulture* 60: 199–204.

Slade, N. and G. Lane. 2018. *Dahlias: Beautiful Varieties for Home & Garden*. Gibbs Smith, Layton, UT. 240 pp.

Sørensen, P.D. 1969. Revision of the genus *Dahlia* (Compositae, Heliantheae-Coreopsidinae). *Rhodora* 71: 309–365, 367–416.

Sørensen, P.D. 1970. The dahlia: An early history. *Arnoldia* 30: 121–136. http://arnoldia.arboretum.harvard.edu/pdf/articles/1970-30-4-the-dahlia-an-early-history.pdf.

Thakur, P., A.H. Shah, Y. Adhikari, M. Kumar, and S. Verma. 2022. Dahlia cultivation in India and abroad: A review. *International Journal of Plant & Soil Science* 34(19): 240–251.

Vernon, A. 2016. Dahlia in Mexico. *Plantsman* 15: 244–249.

Whitley, G.R. 1985. The medicinal and nutritional properties of *Dahlia* spp. *Journal of Ethnopharmacology* 14: 75–82.


### KEY CULTIVATION WEBSITES


American Dahlia Society. www.dahlia.org/.

Bates, D. 2022. *The Growing World of Dahlias*. www.dahliaworld.co.uk/index.htm.

National Dahlia Society of New Zealand. www.nzdahliasociety.50megs.com/.

National Dahlia Society (U.K.). www.dahlia-nds.co.uk/about-dahlias/.

The Spruce. *How to Grow and Care for Dahlias*. www.thespruce.com/growing-and-caring-for-dahlias-1402255.

The Spruce. *How to Grow Dahlias in Pots*. www.thespruce.com/growing-dahlias-in-pots-847837.

University of Georgia Extension. 2015. *Dahlias. Circular 576*. https://extension.uga.edu/publications/detail.html?number=C576&title=dahlias.

U.S. Department of Agriculture. *Home and Garden Bulletin 131.1968. Growing Dahlias*. https://naldc.nal.usda.gov/download/CAT87212971/PDF.

WikiHow. 2020. *How to Care for Dahlias*. www.wikihow.com/Care-for-Dahlias.

WikiHow. 2021. *How to Grow Dahlias*. www.wikihow.com/Grow-Dahlias.

WikiHow. 2022a. *How to Cut Dahlia Tubers*. www.wikihow.life/Cut-Dahlia-Tubers.

WikiHow. 2022b. *How to Grow Dahlias in Pots*. www.wikihow.com/Grow-Dahlias-in-Pots.

WikiHow. 2022c. *How to Prune Dahlias*. www.wikihow.com/Prune-Dahlias.

WikiHow. 2022d. *How to Store Dahlias Over Winter*. www.wikihow.com/Store-Dahlia-Bulbs.

# 36 Daylily (*Hemerocallis* Species and Hybrids)

**Scientific Names:** The genus name *Hemerocallis* is based on the Greek *hemera*, "day," and *kalos*, "beautiful."

**Common Names:** The common name *daylily* (less frequently, "day lily") is descriptive of the short blooming time of individual flowers, which typically are in full bloom for only a day before the petals wilt. The "lily" in the name points out the similarity of the flowers to those of true lilies (species of *Lilium*). Similarly, numerous other unrelated plants have "lily" as part of their names.

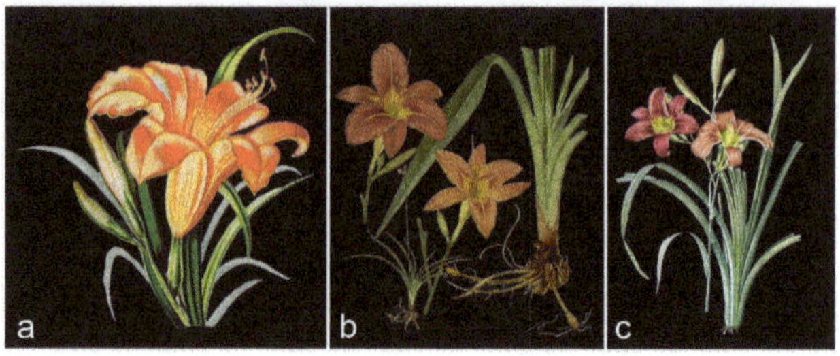

FIGURE 36.2 Illustrations (public domain) of orange daylily (*Hemerocallis fulva*) from early botanical publications.

*Source:* **(a) Vick's 1898 Garden and Floral Guide. (b) Kops, J. 1901–1906. Flora Batava 22. (c) Redouté, P. J. 1805. Les liliacée.**

## 36.1 INTRODUCTION

Daylilies are very popular garden ornamentals, valued for very attractive, large, lily-like flowers available in an extensive range of colors and for hardiness and limited need for care (this is decidedly a great plant for lazy gardeners). There are many societies and numerous breeders dedicated to daylilies, which are one of the world's most frequently cultivated herbaceous ornamentals grown mainly for their blooms.

## 36.2 OVERVIEW

The species of daylilies (estimates varying from 16 to 30, depending on authority) are members of the genus *Hemerocallis*, which is native to Asia, mostly the eastern part, including China, Japan, Korea, and southern Siberia. The plants are herbaceous perennials regrowing annually from underground parts (most species have fibrous roots, but in some the roots have swollen, that is, tuberous, areas). The widespread *H. fulva*, in addition to producing tuberous roots, is rhizomatous, that is, it spreads by underground stems. Daylily plants typically form clumps. The long, strap-shaped leaves occur in two opposite sets (called "fans"), within which newer leaves develop directly above older ones. In warm climates, the foliage may be semi-evergreen or evergreen. The flowering stalks, which are leafless (and termed "scapes"), bear one or (usually) several large trumpet-shaped flowers, each with two sets of three similar "tepals" (the term for similar sepals and petals). Some of the cultivars are

FIGURE 36.1 Displays of beautiful daylilies (*Hemerocallis* species and hybrids).

*Source:* **(a) Photo by Normanack (CC BY 2.0). (b) Photo by VasenkaPhotography (CC BY 2.0). (c) Photo by F. D. Richards (CC BY SA 2.0). (d) Photo by Carl Lewis (CC BY 2.0).**

FIGURE 36.3 Orange daylily (*Hemerocallis fulva*), the main ancestor of modern daylily cultivars.

*Source:* **Photo by Manuel M. V. (CC BY 2.0).**



DOI: 10.1201/9781003586227-36

attractively scented, although many have limited fragrance. Individual flowers are very short-lived (typically opening at sunrise and wilting by sunset), but the flowers on a scape open sequentially, one after another, day after day. A clump of plants often produces flowers for several weeks, the seasonal timing differing among cultivars. Some cultivars, grown in appropriate regions, have a second blooming period.

## 36.3 HISTORY

Daylilies have been grown in Asia for millennia but were imported into Europe no earlier than the 16th century, and to North America in the 17th century. The main ancestors of modern daylily cultivars, which are mostly hybrids, are *H. fulva* (ditch lily, orange daylily, tawny daylily) and *H. lilioasphodelus* (also called *H. flava*, lemon daylily). About 100,000 cultivar names have been published, mostly during the last 100 years, and about 1,000 new cultivars are released annually. *Hemerocallis fulva*, grown commonly for centuries, frequently persists on abandoned homesteads and old garden sites (a testimony to the plant's toughness), and so its light-orange flowers are widely observed. However, garden centers and seed suppliers now usually sell much more colorful modern cultivars. New cultivars have especially colorful flowers, often doubled (with extra petals) or with fringed petals. Some are dwarf forms; others are bred for fragrance. Some recent pleasant-smelling, "night-blooming" varieties have flowers that open in the evening and remain open until the evening of the following day. *Hemerocallis* 'Stella de Oro' (often changed to *Stella d'Oro*, Italian for "gold star"), a dwarf hybrid released to the public in 1975, may be the most popular daylily cultivar, admired for flowering from late spring to autumn, longer than almost all other cultivars.

## 36.4 ECONOMIC IMPORTANCE

Daylilies are one of the garden standards with a very large associated horticultural industry. In Asia, they are also a minor vegetable crop.

## 36.5 ORNAMENTAL USES

Daylilies are widely employed in borders and mass plantings. Dwarf cultivars are suitable for rock gardens and containers. Daylilies are not suitable as houseplants, but the flower stalks look excellent in vases. Their beauty

FIGURE 36.4 Variation of colors and structure of cultivars of daylilies.

*Source:* The following photo sources provide cultivar identification if available: (a) Photo by Famartin (CC BY SA 4.0). (b) Cultivar Prestor John, original photo by Salicyna, modified by James Steakley (CC BY SA 4.0). (c) Cultivar Macbeth, photo by B. Brookes. (d) Cultivar Paprika Flame, original photo by Salicyna, modified by James Steakley (CC BY SA 4.0). (e) Cultivar Double Pompon, original photo by Agnieszka Kwiecień, Nova, modified by James Steakley. (f) Cultivar Spacecoast Blast Off, original photo by Salicyna, modified by James Steakley (CC BY SA 4.0). (g) Photo by Kudelina Larisa (public domain). (h) Photo by Shutterstock/Jerry Gantar. (i) Cultivar Schnickelfritz, photo by Shutterstock/Jerry Gantar. (j) Cultivar Bold Courtier, original photo by Salicyna, modified by James Steakley (CC BY SA 4.0). (k) Cultivar Prissy Peekaboo, original photo by Salicyna, modified by James Steakley (CC BY SA 4.0). (l) Photo by Muhammad Habibullah (CC BY SA 4.0).

FIGURE 36.5 Daylily cultivar Stella d'Oro, probably the most widely purchased variety in recent decades, admired for its short stature and its ability to flower continuously from spring to autumn.

*Source:* (a) Photo (from Maine) by Paul VanDerWerf (CC BY 2.0). (b) Photo by David J. Stang, altered by James Steakley (CC BY SA 4.0). (c) Photo by Shutterstock/NMLenses.

FIGURE 36.6 Attractive daylilies planted in borders. (a) *Hemerocallis* 'Little Tyke'. (b) Mixed cultivars (Colorado blue spruce in background).

*Source:* (a) Photo by F. D. Richards (CC BY SA 2.0). (b) Photo by Berellian (CC BY 2.0).

FIGURE 36.7 Roadside beautification in North Carolina employing daylilies.

*Source:* CDOT Communications (CC BY 2.0).

FIGURE 36.8 Production and consumption of daylily flowers as a vegetable: (a) Hand harvest in Taiwan. Note that daylilies are naturally adapted to mountainsides, as shown here, which are otherwise unsuitable for most crops. (b) Harvested flowers being sun-dried in Taiwan. (c) A commercial package of dried flower buds. (d) Rehydrated flower buds, ready for culinary usage. (e) A dish with fresh flowers.

*Source:* (a) Photo by Rita2253419 (CC BY SA 4.0). (b) Photo by Chen Qixuan (CC BY SA 4.0). (c) Photo by Nikolaj Potanin (CC BY SA 2.0). (d) Photo by Benjwong (public domain). (e) Photo by Michal Orela (CC BY 2.0).

FIGURE 36.9 A cat beside daylilies that could kill it, even from slight contact. Daylilies cause severe kidney disease in cats, but not dogs.

*Source:* Photo by Shutterstock/Daykiney.

combined with their ability to outcompete weeds and grow without care after they are established make them ideal for roadside beautification.

## 36.6 OTHER USES

Certain species of *Hemerocallis* are employed as vegetables in Asia, particularly China, notably the flowers of *H. citrina* and the tuberous roots and young leaves of *H. fulva*. The flowers can be eaten raw (especially the petals) and have a taste like green beans. In Asia, the flower buds are usually dried or pickled and used in soups as flavoring. The foliage is employed as a salad ingredient, and the small white tubers sometimes associated with the roots are eaten like water chestnuts.

Daylilies hold soil well and have been planted for erosion control, especially to stabilize steep slopes and the sides of waterways.

## 36.7 PROBLEMS

A neurotoxic principal found in *Hemerocallis* species has been termed hemerocallin (identical to stypandrol, a name which has priority). The chemical seems to be present particularly in the roots, and since the roots are consumed in Asia, concern has been expressed about possible toxicity if they are eaten raw. Additionally, various reports on the internet suggest that the stalks and leaves (which are sometimes eaten fresh in salads) should be avoided because the hemerocallin content has hallucinogenic potential. Similar concern has not been expressed for the flowers, which are consumed in large amounts in Asia. As noted in the textbox, a different (but unidentified) chemical can be deadly for cats.

**Textbox:**

Lilies in the "true lily" and "daylily" families are very dangerous for cats. The entire lily plant is toxic: the stem, leaves, flowers, pollen, and even the water in a vase. Eating just a small amount of a leaf or flower petal, licking a few pollen grains off its fur while grooming, or drinking the water from the vase can cause your cat to develop fatal kidney failure in less than 3 days. The toxin, which only affects cats, has not been identified. Dogs that eat lilies may have minor stomach upset but they don't develop kidney failure.

– US Food and Drug Administration. 2021. "Lovely lilies and curious cats: A dangerous combination." www.fda.gov/animal-veterinary/animal-health-literacy/lovely-lilies-and-curious-cats-dangerous-combination

The two most widely planted daylilies, orange and lemon daylilies, are considered to be potentially invasive. They have become naturalized (established) outside their native areas in various parts of the world and are considered to be significant threats to the ecology of some regions.

## 36.8 CULTIVATION

### 36.8.1 Climate and Site

Daylilies are rated as tolerant of climate zones 3 to 9 (or 2 to 10 according to some websites). However, cultivars differ in their suitability to the different climatic regions. Those adapted to colder regions are deciduous (lose their foliage) with the onset of winter, becoming dormant, and this stimulates vigorous spring regrowth. Those adapted to the hottest regions are evergreen, capable of growing through mild winters. The plants are adapted to full sun and tolerate part shade (but flower colors may not be as brilliant, becoming more pastel). Daylilies should not be planted directly under trees and shrubs, with which they would compete for water and nutrients, as well as light.

### 36.8.2 Propagation and/or Purchasing

Daylilies can be grown from seed, although today almost all superior cultivars are reproduced vegetatively. Many modern cultivars are sterile and do not produce seeds. (In North America, the orange daylily rarely sets seed, while the lemon daylily often does). Daylilies multiply naturally by forming new shoots from the roots, and dividing the clumps by pulling or cutting them apart is the most common form of daylily propagation. Tuberous roots (usually termed "bulbs") which form on some of the species can be purchased from suppliers. "Proliferations" are small plants that sometimes grow on the scapes (flower stalks) of daylilies. Such plantlets can be cut off and planted.

### 36.8.3 Planting

Daylily "bulbs" are usually planted 30–60 cm apart, depending on cultivar. Seedlings started indoors from seeds can also be planted but will not produce flowering plants for more than a year.

### 36.8.4 Soil and Moisture Conditions

Daylilies do best in soil that is moist but well-drained, organically rich, and neutral or slightly acidic. They prefer loams but will grow in clay, and indeed, these tough plants can survive in poor soils.

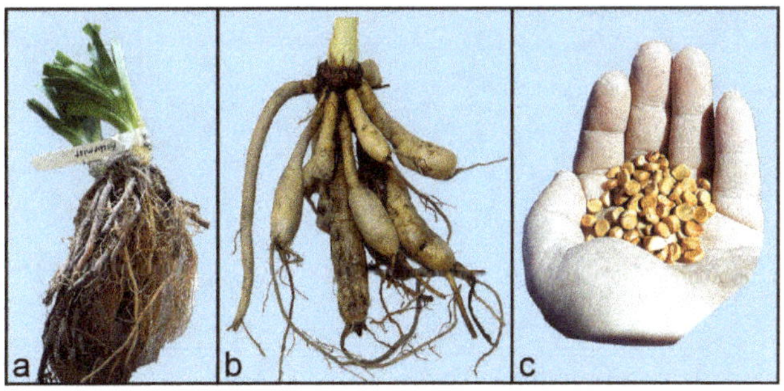

FIGURE 36.10 Materials for reproducing daylilies: (a) A fibrous-rooted clump. Larger clumps than this are divided to produce new plants. (b) A plant with tubers and a rhizome, both of which can be separated and replanted. (c) Seeds.

***Source:*** **(a) Photo by F. D. Richards (CC BY SA 2.0). (b) Photo by Hardyplants (CC BY SA 4.0). (c) Photo by Retired electrician (public domain).**

### 36.8.5 Maintaining Plants

Most daylilies tolerate neglect, even growing so thickly that they make it difficult for weeds to invade. However, unless divided every 4 or 5 years, flowering will be less attractive. The plants usually have limited need of fertilization, but occasional applications will enhance flowering. Similarly, irrigation is usually not required except in prolonged dry spells. Deadheading (removing spent flowers) is sometimes recommended, especially for older cultivars.

## 36.9 FASCINATING FACTS, FABLES, FOLKLORE, AND FRIVOLITY

- In ancient China, the belief was prevalent that daylilies were an antidote for poisons. It was even thought that deer deliberately ate the plants to nullify the effects of poisonous material, and as a result, the name "deer-onion" was coined for daylilies.
- Perennial herbs like daylilies grow anew each year from an underground "crown" (the condensed short stem from which leaves and floral shoots grow upward and roots descend downward). Nature has designed the crowns to be under the ground, where they are protected from animals and the elements overwinter. However, the new crown that develops is situated directly above the older (parental) crown so that season after season the newer crowns are higher and could pop out of the ground, where they would not be safe. In daylilies and many other similar plants (like dandelions), the roots under the new crowns are "contractile," shortening to deliberately pull the new crowns back into the soil for the coming winter.
- As noted in the introduction, the genus name *Hemerocallis* is based on Greek *hemera*, "day," and *kalos*, "beautiful," coined for the ephemeral blooming period of individual flowers. The ancient Greek personification of day was Hemera (the bringer of light), and she was paired with an opposite personality, Nyxt (night, the bringer of darkness). These goddesses of night and day were not allowed to be in the same room, and should one be present when the other tried to enter, the occupant had to leave. However, while passing, they greeted each other cordially.
- Daylilies are the official flower of numerous cities in the United States, including Ogden, New York; Amarillo, Texas; Hattiesburg, Mississippi; West Bend, Wisconsin; and Brunswick, Maryland.

## KEY PUBLICATIONS

American Hemerocallis Society. 1991. *Daylilies: The Beginners Handbook*. AHS Publications, Bells, TN. 159 pp.

Erhardt, W. 1992. *Hemerocallis: Daylilies.* Timber Press, Portland, OR. 160 pp.

Gatlin, F.L. 2002. *The New Daylily Handbook (for 2002).* American Hemerocallis Society, Palm Coast, FL. 320 pp. [The 1968 edition on which this is based is available at: https://ahsgardening.org/wp-content/pdfs/1968-02r.pdf]

Gulia, S.K., B.P. Singh, J. Carter, and R.J. Griesbach. 2009. Daylily: Botany, propagation, breeding. *Horticultural Reviews* 35: 193–220.

Hill, L. and N. Hill. 1991. *Daylilies: The Perfect Perennial.* Storey Communications, Inc., Pownal, VT. 208 pp.

Hu, S.Y. 1968. The species of *Hemerocallis. American Horticultural Magazine* 47: 86–111.

Li, X., S. Jiang, J. Cui, X. Qin, and G. Zhang. 2022. Progress of genus *Hemerocallis* in traditional uses, phytochemistry, and pharmacology. *The Journal of Horticultural Science and Biotechnology* 97: 298–314.

McGarty, T.P. 2009. *Hemerocallis: Species, Hybrids and Genetics.* Режим доступу. 437 pp. www.telmarcgardens.com; www.researchgate.net/profile/Terrence-Mcgarty/publication/265206495_Hemerocallis_Species_Hybrids_and_Genetics/links/54047e1b0cf2bba34c1c8ed2/Hemerocallis-Species-Hybrids-and-Genetics.pdf.

Munson, R.W., Jr. 1989. *Hemerocallis, the Daylily.* Timber Press, Portland, OR. 160 pp.

Peat, J.P. and T.L. Petit. 2000. *The Color Encyclopedia of Daylilies.* Timber Press, Portland, OR. 296 pp.

Peat, J.P. and T.L. Petit. (eds.). 2004. *The Daylily: Guide for Gardeners.* Timber Press, Portland, OR. 200 pp.

Rodriguez-Enriquez, M.J. and R.T. Grant-Downton. 2013. A new day dawning: *Hemerocallis* (daylily) as a future model organism. *AoB PLANTS,* 5: pls055. https://doi.org/10.1093/aobpla/pls055.

Schabell, J. 1990. The daylily: 5000 years of glory. *Daylily Journal* 45: 348–353.

Stout, A.B. 1934. *Daylilies.* Sagapress, Inc., Miliwood, NY. 145 pp. [Reprinted 1986, J.M. Dent and Sons, London.]

## Key Cultivation Websites

American Daylily Society. https://daylilies.org/.

Bill's Hemerocallis – the Daylily. www.ofts.com/bill/daylily.html.

Cornell University, Extension. *Daylilies.* https://ccenassau.org/resources/-daylilies.

The Spruce. *How to Grow and Care for Daylilies.* www.thespruce.com/how-to-grow-and-care-for-daylilies-5075400.

The Spruce. *How to Grow and Care for Orange Daylily.* www.thespruce.com/growing-orange-daylily-5116342.

University of Minnesota, Extension. *Daylilies.* https://extension.umn.edu/flowers/daylilies#plant-hardiness-586960.

WikiHow. *How to Divide Daylilies.* www.wikihow.com/Divide-Daylilies.

WikiHow. *How to Eat Daylilies.* www.wikihow.com/Eat-Daylilies.

WikiHow. *How to Grow Daylilies.* www.wikihow.com/Grow-Daylilies.

WikiHow. *How to Plant Daylilies.* www.wikihow.com/Plant-Daylilies.

WikiHow. *How to Prune Daylily.* www.wikihow.com/Prune-Daylily.

# 37 Delphinium (*Delphinium* Species and Hybrids)

**Scientific Names:** The genus name *Delphinium* traces to the ancient Greek *delphínion*, meaning "dolphin," presumed to refer to a species of the genus and alleged to reflect an old interpretation that the flowers are dolphin-shaped.

**Common Names:** In the garden trade, the Consolida (annual) group (discussed later) is frequently termed *larkspur* and *annual delphinium*, while the name "delphinium" is usually restricted to the perennial species. The alternative common name *larkspur* is based on the prominent spur of the flowers, which allegedly resembles a lark's spurs.

## 37.1 INTRODUCTION

Modern delphinium cultivars include a wide range of very colorful varieties ranging from dwarfs to giants approaching 3 m in height. Tall delphiniums tower over many other flowering perennial herbs, and their stately appearance has earned them their reputation as aristocrats of the garden.

## 37.2 OVERVIEW

The genus *Delphinium* contains about 350 species of herbaceous plants, mostly native to the north temperate zone, but there are also some in the mountains of tropical Africa. Traditionally, about 40 species now understood to belong to *Delphinium* were placed in the genus *Consolida*, and most literature produced until recently was based on the assumption that the two belong to separate genera. Both *Delphinium* and *Consolida* have been called "larkspur." In the horticultural trade, the name *Consolida* persists and is applied mostly to annual species, while *Delphinium* is applied mostly to perennial species. There are anatomical differences between the flowers and the fruits that distinguish species once placed in *Consolida* from other species of *Delphinium*. Some species are as short as 10 cm; others (some perennials) grow to a height of over 2 m. The flowers,

FIGURE 37.1 A bed of mixed delphiniums.

*Source:* **Photo by Andrew Stawarz (CC BY ND 2.0).**

FIGURE 37.2 Examples of annual delphinium cultivars (group Consolida, "larkspur" in the limited horticultural sense). (a) *Left*: *Delphinium orientale*; *right*: *D. ajacis.* (b) Cultivar Rosamond. (c) *Delphinum ajacis.* Painted by German artist Hans Simon Holtzbecker (1610–1671).

*Source:* **Public domain illustrations. (a) Guillaumin, A. 1929. *Les Fleurs de Jardins*, vol. I. Les Fleurs de Printemps. Published by Paul Lechevalier. (b) Cover of Beckert Seed & Bulb Co. (Pittsburgh), spring 1935 catalogue. (c) Photo from National Gallery of Denmark.**

FIGURE 37.3 Examples of perennial *Delphinium* species ("delphinium" in the narrow sense, as generally understood in the horticultural trade): (a) Delphiniums on the front cover of Currie Bros. (Milwaukee) Farm and Garden 1906 catalogue. (b) Menzies' larkspur (*D. menziesii*). (c) A garden hybrid.

*Source:* **Public domain illustrations. (a) Public domain. (b) *The Botanical Register*, vol. 14 (1828). (c) Step, E. 1896. *Favourite flowers of Garden and Greenhouse*, vol. 1.**

DOI: 10.1201/9781003586227-37

FIGURE 37.4 *Delphinium* flower clusters: (a) *Delphinium ajacis* (annual, Consolida group). (b) *Delphinium elatum* cultivar Green Twist (perennial). (c) *Delphinium regalis* cultivar Carmine King (annual, Consolida group).

***Source:*** **(a) Photo by HQ (CC BY SA 2.0). (b) Photo by Jacinta Iluch Valero (CC BY SA 2.0). (c) Photo by Shutterstock/Julia Gardener.**

usually in prominent "spires" or "spikes" (botanically, they are "racemes"), are usually blue or purple, but horticultural forms have been selected with white, pink, red, and yellow flowers. One of the sepals is prominently spurred. The contrasting centers of the flowers are called "bees" by horticulturalists; in cultivars, they range in color from white to tan, brown, and black and can be striped. Some perennial cultivars are early-flowering; others grow into the autumn.

## 37.3 HISTORY

*Delphinium* has been cultivated as ornamental plants in Europe since the 17th century. By the 19th century, interest in breeding delphinium increased greatly, pioneered by the prominent horticulturalist Victor Lemoine in France, and later by breeders in other countries, including the United States. Thousands of cultivars, especially of hybrid origin, have been bred as ornamental plants, predominantly from *D. elatum*, but also with parentage from *D. bruninianum*, *D. cardinale*, *D. cheilanthum*, and *D. formosum*. Cultivars of the Consolida group originate from different species, notably *C. ajacis* and *C. regalis* (= *D. consolida*). Breeding for different flower color from the usual blue has resulted in flowers that are white, pink, red, and sometimes yellow. Also, doubled flowers have been created, as well as forms suitable for cut flowers.

## 37.4 ECONOMIC IMPORTANCE

Delphinium is a frequently encountered ornamental plant, but of somewhat lesser importance than the top-tier commercial offerings of garden plants. There are several other popular flowering ornamentals (such as lupin and foxglove) that produce tall spikes of colorful flowers, and these similar plants provide competition for attention. Nevertheless, in various countries, delphinium farms grow the plants on an industrial scale to supply the horticultural industry.

FIGURE 37.5 Roadside beautification in North Carolina with rocket larkspur (*Delphinium ajacis*, an annual species of the Consolida group).

***Source:*** **Photos by NCDOTcommunications (CC BY 2.0).**

FIGURE 37.6 Exhibitions of delphinium cultivars at flower shows in the UK: (a) Potted plants. (b) Cut flowering branches in vases.

***Source:*** **(a) Photo by James (CC BY SA 2.0). (b) Photo by Verity Cridland (CC BY 2.0).**

FIGURE 37.7 A woman holding a bouquet of recently harvested delphinium flowers.

***Source:*** **Photo by Shutterstock/Stakhov Yuriy.**

## 37.5 ORNAMENTAL USES

Delphiniums are common, popular garden ornamentals that are employed in mass plantings and borders. Tall delphiniums are often employed at the back of ornamental displays. Cut flower are widely employed for bouquets. The plant is rarely employed as a houseplant.

## 37.6 OTHER USES

*Delphinium* species have been employed historically for various purposes. The flowers have been used in both Eurasia and North American as sources of natural dyes. Similarly, they have been employed as herbal medicines in different areas of the world.

FIGURE 37.8 Delphiniums in flower at the Real Flower Petal Confetti Company, a flower farm in Wick, Pershore, Worcestershire, UK, specializing in employing the flowers to produce natural wedding confetti.

***Source:* Photo by Shutterstock/Golden Shark 2.**

FIGURE 37.9 A female bison grazing on delphinium in Grand Teton National Park, Jackson Hole, Wyoming. Delphinium is very toxic to domesticated cattle, but wild grazers have often become relatively tolerant of poisonous wild plants that occur naturally in their environment.

***Source:* Photo by Shutterstock/Alan Lipkin.**

A large production facility in Wick, Pershore, Worcestershire, UK, specializes in employing delphinium flowers to produce natural wedding confetti and has become a tourist site, attractive for its spectacular colorful fields.

## 37.7 PROBLEMS

All *Delphinium* species are toxic, and wild plants are frequently responsible for poisoning livestock. Cattle poisoning from cattle grazing on wild delphinium species is a major concern in the western United States (sheep are relatively resistant, and horses usually refuse to eat significant amounts). Some people are contact-sensitive to the plants, and to avoid developing dermatitis, they should wear gloves and long sleeves when working with them.

## 37.8 CULTIVATION

### 37.8.1 Climate and Site

Modern cultivars differ significantly in their appropriateness for different climates. Annual delphiniums (Consolida group) are adapted to climate zones 3 to 9. Generally, perennial delphiniums are appropriate for climate zones 3 to 7, and in warmer areas, they are preferably grown as annuals. The plants prefer cool, mild summers with low humidity. They grow best in full sun, although partial or afternoon shade is beneficial in hot climates. Tall varieties are best positioned where they are protected from wind.

### 37.8.2 Propagation and/or Purchasing

Delphiniums are reproduced from seed and basal cuttings. Seeds are best started indoors, and the seedlings transplanted. It can be difficult to germinate the seeds, but seedlings in cell packs can often be purchased early in the growing season. When planted from seeds (or seedlings), perennials frequently do not produce flowers until the second year. Seeds collected from hybrids may not produce offspring with the same flower color. Larger potted plants intended for transplanting are often available from garden centers. Established mature plants can be divided.

### 37.8.3 Planting

Annual (Consolida group) cultivars are planted 25 to 60 cm apart, depending on the size of the cultivar. Perennial delphiniums are usually spaced 60 cm or more apart. Some gardeners recommend that tall delphiniums be especially well-spaced (1 m apart when planted), to avoid problems of poor air circulation, which contributes to fungal diseases. Dwarf varieties can be established more closely.

### 37.8.4 Soil and Moisture Conditions

The soil should be moist but well-draining, neutral to slightly alkaline, fertile/rich, and humusy (if not, organic matter should be added). A thin layer of compost can be added to beds annually; mulching in autumn with well-rotted manure or leaf mold is a good idea. The plants respond well to fertilization, but high-nitrogen fertilizers encourage vegetative growth rather than flowering. Ideally, slow-release organic fertilizer should be added to the soil preceding planting, but a balanced liquid fertilizer can be added every 3 weeks or so during the growing period. Some gardeners try to water plants at their base to avoid wetting the foliage, since delphinium is susceptible to fungi in humid areas.

### 37.8.5 Maintaining Plants

Perennial delphiniums, especially older varieties, are often considered to be relatively fussy (they don't like hot weather or drought, and too much shade encourages fungi), difficult to grow (they often are short-lived), and high-maintenance. Taller varieties frequently require staking, especially in windy areas, as the hollow stems can be fragile. Supporting stakes should be inserted at planting time to avoid spearing roots later. Massed plantings require thinning when they grow thickly (dividing every 3 or 4 years is recommended). Some garden specialists advise allowing first-year plants to develop only one main shoot the initial year and limiting the number of main shoots for the first several years. Some gardeners cut away the top of the main stalk (or even cut to the ground) after flowering to encourage side shoots to produce a second flowering period. Removal of spent flowers prolongs the flowering display. After the perennials wither in the autumn, they should be cut back to the ground.

## 37.9 FASCINATING FACTS, FABLES, FOLKLORE, AND FRIVOLITY

- In the 19th and early 20th centuries, before photography became common, illustrations were usually prepared by artists. Young children, of course, are naturally attractive subjects. And so are flowering plants. Children and flowers (including delphinium) were often portrayed in imaginative situations.
- In Transylvania (the home of many superstitions), delphinium was kept in stables to protect animals from witchcraft.
- In different parts of Eurasia, people employed delphinium to deter pests. During the Battle of Waterloo (1815, the site where Napoleon and the French were defeated, now in Belgium), the British army used powdered delphinium plants as pesticides to prevent and kill lice. Application to humans of such toxic material seems drastic today, but recall that DDT was employed the same way.
- Since ancient times, herbal medicinal practitioners in the Mediterranean area have employed the seeds of "*Delphinium staphisagria*" (now reclassified as *Staphisagria macrosperma*) to treat hair loss, to control lice, and even to remedy erectile dysfunction. The seeds are available on the web and occasionally still employed medicinally but are dangerously toxic and have resulted in serious poisoning by people who tried to self-medicate. (Note Koparal and Bostancıoğlu 2016; Baby 2023.)
- In birds, spurs are thorny extensions, usually on feet or legs (sometimes on wings). Spurs occur predominantly in males and serve particularly as weapons against other males. However, spurs sometimes occur in both sexes and presumably serve for defense. Most of the approximately 100 "lark" species (including 17 genera) have a

FIGURE 37.10 Vintage portrayals (public domain) of innocent young children in combination with attractive flowers: (a) A charming interaction of a young girl holding delphiniums and foxgloves, being offered pansies by a young boy, with gladiolus behind him. (b) Two "flower children" costumed as pansy and delphinium.

*Source:* **(a) Vaughan's (Chicago) 1914 seed catalogue. (b) Gordon, E. 1910. *The Flower Children*. P. F. Volland, Chicago.**

FIGURE 37.11 Spurs of a "larkspur" (delphinium) flower and a lark bird: (a) Flower of *Delphinium nuttallianum* showing prominent spur in which nectar is present at the base. (b) White-tailed lark (*Mirafra albicauda*). Like most birds, the feet have three forward-facing toes and one rear-facing toe, all of which terminate in a claw. The "larkspur" is the claw of the rear toe, which is noticeably more elevated than the other claws of the foot.

*Source:* **(a) Photo by Walter Siegmund (CC BY SA 3.0). (b) Shelley, G. E. 1902. *The Birds of Africa*, vol. 3. R. H. Porter, London.**

"larkspur," a long straight claw on the hind toe (not like roosters, which have spurs much higher up on their legs). In flowers, spurs have a quite different purpose. They are a hollow projection on a flower containing nectar at the base. Frequently, the length and shape of the spur allow only a particularly adapted pollinator (with a long narrow beak or tongue) to reach the nectar, simultaneously dusting it with pollen to be carried to other flowers.

## KEY PUBLICATIONS

Baby, D.P. 2023. What is staphysagria used for? *MedicineNet.* www.medicinenet.com/what_is_staphysagria_used_for/article.htm.

Bassett, D. and S. Bassett. 2007. *Delphiniums.* Timber Press, Portland, OR. 160 pp.

Booth, P. and L. Cooper. (eds.). 2003. *The Delphinium Garden: A Comprehensive Guide to Growing Delphiniums.* The Delphinium Society, London. 112 pp.

Edwards, C. 1990. *Delphiniums: The Complete Guide.* Crowood Press, Marlborough, Wiltshire, UK. 151 pp.

Jabbour, F. and S.S. Renner. 2011. *Consolida* and *Aconitella* are an annual clade of *Delphinium* (Ranunculaceae) that diversified in the Mediterranean basin and the Irano-Turanian region. *Taxon* 60: 1029–1040.

Koparal, A.T. and R.B. Bostancıoğlu. 2016. Promotion of hair growth by traditionally used *Delphinium staphisagria* seeds through induction of angiogenesis. *Iranian Journal of Pharmaceutical Research* 15: 551–560.

Lucas, H.R. (ed.). 1974. *The Delphinium Society Year Book, 1964.* The Delphinium Society, London. 148 pp.

Munz, P.A. 1967. A synopsis of the Asian species of *Delphinium*, sensu stricto. *Journal of the Arnold Arboretum* 48: 476–545.

Welch, K.D., D. Cook, B.T. Green, D.R. Gardner, J.A. Pfister, T.G. McDaneld, et al. 2015. Adverse effects of larkspur (*Delphinium* spp.) on cattle. *Agriculture* 5: 456–474.

### KEY CULTIVATION WEBSITES

The Delphinium Society (U.K.). https://delphinium-society.co.uk/.

Garden Design. *Delphiniums: How to Grow & Care for Perennial Delphinium.* www.gardendesign.com/flowers/delphinium.html.

GardenersWorld.com. *Complete Guide to Growing Larkspur Flowers.* www.gardenersworld.com/plants/larkspur/.

GardenersWorld.com. *How to Grow Delphiniums.* www.gardenersworld.com/how-to/grow-plants/how-to-grow-delphiniums/.

The Spruce. *How to Grow and Care for Delphinium.* www.thespruce.com/delphinium-care-guide-7105084.

The Spruce. *How to Grow and Care for Larkspur.* www.thespruce.com/popular-larkspur-types-5090142.

WikiHow. *How to Grow Delphiniums.* www.wikihow.life/Grow-Delphinium.

# 38 Echinacea (*Echinacea* Species and Hybrids)

**Scientific Names:** The genus name *Echinacea* is based on the Greek *echinos*, "hedgehog" or "sea urchin," prickly animals that are suggested by the ray flowers (i.e., the central tiny flowers of the daisylike flower heads), which have a decidedly spiny appearance. Alternatively, some authors state that the prickly leaf-bracts (scales) under the flower heads are the basis of the name.

**Common Names:** The common name "echinacea," based on the genus name, is unambiguous. *Echinacea* species are also called coneflowers, in reference to the conical shape of the flower heads. However, other species in the daisy family (Asteraceae or Compositae), notably in the genera *Dracopsis*, *Ratibidia*, and *Rudbeckia*, are also called coneflower because the flowers have the same cone-like appearance. Garden cultivars of these and some others in the same family, such as *Helenium* and *Gaillardia*, are often misidentified as echinacea.

## 38.1 INTRODUCTION

Echinacea ornamentals include stunningly attractive hybrids with showy, intriguingly constructed flowers in almost all colors of the rainbow. The plants are very hardy, low-maintenance perennial herbs that have justly earned their reputation as essential garden staples.

FIGURE 38.1 Spectacular massed plantings of echinacea cultivars showing the range of colors available.

*Source:* (a) Photo by Leonora (Ellie) Enking (CC BY SA 2.0). (b) Photo by Bennilover (CC BY ND 2.0). (c) Photo by Alicja from Pixabay (public domain). (d) Photo by Leonora (Ellie) Enking (CC BY SA 2.0).

## 38.2 OVERVIEW

*Echinacea* is a North American genus of nine perennial herbaceous species. These produce daisylike flower heads in mid- and late summer. Many of the cultivars have extremely attractive large, colorful flowers. The principal garden cultivars are selections and hybrids of *E. angustifolia*, *E. pallida*, and *E. purpurea* (the main cultivated species). *Echinacea angustifola*, *E. pallida*, and *E. tennesseensis* are closely related, and some authorities combine them into one species (called *E. pallida*). Flower colors are commonly shades of white, pink, and purple, but also red, orange, and yellow, as well as combinations of these within the flower heads. Yellow flowers usually indicate parentage from yellow echinacea (*E. paradoxa*). Doubled flower heads are also available, and the plants vary from dwarf forms to heights of over 2 m.

## 38.3 HISTORY

In comparison with those of other ornamental plants, the history of echinacea is short. The plants originated from North America and were employed medicinally and culturally by the Indigenous Indians, but not as ornamentals. Native Americans used echinacea topically for wounds, burns, and insect bites; chewed the roots for toothache and throat infections; and consumed the plant internally for pain, coughs, stomach cramps, and snakebites. The first marketed

FIGURE 38.2 Purple coneflower (*Echinacea purpurea*), the main cultivated species.

*Source:* (a) Photo by Shutterstock/Ben Schonewille. (b) Photo by EddyCubausa (CC BY SA 4.0). (c) Photo by F. D. Richards (CC BY SA 2.0).


DOI: 10.1201/9781003586227-38

FIGURE 38.3 *Echinacea* cultivars with doubled flowers (i.e., with extra petals): (a) Cultivar Pink Double Delight. (b) Cultivar Hot Papaya.

*Source:* **(a) Photo by Ltshears (CC BY SA 3.0). (b) Photo by Daderot (public domain).**

echinacea preparations appeared in the late 19th century, alleged to be cure-all tonics for treatment of a very wide variety of conditions. By the start of the 20th century, echinacea was one of the most frequently used herbal preparations in the United States (some claim it is the most widely employed herbal in both North America and Europe). Consistent with a huge market that has developed in Europe for herbal products, most cultivation and consumption of medical echinacea now occur in Europe. For the last century, there has been extensive selection of ornamental cultivars.

## 38.4 ECONOMIC IMPORTANCE

Echinacea is one of the best-selling herbal preparations in many developed countries. Echinacea has also become an ornamental plant favorite, although its economic horticultural significance is much less than its medical value.

## 38.5 ORNAMENTAL USES

Echinacea is widely employed as tall background plants in the rear of perennial garden beds, in borders, in commercial landscapes, and as mass plantings in public naturalized areas. The blooms are used for the fresh-cut floral industry, as well as for dried-flower arrangements. The flowers are pollinator magnets, the nectar and pollen attracting bees and butterflies, and the seeds enjoyed by birds. Echinacea is a principal plant recommended for "butterfly gardens" and "wildlife gardens."

## 38.6 OTHER USES

Echinacea plants were employed by Native Americans of the Great Plains region as traditional medicines. In recent times, echinacea extracts have been promoted as dietary supplements, allegedly effective against the common cold and other respiratory infections, based on the hypothesis that the extract stimulates the immune system to more effectively fight infection. Echinacea preparations have also been promoted for use on skin problems, as an anti-inflammatory, and as an aid in healing wounds. Most of

FIGURE 38.4 Commercial fields of echinacea: (a) In the United Kingdom. (b) *Echinacea pallida*, in Ukraine.

*Source:* **(a) Photo by RobsAllotment from Pixabay (public domain). (b) Photo by Oris2012 (CC BY SA 3.0).**

FIGURE 38.5 Border planting of mixed echinacea cultivars.

*Source:* **Photo by Shutterstock/Shriesik.**

FIGURE 38.6 Wildlife feeding on purple coneflower (*Echinacea purpurea*): (a) Monarch butterfly (*Danaus plexippus*). (b) Hummingbird. (c) Painted Lady butterfly (*Vanessa cardui*) in Belgium. (d) American goldfinch (*Spinus tristis*). (e) Bumblebees.

*Source:* **(a) Photo by Jim Hudgins/US Fish and Wildlife Service (public domain). (b) Photo by Shutterstock/Yanik Chauvin. (c) Photo by Jean-Pol Grandmont (CC BY 3.0). (d) Photo by Joseph Gage (CC BY SA 2.0). (e) Photo by Lance Cheung, US Department of Agriculture/Forest Service (public domain).**

the market is based on extracts from the tops and flowers from cultivated fields of *E. purpurea*, but significant quantities are also obtained from the roots of *E. angustifola*. (*E. purpurea* has fibrous roots, unlike the taproots of the other species, making root extracts more difficult.) Products include teas, ointments, pills, tinctures, creams, and powders. The evidence is unclear whether echinacea is significantly helpful for health conditions. Some people have allergic reactions to echinacea, and there is some hesitancy about consuming it under some circumstances, such as pregnancy. Nevertheless, echinacea over-the-counter products are widely touted and marketed, and most authorities have concluded that whether or not echinacea is medically effective, at least it is minimally, if at all, harmful. There are hundreds of publications stating that echinacea is medically valuable, and a comparable number reporting that it is not or at least its merit has been hugely inflated, exemplifying the unreliability of numerous medical reports (see Paulos, J. A. 2005. *Why Medical Studies Are Often Wrong*. https://abcnews.go.com/Technology/WhosCounting/story?id=997688).

## 38.7 PROBLEMS

Echinacea plants are tough and hardy and generally about as problem-free as ornamental plants can be. They are rich in chemicals that can be poisonous if consumed in large amounts, which is helpful in making them resistant to deer, but not necessarily good for children, pets, and livestock. However, their toxic potential is considered quite limited, with possible digestive upsets if consumed and dermatitis for sensitive people.

## 38.8 CULTIVATION

### 38.8.1 Climate and Site

Wild echinacea species are adapted to the dry open woodlands and prairies of North America, and so they are quite tolerant of summer heat, winter temperatures as low as –20°C, high humidity, drought, and partial shade. The plants are recommended for climate hardiness zones 3 to 9, depending on cultivar.

### 38.8.2 Propagation and/or Purchasing

Echinacea is propagated by seeds and by division of the underground portions. The plants can be grown from cuttings, but few people are likely to do this. Seeds can be collected from plants growing in a personal garden, or a neighbor's, or from garden seed suppliers. Commercial seeds often can be planted directly, but echinacea seeds may not have been pre-stratified (cold-treated to overcome dormancy). This can be done by refrigerating the seeds for a few weeks before planting them. When started from seed, it may require 2 or more years to develop into sizable plants. Most people will opt to simply purchase potted plants intended for transplanting from nurseries or online suppliers.

### 38.8.3 Planting

Echinacea are best planted in open (sunny) places, well-spaced from competing species, and 0.3 to 0.9 m from each other (0.5 m apart has been recommended, even if it seems excessive), depending on size. Seedlings are best established indoors in small pots (which may require starting 8 weeks in advance) and transplanted. Existing mature plants (about 4 years of age) can be divided in the fall or spring.

### 38.8.4 Soil and Moisture Conditions

The species prefer well-drained, moist loams and sandy loams with pH 6.0 to 7.0 but are tolerant of various soil types. They are drought-tolerant once established but benefit from watering during dry spells.

### 38.8.5 Maintaining Plants

Echinacea stems are sturdy and rarely require staking. Removing spent flowers (deadheading) will prolong flowering. However, some gardeners prefer to let seed heads form, for winter interest, to feed birds, and to promote self-seeding. Whether in the late fall or early spring, the dead stubble needs to be removed to promote regrowth. Dividing mature plants every 4 or so years is recommended.

## 38.9 FASCINATING FACTS, FABLES, FOLKLORE, AND FRIVOLITY

- The Meskwaki tribe of Native Americans called echinacea "the hairs of Grandmother Earth's head." The Lakota employed the dried, prickly flower heads to brush hair, presumably inspired by the hairlike appearance of the flowers.
- Western American Indian tribes, like the Ute, associated echinacea with elk and coined the name "elk root" for it in the belief that wounded elk seek the plant as medicine.
- Some Native Americans thought that echinacea was protective from heat. Claims were made that after chewing echinacea (known to cause an unusual, acrid, tingling sensation on the tongue, which has been claimed to represent numbing), a hot coal could be put in one's mouth. The Sioux put its juice on their hands to prevent burns while handling very hot meat, and the Pawnee employed it to make the heat of steam baths more endurable.
- The medical and other traditional uses of plants by Indigenous cultures often seem peculiar, indeed primitive today, but this perspective fails to appreciate the state of knowledge not just by Native peoples but, indeed, everywhere. Europeans took up the use of the plant for equally questionable applications,

such as treating sexually transmitted diseases, dog bites, and numerous other afflictions. Probably the majority of herbal medicines have been considered to be panaceas, leading to disrespect for them today, although they usually do have metabolic effects, and the potential to be employed medicinally should be researched to justify applications.

- In North America in the late 19th and early 20th centuries, some medical practitioners called "Eclectics" advocated a resurgence of the use of medical substances and practices employed by Indigenous peoples. Although denounced by conventional doctors, they promoted a resurgence in the use of medicinal herbs, such as echinacea.
- Echinacea was widely believed to be a treatment for snakebite by Native Americans (and subsequently by European colonists, as indicated by the occasional names "Missouri snakeroot" and "Kansas snakeroot"). H. C. F. Meyer was a (German) Nebraskan lay physician who created an echinacea mystique by widely publicizing an offer to allow a rattlesnake to bite him (which he never permitted) to demonstrate the curative power of his miraculous medicine, called "Meyer's blood purifier." He has been credited as the source of the phrase "snake oil salesman."
- The state of Tennessee has three official state flowers. The Tennessee coneflower (*Echinacea tennesseensis*) and the passionflower are the designated state wildflowers, while the iris is the designated official state cultivated flower. *Echinacea tennesseensis* was considered in danger of extinction and classified as endangered for decades, but conservation efforts have led to recovery.

FIGURE 38.7 A snake oil salesman. An 1850 lithograph by G. Frison (public domain).

*Source:* Wellcome Trust.

FIGURE 38.8 Tennessee coneflower (*Echinacea tennesseensis*).

*Source:* Photo by Funke (CC BY SA 4.0).

- An aphid, *Aphis echinaceae*, cannot survive on any other host species except *Echinacea angustifolia*. Sometimes over 1,000 of these insects occur on a single plant. It has been observed that ants protect the aphids so that they can consume the sweet liquid that the insects excrete. Plant species that similarly are hosts to such ant–aphid symbiotic relationships often benefit indirectly by the ants not only guarding the aphids but also protecting the plants from herbivores.
- An embarrassing issue confronted by those interested in the names of species is how often they were coined in pastimes using racist language which was much more accepted than today. Frequently obviously racist names (which exist for echinacea) are simply not reported. The names Samson and black Samson (the root of echinacea is often black) seem harmless and so are often listed as among the many English synonyms. The names, in fact, are tinged with racism. Black Americans came to regard the biblical character Samson as an iconic hero (sometimes represented as a Black man) who defended their rights (see Junior, N., and J. Schipper. 2020. *Black Samson: The Untold Story of an American Icon*. Oxford University Press, Oxford).

## KEY PUBLICATIONS

Binns, S.E., B.R. Baum, and J.T. Arnason. 2002. A taxonomic revision of *Echinacea* (Asteraceae: Heliantheae). *Systematic Botany* 27: 610–632.

Fish and Wildlife Service. 2011. Endangered and threatened wildlife and plants; removal of *Echinacea tennesseensis* (Tennessee purple coneflower) from the federal list of endangered and threatened plants. *Federal Register, the Daily Journal of the United States Government*. www.federalregister.gov/documents/2011/08/03/2011-19674/endangered-and-threatened-wildlife-and-plants-removal-of-echinacea-tennesseensis-tennessee-purple.

Foster, S. 1984. *Echinacea Exalted!: The Botany, Culture, History and Medicinal Uses of the Purple Coneflowers*. Ozark Beneficial Plant Project. 32 pp. www.amazon.com/exec/obidos/ASIN/B00072EHGE/natilangofthe-20.

Foster, S. 1991. *Echinacea: Nature's Immune Enhancer*. Healing Arts Press, Rochester, NY. 150 pp.

Hobbs, C. 1994. Echinacea: A literature review: Botany, history, chemistry, pharmacology, toxicology, and clinical uses. *HerbalGram* 30: 33–48. www.herbalgram.org/resources/herbalgram/issues/30/table-of-contents/article702/.

Kindscher, K. 1989. Ethnobotany of purple coneflower (*Echinacea angustifolia*, Asteraceae) and other *Echinacea* species. *Economic Botany* 43: 498–507.

Kindscher, K. 2006. *The Conservation Status of Echinacea Species*. www.researchgate.net/profile/Dana-Price-3/publication/226337724_One_Hundred_Years_of_Echinacea_angustifolia_Harvest_in_the_Smoky_Hills_sof_Kansas_USA/links/55fc4af308aec948c4b2f18a/One-Hundred-Years-of-Echinacea-angustifolia-Harvest-in-the-Smoky-Hills-sof-Kansas-USA.pdf#page=162.

Kindscher, K. (ed.). 2016. *Echinacea: Herbal Medicine with a Wild History*. Springer, Cham. 238 pp.

Lemaire, E., M. Trepanier, S. Simard, and M. Comtois. 2014. Evaluation of different crop management and overwintering techniques on winter survival of different *Echinacea* cultivars. *Landscape Ontario*. https://horttrades.com/evaluation-of-different-crop-management-on-winter-survival-of-echinacea-cultivars.

Li, T. 1998. Echinacea: Cultivation and medicinal value. *HortTechnology* 8: 122–129.

McGregor, R.L. 1968. The taxonomy of the genus *Echinacea* (Compositae). *University of Kansas Science Bulletin* 48: 113–142.

McKeown, K.A. 1999. A review of the taxonomy of the genus *Echinacea*. In: *Perspectives on New Crops and New Uses*, edited by J. Janick, 482–489. ASHS Press, Alexandria, VA.

Miller, S.C. and H. Yu. (eds.). 1994. *Echinacea. The Genus Echinacea*. CRC Press, New York. 296 pp.

Muller, K.E. and S. Wagenius. 2016. *Echinacea angustifolia* and its specialist ant-tended aphid: A multi-year study of manipulated and naturally-occurring aphid infestation. *Ecological Entomology* 41: 51–60.

Parmenter, G.A., L.C. Burton, and R.P. Littlejohn. 1996. Chilling requirement of commercial echinacea seed. *New Zealand Journal of Crop Horticultural Science* 24: 109–114.

Parsons, J.L., S.I. Cameron, C.S. Harris, and M.L. Smith. 2018. Echinacea biotechnology: Advances, commercialization and future considerations. *Pharmaceutical Biology* 56: 485–494.

Urbatsch, L.E., K.M. Neubig, and P.B. Cox. 2005. Echinacea. In: *Flora of North America North of Mexico*, Vol. 21, edited by Flora of North America Editorial Committee, 88–92. Oxford University Press, New York.

## Key Cultivation Websites

Kaiser, C., R. Geneve, and M. Ernst. 2016. *Echinacea*. Center for Crop Diversification, University of Kentucky College of Agriculture, Food and Environment, Lexington, KY. www.uky.edu/ccd/sites/www.uky.edu.ccd/files/echinacea.pdf.

King, C. (ed.). 2005. *Commercial Echinacea Production (Agdex 188/20-1)*. Alberta Agriculture, Food, and Rural Development, Edmonton, Canada. 120 pp. https://www1.agric.gov.ab.ca/$Department/deptdocs.nsf/all/agdex10531/$FILE/188_20_1.pdf.

The Spruce. *How to Grow Purple Coneflower*. www.thespruce.com/echinacea-purple-coneflowers-1402840.

The Spruce. *Top 35 Coneflower Varieties and Care Tips*. www.thespruce.com/top-coneflower-varieties-1315807.

WikiHow. *How to Care for an Echinacea Plant*. www.wikihow.com/Care-for-an-Echinacea-Plant.

WikiHow. *How to Divide Coneflowers*. www.wikihow.com/Divide-Coneflowers.

WikiHow. *How to Grow Coneflowers*. www.wikihow.com/Grow-Coneflowers.

WikiHow. *How to Harvest Echinacea*. www.wikihow.com/Harvest-Echinacea.

WikiHow. *How to Prune Coneflowers*. www.wikihow.com/Prune-Coneflowers.

# 39 Forget-Me-Not (*Myosotis* Species and Hybrids)

**Scientific Names:** The genus name *Myosotis* originates from the Greek *mus*, "mouse," and *ous*, "ear," literally "mouse's ear," for the perceived similarity to the leaves.

**Common Names:** The common name *forget-me-not* is recorded in English dating to before the mid-1500s, but its origin is lost in history. Nevertheless, there are many tales accounting for the name. A popular legend relates that a 15th-century German knight picking riverside *Myosotis* blossoms for his beloved fell into the water. While drowning, he tossed the bouquet to her, crying, "Forget me not!" The name forget-me-not is mostly employed in the north temperate world. The name "scorpion grass" is also sometimes applied to many of the *Myosotis* species, but often specifically to *M. scorpiodes*. Chinese forget-me-not is *Cynoglossum amabile*. Chatham Island forget-me-not (giant forget-me-not) is *Myosotidium hortensium*.

## 39.1 INTRODUCTION

Forget-me-nots are dainty plants with sprays of delicate flowers. Reminiscent of lily of the valley, these diminutive, beautiful flowering herbs have acquired widespread popularity and a loyal following.

FIGURE 39.1 Illustration (public domain) by J. S. Eland (1872–1933) showing the story of a medieval knight about to drown while attempting to collect waterside flowers, during which he shouted, "Forget-me-not!"

*Source:* **Public domain. Murray, H., and J. S. Eland. 1890. *Flower Legends for Children*. Longmans, Green & Co., London.**

## 39.2 OVERVIEW

*Myosotis* contains about 80 annual or perennial herbaceous species. They are native to Eurasia, Australasia, and the Americas, mostly in the temperate zones of both hemispheres. Centers of distribution and evolution occur in western Eurasia and New Zealand. Many species have escaped from cultivation and have become naturalized in much of

FIGURE 39.2 Garden displays of forget-me-nots (*Myosotis* species).

*Source:* **(a) Photo by Shutterstock/Lialina. (b) Photo by Leonora (Ellie) Enking (CC BY SA 2.0).**

FIGURE 39.3 "Scorpioid cyme," a type of flower arrangement of *Myosotis sylvatica*. Note that the flowers are borne on a curving stalk reminiscent of a scorpion's tail, the flowers are all on the convex side of the stalk, and the youngest ones are at the distal end of the stalk.

*Source:* **Photo by Thomas Ebner (CC BY 4.0).**

DOI: 10.1201/9781003586227-39

the temperate world. Some occur in alpine places, many in woodlands. The flowers are small, usually 1 cm or less in diameter; blue (predominantly), pink, purple, or white; and sometimes with flowers of different colors (some changing color as they mature). A central "eye" of the flower is often yellow or white. The blooms are often borne in a "scorpioid cyme" (a type of flower cluster in which the axis is curved or coiled, like the tail of a scorpion, and the flowers open successively downward from the apex, always on the convex side). *Myosotis sylvatica*, *M. scorpiodes*, and *M. alpestris* are the most frequently cultivated species. The species grown are rather similar in appearance.

*Myosotis sylvatica* (garden forget-me-not, woodland forget-me-not), a native of Europe, is the most popular species, with numerous cultivars. It is a biennial or short-lived perennial, often grown as an annual or biennial. Its foliage is quite hairy. Established (biennial) plants bloom in the spring. Plants started from seed outdoors may bloom late in the same season, but often the flower display is limited.

*Myosotis scorpiodes* (water forget-me-not, true forget-me-not, scorpion grass), a native of Eurasia (and naturalized in North America), is a perennial, with several cultivars. It is adapted to the damp margins of streams and ponds and also grows well in water as deep as 10 cm. The plants bloom during the summer.

FIGURE 39.4 Flower color variation in *Myosotis sylvatica*.

*Source:* (a) Cultivar Bluesylva. Photo by Agnieszka Kwiecień, Nova (CC BY SA 4.0). (b) Photo by Tim Green (CC BY 2.0). (c) Cultivar Rosylva. Photo by David Monniaux (CC BY SA 3.0). (d) Photo by HQ (CC BY SA 2.0). (e) Cultivar Snowsylva. Photo by David Monniaux (CC BY SA 3.0). (f) Photo by Salicyna (CC BY SA 4.0).

FIGURE 39.5 *Myosotis scorpioides.*

*Source:* (a) Photo by Bernt Fransson (CC BY SA 4.0). (b) Photo by Jörg Hempel (CC BY SA 2.0).

FIGURE 39.6 *Myosotis alpestris.*

*Source:* (a) Photo by Meneerke bloem (CC BY SA 4.0). (b) Photo by US National Park Service (public domain).

*Myosotis alpestris* (alpine forget-me-not) is a clump-forming, short-lived perennial native to northern Eurasia and North America. The plant is very small, usually no more than 15 cm in height, and has blue flowers with yellow centers. Flowering occurs from spring to midsummer. The identity of plants sold commercially under the name *M. alpestris* is sometimes problematic. Moreover, the name "alpine forget-me-not" is ambiguous. It also refers to several other plants in the borage family (Boraginaceae), notably *Eritrichium aretioides*, which, like *M. alpestris*, is also cultivated as a rock garden ornamental. Another nomenclatural issue concerns the name "*Myosotis alpina*," which some people consider to be a different species endemic to the Pyrenees and others rank as a subspecies of *M. alpestris*.

## 39.3 HISTORY

Wild forget-me-nots were grown in "cottage gardens" (informal gardens maintained by common people everywhere) many centuries ago in Europe and are the ancestral material from which modern cultivars have been selected.

## 39.4 ECONOMIC IMPORTANCE

Forget-me-not is a popular but minor commodity in horticultural industries. It is of commercial interest to suppliers of seeds and starter plants for garden supply centers and to florists and the cut flower industry.

FIGURE 39.7 Scientifically accurate paintings (public domain) of cultivated forget-me-not (*Myosotis*) species by professional European botanical illustrators: (a) *Myosotis sylvatica*. (b) *Myosotis scorpiodes*. (c) *Myosotis alpestris*.

*Source:* **(a) Guillaumin, A. 1929. *Les Fleurs de Jardins*, vol. I. Paul Lechevalier, Paris. (b) *Billeder af nordens flora*, vol. 1, 1917–1927, Gad's forlag, København. (c) Step, E., D. Bois, B. Herincq, and W. Watson. 1896. *Favourite Flowers of Garden and Greenhouse*, vol. 3. Frederick Warne, New York.**

FIGURE 39.8 Large-scale production of forget-me-not potted plants in a Dutch greenhouse.

*Source:* **Photo by Shutterstock/Martin Bergsma.**

## 39.5 ORNAMENTAL USES

Forget-me-nots are grown in woodland gardens, in rock gardens, in open areas, as groundcover, and in containers. There is also some use as a cut flower. As they are very small plants (most cultivars are 15 to 30 cm in height), the plants are often positioned at the front in borders and flower gardens. They are often intermixed with taller bulb plants, such as tulips and daffodils. *Myosotis sylvatica* is sometimes grown as a houseplant.

## 39.6 OTHER USES

True to its name, forget-me-not flowers have become widespread symbols of commemoration, love, and fidelity, especially of valued people who have passed away, often tragically. In old paintings, the flowers are often included as symbols of remembrance. At funerals, seeds are sometimes given out. Forget-me-nots are especially symbolic of the widespread death resulting from the two world wars and the Armenian genocide. The flower has been employed as an emblem of Masonic organizations and has been adopted as the symbol of Alzheimer and dementia advocate societies, and the symbol of International Missing Children's Day (May 24).

FIGURE 39.9 Landscape uses of forget-me-nots.

*Source:* **(a) Photo by Shutterstock/Fortish. (b) Photo by Shutterstock/Min C. Chiu. (c) Photo by Mim (CC BY 2.0). (d) Photo by Mark (CC BY 2.0).**

FIGURE 39.10 Forget-me-nots inside the home: (a) Potted plant. (b) Cut flowers in a vase.

*Source:* **(a) Photo by Shutterstock/Vilax. (b) Photo by Shutterstock/LiudmylaChe.**

FIGURE 39.11 Commemorative use of forget-me-nots in the UK: (a) In a graveyard. (b) Paper forget-me-not flowers at a charity event.

*Source:* **(a) Photo by Nottingham General Cemetery (CC BY 2.0). (b) Photo by Peter Lowe (CC BY ND 2.0).**

FIGURE 39.12 Objets d'art with forget-me-not themes: (a) A German porcelain vase dating to 1740, hand-painted mostly with forget-me-not flowers. (b) Forget-me-not earrings. (c) A forget-me-not brooch fashioned from metal harvested from a Russian bomb launched against the Ukraine in 2022. (d) Beads in the shape of forget-me-not flowers.

*Source:* **(a) Courtesy of the Honolulu Museum of Art (public domain). (b) Photo by Olga Dudenko (CC BY SA 2.0). (c) Photo by Iryna Udovychenko (CC BY SA 4.0). (d) Photo by Preciosa Ornella (CC BY 2.0).**

Forget-me-nots are tiny and beautiful, considerations that make them ideal decorative images for jewelry and craft items.

## 39.7 PROBLEMS

*Myosostis* species have been naturalized in many areas throughout the world and are significantly invasive in some regions. In some places, the sale of the plant for cultivation is regulated. Generally, however, the plants are not ranked as a significant threat to native ecosystems.

The flowers of *M. sylvatica* are sometimes consumed. Extracts of some species of *Myosotis* are employed in herbal medicine, suggesting that they may contain potent chemicals and might be somewhat toxic. The chemical pyrrolizidine is present in concerning levels in some of the species. Some *Myosotis* species have been reported to be mildly toxic to grazing animals. The Chinese forget-me-not (*Cynoglossum amabile*), which could be mistaken for *Myosotis* species, is mildly poisonous.

## 39.8 CULTIVATION

### 39.8.1 Climate and Site

Forget-me-nots are suited to cool weather. *Myosotis sylvatica* is adapted to climate zones 4 to 8 (marginally in zones 3 and 9). *Myosotis scorpioides* can grow in zones 5 to 9 (marginally in zone 4). *Myosotis alpestris* has been rated for zones 3 to 8. *Myosotis* cultivars usually grow well in full sun (provided moist soil is available) but thrive in part shade.

### 39.8.2 Propagation and/or Purchasing

Starter plants are often available from garden centers. Once planted, forget-me-nots self-sow their seeds and tend to spread in the area. *Myosotis scorpiodes* spreads by creeping rhizomes, and it can be divided and replanted (preferably in the spring).

FIGURE 39.13 Introduced *Myosotis sylvatica* growing as a naturalized invasive weed in a beech forest in Wisconsin.

*Source:* **Photo by Joshua Mayer (CC BY SA 2.0).**

### 39.8.3 Planting

Seeds of forget-me-nots are often planted directly in the ground – several weeks in advance of freeze-up in cold climates or in the spring in warm regions. *Myosotis scorpiodes* can be planted in the soils of muddy banks near the waterline. Forget-me-nots are frequently grown as annuals by starting seeds indoors 8 to 10 weeks in advance of the spring planting season. A spacing of 12 to 20 cm is appropriate for most cultivars.

### 39.8.4 Soil and Moisture Conditions

Forget-me-nots grow best in organically enriched, moist soil with good drainage (except for *M. scorpioides*, which is adapted to waterlogged soil). Soil pH should be 5.5 to 7.5. Addition of fertilizer once or twice a season (once in the spring and again in the autumn) is recommended by some (others state little or no fertilization is needed).

### 39.8.5 Maintaining Plants

To prevent spread by seeds, the plants can be cut back after the flowers senesce (unfortunately, a laborsome task). *Myosotis scorpioides* (which is listed as a prohibited weed in some jurisdictions) can be placed in containers (set in the water or mud) to control spread by the rhizomes.

## 39.9 FASCINATING FACTS, FABLES, FOLKLORE, AND FRIVOLITY

- For more than two centuries, talented artists illustrated "flower fairy" books, seemingly written for children but of considerable interest to their parents as well. These works showed female figures elaborately dressed as familiar garden flowers, such as forget-me-nots. Printer's plates and original prints have become collector items.
- In England, various superstitions were associated with forget-me-nots. Wearing the flower was thought to be protective against witches. The juice of the plant was believed to make steel blades so sharp that they could cut through stone.
- Henry Bolingbroke (Duke of Lancaster) clashed with his cousin King Richard II of England (who reigned from 1377 to 1399) and was exiled in 1397. During his banishment in France, he chose the forget-me-not and its French name, "souvenez-vous de moi," as his emblem and motto. The flower and slogan were woven into his knight's collar, and his followers wore forget-me-nots as signs of support. Henry returned to power and reigned in England from 1399 until his death in 1413.
- Madame Pompadour (Jeanne Antoinette Poisson, 1721–1764), the mistress of King Louis XV of France, wore blue forget-me-not flowers in her hair and on her gowns as a symbol of faithfulness to the king. Indeed, for centuries in Europe, the flower was often worn by ladies as a sign of enduring love.

FIGURE 39.15 Portrait of Madame Pompadour, mistress of King Louis XV of France, with a twig of forget-me-not, which she regarded as a symbol of her loyalty to the king. Painting by Charles-André van Loo (1705–1765).

*Source:* **Photo by Château de Versailles.**

FIGURE 39.14 "Flower fairy" figures based on forget-me-nots.

*Source:* **(a) Grandville, J. J., in Delord, T., and J. J. Grandville. 1847. *Les fleurs animées* (the flowers personified). Gabriel de Gonet, Paris. (b) Gordon, E., and M. T. Rose. 1910. *The Little Cousins of the Field and Garden*. Volland, Chicago.**

- The favorite flower of Diana, Princess of Wales, was forget-me-not. After her passing, the flowers were planted in a memorial garden dedicated to her at her former residence, Kensington Palace, in London. When Prince Harry, Duke of Sussex, married Meghan, Dutchess of Sussex, her wedding bouquet included forget-me-nots from the memorial garden. Prince Harry named his charity benefitting African young people who are HIV-positive "Sentebale," Sesotho language for "forget me not."
- *Myosotis alpestris* became the territorial flower of Alaska in 1917, and the state flower in 1959, when Alaska became the 49th state of the Union. It is also the floral emblem of the old Swedish province of Dalsland.
- Grandparents Day is a holiday recognized in various countries, occurring on various days of the year depending on nation. In the United States, the forget-me-not was adopted in 1978 as the official flower of National Grandparents Day.

FIGURE 39.16 Postcards produced from 1900 to 1910 showing scenes based on forget-me-nots.

*Source:* **Public domain.**

- In the early 20th century, postcards (often printed on linen) became extremely popular and were widely exchanged, especially during holidays. Talented artists created remarkable scenes for these, and many have become collector's items. Many postcards of the time were highly sentimental, and forget-me-nots were frequently shown.

## KEY PUBLICATIONS

Elkington, T.T. 1964. *Myosotis alpestris* F.W. Schmidt. *Journal of Ecology* 52: 709–722.

Grau, J. and H. Merxmuller. 1972. *Myosotis* (Boraginaceae). In: *Flora Europea*, Vol. 3, edited by T.G. Tutin, V.H. Heywood, N.A. Burges, D.M. Moore, D.A. Valentine, S.M. Walters, et al., 111–117. Cambridge University Press, Cambridge.

Karlsson, M.G. and J.W. Werner. 2003. Cold requirements for flowering of forget-me-not. *Acta Horticulturae* 624: 223–227.

Resch, J.F., D.F. Rosberger, J. Meinwald, and J.W. Appling. 1982. Biologically active pyrrolizidine alkaloids from the true forget-me-not, *Myosotis scorpioides*. *Journal of Natural Products* 45: 358–362.

Russell, C. 2022. *Forget Me Not Day 2023: When, History, Significance & Activities*. Next National Holiday (Blog). https://nextnationalday.com/forget-me-not-day/.

Sherafati, M., S. Kazempour Osaloo, M. Khoshsokhan Mozaffar, S. Esmaeilbagi, and N. Saadati. 2014. Molecular phylogeny of the genus *Myosotis* (Boraginaceae) based on nrDNA ITS sequences. *Taxonomy and Biosystematics* 6: 85–96.

Varopoulos, A. 1979. Breeding systems in *Myosotis scorpioides* L. (Boraginaceae). *Heredity* 42: 149–157.

Winkworth, R.C., J. Grau, A.W. Robertson, and P.J. Lockhart. 2002. The origins and evolution of the genus *Myosotis* L. (Boraginaceae). *Molecular Phylogenetics and Evolution* 24: 180–193.

## Key Cultivation Websites

Global Flowers. *Myosotis – Care and Cultivation, Watering, Reproduction*. https://global.flowers/en/plants/myosotis/.

Horticulture Magazine. *How to Plant Myosotis 'Forget-Me-Nots'*. https://horticulture.co.uk/forget-me-nots/planting/.

The Spruce. *Forget-Me-Not: Plant Care, Growing Tips, and Symbolism*. www.thespruce.com/forget-me-not-plant-profile-4799948.

WikiHow. *How to Grow Forget-Me-Nots Indoors*. www.wikihow.com/Grow-Forget-Me-Nots-Indoors.

WikiHow. *How to Plant Forget-Me-Not Seeds*. www.wikihow.com/Plant-Forget-Me-Not-Seeds#:~:text=Forget%2Dme%2Dnots%20tend%20to,than%206%20hours%20of%20sunlight.

# 40 Foxglove (*Digitalis* Species and Hybrids)

**Scientific Names:** The genus name *Digitalis* traces to the Latin *digitus*, "finger." The German botanist Leonhart Fuchs translated the German name of the plant, *fingerhut*, meaning "finger hat" and "thimble," to the Latin *digitalis*, to indicate the thimble-shaped flowers.

**Common Names:** The name "digitalis" is based on the genus name. The name *foxglove* is speculated to have arisen in England, where foxes were observed to frequently make their dens on woody hillsides covered with the plants. Chinese foxglove denotes species of *Rehmannia*, especially *R. elata*. Madeiran or bush foxglove is *Isoplexis sceptrum*. Foxglove beardtongue and foxglove penstemon denote certain species of *Penstemon*. False foxglove refers to *Agalinis* and *Aureolaria*. Foxglove tree is a name for *Paulownia tomentosa*.

FIGURE 40.1 Foxglove (*Digitalis purpurea*) flowers.

*Source:* (a) A Polkadot Series cultivar by Swallowtail Garden Seeds (CC BY 2.0). (b) Photo by Beverly Buckley from Pixabay (public domain). (c) Photo by Keith Double (public domain). (d) Photo by Shutterstock/DW0318. (e) Photo by Shutterstock/Olga Lipatova.

## 40.1 INTRODUCTION

Foxgloves are tall, stately plants with elegant spires of colorful bell-shaped blossoms that have made the plants favorite garden ornamentals. Indeed, foxglove is one of the most beloved of all garden flowers, despite being quite poisonous.

## 40.2 OVERVIEW

The genus *Digitalis* has about 20 species, native to Europe, western Asia, and northwestern Africa. The plants are biennials (which sometimes become short-lived perennials) or perennials (or rarely shrubs), with tall spikes of distinctive bell-shaped flowers. A few hybrids produce flowers in just 1 year but need to be started indoors months in advance of transplanting. Depending on species and cultivar, the flowers are cream, lavender, purple, pink, red, white, or yellow, frequently spotted, and sometimes bicolored. *Digitalis purpurea* (common foxglove, purple foxglove for the most common flower color, or just foxglove), a biennial herb, is the most frequently cultivated ornamental, but other species grown in gardens include *D. ferruginea* (rusty foxglove), *D. grandiflora* (yellow foxglove), *D. lutea* (straw, or small yellow foxglove), and *D. parviflora* (chocolate, or small-flowered foxglove). The biennials develop a rosette of large leaves in the first year. In the second year, the plants produce an upright stem with smaller leaves, originating from the center of the basal clump. The flowering stem typically grows 50 to 150 cm in height, sometimes up to 2 m, with blooms in most varieties developing progressively up one side of the stem. The bell-shaped flowers face downward.

FIGURE 40.2 Botanical illustrations (public domain) of *Digitalis purpurea*.

*Source:* (a) Sturm, J. G. 1796. Deutschlands Flora in Abbildungen, Stuttgart, Germany. (b) Köhler, F. E. 1887. Medizinal-Pflantzen. F. E. Köhler, Gera-Untermhaus, Germany. (c) Cultivar Beauty of Dorking. *The Floral Magazine*, vol. 6 (1867).

DOI: 10.1201/9781003586227-40

Blooming occurs from late spring through midsummer. The fruits are dry capsules with many small seeds.

## 40.3 HISTORY

Foxgloves have been cultivated at least since the 1400s in England. Breeding in many countries has produced hundreds of cultivars, varying particularly in floral displays. In wild plants, the flowers appear only on one side, but in some cultivated forms, flowers completely surround the flowering stem. There have also been attempts to increase the rather limited flowering season of foxgloves. Some cultivars have variegated leaves, although the foliage is not of particular importance to gardeners.

## 40.4 ECONOMIC IMPORTANCE

Foxglove species are of significant value as sources of medicinal digoxin and related compounds, although in recent years alternative pharmaceuticals are replacing these medicinals to some degree. As ornamentals, foxgloves have long been garden favorites, but even with the production of new cultivars of *Digitalis*, foxglove is just one of a very large number of offerings of the horticultural trade.

## 40.5 ORNAMENTAL USES

Foxgloves are employed primarily in flower borders and beds. Tall varieties are excellent in the background of shorter plantings. Foxgloves are particularly attractive against buildings and fences. Mass plantings are also especially effective when in flower.

The flower spikes can be used for cut flowers, but it should be kept in mind that handling them requires caution, and pets and children are potential poison victims – even simply from drinking water from the vases in which foxgloves may be kept.

FIGURE 40.3 Garden displays of foxglove: (a) Border beside a fence. (b) Massed planting. (c) Another massed planting. The yellow cultivar is Lemoncello.

*Source:* **(a) Photo by Tsc Traveler (CC BY 2.0). (b) Photo by Shutterstock/Leungchopan. (c) Photo by Shutterstock/Andrew Fletcher.**

The flowers are visited especially by bees – primarily bumblebees. The projecting lower lip of the flower is a landing platform facilitating access by the bees to the nectar at the base of the tubular flower. Occasionally, butterflies and hummingbirds also visit the flowers. Plants left to develop seeds will attract a variety of wild birds (presumably immune to the toxins, although poultry are susceptible). Accordingly, foxgloves are useful in pollinator gardens.

## 40.6 OTHER USES

The word *digitalis* is employed not just for the plants but for extracted drugs that contain cardiac glycosides, especially

FIGURE 40.4 Pollinators visiting foxglove flowers: (a) Hummingbird in Costa Rica. (b) American painted lady butterfly (*Vanessa virginiensis*). (c) Bumblebee at the entrance of a flower. (d) Buff-tailed bumblebee (*Bombus terrestris*) inside the throat of a flower.

*Source:* **(a) Photo by Shutterstock/Crbellette. (b) Photo by Shutterstock/John A. Anderson. (c) Photo by Shutterstock/Mr.Nikon. (d) Photo by Netopyr (public domain).**

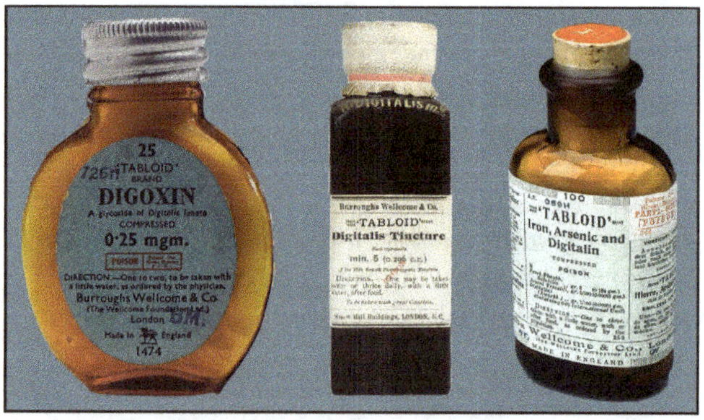


FIGURE 40.5 Digitalis medicinal preparations manufactured in London, United Kingdom, and available in the late 19th and early 20th centuries.

*Source:* **Wellcome Images (CC BY 4.0).**

digoxin, employed in medicine. Digoxin, the most important extract, is obtained from *D. lanata* (Grecian foxglove) and, to a limited extent, from *D. purpurea*. It is used to control heart rate, particularly irregular atrial fibrillation. Different preparations are also employed for the same purposes in modern medicine. As noted next, foxglove not only has the power to cure but also can kill.

## 40.7 PROBLEMS

Foxglove is very toxic, and consumption can lead to the death of humans and other animals. All parts of the plants are toxic, not only when consumed, but also simply from skin contact. However, usually, contact can just irritate sensitive skin. Nevertheless, gardeners are advised to wear gloves and protective clothing when handling the plants. There are examples of people confusing foxglove for comfrey (*Symphytum*), which looks very similar before the plants flower and is occasionally employed to brew tea. Death has resulted from mistaken identification. More rarely, foxglove is confused with a relative of comfrey, borage (*Borago officinalis*), employed as food. All mammals and poultry are susceptible. Goats, among the most resistant of animals to toxic plants, usually find the taste so objectionable that they rarely persist in eating foxglove. The same is true for deer and rabbits, which rarely feed on the plants.

FIGURE 40.6 Similarity of the pre-flowering rosette of foxglove (*Digitaria*) and the pre-flowering rosette of comfrey (*Symphytum*). People have died from mistakenly employing foxglove to make herbal tea instead of comfrey: (a) Toxic *Digitalis purpurea*. (b) Comfrey, commonly employed for herbal tea.

*Source:* **(a) Photo by Jean-Marc Pascolo (CC BY SA 3.0). (b) Photo by Nancy Buron from Pixabay (public domain).**

FIGURE 40.7 Livestock contemplating eating poisonous foxglove in the UK: (a) Sheep. (b) Goat.

*Source:* **(a) Photo by Ashley Coates (CC BY SA 2.0). (b) Photo by Shutterstock/Sarah2.**

FIGURE 40.8 Foxglove (*Digitalis purpurea*) growing as an aggressive weed in the United Kingdom.

*Source:* **Photo by Mark AC (CC BY SA 2.0).**

Children and pets should be kept away from foxgloves. While this may deter some from growing the plant, it should be kept in mind that many ornamental household and garden plants are toxic, and the same recommendation applies.

*Digitalis purpurea* is widely naturalized outside its native area, often growing as a weed along roads, and in some regions, it is considered to be an invasive plant. A single plant can produce thousands of seeds, so the species has considerable ability to spread. If the plants aren't deadheaded after they flower, they may produce so many seedlings they will dominate the area. The seeds may remain dormant for years, so eradicating the plant might require years of effort.

Most foxglove cultivars are biennials, and most gardeners do not like biennials because they require 2 years to produce flowers. As noted earlier, some cultivars are annual but require starting indoors, and established plants which will flower in short order can simply be purchased and transplanted.

## 40.8 CULTIVATION

### 40.8.1 Climate and Site

*Digitalis purpurea* is rated as hardy in climate zones 4 to 9 or 10. In nature, it sometimes occurs in deep shade, but it grows best in full sun and light shade.

### 40.8.2 Propagation and/or Purchasing

Foxgloves generally are planted as seeds (often sold in packets in a mixture of colors) but are available from plant nurseries and garden centers as potted plants intended for transplanting. Because they are biennial, they will not flower from seed until the second year. However, once established, plantings typically self-seed, and in a bed, flowering plants develop annually.

### 40.8.3 Planting

Depending on cultivar, purchased young plants should be placed 30 to 60 cm apart. Foxglove seeds are tiny and not easy to control. They are best started indoors in flats or small pots. Alternatively, seeds can simply be scattered on the ground and lightly pressed down (some of the seeds may be dormant and may not germinate until future years). Some authorities recommend covering very shallowly with fine soil; others state that covering with soil is not recommended, as light is thought to aid in germination. Seedlings need to be thinned to 30 to 60 cm apart.

### 40.8.4 Soil and Moisture Conditions

Foxgloves prefer light, moist, well-drained soils high in organic matter but are quite tolerant of soil texture (including sand and clay). In nature, foxgloves can grow well in acidic soils, but slightly acidic to neutral soils are usually recommended.

### 40.8.5 Maintaining Plants

Tall varieties may require staking to keep them upright. Plants that complete their life cycle (usually in the second year) need to be cleaned away (indeed, this can be done as soon as the flowering display fades significantly). Some gardeners cut off the flowering stalks when the flowers fade in the hope that a second round of flowering will occur. Seedlings growing up from past plantings need to be thinned.

## 40.9 FASCINATING FACTS, FABLES, FOLKLORE, AND FRIVOLITY

- Many old names for foxglove are based on the idea that it was a favorite flower of fairies: *fairy bells*, *fairy fingers*, *fairy gloves*, *fairy's cap*, *fairy herb*, *fairy's petticoat*, *fairy's thimbles*, and *goblin gloves*. Fairies are said to have saved foxes by showing them how to ring the foxglove bells to warn of approaching hunters.

FIGURE 40.9 Old illustrations (public domain) showing the association of foxgloves with fairies.

***Source:*** **(a) The vintage book *How Nice*. Photo by In Pastel (CC BY 2.0). (b) Cover from Kay, G. A. 1918. *The Fairy Who Believed in Human Beings*. Moffat Yard and Company.**

- Other old names for foxglove indicate its poisonous nature: *bloody bells*, *bloody fingers*, *dead man's thimbles*, *dead men's bells*, *witches' bells*, *witches' gloves*.
- Consumption of foxglove can induce xanthopsia, a condition that alters color perception, making objects appear yellowish or green and producing blue halos around light sources. The possibility has been extensively debated in medical circles that Vincent van Gogh's "yellow period" of 1886–1890 (his paintings such as *Starry Night* and *The Night Café* of that time were yellowish and showed conspicuous halos) was influenced by xanthopsia (although malnutrition and absinthe may have been causes). Coincidentally, van Gogh was prescribed digitalis by his physician, Dr. Paul-Ferdinand Gachet, while he was in an asylum in 1889, perhaps to alleviate possible epilepsy (van Gogh suffered from seizures), which was treated with foxglove at that time. Van Gogh painted two portraits of Dr. Gachet in 1890 in which the doctor is holding a stem of foxglove.
- Agatha Christie (1890–1976), the world's most famous mystery writer, was a volunteer nurse in a military dispensary during World War I, giving her a knowledge of medicine. This likely helped her write her 1938 novel in which foxglove extract was employed as poison. Over her career, Christie employed foxglove poisoning six times in her stories.
- A 3.7 m tall foxglove plant grown in Victoria, British Columbia, in 2021 established a record for height (www.darlingtonandstocktontimes.co.uk/history/lookingback/19542222.a-flower-steroids-world-record-foxglove-tops-11-feet-tall/).

## KEY PUBLICATIONS

Ardelean, M., A.M. Costea, and M. Cordea. 2006. Breeding foxglove (*Digitalis* sp.) for ornamental and/or medical purposes. *Bulletin of University of Agricultural Sciences and Veterinary Medicine, Cluj-Napoca. Horticulture* 63: 22–31.

Aronson, J.K. 1986. *An Account of the Foxglove and Its Medical Uses, 1785–1985*. Oxford University Press, UK. 420 pp.

Burchell, H.B. 1983. Digitalis poisoning: Historical and forensic aspects. *Journal of the American College of Cardiology* 1: 506–516.

Kreis, W. 2017. The foxgloves (*Digitalis*) revisited. *Planta Medica* 83: 962–976.

Nebauer, S.G., L. Del Castillo-Agudo, and J. Segura. 2000. An assessment of genetic relationships within the genus *Digitalis* based on PCR-generated RAPD markers. *Theoretical and Applied Genetics* 100: 1209–1216.

Nicole, G. 2022. *Growing Foxglove: The Complete Guide for Beginners on Growing, Caring, Pruning, Propagation, Maintenance and More*. Author-Published. 39 pp.

Rady, M.R. 2019. Plant biotechnology and foxglove. In: *Plant Biotechnology and Medicinal Plants*, edited by M.R. Rady, 149–197. Springer, Cham.

Smith, G. 2023. *Cultivating Foxglove Plant*. Amazon Digital Services. 90 pp.

Verma, S.K., A.K. Das, G.S. Cingoz, and E. Gurel. 2016. In vitro culture of *Digitalis* L. (foxglove) and the production of cardenolides: An up-to-date review. *Industrial Crops and Products* 94: 20–51.

Verma, S.K., S. Gantait, B.R. Jeong, and S.J.Hwang. 2018. Enhanced growth and cardenolides production in *Digitalis purpurea* under the influence of different LED exposures in the plant factory. *Scientific Reports* 8(1): 18009. https://doi.org/10.1038/s41598-018-36113-9.

## Key Cultivation Websites

BBC Gardeners' World Magazine. *How to Grow Foxgloves.* www.gardenersworld.com/how-to/grow-plants/how-to-grow-foxgloves/.

The Botanic Nursery, Atworth. *Foxglove Cultivation and Propagation.* https://static1.squarespace.com/static/5faeaa54a363746603cf39d9/t/6061b47b0598c25f7a9b2240/1617015932403/FOXGLOVE+CULTIVATION+AND+PROPAGATION.pdf.

Flower Magazine. *The Basics of Growing Foxglove Plants.* https://flowermag.com/growing-foxglove-plant-and-foxglove-flowers/.

Gardening. 2023, November 24. *Growing Foxglove Plant Guide: Growing Zones, Care Tips, Foxglove Flowers.* https://8billiontrees.com/gardening/foxglove-plant/.

Martha Stewart. *Everything You Need to Know about Foxgloves.* www.marthastewart.com/8038347/how-to-grow-foxgloves.

The Spruce. *How to Grow and Care for Common Foxglove.* www.thespruce.com/tall-toxic-foxglove-plants-2132588.

WikiHow. *How to Grow Foxglove.* www.wikihow.com/Grow-Foxglove.

Wisconsin Horticulture, Division of Extension. *Common Foxglove, Digitalis purpurea.* https://hort.extension.wisc.edu/articles/common-foxglove-digitalis-purpurea/.

# 41 Freesia (*Freesia* Species and Hybrids)

**Scientific Names:** The name *Freesia* honors German botanist and physician Friedrich Heinrich Theodor Freese (1795–1876), who was a teacher and friend of Christian Friedrich Ecklon, the Danish botanist who first named the genus.

**Common Names:** The common name "freesia," like so many other important ornamental plants, simply repeats the genus name. The phrase "florists' freesia" refers to the hybrid species commonly employed as cultivars and grown commercially to supply the cut flower trade. "Blue freesia" refers to *Babiana stricta*, better known as *baboon flower*. Like freesias, it also is native to South Africa, is a member of the Iris family (Iridaceae), and is grown as an ornamental. The genus name *Babiana* is derived from the Dutch word *baviaan*, referring to the Chacma baboon, *Papio ursinus*, which consumes the corms of plants in the genus.

## 41.1 INTRODUCTION

Freesias have gorgeous flowers, often with heavenly scents. They are primarily encountered as cut flowers obtained from florists and are one of the world's leading ornamentals in this category.

## 41.2 OVERVIEW

The genus *Freesia* is made up of about 15 species of perennial herbs native to Southern Africa. The plants are perennial, most surviving the dry winter season as underground swollen stems called corms. The interior of corms is solid tissue, in contrast to true bulbs, which are mostly made up of layered fleshy, scale-like leaves, such as an onion. On the outside, corms look much like bulbs, and in the horticultural trade, freesia corms are often called bulbs. The corms of freesia are conical, 1–2.5 cm in diameter. Some freesia species (especially *F. laxa*) are cultivated as ornamental plants, but most cultivars are hybrids of *F. refracta* and *F. leichtlinii*. The plants are typically 30 to 60 cm in height. The leaves are long and reminiscent of grass blades (one of the names of *F. laxa* is "flowering grass"). Ornamental freesias usually have fragrant, funnel-shaped flowers, but some (such as *F. laxa*) have flattish flowers. Flower colors include blue, mauve, pink, orange, purple, red, white, and yellow. Some cultivars have bicolored flowers; other cultivars have doubled flowers (with extra petals). Five to ten flowers occur on the leafless, arching, wiry stems on one side, opening sequentially back from the tip of the stem. The stems turn at right angles just below the lowest flower, and the flowers face upward, a rather unique architecture.

FIGURE 41.1 Flowering displays of freesia (*Freesia* hybrids): (a) Cut flowers on sale. (b) In a garden. (c) In a garden.

*Source:* **(a) Photo by Shutterstock/Neirfy. (b) Photo by Manuel (CC BY 2.0). (c) Photo by Joanbanjo (CC BY SA 3.0).**

## 41.3 HISTORY

Freesia was first grown in Europe in The Netherlands in the mid-18th century. Today, The Netherlands remains the most important center of freesia breeding, followed by Japan. The popularity of freesias has developed particularly since the 1970s. Hundreds of cultivars have been created. Newer cultivars include double-flowered forms. Although greatly valued for their attractive scent, many recent cultivars, bred primarily for appearance, have greatly reduced fragrance.

## 41.4 ECONOMIC IMPORTANCE

Freesia is one of the most important commercial cut flowers, and indeed, this is its principal horticultural value.

FIGURE 41.2 Freesia (*Freesia* hybrids) flower clusters, illustrating the characteristic upward arrangement of the flowers on the floral stalk.

*Source:* **(a) Photo by Kevin Thiele (CC BY 2.0). (b) Photo by Shutterstock/Neirfy.**



DOI: 10.1201/9781003586227-41

FIGURE 41.3 Old botanical illustrations of *Freesia* (public domain): (a) Garden hybrids. (b) *Freesia laxa*.

*Source:* **(a)** ***Revue horticole*****. 1907.** ***Librairie agricole de la maison rustique*****. Paris. (b) Paxton, J. 1834.** ***Magazine of Botany and Register of Flowering Plants*****, vol. 1.**

FIGURE 41.4 Commercial greenhouse production of freesia.

*Source:* **Photo by Shutterstock/Lidian Neeleman.**

Growers often produce a crop of corms outdoors, then manipulate temperature to shorten the natural dormancy period so that plants raised in greenhouses from the corms will produce flowers to meet seasonal demand. In the Northern Hemisphere, outdoor production of corms is possible in very mild areas, such as the Channel Islands, islands of southern Japan, and parts of California and The Netherlands.

## 41.5 ORNAMENTAL USES

Freesias are primarily cultivated as commercial cut flowers and are widely employed in mixed flower arrangements and wedding bouquets. They can be grown as outdoor annuals in temperate climates, in beds, borders, and containers, although they are much more employed for these purposes (especially as perennials) in subtropical and tropical areas.

FIGURE 41.5 Cut flower uses of freesia: (a) Bouquet of mixed flower colors in a vase. (b) Flowers in a basket. (c) Flowers in a vase. (d) Wedding bouquet.

*Source:* **(a) Photo by Shutterstock/Neirfy. (b) Photo by Shutterstock/Alla vector. (c) Photo by M. Z. Wojalski (CC BY SA 4.0). (d) Photo by Shutterstock/AlexanderLipko.**

FIGURE 41.6 Outdoor plantings of freesia hybrids: (a) Growing in a bed. (b) Growing in a border.

*Source:* **(a) Photo by Shutterstock/Balazovic Lubos. (b) Photo by Bgwashburn (CC BY 2.0).**

Freesia corms can be forced to bloom indoors at any time of the year, and the resulting plants treated as short-term houseplants.

## 41.6 OTHER USES

Freesia extracts are often used to provide scent for cosmetics, such as hand creams and shampoos, and to products such as fragrancy candles. Indeed, the flowers can simply be employed as an air freshener.

## 41.7 PROBLEMS

Freesias have exacting requirements with respect to temperature in order to flower. As described later, both the corms and seeds need cool temperatures at both the early and flowering stages, which are often difficult to provide. Because they are so demanding ("fiddly," as some would say), they are mostly employed as cut flowers obtained from florists. The sap from freesia can irritate the skin of sensitive individuals.

**FIGURE 41.7** Freesia and perfume: (a) Illustration featuring freesia flowers prepared for perfume advertisements. (b) Vintage perfume illustration.

*Source:* **(a) Photo by Shutterstock/HstrongART. (b) Photo by Amy from Pixabay (public domain).**

**FIGURE 41.8** Freesia corms being started in a cell tray.

*Source:* **Photo by Shutterstock/Dean Clarke.**

## 41.8 CULTIVATION

### 41.8.1 Climate and Site

Freesias are quite frost-sensitive. They can be grown as perennials in climate zones 9 and 10. In temperate climates, they are often placed in sheltered locations but need substantial sun during much of the day. Some gardeners grow them in pots, sunken in the soil so that they can be brought indoors when cold weather approaches. Freesias are also sensitive to hot temperatures and may not flower if temperatures are significantly above 21°C.

### 41.8.2 Propagation and/or Purchasing

Freesias are sometimes grown as annuals established from seed, but they can take 8 months to mature, and some plants may not flower for several years. Most popular cultivars are reproduced commercially by cormlets (small corms produced by the plants; these are often called offsets), and most gardeners will plant these rather than seeds. Corms (or cormlets) are available from some nurseries and online sources, but it is necessary to obtain good information on when these can be planted, since they may or may not have been given a dormancy-requiring treatment. Typically, properly treated corms will produce blooming plants in 3 to 4 months when revived, so if earlier flowering is desired, the corms should be started indoors well ahead of transplanting. Cool temperatures (no more than 15°C) are required to keep indoor corms from producing spindly plants. Corms won't survive cold winters, but their offspring (cormlets) can be salvaged in the autumn and stored for the next season. Excavated corms should be kept warm for several months to ripen them (as would happen naturally in South African summers), then stored in a paper bag in a cool, frost-free area to plant the following year. However, most freesia plants are discarded after flowering because the corms are relatively cheap.

### 41.8.3 Planting

When grown from the conical corms, the pointy end is planted upward, with 2 to 3 cm of soil above the top of the corms. In beds, planting distance is 7.5 to 10 cm apart. In pots, they may be planted 5 cm apart. For visual impact, at least six plants should be established in containers, and whether these should be one or different colors is a matter of personal choice.

### 41.8.4 Soil and Moisture Conditions

Recommended soils should incorporate compost, be moisture-retaining but well-drained, and have a pH of 6.3 to 7.5. Freesias are sensitive to fluoride, a concern mainly for professional growers.

### 41.8.5 Maintaining Plants

Freesia plants (especially the slim flower stalks) tend to flop over (note Figure 41.1c), and stakes are recommended (even for container plants). For outdoor beds, netting can be used.

## 41.9 FASCINATING FACTS, FABLES, FOLKLORE, AND FRIVOLITY

- Freesia produces one of the most attractive of all floral scents, and perfumers have tried in vain to discover natural constituents that could be combined to produce the fragrance. Freesia contains linalool, which is a known sensitizer that some people need to avoid, and this complicates extracting scent constituents from the flowers. In practice, in perfumery, freesia scent is produced synthetically, employing various chemicals. Freesia scent is present in such popular perfumes as Calvin Klein Eternity, Elizabeth Arden Red Door, and Estée Lauder Pleasures.
- Freesia is one of the strongest of scents, but some people cannot smell the fragrance. It has been found that a genetically recessive gene responsible for this inability occurs in about 10% of Eastern

FIGURE 41.9 The Freesia Festival on Hachijo-jima Island in Japan. Mt. Hachijo-Fuji is in the background.

*Source:* **Photo by Geomr~commonswiki (CC BY SA 3.0).**

Europeans and people of British Celtic descent, but in only about 4% of people predominantly of Scandinavian, English, Dutch, and German descent.

- The scent of freesia has been reported to be effective in encouraging gambling patrons in Las Vegas to continue gambling.
- Hachijo-jima Island in Japan celebrates an annual freesia festival from late March to early April. Visitors walk among huge fields of blooming freesias.

## KEY PUBLICATIONS

Brown, N.E. 1935. *Freesia* Klatt and its history. *Journal of South African Botany* 1: 1–31.

Gilbertson-Ferriss, T.L. 1985. *Freesia × hybrida*. In: *Handbook of Flowering*, Vol. 3, edited by A.H. Halevy, 34–37. CRC Press, Boca Raton, FL.

Goemans, R.A. 1980. The history of the modern freesia. In: *Petaloid Monocotyledons: Horticultural and Botanical Research*, edited by C.D. Brickell, D.F. Cutler, and M. Gregory, 161–170. Academic Press, London.

Goldblatt, P. 1982. Systematics of *Freesia* Klatt (Iridaceae). *Journal of South African Botany* 48: 39–91.

Goldblatt, P. and J.C. Manning. 1995. Phylogeny of the African genera *Anomatheca* and *Freesia* (Iridaceae – Ixioideae), and a new genus *Xenoscapa*. *Systematic Botany* 20: 161–178.

Hancock, G.D. 2009. *The Efficacy of Fragrance Use for Enhancing the Slot Machine Gaming Experience of Casino Patrons*. Master's thesis, Dissertations, Professional Papers, and Capstones. University of Nevada, Los Vegas. 85 pp. https://doi.org/10.34917/1383413.

James, L. 2021. *Growing Freesia Bulbs: The Gardeners Guide on How to Grow and Care for Freesia Bulbs*. Self-Published. 24 pp.

Manning, J.C. and P. Goldblatt. 2010. *Botany and Horticulture of the Genus Freesia (Iridaceae)*. South African National Biodiversity Institute, Pretoria. 114 pp. Free Download, Borrow, and Streaming: Internet Archive.

McWhirter, K.G. 1969. Ethnography of specific anosmia. *Canadian Journal of Genetics and Cytology* 11: 479.

Sennels, N.J. 1951. *The Cultivated Species of Freesia*. J.E. Ohlsens Enke, Copenhagen. 64 pp. (In Danish & German).

Startek, L., K. Mynett, and P. Zurawik. 2002. The effect of the method of cultivation on the development of freesia (*Freesia* Eckl. ex Klatt). *Acta Horticulturae* 570: 377–383.

Van Toorenenbergen, A.W. and P.H. Dieges. 1984. Occupational allergy in horticulture: Demonstration of immediate-type allergic reactivity to freesia and paprika plants. *International Archives of Allergy and Immunology* 75: 44–47.

Wang, L. 2007. Freesia: *Freesia × hybrida*. In: *Flower Breeding and Genetics*, edited by N.O. Anderson, 665–693. Springer, Dordrecht.

### KEY CULTIVATION WEBSITES

Dutch Grown. *Growing Guide: How to Grow Freesias*. www.dutchgrown.com/blogs/the-dutchgrown-blog/growing-guide-how-to-grow-freesias.

Pacific Bulb Society. *Freesia*. www.pacificbulbsociety.org/pbswiki/index.php/freesia.

Royal Horticultural Society. *How to Grow Freesias*. www.rhs.org.uk/plants/freesia/growing-guide.

The Spruce. *How to Grow and Care for Freesia*. www.thespruce.com/freesia-flowers-1315761.

University of Mississippi, Extension. *Freesia (Freesia × hybrida) for the Farmer Florist*. http://extension.msstate.edu/publications/freesia-freesia-%C3%97-hybrida-for-the-farmer-florist.

WikiHow. *How to Care for a Potted Freesia after Blooming*. www.wikihow.com/Care-for-a-Potted-Freesia-After-Blooming.

WikiHow. *How to Grow Freesias*. www.wikihow.com/Grow-Freesias.

# 42 Fuchsia (*Fuchsia* Species and Hybrids)

**Scientific Names:** The genus *Fuchsia* was named for German botanist Leonhart Fuchs (1501–1566; "Fuchs" is German for *fox*), considered to be one of the most important historical figures in botany. Fuchs had been dead for a century when he was honored with the plant name by Charles Plumier (1646–1704), a missionary French monk and botanist, who discovered *Fuchsia triphylla* in the Dominican Republic. Plumier was also the discoverer of *Begonia*, *Lobelia*, and *Magnolia* (and, in turn, is honored by the genus name *Plumeria*, best known for the popular ornamental shrub frangipanis). The pronunciation of fuchsia has shifted from historical, awkward, hard German pronunciations (fuk-si-a; fook-sy-a) to a soft English (fyu-sha, foo-shia), the same way the color fuchsia is pronounced.

**Common Names:** The common name *fuchsia* is well-established (although frequently misspelled as *fuschia*), but the plants are sometimes called *lady's eardrops* (*ladies' eardrops*, that is, earrings that hang down from the earlobes), in reference to the pendent, teardrop-shaped flowers, reminiscent of earrings. "California fuchsia" refers to the genus *Zauschneria*. "Cape fuchsia" is a Southern African shrub (*Phygelius capensis*). "Tree fuchsia" may refer to the very few fuchsia species that develop a tree form or to fuchsia standards (plants trained to look like small trees) or to *Schotia brachpetala*, an African tree cultivated in warm climates. "Fuchsia heath" is an Australian shrub, *Epacris longifolia*, with flowers reminiscent of fuchsia.

FIGURE 42.1 Cultivated fuchsias on display at a flower show in England.

*Source:* **Photo by Andy Mabbett (CC BY SA 3.0).**

## 42.1 INTRODUCTION

Fuchsias are widely acknowledged to be among the most attractive of all flowering ornamental plants. The eye-catching flowers have been likened to hanging jewels, and the plant has been called the champion of hanging baskets.

## 42.2 OVERVIEW

The majority of the more than 110 species of *Fuchsia* are native to South America; a few are in Central America, Mexico, New Zealand, or Tahiti. Most are tropical or subtropical shrubs, and one New Zealand species is a tree. The numerous herbaceous hybrid cultivars for which fuchsia is mainly known are tender evergreens, most often grown as annuals. Cultivars suited to planting in gardens often originate mainly from *F. magellanica*, which is relatively cold-hardy. Dwarf and trailing forms of *F.* × *hybrida* are commonly grown in hanging baskets. *Fuchsia procumbens*, which grows in a prostrate fashion, is also cultivated in pots and hanging baskets. The flowers have four long sepals, which are often bright red, and four short petals, which are often purple (thus, they are usually "bicolors" horticulturally). Colors vary through white, blue, orange, pink, purple, red, violet, and (infrequently) yellow, and the flowers may be single or doubled (i.e., with extra petals and/or sepals). The very distinctive, attractive flowers are naturally adapted to being pollinated by hummingbirds. Varieties with variegated foliage are available, although fuchsia is grown essentially for its flowering display.

## 42.3 HISTORY

Fuchsias were first grown in England in the late 18th century and became increasingly popular as hardy garden cultivars

FIGURE 42.2 Shrub species of *Fuchsia*: (a) Fuchsia cultivar Heron, an upright, bushy, deciduous shrub growing in New Zealand. (b) *Fuchsia regia*, a species native to Brazil.

*Source:* **(a) Photo by Krzysztof Ziarnek, Kenraiz (CC BY SA 4.0). (b) Photo by André Karwath (CC BY SA 2.5).**



DOI: 10.1201/9781003586227-42

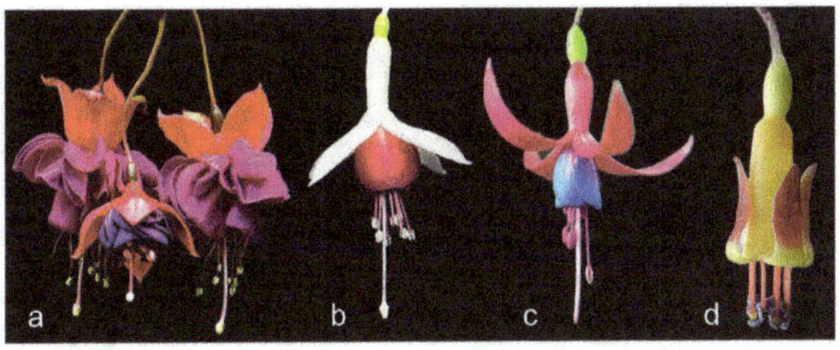


FIGURE 42.3 Examples of flowers of *Fuschia.*

*Source:* (a) Photo by Dave Shafer (CC BY 2.0). (b) Cultivar Triantha. Photo by Dominicus Johannes Bergsma (CC BY SA 4.0). (c) Photo by William Warby (CC BY 2.0). (d) *Fuchsia procumbens* "Variegata." Photo by Schurdl (CC BY SA 3.0).

FIGURE 42.4 Nineteenth-century botanical illustrations (public domain) of old fuchsia cultivars (except for *a*, several cultivars are shown in each painting).

*Source:* (a) Cultivar Abel Carriere. *Revue horticole* (Paris), Series 4, vol. 55 (1883). (b) Deutsches Magazin für Garten- und Blumenkunde; Stuggart, G. W. (1855). (c) *Journal d'horticulture pratique de la Belgique, ou Guide des amateurs et jardiniers*, vol. 13 (1855–56). (d) *La Belgique horticole, Journal des Jardins, des Serres, et des Vergers*, vol. 9 (1859).

were selected. There was a craze in Britain for fuchsia, but by the late 19th century, other ornamentals became more popular. Almost all the thousands of cultivars are hybrids originated from a few species. Hybrids are often assigned to *F.* × *hybrida*, and its likely parents are *F. fulgens* and *F. magellanica*.

## 42.4 ECONOMIC IMPORTANCE

Fuchsia is a popular offering at plant nurseries and garden centers and is one of the standard ornamental plants of horticultural industries in numerous countries.

## 42.5 ORNAMENTAL USES

In warm climates (like much of California), depending on cultivar, fuchsia is grown as an herbaceous bedding plant or a shrub. In such hospitable climates, the plant can be employed in many ways – from dwarf foreground plantings to specimens which can climb against arbors or even a two-story house. Frequently, fuchsia are grown as "standards," that is, trained to look like a short tree (2 m or so tall) with a bare trunk and a canopy on top. By contrast, in colder areas, fuchsia is usually grown as a pot plant or in hanging

FIGURE 42.5 Fuchsia plants cultivated in a commercial Dutch greenhouse.

*Source:* Photo by Shutterstock/T. W. van Urk.

FIGURE 42.6 Fuchsia trained to grow as "standards," that is, to look like miniature trees.

*Source:* (a) Photo by Shutterstock/Guentermanaus. (b) Photo by Shutterstock/Geraldick. (c) Photo by Stas Kirenkov (CC BY ND 2.0). (d) Photo by Raul654 (CC BY SA 3.0).

FIGURE 42.7 Fuchsia grown in small containers: (a) Hanging basket. (b) Potted. (c) Bonsai.

*Source:* (a) Photo by Shutterstock/Clarst5. (b) Photo by Shutterstock/Scisetti Alfio. (c) Photo by Shutterstock/Pytralona.

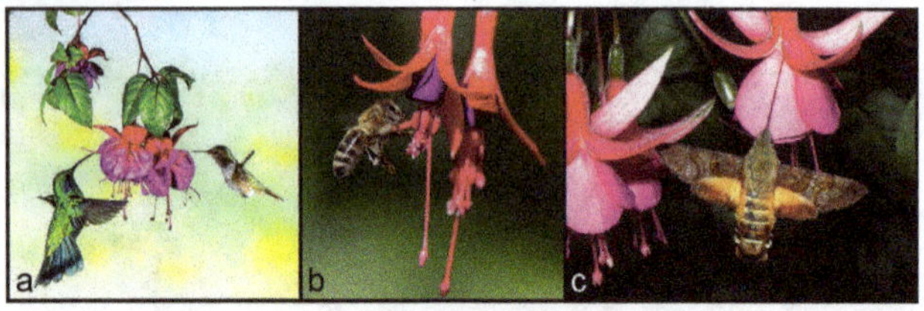

FIGURE 42.8 Pollinators attracted to fuchsia flowers: (a) Hummingbirds. (b) Honeybee. (c) Hummingbird hawk moth.

*Source:* **(a) Photo by Shutterstock/Raul Cole. (b) Photo by Andreas Hoja from Pixabay (public domain). (c) Photo by Luzonne from Pixabay (public domain).**

baskets, outdoors during the summer, and optionally also as a houseplant.

Fuchsia flowers are very attractive to pollinators, including bees, butterflies, and hummingbirds, which employ them as a source of nectar. The plants are excellent additions to butterfly, pollinator, and wildlife gardens.

## 42.6 OTHER USES

Fuchsia plants are not believed to be toxic to people or pets. The flowers and berries are considered to be edible and are employed to a minor extent in jams, chutneys, and baked goods. The fruits are slightly tart. Like most edible flowers, their chief culinary value is usually just as a decorative garnish.

## 42.7 PROBLEMS

Fuchsias are fairly easily grown outdoors as annuals, although demanding when cultivated as houseplants or as permanent garden transplants in cool climates. They do not cause some of the more concerning problems associated with many other ornamentals, such as toxicity and invasiveness.

## 42.8 CULTIVATION

### 42.8.1 Climate and Site

In climate zones 10 and 11, fuchsia can be grown outdoors as a perennial (debatably, some cultivars will survive overwinter as far north as zone 8). The plants grow best in cool weather (they originate from species native to cool mountains) and respond poorly to summer temperatures above 26°C. Relatively heat-tolerant varieties are available but tend not to be as attractive. In colder areas, the frost-tender plants are usually grown as annuals, discarded when the weather turns cold. Most fuchsia species are native to damp, shady woods, and this is reflected in their needs for shade and humidity. Fuchsias are suited to growing in part shade but often tolerate full shade. When grown as houseplants, they need bright window light in winter (or filtered light in very sunny climates) and high humidity (minimally 40%).

FIGURE 42.9 Fuchsia with berries: (a) Cultivar Howlett's Hardy. (b) *Fuchsia procumbens*.

*Source:* **(a) Photo by Agnieszka Kwiecień, Nova (CC BY SA 4.0). (b) Photo by Martin Reith (CC BY SA 4.0).**

Outdoors, windy places should be avoided (they dry out the plant and can harm the flowers); indoors, drafty locations should be avoided.

### 42.8.2 Propagation and/or Purchasing

Hybrid cultivars (most offerings in the horticultural trade) do not come true from seed. Plant nurseries and garden centers commonly provide starter plants in pots – usually trailing varieties established in hanging baskets. Enthusiasts sometimes take stem cuttings in late summer and maintain them indoors in pots overwinter. Cuttings can be started in water.

### 42.8.3 Planting

Weeping or trailing types are appropriate for planting in hanging baskets; upright varieties are best for containers suited to ground level.

### 42.8.4 Soil and Moisture Conditions

Soils for fuchsia need to be kept moist, but not waterlogged. In many dry, warm climates, this often requires daily watering, or at least daily checking. The soil should be rich, with a recommended pH of 6.5 to 7.0. Applications of a balanced fertilizer is recommended. If outdoor plants grown in garden soil are transferred indoors to be employed as houseplants, they should be placed in pots with mostly potting mixtures intended for houseplants (garden soil tends to be too poorly draining for houseplants).

### 42.8.5 Maintaining Plants

Removing spent flowers and "pinching" or cutting back some of the stems in early season (by one-quarter to half) tends to stimulate and prolong blooming. Most residences in cold areas are too dry and too dark to grow fuchsia as a houseplant, but for gardeners willing to try to maintain their beautiful potted plants overwinter, the containers can be brought indoors before the last frost and forced into dormancy by cutting them back (recommendations vary from cutting to near soil level to cutting away just one-third) and maintaining them in a cool (ca. 5–7°C) dark place until spring. Just enough water should be applied to keep the soil from completely drying out.

## 42.9 FASCINATING FACTS, FABLES, FOLKLORE, AND FRIVOLITY

- Colors are often named for the predominant hue of well-known flowers, but they can be difficult to define precisely, as indicated by the following. Fuchsia, named for the pink-purple flower, has variously been described as hot pink, reddish-purple, vivid pink, and light purple. Fuchsia is reminiscent of magenta. Indeed, an aniline dye originally patented by a French chemist in 1859 as fuchsine (for its similarity to the color fuchsia) was renamed the same year as magenta, to commemorate a French victory at the battle of Magenta, a city in Italy. Many shades of fuchsia have been recognized. "Antique fuchsia" is a lavender shade of fuchsia. "Fuchsia pink" is a light magenta. The Pantone color system is a standardized color matching system widely used around the world. "Fuchsia Rose," chosen by Pantone as its "color of the year for 2001," is defined as "a variable color averaging a moderate purplish red that is bluer and deeper than average rose, redder and deeper than violine pink, redder, lighter, and stronger than magenta rose, and bluer and stronger than solferino" (solferino is a brilliant deep-pink-colored aniline dye with a purplish tinge).
- Today, flowers are not considered manly, but in the past, there were exceptions, and ships were often named after well-known flowers. *Fuchsia*, a steam tug built in 1863 in New York, became a military vessel of the Union forces, patrolling eastern rivers against the Confederate Army. In 1864, *Fuchsia* encountered the abandoned Confederate Army tug *Titan* and disabled it.
- The Isle of Man is a self-governing British Crown Dependency in the Irish Sea between Great Britain and Ireland. It has a population of about 90,000 and an area of 574 km$^2$. The Manx ethnic group originated on the Isle of Man, but its unique language has mostly been replaced by English. *Fuchsia magellanica* hybrids were introduced to the Isle of Man from South America in the 1820s and now represent a prominent part of the landscape. *Jacobaea vulgaris* (*Senecio jacobaea*), known as ragwort and cushag, has often been interpreted as the Manx/Isle of Manx national flower emblem but is disliked by many for its toxicity (www.gov.im/media/402197/cushag_code_july_2013.pdf). "Bollan bane" (mugwort, usually interpreted as *Artemisia vulgaris*) is pinned to chests on Tynwald, the national day of the Isle of Man. However, fuchsia is now often preferred as the representative plant of the Isle of Man and of Tynwald. The Manx name for fuchsia is *jeiryn yee*, which means "God's tears."

FIGURE 42.10 Fuchsia flowers intergrade in color between shades of pink, purple, and red, and like the color fuchsia, specifying the hue is imprecise and difficult.

*Source:* **Photo by Shutterstock/Viktoriia Protsak.**

## KEY PUBLICATIONS


Berry, P.E. 1982. The systematics and evolution of *Fuchsia* sect. *Fuchsia* (Onagraceae). *Annals of the Missouri Botanical Garden* 69: 1–198.

Berry, P.E. 1985. The systematics of the apetalous fuchsias of South America, *Fuchsia* sect. *Hemsleyella* (Onagraceae). *Annals of the Missouri Botanical Garden* 72: 213–251.

Berry, P.E. 1989. A systematic revision of *Fuchsia* sect. *Quelusia* (Onagraceae). *Annals of the Missouri Botanical Garden* 76: 532–584.

De Cooker, M.G.R.T., E.J. Goulding, J.H. Waldenmaier, and P.E. Berry. 1985. Fuchsia. In: *Ornamental Crops. Handbook of Plant Breeding*, Vol. 11, edited by J. Van Huylenbroeck, 383–485. Springer, Cham. [Updated online version: https://link.springer.com/chapter/10.1007/978-3-319-90698-0_16].

Goulding, E.J. (ed.). 2002. *Fuchsias: The Complete Guide*. Batsford, London. 192 pp.

Green, D. 2011. *Fuchsia Species*. AuthorHouse (Self-Publisher), UK. 444 pp.

Lim, T.K. 2014. *Fuchsia × hybrida*. In: *Edible Medicinal and Non Medicinal Plants*, edited by T.K. Lim, 548–551. Springer, Dordrecht.

Margelony, R.T. 2012. *The Fuchsia+Blog. The Color Fuchsia*. https://fuchsietum.com/blog/files/color-fuchsia.php.

Margelony, R.T. 2015. *The Fuchsia+Blog. Eat Your Fuchsias*. https://fuchsietum.com/blog/files/eat-your-fuchsias.php.

Margelony, R.T. 2020. *The Fuchsia+Blog. The Inca and the Fuchsia.* https://fuchsietum.com/blog/files/Inca-and-the-fuchsia.php.

Munz, P.A. 1943. A revision of the genus *Fuchsia* (Onagraceae). *Proceedings of the California Academy of Sciences (Series 4)* 25(1): 1–138.

Nicholass, J. 2022. *The Complete Guide to Growing Fuchsias: How to Cultivate Fuchsias with Practical Gardening Advice and an Illustrated Directory of 500 Varieties.* Anness, London. 256 pp.

Rosenfeld, R. 2004. *Fuchsias.* DK, London. 320 pp.

Thorne, T. (ed.). 1964. *Fuchsias for All Purposes.* Collingridge, London. 175 pp.

Wilkins, H.F. 1985. *Fuchsia* × *hybrida.* In: *Handbook of Flowering*, Vol. 3, edited by H.F. Wilkins, 38–41. CRC Press, Boca Raton, FL.

## Key Cultivation Websites

British Fuchsia Society. www.thebfs.org.uk/index.asp.

Clemson University, Extension. *Fuchsia.* https://hgic.clemson.edu/factsheet/fuchsia/.

The Fuchsietum: Fuchsia Societies & Groups around the Globe. https://fuchsietum.com/about-fuchsias/societies/societies.html.

Garden Geek. *How to Grow Fuchsia in Hot Climate (Lahore).* www.lahoregardening.com/2020/07/how-to-grow-fuchsia-in-hot-climate-lahore.html.

Plantura Magazine. *Overwintering Fuchsias & Hardy Fuchsia Varieties.* https://plantura.garden/uk/trees-shrubs/fuchsias/overwintering-fuchsias.

Royal Horticultural Society. *How to Grow Hardy Fuchsias.* www.rhs.org.uk/plants/fuchsia/hardy/growing-guide.

The Spruce. *How to Grow and Care for Fuchsia.* www.thespruce.com/fuchsia-great-container-plant-for-shade-847937.

The Spruce. *How to Grow and Care for Fuchsias Indoors.* www.thespruce.com/grow-fuchsia-as-houseplants-1902598.

WikiHow. *How to Care for Fuchsia.* www.wikihow.com/Care-for-Fuchsia.

WikiHow. *How to Overwinter Fuchsias.* www.wikihow.com/Overwinter-Fuchsias.

WikiHow. *How to Prune Fuchsias.* www.wikihow.com/Prune-Fuschias.

# 43 Gardenia (*Gardenia* Species)

**Scientific Names:** The genus name *Gardenia* commemorates Scottish-born naturalist and physician Alexander Garden (1730–1791), who resided in Charleston, South Carolina.

**Common Names:** The most widely grown species, *G. jasminoides*, is usually referred to simply as gardenia, but also as Cape jasmine ("Cape," a reference to the Cape of South Africa; "jasmine," an allusion to jasmine, that is, the highly perfumed flowers of *Jasminum* species). *Kopsia fruticosa*, an Asian shrub sometimes grown as an ornamental, is called pink gardenia.

## 43.1 INTRODUCTION

Gardenias are gorgeous plants with heavenly fragrances, popular as outdoor ornamental shrubs and small trees in climates that don't experience freezing temperatures. When grown as houseplants, however, they have a well-earned reputation for being difficult, demanding, and frustrating, even for experienced gardeners. Although beautiful as cut flowers, they are short-lived and expensive and so primarily employed for special occasions. Although challenging, gardenias rank among the world's leading ornamental plants.

The genus *Gardenia*, with as many as 200 species, is native to the tropical and subtropical regions of the Old World, including Africa and Asia, as well as the Pacific Islands and Australia. The plants are evergreen shrubs and small trees, varying in height from 1 to 15 m. The flowers are usually white or yellow (white flowers often yellow as they age) and occur singly or in small clusters. In the ornamental species, they are large and extremely sweet-scented. *Gardenia jasminoides* is the leading ornamental species. It is a native of Asia, including Vietnam, southern China, Taiwan, Japan, Myanmar, and India. This species is cultivated as an outdoor multi-trunked shrub or small tree and as a houseplant. *Gardenia thunbergia*, white gardenia, is often grown as a rootstock (to which other plants are grafted; this practice is useful for dwarfing and for increasing hardiness).

FIGURE 43.1 Nineteenth-century illustrations (public domain) of *Gardenia jasminoides*.

*Source:* (a) **Revue horticole, Journal d'Horticulture Pratique. 1886, Librairie Agricole de la maison rustique, Paris. (b) Blanco, F. M. 1880–1883. *Flora de Filipinas*. Establecimiento tipográfico de Plana y C, Manila. (c) Van Houtte, L. (ed.). 1846. *Flore des Serres et des Jardins de l'Europe*, vol. 2.**

## 43.2 OVERVIEW

*Gardenia jasminoides* grown outdoors may reach a height of 2.5 m but, as a houseplant, is usually no taller than 0.5 m. The leaves are glossy-lustrous, dark emerald-green, thick, leathery, and arranged in whorls (sometimes opposite pairs) on the stems. The flowers are 8 to 13 cm wide, usually white or yellowish, waxy, extremely fragrant, and borne singly in the leaf axils. Flowering usually occurs from spring to early summer. The flowers in many of the most popular cultivars are doubled (i.e., with extra petals). The fruits produced are fleshy, orange or yellow, ovoid or ellipsoid capsules, 3–6 cm long, containing many seeds. Breeding has produced cultivars with variegated foliage and dwarf forms.

FIGURE 43.2 Views of *Gardenia jasminoides*: (a) Tree form. (b) Shrub form. (c) Flowering branches. (d) Flower.

*Source:* **(a) Photo by Shutterstock/Alan Parker. (b) Photo by Natalie Maynor (CC BY 2.0). (c) Photo by Alpsdake (CC BY SA 3.0). (d) Photo by Beverly Buckley from Pixabay (public domain).**

DOI: 10.1201/9781003586227-43

FIGURE 43.3 Transformation of gardenia flowers from white to yellow as they mature: (a) Common gardenia (*Gardenia jasminoides*). Note how three of the older flowers have yellowed. (b) Transvaal gardenia (*G. volkensii*). The flowers were white when young but turned yellow as they matured.

*Source:* **(a) Photo by Alpsdake (CC BY SA 3.0). (b) Photo by Wendy Cutler (CC BY 2.0).**

## 43.3 HISTORY

Gardenias have been cultivated in Asia, particularly China, for over a thousand years. *Gardenia jasminoides* was introduced from China into Britain in the mid-18th century in its sterile double-flowered form. It became popular because of its large white jasmine-like scented flowers. Gardenias subsequently became widespread in the Western world and the subject of breeding to produce superior cultivars.

## 43.4 ECONOMIC IMPORTANCE

Gardenias are popular landscape and residential shrubs in warm temperate and subtropical regions throughout the world. Extracts from gardenias, especially from the fruits, are employed in traditional medicine, particularly in Asia, and there are very extensive studies of their potential pharmacological and industrial applications.

## 43.5 ORNAMENTAL USES

Gardenias are often employed as flowering shrubs, sometimes as trees. In hot (tropical and subtropical) climates, they are cultivated for a variety of landscape purposes in both public and residential areas, including use as hedges and even as groundcover. They are frequently grown in containers, often as houseplants, although considered to be demanding and difficult. In warmer areas, they are also grown in large tubs that can be protected indoors in cold seasons. They are also used as a cut flower but do not last long (2 days, or up to 4, with special care). Because they are expensive when purchased from florists, they are often reserved for special occasions, such as marriages. Gardenias are employed as very expensive corsage flowers worn by women, and sometimes (particularly in France) as boutonnières for men. In Hawaii, where they are commonly available, they are used in leis. The large flowers make a beautiful display when floating in attractive vessels in the center of a table.

FIGURE 43.4 Pre–World War II postcard titled "Gardenia time in Florida Cypress Gardens," showing spectacular landscaping with gardenia. At that time, special postcards were often printed on linen, as reflected by the diffuse texture of the image.

*Source:* **Boston Public Library (public domain).**

FIGURE 43.5 Indoor use of gardenia as an ornamental: (a) A houseplant. (b) Cut flowers in a vase.

*Source:* **(a) Photo by Shutterstock/Madlen. (b) Photo by Shutterstock/Fancy.**

## 43.6 OTHER USES

The fruits of gardenia are employed in traditional Chinese medicine for treating inflammation, jaundice, headache,

FIGURE 43.6 A bride holding a gardenia bouquet.

*Source:* **Photo by Shutterstock/Lapha.R.**

FIGURE 43.7 Women decorated with gardenia flowers.

*Source:* **(a) Photo by Shutterstock/Oceanoart. (b) Photo by Shutterstock/Missha Vara.**

edema, fever, hepatic disorders, hypertension, and many other conditions. Also in China, the fruits of some species are employed as food. "Gardenia yellow dye" extracted from the fruit has minor usage for coloring clothing and food. "Gardenia blue" is a natural food colorant produced by treating extracts from the fruits of *G. jasminoides* or *G. augusta*. It has been used for decades to color food in Asia. Gardenia flowers are considered edible for humans, and the green flower buds are sometimes pickled.

## 43.7 PROBLEMS

Online sources are unclear regarding the toxicity of gardenias to pets, often stating that "certain species" are at least mildly or potentially poisonous for dogs, cats, and horses. As noted earlier, *G. jasminoides* is sometimes consumed by people.

FIGURE 43.8 Dried fruit of gardenia, extensively used in traditional Chinese medicine.

*Source:* **Photo by Shutterstock/Sasimoto.**

## 43.8 CULTIVATION

### 43.8.1 Climate and Site

*Gardenia jasminoides* is native to warm, humid tropical regions, and when grown as a houseplant, it needs high humidity and temperatures of 20–23°C (preferably cooler night: 15–16°C) and bright but not direct sunlight (except in Northern cold areas during dark winters). Some cultivars are rated as tolerating climate zone 7, or even zone 6. In hot climates, gardenia is planted outdoors, in semi-shade, where temperatures are high, but often in full sun in cooler areas. Sheltering from the cold and from strong winds is recommended. Gardenias are not only difficult to grow in cold climates but, in very hot climates (such as in Phoenix), also have been described as short-term, high-maintenance plants. Gardenias are shallow-rooted, do not compete well with tree roots, and so should not be located near trees. To enjoy the fragrance of the flowers, outdoor plants are preferably located near patios, balconies, windows, entrances to buildings, and pathways. Container-grown plants can be moved outdoors for the summer, preferably in protected locations.

### 43.8.2 Propagation and/or Purchasing

Gardenia is usually professionally propagated by taking cuttings from cultivars and grafting them to hardy rootstocks. Layering is also practiced (bending branches so they are partly in soil, where they are encouraged to root, or in air layering, a portion of the living stem is covered and kept humid and induced to root). The plants can be grown from seeds, but not most cultivars, which won't reproduce to type. Cultivars with doubled flowers tend to be sterile (they won't produce seeds). The average gardener can simply purchase young plants established in containers.

### 43.8.3 Planting

Depending on cultivar, specimen trees or large shrubs are planted outdoors 1 to several meters apart. Standard recommendations for transplanting containerized specimens should be followed.

### 43.8.4 Soil and Moisture Conditions

Gardenia grows well in well-drained, acidic soils (pH 5.0–6.5) high in organic matter. Some authorities recommend avoiding irrigation with alkaline water (frequent in tap water), which can raise the pH and result in chlorosis from iron deficiency. Fertilizing outdoor plants in the fall in cool climates is inadvisable, as this produces frost-tender growth. Commercial potting soils for container growing of gardenias are available. The soil should be slightly moist, but not overwatered. Standard methods of increasing humidity for houseplants include running a humidifier and grouping plants together on trays of wet pebbles. Some indoor gardeners mist their plants, although this is not considered sufficient for gardenia, and water contacting the flowers may turn them brown. Fertilization with a preparation formulated for acid-loving, blooming plants, such as azalea, should be carried out according to label instructions.

### 43.8.5 Maintaining Plants

Spent flowers should be removed to encourage further blooming.

## 43.9 FASCINATING FACTS, FABLES, FOLKLORE, AND FRIVOLITY

- According to legend, in biblical times, Solomon, said to have a thousand wives, gave each a gardenia flower to show he loved all of them equally and they had no reason to be jealous.
- Transvaal gardenia (*G. volkensii*) is a multi-stemmed shrub or small tree indigenous to Southern Africa. In its native area, it is often planted near houses as protection from lightning and evil spirits.
- The British Royal Navy frequently named ships after flowers. HMS *Gardenia* (1917) was launched in 1917 and sold in 1923. HMS *Gardenia* (K99) was launched in 1940 and accidentally sunk in 1942 when rammed by another British ship.
- *Gardenia thunbergia*, as noted earlier, is commonly used as a hardy rootstock, on top of which ornamental species are grafted. In its native South Africa, it is a large shrub or small tree up to 5 m tall. It produces a woody fruit, which protects the seeds. Several large South African antelopes, including the greater kudu, the lesser kudu, and the Cape buffalo, chew open the woody fruits and eat the seeds, which pass unharmed through the digestive system of the animals and become dispersed in their dung, which provides fertilizer for the resulting seedlings.
- The Principality of Andorra is a landlocked microstate (with fewer than 100,000 citizens) bordered by France to the north and Spain to the south. The gardenia is its national flower.
- *Gardenia taitensis*, called Tahitian gardenia, although it originated from Melanesia and western Polynesia, is the national flower of Tahiti, French Polynesia, and the Cook Islands. Polynesians in the Pacific islands use its flowers to make fragrant necklaces. Sometimes they wear a flower on the left ear to indicate that they are married, but if worn on the right ear, it means the person is available.

FIGURE 43.9 Use of *Gardenia taitensis* for personal floral decorations: (a) Wreaths made from gardenia (white) and bougainvillea (red) flowers, to be worn on the head, in the Cook Islands. (b) Fragrant flower necklace made in Tahiti.

***Source:*** **(a) Photo by Dustinpsmith (CC BY 2.0). (b) Photo by Shutterstock/EQRoy.**

## KEY PUBLICATIONS


Armstrong, W.P. 1986. The gardenia, the kudu, and a long-tongued moth. *Zoonooz* 59(1): 17–19.

Bradshaw, J. and S.P. Brown. 2012. *Growing Gardenias in Florida*. University of Florida, Extension, Circular 1098. 7 pp.

Çelikel, F.G., M.S. Reid, and C.-Z. Jiang. 2019. Water uptake and vase life of cut *Gardenia jasminoides* flowers. *Acta Horticulturae* 1263: 335–342.

Chen, L., M. Li, Z. Yang, W. Tao, P. Wang, X. Tian, et al. 2020. *Gardenia jasminoides* Ellis: Ethnopharmacology, phytochemistry, and pharmacological and industrial applications of an important traditional Chinese medicine. *Journal of Ethnopharmacology* 257: 112829. https://doi.org/10.1016/j.jep.2020.112829.

Jarvis, C.E., A. DuVal, and P.R. Crane. 2014. *Gardenia jasminoides*: A traditional Chinese dye plant becomes a garden ornamental in Europe. *Curtis's Botanical Magazine* 31: 80–98.

Jin, C., A.W.-S. Zongo, H. Du, Y. Lu, N. Yu, X.A. Nie, et al. 2023. Gardenia (*Gardenia jasminoides* Ellis) fruit: A critical review of its functional nutrients, processing methods, health-promoting effects, comprehensive application and future tendencies. *Critical Reviews in Food Science and Nutrition*. https://doi.org/10.1080/10408398.2023.2270530.

Lim, T.K. 2014. *Gardenia jasminoides*. In: *Edible Medicinal and Non-Medicinal Plants*, Vol. 8, edited by T.K. Lim, 705–729. Springer, Dordrecht.

Nower, A.A. and E.M. Hamza. 2013. Production of *Gardenia jasminoides* scions via tissue culture for grafting on *Gardenia thunbergia* under greenhouse conditions. *Journal of Applied Sciences Research* 9: 3118–3128.

Reuben, P. 2022. *Growing Gardenia in Your Home Garden: The Expert Farmers Guide to Growing, Caring, Maintaining, Pruning and Simple Tips to Keeping Your Gardenia Flourishing*. Self-Published. 47 pp.

Rodríguez-Deméneghi, M.V., F. Gallardo-López, H.E. Lee-Espinosa, M.E. Galindo-Tovar, N. Aguilar-Rivera, O.R. Leyva-Ovalle, et al. 2016. Cultivation of *Gardenia jasminoides* Ellis in the central region of Veracruz, Mexico. *Agroproductividad* 9(6): 39–43.

Wilkins, H.F. 1986. *Gardenia jasminoides*. In: *Handbook of Flowering*, Vol. V, edited by A.H. Halevy, 127–131. CRC Press, Boca Raton, FL.

## Key Cultivation Websites

Clemson Cooperative Extension. *Gardenia Diseases & Other Problems*. https://hgic.clemson.edu/factsheet/gardenia-diseases-other-problems/.

Epic Gardening. *How to Plant, Grow and Care for Gardenias*. www.epicgardening.com/gardenia/.

Gardenia.net. *Learn How to Grow and Care for Your Gardenia*. www.gardenia.net/guide/learn-how-to-grow-and-care-for-your-gardenia.

Hodgson, L. *Why I Don't Grow Gardenias*. https://laidbackgardener.blog/2017/10/14/why-i-dont-grow-gardenias/?doing_wp_cron=1667197890.4661281108856201171875.

Kobayashi, K.D. and A.J. Kaufman. 2006. *Common Gardenia*. University of Hawaii Publication. www.ctahr.hawaii.edu/oc/freepubs/pdf/of-32.pdf.

The Spruce. *How to Grow and Care for Gardenia*. www.thespruce.com/grow-gardenia-indoors-1902758.

WikiHow. *How to Care for a Gardenia*. www.wikihow.com/Care-for-a-Gardenia.

WikiHow. *How to Grow Gardenia from Cuttings*. www.wikihow.com/Grow-Gardenia-from-Cuttings.

WikiHow. *How to Grow Gardenias*. www.wikihow.com/Grow-Gardenias.

WikiHow. *When & How to Prune Gardenias for Healthy, Happy Blooms*. www.wikihow.com/Prune-a-Gardenia-Bush.

# 44 Gazania (*Gazania* Species and Hybrids)

**Scientific Names:** The genus name *Gazania* is usually interpreted as commemorating Theodorus Gaza (about 1398–1475), born in Greece and subsequently a scholar in Italy who was one of the leaders in reviving learning in Europe. An occasionally proposed alternative explanation of the name is that it originated from the Latin and Greek *gaza*, "treasure" or "riches," consistent with the common name *treasure flower*.

**Common Names:** Gazanias are also called treasure flowers, a reference to the jewel-like appearance of the blooms. Like several other ornamentals from South Africa with daisylike flowers, they are also known as African daisies.

## 44.1 INTRODUCTION

Gazanias are low-growing plants that produce numerous extremely attractive, daisylike, long-lasting flowers. They are very popular for annual beds, borders, and containers and are so low-maintenance that they are recommended for lazy gardeners.

## 44.2 OVERVIEW

*Gazania* is a genus of seven species native to Southern Africa (in the past, many more species were recognized). The plants are low-growing (typically 15–30 cm high) herbaceous or shrubby perennials typically grown as annuals in temperate regions, but also as evergreen perennials in semi-tropical and tropical areas. The species of *Gazania* are hard to distinguish from each other and, indeed, resemble allied genera. *Gazania rigens* (= *G. splendens*, coastal gazania), currently interpreted as a very variable plant, is the most popular species, but most garden cultivars are interspecific hybrids. The flowers are often brilliantly colored, in shades of orange, pink, red, white, and yellow. Bicolored flowers (often striped) are common, and cultivars with extra petals have been selected. Flowering is often prolonged over much of the summer. However, the flowers do not open fully on cloudy days, and they tend to close at night. Some recent cultivars have huge flower heads (up to 15 cm in diameter)

FIGURE 44.1 Gazania in flower.

*Source:* (a) Photo by Denisbin (CC BY ND 2.0). (b) Photo by Unit20072007 (CC BY 2.0). (c) Photo by Symphori from Pixabay (public domain). (d) Photo by Shutterstock/Ulkastudio.

FIGURE 44.2 Gazania flowers.

*Source:* (a) Photo by Alvesgaspar (CC BY SA 4.0). (b) Photo by HQ (CC BY SA 2.0). (c) Photo by M. Fitzsimmons (CC BY 3.0). (d) Photo by Shirley Hirst from Pixabay (public domain). (e) Photo by Hedwig Storch (CC BY 3.0). (f) Photo by Manfred Richter from Pixabay (public domain).

FIGURE 44.3 Gazania flowers that have closed up in response to darkness.

*Source:* Photo by Shutterstock/NataliaVo.



DOI: 10.1201/9781003586227-44

FIGURE 44.4 Gazania growing in commercial greenhouses.

*Source:* (a) Photo by Shutterstock/Nataly Mayak. (b) Photo by Leonora (Ellie) Enking (CC BY SA 2.0).

FIGURE 44.5 Garden displays of gazania cultivars.

*Source:* (a) Photo by Leonora (Ellie) Enking (CC BY SA 2.0). (b) Photo by Shutterstock/Kim Britten.

with semi-doubled flowers that have so many petals in the center that they cannot close up completely in the evening like most other varieties.

## 44.3 HISTORY

Although the scientific name *Gazania* was coined in 1791, the ornamental usage of the species was quite limited until recently, and the attractive cultivars currently available are primarily products of recent decades. Most cultivars of gazania are hybrids, often involving *G. rigens* in their parentage.

## 44.4 ECONOMIC IMPORTANCE

Gazania is a very attractive flowering ornamental currently under extensive development for new cultivars. It competes with many other daisylike flowers but is certainly among the most popular.

## 44.5 ORNAMENTAL USES

Gazania is suitable for borders, garden edging, mass planting, and containers, including hanging baskets. The ground-hugging cultivars are superb as groundcover and are suitable for rock gardens. Gazania combines well with other low-growing ornamentals. Blooms that tend to close when shaded, as in gazania, are not suitable for use as cut flowers.

FIGURE 44.6 Gazania in containers.

*Source:* All photos from Shutterstock: Contributors: (a) KanphotoSS, (b) P. Nova, (c) KanphotoSS, (d) Sarawut Muensang.

FIGURE 44.7 Gazania growing as a groundcover in hot climates.

*Source:* (a) Photo by Sheba (CC BY 2.0). (b) Photo by Mary Gillham (CC BY 2.0).

FIGURE 44.8 Gazania stabilizing a slope in southern California.

*Source:* Photo by Sharon Tate Soberon (CC BY ND 2.0).

## 44.6 OTHER USES

In hot climates, gazania is employed for erosion control, especially on sloping banks.

## 44.7 PROBLEMS

Gazania has escaped from cultivation and established in the Mediterranean, parts of Australia, the United Kingdom, France, Spain, Portugal, and parts of the United States. In some areas (notably Australia and California), it is a significantly invasive weed.

## 44.8 CULTIVATION

### 44.8.1 Climate and Site

Gazania is grown as an annual in climate zones 8 or 9 and northwards. The plant is drought-tolerant, growing well in hot, dry areas, but the flowers may close in response to drought. Full sunlight is strongly recommended. In cold climates, container-grown plants can be brought indoors in the autumn and maintained in a sunny windowsill.

FIGURE 44.9 Weedy *Gazania rigens* near San Diego, California.

*Source:* **Photo by Cultivar413 (CC BY 2.0).**

FIGURE 44.10 A planting of gazania beside pavement. "Sidewalk gardens" or ornamental borders beside paved or concrete driveways can be hostile to plants because of the heat absorbed from the sun. Gazania is ideal for plantings beside pavement, as shown here, because it is very tolerant of heat and drought.

*Source:* **Photo by Sydh2010 (CC BY ND 2.0).**

FIGURE 44.11 Appropriately spaced young plants of gazania.

*Source:* **Photo by Jim, the Photographer (CC BY 2.0).**

### 44.8.2 Propagation and/or Purchasing

Gazania is frequently grown from seeds, but many recent attractive cultivars are established from cuttings, as they will not breed true. In hot climates, in which gazania can be grown as a perennial, it can be divided. Creeping forms will expand naturally, the spreading stems establishing new plants. In cold climates, cuttings can be taken in the fall and overwintered indoors.

### 44.8.3 Planting

Suggested spacing in beds is 30 to 45 cm. In hanging baskets, plants can be placed close to the edge of the container to encourage the flowering stems to trail down.

### 44.8.4 Soil and Moisture Conditions

In their native habitat (rocky cliffs of South Africa), gazanias grow in low-fertility soils. Although gazania tolerates infertile, sandy soils, for better flowering displays, it should be established in more fertile soils or fertilized regularly. Circumneutral soil is best, but the plant withstands pH as acidic as 5.8. Although it tolerates dryness, it grows better if watered moderately. Watering in the morning, allowing the moisture to dry during the day, reduces exposure to fungal diseases affecting gazania.

### 44.8.5 Maintaining Plants

Compared to most flowering ornamentals, gazania requires much less attention. Deadheading of wilted flowers is recommended. In hot areas where gazania is grown as a groundcover, it should be pruned annually in late winter or early spring, starting in the second season, cutting the foliage back as low as 5 to 8 cm above the ground to encourage new growth.

## 44.9 FASCINATING FACTS, FABLES, FOLKLORE, AND FRIVOLITY

- All gazanias have white woolly hairs on the lower sides of the leaves. In 1883, John Wood, a South

African botanist, recorded that the only dress worn by many native Zulu girls was a feltlike garment fashioned from the cottony hairs of the foliage of gazania.

- In Africa, gazania is called *botterblom*, meaning "butter flower," supposedly in reference to the butter-like taste of its petals. The plants are not considered toxic but can cause significant gastrointestinal upset, so they should not be consumed.
- *Nyctinasty* is movement by plants (especially leaves and flowers) in response to light being reduced, such as occurs at dusk, or being increased, such as at dawn. Many plants exhibit nyctinasty, but the majority do not. Gazania flowers are light-sensitive, closing up at night and on especially cloudy days. It is thought that when the petals cover the flower head, the inner parts, particularly the pollen, are protected from dew, rain, and pest insects.
- In 1999 in Australia, a horse was registered for racing although it was 24 years of age (most racehorses cease competition at 3 years of age), had never won a race, and its lifetime earnings amounted to nothing. Additionally, demonstrating the optimism of the owners, the horse's name was Gazania, which, as noted earlier, translates from Latin and Greek as "riches."

## KEY PUBLICATIONS


Funk, V.A. and R. Chan. 2008. Phylogeny of the spiny African daisies (Compositae, tribe Arctotideae, subtribe Gorteriinae) based on trnL-F, ndhF, and ITS sequence data. *Molecular Phylogenetics and Evolution* 48: 47–60.

Giuffrida, F., M. Agnello, and D. Gangi. 2017. Compost-based substrate for *Gazania rigens* (L.) cultivation. *Acta Horticulturae* 1168: 237–244.

Howis, S. 2007. *A Taxonomic Revision of the Southern African Endemic Genus Gazania Asteraceae) Based on Morphometric, Genetic and Phylogeographic Data.* Doctoral dissertation. Rhodes University, South Africa. 293 pp. https://core.ac.uk/download/pdf/145040966.pdf.

Howis, S. and N.P. Barker. 2006, May. Phylogeographic analysis suggests a recent radiation of the southern African endemic genus *Gazania* (Gorteriinae; Asteraceae). *South African Journal of Botany* 72: 322–323.

Howis, S., N.P. Barker, and L. Mucina. 2009. Globally grown, but poorly known: Species limits and biogeography of *Gazania* Gaertn. (Asteraceae) inferred from chloroplast and nuclear DNA sequence data. *Taxon* 58: 871–882.

Justiss, E.S. 1985. *The Performance of* Gazania splendens *Cultivars under Kentucky Conditions.* Master's thesis. Western Kentucky University, Bowling Green, KY. 44 pp. https://digitalcommons.wku.edu/cgi/viewcontent.cgi?article=2739&context=theses.

Zulfiqar, F., A. Younis, Z. Abideen, A. Francini, and A. Ferrante. 2019. Bioregulators can improve biomass production, photosynthetic efficiency, and ornamental quality of *Gazania rigens* L. *Agronomy* 9(11): 773. https://doi.org/10.3390/agronomy9110773.


### Key Cultivation Websites


Better Homes & Gardens. *How to Plant and Grow Gazania.* www.bhg.com/gardening/plant-dictionary/annual/gazania/.

The Garden Helper. *How to Grow and Care for Your Treasure Flower.* www.thegardenhelper.com/gazania.html.

GardenersWorld.com. *How to Grow Gazanias.* www.gardenersworld.com/how-to/grow-plants/how-to-grow-gazanias/.

Gardenia. *How to Plant and Grow Gazanias.* www.gardenia.net/genus/gazania-treasure-flower.

Horticulture Magazine. *Gazania – Treasure Flower.* https://horticulture.co.uk/gazania/.

The Spruce. *How to Grow and Care for Gazania (Treasure Flower).* www.thespruce.com/gazania-flowers-1315701.

# 45 Geranium (*Pelargonium* Species and Hybrids)

**Scientific Names:** The genus name *Pelargonium* is derived from the Greek *pelargos*, a stork, the long slender fruit reminiscent of a stork's bill. As discussed in this chapter, the two most important classes of geranium cultivars are the "zonals" and the "regals." The name *P.* × *hortorum* has been applied to the zonals, ands the name *P.* × *domesticum* to the regals.

**Common Names:** The species of the genus *Pelargonium* have the common name "geranium." However, there is also a related genus, *Geranium*, many of which are also garden ornamentals. This regularly results in confusion, which could be avoided by using the vernacular name "pelargonium" for species of the genus *Pelargonium*. Unfortunately, the common name geranium is entrenched in North America for species of *Pelargonium*. In England, New Zealand, Australia, and South Africa, geraniums are often referred to as pelargoniums. "Hardy geranium" generally refers to *Geranium* species. Species of *Pelargonium* are frequently called storkbill or stork's bill, while species of *Geranium* are often called cranesbill. "East Indian geranium" and "Turkish geranium" are names for the grass *Cymbopogon martinii* and its essential oil, used as a flavorant.

## 45.1 INTRODUCTION

Geraniums (*Pelargonium* species) are one of the world's leading ornamentals grown as annual garden and container plants. Their popularity is due to the production of spectacular masses of blooms which look good from summer to a killing frost, ease of cultivation, and the availability of thousands of cultivars providing a dazzling kaleidoscope of colors and forms. There are numerous societies devoted to *Pelargonium*.

The genus *Pelargonium* is native mainly to the Southern Hemisphere, mostly in South Africa, where the majority of cultivated forms originated. A few species are also indigenous to Australia, eastern Africa, New Zealand, Madagascar, Yemen, Asia Minor, and the Atlantic Ocean islands of Saint Helena and Tristan de Cunha. In some areas of the world, notably Australia and California, domesticated geraniums have escaped to the wild and grow as weeds. Geraniums grow in quite varied habitats but are usually adapted to low-humidity, cool areas with limited seasonal rainfall.

## 45.2 OVERVIEW

*Pelargonium* includes annual and perennial herbs and shrubs. Some species develop underground tubers. The foliage may be evergreen or deciduous, lobed or dissected; the margins smooth, curly, or crisped; and the leaves may have a rough, sticky, or velvety texture. Some cultivars have variegated leaves or foliage with distinct color along the veins. Stems may be erect or trailing, succulent, spiny, or initially herbaceous but later woody. The blooms are arranged in clusters. The flowers may be shaped like butterflies, funnels, saucers, stars, or trumpets and may resemble cactus flowers, carnations, tulips, or pansies. One of the five sepals is modified into a nectary, which is fused to the flower stalk. The five petals are arranged into two upper ones and three lower ones. At maturity, the five fruit segments (each with a single seed) split apart, dispersing their plumed seeds.

The two so-called geranium genera both provide many cultivars and are widely confused. The easiest way to differentiate *Pelargonium* species from *Geranium* species is by floral shape. In *Pelargonium*, the flower has two upper petals which are different in shape and size from the three lower petals, but in *Geranium*, all five petals are more or less the same in shape and size. (However, some zonal and regal *Pelargonium* cultivars develop flowers resembling *Geranium* flowers, and some cultivars have doubled flowers with extra petals.) *Geranium* species frequently fling their seeds away by an explosive mechanism, while *Pelargonium* seeds float away on the breeze with the aid of a feathered end (lacking in *Geranium*). The foliage of cultivated *Geranium* species is often more divided than in commonly cultivated *Pelargonium* species. *Pelargonium* tends to have rather thick, succulent stems, unlike cultivated *Geranium* species.

FIGURE 45.1 Displays of ornamental geranium (*Pelargonium*) cultivars.

*Source:* **All photos from Shutterstock. Contributors: (a) Helen J. Davies, (b) Josie Gilbert, (c) Marcelino Pozo Ruiz.**



DOI: 10.1201/9781003586227-45

FIGURE 45.2 Seed dispersal of *Pelargonium aestivale*: (a) Seeds with an awn (extension) that is spirally twisted at the base and has fluffy hairs on the end. The spirally twisted portion is hygroscopic, winding and unwinding in response to moisture, drilling the seed into the ground. (b) Geranium seeds being dispersed by the wind blowing against the hairs on the awn.

***Source:* (a) Photo by Michael Wolf (CC BY 3.0). (b) Photo by P. Jeremy (CC BY 2.0).**

There are thousands of cultivars of *Pelargonium*, in a very wide range of shapes, sizes, floral colors, and scents. Most of the common cultivars produce excellent, long-lasting floral displays; many have decorative foliage; and some are trailing, suitable for hanging baskets and window boxes. Mature geraniums are usually 30 to 50 cm tall and 40 to 60 cm wide, depending on variety. *Pelargonium* cultivars are widely classified into the following informal groups, which only partly reflect the technical taxonomic divisions recognized in the genus.

### 45.2.1 Zonal Geraniums

These are mainly hybrids derived from *P. zonale*, *P. inquinans*, and a few other species in section *Ciconium*. Zonals are the most common kind of geranium, with over 2,000 recognized cultivars, including the ill-defined *P.* × *hortorum*. They are upright, bushy, succulent-stemmed perennials grown for their single or double flowers. The group includes large plants as well as "dwarfs" (13–20 cm tall) and "miniatures" (not exceeding 13 cm). As the name *zonal* suggests, there is a dark (pigmented) crescent or horseshoe-shaped zone on the leaf (although not in all selections), and various cultivars are noted for their attractive, rounded foliage. This class is mostly used for bedding displays and containers.

### 45.2.2 Regal Geraniums

This group of hybrids is derived from *P. cucullatum* and perhaps a few other species, including *P. graveolens*, *P. angulosum*, *P. fulgidum*, and *P. grandiflorum*. They are also known as Martha Washington, Lady Washington, and show pelargoniums. The plants are bushy evergreen perennials and shrubs with rounded, sometimes lobed or deeply dissected, leaves producing single or rarely double flowers in shades of mauve, pink, purple, or white and grown both for outdoor and indoor display. The flowers are much larger than in the zonals, and individual potted plants can be rather spectacular. Regal geraniums have beautifully colored petals in

FIGURE 45.3 Paintings (public domain) by European botanical illustrators showing geraniums (*Pelargonium*) cultivated in the 19th century.

***Source:* (a) Step, E., and D. Bois. 1896. *Favourite Flowers of Garden and Greenhouse*, vol. 1. Frederick Warne, London. (b) *Flore des serres et des jardins de l'Europe* (Ghent, Belgium), vol. 15, 1851. (c) Andrews, J., and V. Brooks. 1867. *The Floral Magazine*. Reeve & Co., London.**

FIGURE 45.4 Regal geranium cultivars: (a) Cultivar Regal. (b) Cultivar Tornado.

***Source:* (a) Photo by Habib M'henni (CC BY 3.0). (b) Photo by Andy Mabbett (CC BY 3.0).**

a wide range of hues but tend to have a restricted flowering period, do not do well in wet summers, attract whiteflies, and tend to be less successful outdoors than the zonals.

### 45.2.3 Angel Geraniums

Angels are mostly derived from *P. crispum* with input also from *P. grossulariodes*. These are similar to regals but are more compact and bushy and are rarely taller than 30 cm. They are said to have a distinctive "pansy-face."

### 45.2.4 Ivy-Leaved Geraniums

Derived especially from *P. peltatum*, this class consists of trailing evergreen perennials with stiff, fleshy, shiny, five-angular-lobed leaves (reminiscent of ivy, *Hedera helix*) and single or double flowers like zonal geraniums. Ivy-leaved geraniums are frequently grown in hanging baskets.

### 45.2.5 Unique Geraniums

This is a class of shrubby plants, woody at the base, mostly cultivated in the early 19th century, probably from crosses with *P. fulgidum*.

**FIGURE 45.5** *Pelargonium* cultivar Angeleyes Randy.

*Source:* **Photo by Gothika (CC BY 3.0).**

**FIGURE 45.6** Paintings (public domain) by European botanical illustrators showing "angeleye" geraniums (*Pelargonium*) cultivated in the 19th century.

*Source:* **(a) Andrews, J., and V. Brooks. 1867. *The Floral Magazine*. Reeve & Co., London. (b) *Neerland's Plantentuin* (Groningen, Netherlands), vol. 1, 1865–1867. (c) *The florist, fruitist, and garden miscellany*, 1859, Chapman and Hall, London.**

### 45.2.6 Scented-Leaved Geraniums

These are shrubby evergreen perennials and shrubs, which are mainly cultivated for their scented and often distinctly lobed, toothed, or incised or variegated leaves. These plants have a wide range of kinds of foliage and habits. These are excellent houseplants, although the flowers are usually not showy. This is a very heterogenous group.

## 45.3 HISTORY

The first *Pelargonium* species exported to Europe was *P. triste*, transported from Cape Province to The Netherlands and England at least by the early 16th century. Subsequently, European gardeners created many strains by hybridization. Although some modern cultivars of *Pelargonium* are derived from one species, most have originated by hybridization from two or frequently several species. Two dozen or so wild species have contributed the parentage of most cultivated geraniums. Much of the hybridization occurred during Victorian times, when rich landowners included unusual specimens in their conservatories and glasshouses.

**FIGURE 45.7** Ivy-leaved cultivars of geranium (*Pelargonium*). (a) A 1917 postcard featuring displays of ivy geranium in Berkeley, California. (b) *Pelargonium* × *lateripes* = *P. peltatum*. (c) *Pelargonium peltatum*.

*Source:* **Public domain figures. (a) Published by Edw. H. Mitchell. (b) Step, E., and D. Bois. 1896. *Favourite Flowers of Garden and Greenhouse*, vol. 1. Frederick Warne, London. (c) Dutch botanical artist Jan Moninckx (ca. 1656–1714), from his nine-volume *Moninckx Atlas* (1686–1756).**

**FIGURE 45.8** Geranium (*Pelargonium*) cultivars of the unique class: (a) Cultivar Crimson Unique. (b) Cultivar White Unique. (c) Cultivar Purple Unique.

*Source:* **(a) Photo by Paolo Harlock (CC BY 2.0). (b) Photo by HQ (CC BY SA 2.0). (c) Photo by HQ (CC BY SA 2.0).**

FIGURE 45.9 Rose geranium (hybrids of *Pelargonium graveolens*), the most popular scented-leaved geranium, and the source of rose geranium oil: (a) Drawing of the plant. (b) Illustration of essential oil extract from the plant.

*Source:* **(a) Step, E., and D. Bois. 1896. *Favourite flowers of garden and greenhouse*, vol. 1. Frederick Warne, London. (b) Shutterstock/Daria Ustiugova.**

FIGURE 45.10 Commercial production of geraniums (*Pelargonium*): (a) An Illinois greenhouse full of plants. (b) A greenhouse worker and her geraniums.

*Source:* **(a) Photo by Wendy Piersall (CC BY 2.0). (b) Photo by Shutterstock/Andriiii.**

## 45.4 ECONOMIC IMPORTANCE

Geraniums are very widely sold in the nursery trade, principally in temperate climates. Rose geranium oil is a significant industrial extract employed in the perfumery industry.

## 45.5 ORNAMENTAL USES

Geraniums are very popular bedding plants, used as staples in public and residential gardens, in beds and borders, and in containers maintained indoors and outdoors. Geraniums are one of the most widely employed plants on interior windowsills and exterior window boxes.

## 45.6 OTHER USES

All *Pelargonium* species have gland-tipped hairs containing essential oils, some species producing much more than others. The "scented-leaved" or "fragrant-leaved" geraniums are a mixed group of odoriferous species, including several dozen used for culinary purposes. The species smell like fruits (apricot, apple, peach, strawberry), citrus (lemon, lime, orange), "nuts" (coconut, filbert), spices (allspice, cinnamon, ginger, nutmeg), mint, pine, camphor, and especially roses. It should be noted that the perception of smell is at least somewhat unique to individuals, and one person's apricot-scented plant may be another's sour milk. The taste does not necessarily reflect the smell. So-called chocolate geraniums have a brownish spot on the leaves, but imagination is required to perceive a chocolate taste. Scented-leaved geraniums are widely grown in flower gardens and as potted

FIGURE 45.11 Landscape uses of geraniums (*Pelargonium*): (a) A bed of mixed cultivars. (b) The front of a house in Italy decorated with pots of geraniums. (c) Window boxes. (d) Puente de las Flores (flower bridge) in Valencia, Spain, decorated with flowers, especially pelargoniums. (e) Buckingham Palace and Victoria Memorial in London, UK.

*Source:* **(a) Photo by Shutterstock/Ailisa. (b) Photo by Shutterstock/JeniFoto. (c) Photo by Shutterstock/Anna Nahabed. (d) Photo by Diego Delso (CC BY SA 3.0). (e) Photo by Shutterstock/Pajor Pawel.**

FIGURE 45.12 A cake decorated with edible geranium flowers.

*Source:* **Photo by Shutterstock/Mostovyi Sergii.**

plants, although their flowers are often small and not as attractive as species grown mainly for floral display. Plants that lack an appetizing aroma should not be used as culinary herbs, but an attractive odor does not guarantee an attractive taste. The wide range of geranium tastes and scents allows for an extensive range of culinary applications. For edible purposes, fresh or dried leaves are used to flavor gelatins, puddings, preserves, and the like. The leaves can be infused in liquid, and the filtered liquid employed to impart aroma to sauces, custards, jellies, buns, water ices, butters, jams, sugar, syrups, vinegars, meats, and soups. Cakes can be flavored simply by putting a leaf in the bottom of a cake pan. The flowers can be tossed in salads. Geranium ice cubes can be prepared by freezing sprigs (especially of lemon- or mint-flavored forms) in an ice cube tray. Geranium oil is used in baked goods and fruit desserts. Except for some utilization of geranium oil as a flavoring in commercial food preparations, the culinary use of geranium is insignificant and is essentially confined to occasional home use.

Geranium oil is extensively used in manufacturing perfumes and for providing scent for cosmetics, soap, and other toilet products. Hybrids of *Pelargonium* species with rose geranium (*P. graveolens*) as one of the parents are employed as a source of geranium oil. These hybrids are especially a source of "rose geranium oil," which is almost indistinguishable from the more expensive oil of roses. Geraniums are grown commercially as a source of geranium oil in many countries, especially China, Egypt, Morocco, and the South African island of Réunion, but also in Algeria, El Salvador, France, India, Madagascar, and Spain. The annual world production of geranium essential oil is about 250 t, with a value exceeding $10 million.

Extracts from *Pelargonium* species are occasionally employed in modern Western medicine. There have been many folk uses, some persisting to the present. (*Geranium* species are much more commonly used in herbal medicine than are *Pelargonium* species, and accounts of "geraniums" employed medicinally often confuse the two genera.) Geranium oil from *Pelargonium* is used in aromatherapy. In their native area of Africa, the main region where *Pelargonium* species were used medicinally, the plants have been regarded as beneficial for treating diarrhea and dysentery, with the root of several species employed to treat syphilis. In Middle America, geranium decoctions (teas) are added to baths to relieve skin problems. Rose geranium leaves mashed in vinegar and salt were once applied in New Mexico to the forehead for headache, and warmed leaves were stuffed into the ear for earache. In the West, geranium has been used as a folk remedy in the treatment of ulcers. In Chinese medicine, the essential oil of geranium has been applied locally to treat cervical cancer.

## 45.7 PROBLEMS

While consumption of *Pelargonium* species is not thought to be a hazard for humans, it is quite dangerous for dogs and cats. Geraniol and linalool are responsible for this toxicity

FIGURE 45.13 Well-known paintings by European artists showing cats and geraniums on a windowsill. The geraniums are capable of poisoning the cats: (a) *Cat in a Cottage Window* (1881), by British painter Ralph Hedley (1848–1913). (b) *Flowers and cats* (also 1881), by French artist Pierre-Auguste Renoir (1841–1919).

and can also cause skin irritation and skin allergies in susceptible people.

## 45.8 CULTIVATION

### 45.8.1 Climate and Site

In warm climates (zones 9 to 11, debatably if protected in zone 8), geraniums can be grown outdoors as perennials, but in colder areas, they are normally treated as annuals. Although many geraniums are evergreen perennials, almost all the species are tender. *Pelargonium* does not tolerate frost. In cold climates, geraniums can be brought indoors during the winter as potted houseplants, preferably in a cool location, and lightly watered. Almost all *Pelargonium* species grow and flower best under cool conditions. Curiously, they are poor ornamentals in their native South Africa because the spring season, although ideal, is only a few weeks long and is followed by a summer which is far too hot. By contrast, the whole summer is ideal in Europe, where one sees splendid displays in flower beds. Areas of the tropics and subtropics where geranium oil is produced are all at high altitudes, with a temperate climate all year round. A high light level is very important, and in nature, practically all species occur in direct sunlight (although some also grow well in semi-shade).

### 45.8.2 Propagation and/or Purchasing

Some cultivated geraniums can be grown from seed (which can be obtained commercially or from the capsules). Seeds are sown shallowly and preferably kept at no lower than 13°C. Germination usually occurs in several weeks or, in some cases, in several months. Many geranium plants are of hybrid origin and need to be propagated vegetatively. Most gardeners today buy geranium plugs (young plants or seedlings in small fiber pots). Most geraniums can be reproduced by root cuttings, but stem cuttings are universally employed. Reproducing plants by cuttings rather than dividing the plants tends to result in healthier stock. The best time to take stem cuttings is soon after the flowering season. Typically, cuttings are made in early fall, before frost. Unlike with some other houseplants, cuttings of geraniums tend to rot if the attempt is made to root them in water. (Geranium cuttings are widely started in water, but planting them in moist soil tends to be more successful.) Cuttings (7–13 cm long) from short-jointed young shoots can be placed in a sand-and-peat mixture to root. Using a sharp knife, cleanly cut a stem just below a node. Large varieties require longer cuttings 10–13 cm long, while cuttings of smaller varieties can be about 7 cm long. Remove all but three leaves. Cut ends may be dipped in root-stimulating hormone. The cuttings should be inserted sufficiently deep in sand that they will stay erect. These are watered and kept shaded for a few days, then gradually exposed to sunlight. Subsequently, the cuttings should be kept on the dry side and not overwatered. Cuttings are potted when the roots are 1.5–2 cm long.

### 45.8.3 Planting

In beds and borders, a planting distance of 30 cm is appropriate for most cultivars, as little as 15 cm for dwarf varieties and as much as 60 cm for the largest varieties.

### 45.8.4 Soil and Moisture Conditions

In nature, all species of *Pelargonium* grow well on nutrient-poor sand, but geraniums also thrive in well-rotted, well-drained compost and fertile soils. A pH of 6.0–6.8 is recommended. For pot culture, three parts loam, one part coarse sand, and one part leaf mold (or one-half part leaf mold and one-half part peat moss) has been recommended. A commercial grower's mix prepared specifically for geranium can often be purchased. Little addition of fertilizer is necessary if soils in beds have been well prepared (but plants in containers should be fertilized regularly, and fertilizers formulated for geraniums may be available). A slow-release, granular fertilizer can be added in the spring or a liquid 20–20–20 fertilizer employed about three times during the growing season. In their native habits, geraniums generally grow on dryish, well-drained loamy soils, and the key to successful cultivation is a very well-drained soil. Growing geranium in the same soil year after year is inadvisable, as diseases may build up. Many of the species endure some drought and, indeed, are damaged by heavy rains. In general, plants should be watered sparingly, the soil allowed to become barely dry before watering. If the soil becomes too dry, the lower leaves tend to turn yellow and fall off. Pots of geraniums should be allowed to dry out a little between thorough soakings, but water should never be allowed to collect in pots or saucers under pots, and the foliage should not be misted.

### 45.8.5 Maintaining Outdoor Plants

Pinching stem leaders is often employed to encourage branching and avoid legginess. Deadheading spent flowers encourages continuous bloom throughout the summer. Pruning is only necessary for broken branches. Yellowing leaves are very common and can be removed as they appear.

### 45.8.6 Maintaining Indoor Plants

Potted plants that have been sunken in the garden during the summer can simply be brought indoors before threat of frost. Keeping the root systems of potted specimens slightly pot-bound usually prevents the plants from growing large and taking excessive room. Regular pruning of some species encourages branching, but many cultivars do not respond well to pruning. (However, some authorities recommend cutting branches back by two-thirds when they are brought indoors in the autumn.) Indoor plants should be kept fairly cool if possible (12°C at night and 18°C during the day are desirable), best achieved in a house by placing the plants close to a sunny but cool window. When overwintered indoors, geraniums should be given at least 4 hr of direct sunlight daily, or 14 to 18 hr of artificial light. A half-strength houseplant fertilizer can be used every 2 weeks during the flowering season, and monthly during the remainder of the year. Spent flower heads should be removed. It is difficult to maintain garden-grown plants disease- and insect-free indoors overwinter. Whiteflies, aphids, and other insects may weaken indoor plants considerably, but once taken outside again in the spring, they recover.

## 45.9 FASCINATING FACTS, FABLES, FOLKLORE, AND FRIVOLITY

- Geraniums have had some interesting symbolic associations. In Poland, pelargoniums are considered to be symbols of hope and protectors of the home. In traditional Bulgarian weddings, every bouquet and wreath should have a red carnation – symbol of manliness – and a geranium – symbol of health ("geranium" in Bulgarian is "zdravetz," and the word *health* is the very similar "zdrave"). In 1964 in Chicago, a geranium on a white person's windowsill signified that they opposed racism.
- According to a Muslim legend, the first geranium was created when the prophet Mohammed washed his shirt and hung it over a mallow plant to dry. The humble mallow was transformed into a magnificent geranium.
- English chemist and physicist John Dalton (1766–1844), who pioneered modern atomic theory, first realized he was color-blind when he heard others describe some geraniums as red while he perceived them as blue. He went on to conduct research on color blindness, which is sometimes called Daltonism in his honor.

- A mixture of the fat of a hippopotamus or a python and the powdered root of *Pelargonium luridum*, which has flowers that release their scent at night, has been rubbed on the faces of young men during Zulu courtship rituals in South Africa, in the belief that this charmed the opposite sex.
- South African Xhosa warriors fighting European settlers in the 1850s thought that the roots of *P. pulverulentum* warded off the bullets of their enemies.
- In South Africa, the foliage of *P. papilionaceum* has been smoked and employed as a tobacco substitute.
- In the Victorian era (1837–1901), "floriography" or "the language of flowers" was a coded system of communicating private emotions by mentioning plants or flowers in letters or placing them in small bouquets called tussie-mussies, which were very fashionable at that time. "Geranium" unqualified meant gentility. Some more specific geranium meanings are: lemon geranium, unexpected meeting; nutmeg geranium, expected meeting; oakleaf geranium, true friendship; rose-scented geranium, preference; fish (zonal) geranium (which can be unpleasantly scented), disappointed expectation.
- In Victorian England, pots of scented geraniums were strategically positioned indoors so that the long skirts of ladies would brush against them and spread the scent throughout the room.
- Scented geraniums are very popular as houseplants but generally do not thrive in the low light levels in many interiors. With the invention of Edison's electric light in the late 1800s, one of the first houseplants suggested as benefitting from the extra illumination was the scented geranium.
- The *Guinness Book of World Records* reported a record length of 3.4 m for a geranium plant grown in 1997.
- The Japanese beetle (*Popillia japonica*), an alien invader that arrived in North America about a century ago, is one of the most destructive pests of ornamental and turf plants, feeding on hundreds of different species. Established populations are in the eastern United States and adjacent Canada. Within half an hour of eating the petals (but not the foliage) of zonal or some other geraniums, the insect becomes paralyzed for several hours but will usually recover (unless discovered by a predator) within 24 hr and may resume feeding on geraniums. The amino acid quisqualic acid is responsible for the paralysis, through to affect neuromuscular transmission in the insect.
- Methylhexanamine (also known as 1,3-dimethylamylamine and DMAA) has been extensively marketed as a dietary supplement, employed by athletes and body builders. It has also been called "Geranamine," and according to an early report (which has been discounted), it is present in "geranium oil." The chemical has become widely known as "geranium." Deaths have been reported from its use, which was restricted (essentially banned as a supplement) in Canada in 2011, and in the United States in 2013. It has been forbidden by numerous sports organizations.
- "Geranium" (the plant) is often confused with "germanium," a chemical element with the symbol *Ge*. This is a lustrous, hard grayish-white metalloid in the carbon group, chemically similar to tin and silicon. Purified germanium is a semiconductor. John Bardeen and Walter Brattain invented the transistor in 1947 by inserting two electrodes into a sliver of germanium. Consequently, they and their colleague William B. Shockley received the Nobel Prize for Physics in 1956. The transistor effectively initiated the information age and has been characterized as the most important invention of the 20th century. There are many accounts on the web that mistakenly claim the invention of the transistor was based on "geranium."
- Because of its citronellol content, geraniums have been reported to have some value in repelling insects, and indeed, some scented geraniums are actually used as intercrops to repel insects. However, a hoax involved the marketing of a scented geranium called 'Mosquito Fighter'. It was claimed to be the result of genetic engineering that transferred to it a gene from citronella grass (*Cymbopogon nardus*) producing citronella oil, a common ingredient in insect repellents. Said to keep mosquitoes, black flies, and other insects away (tests have shown that these claims are untrue – indeed, mosquitoes were observed happily landing on the plants), the plants are still widely sold today. 'Mosquito Fighter' has produced allergic reactions. However, such "citrosa geraniums" (often identified by the invalid name *P. citrosum*) can be grown simply for their attractive, lemony scent.

FIGURE 45.14 *Pelargonium* 'Mosquito Fighter' on sale in a nursery.

*Source:* Photo by Mokkie (CC BY SA 4.0).

## KEY PUBLICATIONS


Armitage, A.M. 1992. *Seed-Propagated Geraniums and Regal Geraniums: Production Guidelines and Future Concerns.* Timber Press, Portland, OR. 136 pp.

Bagust, H. 1988. *Miniature and Dwarf Geraniums (Pelargoniums)*, Revised edition. C. Helm, London, UK. 163 pp.

Bakker, F.T., A. Culham, P. Hettiarachi, T. Touloumenidou, and M. Gibby. 2004. Phylogeny of *Pelargonium* (Geraniaceae) based on DNA sequences from three genomes. *Taxon* 53: 17–28.

Becker, J. and F. Brawner. 1996. *Scented Geraniums: Knowing, Growing, and Enjoying Scented Pelargoniums.* Interweave Press, Loveland, CO. 96 pp.

Bennett, M. 1972. *Geraniums: The Successful Grower's Guide.* Hippocrene Books, New York. 141 pp.

Blerot, B., S. Baudino, C. Prunier, F. Demarne, B. Toulemonde, and J.-C. Caissard. 2016. Botany, agronomy and biotechnology of *Pelargonium* used for essential oil production. *Phytochemistry Reviews* 15: 935–960.

Boddy, K. 2013. *Geranium.* Reaktion Books, London, UK. 216 pp.

Brawner, F. 2003. *Geraniums: The Complete Encyclopedia.* Schiffler, Atglen, PA. 176 pp.

Clark, D.W.H. 1988. *Pelargoniums.* Timber Press, Portland, OR. 124 pp.

Clifford, D.P. 1970. *Pelargoniums, Including the Popular 'Geranium'*, 2nd edition. Blandford Press, London. 350 pp.

Conder, S. 1992. *The Complete Geranium: Cultivation, Cooking, and Crafts.* Crown Publishers, New York. 89 pp.

Cross, J.E. 1951. *The Book of the Geranium.* Saturn Press, London. 184 pp.

Esser, M. 1998. *Pelargoniums.* Rebo, Lisse, The Netherlands. 144 pp.

Feltwell, J. 2001. *Geraniums and Pelargoniums.* Firefly Books, Buffalo, NY. 128 pp.

Fogg, H.G.W. 1975. *Geraniums and Pelargoniums.* J. Gifford, London. 158 pp.

Horst, R.K. and P.E. Nelson. 1985. *Diseases of Geraniums.* Cornell University, Ithaca, NY. 31 pp.

Key, H. 2000. *1001 Pelargoniums.* Batsford, London. 184 pp.

Krauss, H.K. 1955. *Geraniums for Home and Garden.* Macmillan, New York. 194 pp.

Lis-Balchin, M. (ed.). 2002. *Geranium and Pelargonium: The Genera Geranium and Pelargonium.* Taylor & Francis, New York. 318 pp.

Mastalerz, J.W. and E.J. Holcomb. (eds.). 1982. *Geraniums: A Manual on the Culture of Geraniums as a Greenhouse Crop*, 3rd edition. Pennsylvania Flower Growers, University Park, PA. 410 pp.

Meve, U. and F. Albers. 2022. *Pelargonium*, Geraniaceae. In: *Dicotyledons: Rosids. Illustrated Handbook of Succulent Plants*, edited by U. Eggli and R. Nyffeler, 1–86. Springer, Cham.

Miller, D. 1996. *Pelargoniums, a Gardener's Guide to the Species and Their Cultivars and Hybrids.* Batsford, London. 175 pp.

Miller, D. 2002. The taxonomy of *Pelargonium* species and cultivars, their origins and growth in the wild. In: *Geranium and Pelargonium: The Genera Geranium and Pelargonium*, edited by M. Lis-Balchin, 49–79. Taylor & Francis, New York.

Peddie, M., J. Lewis, and J. Lewis. 1991. *Growing & Using Scented Geraniums.* Storey Communications, Pownal, VT. 32 pp.

Plaschil, S., H. Budahn, M. Wiedemann, and K. Olbricht. 2017. Genetic characterization of *Pelargonium* L'Hér. *germplasm. Genetic Resources and Crop Evolution* 64: 1051–1059.

Ranger, C.M., R.E. Winter, A.P. Singh, M.E. Reding, J.M. Frantz, J.C. Locke, et al. 2011. Rare excitatory amino acid from flowers of zonal geranium responsible for paralyzing the Japanese beetle. *PNAS* 108: 1217–1221.

Schulz, P. 1965. *All about Geraniums.* Doubleday, Garden City, NY. 175 pp.

Shellard, A. 1981. *Geraniums for Home and Garden.* David & Charles, Newton Abbot, UK. 232 pp.

Struck, M. 1996. Floral divergence and convergence in the genus *Pelargonium* (Geraniaceae) in Southern Africa: Ecological and evolutionary considerations. *Plant Systematics and Evolution* 208: 71–97.

Taylor, J. 1988. *Geraniums and Pelargoniums: The Complete Guide to Cultivation, Propagation, and Exhibition.* Crowood Press, Ramsbury, Marlborough, UK. 176 pp.

Taylor, J. 1990. *Pelargoniums: For Colour and Variety.* Crowood Press, Ramsbury, Marlborough, UK. 128 pp.

Tucker, A.O. and M.J. Maciarello. 1993. Is this plant a hoax? A lecture in honor of Otto Richter. In: *Herbs 93: Eighth National Herb Growing and Marketing Conference*, edited by A. Kestner and J.E. Simon, 177–182. International Herb Growers & Marketers Association, Bellevue, WA.

Van der Walt, J.J.A, P.J. Vorster, and E. Ward-Hilhorst. 1981. *Pelargoniums of Southern Africa.* Juta, Cape Town, South Africa. 148 pp.

Weng, M.-L., T.A. Ruhlman, M. Gibby, and R.K. Jansen. 2012. Phylogeny, rate variation, and genome size evolution of *Pelargonium* (Geraniaceae). *Molecular Phylogenetics and Evolution* 64: 654–670.

White, J.W. (ed.). 1993. *Geraniums IV: The Grower's Manual*, 4th edition. Ball Publications, Geneva, IL. 412 pp.

Wilkinson, A. 2007. *The Passion for Pelargoniums: How They Found Their Place in the Garden.* Sutton Publishers, Stroud, UK. 302 pp.

Wilson, H. Van Pelt. 1965. *The Joy of Geraniums; The Standard Guide to the Selection, Culture, and Use of the Pelargonium.* M. Barrows, New York. 364 pp.

Wood, H.J. 1966. *Pelargoniums: A Complete Guide to Their Cultivation*, 2nd edition. Faber, London. 168 pp.


### KEY CULTIVATION WEBSITES


Clemson University, Extension. *Growing Geraniums Indoors.* https://hgic.clemson.edu/factsheet/growing-geraniums-indoors/.

Geraniums Online. www.geraniumsonline.com/index.htm.

Pelargonium & Geranium Society (U.K). https://thepags.org/about-us/.

Pelargonium Page. www.pelargonium.si/.

The Spruce. *How to Grow Regal Geraniums.* www.thespruce.com/growing-and-caring-for annual geranium-plants-1402908.

WikiHow. *How to Grow Geraniums.* www.wikihow.com/Grow-Geraniums.

WikiHow. *How to Grow Geraniums in Pots.* www.wikihow.com/Grow-Geraniums-in-Pots.

WikiHow. *How to Propagate Geraniums from Cuttings.* www.wikihow.com/Propagate-Geraniums-from-Cuttings.

WikiHow. *How to Prune Geraniums.* www.wikihow.com/Prune-Geraniums.

# 46 Gerbera (*Gerbera* Species and Hybrids)

**Scientific Names:** The genus *Gerbera* was named for the German medical doctor and naturalist Traugott Gerber (1710–1743), director of a public garden in Moscow and collector of numerous plants throughout Russia.

**Common Names:** Gerbera, gerber daisy; African daisy (other members of the Asteraceae are also called African daisy, particularly cultivars of *Osteosperum* and *Arctotis*, which have flowers resembling those of *Gerbera*; flowers of *Gazania* also can look like those of gerbera). Often, common names for *G. jamesonii*, the main ancestor of domesticated forms, are employed for *Gerbera* cultivars, regardless of their parental species. Names for *G. jamesonii* include Barberton daisy (Robert Jameson, a Scotsman, for whom the species was named, was a prominent Durban merchant who collected specimens in the area of Barberton, a mining town in South Africa), Transvaal daisy (the species is endemic to the Transvaal region of South Africa), flame-ray daisy, and veldt daisy.

FIGURE 46.1 Colorful displays of gerbera flowers.

*Source:* **(a) Fotosforyou-rk from Pixabay (public domain). (b) Manuela724 from Pixabay (public domain). (c) Photo by Shutterstock/Makasana photo.**

## 46.1 INTRODUCTION

Gerbera is one of the stars of the ornamental floral industry, ranking among the top ten cut flowers in the world and also often sold as a potted plant. The flowering tops are very beautiful, available in a wide range of spectacular colors, and are long-lasting both in vases and pots. As a sales commodity, gerberas are one of the leading supermarket plants, appealing strongly to impulse buyers as well as to more experienced purchasers who have become familiar with these attractive flowers.

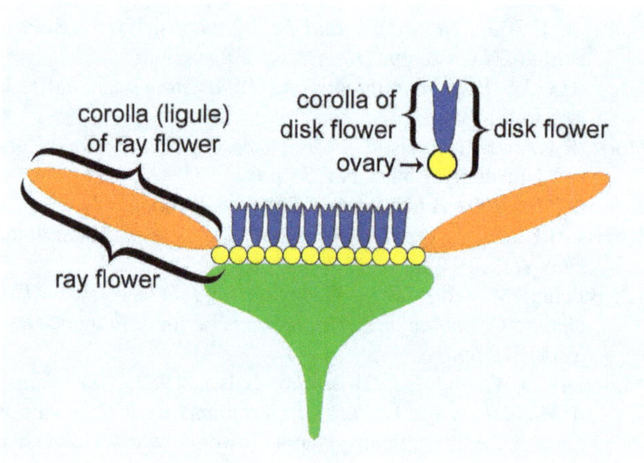


FIGURE 46.2 Simplified sectional view of a flower of the daisy (Asteraceae) family.

*Source:* **Prepared by B. Brookes.**

## 46.2 OVERVIEW

Thirty to forty species of *Gerbera* have been recognized, but the delimitation of the genus requires reassessment. The species are native to South Africa (at least 15 species) and Asia. Two South American species were recently transferred to *Gerbera*, but this is questionable. Gerberas mostly inhabit temperate and mountainous regions. The plants are herbaceous, developing from perennial rootstocks (which can become somewhat woody with age). They produce a basal rosette of leaves and one to several long leafless "stems" each bearing a solitary flower head. *Gerbera* does not produce significantly tall true stems and is often described as stemless.

*Gerbera* is a member of the Asteraceae, the flowers of which are arranged in "daisylike" inflorescences ("heads," technically "capitula"; singular: *capitulum*). The flowering heads are usually simply called "flowers" by non-botanists, and indeed by the floral industry. The *ray flowers* on the periphery (outermost whorl) of the head have a prominent asymmetrical ("zygomorphic") corolla which mostly appears as a single strap-like petal (technically, the "ligule," consisting of three fused true tiny petals). The ray flowers of *Gerbera* are large – up to 8 cm in length. The central (*disk*) flowers have comparatively inconspicuous symmetrical ("actinomorphic") corollas with five evident petals. *Trans* flowers are located between ray and disk flowers. They have shorter,



DOI: 10.1201/9781003586227-46

less-conspicuous corollas than the ray flowers, and toward the center of the inflorescence, their corollas gradually become less prominent and more symmetrical. In *Gerbera*, the central disc flowers are hermaphroditic, while the marginal ray and intermediate trans flowers are female (the anthers aborting during flower development). Although the large "petals" (ligules) of the ray flowers serve to attract pollinators, only the central disk flowers produce functional pollen, and pollinators congregate only on the disk flowers.

The South African *G. jamesonii* was formally recognized by the botanist J. D. Hooker in 1889 and is considered to be the most important ancestor of modern gerbera cultivated varieties. Numerous horticultural publications today identify gerbera cultivars by the name *G. jamesonii*, although other species are usually involved in their parentage. *Gerbera jamesonii* is endemic to Transvaal and Swaziland in Southern Africa. It occurs in bushveld (mixed woodland) and steep slopes, on dolomitic and stony clay soils, on burned ground, and in dry, shaded habitats.

*Gerbera viridifolia* is considered to be the second most important ancestral species of modern cultivars. It is rather variable, a number of alleged species now included in it. *Gerbera viridifolia* ranges along eastern Africa, occurring in open grassland with stony soil.

Houseplant potted gerberas are usually less than 60 cm high and spread laterally no more than 45 cm. Some gerbera cultivars produce heads up to 15 cm in diameter, but "mini" cultivars with small (less than 10 cm wide) but numerous heads have recently become very popular. Cultivars used for cut stems have long stems and large foliage (both characteristics considered undesirable in pot cultivars).

Gerbera flowering heads consist of one to several whorls of ray flowers and several whorls of trans and disk flowers. In wild gerbera flower heads, there is just one whorl of large peripheral ray flowers. Some cultivars have acquired "doubled" or "crested" heads. Most so-called doubled ornamental flower species (outside of the Asteraceae family) simply have extra petals in each flower. However, in doubled gerbera heads (and doubled flowers of other Asteraceae), individual flowers retain the same number of petals, but the corollas of additional whorls of flowers (particularly trans flowers) have become as large as the outermost whorl of ray flowers. Some doubled gerbera cultivars produced heads with as many as 500 of the large ray flowers. "Spider flowers" (the plants are sometimes called "spider daisies") have a laciniated (split or notched) corolla, so that the heads appear lacy or fringed.

FIGURE 46.3 *Gerbera jamesonii* hybrid cultivars: (a) Cultivar Festival Apricot with Eye. (b) Cultivar Revolution.

*Source:* **(a) Photo by Jim, the Photographer (CC BY 2.0). (b) Photo by Satin66Flower (CC BY SA 4.0).**

FIGURE 46.4 Wild *Gerbera viridifolia* growing in a grassland in Pretoria, South Africa.

*Source:* **Photo by SAplants (CC BY SA 4.0).**

FIGURE 46.5 Pots of "mini" gerberas (with relatively small but numerous flowering heads).

*Source:* **Photo by Serres Fortier (CC BY 2.0).**

**FIGURE 46.6** Highly "doubled" form of gerbera flowering head (i.e., the number of whorls of flowers with large rays has been increased).

*Source:* **(a) Photo by Monica Loirinha (CC BY SA 2.0). (b) Photo by Miglena from Pixabay (public domain). (c) Photo by ThomasWolter from Pixabay (public domain). (d) Photo by Ben A. Bergmann (CC BY 2.0). (e) Photo by Sonyuser from Pixabay (public domain). (f) Photo by Simplysabi0116 from Pixabay (public domain).**

**FIGURE 46.7** Old paintings (public domain) of "spider" forms of gerbera flowering head (i.e., the ligules (petals) of the rays are slim and long, like a spider's legs).

*Source:* **(a) *Revue horticole* (Paris), Sér. 4, vol. 81, 1909. (b) *Flora and Sylva*, a monthly review for lovers of garden, woodland, tree, or flower; London, 1903–1905. (c) Armstrong Nurseries (Ontario, California) 1941 fall bulb catalog.**

**FIGURE 46.8** Photos of "spider" forms of gerbera flowering head: (a) Plants growing in a garden in Japan. (b) Flower with very slim "petals."

*Source:* **(a) Photo by M-Louis (CC BY SA 2.0). (b) Photo by Rick Ligthelm (CC BY 2.0).**

Gerbera cultivars are available in a wide array of brilliant flower colors, including shades of cream, lavender, orange, pink, purple, red, salmon, scarlet, violet, white, and yellow. True blue and green flowers are not yet available, and in stores, flowers with such colors have been dyed. The center of the heads (often called the "eye" of the flower) is often a contrasting darker color (commonly greenish, cream, or dark brown). Since the disk flowers have very small corollas, the color is usually due to the accumulation of anthocyanins in the bristles (collectively the "pappus") at the base of the disk flowers. Bicolored flowers have also been selected based on differently colored whorls of ray flowers or on different coloration of the ligules.

**FIGURE 46.9** Gerbera flowering heads with bicolored petals.

*Source:* **(a) Photo by Andre Rau from Pixabay (public domain). (b) Photo by Ben A. Bergmann (CC BY 2.0).**

## 46.3 HISTORY

Gerberas have a relatively limited history as cultivated flowers. Breeding of gerbera started at the end of the 19th century in England, with the crossing of two South African species, *G. jamesonii* and *G. viridifolia.* The hybrid was originally called *Gerbera cantebrigiensis* ("of Cambridge," where the cross was made) but is often known today as *G.* × *hybrida*. Most cultivars originate from crosses of these two species. By the beginning of the 20th century, gerbera was grown in much of Europe and North America. Today, The Netherlands is the center of breeding of gerbera cultivars, but the United States, Canada, Israel, Japan, Korea, Brazil, China, India, and other countries are also involved in creating new cultivars. It has been estimated that thousands of cultivars have been released in the last century. Currently, more than 1000 gerbera cultivars are produced annually, about a quarter of them on a large scale.

## 46.4 ECONOMIC IMPORTANCE

On a world basis, gerbera frequently ranks fifth in the cut flower industry, exceeded in importance only by rose, carnation, chrysanthemum, and tulip. Gerbera is now produced commercially in over 80 countries. In the United States, cut flower sales of gerbera have exceeded $30,000,000 annually, for about 100 million stems. Gerbera is well adapted to be a cut flower, as the harvested stems have excellent keeping qualities, rehydrate rapidly, and withstand the rigors of transportation well.

FIGURE 46.10 Old paintings (public domain) of the two parental species of most *Gerbera* cultivars: (a) *Gerbera jamesonii*. (b) *Gerbera viridifolia*.

**Source:** **(a) Hooker, J. D. (ed.). 1889. *Curtis's Botanical Magazine, Plate 7087*. L. Reeve & Co., London. (b) Pole-Evans, I. B., E. P. Phillips, and R. A. Dyer. 1923. *Flowering Plants of (South) Africa*, vol. 3: t. 85. South African National Biodiversity Institute, Pretoria, South Africa.**

FIGURE 46.11 Commercial production of cut flower gerberas in The Netherlands: (a) A greenhouse full of gerberas. (b) Apparently identical greenhouse workers preparing bouquets. Note the very long stems typical of cut flower cultivars.

**Source:** **(a) Photo by Alias 0591 (CC BY 2.0). (b) Photo by Sander van der Wel (CC BY SA 2.0).**

## 46.5 ORNAMENTAL USES

Gerberas are predominantly employed as cut flowers, and there are large production industries providing the plants to florists. The plants are occasionally grown as potted plants, both indoors and outside, and in warm climates, they are occasionally treated as garden ornamentals.

FIGURE 46.12 Cut flowers of gerbera, the main use of the plant.

**Source:** **(a) Photo by Shutterstock/Kozak Photo. (b) Photo by Shutterstock/Brian Chase. (c) Cut flowers on sale in Rio de Janeiro, Brazil. Photo by Anna Carol (CC BY 2.0). (d) Photo by Shutterstock/Viacheslav Lopatin.**

FIGURE 46.13 Cut flowers employed as hair decorations.

**Source:** **(a) Photo by Vera Kratochvil (public domain). (b) Photo by Shutterstock/Irina Bg.**

## 46.6 OTHER USES

Cultivars of gerbera are not considered to pose a poison threat to children or pets but are not generally viewed as edible. Nevertheless, in their native areas, species of *Gerbera* are sometimes employed in cooking and medicine. In Indigenous communities of Southwest China, *G. piloselloides* is consumed as a spice for meat and distilled beverages, and is used as a traditional herbal remedy for various conditions.

## 46.7 PROBLEMS

Gerberas are subject to various diseases and pests, but no more so than other leading flowering ornamentals. The chief limitation is their narrow climate tolerance – unsuitable for growing as outdoor perennials in cold climates, but also fairly intolerant of high temperatures.

## 46.8 CULTIVATION

### 46.8.1 Climate and Site

Gerberas are grown outdoors as perennials in public landscape and private residential flowerbeds, borders, and rock gardens, but only in frost-free climates (although the plants may withstand a short, mild freeze). Gerberas are rated to grow as perennials in climate zones 8 to 10 (also 11, but the higher temperature is a concern). In quite warm climates, the plants might stay evergreen, but in cooler areas, the plants go dormant over the winter and regrow in the spring. Gerberas need a full sun position that is sheltered from strong winds and frost. The plants are occasionally grown as annuals planted in garden beds in cold climates but are more often grown in containers, which allow the plants to be protected indoors overwinter. Gerberas rarely last longer than 3 years, unless divided. The plants are adapted to full sun, but because they are not tolerant of high heat, afternoon shade is recommended in hot climates.

### 46.8.2 Propagation and/or Purchasing

Those wishing to grow gerberas from seeds should realize that they will require 4 to 5 months for flowers to develop, and up to a year to produce a plant the size of those that can be purchased relatively cheaply. The seeds are sensitive to dryness, and while germinating, they should be kept moist, for example, by placing plastic wrap over the pot of soil containing the seeds. Often, attractive plants are clones, and their seeds will not reproduce true to type.

Although several decades ago gerberas were mostly reproduced commercially by seed, they are now usually propagated vegetatively. Gerberas can be grown from cuttings, but clumps are more commonly simply divided. Micropropagation (using tissue culture) has been the main approach to establishing cut flower gerberas since the 1970s.

FIGURE 46.14 A bed of gerberas. As noted in the text, the plants are grown as perennials in frost-free climates, but rarely as outdoor annuals in temperate regions.

*Source:* **Photo by Shutterstock/RKMT1.**

### 46.8.3 Planting

In outdoor beds, gerbera is planted 30 to 45 cm apart. Compact varieties are preferable for containers.

### 46.8.4 Soil and Moisture Conditions

When planted in beds, the soil should be enriched with compost and well-draining. Soil pH ideally is 5.5 to 6.2. Unamended clay soil is unsuitable. Regular fertilization (e.g., a soluble fertilizer every 2 weeks) or addition of a slow-release fertilizer is recommended. Excessive nitrogen can cause the foliage to overgrow at the expense of flowering, and some authorities recommend fertilizers with relatively limited nitrogen.

### 46.8.5 Maintaining Outdoor Plants

Deadheading is recommended to prolong flowering.

### 46.8.6 Maintaining Potted Plants Indoors

Gerberas can be employed as potted patio plants so long as the weather remains warm. Most container-grown gerberas are purchased for short-term indoor residential use and discarded in a few months, after the current flowering display is completed. Gerberas will last longer if kept in bright light (although they are shade plants in their native habitats), and cool temperatures (under 24°C, preferably 7–18°C) are best. Spent blooms should be removed (cut off at the flowering stem base) to prevent energy being diverted into seed production, and to improve appearance. Gerberas are considered to be relatively difficult indoor plants, but with careful treatment and good exposure to sun, they can last for several years. If transplanting to a larger pot, maintain a soil pH of 5.5 to 6.5. Gerberas benefit from slow-release fertilizers, preferably balanced or high in potassium, and prefer a moist but not waterlogged substrate. As with African violets, avoid placing water on the foliage.

FIGURE 46.15 Potted gerberas. (a) Girl and plant. (b) Cultivar Durora, a compact form often employed indoors.

*Source:* **(a) Photo by Shutterstock/Aldegonde. (b) Photo by Shutterstock/Madlen.**

FIGURE 46.16 Vases of gerbera.

*Source:* **Photo by Shutterstock/AnnaPustynnikova.**

### 46.8.7 Maintaining Cut Flowers

Stem end blockage of water transport ("vascular occlusion") due primarily to bacterial contamination is the chief cause of short vase life of most cut flower species, including gerbera. Cut flowering gerbera stems typically deteriorate within 5 days, but properly managed, they may last 10 days or so. Cut stems should be recut 2.5–5 cm from their bases at an angle with a sharp knife before being placed in a clean vase with lukewarm water, along with a preservative (which can be purchased from floral shops if not provided with the bouquet). Recutting the stems every 2 or 3 days and changing the water will prolong the life of the flowers. The level of water should be shallow. Bright light (but not direct sunlight) and cool temperatures (13–24°C) are recommended.

## 46.9 FASCINATING FACTS, FABLES, FOLKLORE, AND FRIVOLITY

- Tracing particularly to a 1989 US National Aeronautics and Space Administration study (https://ntrs.nasa.gov/search.jsp?R=19930073077) seeking to find plants that could reduce air pollution in space stations, several common houseplants, including gerbera, have been claimed to have the ability to reduce indoor air pollutants, such as benzene, formaldehyde, and trichloroethylene. However, evidence for this is weak (Dela Cruz et al. 2014).
- A study (Liu et al. 2006) of the effect of spaceflight on *Gerbera* seeds indicated that they germinated more quickly in space than on land.
- In southwestern China, the leaves of *G. delavayi* are traditionally employed as a source of fiber for weaving and clothing by ethnic minority tribes.
- Gerbera is a "daisy" (African daisy, Transvaal daisy), a term tracing to the Old English *dæges ēage*, interpreted as "day's eye," because the petals of some species open at dawn and close at dusk. The flower head of Asteraceae is so recognizable that numerous species have been termed daisies, and as a result, the word has become part of many phrases, such as "Oops-a-daisy" and "Whoops-a-daisy" (an expression of encouragement, such as to a child who has fallen or is being lifted), "fresh as a daisy" (to be eager and enthusiastic), "pushing up daisies" (to be dead), "daisy chain" (a series of connected events, activities, or experiences; also a pornographic term), and "daisy cutter" (a type of fuse designed to detonate an aerial bomb at or above ground level, also a bomb used in the Vietnam War).
- The flag and coat of arms of Mpumalanga (a province of South Africa) depict a stylized *Gerbera jamesonii*, which is endemic to the region.

## KEY PUBLICATIONS


Brisco-McCann, E.I. and M.K. Hausbeck. 2018. Diseases of *Gerbera*. In: *Handbook of Florists' Crops Diseases, Handbook of Plant Disease Management*, edited by R.J. McGovern and W.H. Elmer, 533–599. Springer International.

Codd, L.E. 1979, December. The story of Barberton daisy, *Gerbera jamesonii*. *Veld & Flora*: 114–115.

Dela Cruz, M., J.H. Christensen, J.D. Thomsen, and R. Müller. 2014. Can ornamental potted plants remove volatile organic compounds from indoor air? – a review. *Environmental Science and Pollution Research International* 21: 13909–13928.

Deng, Z. and K. Bhattarai. 2018. Gerbera. In: *Ornamental Crops, Handbook of Plant Breeding 11*, edited by J. Van Huylenbroeck, 407–438. Springer International.

Dufault, R.J., T. Phillips, and J.W. Kelly. 1990. Gerbera daisies: A potential field-produced cut flower crop for coastal South Carolina. In: *Advances in New Crops*, edited by J. Janick and J.E. Simon, 457–459. Timber Press, Portland.

Elomaa, P. and T.H. Teeri. 2001. Transgenic *Gerbera*. In: *Transgenic Crops III. Biotechnology in Agriculture and Forestry*, Vol. 48, edited by Y.P.S. Bajaj, 139–154. Springer, Berlin.

Hansen, H.V. 1985. *A Taxonomic Revision of the Genus Gerbera (Compositae, Mutisieae) Sections Gerbera, Parva, Piloselloides (in Africa), and Lasiopus (Opera Botanica No. 78)*. Council for Nordic Publications in Botany, Copenhagen. 36 pp.

Hansen, H.V. 1999. A story of the cultivated gerbera. *New Plantsman* 6: 85–95.

Kanwar, J.K. and S. Kumar. 2008. In vitro propagation of *Gerbera* – a review. *Horticultural Science (Prague)* 35: 35–44.

Katinas, L. 2004. The *Gerbera* complex (Asteraceae: Mutisieae): To split or not to split. *SIDA* 21: 935–940.

Keditsu, R. 2013. Gerbera: Soil fertility and plant nutrition. *Scientific Journal of Agriculture* 2: 97–114.

Keditsu, R. 2014. Gerbera: Crop phenology and plant nutrition. In: *Physiology of Nutrition and Environmental Stresses on Crop Productivity*, edited by A. Hemantaranjan, 74–110. Scientific Publishers, Jodhpur, India.

Kumar, R., T.N. Saha, and S. Saha. 2021. Gerbera (*Gerbera jamesonii* Bolus ex. Hooker f.). In: *Floriculture and Ornamental Plants. Handbooks of Crop Diversity: Conservation and Use of Plant Genetic Resources*, edited by S.K. Datta and Y.C. Gupta, 1–25. Springer, Singapore.

Liu, Q., X. Wang, F. Liao, M. Zhang, and B. Wang. 2006. Effect of spaceflight treatment on seeds germination and culture of *Gerbera jamesonii* in vitro. *Chinese Agricultural Science Bulletin* 22: 281.

Manning, J.C., B. Simka, J.S. Boatwright, and A.R. Magee. 2016. A revised taxonomy of *Gerbera* sect. *Gerbera* (Asteraceae: Mutisieae). *South African Journal of Botany* 104: 142–157.

Mehdikhah, M., R. Onsinejad, and D. Hashemabadi. 2016. Postharvest life of cut gerbera (*Gerbera jamesonii*) flowers as affected by salicylic acid, citric acid and ascorbic acid. *ARPN Journal of Agricultural and Biological Science* 11: 170–174.

Mercurio, G. 2002. *Gerbera Cultivation in Greenhouse*. Schreurs, The Netherlands. 206 pp.

Nesom, G.L. 2004. Response to "the *Gerbera* complex (Asteraceae, Mutisieae): To split or not to split" by Liliana Katinas. *SIDA* 21: 941–942.

Pasini, E., V.A. Funk, T.T. de Suza-Chies, and S.T.S. Miotto. 2016. New insights into the phylogeny and biogeography of the *Gerbera*-complex (Asteraceae: Mutisieae). *Taxon* 65: 547–562.

Reddy, P.P. 2016. Gerbera. In: *Sustainable Crop Protection under Protected Cultivation*, edited by P.P. Reddy, 355–362. Springer, Singapore.

Rogers, M.N. and B.O.S. Tjia. 1990. *Gerbera Production for Cut Flowers and Pot Plants*. Timber Press, Portland, OR. 116 pp.

Singh, S. and D. Dhyani. 2016. Domestication of wild gerbera (*Gerbera gossypina* (royle) beauved). *International Journal of Plant Science and Ecology* 2: 10–14.

Tourjee, K.R., J. Harding, and T.G. Byrne. 1994. Early development of gerbera as a floricultural crop. *HortTechnology* 4: 34–40.

Xu, X., W. Zheng, V.A. Funk, K. Li, J. Zhang, and J. Wen. 2018. Home at last III: Transferring *Uechtritzia* and Asian *Gerbera* species into *Oreoseris* (Compositae, Mutisieae). *PhytoKeys* 96: 1–19.

Zheng, W., X. Xu, and J. Wen. 2017. The ethnic textile use of natural fibers from fireweed (*Gerbera delavayi*) in Southwest China. *Economic Botany* 71: 380–386.

## Key Cultivation Websites

Agriculture Guruji. *Gerbera Cultivation Guide*. https://agricultureguruji.com/gerbera-cultivation-guide/.

Clemson University, Extension. *Gerbera Daisy*. https://hgic.clemson.edu/factsheet/gerbera-daisy/.

Moorman, G.W. 2016. *Gerbera Diseases*. https://extension.psu.edu/gerbera-diseases.

Murphy, G. 2018. *Taking Control: Using Biocontrol in Cut Gerbera*. www.greenhousecanada.com/news/taking-control-using-biocontrol-in-cut-gerbera-32410.

Royal Horticultural Society. *How to Grow Gerberas*. www.rhs.org.uk/plants/gerbera/growing-guide.

The Spruce. *How to Grow and Care for Gerbera Daisies*. www.thespruce.com/gerber-daises-4121360.

WikiHow. *How to Care for a Gerbera Daisy*. www.wikihow.com/Care-for-a-Gerbera-Daisy.

WikiHow. *How to Divide Gerbera Daisies*. www.wikihow.com/Divide-Gerbera-Daisies.

WikiHow. *How to Grow Gerbera Daisies*. www.wikihow.com/Grow-Gerbera-Daisies#:~:text=Water%20the%20soil%20around%20the,will%20dry%20during%20the%20day.

# 47 Gladiolus (*Gladiolus* Species and Hybrids)

**Scientific Names:** The genus name *Gladiolus* is the Latin diminutive of *gladius*, meaning "sword," referring to the shape of the leaves. Most modern cultivars are hybrids, sometimes identified as *G.* × *hybridus*, *G.* × *grandiflorus*, *G.* × *hortulanus*, *G. gandavensis*, and *G. lemoinei*. (The multiplication sign, ×, is a convention indicating that the taxon is of hybrid origin; in non-technical literature, the × is frequently omitted.)

**Common Names:** *Gladiolus* (the plural is usually "gladioli," or less formally "glads"; "gladioluses" is correct as a plural but is quite rare; "gladiola" is also rare and is encountered both as singular and as plural), *garden gladiolus*, *sword lily*. In human anatomy, *gladiolus* refers to the middle of the three parts of the sternum.

FIGURE 47.2 Great paintings featuring gladiolus: (a) *Gladioluses in a vase*, by German artist Bernd Schuchert (1886–after 1936). (b) *Gladioli*, by French painter Claude Monet (1840–1926), housed in Detroit Institute of Arts.

*Source:* **(a) Photo by Irina, based on a private collection (CC BY 2.0). (b) Public domain image.**

## 47.1 INTRODUCTION

Gladiolus is one of the most beautiful and popular of decorative plants. It is honored by numerous horticultural societies, flower shows, amateurs dedicated to its spectacular blooms, and many paintings by the world's greatest artists. As a cut flower employed in bouquets and floral arrangements, it is among the leading commercial ornamentals in international trade. Gladiolus is also admired as a majestic garden and landscape plant, although insufficiently hardy to overwinter outdoors in much of the cold temperate world. Several decades ago, gladiolus became so common that it seemed old-fashioned, and demand diminished. However, the dazzling floral displays of recent cultivars have resurrected the popularity of this long-admired, elegant classic.

FIGURE 47.3 An artistic display of flowering gladiolus cultivars.

*Source:* **Photo by Shutterstock/Kanea.**

## 47.2 OVERVIEW

*Gladiolus* species are herbs. The height of garden varieties varies from 50 cm to 2 m, and "miniature" cultivars are

FIGURE 47.1 A display of flowering gladiolus cultivars.

*Source:* **Photo by Shutterstock/LiliGraphie.**

available for indoor cultivation as houseplants. The plants produce up to nine or so narrow, deciduous, sturdy leaves with overlapping bases. The foliage is displayed in a vertical fan, with the opposite leaves in two ranks. The longest leaves may exceed 1 m in some cultivars. The flowers usually occur in a double row on one side of a terminal spike, blooming occurring successively from the base to the top. The flowers have six "tepals" (petal-like parts) forming a tube at the base, while at the open end the tepals may be plain, cut, frilled, or ruffled. The flowers are typically 5–10 cm across but may range up to 20 cm in diameter and are often trumpet-shaped in the large-flowered cultivars. The spikes sometimes develop several dozen flower buds; as many as half of them open simultaneously. Common flower colors include white and shades of yellow, pink, red, and purple. Greenish, blue, lavender, orange, apricot, burgundy, gold, and multicolored flowers are also well-known. Some cultivars have doubled flowers, that is, the number of tepals has been increased. Modern cultivars usually lack significant odor, but there are recent efforts to breed attractive

DOI: 10.1201/9781003586227-47

scents (for example, the cultivar Limoncello, which is lemony). The fruit is a many-seeded capsule.

Typically, *Gladiolus* is adapted to a subtropical climate. Most of the approximately 300 species are native to sub-Saharan Africa, and the main distribution area is South Africa. About ten species are indigenous to Eurasia. Some occur only in very restricted areas; others are widespread noxious weeds. Although wild species of *Gladiolus* are much humbler in appearance than modern cultivars, they are often grown in conservatories, parks, and rock gardens.

In addition to sexual reproduction by seeds, gladiolus reproduces vegetatively by corms, and indeed, most cultivation is by corms. The mature corm is enveloped in several layers (collectively termed "tunic") of brownish, fibrous husks, each composed of the petiole base of a deteriorated old leaf. The tunic is protective and, in some genera related to *Gladiolus*, is more substantial. The corm produced in the previous season serves as a "mother corm" to produce a "daughter corm" immediately on top of it. As the energy reserves are consumed from the mother corm, it shrivels and shrinks, and conversely, the daughter corm expands. This system of having a daughter corm (sometimes more than one) develop on top of its mother corm every season would eventually result in the elevation of the newest corm out of the ground. However, young gladiolus corms produce some contractile roots, which pull them down into the soil. Depending on cultivar, corms may be as large as 15 cm in diameter. Also produced are much smaller offsets termed *cormels* (also *cormlets*) at the tips of short, branched stolons (shoots) arising between the new and old corms. Cormels are miniature corms, varying in size from a pinhead to a hazelnut. Wild species produce one or occasionally two corms and a few cormels, but modern cultivars can produce 2–4 corms and sometimes over 400 cormels. The fleshy tissue of a cormel is surrounded by a single loosely fitting hard shell. At maturity, corms and cormels are easily detached. At the bottom of the corm, a fibrous root system is generated, while above a leafy shoot is produced. If the corm is large enough to provide sufficient energy, the shoot produces a flowering spike.

FIGURE 47.4 Gladiolus corms: (a) Sprouting corms. (b) Base of a gladiolus plant showing a new corm, an old corm, and cormlets. (c) Base of gladiolus plants showing many cormlets developed between the lower old corm (which has shrunk considerably) and the upper new corm. (d) Corms covered by brown tunics (made up of husks, which are often removed when offered for sale).

*Source:* **(a) Photo by Shutterstock/AnkaFed. (b) Photo by Emőke Dénes (CC BY SA 4.0). (c) Photo by Shutterstock/ARTamonovy Studio. (d) Photo by Legirons (CC BY SA 3.0).**

The corms of gladioli serve to survive (in a dormant state) the dry season found in most of their Southern African habitats, and to reproduce the plants when spring rains arrive. Corms are designed by nature for reproduction in the same location as the mother plant. In their native areas, *Gladiolus* corms are consumed by many animals, including baboons, porcupines, naked mole rats and other rodents, and some birds, such as guinea fowl. The smaller and more numerous cormels tend to break off and become scattered during attacks by herbivores, and this serves to spread genetically identical daughter plants locally in the vicinity of the mother corm. Corms and cormels exhibit a period of dormancy, which may be overcome in weeks at either high or low temperature. The gladiolus industry has conducted extensive research on environmental and chemical stresses that are useful commercially to break dormancy.

## 47.3 HISTORY

Gladioli were cultivated in ancient Greece and Rome, but the plants could not have been nearly as impressive as today's garden selections. Extensive hybridization, especially by the Dutch, began after several species were brought from South Africa to Europe in the early 18th century. Modern cultivars are the product of hybridization, mostly from southern and eastern African species. Species considered to be progenitors include *G. cardinalis*, *G. carneus*, *G. cruentus*, *G. dalenii* (*G. natalensis*), G. *oppositiflorus*, *G. papilio*, *G. saundersii*, and *G. tristis*. An estimated 70,000 cultivars of *Gladiolus* have been bred to date.

FIGURE 47.5 Illustrations (public domain) of gladiolus cultivars in old garden catalogues.

*Source:* **(a) John Lewis Childs 1900 gladiolus catalogue. (b) Vaughan's Seed Company (Chicago) 1910 catalogue. (c) Germain Seed and Plant Company (California) 1940s catalogue.**

FIGURE 47.6 Commercial gladiolus fields: (a, b) Production of gladiolus south of Santa Maria, California. (c) A woman harvesting flowers.

*Source:* **(a) Photo by Anita Ritenour (CC BY 2.0). (b) Photo by Harold Litwiler (CC BY ND 2.0). (c) Photo by Shutterstock/Anna Nahabed.**

FIGURE 47.7 Landscape uses of gladioli: (a) A cutting garden (employed as a source of vase flowers for a home). (b) A massed planting. (c) A planted border beside a house. (d) A planted bed.

*Source:* **(a) Photo by Shutterstock/Anna Nahabed. (b) Photo by Carl Lewis (CC BY 2.0). (c) Photo by Shutterstock/New Africa. (d) Photo by Fredrik Bruno (public domain).**

Breeders have created new varieties with attractive flower colors and an increased number of flowers that are well displayed on long spikes. The blooms of modern *Gladiolus* hybrids are much larger than those of unmodified wild species. The ability of corms to generate many new corms and cormels has also been selected, since this is highly advantageous commercially. There are ongoing efforts to breed cultivars which produce cut flowering spikes that have longer shelf life.

## 47.4 ECONOMIC IMPORTANCE

For many years, gladiolus has been considered to be a traditional flowering ornamental and the most desirable of cut flowers, loved particularly by florists for preparing bouquets. After the mid-1970s, sales of gladiolus waned somewhat because it had become cheap, common, and old-fashioned, but new cultivars, diverse in color, shape, and size, revived its popularity. In recent years, gladiolus has often been ranked among the five most important cut flowers in international trade. Major producers include the United States (Florida and California), The Netherlands, Italy, France, Poland, Bulgaria, Brazil, India, Australia, and Israel.

## 47.5 ORNAMENTAL USES

Although gladioli are primarily employed as cut flowers, they also have some usage in landscaping. Gladioli can be used as ornamentals in many circumstances – in herbaceous borders, rock gardens, and containers for patios and decks.

## 47.6 OTHER USES

Some species of *Gladiolus* are employed for food and medicine by Indigenous peoples in Africa. Gladiolus corms are consumed in Ghana, Nigeria, Cameroon, and Botswana, while in Nigeria, together with cereals, they are used in the preparation of a non-alcoholic drink. *Gladiolus dalenii*, one of the primary parents of many modern cultivars, is widely distributed, from eastern South Africa and Madagascar throughout tropical Africa and into western Arabia. The plant is used by Indigenous groups both as food and as medicine. The flowers are eaten raw or cooked, and the corms are consumed cooked. Extracts from the corms have been used in traditional medicine to treat a variety of conditions, including fungal infections, convulsions, seizures, and insomnia.

## 47.7 PROBLEMS

Viral, microbial, and fungal pathogens cause significant damage to gladiolus. Indeed, viruses are a major problem for most bulb crops because, once infected, all subsequent propagating material may harbor the virus. Viruses affecting gladiolus are a particular economic concern of commercial growers.

A few *Gladiolus* species are significant weeds. *Gladiolus angustus* has become invasive in many places, especially Australia, and *G. undulatus* is considered a threat in New Zealand. *Gladiolus italicus* (*G. segetum*), a native of much of Eurasia and North Africa, is occasionally grown as an ornamental and has become a common weed on other continents, particularly in cultivated fields. It is employed in North Africa as a traditional medicine. The species has been responsible for the death of grazing livestock in Algeria, where the plant is widespread. Nevertheless, Burrows and Tyrl (2013), the most authoritative survey of toxic plants affecting livestock, states, "Intoxication problems are only occasionally caused by species of *Gladiolus*."

There is very limited information on either the edibility or toxicity of gladiolus cultivars. Some reports on the internet (none from authoritative sources) claim that the flowers (at least the petals) are edible, but that the corms, and perhaps other parts of the plant, are toxic to cats, dogs, and possibly some wild rodents. Garden gladiolus has been so widely cultivated that if it constituted a significant threat of poisoning, this should be well-known by now. Nevertheless, it is premature to assess its toxicity, so deliberate consumption is unwise.

## 47.8 CULTIVATION

### 47.8.1 Climate and Site

Gladiolus is frost-intolerant, and the corms will not withstand freezing. Most cultivars are rated for growth as perennials in climate zones 8 to 10, but some can survive in zone 5. For most cultivars, optimal growth occurs when daylight temperatures are 10–25°C and night temperatures are 16–18°C. In the colder parts of the range, where gladiolus can be grown as a perennial outdoors, the corms of some cultivars can survive overwinter in the ground if protected by a mulch. In cold climates, gladioli are usually treated as a dispensable annual flowering ornamental, and new starting stock is purchased each year. Alternatively, as detailed in the following, the corms produced can be dug up in the autumn, stored, and replanted the following spring. Gladiolus should not be planted near trees, shrubs, or other large plants, as it does not compete well. The plants grow best in full sun. A location that is sheltered from wind is advisable, especially to prevent windthrow of tall varieties.

### 47.8.2 Propagation and/or Purchasing

Home gardeners typically buy large corms in order to rapidly produce large flowering plants. Corm diameter of 2.5 cm is minimal for most cultivars to produce flowers. Larger corms ("Jumbo" class is at least 5 cm) will produce better plants. Corms may be in sealed packages, making evaluation difficult; nevertheless, it is wise to check that the corms have a smooth tunic, showing no signs of damage or discoloration, and when the tunics are removed, the corms should be firm and without dark areas or spots. While chain stores offer a selection of cultivars, a far greater and more interesting range of gladioli are offered in the online catalogues of specialty dealers. Local nurseries are likely to be familiar with cultivars that perform well in the region and should be able to provide advice on how to minimize problems.

### 47.8.3 Planting

Gladiolus corms are best established in the spring, in well-tilled soil, when danger of frost is minimal. Planting distance varies according to variety, generally 15 to 20 cm apart. The bulbs are frequently planted in a trench, pointed end up. Large bulbs can be planted 10–15 cm deep, initially covered with only 5 cm of soil, but the trench progressively filled over a period of weeks. Smaller bulbs are best planted more shallowly. Rows may be spaced from 50 to 90 cm apart. Taller cultivars (up to 2 m) are best located at the back of a bed, shorter ones (less than 1 m) in the foreground. Staking is recommended for tall cultivars, as these could be blown over by wind. Home growers are often advised to apply a combination fungicide and insecticide to corms when they are being planted (which is inconsistent with modern attempts to grow ornamentals without pesticides).

As an outdoor plant, gladiolus blooms in midsummer, but given plants are in good display only for 2 or 3 weeks. A flowering bed can be arranged to have continuing displays by staggering planting times of early-, mid-, and late-season cultivars at 2- or 3-week intervals. In short-season climates, starting corms indoors is a technique that encourages earlier flowering. Generally, 70 to 90 days are required from planting to maturation of corms, depending on cultivar and weather, and if the objective is to produce new corms, the local date of frost needs to be considered. As noted in the following, to save the corms, they must be dug up after they have matured in late summer or fall, dried, and stored during the winter months.

### 47.8.4 Soil and Moisture Conditions

Gladiolus thrives in well-drained, rich, fertile sandy-loam soils but is tolerant of a range of soil conditions. However, heavy clay promotes waterlogging, which is harmful. Soil pH of 6–7 is ideal. Ample irrigation should be provided to maintain the soil in an evenly moist condition. A fertilizer with extra phosphorus (such as 5–10–5) to encourage flowering may be added as a side-dressing once the plants are established.

### 47.8.5 Harvesting Flower Stalks

To harvest the flowers for display in a vase, cut the spike when the first (lowest) flower is open (remaining flowers will subsequently open), using a sharp knife to make a slanting cut so that a few leaves are included. Immerse

FIGURE 47.8 Gladioli in vases.

*Source:* **(a) Photo by Garten-gg from Pixabay (public domain). (b) Photo by Kelly Cookson (CC BY SA 2.0). (c) Photo by Shutterstock/Flower Studio.**

immediately in water. Spent flowers need to be removed to maintain appearance.

### 47.8.6 Harvesting Corms

Gladiolus is propagated commercially by natural multiplication of new corms and cormels, although occasionally by in vitro (micropropagation) techniques. A mother corm may produce about 25 cormels, but these require three to four seasons to produce full-size daughter corms. The home gardener can dig up the new corms 4–6 weeks after the flowers have faded or when the aboveground plant has died, leaving about 1.3 cm of stem attached to the new corm. Corms should be dried well for several weeks, and the old, shriveled corm cut away. Cormels may be saved, but replanting these will require 2–4 years to produce a corm large enough to produce a flowering spike. Corms should be stored during the winter at 2–7°C in a well-ventilated, dry area. Dusting the corms before storage with a garden fungicide may prevent both insect and disease issues in storage. In principle, harvesting corms can maintain a gladiolus bed indefinitely, but after 6 years or so, the vigor of corms that have been harvested year after year may decrease.

## 47.9 FASCINATING FACTS, FABLES, FOLKLORE, AND FRIVOLITY

- In the King James Version of the Bible, Matthew 6:28 reads: "And why take ye thought for raiment? Consider the lilies of the field, how they grow; they toil not, neither do they spin." The identity of the "lilies" mentioned has been debated, since the biblical names applied to the many flowering plants in Galilee are uncertain. It has been suggested that they are, in fact, gladioli.
- "Gladiolus" was correctly spelled by 11-year-old Frank Neuhauser to win the first US National Spelling Bee in 1925. Subsequently, words to be spelled became more challenging. The 2018 second-place finisher failed to spell "bewusstseinslage" (a state of consciousness or a feeling devoid of sensory components), while the winner correctly spelled "koinonia" (the Christian fellowship or body of believers).
- *Corvettes* are small warships designed for convoy escort duty. The "flower-class corvette," also called the gladiolus-class after the lead ship, was a British set of 267 ships used during World War II by the Allied navies (mostly Great Britain and Canada) as anti-submarine guards during the Battle of the Atlantic. Most of the ships were named after flowers. The flower class represented half of all Allied convoy escort vessels in the North Atlantic during the war. Thirty-six flower-class ships were lost, while 51 enemy submarines were sunk. The *Gladiolus* was torpedoed and sunk in 1941 by a U-boat while escorting a convoy south of Iceland. All 65 on board were lost.
- Ornamental flowers have been assigned meanings in the "language of flowers" (floriography), especially popular in Victorian England, although attributing particular significance to given plants has been practiced for millennia. Today, internet florists repeat many of the old associations and often invent additional meanings, since they are interested in selling bouquets, and gifts of flowers are particularly associated with holidays, marriages, funerals, and anniversaries. The symbolisms alleged for gladiolus (all quite positive) include strength of character, infatuation, faithfulness, sincerity, integrity, and honor. Gladiolus has been represented as the birth month flower for August and the 40th wedding anniversary flower. The alternative name "sword lily" has likely been influential in creating symbolism: Lilies have long been associated with nobility of character, positive aspects of Christianity and tradition, and the swords reflect military and national honor.
- In The Netherlands in the 21st century, gladiolus became symbolic of athletic prowess, military strength, and victory, an association attributed to the origin of the word *gladiolus* from the Latin word for swords, such as employed by ancient Roman gladiators. The largest civilian and military marching event in the world, "International Four-Day Marches," has been held annually since 1909 in July in Nijmegen. The participants are showered with gladioli near the finish on a street called the Via Gladiola.
- *Gladiolus flanaganii* is known as "suicide lily," so-named because photographers need to perform death-defying climbs to get near it in the wild, because it grows on dangerous cliff faces in the Drakensburg Mountains of South Africa. The

FIGURE 47.9 Arriving participants of the 2019 Nijmegen (Netherlands) Four-Day March, carrying gladioli awarded to them for surviving the grueling event.

*Source:* **Photo by Roger Veringmeier (CC BY SA 4.0).**

FIGURE 47.10 Suicide lily (*Gladiolus flanaganii*): (a) Growing on a steep mountainside in Drakensburg, South Africa. (b) A cultivated plant.

***Source:*** **(a) Photo by Chris Vynbos (CC BY SA 4.0). (b) Photo by Peganum (CC BY SA 2.0).**

flowers of wild plants last little more than a day before wilting, increasing the difficulty of finding the species in flower. However, the plant is acquiring popularity in cultivation.

## KEY PUBLICATIONS

Abdessemed, D., N. Alloui, and A. Dibi. 2010. Phytochemical studies on the toxic compounds of *Gladiolus segetum*. *Asian Journal of Chemistry* 23: 609–613.

Anderton, E. and R. Park. 1989. *Growing Gladioli*. Timber Press, Portland. 166 pp.

Burrows, G.E. and R.J. Tyrl. 2013. *Toxic Plants of North America*, 2nd edition. Wiley-Blackwell, Oxford. 1383 pp.

Cantor, M. and J. Tolety. 2011. Gladiolus. In: *Wild Crop Relatives: Genomic and Breeding Resources*, edited by C. Kole, 133–159. Springer, Berlin.

Datta, S.K. 2020. Breeding of ornamentals: Gladiolus. *International Journal of Life Sciences* 9: 115–133.

Dhiman, M.R., N. Thakur, Y.C. Gupta, and N. Sharma. 2022. Gladiolus. In: *Floriculture and Ornamental Plants, Handbooks of Crop Diversity: Conservation and Use of Plant Genetic Resources*, edited by S.K. Datta and Y.C. Gupta, 47–79. Springer, Singapore.

Elmer, W.H. and K.K. Kamo. 2018. Diseases of gladiolus. In: *Handbook of Florists' Crops Diseases*, edited by R.J. McGovern and W.H. Elmer, 1289–1307. Springer, Berlin.

Garrity, J.B. 1975. *Gladioli for Everyone*. Devon & Charles, London. 196 pp.

Goldblatt, P. 1996. *Gladiolus in Tropical Africa. Systematics, Biology and Evolution*. Timber Press, Portland, OR. 338 pp.

Goldblatt, P. and J. Manning. 1998. *Gladiolus in Southern Africa*. Missouri Botanical Garden, St. Louis, MO. 320 pp.

González, A., J. López, S. Bañón, J. Ochoa, and R. Rodríguez. 2003. Ornamental use of wild species of genus *Gladiolus*. *Acta Horticulturae* 598: 59–63.

Hamilton, A.P. 1976. A history of the garden gladiolus. *The Garden* 101: 424–428.

Huxley, A., M. Griffiths, and M. Levy. (eds.). 1992. *Gladiolus*. In: *The New Royal Horticultural Society Dictionary of Gardening*, Vol. 2, 413–422. MacMillan, London.

Kamenetsky, R. and H. Okubo. 2012. *Ornamental Geophytes: From Basic Science to Sustainable Production*. CRC Press, Boca Raton. 597 pp.

Kamo, K., M. Roh, A. Blowers, F. Smith, and J. van Eck. 2001. Transgenic gladiolus. In: *Transgenic Crops III*, edited by Y.P.S. Bajaj, 155–170. Springer-Verlag, Berlin.

Katoch, M., M.Z. Abdin, R. Ram, and A.A. Zaidi. 2003. An overview of diagnostics for viruses infecting gladiolus. *Crop Protection* 22: 53–56.

Koenig, N. and W. Crowley. (eds.). 1972. *The World of the Gladiolus*. Edgewood Publishing Company, Edgewood, MA. 241 pp.

Lim, T.K. 2014a. *Gladiolus dalenii*. In: *Edible Medicinal and Non Medicinal Plants*, Vol. 8, 141–143. Springer, Dordrecht.

Lim, T.K. 2014b. *Gladiolus grandiflorus*. In: *Edible Medicinal and Non Medicinal Plants*, Vol. 8, 144–150. Springer, Dordrecht.

Memon, N.-U.-N., N.A. Wahocho, T.F. Miano, and M.H. Leghari. 2016. Propagation of *Gladiolus* corms and cormels: A review. *African Journal of Biotechnology* 15: 1699–1710.

Mukhopadhyay, A. 1995. *Gladiolus*. Publications and Information Division, Indian Council of Agricultural Research, Krishi Anusandhan Bhavan, New Delhi. 82 pp.

Ohri, D. 2013. Cytogenetics of domestication and improvement of garden *Gladiolus* and *Bougainvillea*. *The Nucleus* 56: 149–153.

Ohri, D. and T.N. Khoshoo. 1983. Cytogenetics of garden gladiolus: 4. Origin and evolution of ornamental taxa. *Proceedings of the Indian National Science Academy Part B Biological Sciences* 49(3): 279–294.

Singh, N., K.S. Mahar, S. Verma, B. Meena, R.K. Roy, S.K. Tewari, et al. 2018. Molecular analysis of genetic variability and relationship among *Gladiolus* cultivars. *Indian Journal of Biotechnology* 17: 118–127.

### Key Cultivation Websites

British Gladiolus Society. www.britglad.com/index.html.

Dutch Grown. *Growing Guide: How to Grow Gladiolus*. www.dutchgrown.com/blogs/the-dutchgrown-blog/how-to-grow-gladiolus.

North American Gladiolus Council. www.gladworld.org/.

The Old Farmer's Almanac: Growing Gladiolus, How to Plant, Grow, and Care for Gladioli. www.almanac.com/plant/gladiolus.

The Spruce. *How to Grow Gladiolus*. www.thespruce.com/gladiolus-flowers-1315704.

Trinklein, D. *Summer Flowering Bulbs: Gladiolus*. https://extension2.missouri.edu/g6620.

WikiHow. *How to Cut Gladiolus*. www.wikihow.com/Cut-Gladiolus.

WikiHow. *How to Grow Gladiolus*. www.wikihow.life/Grow-Gladiolus#:~:text=Care%20for%20Gladiolus,-%7B%22smallUrl%22%3A&text=Glads%20often%20lose%20ground%20to%20weeds.&text=Water%20gladiola%20plants%20with%201,during%20periods%20of%20insufficient%20rainfall.&text=Spray%20or%20dust%20plants%20with,to%20protect%20them%20from%20thrips.f.

Zwaal, P. 2016. *A Short Cultural History of Gladioli in The Netherlands*. https://peterzwaal.nl/a-short-cultural-history-of-gladioli-in-the-netherlands/#:~:text=In%20the%20last%20quarter%20of,were%20exported%20to%20all%20continents

# 48 Gloxinia (*Sinningia speciosa*)

**Scientific Names:** The genus name *Sinningia* honors Wilhelm Sinning (1792–1874), Prussian horticulturalist and botanist.

**Common Names:** The common name *gloxinia* reflects the initial placement of *Sinningia speciosa* in the genus *Gloxinia* (as *Gloxinia speciosa*) in 1817, after its discovery, and subsequently for the next 60 years, after which it was transferred to *Sinningia*. The genus name *Gloxinia* commemorates Benjamin Peter Gloxin (1765–1794), a German physician and botanical writer who lived in France. Such nomenclatural re-assignments are common in biology as a result of new information. Gloxinia is also known as florist's gloxinia, and sometimes as Brazilian gloxinia. It is clear what gloxinia means in this chapter, but in general, to avoid confusion, the best English name is florist's gloxinia. Peruvian gloxinia is *Gloxinia perennis*. Garden gloxinia (hardy gloxinia) is *Incarvillea delavayi*. Tree gloxinia is *Kohleria*. Creeping gloxinia is *Asarina erubescens*. The name "Canterbury bell(s)" is occasionally mistakenly applied to gloxinia but is usually applied to several unrelated species with bell-shaped flowers, often to *Gloxinia perennis*, but principally to *Campanula medium*.

## 48.1 INTRODUCTION

The gloxinia is a small flowering houseplant reminiscent of the African violet, which is also in the gesneriad family (Gesneriaceae). Its exquisite blooms are much larger and more colorful, but the plants are harder to maintain and are usually discarded after flowering. These gorgeous, dainty plants are ideal as gift plants.

FIGURE 48.1 Various gloxinia (*Sinningia speciosa*) cultivars.

*Source:* Photo by Shutterstock/Anaion.

## 48.2 OVERVIEW

*Sinningia* is a genus of about 70 species of tuberous herbaceous perennials of Central and South America, especially in southern Brazil. The ornamental species of principal interest is *S. speciosa*, known as gloxinia. It is native to the Atlantic coastal forests of southeastern Brazil, a tropical region known for concentrations of biodiversity and endemic species (ones that are not found elsewhere). This perennial herb grows from a tuberous storage root (often called a bulb), which the plant employs to survive the local drought season in a dormant condition. Gloxinia is 15 to 30 cm high, with a basal rosette of evergreen, hairy ("velvety" or downy) leaves, with leaf blades that seem overly large for such a small plant. Gloxinia flowers are about 5 cm in diameter (often twice or more as wide at the spreading mouth), tubular or bell-shaped, velvety, and often ruffled. They may be blue, pink, purple, red, white, or bicolored, and many cultivars have doubled flowers. The blooming period can last for several weeks.

## 48.3 HISTORY

The florist's gloxinia (*Sinningia speciosa*) was first sent to England from Brazil in 1815. In 1845, John Fyfe, a gardener in Scotland, discovered a mutation with symmetrical, bell-shaped, upright flowers, unlike the slipper-shaped flowers that hung down and are usual in wild plants. The mutated form is now standard. Hundreds of cultivars that were named and circulated in the 19th and early 20th centuries are no longer available. Until recently, it has been claimed that most marketed plants are hybrids of *S. speciosa* and

FIGURE 48.2 Variation of flowers of gloxinia (*Sinningia speciosa*).

*Source:* (a) Photo by Mokkie (CC BY SA 4.0). (b) Photo by Burkhard Mücke (CC BY SA 4.0). (c) Photo by Hartono Subagio from Pixabay (public domain). (d) Photo by HQ (CC BY SA 2.0). (e) Photo by Alia1906 from Pixabay (public domain). (f) Photo by HQ (CC BY SA 2.0).

DOI: 10.1201/9781003586227-48

other species of *Sinningia* or related genera, and that hybridization has been frequent (note "Further Introductions and Early Hybrids" in the Gesneriad reference website: https://gesneriads.info/articles/the-sinningia-alliance/sinningia-speciosa-history-in-horticulture/introductions-early-hybrids/). However, recent evidence (note Zaitlin 2012) suggests that most, if not all, modern cultivars trace only to *S. speciosa*, indeed, particularly to the Rio de Janeiro region.

## 48.4 ECONOMIC IMPORTANCE

The gift potted plant industry is based to a considerable extent on throwaway plants, and gloxinia is important in this respect. Unlike cut flowers, which rarely last 2 weeks in vases, potted flowering plants have a considerably longer life expectancy. Gloxinia is of relatively minor horticultural importance, but to plant aficionados, it is one of the world's most attractive flowers, despite its diminutive size.

## 48.5 ORNAMENTAL USES

Gloxinias are normally grown as small potted houseplants. They are considered to be safe for pets and children, unlike many other houseplants. The potted plants are so small, compact, and light that they can be easily employed temporarily in very shady locations for social events and moved back to better-lighted conditions afterward.

## 48.6 OTHER USES

Before the age of photography, floral art was extremely popular, and up to the early 20th century, many artists made their living by creating paintings of plants. The exceptional beauty of gloxinias made these a favorite subject. Although copies of most of the best of these are now copyright-free, the original paintings sell for high prices.

FIGURE 48.3 A commercial greenhouse-growing gloxinia (*Sinningia speciosa*).

*Source:* **Photo by Shutterstock/Barmalini.**

FIGURE 48.4 Potted gloxinia (*Sinningia speciosa*) houseplants. Note that some of the cultivars have doubled flowers.

*Source:* **(a) Photo by Shutterstock/MaskaRad. (b) Photo by Shutterstock/Audrius Merfeldas. (c) Photo by Kor!An (CC BY SA 3.0). (d) Photo by Anne Beaumont (CC BY SA 2.0). (e) Photo by Antoninhocd (CC BY SA 4.0).**

FIGURE 48.5 Old paintings (public domain) of gloxinia (*Sinningia speciosa*) by botanical illustrators. Note how the oldest cultivars have drooping flowers, while the newer ones have erect blooms, due to the peloric mutation discussed in the text.

*Source:* **(a) *L'Illustration horticole* (Gand, Belgium), vol. 21–22, 1874–1875. (b) Sydenham, E., J. Lindley, and J. Ridgway. 1829–1847. *Edwards's Botanical Register.* (c) Hulme, F. E., and S. Hibberd. 1879. *Familiar Garden Flowers, Fourth Series.* Cassell, London. (d) Harrison, J. 1841. *Floricultural Cabinet.* Gesneriad Reference Web. (e) Lemaire, C. A. 1854. *L'Illustration horticole.* (f) Morren, C. (ed.). 1846. *Annales de la Société royale d'agriculture et de botanique de Gand*, Tome 2. (g) Van Houtte, L. (ed.). 1869–1870. *Flore des serres et des jardins de l'Europe*, vol. 18. (h) Van Houtte, L. (ed.). 1867–1868. *Flore des Serres et des Jardins de l'Europe*, vol. 17.**

## 48.7 PROBLEMS

The work necessary to rejuvenate gloxinia after it flowers is its chief drawback, since most people today are accustomed to time-saving, convenient consumer items. Online garden writers and online seed catalogues sometimes confuse gloxinia (*Sinningia speciosa*) and the genus *Gloxinia* and provide misleading instructions.

## 48.8 CULTIVATION

### 48.8.1 Climate and Site

Gloxinias are adapted to semi-shade and shade, and as houseplants, they should be maintained in light that is bright but indirect. They are typically placed on interior windowsills, which can, in some climates and seasons, produce more light than desirable, and curtains or shading may be required. The plants are adapted to standard room temperatures (18 to 24°C), reflecting the climate of their native Brazilian mountains.

Some authorities state that the plants can be summered outside or even maintained permanently outdoors in hardiness zones 11 and 12. However, gloxinia cultivars are not tolerant of cool night temperatures (below 12°C) while in bloom, and the upward-facing trumpet flowers of most modern cultivars are very susceptible to rain damage.

### 48.8.2 Propagation and/or Purchasing

Gloxinia can be propagated (as described later) by their tiny, dust-like seeds. However, establishing new plants is a lengthy process. Leaf cuttings and division of tubers are also possible, but these procedures are best left to professionals or dedicated amateurs. The sprouts that arise from the tubers, when 3 to 4 cm long, can be cut off and planted in soil to produce new plants. However, in most cases, gloxinias are purchased as potted plants from plant outlets, or tubers are obtained from online sources. The best plants to buy should have buds that are just beginning to open.

FIGURE 48.6 Gloxinia (*Sinningia speciosa*) growing on a shaded windowsill.

***Source:*** **Photo by Shutterstock/Alena Dafna.**

### 48.8.3 Planting

Seeds are usually available from online catalogs, along with instructions. Seeds are typically sown in a pot of moistened soil covered with a clear plastic bag to promote humidity, maintained in good indirect light at a temperature of 21 to 24°C. The seeds should sprout in 1 to 3 weeks, and the seedlings are then transplanted to individual pots. Blooming usually takes 5 to 7 months from seed. Larger cultivars are usually grown as individual houseplants in 5 to 6 in (13–15 cm) pots, compact types in 4 in (10 cm) pots.

### 48.8.4 Soil and Moisture Conditions

Plants are best kept in potting soil, kept evenly moist. In their native areas, the wild plants often grow in soil rich in humus. The soil provided with purchased potted plants should be loose, well-draining, and slightly acidic (pH 5.5 to 6.5). Watering the foliage should be avoided, as this can result in permanent brown spots. Water added to the pots should be at room temperature. Bottom watering is recommended (but after the base of the pot is immersed in water, the container should be allowed to drain well). Gloxinia is native to South American rainforests, where the relative humidity is above 70%. Accordingly, humidity around the plants is strongly recommended during the winter dry season (provided, for example, by placing the pot on a bed of pebbles in a tray of water). However, the leaves should not be misted. If the plant is to be rejuvenated for reflowering, it can be fertilized, but not after the flowers fade. A high-phosphorus, water-soluble fertilizer, not a balanced fertilizer, is recommended (high nitrogen seems to distort leaf growth).

### 48.8.5 Maintaining Plants

Senescent leaves and wilted flowers should be removed. Diseased plants are best discarded, as any kind of control to kill fungi, bacteria, or insects tends to damage the flowers.

Gloxinia is often treated as a throwaway plant after blooming, since effort is required to get the plants to reflower. Restoring vigorous growth to many modern cultivars is very difficult (it appears that such cultivars devote their energy to flowering well just once, not to producing an energy-rich tuber). For those who wish to try, the following summary of steps to take is presented. After flowering, to rejuvenate the plants, they should be induced to become dormant by gradually reducing watering. Once the leaves die back, they should be cut off. The pot should be transferred to a cool dark location (10 to 15°C, although tubers kept at room temperature often survive), and the tuber rested for 2 to 3 months. When new growth appears, the tuber should be removed from the old potting soil and repotted in a slightly larger pot, preferably in a potting soil recommended for African violets. The top of the tuber should be at soil level. The plant should be returned to the growth conditions previously described.

FIGURE 48.7 Tuberous root ("bulb") of gloxinia (*Sinningia speciosa*).

*Source:* **Photo by Shutterstock/Andrei Dubadzel.**

## 48.9 FASCINATING FACTS, FABLES, FOLKLORE, AND FRIVOLITY

- *Pelorism* is a term thought to have been coined by the father of evolution, Charles Darwin. It refers to the formation of "peloric" flowers – mutations of bilaterally symmetrical (zygomorphic) flowers which are normal in a species – to radially symmetrical (actinomorphic) flowers. As noted earlier, in 1845, John Fyfe, a gardener in Scotland, discovered a peloric mutation with symmetrical, bell-shaped, upright flowers, unlike the slipper-shaped flowers that hung down and are usual in wild plants, and the mutation is retained in all modern gloxinia cultivars. In some academic publications, these cultivars of *S. speciosa* are sometimes termed "Darwin's gloxinia." (See "The first peloric flowers" in the Gesneriad reference website: https://gesneriads.info/articles/the-sinningia-alliance/sinningia-speciosa-history-in-horticulture/introductions-early-hybrids/.)
- Gloxinia is a brand of pharmaceutical tablets containing ciprofloxacin, employed in the treatment of a wide variety of bacterial diseases (including plague and anthrax). The coincidental name is unrelated to the plants called gloxinia.
- As noted in the text, the phrase "Canterbury bell(s)" is applied to the bell-shaped flowers of several species with bell-shaped flowers like those of gloxinia. Canterbury Cathedral, in Canterbury, England, was founded in 597 and rebuilt several times, especially between 1070 and 1077. To this day, it is famous for its many bells. Pilgrims to the site often have purchased miniature bells, and in the late 16th century, the practice became common for pilgrims on horseback to Canterbury to place bells on their horses (the word "canter," meaning a gait between a trot and a gallop, arose from this association). Since most English vernacular names of plants originated in Great Britain, it is understandable that those with bell-shaped flowers were often called Canterbury bells.

## KEY PUBLICATIONS


Chautems, A., T.C. Costa Lopes, M. Peixoto, and J. Rossini. 2010. Taxonomic revision of *Sinningia* Nees (Gesneriaceae) IV: Six new species from Brazil and a long overlooked taxon. *Candollea* 65: 241–266.

Citerne, H. and Q.C.B. Cronk. 1999. The origin of peloric *Sinningia*. *The New Plantsman* 6: 219–222.

Clayberg, C.D. 1996. Interspecific hybridisation in *Sinningia* (Gesneriaceae). *Baileya* 23: 184–194.

De Worth, A.F. 1965. *Gloxinia Culture*. Texas A&M University, Texas Agricultural Extension Service Bulletin L628. https://oaktrust.library.tamu.edu/bitstream/handle/1969.1/162758/628/Leaf0628.pdf?sequence=6.

Dong, Y., J. Liu, P.-W. Li, C.-Q. Li, T.-F. Lü, X. Yang, et al. 2018. Evolution of Darwin's peloric gloxinia (*Sinningia speciosa*) is caused by a null mutation in a pleiotropic TCP gene. *Molecular Biology and Evolution* 35: 1901–1915.

Ferreira, G.E., A. Chautems, and J.L. Waechter. 2015. Taxonomy of *Sinningia* Nees (Gesneriaceae) in Rio Grande do Sul, southern Brazil. *Acta Botanica Brasilica* 29: 310–326.

Hasing, T., E. Rinaldi, M. Silvia, L. Colombo, D.C. Haak, D. Zaitlin, et al. 2019. Extensive phenotypic diversity in the cultivated florist's gloxinia, *Sinningia speciosa* (Lodd.) Hiern, is derived from the domestication of a single founder population. *Plants, People, Planet* 1: 363–374.

Kessler, J.R. 2004. *Greenhouse Production of Gloxinias*. Alabama Cooperative Extension System, ANR-1258. www.scribd.com/document/462807831/instructiuni-gloxinia.

Sudthai, S. and S. Sakhonwasee. 2022. Reproductive development, seed yield and seed quality of gloxinia (*Sinningia speciosa*) in a closed plant production system as influenced by LED light quality and intensity. *The Horticulture Journal* 91: 24–32.

Zaitlin, D. 2011. A history of the florist gloxinia in pictures and words. *Gesneriads* 61(3): 14–27.

Zaitlin, D. 2012. Intraspecific diversity in *Sinningia speciosa* (Gesneriaceae: Sinningieae), and possible origins of the cultivated florist's gloxinia. *AoB Plants* 2012: pls039. https://doi.org/10.1093/aobpla/pls039.


### KEY CULTIVATION WEBSITES


Alabama, A&M, and Auburn Universities Bulletin ANR-1258. *Greenhouse Production of Gloxinias*. https://ssl.acesag.auburn.edu/pubs/docs/A/ANR-1258/ANR-1258-archive.pdf.

Chambers, K. 2021. *8-Step Guide on How to Grow and Take Care of Gloxinia Plant (Blog)*. www.homify.co.uk/diy/14472/8-step-guide-on-how-to-grow-and-take-care-of-gloxinia-plant.

Gesneriad Reference Web. *Sinningia speciosa – History in Horticulture*. https://gesneriads.info/articles/the-sinningia-alliance/sinningia-speciosa-history-in-horticulture/.

Hellendooorn, B. *28 Impressive and Easy Gloxinia Plant Care Tips (Blog)*. https://twopeasinacondo.com/growing-guides/gloxinia-care/.

Homify.co.UK. *8-Step Guide on How to Grow and Take Care of Gloxinia Plant*. www.homify.co.uk/diy/14472/8-step-guide-on-how-to-grow-and-take-care-of-gloxinia-plant.

Laidback Gardener. *How to Grow a Florist's Gloxinia*. https://laidbackgardener.blog/2021/03/10/how-to-grow-a-florists-gloxinia/.

The Spruce. *How to Grow and Care for Gloxinia Hybrids*. www.thespruce.com/grow-senningia-hybrids-indoors-1902487.

# 49 Heath (*Erica* Species and Hybrids) and Heather (*Calluna vulgaris*)

**Scientific Names:** The genus name *Calluna* is based on the Greek *kallunein*, "to cleanse or clean," interpreted as beautifying a place by sweeping it (the plants were once employed to make brooms). The genus name *Erica* traces to the Greek *ereiko*, "to break," based on the belief that an infusion (a preparation made by steeping in a solvent such as water or alcohol at room temperature) of the leaves could break bladder stones.

**Common Names:** *Calluna vulgaris* (the only species of the genus) is usually called heather, Scots (Scotch) heather, or common heather, sometimes also ling (speculated to trace to the old Norse word *lyng*, meaning "light in weight," or the Anglo Saxon *lig*, meaning "fire" and referring to use as a fuel). *Erica* species are usually called heath, but some are termed *heather*. *Calluna vulgaris* has also been termed "summer heather," in contrast to cold-hardy *Erica* species, termed "winter heather" and "spring heath" (phrases reflecting the flowering periods of these popular ornamentals). In ecology (particularly in Great Britain), a "heath" is a landform with low-growing shrubs on free-draining, infertile, acidic soils. Confusingly, the "heather" *C. vulgaris* is the predominant plant growing on "heaths" in Great Britain. The word *heath* applied to plants is speculated to have originated as a reference to plants that occupy heaths, and the same may be true for *heather* ("heath-er" meaning "heath-loving"). However, the word *heather* is also speculated to be derived from the ancient Scots word *hadddre*. *Cuphea hyssopifolia* is known as Mexican heather, false heather, and Hawaiian heather.

## 49.1 INTRODUCTION

"Heaths and heathers" is an established phrase that has been applied collectively to the small shrub species of *Erica* and *Calluna vulgaris* that are hardy enough to grow in Europe. Sometimes also included is the related small genus *Daboecia* (notably *D. cantabrica*, the Irish heath [a name also applied to *E. erigena*], *D. azorica*, and hybrids of the two). Contributing to semantic difficulties, the word *heather* is often also employed collectively to include all these species. As indicated earlier, plants often called "heather" belong both to *Erica* and *Calluna*, although the term is much more strongly associated with *Calluna* (the "true heather") than with *Erica*. Most *Erica* species are tropical or subtropical (these South African species are

FIGURE 49.1 Heaths and heathers growing as wild plants in Europe: (a) *Calluna vulgaris* in Lueneburg Heath, Germany. (b) *Erica carnea* in the Czech Republic. (c) *Erica cinerea* in Italy.

***Source:*** **(a) Photo by Shutterstock/LianeM. (b) Photo by Lubor Ferenc (CC BY SA 4.0). (c) Photo by Umberto Ferrando (CC BY 2.0).**

known as "Cape heaths") and are unsuitable for permanent outdoor cultivation in freezing climates (they need to be moved inside or treated as annuals). The Cape heaths are sold as short-lived houseplants (often inappropriately called "heathers") in temperate climates, frequently at Christmas. This chapter treats *Calluna vulgaris* (which was once placed in *Erica*) and the cultivated European species of *Erica*, all of which look very similar and require much the same conditions for cultivation.

## 49.2 OVERVIEW

*Calluna vulgaris* is a low evergreen shrub growing to 60 cm in height (rarely to 1 m). It produces multiple-stemmed ascending branches, very small leaves, and small flowers displayed on long flowering floral stems. The seeds are tiny, and a single plant can produce more than 100,000. The species is native to Europe and the Azores but has spread widely across Asia, Australia, North America, and New Zealand. It is much like *Erica* species, but its sepals (which look like petals and are colored) enclose and conceal the remainder of the flowers (including the true petals), and its short leaves (1–2 mm long) are scale-like rather than needle-like. Wild plants are usually purple-flowered, but the more than 1,000 cultivars exhibit an impressive range of flower colors and, occasionally, doubled flowers. Blooming occurs in late summer and autumn (the cultivated *Erica* species bloom earlier). Complementing the many colors of the flowers, the foliage of cultivars also exhibits a considerable range of shades of bronze, gold, green, and silver, occasionally with variegations of pink or white. Often, foliage color changes seasonally.

DOI: 10.1201/9781003586227-49

*Erica* is a huge genus of at least 800 species of woody plants, ranging from low shrubs to a tree (*E. arborea*) as tall as 7 m. Over 90% are native to South Africa, but several of the European species are widely cultivated. The leaves are small and short, needle-like, and evergreen. The flowers are bell- or trumpet-like or tubular, small but numerous, usually pink or white, but available in many colors in cultivars. *Erica carnea* (spring heath, snow heath, winter heath, mountain heath, alpine heath), with over 150 cultivars, is the principal cultivated species. Also popular, with many cultivars, is *E. cinerea* (bell heather, Scots heath, Scotch heath).

FIGURE 49.2 (a) Colorful shrubs of common heather (*Calluna vulgaris*). (b) Cultivar Winter Chocolate. (c) Cultivar Dark Beauty. (d) Cultivar Kinlochruel. (e) Cultivar Annemarie.

***Source:*** **(a) Photo by Shutterstock/Vilor. (b–e) Photos by Krzysztof Ziarnek, Kenraiz (CC BY SA 4.0).**

FIGURE 49.3 Colorful massed pots of common heather (*Calluna vulgaris*) showing the impressive range of colors of the flowers and the foliage.

***Source:*** **(a) Photo by Shutterstock/Barmalini. (b) Photo by Alessio Maffeis (CC BY 2.0).**

FIGURE 49.4 *Erica carnea* (winter heath), the most popular cultivated species of *Erica*: (a) Close-up of the flower and leaf. (b) Cultivar Whisky. (c) Cultivar Pink Spangles. (d) Cultivar Whitehall.

***Source:*** **(a) Photo by Captainpixel (public domain). (b) Photo by Krzysztof Ziarnek, Kenraiz (CC BY SA 4.0). (c) Photo by cultivar413 (CC BY 2.0). (d) Photo by Andy Morffew (C BY 2.0).**

## 49.3 HISTORY

The European heathers and heaths (*Calluna* and *Erica*) are so common in Europe that there was little motivation to grow them for most of history. However, in the late 18th and early 19th century, the Cape heaths (South African ericas)

FIGURE 49.5 Some of the cultivated European species of *Erica*: (a) *Erica vagans*. (b) *Erica gracilis*. (c) *Erica terminalis*. (d) *Erica cinerea*. (e) *Erica × darleyensis* (*E. carnea × E. erigena*) cultivar Ghost Hills. (f) *Erica manipuliflora*.

***Source:*** **(a) Photo by Manuel M. V. (CC BY 2.0). (b) Photo by Assianir (CC BY SA 3.0). (c) Photo by Réginald Hulhoven (CC BY SA 3.0). (d) Photo by Ligurian Vascular Flora (CC BY 2.0). (e) Photo by Babij (CC BY SA 2.0). (f) Photo by Abu Shawka (CC BY SA 3.0).**

FIGURE 49.6 Nineteenth-century paintings (public domain) of cultivated heath and heather species by European botanical illustrators: (a) *Calluna vulgaris*. (b) *Erica tetralix*. (c) *Erica carnea*.

***Source:*** **(a, b) Thomé, O. W. 1885. Flora von Deutschland, Österreich und der Schweiz, Gera, Germany. (c) Club alpin allemand et autrichien and H. Correvon. 1899. *Atlas de la flora alpine*. Georg & Co., Genève.**

were imported, grown in conservatories, and stimulated interest in selecting cultivars of the European species.

## 49.4 ECONOMIC IMPORTANCE

Heaths and heathers are popular small ornamental shrubs cultivated in many countries, but due to their limited stature, they compete with a large number of herbaceous species for the attention of gardeners. The harvested beautiful flowering stems are widely marketed.

FIGURE 49.7 Commercial activities related to heaths and heathers: (a) A huge *Erica* nursery in Geldern, Germany. (b) Starter potted plants on sale at a nursery.

***Source:*** **(a) Photo by Frank Vincentz (CC BY SA 3.0). (b) Photo by Arune from Pixabay (public domain).**

FIGURE 49.8 Examples of landscape use of heaths and heathers: (a) A large garden of common heather (*Calluna vulgaris*). (b) A planting of winter heath (*E. carnea*) elegantly situated in a pit surrounded by stones.

***Source:*** **(a) Photo by David Short (CC BY 2.0). (b) Photo by Chris.urs-o (CC BY SA 3.0).**

FIGURE 49.9 Heaths and heathers in containers: (a) A stone trough with mixed cultivars. (b) A kitten playing with a pot of heather. (c) Plants growing in a basket. (d) Plants growing on a windowsill.

***Source:*** **(a) Photo by Shutterstock/Peter Turner Photography. (b) Photo by Shutterstock/Veronika Nizova. (c) Photo by Heike from Pixabay (public domain). (d) Photo by Shutterstock/Q77photo.**

## 49.5 ORNAMENTAL USES

Heaths and heathers are primarily grown as low-growing flowering shrubs in public landscapes and in the gardens and borders of private residences. They are excellent as groundcover, in rock gardens, and in tubs and window boxes. They are also extensively employed as cut flowers, both for fresh bouquets and dried floral arrangements.

## 49.6 OTHER USES

Specialists in the ecology of mammalian herbivores distinguish "grazing" (feeding mainly on low-growing, generally herbaceous vegetation, especially grasses) from "browsing" (feeding mainly on the foliage and woody twigs of taller shrubs and trees). For example, goats are primarily

FIGURE 49.10 Heath and heather cut flowers.

***Source:*** **(a) Photo by Shutterstock/Zelma Brezinska. (b) Photo by Shutterstock/Lisavo.**

FIGURE 49.11 Livestock grazing on heather: (a) Sheep in North Yorkshire, UK. (b) Highland cow in Scotland. (c) Highland ponies in Scotland. (d) Goat in Belgium.

***Source:*** **(a) Photo by Anthony Parkes (CC BY SA 2.0). (b) Photo by Shutterstock/DaveHawtree. (c) Photo by Shutterstock/Ellen Thornell. (d) Photo by Shutterstock/NAPA.**

browsers, and sheep are mostly grazers. Heaths and heathers provide low-growing herbaceous plant material (the younger green stems), but also woody twigs (the older stems), so both terms could be applied to animals consuming *Calluna* and *Erica*. *Calluna vulgaris* is widely grazed in Europe, especially by sheep, although its nutritional value is controversial, and some diseases of sheep have been attributed to overconsumption.

Heather honey is widely produced in Europe, although because of its strong taste and tendency to be a jelly until stirred, it is less valued than in past times.

## 49.7 PROBLEMS

*Calluna* and *Erica* species have become aggressively spreading weeds in some countries. *Calluna* is listed as invasive in parts of Australia, Canada, New Zealand, and the United States.

## 49.8 CULTIVATION

### 49.8.1 Climate and Site

*Calluna vulgaris*, probably the most frost-tolerant of all plants termed *heather* or *heath*, is adapted to climate zones 4 to 7. Some cultivars can grow in zone 3, and with winter protection, some sensitive varieties can also be grown as outdoor perennials. Many cultivars have difficulty with heat in zones 8 and southward. *Erica* species differ greatly in climate adaptation. Some cultivars of the European species are able to survive as far north as zone 4, but the tropical species are not able to survive frost. The popular *E. carnea* and *E. cinerea* are rated for zones 5 to 7; *E. tetralix* (cross-leaved heath) is suited to zones 4 and 5. *Erica carnea*, appropriately termed *winter heather*, comes from the mountains of Central Europe, where it thrives under a blanket of snow, with flowers emerging as the snow melts. Heaths and heathers are adapted to full sun but tolerate partial shade. Many do well in seashore locations, partly because most species are tolerant of wind.

### 49.8.2 Propagation and/or Purchasing

The widest range of cultivars is available from specialist mail-order nurseries, but popular varieties are often sold in walk-in retail outlets, nurseries, and garden centers. Cultivars need to be propagated vegetatively (by cuttings or layering, or by division) to remain true-to-the-parent plants. Hybrids of the European heathers (such as the popular *E.* × *darleyensis*) are normally sterile, not producing seed. Many gardeners simply purchase starter plants from nurseries. Large areas requiring groundcovers are usually planted by professional landscapers, employing seeds.

FIGURE 49.12 Heath and heather in snow: (a) Common heather (*Calluna vulgaris*). (b) Winter heath (*Erica carnea*).

*Source:* **(a) Photo by Olga Fil from Pixabay (public domain). (b) Photo by H. Zell (CC BY SA 3.0).**

### 49.8.3 Planting

Early spring or early fall are usually appropriate planting times. Most ericas and dwarf *Calluna* cultivars can be planted at least 25 to 30 cm apart. Larger *Erica* cultivars and most *Calluna* varieties should be at least 30 to 50 cm apart.

### 49.8.4 Soil and Moisture Conditions

*Calluna vulgaris* and most species of *Erica* prefer acidic (pH 6.5 or less; 4.5 to 5.0 is often recommended), sandy, organically rich, moist, but very well-drained soil. Peat moss has been a widely employed soil amendment, but its use is being discouraged because it is now considered a limited resource that should be left in nature to support natural ecosystems. *Erica carnea* and some other *Erica* species are noted for tolerating some alkalinity, although thriving in acidic soils. Relatively little fertilizer is required (the claim is often made that none is needed), but the species usually respond well to rhododendron/azalea fertilizers. The plants are drought-resistant once established.

### 49.8.5 Maintaining Plants

Pruning is very important for maintaining the appearance of heathers and heaths, especially of the faster-growing cultivars. In colder areas, flower stalks of *Calluna* are best pruned off in the spring, but in warmer climates, pruning can also take place shortly after blooming is finished. In relatively mild but snowy areas (as in Great Britain), flower stalks of *Calluna* are often deliberately allowed to persist overwinter (partly covered with snow) for their attractive appearance. Ericas (which are "winter-blooming," which usually means "spring-blooming") are pruned in the early spring, not in the summer or fall, which would destroy the buds for the next season. Flower stems of *Calluna* should be cut off just below where flowers start at the lower part of the stem, avoiding cutting into the woody part of the stem just below the flowers. The cultivated European ericas require lighter trimming of flowering branches (not necessarily to the base of the flowering part of stems). Pruning heathers and heaths should be restricted to cutting (upper) green parts of stems, not (lower) woody parts, except to cut off stems at their base that are spreading the plant in an unwanted direction. Vigorous cultivars often require strong pruning to prevent the center of the shrub from going bare.

FIGURE 49.13 Pruning heather.

*Source:* Photo by Shutterstock/ArtCookStudio.

## 49.9 FASCINATING FACTS, FABLES, FOLKLORE, AND FRIVOLITY

- Heather (predominantly *Calluna vulgaris*) is iconic of Scotland, where it is extremely widespread, occupying great tracts of land, and has been utilized historically for many practical purposes. Nevertheless, "Scotch thistle" – either *Cirsium vulgare* or *Onopordum acanthium* – is accepted as the national flower. Heather was widely employed in building construction, especially in walls and thatched roofs. Implements made from heather included brooms, hoes, and rakes. Heather mattresses were common. Heather was extensively employed as fuel and occasionally utilized as a yellow dye. Various drinks, potions, ointments, and salves made with heather were used medicinally.
- Purple heather and pyramidal saxifrage (*Saxifraga cotyledon*) are the national flowers of Norway.
- Sticky heaths (several *Erica* species of South Africa) have flowers that are sticky on the outside, which appears to protect them from insects boring into the flower to steal nectar. Insects sometimes get stuck permanently on the flowers, attracting birds to consume them.
- The popular liqueur Drambuie is made from Scotch whisky, heather honey, herbs, and spices. In old Scotland, ales were bred employing heather, but recipes do not seem to have survived. By tradition, only the king and his firstborn son were allowed to know how heather ale was prepared. In the poem titled "Heather Ale: A Galloway Legend" by Robert Louis Stevenson, a Pictish king (*Picts* were a tribal confederation of Celtic peoples of ancient Scotland) sacrificed both his life and that of his son to protect the secret recipe.
- Heather has given its name to a class of colors, but identifying the shades can be difficult. "Heather gray" (described as a light-weathered, warm gray) is the most common "heathered color," but other colors can be heathered. In clothing, *heather* refers to interwoven yarns of mixed colors resulting in flecks of an alternate color (often, the result of such mixing is an apparent color different from the components). Generally, clothing that has been "heathered" has muted coloration. Heathered colors (for clothing, painted surfaces, and home furnishings) are considered sophisticated and are employed by designers and the fashion-conscious. The derivation of this usage of heather has been explained as follows: The small flowers of heather range naturally from purple to pink, the two colors often bleeding into each other on the same petal. Accordingly, *heather* came to describe things that had flecks of various colors in them.

## KEY PUBLICATIONS

Cucu, A.-A., G.-M. Baci, A.-B. Cucu, Ş. Dezsi, C. Lujerdean, I.C. Hegeduş, et al. 2022. *Calluna vulgaris* as a valuable source of bioactive compounds: Exploring its phytochemical profile, biological activities and apitherapeutic potential ["apitherapy" is a branch of alternative medicine that uses honey bee products.] *Plants (Basel)* 11(15): 1993 https://doi.org/10.3390/plants11151993; www.ncbi.nlm.nih.gov/pmc/articles/PMC9370339/.

Davies, G.M., V. Vandvik, R. Marrs, and L.G. Velle. 2022. Fire management in heather-dominated heaths and moorlands of North-West Europe. In: *Global Application of Prescribed Fire*, edited by J.R, Weir and D. Scasta, 194–229. CRC Press, Boca Raton, FL.

Gimingham, C.H. 1960. Biological flora of the British Isles. *Calluna* Salisb. A monotypic genus. *Calluna vulgaris* (L.) Hull. *Journal of Ecology* 48: 455–483.

Matthews, R.F. 1993. *Calluna vulgaris*. In: *Fire Effects Information System [Online]*. U.S. Department of Agriculture, Forest Service, Rocky Mountain Research Station, Fire Sciences Laboratory. www.fs.usda.gov/database/feis/plants/shrub/calvul/all.htmlM.

McGuire, A.F. and K.A. Kron. 2005. Phylogenetic relationships of European and African ericas. *International Journal of Plant Sciences* 166: 311–318.

Moise, A., L. Al Mărghitaş, D.S. Dezmirean, O. Bobiş, and O. Maghear. 2011. Theoretical study regarding the heather honey (*Calluna vulgaris*). *Bulletin University of Agricultural*

*Science and Veterinary Medicine, Animal Science and Biotechnologies* 68: 233–237.

Nelson, E.C. 2012. *Hardy Heathers from the Northern Hemisphere: Calluna – Daboecia – Erica.* Royal Botanical Gardens, Kew, UK. 442 pp.

Nelson, E.C. and M.D. Pirie. 2022. Where have all the heathers gone? *Sibbaldia.* file:///C:/Users/ottaw/Downloads/1887.pdf.

Small, D. and E.M.T. Wulff. 2008. *Gardening with Hardy Heathers.* Timber Press, Portland, OR. 296 pp.

Wallace, A. 1903. *The Heather in Lore, Lyric and Lay.* A.T. de la Mar, New York. 245 pp. https://books.google.ca/books?id=dugpAAAAYAAJ&pg=PA6&redir_esc=y#v=onepage&q&f=false.

## Key Cultivation Websites

Florgeous. *Heather and Heath Flowers: What's the Difference*? https://florgeous.com/heather-and-heath/.

Gardenia. *Heaths and Heathers.* www.gardenia.net/plant-variety/about-heaths-and-heathers.

Heather World (Replaces British Heather Society). www.heatherworld.org/.

Homes & Gardens. *How to Grow and Care for Winter Heather.* www.homesandgardens.com/gardens/how-to-grow-and-care-for-winter-heather.

Horticulture Magazine. *5 Tips to Rejuvenate Heather.* https://horticulture.co.uk/heathers/rejuvenation/.

Horticulture Magazine. *Heather Growing in Pots.* https://horticulture.co.uk/heathers/containers/.

The Spruce. *How to Grow and Care for Heather Shrubs.* www.thespruce.com/heather-shrub-growing-profile-3269229.

Virginia Cooperative Extension. *Heaths (Several Species of Erica) and Heathers (Calluna vulgaris).* https://vtechworks.lib.vt.edu/bitstream/handle/10919/48735/3010-1489_pdf.pdf?sequence=1.

WikiHow. *How to Prune Heather.* www.wikihow.com/Prune-Heather.

# 50 Hibiscus (*Hibiscus* Species and Hybrids)

**Scientific Names:** The genus *Hibiscus* is said to be based on the Greek *hibiscus* or *ibiscum*, an allusion to associating with storks (*ibis*) in marshes. Alternatively, the Greek word is claimed to simply designate marsh-dwelling plants of the genus.

**Common Names:** The common name hibiscus, of course, is based on the Latin name. As noted next, hibiscus is sometimes called Chinese rose. The English names of the principal cultivated species are discussed later.

## 50.1 INTRODUCTION

Ornamental plants are grown for their beauty – in most cases, for the appearance of their flowers – so plants that produce long-lasting, colorful displays are favored. A simple principle employed by horticulturalists is to exhibit them, whether planted in the ground or in pots, in groups (so-called "mass planting") so as to maximize the impact of a large, continuous presentation of color. Roses have achieved the status of "queen of the flowers" because individual plants are capable of generating large displays of color that attract the eye. But roses have a competitor – hibiscus – which significantly is sometimes termed "Chinese rose" and "queen of tropical flowers." Hibiscus ornamentals include cultivars that bloom copiously for months, with eye-popping flowers of all colors as large as 30 cm across. Visitors to tropical countries are invariably impressed with the stunning giant flowers of hibiscus shrubs and trees, which are unable to survive cold winters, so they aren't grown in cooler climates. However, for the past century, breeders have been creating gorgeous "hardy hibiscus" cultivars which are well suited to temperate climates.

FIGURE 50.1 A collection of colorful hibiscus flowers.

*Source:* **Photo by Shutterstock/Irabel8.**

## 50.2 OVERVIEW

There are perhaps as many as 400 species (or as few as 200) of *Hibiscus*, including annual or perennial herbs, vines, subshrubs, shrubs, and trees. Study of the genetic relationships of species currently placed in *Hibiscus* has led to debates about the wisdom of including traditionally separated but phylogenetically related small genera. *Hibiscus* is indigenous to all continents except Antarctica and also occurs on islands of the Indian and Pacific Oceans. The plants are found

FIGURE 50.2 Little girls and big hibiscus flowers: (a) *Hibiscus moscheutos*. (b) *Hibiscus rosa-sinensis*. (c) *Hibiscus* 'Bordeau'.

*Source:* **(a) Photo by SB Johnny (CC BY SA 3.0). (b) Photo by Christina Mossaad (CC BY SA 2.0). (c) Photo by Famille Wielosz-Caron (CC BY SA 2.0).**

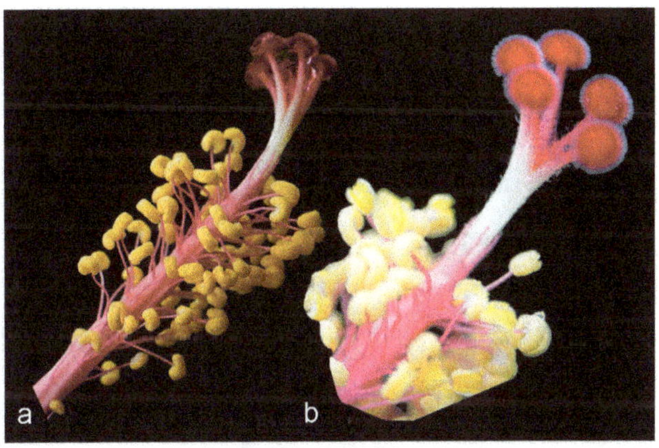


FIGURE 50.3 (a) Sexual column (a cylinder of male flowers surrounding a single female flower) of *Hibiscus*. (b) End of the column showing the five-stigma lobes of the female flower.

*Source:* **(a) Photo by Dinuraj K (CC BY 2.0). (b) Photo by Mondberg25 (CC BY SA 3.0).**

DOI: 10.1201/9781003586227-50

mostly in warm-temperate, subtropical, and tropical regions. *Hibiscus* flowers are characterized by a very prominent staminal column which surrounds the pistil. The flowers are very attractive to pollinators, including hummingbirds and butterflies. *Hibiscus* flowers of the tropical species generally last only 1 day (often senescing by dusk), occasionally persisting up to 3 days, but the plants compensate by flowering profusely and continuously producing new blooms.

Several other cultivated species in the Malvaceae family are similar in appearance to hibiscus. These include hollyhocks (*Alcea*), mallows (*Malva* and *Kosteletzkya*), and okra (*Abelmoschus esculentus*).

*Hibiscus rosa-sinensis* (tropical hibiscus, Chinese hibiscus) is a shrubs or tree 1–3 m (sometimes up to 5 m) in height. It appears to be a cultigen – that is, it seems to have originated in cultivation and does not have a natural indigenous range. The plant has been extensively cultivated and hybridized for centuries. Its ancestors may be traced to China, Africa, and/or the New World tropics, and different cultivars likely arose as hybrids from different parents. Most plants are sterile and do not set seed, so they are propagated by cuttings. Nevertheless, in the tropics and subtropics, where it is principally cultivated, the plant has become somewhat naturalized. There are thousands of cultivars, a very wide range of flower sizes, and huge variation of colors and color combinations. Modern cultivars are mainly hybrids of about eight tropical species, natives of Hawaii (perhaps the most important center of selection), Madagascar, Mauritius, China, and Fiji. These probably include the African *H. schizopetalus* and the Hawaiian species *H. arnottianus*, *H. kokio*, and *H. waimeae*. Australia has also become an important breeding center. *Hibiscus rosa-sinensis*, the most important ornamental species in the genus, is distantly related to *H. syriacus*, the second most important species, and breeders have been unable to cross them. *Hibiscus rosa-sinensis* is grown as a perennial shrub or tree in warmer areas of the world, often as a hedge plant. It is extremely popular outdoors in tropical and subtropical areas. In hot climates, the foliage is evergreen. The plant freezes and dies at about −3°C, so it is unsuitable as an outdoor garden perennial in areas with cold winters. However, it makes an excellent potted specimen that can be moved to the patio or deck in the summer and sheltered indoors in the winter.

*Hibiscus syriacus* (common hibiscus, garden hibiscus, rose of Sharon, shrub althaea, shrub althea) is the most widely grown ornamental hibiscus in temperate regions. The species is not the biblical "rose of Sharon," which has been interpreted as a species of lily, tulip, or any of several other species. Nor is it *Hypericum calycinum*, another ornamental plant which is also well-known as rose of Sharon. *Hibiscus syriacus* is a native of China and India but is not indigenous to Syria, despite its name. It is naturalized in many areas of the world. This winter-hardy deciduous shrub is 1–4 m in height and is sometimes trained to look like a small tree.

FIGURE 50.4 Illustrations (public domain) of pollinators visiting hibiscus flowers: (a) A 1960 Trinidad and Tobago stamp showing a hummingbird and a flower. (b) Queen swallowtail (*Papilio androgeus*) on Confederate rose (*Hibiscus mutabilis*).

**Source:** **(a) Public domain. (b) Merian, M. S. 1705. *Metamorphosis insectorum Surinamensium [The Metamorphosis of the Insects of Suriname]*. Amsterdam.**

FIGURE 50.5 *Hibiscus rosa-sinensis*: (a) A mature shrub. (b) Painting (public domain).

**Source:** **(a) Photo by John (CC BY SA 2.0). (b) Step, E., D. Bois, B. Herincq, and W. Watson. 1896. *Favourite Flowers of Garden and Greenhouse*, vol. 3. Frederick Warne, New York.**

FIGURE 50.6 Examples of colorful flowers of *Hibiscus rosa-sinensis*.

**Source:** **(a) Photo by Xemoth1 (CC BY SA 2.0). (b, c) Photos by John (CC BY SA 2.0). (d) Photo by Franz Winter (CC BY SA 3.0).**

FIGURE 50.7 Illustrations (public domain) of rose of Sharon (*Hibiscus syriacus*).

**Source:** **(a) Photo by I. Sáček. (b) Chromolithograph (1868) of old cultivars by A. J. Wendel (1826–1915).**

The flowers are typically pink (cultivars with many other colors are available) and usually last for just 1 day, but the plant blooms profusely from midsummer and into fall. Some cultivars produce semi-doubled or doubled flowers.

So-called "hardy hibiscus" species are cold-tolerant but not necessarily able to survive in significantly cold areas. These include *H. syriacus*, discussed earlier, and some members of *Hibiscus* section *Muenchhusia*, a small North American indigenous group of five or six herbaceous perennial plants which die back to the ground in which they overwinter. These are known as "rose mallows" (rosemallows). They include *H. coccineus* (scarlet rose mallow), *H. dasycalyx* (Neches River rose mallow), *H. grandiflorus* (swamp rose mallow), *H. lasiocarpos* (hairy or wooly rose mallow, often considered part of *H. moscheutos*), *H. laevis* (halberd-leaf rose mallow), and *H. moscheutos* (common rose mallow, crimson-eyed rose mallow). The latter two are quite cold-hardy. *Hibiscus moscheutos* is the most important of these. Offerings labelled *H. moscheutos* in nurseries are likely to be hybrids with related species. The species grows as tall as 2.5 m. The dwarf cultivar 'Luna Red' has become popular in recent years. It grows to about 1 m in height, has huge flowers, and is easy to cultivate.

FIGURE 50.8 Illustrations of rose mallow (*Hibiscus moscheutos*): (a) Painting (public domain) by Mary Vaux Walcott (1860–1940). (b) A modern cultivar.

*Source:* **(a) From Smithsonian American Art Museum. (b) Photo by André Alliot (CC BY SA 3.0).**

FIGURE 50.9 *Hibiscus acetosella* 'Mahogany Splendor', grown as an annual near Detroit.

*Source:* **Photo by F. D. Richards (CC BY SA 2.0).**

FIGURE 50.10 *Hibiscus schizopetalus*, a native of Africa, growing in a city square in Ecuador.

*Source:* **Photo by David J. Stang (CC BY SA 4.0).**

In warm countries, a wide variety of *Hibiscus* are cultivated, including some that grow into extremely attractive large shrubs and small trees. Red-leaf hibiscus (*H. acetosella*) is a tropical shrub, often grown as an annual in temperate regions. Popular cultivars develop maple-like leaves that are burgundy red.

## 50.3 HISTORY

As with other major ornamental plants, most cultivars grown today are hybrids, often with parentage tracing to several wild species. Malinowski et al. (2019) provide an extensive account of the probable ancestors of modern cultivars.

## 50.4 ECONOMIC IMPORTANCE

The edible flowers of *H. sabdariffa* (roselle) and the fiber from *H. canabinus* (kenaf) are important economic products, rivaling the value of the ornamental forms of *Hibiscus*. *Hibiscus* flower cultivars contribute significantly to the horticultural industries of many countries, especially those with warm and hot climates.

## 50.5 ORNAMENTAL USES

The larger, shrubby hibiscus species are employed as lawn and garden specimens and as hedges, especially in subtropical and tropical climates. Species grown as annuals or as perennials that die back to the ground in the fall are used in garden beds and borders. The large flowers are often employed for personal beautification or as decorations.

FIGURE 50.11 Pots of hibiscus on sale in a garden center.

*Source:* **Photo by Shutterstock/New Africa.**

FIGURE 50.12 Rose of Sharon (*Hibiscus syriacus*) in front of a house.

*Source:* **Photo by Shutterstock/Spiroview Inc.**

FIGURE 50.13 Chinese rose (*Hibiscus rosa-sinensis*) hedges in Hawaii.

*Source:* **Photos by Forest and Kim Starr (CC BY 3.0).**

## 50.6 OTHER USES

Most economically significant species of *Hibiscus* are grown as ornamentals, but *H. cannabinus* (kenaf) is an important stem fiber crop, and *H. sabdariffa* (roselle, Jamaica sorrel) is a source of fleshy calyces (the sepals of the flowers) employed to flavor beverages. Both of the latter species are adapted to warm climates. These plants, as well as those grown for ornamentals, have been extensively employed in herbal medicine to treat a variety of illnesses and are currently being researched for possible modern medicinal applications.

FIGURE 50.14 A planter with hibiscus in Vietnam.

*Source:* **Photo by Shutterstock/Phetsamay Philavanh.**

FIGURE 50.15 A Hawaiian lady wearing a hibiscus lei and hibiscus headband.

*Source:* **Photo by Shutterstock/Chiyacat.**

## 50.7 PROBLEMS

Several *Hibiscus* species are invasive weeds. *Hibicus syriacus* is weedy in much of the eastern United States.

FIGURE 50.16 *Hibiscus sabdariffa*, the roselle or Jamaica sorrel plant, grown for its edible flowers, widely employed to prepare sweet beverages and desserts: (a) Plant in flower. (b) A farmer in a field of cultivated roselle in Argentina. (c) Calyxes (sepals), the edible parts of the flowers. (d) Hibiscus tea (made with roselle).

*Source:* **(a) Delaunay, M., and J. L. A. Loiseleur-Deslongchamps. 1821. *Herbier général de l'amateur*, vol. 5, Paris. (b) Photo by Agro Emprendedores (CC BY 2.0). (c) Photo by Mohamad bin Osman (public domain). (d) Photo by MRG90 (CC BY SA 4.0).**

## 50.8 CULTIVATION

### 50.8.1 Climate and Site

Hibiscus species differ greatly in climate suitability, and even the hardy cultivars differ in overwintering cold tolerance. New cultivars are being generated, and local advice from horticulturalists is advantageous. *Hibiscus syriacus* and many cultivars of *H. moscheutos* will usually survive as far north as climate zone 5, but some can tolerate zone 4. Mulching the base of the shrubs in late fall will promote survival. *Hibiscus rosa-sinensis* and many other species do not tolerate frost, and in areas with freezing winters, they may be grown as potted plants that can be overwintered indoors. *Hibiscus* species prefer sunny sites but tolerate light shade. Tall cultivars should be situated where they are not exposed to strong winds to avoid breaking of the long stems.

### 50.8.2 Propagation and/or Purchasing

The herbaceous perennials such as *H. moscheutos* are propagated by seeds or division. *Hibiscus rosa-sinensis* often does not produce seeds, and in any event, named cultivars are frequently hybrids, which will not grow true from

FIGURE 50.17 Kenaf (*Hibiscus cannabinus*), an important tropical source of stem fiber: (a) Plants with very tall stems, ideal for producing quality stem fiber. (b) Flowering stem.

*Source:* **(a) Photo by US Department of Agriculture (public domain). (b) Photo by Lalithamba (CC BY 2.0).**

FIGURE 50.18 Indoor cultivation of *Hibiscus rosa-sinensis.*

*Source:* **Photos from public domain: (a) Pipsels, (b) Maxpixel.**

seeds, so they are reproduced vegetatively (by cuttings, grafting, or layering; sometimes by marcottage, which is air layering – that is, the rooting medium is bound to the plant rather than enclosed in a pot). Although *Hibiscus* species can often be grown from seed started indoors, it is much faster to buy potted plants, and most of the ornamental cultivars are reproduced only vegetatively. Larger nurseries and specialty mail-order businesses have a greater range of plants (which are often healthier) than big-box chain stores.

### 50.8.3 Planting

Planting distance needs to be based on the anticipated size of the mature plants. Some of the species are capable of

FIGURE 50.19 Hibiscus grown as bonsai.

*Source:* **Photo by Shutterstock/Walter Pall.**

FIGURE 50.20 Hibiscus flowers worn on the right indicates availability for marriage, on the left indicates attached status.

*Source:* **(a) From DLPNG public domain image database. (b) "Tahitian women on the beach" wearing hibiscus flowers, by French artist Paul Gaugin (1848–1903). Public domain image.**

living for decades, and so large shrubs need to be spaced 1–2 m or more apart.

### 50.8.4 Soil and Moisture Conditions

*Hibiscus moscheutos* and many of its herbaceous, hardy perennial relatives are native to swamps, so they need ample moisture. *Hibiscus rosa-sinensis* requires regular watering and good drainage. Hibiscus plants prefer soil that has appreciable organic matter. Ideal pH for the soil is 6.0 to 7.0. Moderate fertilization is recommended, with a slow-release, balanced fertilizer or one formulated for hibiscus plants.

### 50.8.5 Maintaining Plants

To encourage blooming, spent flowers should be removed (advice that applies to most of the plants discussed in this book).

## 50.9 FASCINATING FACTS, FABLES, FOLKLORE, AND FRIVOLITY

- The expression "rose of Sharon" traces to the translation of a Hebrew passage in the Old Testament (Song of Solomon 2:1) by the editors of the King James Version of the Bible (published in 1611). "Sharon" refers to a fertile valley plain in northern Palestine, but what the "rose" was is doubtful. Although botanists today primarily employ the phrase for both *Hibiscus syriacus* and *Hypericum calycinum* (sometimes also for other species), biblical scholars strongly disagree on what material this originally designated (it apparently was a metaphor for the beauty of one of King Solomon's lovers/brides). The English phrase "rose of Sharon" has been a popular designation for characters in literature, notably for a young woman in John Steinbeck's novel *Grapes of Wrath*, where after her baby was stillborn, she provided her milk to a starving stranger. "Rose of Sharon" has also become a favored name for foundations, businesses, movements, and buildings.
- The red flowers of some forms of *H. rosa-sinensis* are so dark they appear to be black. Preparations of such flowers have been employed as shoe blackening in Asia, and this gave rise to the alternative names "shoe plant" and "shoe black plant."
- Of all the world's navies, the British have named ships after plants more than any other nation. Two British Royal Navy vessels were named HMS *Hibiscus*: HMS *Hibiscus* (1917), a sloop used until 1923, and HMS *Hibiscus* (K24), a corvette launched in 1940 and retired in 1946. Several non-naval British ships have also been named *Rose of Sharon*.
- *Hibiscus rosa-sinensis* is the national flower of Malaysia, the Solomon Islands, and Niue, and the national symbol of Haiti. The state flower of Hawaii is *H. brackenridgei*, which is native there. *Hibiscus syriacus* is the national flower of South Korea.
- In Tahiti and Hawaii, hibiscus flowers are worn by women as an indication of their marriage eligibility. Wearing the flower behind the right ear signals availability, behind the left ear indicates the woman is not available.

## KEY PUBLICATIONS

Akpan, G.A. 2007. *Hibiscus: Hibiscus rosa-sinensis*. In: *Flower Breeding and Genetics: Issues, Challenges and Opportunities for the 21st Century*, edited by N.O. Anderson, 479–490. Springer, Dordrecht.

Al-Snafi, A.E. 2018. Chemical constituents, pharmacological effects and therapeutic importance of *Hibiscus rosa-sinensis* – a review. *IOSR Journal of Pharmacy* 8: 101–119.

Bates, D.M. 1965. Notes on the cultivated Malvaceae, 1. *Hibiscus*. *Baileya* 13: 56–130.

Blanchard, O.J., Jr. 2015. *Hibiscus*. In: *Flora of North America*, Vol. 6, edited by Flora of North American Editorial Committee, 252–267. Oxford University Press, New York.

Gilman, E.F. and D.G. Watson. 2014. *Hibiscus syriacus: Rose-of-Sharon*. University of Florida Extension Bulletin ENH454. 3 pp.

Khan, I.M., R. Rahman, A. Mushtaq, and M. Rezgui. 2017. *Hibiscus rosa-sinensis* L. (Malvaceae): Distribution, chemistry and uses. *International Journal of Chemical and Biochemical Sciences* 12: 147–151.

Kuligowska, K., H. Lutken, B. Christensen, and R. Muller. 2016. Interspecific hybridization among cultivars of hardy *Hibiscus* species section *Muenchhusia*. *Breeding Science* 66: 300–308.

Kuligowska, K., M. Simonsen, H. Lutken, and R. Muller. 2013. Breeding of *Hibiscus rosa-sinensis* for garden use in Denmark. *Acta Horticulturae* 990: 235–242.

Lawton, B.P. 2004. *Hibiscus: Hardy and Tropical Plants for the Garden*. Timber Press, Portland, OR. 184 pp.

Lim, T.K. 2014a. *Hibiscus rosa-sinensis*. In: *Edible Medicinal and Non-Medicinal Plants*, edited by T.K. Lim, 306–323. Springer, Dordrecht.

Lim, T.K. 2014b. *Hibiscus syriacus*. In: *Edible Medicinal and Non-Medicinal Plants*, edited by T.K. Lim, 374–380. Springer, Dordrecht.

Magdalita, P.M. and A.O. San Pascual. 2022. Hibiscus (*Hibiscus rosa-sinensis*): Importance and classification. In: *Floriculture and Ornamental Plants. Handbooks of Crop Diversity: Conservation and Use of Plant Genetic Resources*, edited by S.K. Datta and Y.C. Gupta, 483–522. Springer, Singapore.

Malinowski, D.P., W.E. Pinchak, and K. Yanker-Hansen. 2019. Phenotypic similarities in flower characteristics between novel winter-hardy hibiscus hybrids and their tropical relatives. *Frontiers in Plant Science* 10: 1528. Frontiers | Phenotypic Similarities in Flower Characteristics Between Novel Winter-Hardy Hibiscus Hybrids and Their Tropical Relatives.

Merritt, P.B. 2003. *The Tropical Hibiscus Handbook*. American Hibiscus Society, Cocoa Beach, FL. 120 pp.

Nakasone, H.Y. and F.D. Rauch. 1980. *Ornamental Hibiscus Propagation and Culture. Hawaii Agricultural Experiment Station Research Bulletin 175*. University of Hawaii, Honolulu. 15 pp.

Pfeil, B.E., C.L. Brubaker, L.A. Craven, and M.D. Crisp. 2002. Phylogeny of *Hibiscus* and the tribe Hibisceae (Malvaceae) using chloroplast DNA sequences of ndhF and the rpl16 intron. *Systematic Botany* 27: 333–350.

Small, R.L. 2004. Phylogeny of *Hibiscus* sect. *Muenchhusia* (Malvaceae) based on chloroplast rpL16 and ndhF, and nuclear ITS and GBSSI sequences. *Systematic Botany* 29: 385–392.

Sparnaaij, L.D. 1973. The introduction of new hibiscus cultivars. *Acta Horticulturae* 31: 133–136.

Walker, J. 2001. *Hibiscus*. Firefly Books, Richmond Hill, ON. 96 pp.

Winters, H.F. 1970. Our hardy *Hibiscus* species as ornamentals. *Economic Botany* 24: 155–164.


## Key Cultivation Websites


American Hibiscus Society. www.americanhibiscus.org/.

Gardener's Path. *How to Grow and Care for Hardy Hibiscus*. https://gardenerspath.com/plants/flowers/grow-hardy-hibiscus/.

Hibiscusmania. www.hibiscusmania.com/.

International Hibiscus Society. www.internationalhibiscussociety.org/.

The Spruce. *How to Grow and Care for Hibiscus*. www.thespruce.com/hibiscus-growing-guide-5192541#:~:text=Hibiscus%20tree%20is%20an%20annual,moist%2C%20well%2Ddrained%20soil.

The Spruce. *How to Grow and Care for Tropical Hibiscus*. www.thespruce.com/caring-for-hibiscus-847914.

Today's Gardener. *How to Grow Hibiscus Indoors*. https://indoorgardenook.com/hibiscus-from-a-to-z/.

WikiHow. *How to Care for a Chinese Hibiscus*. www.wikihow.life/Care-for-a-Chinese-Hibiscus.

WikiHow. *How to Care for a Hibiscus Plant*. www.wikihow.com/Care-for-a-Hibiscus-Plant.

WikiHow. *How to Grow Hibiscus Outdoors*. www.wikihow.com/Grow-Hibiscus-Outdoors.

WikiHow. *How to Plant Hibiscus*. www.wikihow.com/Plant-Hibiscus#:~:text=Try%20to%20keep%20your%20hibiscus,soil%20gets%20very%20dried%20out.

WikiHow. *How to Propagate Hibiscus*. www.wikihow.com/Propagate-Hibiscus.

WikiHow. *How to Prune Hibiscus*. www.wikihow.com/Prune-Hibiscus.

WikiHow. *How to Winterize a Perennial Hibiscus*. www.wikihow.com/Winterize-a-Perennial-Hibiscus.

The World of Hibiscus: Hibiscus Mail Lists. www.hibiscusworld.com/Mail-Lists.html.

# 51 Hollyhock (*Alcea rosea* and Hybrids)

**Scientific Names:** The genus name *Alcea* traces to the ancient Greek *alkea* (mallow), through the Latin *alcea* (for the garden mallow, *Malva alcea*). Species in the closely related genus *Althaea* are known as "marshmallows," best known of which is *A. officinalis*, the common marshmallow, once often used in medicine and in confectionary. The genus *Alcea* has often been considered as part of *Althaea*, and *Alcea rosea* is often called *Althaea rosea*.

**Common Names:** The name hollyhock originated from Middle English *holy* + the obsolete *hock*, "mallow" (a reference to the marshmallow plant, *Malva officinalis*).

FIGURE 51.2 Hollyhock flowering stalks.

*Source:* **(a) Photo by Angie Six (CC BY ND 2.0). (b) Photo by Biswarup Ganguly (CC BY SA 3.0). (c) Cultivar Fiesta Time. Photo by Cephas (CC BY SA 4.0).**

## 51.1 INTRODUCTION

Hollyhocks are the opposite of shrinking violets: They are giants of the garden, with very large, extremely colorful flowers. Hollyhocks are not the easiest of plants to grow but are among the most rewarding. They are traditional "stalwarts of the cottage garden," which is a euphemistic way of pointing out that many consider them old-fashioned.

## 51.2 OVERVIEW

The genus *Alcea* is made up of over 80 species, most biennial and short-lived perennial herbs, native to the eastern Mediterranean to Central Eurasia. *Alcea rosea* (common hollyhock), the principal cultivated species, is speculated to have originated in ancient times from the Anatolian region of Turkey (Asia Minor), but today its presence outside of cultivation appears to be due to escapes from cultivation. The species has become naturalized (established) in numerous countries around the world but is not an aggressive invader. Many hollyhock cultivars are hybrids of *Alcea rosea* and other species of *Alcea*. Garden hollyhocks are mostly short-lived perennials, usually grown as biennials. A basal rosette is developed the first year, from which, in the second year, a sturdy unbranched or sparsely branched flowering stem arises. Sometimes the plant persists and flowers in a third year. Some cultivars, especially dwarf forms, flower the first year and can be grown as annuals. The flowers are up to 15 cm in diameter when open, on main flowering stems that are as tall as 2.5 m. Flower colors in cultivars include lavender, pink, maroon, purple (including a shade that looks black), red, white, and yellow. Bicolor (two-color) flowers and doubled flowers are also available. The male and female parts of the flower are located in a columnar structure in the center, numerous stamens surrounding and hiding the pistil, except for its threadlike stigmas. In doubled flowers (i.e., with extra petals), the sexual parts have been transformed into petals. Plants typically bloom from midsummer to autumn.

FIGURE 51.1 Hollyhock gardens.

*Source:* **(a) Photo by Shutterstock/Wei Ling Chang. (b) Photo by Shutterstock/Alex Manders. (c) Photo by Akiyoshi Matsuoka (CC BY SA 4.0). (d) Photo by Maxwell Hamilton (CC BY 2.0).**

## 51.3 HISTORY

Hollyhock was introduced to Great Britain in the 16th century. Cultivars with doubled flowers were well developed by the 18th century. In the late 19th century, rust disease attacked hollyhocks in Europe, leading to a considerable



DOI: 10.1201/9781003586227-51

FIGURE 51.3 Botanical illustrations of flowering hollyhocks (public domain).

*Source:* **(a) Painting by (female) Danish artist Anthonore Christensen (1849–1926). (b) Chromolithograph by E. T. Fisher, published by Prang & Co. in 1885. Photo from Boston Public Library. (c) 1900 John Lewis Childs seed catalogue. (d) Köhler, F. E. 1887. Medizinal Pflanzen, vol. 1.**

FIGURE 51.4 Paintings (public domain) of cottage gardens with hollyhocks by Danish artist Peder Mørk Mønsted (1859–1941).

*Source:* **Photo from Bbruun-rasmussen.dk.**

decrease in their popularity. In the early 20th century, there was a revival of interest in the plant, and recent cultivars tend to be relatively rust-resistant. An extensive historical account of hollyhocks by the Garden Trust is available at: https://thegardenstrust.blog/2020/09/05/hollyhocks/.

## 51.4 ECONOMIC IMPORTANCE

Hollyhock is economically important to the world's horticultural trade, mostly for sale of seeds and starter plants for cultivation in private gardens. It is of relatively minor importance compared to most of the other well-known standard garden ornamentals.

## 51.5 ORNAMENTAL USES

"Cottage gardens" are small, informal plantings of mixed ornamental plants (and sometimes vegetables) traditionally associated with modest houses in England. These very picturesque gardens often were enclosed with a rose-bowered gateway. Hollyhocks are one of the premier examples of cottage garden plants. Tall hollyhocks are frequently grown against walls and fences and used as backdrops in mixed beds and borders.

Hollyhock flowers are extremely attractive to hummingbirds and bees. Other birds happily consume the seeds. Hollyhocks are, accordingly, often planted in pollinator and wildlife gardens.

The term "mallow" may refer to any of the more than one thousand plants in the mallow family (Malvaceae), but it often denotes only species of *Malva*, of which common mallow

FIGURE 51.5 Use of hollyhocks beside fences and houses.

*Source:* **(a) Photo by Chic Bee (CC BY 2.0). (b) Photo by Shutterstock/a40757. (c) Photo by Shutterstock/Anna50. (d) Photo by Marja van Bochove (CC BY 2.0).**

FIGURE 51.6 Hollyhock flower visitors: (a) Hummingbird in New York state. (b) Ladybug. (c) Bumblebee.

*Source:* **(a) Photo by All Things Colorful CC BY ND 2.0). (b) Photo by Jeri Leandera (CC BY 2.0). (c) Photo by Alh1 (CC BY ND 2.0).**

FIGURE 51.7 Commonly cultivated *Malva* ("mallow") species, which are often misidentified as hollyhocks (public domain illustrations): (a) Common mallow (*M. sylvestris*). (b) Rose mallow (*M. alcea*).

*Source:* **(a) Hall, C. A. 1915. Plant-life. A. & C. Black, London. (b) Witte, H., and A. J. Wendel: 1868. Flora: Afbeeldingen en beschrijvingen van boomen, heesters, éénjarige planten, enz. voorkomende in de Nederlandsche Tuinen. Wolters, Groningen.**

(*Malva sylvestris*) is especially cultivated as an ornamental. Some garden cultivars of *Malva* resemble hollyhocks in flower and foliage, albeit these are usually shorter and have smaller flowers with mostly separate petals. Some garden catalogues even label some *Malva* varieties as "hollyhocks." The Latin name of rose mallow is *Malva alcea*, meaning "the mallow that looks like hollyhock." Because they are smaller, mallows may be more suitable for containers and small gardens.

## 51.6 OTHER USES

Hollyhock, like most cultivated plants, has a long history of usage in herbal medicine, and there still exists a small market for such uses. There are efforts underway to find modern medicinal applications.

Many members of the mallow family (Malvaceae), including several *Malva* species (notably marshmallow, *Althaea officinalis*), are edible. Although the young foliage and petals can be consumed, they are bland with little flavor. The plant is not considered to be toxic (but as is the case for all plants, occasional people can experience contact irritation). Livestock will graze on the flowers and foliage.

## 51.7 PROBLEMS

Fashion affects the popularity of ornamentals, just as it does many other aspects of life. Hollyhocks are considered extremely old-fashioned, and this does seem to affect the choices of gardeners. Hollyhocks are generally among the very tallest of herbaceous ornamentals, and the trend to smaller gardens makes them less desirable. Most cultivars are biennials, and waiting for plants to flower in the second year does not please many gardeners. Although resistant varieties are becoming available, susceptibility to hollyhock rust disease also discourages cultivation.

## 51.8 CULTIVATION

### 51.8.1 Climate and Site

Hollyhock can be grown in climate zones 2 through 10 (ideally in zones 3 to 8). The plants grow best in full sun but can tolerate part shade. Because they are tall, sites protected from wind are preferable.

FIGURE 51.8 Hollyhock rust (*Puccinia malvacearum*), a fungus producing pustules containing numerous spores on the lower leaf surface (shown on the left). Heavy infestations can kill the plant, as shown on the right.

*Source:* **Photos by Whitney Cranshaw (CC BY 3.0).**

FIGURE 51.9 Extreme climates which hollyhocks tolerate: (a) Very hot. (b) Very cold.

*Source:* **(a) Photo by Shutterstock/Aimful. (b) Photo by Dicho-hecho (CC BY 2.0).**

### 51.8.2 Propagation and/or Purchasing

Hollyhock is usually started from seeds. Cultivars with doubled flowers are usually propagated vegetatively. Mature plants can be divided, and pieces of the larger roots ("root cuttings") from the root crown (the root system attached to the base of the central stalk) can be employed to grow new plants. Young starter plants may be available from garden centers. Older plants develop deep taproots, which are not easily transplanted.

### 51.8.3 Planting

Whether established from seeds, root cuttings, or starter plant, individuals should be sufficiently separated – about 50 cm apart – to promote good air circulation and reduce the fungal disease rust associated with higher humidity.

### 51.8.4 Soil and Moisture Conditions

Hollyhock is a heavy feeder and benefits from fertile, organically rich soils. It does well in loams and clay, but the soil must be well-draining. The pH can be 6.0 to 8.0. When irrigating, attempt should be made to apply water to the base of the plant and avoid wetting the foliage, since this encourages fungal disease.

### 51.8.5 Maintaining Plants

Hollyhocks often self-seed and establish long-lived colonies, with older plants in flower and younger ones growing a basal rosette of leaves. Hollyhock rust can be limited by removing lower, senescing leaves and plant debris (both can harbor rust spores), deadheading wilted blooms, and cutting the plants back in the fall. Taller plants may need staking to remain upright.

## 51.9 FASCINATING FACTS, FABLES, FOLKLORE, AND FRIVOLITY

- Many of the best artists in the past produced paintings associating beautiful flowers together with beautiful women. Hollyhocks were frequently among the flowers shown.

FIGURE 51.10 Examples of hollyhock art (public domain): (a) *Hollyhocks*, by American artist Eastman Johnson (1824–1906), housed in the New Britain Museum of American Art. (b) *Hollyhocks and Sunlight*, by American artist Charles Courtney Curran (1861–1942).

*Source:* **(a) Photo by Daderot. (b) Photo by Photographer Sotheby's.**

FIGURE 51.12 Independent "Hollyhock Lanes" in early 20th-century coastal Massachusetts, shown in old-fashioned linen postcards (notice the diffuse texture): (a) In Nantucket. The structure on top of the house is a "captain's walk," a lookout to view incoming ships. (b) On the Cape Cod Peninsula. Notice the lookout tower to view ships.

*Source:* **Public domain illustrations. Photo from Boston Public Library.**

FIGURE 51.11 Hollyhocks growing at Frank Lloyd Wright's "Hollyhock House" in Los Angeles, a UNESCO World Heritage Site. The patterned figures decorating the building are abstract representations of the flowers.

*Source:* **Photo by Jeremy Thompson (CC BY 2.0).**

- "Hollyhock House" (Aline Barnsdall Hollyhock House) is a mansion in Los Angeles designed by American architect Frank Lloyd Wright (1867–1959) for oil heiress Aline Barnsdall. The innovative building was declared a UNESCO World Heritage Site in 2019. The hollyhock, Aline Barnsdall's favorite flower, is a central theme of the house, with geometric interpretations of the hollyhock flower throughout the house, planters with hollyhocks, and stained-glass windows with stylized hollyhock patterns.
- Frank Lloyd Wright is famously quoted as saying, "The physician can bury his mistakes, but an architect can only advise his clients to plant vines." In a similar vein, hollyhocks were often found on rural farms, where they often served to hide eyesores, such as outhouses, manure piles, pig sheds, and decrepit buildings. Indeed, hollyhocks were once given the nickname "outhouse plant," and the very tall plants served as unspoken signage to the location of outdoor toilets. Asking where one could find the hollyhocks was a euphemistic way of asking for directions to the commode.
- A record-setting hollyhock 7.4 m high was grown in Israel in 2021 (www.guinnessworldrecords.com/world-records/72269-tallest-hollyhock).
- Curiously, there were once two different roads named "Hollyhock Lane" in the state of Massachusetts in pre–World War II times, one in Provincetown (the northern tip of Cape Cod Peninsula), the other on the island of Nantucket. Both, the sites of considerable cultivation of hollyhocks, were frequently shown in postcards.

## KEY PUBLICATIONS

Aly Maher Arfa, N. 2022. Utilization of hollyhock (*Althaea officinalis* L.) roots in manufacture of juices and jams. *International Journal of Family Studies, Food Science and Nutrition Health* 3: 193–211.

Celka, Z., K. Buczkowska, A. Baczkiewicz, and M. Drapikowska. 2010. Genetic differentiation among geographically close populations of *Malva alcea*. *Acta Biologica Cracoviensia Series Botanica* 52(2): 32–41.

Celka, Z. and M. Drapikowska. 2008. Relics of cultivation in Central Europe: *Malva alcea* L. as an example. *Vegetation History and Archaeobotany* 17: 251–255.

Escobar García, P., M. Pakravan, P. Schönswetter, J.F Aguilar, and G.M. Schneeweiss. 2012. Phylogenetic relationships in the species – rich Irano – Turanian genus *Alcea* (Malvaceae). *Taxon* 61: 324–332.

Friedman, A.T. 1992. A House is not a home: Hollyhock House as art-theater garden. *The Journal of the Society of Architectural Historians* 51: 239–260.

Hoffmann, D. 1992. Frank *Lloyd Wright's Hollyhock House: The Illustrated Story of an Architectural Masterpiece*. Dover, Mineola, NY. 118 pp.

Li, Q., C.-J. Ruan, J.A. Teixeira da Silva, and X.-Y.Wang. 2012. Floral morphology and mating system of *Alcea rosea* (Malvaceae). *Plant Ecology and Evolution* 145: 176–184.

Lim, T.K. 2014. *Alcea rosea*. In: *Edible Medicinal and Non Medicinal Plants*, edited by T.K. Lim, 292–299. Springer, Dordrecht.

Shehzad, M.R., M.A. Hanif, R. Rehman, I.A. Bhatti, and A. Hanif. 2020. Hollyhock. In: *Medicinal Plants of South Asia*, edited by M.A. Hanif, H. Nawaz, M.M. Khan, and H.J. Byrne, 381–391. Elsevier, Amsterdam.

## Key Cultivation Websites

Better Homes & Gardens. *Hollyhock*. www.bhg.com/gardening/plant-dictionary/perennial/hollyhock/.

Harvesting History. *Hollyhock*. A little history and some growing instructions. https://harvesting-history.com/hollyhock/.

Royal Horticultural Society. *How to Grow Hollyhocks*. www.rhs.org.uk/plants/hollyhocks/growing-guide.

The Spruce. *How to Grow and Care for Hollyhock (Alcea)*. www.thespruce.com/hollyhock-alcea-growing-guide-5200999.

University of Illinois Urbana-Champaign, Extension. *How to Grow and Care for Hollyhocks*. https://extension.illinois.edu/blogs/good-growing/2023-05-26-how-grow-and-care-holly hocks.

WikiHow. *4 Easy Ways to Propagate Hollyhocks from Root Cuttings Or Seeds*. www.wikihow.com/Propagate-Hollyhocks.

WikiHow. *How to Grow Hollyhocks*. www.wikihow.com/Grow-Hollyhocks.

WikiHow. *How to Prune Hollyhocks (When & How to Cut Back Hollyhocks)*. www.wikihow.com/Prune-Hollyhocks.

WikiHow. *How to Transplant Hollyhocks*. www.wikihow.com/Transplant-Hollyhocks.

# 52 Hyacinth (*Hyacinthus orientalis*)

**Scientific Names:** *Hyacinthus orientalis*. The genus name *Hyacinthus* is thought to trace to the classical Greek and Latin names for hyacinth or (more likely) some other blue-flowered plant (larkspur and iris have been suggested). *Orientalis* in the Latin name refers to "eastern" but is sometimes translated as "from the Orient" (understood as east of Europe). *Hyacinthus* is a member of the Hyacinthaceae or the Asparagaceae (see www.pacificbulbsociety.org/pbswiki/index.php/Hyacinthaceae for a justification of placement in the latter).

A pre-Hellenic mythological figure who came to be called Hyakinthos by the Greeks is often interpreted as the basis of the names *Hyacinthus* and hyacinth. In one of many imaginative accounts, a Spartan prince with this name was killed by Apollo in an unfortunate throw of his discus. The hyacinth plant grew from the youth's blood, planted by the regretful Apollo. The event is portrayed in several paintings by European masters. Apollo and Hyacinth are often alleged to be lovers and so representatives of homosexual concerns. By contrast, "[t]he hyacinth is sometimes regarded as a Christian symbol of prudence, peace of mind, and the desire for Heaven. This was derived from the myth of Hyacinthus, upon whose death the flower sprung forth" (www.catholictradition.org/Saints/signs4.htm).

FIGURE 52.1 *The Death of Hyacinth*, public domain painting by Russian artist Alexandre Kisseliov (1838–1911). The Greek God Apollo is holding the fallen Hyacinth, whom he has just accidentally killed with his discus.

**Common Names:** Hyacinth, common hyacinth, Dutch hyacinth, garden hyacinth. Rarely called *jacinth* (a word of Anglo-French origin). One needs to be careful when reading information about plants called "hyacinth," because they could belong to other genera. For example:

Grape hyacinth (*Muscari* species; Liliaceae or Asparagaceae family)

Hyacinth bean (*Lablab purpureus*, *Dolichos lablab*; Fabaceae)

Water hyacinth (*Eichhornia crassipes*, *Pontederia crassipes*; Pontederiaceae)

Wild hyacinth (*Camassia scilloides*; Liliaceae or Asparagaceae)

Wild hyacinth (*Brodiaea* species; Liliaceae, Asparagaceae, or Themidaceae)

Wood hyacinth (*Hyacinthoides hispanica*; Asparagaceae)

## 52.1 INTRODUCTION

Hyacinths are iconic, easily grown, miniature early-season plants with gorgeous flower clusters and unforgettable intense perfume. They are standard ornamentals for containers, gardens, landscapes, and short-term houseplants.

## 52.2 OVERVIEW

The genus *Hyacinthus* has three species (more were recognized in the past, but they were transferred to other genera).

FIGURE 52.2 Large beds of hyacinth (*Hyacinthus orientalis*) displaying the range of colorful flowers available.

*Source:* **(a) Photo by Shutterstock/Bruce Baron. (b) Photo by Jennifer Kramer (CC BY 2.0). (c) Photo by Marianne Bevis (CC BY ND 2.0). (d) Photo by Dguendel (CC BY SA 4.0).**

DOI: 10.1201/9781003586227-52

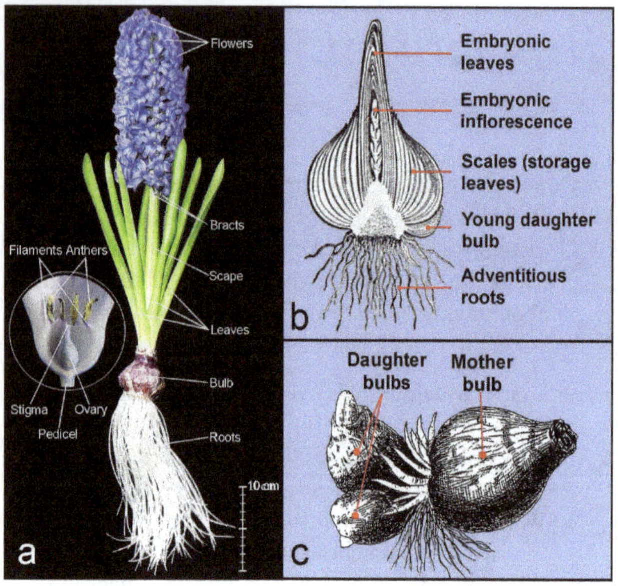


FIGURE 52.3 (a) Parts of the plant. (b) Longitudinal section of bulb. (c) Mature offsets from a hyacinth bulb.

*Source:* **Public domain diagrams. (a) QuazDelaCruz. (b) Bergen, J. Y. 1896.** ***Elements of botany.*** **Ginn, Boston.** ***Text added.*** **(c) Weathers, J. 1867.** ***The bulb book.*** **J. Murray, London.** ***Text added.***

FIGURE 52.4 Early 20th-century garden catalogues featuring hyacinths (*Hyacinthus orientalis*). Public domain reproductions. Note in (a) double-flowered cultivar at left, normal-flowered cultivar at right.

*Source:* **(a) Photo from Swallowtail Garden Seeds. (b) Photo from Internet Archive.**

*Hyacinthus litwinovii* and *H. transcapsicus* occur in Turkmenistan and Iran, and some authorities place them in the related genus *Hyacinthella*. These do not appear to have contributed to the ancestry of domesticated hyacinths. Wild *H. orientalis*, the ancestor of domesticated hyacinths, is native to the eastern Mediterranean region from Turkey to Iran, and Syria to northern Israel. The species occasionally occurs elsewhere as a feral introduction. It is a perennial, long-lived, bulbous herb with a few basal, fleshy, narrow, strap-shaped leaves. The very fragrant flowers are produced in the spring, usually singly on an erect stalk 20–30 cm (sometimes as much as 40 cm) tall. The six-petalled, bell- or funnel-shaped florets range in length from 0.5 to 2.5 cm long, with 5–40 densely packed on the upper part of the stalk. The wild plants usually have blue flowers. Flower colors include shades of blue, purple, pink, red, yellow, and white. Blue- or purple-flowered cultivars are frequent. Double-flowered forms are also available in the trade, although they were more popular in past times. A few cultivars produce plants with more than a single flowering stem. Varieties differ notably in how closely the flowers are packed together.

The bulbs are overwintering, starch-storing reproductive organs which, when mature, are 3–7 cm in diameter. Hyacinth bulbs are "tunicate," that is, they have an outer layer of dry, protective scales. The original (mother) bulbs produce smaller adventitious bulbs from the base, and these vegetatively reproduce the species. As in many other bulb-producing perennials, some contractile roots are present, which pull the new bulbs downward to prevent them from popping out of the soil. Hyacinths also reproduce sexually, and in nature, the seeds are ant-dispersed, employing an edible portion (i.e., an elaiosome) to attract the ants.

## 52.3 HISTORY

Wild hyacinths allegedly were grown in classical Greece and Rome, but interpreting names used in ancient literature is hazardous. Hyacinth, like tulip and other popular bulb ornamentals, was domesticated in Asia Minor. The plant was valued in the gardens of the rulers of the Ottoman Empire in the 15th century. By the 1550s, it was imported into Europe, particularly The Netherlands, which subsequently developed the world-famous Dutch bulb industry. In the 1730s, a market frenzy was based on double-flowered hyacinth clones, which sold for preposterously inflated prices, just as had occurred with tulips a century earlier.

Unlike numerous garden plants, ornamental hyacinths originated from just one species, *H. orientalis*. "Hybrids" are intraspecific combinations of selections, not interspecific. Thousands of cultivars have been named, but most are extinct, and only a few dozen clones dominate current markets.

In the earlier part of the 20th century, the low-growing bulb ornamentals, especially hyacinths, were often laid out in elaborate geometric patterns in lawns and in front of large buildings. This labor-intensive practice is still sometimes observed in large commercial and public spaces.

## 52.4 ECONOMIC IMPORTANCE

Hyacinths are one of the world's major ornamentals, ranking behind tulips and daffodils in the "bulb" class (which includes not just true bulbs but also corms, rhizomes, and other underground storage organs that serve for annual regeneration). Hyacinth bulbs and potted plants represent important commodities in international trade. The Netherlands, the chief trader, regularly exports about 30 million euros worth of "hyacinth" species (mostly Dutch

FIGURE 52.5 Elaborate landscape layouts of hyacinths (*Hyacinthus orientalis*). Such artistic patterns used to be very popular but are uncommon today: (a) Long row of blue and white hyacinths at Keukenhof Gardens in The Netherlands. (b) Hyacinth garden at Anglesey Abbey, UK. (c, d) Early 20th-century hyacinth catalogues featuring hyacinths.

*Source:* **(a) Photo by Will Bakker (CC BY SA 2.0). (b) Photo by Andrew Stawarz (CC BY ND 2.0). (c, d) Public domain.**

hyacinth; www.statista.com/statistics/581418/value-of-the-import-and-export-of-hyacinth-bulbs-in-the-netherlands/).

## 52.5 ORNAMENTAL USES

For the most part, hyacinths are grown in public landscapes, in residential gardens, and as houseplants, but occasionally they are used as cut flowers. Hyacinths are widely employed in landscaping as massed beds, in borders, and occasionally in rock or alpine gardens. The bulbs are frequently grown in containers, especially when forced for interior winter bloom, as described later.

## 52.6 OTHER USES

As early as the 17th century, the French began making perfumes from the flowers, and extraction of essential oil for scent still occurs in France. Flowering potted plants are so pungent that they are sometimes marketed as natural plant deodorizers for the home kitchen or bathroom. In traditional (folk or herbal) medicine, foliage and bulbs were used as a hemostat (to control bleeding), promote healing, treat prostate disease and hemorrhoids, and alleviate other conditions. There are no current approved medicinal applications, but as with numerous other plants, experiments are underway to assess the value of the chemical constituents of hyacinth.

FIGURE 52.6 Commercial fields of hyacinths (*Hyacinthus orientalis*) in The Netherlands.

*Source:* **(a) Photo by Shutterstock/Muratart. (b) Photo by Kyoung87 from Pixabay (public domain).**

FIGURE 52.7 Examples of landscape use of hyacinth (*Hyacinthus orientalis*) at the world-famous Keukenhof Park in The Netherlands (32 ha, planted with about 7 million flower bulbs annually).

*Source:* **(a) Photo by Shutterstock/T. Irina. (b) Photo by Shutterstock/Rob3rt82. (c) Photo by Shutterstock/Sriya Pixels. (d) Photo by Michal Osmenda (CC BY SA 2.0).**

## 52.7 PROBLEMS

Hyacinths contain quite toxic alkaloids (notably lycorine), especially in the bulbs, which are the most toxic part of the

**FIGURE 52.8** Hyacinth (*Hyacinthus orientalis*) in indoor containers.

*Source:* **(a) Photo by Rampa55 from Pixabay (public domain). (b) Photo by Shutterstock/Sarycheva Olesia. (c) Photo by Geolina 163 (CC BY SA 4.0). (d) Photo by Jill Wellkgton from Pixabay (public domain).**

**FIGURE 52.9** Hyacinth (*Hyacinthus orientalis*) in outdoor containers.

*Source:* **(a) An exhibit at the Chelsea Flower Show in England. Photo by Magnus D (CC BY 2.0). (b) Photo by Dguendel (CC BY 4.0). (c) Photo by Shutterstock/Sun Shock. (d) Photo by BKP (CC BY SA 3.0).**

**FIGURE 52.10** Hyacinth (*Hyacinthus orientalis*) employed as cut flowers.

*Source:* **(a) Photo by Shutterstock/Julia Nikitina. (b) Photo by Shutterstock/Pixel-Shot.**

plant. The outer bulb scales contain calcium oxalate crystals in the form of bundles of raphides (needle-like crystals), which act as mechanical irritants. Up to 6% of the bulbs may be calcium oxalate. Ingestion could cause severe gastrointestinal problems. Contact can cause allergic skin reactions (known as "hyacinth itch" and "hyacinth fingers") and eye irritation. Gloves are advisable when handling.

**FIGURE 52.11** The potential toxicity of hyacinth (*Hyacinthus orientalis*) to pets: (a) Illustrations of houseplants, including hyacinth (*Hyacinthus orientalis*), poisonous to cats. (b) Ginger cat and a hyacinth. (c) English cocker spaniel and a hyacinth.

*Source:* **(a) Photo by Shutterstock/New Africa. (b) Photo by Shutterstock/Kamyshnikova Viktoria. (c) Photo by Shutterstock/Maria Sidelnikova.**

Breathing in dust from the bulbs can irritate the lungs and trigger an asthma attack in some people. Workers who professionally handle hyacinth bulbs are at special risk. Dogs and cats are also susceptible to being poisoned from both indoor and outdoor plants. Nevertheless, squirrels, chipmunks, voles, and other rodents will excavate and consume the bulbs and also eat the flowers (the plants are claimed to be resistant to deer and rabbits). The intense cloying scent is off-putting to some people when potted plants are in a confined space.

## 52.8 CULTIVATION

### 52.8.1 Climate and Site

Hyacinth is considered to be moderately heat-tolerant in the summer. The plants require an annual period of dormancy (4–7°C, 12–14 weeks) to reflower. In hot climates, where adequate winter vernalization cannot occur, hyacinths are often grown outdoors as annuals or lifted and replanted every year after artificial chilling. In areas with freezing winters, an adequate cold period is usually available. The literature is ambiguous regarding how much freezing the bulbs can tolerate, or the lowest temperatures that can be survived, some authorities recommending that freezing should be avoided. Hyacinths are cold-tolerant, but not as winter-hardy or resistant to spring frosts as tulips, daffodils, or crocuses. In colder areas, mulching of beds to provide insulation overwinter may be necessary. Various references to the cold-hardiness of hyacinth state that its Northern limit is climate zone 3 (defined as average annual minimum air temperature no lower than −40°C) or zone 4 (average annual minimum air temperature no lower than

FIGURE 52.12 Hyacinth (*Hyacinthus orientalis*) exposed to early-season snow.

*Source:* (a) Photo by Shutterstock/Colin Grice. (b) Photo by Bru-nO from Pixabay (public domain).

–34.4°C), but neither measure accurately indicates freezing tolerance. At the Northern limit of hardiness, the bulbs may be planted 5 cm deeper than normal to promote survival. Hyacinth is heliophilous (adapted to full sunlight) but also grows well in light shade. Outdoor plants are sometimes strategically located (easier when they are in containers) near doors, paths, patios, decks, and picnic tables, where their fragrance will be easily noticed.

### 52.8.2 Propagation and/or Purchasing

Hyacinth is mostly reproduced by bulb segments commercially (by skillful cutting, professionals can obtain up to 40 segments) or by bulbs by home growers. The mature bulbs naturally produce one to three daughter bulbs (termed *bulblets* or *offsets*), and if planted, these will produce flowers in 2 or 3 years (the mother bulb itself may also regenerate annually). Seeds are available but often won't breed true to type, and it will require years before flowering occurs. Hyacinth bulbs are commonly sold in chain stores and garden outlets. A wider choice is available online, where one is more likely to obtain hardier cultivars. Out-of-season, discounted ornamental bulbs may not be of good quality.

### 52.8.3 Planting

Bulbs are best planted in late fall, preferably between first frost and soil freezing. The pointy end should face up, the flatter side down. Outdoors, grouped or massed plantings are strongly recommended because the plants are small. Suggested spacing in a garden: about 15 cm apart, 10–15 cm deep. A general recommendation is that the bulbs should be covered with soil three times their vertical diameter, so the smaller bulbs are planted more shallowly. To avoid buildup of diseases, commercial growers recommend against replanting the bulbs in the same place, unless about 4 years have elapsed.

### 52.8.4 Soil and Moisture Conditions

Hyacinth does best in moderately rich soils, preferably with appreciable organic content. The soils should be well-drained and maintained medium-moist during the growth period. Loose-textured soil is desirable, but not essential. A circumneutral soil is preferable, although the plants grow in a range of pH. However, strongly acidic substrates are not tolerated. Once established, hyacinth can endure summer drought.

### 52.8.5 Maintaining Outdoor Plants

Spent flower spikes should be removed so the plants do not use their reserves on seed production. Following flowering, as the plants go into dormancy, reduce watering. The foliage should not be cut down until the leaves have senesced naturally, in order to allow starch reserves to build up in the bulbs. Outdoor plants maintained for several years may produce less-vigorous plants with smaller flowers, requiring replanting with new stock. Shorter cultivars are better suited to containers than to culture in the ground.

### 52.8.6 Care of Indoor Plants

Hyacinth bulbs are among the easiest to force. High-quality, large bulbs of cultivars known to be suitable for forcing should be chosen (these are often available in stores). The bulbs require 8 to 16 weeks of chilling at 4–7°C to flower. For most people, a refrigerator is the most available suitable place. If stored in a refrigerator, the bulbs should be kept slightly moist in a plastic bag to prevent ethylene gas (from any ripening fruit also present) from inhibiting development of the bulbs. Pre-cooled bulbs are sometimes available, but these may nevertheless require at least a shortened chilling. After the cold period, the bulbs can be placed in a well-drained (porous) commercial potting mix, the tops of the bulbs remaining uncovered by soil. The container should have drainage holes. Alternatively, the bulbs can be grown in water. "Forcing vases" designed for the large bulbs of hyacinths are often available from florist shops, garden centers, and online. "Hyacinth glasses" hold the bulb (generally just a single bulb) so that just the very bottom of the base touches the water. These containers are frequently transparent so the roots can be viewed. (Narrow-topped jam jars can be employed as a cheap alternative.) Because changing the water is difficult, charcoal is typically added

FIGURE 52.13 Closely packed bulbs developing into new plants: (a) Massed hyacinth bulbs forced in a pot of soil. (b) Hyacinth bulbs.

*Source:* (a) Photo by Tiger (CC BY 2.0). (b) Photo by Pixel Addict (CC BY 2.0).

FIGURE 52.14 Special narrow-necked containers ("hyacinth glasses," "hyacinth vases") for forcing bulbs. The hourglass constrictions are designed to hold the bulbs above water while the roots grow into the water.

*Source:* **(a) Photo by Ekochin from Pixabay (public domain). (b) Photo by Vitavia (CC BY SA 4.0). (c) Photo by Paul VanDerWerf (CC BY 2.0).**

FIGURE 52.15 Floral-themed posters by Alphonse Mucha.

*Source:* **Public domain.**

to the water to limit bacteria. Chilled bulbs initiate flowering about 5 weeks after being moved to room temperature. After flowering, hyacinths and, indeed, most other spring-flowering bulbs that have been forced indoors are usually discarded because they have been so significantly weakened that they are unlikely to perform well if planted outdoors. Nevertheless, replanting takes little effort.

## 52.9 FASCINATING FACTS, FABLES, FOLKLORE, AND FRIVOLITY

- The British Navy named many of its boats after flowers. The HMS *Hyacinth* was an 18-gun Royal Navy combat ship launched in 1829, which destroyed dozens of Chinese junks during the First Opium War and was eventually broken up in 1871. Another HMS *Hyacinth* was a cruiser built for the Royal Navy in the 1890s which was involved in hostilities during the First World War before being sold for scrap in 1923.
- The US National Garden Bureau, a non-profit organization, annually highlights one annual, one perennial, one vegetable, one shrub, and one bulb. The hyacinth was chosen as the bulb for 2021.
- Words can be remarkably ambiguous. The adjective "hyacinthine" can refer to the purplish-blue color typical of numerous hyacinth flowers. However, it also has unusual literary meanings tracing to translations from classical Greek literature pertaining to descriptions of the hair of Odysseus (also known by the Latin Ulysses), the hero of Homer's epic poem the *Odyssey* (Irwin 1990). Thus, *hyacinthine* in reference to hair sometimes means "golden," "reddish-gold," "black," or "curly" (supposedly like the inflorescence of hyacinth – a rather obscure description).
- Cyril Darlington, professor of botany in Oxford's Department of Botany, conducted pioneering investigations of the fundamental behavior of plant chromosomes during the 1930s and 1940s. He employed the large, distinctive chromosomes of *Hyacinthus orientalis* as his most important model (note Darlington et al. 1951).
- Czech artist Alphonse Mucha (1860–1939) is known for his distinctly stylized and decorative posters, many of which featured flowers, including hyacinths.

## KEY PUBLICATIONS

Addai, I.K. 2010. Growth and Biochemistry of the Common Hyacinth (*Hyacinthus orientalis* L.) and the Lily (*Lilium longiflorum* L.). Ph.D. thesis. School of Life Sciences, Biology and Environmental Science Department, University of Sussex, England. 306 pp.

Amigues, S. 1992. Hyakinthos: Fleur mythique et plante réelle. [The hyacinth: Mythical flower and real plant]. *Revue des Études Grecques* 105: 19–36.

Bates, N. 2012. Exposure to cut flowers and spring flowering plants in cats and dogs in the UK. *The Veterinary Nurse.* www.theveterinarynurse.com/review/article/exposure-to-cut-flowers-and-spring-flowering-plants-in-cats-and-dogs-in-the-uk.

*Botanical Dermatology Database.* www.botanical-dermatology-database.info/BotDermFolder/HYAC.html.

Bożek, M. 2019. Nectar secretion and pollen production in *Hyacinthus orientalis* 'Sky Jacket' (Asparagaceae). *Acta Agrobotanica* 72(4): 1–11.

Darlington, C.D., J.B. Hair, and R. Hurcombe. 1951. The history of the garden hyacinths. *Heredity* 5: 233–252.

Doerflinger, F. 1989. *The Hyacinth Story.* https://oldhousegardens.com/HyacinthHistory.

Dwarswaard, A. 2015. Hyacinth breeding since the 19th century. *Plantsman* 14: 98–101.

Elliott, C. 2017. *Hyacinths*. https://thegardenstrust.blog/2017/03/25/hyacinths/.

Hjorth, N. and D.S. Wilkinson. 1968. Contact dermatitis – IV. Tulip fingers, hyacinth itch and lily rash. *British Journal of Dermatology* 80: 696–698.

Ina, V. and G. Svetlana. 2015. Vegetative propagation of *Hyacinthus orientalis* L. In: *Conservation of Plant Diversity (symposium, Moldova)*, organized by G. Duca, Academy of Sciences of Moldova, 4: 112.

Irwin, M.E. 1990. Odysseus' "hyacinthine hair" in "Odyssey" 6.231. *Phoenix* 44: 205–218.

James, L. 2020. *Growing Hyacinth Bulbs: The Gardeners Guide on How to Grow and Care for Hyacinth Bulbs*. Published by the author. 30 pp.

Keeler, K. 2019. *Plant Story: Hyacinth, Beautiful and Memorable*. http://khkeeler.blogspot.com/2019/03/plant-story-hyacinth-beautiful-and.html.

Kury, L., Z. Taha, and W.H. Talib. 2021. Immunomodulatory and anticancer activities of *Hyacinthus orientalis* L.: An in vitro and in vivo study. *Plants (Basel)* 10(4): 617. https://doi.org/10.3390/plants10040617.

Li, X. and S. Wang. 2018. Research advances on flower bulb *Hyacinthus orientalis*. *Jiansu Agricultural Sciences* 46(11): 5–9.

Martin, R. 2017. *Alan Shipp: The Noah of Hyacinth Bulbs*. www.motherearthgardener.com/profiles/heirloom-bulbs-zm0z17fzmar/.

Mathew, B. 1987. *The Smaller Bulbs*. Batsford, London. 190 pp.

Mulholland, D.A., S.L. Schwikkard, and N.R. Crouch. 2013. The chemistry and biological activity of the Hyacinthaceae. *Natural Products Reports* 30: 1165–1210.

Pfosser, M. and F. Speta. 1999. Phylogenetics of Hyacinthaceae based on plastid DNA sequences. *Annals of the Missouri Botanical Garden* 86: 852–875.

Rees, A.R. 1985. *Hyacinthus orientalis*. In: *CRC Handbook of Flowering*, Vol. 1, edited by A.H. Halevy, 278–281. CRC Press, Boca Raton.

Stebbings, G. 2005. *Spring Bulbs: Daffodils, Tulips and Hyacinths*. Batsford, London. 208 pp.

Van Gelder, K. 2021. *Flower Market in the Netherlands: Statistics and Facts*. www.statista.com/topics/3732/flower-industry-in-the-netherlands/.

Wilford, R. and Kew Royal Botanic Gardens. 2019. *Kew Gardener's Guide to Growing Bulbs: The Art and Science to Grow Your Own Bulbs*. White Lion, London. 144 pp.

Wilson, K. and J.N. Honey. 1966. Root contraction in *Hyacinthus orientalis*. *Annals of Botany* 30: 48–61.

## Key Cultivation Websites

GardenersWorld.com. *How to Grow Hyacinths*. www.gardenersworld.com/how-to/how-to-grow-hyacinths/#indoors.

Hodgson, L. ("Laidback Gardener"). 2018. *Forcing Bulbs over Water: A Project for the Whole Family*. https://laidbackgardener.blog/2018/11/10/forcing-bulbs-over-water-a-project-for-the-whole-family/.

Royal Horticultural Society. *How to Grow Hyacinths*. www.rhs.org.uk/plants/hyacinth/growing-guide.

The Spruce. *How to Plant, Grow, and Care for Hyacinth*. www.thespruce.com/growing-and-caring-for-hyacinth-plants-1402248.

The Spruce. *Hyacinth Care Guide (before and after Flowering)*. www.thespruce.com/grow-hyacinth-indoors-1902762.

WikiHow. *How to Grow a Hyacinth Bulb in Water*. www.wikihow.com/Grow-a-Hyacinth-Bulb-in-Water.

WikiHow. *How to Plant Hyacinths*. www.wikihow.life/Plant-Hyacinths.

WikiHow. *How to Prevent Hyacinth Flowers from Flopping*. www.wikihow.com/Prevent-Hyacinth-Flowers-from-Flopping.

# 53 Hydrangea (*Hydrangea* Species and Hybrids)

**Scientific Names:** The genus name *Hydrangea* is based on Greek *hydor*, "water," + *angeion*, "vessel or container," alluding to the shape of the mature, opened (dehisced) fruit capsule.

**Common Names:** Like many popular ornamentals, the common name (hydrangea) is the same as the genus name. The occasional name "hortensia" (a standard name in Europe and the French name) has the same etymological origin as *Hydrangea*. Common names for the five species highlighted are presented later. In warmer climates, "hydrangea" unqualified usually refers to *H. macrophylla* (which is usually called "hortensia" in Europe). *Hydrangea paniculata* is the most widely grown species in cold-temperate countries, where "hydrangea" is likely to refer to this species.

## 53.1 INTRODUCTION

Hydrangeas are admired for their large, long-lasting flower clusters in attractive colors. They are among the most popular of ornamental shrubs, some cultivars producing spectacular floral displays. In recent years, hydrangea has also become a major potted plant, marketed in stores for holiday gifts and for use as short-term houseplants. Worldwide, hydrangeas are among the leading floral retail commodities, generating hundreds of millions of dollars annually. As will be noted, the determination of flower color in some species by soil pH represents one of the most curious phenomena in nature.

Some botanists consider *Hydrangea* to include as many as 1,000 species, but most of these are usually placed in seven other genera. *Hydrangea*, in a narrow sense, only includes several dozen species (the exact number recognized depending on authority). Interpreting the genus in this restricted sense, it is made up of shrubs, small trees, and vines. The species native to temperate areas of eastern Asia and North America are deciduous and the main source of the cultivars. The species native to subtropical Central America and western South America are evergreen. Five species (all deciduous) are indigenous in the United States, and several of the cultivated species have escaped from cultivation and become established outside of their native areas.

## 53.2 OVERVIEW

Hydrangeas grown outdoors usually form rounded shrubs 1–3 m tall. The most common cultivated *Hydrangea* species are four deciduous shrubs: *H. arborescens*, *H. macrophylla*, *H. paniculata*, and *H. quercifolia*. Cultivars of several other species are available but are much less significant commercially. However, the deciduous vine *H. anomala* subsp. *petiolaris* (also known as *H. petiolaris*) has also acquired appreciable popularity. The genus includes at least 1,000 cultivars (the majority in *H. macrophylla*), and some interspecific hybrids have been generated for garden cultivation. The five species mentioned here are popular in temperate climates.

FIGURE 53.1 Massed blooms of *Hydrangea macrophylla*, the most widely grown species of *Hydrangea*: (a) In China. (b) In Thailand.

*Source:* **(a) Shutterstock/Meiqianbao. (b) Photo by Shutterstock/Guitar photographer.**

FIGURE 53.2 The two basic flower cluster forms in *Hydrangea*: (a) Mophead (all florets sterile, large, and showy). (b) Lacecap (central florets fertile, small, and inconspicuous; outermost florets sterile, large, and showy).

*Source:* **Photos from Pixabay (public domain).**

Hydrangeas are best known for their attractive inflorescences (flower clusters). Like the Asteraceae (Compositae; daisy family), these are usually made up of a combination of small, inconspicuous fertile flowers in the center of the cluster and large, showy (pollinator-attracting) infertile flowers surrounding the central sterile ones. (Small flowers such as in the Asteraceae and Hydrangeaceae are often informally termed "florets.") The showiness of the sterile flowers resides mostly in the petaloid sepals (an enlarged calyx is present, a corolla of true petals is not, and sexual



DOI: 10.1201/9781003586227-53

organs, if present, are rudimentary). The fertile flowers produce nectar; the infertile ones do not. Wild populations have inflorescences composed mostly or entirely of small fertile flowers. Domestication has produced many cultivars in which most of or all the flowers are of the large, showy, sterile type. Inflorescences in which most of the flowers are large, showy, and sterile and clustered into a pom-pom shape are referred to as "mopheads" and "hortensias," in contrast to the "lacecap" category, which has retained central small fertile flowers. Large, globe-shaped, puffy mophead blooms (named for the alleged resemblance to a mophead) are what most people associate with hydrangeas. Lacecap inflorescences tend to be flattened rather than spherical, and many lacecap cultivars have differently colored fertile and sterile flowers. Lacecaps have been claimed to be named for their likeness to colonial ladies' head coverings. The outer flowers are often much larger than the sterile flowers, primarily made up of four or five sepals. The flowers of some cultivars have been doubled – that is, in addition to the petaloid sepals, the true petals and stamens have been mutated to increase the number of petaloid structures. While white is the most common floral color in wild *Hydrangea* species, in cultivars, flowers may also be pink, blue, purple, red, and occasionally yellow. In some cultivars, flower color (or at least depth of shade) changes with maturity. "Remontant" cultivars rebloom (usually in later summer).

### 53.2.1 *Hydrangea macrophylla*

This, by a considerable margin the most widely cultivated species of *Hydrangea* in the world, is known as bigleaf hydrangea, common hydrangea, florist's hydrangea (potted forms used as houseplants), French hydrangea (reflecting the importance of France in the past as a source of cultivars), garden hydrangea, hortensia (the predominant name in Europe), house hydrangea, and snowball hydrangea. It is native to Japan and Korea, commonly growing 1–2 m in height (occasionally more than 3 m), and spreading to 2.6 m. *Hydrangea macrophylla* subsp. *serrata* (mountain hydrangea, often separated as the species *H. serrata*) has somewhat smaller flowers and leaves. *Hydrangea macrophylla* is popular because it is easily cultivated, requires little maintenance, has few pests and diseases, is tolerant of shade and sun, can be grown in acid and slightly alkaline soil, and provides outstandingly beautiful flowers. There are many seaside biotypes in Japan that are very salt-tolerant. Most cultivars are insufficiently cold-hardy to produce a good flowering display outdoors in colder parts of the temperate region. The flower buds (produced in late summer and fall) are susceptible to low winter temperatures and are too tender when they sprout in the spring to survive heavy frosts. Planting in a sheltered location may help. Where it can be grown, pruning should be avoided or limited once flower buds begin to develop, or none will be available to bloom the next season. Several new cultivars, most notably 'Endless Summer', develop their flowers not only on old wood but also on new wood (the current year's growth) and so are not completely dependent on mild weather during both winter and spring to produce good floral displays.

FIGURE 53.3 Flowering shrubs of *Hydrangea macrophylla*.

*Source:* (a) Photo by Olive Titus (public domain). (b) Photo by A. Knop from Pixabay (public domain).

### 53.2.2 *Hydrangea paniculata*

Known as panicle hydrangea, panicled hydrangea, PG hydrangea, and hardy hydrangea, this species is native to

FIGURE 53.4 Cultivars of *Hydrangea paniculata*: (a) 'Vanilla Fraise'. (b) 'Limelight'.

*Source:* (a) Photo by Leonora (Ellie) Enking (CC BY SA 2.0). (b) Photo by Michelle Oshen (CC BY SA 2.0).

FIGURE 53.5 *Hydrangea paniculata* growing in a cemetery (hence the common name "cemetery flower") in Berlin, New York State.

*Source:* Photo by Famartin (CC BY SA 4.0).

cool-temperate and subtropical regions of Japan, Sakhalin (a Russian island north of Japan), and eastern and southern China, where it grows as a shrub or small tree as high as 8 m. Cultivars of *H. paniculata* are usually 1–2.5 m in height and tend to have cone-shaped inflorescences, unlike the usually spherical inflorescences of *H. macrophylla*. *Hydrangea paniculata* was collected from Japan and named in 1829 by the German physician and botanist Philipp Franz von Siebold, who was subsequently banned from the country. Thirty years later, he was allowed to return and collected the cultivar 'Grandiflora', now generally known as 'PeeGee' (forPaniculataGrandiflora). Although now old-fashioned, it is still the world's most popular cultivar of *H. paniculata* and the most widespread hydrangea grown outdoors in many cold climates. (Because it is often grown in graveyards, it has been called "cemetery flower.") Panicle hydrangea is frequently trained as a small tree. *Hydrangea paniculata* flowers on new wood developed in the spring and may be pruned in winter or early spring. Compared to the more popular *H. macrophylla*, it is more drought-tolerant, thrives better in both full sun and partial shade, and also has a broader tolerance to soil conditions, withstanding not only acidic but also somewhat-alkaline soil. Most importantly, insofar as cold-temperate regions are concerned, *H. paniculata* is much more cold-tolerant than *H. macrophylla*, capable of surviving –30°C, and unlike for *H. macrophylla*, its buds are not susceptible to early-spring frosts.

### 53.2.3 *Hydrangea quercifolia*

Oakleaf hydrangea (so named for its leaves, which are lobed somewhat like red oaks) is native to the southeastern United States and is marginally cold-hardy. Unlike most cultivated hydrangeas, which have uninteresting foliage color in the autumn, *H. quercifolia* develops a striking range of reds, oranges, and deep maroon leaf hues in the fall. The shrubs commonly grow to a height of 2 to 2.5 m but tend to spread wider than their height from root suckers. Large white to cream-colored flowers are produced in early to midsummer, these sometimes turning pink to rose with age. The flowers develop on old wood, and pruning, if necessary, is best done after flowering, not in late summer or fall, which would remove potential buds.

FIGURE 53.6 Oakleaf hydrangea (*Hydrangea quercifolia*): (a) Cultivar Snowflake. (b) Colorful foliage in autumn.

***Source:*** **(a) Photo by Krzysztof Ziarnek Kenraiz (CC BY SA 4.0). (b) Photo by University of Washington Botanic Garden (CC BY 2.0).**

### 53.2.4 *Hydrangea arborescens*

This species is known as seven-bark (for the layers of bark which peel away easily), smooth hydrangea, tree hydrangea, and wild hydrangea. It is native to eastern US woodlands and has become introduced in eastern Canada. The plants are usually 1–2 m tall, occasionally 3 m. Flowering is on new wood of the current season, and so the plants can be pruned in early spring without fear of removing flower buds. The species can be grown in the warmer regions of temperate climates. Most cultivars are white-flowered, but pink forms are available.

### 53.2.5 *Hydrangea anomala* subsp. *petiolaris*

Climbing hydrangea is a woody deciduous vine native to woodlands and forests of eastern Siberia, Japan, China, and South Korea. It grows over shrubs and trees by twining and by aerial rootlets that develop along the stems. The rootlets can adhere to brick and vinyl siding so tenaciously that they are impossible to remove. Although slow to develop, this vine produces vigorous, sprawling growth that can be almost shrub-like. Climbing hydrangea is employed as an ornamental vine and as a groundcover. Blooming occurs on old wood, so to promote good floral displays, pruning is best done after flowering. However, when growing vigorously, it is advisable to prune whenever necessary to prevent stray stems from developing where they are unwanted. Climbing hydrangea is hardy in the warmer parts of temperate climates. It is tolerant of different soil types but does best in well-drained substrates.

FIGURE 53.7 *Hydrangea arborescens.*

***Source:*** **(a) Photo by Kor!An (CC BY SA 3.0). (b) Curtis, W. 1799. *Botanical Magazine* (London), vol. 3, plate 437 (public domain).**

FIGURE 53.8 Climbing hydrangea (*Hydrangea anomala* subsp. *petiolaris*): (a) Vine climbing on a brick wall. (b) Close-up of flowering stems.

***Source:*** **(a) Photo by Leonora (Ellie) Enking (CC BY SA 2.0). (b) Photo by A. Barra (CC BY SA 3.0).**

## 53.3 ENVIRONMENTAL CONTROL OF FLOWER COLOR

One of the first chemistry lessons that students learn is that litmus paper becomes blue under alkaline conditions (pH $>7.0$), and red in an acidic environment (pH $<7.0$). Curiously, the reverse often occurs with respect to *Hydrangea* flower hues in respect to soil pH. Blue and pink are principal floral colors of *Hydrangea* cultivars, some of which are famous for having the capacity to develop one or the other color, depending on the soil in which they grow. The color produced in such species (most notably *H. macrophylla*) has been found to be determined by (1) a cultivar's inherent ability to produce pigment, (2) the aluminum (aluminium) availability in the soil or substrate, and (3) the biotype's ability to uptake aluminum. Blue color results when aluminum complexes with a pigment (the anthocyanin delphinidin 3-glucoside) in the sepals. When the pH of the growth medium is acidic (4.5–5.5), flower color tends to become blue because aluminum is generally highly available at a low pH. At pH $\geq 6.0$, flower color tends to be pink. Between pH 5.5 and 6.0, the color could be pink, blue, lavender, or a mixture of pink and blue on the same plant. However, some species, and some cultivars, simply cannot be made to develop blue flowers regardless of soil pH; some will only develop scattered blue flowers within an inflorescence, and some will produce rather unattractive floral shades.

Blue is the world's most popular color, the preference for it showing up across cultures. It is no accident that the sexist preference for baby boys is associated with blue garments, and the traditional powder blue of the majority of cultivated hydrangeas is consistent with this favoritism. Psychologists have explained the preference for blue as perhaps reflecting the calmness of blue skies and blue oceans. Nevertheless, red is the world's most "attractive" color, consistent with lips and nipples being reddish, but so is blood, the sight of which is alarming, which may explain why pink (a softened red: attractive, but not overly so) has become traditionally associated with baby girls (indeed, with women).

Nursery growers have reported that they sell, on average, ten blue hydrangeas for every pink one, so controlling flower color is crucial. (Nevertheless, in Asia, pink hydrangeas seem to be increasing in popularity.) For many varieties of *H. macrophylla* grown in containers, the development of blue flowers requires particular substrate conditions. Soil pH is usually maintained at 4.5 to 5.5. Some growth media may have very little aluminum, and so just influencing pH may not affect flower color in plants potted in such material. For some biotypes, care must be taken to regulate fertilizers and water quality to avoid deficiencies of minerals such as calcium, magnesium, and aluminum, and to also avoid metal toxicities. For home growers, to produce pink flowers, it may suffice to simply add lime to raise soil pH, or to produce blue flowers, add sulfur to lower pH. A quick-and-dirty procedure to produce blue flowers in susceptible cultivars is to drench the soil around the plant with a gallon (4 L) of water in which 1 tbsp of alum (aluminum sulphate) has been dissolved. Or to make flowers pink, dissolve 1 tbsp of hydrated lime instead. Fertilizers low in phosphorus will tend to produce pink flowers, since this element ties up aluminum. Those expecting to see color changes overnight will be disappointed: Generally, soil conditions must be provided well before and sometimes during flower development.

## 53.4 HISTORY

*Hydrangea* species have been cultivated and domesticated as ornamentals in China and Japan at least for centuries. In 1789, the famous British naturalist Sir Joseph Banks imported *H. macrophylla* into England from China, this species destined to become the predominant source of cultivars in the genus. Subsequent 18th- and 19th-century plant collectors brought other Asian species and garden forms to Europe and North America. About the same time, the eastern North American species *H. arborescens* and *H. quercifolia* were brought to the Old World. Breeding has been continuing in both the Old and New Worlds, but to date, the majority of cultivars arose in Europe and Asia.

## 53.5 ECONOMIC IMPORTANCE

Hydrangea is one of the world's leading ornamental crops, familiar in most of the temperate and subtropical world, where it is produced and sold in large quantities.

FIGURE 53.9 Commercial production and sale of hydrangea: (a) Greenhouse scene. (b) Colorful hydrangeas in a store.

*Source:* **(a) Photo by Shutterstock/Iryna Inshyna. (b) Photo by Shutterstock/NNattalli.**

## 53.6 ORNAMENTAL USES

Hydrangea shrubs are extremely widely grown as large lawn specimens and hedges, and as smaller plants in beds and containers.

Potted hydrangeas are increasingly being sold in supermarkets and other chain stores, especially for holidays. In both Europe and North America, there is high demand for potted hydrangeas at Christmas, Easter, and Mother's Day. In Europe, more than 60 million pots are sold annually, with production centered on The Netherlands, Germany, and

FIGURE 53.10 Large beds of flowering *Hydrangea macrophylla*: (a) In Thailand. (b) In Yangmingshan National Park, Taiwan.

*Source:* **(a) Photo by Shutterstock/TualekPhoto. (b) Photo by Shutterstock/Jack Hong.**

FIGURE 53.11 Hydrangea flowering hedges in front of houses: (a) *Hydrangea macrophylla* in South Korea. (b) *Hydrangea paniculata* cultivar Vanilla Fraise in Russia.

*Source:* **(a) Photo by Shutterstock/Sanga Park. (b) Photo by Shutterstock/Sergey Bezgodov.**

FIGURE 53.12 Potted *Hydrangea macrophylla*.

*Source:* **(a) Photo by Johanna84 from Pixabay (public domain). (b) Photo by Dalibri (CC BY SA 2.0). (c) Photo by Freak 222 (CC BY SA 3.0).**

FIGURE 53.13 Hydrangea cut flowers: (a) A luxury restaurant, the tables decorated with spectacular bouquets dominated by hydrangea. (b) Bride with a bouquet. (c) A basket of flowers. (d) A single harvested flower cluster in a container of water.

*Source:* **(a) Photo by Shutterstock/IvashStudio. (b) Photo by Shutterstock/Bloom. (c) Photo by Choiceflowers from Pixabay (public domain). (d) Photo by Initsyk from Pixabay (public domain).**

France. Most of these so-called "florist hydrangeas" are *H. macrophylla*, grown to bloom very early and abundantly. They may be dwarf forms or simply cultivated to be small. In warm regions, they can be planted outdoors once the flowers fade, but in cold climates, they may not survive winter.

Hydrangea is employed as a cut flower crop popular because of the large size of the flower heads and the variety of colors available. The flowers are often used in bouquets and floral decorations for weddings, parties, and other events. The flowering stems have excellent vase life and preserve well for dried flower arrangements. The most important countries of production are Colombia and The Netherlands.

## 53.7 OTHER USES

Amacha is a Japanese beverage prepared from fermented leaves of *H. macrophylla*. A similar herbal tea, sugukcha, is available in Korea. These preparations are sweet because of the presence of phyllodulcin, which is hundreds of times as sweet as table sugar and is under investigation for control

FIGURE 53.14 Hydrangea tea in relation to celebrations of the birthday of Buddha: (a) Worshippers pouring amacha (hydrangea) tea on the head of a Buddha statue, part of an annual celebration of Buddha's birthday in Japan. (b) Celebration of Buddha's birthday in Borobudur in Indonesia, the world's largest Buddhist temple.

*Source:* **(a) Photo by Maarten Heerlien (CC BY 2.0). (b) Photo by PWBaker (CC BY 2.0).**

of obesity and diabetes. Amacha tea is frequently used in ceremonies during Kambutsue (Kanbutsue), the holiday celebrating the birthday of Buddha (thought to be April 8) in Japanese Buddhism. In other Asian countries, the comparable celebration is known as Vesak or Buddha Day. According to legend, when Buddha was born, it was raining amacha or (alternatively) nine dragons poured amacha over him. In commemoration of this, amacha tea is ceremoniously poured over a statue of Buddha. Some temples display a small statue of the Buddha in front of the altar and provide a basin of sweet tea to pour over the statue.

Preparations of hydrangea roots and rhizomes are sometimes employed to treat kidney and bladder conditions, allegedly because of diuretic properties (which increase the flow of urine). As well, such preparations have been used to treat kidney stones and enlarged prostate. Foliage decoctions have been employed to treat intestinal ailments. These traditional herbal medicinal treatments have not been validated and should be avoided. The antimalarial potential of leaf compounds in *Hydrangea* is under investigation, as is the potential of root compounds to treat autoimmune diseases like multiple sclerosis.

## 53.8 PROBLEMS

Some species of *Hydrangea* are considered to be toxic because of the presence of cyanogenic glycosides. Hydrangeas have a very minor reputation as an intoxicant when smoked, which could simply be the result of cyanide poisoning. Hydrangeas are reportedly poisonous to dogs, cats, horses, and cows.

## 53.9 CULTIVATION

### 53.9.1 Climate and Site

Where winter temperatures permit, hydrangeas are not difficult to grow. The following climate zone ratings have been assigned to the species discussed in detail in this chapter: *Hydrangea anomala* subsp. *petiolaris*, 4 to 9; *H. arborescens*, 4 to 9; *H. macrophylla*, 6 to 9; *H. paniculata*, 4 to 8; *H. quercifolia*, 5 to 9. Like most other shrubs, hydrangeas should be transplanted to spaces large enough to allow maximum development and to minimize pruning. In temperate regions, sunny or partly shady situations are suitable (shade will reduce flower displays). In very cold regions, locations providing protection from wind and temperature fluctuations are desirable.

### 53.9.2 Propagation and/or Purchasing

Most commercial propagation of hydrangea cultivars is by stem cuttings or tissue culture. Accordingly, most plants will be purchased in the form of pot-grown specimens ready for transplanting.

### 53.9.3 Planting

The hole for starter pot plants should be excavated to be as deep as the root ball (place the root ball at the same depth as in the pot) and three times its width. Backfill with soil amended with organic matter to promote good drainage.

### 53.9.4 Soil and Moisture Conditions

In gardens, the best soils for hydrangeas are fertile sandy or silt loams with more than 1% organic matter, although with proper watering, they will grow well in clay soils. A soil pH of 5.0–6.5 is usually recommended. Soils should be well-drained but maintained moist.

### 53.9.5 Maintaining Plants

Because of the popularity of hydrangea, extensive information is available online (see key cultivation websites).

## 53.10 FASCINATING FACTS, FABLES, FOLKLORE, AND FRIVOLITY

- Hydrangeas bloom abundantly during the rainy season in Japan (June and early July), and accordingly, they often symbolize early summer in the

FIGURE 53.15 Planting a potted starter plant of hydrangea in a garden.

*Source:* **Photo by Shutterstock/Ajlatan.**

FIGURE 53.16 *Hydrangea macrophylla* growing on a hillside by a shrine in Japan.

*Source:* **Photo by Aiko Matsuoka (CC BY SA 3.0).**

country. In Japanese tradition, hydrangea flowers represent deep or heartfelt emotion as well as a fickle or changeable heart (reflecting the instability of flower color). In feudal Japan, warriors despised hydrangeas because changing colors came to symbolize unstable loyalty.

- In Western countries, hydrangeas have come to symbolize gratitude, grace, and beauty on the positive side, and on the negative side, vanity and boastfulness (perhaps reflective of the abundance and beauty of the flowers), and frigidity when the flowers are blue (allegedly because in medieval times, it was believed that young women who grew or picked hydrangeas would never find a husband).
- Alabama designated the oakleaf hydrangea (*H. quercifolia*) as its "official state wildflower" in 1999 (camellia is the "state flower" of Alabama). Unofficially, hydrangea is considered to be the floral symbol of Brittany (France).
- Philipp Franz Balthasar von Siebold (1796–1866), a German physician and botanist who taught Western medicine in Japan, coined the scientific name of the bigleaf hydrangea as "*Hydrangea otakusa*" (although the correct name is now considered to be *H. macrophylla*). The epithet *otakusa*

FIGURE 53.17 Examples of hydrangea flowers in popular art: (a) Vintage advertising poster. (b) Heirloom postcard. (c) Painting titled *Young woman with hydrangeas* by the French artist Elisabeth Sonrel (1874–1953). (d) *Two Little Fraid Cats*, a hand-colored lithograph published by Currier & Ives in 1863 showing two cats in hydrangea blooms, cowering in fear of a mouse.

*Source:* **(a) Photo by Shutterstock/AXpop. (b, c, d) Public domain.**

has been interpreted as a reference to the name of the courtesan known as Kusumoto Taki, who became von Siebold's mistress (marriage with foreigners was not permitted in Japan at that time). Their daughter, Kusumoto Ine, became the first female doctor of Western medicine in Japan.

- Portugal's Faial Island in the Azores became widely known as "Blue Island" after Portuguese poet Raul Brandão in the early 20th century marvelled at the abundant flowering hydrangeas established there.
- In 2009, the chef of a pricey restaurant in Japan seasoned a dish with hydrangea leaves, which, as noted earlier, are toxic, producing cyanide when ingested. The diners experienced severe vomiting, and the chef was prosecuted.
- Because hydrangea flower heads are so large and impressive, they have frequently been employed in popular art.

## KEY PUBLICATIONS

Bailey, D.A. 1989. *Hydrangea Production*. Timber Press, Portland, OR. 91 pp.

Boebel, T. 2011. *Hydrangeas in the North: Getting Blooms in the Colder Climates*. CreateSpace Independent Publishing Platform (self-published by author). 120 pp.

Bruynzeel, D.P. 1986. Allergic contact dermatitis to hydrangea. *Contact Dermatitis* 14: 128.

Bowman, D.P. 2000. *Hydrangeas*. Friedman/Fairfax (international). 112 pp.

Church, G. 2007. *Complete Hydrangeas*. Firefly Books, Richmond Hill, ON. 144 pp.

De Smet, Y. 2020. *Through the Fog: Evolutionary Insights Provide Novel Genus- and Species-Level Boundaries in* Tribe *Hydrangeeae and* Genus *Hydrangea*. Doctoral dissertation. Ghent University, Belgium. 239 pp.

De Smet, Y., C. Granados Mendoza, S. Wanke, P. Goetghebeur, and M.-S. Samain. 2015. Molecular phylogenetics and new (infra)generic classification to alleviate polyphyly in tribe Hydrangeeae (Cornales: Hydrangeaceae). *Taxon* 64: 741–752.

Dirr, M. 2004. *Hydrangeas for American Gardens*. Timber Press, Portland, OR. 236 pp.

Freeman, C.C. 2017. *Hydrangea*. Flora North America (online). Vol. 12. www.efloras.org/florataxon.aspx?flora_id=1&taxon_id=115977.

Granados Mendoza, C., S. Wanke, P. Goetghebeur, and M.-S. Samain. 2013. Facilitating wide hybridization in *Hydrangea* s. l. cultivars: A phylogenetic and marker-assisted breeding approach. *Molecular Breeding* 32: 233–239.

Harrison, J. 2013. *Heavenly Hydrangeas: A Practical Guide for the Home Gardener*. Schiffer, Atglen, PA. 144 pp.

Haworth-Booth, M. 1984. *The Hydrangeas*, 5th edition. Constable, Edinburgh. 217 pp.

Hufford, L., M.L. Moody, and D.E. Soltis. 2001. A phylogenetic analysis of Hydrangeaceae based on sequences of the plastid gene *mat*K and their combination with *rbc*L and morphological data. *International Journal of Plant Sciences* 162: 835–846.

Lawson-Hall, T. and B. Rothera. 2005. *Hydrangeas: A Gardener's Guide*. Timber Press, Portland, OR. 176 pp.

Li, Y., M.T. Mmbaga, B. Zhou, J. Joshua, E. Rotich, and L. Parikh. 2018. Diseases of hydrangea. *Handbook of Florists' Crops Diseases* 2018: 987–1006.

Mallet, C. 2008. *Hortensienatlas*. Ulmer Eugen Verlag, Stuttgart. 208 pp. (In German, with English portions; presents 1,000 illustrations.)

Mallet, C., R. Mallet, and H. van Trier. 1992–1994. *Hydrangeas: Species & Cultivars*. Centre d'Art Floral, Varengeville, France, 2 vols.

McClintock, E. 1957. A monograph of the genus *Hydrangea*. *Proceedings California Academy of Sciences*, ser. 4, 29: 147–256.

Reed, S.M. and T.A. Rinehart. 2006. *Hydrangea macrophylla* and *serrata* – should we lump 'em or split em? *SNA Research Conference* 51: 573–576.

Samain, M.-S., S. Wanke, and P. Goetghebeur. 2010. Unraveling extensive paraphyly in the genus *Hydrangea* s.l. with implications for the systematics of tribe Hydrangeae. *Systematic Botany* 34: 593–600.

Schreiber, H. 2014. Curious chemistry guides hydrangea colors. *American Scientist* 102(6): 444. www.americanscientist.org/article/curious-chemistry-guides-hydrangea-colors.

Schreiber, H.D., A.H. Jones, C.M. Lariviere, K.M. Mayhew, and J.M. Cain. 2011. Role of aluminum in red-to-blue color changes in *Hydrangea macrophylla* sepals. *Biometals* 24: 1005–1015.

Sherwood, A., M.D. Clark, and S.C. Hokanson. 2022. Oakleaf hydrangea (*Hydrangea quercifolia* Bartr.) horticulture, genetics, breeding, and conservation. *Horticultural Reviews* 49: 1–33.

Uemachi, T., Y. Mizuhara, K. Deguchi, Y. Shinjo, E. Kajino, and H. Ohba. 2014. Phylogenetic relationships of *Hydrangea macrophylla* (Thunb.) Ser. and *H. serrata* (Thunb.) Ser. evaluated using RAPD markers and plastid DNA sequences. *Journal Japanese Society for Horticultural Science* 83: 163–171.

Van Gelderen, C.J. and D.M. van Gelderen. 2004. *Encyclopedia of Hydrangeas*. Timber Press, Portland, OR. 280 pp.

Yoshida, K., K.-I. Oyama, and T. Kondo. 2021. Insight into chemical mechanisms of sepal color development and variation in hydrangea. *Proceedings of the Japan Academy, Series B* 97(2): 51–68.

## Key Cultivation Websites

American Hydrangea Society. http://americanhydrangeasociety.org/.

Bailey, D.A. 1998. *Commercial Hydrangea Forcing*. North Carolina Cooperative Extension Service, Leaflet 524. www.clamerinforma.it/EnglishFiles/PDF/HIL524.pdf.

Fulcher, A., J. Owen, and A. Lebude. 2016. *Hydrangea Production: Species-specific Production Guide*. www.researchgate.net/profile/James_Owen5/publication/311558337_Hydrangea_Production_Species-Specific_Production_Guide/links/584c472208aecb6bd8c2e37c/Hydrangea-Production-Species-Specific-Production-Guide.pdf.

Lancaster, N. 2008. *RHS Plant Trials Bulletin: Hydrangea paniculata*. www.rhs.org.uk/plants/pdfs/plant-trials-and-awards/plant-bulletins/hydrangea-paniculata.pdf.

Owen, J.S., A. Fulcher, and A. LeBude. 2016. *Hydrangea Production: Cultivar Selection and General Practices to Consider When Propagating and Growing Hydrangea*. https://extension.tennessee.edu/publications/Documents/PB1840-A.pdf.

The Spruce. How to grow and care for hydrangeas. www.thespruce.com/growing-hydrangeas-1402684.

The Spruce. 14 Beautiful types of hydrangea to grow. www.thespruce.com/top-hydrangea-varieties-4129245.

WikiHow. *How to Care for Limelight Hydrangeas*. www.wikihow.com/Care-for-Limelight-Hydrangeas.

WikiHow. *How to Cut Hydrangea Blooms*. 3 Ways to Cut Hydrangea Blooms – wikiHow.

WikiHow. *How to Get Hydrangeas to Bloom*. www.wikihow.com/Get-Hydrangeas-to-Bloom.

WikiHow. *How to Grow Hydrangea from Cuttings*. www.wikihow.com/Grow-Hydrangea-from-Cuttings.

WikiHow. *How to Grow Hydrangeas*. www.wikihow.com/Grow-Hydrangeas.

WikiHow. *How to Grow Hydrangeas in a Pot*. www.wikihow.com/Grow-Hydrangeas-in-a-Pot.

WikiHow. *How to Keep Blue Hydrangeas Blue*. www.wikihow.com/Keep-Blue-Hydrangeas-Blue.

WikiHow. *How to Keep Hydrangeas Small*. www.wikihow.com/Keep-Hydrangeas-Small.

WikiHow. *How to Mulch Hydrangeas*. How to Mulch Hydrangeas: 12 Steps (with Pictures) – wikiHow.

WikiHow. *How to Plant Hydrangeas*. www.wikihow.com/Plant-Hydrangeas.

WikiHow. *How to Prepare Hydrangeas for Winter*. www.wikihow.com/Prepare-Hydrangeas-for-Winter.

WikiHow. *How to Prune Limelight Hydrangeas*. www.wikihow.com/Prune-Limelight-Hydrangeas.

WikiHow. *How to Revive Hydrangeas*. www.wikihow.com/Revive-Hydrangeas.

WikiHow. *When & How to Prune Hydrangeas So They Stay Happy and Healthy*. www.wikihow.com/Prune-Hydrangeas.

# 54 Impatiens (*Impatiens* Species and Hybrids)

**Scientific Names:** The genus name *Impatiens* is based on the idea that the plants are explosively impatient to distribute their seeds (see "touch-me-nots," explained later).

**Common Names:** The most frequent general English name for the genus *Impatiens* is "impatiens," followed by "balsam." The species are also known as "touch-me-nots" for the way ripe pods explosively disperse their seeds when touched. For the same reason, they are also often termed "snapweed." The name "balsam" (commonly applied to plants used as topical balms) reflects widespread historical use of the plants as skin ointments. Some species are called "jewel weed" (for rationale, see "Fascinating Facts, Fables, Folklore, and Frivolity").

*I. hawkeri*: New Guinea impatiens.

*I. walleriana*: Patience (plant), garden impatiens, busy Lizzie (a popular name in the UK; "busy" because the plant is "busily" setting flowers for a very long season; "Lizzie" is a variant of "Lucy," explained next), patient Lucy ("Lucy" for the scientific name *I. lujai*), Japanese balsam, shady lady, sultana (for the scientific synonym *I. sultanii*, coined in honor of the sultan of Zanzibar), sultan's balsam, sultan's flower, Zanzibar balsam.

*I. balsamina*: bush balsam, garden balsam, garden jewelweed, rose balsam, spotted snapweed.

FIGURE 54.1 Explosive fruit of *Impatiens*, the basis of the name "touch-me-not." A slight touch of the ripe fruits of *Impatiens* results in contraction of elastic valves of the capsule and an ejection of their seeds with great force. A study of explosive seed dispersal in *Impatiens capensis* showed that the seeds could travel at up to 4 m/sec (9 mi/hr) and can be dispersed up to 1.8 m from the plant: (a) A finger about to touch and trigger the explosion of a mature fruit of western touch-me-not (*Impatiens ecornuta*, previously called *I. ecalcarata*). (b) Children in Indonesia mischievously popping the pods of garden balsam (*Impatiens balsamina*).

*Source:* **(a) Photo by Matt Lavin (CC BY SA 2.0). (b) Photo by David E Mead (public domain).**

## 54.1 INTRODUCTION

Until about 2012, the garden patience (*Impatiens walleriana*) was one of the most important bedding plants in the world and the premier floral annual for shady locations. As described in this chapter, a devastating, virulent mildew has substantially reduced its popularity, although there are efforts underway to breed resistant cultivars. Today, New Guinea impatiens (*I. hawkeri*) has become the most important *Impatiens*, while many of the other species (most notably balsam – *I. balsamina*) are occasionally grown. Even with the current difficulties experienced by *I. walleriana*, the genus *Impatiens* remains a leading source of easily grown, economical, colorful garden ornamentals suited for shadier areas.

*Impatiens* is one of the largest plant genera, with over 1,000 species. Together with the genus *Hydrocera*, which has just one species, *Impatiens* makes up the family

FIGURE 54.2 Spectacular gardens displaying mixed *Impatiens* hybrid cultivars.

*Source:* **(a) Photo by Shutterstock/Suratn Sridama. (b) Photo by Shutterstock/Daimond Shutter.**



DOI: 10.1201/9781003586227-54

Balsaminaceae. *Impatiens* species are predominantly native to the tropics and subtropics of Asia and Africa, and most are confined to small regions. The distribution is mainly Northern Hemisphere, nearly all species occupying elevations between 1000 and 3000 m, where cooler temperatures prevail, albeit in warmer regions of the world. Major centers of diversity include the Indian subcontinent, Southeast Asia, and the highlands and mountains of tropical and subtropical Africa. The plants are annual or perennial herbs or subshrubs. Most species occur in damp, shady environments near rivers or streams. Although very few species of *Impatiens* are grown, the huge number represents an invaluable store of breeding material for future development of improved cultivars.

## 54.2 OVERVIEW

The following three species are the most important ornamentals in *Impatiens*. Although traditionally identified by the scientific names *I. balsamina*, *I walleriana*, and *I. hawkeri*, advanced cultivars are hybrids, with at least some parentage of other species.

### 54.2.1 Garden Balsam (*Impatiens balsamina*)

Garden balsam is an annual native to southern Asia, including India, Bangladesh, and Myanmar, and has become widely naturalized in tropical, subtropical, and warm-temperate regions. This is the first *Impatiens* species to have been cultivated and has been grown for centuries. It was introduced into Europe in the 16th century, where it was once a cottage-garden favorite. It grows to 1 m in height, although dwarf varieties have been bred. The species is available in a wide variety of flower colors and double-flowered forms. On the whole, garden balsam has limited popularity today.

### 54.2.2 Garden Impatiens (*Impatiens walleriana*)

*Impatiens walleriana* (frequently misspelled *I. wallerana*) is a perennial native to Eastern Africa (Kenya, Tanzania) and Equatorial (South Tropical) Africa (Malawi, Mozambique, Zambia, and Zimbabwe). It has also been widely naturalized in tropical, subtropical, and warm-temperate regions. In its indigenous areas, it occurs on the margins of moist, shady forests and along river edges, while in introduced locations, it is often found along roadsides and in disturbed sites. The species was named for Horace Waller, a missionary, member of the Royal Geographic Society, and publisher of the journals of David Livingston, the famous explorer of Africa. Physician and naturalist John Kirk, who travelled with Livingston, appears to have first brought the species to England in 1896 from the island of Zanzibar. The species was grown primarily as a potted plant until the 1970s, when breeding of cultivars expanded considerably. A very wide range of flower colors became available, as well as striped flowers and doubled flowers, and forms with different sizes and habits. Sales of *I. walleriana* underwent a spectacular increase in the early 1970s, and by 1977, it ranked third as a bedding ornamental in North America. In some recent years, annual sales of the species exceeded $100 million in the United States alone. However, because of a serious disease problem, discussed next, *I. walleriana* has recently undergone a catastrophic reduction in popularity. *Impatiens hawkeri* (New Guinea impatiens) has largely replaced *I. walleriana*, although the overall market value of *Impatiens* has decreased.

FIGURE 54.3 Garden balsam (*Impatiens balsamina*) in flower.

***Source:* (a) Photo by Jalaja Puzhankara (CC BY SA 4.0). (b) Photo by Rameshng (CC BY SA 3.0).**

FIGURE 54.4 (a) Garden impatiens (*Impatiens walleriana*) in flower. (b) Cultivar Rockapulco Pink.

***Source:* (a) Photo by Satin66Flower (CC BY SA 4.0). (b) Photo by ksblack99 (public domain).**

#### 54.2.2.1 Downy Mildew and the Demise of *Impatiens walleriana*

Although recorded for centuries, impatiens downy mildew emerged as a devastating foliar disease of *I. walleriana* in the 21st century on different continents (in 2011 in North America). The disease is caused by *Plasmopara obducens*, which is not a true fungus but a member of the "water molds" (Oomycota, which affect numerous economically important plants around the world; most famously, the group includes the pathogen that caused the Irish potato famine). Some pathologists suspect that the sudden recent emergence of virulence of *P. obducens* is due to mutation. Reminiscent of true fungi, branched threadlike structures (sporangiophores) bearing reproductive sacs (sporangia, in which spores are generated) develop inside the foliage. These sporangiophores emerge through stomates on the lower surface of the leaves, producing a downy white growth on the leaf undersides. The sporangia on the sporangiophores are easily dislodged and can be distributed by air currents or splashed by rain to other plants. Zoospores released from the sporangia swim in moisture on leaf surfaces, and they penetrate leaves by means of germ tubes which enter through

**FIGURE 54.5** Lower leaf surfaces of *Impatiens walleriana* infected by impatiens downy mildew: (a) Early development of sporangiophores and sporangia (spore-bearing structures). (b) Advanced development of infection.

*Source:* **(a) Photo by Don Ferrin (CC BY 3.0). (b) Photo by Bruce Watt (CC BY NC 3.0).**

the stomata. In addition to the zoospores, *Plasmorpara* produces "oospores" inside infected plants. As the plants decay, these thick-walled, long-lived resting spores are released into the soil. Accordingly, infected plants should be discarded in garbage, not placed in compost.

Impatiens downy mildew thrives in cool, wet weather. Initially, leaves may curl downward at the edges and become yellow or stippled, and eventually, plants become stunted. A light-gray fluffy coat of sporangiophores appears on the underside of the foliage, and many of the leaves and flowers fall away. Several fungicides can protect the plant from infection, but no cure is available once a plant is infected. The disease appears confined to *Impatiens*. All cultivars of *I. walleriana* as well as some wild species of *Impatiens* are susceptible. *Impatiens balsamina* is also considered susceptible, but significantly less so than *I. walleriana*, usually only exhibiting a yellow leaf spot. (In any event, *I. balsamina* is much less popular than the other two major ornamental species of *Impatiens*.) Some hybrid combinations between *I. walleriana* and other species of *Impatiens* are relatively resistant, but a large percentage of *I. walleriana* in the parentage of a hybrid generally is associated with susceptibility.

Impatiens downy mildew changed the bedding plant business in North America by drastically reducing the most popular bedding plant, *I. walleriana*, from the market. New Guinea impatiens (*I. hawkeri*) is very resistant to the disease and so has replaced *I. walleriana* as the leading ornamental species of *Impatiens*. However, it is not as shade-tolerant as *I. walleriana* and has not become as popular. Alternative shade-tolerant ornamentals such as begonias and coleus are often recommended, but none of these has proven as satisfactory to consumers. Breeding is underway to create hybrids of *I. walleriana* with other *Impatiens* species (particularly *I. hawkeri*) that are resistant to impatiens downy mildew. "Bounce" impatiens and "SunPatiens" (which is tolerant of full sun) are relatively resistant hybrids that have recently appeared on the market.

### 54.2.3 New Guinea Impatiens (*Impatiens hawkeri*)

New Guinea impatiens (*I. hawkeri*) is a perennial native to Papua New Guinea and the Solomon Islands. It was

**FIGURE 54.6** "Sunpatiens Compact Orange," a recent *Impatiens* hybrid with *I. walleriana* and *I. hawkeri* parentage. Such varieties are resistant to downy mildew, but not as tolerant of deep shade as older *I. walleriana* cultivars that are susceptible to the disease.

*Source:* **Photo by Serres Fortier (CC BY 2.0).**

**FIGURE 54.7** (a) New Guinea impatiens (*Impatiens hawkeri*). (b) Cultivar Florifica Sweet Orange.

*Source:* **(a) Photo by Globetrotter19 (CC BY SA 3.0). (b) Photo by F. D. Richards (CC BY SA 2.0).**

collected in Papua New Guinea by Lt. Hawker in 1884 and brought into cultivation. However, it did not become popular until 1970, after which extensive hybridization produced numerous attractive cultivars. In very recent times, it has become the principal commercial species of *Impatiens*.

The species is generally larger than *I. walleriana*, up to 1 m in height, with leaves arranged in distinct whorls. New Guinea impatiens is available in about as wide a range of flower colors as was *I. walleriana* (including white, purple red, pink, salmon, and orange), and many of the new cultivars have flowers that are larger (sometimes 7 cm across) and more numerous. Moreover, the foliage of New Guinea impatiens is available in a wide range of variegated patterns, exceeding what was available in *I. walleriana*. However, New Guinea impatiens has proven to be more expensive than *I. walleriana* (particularly because hybrids are reproduced by cuttings, not by seeds) and not as tolerant of deep shade.

## 54.3 HISTORY

Detailed information on the history of each of the three principal cultivated species is provided in the preceding. New Guinea impatiens (*I. hawkeri*) was first introduced in Europe in 1886 and quickly became popular. However, by the early 1900s, it had virtually disappeared from cultivation because of begonia mite infestations. Garden impatiens (*I. walleriana*) was also introduced to Europe in the late 19th century and became the dominant *Impatiens* until the 1970s, when downy mildew made it too difficult to grow for most gardeners, leading to the resurrection of New Guinea impatiens as the predominant species. Breeding of hybrids in the last several decades has somewhat obscured the differences between the two competing species.

## 54.4 ECONOMIC IMPORTANCE

Impatiens is an important horticultural crop with significant sales by plant nurseries for the gardening trade in both temperate and semi-tropical regions.

## 54.5 ORNAMENTAL USES

Impatiens is widely employed as a bedding and border ornamental in private residential gardens and public areas. It is

FIGURE 54.8 Paintings (public domain) of early cultivars of *Impatiens*: (a) *Impatiens hawkeri*. (b) *Impatiens walleriana*. (c) *Impatiens balsamina*. (d) A garden hybrid with doubled flowers.

***Source:*** **(a) *Curtis's Botanical Magazine* (London), vol. 135 (= ser. 4, vol. 5), 1909. (b) *Revue horticole* (Paris), vol. 56, 1884. (c) *Edwards's Botanical Register* (London), vol. 27, 1841. (d) *Flore des serres et des jardins de l'Europe* (Ghent, Belgium), vol. 12, 1857.**

FIGURE 54.9 Impatiens growing in commercial greenhouses.

***Source:*** **(a) Photo by Shutterstock/Yuliasis. (b) Photo by Kevin Marsh (CC BY 2.0).**

FIGURE 54.10 Garden beds of impatiens: (a) New Guinea impatiens (*Impatiens hawkeri*). (b) Garden impatiens (*I. walleriana*). (c) New Guinea impatiens (*I. hawkeri*).

***Source:*** **(a) Photo by Leonora (Elle) Enking (CC BY SA 2.0). (b) Photo by Rameshng (CC BY SA 3.0). (c) Photo by Leonora (Elle) Enking (CC BY SA 2.0).**

FIGURE 54.11 Impatiens in containers: (a) New Guinea impatiens (*Impatiens hawkeri*). (b) New Guinea impatiens (*I. hawkeri*). (c) Garden impatiens (*I. walleriana*) grown in boots. (d) Garden impatiens (*I. walleriana*).

***Source:*** **(a) Photo by Shutterstock/Ole Schoener. (b) Photo by Shutterstock/Totokzww. (c) Photo by Forest and Kim Starr (CC BY 2.0). (d) Photo by Shutterstock/Patrick Hatt.**

also very popular as an annual container plant in temperate climates and is occasionally employed as a houseplant and cut flower.

## 54.6 OTHER USES

In Asia, *I. balsamina* has been employed as a folk medicine to treat numerous conditions, such as rheumatism, fractures, inflammation, tuberculosis, snakebite, and menstruation. The plant is cathartic (purgative), diuretic (promoting

FIGURE 54.12 A peacock fashioned from potted garden impatiens (*Impatiens walleriana*).

*Source:* **Photos by Shutterstock/Peter Maerky.**

FIGURE 54.13 Cut flowers of impatiens: (a) A display of cut flowers of *Impatiens walleriana* stuck to a sphere representing a baseball. (b) A bouquet of cut flowers of *Impatiens hawkeri* in a fancy container.

*Source:* **(a) Photographed in Tokyo, Japan, in 2005 by Mom the Barbarian (CC BY 2.0). (b) Photo by Shutterstock/Ted Huizinga.**

urination), emetic (causing vomiting), and antibiotic, and these properties may provide a rationale for some of the traditional usages. Moreover, a variety of chemical constituents are under investigation for possible medical applications.

Throughout the indigenous areas of *Impatiens* species, the plants have often been employed as topical salves to treat skin irritation. In particular, North American Indians widely employed a plant mash of the jewelweed *I. capensis*

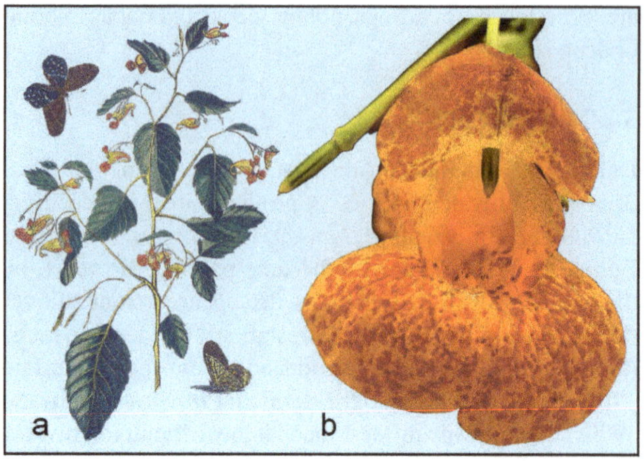


FIGURE 54.14 *Impatiens capensis* (common jewelweed, spotted jewelweed), often employed to treat poison ivy dermatitis.

*Source:* **(a) Meerburgh, N. 1775, *Afbeeldingen van zeldzaame gewassen*, vol. 10 (public domain). (b) Photo of flower by Helen Lowe Metzman/US Geological Survey (public domain).**

(often called spotted touch-me-not) to reduce development of poison ivy dermatitis. It has long been known that soaps are useful for this purpose by acting as a detergent to remove oily urushiol, the causative agent of poison ivy inflammation. It has recently been demonstrated that saponins (soap-like chemicals) in this jewelweed act like soaps to alleviate poison ivy rash. In various parts of the world, similar preparations of *Impatiens* species have been employed to treat comparable skin conditions, but chemical constituents that could explain therapeutic effects have not been reliably identified.

*Impatiens balsamina* flowers are employed to prepare a red dye which is used as a substitute for henna (*Lawsonia inermis*) for coloring finger- and toenails and in cosmetics. Tubers of a species termed *I. tinctoria* were allegedly employed as a source of a black pigment used in Ethiopia for the ritual dying of hands and feet. In North America, Amerindians used the floral pigments of *I. capensis* as a food dye (to color dried salmon), and they, as well as European colonists, also employed the plant to dye fabrics. The seed oil of *Impatiens* species has been used as a lamp fuel and in the surface-coating industry.

*Impatiens walleriana* has been alleged to have sweet-tasting, edible flowers which can be added to salads or mixed into fancy drinks. The flowers, seeds, foliage, and young shoots of *I. balsamina* have been consumed in Asia. In Nepal, the seeds of *I. glandulifera* are harvested for food and also pressed for their edible oil.

## 54.7 PROBLEMS

Despite the occasional food uses previously described, *Impatiens* plants are generally bitter-tasting and should be considered to be slightly poisonous. Preparations of the leaves and roots of *I. walleriana* have been employed as an abortifacient, suggesting that the plants are potentially

FIGURE 54.15 Himalayan balsam (*Impatiens glandulifera*), an invasive weed in many countries, especially in wetlands.

*Source:* **(a) Photo by Oatsy40 (CC BY 2.0). (b) Photo by Maja Dumat (CC BY 3.0).**

toxic. Consumption of the species can cause vomiting, diarrhea, and other intestinal disturbances, so caution is necessary in using the plants as food. The ornamental species of *Impatiens* are usually not considered to be hazardous for dogs and cats, and wild species are not thought to be significant hazards for most livestock.

A few cultivars of so-called "ornamental jewelweed" (*I. glandulifera*) are available. This species was indeed once cultivated as a garden ornamental, until it became evident that it is extremely invasive. This native of northern India is a serious weed of North America. To avoid confusion with the desirable ornamental species, *I. glandulifera* is better termed by its alternative names, "Himalayan balsam" or "policeman's helmet."

## 54.8 CULTIVATION

### 54.8.1 Climate and Site

The species mentioned in this review are frost-intolerant (they are said to be "half-hardy") and are suited to grow outdoors as perennials in climate zones 10 and 11. They are usually grown as annuals in colder areas. These tropical plants succumb to the first light frost. Impatiens grow well as houseplants but requires more light than most other plants grown indoors. Outdoors, New Guinea impatiens is the most popular choice for semi-sunny locations, and some cultivars are tolerant of full sun (and are sometimes termed "sun impatiens"). *Impatiens walleriana*, until recently, was the chief impatiens grown as a bedding annual in quite shady locations, but because of its susceptibility to downy mildew, it is no longer recommended. Hybrids between *I. walleriana* and *I. hawkeri*, based primarily on (and often looking like) New Guinea impatiens, are the wisest choice for shady areas, but the plants do not grow as well in strong shade as *I. walleriana*. The latter is still available from some nurseries and as seed, and those interested in cultivating it are more likely to be successful if it is grown in a hanging basket or pot in a protected situation, where the mildew is less likely to infect the plants.

### 54.8.2 Propagation and/or Purchasing

Seeds are available for the dominant ornamental species, but for the most part, the plants are established from young established specimens purchased from nurseries or garden centers. Commercially grown plants are frequently propagated from cuttings. *Impatiens* are often treated as houseplants and can be obtained as potted plants to be maintained indoors as perennials.

### 54.8.3 Planting

Young plants are usually set out in beds or borders after danger of frost has passed. A planting distance of 25 to 30 cm suffices for most cultivars. The plants are also sometimes employed as a groundcover in shady locations, particularly at the base of trees. They are also popular in hanging baskets, patio containers, window boxes, and pots.

### 54.8.4 Soil and Moisture Conditions

The species discussed here are drought-intolerant and grow best in moist, well-draining soil rich in humus, ideally with a slightly acidic pH of 5.8–6.5. *Impatiens hawkeri* requires regular watering to keep the soil moist, and regular fertilizing (or the use of a slow-release fertilizer). Overfertilization can encourage the plants to become leggy.

### 54.8.5 Maintaining Plants

Plants grown outdoors, whether in beds or containers, should be discarded once they become frost-damaged. If left in place, pathogens are encouraged to become established in the soil.

## 54.9 FASCINATING FACTS, FABLES, FOLKLORE, AND FRIVOLITY

- The flowers of jewelweeds are very attractive (some sparkle in sunlight like jewels) and have been interpreted as the inspiration for the name jewelweed. However, the term is based on the foliage, which, in the morning, often has beads of water on the leaf edges, which sparkle like jewels. At night, the leaf edges expel excess water by guttation (exudation of xylem sap through special pores called hydathodes, or water glands). This secretion of drops of water is an adaptation to the moist habitat and compensates for the lack of transpiration at night, when the stomates are closed. The drops of water are present in early morning but evaporate quickly in sunlight.
- Unlike almost all other plant species, the number of chromosomes of *Impatiens* can be quickly determined from pollen grains on herbarium sheets. Pollen typically contains a vegetative cell, responsible for producing an elongated tube in a stigma on which the pollen grain is deposited, and a generative cell, which divides by mitosis to produce two sperm cells which migrate down the tube to achieve the characteristic double fertilization of

FIGURE 54.16 Guttation in western jewelweed (*Impatiens noli-tangere*). Notice the exuded drops of liquid on the leaf margins.

*Source:* **Photo by Hans Schneider (CC BY SA 2.0).**

FIGURE 54.17 Parrot plant (*Impatiens niamniamensis*): (a) Houseplant. (b) Flower. (c) Garden plant.

*Source:* **(a) Photo by Maja Dumat (CC BY 2.0). (b) Photo by Lark Ascending (public domain). (c) Photo by Darwinius (CC BY SA 3.0).**

flowering plants. In *Impatiens*, the chromosomes go into a prolonged resting stage at the first mitotic metaphase of the generative cell, and in this state, the chromosomes can be easily visualized.

- One of the most amusingly labelled plants is a cultivar of *Impatiens walleriana* named "Izzy Wizzie," otherwise called "Busy Lizzie 'Izzy Wizzy' " (www.shootgardening.com/plants/impatiens-walleriana-izzy-wizzy).
- Like many other genera, species of *Impatiens* often have extrafloral nectaries (i.e., nectaries not located in the flowers), located on the stipules (small leaflike bracts typically borne in pairs at the base of the leafstalk). The nectar produced is especially attractive to ants, which have been hypothesized to symbiotically protect the plants against other insects in return for the sweet liquid.
- Parrot plant (*Impatiens niamniamensis*, also called Congo cockatoo plant and parrot beak plant) is a small evergreen bush (30–100 cm tall) native to tropical Central Africa. The striking, cornucopia-shaped, greenish-yellow flowers are said to resemble birds (hence the reference to parrots and cockatoos). The blooms have long reddish or purple spurs in which nectar is produced. Like other species of *Impatiens*, it can be grown as a perennial outdoors in climate zones 10 and 11 but is usually cultivated as a houseplant (the foliage can be toxic if consumed).

## KEY PUBLICATIONS

Bailey, D.A. 1999. *Commercial Production of New Guinea Impatiens*. Horticulture Information Leaflet 526. Department of Horticultural Science, North Carolina State University. 10 pp. www.clamerinforma.it/EnglishFiles/PDF/HIL526.pdf.

Banner, W. and M. Klopmeyer (eds.). 1995. *New Guinea Impatiens: A Ball Guide*. Chicago Review Press, Chicago. 288 pp.

Carlson, W.H. and C.L. Crankshaw. 1981. *Producing Impatiens [walleriana] for Profit*. Michigan State University Cooperative Extension Service Bulletin E-1580.4 pp. http://archive.lib.msu.edu/DMC/Ag.%20Ext.%202007-Chelsie/PDF/e1580-1981-print1.pdf.

Carr, J., M. Xu, J.W. Dudley, and S.S. Korban. 2003. AFLP analysis of genetic variability in New Guinea impatiens. *Theoretical and Applied Genetics* 106: 1509–1516.

Clements, D.R., K.R. Feenstra, K. Jones, and R. Staniforth. 2008. The biology of invasive alien plants in Canada. 9. *Impatiens glandulifera* Royle. *Canadian Journal of Plant Science* 88: 403–417.

Coakley, S. and P. Carloalberto. 2021. Impacts of the invasive *Impatiens glandulifera*: Lessons learned from one of Europe's top invasive species. *Biology* 10(7): 619; https://doi.org/10.3390/biology10070619.

Getter, K.L. and B.K. Behe. 2013. Consumer willingness to purchase *Impatiens walleriana* alternatives. *HortScience* 48: 1370–1377.

Gill, L.S. and C.C. Chinnappa. 1977. Chromosome numbers from herbarium sheets in some Tanzanian *Impatiens* L. (Balsaminaceae). *Caryologia* 30: 375–379.

Grey-Wilson, C. 1980. *Impatiens of Africa. Morphology, Pollination, and Pollinators, Ecology, Phytogeography, Hybridisation, Keys and a Systematic Treatment of all African Species with a Note on Collecting and Cultivation*. Balkema, Rotterdam, Netherlands. 235 pp.

Hayashi, M., K.L. Feilich, and J. Ellerby. 2009. The mechanics of explosive seed dispersal in orange jewelweed (*Impatiens capensis*). *Journal of Experimental Botany* 60: 2045–2053.

Janssens, S., K. Geuten, Y.-M. Yuan, Y., Song, P. Küpfer, and E. Smets. 2006. Phylogenetics of *Impatiens* and *Hydrocera* (Balsaminaceae) using chloroplast atpB-rbcL spacer sequences. *Systematic Botany* 31: 171–180.

Janssens, S.B., Y.S. Wilson, Y.-M. Yuan, A. Nagels, E.F. Smets, and S. Huysmans. 2012. A total evidence approach using palynological characters to infer the complex evolutionary history of the Asian *Impatiens* (Balsaminaceae). *Taxon* 61: 355–367.

Morgan, R.F. 2007. *Impatiens: The Vibrant World of Busy Lizzies, Balsams, and Touch-Me-Nots*. Timber Press, Portland, OR. 219 pp.

Motz, V.A., C.P. Bowers, A.R. Kneubehl, E.C. Lendrum, L.M. Young, and D.H. Kinder. 2015. Efficacy of the saponin component of *Impatiens capensis* Meerb. in preventing urushiol-induced contact dermatitis. *Journal of Ethnopharmacology* 162: 163–167.

Qian, H.-Q., B. Wang, J. Ma, C. Li, Q. Zhang, and Y. Zhao. 2023. *Impatiens balsamina*: An updated review on the ethnobotanical uses, phytochemistry, and pharmacological activity. *Journal of Ethnopharmacology* 303: 115956, https://doi.org/10.1016/j.jep.2022.115956.

Salgado-Salazar, C., Y. Rivera, D. Veltri, and J.A. Crouch. 2015. Polymorphic SSR Markers for *Plasmopara obducens* (Peronosporaceae), the newly emergent downy mildew pathogen of *Impatiens* (Balsaminaceae). *Applications in Plant Sciences* 3(11): 1500073. http://dx.doi.org/10.3732/apps.1500073.

Uchneat, M.S. 2007. Impatiens: *Impatiens wallerana*. In: *Flower Breeding and Genetics: Issues, Challenges and Opportunities for the 21st Century*, edited by N.O. Anderson, 277–299. Springer, Netherlands.

Yu, S.-X., S.B. Janssens, X.-Y. Zhu, M. Lidén, T.-G. Gao, and W. Wang, W. 2016. Phylogeny of *Impatiens* (Balsaminaceae): Integrating molecular and morphological evidence into a new classification. *Cladistics* 32: 179–197.

## Key Cultivation Websites

Greenhouse Grower. *Production of New Guinea Impatiens*. www.canr.msu.edu/uploads/resources/pdfs/new-guinea-impatiens.pdf.

*Horticulture Magazine*. How to grow impatiens 'busy lizzies'. https://horticulture.co.uk/impatiens/.

Iowa State University Extension. *Growing Impatiens in the Home Garden*. https://hortnews.extension.iastate.edu/growing-impatiens-home-garden.

North Carolina State University Extension. *Impatiens walleriana*. https://plants.ces.ncsu.edu/plants/impatiens-walleriana/.

The Spruce. *How to Grow and Care for Impatiens*. www.thespruce.com/shade-tolerant-impatiens-flowers-2132141.

University of Michigan Extension. *Impatiens*. https://extension.umn.edu/flowers/impatiens.

WikiHow. *How to Grow Impatiens*. www.wikihow.life/Grow-Impatiens.

WikiHow. *How to Fertilize Impatiens*. www.wikihow.life/Fertilize-Impatiens.

# 55 Iris (*Iris* Species and Hybrids)

**Scientific Names:** The genus name *Iris* is based on *iris*, Latin and Greek names usually interpreted as "rainbow," but also as "a sweet-smelling plant." The name is frequently explained as a reference to the many attractive colors of the flowers of the species being reminiscent of the colors of a rainbow.

**Common Names:** The common name "iris" has generally been employed to designate all the species in the genus *Iris*. A few unrelated species have "iris" as part of their name, and several of the debatably included species are known by other common names. "Iris" is also the name in numerous other languages, including Danish, Dutch, French, German, Italian, Norwegian, Portuguese, Spanish, Swedish, and Turkish.

## 55.1 INTRODUCTION

The iris is one of the leading ornamentals in the gardens of temperate regions. Its gorgeous flowers are unsurpassed in the plant kingdom, with an astonishing range of colors, sizes, and interesting forms. The plants can be grown in a wide variety of situations; many are easily cultivated, and some are widely sold by florists as cut flowers for bouquets.

## 55.2 OVERVIEW

The genus *Iris* has close to 300 species (or much fewer, depending on taxonomic concepts). Specialists have noted that *Iris* can be recognized as a large genus composed of several groupings (subgenera) or as a smaller genus (of about 100 species), while what were previously classified as subgenera are instead recognized as allied separate genera. The simplified scheme shown in Figure 55.2 is roughly reflective of current taxonomic opinion regarding relationships in *Iris* and is also basically the grouping system used horticulturally (rarely cultivated groups have been excluded). There are many genera related to *Iris*, sometimes included in the genus, and a few of these (not mentioned here) are cultivated.

*Iris* species are herbs, deciduous, or less commonly evergreen (in warm climates). They reproduce by seeds, but also by rhizome or bulb replication. (Plants of a third category, with swollen or "tuberous" roots, are rarely cultivated.) Irises are always referred to as "perennials," but the aboveground (or above-water) shoots are usually annual (evergreen shoots might endure longer than a year), and the tissues of the below-ground-level storage organs (rhizomes or bulbs) may often live for only one season before reproducing themselves. Most horticultural iris specialists employ the presence of rhizomes on the one hand or bulbs on the other as a principal basis for recognizing two broad groups. Fibrous roots are attached to the storage organs (whether rhizomes or bulbs). Most irises grown in gardens are rhizomatous species. Bulb-bearing species, by contrast, are much less popular because they mostly originate from relatively cool areas of Asia that are well-supplied with water in early season but the shoots dry up to avoid

FIGURE 55.1 Spectacularly colored iris flowers. These are all rhizomatous bearded irises, the dominant class of garden irises, as discussed in the text.

*Source:* **(a) Photo by Lady Dragonfly (CC BY 2.0). (b) Photo by Mike Ball (CC BY 2.0). (c) Cultivar Wild West, photo by Kor!An (CC BY 3.0).**

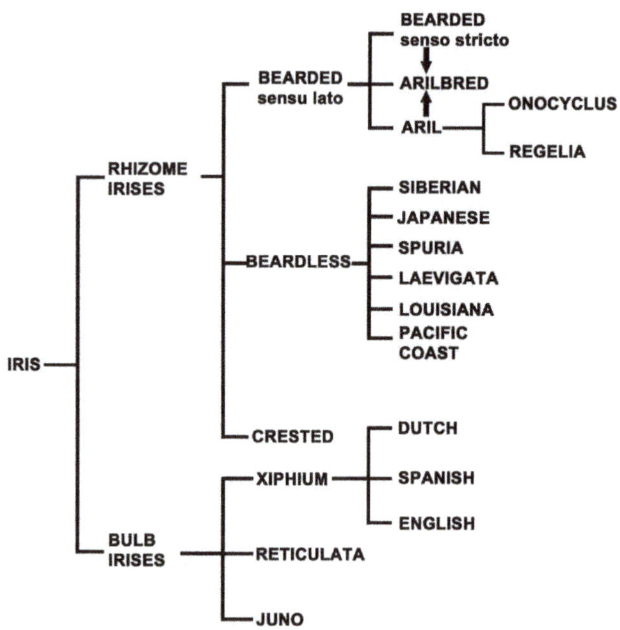


FIGURE 55.2 Simplified classification of the groups of cultivated irises discussed in this chapter. The groups are horticultural (i.e., employed in the ornamental industry), but the hierarchy shown substantially also reflects their evolution and taxonomic classification. Bearded irises are dominant in temperate climates, followed by Siberian and Japanese irises. *Sensu lato* means "in the wide sense" (with more species); *sensu stricto* means "in the narrow sense" (with fewer species).


DOI: 10.1201/9781003586227-55

FIGURE 55.3 Contrast of bulbous and rhizomatous irises: (a) *Iris persica*, showing basal bulb. (b) *Iris pumila*, showing basal rhizome.

*Source:* **Illustrations from Sowerby, J. 1787. *Botanical Magazine*, vol. 1. Public domain figures.**

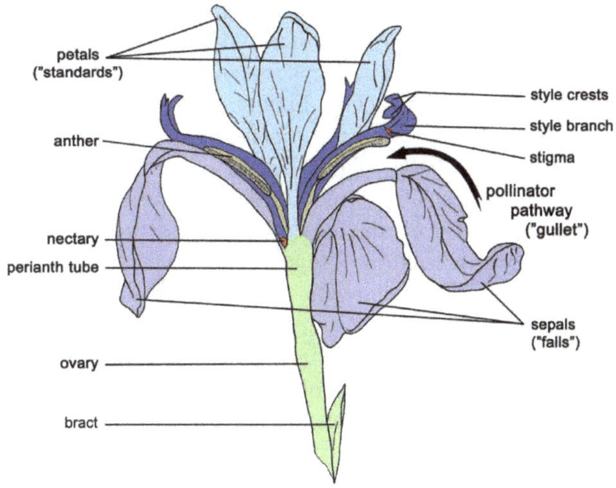


FIGURE 55.4 Diagram of the basic parts of an iris flower, based on *Iris versicolor.* Pollinators land on one of the large sepals and are guided toward the basal nectary through a tunnel ("gullet") formed by the base of the sepal and the overhead style branch. During the inward trip, the insect deposits previously collected pollen on the upper (receptive) part of the stigma flap and then picks up pollen from the anther; during the outward trip, it brushes against the lower part of the stigma flap, but this is non-receptive, to prevent self-fertilization.

*Source:* **Prepared by B. Brookes.**

midsummer droughts, so the floral display period is limited. Moreover, most of the bulbous species are relatively difficult to cultivate. However, many of the bulb-bearing species are suitable for rock gardens, indoor cultivation, and the cut flower industry. In most *Iris* species, the foliage is narrow and basal. The species with rhizomes usually have a clump (or fan) of sword-shaped leaves, while the species with bulbs often have cylindrical leaves.

*Iris* species occur primarily in the Northern Hemisphere, with the greatest diversity in Eurasia (including 60 or so in China). In North America, 34 species have been recognized north of Mexico. Habitats mainly include open grasslands, dense woodlands, mountainsides, deserts, and sandy coastal areas. *Iris* species are found from sea level to alpine meadows over 4,000 m and from the Arctic tundra to the semi-tropics. Most irises (particularly the bearded and bulbous kinds, discussed later) are found in circumneutral soils, but some (particularly the beardless kinds) occupy acidic ground. Species are variously adapted to substrates that are poor, fertile, or stony. Species also differ in adaptation to wetness of soils, which may be dry, damp, or covered by shallow water. Most irises prefer open, sunny locations, some tolerate semi-shade, and a few are adapted to shade.

With the exception of orchids, no other popular group of ornamentals has flowers that are as structurally complex as those of iris species. Although the vegetative features of irises are considered attractive, as with the majority of the world's most important ornamentals, the plants are cultivated mostly for their large colorful flowers, particularly the "tepals" (sepals and petals collectively), which are divided into three outer sepals ("falls" in iris terminology) that flex downward and three inner petals ("standards") which remain upright. In some cultivars (especially in the Japanese group), doubling has occurred (i.e., extra petals have been generated). The style (the portion of the pistil between the ovary and the stigma) of irises is complex: It is divided into three "style arms," or style branches, each of which terminates in an upturned crest (or crest lobes), and at the base of the crest there is a small shelf termed the *stigmatic lip*. An anther is located under each style arm. The sepals, petals, and style arms are often differently colored, and indeed, each of these may be variegated, and (although purple is the dominant hue) an astonishing range of color combinations has been generated by hybridization and selection.

Like the flowers of many insect-pollinated plant species, the stigma of *Iris* flowers receives pollen from visiting pollinators (especially bees), and the anthers provide a fresh load of pollen to the visitors to fertilize other flowers. However, the *Iris* flower is peculiar in the architectural way it separates these two events. The large petal-like sepals of the iris flower function as a landing platform for flying insects, and markings (guides) direct the visitors to the nectary at the base of the flower. The petal-like style branch (above) and the petal-like sepal (below) provide a channel ("gullet") toward the basal nectary. This channel is partially blocked by the stigma, in the form of a flap, which is cleverly arranged so that the receptive side faces the incoming pollinator. As the pollinator pushes against the stigma flap, the receptive surface removes pollen that was acquired by the insect from previous floral visits. The insect next contacts an anther and acquires a new load of pollen grains before reaching the nectary. Finally, while backing out of the flower, the insect comes in contact only with the non-receptive face of the stigma flap. Thus, the insect is prevented from depositing its newly collected batch of pollen on the stigma of the same flower. Each of the three divisions of the flower (including a sepal, petal, style, stigma, and ovary "locule" [segment with seeds]) effectively functions as a separate flower.

Although the flowers of *Iris* are extremely well-adapted for sexual reproduction, the plants (especially the rhizomatous species) spread very extensively by vegetative extension. Clones often persist for decades in abandoned gardens and other locations where they have become established.

Breeding has greatly modified the flowers of *Iris*, resulting in cultivars with unusual floral colors and structure. Some cultivars have been selected for long blooming period, and also for reblooming (although in cold climates a second blooming in the autumn can reduce winter-hardiness).

### 55.2.1 Rhizomatous Irises

Irises with rhizomes, the most popular kind of iris, are divisible into three basic classes, based on differences in nectar guides on the sepals, which assist pollinators in finding nectar. The three classes are bearded, beardless, and crested. Most garden irises are hybrids produced between bearded irises or between beardless irises (bearded irises rarely cross with beardless irises).

### 55.2.2 Bearded Irises

Bearded (or "pogon," based on Greek for *bearded*) irises constitute the most popular, indeed the dominant class, of rhizomatous irises. There are perhaps about 30 species, but additionally many hybrids have been given species names. Bearded iris species are mostly native to Central and southern Europe east to Turkey, as well as a few from

FIGURE 55.5 The different sepal nectar guide systems characterizing the three major horticultural classes of rhizomatous irises: (a) Bearded guide, representing the bearded group (species not identified). (b) Simple distinctly colored spot guide in *I. versicolor*, representing the beardless group. (c) Crested guide (with raised areas) in *I. japonica*, representing the crested group.

*Source:* **(a) Photo by Sid Mosdell (CC BY 2.0). (b) Photo by D. Langlois (CC BY 3.0). (c) Photo by Sigusr0 (CC BY 2.0).**

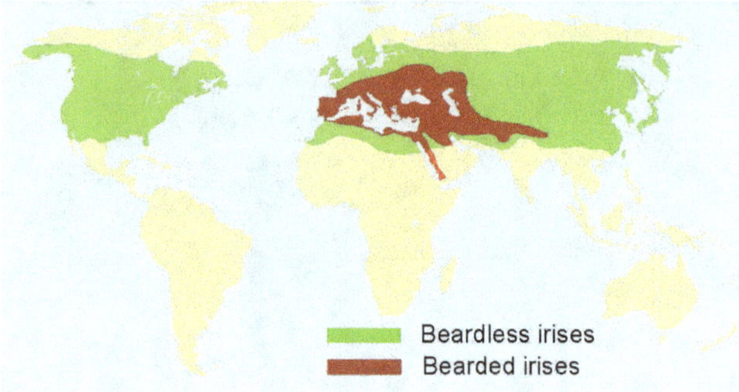


FIGURE 55.6 Geographical distribution of species of the two major classes of garden irises, bearded (pogon) and beardless (apogon). Bearded irises occur only in part of the range of beardless irises, and although a minority of iris species, they are a majority of garden irises.

*Source:* **Based on information in Köhlein (1987).**

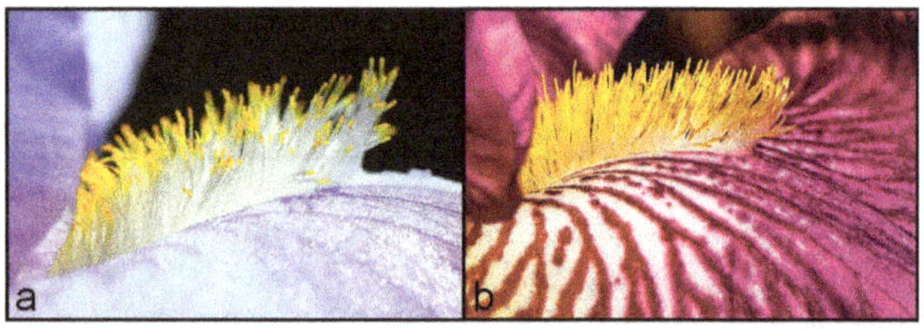


FIGURE 55.7 Floral "beards" of bearded irises.

*Source:* **(a) Photo by Romavor from Pixabay (public domain). (b) Photo by Shutterstock/Gkordus.**

FIGURE 55.8 Flower of cultivar 'Thornbird', a "horned iris" (note upturned tip of the beard).

*Source:* **Photo by Hope Abrams (CC BY 2.0).**

the Caucasus, North Africa, and western Asia. The leaves develop in a fan (i.e., in a horizontal plane) and are usually sword-shaped and broader than in most other irises. Over 30,000 cultivars have been recognized. Bearded irises are characterized by a hairy, caterpillar-like row of hairs (plant hairs are called trichomes) on the distal half of the upper (adaxial) side of the sepals (rarely, also elsewhere). *Iris × germanica*, a complex species with hundreds of cultivars, is perhaps the most widely grown. Bearded irises are very variable and, horticulturally, have been divided into several height classes. Most garden irises are "bearded irises" in the narrow sense, that is, excluding the aril irises mentioned next. The flowers of bearded irises have undergone considerable selection. One of the interesting changes has been the selection of so-called "horned irises," which have a distinctive upturned horn at the tip of the beard. More elaborate, leafy mutations of the beard have been selected and referred to as "space-age irises," a controversial development intensely disliked by some iris specialists.

### 55.2.3 Aril Irises

"Aril irises" are a class of irises with true arils – appendages on the seeds which serve as food attractants to insects (particularly ants) for propagule distribution. Aril iris flowers have beards, but these are so sparse that the group is often not considered to be truly bearded. Two small groups of aril irises are relatively popular, Oncocylus irises (mostly Middle Eastern, extending from southwestern Europe to western Asia) and the closely related Regelia irises (native to open rocky hillsides and high mountains in Afghanistan and Turkestan). There are very attractive hybrid cultivars ("Regeliocyclus" irises) between Regalia and Oncocyclus irises. Because aril irises are adapted to warm, dry areas (in summer), in cold-temperate climates, they are usually grown in greenhouses. However, very attractive hybrids of aril irises and bearded irises (in the strict sense) have been generated (these are referred to as "arilbreds"). These are easier to grow than aril irises (they are often adapted to temperate areas with low rainfall), are generally tall with large flowers, and are occasionally cultivated in temperate gardens.

FIGURE 55.9 Examples of the aril group: (a) *Iris iberica*. (b) *Iris hoogiana*. (c) *Iris gatesii*.

*Source:* **Public domain illustrations. (a) The florist and pomologist, London, 1873. (b) *Curtis's Botanical Magazine* (London), vol. 146 (= ser. 4, vol. 16), 1920. (c) *Curtis's Botanical Magazine* (London), vol. 128 (= ser.3, vol. 58), 1902.**

### 55.2.4 Beardless Irises

Much less dominant but still very popular in gardens is the beardless (or "apogon," Greek for *beardless*) class, in which instead of a fuzzy "beard" to guide pollinators, there is simply a distinctive colored area (generally yellow) signalling pollinators toward the nectary. Beardless irises, on the whole, do not produce as glamorous a flower as do bearded irises, but nevertheless, many cultivars are very attractive. All native North American irises belong to this group. The main subgroupings of beardless irises are noted in the following.

### 55.2.5 Siberian Irises

These are the best-known and most popular of the beardless irises. Most Siberians come from Europe, eastern Russia, China, and Japan. There are about a dozen species, of which *I. sibirica* and *I. sanguinea* and their hybrids are the most widely grown. Siberians grow well in climates with cold winters. Some *Iris* species of Oregon and California are closely related and have been hybridized with the Eurasian species to produce cultivars. Siberians produce smaller but more numerous flowers than the more popular bearded irises. The flowers are available in a smaller but still interesting range of colors. Siberian irises are fairly resistant to environmental and biotic stresses.

### 55.2.6 Japanese Irises

Japanese irises are largely based on *I. ensata* (= *I. kaempferi*) and have been bred in Japan for at least the last eight

FIGURE 55.10 Examples of the Siberian group: Siberian iris, *Iris sibirica*, showing purple and white selections.

*Source:* **(a) Photo by Derek Ramsey (GNU Free Documentation License). (b) Cultivar King of Kings. Photo by F. D. Richards (CC BY SA 2.0).**

FIGURE 55.11 (a) Examples of the Japanese group (Japanese iris, *I. ensata*). (b) Japanese iris in a park in Japan.

*Source:* **(a) Photo by Cheryl Marland (CC BY 2.0). (b) Photo by Bariston (CC BY SA 4.0).**

centuries. Many cultivars produce spectacularly large, doubled, and often ruffled flowers. The largest iris flowers are produced by Japanese irises, some exceeding 25 cm in width. The species are moisture-loving and are adapted to areas near ponds and bogs. In warm regions, they can be grown submersed – no more than 8 cm has been recommended. They will die overwinter in frozen water, and in freezing areas, they should be grown terrestrially, providing adequate irrigation. The color range is limited (most cultivars are shades of violet), but wide ranges of color variegation and floral structure are available.

### 55.2.7 Laevigata Irises

These are wetland species of Eurasia and North America typically occurring in standing water in their native locations but capable of being grown in well-watered gardens. Many cultivars have been selected of *I. laevigata* by the Japanese. Blue flag (*I. versicolor*) and Virginia iris (*I. virginica*) of North America, and yellow iris (*I. pseudacorus*) of Europe, belong to this group, and hybrid cultivars of these species have been bred. These species are suitable for water gardens and for naturalizing in privately owned marshy areas.

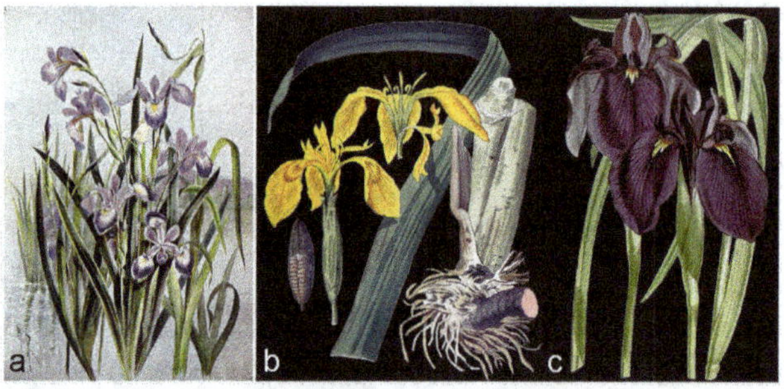

FIGURE 55.12 Examples of the laevigata group: (a) *Iris versicolor.* (b) *Iris pseudacorus.* (c) *Iris laevigata.*

***Source:*** **Public domain illustrations. (a) Lounsberry, A. 1899. *A guide to the wild flowers.* (b) Kops, J., H. C. van Hall, and C. Sepp. 1828. *Flora Batava*, vol. 5. (c) *Curtis's botanical magazine* (London), vol. 100 (ser. 3: vol. 30), 1874.**

FIGURE 55.13 (a) Examples of the spuria group (*I. spuria*). (b) Cultivar Cinnabar Red.

***Source:*** **(a) Photo by Manuel (CC BY 2.0). (b) Photo by Derek Ramsey, courtesy of Chanticleer Garden (CC BY SA 3.0).**

### 55.2.8 Spuria Irises

Spuria irises are native from western Europe through the Middle East to western China. They are not very popular but have attractive, orchid-like blooms sometimes reminiscent of bulb irises. The name *I. spuria* has been applied to a complex of species, and the term "spuria irises" is more broadly employed to include over a dozen cultivated species and hybrids. Spuria irises range from dwarfs to quite tall plants. Some of the species are very vigorous, but most are unsuitable for cold climates.

### 55.2.9 Louisiana Irises

"Louisiana irises" include a group of five wetland species native to swamps of the American Gulf Coast (particularly concentrated in Louisiana). They can be grown on dry land but need acidic soil that is wet in the spring. Hundreds of attractive cultivars, some with flowers up to 18 cm wide, are available. To overwinter, they require a relatively mild climate.

### 55.2.10 Pacific Coast Irises

"Pacific Coast irises," or Californian irises, include a dozen or so species adapted to the Pacific West Coast of the United States, particularly California (most are plants of mountainous and forested areas, often found in light shade). As for the previous group, to overwinter outdoors, they require a relatively mild climate.

### 55.2.11 Crested Irises

The least-popular class of rhizomatous irises, crested irises, includes about a dozen species, mostly in eastern Asia, with three in eastern North America. These are also called orchid irises (several of the Asian species look remarkably like orchids) and Evansia irises (after Thomas Evans, who introduced them to England in 1794). Crested irises usually have an elevation (crest), ridge, or occasionally, a set of colored bumps, where the beard or color signal is found in other rhizomatous irises. Although some species are fairly tall, the group includes many miniature cultivars, which are very small and produce tiny blooms. Crested irises are fairly demanding, the half-dozen or so cultivated species differing in their growth requirements.

### 55.2.12 Bulbous Irises

Most irises with bulbs are native to high altitudes from Israel and the Russian Caucasus to Turkey, Iran, and Central Asia. They are adapted to flowering in the high rainfall of spring and go dormant during the very dry summers. They produce their foliage before flowering. Bulbous irises should be fertilized in the spring while they are growing, and split for replanting when they are dormant. A few of the hardier species can be grown outdoors as perennials in rockeries, borders, and protected areas, but most are better suited for raised beds, containers, and indoor situations. The bulbs can be lifted and stored overwinter (most, but not all, cultivars can be maintained dry). Bulb irises are not popular in North America, and the selection of cultivars available is quite limited.

### 55.2.13 Xiphium Irises

Xiphium irises (named for the most common species, *I. xiphium*) include about eight species native to southern Europe and the Mediterranean area, mostly occupying well-drained soils in hot, dry climates. These irises are relatively tall and slender, producing flowers that persist for many days and leaves that are narrow, are deeply channelled, and tend to be semi-circular in cross section. They are not suited to cold climates. The group is divided into "Dutch," "English," and "Spanish" irises, although these terms are misleading. The English irises are from *I. latifolium*, from the marshy areas in the High Pyrenees of Spain and France; Spanish irises are selections of *I. xiphium* which occur in Europe (including Spain) and North Africa; the Dutch irises are hybrids of *I. xiphium* and other xiphium irises which arose in cultivation (at first in The Netherlands, although none of the xiphium irises are native there). English and Spanish irises were once popular but are rarely grown today. They can be cultivated outdoors in mild climates, and the English selections tolerate shade. Spanish irises are sometimes grown as annuals in cold areas. Dutch irises can also be raised outdoors in mild climates but are now chiefly grown for the cut flower florist's market, because the bulbs can easily be forced to come into flower. The bulbs are discarded after the flowering stems are harvested. Because they are mass-produced, the bulbs are very cheap and can be used to establish potted plants for indoor use.

### 55.2.14 Reticulata Irises

The word "reticulata" is Latin for *netted*, a description of the appearance of the outer coat of the bulbs when stretched out. The dozen or so species are found from the eastern Mediterranean to Central Asia. The species of this group are small and have small bulbs. A few species have been grown as rock garden plants but are scarcely frost-tolerant. However, the bulbs can be started in pots and cold frames and maintained outdoors during the summer.

FIGURE 55.14 Xiphium class of bulbous irises. "Dutch iris" (*Iris xiphium*).

*Source:* **(a) Photo by Wylie (CC BY 2.0). (b) Illustration from Witte, H. 1868. *Flora: Afbeeldingen en beschrijvingen van boomen, heesters, éénjarige planten, enz. voorkomende in de Nederlandsche Tuinen*. Wolters, Groningen (public domain).**

### 55.2.15 Juno Irises

Junos include 60 or so species of exotic-looking plants of limited stature, but only about four are cultivated as ornamentals, and only infrequently. The bulbs tend to have several long, fairly thick roots, reminiscent of daylilies, and these can be used to grow new plants. Junos are natives of semi-arid areas of Asia, occupying steppes and mountainsides, and climates with hot, dry summers and cold winters. They are usually grown in pots indoors, and because they are very difficult to maintain and most are not particularly attractive, they are mostly grown by specialists, although a few (such as *I. bucharica*) are occasionally cultivated in gardens.

## 55.3 HISTORY

Irises have been valued since ancient times. A fresco in Crete dating back 4 millennia has been claimed to picture the Xiphium category of bulbous irises (however, the plants have also been accepted as lilies, as discussed in

FIGURE 55.15 Reticulata class of bulbous irises: *Iris reticulata*.

*Source:* **(a) Photo by Hedwig Storch (CC BY SA 3.0). (b) Photo by Averater (C BY SA 4.0).**

FIGURE 55.16 Juno class of bulbous irises: (a) *Iris magnifica*. (b) *Iris bucharica*.

*Source:* **(a) Photo by Denis Prévôt (CC BY 3.0). (b) Photo by Tejvan Pettinger (CC BY 2.0).**

FIGURE 55.17 Old iris cultivars illustrated in vintage seed catalogues: (a, b) Bearded irises. (c) Japanese (*Iris ensata* = *I. kaempferi*) cultivars.

***Source:*** **John Lewins Childs, Floral Park, New York (public domain).**

FIGURE 55.18 Art showing beautiful women associated with beautiful irises.

***Source:*** **(a) Illustration by Shutterstock/NemesisINC. (b) *Fleurs des Eaux* (1923, public domain), by French artist Elisabeth Sonrel (1874–1953). (c) Illustration by Shutterstock/HstrongART.**

the "History" section of the chapter on lily). To the early Egyptians, the iris represented power and majesty. The flower was depicted on the brow of the Sphinx and on regal scepters. In classical Greece, irises were planted on the tombs of women. By Islamic tradition, white-flowered irises were grown on the graves of soldiers. In Christianity, it has been claimed that many biblical "lilies" were actually irises. Irises have represented the birth of Christ in several well-known paintings, including Dürer's *Madonna with the Iris* and da Vinci's *Madonna of the Rocks*. The three-parted flower has been interpreted as reflecting the three virtues of faith, wisdom, and courage. For centuries, artists have celebrated the beauty of women by showing them associated with beautiful flowers, and irises have been widely employed for the purpose.

## 55.4 ECONOMIC IMPORTANCE

Iris species are among the most commonly grown ornamental plants and are important in the nursery stock industry. There are numerous books dedicated to horticultural aspects, reflective of the popularity of irises.

## 55.5 ORNAMENTAL USES

Irises are extensively grown in public and residential gardens, ranging from a few plants in a border to immense displays in parks. Smaller species are also grown in containers.

FIGURE 55.19 Commercial iris fields. (a) Young girl in a field. (b) Huge field in Salem, Oregon.

***Source:*** **(a) Photo by Shutterstock/Victoria Ditkovsky. (b) Photo by Shutterstock/Danita Delimont.**

FIGURE 55.20 Huge beds of irises planted for public display: (a) Iris garden at Royal Botanical Gardens, Burlington, Ontario, Canada. (b) Iris festival in Hiraizumi, Japan. (c) Iris garden of Vullierens Castle, Switzerland.

***Source:*** **(a) Photo by RichardBH (CC BY 2.0). (b) Photo by Shutterstock/Yspbqh14. (c) Photo by Shutterstock/Mariepitt (CC BY SA 4.0).**

FIGURE 55.21 Iris cut flowers: (a) Bride with bouquet. (b) Vase with bouquet of flowering stems. (c) Girl with long-stemmed bouquet.

***Source:*** **(a) Photo by Shutterstock/Alena Ozerova. (b) Photo by Dave Shafer (CC BY 2.0). (c) Photo by Shutterstock/Yana Tkachenko.**

Bulbous irises are predominantly used in the cut flower industry. So-called "Dutch irises" (Xiphium group) are the primary commercial cut flower irises. These are hybrids of *I. xiphium* and other species, particularly *I. filifolia* and *I. tingitana*. They have been bred specifically for the cut flower market. Some of the other bulbous irises are also employed as commercial cut flowers. Rhizomatous irises from private gardens are often employed in bouquets of cut flowers maintained indoors in residences, although individual flowers of most of the popular bearded irises last only a day or two. Many of the beardless irises also make good cut flowers, and the Japanese irises are very popular in Japan for display in vases. Because of their brief life, stems bearing buds rather than opened flowers are preferable.

FIGURE 55.22 Orris root: iris species employed for rhizome extracts, and examples of use in perfumes and gin. The two irises shown here belong to the rhizomatous beardless iris class: (a) *Iris germanica*. (b) An array of perfumes. (c) An array of gins. (d) *Iris pallida*.

***Source:*** **(a) Thomé, O. W. 1885. Prof. Dr. Thomé's Flora von Deutschland, Österreich und der Schweiz. Gera, Germany. (b) Photo by Joeannenah (CC BY 2.0). (c) Photo by JonWorth (CC BY 2.0). (d) Köhler, F. E. 1890. *Köhler's Medizinal-Pflanzen*, vol. 2.**

## 55.6 OTHER USES

"Orrisroot powder," prepared by drying and pulverizing the rhizomes of certain *Iris* species (especially the closely related *I. florentina*, *I. pallida*, and *I. germanica*), has been used in spices since the eras of the classical Greeks and Romans. Powdered orris, known as "violet powder," was once widely employed in Europe to scent clothing and linens. The juice of steeped *Iris* rhizomes has imparted a flowery bouquet to Chianti wine. Today, orris is still employed to flavor alcoholic beverages, particularly gin, vermouth, and brandy. Orris is also widely used in cosmetics, and especially in perfumes. Irone, an ephemeral fragrance in the rhizomes, produces a pleasant violet odor, and orris also acts as a fixative for other scents. Much of orrisroot today is produced in southern Europe.

Iris rhizomes and orrisroot have been used medicinally in Europe, since the times of the classical Greeks and Romans, for a wide variety of conditions. Similarly, North American Indians used *I. versicolor* rhizome for a range of medicinal purposes, especially topically to treat wounds, sores, and burns, and internally as a purgative, and to treat rheumatism, as well as disorders of the kidney and the liver. Iridin (irisin) is a bitter extract of *I. versicolor* and *I. pseudacorus* that has been used in herbal medicine for its diuretic (urine-stimulating), laxative, and blood-cleansing properties. Rhizomes of *I. versicolor* were even employed as a cure for syphilis. Virtually all medical usages of iris are considered obsolete today and are potentially dangerous, given the toxicity pointed out next. The rhizomes or their extracts have been used as an abortifacient, indicative of the need for caution.

## 55.7 PROBLEMS

*Iris* species are well-known for their toxicity. Small amounts of the rhizomes, or large amounts of the leaves, even if dried, are quite purgative and can also cause depression and respiratory problems. Consumption of iris bulbs can also result in severe digestive upset. The juice of irises can produce dermatitis in sensitive individuals. Livestock avoid eating iris foliage. Cases of death of humans have been recorded, and ornamental rhizomes that are stored for planting are a potential risk to dogs. Irisin (mentioned in the previous section) may be responsible for some of the poisonous effects, but other compounds are suspected of being more toxic.

## 55.8 CULTIVATION

### 55.8.1 Climate and Site

Irises differ considerably in ability to survive winter temperatures, tolerance varying depending on species through zones 3 to 11. Information was provided earlier on the ecological adaptations of the many different classes. When purchasing iris cultivars, it is important that their ecological requirements be appropriate for the sites where they will be cultivated. Irises usually grow best in an open, sunny location (at least 6 hr of direct sunlight daily). However, many Siberian iris cultivars will tolerate light shade. Avoiding areas crowded with other plants is a way of increasing air circulation and lessening the humidity that promotes fungal infections. Unlike most large rhizomatous irises, Japanese irises grow well in containers.

### 55.8.2 Propagation and/or Purchasing

Irises can be established from seeds, but this will usually delay the development of flowers for 2 or 3 years, and the hybrids that now dominate garden irises will not reproduce true to type from seeds (some cultivars will not even produce viable seeds). Most irises are started in the form of rhizomes with clipped leaves and roots attached (purchased or obtained by donation or trading). These need to be planted soon, preferably within a week. Depending on how well the plants grow and become overcrowded (3–5 years, generally) or if rhizomes are frost-heaved out of the ground, new rhizomes should be cut off and used to establish new plantings or to replace old clumps that have become overcrowded. Such new rhizomes should be harvested 1–2 months after flowering (cutting away the upper two-thirds of the leaves traditionally so that the remaining fan of leaves is pointed), and also trimming the fibrous roots. It is also possible to transplant entire clumps or portions of clumps (again, trimming the foliage). Transplanting should be carried out in time to allow 6 weeks or more for roots to establish before a hard freeze. Rhizomes transplanted in late summer should produce flowers the next season. Rhizomes transplanted in the spring may flower the same season, but flower production will be limited.

### 55.8.3 Planting

Bearded iris rhizomes should be planted shallowly, the top barely below the surface (after soil settling). However, the earth surrounding the rhizome should be firmly compacted to keep the growing plant from toppling over. Beardless irises are cultured somewhat different from bearded irises. As most require higher soil moisture, planted rhizomes should be maintained moist and well-watered when planting. Allowing the young plants to become dry during early growth is a common cause of mortality. The rhizomes should be buried shallowly, but deeper than for bearded irises (at least 2.5 cm).

Depending on ultimate size, plants should be about 50 cm apart (or more, to lengthen the time necessary to replenish the bed), and several rhizomes can be planted, with the growing tips pointed in the same direction so that an area will be filled in during the next few years. Iris specialists commonly plant iris beds in patterns that anticipate even filling of the bed in the future, particularly rows and triangles (one plant at each of the three apices).

Transplanting rhizomes is commonly done in the fall (allowing at least a month before freeze-up to allow root establishment), but Siberian irises also transplant particularly well in the spring.

### 55.8.4 Soil and Moisture Conditions

Rhizomes should be established in well-draining circumneutral soil (ideally, 6.8 for bearded irises, slightly more acidic for beardless irises) which preferably has been well worked, with competing nearby grass and weeds removed. Water should not accumulate in the planting area, and raised beds and slopes can be employed to promote drainage. Most irises thrive with light fertilization, and high-nitrogen fertilizers should be avoided as they promote soft growth susceptible to stresses.

Japanese irises grow best in slightly more acidic substrates (5.0–6.5). They also grow better in richer soils than most irises and benefit from spring and post-flowering fertilization and the incorporation of organic matter into the soil. Also, rhizomes should be planted deeper (5–8 cm or 2–3 in).

As noted earlier, some irises are adapted to moist soil and thrive in shallow water. These are sometimes termed "bog irises."

### 55.8.5 Maintaining Plants

Occasionally, iris beds are "contaminated" by seeds formed on the plants that have developed as a result of hybridization with plants outside the bed, and when they grow in the bed, the resulting plants are often not as attractive. This can be prevented by removing fruits.

## 55.9 FASCINATING FACTS, FABLES, FOLKLORE, AND FRIVOLITY

- In ancient Greece, Iris was the Greek goddess of the rainbow, daughter of the ocean spirit Electra, and a messenger of the gods. Depicted with rainbow wings, she is said to have used the rainbow as a pathway through the sky, and wherever her

FIGURE 55.23 Illustrations of the mythological Greek goddess Iris.

*Source:* (a) 1904 yearbook of Ward Seminary, Nashville, Tennessee (public domain). (b) *The Iris: An Illuminated Souvenir for 1852*, published by Lippincott, Grambo & Co., Philadelphia, 1852 (public domain). (c) Prepared and published in 2021 by Randomactofflower (CC BY SA 4.0).

FIGURE 55.24 (a) The Taj Mahal. (b) Limestone wall in the Taj Mahal showing floral carvings (below) and inlays of semi-precious stones in the form of stylized flowers (top).

*Source:* (a) Photo by Yann and Jim Cartar (CC BY SA 3.0). (b) Photo by Buddhika Siddhisena (CC BY 2.0).

feet touched earth, there appeared iris flowers in a rainbow of colors.

- Shah Jahan had the world-famous Taj Mahal erected in India after the death of his favorite wife in 1631. Some of the rooms are decorated with stone inlay based on an iris motif (daffodils, narcissus, tulips, and poppies are also illustrated). There are old bearded iris cultivars named 'Shah Jahan' and 'Taj Mahal'.
- The Ojibwa carried a piece of blue flag iris as a charm against snakes. The Arizona Indians are believed to have chewed blue flag before holding rattlesnakes in their mouths during snake dances, the odor said to be protective against bites.
- A solution of the flowers of blue flag produces a blue dye that has been employed like litmus paper to test for pH.
- Adding a cut daffodil stem to a bouquet of cut iris flowering stems in water delays senescence of the iris flowers due to the presence of the compound narciclasine in daffodil mucilage. By contrast, adding a daffodil stem to other common flowers (tulips, for example) greatly hastened their senescence (Van Doorn et al. 2004).
- While most irises have nectar guides on the sepals and, indeed, produce nectar for pollinators, some Asian species lack both nectar guides and nectar. These nevertheless lure pollinators (large male bees), apparently by providing early-morning warmth. The flowers are quite dark, absorbing heat in the cool early morning, allowing the bees which shelter overnight inside the flowers to emerge earlier (Sapir et al. 2005, 2006).
- The leaves of *Iris florentina* yield iris green, a natural coloring once popular with artists. Coincidentally, irises have also been popular subjects for artists, particularly Vincent van Gogh (1853–1890).

FIGURE 55.25 Photo of one of Vincent van Gogh's several paintings titled *Irises*, currently owned by the Getty Museum in Los Angeles. This version was painted in 1889, while van Gogh was in an asylum in France. In 1987, it sold for $53,900,000, at that time the highest selling price ever obtained for a painting.

*Source:* Public domain.

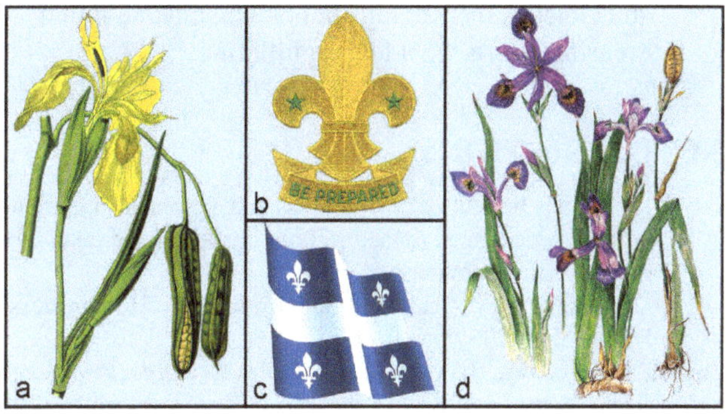


FIGURE 55.26 Symbolic (particularly French) connections of irises: (a) European yellow iris (*I. pseudacorus*), often interpreted as the basis of the fleur-de-lis. (b) The fleur-de-lis backpatch, used as the emblem of Scouts Canada up until 1977. (c) The flag of Quebec, showing four images of the fleur-de-lis and a white cross. (d) The floral emblem of Quebec, blue flag (*I. versicolor*). The two emergent aquatic species shown here are allied to the rhizomatous beardless Japanese iris group discussed in the text.

*Source:* (a) Thomé, O. W. 1903. Prof. Dr. Thomé's Flora von Deutschland, Österreich und der Schweiz. Band 1, Fig. 139. H. V. Verlag, Berlin-Lichterfelde, Germany. (b) Photo by Chris (CC BY 2.0). (c) Image courtesy of the Government of Quebec. (d) Painting by Lavonia R. Stockelbach (1874–1966), housed by the herbarium of Agriculture and Agri-Food Canada in Ottawa.

- *Iris* species are fairly widely afforded official symbolic status. *Iris versicolor* has been the floral emblem of Quebec since 1999. *Iris giganticaerulea* (giant blue iris, or giant blue flag), a native of the southeastern United States, was declared to be the official state wildflower of Louisiana in 1990. *Iris lacustris* (dwarf lake iris), a tiny plant endemic to the northern shores of Lake Michigan and Lake Huron, was designated the state wildflower of Michigan in 1998. It occurs in the Bruce Peninsula and Manitoulin Island in Ontario and has been designated a threatened species in the United States and Canada. Iris (no particular species) was recognized as the "state cultivated flower" of Tennessee in 1933. *Iris nigricans* (black iris) is the national flower of Jordan. *Iris croatica* is the national flower of Croatia. The "iris" is often claimed to be the national flower of France. Several cities have designated the iris as their official flower, including Kansas City, Brussels, and Athens.
- The fleur-de-lis, interpreted as "the iris," was the royal badge of France until the end of the 19th century, when the French Revolution dissolved the monarchy. The French revolutionaries adopted the violet as their floral symbol, and anyone found wearing an iris went to the guillotine.

## KEY PUBLICATIONS

Note: Irises may be second only to roses in the number of gardening books dedicated to them. Those cited here are relatively recent and comprehensive.

Austin, C. 2005. *Irises: A Gardener's Encyclopedia*. Timber Press, Portland, OR. 339 pp.

British Iris Society, Species Group. 1997. *A Guide to Species Irises: Their Identification and Cultivation*. Cambridge University Press, Cambridge. 371 pp.

Caillet, M. 2000. *The Louisiana Iris: The Taming of a Native American Wildflower*, 2nd edition. Timber Press, Portland, OR. 211 pp.

Goldblatt, P. and J. Manning. 2008. *The Iris Family: Natural History & Classification*. Timber Press, Portland, OR. 290 pp.

Henderson, N.C. 2002. *Iris*. In: *Flora of North America North of Mexico*, Vol. 26, edited by Flora of North America Editorial Committee, 371–395. Oxford University press, Oxford, UK.

Köhlein, F. 1987. *Iris*. Timber Press, Portland, OR. 370 pp.

Mathew, B. 1989. *The Iris*, 2nd edition. Timber Press, Portland, OR. 215 pp.

Mavrodiev, E.V., M. Martínez-Azorín, P. Dranishnikov, and M.B. Crespo. 2014. At least 23 genera instead of one: The case of *Iris* L. s.l. (Iridaceae). *PLoS ONE* 9(8): e106459. https://doi.org/10.1371/journal.pone.0106459.

McEwen, C. 1990. *The Japanese Iris*. University Press of New England, Hanover, NH. 153 pp.

McEwen, C. 1996. *The Siberian Iris*. Timber Press, Portland, OR. 206 pp.

Norris, K.D. 2012. *A Guide to Bearded Irises: Cultivating the Rainbow for Beginners and Enthusiasts*. Timber Press, Portland, OR. 348 pp.

Randolph, L.F. (ed.). 1959. *Garden Irises*. American Iris Society, St. Louis. 575 pp.

Sapir, Y., A. Schmida, and G. Ne'eman, G. 2005. Pollination of *Onocyclus* irises (*Iris*: Iridaceae) by night sheltering male bees. *Plant Biology* 7: 417–424.

Sapir, Y., A. Schmida, and G. Ne'eman, G. 2006. Morning floral heat as a reward to the pollinators of the *Onocyclus* irises. *Oecologia* 147: 53–59.

Shear, W. 1998. *The Gardener's Iris Book*. Tauten Press, Newtown, CT. 170 pp.

Stebbings, G. 1997. *The Gardener's Guide to Growing Irises*. Timber Press, Portland, OR. 160 pp.

Tillie, N., M.W. Chase, and T. Hall, T. 2000. Molecular studies in the genus *Iris* L.: A preliminary study. *Annali di Botanici n.s. (Roma)* 58: 205 112.

Van Doorn, W.G., A. Sinz, and M.M. Tomasssen. 2004. Daffodil flowers delay senescence in cut *Iris* flowers. *Phytochemistry* 65: 571–577.

Waddick, J.W. and Y. Zhao. 1992. *Iris of China: Chinese Iris in the Wild and in the Garden*. Timber Press, Portland, OR. 192 pp.

Warburton, B.A. and M. Hamblen. (eds.). 1978. *The World of Irises*. American Iris Society, Wichita, KS. 494 pp.

Wilson, C.A. 2004. Phylogeny of *Iris* based on chloroplast matK gene and trnK intron sequence data. Molecular *Phylogenetics and Evolution* 33: 402–412.

Wilson, C.A. 2009. Phylogenetic relationships among the recognized series in *Iris* section *Limniris*. *Systematic Botany* 34: 277–284.

Wilson, C.A. 2011. Subgeneric classification in *Iris* re-examined using chloroplast sequence data. *Taxon* 60: 27–35.

### Key Cultivation Websites

American Iris Society. www.irises.org/.

BBC Gardeners' World Magazine. *How to Grow Iris*. www.gardenersworld.com/how-to/grow-plants/how-to-grow-iris/.

British Iris Society. www.britishirissociety.org.uk/.

Iris in Canada. www.cdn-iris.ca/.

Pacific Bulb Society. www.pacificbulbsociety.org/pbswiki/index.php/HomePage.

The Spruce. *How to Grow and Care for Iris Flowers*. www.thespruce.com/iris-flowers-growing-guide-5120188.

The Spruce. *How to Grow and Care for Bearded Iris*. www.thespruce.com/iris-plant-profile-4589585.

WikiHow. *How to Deadhead an Iris*. www.wikihow.com/Deadhead-an-Iris.

WikiHow. *How to Divide Bearded Irises*. www.wikihow.life/Divide-Bearded-Irises.

WikiHow. *How to Get Irises to Bloom*. www.wikihow.com/Get-Irises-to-Bloom.

WikiHow. *How to Grow Bearded Irises*. www.wikihow.life/Grow-Bearded-Irises.

WikiHow. *How to Grow Irises*. www.wikihow.com/Grow-Iris#:~:text=Plant%20in%20late%20summer.,the%20best%20months%20for%20planting.

WikiHow. *How to Take Care of Irises When Their Flowers Die*. www.wikihow.com/Take-Care-of-Irises-when-Their-Flowers-Die.

# 56 Jacaranda (*Jacaranda mimosifolia*)

**Scientific Names:** The genus (and common) name *Jacaranda* originated from native (Tupi-Guarani) South American words meaning "fragrant." (The language is spoken by millions in Argentina, Bolivia, Brazil, and Paraguay.) The species term *mimosifolia*, Latin for *mimosa-leaved*, reflects similarity to the compound, fernlike leaves of species of the genus *Mimosa*.

**Common Names:** The common name jacaranda could refer to any of the other 50 or so species of the genus *Jacaranda* but normally refers to *J. mimosifolia* (which is often mislabelled *J. acutifolia*). The alternate name *blue jacaranda* also refers to this species.

## 56.1 INTRODUCTION

Jacaranda is a subtropical tree indigenous to south-central South America. The species produces a spectacular flowering display that few other plants can match. Its blooms are extraordinarily striking and attractive, and as a result, jacaranda has been planted in much of the subtropical and warm-temperate world and the highland tropics. Ironically, while the plant is in no danger of extinction because it is so widely cultivated and, indeed, has become naturalized in many countries, in its native area (Argentina, Bolivia to Brazil, and Paraguay), it is considered in need of protection because its habitat is being reduced by development.

FIGURE 56.1 Jacaranda (*Jacaranda mimosifolia*) trees: (a) In Australia. (b) A carpet of flowers underneath a tree at Sydney University, Australia. (c) In Australia. (d) Uncommon white-flowered jacaranda trees in Groenkloof, South Africa.

*Source:* **(a) Photo by John (CC BY SA 2.0). (b) Photo by Mary and Andrew (CC BY 2.0). (c) Photo by John (CC BY SA 2.0). (d) Photo by Deanerasmus2006 (CC BY SA 4.0).**

## 56.2 OVERVIEW

Jacaranda trees grow to a height of up to 20 m and can cover at least as wide an area. The large leaves (up to 45 cm long) are compound (with many leaflets about 1 cm long). The fernlike foliage is basically deciduous, although in some places, the old leaves are retained in winter (most trees drop their leaves after fruits appear, and they become partly dormant). The new leaves appear before or while the flowers open. The flowers are up to 5 cm long, occurring in dense flowering clusters in the spring and early summer.

FIGURE 56.2 Paintings of jacaranda (*Jacaranda mimosifolia*): (a) *Under the Jacaranda*, by R. Godfrey Rivers (1858–1925), English artist active in Australia. (b) Vintage Florida postcard. (c) Vintage (1930s) Florida postcard printed on linen (note diffuse texture).

*Source:* **(a) Photo of public domain painting in Queensland Art Gallery by Joyofmuseums (CC BY SA 4.0). (b) Photo by Dave (CC BY ND 2.0). (c) Public domain.**

FIGURE 56.3 Technical 19th-century botanical illustrations of the flower clusters of jacaranda (*Jacaranda mimosifolia*).

*Source:* **Public domain. (a) *Curtis's Botanical Magazine*, 1822, vol. 49, plate 2327. (b) *Revue horticole*, series 4, 1897, Librairie agricole de la maison rustique, Paris. (c) Paxton, J., 1844, *Magazine of Botany and Register of Flowering Plants*, vol. 11.**

DOI: 10.1201/9781003586227-56

FIGURE 56.4 Foliage of jacaranda (*Jacaranda mimosifolia*).

*Source:* **Photo by Crusier (CC BY 3.0).**

FIGURE 56.5 Fruit and seeds of jacaranda (*Jacaranda mimosifolia*): (a) Fruit capsules hanging on a tree. (b) Seeds.

*Source:* **(a) Photo by Philmarin (CC BY SA 3.0). (b) Photo by Blumenbiene (CC BY 2.0).**

The lightly fragrant, trumpet-shaped flowers are almost always blue-purple, lavender, or lilac-blue, although a white-flowered form is available. The blooms last up to 2 months and transform into woody seedpods 3 to 8 cm in diameter. The pods contain many flat, winged seeds. When the pod dries, it splits open, releasing the papery, winged seeds, which are blown away by the wind.

## 56.3 HISTORY

Following European colonization of South America, jacaranda was imported to numerous countries throughout the world. It is so well-established in areas like Florida, Australia, and South Africa that local inhabitants often assume it is indigenous there. Some cultivars have been selected, but for the most part, the striking blue-purple appearance of the flowers has been maintained.

## 56.4 ECONOMIC IMPORTANCE

The main value of jacaranda is as an ornamental tree, widely grown in urban areas throughout the world.

## 56.5 ORNAMENTAL USES

Jacaranda is primarily grown as a specimen tree around buildings and as a street and park tree.

The arching branches are ideal for creating streets covered by a canopy. Dwarf varieties can be grown in pots, kept outdoors during the summer, and moved under cover over winter in a cool, frost-free place. Jacaranda is often grown as bonsai or at least as potted plants maintained indoors. The tough, unusually shaped pods are employed in dried flower arrangements and as Christmas tree ornaments.

## 56.6 OTHER USES

Jacaranda, like numerous other plants, is employed for some herbal medicinal uses and is currently the subject

FIGURE 56.6 Container-grown jacaranda (*Jacaranda mimosifolia*). Such saplings can be planted in the ground, transferred to larger containers, or employed as patio and houseplants.

*Source:* **Photo by Blumenbiene (CC BY 2.0).**

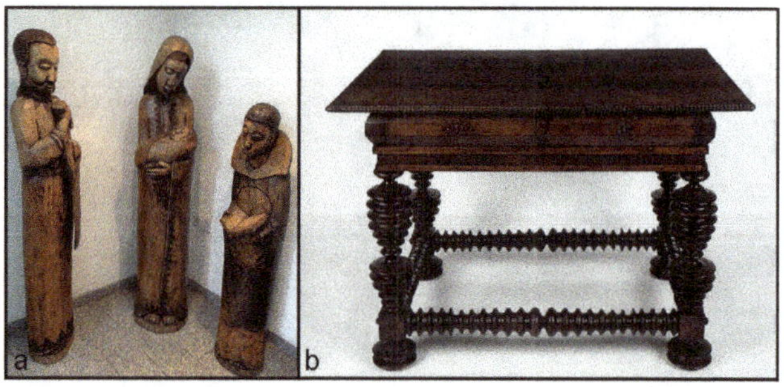


FIGURE 56.7 Exquisite jacaranda wood carvings: (a) Three sculpted figures (an unknown man, Mary holding Jesus, and Joseph) by Indigenous African artist John Chikerema. Originally from Zimbabwe, now in St. Martin-in-the-Fields Church in London. (b) Elaborately carved table.

*Source:* **(a) Photo by Andy Scott (CC BY SA 4.0). (b) José Rosael/Hélio Nobre/Museu Paulista da USP, Brazil (CC BY SA 4.0).**

of research to examine modern potential medical applications. "Jacaranda wood" is employed in fine furniture and to make various carved objects, such as bowls and objets d'art. However, identifying exotic woods can be difficult because the timber industry often employs names that are different from standard botanical nomenclature. "Jacaranda" wood can be based on *Dalbergia nigra*, and wood from *Jacaranda* species has sometimes been called South American rosewood and tulipwood. Regrettably, the wide use of jacaranda species for wood contributed to the devastation of their indigenous habitats.

## 56.7 PROBLEMS

Jacaranda is considered an invasive, weedy species in parts of Africa, Australia, Hawaii, and Chile. It can outcompete many native plants, often simply by shading them out. Although most people are entertained by the carpet of fallen flowers under jacarandas, some dislike cleaning them away. The flowers can be sticky and difficult to wash away (recommended, as they may turn into slime, the slipperiness representing potential lawsuits).

## 56.8 CULTIVATION

### 56.8.1 Climate and Site

Jacarandas are drought- and heat-tolerant, adapted to frost-free climates, especially climate zones 9 to 11, although established trees tolerate brief spells of freezing, down to –7°C. Trees that die back from frost may regrow from the roots but rarely produce attractive trees. The plants grow best in full sun. In the United States, the species is cultivated in the Southern states, including coastal California and Florida (but maritime winds can inhibit flowering, and the tree is intolerant of salt spray). Jacaranda trees should not be planted where its falling accumulating flowers (and leaves) will be a problem (for example, near a swimming pool, where the water filter could be clogged).

FIGURE 56.8 Dogs in carpets of fallen jacaranda flowers: (a) Cavalier King Charles spaniel. (b) Lhasa apso.

*Source:* **(a) Photo by Shutterstock/Rian Scheepers. (b) Photo by John (CC BY SA 2.0).**

### 56.8.2 Propagation and/or Purchasing

Jacaranda cultivars can be propagated by grafting and softwood cuttings. The trees can also be grown from seeds, although they take many years to reach blooming size and will not reproduce clonal forms well. Most people will purchase saplings established in containers, intended for transplanting.

### 56.8.3 Planting

Trees planted directly into the ground should be located at least 5 m from other trees and from buildings. Alternatively, young trees can be transplanted into large containers.

### 56.8.4 Soil and Moisture Conditions

The species grows best in well-drained, sandy or loamy, neutral or slightly acidic soil.

### 56.8.5 Maintaining Plants

Pruning of branches, unless dead, should be avoided, as this may promote additional branching. Jacarandas sometimes develop multiple trunks, and most horticulturalists recommend removing all but one, which usually results in a desirable, umbrella-shaped habit.

## 56.9 FASCINATING FACTS, FABLES, FOLKLORE, AND FRIVOLITY

- In the Amazon of South America, jacaranda is associated with the goddess of the moon. Folktales tell of her flying to Earth, landing on top of a jacaranda tree, and sharing divine knowledge with the common people.
- Jacaranda is often planted on the campuses of schools and universities, giving rise to the belief that the tree is associated with wisdom.
- In Queensland, Australia, jacaranda is in flower in late spring and early summer, when students are under stress to complete assignments and prepare for exams. As a result, students refer to the time as "purple panic." Similarly, at the University of Pretoria in South Africa, the flowering jacaranda is known as the exam tree. A flower falling on a student's head has been interpreted as a warning that they should start studying hard or as an omen (either good or bad).
- The jacaranda is the official urban tree of the city of San Diego, California.
- Many cities have claimed that they are a "jacaranda capital" (of their country or of the world), including Brisbane and Grafton, Australia; Tauranga, New

Zealand; Pretoria, South Africa; Lisbon, Portugal; and Joburg, Pretoria, and Robertson, South Africa.

- The color of jacaranda flowers is often debated – is it blue, purple, lavender, lilac, or indigo? The color simply illustrates biological and environmental variation, depending on the specimen, soil, and climate.

## KEY PUBLICATIONS

Brett, M.R. 2021. Would you like your carving in mahogany or jacaranda? Woodcarvers of Nyalazi State Forest. *Journal of African Films and Diaspora Studies* 4(107). https://journals.co.za/doi/abs/10.31920/2516-2713/2021/4n3a6.

Gachet, M.S. and W. Schühly. 2009. Jacaranda – An ethnopharmacological and phytochemical review. *Journal of Ethnopharmacology* 12: 14–27.

Gilman, E.F. and D.G. Watson. 1993. *Jacaranda mimosifolia*, Jacaranda. U.S. Forest Service Fact Sheet Fact Sheet ST-317.

Henderson, L. 1990. Jacaranda. *Farming in South Africa: Weeds A* 30: 9–10.

Martin, S.K. 2023. *How the South American Jacaranda became a Symbol of Australian Spring*. Australian Geographic (blog). www.australiangeographic.com.au/travel/2023/11/the-way-of-the-water/.

Mostafa, N.M., O.A. Eldahshan, and A.N.B. Singab. 2014. The genus *Jacaranda* (Bignoniaceae): An updated review. *Pharmacognosy Communications* 4: 31–39.

Olowoyo, J.O., E. van Heerden, and J.L. Fischer. 20101. Investigating *Jacaranda mimosifolia* tree as biomonitor of atmospheric trace metals. *Environmental Monitoring and Assessment* 164: 435–443.

Pasiecznik, N. 2022. *Jacaranda mimosifolia* (jacaranda) Invasive Species Compendium. www.cabi.org/isc/datasheet/29212.

Serra, M.B., W.A. Barroso, F.F.S. Filho, S. do Nascimento Silva, A.C.R. Borges, M.B. Campos, et al. 2020. Pharmacological evidence from plants of genus *Jacaranda*. *Biomedical Journal of Scientific & Technical Research* 28: 21730–21734.

Thakur, S.S. 2006. Jacaranda – an excellent source of nectar for African honey bees in Ethiopia. *Insect Environment* 12(2): 75–77.

van Vollenhoven, A.C. 2020. The cultural historical significance of Pretoria's jacaranda trees. *New Contree* 85: 60–85.

Wang, M., L. Zhang, and Z. Wang. 2021. Chromosomal-level reference genome of the neotropical tree *Jacaranda mimosifolia* D. Don. *Genome Biology and Evolution* 13(6): evab094.

## KEY CULTIVATION WEBSITES

GardenersWorld.com. *How to Grow a Jacaranda Tree*. www.gardenersworld.com/how-to/grow-plants/how-to-grow-a-jacaranda-tree/.

The Spruce. *How to Grow and Care for a Jacaranda Tree*. www.thespruce.com/growing-jacaranda-mimosifolia-3269356.

University of Florida, Extension. *Jacaranda mimosifolia*. https://citeseerx.ist.psu.edu/document?repid=rep1&type=pdf&doi=2440f7d85c7459029279214efe1937a790c96206.

WikiHow. *How to Grow a Jacaranda Tree*. www.wikihow.com/Grow-a-Jacaranda-Tree#References.

# 57 Jasmine (*Jasminum* Species and Hybrids)

**Scientific Names:** The genus name *Jasminum* and the corresponding English name *jasmine* (as well as the variant *jessamine*) traces to the Arabic and Persian *yasmin* or *yasaman*, which probably originally referred to Arabian jasmine (*J. sambac*).

**Common Names:** Numerous unrelated species have *jasmine* as part of at least one of their names, such as Brazilian jasmine (*Mandevilla sanderi*), Cape jasmine (*Gardenia*), Carolina jasmine (*Gelsemium sempervirens*), crape jasmine (*Tabernaemontana divaricata*), Chilean jasmine (*Mandevilla laxa*), Madagascar jasmine (*Stephanotis floribunda*), New Zealand jasmine (*Parsonsia capsularis*), night-blooming jasmine (*Cestrum nocturnum*), night-flowering jasmine (*Nyctanthes arbor-tristis*), orange jasmine (*Murraya paniculata*), red jasmine (*Plumeria rubra*), star/Confederate jasmine (*Trachelospermum*), and tree jasmine (several genera). For more information on the wide variety of plants called jasmine, see Green (1967) and Weaver and Anderson (2012).

## 57.1 INTRODUCTION

Jasmines are members of the genus *Jasminum*, which is made up of about 200 species of shrubs and vines (both trailing and climbing) native to tropical and warm temperate regions of Eurasia, Africa, and Oceania. Depending on species, cultivar, and climate, the foliage is evergreen, semi-evergreen, or deciduous. In frost-free areas, most cultivars are evergreen. The flowers are usually white or yellow, not much larger than 2.5 cm in diameter. Although their flowering display can be impressive, the chief claim to fame of jasmines is that the flowers of many of the species are exquisitely fragrant. The three species featured here are often encountered as ornamentals, but many others are cultivated in semi-tropical and tropical regions.

## 57.2 OVERVIEW

The most important species is *J. officinale*, known as common jasmine, poet's jasmine, summer jasmine, true jasmine, white jasmine, and jessamine. The foliage is deciduous, semi-evergreen, or evergreen, depending on climate. The plant is a twining vine or shrub (best described as a semi-vining shrub), with clusters of small (up to 2.5 cm in diameter) white very fragrant flowers that bloom all summer. As a vine, the plant can grow to a height of over 5 m (sometimes as high as 9 m) and needs pruning to be grown as a shrub. The plant is hardy northward to climate zone 7.

FIGURE 57.1 Common jasmine (*Jasminum officinale*): (a) Jasmine vines climbing up street-side trees in Switzerland. (b) Painting of the plant. (c) Flowering branches hanging over a fence.

*Source:* **(a) Photo by Chris.urs-o (CC BY SA 3.0). (b) Figure prepared by AJC1 (CC BY 2.0). (c) Photo by Jim Evans (CC BY SA 4.0).**

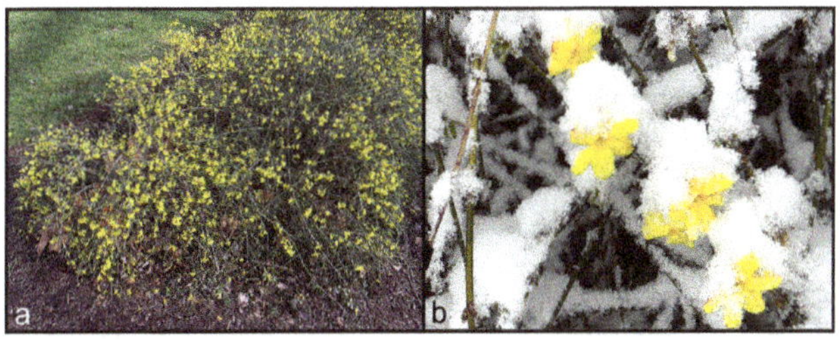


FIGURE 57.2 Winter jasmine (*Jasminum nudiflorum*). (a) Shrub in flower in Oak Hill, Virginia, United States, on December 15, 2015. (b) The same plant in a snowstorm on March 12, 2022.

*Source:* **Photos by Famartin (CC BY SA 4.0).**

Winter jasmine (*J. nudiflorum*) is widely planted in temperate regions as an ornamental deciduous, viny shrub. It can grow to 5 m on a trellis, much less as a shrub, and makes a good cover for slopes, spreading by arching branches that root where they touch the soil. With winter protection, it sometimes survives as far north as hardiness zone 6 and is considered the most cold-hardy of the ornamental jasmines. Winter jasmine produces yellow unscented flowers in the spring, before the foliage unfolds, the flowers sometimes poking through a covering of snow. However, unlike forsythia shrubs, well-known for their flush of yellow flowers in the very early spring, when other shrubs are bare, winter jasmine flowers appear more sporadically over a longer interval.

Arabian jasmine (*J. sambac*), a native of southern and eastern Asia, is widely cultivated around the world. This evergreen shrub grows as tall as 3 m, although it is usually 1 to 1.5 m in height. It is winter-hardy in climate zones 9 to 12. Some of the numerous cultivars produce fully double flowers resembling small gardenias or roses. The flowers are intensely fragrant and are very extensively used (especially dried) to flavor jasmine tea.

DOI: 10.1201/9781003586227-57

FIGURE 57.3 Arabian jasmine (*Jasminum sambac*) cultivars with highly doubled flowers resembling gardenias and roses.

*Source:* **(a) Photo by Shutterstock/Praphaporn Chakaew. (b) Photo by Jayachandranjay (CC BY SA 4.0). (c) Flowers collected for providing fragrance and for flavoring tea. Photo by Shutterstock/MRS Siwaporn.**

## 57.3 HISTORY

*Jasminum officinale* has been cultivated since ancient times in Asia. It was introduced into Europe in the Middle Ages, and by the 16th century, cultivars with doubled flowers were available. Winter jasmine was introduced to the West from China in 1844, when the English plant collector Robert Fortune purchased plants in Shanghai and sent them to the Royal Horticulture Society. *Jasminum sambac* (Arabian jasmine) and several other species of the genus were cultivated in early times in the gardens of Arabia and Persia. The Moors may have introduced Arabian jasmine to Spain during the 1600s.

## 57.4 ECONOMIC IMPORTANCE

Jasmines are relatively minor ornamental shrubs and vines, but thanks to the alluring floral fragrance of many of the species, they are widely planted and extraordinarily respected in many cultures. Jasmines are also significant semi-tropical and tropical crops, grown for flower buds employed for culinary flavoring, and for extracts used for perfumery, aromatherapy, and sometimes medicines.

## 57.5 ORNAMENTAL USES

Aromatic jasmines are often planted near a house (especially bedrooms) or walkway so their fragrance can be enjoyed, and it is easy to watch hummingbirds and butterflies visiting

FIGURE 57.4 Old paintings (public domain) of jasmine (*Jasminum*) species by botanical illustrators: (a) Italian jasmine (*J. humile*). (b) Royal/Spanish jasmine (*J. grandiflorum*). (c) Downy/star jasmine (*J. multiflorum*). (d) Winter jasmine (*J. nudiflorum*). (e) Common/poet's jasmine (*J. officinale*). (f) Arabian jasmine (*J. sambac*). (g) Lemon-scented jasmine (*J. azoricum*). (h) South African/wild/angular jasmine (*J. angulare*).

*Source:* **(a) Denisse, E. 1843–1846. *Flore d'Amérique*, vol. 107. (b) Redouté, P. J. 1833. *Choix des plus belles fleurs*. (c) Lemaire, C. (ed). 1841. *Herbier général de l'amateur* (Paris), 2nd series, vol. 2. (d) Van Houtte, L. (ed). 1852–1853. *Flore des serres et des jardins de l'Europe*, vol. 8. (e) Step, E., D. Bois, B. Herincq, and W. Watson. 1897. *Favourite flowers of garden and greenhouse*, vol. 3. Frederick Warne, New York. (f) Blanco, F. M. 1837. *Flora de Filipinas*. Manila. (g) Sydenham, E., and J. Ridgway. 1815. *The Botanical Register* (London), vol. 1. (h) Curtis, W. 1886. *Botanical Magazine*, vol. 112.**

FIGURE 57.5 Great art showing jasmine: (a) Painting, by French-born British artist Sophie Gengembre Anderson (1823–1903), titled *No Walk Today* showing a little girl staring at the bad weather outside. A jasmine vine is growing outside of her window. (b) Painting of a vase of jasmine by Viennese artist Georg Daumer (1845–ca. 1911). (c) A flowering jasmine branch painted on silk, attributed to the early 12th-century Chinese artist Zhao Chang, now located in the Shanghai Museum in China.

*Source:* **Public domain paintings.**

the flowers. Depending on cultivar, the plants make good hedges, groundcovers, and houseplants. Jasmine flowers are widely employed in garlands and leis, especially in religious rituals and marriages in Asia. The flowering branches

FIGURE 57.6 Commercial production and retail sale of jasmine: (a) Workers handpicking flowers in India. (b) Another jasmine field in India. (c) Potted plants of Arabian jasmine (*Jasminum sambac*) on sale. (d) Flower shop in Hong Kong selling bouquets of Arabian jasmine.

***Source:* (a) Photo byMcKay Savage (CC BY 2.0). (b) Photo by Shutterstock/Porjai Kittawornrat. (c) Photo by Shutterstock/Amornie. (d) Photo by Darifengm 092 Shtiah (CC BY SA 4.0).**

are employed as cut flowers. Jasmine flowers are sometimes expensive in northern regions, where they need to be shipped.

## 57.6 OTHER USES

Jasmine is cultivated commercially for flowers employed as a flavoring for tea, rice, and other food preparations, and extracts (especially essential oils) for the perfume industry. Among perfumers, jasmine is known as the king of flowers, with rose being the queen. "Jasmine flower tea" is Chinese tea flavored with jasmine flowers (preferably mature but unopened flower buds), usually from Arabian jasmine (*J. sambac*). Sometimes, tea connoisseurs buy jasmine flowers from florists so they can prepare jasmine tea using fresh rather than dried flowers.

## 57.7 PROBLEMS

Several species of *Jasminum* are invasive weeds in subtropical and tropical regions.

## 57.8 CULTIVATION

### 57.8.1 Climate and Site

Most *Jasminum* species require semi-tropical, warm climates. The climate zone limits of the three species discussed in detail were presented earlier. Jasmines grow well in full sun to partial shade. Twining jasmines require supports on an arbor or trellis. When they are grown in containers, a heavy support should be provided. Vine forms can benefit from being sheltered from exposure to wind.

FIGURE 57.7 Landscape uses of jasmine (*Jasminum* species): (a) A blooming jasmine vine climbing very high on a building in Bordeaux, France. (b) White jasmine (*J. polyanthum*) in a pot. (c) A jasmine-covered wall. (d) Flowering hedges lining the driveway entrance to an estate. (e) Fragrant hedge lining a stairway.

***Source:* Photos from Shutterstock, contributed by (a) Redzen2, (b) Totokzww, (c) ErreCh, (d) Fotografiche, and (e) Iryna Bozhenkova.**

FIGURE 57.8 Jasmine garlands in India: (a) A woman wearing a garland in her hair. (b) A woman selling garlands. (c) Garlands for sale on a street-side table.

***Source:* (a) Photo by Shutterstock/Sivaleela-V. (b) Photo by PlaneMad (CC BY SA 3.0). (c) Photo by Parvathisri (CC BY SA 3.0).**

### 57.8.2 Propagation and/or Purchasing

Although jasmines can be started from seeds and cuttings, in most cases, potted saplings are purchased. Many cultivars

FIGURE 57.9 Use of jasmine to flavor tea: (a) Basket of jasmine blossoms collected in China for combining with green tea. (b) A commercial container of jasmine tea. (c) Jasmine flower tea served in Shanghai.

*Source:* **(a) Photo by Daqve Dahl (public domain). (b) Photo by National Museum of Korea (CC BY 2.0). (c) Photo by Wesley Fryer (CC BY SA 2.0).**

FIGURE 57.10 A monster ambushing a jasmine flower-picker.

*Source:* **Prepared by H. J. Ford. Colorized by B. Brookes. From Lang, A. 1900. *The Grey Fairy Book*. Green, London (public domain).**

will not breed true from seed. In northern climates, where the plants need to be grown indoors, at least much of the year, potted plants may be available.

### 57.8.3 Planting

Plants should be separated by at least 2.5 m when grown as a shrub border.

### 57.8.4 Soil and Moisture Conditions

Jasmines grow best in well-drained, humus-enriched garden soil with moderate levels of soil fertility and moisture and an optimal pH of 5.5 to 6.0.

### 57.8.5 Maintaining Plants

Jasmine shrubs are often semi-vining and may need frequent pruning to keep them from looking sloppy. Some cultivars are sometimes grown as groundcover, requiring pruning of upward-growing stems.

## 57.9 FASCINATING FACTS, FABLES, FOLKLORE, AND FRIVOLITY

- Picking beautiful flowers of plants so that their beauty can be appreciated at one's home is a natural behavior, but sometimes it's a bad idea. Henry Justice Ford (1860–1941), a British illustrator, prepared a frightening illustration showing a lurking monster waiting to ambush anyone daring to pick jasmine flowers.
- Jasmine (or Jasmin) is a common female given name, often considered to be rather exotic. Perhaps the best-known Jasmine today is Princess Jasmine, the fictional character who played the role of girlfriend in the 1992 Walt Disney animated film *Aladdin*. She is loosely based on Princess Badroulbadour from the Aladdin fairy tale in the ancient *One Thousand and One Nights* collection of stories. The original name Badroulbadour had nothing to do with jasmine but is related to "full moon," an Arabic metaphor for extraordinary female beauty.
- Two independent historical events have been called "Jasmine Revolutions": a change in presidency in Tunisia in 1987, and a revolution in Tunisia in 2011.
- Because it is in great demand for luxury perfumes, a kilogram of jasmine oil can cost $5,000.

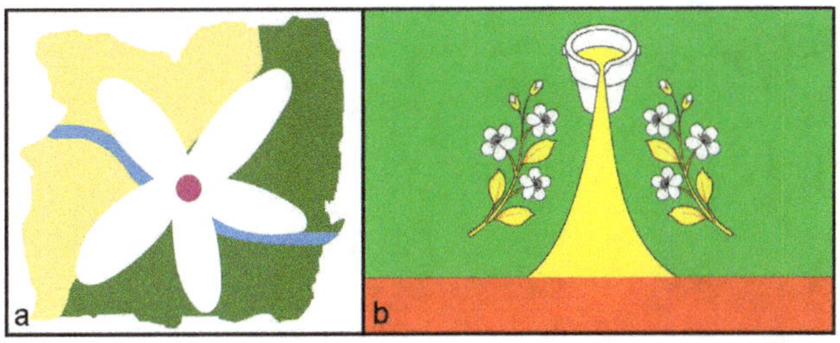


FIGURE 57.11 Regional flags featuring jasmine: (a) City flag of Damascus, Syria, with a stylized jasmine flower. (b) Flag of the Lvovsky Municipality of the Moscow Oblast (province) of Russia. It shows a silver bowl pouring out molten gold, accompanied on the sides by golden branches of jasmine with silver flowers and buds.

***Source:*** **(a) Prepared by Siirski (CC BY SA 4.0). (b) Union of Heraldists of Russia (public domain).**

In India, harvesters need to handpick over 5,000 buds to make just 1 kg.

- Countries that consider jasmine as the national flower include Indonesia, Pakistan, Philippines, and Tunisia.
- Damascus, the capital city of Syria, is called the "City of Jasmine" because of the widespread cultivation of the plant there for many generations. In pre-war times, a jasmine festival was held every April.

## KEY PUBLICATIONS

Al-Snafi, A.E. 2018. Pharmacology and medicinal properties of *Jasminum officinale* – A review. Indo American *Journal of Pharmaceutical Sciences* 05: 2191–2197.

Green, P.S. 1967. The name jasmine. *Arnoldia* 25: 71–74.

Green, P.S. and D. Miller. (eds.) 2009. *The Genus* Jasminum *in Cultivation*. University of Chicago Press, Chicago. 150 pp.

Hetman, J. 2016. A dandy for winter: *Jasminum nudiflorum*. *Arnoldia* 73: 44.

Lim, T.K. 2014. *Jasminum sambac*. In: *Edible Medicinal and Non Medicinal Plants*, edited by T.K. Lim, 529–540. Springer, Dordrecht.

Rojas-Sandoval, J. 2016. *Jasminum sambac* (Arabian jasmine). *CABI Compendium*. https://doi.org/10.1079/cabicompendium.28388.

Sahu, R., V, Arya, R. Joshi, P. Kaushik, and M. Chauhan. 2022. Pharmacological and therapeutic Properties of *Jasminum officinale*. L: A review. *Indian Journal of Ecology* 49.1122–1128.

Usha, S., M. Ganga, K. Rajamani, S. Manonmani, and R. Gnanam. 2022. Reproductive biology analyses in jasmine (*Jasminum* spp.). *International Journal of Plant & Soil Science* 34(19): 149–160.

Weaver, R.E. and P.J. Anderson. 2012. *Jasmines, a Diversity of Plants with Fragrant Flowers*. Florida Department of Agriculture and Consumer Services, Botany Circular No. 38.6 pp.

## Key Cultivation Websites

GardenersWorld.com. *How to Grow Jasmine*. www.gardenersworld.com/how-to/grow-plants/how-to-grow-jasmine/.

Horticulture Magazine. *Jasmine*. https://horticulture.co.uk/jasmine/.

Plantin (blog). *Most Popular Types of Jasmine Flowers*. https://myplantin.com/blog/most-popular-types-of-jasmine-flowers.

The Spruce. *How to Grow and Care for Vining Jasmine*. www.thespruce.com/grow-jasmine-indoors-1902621.

WikiHow. *How to Grow Jasmine*. www.wikihow.com/Grow-Jasmine.

WikiHow. *How to Grow Jasmine from Cuttings*. www.wikihow.com/Grow-Jasmine-from-Cuttings.

WikiHow. *How to Grow Jasmine in a Pot*. www.wikihow.com/Grow-Jasmine-in-a-Pot.

WikiHow. *How to Make Jasmine Tea*. www.wikihow.com/Make-Jasmine-Tea.

WikiHow. *How to Prune Jasmine*. www.wikihow.com/Prune-Jasmine.

# 58 Kalanchoe (*Kalanchoe* Species and Hybrids)

**Scientific Names:** The genus name *Kalanchoe* is a Latinization of the Chinese *Kalan Cauhuy*, "that which falls and grows," a reference to the plantlets produced on leaf margins which fall off and grow on the ground. As a scientific name, *Kalanchoë* with a diaeresis (dieresis, diëresis) is incorrect, although occasionally encountered. In the past, *Kalanchoe* was separated from the genera *Bryophyllum* and *Kitchingia*, but they are now united.

**Common Names:** Kalanchoe. It is generally recommended that the word be pronounced with four syllables, accenting the third syllable and pronouncing the *ch* as *k*, that is, kal-un-KOE-ee. Several other names are applied to the cultivated species, as presented later. A group of very attractive double-flowered cultivars under the trademarked name "calandiva" is currently widely marketed (calandiva is kalanchoe, although sold as a different kind of plant). The name "mother-of-thousands" is applied to some species of kalanchoe, but also to *Saxifraga stolonifera*, which also generates daughter plantlets attached to the mother plant. "Mother-of-millions" refers to *K. delagoensis*.

## 58.1 INTRODUCTION

Kalanchoes are popular ornamental houseplants most widely encountered as small potted specimens, like African violets. These beautiful succulents are relatively hardy, require limited water, are easily propagated and maintained, and are available in an increasing variety of cultivars with very attractive floral and foliage displays. Kalanchoes are one of the most important indoor potted plants produced commercially. They are very popular gift plants and can last for years. The species frequently exhibit "vivipary": the vegetative (non-sexual) production of embryos (young plantlets) that remain attached to the mother plant for a period. (In animals, *vivipary* has a different meaning: "live birth," as opposed to hatching external to the mother from an egg.)

FIGURE 58.1 *Kalanchoe blossfeldiana* cultivars with different flower colors.

*Source:* Photo by Quadell (CC BY SA 3.0).

## 58.2 OVERVIEW

There are about 140 species of *Kalanchoe*, although some authorities recognize as many as 200. They are tropical natives of South and East Africa, Arabia, and Southeast Asia. Most are indigenous to Madagascar and tropical Africa. A few have been introduced and grow wild in the New World. Most are shrubs or perennial herbs, but several are annual or biennial. Kalanchoes are leafy succulents, producing clusters of small, usually brightly colored flowers – frequently yellow, orange, red, pink, or magenta. Most species are less than 1 m in height, but the largest, *K. beharensis* of Madagascar, reaches 6 m.

Cultivated kalanchoes are mostly hybrids of *K. blossfeldiana* and are mainly grown as flowering houseplants. Many are compact and will grow well in very small containers, even in teacups, but modern varieties have also been bred that branch and spread extensively. Hybridization

FIGURE 58.2 *Kalanchoe beharensis*, the largest species of the genus. This tree is growing in the city nursery in Ellfenau Park, Bern, Switzerland.

*Source:* Photo by Sandstein (CC BY 3.0).



DOI: 10.1201/9781003586227-58

has generated cultivars with attractive foliage and a range of flower colors. Cultivated varieties usually have fleshy, thick, succulent leaves with scalloped margins and clusters of small flowers with four petals (most other members of the Crassulaceae family have flowers with five petals). The blooms are available in many colors, most popular of which are red, yellow, pink, orange, and white. Some cultivars have bicolored flowers, and some have doubled, larger flowers (especially the calandiva selections). Some cultivars are grown specifically for their attractive, often variegated foliage. Several new cultivars are asexually propagated. Dozens of species of *Kalanchoe* are cultivated in specialty collections, but only a few have been domesticated. The most widely grown are mentioned in the following.

*Kalanchoe blossfeldiana* (florist kalanchoe, Christmas kalanchoe, flaming Katy, Madagascar widow's thrill), the most widely grown species of *Kalanchoe*, is named for Robert Blossfeld (1882–1945), a German nurseryman who

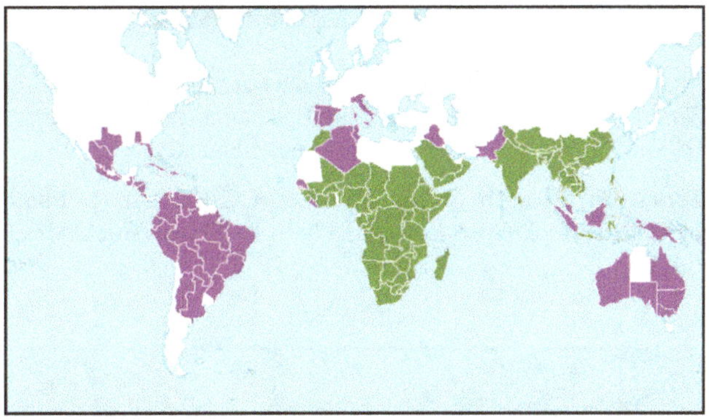

FIGURE 58.3 Geographical distribution of the genus *Kalanchoe*. The original (native) range is shown in green; the introduced areas are shown in purple.

*Source:* **Board of Trustees of the Royal Botanic Gardens, Kew (CC BY SA 4.0).**

FIGURE 58.4 *Kalanchoe blossfeldiana* cultivars developing flowers with just four petals (typical of the species).

*Source:* **(a) Photo by HitroMilanese (CC BY SA 3.0). (b) Photo by Bodie Pennisi, University of Georgia, Bugwood.org (CC BY 3.0).**

FIGURE 58.5 *Kalanchoe blossfeldiana* cultivars: (a) Cultivars with flowers with slightly more than the normal four petals. (b) Cultivars with doubled flowers (i.e., with many petals).

*Source:* **(a) Photo by Britchi Mirela (CC BY SA 4.0). (b) Photo by Shutterstock/9Wolachode.**

FIGURE 58.6 Calandiva® series of double-flowered cultivars of *Kalanchoe blossfeldiana*.

*Source:* **(a) Photo by Jonathan Cardy (CC BY SA 3.0). (b) Photo by JC7001 (CC BY SA 2.0).**

FIGURE 58.7 *Kalanchoe pinnata*: (a) A well-developed flowering plant. (b) Plantlets developed on the margin of a leaf.

*Source:* **(a) Photo by SKsiddhartthan (CC BY SA 4.0). (b) Photo by Navaho (Stefan Walkowski) (CC BY SA 3.0).**

FIGURE 58.8 *Kalanchoe daigremontiana*: (a) Flowering plants. (b) Vegetative plant with bluish plantlets growing on leaf margins.

*Source:* **(a) Photo by Chic Bee (CC BY 2.0). (b) Photo by Gmihail (CC BY SA 3.0).**

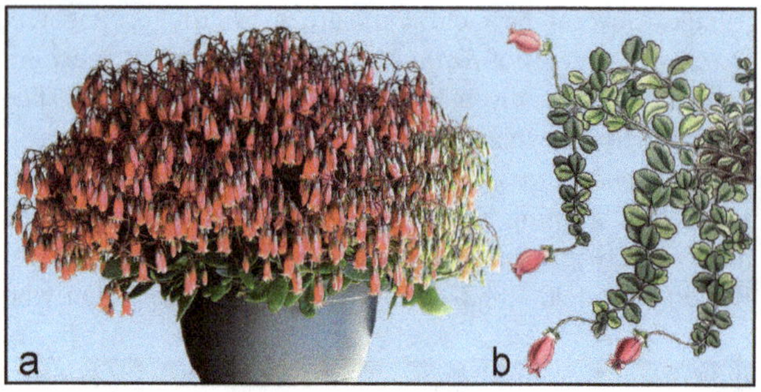


FIGURE 58.9 *Kalanchoe uniflora.*

*Source:* (a) Photo by Shutterstock/Dorothy Chiron. (b) Illustration from *Curtis's Botanical Magazine* (London), vol. 135 (= ser. 4, vol. 5), 1909 (public domain).

introduced it to Europe in 1932 from its native Madagascar. The wild species is a succulent herb or subshrub which can develop woody tissues with age.

*Kalanchoe pinnata* (air plant, Mexican love plant, cathedral bells) is another native of Madagascar. It is often grown as a curiosity, because a leaf cut from the plant is able to produce new plantlets along the margin.

*Kalanchoe daigremontiana* (mother-of-thousands, Mexican hat plant, Devil's backbone, alligator plant) is another curiosity, producing numerous plantlets on the leaf edges, which will develop into new plants if they land in soil when they fall off. This native of Madagascar has lost the ability to reproduce sexually: It cannot generate viable seeds.

*Kalanchoe uniflora* is a trailing plant ideal for hanging baskets. 'Freedom Bells' is a popular cultivar.

## 58.3 HISTORY

After German botanist Robert Blossfeld discovered *K. blossfeldiana* in Madagascar and brought it to Germany in 1932, the popularity of the species expanded greatly. Breeding, especially in the 1960s and 1970s, generated colorful varieties that led to kalanchoe becoming one of the world's most popular houseplants.

## 58.4 ECONOMIC IMPORTANCE

There have been some efforts to develop a cut flower industry based on kalanchoe, but this is currently very limited. The plants are popular with florists, in part because they can be forced to come into bloom at any time of the year and so are available at peak selling seasons, such as holidays (the name "Christmas kalanchoe" reflects the plant's popularity at this season).

## 58.5 ORNAMENTAL USES

Kalanchoe is grown primarily as a potted indoor plant in temperate regions but is also employed in garden beds in tropical and subtropical areas.

FIGURE 58.10 Commercial production and marketing of *Kalanchoe blossfeldiana*: (a) Potted plants on sale in a market. (b) Growth facility. (c) Florist with a potted plant.

*Source:* (a) Photo by Richard Parker (CC BY 2.0). (b) Photo by Shutterstock/Grafvision. (c) Photo by Shutterstock/Africa Studio.

FIGURE 58.11 Potted houseplants of *Kalanchoe blossfeldiana.*

*Source:* (a) Photo by Leonora (Ellie) Enking (CC BY SA 2.0). (b, c) Photos by Vinicius Nunes (public domain).

FIGURE 58.12 Landscape uses of kalanchoe.

*Source:* (a) Photo by Shutterstock/Topimages. (b) Photo by Shutterstock/Fotoaddicta. (c) Photo by Shutterstock/Anney Lier. (d) Photo by Shutterstock/Max_555.

## 58.6 OTHER USES

Kalanchoe has been extensively employed in Indigenous herbal medicine in tropical regions and is being examined for modern pharmacological applications.

## 58.7 PROBLEMS

All parts of kalanchoes are poisonous, especially the flowers. Several species are known for the presence of toxic cardiac glycosides known as bufadienolides (these are of the same class as the cardiotoxins secreted by the skin of toads of the genus *Bufo*). Sheep and cattle have been poisoned by consuming introduced species in South Africa and Australia. Cats, birds, iguanas, and especially dogs have also been poisoned, and it has been recommended that the plants be kept at a safe distance from household pets and young children. Common gastrointestinal symptoms include vomiting, bloody diarrhea, and cramping, indicating that these are definitely not edible plants. Despite the toxicity, as noted earlier, several kalanchoes are employed both internally and externally as traditional herbal medicines in regions where they are indigenous, and there is considerable research underway to examine potential applications as modern pharmaceuticals.

## 58.8 CULTIVATION

### 58.8.1 Climate and Site

Kalanchoes originate predominantly from tropical climates, and most species require frost-free conditions to grow outside. In regions where day temperatures fall below 12°C, it is recommended that kalanchoe be grown as an indoor plant. In quite warm climates (zones 10 to 12), kalanchoes are grown outdoors, especially in rock gardens, often in semi-shade in tropical regions. While primarily a houseplant, kalanchoe is also a popular balcony or patio plant for summer use, tolerant of both sun and shade. Growing the plants in hanging baskets allows them to be placed outdoors but quickly brought inside when the temperature becomes too low.

### 58.8.2 Propagation and/or Purchasing

Some cultivars are grown from seed, but most are reproduced from terminal (tip) stem or petiole cuttings. Home growers can harvest stem-tip cuttings 5–7.5 cm long, removing any flowers. Preferably, use portions of vegetative stems with two pairs of leaves, denuding leaves from the bottom part that is planted. The cuttings should be placed in a porous germination medium like a mixture of peat and vermiculite, and a high humidity should be maintained (for example, as developed in a plastic bag). A rooting hormone is unnecessary. After the cuttings root in 2 to 3 weeks, they can be transplanted to a potting soil.

Although *Kalanchoe* species are easily propagated from leaf and stem cuttings, many produce plantlets (miniature plants; more technically known as bulbils or gemmae) in the notches of the leaf margins, sometimes so prolifically that the plants easily escape to the wild in hot climates and can even become pests inside a greenhouse.

Acquiring a plant as a gift usually means that it is in full bloom or at least many of the flowers are mature. Similarly, when buying a plant to give as a gift, many of the flowers will be attractively open. Since gift pots are often purchased in the winter, care should be taken that the plant is protected from cold stress, as kalanchoes are damaged at temperatures below 12°C. When obtaining a plant for personal use, it is preferable to buy one when the flowers are all in bud but showing some color (so that the blooming period in the new location will be long).

FIGURE 58.13 Plantlets of *Kalanchoe daigremontiana* being established.

*Source:* **Photos by Maja Dumat (CC BY 2.0).**

### 58.8.3 Planting

In tropical regions, larger kalanchoe shrubs and trees are spaced according to their size. Information regarding the indoor species described in this chapter is provided in the following.

### 58.8.4 Soil and Moisture Conditions

In tropical regions, kalanchoe is grown outdoors in well-drained, humus-rich, moderately fertile, circumneutral or slightly acidic soils. Information regarding potted indoor plants is provided next.

### 58.8.5 Maintaining Indoor Potted Plants

Houseplant kalanchoes should be given as much light as possible, or they will become spindly. Plants that have become too tall and stretched should be cut back to about half their height and provided with better growth conditions. Pinching the stem tips encourages branching and the production of a compact form. The leaves can be brittle, so plants should be handled carefully. Like most succulents, kalanchoes do well in dry conditions and thrive in a well drained growing medium, such as specifically formulated for succulents (preparations for cacti will suffice but are less suitable). Recommended soils are slightly acidic or neutral. The substrate is best allowed to become slightly dry before watering. Overwatering can result in crown rot and the death of the plant. Fertilization with a houseplant fertilizer (such as diluted 20–20–20, or a preparation higher in P to promote flowering) is best done when the plant is growing actively, applied at 2- to 4-week intervals.

Night temperatures of about 15°C and day temperatures about 10°C higher are recommended. Temperatures exceeding 24°C tend to delay or reduce flowering.

Kalanchoes bloom for weeks, often several months, but eventually lose their attractive floral display, which should be cut away ("deadheaded") when senesced to encourage natural reblooming. In the past, kalanchoes were usually regarded only as flowering gift plants, to be discarded after the flowers died. However, many modern cultivars can bloom repeatedly indoors, at any time of the year.

After the initial blooming, kalanchoes may not subsequently produce an impressive flowering display unless rejuvenated. Like poinsettias, kalanchoes are short-day plants, requiring a period of long nights to induce flower development. Poinsettias are usually tossed out after blooming, and too many people do the same thing for kalanchoes, which, however, are much easier to rejuvenate thanks to their small size. It is generally recommended that they be provided with about 6 weeks of alternating daily periods (about 10 hr) of good light and uninterrupted periods (about 14 hr) of total darkness, starting in October. (Some authorities state a shorter period will suffice.) A cool closet with a tightly closed door is suitable for the dark period. The plant may be repotted into a slightly larger container. In cold Northern climates, the naturally occurring daylight regime (long hours of light in the summer followed by short daylight hours in the winter) could result in excellent flowering of kalanchoes in midwinter, but it would be necessary to ensure that the rooms in which the plants are kept are not exposed to light during the nights.

## 58.9 FASCINATING FACTS, FABLES, FOLKLORE, AND FRIVOLITY

- In China, red-flowered plants are popular, because red is considered to be the color of joy, symbolizing virtue, truth, and sincerity and having the power to scare away bad fortune. Red-flowered kalanchoes are especially popular and are commonly purchased during the Chinese Lunar Calendar New Year. The plants are known as *wan zi qian hong* (literally "thousands of red flowers") and are thought to bring wealth and prosperity. Western flower merchants have assigned the symbolic meanings "endurance" and "lasting affection" to kalanchoe, presumably because the plants are long-lasting and attractive.
- Kalanchoe was one of the first plants to voyage in space, transported to the Soviet Salyut 1 space station in 1971. A mature plant was sent to comfort the cosmonauts, who were experiencing stress (www.airspacemag.com/space/growing-pains-4148507/). Cultivation of vegetables in the microgravity of space, as a source of food for interplanetary travel, has been a primary research area, but several ornamental plants have also been grown in space, reflecting the basic human admiration of beautiful flowers.
- Kalanchoes open their flowers by growing new cells on the inner surface of the petals, forcing them outwards, and similarly close the flowers by cell division on the outside of the petals.
- Kalanchoe is notable for an extraordinary system of photosynthesis called crassulacean acid metabolism (CAM). Three metabolic systems of photosynthesis are commonly recognized. *The C3 pathway*, employed by about 85% of plants, was named for the first molecule produced in the cycle, the 3-carbon 3-phosphoglyceric acid. The presence of oxygen reduces the efficiency of uptake of $CO_2$ by about 25%. The carbon fixation pathways used by C4 and CAM plants, noted next, are more efficient than the C3 pathway, because their chloroplasts function in an oxygen-reduced environment, but this is achieved by investment in special adaptations. These alternative photosynthetic mechanisms have each evolved independently in over two dozen groups of plants. *The C4 pathway*, employed by about 3% of plants, is named for 4-carbon molecules produced during the process. Carbon dioxide from the atmosphere is chemically fixed and transported to special cells surrounding the veins. These "bundle sheath cells," which are a characteristic feature of the "Kranz" leaf anatomy of C4 plants, are low in oxygen and so increase the efficiency of $CO_2$ utilization. C4 plants are common in hot habitats. *The CAM pathway* occurs in about 7% of plants. CAM plants are not only adapted to hot environments like C4 plants but are also adapted to dry environments by succulence (not all succulents are CAM plants, and CAM has been observed in non-succulent organs). Succulents have thick, fleshy leaves and stems that retain moisture, along with adaptations to reduce evaporation through the epidermis. Terrestrial plants usually take up $CO_2$ through closable pores called stomata (= *stomates*; singular: *stoma*, *stomate*). CAM plants keep their stomata closed during the day to prevent water loss by transpiration. Instead, the stomata are opened at night to take in $CO_2$. Like C4 plants, they transform $CO_2$ into a chemically fixed form (malate). But unlike C4 plants, the fixed carbon is not quickly transformed into carbohydrates but is stored until daylight, when the stomates are closed and once again an oxygen-poor environment exists that promotes more efficient photosynthesis. The advantage of CAM is that temperatures are generally lower at night, when stomates are open, so water loss is reduced. Some plants are flexible in the photosynthetic pathways they use, and this is especially true of *Kalanchoe*. Light intensity, water availability, temperature, and stage of reproduction have all been found to be influential. *Kalanchoe* species under stress

FIGURE 58.14 A garden in San Diego, California, dedicated to desert plants tolerant of heat and drought. Several of the species shown here, including *Kalanchoe luciae*, are succulents adapted to hot, dry conditions by Crassulacean acid metabolism photosynthesis.

*Source:* **Photo by Cultivar 413 (CC BY 2.0).**

tend to shift from conventional C3 metabolism to the CAM pathway, which is doubtless adaptive for the wild species, most of which occur in dry environments. However, the tight closure of stomates during the day, characteristic of CAM plants, is sometimes poorly developed in *Kalanchoe*, particularly in cultivars.

## KEY PUBLICATIONS


Akulova-Barlow, Z. 2009. Kalanchoes: Beginner's delight, collector's dream. *Cactus Succulent Journal* 81: 268–276.

Assis de Andrade, E., I. Machinski, A.C.T. Ventura, S.A. Barr, A.V. Pereira, F.L. Beltrame, et al. 2023. A review of the popular uses, anatomical, chemical, and biological aspects of Kalanchoe (Crassulaceae): A genus of plants known as "miracle leaf". *Molecules* 28(14): 5574. https://doi.org/10.3390/molecules28145574.

Boiteau, P. and L. Allorge-Boiteau. 1995. Kalanchoe *(Crassulacées) de Madagascar: Systématique, Écophysiologie et Phytochimie*. Editions Karthala, Paris. 252 pp.

Descoings, B., 2003. Kalanchoe. In: *Illustrated Handbook of Succulent Plants, Crassulaceae*, edited by U. Eggli, 143–181. Springer-Verlag, Berlin.

Gehrig, H., O. Gaußmann, H. Marx, D. Schwarzott, and M. Kluge. 2001. Molecular phylogeny of the genus *Kalanchoe* (Crassulaceae) inferred from nucleotide sequences of the ITS-1 and ITS-2 regions. *Plant Science* 160: 827–835.

Guerra-García, A., J. Golubov, and M.C. Mandujano. 2015. Invasion of *Kalanchoe* by clonal spread. *Biological Invasions* 17: 1615–1622.

Kahraman, M.U., Y.Y. Mendi, Ş. Karabıyık, H.V. Lütken, and B.T. Favero. 2022. Kalanchoe breeding: Past, present and future. *Ornamental Horticulture* 28: 19–35.

Kawade, R.M., N.B. Ghiware, M.H. Ghante, S.M. Malwatkar, S.M. Vadvalkar, A.K. Dhadwe, et al. 2014. A review on pharmacognostical, phytochemical and pharmacological potentials of *Kalanchoe pinnata* (Crassulaceae). *America Journal of PharmTech Research* 4(1): 1–15.

Kolodziejczyk-Czepas, J. and A. Stochmal. 2017. Bufadienolides of *Kalanchoe* species: An overview of chemical structure, biological activity and prospects for pharmacological use. *Phytochemistry Reviews* 16: 1155–1171.

Love, J.W. 1976. Kalanchoe production. *North Carolina Flower Growers Bulletin* 20(2): 1–3. https://hortscans.ces.ncsu.edu/library/all/doc_id/3518/.

Milad, R., S. El-Ahmady, and A.N. Singab. 2014. Genus *Kalanchoe* (Crassulaceae): A review of its ethnomedicinal, botanical, chemical and pharmacological properties. *Journal of Medicinal Plants* 4: 86–104.

Pertuit, A.J., Jr. 1992. Kalanchoe. In: *Introduction to Floriculture*, 2nd edition, edited by R.A. Larson, 429–450. Academic Press, New York.

Raadts, E. 1977. The genus *Kalanchoe* (Crassulaceae) in Tropical East Africa. *Willdenowia* 8: 101–157.

Smith, G.F. and E. Figueiredo. 2018. The infrageneric classification and nomenclature of *Kalanchoe* Adans. (Crassulaceae), with special reference to the southern African species. *Bradleya* 36: 162–172.

Smith, G.F., E. Figueiredo, and A. van Wyk. 2019. Kalanchoe *(Crassulaceae) in Southern Africa: Classification, Biology, and Cultivation*. Academic Press. 338 pp.

Smith, G.F. and R. Shtein. 2022. A Review of horticulturally desirable characters in *Kalanchoe* (Crassulaceae subfam. Kalanchooideae): Variable and deviating vegetative and reproductive morphologies useful in breeding programmes. *Haseltonia* 28: 106–119.

Voorst, A. van and J.C. Arends. 1982. The origin and chromosome numbers of cultivars of *Kalanchoe blossfeldiana* Von Poelln.: Their history and evolution. *Euphytica* 31: 573–584.


### Key Cultivation Websites


Clemson University, Extension. *Kalanchoe*. https://hgic.clemson.edu/factsheet/kalanchoe/.

Horticulture Magazine. *Kalanchoe*. https://horticulture.co.uk/kalanchoe/.

Houseplants UK. *Kalanchoe*. www.ukhouseplants.com/plants/kalanchoe-1.

Kalanchoe Culture Guide. https://siteadmin.dummenorange.com/site/documents/7737/kalanchoe-culture-guide-engels-march-2022.pdf.

Kalanchoe Garden. www.kalanchoe.nl/en/.

Smith, G. 2004. *Kalanchoe* species poisoning in pets. *Veterinary Medicine* (November): 933–936. www.aspcapro.org/sites/default/files/v-vetm1104_933-936.pdf.

Stein, G. 2012. Introduction to kalanchoes. *Dave's Garden*. https://davesgarden.com/guides/articles/view/2673.

The Kalanchoe Page. http://succulent-plant.com/families/crassulaceae/kalanchoe.html.

The Spruce. *How to Grow and Care for Kalanchoe*. www.thespruce.com/growing-kalanchoe-plants-1902982.

WikiHow. *How to Grow Kalanchoe*. www.wikihow.com/Grow-Kalanchoe#:~:text=Kalanchoe%20need%20plenty%20of%20sunlight,indirect%20sunlight%20in%20the%20summer.

WikiHow. *How to Prune Kalanchoe*. www.wikihow.com/Prune-Kalanchoe.

# 59 Lantana (*Lantana* Species and Hybrids)

**Scientific Names:** The genus name *Lantana* originates from the Latin name of the wayfaring tree (*Viburnum lantana*), a hardy Old World shrub or small tree often grown as an ornamental in temperate regions. Its flowers are displayed in similar clusters but do not develop the amazing range of colors within the flower heads of lantana.

**Common Names:** Lantana plants are sometimes called *verbena bush* or *shrub verbena* (*Lantana* and *Verbena* are both in the Verbenaceae family and look somewhat similar). *Lantana camera*, the most important ornamental species of the genus, is called Spanish flag, West Indian lantana, and lava. These common names are also applied to ornamental cultivars, which are usually hybrids of *L. camera* and other *Lantana* species. Names with "sage" (*big sage*, *Jamaica mountain sage*, *yellow sage*, *white sage*, *wild sage*) are sometimes encountered but are more often employed for different plants. Trailing lantana (*L. montevidensis*, also known as trailing shrubverbena, creeping lantana, purple lantana, small lantana, and weeping lantana) is also frequently cultivated. The name "ham-and-egg plant" is based on forms with pink (ham-like) and yellow (egg-like) flower colors. The name "convertible rose" reflects the changing flower colors.

## 59.1 INTRODUCTION

Lantanas are flowering shrubs, frequently grown as annuals, which often produce small flowers in heads, each of which features a rainbow of hues. The color effect is stunning and contributes to the unique charm of the plant. Less charming are the facts that the plant is ranked among the world's worst environmental weeds and is quite poisonous.

FIGURE 59.1 Flowering displays of common lantana (*Lantana camera* and hybrids).

*Source:* (a) Photo by Hans from Pixabay (public domain). (b) Photo by Ryan Vaarsi (CC BY 2.0). (c) Photo by Megan Zopf from Pixabay (public domain). (d) Photo by Dick Thompson Sandian (CC BY SA 2.0). (e) Photo by Josch13 from Pixabay (public domain). (f) Photo by Natasha de Vere & Col Ford (CC BY 2.0).

## 59.2 OVERVIEW

Lantanas include over 150 species of woody, evergreen perennials native to Africa and the tropical Americas. Some of the species and cultivars are capable of growing well over 3 m (even over 5 m) in hot climates, but typically not over 1 m in height when grown as annuals. The foliage is abrasive (due to small rough hairs) and usually has a disagreeable odor when rubbed or crushed. The angular stems may have small spines (frequent in wild material, rare in cultivated varieties). Lantanas produce stalked, rounded (spherical), or flat-topped clusters (2 to 5 cm in diameter) of small, tubular, brightly colored flowers (small

FIGURE 59.2 Creeping lantana (*Lantana montevidensis*).

*Source:* (a, b) Photos by Forest and Kim Starr (CC BY 3.0). (c) Photo by Mathew Townsend (CC BY SA 3.0).



DOI: 10.1201/9781003586227-59

flowers are often called florets). Some cultivars produce flowers of a single, uniform color, but in many varieties, there are flowers of different colors, including two or three of blue, lavender, orange, pink, purple, red, yellow, and/or white. The flowers may have a slight fragrance. In varieties with flowers of multiple colors, the blooms change color as they mature. The newest flowers, opening in the center of the umbel, are one color, but the shade changes as the flowers mature and move to the outer edges of the cluster. The flowers tend to be a light color when they open, and darken with age. The multiple hues of many cultivars provide a spectacular kaleidoscope of colors, with blooming lasting for much of the season (almost year-round in hot climates). Pollinated flowers develop into fleshy, berry-like green fruits that become bluish black and, in many popular selections, contain one or two seeds.

## 59.3 HISTORY

Lantana was imported to Europe in the 1700s from Brazil and became a popular ornamental there, with hundreds of cultivars bred through the 19th and 20th centuries. Most cultivated lantanas trace to *L. camera*. Related species (such as *L. depressa*, *L. hirsuta*, *L. horrida*, *L. splendens*, and *L. strigocamara*) are also grown, and there are many cultivars of hybrid origin. The taxonomic identification of both cultivars and invasive populations is often difficult to determine because of hybridization.

FIGURE 59.3 Paintings (public domain) by botanical illustrators of old cultivars of common lantana (*Lantana camera* and hybrids).

***Source:*** **(a) Lithograph by French artist Charles Lemaire (1800–1871). (b) Moninckx, J. 1682–1701. *Moninckx Atlas*, vol. 2. (c) Regel, E. A. von. 1865. Gartenflora, vol. 14. (d) *The Floral Magazine*, 1867, vol. 6. (e) Step, E., D. Bois, B. Herincq, and W. Watson. 1897. *Favourite flowers of garden and greenhouse*, vol. 3. Frederick Warne, New York. (f) *Revue horticole*, vol. 81, 1909.**

FIGURE 59.4 Common lantana (*Lantana camera*) in flower pots for sale in Thailand.

***Source:*** **Photo by Shutterstock/Joloei.**

## 59.4 ECONOMIC IMPORTANCE

Lantana is one of the world's leading ornamental flowering shrubs, important mainly in subtropical and tropical regions. Nevertheless, the ecological and economic damage caused by invasive lantana greatly outweighs its value as an ornamental. Although it is an extremely harmful weed, efforts are underway in many countries to find ways of employing the plant as a source of useful materials.

## 59.5 ORNAMENTAL USES

In subtropical and tropical areas, many lantana cultivars can grow into large shrubs, useful as isolated specimens and hedges, while short, sprawling forms can be employed as groundcover.

Trailing lantana is often used in hot climates as a groundcover for erosion control and is planted on top of retaining walls so that the stems hang down gracefully. Also known as purple lantana, this species is a good choice for erosion control on slopes. In cold climates, lantana plants are usually grown as annuals, often in containers, especially in hanging baskets (employing spreading cultivars). Plants in pots are sometimes utilized as houseplants (although they usually do not thrive in the winter because of limited light) or overwintered indoors in a dormant condition in a cool place.

## 59.6 OTHER USES

Lantana is one of the world's most widely employed flowering plants to attract flying wildlife to gardens, especially butterflies, bees, and hummingbirds.

In parts of the world, notably India, the stems of *L. camera* can become significantly thick (up to 15 cm in diameter) and furnish wood that is very resistant to decay and is useful for furniture and handicrafts.

FIGURE 59.5 Landscape usage of common lantana (*Lantana camera*): (a) A hedge. (b) Lantana bushes beside Santa Catalina Monastery at Arequipa in the Peruvian Andes.

***Source:*** **(a) Photo by Shutterstock/Alybaba. (b) Photo by Shutterstock/Franck Camhi.**

FIGURE 59.6 Common lantana (*Lantana camera*) in containers.

***Source:*** **(a) Photo by Shutterstock/Nuk2013. (b) Bonsai. Photo by ShutterstockOnO Gallery01. (c) Bonsai. Photo by This-feministrox (CC BY 2.0). (d) Photo by Shutterstock/Manfred Ruckszio.**

## 59.7 PROBLEMS

All parts of lantana are toxic. The plants should not be positioned where small children, pets, and livestock can

FIGURE 59.7 Pollinators attracted to lantana flowers: (a) Antillean crested hummingbird (*Orthorhyncus cristatus*) feeding in Dominica. (b) Leopard lacewing (*Cethosia cyane*). (c) Honeybee.

***Source:*** **(a) Photo by Charles J. Sharp (CC BY 3.0). (b) Photo by Quartl (CC BY SA 3.0). (c) Photo by John Box (CC BY SA 2.0).**

FIGURE 59.8 Invasive weedy lantana infestations: (a) Carpeting a seaside hill in Hawaii. (b) In Australia.

***Source:*** **(a) Photo by Forest and Kim Starr, Starr Environmental, Bugwood.org (CC BY 3.0). (b) Photo by Harry Rose (CC BY2.0).**

contact them. The unripe berries are particularly poisonous (fatalities have been recorded), and contact with the foliage can cause irritation in sensitive individuals. The internet presents a lively debate over just how poisonous the ripe berries are, with many anecdotal statements indicating that the berries have been consumed without much, if any, ill effects. Lantana causes liver failure and even death in livestock animals, such as cattle, sheep, goats, and horses. Some researchers have claimed that the toxic potential for livestock has been exaggerated, but there are many reports of serious poisoning.

Introduced *Lantana* species have become invasive weeds in subtropical and tropical areas of the world. Large plants can produce more than 10,000 seeds, which are easily dispersed by birds and other animals. Some species of *Lantana* are rare and endangered, and hybridization with introduced species (including the popular *L. camera*) further endangers their genetic survival.

## 59.8 CULTIVATION

### 59.8.1 Climate and Site

Lantanas are mostly tropical plants and tolerate high heat. Depending on cultivar, they can be grown as a perennial from climate zones 8 to 11, some cultivars rated for the milder areas in zone 7. In colder areas, the plant is cultivated as an annual. Lantana is not frost-tolerant but can survive a brief, mild freeze (a hard freeze will kill the roots, preventing survival). The plants should be grown in full sun, but in very hot regions, some afternoon shade may be beneficial. Lantanas can exhibit an odor that some find objectionable, and sometimes the recommendation is made to avoid planting it beside walkways.

### 59.8.2 Propagation and/or Purchasing

In some very warm regions where lantana is invasive, it (or at least most cultivars) cannot be purchased (this should not apply to sterile varieties – these produce few, if any, seeds). The plants can be grown from seed, started well in advance of the planting season in cold climates, but leaving this task to plant nurseries is simpler, and in any event, there is very little deliberate reproduction of cultivated plants by seeds. Most cultivars won't breed true to type and are reproduced by cuttings or division of root balls. Most gardeners buy lantanas as bedding plants from garden centers.

### 59.8.3 Planting

When planted permanently (as perennials) outdoors employing starter nursery stock, the standard recommendations for planting shrubs and trees should be followed. Single plants (of the kinds that form mounds) when planted in beds should be spaced 30 cm apart (although, often, plants are separated 60 cm, especially in warmer areas, where they will grow more vigorously). A bed of plants can be established with a spacing of at least 25 cm apart in rows, the rows at least 35 cm apart. Plants of the sprawling kind (notably *L. montevidensis*) can spread over a diameter of 1 m and may be given more space. Large shrub-type cultivars should also be afforded more space. In colder climates, containers are most often employed, and their size will determine how large the plants grow.

### 59.8.4 Soil and Moisture Conditions

Lantana is adapted to well-drained (usually sandy or loamy), slightly acidic soil, ideally with pH of 6.0 to 6.5 (5.0 to 8.0 is tolerated). The plants withstand both salt spray and high humidity and so are popular near seashores. Limited addition of fertilization is recommended, because too much can decrease flower production. While high soil quality is not essential, adding compost to outdoor soil is desirable, and plants grown in containers should be provided with commercial soil mixes, not simply earth from one's garden.

### 59.8.5 Maintaining Plants

When grown as a perennial, pruning is recommended to keep the plant in check and promote branching, which, in turn, encourages flowering. In areas where lantana is winter-hardy but dies to the ground (climate zones 7 and 8), in the early spring, pruning the remaining stems back to about

FIGURE 59.9 Indian elephants fashioned by covering steel scaffolds with lantana wood. The creations have been employed on behalf of biodiversity conservation campaigns.

*Source:* **Photo by London Road (CC BY 2.0).**

20 cm above the soil is recommended (and old, dead growth should be cleared away). Plants are sometimes grown as standards (tree form), requiring careful pruning so only one bare stem persists. To prevent seed distribution, the berries should be removed before they mature (or flower clusters should be pruned away when they are spent). In warm climates, lantanas typically live for 2 to 5 years, sometimes as long as 10 years. Using pesticides on lantanas can have the unfortunate effect of harming the many pollinators attracted to the flowers.

## 59.9 FASCINATING FACTS, FABLES, FOLKLORE, AND FRIVOLITY

- A "herd" of wooden Indian elephants sculpted by Indigenous people in India from lantana wood has toured Europe as part of fundraising efforts to conserve biodiversity (https://vikalpsangam.org/article/south-indias-famous-lantana-elephants-to-raise-their-trunks-for-conservation-funds/).
- Lantana is a town of about 12,000 in Florida, part of the Miami metropolitan area. It was named for the many lantana plants growing wild there. Under the auspices of the tabloid magazine the *National Enquirer*, which moved there in 1971, what was advertised as "the world's largest Christmas tree" was exhibited annually until 1988.
- Some male *Ploceus* species (weaverbirds) employ colorful lantana flowers to decorate their nests to attract females.

## KEY PUBLICATIONS


Barrows, E.M. 1976. Nectar robbing and pollination of *Lantana camara* (Verbenaceae). *Biotropica* 8: 132–135.

Binggeli, P. 2003. Verbenaceae, *Lantana camara*. In: *The Natural History of Madagascar*, edited by S.M. Goodman and J.P. Benstead, 415–417. Princeton University Press, Princeton, NJ.

Chaudhary, N., 2011. Lantana – A menace or friend? *Science Reporter* (November). https://nopr.niscpr.res.in/bitstream/123456789/13019/1/SR%2048(11)%2054-57.pdf.

Howard, R.A. 1969. A check list of cultivar names used in the genus *Lantana*. *Arnoldia* 29: 73–109.

Howard, R.A. 1970. *Lantana camara* – a prize and a peril. *American Horticultural Magazine* 49: 31–36.

Lu-Irving, P., A.M. Bedoya, F.R.G. Salimena, T.R. dos Santos Silva, L.F. Viccini, C. Bitencourt, et al. 2021. Phylogeny of *Lantana*, *Lippia*, and related genera (Lantaneae: Verbenaceae). *American Journal of Botany* 108: 1354–1373.

Maschinski, J., E. Sirkin, and J. Fant. 2010. Using genetic and morphological analysis to distinguish endangered taxa from their hybrids with the cultivated exotic pest plant *Lantana strigocamara* (syn: *Lantana camara*). *Conservation Genetics* 11: 1607–1621.

Mohan Ram, H.Y. and G. Mathur. 1984. Flower colour changes in *Lantana camara*. *Journal of Experimental Botany* 35: 1656–1662.

Negi, G.C.S., S. Sharma, S.C. Vishvakarma, S.S. Samant, R.K. Maikhuri, R.C. Prasad, et al. 2019. Ecology and use of *Lantana camara* in India. *The Botanical Review* 85: 109–130.

Patel, S. 2011. A weed with multiple utility: *Lantana camara*. *Reviews in Environmental Science and Bio/Technology* 10: 341–351.

Sanders, R.W. 2006. Taxonomy of *Lantana* sect. *Lantana* (Verbenaceae): I. Correct application of *Lantana camara* and associated names. Sida, *Contributions to Botany* 22: 381–421.

Sanders, R.W. 2012. Taxonomy of *Lantana* sect. *Lantana* (Verbenaceae): II. Taxonomic revision. *Journal of the Botanical Research Institute of Texas* 6: 403–441.

Sharma, G.P., A.S. Raghubanshi, and J.S. Singh. 2005. Lantana invasion: An overview. *Weed Biology and Management* 5: 157–165.

Sharma, O.P., H.P.S. Makkar, and R.K. Dawra. 1988. A review of the noxious plant *Lantana camara*. *Toxicon* 26: 975–987.

Sinha, S. and A. Sharma. 1985. *Lantana camara* L. – A review. *Feddes Repertorium* 95: 621–633.

Thomas, S.E. *Lantana camara* (lantana). CABI Compendium. https://doi.org/10.1079/cabicompendium.29771.

## Key Cultivation Websites

Clemson University Home & Garden Information Centre. *Lantana*. https://hgic.clemson.edu/factsheet/lantana/.

The Spruce. *How to Grow and Care for Lantana*. www.thespruce.com/add-a-florida-touch-with-lantana-plants-2132142.

WikiHow. *How to Fertilize Lantana*. www.wikihow.life/Fertilize-Lantana.

WikiHow. *How to Grow Lantana Plants*. www.wikihow.com/Grow-Lantana-Plants.

WikiHow. *How to Prune Lantana*. www.wikihow.com/Prune-Lantana.

WikiHow. *How to Winterize Your Lantana Plant*. www.wikihow.life/Winterize-Your-Lantana-Plant.

Wisconsin Horticulture, Division of Extension. *Lantana*. https://hort.extension.wisc.edu/articles/lantana/.

# 60 Lavender (*Lavandula* Species and Hybrids)

**Scientific Names:** The genus name *Lavandula* originates from the Latin *lavo*, "I wash," in reference to the frequent use of lavender water, made from oil of lavender, and used as a fragrant wash.

**Common Names:** The common name lavender has the same origin as the genus name mentioned earlier. The name is employed in many names unrelated to *Lavandula*. Some examples follow. Desert lavender is *Hyptis emoryi*, a cultivated ornamental that originated in the southwestern United States and Mexico. Sea lavender (or lavender thrift) includes many of the ornamental species of *Limonium*. Lavender cotton is the strong-scented ornamental shrub *Santolina chamaecyparissus* originating from the Mediterranean region. Lavender thyme is *Thymus thracicus*. Lavender Gem is a citrus fruit, the product of a grapefruit–tangelo cross, available only in specialty stores; this should not be confused with Imperial Gem, a cultivar of English lavender. The name Spanish lavender, noted in this chapter, is also employed for the fragrant geranium species *Pelargonium cucullatum*.

FIGURE 60.1 Illustrations of the principal lavender (*Lavandula*) species: (a) *Lavandula angustifolia* (English lavender). (b) *Lavandula latifolia* (spike lavender). (c) *Lavandula stoechas* (Spanish lavender, French lavender). (d) *Lavandula dentata* (Spanish lavender, fringed lavender).

***Source:*** **(a) Köhler, F. E. 1887. Köhler's Medizinal-Pflanzen. Gera-Untermhaus, F. E. Köhler. (b) Du Monceau, H.-L. D. 1801–1919. *Traité des arbres et arbustes que l'on cultive en France en pleine terre.* H. L. Guerin et L. F. Delatour, Paris. (c) Hortus Romanus Juxta Systems Tournefortianum Paulo, 1772 (https://digitalcollections.nypl.org/items/510d47dd-cb5d-a3d9-e040-e00a18064a99). (d) Curtis, W. 1798. *The Botanical Magazine*, vol. 12.**

FIGURE 60.2 White-flowered cultivars of lavender: (a) *Lavandula angustifolia* 'Rosea', a whitish variety. (b) White and purple varieties of lavender.

***Source:*** **(a) Photo by Nova (CC BY SA 4.0). (b) Photo by Saffron Blaze (CC BY SA 3.0).**

## 60.1 INTRODUCTION

True lavenders are species of the genus *Lavandula*. The approximately 40 species are aromatic shrubs and perennial (sometimes annual) herbs found from the Atlantic islands and Mediterranean region to northern tropical Africa, western Asia, Arabia, and India. About half of the species are commonly grown as ornamentals. In warm climates, some of the species are almost evergreen. Lavender is of principal importance as a source of essential oil.

## 60.2 OVERVIEW

Many of the lavender species are quite similar, and garden guides sometimes inadequately distinguish them. The names French lavender, Spanish lavender, and Dutch lavender are often unreliable but usually refer to one of the species highlighted here. The principal species employed as ornamentals are mentioned in the following.

English lavender (*L. angustifolia*), widely grown in gardens, is the chief species grown for essential oil, employed in cosmetics and many other manufactured products. This native of the Mediterranean region is a much-branched, pubescent shrub growing 0.3–2 m in height. The bluish-green leaves are narrow, often whitish when young, giving a silvery or silvery-gray cast to the plant. The tiny flowers are typically mauve or gray-blue, growing in whorls of 6–10 flowers along spiky inflorescences. Cultivars have been selected which have purple, pink, and white flowers. Flowering occurs in midsummer to early fall. English lavender is an intensely and very pleasantly perfumed plant.

Spike lavender (*L. latifolia*, also called sweet lavender) is much like English lavender, but the leaves are more pubescent and less green. It is commonly grown as an ornamental. Essential oil from its flowers is used in the manufacture of lacquers, varnishes, and cheaper grades of perfume

DOI: 10.1201/9781003586227-60

and sometimes employed to flavor salads and jellies, but its oil is inferior to that from English lavender. Hybrids of English lavender and spike lavender occur naturally in the Mediterranean region but have also been widely produced in cultivation. Known as lavandin (and sometimes as Dutch lavender), these hybrids are very vigorous and easy to grow, and because they are very productive of essential oil of reasonable quality, they are widely grown for this purpose.

*Lavandula stoechas*, known as Spanish lavender, Italian lavender, and French lavender, is cultivated extensively as an ornamental and for extraction of low-grade essential oil. It is often confused with *L. dentata*, also widely known as French lavender (although like *L. stoechas* it occurs wild predominantly in Spain) and is also cultivated both as an ornamental and for essential oil. *Lavandula dentata* is sometimes called fringed lavender because of its fernlike foliage with deeply indented edges.

## 60.3 HISTORY

Lavenders were well-known to the ancient Greeks and Romans (precisely which species is debatable). The Romans and Carthaginians put lavender in their bathwater for the fragrance and apparent therapeutic properties. Lavenders were cultivated in Elizabethan England, 1568 commonly cited as the earliest known date. Lavender was a very popular strewing and cosmetic herb since at least the 12th century. It was believed to have disinfectant qualities and was liberally used to mask household smells and reduce disease. Lavender was also thought to be antiseptic, for example, being used to treat snakebite. The herb was widely used medicinally, and when the patient died, it was frequently used for embalming. Lavender was brought to North America by the Pilgrim Fathers. In recent times, the lavenders have been extensively bred as ornamentals and for yield of essential oil, especially in Europe. Although lavender is not currently considered to be of much commercial significance as a culinary herb, several centuries ago, such sweet-scented flowers were commonly and extensively employed in the kitchen.

## 60.4 ECONOMIC IMPORTANCE

Lavender is chiefly a source of a commercial fragrant oil which has many uses. The oil is employed in perfumes, lotions, cosmetics, soaps, colognes, shaving creams, toilet water, and other grooming products. "Lavender water" is a perfume consisting primarily of an alcoholic solution of lavender oil. English lavender oil is of the highest quality, but "lavender oil" is often adulterated with spike oil (i.e., from spike lavender) or oil from other inferior species which can be produced more cheaply. Because they are not only very productive of oil but also produce a reasonably high quality of oil, hybrids of the English and spike lavenders (i.e., lavandin) produce most of the "lavender oil" marketed today.

## 60.5 ORNAMENTAL USES

Lavender is a widely grown ornamental, especially useful as a sweet-smelling border plant along pathways and

FIGURE 60.3 Commercial lavender fields in France.

*Source:* **(a) Photo by Santiago Puig Vilado (CC BY SA 3.0). (b) Photo (public domain) by Andyblind.**

FIGURE 60.4 Harvesting lavender: (a) Mechanized harvest with a combine. (b) Bundles of harvested lavender drying in a commercial factory. (c) Traditional harvest with a sickle. (d) Harvesting by hand. A Bulgarian girl in traditional folklore costume picking lavender by hand.

*Source:* **(a) Photo by Celeda (CC BY SA 4.0). (b) Photo by Shutterstock/Ppa. (c) Photo by Shutterstock/Photolona. (d) Photo by Shutterstock/Valentin Valkov.**

FIGURE 60.5 Lavender plants in different settings: (a) Garden plant. (b) A border of lavender in front of Clare College, Cambridge. (c) A large outdoor pot of lavender. (d) A pot of lavender strategically placed in a bathroom to control odor.

*Source:* **(a) Hans from Pixabay (public domain). (b) Roberto-M from Pixabay (public domain). (c) Photo by Allan Henderson (CC BY 2.0). (d) Photo by Shutterstock/Ground Picture.**

in rockeries. In warm climates, tall-growing forms can be used as hedges. Dwarf cultivars make excellent potted patio plants.

## 60.6 OTHER USES

Since classical times, the smell of lavender has been considered to be therapeutic for many nervous diseases, and today it is perhaps the premier herb employed in aromatherapy. *Aromatherapy* has been defined as "the therapeutic use of fragrances or volatile substances to cure and to mitigate or to prevent diseases, infections and indispositions only by means of inhalation." Lavender has long had a reputation for calming the nerves and relaxing tensions, typically administered as a bath at night with a few drops of lavender oil or as a sedative tea. Like hop (*Humulus lupulus*), it has been recommended in "sleep pillows" to induce tranquil slumber. A number of reports on the internet claim that the scent of lavender is aphrodisiac. The extent to which aromatherapy offers a direct therapeutic pharmacological interaction of fragrance molecules with body tissues, rather than simply inducing a pleasant feeling, remains to be determined. In modern medicine, lavender drugs are occasionally used (mostly in Europe) as a mild sedative to treat excitement, nervous exhaustion, disturbances of sleep, and functional and intestinal disorders related to nervousness and difficulties in sleeping. Lavender is marketed commercially in a number of patent medicines and calmative teas (again, mostly in Europe, although some medicines in North America are formulated with lavender).

Lavender is also employed in a variety of other ways, albeit minor. It is used in crafts, sachets (small bags containing scented material, used especially to scent clothing), and potpourris (mixtures of materials, especially botanicals, placed in a bowl or small sack to perfume clothing or a room). The fragrant leaves and blossoms have been employed for centuries in scented soaps, perfumes, and aromatic tobaccos. The dried flower spikes continue to give off their pleasant odor for many years. Lavender has been used since ancient times to scent clothing and linens. While the aroma is very attractive to humans, it appears to dissuade many insects and so has been used in clothes chests as a moth repellent and simply hung in a room to drive away flies and mosquitoes. On the other hand, the nectar in the lavenders is extremely attractive to bees and butterflies, and often, lavender is planted deliberately to attract these desirable insects. Lavender honey has an outstanding reputation and is widely sold as a gourmet item.

FIGURE 60.6 Commercial lavender products: (a) A specialty lavender products store in Italy. (b) Dried flowers. (c) Sachets of lavender. (d) Lavender shortbread cookies.

*Source:* **(a) Photo by Son of Groucho (CC BY 2.0). (b) Photo by Shutterstock/Fotokon. (c) Photo by Véronique Pagnier (public domain). (d) Photo by Turku Gingerbread from Finland (CC BY 2.0).**

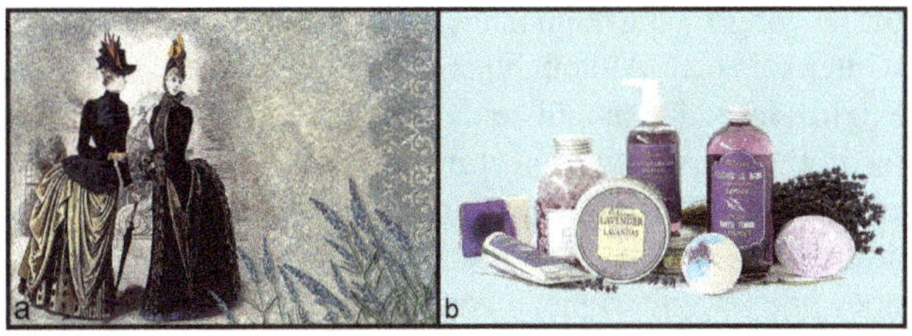

FIGURE 60.7 The association of lavender with femininity: (a) Vintage linen postcard. (b) Lavender commercial cosmetics.

*Source:* **(a) Photo by Dorothe from Pixabay (public domain). (b) Photo by Andyblind (public domain).**

Lavender has other minor culinary usages. The essential oil is used commercially to a small extent as a food flavorant. The flowers are the chief part of the plants employed as a culinary herb, although sometimes the foliage is also employed. The flowers (fresh or dried) are often brewed to make tea. Lavenders are sometimes grown near highways for stabilization of soil and beautification.

## 60.7 PROBLEMS

While lavender plants are not hazardous for most people, some individuals experience dermatitis from contact. Lavender oil can cause dermatitis and can be fatal if taken internally in large doses.

## 60.8 CULTIVATION

### 60.8.1 Climate and Site

In the wild, species of *Lavandula* occur in exposed, sunny, usually parched, hot situations. Lavender should not be planted in a shady, damp location, as it will not develop sturdily or flower well. English lavender, in particular, should be provided with full sun. To take advantage of the delightful odor of the plants, they may be planted beside pathways and benches. English lavender is generally listed as tolerating climate zones 3 to 5. Spike lavender is comparatively tender and will not grow as a perennial farther north than zones 8 or 9, while lavandin is slightly less hardy than English lavender. *Lavandula stoechas* and *L. dentata* are typically only winter-hardy as far north as zones 7 to 9. Improving the drainage of *Lavandula* species by amending the soil or growing them in inclined or raised beds enhances winter survival. Although hardy in climate zones 5 and 6, English lavender should be covered with mulch in the winter in these areas to protect it from the cold. In the northern regions of its climate range, English lavender will not attain the considerable size and floral profusion that it does in its Mediterranean homeland. Dwarf cultivars such as Munstead and Hidcote winter well in northern regions when covered with straw or evergreen boughs.

### 60.8.2 Propagation and/or Purchasing

Some lavender cultivars, such as Lavender Lady and Munstead, can be grown efficiently from seeds germinated at about 21°C, with emergence in 15–25 days. Moist prechilling, at 3–5°C for 1 to several weeks, can increase speed of germination. A planting depth of no deeper than 6 mm has been recommended. Because the seeds of many cultivars may take 6 weeks (sometimes 3 months) to germinate and the seedlings grow very slowly, in cold climates it may be necessary to overwinter the seedlings indoors. Plants produced from seeds are often variable and are unlikely to reproduce the distinctive characteristics of the parental cultivar, while cuttings have the advantage of propagating desired traits well.

Softwood stem cuttings taken in spring or autumn are usually preferable to propagation by seeds. However, growth of lavenders is slow, and for most people, purchasing small plants from nurseries is the easiest way to establish plants. If access to large plants is available, they can simply be divided, and the portions transplanted.

Established commercial plantings may last as long as 30 years, although ornamental forms usually require replacing at least by 10 years of age and, if not trimmed annually, are probably best discarded by the age of 5.

### 60.8.3 Planting

Most lavender planting stock should be established 60 to 90 cm apart. Some cultivars are prone to fungus diseases. Plants showing symptoms of disease should, of course, not be used for propagation, and it would be prudent not to replant lavender for several years in a location where disease has developed.

### 60.8.4 Soil and Moisture Conditions

Wild species of *Lavandula* frequently occur in rocky situations, especially on calcareous soils. A pH tolerance range of 5.8–8.3 has been reported, but alkaline soil is best, and acidic soils should be neutralized, for example, with ground limestone. Although tolerant of various types of soil, the lavenders prefer well-drained substrates, and waterlogging is detrimental during the winter. Dry, sandy, well-drained soil is best. Wet conditions encourage root rot. It has been claimed that infertile soil seems to improve the fragrance of lavender (indeed, in a similar vein, it is often alleged that the aroma of culinary herbs of Mediterranean origin is improved when grown with no soil fertilization).

### 60.8.5 Maintaining Plants

English lavender tends to become unkempt after several years and should be trimmed for an attractive shape or replaced with young plants. Cutting away old woody growth every year in the spring is advisable. Nipping the flower spikes off before they bloom during the first year makes the bushes more compact. In cold areas, lavender can be maintained in pots, overwintered indoors in a dormant state.

## 60.9 FASCINATING FACTS, FABLES, FOLKLORE, AND FRIVOLITY

- The color lavender, of course, is named for the flower, but like many colors, it is difficult to define exactly. Webster's (1993) *Third New International Dictionary* defines *lavender* as "a variable color averaging a pale purple that is bluer and deeper than wisteria, flossflower blue, or mauvette and bluer, darker, and slightly stronger than phlox pink." Other colors defined in *Webster's*: lavender blue, "a light to brilliant purplish blue"; lavender gray, "a light bluish gray that is darker and slightly redder than sky gray or chicory"; lavender mist, "a pale purple that is paler than average lavender, bluer and duller than wisteria, and bluer, stronger, and slightly lighter than flossflower blue."
- Ancient Egyptians created mummification casts that would last indefinitely by soaking linen in oil of lavender containing asphalt, wrapping the bodies with these, and drying them in the sun until the casts were hard. The use of lavender in mummification was also practiced by ancient Phoenicians and peoples of Arabia.
- It was once believed that the asp, a dangerous kind of viper, lived where lavender grew, so that the plant had to be approached with great caution. Lavender growers are said to have once spread the idea that snakes live in lavender fields in the belief that the danger of growing and harvesting the herb would drive the price of lavender higher. Ironically, lavender has been used to treat snakebite. In the Victorian language of flowers, a way of communicating coded messages within letters that was popular during Victorian times, lavender signifies distrust. This is thought to have come about because of the belief that snakes lurked within lavender bushes and one had to be extremely wary when picking the flowers.
- Lavender is often mentioned in the Bible, not by this word, but rather by the name used at that time – spikenard. *Nard* (the name originating from Nardus in Syria), or spikenard, was freely used in biblical times for anointing the feet. In the Gospel

of Luke, the writer reports: "Then took Mary a pound of ointment of spikenard, very costly, and anointed the feet of Jesus, and wiped his feet with her hair; and the house was filled with the odor of the ointment." According to legend, lavender was taken from the Garden of Eden by Adam and Eve. Another Christian legend explains how lavender acquired its wonderful scent. This happened when the clothing of baby Jesus was laid upon a lavender bush to dry by Mother Mary. This may explain why the plant was also regarded as a holy safeguard against evil. In many Christian houses, a cross of lavender was hung over the door for protection. In one of the apocryphal books of the Bible, Judith anointed herself with perfumes including lavender before seducing Holofernes, the enemy commander. Once he was under her heavenly scented influence, she murdered him and saved the city of Jerusalem.

- The seductive scent of lavender has been credited with assisting Egyptian Queen Cleopatra (69–30 BC) to seduce the Romans Julius Caesar (100–44 BC) and Mark Antony (Mark Anthony, 83?–30 BC). In a similar vein, the Queen of Sheba offered lavender with frankincense and myrrh to King Solomon (10th century BC) of Israel.
- In 16th- and 17th-century England, it was widely thought that lavender had soothing qualities, a belief that has persisted to today. Brides, often the subject of arranged marriages, sometimes placed a sprig of lavender in the house on their wedding day for protection against possible verbal or physical abuse from their new husbands. In North Africa, lavender has similarly been used to protect the Kabyle women from being mistreated by their husbands. Another old European practise was perfuming a new bride with lavender on her wedding night before she met her groom. It has been suggested that this was to calm her fears, rather than make her more attractive to the groom.
- In medieval (i.e., the period 500–1500) and Renaissance (14th–17th centuries) Europe, the washing women were known as "lavenders," and they used lavender to scent laundry, sometimes simply by drying clothes on lavender bushes.
- Lavender was worn by prostitutes several centuries ago to advertise their profession as well as to attract customers.
- Lavender has a long association with European royalty. Queen Elizabeth I (1533–1603) of England liked it as a conserve to accompany meat roasts, and in lavender tea. She also demanded that fresh lavender flowers be available every day of the year, a formidable task for a gardener in the climate of England. Queen Victoria (1819–1901) found it attractive for bathing and deodorising. Emperor Napoleon Bonaparte (1769–1821) is said to have used it to calm his frayed nerves before going into battle. King Louis VI (Louis the Mad, 1368–1422) of France had all the cushions in his palace filled with lavender so that when he sat down, he would smell the fragrance. King Louis XIV of France (the Sun King, 1638–1715) also loved lavender and bathed in water scented with it.
- Color has long played an important role in the identification of the homosexual community. In Victorian England, the color green was associated with homosexuality. The color purple (or, more accurately, lavender) was associated with homosexuality at least by the early 20th century and became popularized as a symbol for gay pride in the late 1960s. It has been suggested that the association traces to the 16th-century practice in England of indicating that one would not marry by wearing a sprig of purple flowers. A pink triangle was first used in Nazi Germany to identify gay males in concentration camps, and a black triangle was similarly used to identify lesbians (and others deemed "asocial"), but these were adopted by the homosexual community in the early 1980s. The multicolored "rainbow flag" has become emblematic of the diversity of the community.
- In Britain during the First World War, there was a scarcity of drugs normally obtained from continental Europe. At the request of the government, citizens assisted the war effort by gathering tons of various medicinal herbs, including lavender from gardens, which was used to produce an antiseptic oil. During both world wars, wild-growing lavender in the Mediterranean area was sometimes used to dress wounds, when proper surgical supplies were lacking.
- Yardley and Company was established in London, England, in 1770, and is famous for its advertising slogan: "Yardley lavender soap. The soap

FIGURE 60.8 Lavender mazes: (a) The stone circle of the Michigan Lavender Labyrinth, a 1.2 ha garden near Shelby. This central area is surrounded by a labyrinth primarily of lavender. (b) Aerial view of a bride and groom holding hands in the middle of a lavender maze in Croatia.

*Source:* **(a) Photo by K. E. Murphy, SandstormHART (CC BY SA 4.0). (b) Photo by Shutterstock/Nadtochiy.**

that's kept women in hot water for 200 years – and they've loved every minute."

- Lavender mazes are popular. A huge lavender labyrinth requiring about an hour to reach the center is a popular attraction in Michigan, and similar mazes occur throughout Europe.

## KEY PUBLICATIONS

Ashraf, A., P. Sultan, P. Qazi, and S. Rasool. 2019. Approaches for the genetic improvement of lavender: A short review. *Journal of Pharmacognosy and Phytochemistry* 8: 736–740.

Bader, S.B. 2012. *The Lavender Lover's Handbook: The 100 Most Beautiful and Fragrant Varieties for Growing, Crafting, and Cooking*. Timber Press, Portland, OR. 192 pp.

Fismer, K.L. and K. Pilkington. 2012. Lavender and sleep: A systematic review of the evidence. *European Journal of Integrative Medicine* 4: e436–e447.

Héral, B., E. Stierlin, X. Fernandez, and T. Michel. 2021. Phytochemicals from the genus *Lavandula*: A review. *Phytochemicals Reviews* 20: 751–771.

Kang, H.-J., E.S. Nam, Y. Lee, and M. Kim. 2019. How strong is the evidence for the anxiolytic efficacy of lavender?: Systematic review and meta-analysis of randomized controlled trials. *Asian Nursing Research* 13: 295–305.

Lis-Balchin, M. (ed.). 2002. *Lavender: The Genus Lavandula*. CRC Press, London. 296 pp.

Mason, J.L. 2014. *Growing & Knowing Lavender*. ACS Distance Education, Nerang MDC, Queensland, Australia. 117 pp. (e-book). https://archive.org/details/Growing_and_Knowing_Lavender_by_John_Mason; https://dl.acsedu.com/ebook/Samples/Growing_Knowing_Lavender_Sample.pdf.

McNaughton, V. 2010. *Lavender: The Grower's Guide*. Timber Press, Portland, OR. 180 pp.

Platt, E.S. 2009. *Lavender: How to Grow and Use the Fragrant Herb*, 2nd edition. Stackpole Books, Mechanicsburg, PA. 160 pp.

Salehi, B., D. Mnayer, B. Özçelik, G. Altin, K.N. Kasapoğlu, C. Daskaya-Dikmen, et al. 2018. Plants of the genus *Lavandula*: From farm to pharmacy. *Natural Product Communications* 13(10): 1934578X1801301037. https://journals.sagepub.com/doi/abs/10.1177/1934578X1801301037.

Tucker, A. and K. Hensen. 1985. The cultivars of lavender and lavandin (Labiatae). *Baileya* 22(4): 168–177.

Upson, T. and S. Andrews. 2004. *The Genus Lavandula*. Royal Botanic Gardens, Kew. 456 pp. [Upson, T., S. Andrews, and G. Harriott. 2004. The Genus *Lavandula*. Timber Press, Portland. 442 pp.]

Webster's Third New International Dictionary. 1993. *Webster's Third New International Dictionary*, edited by P. B. Gove and the Merriam-Webster editorial staff. Merriam-Webster Inc., Springfield, MA. 2662 pp.

### Key Cultivation Websites

Fine Gardening. *The Allure of Lavender*. www.finegardening.com/article/the-allure-of-lavender.

McCoy, J.-A. and J.M. Davis. 2021. *Lavender: History, Taxonomy, and Production*. https://newcropsorganics.ces.ncsu.edu/herb/lavender-history-taxonomy-and-production/.

Ontario, Canada. *Growing Lavender in Ontario: An Introduction for Prospective growers*. www.ontario.ca/page/growing-lavender-ontario-introduction-prospective-growers.

The Spruce. *How to Grow Lavender: Planting and Care*. www.thespruce.com/growing-lavender-1402779.

WikiHow. *How to Divide a Lavender Plant*. www.wikihow.com/Divide-a-Lavender-Plant.

WikiHow. *How to Dry Lavender*. www.wikihow.com/Dry-Lavender.

WikiHow. *How to Easily Plant Lavender Seeds*. www.wikihow.com/Grow-Lavender-from-Seed.

WikiHow. *How to Grow Lavender*. www.wikihow.com/Grow-Lavender.

WikiHow. *How to Grow Lavender from Cuttings*. www.wikihow.com/Grow-Lavender-from-Cuttings.

WikiHow. *How to Harvest Lavender*. www.wikihow.com/Harvest-Lavender.

WikiHow. *How to Make Lavender Oil & Lavender Essential Oil*. www.wikihow.com/Make-Lavender-Oil.

WikiHow. *How to Plant Lavender*. www.wikihow.com/Plant-Lavender.

WikiHow. *How to Plant Lavender in Pots*. www.wikihow.com/Plant-Lavender-in-Pots.

WikiHow. *How to Trim Lavender*. www.wikihow.life/Trim-Lavender.

# 61 Lilac (*Syringa* Species and Hybrids)

**Scientific Names:** The genus name *Syringa* is based on the Greek *syrinx*, meaning "hollow tube or pipe," a reference to the pith-containing stems of lilac, which are easily hollowed out and were once used as primitive musical instruments. Ancient Greek legend holds that the nymph Syrix, pursued by the god Pan, was transformed into a hollow reed which Pan used as his first flute or panpipe. The common name "syringa" is sometimes applied in western North America to the genus *Philadelphus* (more appropriately termed mock-orange). The "syringa tree" is *Melia azedarach* (also called as Chinaberry, Persian lilac, and many other names) and is known for producing quite poisonous berries.

**Common Names:** Lilac (from the Persian *nilak*, blue, a reference to the flower color). Lilac almost always refers to *Syringa*, but the word *lilac* is sometimes part of the names of some species of *Callicarpa*, *Ceanothus*, *Hardenbergia*, *Hesperis*, *Melia*, and *Veronica*. Butterfly bush (*Buddleja davidii*), also known as summer lilac, is one of the chapters in this book.

## 61.1 INTRODUCTION

Of all woody ornamentals, lilacs are an exceptionally attractive combination of eye candy and nose candy, the result of their spectacular, colorful flower clusters and exquisite fragrance. Lilacs are adapted to regions with cold winters, because they must have a period of cold weather to induce good flowering. They are low-maintenance, easily grown, durable shrubs and small trees which are excellent for a range of landscaping applications, including hedges, foundation plantings, small and large gardens, and street trees. There may be no other common woody plant that is so ideal for regularly producing cut flower bouquets. Next to roses, no other woody ornamental is as celebrated by great public gardens and festivals.

The genus *Syringa* consists of about 20 Eurasian species of deciduous shrubs or small trees, most native in eastern Asia (16 in China, a few in Japan and Korea), some extending west to Afghanistan, Pakistan, Kashmir, Himalayan India, and Nepal, and two separated in southeastern Europe. The species grow naturally in woodland and scrub, almost always in cool-temperate climates.

The relationships of *Syringa* and *Ligustrum* (privet) are unsettled. *Ligustrum*, until recently, has been considered to comprise 40–50 species of Asia, Europe, Malaysia, and Australia. Their flowers are not as showy or fragrant as those of *Syringa*, but some of the species are commonly employed as hedge plants. Evidence (Li et al. 2012) has suggested that *Ligustrum* is best placed within *Syringa*.

## 61.2 OVERVIEW

Lilacs are grown mostly for their display of flowers, complemented by attractive odor. It is said that lilac colors are most intense during cool, damp springs, and that the color

FIGURE 61.1 Beautiful paintings of lilacs (public domain).

*Source:* (a) Larousse du XXème siècle 1932, vol. 5. (b) Vintage greeting card. (c) Houtte, L. van. 1845. *Flores des serres et des jardins de l'Europe*, vol. 14. (d) Charles François Antoine Morren, C. F. A., and Moren, C. J. É. 1854. *La Belgique horticole, journal des jardins et des vergers*, vol. 4.

FIGURE 61.2 Common lilac (*Syringa vulgaris*): (a) Cultivar Indiya. (b) Cultivar Mme Lemoine. (c) Unidentified cultivar.

*Source:* (a) Photo by Kor!An (CC BY SA 3.0). (b) Photo by Salicyna (CC BY SA 4.0). (c) Photo by Joseph Berger/Bugwood.org (CC BY 3.0).

DOI: 10.1201/9781003586227-61

FIGURE 61.3 Advanced kinds of flowers in domesticated common lilac (*Syringa vulgaris*): (a) Variegated flowers. (b) Doubled flowers.

***Source:*** **(a) Photo by Angel Caboodle (CC BY 3.0). (b) Photo by Rex (public domain).**

FIGURE 61.4 Several of the frequently grown species of *Syringa*: (a) Korean lilac (*S. pubescens* subsp. *patula*), cultivar Miss Kim. (b) Meyer lilac (*S. meyeri*). (c) Japanese tree lilac (*S. reticulata*). (d) Littleleaf lilac (*S. pubescens* subsp. *microphylla*).

***Source:*** **(a) Photo by TownePost Network (CC BY 2.0). (b) Photo by Daderot (public domain). (c) Photo by TheKohser – Gregory Kohs (CC BY 3.0). (d) Photo by Sten Porse (CC BY 3.0).**

of given cultivars sometimes is altered by the soil in which the plant is grown. Flowers have been selected for different colors, size, variegation, and doubling. Modern cultivars are often (artificially) classified by color (blue, lilac, magenta, pink, purple, violet, and white) and whether the flowers are single or double.

The common lilac is in good flower for only about 10 days, but cultivars are available with different flowering periods, and it is possible to establish a lilac garden with staggered blooming times so at least some plants will be in flower over a 6-week period, usually from April through June in North America. A few reblooming cultivars are available, which have the usual main blooming season in the spring, and a less-pronounced second flowering period in the summer. The cultivar Bloomerang, advertised as "continuously blooming," produces the usual large flush of flowers in the spring, sporadically flowers during the summer, and develops a second flush of flowers in the autumn.

Most lilacs are scented. It is claimed that lilacs are most fragrant on a warm, sunny afternoon, when the flowers are completely open. The common lilac (*S. vulgaris*) is generally considered to have the most pleasant odor, although cultivars differ in smell, and some recently bred cultivars with large florets are quite deficient in odor. The tree lilacs (*S. reticulata*) have a different, musk-like scent which is unpleasant to some noses. Doubled flowers of lilac (as is frequently the case for ornamentals) tend to have less scent than normal flowers.

Most species of lilacs are cultivated (perhaps half of these only in special collections), and a dozen or so have given rise to cultivars, either directly or in hybrids. Information for the most frequently cultivated species is given in Table 61.1, and for commonly cultivated hybrid species in Table 61.2. *Syringa* species are naturally adapted to continental cool-temperate regions and usually do not thrive in humid coastal areas or in hot climates. However, some recent cultivars are relatively tolerant of high humidity and/or warmer temperatures.

### 61.2.1 Tree Lilacs

After the common lilac (*S. vulgaris*), tree lilac (*S. reticulata*) is the second most frequently cultivated lilac species. It is divided into Japanese tree lilac (subsp. *reticulata*, native of Japan), Amur tree lilac (subsp. *amurensis*, native of northeastern China, Korea, and southeastern Russia), and Peking tree lilac (subsp. *pekinensis*; native of north-central China). Japanese tree lilac is the most widely cultivated of these. The most widespread cultivar is Ivory Silk, an early-flowering form which was bred by Sheridan Nurseries (Toronto) and released in 1973. Tree lilacs are often shrubs, but some cultivars develop into short trees, sometimes as tall as 9 m. Tree lilacs are late-blooming (often into midsummer), developing creamy-white or yellowish fragrant flowers, and the bark, reminiscent of cherry trees, has horizontal lenticels. Japanese tree lilacs are resistant to most pests and diseases, grow well in a wide range of soil conditions (including clay and infertile soils), transplant well, are not invasive, and are moderately resistant to pollution, drought, and de-icing salt. Their relatively short stature makes them ideal as urban (street and lawn) trees.

### 61.2.2 Dwarf Lilacs

Most lilacs are large shrubs, which are unsuitable for small yards and gardens. To address this limitation, small shrub

**TABLE 61.1**
**Commonly Cultivated Lilac Species**

| *Syringa* Taxon | Common Name(s) | Origin | Height[1] | Dominant Flower Colors | Coldest Hardiness Zone Recommended |
|---|---|---|---|---|---|
| *S. meyeri* | Meyer lilac (often inappropriately called "Korean lilac") | China (found only in cultivation) | 1.2–2.4 m (compact garden forms often 1.2–1.5 m) | Lilac, pink | 5 |
| *S. oblata* subsp. *oblata* | Early-blooming lilac, large-leaf lilac | China, Korea | Up to 5 m, but mostly 2.4–3.6 m | Lilac, white | 3 |
| *S. pekinensis* | Peking lilac, Chinese tree lilac | China | 2–5 (exceptionally to 10) m | Yellow, white, yellowish-white | 5/4 |
| *S. pubescens* subsp. *microphylla* var. *microphylla* | Littleleaf lilac | Northwestern China | 1.5–2.4 m | Pink | 4 |
| *S. pubescens* subsp. *patula* | Korean lilac, Manchurian lilac (a name also applied to *S. reticulata* subsp. *amurensis*) | Northern China, Korea | 1.8–3.6 m; ('Miss Kim', the most popular cultivar, usually not higher than 2 m) | Lilac | 4 |
| *S. reticulata* subsp. *reticulata* | Japanese tree lilac | Japan, China, Korea | 6–9 m | Yellowish-white | 3 |
| *S. villosa* | Late lilac | China | Up to 6 m | Lilac, pinkish-white, yellow | 2 |
| *S. vulgaris* | Lilac, common lilac, French lilac | Balkans, southeastern Europe | 1.5–6 m | Lilac, white, yellow, pink, burgundy | 4 |

[1] Height in colder areas tends to be smaller. For several of the species, dwarf forms are available, which are much smaller than the indicated range.

FIGURE 61.5 Four of the commonly grown hybrid species of *Syringa*: (a) Cut-leaved lilac (*S.* × *laciniata*). (b) Chinese lilac (*S.* × *chinensis*). (c) Hyacinth lilac (*S.* × *hyacinthiflora*), cultivar Esther Stanley. (d) Preston lilac (*S.* × *prestoniae*) cultivar Oberon.

**Source: (a) Photo by Linél (CC BY 3.0). (b) Photo by A. Barra (CC BY 3.0). (c) Photo by Patche99z (public domain). (d) Photo by Kor!An (CC BY SA 3.0).**

cultivars have been selected, particularly of *S. meyeri* (Meyer lilac) and *S. pubescens* subsp. *patula* (Manchurian lilac). These are particularly useful as foundation specimens and develop relatively small but very abundant floral clusters. Both of the preceding species do not require pruning and are resistant to powdery mildew. 'Charisma' (identified as a cultivar of *S.* × *prestoniae*) and 'Prairie Petite' (a cultivar of *S. vulgaris*) grow only to about a meter in height.

"Witches' broom" is a condition in which plants form thick, twiggy growths at the ends of branches. This is an occasional disease of lilacs, usually caused by a phytoplasma (formerly called a mycoplasma-like organism), although sometimes the response is due to environmental stresses, pollution, or invertebrates. Some dwarf lilac cultivars may simply be phytoplasma-infected plants (i.e., dwarfing is due to the disease). This was found to be the case for 'Charisma', which accordingly was withdrawn from the market. Since the disease can be transferred by insects to other lilacs, dwarf lilacs that are the result of phytoplasmas should not be propagated.

## 61.3 HISTORY

Lilacs appear to have been known for more than 500 years. Several spontaneous interspecific hybrids (i.e., not deliberately created) became established before modern breeding began. One of these, the Persian lilac (*S.* × *persica*), was recorded over 400 years ago. In the 16th century, the common lilac (*S. vulgaris*) was transported to Istanbul, center of the Ottoman Empire at that time. Although the common

**TABLE 61.2**
**Popular Lilac Hybrids**

| *Syringa* Taxon | Common Name(s) | Origin | Height | Flower Colors | Coldest Climate Hardiness Zone Recommended |
|---|---|---|---|---|---|
| *S.* × *chinensis* | Chinese lilac | *S. protolaciniata* × *S. vulgaris* | 3.6–4.5 m, mostly 2.4–3.6 m | Lilac, red, white | 3 |
| *S.* × *hyacinthiflora* | Hyacinth lilac, early-flowering lilac, American hybrid lilac | *S. oblata* × *S. vulgaris* | 2–3.6 m | Lilac, blue, red, pink, white | 4/3 |
| *S.* × *laciniata* | Cut-leaf lilac, cut-leafed lilac | A hybrid lilac of unknown, old origin, often cited as from China, but possibly southwestern Asia | 1.5–3.6 m | Lilac | 4 |
| *S.* × *persica* | Persian lilac | *S.* × *laciniata* × *S. afghanica* | 1.2–2.4 (3) m | Lilac, red, white | 4 |
| *S.* × *prestoniae* | Preston lilac, Canadian lilac | *S. villosa* × *S. komarowii* subsp. *reflexa* | 1.8–3.6 m | Violet, pink, pinkish-violet | 2 |

**FIGURE 61.6** Lilac in great European art paintings (public domain): (a) *The Bunch of Lilacs*, by Jacques Joseph Tissot (French artist better known as James Tissot, 1836–1902). (b) *The Time of the Lilacs*, by Sophie Gengembre Anderson (French-born British artist, 1823–1903). (c) *Lilac*, by Edmund Leighton (English artist, 1853–1922).

lilac is native to Europe, cultivars were imported from Istanbul to Europe (as early as 1563). By the mid-1600s, common lilac was growing in the European overseas colonies. Today there are more than 2,000 cultivars of *Syringa*, mostly of the common lilac.

The common lilac, *S. vulgaris*, is the dominant cultivated lilac species. The earliest cultivars mostly have purple or white flowers and are quite hardy and drought-tolerant. They propagate readily by suckers (also sometimes called offshoots and runners) and so are often encountered persisting on abandoned homesteads. Suckers are the plant's method of rejuvenating itself, indeed potentially for centuries. The Lemoine family of Nancy, France, was instrumental in breeding (mainly by hybridization) over 200 cultivars (so-called "French hybrids") with a wide spectrum of plant size and shape and flower color, fragrance, and floral doubling. Many modern cultivars tend to sucker less frequently than the older cultivars, and although more attractive, they are often less hardy.

## 61.4 ECONOMIC IMPORTANCE

Lilacs are one of the most popular flowering shrubs in the world, utilized in both public and private spaces and contributing substantially to the nursery trade.

**FIGURE 61.7** Lilacs in open landscapes: (a) A long hedge in a park. (b) Tree-form lilacs in a park.

*Source:* **(a) Photo by Shutterstock/Magnus Binnerstam. (b) Photo by Shutterstock/Svetlana Mahovskaya.**

## 61.5 ORNAMENTAL USES

Lilacs are versatile ornamentals, serving in both home and public gardens as border and foundation plants, street-side and highway median specimens, and flowering hedges. They have also found application as windscreens, notably in areas with snowy winters. Several lilacs are either so naturally small or capable of being dwarfed that they can be raised as bonsai or grown as container (patio) plants or

FIGURE 61.8 Lilacs on hillsides beside large buildings: (a) In Moscow, Russia. (b) Beside St. Basil's Cathedral in Moscow.

*Source:* (a) Photo by Timur Lazarenko (CC BY 2.0). (b) Photo by Shutterstock/Maryna Kulchytska.

FIGURE 61.9 Artistic modification of lilacs: (a) An old common lilac trained into a tree. Notice the spirally contorting bark, characteristic of very old lilac stems. (b) Lilac bonsai.

*Source:* (a) Public domain photo. (b) Photo by Ryan Somman (CC BY 2.0).

FIGURE 61.10 Harvesting lilac flowers.

*Source:* (a) Photo by Lovelydress0 from Pixabay (public domain). (b) Photo by Shutterstock/Tatiana Bobkova. (c) Photo by Shutterstock/Palchik Kseniya.

in rock gardens. In Europe, lilacs are forced in greenhouses and are harvested for a substantial cut flower market at Christmas.

## 61.6 OTHER USES

Lilac flowers (at least the petals) are edible. They can be candied, added to dishes as a decoration, or incorporated in salads or yogurt.

FIGURE 61.11 Cut flowering stems of lilacs in vases.

*Source:* (a) Photo by Shutterstock/Kateryna Upit. (b) Photo by Shutterstock/Yuliya Evstratenko. (c) Photo by Neelam279 from Pixabay (public domain). (d) Photo by George Chernilevsky (public domain).

FIGURE 61.12 Edible lilac flower dishes. *Left*: Sugared lilac corollas on a cupcake. *Right*: Toast with lilac petal jelly and goat cheese, along with decorative garnishes of lilac flowers.

*Source:* Photos courtesy of Valerie Lugonja (www.acanadianfoodie.com).

## 61.7 PROBLEMS

Lilacs are subject to various diseases and physiological problems, but no more so than most ornamental plans. A special issue that affects lilacs is simply that they are so attractive that pedestrians often break off flowering branches to take home from shrubs that they do not own. Many have argued that stealing lilacs is a harmless pastime that should be tolerated, but some "lilac thieves" have been prosecuted.

## 61.8 CULTIVATION

### 61.8.1 Climate and Site

Depending on species and cultivar, lilacs grow best in hardiness zones 3 to 5 (average lowest winter temperatures ranging between −40°C and −23°C) but can survive in zone 2

FIGURE 61.13 Lilac thieves.

*Source:* **Prepared by Jessica Hsiu.**

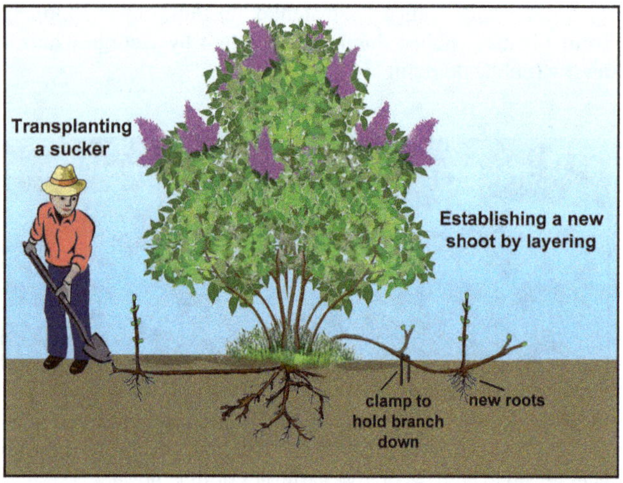


FIGURE 61.14 Simple methods of propagating lilacs vegetatively. *Left*: Digging out a sucker for replanting. *Right*: Layering.

*Source:* **Prepared by B. Brookes.**

(where average lowest temperatures can go down to −46°C) and in zones 6 and 7 (–23°C to −12°C). Tables 61.1 and 61.2 provide information on climate zone tolerance of the most commonly grown species and hybrids. Particular cultivars may survive but not flower in regions that are either too cold or too warm.

In poorly drained sites, roots may be frozen to death in ice during the winter. Planting in "frost pockets" (areas where cold air accumulates) is inadvisable. Reflection of sunlight from light-colored buildings can force the buds of foundation plantings to open early and be damaged by frost. Cultivars of *S.* × *hyacinthiflora* are among the earliest-flowering lilacs (usually 1–2 weeks before the common lilac) and so are relatively susceptible to early-spring frosts.

An open location (with at least 6 hr of direct sun if in cold northern climates) is strongly recommended. Lilacs can tolerate considerable shade, but at the cost of suppressed flowering. Sometimes, after lilacs are established, surrounding trees grow tall or buildings arise, which produces too much shade for good lilac growth. Some of the dwarf shrub cultivars are relatively tolerant of semi-shade, but generally, lilacs and shade are incompatible.

### 61.8.2 Propagation and/or Purchasing

Almost all cultivated lilacs are of hybrid origin and have been maintained vegetatively, so seeds (if generated) will not reliably reproduce the mother plant's characteristics. Professional lilac producers employ several reproductive methods: root sprouts, layering, cuttings, and cleft and bud grafts. Grafts, sometimes to root systems of different cultivars or different species of *Syringa* and *Ligustrum*, are the chief commercial way of propagating lilacs, but this requires knowledge, skill, and greenhouse resources. (Grafts to foreign rootstocks are often intended to be short-term, in which case they are discarded after the cutting develops its own roots.) The easiest way for home gardeners to reproduce an attractive plant is by cutting off and establishing root sprouts (i.e., suckers, if produced; this won't work if the plant is grafted and the suckers originate from the foreign root system) and by layering (bending over a lower stem, fastening part of it underground, wounding it below a leaf bud, and dusting the wound with rooting hormone to stimulate root formation; this may take a year or more and is only advisable if suckers are unavailable).

To establish lilac cuttings, professional gardeners may graft cuttings to the roots of different species or even genera (especially *Ligustrum* and *Fraxinus*), but this should be temporary, allowing the cutting to develop its own roots by burying the graft juncture below the level of the substrate to encourage rooting. Sometimes young grafted lilacs are marketed with the warning to plant the graft juncture below the ground line to encourage development of the desired cultivar's own root system. Lilacs are usually more cold-tolerant on their own roots, so it is inadvisable to purchase grafted lilacs (indeed, in cold climates, they should not be marketed). Moreover, should the root system of a grafted lilac survive so that the canes of the desired cultivar are growing on it rather than on their own roots, it is almost impossible to successfully prune the plant (pruning the canes will kill them, and new suckers arise only from the undesirable foreign root system).

Because lilacs can be lifelong companions, the characteristics of a desired lilac cultivar (size, tendency to sucker, color, flowering period, tolerance of diseases, cold-hardiness, requirements for pruning) should be appreciated before purchase. Lilacs are usually bought as containerized or bare-root plants 0.6–1.2 m in height, although larger, balled, and burlapped plants are also available.

### 61.8.3 Planting

Container-grown plants should be planted in early spring or in the fall, while bare-root plants need to be put in the ground in early spring, before budbreak. Spacing lilacs depends on the natural size of the cultivar: 3–5 m for large specimens and trees, 2–2.4 m for informal hedges. Interplant distances for clipped hedges depend on anticipated (clipped) height: 1–1.2 m tall, 45–60 cm on center; 2–2.4 m tall, 0.6–1 m on center. Most lilacs spread by suckers, so it is important not to establish specimen plants too close together. When transplanting a lilac, replant at the original soil depth.

### 61.8.4 Soil and Moisture Conditions

Loam soils that are not overly rich are best. Heavy (clay) soils should be amended. A soil pH of 6.5–7.5 (circumneutral or slightly alkaline) has been recommended, but 6.5 to 8.5 is tolerated. Compost and well-decomposed manure may be added to the soil. Light chemical fertilization (high in phosphorus) may be applied in the early spring, but some authorities recommend that if plants are growing well, they should not be fertilized. Excessive application of high-nitrogen fertilizers can reduce or even prevent the formation of flowers. Although lilacs require a regular supply of water (especially while they are establishing), the substrate should be well-drained. Raised beds or mounds may be employed to avoid naturally damp ground, or lilacs can be located on the upper part of a hill, where natural drainage occurs. Mulch will aid in conserving soil moisture.

### 61.8.5 Maintaining Plants

Lilacs are tough landscape plants living for a century or more in sites with full sunlight, good drainage, and space for air movement. Normally, the only care needed is pruning. This is recommended to maintain vigor of most cultivars, to allow sunlight to reach leaves in the lower and central parts of the plant, to control size, and to promote symmetrical shape. Pruning is also desirable so that vigorous growing stems and flowers occur at different heights, resulting in flowers all over the bush, not just at the top. Pruning is usually unnecessary for the first 3 to 5 years of growth. As with all woody plants, pruning away misshapen, broken, diseased, declining, and undesirably oriented stems should be routinely practiced (branches should not be allowed to cross). Some cultivars of some species (especially dwarf forms) do not require much, if any, pruning, and this information should be available (either from the nursery where purchased or on literature accompanying the plant). Two kinds of pruning, or "thinning," need to be distinguished: removing entire canes of shrubs down to the base and clipping off side branches or the tops of main branches. Removing main canes is best done during the dormant period (winter or early spring), preferably from the center of the bush. A second mode of pruning is the annual removal of selected branches. Thinning away some of the youngest branches promotes air circulation, discouraging diseases. Such pruning is best done immediately after flowering, for the following reason. Flower buds for the next year are formed after the early-season flush of flowering, so pruning lilacs in the autumn, winter, or early spring could reduce stems that would have produced flowers. (Although sometimes a second blooming period occurs in the same season, most cultivars flower just once.) Spent flower clusters should be removed as soon as they wither, to prevent the plant from directing its energy into seed formation (some hybrids will not produce seeds); moreover, persisting clusters reduce the attractiveness of the plant. Such deadheading of younger lilacs does seem desirable, but doing so for large, tall specimens can be very labor-intensive, and some authorities contend it does not improve flowering significantly. Suckers can be removed at any time.

Probably, most cultivated lilacs never receive any care whatsoever and still manage to produce impressive floral displays for decades. Nevertheless, the uncontrolled rampant growth of old bushes reduces their overall attractiveness and flower clusters. It has been recommended that such plants be renovated by removing one-third of the major (largest-diameter) canes for three consecutive years. This will, of course, completely eliminate the old aboveground plant, allowing new canes to develop from the ground. Cutting away so much old wood will reduce flowering for a period but will eventually produce many more flowers.

In rural areas, deer commonly browse the tips of young branches, and rutting bucks use their antlers to strip trunks of their bark. Some cultivars are resistant to browsing, but in areas where the problem develops, fences and hunting need to be considered. Rabbits will damage canes that are less than 3 years old, and where snow accumulates, they

FIGURE 61.15 Powdery mildew and common lilac (*Syringa vulgaris*): (a) Powdery mildew (*Erysiphe syringae*) on foliage of several branches. (b) An affected leaf. (c) 'Betsy Ross', a cultivar of common lilac which is resistant to powdery mildew.

**Source:** **(a) Photo by Whitney Cranshaw (CC BY 3.0). (b) Photo by Sarah Pethybridge (CC BY 2.0). (c) Photo by Margaret Pooler, from Bugwood.org (CC BY 3.0).**

will also consume young twigs that can be reached. Field mice and moles can girdle canes below snow level. Canes can be protected with wire and other barriers, although most lilacs seem to survive with no protection.

Powdery mildew is the most common disease of lilacs, causing infected leaves to turn yellow and fall off and often stunting new growth. There are over 1000 species of powdery mildew fungus, but the two most affecting lilacs are *Erysiphe syringae* (previously *Microsphaera syringae*) and *Phyllactinia syringae*. Although unsightly, powdery mildew usually develops in late season and so does little permanent harm to the plants. Fungicides can be applied as a preventative, before the grayish-white powdery mat appears on the foliage, but this is rarely recommended. Affected tissues should be pruned away and burned, sterilizing the shears with rubbing alcohol or diluted bleach. Powdery mildew is promoted by high humidity, poor air circulation, low light (due to shady sites or overcast skies), and warm days coupled with cool nights. Most cultivars of common lilac are susceptible, although some resistant cultivars are available. Japanese tree lilacs, Manchurian lilacs, and Meyer lilacs are quite resistant.

FIGURE 61.16 Swallowtail butterflies visiting "lilacs": (a) A true lilac (*Syringa vulgaris*). (b) A "summer lilac" (*Buddleja davidii*).

*Source:* **(a) Photo by Spangles 44 (CC BY 2.0). (b) Photo by Cat-is from Pixabay (public domain).**

## 61.9 FASCINATING FACTS, FABLES, FOLKLORE, AND FRIVOLITY

- Allegedly, hollow lilac stems were used in ancient times as primitive hypodermic devices to inject medicine or to bleed patients. Curiously, because the scientific name of lilacs is "*Syringa*," computer searches for lilacs sometimes returns information on "syringes."
- Color definitions are often based on well-known plants (e.g., lemon, orange, peach, and plum are accepted color names), and variations of purplish flowers illustrate this well: heliotrope, lilac, mauve (for the mallow flower), phlox, "orchid," and thistle are all employed as color shades.
- Cloned lilac cultivars have been employed in an American program to examine how flowering dates have varied in response to climate change since the 1950s. Cloned plants, of course, control for possible ecogeographical adaptation which could influence phenology, and the popularity of lilacs facilitated distribution of the lilacs to volunteer participants over a wide geographical range. (Note Nummer and Qian 2020.)
- Trunks of very old common lilacs (*S. vulgaris*) have been measured at 60 cm in diameter at ground level.
- Although bouquets of cut lilac flowers are very widely displayed in houses, many in the midland counties of England feared bringing cut lilacs (especially white ones) indoors, in the belief that it would result in death or misfortune. On the other hand, in England, finding a five-petalled lilac flower was thought to produce good luck.
- It is well-established that blue is the most preferred color, regardless of cultural background. However, there is evidence that toward the red end of the spectrum (such as pink and lilac), women exhibit greater preferences than men. It has been contended that this is genetically intrinsic, not learned (http://people.howstuffworks.com/gender-color.htm).
- The larvae of several Lepidopteran species employ lilacs as food plants, and lilacs are excellent as attractors of butterflies, such as swallowtails, to a garden. However, on the web, photos of butterflies visiting "lilac flowers" frequently confuse true lilacs (*Syringa*) with the similar butterfly bush (*Buddleja*), also known as summer lilac.

## KEY PUBLICATIONS

Cathey, H.M. 1980. *Growing Lilacs*, revised edition. United States Department of Agriculture, Washington, DC. 8 pp.

Chang, M.-C. and P.S. Green. 1996. *Syringa*. In: *Flora of China*, Vol. 5, edited by Z.Y. Wu and P.H. Raven, 280–286. Missouri Botanical Garden, St. Louis.

DeBard, M.L. 2019. An updated taxonomy of genus *Syringa* (lilacs). *Lilacs – Quarterly Journal of the International Lilac Society* 48(4): 153–167.

Fiala, J.L. 2008. *Lilacs: A Gardener's Encyclopedia*. Timber Press, Portland, OR. 416 pp. (This version is an update, by Freek Vrugtman, of the highly respected original published in 1988.)

Green, P.S. and H.J. Fliegner. 1991. When is a privet not a lilac? *Kew Magazine* 8: 58–63.

Kang, K.H. and C.S. Chang. 1998. Systematic study of *Syringa* (Oleaceae) in Korea. *Korean Journal of Plant Taxonomy* 28: 249–279.

Li, J., B. Goldman-Huertas, J. DeYoung, and J. Alexander III. 2012. Phylogenetics and diversification of *Syringa* inferred from nuclear and plastid DNA sequences. *Castanea* 77: 82–88.

Lim, T.K. 2014. *Syringa vulgaris*. In: *Edible Medicinal and Non Medicinal Plants*, Vol. 8, Flowers, edited by T.K. Lim, 541–547. Springer Netherlands, Dordrecht.

McKelvey, S.D. 1928. *The Lilac: A Monograph*. Macmillan, New York. 581 pp.

Nummer, S.A. and S.S. Qian. 2020. A hierarchical threshold modeling approach for understanding phenological responses to climate change: When did North American lilacs start to bloom earlier? *SN Applied Sciences* 2(12): 2060. https://link.springer.com/content/pdf/10.1007/s42452-020-03847-z.pdf.

Pringle, J.S. 1997. An updated summary of classification in *Syringa* at the ranks of species, subspecies, and variety. *Lilacs* 26(1): 19–26.

Upton, E.A. 1980, 1986. *The Edward A. Upton Scrapbooks of Lilac Information*. International Lilac Society, Rumford, ME. 2 vols.

Zhu, W., Z. Wang, Y. Sun, B. Yang, Q. Wang, and H. Kuang. 2021. Traditional uses, phytochemistry and pharmacology of genus *Syringa*: A comprehensive review. *Journal of Ethnopharmacology* 266: 113465.

## Key Cultivation Websites

Cornell University, Extension. *Growing Lilacs*. https://chemung.cce.cornell.edu/resources/lilacs.

*Friends of the Farm: Lilacs at the Central Experimental Farm, Ottawa, Canada*. www.friendsofthefarm.ca/lilacs/.

Horticulture Magazine. *How to Plant Lilacs in Pots*. https://horticulture.co.uk/lilac/containers/.

International Lilac Society. www.internationallilacsociety.org/.

International Lilac Society. 2021. *International Register and Checklist of Cultivar Names in the Genus Syringa L*. www.internationallilacsociety.org/wp-content/uploads/2021/03/ILS-register-update-March-5-2021.pdf.

Jull, L. 2006. *Lilacs for Cold Climates*. http://learningstore.uwex.edu/assets/pdfs/A3825.pdf.

New York Botanical Garden. *Trouble-free Lilacs*. https://libguides.nybg.org/c.php?g=655037&p=4597634.

The Spruce. *How to Grow and Care for Lilac Bushes*. www.thespruce.com/lilac-bushes-growing-tips-2132722.

The Spruce. *How to Prune Lilac Bushes*. www.thespruce.com/pruning-lilacs-how-and-when-to-prune-lilacs-1403004.

The Spruce. *Tips for Growing Common Lilac (Syringa vulgaris) in Your Garden*. www.thespruce.com/common-lilac-growing-tips-3269207.

University of Nebraska, Extension. *Lilacs*. https://extension.unl.edu/statewide/douglas-sarpy/pdfs/ce/resources/ce-g1860-lilacs.pdf.

Virginia State University, Extension. *Lilacs, Syringa spp*. https://vtechworks.lib.vt.edu/server/api/core/bitstreams/12b1caa0-8f6f-49a2-a5d5-b946a6f2f81f/content.

WikiHow. *How to Grow Lilacs*. www.wikihow.com/Grow-Lilacs.

WikiHow. *How to Prune Lilacs*. www.wikihow.com/Prune-Lilacs.

WikiHow. *How to Trim an Overgrown Lilac Bush*. www.wikihow.life/Trim-an-Overgrown-Lilac-Bush.

# 62 Lilies (*Lilium* Species and Hybrids)

**Scientific Names:** The genus name *Lilium* is derived from the Greek *leírion*, thought to be based on white lilies.

**Common Names:** The vernacular name "lily" traces to the Latin *Lilium*, explained earlier.

The "true lilies" are species of *Lilium*. Some species formerly included within this genus have now been placed in other genera. These segregated genera include *Cardiocrinum*, *Notholirion*, *Nomocharis*, and *Fritillaria*. Species in dozens of unrelated genera have "lily" in their names. For example: African lily (*Agapanthus*), arum or calla lily (*Zantedeschia*), belladonna lily (*Amaryllis*), daylily (*Hemerocallis*), glory lily (*Gloriosa*), lily of the valley (*Convallaria*), Mariposa lily (*Calochortus*), plantain lily (*Hosta*), peace lily (*Spathiphyllum*), Peruvian lily (*Alstroemeria*), spider lily (*Crinum*, *Hymenocallis*, *Lycoris*), St. Bernard's lily (*Anthericum*), St. Bruno's lily (*Paradisia*), torch lily (*Kniphofia*), water lily (*Nymphaea* and *Nuphar*), and wood lily (*Trillium*). The phrase "tree lily" has been used in various ways (e.g., for species of the Velloziaceae and *Dracaena*), most recently to refer to new, robust, true lily hybrids with stems up to 5 cm in diameter and up to 2.4 m in height. Lilly pillies (lillypillies) are native trees of Australia and Southeast Asia, of the genus *Syzygium*, used as evergreen hedges and ornamentals.

FIGURE 62.1 Collections of lily cultivars.

*Source:* **(a) Photo by Shutterstock/Kisialiou Yury. (b) A display at the 2009 Chelsea (UK) flower show. Photo by Foshie (CC BY 2.0). (c) Photo by Shutterstock/zzz555zzz. (d) Photo by Lolame from Pixabay (public domain).**

FIGURE 62.2 Lily bulbs: (a) A group of bulbs ready for planting. (b) Bulbs of *Lilium candidum* that have started to grow.

*Source:* **(a) Photo by Shutterstock/Damian Lugowski. (b) Photo by Shutterstock/Sarycheva Olesia.**

## 62.1 INTRODUCTION

Lilies are among the most beautiful, regal, enchanting, popular, and important ornamental plants. They exhibit an astonishing range of variation, providing for flexibility of applications. The plants differ greatly in habit and height, and the flowers range widely in size, form, color, flowering time, and fragrance. Indeed, in the last half century, breeders have created thousands of extraordinarily attractive cultivars.

## 62.2 OVERVIEW

There are about 100 species of *Lilium*, which is centered on Eastern Asia (with about 60 native species) and North America (with 21 species). Most lilies are native to the temperate Northern Hemisphere, although some reach the northern subtropics. The species are frequent in woodlands, often in montane habitats, sometimes in grasslands, and rarely in marshy or boggy areas. A few are epiphytes in tropical Southeast Asia. Lilies have suffered from excessive removal from the wild. Transplanting wild lilies is inadvisable, because most are adapted to their natural ecological conditions and tend to grow poorly in cultivation. Modern hybrid garden cultivars, by contrast, have a greater range of tolerance to soils, climates, and diseases.

Lilies are herbaceous perennials, typically 60–180 cm in height. The aboveground portions are usually annual, the plants overwintering from scaly bulbs that are usually deeply buried. The bulbs are annual or perennial (sending up a shoot in the spring for several years), depending on species. The bulbs are composed of fleshy scales (which are leaves modified for storage of starch) and lack an external protective coating (like the membranous tunic of an onion). Some species reproduce from new bulbs (often called "bulblets") generated on the sides of the old bulb, or above



DOI: 10.1201/9781003586227-62

the old bulb, sometimes aboveground. So-called "rhizomes" bearing many small bulbs develop from the base of the bulb in some North American lilies. Several species develop so-called "stolons." *Stolons* are aboveground trailing stems (runners) or laterally spreading stems at or just below ground level. Rhizomes, which overlap definitionally with stolons, are stems that also grow horizontally, either underground or at ground level. The lily literature mostly uses the term *stolon* (or "underground stolon") for spreading underground stems and refers to bulbs proliferating off but attached to the base of the main bulb as "rhizomatous bulbs." Basal roots develop at the bottom of the main bulb. These are often contractile, pulling the bulb down so that it doesn't reach the soil surface (bulbs of some species, such as *L. polyphyllum*, can be more than 60 cm underground). Many species also produce "stem-roots" – adventitious roots that emerge from the growing stem above the bulb, but under the soil. The flowering stem is leafy, and basal leaves are usually absent. However, some species growing in relatively mild areas, after the leaves senesce, develop a rosette (a ring of leaves at ground level) in early winter, and this grows into a new stem with the resumption of warm weather.

Flowering occurs in late spring or summer, the flowers developing singly or in clusters at the top of unbranched stems. The flowers are large – those of *L. auratum*, perhaps the largest in the genus, can be 30 cm across. The three sepals are usually different from the three petals. The six tepals (i.e., sepals and/or petals) are free from each other and are spreading or reflexed. The flowers are attractive, often fragrant, in all colors except blue, notably white, red, orange, pink, purple, and yellow, sometimes marked with spots and streaks. A nectar gland occurs at the base of each tepal. The three-celled fruit capsule contains 60 to over 300 seeds, depending on the species. The seedpod opens at the top and splits down the middle of each compartment, releasing the flattened, very light, papery seeds to be scattered gradually by the wind.

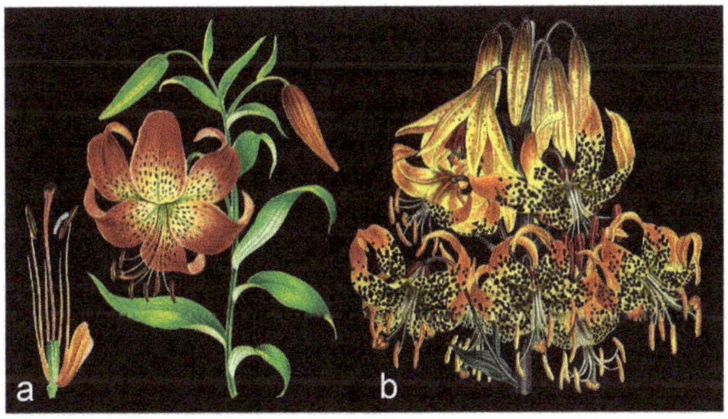


FIGURE 62.3 Botanical illustrations of *Lilium superbum* (public domain): (a) A wild form. (b) A garden hybrid cultivar.

***Source:*** **(a) A chromolithograph published in 1872 by L. Prang & Co., Boston. Photo from Library of Congress Prints and Photographs Division, Washington, DC. (b) *Curtis's Botanical Magazine* (London), vol. 24, 1806.**

FIGURE 62.4 *Lilium auratum* (division VII, as explained in the text), the goldband lily (also *goldbanded lily* and *golden-rayed lily*). This species produces exceptionally large and colorful flowers, and in its Japanese homeland, it is one of the species with bulbs employed as food.

***Source:*** **(a) Photo by Tanaka Juuyoh (CC BY 2.0). (b) Painting by John Frederick Lewis (1804–1876), an English artist specializing in Oriental scenes.**

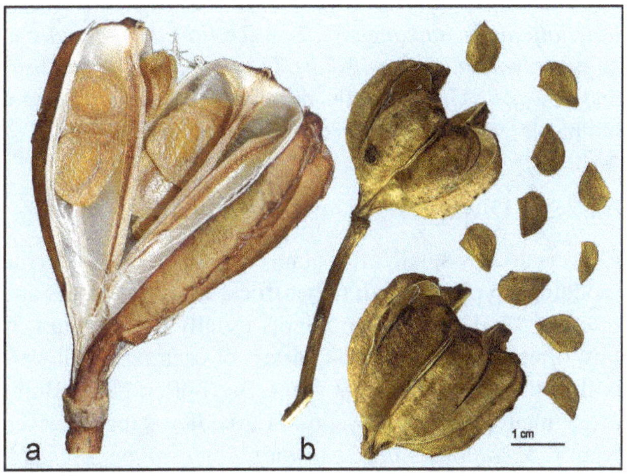


FIGURE 62.5 Mature fruit and seeds of *Lilium* species: (a) A capsule of *L. japonicum* that has split open to reveal the seeds. (b) Seeds dispersing from capsules of Madonna lily (*L. candidum*).

***Source:*** **(a) Photo by Alpsdake (CC BY SA 4.0). (b) Photo by Roger Culos (CC BY 3.0).**

Modern lily cultivars are of principal interest because of the spectacular beauty of the flowers. The genus *Lilium* presents an extraordinary variety of floral forms. As noted in the following presentation, horticultural classification is very heavily based on flower characteristics. The flowers may facc upward, flare horizontally, or hang in inversed tiers. A flaring trumpet shape is particularly popular, exemplified by the cultivar 'Sentinel' and the Easter lily. The flowers of many cultivars are shaped like saucers, facing either upward or outward. Cultivars with semi-doubled (often with two sets of six tepals) or fully doubled flowers are available.

The Royal Horticultural Society in London maintains an international register for lily cultivar names and records descriptive information for over 13,000 entries. To manage the huge number of cultivars of *Lilium*, in 1963, the Royal Horticultural Society Lily Committee created a horticultural classification of lilies, recognizing nine "divisions" of cultivars. Almost all lily cultivars are hybrids, and the first eight divisions include only hybrids. The divisions are quite independent of the "natural" (evolutionary) taxonomy of *Lilium*. This "artificial" system, described in the following, remains in general use by the floral industry, although many recent hybrids have obscured the distinctions between the divisions. Divisions VII and VIII dominate the marketplace.

### 62.2.1 Division I: Asiatic Hybrids

Flowers usually small to medium-sized. Flower color often uniform, or with contrasting tepal tips and/or throat. Spots absent or, when present, well-defined and often rounded. Conspicuous brush marks sometimes present. Tepal margins usually smooth or slightly ruffled. Flowers usually with little or no scent (most of the unscented lilies). Derived from *L. amabile*, *L. bulbiferum*, *L. callosum*, *L. cernuum*, *L. concolor*, *L. dauricum*, *L. davidii*, *L. ×hollandicum*, *L. lancifolium*, *L. lankongense*, *L. leichtlinii*, *L. maculatum*, *L. pensylvanicum* (*L. wilsonii*), *L. pumilum*, *L. × scottiae*, and *L. wardii*. Until recently, this group was very important but has largely been replaced by division VIII.

### 62.2.2 Division II: Martagon Hybrids

Flowers usually small, often numerous, mostly down-facing (nodding), typically with rather thick, recurved tepals and often of "Turk's cap" form. Spots usually present, numerous, often on at least three-quarters of each tepal. Flowers with little or an unpleasant scent. Buds often hairy. Bulbs often mauve or orange-yellow. Early flowering. Derived from *L. × dalhansonii*, *L. hansonii*, *L. martagon*, *L. medeoloides*, and *L. tsingtauense*.

### 62.2.3 Division III: Euro-Caucasian (*Candidum*) Hybrids

Flowers mostly small to medium-sized, often bell-shaped to Turk's cap-shaped, often downfacing. Flower color often in rather pale muted shades. Spots absent to numerous. Tepal-margins smooth, often gently reflexed. Flowers often scented. Many are lime-tolerant. Infrequently marketed. Derived from *L. candidum*, *L. chalcedonicum*, *L. kesselringianum*, *L. monadelphum*, *L. pomponium*, *L. pyrenaicum*, and *L. × testaceum*.

### 62.2.4 Division IV: American Hybrids

Mostly tall. Rootstocks mostly rhizomatous. Flowers small to medium-sized, mostly down-facing. Flower color often strong yellow to orange or orange-red, often with contrasting center and tepal tips. Spots often very conspicuous over at least half of each tepal, often surrounded by a paler halo. Tepals rather narrow, margins smooth, usually gently to strongly reflexed. Flowers with little scent. Derived from native North American species, including *L. bolanderi*, *L. × burbankii*, *L. canadense*, *L. columbianum*, *L. grayi*, *L. humboldtii*, *L. kelleyanum*, *L. kelloggii*, *L. maritimum*, *L. michauxii*, *L. michiganense*, *L. occidentale*, *L. × pardaboldtii*, *L. pardalinum* sensu lato, *L. parryii*, *L. parvum*, *L. philadelphicum*, *L. pitkinense*, *L. superbum*, and *L. washingtonianum*.

### 62.2.5 Division V: *Longiflorum* Hybrids

Flowers medium-sized to large, often few; trumpet-shaped. Flower color typically uniform. Spots, papillae, and brush marks absent. Tepal margins smooth. Flowers usually scented. Flowers usually white, trumpet-shaped. Mostly grown for cut flowers. Derived from *L. formosanum*, *L. longiflorum*, *L. philippinense*, and *L. wallichianum*.

### 62.2.6 Division VI: Trumpet and Aurelian (*L.* × *aurelianse*) Hybrids

Flowers medium-sized to large. Flower color white, cream, yellow to orange or pink, often with a contrasting star-shaped throat and/or strong bands of color outside. Trumpet shape common, usually scented, without markings; other types often with spotting as small streaks in two bands at the base. Tepal margins smooth or twisted, with irregular ruffling; tips often reflexed. Derived from *L. × aurelianense*, *L. brownii*, *L. × centigale*, *L. henryi*, *L. × imperiale*, *L. × kewense*, *L. leucanthum*, *L. regale*, *L. rosthornii*, *L. sargentiae*, *L. sulphureum*, and *L. × sulphurgale* (but excluding hybrids of *L. henryi* with all species listed in division VII). Aurelian hybrids are derived from a combination of *L. henryi* and trumpet lilies.

### 62.2.7 Division VII: Oriental Hybrids

Flowers usually medium-sized to very large. Inner tepals often very broad and overlapping at the base, producing a "closed" center. Flower color mostly white to pink to purplish red, some golden yellow; ground color often white, with a contrasting central ray. Spots absent to numerous, sometimes over half of each tepal. Papillae and nectaries often conspicuous. Flowers usually scented. Often late-flowering.

FIGURE 62.6 Stargazer lilies.

*Source:* (a) Photo by Leimenide (C BY 2.0). (b) Photo by Pro-Flowers (CC BY 2.0).

Derived from *L. auratum*, *L. japonicum*, *L. nobilissimum*, *L.* × *parkmanii*, *L. rubellum*, and *L. speciosum* (excluding all hybrids of these with *L. henryi*). (Upward appearance of flowers has led to the name "stargazers.")

### 62.2.8 Division VIII: Interdivisional Hybrids

Any other hybrids, mostly derived as crosses of species between two or more of the previous groups. Interdivisional hybrids between Asiatic and Trumpet are "Asiapets" or AT hybrids, Longiflorum/Asiatic crosses are LA hybrids, Longiflorum/Oriental crosses are LO hybrids, Longiflorum/Trumpet crosses are Longipet or LT hybrids, Oriental/Asiatic crosses are OA hybrids, and Oriental/Trumpet crosses are "Orienpets" or OT hybrids. Hybrids of *L. henryi* with *L. auratum*, *L. japonicum*, *L. nobilissimum*, *L.* ×*parkmanii*, *L. rubellum*, and *L. speciosum* (excluded from divisions VI and VII) are placed here. Most commercial lily cultivars belong to this group.

### 62.2.9 Division IX: Wild Species

Wild species and cultivars of them (but excluding those derived exclusively from *L. formosanum*, *L. longiflorum*, *L. philippinense*, and *L. wallichianum*, which are placed in division V). The "species lilies" tend to be susceptible to diseases and are usually difficult to maintain in gardens. However, the tiger lily (*L. lancifolium*) from China is widely grown, and several North American natives are occasionally grown, including *L. pardalinum*, *L. canadense*, and *L. superbum*.

### 62.2.10 Madonna Lily

*Lilium candidum*, Madonna lily (often also referred to as the white lily, although other lilies are white), is the most famous species of *Lilium*. It is thought to originate from the Balkans and western Asia but has been cultivated for so long its native area is difficult to establish. It was painted on the walls of the tombs of the Egyptian pharaohs, and the Egyptians used the plant as a source of perfume. The classical Romans consumed the bulbs and used them as a medicinal ointment. At least since the AD 7th century, the species has been associated with the Virgin Mary, and medieval depictions of the Blessed Virgin frequently show her holding the flower.

FIGURE 62.7 Tiger lily (*Lilium lancifolium*), one of the most popular lilies: (a) A large outdoor container. (b) Two monarch butterflies (*Danaus plexippus*) on one flower in California. (c) Photo of a flower.

**Source:** **(a) Photo by Shutterstock/PT Pictures. (b) Photo by Lisa Redfern (public domain). (c) Photo by Ron Conlon (CC BY 2.0).**

FIGURE 62.8 Madonna lilies (*Lilium candidum*), symbolic of innocence, purity, chastity, and humility (public domain illustrations): (a) Plants in a pot. (b) Illustration. (c) Stained-glass church window showing the Madonna with a Madonna lily.

**Source:** **(a) Photo by Dobre Wzory from Pixabay. (b) Eeden, A. C. van. 1872–1881. *Album van Eeden, Haarlem's flora, afbeeldingen in kleurendruk van verschillende bol-en knolgewassen.* (c) Photo by Falco from Pixabay.**

### 62.2.11 Easter Lily

The "Easter lily," *L. longiflorum* Thunb., is native to the southern islands of Japan, Korea, and Taiwan. During the late 1800s, most production was in Bermuda, and the name "Bermuda lily" is still sometimes encountered. The tradition of marketing lily species for Easter first developed sometime before 1930 in North America and remains predominantly a North American phenomenon. *Lilium longiflorum* became symbolic of Jesus's resurrection at Easter, a curious parallel to *L. candidum*'s symbolic association with Mary, mother of Jesus. Prior to the Second World War, the Easter lily was usually imported to North America from Japan. However, when this source was halted during the Second Word War, production began in the California–Oregon border area, which today produces many of the bulbs grown in the world for the potted Easter lily market. In the US potted plant market, the Easter lily is the fourth largest crop, after poinsettia, chrysanthemum, and azalea. However, the sales period for Easter lily lasts only about 2 weeks, whereas poinsettias are sold for 6 weeks, and chrysanthemums and azaleas are sold year-round. Although the demand for Easter lilies collapses after Easter, there is still a market for flowering, potted lilies at other times of the year. The "florist's lily" is usually a hybrid with *L. longiflorum* as one of the parents.

Lilies do not bloom naturally during the spring, so marketing flowering plants for Easter requires greenhouse forcing technology. The vernalization required to induce flowering is a period of cold, moist treatment, sometimes

FIGURE 62.9 Easter lily (*Lilium longiflorum*). (a) Potted plant. (b) Interior of the Cathedral of Saint Joseph in Burlington, Vermont, at Easter. (c) Woman with Easter lily. (d) Vintage Easter greeting illustration, circa 1910.

***Source:*** **(a) Photo by Shutterstock/Mountain Studio. (b) Photo by Farragutful (CC BY SA 4.0). (c) Photo by Dobre Wzory from Pixabay (public domain). (d) Photo by Shutterstock/Victorian Traditions.**

sped up by a photoperiodic treatment (long days). The cultivar ‘Nellie White’ accounts for most production. Ideally, the first blooms should open on Easter Saturday to maximize the display on Sunday, but for sale in stores, flowering is often scheduled sooner. Because Easter can vary from March 22 to April 25, flowering time needs to be scheduled accordingly. To compensate for plants that are coming into flower too quickly or too slowly, growers have sometimes used wheelbarrows to move plants between hot and cold greenhouses, leading to the expression “wheelbarrow crop.”

FIGURE 62.10 The “Prince of the Lilies,” or the “Lily Prince,” a famous wall fresco showing a young man (possibly a priest king) with lilies, in the Minoan palace of King Minos at Knossos on the Greek island of Crete, painted about 1550 BC and excavated in the early 20th century. The lilies depicted have been interpreted as the Madonna lily, *Lilium candidum*. (However, for an argument that the plants shown look like the *Xiphium* category of bulbous irises, see the Bulb Newsletter, 2001, no. 36, www.pacificbulbsociety.org/pbswiki/files/TBN/BN.php?v=36.)

## 62.3 HISTORY

Lilies have been significant ornamentals for thousands of years. Drawings of Madonna lilies (*L. candidum*) on Cretan vases and frescoes have been dated to 3,600 years ago. Lilies were also grown in the gardens of ancient Mesopotamia and Rome. It has been claimed that Madonna lily may be the oldest domesticated ornamental plant, although it is uncertain when sterile plants of the species, that can only be propagated vegetatively, arose. Until the middle of the 15th century, the Madonna lily was the only lily grown in European gardens. Madonna lilies were immortalized as religious icons in paintings by some of the finest artists of Renaissance Europe. *Lilium lancifolium* (*L. tigrinum*), the species most commonly labelled “tiger lily,” has been grown since ancient times in China and may also have been domesticated thousands of years ago. Several Asian and North American lilies were introduced to Europe in the 17th and 18th century, but very few species were cultivated until the very early 20th century. Then, several more Asian lilies were introduced to the West, and their use in generating hybrids substantially increased the popularity of the flower.

A very large majority of modern garden lilies are artificial hybrids that were generated during the 20th century. Most of the species cross easily with only a few others. In vitro (tissue culture) techniques have been applied to lilies by plant breeders to overcome difficulties in generating hybrids. As is common among distantly related species, hybrid embryos are often inviable but sometimes can be saved by embryo rescue culture. Indeed, lilies were among the first floricultural crops employing tissue culture methods. Many new hybrids are sterile triploids (with three sets of chromosomes, hence sterile), but polyploidization (doubling of chromosome sets) has been used to restore fertility,

FIGURE 62.11 Examples of paintings (public domain) of old European lily (*Lilium*) varieties by Belgian painter and botanist Pierre-Joseph Redouté (1759–1840). Redouté is best known for his outstanding portraits of roses, but he also specialized on plants of the lily family: (a) *Lilium bulbiferum* (a parental species of horticultural division I). (b) *Lilium superbum* (parental species of div. IV). (c) *Lilium canadense* (as "*L. penduliflorum*"; parental species of div. IX).

*Source:* **Redouté 2000).** ***Les liliacées*** **(reprint).**

FIGURE 62.12 Examples of paintings (public domain) of old European lily (*Lilium*) varieties by Scottish botanical illustrator Walter Hood Fitch (1817–1892). Fitch prepared some 10,000 drawings for various publications, and those of lilies were among his best (see Synge 1980): (a) *Lilium hansonii* (parental species of horticultural division II). (b) *Lilium martagon* (parent of div. II). (c) *Lilium monodelphum* (parent of div. III). (d) *Lilium longiflorum* (parent of div. V). (e) *Lilium speciosum* (parent of div. VII). (f) *Lilium pardalinum* (parent of div. IX).

*Source:* **Elwes (1877–1880).**

FIGURE 62.13 Old garden catalogues (public domain) featuring lilies (*Lilium*).

*Source:* **(a) John Lewis Childs, Floral Park, NY, 1925. (b) Currie Bros., Milwaukee, WI, 1908. (c, d) Yokohama Nursery Co., Yokohama, Japan, 1909–1910. (e) James Vick, Rochester, NY, 1878. (f) H. A. Dreer, Philadelphia, PA, 1869–1870.**

FIGURE 62.14 Commercial production and marketing of lilies (*Lilium*): (a) A lily farm in 1947 in Oregon. (b) Lilies on sale in a Taiwan market. (c) A florist with cut flowering stems.

*Source:* **(a) Photo by Oregon State University Special Collections and Archives (public domain). (b) Photo by Fēngniǎo (CC BY ND 2.0). (c) Photo by Shutterstock/Liza888.**

as well as to create plants with larger flower and more robust stems. Micropropagation (using very small amounts of tissue rather than seeds or bulbs) is now often employed to produce large numbers of plants of hybrid cultivars. Meristem-tip cultures have also been employed to eliminate viruses from clones. Genetic engineering of lilies is under experimental development, but the future of transgenic ornamentals faces public opposition. Nevertheless, genetic engineering is being used today to create disease-resistant Easter lilies for the marketplace.

## 62.4 ECONOMIC IMPORTANCE

Lilies are important as outdoor plants in gardens and borders, but their economic value is due primarily to their appeal in the hugely profitable cut flower, bouquet, and short-lived potted plant industries. They are a mainstay of the world's bulb trade, very widely marketed in the form of bulbs to be used as planting stock. Lilies are also one of the most important plants grown as temporary potted houseplants and as cut flowers. They are often ranked fourth in world importance of cut flowers, after roses, chrysanthemums, and tulips. Both cut and potted lilies are widely used in floral arrangements for holiday celebrations, weddings, and funerals.

The Netherlands is the leading grower of lily bulbs, with an acreage exceeding 5,000 ha, producing over 2 billion lily bulbs annually, three-quarters of which are exported. Much smaller amounts are grown in other countries, including Australia, China, France, Japan, Israel, Netherlands, South Africa, and the United States. Several hundred hectares have been grown in each of Chile and New Zealand.

## 62.5 ORNAMENTAL USES

Lilies are prominent in horticulture as a cut flower, potted plant, and garden ornamental.

FIGURE 62.15 Huge lily (*Lilium*) patterned display gardens: (a) In Switzerland. (b) In Japan. (c) In China.

*Source:* (a) Photo by Shutterstock/Natali Glado. (b) Photo by Shutterstock/yspbqh14. (c) Photo by Shutterstock/Svilen G.

FIGURE 62.16 Lily (*Lilium*) potted plants.

*Source:* (a) Photo by Shutterstock/Wilm Ihlenfeld. (b) Photo by Michelle Raponi from Pixabay (public domain). (c) Photo by Shutterstock/Aleksandrs Samuilovs.

FIGURE 62.17 Use of lily (*Lilium*) as an ornamental cut flower.

*Source:* (a) Photo by Shutterstock/FamVeld. (b) Photo by MemoryCatcher from Pixabay (public domain). (c) Photo by Shutterstock/SvetlanaFedoseyeva.

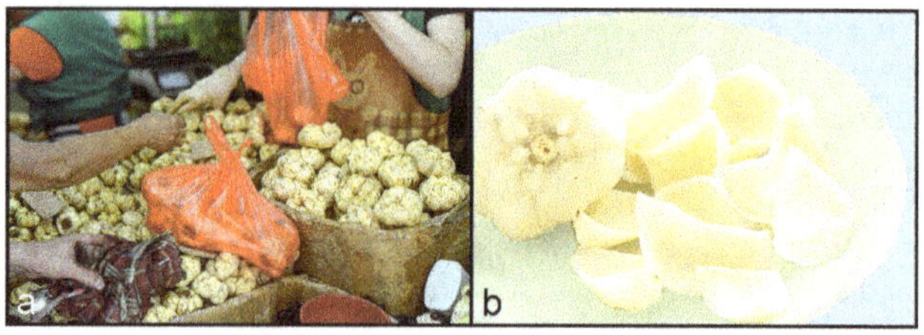

FIGURE 62.18 Edible lily bulbs: (a) A vendor selling edible lily bulbs to a customer in a food market in China. (b) An intact bulb at *left*, a bulb broken into its component scales at *right*.

*Source:* (a) Photo by Shutterstock/Yuriy Golub. (b) Photo by Shutterstock/Jreika.

## 62.6 OTHER USES

For more than a thousand years, lily bulbs have been consumed in China. Today, large acreages are also cultivated for food in Japan, Korea, and Vietnam. Although the bulbs of some species are very bitter, those of several are tasty. In China, species that are eaten include *L. candidum*, *L. lancifolium*, *L. pumilum*, and especially *L. brownii* and *L. davidii* var. *unicolor*. The Chinese usually consume dried bulbs, which are stir-fried, grated, and employed to thicken soup. The bulbs are starchy and are sometimes a source of starch. In Japan, lily bulbs are a common ingredient of chawan-mushi (savory egg custard). Flower buds from the daylily, *Hemerocallis fulva*, not from true lilies, are consumed in Chinese cuisine. North American Indians used the bulbs of many species of *Lilium* for food. The Cherokee employed the bulbs of *L. canadense* as a famine source of flour for bread. The Thompson Indians in British Columbia combined the bulbs of *L. columbianum* and salmon roe. The Cree of Saskatchewan ate the bulb scales of *L. philadelphicum* as a snack. Indigenous North Americans also steamed the bulbs, dried them into cakes for winter use, baked them in an earth oven, or used them in soup.

In Asia, species of *Lilium* have been used medicinally for at least 2,000 years to treat a variety of conditions. The Persian physician Avicenna (980–1037) recommended bulbs, leaves, and oil from the flowers for injuries, burns, and inflammation. In China, lilies have been used to treat cough, blood disorders, and insomnia. North American Indians also used the bulbs of many species of *Lilium* to treat illnesses. Mashed bulbs were applied to spider bites, cuts, and bruises and consumed for fever, coughs, tuberculosis, stomach ailments, and rheumatism. Today, *Lilium* is used to some degree as a source of medicinal steroids, and the medical value of a variety of constituents is under investigation.

## 62.7 PROBLEMS

Although humans sometimes eat the bulbs of true lilies, cats are easily poisoned by the plants. As little as two leaves or part of a flower has resulted in death. Toxicity is generally attributed to a water-soluble toxin thought to produce renal failure. Symptoms include salivation, vomiting, anorexia, and depression. The toxin acts rapidly, and treatment needs to be administered quickly. It is recommended that lilies (notably Easter lilies) and cats should not be kept indoors together. Daylilies (*Hemerocallis*) also cause nephrotoxicity in cats. Several other "lilies" can be toxic to cats (and other mammals). Bulbs of the Peruvian lily (*Alstroemeria*) contain tulipalin A, which can cause diarrhea, mouth irritation, vomiting, salivation, and digestive irritation in cats. Lily of the valley (*Convallaria*) is toxic because it possesses potent cardiac glycosides similar to those of *Digitalis*. "Lilies" of the arum family (peace lily, *Spathiphyllum*; calla lily, *Zantedeschia*) possess insoluble oxalate crystals that can irritate tissues of the mouth, tongue, pharynx, and esophagus.

## 62.8 CULTIVATION

### 62.8.1 Climate and Site

Lilies are hardy in climate zones 3 to 9, depending on species and cultivar. Lilies prefer sun or part shade, especially a south-facing exposure. A sheltered site is preferable for tall cultivars, which are susceptible to wind.

### 62.8.2 Propagation and/or Purchasing

Lilies are easily propagated by bulbs so that spectacular but sterile hybrids among the species can be easily perpetuated, sold, and established. Most lilies are grown from well-developed bulbs, although small offshoot bulbs ("bulblets") are also employed. Sometimes bulbs are simply split. Seeds can be planted but, of course, require more time to develop plants and will not reproduce hybrids accurately. Some hybrids are triploids (with three sets of chromosomes, resulting in sterility) and will not produce viable seed. Commercial growers often use individual bulb scales to grow plants. A few species produce tiny aerial bulbs ("bulbils") resembling peas, which develop in the axils of the leaves, and these can be used to propagate the plants. Some species can be reproduced easily using stem cuttings. Growers today often employ micropropagation (tissue culture based on growing small amounts of tissue in nutrient media).

FIGURE 62.19 Cats and *Lilium* lilies: an incompatible pair because the lilies can kill the cats.

*Source:* **(a) Photo by Shutterstock/Daykiney. (b) Photo by Shutterstock/Klevers.**

### 62.8.3 Planting

Bulbs are usually planted in late fall (but often also in early spring). Planting distance varies with cultivar, but a rough rule of thumb is to keep the bulbs 15 cm apart (sometimes several bulbs are planted together in a hole for a concentrated display, and sometimes exceptional plants are separated by as much as 50 cm). A depth of 2 1/2 times the height of the bulb is recommended (but Madonna lilies, *L. candidum*, should have no more than 2.5 cm of settled soil over the tops of the bulbs).

Some ornamentals, such as tulips, are best planted in large formal beds or set out in regular rows or geometrical patterns, but lilies are not one of these, because they are not particularly attractive when out of bloom. Lilies are best planted in association with other ornamentals (it is said that "lilies love mixed company"). Lilies are sometimes laid out in large blocks but show at their best in small isolated groups of five or more plants, each clump of a uniform color of a given cultivar. It is wise to grow lilies in association with companion species that do not compete seriously with their roots. Lilies have been described as the aristocrats of the summer border, a role they serve well because many are tall, with impressively large, colorful flowers. Provided they are not crowded, they display well in a mixed border with low-growing or medium-sized evergreens and deciduous shrubs, producing a needed color accent. The lilies show to advantage between or in front of shrubs, their colorful flowers complementing the green foliage of their companions. Few lilies flower in the spring, so it is advantageous to intermix them with spring-blooming perennials. Established clumps of lilies can provide color over an extended period lasting into late summer. Low-growing plants and groundcovers combine well with lilies. Clumps of lilies interspersed with short annuals can also be very attractive.

### 62.8.4 Soil and Moisture Conditions

Most lilies grow naturally in acidic soil, but a few prefer somewhat-alkaline conditions (notably Turk's cap lily, *L. martagon*, and Easter lily, *L. longiflorum*). For most species, a slightly acidic soil (pH 6.0 to 7.0) which is rich in compost is best (note that mushroom compost may have a high lime content and may change pH). The most essential requirement for lilies is good drainage, which can be encouraged by growing on a slight slope or in a raised bed. Regular moisture is needed during the growing season.

### 62.8.5 Maintaining Plants

Some cultivars with heavy inflorescences should be staked (with caution so the bulbs are not damaged). Wilted flowers should be removed. In the fall, remove senesced leaves (taking care to not pull up the bulb). Over winter, a protective mulching is recommended in cold areas. After several years, should crowding be evident by the appearance of spindly plants, the clumps should be lifted, divided, and replanted, preferably in a different location.

The lily leaf beetle (*Lilioceris lilii*), a native of Europe and Asia, seems to have first immigrated to North America in Montreal, in 1943, and subsequently spread to eastern Canada and northeastern United States. It is a devastating defoliator of all lilies but affects some other plants. The insect is 6–8 mm long, red above except for the black head, and often produces distinct chirping sounds when picked up. The larvae cover themselves with a "fecal shield" (their own excrement) which seems to repel predators, including gardeners (although simply picking them off is recommended). The beetle is well-controlled by natural predatory insects in Europe, and several of these have been released in North America in an effort to reduce the impact of the pest.

### 62.8.6 Advice Regarding Easter Lilies

*Purchasing.* Ideal plants should have flowers in various stages of bloom (from buds to open or partially opened flowers), the foliage should be dense and dark green, the plant should be well-proportioned (about twice as high as the pot), and it should be free of yellowing, insects, or signs of disease. Avoid waterlogged plants and plants displayed in shipping sleeves.

FIGURE 62.20 Lily leaf beetle (*Lilioceris lilii*).

*Source:* **Photo by dave-pemcoastphotos.com (CC BY 2.0).**

*Maintenance.* Place plant in indirect light (or light filtered through a curtain), with daytime temperatures of 15–18°C and night temperatures of 5–10°C. Keep away from drafts and drying heat sources, such as heating ducts. Water soil well when dry, but do not overwater (aluminum foil wrapping can hide waterlogged soil). Remove flowers when they wither (allegedly, flowers can be prolonged by removing anthers). Turning the pot every few days can prevent the stalk from leaning toward light. Do not fertilize.

*After Easter.* Attempting to grow an Easter lily indoors as a regular houseplant or to get it to rebloom indoors is unlikely to be successful. However, it can often be transplanted outdoors. If the plant needs to be maintained indoors because of possible frost outside, place it in bright light, remove flowers and browning or yellowed leaves, and add a slow-release fertilizer. After danger of frost, Easter lilies can be planted outside, setting the bulb 15 cm below the soil surface. Loosen clumped and matted roots, and spread the root ball. Leave the stem and leaves, but cut off remaining flowers. Blooming might occur in mid- or late summer, but probably not until next season. In cold areas (the species has been rated as capable of surviving climate zone 3, although it is generally evaluated as semi-hardy), consider planting near the house foundation, and mulch well overwinter.

## 62.9 FASCINATING FACTS, FABLES, FOLKLORE, AND FRIVOLITY

- In ancient times, lilies were frequently considered to be signs of fertility and a pure life and were used as offerings to appease the gods. Lilies have also been associated with death, perhaps because the dead are no longer capable of sin and so regain lost innocence and purity. At one time, lilies were placed on the graves of the young (assumed to be innocent).
- *Lilium candidum* (Madonna lily, white lily) is especially significant as a Christian symbol. It is commonly associated with St. John the Baptist, St. Joseph, St. Dominic, St. Francis of Assisi, and St. Anthony of Padua. Most of all, as the "Madonna lily," it is the symbol of the Virgin Mary, symbolizing purity and chastity. This is ironic, as the Church adopted the lily from the pagan Greek goddess Aphrodite (later adopted by the Romans as Venus), who symbolized love, beauty, pleasure, and procreation and had many lovers. The ancient Greeks also associated *L. candidum* with Hera (later adopted by the Romans as Juno), a competitor of Aphrodite and goddess of women and marriage, who might, at first blush, seem to be a superior figure since she was worshipped as a virgin, but she spent most of her time in a jealous

rage, plotting vengeance against people who innocently crossed her path.

- The "white lily" was associated with slavery in the American South in the early 1850s, before the Civil War. Slaves were sometimes branded with a lily symbol. In the early 1880s, the white lily was adopted as a symbol by racist groups in the Deep South, especially Georgia. The emblem was placed on buildings to indicate sympathy for racism. The Stem of the White Lily, an organization parallel to the Ku Klux Klan, was responsible for over 200 lynchings. The white lily is still considered a racist emblem in the Southern United States.
- Dating to the 1916 uprising by Irish republicans, the Easter lily has been used as a badge worn at Easter by Irish republicans as a symbol of remembrance for Irish combatants who died during the hostilities with the British.
- A trend (promoted by the flower industry) is to use a gift of lilies as symbolic of a 30th wedding anniversary.
- According to an old wives' tale, approaching an expectant mother with a lily in one hand and a rose in the other could be used to tell the sex of the child. If the mother chooses the rose, it will be a girl, but picking the lily indicated that it will be a boy (because of the phallic appearance of the long pistil).
- In the Victorian era, "floriography," or the language of flowers, was the basis of sending messages by careful choice of the composition and color of the flowers included. A white lily conveyed the sentiment that the receiver was pure and sweet. A yellow lily could mean gratitude or jubilation; an orange lily, however, indicated hatred.
- The rose is the leading flower for the names of people (Rose, Rosie, Rosalyn, etc.). The lily is the second most popular basis for floral names (Lily, Lil, Lilliane, etc.).
- Reflecting the importance of the plants, the word *lily* appears in several idiomatic phrases. "Lily-white" can mean (1) white as a lily (especially pertaining to skin); (2) above reproach, untouched by corruption or imperfection; or (3) excluding Blacks (especially in the Southern United States). "Lily-livered" means cowardly. "To gild the lily" means to give an often deceptively attractive or improved appearance to something.
- The English plant collector E. H. Wilson (1876–1930) discovered the regal lily (*L. regale*) in 1903 in western Szechuan, China. In 1910, he returned to the site to obtain more bulbs, and while he was being carried in a sedan chair, an avalanche of boulders crushed his leg. He set his broken leg with the tripod of his camera and endured a three-day forced march, along with his collected bulbs, which subsequently were used to introduce the species into commercial cultivation in much of the world. After his trip, Wilson walked with what he called his "lily limp."
- *Lilium candidum* (called the white garden lily) was designated as the floral emblem of the province of Quebec on January 23, 1963, the result of interpreting the fleur-de-lis emblem, symbolic of France and, consequently, of Quebec, as a lily. However, this came to be appreciated as an erroneous interpretation (the fleur-de-lis is probably not based on a flower), and on November 5, 1999, the blue flag (*Iris versicolor*) was substituted as the official provincial flower of Quebec.
- The Western red lily (*L. philadelphicum*) was chosen as the official floral emblem of Saskatchewan in 1941. Aside from Saskatchewan, no species of *Lilium* appears to have official designation as representative of a political region. *Lilium bosniacum* has been claimed to be the national flower of Bosnia and Herzegovina, but this seems to be based on a tradition of it being a national emblem among the Bosniacs. Several so-called "lilies" are emblems of regions, but not species of *Lilium* (e.g., the flame lily, *Gloriosa rothschildiana*, is the national flower of Zimbabwe).
- "Sea lilies" are marine animals called crinoids, which make up the class Crinoidea. Crinoids are attached to the seafloor by a stalk when young, and the forms that remain stalked when mature are called sea lilies because of their similarity to lily flowering stalks. Crinoids are allied to starfish, brittle stars, sea urchins, and sea cucumbers. There are about 700 living species, a group which was much more diverse in ancient times.

FIGURE 62.21 *Sea lilies*, by H. Harder (1858–1935).

*Source:* **Hader, H. 1920. *The Wonderful Paleo Art of Heinrich Harder* (public domain).**

## KEY PUBLICATIONS


Allen, T., D.C. Kiplinger, and R.W. Langhans. (eds.). 1967. *Easter Lilies. The Culture, Diseases, Insects and Economics of Easter Lilies*. The New York and Ohio Lily Schools, New York, NY and Wooster, OH. 158 pp.

Cohen, A. 2011. Biotechnology in lilies – dreams vs. reality. In: *Proceedings 2nd International Symposium on the Genus Lilium*, edited by A. Grassotti and G. Burchi, 149–159. International Society of Horticultural Science, Leuven, Belgium.

Comber, H.F. 1947. A new classification of the *Lilium*. *Lily Yearbook, Royal Horticultural Society (London)* 15: 86–105.

De Graaff, J. and E. Hyams. 1967. *Lilies*. Thomas Nelson and Sons, London. 142 pp.

Dhiman M.R., S. Moudgil, C. Parkash, R. Kumar, and S. Kumar. 2018, Biodiversity in *Lilium*: A review. *International Journal of Horticulture* 8(8): 83–97.

Elwes, H.J. 1877–1880. *A Monograph of the Genus Lilium*. Taylor & Francis, London, U.K. (Illustrated by W.H. Fitch. 7 parts in one volume. One of the great flower encyclopedias of the Victorian age.)

Feldmaier, C. 1970. *Lilies*. Arco, New York. 228 pp.

Fitzgerald, K.T. 2010. Lily toxicity in the cat. *Topics in Companion Animal Medicine* 25: 213–217.

Fox, D. 1985. *Growing Lilies*. Dover, London. 264 pp.

Grassotti, A. and F. Gimelli. 2011. Bulb and cut flower production in the genus *Lilium*: Current status and the future. In: *Proceedings 2nd International Symposium on the Genus Lilium*, edited by A. Grassotti and G. Burchi, 21–35. International Society of Horticultural Science, Leuven, Belgium.

Hayashi, K. and Kawano, S. 2000. Molecular systematics of *Lilium* and allied genera (Liliaceae): Phylogenetic relationships among *Lilium* and related genera based on rbcL and matK gene sequence data. *Plant Species Biology* 15: 73–93.

Hiramatsu, M., K. Ii, H. Okubo, K.L. Huang, and C.W. Huang. 2001. Biogeography and origin of *Lilium longiflorum* and *L. formosanum* (Liliaceae) endemic to the Ryukyu Archipelago and Taiwan as determined by allozyme diversity. *American Journal of Botany* 88: 1230–1239.

İkinci, N., C. Oberprieler, and A. Güner. 2006. On the origin of European lilies: Phylogenetic analysis of *Lilium* section *Liriotypus* (Liliaceae) using sequences of the nuclear ribosomal transcribed spacers. *Willdenowia* 36: 647–656.

Jefferson-Brown, M.J. and A. Lawson. 2003. *Lilies: A Guide to Choosing and Growing Lilies*. Quadrille, London. 160 pp.

Lim, K-B., R. Barba-Gonzalez, S. Zhou, M.S. Ramanna, and J.M. van Tuyl. 2008. Interspecific hybridization in lily (*Lilium*): Taxonomic and commercial aspects of using species hybrids in breeding. In: *Floriculture, Ornamental and Plant Biotechnology*, Vol. 5, edited by J.A. Teixeira da Silva, 146–151. Global Science Books, Isleworth, U.K.

Matthews, V. (ed.). 2007. *Lily Register and Checklist*. International Fourth Edition. Royal Horticultural Society, London. 948 pp.

McGeorge, P. 2004. *Lilies*. Firefly Books, Buffalo, NY. 96 pp.

McRae, E.A. 1998. *Lilies: A Guide for Growers and Collectors*. Timber Press, Portland, OR. 392 pp.

Nishikawa, T., K. Okazaki, T. Uchino, K. Arakawa, and T. Nagamine. 1999. A molecular phylogeny of *Lilium* in the internal transcribed spacer region of nuclear ribosomal DNA. *Journal of Molecular Evolution* 49: 238–249.

Redouté, P.J. 2000. *The Lilies*. Taschen, New York. 494 pp. [Reproduction of illustrations from Redouté's book, Les Liliacées, originally published in eight volumes between 1802 and 1816.]

Rockwell, F.F., E.C. Grayson, and J. de Graaff. 1961. *The Complete Book of Lilies: How to Select, Plant, Care for, Exhibit, and Propagate Lilies of All Types*. Doubleday, Garden City, NY. 352 pp.

Skinner, M.W. 2002. *Lilium*. In: *Flora of North America*, Vol. 26, edited by Flora of North America Editorial Committee, 172–197. Oxford University Press, New York.

Synge, P.M. 1980. *Lilies: A Revision of Elwes' Monograph of the Genus Lilium and Its Supplements*. Batsford, London. 276 pp.

Van Tuyl, J.M. and P. Arens. 2011. *Lilium*: Breeding history of the modern cultivar assortment. In: *Proceedings of the 2nd International Symposium on the Genus Lilium*, edited by A. Grassotti and G. Burchi, 223–230. International Society of Horticultural Science, Leuven, Belgium.

Van Tuyl, J.M., P. Arens, M.S. Ramanna, A. Shahin, N. Khan, S. Xie, et al. 2011. *Lilium*. In: *Wild Crop Relatives: Genomic and Breeding Resources. Plantation and Ornamental Crops*, edited by C. Kole, 161–183. Springer-Verlag, Belin.

Woodcock, H.D. and W.T. Stearn. 1950. *Lilies of the World, Their Cultivation & Classification*. Scribner, New York. 431 pp.

Xia, Z, J. Wan, Y. Chen, Y. He, and J. Yu. 2013. Experimental oriental hybrid lilies (*Lilium* hybrids) poisoning in cats. *Journal of Clinical Toxicology* 3(1): 1–4.

Zhou, N., K. Miao, C. Liu, L. Jia, J. Hu, Y. Huang, et al. 2023. Historical biogeography and evolutionary diversification of *Lilium* (Liliaceae): New insights from plastome phylogenomics. *Plant Diversity*. www.sciencedirect.com/science/article/pii/S2468265923001087.


## KEY CULTIVATION WEBSITES

*Gardenia: Lilies: How to Grow and Care with Success*. www.gardenia.net/guide/learn-how-to-plant-and-care-for-your-lilies.

Royal Horticultural Society Lily Group. www.rhslilygroup.org/.

The Genus *Lilium*. www.the-genus-lilium.com/main.htm.

The North American Lily Society. www.lilies.org/.

The Spruce. *9 Types of True Lilies to Grow in Your Garden*. www.thespruce.com/lily-types-to-grow-in-garden-1315809.

The Spruce. *14 Recommended Lily Varieties for Your Garden*. www.thespruce.com/beautiful-lily-varieties-to-grow-4136203.

WikiHow. *How to Care for Peace lilies*. www.wikihow.com/Care-for-Peace-Lilies.

WikiHow. *How to Divide and Transplant Lilies*. www.wikihow.com/Divide-and-Transplant-Lilies.

WikiHow. *How to Grow and Care for Asiatic Lilies*. www.wikihow.com/Grow-and-Care-for-Asiatic-Lilies.

WikiHow. *How to Grow Stargazer lilies*. www.wikihow.com/Grow-Stargazer-Lilies.

WikiHow. *How to Overwinter Lily Bulbs*. www.wikihow.com/Overwinter-Lily-Bulbs.

WikiHow. *How to Plant Easter lilies*. www.wikihow.com/Plant-Easter-Lilies.

WikiHow. *How to Plant Lily Bulbs*. www.wikihow.com/Plant-Lily-Bulbs.

WikiHow. *When & How to Prune Lilies for Healthy, Happy Blooms*. www.wikihow.com/Prune-Lilies.

# 63 Lily of the Valley (*Convallaria majalis*)

**Scientific Names:** The genus name *Convallaria* is based on Latin *convallis*, meaning "deep valley." *Majalis* in the scientific name *C. majalis* is derived from *maius*, meaning "of or belonging to May," in reference to the month of flowering in many regions. (Coincidentally, *majalis* in Latin means "castrated pig.")

**Common Names:** The common name lily of the valley (also lily-of-the-valley), like the genus name, is based on past observations of the species in valleys. The phrase *lily of the valley* has been incorporated into the names of unrelated plants, as illustrated in the following. "False lily of the valley" refers to several species of *Maianthemum* ("Canadian lily of the valley" is *Maianthemum canadense*, but the European form of *C. majalis* has become established in the wild in much of Canada). "Lily of the valley vine" is *Salpichroa origanifolia*. "Chilean lily of the valley tree" is *Crinodendron patagua*. "Lily of the valley bush" (or shrub) is *Pieris japonica*.

## 63.1 INTRODUCTION

Lily of the valley is an unassuming plant with tiny flowers. Despite its inconspicuous nature, it has, for centuries, become a favorite bloom of many, mostly women, including royalty, and has become a staple in bouquets at weddings. It appears to be pre-adapted for the purpose, having delicate white flowers that complement the traditional stereotype of dainty, virginal brides. It seems miraculous that such a wonderful fragrance is produced by its diminutive flowers. In the garden, the natural adaptation of the plant to shade is valued, since most ornamental plants thrive in full sun. Culturally, lily of the valley has attracted the interest of the artistic community and has been featured in paintings and literary works. On the negative side, the plant can be invasive, and in cultivation, it is sometime so aggressive that it has been referred to as a "garden thug."

FIGURE 63.1 Views of lily of the valley (*Convallaria majalis*). All are the "European lily of the valley"): (a) Close-up of a flower (in France). (b) Flowering plants (in Austria). (c) Flowering branch (in Germany). (d) Plants in fruit (in Italy). (e) Fruiting branches (in Europe).

*Source:* (a) Photo by Thomas Bresson (CC BY 3.0). (b) Photo by GT1976 (CC BY SA 4.0). (c) Photo by Björn S (CC BY SA 2.0). (d) Photo by Frank Vincentz (CC BY SA 3.0). (e) Shutterstock, contributed by SariMe.

## 63.2 OVERVIEW

The genus *Convallaria* is usually considered to be made up of just one species, *C. majalis*, but some botanists treat the plants of the eastern United States as *C. pseudomajalis* (*C. montana*; "American lily of the valley"), and the plants of eastern Asia as *C. keiskei*, while restricting the name *C. majalis* to the plants of Europe and western Asia. Regardless of whether species names are assigned, there are genetic differences separating American, European, and Asian plants. Lily of the valley is found in cool woodlands and hillsides of the temperate Northern Hemisphere. The European form of the plant ("European lily of the valley") has been widely employed as an ornamental and has been introduced in North America, often growing about abandoned homesteads, roadsides, and cemeteries. The plants are herbaceous perennials, arising from a branching rhizome (underground stem), and continued spread of the rhizome often produces extensive colonies. Shoots arise from the rhizome, usually growing up to 30 cm in height and bearing one or two leaves. Flowering stems produce 5 to 15 very sweetly scented, pendent (nodding), bell-shaped white (occasionally pink) flowers which develop on one side of an elongated, arching, slender stalk in the spring. The fruits are orange-red berries, 5–7 mm in diameter, containing a few seeds. In many natural colonies (especially in North America), berries are infrequent or absent.

FIGURE 63.2 Old botanical paintings (public domain) of lily of the valley.

*Source:* (a) Prof. Dr. Otto Wilhelm Thomé. 185. Flora von Deutschland, Österreich und der Schweiz. Gera, Germany. (b) Guillaumin, A. 1929. Les Fleurs de Jardins, tome I, Les Fleurs de Printemps, Paul Lechevalier, France. (c) A vintage postcard.

DOI: 10.1201/9781003586227-63

## 63.3 HISTORY

Throughout its distribution areas, lily of the valley has been wildcrafted (harvested from the wild) for use as a cut flower, and sometimes also for medicinal applications. The plant has not been domesticated extensively, but there are selections that have doubled flowers, larger flowers, pink- or rose-colored flowers, larger size, and/or variegated (especially striped) foliage.

## 63.4 ECONOMIC IMPORTANCE

Lily of the valley is a minor garden ornamental, but one that nevertheless has attracted enthusiastic supporters in much of the world. It is sold in specialty niche markets but maintains a presence on the property and in the hearts of many people everywhere.

## 63.5 ORNAMENTAL USES

*Convallaria majalis* is extensively cultivated in gardens for its scented flowers and as a groundcover, particularly in shady locations. The plants are suitable for beds and woodland gardens and are superb for supplying cut flowers for one's residence in the spring (cut flowers last as long as a week). They are also grown in borders and rock gardens, but in shady places, they can become too aggressive (although often used as low-maintenance groundcover for shaded areas). Lily of the valley is commercially cultivated indoors (often, outdoor plants are harvested, kept cool or dormant until needed, and simply forced to flower out of season) to provide cut flowers, traditionally and especially employed for weddings. The plants, which are a favorite of florists, can be expensive, and often flowering branches of other species that mimic lily of the valley are substituted.

FIGURE 63.3 Ornamental selections of lily of the valley: (a) A pink-flowered form (labelled as *Convallaria majalis* var. *rosea*). (b) A variegated form (with striped leaves). (c) A cultivar with doubled flowers (identified as *Convallaria majalis* 'Plena').

***Source:*** **(a) Photo by Hardyplants (CC BY SA 4.0). (b) Photo by Paul Henjum (public domain). (c) Photo by Aha (CC BY SA 3.0).**

## 63.6 OTHER USES

Extracts of lily of the valley have been employed to treat heart failure, mostly more than a century ago, and only in a limited way. The plant is still used, mostly in European herbal medicine, for regulating heart functions, much as digitalis is employed. Extracts have also been used in aromatherapy and have occasionally been incorporated as fragrance in detergents, soaps, and cleansing creams. All these uses require caution in view of the presence of toxic chemicals in the plant. Despite efforts, chemists have been unable to process chemicals from lily of the valley into perfumes, although several commercial perfumes have been created that simulate the fragrance (including the French firm Dior's "Diorissimo, "La Muguet" by Annick Goutal, "Muguet" by Guerlain, "Muguet du Bonheur" by Caron, and "Varens Essential Muguet Secret" by Ulric de Varens).

## 63.7 PROBLEMS

All parts of lily of the valley are toxic, due to the presence of cardiac glycosides, although the roots develop particularly high concentrations. The red berries can be attractive to children. Ingestion can lead to intestinal pain, nausea, vomiting, and irregular heartbeat. The plant is also toxic to both wild and domesticated mammals (even goats), household pets, and birds (including hens, ducks, and geese).

When flowers are not present, lily of the valley foliage resembles the leaves of some wild onion species that are collected by wild food enthusiasts, particularly *Allium ursinum* (known as ramsons and wild garlic) in Europe. Like

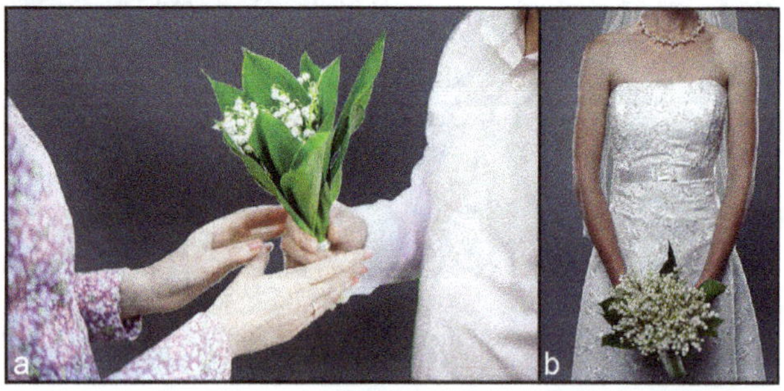


FIGURE 63.4 Lily of the valley as a symbol of love and marriage.

***Source:*** **(a) Photo by Shutterstock/Photokirov. (b) Photo by Shutterstock/Ekaterina Pokrovskaya.**

FIGURE 63.5 Household pets inadvisably posed near lily of the valley, which is capable of severely poisoning them: (a) Pomeranian. (b) Ginger cat.

***Source:*** **(a) Photo by Shutterstock/KristinaSh. (b) Photo by Shutterstock/Avatu.**

*Convallaria*, this species can form extensive colonies in woodland. In North America, collectors of wild leek or ramps (*Allium tricoccum*) have similarly misidentified and consumed lily of the valley, resulting in serious poisoning. *Allium* species may look similar, but they have the characteristic smell of onions and garlic. However, it may be noted that, in general, various poisonous plants mimic edible wild plants, so foraging on wild plants requires some expertise.

Lily of the valley is considered to be invasive in various regions, with potential to negatively affect biodiversity. As with many ornamentals with invasive potential, just how potentially damaging many of the garden forms (which have been weakened by domestication) are is not clear.

## 63.8 CULTIVATION

### 63.8.1 Climate and Site

Lily of the valley is naturally adapted to warm summers and cold winters. This relatively cool-weather plant is not recommended for hot climates. Some authorities rate it as hardy as far north as climate zone 2 (it is known to survive below –20°C); others give the adapted range as 3 to 8. Lily of the valley grows best in partial shade but can be grown in full sun (at least in Northern regions, and if watered regularly), and in full shade. The plants will not be attractive after the flowers senesce in the early summer (although plants bearing red fruit are of interest), and experienced gardeners position them near other flowering plants that will provide ongoing color in the location. Lily of the valley performs well under trees and shrubs.

### 63.8.2 Propagation and/or Purchasing

Lily of the valley is reproduced by division and transplanting portions of the plant in the autumn or the spring (when the plants are dormant is best). The plants can be grown from seeds either purchased or collected from garden plants. The surrounding flesh from the ripe fruit should be removed, and seeds can be planted outdoors in the autumn, preferably in containers in a cold frame, kept semi-moist. However, it can take from 2 months to a full year for seeds to germinate and sprout, and seedlings may require 5 to 10 years before they flower. Florists and online sites often offer pre-chilled potted plants that will bloom when desired as houseplants.

### 63.8.3 Planting

Cool or cold winter temperatures are needed to satisfy the dormancy period needed for flowering, so plants should be established in the ground in the fall, not kept indoors overwinter in heated conditions. (Nevertheless, lily of the valley houseplants that flowered in the winter can be transplanted to the garden in the spring.) Bare-rooted bulbous roots (termed "pips") or container-grown plants should be positioned so the top peeks out over the soil surface. Spacing about 10 cm apart is recommended.

### 63.8.4 Soil and Moisture Conditions

The plants prefer sandy and loamy, well-drained but moist soils enriched with humus, with pH 5.8 to 7.0. Clay soils are tolerated. Summer drought is also tolerated, but the plants benefit from being watered.

### 63.8.5 Maintaining Plants

In nature, lily of the valley can become dominant in a region if conditions are suitable, and similarly, in a garden, the plant can become rather weedy. It is quite long-lived, although like other herbaceous perennials, it may require dividing every few years for rejuvenation (to promote flowering). In a border, it may be necessary to restrict how far the plant spreads. Just how much irrigation, fertilization, and removal of spent materials is required is optional, since it is capable of surviving with limited care.

## 63.9 FASCINATING FACTS, FABLES, FOLKLORE, AND FRIVOLITY

- "Mary's tears" and "Virgin's tears" are occasional old European names for *Convallaria majalis*, reflecting a legend that when Jesus died upon the cross, his mother, Mary, cried, and her tears grew into lily of the valley plants.
- Other comparable folk legends interpret lily of the valley as having originated from Eve's tears after she and Adam were driven from the Garden of Eden, or from the blood shed by Saint Leonard of Noblac during his battles with Malitia, an evil dragon.
- Still more fanciful myths concern lily of the valley. It has been said that fairies use the flowers as steps to the reeds from which they weave their cradles, and that the flowers (sometimes called "fairy bells") ring when the fairies sing. An old fairy tale describes how the nightingale refused to sing unless the lily of the valley agreed to flower every May.
- Lily of the valley is the national flower of Finland and was the national flower of Yugoslavia before it split into independent countries (beginning in 1992).
- Bourgeonal, an odor constituent of lily of the valley, has been found to stimulate mammalian sperm to swim faster.
- In France in the early 20th century, lily of the valley was sold on International Labor Day, May 1 (termed Lily of the Valley Day, traditionally indicating the beginning of spring and often given as a gift at that time), with the sales tax suspended for the day.
- Lily of the valley is one of the wordiest of plant names but is exceeded by such old nicknames as "John go to bed at noon" (salsify, *Tragopogon* species; and others), "Welcome home husband although never so drunk" (yellow stonecrop, *Sedum acre*),

"Kitty come down the lane jump up and kiss me" (cuckoo-pint, *Sedum acre*), and "Meet her in the entry kiss her in the buttery" (pansy, *Viola tricolor*).

- *İnci* is Ottoman Turkish meaning "pearl," "lily of the valley," and all plants with pearl-like white flowers or berries. The perception of lily of the valley flowers as pearls is widespread, and numerous works of art have been created in which genuine pearls represent lily of the valley flowers.
- In this current age of near-instantaneous transfer of electronic messages, it is easy to forget that snail mail was once the standard. In those times, postcards were immensely popular, and in the late 19th and early 20th centuries, artists often created pictorial masterpieces. Today, remaining specimens are collector's items. Ornamental plants were a favorite theme for postcards, and many showed lily of the valley.

FIGURE 63.6 Artistic creations with lily of the valley represented by genuine pearls: (a) A Fabergé jewelled egg showing lily of the valley flowers as pearls. Peter Carl Fabergé created fewer than 60 eggs, mainly as Easter gifts for the Russian aristocracy, from 1885 to 1917. They are considered priceless but have sold for millions of dollars. (b) A jeweled creation by Fabergé in the form of lily of the valley with pearls representing the flowers, housed at the Virginia Museum of Fine Art. (c) Vintage image of a "lily of the valley dress," with pearls representing the flowers.

***Source:* (a) Photo by Pedro Szekely (CC BY SA 2.0). (b) Photo by Chuck Redden (CC BY 2.0). (c) Public domain.**

FIGURE 63.7 Vintage (more than a century old, copyright-free) illustrations of lily of the valley on postcards.

***Source:* Public domain.**

## KEY PUBLICATIONS

Anonis, D.P. 1986. Lily of the valley (muguet) in perfumery. *Perfumer & Flavorist* 11: 31–36.

Boelens, M., H.J. Wobben, and J. Heydel. 1980. Muguet in perfumery: A review of lily of the valley. *Perfumer & Flavorist* 5: 1–8.

Chwedorzewska, K.J., H. Galera, and I. Kosiński. 2008. Plantations of *Convallaria majalis* L. as a threat to the natural stands of the species: Genetic variability of the cultivated plants and natural populations. *Biological Conservation* 141: 2619–2624.

Jones, N. 2003. Sperm get their kicks from lily of the valley. *New Scientist* 178(2389): 19.

Kanchi, G.N., R.L. James, and Z.L. James. 2012. Nomenclatural and taxonomic analysis of *Convallaria majalis*, *C. majuscula*, and *C. montana* (Ruscaceae/Liliaceae). *Phytoneuron* 17: 1–4.

Kondo, T., M. Narita, S.S. Phartyal, S.N. Hidayati, J.L. Walck, J.M. Baskin, et al. 2015. Morphophysiological dormancy in seeds of *Convallaria keiskei* and a proposal to recognize two types of double dormancy in seed dormancy classification. *Seed Science Research* 25: 210–220.

Lu, Q.-X., X. Chang, J. Gao, X. Wu, J. Wu, Z.-C. Qi, et al. 2022. Evolutionary comparison of the complete chloroplast genomes in *Convallaria* species and phylogenetic study of asparagaceae. *Genes* 13(10): 1724. www.mdpi.com/2073-4425/13/10/1724.

Maililie, Z. and Wilkins, H.F., 1985. *Convallaria majalis*. In: *CRC Handbook of Flowering*, Vol. 2, edited by A.H. Halevy, 321–323. CRC Press, Boca Raton, FL.

Ohara, M., A. Kiwako, E. Yamada, and S. Kawano. 2006. Life-history monographs of Japanese plants. 6: *Convallaria keiskei* Miq. (Convallariaceae). *Plant Species Biology* 21: 119–126.

Ramachanderan, R. and B. Schaefer. 2019. Lily-of-the-valley fragrances. *ChemTexts* 5: 1–34.

Reimann-Philipp, R. and G. Reimann-Philipp. 1987. *Convallaria majalis* (lily of the valley) improved as ornamental plant after 35 years of breeding work. *Acta Horticulturae* 205: 65–70.

Rochelle, R. 2022. *Lily of the Valley Cultivation Handbook: The Beginners Guide on Growing, Caring, Pruning, Propagation, Maintenance and Lot More*. Self-published. 31 pp.

Wachsmuth, B. 2017. A historical overview of *Convallaria* cultivars. *Plantsman* 16: 92–99.

Wilkins, H.F. 1985. *Convallaria majalis*. In: *Handbook of Flowering*, Vol. II, edited by A.H. Halevy, 321–323. CRC Press, Boca Raton, FL.

## KEY CULTIVATION WEBSITES

Gardeners World.com. *How to Grow Lily of the Valley*. www.gardenersworld.com/how-to/grow-plants/how-to-grow-lily-of-the-valley-convallaria/.

Longfield Gardens. *All about Lily of the Valley*. www.longfield-gardens.com/article/all-about-lily-of-the-valley/.

Old Farmer's Almanac. *How to Plant, Grow, and Care for Lily of the Valley Flowers*. www.almanac.com/plant/lily-valley.

Sloane & Sons Garden Benches. *The Ultimate Lily of the Valley Growing Guide*. www.gardenbenches.com/blog/lily-of-the-valley/.

The Spruce. *How to Grow and Care for Lily of the Valley*. www.thespruce.com/lily-of-the-valley-2132623.

WikiHow. *How to Plant Lily of the Valley*. www.wikihow.life/Plant-Lily-of-the-Valley.

# 64 Lisianthus (*Eustoma grandiflorum*)

**Scientific Names:** The genus name *Eustoma* originates from the Greek *eu-*, "good" or "pleasing," and *stoma*, "mouth," suggesting that the flowers have a pleasing mouth (possibly a pleasingly colored throat).

**Common Names:** *Lisianthius* (which used to be spelled *Lisianthus*) is a genus of plants native from Mexico to Colombia and to the Caribbean. The name lisianthus is based on the ancient Greek *lysis*, "bitter," and *anthos*, "flower," reflecting the bitter taste of many of the plants in the gentian family (including *Eustoma*). *Eustoma* was once considered to be part of *Lisianthius* or *Lisianthus*, and the common name lisianthus simply persisted for the plants that were reclassified as *Eustoma*. Adding to difficulties of interpretation, the common name lisianthus is also frequently confused with the genus name *Lasianthus*, a quite unrelated group of plants. *Eustoma* specimens have been misidentified as *Lasianthus* (often misspelled as *Lisianthus*), and vice versa. Lisianthus is also called prairie gentian (and less frequently, Texas bluebell, tulip gentian, and bluebells). The name lisianthus is sometimes shortened to "lizzy" (a name also applied to *Impatiens walleriana*, discussed in this book).

## 64.1 INTRODUCTION

Many cultivars of lisianthus produce gorgeous cut flowers which are remarkable mimics of roses but are sometimes produced more cheaply and even last longer in vases. Probably, many who have received gift bouquets featuring lisianthus flowers from florists have thought that they were indeed roses. As a result, lisianthus has become a favorite of florists and one of the world's most admired flowers, even if it has often been mistaken for the rose, the queen of flowers.

FIGURE 64.1 Roselike appearance of lisianthus cultivars with doubled flowers.

*Source:* (a) Photo by Tatters (CC BY SA 2.0). (b) Photo by Swallowtail Garden Seeds (CC BY 2.0). (c) Photo by Marco Verch (CC BY 2.0). (d) Photo by Takeaway (CC BY SA 4.0).

## 64.2 OVERVIEW

The genus *Eustoma* is found in the Southern United States, Mexico, and the Caribbean. The two species occur in disturbed areas, grasslands, and pine forests, frequently in dry environments. The principal cultivated species is *E. grandiflorum* (previously named *E. russellianum* and *Lisianthius russellianus*), often called the florist's lisianthus, since it is primarily obtained from florists (the other species is *E. exaltatum*). The plants are tender herbaceous perennials, and in nature, the wild plants often develop just a low circle of leaves (rosette) which goes dormant over a dry season and flowers the next year. The cultivars are usually grown as annuals, but when planted as seeds outdoors to be grown as perennials in hot climates, they often won't flower until

FIGURE 64.2 Lisianthus gardens.

*Source:* (a) Photo by Shutterstock/Artpritsadee. (b) Photo by Shutterstock/Photo Win1.

DOI: 10.1201/9781003586227-64

the second year. The plants are 30 to 90 cm tall. The bell-shaped flowers are about 5 cm in diameter. The cultivars are available in blue, cream, lavender, mauve, pink, purple, red, white, and yellow; in bicolors; and in doubled forms. The doubled flowers are often very reminiscent of tea roses (i.e., they are large, with many crowded, ruffled petals and a lush appearance) and were called "love roses." The flowers are essentially scentless (many modern rose varieties are also not fragrant). Flowering occurs in midsummer onward, but earlier when purchased in the spring from starter plants raised in nurseries.

## 64.3 HISTORY

Early cultivars of lisianthus grown in Europe were mostly single-flowered. Most modern lisianthus cultivars were bred within the last few decades, especially in Japan (hence the occasional nickname "Japanese rose").

## 64.4 ECONOMIC IMPORTANCE

Lisianthus is ranked in the top ten commercial cut flowers worldwide. It is also employed as a garden ornamental, but this is much less important economically.

## 64.5 ORNAMENTAL USES

Lisianthus is grown primarily as a cut flower, the flowers lasting weeks (at least 2) in vases. Lisianthus stems can carry several flowers that bloom in sequence from the bottom to the top, so a given stem can last up to a month. The plants are typically employed in bridal and gift bouquets and in mixed flower arrangements. Lisianthus is also often used as a flowering bedding and container plant.

FIGURE 64.3 Early paintings (public domain) of garden forms of lisianthus (*Eustoma grandiflorum*), which were predominantly single-flowered (with just one set of five petals).

*Source:* **(a) Painted by Mary Vaux Walcott (1860–1940). Photo from Smithsonian American Art Museum. (b) *Revue horticole*, published by Librairie Agricole de la Maison Rustique, Paris, 1863. (c) Hartinger, A. 1846. *Endlicher's Paradisus vindobonensis*, vol. 54.**

FIGURE 64.4 Field production of lisianthus at Hmong American Farmers Association flower farm in Minnesota: (a) Harvesting plants for market. (b) A bunch of harvested plants.

*Source:* **Photos by Media Mike Hazard (CC BY 2.0).**

FIGURE 64.5 Greenhouse production of lisianthus.

*Source:* **(a) Photo by Shutterstock/Roberta Blonkowski. (b) Photo by Shutterstock/Ruud Morijn.**

## 64.6 OTHER USES

Lisianthus was used in herbal medicine in the past in its Native Mexican and Central American range but is not employed in modern therapies.

## 64.7 PROBLEMS

Lisianthus is considered to be quite difficult to grow properly by home gardeners, but professional growers have mastered its large-scale production.

FIGURE 64.6 Lisianthus as a cut flower.

*Source:* **(a) Photo by Shutterstock/Africa Studio. (b) Photo by Downtowngal (CC BY SA 4.0). (c) Photo by Shutterstock/Soyka.**

FIGURE 64.7 Lisianthus in garden beds.

*Source:* **(a) Cultivar Lisa Pink. Photo by David J. Stang (CC BY SA 4.0). (b) Cultivar Forever Blue. Photo by David J. Stang (CC BY SA 4.0). (c) Photo by Rameshng (CC BY SA 3.0). (d) Photo by Alex Proimos (CC BY 2.0).**

## 64.8 CULTIVATION

### 64.8.1 Climate and Site

For most people, lisianthus is obtained as cut flowers in the form of gifts and flower arrangements, to be maintained as temporary presentations in vases. These should be placed away from drafts or heaters, away from direct sunlight, in a clean container with clean water (and plant food, which is often supplied by florists).

Lisianthus is frost-sensitive, and when treated as a garden plant, it is grown as an annual in temperate regions. It can be cultivated as a perennial in climate zones 8 and southward. The plants develop best in full sunlight (some afternoon shade may be beneficial in very hot areas). They like hot days and dislike humid conditions. Raised beds and containers can be easier to maintain than plants grown in conventional beds and borders. Dwarf varieties (15 to 20 cm tall) can be grown as potted houseplants (but are difficult to maintain). In warm regions where lisianthus is grown as a perennial, it can be divided.

### 64.8.2 Propagation and/or Purchasing

Lisianthus is usually grown as an annual started from seeds. The seeds are tiny and so difficult to hold that they are often pelleted to handle easily. Because the plants grow slowly and it may require months (sometimes over a year!) for flowers to develop, they are usually purchased as plugs from plant nurseries. Because the plants are very slow to flower, it is a good idea to buy specimens that have already produced flower buds.

### 64.8.3 Planting

In hot regions, the seeds can be planted directly in the ground, but as noted earlier, in most regions, transplanting pre-grown seedlings that are relatively mature is recommended. Many authorities encourage close spacing of 15–20 cm apart, which allows the multi-branching stems to support one another.

### 64.8.4 Soil and Moisture Conditions

Lisianthus requires rich, well-drained, silty or loamy soil (preferably high in organic matter) that is kept moist but not overwatered. A circumneutral soil – pH 6.5 to 7.2 – is advisable. Fertilization is also recommended (avoiding high-nitrogen fertilizers, which promote foliage over flowers).

### 64.8.5 Maintaining Plants

Taller varieties may require staking. Flowering stems cut for vases need to be sliced with a clean knife or secateurs before being placed in water. Also, the lower leaves on the stem should be cut off so they won't contaminate the water, and the water level should be kept up.

## 64.9 FASCINATING FACTS, FABLES, FOLKLORE, AND FRIVOLITY

- Lisianthus, sometimes called prairie gentian, is a member of the gentian family (Gentianaceae), which has about 1600 species, mostly in temperate areas worldwide. As exemplified by lisianthus and the genus *Gentiana* (the true gentians), the family is best known for its flowering ornamentals. The English word *gentian* and the groups *Gentiana* and Gentianaceae all trace to Gentius, a second-century BC king in the Balkan region who was initially a cooperative of the Roman Empire but rebelled and was eventually imprisoned.
- As discussed earlier, lisianthus used to be put in the genus *Lisianthus*, although transferred to *Eustoma*. Of the species in the genus *Lisianthius*, *L. nigrescens* of Mexico and Guatemala is especially interesting. It is considered to be the blackest naturally occurring flower in the world. Its Spanish name, flor de muerto, translates to "flower of death," reflecting not just its darkness but also the practice of planting it around graves.
- Aspirin (salicylic acid) prolongs vase life of lisianthus, as indeed it does for many other cut flowers. A common home remedy is simply to crush a 500 mg aspirin tablet and add it to the water. Aspirin lowers the pH of the water, which can help flowers live longer. However, some experts prefer other methods of prolonging vase flowers.

## KEY PUBLICATIONS


Barba-Gonzalez, R., E. Tapia-Campos, T.Y. Lara-Bañuelos, and V. Cepeda-Cornejo. 2017. Lisianthus (*Eustoma*) breeding through interspecific hybridization. *Acta Horticulturae* 1171: 241–244.

Castillo-Gonzâlez, A.M., L.A. Valdez-Aguilar, and E. Avitia-García. 2019. Response of lisianthus (*Eustoma grandiflorum* [Raf.] Shinn) to applications of growth regulators. *Acta Horticulturae* 1263: 241–244

Foti, C., P. Romita, A. Filoni, A. Antelmi, D. Bonamonte, and G. Angelini. 2014. Occupational allergic contact dermatitis caused by *Eustoma exaltatum russellianum* (lisianthus). *Contact Dermatitis* 71(1): 59–60.

Gill, S.A., T. Blessington, E.M. Dutky, R. Balge, K.S. Ross, G. Rosenkranz, et al. 2000. *Production of Lisianthus as a Cut Flower*. University of Maryland Cooperative Extension Fact sheet 770. www.yumpu.com/en/document/view/4558750/production-of-lisianthus-as-a-cut-flower-university-of-maryland-.

Grueber, K.L., B.E. Corr, and H.F. Wilkins. 1984. *Eustoma grandiflorum* (*Lisianthus russellianus*). *Minnesota State Florist Bulletin* 33(6): 10–14.

Halevy, A.H. and A.M. Kofranek. 1984. Evaluation of lisianthus as a new flower crop. *HortScience* 19: 845–847.

Harbaugh, B.K. 1995. Flowering of *Eustoma grandiflorum* (Raf.) Shinn. cultivars influenced by photoperiod and temperature. *HortScience* 30: 1375–1377.

Harbaugh, B.K. 2007. Lisianthus. In: *Flower Breeding and Genetics*, edited by N.O. Anderson, 644–663. Springer, Dordrecht.

Ohkawa, K. and E. Sasaki. 1999. *Eustoma* (lisianthus) – Its past, present, and future. *Acta Horticulturae* 482: 423–428.

Pergola, G. 1992. The need for vernalization in *Eustoma russellianum*. *Scientia Horticulturae* 51: 123–127.

Roh, M.S., A.H. Halevy, and H.F. Wilkins. 1989. *Eustoma grandiflorum*. In: *Handbook of Flowering*, Vol. 6, edited by A.H. Halevy, 322–327. CRC Press, Boca Raton, FL.

Shinners, L.H. 1957. Synopsis of the genus *Eustoma* (Gentianaceae). *Southwestern Naturalist* 2: 38–43.

Struwe, L. 2007. The convoluted history of *Lisianthus*, *Lisianthius*, and *Lisyanthus*. Gentian Research Network. https://gentian.rutgers.edu/LisianthusHistory.htm.

Turner, B.L. 2014. Taxonomic overview of *Eustoma* (Gentianaceae). *Phytologia* 96: 7–11.


### Key Cultivation Websites


Neverland. *How to Plant and Care for Lisianthus*. www.enterneverland.com/blog/how-to-plant-and-care-for-lisianthus.

Sakata. *Lisianathus Cut Flower Production Tutorial*. https://sakataornamentals.com/wp-content/uploads/sites/2/2020/03/Lisisanthus-Cut-Flower-Productuion-Tutorial-SAKATA-1018.pdf.

The Spruce. *How to Grow and Care for Lisianthus*. www.thespruce.com/lisianthus-flowers-1315711.

Utah State University, Extension. *Lisianthus Cut Flower Production in Utah*. https://digitalcommons.usu.edu/cgi/viewcontent.cgi?article=3312&context=extension_curall.

Washington State University/My Thanh Kim. *How to Grow Lisianthus*. https://s3.wp.wsu.edu/uploads/sites/2073/2023/04/Lisianthus-Cool-Hardy-Beauties.pdf.

WikiHow. *How to Grow Lisianthus*. www.wikihow.life/Grow-Lisianthus#:~:text=One%20of%20lisianthus%27%20strengths%20is,the%20base%20of%20the%20plant.

# 65 Lobelia (*Lobelia* Species and Hybrids)

**Scientific Names:** The genus name *Lobelia* was named for Flemish (born in France, active in the UK) physician and botanist Matthias de Lobel (1538–1616), private physician of King James I.

**Common Names:** Most *Lobelia* species are called lobelia. The common name cardinal flower arose because of perceived similarity of the flower color (scarlet, crimson, or vermilion) to the color of the vestments worn by Roman Catholic cardinals. Frequently, cultivated *Lobelia* species are simply called lobelia, which is usually inadequate to identify the species, so the scientific name is often preferable.

## 65.1 INTRODUCTION

Ornamental lobelias are most notable for two species: a very wide variety of popular, colorful, dwarf bedding plant cultivars of the species *L. erinus*, and cardinal flower (*L. cardinalis*) – a rather spectacular wildflower that has become a horticultural favorite.

## 65.2 OVERVIEW

What should be included in the genus *Lobelia* is unsettled, but there are at least 400 species, native to much of the planet, the majority in the Old World, especially Africa. Most have not been grown in the horticultural trade. The main cultivated species are *L. cardinalis* (once called *L. fulgens*; cardinal flower, Indian pink, bog sage), *L. erinus* (annual lobelia, edging lobelia, trailing lobelia, laguna flower), and *L. siphilitica* (blue lobelia, great blue lobelia, great lobelia, blue

FIGURE 65.1 *Lobelia erinus* – principal flower colors: (a) Mixed cultivars. (b) Cultivar White Lady. (c) Cultivar Cambridge Blue. (d) Cultivar Waterfall Blue Ice.

**Source: (a) Photo by Foto-RaBe from Pixabay (public domain). (b) Photo by Shutterstock/Tareka Cheon. (c) Photo by Rob Hille (public domain). (d) Photo by HQ (CC BY SA 2.0).**

FIGURE 65.2 *Lobelia cardinalis* – red and purple flower forms.

**Source: (a) Plant growing at water's edge. Photo by Shutterstock/Meunierd. (b) Thomas G. Barnes, US Fish and Wildlife Service (public domain). (c) F. Delventhal (CC BY 2.0). (d) Photo by Shutterstock/Tom Meaker.**

FIGURE 65.3 *Lobelia siphilitica.*

**Source: (a) Photo by Agnieszka Kwiecień, Nova (CC BY SA 3.0). (b) Photo by Steven Katovich (CC BY 3.0).**

DOI: 10.1201/9781003586227-65

cardinal). Hybrids are also grown, notably *L.* × *speciosa* (*L. cardinalis* × *L. siphilitica*), but these are rarely distinguished from *L. cardinalis*. *Lobelia cardinalis* is often confused with *Salvia splendens* (see the chapter on salvia).

*Lobelia erinus* is native to South Africa. This dwarf species is often less than 30 cm in height, often trailing, although upright varieties are available. It is a tender perennial usually grown as an annual. *Lobelia cardinalis* (native of North and Central America) and *L. siphilitica* (native of North America) are hardy herbaceous perennials (the aboveground parts are killed by frost) that grow in upright clumps, often as tall as 1 m (sometimes up to 2 m) in height. Lobelia flowers are typically tubular at the base, generally with five lobes arranged in two sets, the upper with two upper lobes, and the lower with three. The blooms may be blue, lilac, pink, purple, red, or white, although violet-blue is most common. A much greater range of flower colors are available for *L. erinus* than for other species, although its numerous blooms are quite small. A tiny whitish "eye" is often present in the center of each flower. The species discussed in this chapter bloom from midsummer to autumn.

FIGURE 65.4 Eighteenth- and nineteenth-century paintings (public domain) of cultivated lobelias by European botanical illustrators: (a) *Lobelia cardinalis*. (b) *Lobelia tenuior*. (c) Mixed cultivars. (d) *Lobelia* cultivar lowii. (e) Mixed cultivars. (f) *Lobelia erinus* cultivar Kathleen Maillard.

***Source:*** **(a) Step, E., D. Bois, B. Herincq, and W. Watson. 1896. *Favourite flowers of garden and greenhouse*, vol. 3. Frederick Warne, New York. (b) *Revue horticole: Journal d'horticulture practique* (Paris), vol. 5, 1905. (c) *The garden. An illustrated weekly journal of horticulture* (London), vol. 56, 1899. (d) *Edwards's Botanical Register* (London), vol. 17, 1831. (e) *Revue horticole* (Paris), vol. 63, 1891. (f) Garten flora (Erlangen, Germany), vol. 55, 1906.**

## 65.3 HISTORY

*Lobelia erinus*, the most widely grown species of the genus, was imported from South Africa to Europe, where it has been cultivated since the 17th century. Noted because of its striking scarlet flowers, *L. cardinalis* was imported from Canada to France in the 1620s, and to London in 1626. It became a favorite in the so-called "American gardens" that were popular at that time among wealthy Europeans. *Lobelia erinus* is the most changed from the original wild species, with numerous cultivars. Other lobelias, especially *L. cardinalis* and *L. siphilitica*, have not been as extensively altered, although some selections and hybrids have been bred.

## 65.4 ECONOMIC IMPORTANCE

Lobelias are admired but relatively minor ornamental plants that contribute to the gardening industries of many countries. The hundreds of wild species represent a vast genetic pool for future development.

## 65.5 ORNAMENTAL USES

*Lobelia erinus* is extensively employed in mass plantings, in garden borders, along pathways, in rock gardens, and in containers, including hanging baskets and window boxes. These lobelias are often interplanted with species of similar height in gardens and containers.

The cultivated lobelias are often employed as ornamental groundcover. *Lobelia cardinalis* and *L. siphilitica* are especially suited to damp ground, and along creeks and ponds, and are grown partly submersed in water gardens. Dwarf forms of cardinal flower are available that are employed in aquaria. The large lobelias (*L. cardinalis* and *L. siphilitica*) can be employed as cut flowers, although the stems exude a

FIGURE 65.5 A commercial field of lobelia potted plants in The Netherlands.

***Source:*** **Photo by Shutterstock/Martien van Gaalen.**

FIGURE 65.6 Landscape uses of Lobelias: (a) *Lobelia cardinalis* in a garden. (b) A large bed of mixed *L. erinus* cultivars. (c) Hanging basket of *L. erinus*. (d) *Lobelia erinus* in a decorative wooden chest.

*Source:* **(a) Photo by John Rusk (CC BY 2.0). (b) Photo by Shutterstock/Obraz. (c) Photo by lcm1863 (CC BY ND 2.0). (d) Photo by Shutterstock/Paparazzza.**

FIGURE 65.7 Pollinators visiting lobelia flowers: (a) Ruby-throated hummingbird and *L. cardinalis*. (b) Eastern tiger swallowtail on *L. siphilitica*. (c) European honeybee on *L. siphilitica*.

*Source:* **(a) Photo by Rodger Evans (CC BY ND 2.0). (b) Photo by Michele Dorsey Walfred (CC BY 2.0). (c) Photo by Marta Boroń (CC BY 2.0).**

milky sap. *Lobelia cardinalis* flowers are very attractive to hummingbirds, and most lobelias also attract bees and butterflies. Accordingly, they are often grown in pollinator gardens.

## 65.6 OTHER USES

Lobelias have been employed in traditional herbal medicine and are currently sometimes marketed in homeopathic preparations of doubtful value. *Lobelia siphilitica*, true to its name, was once used as a treatment for syphilis (so were other species of the genus). *Lobelia inflata* (Indian tobacco), an annual plant native to North America, was once smoked to treat asthma. It was also employed as an expectorant and purgative. It is one of several lobelia species, sometimes called pukeweed, gagroot, and vomitwort, used to induce vomiting. The plant (especially the root) contains lobeline, a volatile liquid alkaloid with a tobacco-like odor, and is very dangerous if swallowed. The species is still sometimes employed by modern alternative medicine practitioners, although it has caused coma, convulsions, and even death. *Lobelia tupa* (Devil's tobacco) is an evergreen perennial native to Chile, growing up to 4 m in height, which also has had numerous herbal medicinal uses, although quite toxic. It, too, has been smoked, hence the "tobacco" in its name.

FIGURE 65.8 Lobelias employed in traditional herbal medicine: (a) *Lobelia inflata* (Indian tobacco). (b) *Lobelia tupa* (Devil's tobacco).

*Source:* **(a) Photo by Andrea Moro (CC BY SA 4.0). (b) Photo by Jason Hollinger (CC BY 2.0).**

## 65.7 PROBLEMS

All parts of lobelias are toxic to humans, pets, and livestock. The milky sap can irritate skin, and gloves are recommended when handling the plants.

## 65.8 CULTIVATION

### 65.8.1 Climate and Site

The widely grown *L. erinus* is capable of growing as a perennial in climate zones 10 and 11 but is cultivated as an annual in cooler areas. It grows well in full sun and part shade, but in very hot regions, it is best provided with some shade. The flowering display of *L. erinus* may fade during the hottest time of summer but revive with cooler weather. *Lobelia cardinalis* is rated for zones 3 to 9, and *L. siphilitica* for zones 4 to 9. Both can grow in a range of light from full sun to full shade, and as noted later, both can grow in very wet soils. Because lobelias are quite toxic, consideration should be given to positioning them in areas of gardens that are inaccessible to young children and pets.

### 65.8.2 Propagation and/or Purchasing

Lobelias are often sold as starter seedlings in plugs in garden centers and are also available as seeds (a greater selection is usually available). *Lobelia cardinalis* and *L. siphilitica* are plants of wetlands to a considerable extent and are most likely to be available online or from specialist nurseries that specialize on bog and pond plants. *Lobelia*

FIGURE 65.9 A vertically divided clump of *Lobelia cardinalis.* Notice the offsets (small shoots at right), which are employed to propagate the plant.

*Source:* **Photo by Gilles Ayotte (CC BY SA 4.0).**

FIGURE 65.10 A crimson Catholic Cardinal robe and the species named for it: (a) Male Northern cardinals. (b) Catholic clergy costumes. (c) *Lobelia cardinalis.*

*Source:* **(a) Photo by RoyalTee from Pixabay (public domain). (b) Photo by Shutterstock/Amid999. (c) Walcott, M. V. 1925. *North American wild flowers*, vol. 4. Smithsonian Institution, Washington, DC (public domain).**

*cardinalis* and *L. siphilitica* naturally spread by offsets at the base, but clumps tend to be short-lived. *Lobelia cardinalis* and *L. siphilitica* usually require rejuvenation every 2 or 3 years. Lobelias can be propagated by seed, by stem cuttings, or by division of the plants into sections that contain at least one offset. The cultivated lobelias often self-seed, making it unnecessary to regularly rejuvenate them.

### 65.8.3 Planting

*Lobelia erinus* can be established in gardens at a spacing of 15 to 20 cm. The large species, including *L. cardinalis* and *L. siphilitica*, should be spaced at least 60 cm apart.

### 65.8.4 Soil and Moisture Conditions

*Lobelia erinus* cultivars prefer moist but well-drained, compost-rich soil. *Lobelia cardinalis* and *L. siphilitica* are adapted to wet soil. A pH of 6.0 to 7.0 is recommended. Lobelias benefit from application of a high-phosphorus fertilizer, which promotes blooming.

### 65.8.5 Maintaining Plants

Lobelias are considered to be low-maintenance plants. Deadheading of flowers is usually unnecessary, although sometimes recommended.

## 65.9 FASCINATING FACTS, FABLES, FOLKLORE, AND FRIVOLITY

- Cardinals (also called "Princes") are the top clerics of the Roman Catholic Church. The word is derived from the Latin *cardo*, "hinge," a reference to cardinals being "the hinges on which the Church revolves." It is claimed that the cardinals wear robes that are red (the color of martyrdom) to symbolize their readiness to shed their blood for the Church and Christ, while purple is worn during the Advent and Lent seasons, reflecting sorrow and suffering. The cardinal bird, like the cardinal flower, was named by early European colonists in North America after the bright-red robes worn by Roman Catholic cardinals. The Northern cardinal (*Cardinalis cardinalis*) is the state bird of seven states, and the colorful male (the female is drab) is considered to be one of the world's most recognizable species. There are two other species in the cardinal genus *Cardinalis*: the relatively drab desert cardinal of the American Southwest and northern Mexico, and the highly colored vermilion cardinal of South America.
- Scientific names are sometimes embarrassing, as exemplified by *Lobelia siphilitica.* Common names of plants and animals that are unseemly or racist often simply become obsolete as people stop employing the terms. In the same spirit, scientific codes of nomenclature usually prohibit the coining of new scientific names that are insulting or highly inappropriate. However, established scientific names cannot be changed except for technical reasons relating to procedures. The advantage of this convention is that, unlike the many English names like "yellow daisy" and "red rose" that label many different species, a standard, unique scientific name is permanently available for every species.
- Computer spellcheckers are often misled by botanical names. "Pukeweed," a name employed for several of the lobelias, is usually changed to "pokeweed" (which is *Phytolacca americana*, also a quite toxic species).
- A survey of 1,000 botanists and naturalists throughout North America was conducted in the 1940s by the New York Botanical Garden to determine the

FIGURE 65.11 Giant (tree) lobelias in montane tropical Africa: (a) On the Sanetti Plateau, Ethiopia. (b) Mountain gorilla baby (*Gorilla beringei beringei*) playing on a lobelia branch in Volcanoes National Park, Rwanda.

***Source:* (a) Photo by Rod Waddington (CC BY SA 2.0). (b) Photo by Shutterstock/Joe McDonald.**

"most showy, conspicuous, and interesting wildflowers." Cardinal flower ranked first, with 213 votes. The others: Showy lady's slipper, 155; New England aster, 145; butterfly weed, 141; moccasin flower, 136; large-flowered white trillium, 134; columbine, 131; sweet-scented white water lily, 109; eastern fringed gentian, 107; and marsh marigold, 88.

- Tree-sized lobelia species grow at high elevations of the mountains of tropical Africa.

## KEY PUBLICATIONS

Bergner, P. 1998. Lobelia toxicity: A literature review. *Medical Herbalism* 10: 15–34.

Bowden, W.M. 1982. The taxonomy of *Lobelia* × *speciosa* sl and its parental species, *L. siphilitica* and *L. cardinalis* sl (Lobeliaceae). *Canadian Journal of Botany* 60: 2054–2070.

Ebinger, J.E. 1985. *Lobelia cardinalis* × *L. siphilitica* in Illinois. *Transactions of the Illinois Academy of Science* 78: 29–31.

Felpin, F.-X. and J. Lebreton. 2004. History, chemistry and biology of alkaloids from *Lobelia inflata*. *Tetrahedron* 60: 10127–10153.

Johnson, L.M. and L.F. Galloway. 2008. From horticultural plantings into wild populations: Movement of pollen and genes in *Lobelia cardinalis*. *Plant Ecology* 197: 55–67.

Krochmal, A. and L.O. Wilken. 1970. The culture of Indian tobacco (*Lobelia inflata* L.). *USDA Forest Research Service Paper* NE181.9 pp.

Krochmal, A., L.O. Wilken, and M. Chien. 1972. Plant and lobeline harvest of *Lobelia inflata* L. *Economic Botany* 26: 216–220.

Lammers, T.G. 2011. Revision of the infrageneric classification of *Lobelia* L. (Campanulaceae: Lobelioideae). *Annals of the Missouri Botanical Garden* 98: 37–62.

Lungu, M. 2023. *Lobelia: Plant Overview and Guide*. Self-published. 85 pp.

Máthé, Á. 2020. Indian tobacco (*Lobelia inflata* L.). In: *Medicinal and Aromatic Plants of North America. Medicinal and Aromatic Plants of the World*, Vol. 6, edited by Á. Máthé, 159–186. Springer, New York.

Matthews, V. 1988. *Lobelia tupa*. *Kew Magazine* 5: 157–161.

Vojnich, V.J., Á. Máthé, R. Gaál, and S. Tüű. 2011. Botanical and chemical variability of Indian tobacco (*Lobelia inflata* L.). *Acta Agronomica Óváriensis* 53: 37–48.

## Key Cultivation Websites

Ecofriendly Garden Income (blog). *How to Propagate Trailing Lobelia (Lobelia erinus)*. www.ecofriendlyincome.com/blog/how-to-propagate-trailing-lobelia-lobelia-erinus.

Gardeners World. *How to Grow Lobelia*. www.gardenersworld.com/how-to/grow-plants/how-to-grow-lobelia/.

HGTV. *How to Grow and Care for Cardinal Flower Plants*. www.hgtv.com/outdoors/flowers-and-plants/flowers/cardinal-flower.

Horticulture Magazine. *Lobelia*. https://horticulture.co.uk/lobelia/.

TheGardenMagazine.com. *Lobelia Planting, Uses and Care*. https://thegardenmagazine.com/lobelia-planting-uses-care/.

The Spruce. *How to Grow and Care for Lobelia*. www.thespruce.com/how-to-grow-and-care-for-lobelia-5082044.

Wisconsin Horticulture, Division of Extension. *Cardinal Flower, Lobelia cardinalis*. https://hort.extension.wisc.edu/articles/cardinal-flower-lobelia-cardinalis/.

# 66 Lupin (*Lupinus polyphyllus* Hybrids)

**Scientific Names:** The genus name *Lupinus* and the word *lupin* are based on the classical Latin *lupus*, "wolf." It was once thought that lupins took over land, robbing the soil of its fertility, just as wolves robbed shepherds of their livestock. In fact, lupins, like most members of the pea family, have root nodules housing symbiotic nitrogen-fixing bacteria, resulting in enriched soil. *Polyphyllus* in the scientific name is based on the ancient Greek *polus*, "many," and *phullon*, "leaf," meaning "many-leaved," although a more accurate description is that the leaves have many leaflets.

**Common Names:** "Lupin" is standard in most countries for species of *Lupinus*. "Lupine" is mainly encountered in the United States, but "lupin" is increasing in usage. The name *L. polyphyllus* conventionally collectively designates garden lupins. The garden lupin is also called Russell lupin, large-leaved lupin, big-leaved lupin, and marsh lupin.

## 66.1 INTRODUCTION

Garden lupins are early-flowering, mostly perennial herbs with extremely showy, colorful spires of flowers. They are one of the popular horticultural standards, mainly grown as massed displays in private residential gardens and public places where ornamental plants are found. Their conspicuous beauty has led to distribution to much of the world, and in many places, the plants have escaped to the wild or have been deliberately naturalized. Weedy garden lupins aren't shrinking violets; they are extremely aggressive invaders, displacing and endangering indigenous plants and animals.

FIGURE 66.1 A display of lupin cultivars (*Lupinus polyphyllus* hybrids) at a flower show.

*Source:* **Photo by Shutterstock/Devonpaul.**

## 66.2 OVERVIEW

Estimates of the number of species in the genus *Lupinus* range from about 200 to 300. The plants are annual and perennial herbs and evergreen subshrubs and shrubs. The majority grow wild in the Americas. All the dozen or so Old World species are annuals growing in open and disturbed habitats, mostly in the Mediterranean region and sporadically in Africa north of the equator. *Lupinus* species are common in North Africa, western North America, and South America, where they are found growing on dry, hilly grasslands, on coastal sands and cliffs, and along the banks of streams and rivers.

Indigenous North American *L. polyphyllus* is the principal ancestor of the garden lupin. It is a native of moist meadows in western North America from Alaska to California (some authorities have contended that the Alaskan plants are introduced). Similar plants sometimes assigned to the species also occur in South America. The escaped derivatives of the species have become established in much of the world, often colonizing disturbed habitats. They thrive on sandy soils and tend to avoid heavy soils. The species has

FIGURE 66.2 Extremely colorful Russell hybrid forms of the garden lupin (*Lupinus polyphyllus* hybrids): (a) A display at the Chelsea Flower Show in the UK. (b) A display at Lake Tahoe, California/Nevada. (c) An all-red variety. (d) A variegated white/yellow variety. (e) A variegated, very popular variety: Lupin cultivar 'Manhattan Lights'.

*Source:* **(a) Photo by Ickh710 (CC BY 2.0). (b) Photo by Mitch Lorens (CC BY ND 2.0). (c) Photo by Tony Hisgett (CC BY 2.0). (d) Photo by Hans from Pixabay (public domain). (e) Photo by Hans-Günter Wagner (CC BY SA 2,0).**


DOI: 10.1201/9781003586227-66

FIGURE 66.3 Multifoliolate leaf of *Lupinus polyphyllus* 'My Castle'.

*Source:* **Photo by Vince (CC BY 2.0).**

often been employed to stabilize soils against erosion, its roots capable of descending to 2 m.

*Lupinus polyphyllus*, the chief ancestor of the garden lupin, is an erect perennial herbaceous plant with stout, generally unbranched hollow stems growing up to 1.5 m in height, although miniature forms may be less than 30 cm tall. The underground part of the plant is a rhizome, which may be short- or long-spreading, and from which a deeply penetrating taproot often develops. The leaves are palmately compound, with 5–17 leaflets 3–15 cm long. The basal leaves have long petioles, while those on the upper stems are shorter. In the main ancestral species *L. polyphyllus*, the flowers are usually blue or purple, but the derived cultivars differ considerably in flower color. When the plants escape, after several generations, the progeny very strongly tends to revert to producing blue/purple flowers, presumably because this is adaptive (genes controlling blue are recessive, so they are not extinguished in the cultivars). The flowers are 1–1.5 cm long, produced on a tall, dense spike (up to 40 cm long), especially in early summer. The flowers are nectarless, contradicting some reports. Nevertheless, they are quite attractive to insects. The flowers are sometimes alleged to also attract butterflies and hummingbirds, but this claim may apply only to ornamental *Lupinus* species other than *L. polyphyllus* and to hybrids. Garden lupin flowers open from the lower to the upper part of the spike and last 3 to 5 weeks (sometimes 2 months). The pealike pods are 3–5 cm long, are quite hairy, and typically contain 6–10 mottled seeds. An individual plant may produce hundreds of seeds, which can remain viable in the soil for decades. The seeds are dispersed ballistically up to a few meters from the mother plant (as the pod dries out, it explodes because of uneven tension on opposing sides). *Lupinus polyphyllus* reproduces mainly by seeds, but also vegetatively, by true rhizomes (spreading belowground stems). In cool climates, the aboveground shoot dies back in the autumn, and a new shoot emerges in the spring. Annual cultivated forms are available. Perennials may persist for 20 years or more (or much less, depending on location). Among horticulturalists, garden lupins are considered to be short-term perennials (even biennials), requiring replanting at least by 6 years.

## 66.3 HISTORY

Several lupin species are employed as garden ornamentals, but the most widely grown is the so-called garden lupin, which is a hybrid complex derived particularly from the blue/purple-flowered *L. polyphyllus*. Wild plants of this and other plants from North America were introduced to Europe in the mid-19th century. The garden lupin is the result of hybridization employing several lupin species. This was largely accomplished by George Russell (1857–1951), an English amateur horticulturalist who hybridized *L. polyphyllus* with other lupin perennial species, particularly *L. arboreus* and *L. sulphureus*, and possibly also the annual

FIGURE 66.4 Wild species thought to be the principal wild plants hybridized with *M. polyphyllus* to produce the Russell hybrids (*Lupinus polyphyllus* hybrids) which dominate modern garden lupins: (a) *Lupinus arboreus* (called *tree lupin*, *yellow bush lupin*, and *coastal bush lupin*), a native of California. (b) Sulfur lupin (*L. sulphureus*), photographed in the Blue Mountains of Oregon.

*Source:* **(a) Photo by Krzysztof Ziarnek, Kenraiz (CC BY SA 4.0). (b) Photo by Andrey Zharkikh (CC BY 2.0).**

*L. nootkatensis*. Russell's creations exhibited a range of striking flower colors, including red, pink, purple, orange, yellow, and white. These hybrids were released in 1937 to mark the coronation of King George VI and have come to be known as "Russell lupins." Although Russell lupins are hybrids, they are not first-generation hybrids with equal parentage; *L. polyphyllus* dominates the ancestry of Russell lupins. In recognition of the multi-parental nature of garden lupins, the hybrid names *L.* × *polyphyllus*, *L.* × *russellii*, and *L.* × *hybridus* are sometimes encountered, although *L. polyphyllus* is mostly cited. More recent breeding has resulted in the creation of extremely attractive cultivars.

## 66.4 ECONOMIC IMPORTANCE

*Lupinus* species are economically important primarily as grain crops, the seeds providing considerable food for people and livestock. Lupins are also popular as garden ornamentals, employed mostly as bedding and border plants, and the horticultural industry of many countries provides seeds and starter plants. Although not as popular as many other flowering ornamentals, lupins are a relatively new crop and are increasing in popularity.

## 66.5 ORNAMENTAL USES

The tall, upright flowering spikes of lupins are extremely attractive, and lupins are favorite flowers. Ornamental lupins are primarily employed as bedding plants for residential borders and gardens and landscaping for large commercial and public areas. They can be grown as indoor potted houseplants but usually aren't because it is difficult to supply sufficient light, and the large rootstocks can be challenging to keep within pots. Small cultivars are most suitable for containers. Lupins are used as cut flowers occasionally, especially when they can simply be collected from local naturalized plants. The long spikes look very appealing in cut flower arrangements.

FIGURE 66.5 Lupins (*Lupinus polyphyllus* hybrids) growing on a flower farm in The Netherlands.

***Source:*** **Photo by Shutterstock/Ruud Morijn.**

FIGURE 66.6 Lupins (*Lupinus polyphyllus* hybrids) in beds and borders: (a) A village in the Alps of Slovenia. (b) A house in Christchurch, New Zealand. (c) A woman in a traditional national folk Pavlovo Posad shawl beside an old wooden house in Suzdal, Vladimir region, Russia.

***Source:*** **(a) Photo by Shutterstock/Gaspar Janos. (b) Photo by Angelica Perduta (public domain). (c) Photo by Shutterstock/Sergey Bezgodov.**

FIGURE 66.7 Freshly collected bouquets of lupin (*Lupinus polyphyllus* hybrids): (a) Girl holding flowers. (b) Border collie holding flowers. (c) Woman in folk costume in Russia, with harvested flowers. (d) A bouquet of harvested flowers.

***Source:*** **(a) Photo by Shutterstock/Denis Zai. (b) Photo by Shutterstock/Ksenia Raykova. (c) Photo by N-region from Pixabay (public domain). (d) Photo by Shutterstock/MaskaRad.**

## 66.6 OTHER USES

There has been some consideration given to developing the garden lupin into a food or forage species, but little progress has been made. However, several *Lupinus* species are used for food. Cultivation of lupins began in both the Old and New Worlds between 3,000 and 4,000 years ago, for both livestock forage and human food. The degree of

FIGURE 66.8 Potted garden lupins (*Lupinus polyphyllus* hybrids).

*Source:* **Photo by Shutterstock/Manfredxy.**

FIGURE 66.9 Food aspects of white lupin (*Lupinus albus*): (a) A commercial grain field with lupins in flower. (b) Plants with mature seedpods. (c) Seedpods, some split open to reveal edible seeds. (d) A street vendor with cooked lupins for sale in Madrid, Spain.

*Source:* **(a) Photo by Inra, DIST, Jean Weber (CC BY 2.0). (b) Photo by N-region from Pixabay (public domain). (c) Photo by Inra, DIST, Jean Weber (CC BY 2.0). (d) Photo by Tamorlan (CC BY SA 3.0).**

FIGURE 66.10 Andean lupin (*Lupinus mutabilis*), employed both as an ornamental and a source of edible seeds: (a) Painting by Abraham Jacobus Wendel. (b) Vegetable ceviche, prepared with Andean lupin (a dish popular in South America).

*Source:* **(a) Public domain. From the 1868 Dutch-language book *Flora: Afbeeldingen en beschrijvingen van boomen, heesters, éénjarige planten, enz. voorkomende in de Nederlandsche tuinen* by H. Witte, Groningen: Wolters. (b) Photo by Dtarazona (CC BY SA 4.0).**

domestication (genetic change to meet human needs) was quite limited, however, until about a century ago, when modern breeding was initiated. Four annual species are employed as cool-season grain legumes (grown for their edible seeds) or as forage crops (grown to feed livestock). The Mediterranean Basin is the home of the three most important agricultural species, white lupin (*L. albus*), blue lupin (*L. angustifolius*), and yellow lupin (*L. luteus*). The classical Egyptians, Greeks, and Romans chiefly grew white lupin, both as food for animals and flour for bread. Yellow lupin and blue lupin have probably also been cultivated for thousands of years.

Although there are likely more than 200 *Lupinus* species in the Americas, only Andean lupin (*L. mutabilis*), also known as pearl lupin, is agriculturally important. This is basically an annual (although biennial forms are known) with woody stems and blue, purple, pink, or white flowers. The species is grown not just as a crop but also for its attractive flower colors. Andean lupin has been cultivated for thousands of years in the highlands of South America. It was domesticated in pre-Incan times, at least 1,500 years ago, in the Central Andean region.

## 66.7 PROBLEMS

In parts of the world (including Europe, New Zealand, and eastern North America), plants grown for ornament, erosion control, or soil stabilization have escaped and established wild populations. These "beautiful nuisances" are sometimes so invasive that they are harmful to local vegetation. For example, in New Zealand, an endangered endemic plant, the woodrush (*Luzula celata*), has been shaded out in many areas by lupins. Also in New Zealand, weedy lupins have stabilized river gravels, forcing rivers to form deep, rapid channels unsuitable for wading birds such as the wrybill (*Anarhynchus frontalis*), a threatened plover endemic to New Zealand that has lost about half of its habitat (it is the only bird in the world with a beak that is bent sideways in one direction, always to the right). Invasive lupins can significantly change soils supporting wild plants and their associated fauna: They tend to enrich soil with nitrogen, and their alkaloids leach into the soil, both of these results

**FIGURE 66.11** Weedy Russell lupins (*Lupinus polyphyllus* hybrids) in New Zealand, where the highly invasive plant has caused extensive ecological damage.

*Source:* **Photo by SONY DSC (public domain).**

harming some species. In its introduced range, *L. polyphyllus* outcompetes many local plant species. It can replace other *Lupinus* species, resulting in a reduction in the abundance of butterflies that depend on them (indeed, this has endangered several butterfly species in North America).

Despite the saying "You can't judge a book by its cover," we humans evaluate many things by appearance, and this applies to invasive lupins. In parts of the world, people have deliberately scattered lupin seeds in the wild, sometimes on the theory that beautifying the wilderness will make an area more attractive to tourists and therefore increase business. Undeniably, extensive "infestations" of multicolored garden lupin are very attractive, and so they are sometimes featured in advertisements encouraging tourism. Proposals have been attributed to the Nova Scotia Lupin Society to make invasive lupins the Canadian province's "official weed."

Toxic quinolizidene alkaloids are synthesized in the young, growing parts of the plant and are transferred to the seeds as they develop, making wild lupin foliage and seeds bitter. Both livestock and humans avoid consuming wild lupins, except at times of food scarcity. In past times, human consumption of lupin seeds required reducing the high levels of toxins by a lengthy process of soaking and boiling (still widely practiced in some regions). However, modern agricultural cultivars have been selected for low alkaloid content. Ornamental cultivars are high in alkaloids and should not be employed as food for people or animals.

A small percentage of people (perhaps of the order of 1%) have a food sensitivity to lupins, very likely due to an individual reaction to the seed proteins, not the alkaloids. Of those who have such sensitivity, many are also allergic to peanuts, probably because similar proteins in lupin and peanut cause allergic reactions.

In countries where the wild relatives of lupins occur as native or introduced plants, a special difficulty occurs in maintaining the quality of the related low-alkaloid cultivated varieties. Although most lupins self-pollinate to a considerable degree, sexual crossing occurs readily when the wild and cultivated forms of given lupin species are close together. Also, in some regions, high-alkaloid land races (cultivated forms that are not as domesticated as modern cultivars) are commonly grown. The presence of nearby high-alkaloid plants introduces the genes for alkaloid production back into the seeds of the cultivated forms, making the seeds useless for production of low-alkaloid lupins. In countries such as Spain, Portugal, and Italy, where high-alkaloid forms of *L. albus*, *L. angustifolius*, and *L. luteus* are indigenous, it is difficult to maintain low-alkaloid cultivars. Low-alkaloid varieties of Andean lupin are very unlikely to be grown in Ecuador, Peru, and Bolivia because the local, high-alkaloid forms prevalent there would contaminate the cultivated material.

## 66.8 CULTIVATION

### 66.8.1 Climate and Site

Lupin cultivars have different cold tolerances, and at many locations, microclimates are available that will allow survival as perennials in colder regions. The species usually thrive in relatively cool summers. The popular *L. polyphyllus* hybrid cultivars are best adapted to climate zones 4 to 8. Most species of *Lupinus* occur in open, sunny habitats. Full sun is recommended, partial shade is tolerated, but flower production will likely be reduced. A location sheltered from wind is preferable for tall varieties.

### 66.8.2 Propagation and/or Purchasing

Professional horticulturalists frequently start planting stock from cuttings or divisions of the base ("crown") (the only way of ensuring that most of the attractive cultivars will remain true to type). Occasionally, dormant bare-rooted plants can be obtained. Seeds are the main method of establishing new plants. They are widely available from stores and online sources. Nurseries also usually have potted plants available, which can simply be planted in the spring.

### 66.8.3 Planting

Both seeds and starter plants are employed to plant lupins. Plants should be spaced or thinned to 30 to 60 cm apart. Starter plants can be planted directly in the soil in late spring (especially in stressful climates) or fall (in areas with mild winters). To hasten development, seeds can be started in small fiber pots ("plugs") 4 to 6 weeks in advance of potential spring frost. Perennial forms started from seed won't flower the first year. The seeds are hard-coated and benefit from scarification (scratching the shell to facilitate uptake of water). Seeds collected from the plants tend to produce offspring that revert to developing blue-colored flowers, regardless of the color of the parent plants. Plants allowed to go to seed will self-seed in the location, and the resulting plants also revert to production of blue-purple flowers.

### 66.8.4 Soil and Moisture Conditions

The soils occupied by wild species of *Lupinus* are usually light-textured and acidic to neutral and are usually well drained and fairly deep. Cultivated lupins are quite tolerant of soil conditions. Although the plants are adapted to sandy, nutrient-poor soils, they can tolerate richer soils. Clay soils are unsuitable but can be improved by mixing in compost. Little to no fertilization is necessary. Soil pH is preferably slightly acidic to neutral. Considerable moisture in the soil is tolerated (the plants naturally colonize some wetlands), but well-drained substrates are preferable. Soils should be moist but not waterlogged. Because the roots can penetrate the soil to a depth of 2 m, once plants are established, they are more tolerant of dry topsoil.

### 66.8.5 Maintaining Plants

Staking may be necessary to prevent very tall varieties from flopping over. Plants should be deadheaded (spent flowers removed) to prolong blooming and to encourage a second flowering period in the autumn (which depends on the cultivar). At the end of the season, after the foliage has yellowed, the aboveground stem can be cut down.

## 66.9 FASCINATING FACTS, FABLES, FOLKLORE, AND FRIVOLITY

- In ancient Roman times, larger lupin seeds were used as play money, especially in theatrical productions. This association of lupins and money was well-known in past centuries in England, and the lupin flower was sometimes called "penny bean." In one episode of the classic British comedy series *Monty Python's Flying Circus*, a Robin Hood character played by John Cleese robbed people of their lupins, a subtlety that very few of the audience would understand.
- "Necromancy" is communication with the dead. In classical Greece, it was sometimes believed that lupins had magical powers and could be helpful in communicating with departed relatives. The "Oracle of the Dead" was a shrine on the banks of the Acheron River near the ancient city of Ephyra where pilgrims who wanted to contact the spirits of the dead went, first consuming large quantities of "beans" said to facilitate communication. Since the beans of the time included lupins, which back then were likely toxic (unlike modern cultivars), this may have induced hallucinations that led people to believe they were indeed in touch with ancestors.
- The very bitter Andean lupin (*L. mutabilis*) is used as a protective hedge to prevent damage from grazing animals in South America.
- American Indians used wild American lupins for various purposes. When sulfur lupins (*L. sulphureus*) or many-colored lupins (*L. versicolor*) were in flower, the Okanagan and Colville Indians considered it to be a sign that the groundhogs were fat enough to eat. The Kwakiutl rubbed root ash made from seashore lupin (*L. littoralis*) into a newborn baby's cradle to make it sleep well. The Cherokee rubbed sundial lupin (*L. perennis*) on the hands or body to give a person the power to control horses.
- The British Royal Navy constructed over 200 85-man ships designed for minesweeping and escort operations near ports (they could navigate in waters less than 4 m deep), especially during the Second Word War. These were called "flower-class" ships because all of them were assigned unique flower names. HMS *Lupin* was launched in 1916 and broken up in 1946. For a list of some of the other flower warships, see https://military-history.fandom.com/wiki/Arabis-class_sloop. Canadians were assigned many of the ships, renaming them because floral labels seemed insufficiently aggressive. The only extant flower ship is Canadian Navy HMCS *Sackville*, preserved in a museum in Halifax.
- Seeds of an arctic lupin (*L. arcticus*) from a collared lemming (*Dicrostonyx groenlandicus*) burrow deeply buried in permanently frozen silt of Central Yukon were found to germinate, and it was alleged in a resulting publication in 1967 that they were at least 10,000 years old and thus the oldest seeds ever germinated. Because the paper appeared in the prestigious journal *Science*, it has been widely cited. However, in 2009, seeds from the cache were radiocarbon-dated and found to have been only 10 years of age when originally collected. Lotus (*Nelumbo nucifera*) is often credited with being the record holder for seed longevity at about 1,300 years, but in 2008, a seed was germinated of date palm (*Phoenix dactylifera*) from the ancient Israeli fortress of Masada and was radiocarbon-dated to be about 2,000 years old. In 2007,

FIGURE 66.12 Arctic lupin (*Lupinus arcticus*), a native of Northern Canada and Northwestern North America.

***Source:*** **Photo by Sandy Brown Jensen (CC BY SA 2.0).**

FIGURE 66.13 Bluebonnets (*Lupinus* species of the Southern United States): (a) Bluebonnets in Texas, which is famous for its vast displays of wild-flowering lupins, especially by *L. texensis.* (b) A cow among bluebonnets in Texas. Bluebonnet species are toxic to livestock, especially when large amounts of mature seeds are ingested, but some foraging occurs nevertheless.

***Source:*** **(a) Photo by Shutterstock/Silvio Ligutti. (b) Photo by Shiva Shenoy (CC BY 2.0).**

Russian scientists employed viable placental tissue from inviable seeds of narrow-leaves campion (*Silene stenophylla*), an arctic native of Siberia, to generate viable plants. The seeds were carbon-dated at 32,000 years, so while they were not the oldest seeds to germinate, they represent the oldest preserved plant material to be regenerated.

- Six blue-flowered lupin species of Texas are collectively recognized as its state flower. These are called "bluebonnets" for their resemblance to the bonnets worn by pioneer women to protect themselves from sunburn. Ladies, especially the National Society of Colonial Dames of America, were instrumental in persuading lawmakers more than a century ago to recognize bluebonnets as the state flower. Men at that time favored the hardy cactus and the business-friendly cotton.

## KEY PUBLICATIONS

ABC Science. 2012. Plant grown from 30,000-year-old seeds. *ABC Science* (February 22). www.abc.net.au/science/articles/2012/02/22/3436826.htm.

Ainouche, A.-K. and R.J. Bayer. 1999. Phylogenetic relationships in *Lupinus* (Fabaceae: Papilionoideae) based on internal transcribed spacer sequences (ITS) of nuclear ribosomal DNA. *American Journal of Botany* 86: 590–607.

Australian Government, Office of Health and Ageing. 2013. *The Biology of Lupinus L. (lupin or lupine).* 64 pp. https://bangladeshbiosafety.org/wp-content/uploads/2017/06/Biology_of_Lupine_Au.pdf.

Beuthin, M. 2012. *Plant Guide for Bigleaf lupine (Lupinus polyphyllus).* USDA-Natural Resources Conservation Service, Plant Materials Center, Corvallis. 4 pp. https://view.officeapps.live.com/op/view.aspx?src=https%3A%2F%2Fplants.usda.gov%2FDocumentLibrary%2Fplantguide%2Fdoc%2Fpg_lupo2.docx&wdOrigin=BROWSELINK.

Dunn, D.B. and J.M. Gillett. 1966. *The Lupines of Canada and Alaska.* Canadian Department of Agriculture Research Branch, Ottawa. 89 pp.

Eckstein, R.L., E. Lutz, Y.P. Klinger, T. Lennartsson, J. Wissman, K. Ludewig, et al. 2022. Biological Flora of Central Europe – *Lupinus polyphyllus* Lindl. *Perspectives in Plant Ecology, Evolution and Systematics*: 125715. www.sciencedirect.com/science/article/pii/S1433831922000579.

Gladstones, J.S. 1974. Lupins of the Mediterranean region and Africa. *Western Australia Department of Agriculture, Technical Bulletin* 26: 1–48.

Heller, A.A. 1912. The North American lupines: IV. *Muhlenbergia* 8: 61–82.

International Legume Society. 2022. Lupins: An update on lupin crops and their potential. *Journal of the International Legume Society* (22). www.legumesociety.org/wp-content/uploads/2022/03/legum_perspect_22.pdf.

Kurlovich, B.S. (ed.). 2002. *Lupins. Geography, Classification, Genetic Resources and Breeding.* Intan, St Petersburg. 468 pp.

Kurlovich, B.S., F.L. Stoddard, and P. Earnshaw. 2008. Potential and problems of *Lupinus polyphyllus* Lindl. domestication. In: *Lupins for Health and Wealth Proceedings of the 12th International Lupin Conference, 14–18 Sept. 2008, Fremantle, Western Australia*, edited by J.A. Palta and J.B. Berger, International Lupin Association, Canterbury, New Zealand. www.academia.edu/5964450/POTENTIAL_AND_PROBLEMS_OF_LUPINUS_POLYPHYLLUS_LINDL._DOMESTICATION.

Naganowska, B., B. Wolko, E. Śliwińska, Z. Kaczmarek, and M.T. Schifino-Wittmann. 2005. 2C DNA variation and relationships among New World species of the genus *Lupinus* (Fabaceae). *Plant Systematics and Evolution* 256: 147–157.

Parrett, R. 1959. *The Russell Lupin. Its History and Cultivation.* Holmes & Sons, Andover Hants, U.K. 52 pp.

Pereira, A., F. Ramos, and A.S. Silva. 2022. Lupin (*Lupinus albus* L.) seeds: Balancing the good and the bad and addressing future challenges. *Molecules* 27(23): 8557. www.mdpi.com/1420-3049/27/23/8557.

Phillips, L.L. 1955. A revision of the perennial species of *Lupinus* of North America exclusive of southwestern United States and Mexico. *Research Studies of the State College of Washington* 23: 161–201.

Porsild, A.E., R. Harington, and G.A. Mulligan. 1967. *Lupinus arcticus* Wats. grown from seeds of Pleistocene age. *Science* 158(3797): 113–114.

Prass, M., S. Ramula, M. Jauni, H. Setälä, and D.J. Kotze. 2022. The invasive herb *Lupinus polyphyllus* can reduce plant species richness independently of local invasion age. *Biological Invasions* 24: 425–436.

Ramula, S. 2021. *Lupinus polyphyllus* (garden lupin). *Invasive Species Compendium.* www.cabi.org/isc/datasheet/31710.

Ramula, S. and A. Kalske. 2020. Introduced plants of *Lupinus polyphyllus* are larger but flower less frequently than conspecifics from the native range: Results of the first year. *Ecology and Evolution* 10: 13742–13751.

Ramula, S. and J. Sorvari. 2017. The invasive herb *Lupinus polyphyllus* attracts bumblebees but reduces total arthropod abundance. *Arthropod-Plant Interactions* 11: 911–918.

Small, E. 2012. Blossoming treasures of biodiversity – 38. Lupins – Benefit and harm potentials. *Biodiversity* 13(1): 54–64.

Valtonen, A., J. Jantunen, and K. Saarinen. 2006. Flora and lepidoptera fauna adversely affected by invasive *Lupinus polyphyllus* along road verges. *Biological Conservation* 133: 389–396.

White, J.G.H., P. Jarvis, and R.J. Lucas. 1995. Fertiliser requirements of Russell lupins. *Proceedings Agronomy Society New Zealand* 25: 87–90.

Zazula, G.D., C.R. Harington, A.M. Telka, and F. Brock. 2009. Radiocarbon dates reveal that *Lupinus arcticus* plants were grown from modern not Pleistocene seeds. *New Phytologist* 182: 788–792.

## Key Cultivation Websites

E-How. *How to Transplant Lupine.* www.ehow.com/how_6108482_transplant-lupine.html.

GardenersWorld.com. *How to Grow Lupins.* www.gardenersworld.com/how-to/grow-plants/how-to-grow-lupins/.

Horticulture Magazine. *Lupins.* https://horticulture.co.uk/lupins/.

Plant Propagation Protocol for *Lupinus polyphyllus.* https://courses.washington.edu/esrm412/protocols/2015/LUPO2.pdf.

Royal Horticultural Society. *How to Grow Lupins.* www.rhs.org.uk/plants/lupins/growing-guide.

The Propagator. *Propagating Lupins by Basal Cuttings.* https://thepropagatorblog.wordpress.com/2019/04/02/propagating-lupins-by-basal-cuttings/.

The Spruce. *How to Grow and Care for Lupine (Bluebonnet).* www.thespruce.com/growing-lupine-flowers-1316034.

WikiHow. *How to Grow Lupin Flowers.* www.wikihow.life/Grow-Lupine-Flowers.

# 67 Magnolia (*Magnolia* Species and Hybrids)

**Scientific Names:** The genus name *Magnolia* commemorates Pierre Magnol (1638–1715), a French botanist credited with inventing the modern concept of plant families.

**Common Names:** The word *magnolia* usually refers to *Magnolia* species. The name "tulip magnolia" often refers to saucer magnolia (*Magnolia* × *soulangiana*) but sometimes is applied to the tulip tree (two species of the genus *Liriodendron*, a relative of *Magnolia*). The fruits of schisandra magnolia vines (*Schisandra chinensis*) are employed in Chinese medicine (the "magnolia" in the name reflects an old misunderstanding that *Schisandra* is closely related to *Magnolia*).

## 67.1 INTRODUCTION

Magnolias include extremely attractive ornamental trees and large shrubs, often with large, strongly perfumed flowers which typically appear in the early spring. They are marginally suitable for the colder areas of temperate climates but are widely grown in the warmer temperate and semi-tropical regions of the world.

## 67.2 OVERVIEW

There are over 100 species of magnolias (up to 300 according to some authorities), native to temperate to subtropical areas of the world. A few species occur in high altitudes of the equatorial mountains of tropical America. The plants range from shrubs to tall trees. The foliage may be deciduous, semi-evergreen, or evergreen, depending on species and climate. Some bloom in early spring before leaves appear, others in summer after the foliage has developed. Magnolias are valued for their showy flowers, which are usually white, pink, or purple, and range in size (diameter) from less than 5 cm in some species to over 30 cm in others. The flowers are fragrant, not to attract bees but pollen-consuming beetles, which are the main pollinators. The species with the largest flowers rank among woody ornamentals of the Northern Hemisphere that develop the biggest flowers. Unlike most angiosperms (flowering plants), which have the sexual organs in rings, flowers in the magnolia family (Magnoliaceae) have stamens and pistils in spirals on a conical receptacle, an arrangement believed to be primitive for angiosperms. The tepals (both sepals and petals) occur in two or more whorls of three to six. In some species, the leaves are very large, leathery, and glossy, and the fruit is an aggregate seed cone, sometimes as large as 20 cm long. Many magnolias are insufficiently cold-hardy to grow in Northern climates, and some are suitable for tropical areas.

FIGURE 67.1 Diagram of an ideal magnolia tree.

*Source:* **Larousse du XXème siècle, 1932 (Larousse of the 20th century, 1932; public domain).**

FIGURE 67.2 *Magnolia grandiflora* (southern magnolia): (a) A huge tree, with just a few white flowers, in England. (b) An ideal (photoshopped) tree. (c) Flowers of a cultivar. (d) Painting of a fruit cluster (note red seeds).

*Source:* **(a) Photo by Acabashi (C BY SA 4.0). (b) Photo by Shutterstock/Olsio. (c) Photo by Shutterstock/Olsio. (d) Walcott, M. V. 1925. American wild flowers. Smithsonian Institution, Washington, DC (public domain).**


DOI: 10.1201/9781003586227-67

*Magnolia grandiflora* (Southern magnolia, evergreen magnolia, bull-bay) is native to the southeastern United States. It has large fragrant white flowers up to 20 cm wide, and evergreen leaves, and is a very popular ornamental that has been planted around the world in warm climates (climate zones 7–9).

*Magnolia stellata* (star magnolia), a small deciduous tree native to Japan, can grow to a height of 6 m. The common name *star magnolia* refers to its large white or light-pink star-shaped flowers, 7 to 13 cm long. The species is rated for growth in climate zones 4 to 8. It blooms very early, making the flowers susceptible to frost damage.

*Magnolia* × *soulangiana* (saucer magnolia) is a hybrid of *M. denudata* and *M. liliflora*. It is hardy in climate zones 4–9 but requires protection in zone 4 to avoid flower damage. It is a small, multi-trunked tree, 6–9 m tall and wide. The flowers are white or lavender, up to 25 cm in diameter, and are sometime fragrant.

*Magnolia acuminata* (cucumber tree, a name also used for the tropical *Averrhoa bilimbi*) is one of the largest (up to 21 m tall) and cold-hardiest (climate zones 2–8) of the magnolias. It is a native of forests in the eastern United States and southern Ontario and is often planted as a large deciduous landscape tree. The flowers are slightly fragrant, yellowish-green, and 5–10 cm in diameter. Its cucumber-shaped fruit turns red at maturity. Although its blooms are relatively inconspicuous, horticultural hybrids with other species, producing more attractive floral displays, are available.

FIGURE 67.3 *Magnolia stellata* (star magnolia): (a) Trees in full bloom. (b) Flower.

*Source:* **(a) Photo by Shutterstock/Nancy Kennedy. (b) Photo by Shutterstock/Mitzy.**

FIGURE 67.4 *Magnolia* × *soulangiana* (saucer magnolia).

*Source:* **(a) Photo by Shutterstock/Nature's Charm. (b) Photo by Tatiana Gerus (CC BY 2.0). (c) Photo by Shutterstock/Nature's Charm. (d) Photo by Jean-Pol Grandmont (CC BY 3.0).**

## 67.3 HISTORY

Magnolias have been cultivated in Asia, especially China and Japan, for over a millennium. Asian species were brought to Europe in the 17th century, and American species to Europe in the early 18th century. Breeding involving hybridization of many species has produced hundreds of cultivars.

## 67.4 ECONOMIC IMPORTANCE

Magnolias are popular ornamental trees and shrubs in milder areas of temperate climates. In the United States, they have been ranked as one of the top ten cultivated flowering trees.

## 67.5 ORNAMENTAL USES

Depending on species, magnolias are used as specimen trees and shade trees. Smaller trees and shrubs are useful as hedges, as patio plants, and sometimes as container plants.

## 67.6 OTHER USES

There is a long tradition of employing magnolias for therapeutic applications (especially in China), and today medical studies are in progress to clarify possible modern pharmaceutical uses. The larger trees are employed for lumber. The fruit provides food for wildlife. As one of the most beautiful of flowers, magnolias are the subjects of numerous artistic creations.

FIGURE 67.5 *Magnolia acuminata* (cucumber tree): (a) A large old tree in Connecticut. (b) Flowers of cultivar Ultimate Yellow. (c) Fruit cluster (said to resemble a cucumber).

*Source:* **(a) Photo by Katherine Wagner-Reiss (CC BY SA 4.0). (b) Photo by Shutterstock/J Need. (c) Photo by Sarah Culliton (CC BY 4.0).**

FIGURE 67.6 Magnolias employed as colorful backgrounds for famous tourist sites: (a) Pathways lined with flowering magnolias leading to the Eiffel Tower in Paris, France. (b) White House, Washington, DC.

*Source:* **(a) Photo by Shutterstock/Resul Muslu. (b) Photo by Hillbraith (public domain).**

FIGURE 67.7 Examples of magnolia in artistic creations: (a) Bonsai. (b) A cat nibbling on a sprig of artificial flowers in a vase shaped like a light bulb. (c) "The Magnolia Vase," an American silversmith masterpiece displayed by Tiffany & Co. at the 1893 World's Columbian Exposition in Chicago.

*Source:* **(a) Photo by Shutterstock/Walter Pall. (b) Photo by Shutterstock/Watcher Fox. (c) Photo from the Met (public domain).**

## 67.7 PROBLEMS

As noted next, magnolia flowers are easily damaged by cold temperatures, and unfortunately, the garden forms with the most attractive flowering displays are limited in the extent to which they can be grown in colder climates.

## 67.8 CULTIVATION

### 67.8.1 Climate and Site

Climate zone information was provided earlier for the chief ornamental species. Magnolias are often soft-wooded and prone to breakage in ice storms. The flowers are easily damaged by spring radiation frosts, so planting under the branches of large evergreens may be considered. The plants should not be placed in frost pockets, where cold air cannot drain away. Some of the species (including the large-leaved evergreens and some of the spring-flowering species) have branches that tend to be brittle and benefit from shelter from wind. Magnolias do well in full sun and partial shade.

### 67.8.2 Propagation and/or Purchasing

Local retail nursery or garden centers can be sources of cultivars adapted to the local climate. The majority of cultivars are reproduced vegetatively, but some are grown from seeds.

FIGURE 67.8 Commercial nursery stock of magnolias for transplanting: (a) Young trees in blossom growing in a tree nursery farm in The Netherlands. (b) Magnolias grafted onto hardy rootstocks (the translucent tape covers the joint of the stem of the upper cultivar to the top of the short stem of the rootstock). (c) A well-grown specimen of *Magnolia grandiflora* in a tub ready for transplantation.

*Source:* **(a) Photo by Shutterstock/Barmalini. (b) Photo by Mmmavocado (CC BY 2.0.). (c) Photo by Treeworld Wholesale (CC BY 2.0).**

Because the plants take several to many years to bloom, ornamental plants are best purchased as established rooted specimens in containers that can simply be transplanted.

### 67.8.3 Planting

Standard methods for transplanting young plants in containers or with balled (i.e., wrapped root balls) should be followed. Deciduous magnolias are preferably planted when dormant, typically in late fall or winter in warmer climates, and early spring in cold climates. Evergreen magnolias are best planted in early spring.

### 67.8.4 Soil and Moisture Conditions

Ideal soil for most magnolias is rich, loamy, porous, somewhat acidic (pH 5.5 to 6.5), and well-drained. Most species tolerate moderate drought, and a few tolerate wet soils.

### 67.8.5 Maintaining Plants

Magnolia bark is often thin and easily damaged by lawn mowers, so protection is advisable.

## 67.9 FASCINATING FACTS, FABLES, FOLKLORE, AND FRIVOLITY

- Magnolias are among the most primitive plants in evolutionary history, and indeed, some botanists consider the family to be the oldest of all plant families. Fossils date back over 60 million years ago. Dinosaurs went extinct 65 million years ago, so they probably coexisted with early flowering plants (which may have originated between 250 million and 140 million years ago). Accordingly, some dinosaurs probably munched on early magnolias.
- In ancient China, magnolia specimens were grown in temple gardens, where they symbolized candor, purity, and yin (the female principle).
- In 1900, schoolchildren in Mississippi elected the magnolia to be the state's official emblem, beating out the cotton blossom and the cape jasmine. The nickname of Mississippi is "The Magnolia State," and in 1952, the tree was chosen as the state's official flower. A commemorative quarter for Mississippi issued by the US Mint in 2002 features the magnolia flower. The magnolia flower is also the state flower of Louisiana, chosen in 1900.
- Flowers compete for state honors in North Korea. The current national flower is the Korean mountain magnolia (*Magnolia sieboldii*). Previously, the rose of Sharon was considered to be the national flower. "Kimjongilia," a tuberous begonia cultivar named after the late North Korean leader Kim Jong-il, was used to adorn his body during funeral ceremonies. Similarly, "Kimilsungia," an orchid cultivar, honored Kim Jong-il's father and predecessor, Kim Il-sung.
- According to an old tradition in Europe, a man who had betrayed or abandoned his partner should give her magnolias to indicate that he has repented and wishes to return.
- A Japanese legend suggests that is unwise to sleep under a blooming magnolia tree, because the intense perfume of the flowers could paralyze or even kill.
- The Gullah are an African American ethnic group speaking an English-based creole language and inhabiting the low country region of the southeastern United States. An old Gullah tradition interprets white magnolia flowers as the spirits of the dead and departed.
- The town of Magnolia, Delaware, has a sign that boasts, "This is Magnolia, the center of the universe around which the earth revolves."
- The magnolia warbler (*Setophaga magnolia*) was named by Alexander Wilson (regarded as the greatest American ornithologist prior to John James Audubon), who found the species in magnolias near Fort Adams, Mississippi, in the early 1800s. While the bird overwinters in the southern part of North America, much of its life cycle is spent among conifers in northern Canada. Nevertheless, its inappropriate name has persisted.
- The phrase "steel magnolia," popular in the Southern United States, refers to a woman who exemplifies both traditional femininity and notable fortitude. A 1989 American comedy-drama film titled *Steel Magnolias* featured such women. Rosalyn Carter (1927–2023), wife of Jimmy Carter, the 39th president of the United States, was nicknamed Steel Magnolia for her sweetness and drive. The phrase has inspired artists to prepare metal floral sculptures labelled steel magnolias.

FIGURE 67.9 Triceratops dinosaur feeding on a magnolia tree in primeval times.

*Source:* **Illustration by Shutterstock/Elenarts.**

FIGURE 67.10 Metal floral sculptures titled "Steel magnolias" (Port Lambton, Ontario, Canada).

*Source:* **Photo by Sue Thompson (CC BY ND 2.0).**

## KEY PUBLICATIONS

Alello, T. 2003. Pierre Magnol: His life and works. *Magnolia, the Journal of the Magnolia Society* 38(74): 1–10.

Azuma, H., L.B. Thien, and S. Kawano. 1999. Molecular phylogeny of *Magnolia* (Magnoliaceae) inferred from cpDNA sequences and evolutionary divergence of the floral scents. *Journal of Plant Research* 112: 291–306.

Azuma, H., J.G. García-Franco, V. Rico-Gray, and L.B. Thien. 2001. Molecular phylogeny of the Magnoliaceae: The biogeography of tropical and temperate disjunctions. *American Journal of Botany* 88: 2275–2285.

Barrett, R. and D. Hughes. 2002. *Magnolias*. Firefly Books, Buffalo, NY. 96 pp.

Bunting, A. 2016. *The Plant Lover's Guide to Magnolias*. Timber Press, Portland, OR. 232 pp.

Callaway, D.J. 1994. *The World of Magnolias*. Timber Press, Portland, OR. 260 pp.

Gardiner, J.M. 2000. *Magnolias: A Gardener's Guide*. Timber Press, Portland, OR. 330 pp.

Hunt, D. (ed.). 1998. *Magnolias and Their Allies. Proceedings of an International Symposium*. International Dendrology Society & Magnolia Society. 304 pp.

Kim, S., C.W. Park, Y.D Kim, and Y. Suh. 2001. Phylogenetic relationships in family Magnoliaceae inferred from ndhF sequences. *American Journal of Botany* 88: 717–728.

Liu, Y.H. 2004. *Magnolias of China*. Beijing Science & Technology Press., Hong Kong 392 pp.

Rankin, G. 1998. *Magnolias: A Care Manual*. Laurel Glen, San Diego, CA. 128 pp.

Sauquet, H., M. von Balthazar, S. Magallón, J.A. Doyle, P.K. Endress, E.J. Bailes, et al. 2017. The ancestral flower of angiosperms and its early diversification. *Nature Communications* 8: 16047. https://doi.org/10.1038/ncomms16047.

Treseder, N.G. 1978. *Magnolias*. Faber & Faber, London. 96 pp.

## Key Cultivation Websites

GardenersWorld.com. *How to Grow Magnolia*. www.gardenersworld.com/how-to/grow-plants/how-to-grow-magnolias/#care.

Horticulture Magazine. *Magnolia Tree*. https://horticulture.co.uk/magnolia/.

Magnolia Society International. www.magnoliasociety.org/.

*Register of Magnolia Cultivars*. https://journals.ashs.org/hortsci/view/journals/hortsci/56/12/article-p1614.xml.

Royal Horticultural Society. *How to Grow Magnolias*. www.rhs.org.uk/plants/magnolia/growing-guide.

The Spruce. *12 Popular Types of Magnolia Trees and Shrubs*. www.thespruce.com/twelve-species-magnolia-trees-and-shrubs-3269666.

The Spruce. *How to Grow and Care for Saucer Magnolia*. www.thespruce.com/magnolia-trees-saucer-magnolias-2132135.

The Spruce. *Magnolia Care Tips: Solutions to Common Problems*. www.thespruce.com/magnolia-tree-care-2132812.

WikiHow. *How to Grow a Magnolia Tree*. www.wikihow.com/Grow-a-Magnolia-Tree.

WikiHow. *How to Prune a Magnolia tree*. www.wikihow.com/Prune-a-Magnolia-Tree.

WikiHow. *How to Prune a Tulip Magnolia*. www.wikihow.com/Prune-a-Tulip-Magnolia.

# 68 Mandeville (*Mandevilla* Species and Hybrids)

**Scientific Names:** The genus name *Mandevilla* was named for Sir Henry J. Mandeville (1773–1861), British diplomat in Buenos Aires and gardener who discovered *M. laxa* in Argentina and sent it back to England in the 1800s. The genus *Allamanda* is sometimes confused with *Mandevilla* (both are in the dogbane family, the Apocynaceae, and both have cultivated ornamentals).

**Common Names:** Plants previously placed in the genus *Dipladenia* are now included in *Mandevilla*. Nevertheless, in the horticultural trade, the name "mandevilla" is often employed just for vigorous climbing varieties with large leaves, while the name "dipladenia" is used for more compact and bushy varieties with small leaves. Mandevilla is also called rocktrumpet.

## 68.1 INTRODUCTION

Mandevilla produces gorgeous, long-lasting flowers, but these are relatively unfamiliar to many because the species are intolerant of cold and need to be purchased as relatively expensive potted specimens to be discarded in the autumn. Nevertheless, they are so attractive that they are likely to grow in popularity.

## 68.2 OVERVIEW

The genus *Mandevilla* has over 100 species, including many that were formerly placed in *Dipladenia*. The plants are neotropical, that is, found within Central and South America, including the tropical southern part of Mexico and the Caribbean. Most are in South America. The majority are vines, some growing to a length of over 7 m. A few are herbs or shrubs. Most cultivated plants are hybrids, and varieties differ considerably, ranging from long vines to shrubs that just tend to become viny. The foliage is evergreen in tropical areas but dies back when exposed to freezing temperatures. The large, long-blooming flowers come in apricot, pink, red, and white and are produced in early summer, bloom for a long period, and sometimes rebloom in the fall. Cultivars with variegated foliage and double-petaled flowers are also available. Several cultivated tropical vines (notably "yellow mandevilla," *Urechites lutea* and *Pentalinon luteum*) with yellow trumpet-like flowers resembling those of mandevilla are often misidentified as this plant.

FIGURE 68.2 Botanical paintings of *Mandevilla* (public domain): (a) *Mandevilla splendens.* (b) *Mandevilla laxa.* (c) Horticultural hybrid. (d) *Mandevilla splendens.* (e) *Mandevilla boliviensis.* (f) *Mandevilla splendens.* (g) *Mandevilla sanderi.* (h) *Mandevilla* × *amabilis.*

*Source:* **(a) *L' Illustration horticole*, vol. 30, 1883. (b) *Flore des serres et des jardins de l'Europe*, vol. 11, 1856. (c) *The Garden. An illustrated weekly journal of horticulture*, vol. 20, 1881. (d) Gould, J., and H. C. Richter. 1861. *A monograph of the Trochilidæ, or family of humming-birds*, vol. 2. (e) *The Garden. An illustrated weekly journal of horticulture*, vol. 44, (893). (f) *Curtis's Botanical Magazine*, vol. 69, 1843. (g) *The Garden. An illustrated weekly journal of horticulture*, vol. 51, 1897. (h) Lemaire, C. A. 1880. *L'Illustration horticole*, vol. 27. F. et E. Gyselnyck, Gand, Belgium.**

FIGURE 68.1 Mandevilla flowering displays.

*Source:* **(a) Photo by Shutterstock/SusaZoom. (b) Photo by Sheba (CC BY SA 2.0). (c) Photo by Forest and Kim Starr (CC BY 3.0).**

DOI: 10.1201/9781003586227-68

## 68.3 HISTORY

Mandevilla was first described in 1840, so it has a much shorter history of cultivation than most other ornamental plants. Relatively few species have been domesticated, including *M. boliviensis* (white dipladenia), *M. laxa* (*M. suaveolens*, Chilean jasmine), *M. sanderi*, and *M. splendens*. Most cultivars are hybrids, and most (including plants labelled *M.* × *amoena*) are assignable to *M.* × *amabilis*.

## 68.4 ECONOMIC IMPORTANCE

In tropical regions, mandevilla is one of a large number of attractive ornamentals that are available, which limits its importance. In temperate areas, where the plant has good potential for increasing sales, it is increasingly being marketed as an annual, sold in containers (especially hanging baskets). Efforts are also underway to breed compact, cold-hardy forms that are suitable for growth as perennials in semi-tropical regions. Mandevilla is usually sold at prices that are higher than for plants of similar stature, so it tends to appeal to a smaller clientele.

## 68.5 ORNAMENTAL USES

Cultivars that develop long vines are appropriate for growing on fences, trellises, and arbors. Compact cultivars are better suited for large containers and hanging baskets and for maintaining as houseplants. For viny forms grown in containers, an obelisk can be bought or fashioned from bamboo poles, or support hoops employed for other ornamentals can be employed.

FIGURE 68.3 Commercial aspects of mandevilla: (a) A garden center display. (b) Production in a massive greenhouse in The Netherlands.

*Source:* **(a) Photo by the Greenery Nursery and Garden Shop (CC BY 2.0). (b) Photo by Shutterstock/T. W. van Urk.**

FIGURE 68.4 Mandevilla planted in the ground in hot-climate regions: (a) *Mandevilla* × *amoena*. (b) *Mandevilla sanderi*.

*Source:* **(a) Photo by Shutterstock/RaywattJh. (b) Photo by 4028mdk09 (CC BY SA 3.0).**

FIGURE 68.5 Mandevilla planted in outdoor containers.

*Source:* **(a) Photo by Ivanchay (CC BY SA 3.0). (b) Photo by Smiley Toerist (CC BY SA 4.0). (c) Photo by Shutterstock/ Boryana Manzurova.**

## 68.6 OTHER USES

*Mandevilla* species are employed in herbal medicine in much of their New World range.

## 68.7 PROBLEMS

From a horticultural perspective, the lack of cold tolerance limits mandevilla's popularity in temperate climates. Mandevilla contains a viscous white latex which is toxic if ingested and can be irritating if contacted by bare skin.

## 68.8 CULTIVATION

### 68.8.1 Climate and Site

The majority of *Mandevilla* cultivars are killed by slight frost, so they are grown in temperate areas as annuals or

FIGURE 68.6 Mandevilla grown as a houseplant.

*Source:* **(a) Photo by Shutterstock/RStollner. (b) Photo by Cristina Sanvito (CC BY 2.0). (c) Photo by Jerzy Opioła (CC BY SA 4.0). (d) Photo by Shutterstock/Nnncls.**

maintained indoors, at least during the winter. These tropical plants do well at average room temperatures but are adapted to high humidity. Except in subtropical and tropical areas, the plants are usually grown in containers (often large enough to be called "tubs"). The vine forms need trellises, stakes, or other supports, or they can be allowed to sprawl from hanging baskets. Outdoors, the plants are best provided with partial shade, but bright window light is needed in bleak Northern climate winters.

### 68.8.2 Propagation and/or Purchasing

Most people will acquire mandevilla as potted plants. Professional propagation of the majority of cultivars is done by stem cuttings, but some species are grown from seed.

### 68.8.3 Planting

Mandevilla is usually obtained in pots that may suffice for growth as annuals but are too limiting for longer-term growth, and the plants should be replanted into larger containers. Vine forms may come with small bamboo poles that they will soon outgrow, requiring more substantial supports.

### 68.8.4 Soil and Moisture Conditions

The plants should be provided with rich, humusy, well-drained soil. In containers, a mix of equal parts peat moss, potting mix, and builder's sand is recommended. The soil needs to be maintained moist. In spring and summer, a fertilizer high in phosphorus (e.g., 10–20–10) every 2 weeks is recommended (or alternatively, a slow-release fertilizer).

### 68.8.5 Maintaining Plants

The growing stem tips of young plants should be pinched (destroyed) to induce bushiness. Gloves should be worn to avoid contact with the milky sap.

## 68.9 Fascinating Facts, Fables, Folklore, and Frivolity

- Modern medicine in the Western world has made most herbal medicine obsolete, and indeed, many regard traditional herbal medical practices as mere superstition. In South America, mandevilla has been employed as an herbal treatment for snakebite. Recent research has demonstrated that there are chemicals present which have anti-venom properties, validating the herbal usage.
- The name *Mandevilla sherlockii* was inspired by the fictional character Sherlock Holmes (Alvarado, L. O., and L. Lozada-Pérez. 2017. A new species of *Mandevilla* (Apocynaceae; Apocynoideae) from Mexico. Phytotaxa 319(1): 94).
- Most people are tempted to examine the fragrance of beautiful flowers, in the expectation that the smell will be equally attractive. However, the odor of the flowers of many cultivated species is quite variable, and this is the case for *Mandevilla* species. Some claim that warm weather promotes pleasant smell of mandevilla flowers.

FIGURE 68.7 A young girl testing the fragrance of mandevilla flowers in a closed greenhouse, an environment that concentrates the perfume of blooms.

*Source:* **Photo by Shutterstock/Adrian Eugen Ciobaniuc.**

## KEY PUBLICATIONS


Biondo, R., A.M.S. Pereira, S. Marcussi, P.S. Pereira, S.C. França, and A.M. Soares. 2003. Inhibition of enzymatic and pharmacological activities of some snake venoms and toxins by *Mandevilla velutina* (Apocynaceae) aqueous extract. *Biochimie* 85: 1017–1025.

Biondo, R., A.V. Souza, B.W. Bertoni, A.M. Soares, S.C. França, and A.M.S. Pereira. 2007. Micropropagation, seed propagation and germplasm bank of *Mandevilla velutina* (Mart.) Woodson. *Scientia Agrícola* 64: 263–268.

Endress, M.E. 2018. Proposal to conserve the name *Mandevilla* against *Exothostemon* (Apocynaceae: Mesechiteae). *Taxon* 67: 208–209.

Goyder, D. 1992. *Mandevilla illustris*: Apocynaceae. *The Kew Magazine* 9: 131–134.

Kozak, D., M. Parzymies, A. Świstowska, B. Marcinek, and E. Pogroszewska. 2021. The influence of growth regulators and explant position on the growth and development of *Mandevilla sanderi* (Hemsl.) Woodson in vitro. *Acta Scientiarum Polonorum Hortorum Cultus* 20: 127–138.

Morales, J.F. 1998. A synopsis of the genus *Mandevilla* (Apocynaceae) in Mexico and Central America. *Brittonia* 50: 214–232.

Palumbo, F., S. Draga, F. Scariolo, G. Gabelli, G.B. Sacilotto, M. Gazzola, et al. 2022. First genomic insights into the *Mandevilla* genus. *Frontiers in Plant Science* 13: 983879. https://doi.org/10.3389/fpls.2022.983879.

Simões A.O., M.E. Endress, T. Van Der Niet, L.S. Kinoshita, and E. Conti. 2006. Is *Mandevilla* (Apocynaceae, Mesechiteae) monophyletic? Evidence from five plastid DNA loci and morphology. *Annals of the Missouri Botanical Garden* 93: 565–591.

Wilkins, H.F. 1985. *Mandevilla sanderi*. In: *Handbook of Flowering*, Vol. 3, edited by H.F. Wilkins, 358–359. CRC Press, Boca Raton, FL.


### KEY CULTIVATION WEBSITES


Clemson University Cooperative Extension. *Mandevilla*. https://dc.statelibrary.sc.gov/bitstream/handle/10827/41802/CU_ES_HGIC_1109_1999-06.pdf?sequence=1.

Gardening Know How. *Mandevilla Vine: Tips for Proper Mandevilla Care*. www.gardeningknowhow.com/ornamental/vines/mandevilla/mandevilla-plant.htm.

GardenersWorld.com. *How to Grow Mandevilla*. www.gardenersworld.com/how-to/grow-plants/how-to-grow-mandevilla/.

Horticulture Magazine. *Mandevilla, Rock Trumpet*. https://horticulture.co.uk/mandevilla/.

New York Botanical Garden. *How to Over-winter a Mandevilla Vine*. www.nybg.org/blogs/plant-talk/2011/10/tip-of-the-week/how-to-over-winter-a-mandevilla-vine/.

*Sundaville (Large Commercial Producer)*. https://sundaville.com/.

The Spruce. *Mandevilla: Plant Care & Growing Guide*. www.thespruce.com/growing-dipladenia-or-mandevilla-in-containers-847910.

Thursd. *Mandevilla Plant – All There Is to Know*. https://thursd.com/articles/mandevilla-plant-all-there-is-to-know.

WikiHow. *Yes, Mandevillas Are Easy to Grow. How to Grow and Care for Mandevillas (Even in Winter)*. www.wikihow.life/Grow-Mandevillas.

# 69 Marguerite (*Argyranthemum* Species and Hybrids)

**Scientific Names:** The genus name *Argyranthemum* is based on the Greek *argyros*, "silver," and *anthemos*, "flower."

**Common Names:** The word *marguerite* traces to the Middle French *margarite*, "daisy," a usage tracing at least to the very early 17th century. It has been speculated that the name traces to Margaret of Anjou (1430–1482), wife (and queen) of King Henry VI of England, who displayed daisies on her personal banner. "Paris daisy" is another name reflecting past French usage. Marguerites are also known as argyranthemums, canary marguerites, garden marguerites, Cobbitty daisies (Cobbitty is a district in Australia), and marguerite daisies. "Boston daisy" is also encountered but has been applied to numerous genera with daisylike flowers. Indeed, the same is true for the word *marguerite* (Cape marguerites are *Dimorphotheca*, and marguerite is one of the alternative names applied to cultivars of *Anthemis*, *Leucanthemum vulgare*, and other species). The plants are sometimes called single-flowered chrysanthemums, a reference to their closeness to the genus *Chrysanthemum*. However, many modern cultivars have flower clusters that are no longer single like daisies but have additional petals. In addition to the issue of what name is appropriate, numerous cultivated ornamental plants with daisylike flowers are so similar that identifying them is difficult except for experts. Indeed, chrysanthemums, asters, Shasta daisies, and others in the daisy family can be close mimics of the true marguerites.

FIGURE 69.1 The marriage of Margaret (*Marguerite* in the original French) of Anjou to King Henry VI of England in 1445, officiated by the Bishop of Salisbury. She was 15, and he was 23. She wore a white satin wedding dress embroidered with silver and gold daisies, speculated to have led to the naming of the flowers as marguerites. He experienced frequent bouts of insanity; she became a powerful ruler but was deposed in 1475. This colored etching was created by John Carter in 1793.

*Source:* **Wellcome Images (CC BY 4.0).**

## 69.1 OVERVIEW

Probably the first flower that most children draw is a "daisy," since it is so distinctive and charming and reinforced by the familiar exercise of removing petals while saying, "She

FIGURE 69.2 Daisy petals in answer to the inquiry "She/he loves me. She/he loves me not."

*Source:* **(a) Illustration by Shutterstock/Natvas. (b) Illustration by Nancy Houser (CC BY 2.0).**

FIGURE 69.3 Floral variation of "single-flower" (with a single ring of ray petals) *Argyranthemum* cultivars.

*Source:* **(a) Photo by Shutterstock/Ergfoto. (b) Photo by Hans from Pixabay (public domain). (c) Photo by cTrickY (CC BY ND 2.0). (d) Photo by Elstef from Pixabay (public domain). (e) Photo by Albrecht Fietz from Pixabay (public domain). (f) Photo by Shutterstock/Chuck Hsu.**

DOI: 10.1201/9781003586227-69

FIGURE 69.4 Examples of double-flowered (a, b) and semi-doubled (c, d) blooms of *Argyranthemum* cultivars.

*Source:* **(a) Cultivar Summer Melody. Photo by Serres Fortier (CC BY 2.0). (b) Photo by Forest and Kim Starr (CC BY 3.0). (c) Photo by Chesna from Pixabay (public domain). (d) Photo by Forest and Kim Starr (CC BY 3.0).**

loves, she loves me not." There are numerous plants at least sometimes called daisies (several treated in this book), but the most established as a garden ornamental are the "marguerites" that trace to the genus *Argyranthemum*.

As in other members of the daisy family (Asteraceae), what are often interpreted as single daisylike flowers in *Argyranthemum* are compressed flower clusters or "flower heads" (technically, *capitula*; singular *capitulum*) composed of many florets (tiny flowers). In the wild species, the inner (disc) florets lack evident petals (they are very small), while the outer (ray) florets have conspicuous petals, but selection of cultivars has generated a variety of additional patterns. The wild species are usually perennial shrubs, but some of the garden hybrids are annual or short-lived perennials, and regardless, the garden forms are often grown as annuals. Marguerites produce masses of pretty, daisy-like flowers that may be pink, purple, red, white, yellow, or bicolored (when employed in flower arrangements, the blooms are sometimes dyed other colors). Many have rings of colored petals around a differently colored central eye, and some have doubled flowers (with extra petals). In some, the petals change color as they mature or as the temperature changes. The flowering period is often long, lasting much of the summer. In frost-free climates, many of the plants become branching, woody (at least at the base), evergreen shrubs (often called "subshrubs," as most are small). The foliage typically has deep divisions or lobes. The garden forms are short (usually under 1 m in height), although the genus *Argyranthemum* has some tall shrubs.

## 69.2 HISTORY

Garden marguerites originated as selections and hybrids from wild species of the related genera *Argyranthemum* (with 24 species of the Canary Islands, Madeira, and the Savage Islands), *Glebionis* (two species, native to Europe), *Heteranthemis* (one species, native to the Iberian Peninsula and North Africa), and *Ismelia* (one species of North Africa). Garden selections date back in Europe at least to the early 18th century. Although the cultivars are frequently identified as *A. frutescens* (or often as *Chrysanthemum frutescens* in the past and occasionally also in the present), most are of hybrid origin, often hybrids between different genera. *Argyranthemum* is usually present in the parentage of the several hundred garden cultivars that have been bred.

## 69.3 ECONOMIC IMPORTANCE

Marguerites are an established ornamental in the horticultural industry, long employed as a garden and container plant and as a minor cut flower.

FIGURE 69.5 A nursery with potted plants of *Argyranthemum* cultivars.

*Source:* **Photo by Shutterstock/Fotovale.**

FIGURE 69.6 Marguerites (*Argyranthemum* cultivars) in containers.

*Source:* **(a) Photo by Shutterstock/Chuck Hsu. (b) Photo by Shutterstock/BartTa. (c) Photo by Aart Beijeman from Pixabay (public domain). (d) Photo by Gilbert-Noël Sfeir (CC BY 2.0).**

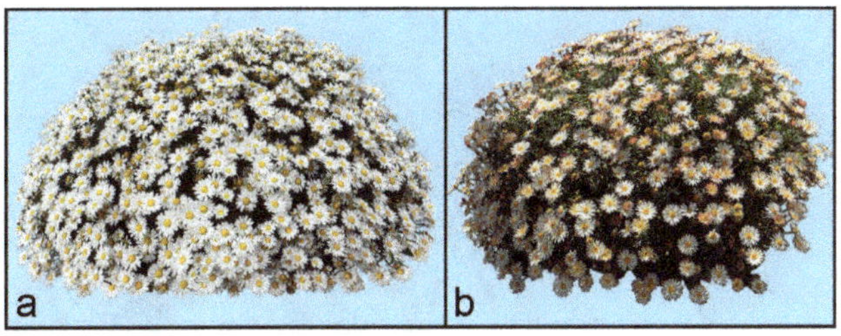

FIGURE 69.7 Natural mound appearance of garden-grown marguerites (*Argyranthemum* cultivars).

*Source:* (a) Photo by Noel Powell from Pixabay (public domain). (b) Photo by Frank Vincentz (CC BY SA 3.0).

FIGURE 69.8 Marguerites (*Argyranthemum* cultivars) trained as "standards" (with a central bare stem so as to resemble miniature trees).

*Source:* (a) Photo by Gerhard from Pixabay (public domain). (b) Photo by Shutterstock/Demamiel62.

FIGURE 69.9 Marguerites for Mother's Day.

*Source:* Photos from Pixabay (public domain): (a) by u_g1c6x-vwjkq, (b) by Dmmagrau, and (c) by Moondance.

## 69.4 ORNAMENTAL USES

Garden marguerites are employed in beds, borders, and containers, and sometimes as cut flowers. When grown as perennials, the plants frequently naturally assume a mound appearance, but they are sometimes trained into shapes such as pyramids or miniature trees (termed "standards" in the horticultural industry).

## 69.5 OTHER USES

Potted marguerites are often given as gift plants.

## 69.6 PROBLEMS

Scientific nomenclature is designed to make the names of species reliable and unambiguous, but unfortunately, many of the Latin names are hard to pronounce, and *Argyranthemum* is no exception. When it is used as a common name, it takes effort. "Marguerites" and "daisies" are easier to remember, but unfortunately, they are employed for many different ornamental plants. The result is that when purchasing plants at a nursery or garden center or when searching for information on the internet, the true identities of the plants are often uncertain.

## 69.7 CULTIVATION

### 69.7.1 Climate and Site

Garden marguerites are mostly frost-tender (they are described as "half-hardy") and in temperate climates are usually grown as annuals. In climate zones 10 and 11 (sometimes also zones 9 and 12), they can be cultivated as perennials. They are best exposed to full sun, although hot days will reduce flowering in these "cool season" plants, and in hotter climates, semi-shade is appropriate. (In their native areas, the plants are adapted to cool sea breezes.)

### 69.7.2 Propagation and/or Purchasing

Marguerite cultivars are usually propagated by cuttings, since they frequently do not breed true from seed, although some can be grown using seed. As for many of the plants described in this book, propagation by given techniques is prohibited by law for some patented varieties. In temperate climates, the most convenient way of establishing plants is simply to buy saplings from a nursery. The plants can be lifted and stored overwinter, but very few consider this worthwhile, as new starter plants can be bought.

### 69.7.3 Planting

Marguerite daisies can spread substantially, and depending on cultivar, spacing in gardens should be 30 to 60 cm (sometimes 90 cm). Compact varieties grow well in containers. Taller varieties (up to 1 m or more) also can be grown in containers and are sometimes trained as standards (all branches removed from a main stem so that the plant looks like a miniature tree).

### 69.7.4 Soil and Moisture Conditions

Well-drained, fertile, sandy or loamy, circumneutral soil is best (a pH of 6.1 to 7.8 is acceptable).

### 69.7.5 Maintaining Plants

At least for older cultivars, pinching (destroying the stem tips) has been recommended to encourage bushiness. Deadheading is sometimes practiced to prolong flowering.

Because there are so many flowers, some gardeners simply cut away all of them when most have wilted.

## 69.8 FASCINATING FACTS, FABLES, FOLKLORE, AND FRIVOLITY

- Gardens often have magical aspects that serve to alleviate the anxieties of people caused by the real world. Beautiful fairies and pixies (and less-attractive gnomes and goblins) are frequently pictured among garden plants, and marguerites (or at least daisylike flowers) are often a theme of imaginative garden illustrations.
- "Marguerite" has a culinary meaning: "A frosted cookie made by spreading a saltine, pastry square, or other base with a mixture of whipped white of egg and boiled sugar syrup and browning in in an oven" (*Webster's Third New International Dictionary*).
- The word "margarita," a cocktail prepared with tequila, triple sec, lime juice, and (sometimes) simple syrup, seems to have originated from the Spanish *Margaret*, daisy. "Daisies" are cocktails that include a base spirit, liqueur, and citrus.
- The pizza margherita is the classic Neapolitan pizza. According to legend, it was invented in 1889 by Raffaele Esposito, chef at Pizzeria Brandi, allegedly created in honor of Italy's unification, with the three toppings – basil, mozzarella, and tomato – respectively representing the green, white, and red of the Italian flag, the pie claimed to be named after the Queen of Italy, Margherita of Savoy.
- The name "Margaret" (the English equivalent of the French name Marguerite) has many equivalent forms, including Maggie, Madge, Daisy, Margarete, Marge, Margo, Margie, Marjorie, Meg, Megan, Rita, Greta, Gretchen, and Peggy (Marg→Meg→Peg). In the United States, the female name Marguerite has been losing popularity for more than a century, whereas the name Margaret has remained fairly stable but applied to less than 1% of the female population (www.ssa.gov/OACT/babynames/).

FIGURE 69.10 Imaginative garden illustrations featuring marguerites.

*Source:* **(a) Grandville, J. J. 1867. *The Flowers Personified*, vol. 1. Garnier Frères, Paris. (b) L. Prang & Co. (publisher), 1895. Photo from Boston Public Library (CC BY 2.0). (c) Photo by Shutterstock/Iren Lukas.**

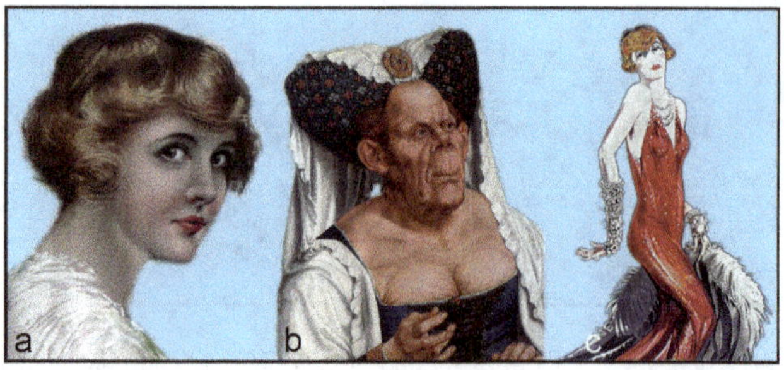

FIGURE 69.11 Some well-known historical personalities named Marguerite (public domain illustrations): (a) American stage and silent film actress Marguerite Clark (1883–1940). (b) Marguerite, Comtesse de Tyrol (1318–1369), satirically depicted as "The Ugly Duchess" by Flemish artist Quentin Matsys (1466–1530). (c) French actress Marguerite Mayane in a 1926 poster advertising a play.

*Source:* **(a) *Theatre Magazine*, vol. 29, 1919. Photo by Hamilton King. (b) Public domain. (c) *Bibliothèque nationale de France*, illustrated by Pol Rab (1898–1933).**

## KEY PUBLICATIONS


Carlson, W.H., C. Bethke, and D. Hopper. 1985. Producing potted marguerite daisies for profit. *Michigan State University Extension Bulletin E-1861*. https://archive.lib.msu.edu/DMC/extension_publications/e1861/e1861.pdf.

Cheek, R. 1993. La belle marguerite. *The Garden* 118: 350–355.

Cunneen, T.M. 1995. Breeding for improvement of the marguerite daisy (*Argyranthemum* spp.). *Acta Horticulturae* 420: 101–103.

Cunneen, T.M. 1996. *The Marguerite Daisy (*Argyranthemum *spp.): Developing an Understanding for Breeding*. Doctoral thesis, University of Sydney (Australia). 352 pp.

Francisco-Ortega, J., D.J. Crawford, A. Santos-Guerra, and R.K. Jansen. 1997. Origin and evolution of *Argyranthemum* (Asteraceae: Anthemideae) in Macaronesia. In: *Molecular Evolution and Adaptive Radiation*, edited by T.J. Givnish and K.J. Sytsma, 407–432. Cambridge University Press, Cambridge.

Francisco-Ortega, J., R.H., Ellis, E., González-Feria, and A., Santos-Guerra. 1994. Overcoming seed dormancy in ex situ plant germplasm collections; an example in the endemic *Argyranthemum* (Asteraceae: Anthemideae) species from the Canary Islands. *Biodiversity & Conservation* 3: 341–353.

Francisco-Ortega, J., A. Santos-Guerra, R. Mosa-Coello, E. González-Feria, and D.J. Crawford. 1996. Genetic resource conservation of the endemic genus *Argyranthemum* Sch. Bip. (Asteraceae: Anthemideae) in the Macaronesian Islands. *Genetic Resources and Crop Evolution* 43: 33–39.

Humphries, C.J., 1976a. A revision of the Macaronesian genus *Argyranthemum* Webb ex Schultz-Bip. (Compositae-Anthemideae). *Bulletin of the British Museum of Natural History (Botany)* 5: 147–240.

Humphries, C.J. 1976b. Evolution and endemism in *Argyranthemum* Webb ex Schultz Bip. (Compositae-Anthemideae). *Botanica Macaronesia* 1: 25–50.

Lim, T.K. 2014. *Argyranthemum frutescens*. In: *Edible Medicinal and Non-Medicinal Plants*, edited by T.K. Lim, 201–203. Springer, Dordrecht.

Morikawa, M., T. Muto, A. Santos-Guerrra, and K. Kondo. 2014. Identifying, discriminating and isolating cultivars of 'marguerites' originated from *Argyranthemum frutescens* parentages and their intergeneric and interspecific hybridities by DNA markers amplified by RAPD (Random Amplified Polymorphic DNA) and ISSR (Inter-Simple Sequence Repeat). *Chromosome Botany* 9: 97–112.

White, O.W. 2018. *Genomics of speciation and hybridisation in the Macaronesian endemic genus* Argyranthemum *(Asteraceae; Anthemideae)*. Doctoral thesis, University of Southampton (U.K.). 250 pp. https://eprints.soton.ac.uk/441933/.

Zhao, Y., D. Qu, and Y. Ma. 2022. Characterization of the chloroplast genome of *Argyranthemum frutescens* and a comparison with other species in anthemideae. *Genes* 13(10): 1720. www.mdpi.com/2073-4425/13/10/1720.

## Key Cultivation Websites

Gardening etc. *17 Types of Popular Argyranthemum Pictorial Guide*. https://asucculent.com/17-types-of-popular-argyranthemum-pictorial-guide/.

Gardening etc. *Marguerites Care and Growing Guide: Expert Tips on These Delightful Daisies*. www.gardeningetc.com/advice/marguerites-care-and-growing-guide.

The Spruce. *7 Species of Daisies for Your Flower Garden*. www.thespruce.com/daisy-types-for-gardens-1316051.

WikiHow. *How to Care for Daisies*. www.wikihow.com/Care-for-Daisies.

WikiHow. *How to Prune Daisies*. www.wikihow.com/Prune-Daisies.

# 70 Marigold (*Tagetes* Species and Hybrids)

**Scientific Names:** The generic name *Tagetes* is generally attributed to Tages, an Etruscan deity (Eutruscia was an ancient country of Central Italy). The Romans changed the myth to state that the deity was a grandson of Jupiter, who is reputed to have taught the Etruscan priests the art of soothsaying.

**Common Names:** The common name "marigold" is part of the name of species in several genera. The word "marigold" is a corruption of the Anglo-Saxon *merso-meargealla*, the marsh marigold (*Caltha palustris*). Aside from *Tagetes* species, the best-known "marigold" is pot marigold (*Calendula officinalis*, sometimes called common marigold), grown as both an ornamental and a culinary herb. Additional so-called marigolds include desert marigold (*Baileya multiradiata*), corn marigold (*Glebionis segetum*), and tree marigold (*Tithonia diversifolia*). All species of *Tagetes* are American in origin. The "French marigold" (*T. patula*) was so-named because it became popular in France. Similarly, the Central American *T. erecta* acquired the name "African marigold" after it was exported to Africa and became popular in monastery gardens there. Although *T. erecta* is usually called African marigold, it is also known as American marigold, big marigold, student flower, and (occasionally) Aztec marigold and saffron marigold.

## 70.1 INTRODUCTION

As noted earlier, there are many "marigolds," but the most widely grown ornamental species belong to the genus *Tagetes*. This has close to 50 species of annual or perennial herbs and shrubs native to the cool tropics of the New World. The genus is particularly diverse in two regions: south-central Mexico and Argentina. African and French marigolds are the species most often cultivated as ornamentals, but several other species are also grown for their showy flowers, notably *T. tenuifolia* (striped Mexican marigold, signet marigold), *T. minuta* (variously known as Muster-John-Henry, stinking Roger, Aztec marigold, dwarf marigold, and Mexican marigold), *T. filifolia* (Irish lace marigold), and *T. lucida* (Mexican mint marigold).

## 70.2 OVERVIEW

Flowers of the sunflower family (Asteraceae, Compositae) are tiny and aggregated into flower heads. The flower heads, not the individual flowers, are commonly referred to as flowers by most people, except by botanists. In the majority of species in the family, the outermost flowers in the flower heads bear a prominent strap-shaped petal, while the petals of the central flowers in the head are inconspicuous. However, in ornamental cultivars, the flowers have been "doubled" by selection – that is, many of or all the inconspicuous central flowers have been converted to have the much larger petals, which often assume different shapes, depending on the variety. A very large number of cultivars are available of the ornamental marigolds, many with doubled flowers.

The *Tagetes* marigolds are among the most familiar, easily recognized, and widely grown of garden plants. The two principal ornamental species are rather similar (indeed, some authorities combine them in the same species), but there are differences. Compared to the French marigold, the African marigold is noted for its larger flowers, ranging from 5 to 10 cm in diameter, and its greater height, ranging from 30 to 120 cm, as well as a tendency to be more erect. The flowers are mostly yellow and orange, sometimes creamy in African marigold, while a greater range of colors is available in French marigold cultivars. In both species, the leaves are pinnate, that is, dissected (fernlike), with leaflets arranged on both sides of the midrib, usually in pairs opposite each other.

The French marigold is an annual herb 20–50 cm tall. Flowers of cultivars differ greatly in color, including yellow, orange, reddish-brown, maroon, and blended mixtures,

FIGURE 70.1 French marigold (*Tagetes patula*).

*Source:* **(a) Cultivar Sparkler. Photo by Jim Evans (CC BY 4.0). (b) Cultivar Jolly Jester. Photo by Addrexoline (CC BY 4.0). (c) Photo by Ben Kerckx from Pixabay (public domain). (d) Photo by Dezidor (CC BY 3.0). (e) Image by Bishnu Sarangi from Pixabay (public domain).**



DOI: 10.1201/9781003586227-70

FIGURE 70.2 African marigold (*Tagetes erecta*).

*Source:* **(a) Photo by Rajesh Balouria from Pixabay (public domain). (b) Photo by JDP90/Joydeep (CC BY SA 3.0). (c) Photo by Yeon Woo Lee from Pixabay (public domain). (d) Photo by Nghia Nguyen from Pixabay (public domain).**

FIGURE 70.3 Hybrids (*T. erecta* × *T. patula*) between African and French marigolds.

*Source:* **(a) Cultivar Vanilla Hybrid. Photo by Don McCulley (CC BY SA 4.0). (b) Photo by Darwinek (CC BY SA 3.0).**

as well as both doubled flowers and what appear to be single flowers with large spreading petals.

Hybrids between the African and French marigolds (*T. erecta* × *T. patula*) are commonly grown as ornamentals. These are often called triploid varieties, since they have a double set of chromosomes from one parent and a single set from the other. They have relatively low seed germination rates. Triploid hybrids combine the large flowers of the African marigold with the more compact size of the French marigold, producing vigorous plants that seem relatively unaffected by high summer heat and generally bloom throughout the summer.

## 70.3 HISTORY

Species of *Tagetes* have been cultivated and extensively used by Native American tribes in Mexico, Central, and South America since pre-Columbian times, and almost every species of the genus has been used by Indigenous Americans in some practical way. Some of the species were domesticated in pre-Columbian times. French marigold was originally bred in Mexico and was first introduced to Europe in 1573. Like other extremely attractive flowering plants, the marigolds have been the subjects of very extensive breeding throughout the world, with the result that a range of amazingly attractive cultivars are now available.

## 70.4 ECONOMIC IMPORTANCE

Marigolds are one of the world's major commercial flowering plants. Aside from their value as ornamental plants, marigolds are very extensively grown in warm-temperate and semi-tropical regions for extracts of industrial chemicals.

## 70.5 ORNAMENTAL USES

French and African marigolds furnish numerous annual cultivars, among the most attractive and popular of all garden ornamentals. They are widely used in massed displays, both in public and private gardens. French marigold and dwarf varieties of African marigold are considered ideal for rockery, edging, hanging baskets, and window boxes. The

FIGURE 70.4 Commercial production of marigolds: (a) In Qatar. (b) In Vietnam. (c) In northern Spain. (d) In Southern California.

*Source:* **(a) Photo by Alex Sergeev (CC BY SA 3.0). (b) Photo by Shutterstock/Phuong D. Nguyen. (c) Photo by Manuel M. V. (CC BY 2.0). (d) Photo by Erica Zabowsk (CC BY ND 2.0).**

FIGURE 70.5 Landscaping of public areas with ornamental marigolds: (a) Marigold-lined path in Kiev, Ukraine. (b) City Botanic Gardens, Brisbane, Australia.

*Source:* **(a) Photo by Mitte27 (CC BY SA 4.0). (b) Photo by Batsv (CC BY SA 4.0).**

FIGURE 70.6 Garlands prepared from African marigolds: (a) An Indian girl in traditional dance costume during the Festival of Color. (b) Honoring a Labrador retriever with a marigold garland and tilaka (red marks painted on the forehead) during the annual Hindu festival of Kukur Tihar, held especially in Nepal. During the period, dogs are considered to be messengers of God. They are fed the choicest foods, and it is considered a sin to act disrespectfully to them. (c) A dalmatian decorated with marigold garlands during the Mexican Day of the Dead celebration.

*Source:* **Photos from Shutterstock, contributed by (a) Soumen-82hazra, (b) Thomas Dutour, and (c) Carlos Tischler.**

larger flowers of African marigold are more extensively employed as cut flowers and in the garland trade.

## 70.6 OTHER USES

The flowers and leaves of all species of *Tagetes* are copiously dotted with glands which produce pungent chemicals, which are extracted extensively for industrial, culinary, and medicinal purposes.

The roots of marigolds contain chemicals which are toxic to insects, fungi, and viruses, but particularly to nematodes (very small, slender worms, sometimes invisible to the human eye). Plant parasitic nematodes can seriously damage or even kill crops, turf grasses, and ornamental plants, and marigold extracts (and even just growing marigolds on a field prior to planting a crop) have proven to be useful in controlling nematodes.

Flowers of African marigold are used to produce a yellow dye sometimes employed to color butter and cheese and to adulterate saffron. In parts of the world, notably in Peru and Mexico, the species is cultivated as a source of pigment for poultry feed. In India and Pakistan, the flowers and roots are employed extensively in home dyes among the poor.

The petals of several species of *Tagetes*, including French marigold, are edible and are consumed in the same manner as pot marigold. The flowers of some cultivars of African marigold are also edible, and the species is used in beverages in parts of the southwestern United States and Latin America.

Marigolds were sacred to Mayan cultures and Aztec cultures, and in modern Mexico and other areas of Latin America, they serve as a symbol in celebrations for Los Dias de Los Muertos (Day of the Dead) on November 2. Marigolds are also commonly associated with the observance of All Saints Day (November 1) and more especially All Souls Day (November 2) in Latin America. Although part of the modern Catholic religious calendar, the celebration coincides with the Indigenous religious practices retained from pre-Columbian times. The dead are believed to come back to their houses to share a meal with their families. The petals are often strewn on graves of relatives and on paths to guide the souls of dead family members to food

FIGURE 70.7 Marigolds in the Day of the Dead celebrations in Latin America.

*Source:* **(a) Photo by MaríaJoséFelgueresPlanells (CC BY SA 4.0). (b) Photo by Isaacvp (CC BY SA 4.0). (c, d, e) Photos by Elenats (CC BY SA 4.0).**

FIGURE 70.8 Marigold decorations for Hindu gods: (a) Vishnu. (b) Ganesh. (c) The goddess Laxmi.

*Source:* **(a) Photo by Springs23 from Pixabay (public domain). (b) Photo by Shutterstock/India Picture. (c) Photo by Shutterstock/Dhananjay Bhagat.**

and offerings placed on home altars. Skeletal- and skull-like figures are often constructed, representing relatives who have passed away; they are surrounded by marigolds and offered food and gifts. People also dress up as skeletons, often decorated with marigolds, and engage happily in events during the holiday.

In Hindu religious ceremonies in India, the French marigold is considered a "holy flower" and is used extensively as an altar decoration, to honor gods and goddesses, and for ritual adornment. Since *Tagetes* marigolds were unknown in pre-Columbian times in the Old World, the use of the North American species probably originated only a few centuries ago.

## 70.7 PROBLEMS

African marigold has escaped from cultivation and become a weed in some semi-tropical countries. The species reputedly also is mildly toxic when consumed by people and pets (although many other species of *Tagetes* are widely used for culinary purposes).

## 70.8 CULTIVATION

### 70.8.1 Climate and Site

The ornamental marigolds are not frost-tolerant but can be grown during summer periods in most temperate and semi-tropical areas. Tall varieties (mostly African marigold) benefit from sheltered locations that provide protection from strong winds and heavy rains. French and African marigolds should be grown in full sun, although in very hot climates, some horticulturalists recommend some late-day shade.

### 70.8.2 Propagation and/or Purchasing

Marigolds are either planted from purchased seeds or seedlings in plugs (small fiber containers) are planted. Seeds can be collected from spent flowers to sow the next spring but may not produce plants with the same appearance. If grown from seeds, they should be started indoors in a warm (20–30°C), bright location. If starter plants in trays are purchased from a nursery, leggy, straggly plants with a flower or two should be avoided, while stocky young plants without open flowers are preferable.

### 70.8.3 Planting

Marigold seedlings should be transplanted outdoors after all danger of frost, when the soil is warm. They best tolerate being transplanted when relatively small (5–10 cm tall).

### 70.8.4 Soil and Moisture Conditions

The marigolds grow well in moderately fertile soils with a pH of 6.5–7.5. African marigold is better adapted than French marigold to hot, dry conditions and poorer soil. For both species, very rich soils can result in excess foliage but fewer blooms. The plants should be watered during periods of drought or very dry conditions. Overwatering encourages disease and suppresses flowering.

FIGURE 70.9 Planting marigolds.

*Source:* **Photo by the New York National Guard Youth Program (CC BY ND 2.0).**

### 70.8.5 Maintaining Plants

Once plants have reached the flowering stage, faded heads of French and African marigold should be removed to prolong flowering. African marigolds tend to become top-heavy, and to prevent this and encourage branching and bushier growth, the young growing tips are often pinched back (destroyed).

## 70.9 FASCINATING FACTS, FABLES, FOLKLORE, AND FRIVOLITY

- Because of the "mari" in *marigold*, Old World "marigolds" (possibly *Calendula*) became associated with the Virgin Mary and Queen Mary. According to legend, the marigold was the flower favored and worn by Mary. Marigold was called "Mary's gold" by early Christians, who placed the flowers around statues of the Virgin. The blossoms were offered in place of coins because it was said that Mary used the blossoms (i.e., "Mary's gold") as coins. Marigolds were said to represent the golden rays of glory that are often shown around Mary's head, and the flowers were commonly associated with Christian celebrations. According to legend, during the flight into Egypt, the Holy Family was accosted by a band of thieves. They took Mary's purse, and when they opened it, marigolds fell out. Because of the association of marigolds and the Virgin Mary, marigolds have been carried by brides to symbolize fidelity.

- During the invasions of the Conquistadores in Mexico, marigolds were known as the "flower of death" (*flor de muerto*), springing up profusely on ground disturbed by war. It was commonly believed that the herb arose from the blood of the Natives killed by the Spanish invaders.
- Another legend explains how marigold became the "flower of death." The Aztec King Tenoch was saddened by all the death during the conquest of Tenochtitlan (what today is Mexico City). He asked Tonatiuh (the sun god) to help him remember the deceased. The next morning, the fields surrounding Tenochtitlan were covered with bright orange marigolds.
- In the language of flowers, popular in Victorian England as a secret code of communicating information within written letters, "African marigold" means "vulgar minds," and "French marigold" means "jealousy." "Garden marigold" means "uneasiness"
- In 1975, the Burpee Seed Company received considerable free publicity by awarding $10,000.00 to Mrs. Alice Vonk of Sully, Iowa, for discovering a white-flowered African marigold.
- Colors are often defined by flowers – for example, "marigold" is a yellow-orange hue.

> For some years I have been carrying the flag in cause, and that cause is to have the American MARIGOLD – *Tagetes erecta* – officially designated by the Congress as the national floral emblem of the United States. . . . I am fully sensible of the fact that all persons, young and old, male and female, rich and poor, have their own preferences when it comes to a flower. But mine goes for the marigold for a variety of reasons. It is actually an American native and really not native to any other place on earth. It is grown in great profusion in every one of the 50 States. . . . Long ago it was acknowledged as a symbol of religious faith and graced the altars both at home and abroad.
>
> One of the outstanding characteristics of the MARIGOLD is its robustness and rugged character. If there is any flower that can resist the onslaught of the insect world better than the MARIGOLD, I would not know what it was. I grow thousands of them every year for my own delight and the delight of my neighbors, and I know what this flower will do.
>
> I class it, therefore, with the American eagle when it comes to a symbol of our country that manifests stamina; and when it comes to beauty, I can think of nothing greater or more inspiring than a field of blooming MARIGOLDS tossing their heads in the sunshine and given a glow to the entire landscape. . . . Like the American eagle and the American flag, [the MARIGOLD is] an exclusively American emblem, unclaimed by any foreign nation.
>
> – Everett McKinley Dirksen (1896–1969), American politician, congressman, and senator, noted for oratory

This passage is taken from the Congressional Record, February 4, 1969, and is misinformed about the nativity of *Tagetes erecta*, the African marigold, probably because it is sometimes called the "American marigold," although it is native to Mexico and more southern areas of the New World.

## KEY PUBLICATIONS

Cicevan, R., A.F. Sestras, M. Plazas, M. Boscaiu, S.Vilanova, P. Gramazio, et al. 2022. Biological traits and genetic relationships amongst cultivars of three species of *Tagetes* (Asteraceae). *Plants (Basel)* March 12;11(6): 760. https://doi.org/10.3390/plants11060760. www.ncbi.nlm.nih.gov/pmc/articles/PMC8951573/.

Kaplan, L. 1960. Historical and ethnobotanical aspects of domestication in *Tagetes*. *Economic Botany* 14: 200–202.

Kumar, S.A. and K. Vijay. 2014. *Genetic Variability in African Marigold (*Tagetes erecta L.*)*. Lap Lambert Academic Publishing, London. 88 pp.

Lim, T.K. *Tagetes*. In: *Edible Medicinal and Non-Medicinal Plants: Volume 7, Flowers*, edited by T.K. Lim, 432–472. Springer, Dordrecht.

Neher, R.T. 1968. The Ethnobotany of *Tagetes*. *Economic Botany* 22: 317–325.

Raihandhany, R. and B. Kurniawan. 2023. The utilization of marigold (*Tagetes erecta* L.) for traditional ceremonies in

Indonesia: A systematic review. *Genbinesia Journal of Biology* 2(3). https://doi.org/10.55655/genbinesia.v2i3.37.

Rodríguez, E. and T.J. Mabry. 1977. Tageteae, chemical review. In: *The Biology and Chemistry of the Compositae*, Vol. 2, edited by V.H. Heywood, J.B. Harborne, and B.L. Turner, 785–797. Academic Press, London.

Rojas-Sandoval, J. 2018. *Tagetes erecta* (Mexican marigold). *Invasive Species Compendium*. www.cabi.org/isc/datasheet/52641.

Soule, J. 1994. Infrageneric systematics of *Tagetes*. In: *Proceedings of the International Compositae Conference, Compositae: Systematics*, edited by J. Beentje Henk, P.D.S. Caligari, and D.J.N. Hind, 435–443. Royal Botanic Gardens, Kew, U.K.

Strother, J.L. 1977. Tageteae, systematic review. In: *The Biology and Chemistry of the Compositae*, Vol. 2, edited by V.H. Heywood, J.B. Harborne, and B.L. Turner, 769–783. Academic Press, London.

Vasudevan, P. 1997. *Tagetes*: A multipurpose plant. *Bioresource Technology* 62: 29–35.

## Key Cultivation Websites

Clemson University, Extension. *Marigold*. https://hgic.clemson.edu/factsheet/marigold/.

Department of Agriculture, Land Reform and Rural Development, South Africa. *Marigold Flower Production Guideline*. www.dalrrd.gov.za/phocadownloadpap/Brochures_and_Production_Guidelines/Marigold%20flower%20production%20guideline.pdf.

Gardeners World.com. www.gardenersworld.com/how-to/grow-plants/how-to-grow-marigolds/.

Horticulture Magazine. *Tagetes – marigolds*. https://horticulture.co.uk/marigolds/.

The Spruce. *How to Grow and Care for Marigolds*. www.thespruce.com/how-to-grow-marigolds-1402909.

University of Minnesota, Extension. *Marigolds*. https://extension.umn.edu/flowers/marigolds.

WikiHow. *How to Plant Marigolds*. www.wikihow.com/Plant-Marigolds.

WikiHow. 2022. *How to Deadhead Marigolds*. www.wikihow.com/Deadhead-Marigolds.

WikiHow. 2022. *How to Grow Marigolds*. www.wikihow.com/Grow-Marigolds.

# 71 Morning Glory (*Ipomoea* Species and Hybrids)

**Scientific Names:** The genus name *Ipomoea* is based on the Greek *ips*, "worm," + *homoios*, "resembling," an allusion to the trailing habit of many of the species.

**Common Names:** The vernacular name *morning glory* is based on the habitual opening of the flowers in the morning. The term "moonflower" is applied to diverse species of plant (including some ipomoeas) that open their usually large white scented flowers in the evening to attract night-flying insect pollinators.

## 71.1 INTRODUCTION

Most ornamental morning glory species are vines with very attractive flowers. With the possible exception of daylilies, they are the most familiar cultivated plants known for flowers that remain open only for a day. Morning glories are very widely cultivated but often have a love–hate relationship with gardeners, lovable for their fast growth and impressive flowering displays, but troublesome because they grow aggressively and, like most ornamental vines, require maintenance.

## 71.2 OVERVIEW

The morning glory family (Convolvulaceae) has about 1,900 species and 60 genera, the largest of which is *Ipomoea*, with about 800 species. Recent classification research has indicated that many other genera should be included in *Ipomoea*, including *Argyreia*, *Astripomoea*, *Blinkworthia*, *Lepistemon*, *Lepistemonopsis*, *Mina*, *Paralepistemon*, *Rivea*, *Stictocardia*, and *Turbina*. Most of the species are native to tropical Asia and America, but some are also found in warm-temperate areas of the world. Many of the plants are very damaging weeds, often called bindweeds. Most ornamentals called morning glories are grown for their attractive flowers, and most of these belong to *Ipomoea*. As in all species in the Convolvulaceae, the large, often funnel-shaped (sometimes trumpet-shaped) flowers of *Ipomoea* open only for one day. The most likely non-*Ipomoea* cultivated ornamental species called morning glory is the dwarf morning glory, *Convolvulus tricolor*. In cooler parts of the world, quick-growing *Ipomoea* species that can be grown as annuals are cultivated as garden ornamentals. These include the annuals *I. nil* (Japanese morning glory), *I. purpurea* (common morning glory, tall morning glory), and *I. tricolor* (Mexican morning glory) and the perennials *I. indica* (blue morning glory) and *I. quamoclit* (Cypress vine). In tropical countries, *Ipomaea* is usually grown as a perennial, particularly *I. cairica* (mile-a-minute vine, railway creeper) and *I. carnea* (bush morning glory), but also *I. alba* (moonflower, tropical white morning glory), *I. horsfalliae* (cardinal creeper, Lady Doorly's morning glory), *I. lobata* (Spanish flag vine, firecracker vine), *I. muricata* (lavender moonvine), and *I. quamoclit*.

FIGURE 71.1 Moonflower (*Ipomoea alba*), one of many night-flowering species. The blooms open in the evening to attract moths. The species has the largest flowers in the Convolvulaceae family but is adapted to hot regions so is difficult to grow in cool climates.

*Source:* **Photo by Shutterstock/Gardens by Design.**

FIGURE 71.2 Paintings (public domain) of attractive morning glory bouquets: (a) *Ipomoea purpurea* cultivars. (b) Various species in the morning glory family (Convolvulaceae). (c) Mixed cultivars.

*Source:* **(a) D'Orbigny, C. D. 1892. Dictionnaire Universel d'histoire Naturelle. (b) Twining, E. 1849. Natural Order of Plants. (c) 1889 lithograph prepared for D. M. Ferry & Co., Detroit, MI. From the Boston Public Library (CC BY 2.0).**


DOI: 10.1201/9781003586227-71

FIGURE 71.3 Dwarf morning glory (*Convolvulus tricolor*, not to be confused with *Ipomoea tricolor*), the most likely non-*Ipomoea* cultivated "morning glory."

***Source:*** **(a) Photo by Andrew Butko (CC BY SA 3.0). (b) Photo by Shutterstock/Calmn8ss.**

FIGURE 71.4 Examples of cultivated *Ipomoea* species in flower: (a) *Ipomoea lobata* (firecracker vine, Spanish flag). (b) *Ipomoea tricolor* (Mexican morning glory). (c) *Ipomea indica* (blue morning glory). (d) *Ipomoea horsfalliae* (cardinal creeper).

***Source:*** **(a) Photo by Cultivar413 (CC BY 2.0). (b) Photo by Dwight Sipler (CC BY 2.0). (c) Photo by Shutterstock/F. Neidl. (d) Photo by Tatters (CC BY 2.0).**

Most species of *Ipomoea* are twining (or at least scrambling), perennial, herbaceous vines, but some are annual and some are woody (shrubs or trees). Some grow erectly; others are leaning or scramble over the ground. On supporting structures in long seasons, some ornamental species can climb to 6 m. Clumps may mound on themselves if not provided with supports to climb on.

## 71.3 HISTORY

Ornamental *Ipomoea* species have been cultivated in China and Japan for over 1,000 years. In Japan, morning glories have been extremely popular, with hundreds of ornamental cultivars created. *Ipomoea nil*, a native of the Americas, represents the imperial flower in Japan and, indeed, is often termed "*I. imperialis*."

FIGURE 71.5 Eighteenth- and nineteenth-century paintings (public domain) of the most extensively cultivated ornamental species of *Ipomoea* by European botanical illustrators: (a) *Ipomoea nil* (Japanese morning glory). (b) *Ipomoea horsfalliae* (cardinal creeper). (c) *Ipomoea purpurea* (common morning glory). (d) *Ipomoea quamoclit* (Cypress vine). (e) *Ipomoea tricolor* (Mexican morning glory). (f) *Ipomoea batatas* (sweet potato vine). (g) *Ipomoea lobata* (firecracker vine, Spanish flag). (h) *Ipomoea indica* (blue morning glory).

***Source:*** **(a) Edwards, S. 1815. *Botanical Register*, vol. 1. (b) Paxton, J. 1839. *Magazine of botany and register of flowering plants*, vol. 3. (c) Lindley, J. 1837. *Edwards's Botanical Register*, vol. 23. (d) Lowis, L. 1881. *Familiar Indian flowers*. (e) *Revue horticole* (Paris), vol. 94, 1922. (f) Denisse, E. 1843–1846. *Flore d'Amérique*. Gihaut Frères, Paris. (g) Lindley, J. (ed). 1842. *Edwards's Botanical Register*, vol. 28. (h) Lindley, J. (ed.) 1841. *Edwards's Botanical Register*, vol. 27.**

Species from the Americas were brought to Europe, primarily Italy and Spain, in the 1500s, and subsequently attractive cultivars were bred. As with perhaps the majority of common garden ornamentals, once the imported plants became established in Europe, they were subsequently popularized in other parts of the world.

Sweet potato (*I. batatas*) was domesticated in the Americas over 5,000 years ago, and there are now over 7,000 cultivars. The species epithet *batatas* is the ancient Taino name for the plant. (The Taíno were an Indigenous people of the Caribbean, from which there are modern descendants.)

## 71.4 ECONOMIC IMPORTANCE

Ornamental morning glory plants have been valued for centuries and are a significant commodity provided by the horticultural industries of many nations. The food species add greatly to the value of *Ipomoea*, and (conversely) the invasive weeds cause considerable damage.

## 71.5 ORNAMENTAL USES

Most cultivars of *Ipomoea* are medium-sized vines employed as climbing flowering ornamentals on frameworks placed

FIGURE 71.6 Sale of potted morning glory plants: (a) Scene in Japan, where the plants are very popular. (b) Morning glory in pots.

*Source:* **(a) Photo by Shutterstock/ST-photo. (b) Photo by Shutterstock/Sleepy Mandy.**

FIGURE 71.7 Landscape uses of ornamental morning glory: (a) Vine growing on an 8 m tall pole. (b) Vines growing on a house in Provence, France. (c) Vines climbing a 3-story building. (d) A dense wall of vines providing privacy.

*Source:* **(a) Photo by Dwight Sipler (CC BY 2.0). (b) Photo by Shutterstock/Emily Marie Wilson. (c) Photo by Ma Shimaro (CC BY 2.0). (d) Photo by Tracy (CC BY 2.0).**

FIGURE 71.8 Ornamental sweet potato vines (*Ipomoea batatas*): (a) Young vines of commercial edible tubers growing as temporary houseplants. (b) A dark-leaved cultivar, Black Tone. (c) A variegated (green-pink-white) cultivar, Tricolor.

*Source:* **(a) Photo by Shutterstock/Ilovehz. (b) Photo by Agnieszka Kwiecień, Nova (CC BY SA 4.0). (c) Photo by David J. Stang (CC BY SA 4.0).**

against houses or grown on fences and other vertical supports. *Ipomoea* species are also employed as ornamental groundcover, but often mainly for stabilization of slopes and beaches. Unfortunately, several species have the potential to be invasive.

Ornamental "sweet potato vines" are usually understood to refer to cultivars of *I. batatas* grown for their attractive foliage (often blackish, bronze, copper, chartreuse, purple, or purple, sometimes variegated, especially with white regions; they often do not produce flowers). The tuberous roots are usually too bitter to consume. The ornamental species *I. pandurata*, which has a large but inedible root, is called "wild sweet potato vine" and could be confused with the ornamental cultivars of *I. batatas*.

## 71.6 OTHER USES

The tuberous roots of several wild species of *Ipomoea* have been employed as food by various Indigenous people. Sweet potato (*I. batatas*), a starchy "root vegetable," has been ranked as the world's seventh most important staple food crop.

Confusion between sweet potatoes and "yams" in the United States began in colonial times, when African slaves thought the sweet potatoes they were fed were like their familiar African vegetables called nyami, which look like

FIGURE 71.9 Harvesting sweet potato (*Ipomoea batatas*).

***Source:*** **(a) Photo by U-ichiro Murakami (CC BY SA 2.0). (b) Photo by Chang Min SHIN from Pixabay (public domain).**

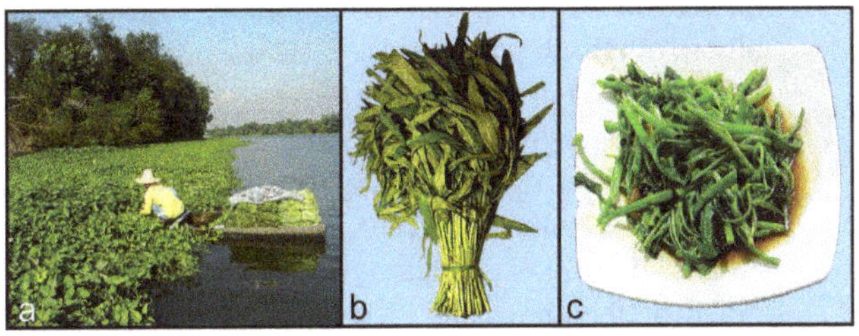

FIGURE 71.10 Water spinach (*I. aquatica*): (a) Harvest in the Philippines. (b) A typical commercial bundle. (c) Stir-fried water spinach.

***Source:*** **(a) Photo by Judgefloro (public domain). (b) Photo by Yamaguchi (public domain). (c) Photo by Midori (CC BY SA 3.0).**

some varieties of sweet potatoes but are produced by species of *Dioscorea*. The African name was eventually shortened to *yam*. To avoid confusion, the term *yam* should be reserved for the tuberous vegetables from *Dioscorea* and not applied to *Ipomoea* sweet potato. However, most people today probably cannot clearly explain the difference between sweet potatoes and yams.

"Water spinach," or kangkong (*I. aquatica*), is a rooted, semi-aquatic, tropical species occurring widely in southern Asia. It is harvested from natural waterways and is also cultivated in shallow water. It is an important green vegetable in Asia, extensively employed in the cuisine of many Asian countries.

*Ipomoea* has been extensively employed in folk medicine by Indigenous people. The roots of several species, especially *I. purga* (known as jalap), a native of Central Mexico, have been employed as a purgative (essentially a laxative). Overconsumption can produce vomiting. The seeds of several species, including *I. alba*, *I. corymbosa* (*Turbina corymbosa*), *I. nervosa*, *I. purpurea*, and *I. tricolor*, are hallucinogenic due to small amounts of LSA (D-lysergic acid amide), which produces similar effects to the usually illicit drug LSD (D-lysergic diethylamide). Cultivars Heavenly Blue, Pearly Gates, and Flying Saucer of *I. tricolor* or hybrids were named to indicate their psychoactive properties. Ancient Aztec, Mayan, and Native American tribes used the seeds to help in spiritual communion. The seeds are legally available in most jurisdictions (but sometimes are specifically prohibited for human consumption) and are frequently sold as a

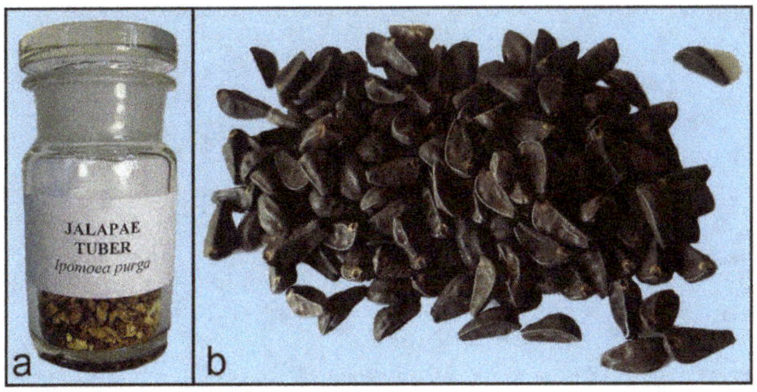


FIGURE 71.11 Potentially dangerous drugs from morning glories: (a) An obsolete medicinal preparation of the tuberous root of *Ipomoea purga* employed as a laxative. (b) Seeds of *I. violacea* employed as a recreational and spiritual hallucinogen.

***Source:*** **(a) Photo by Maša Sinreih (CC BY SA 3.0). (b) Photo by Dominic Milton Trott/DrugUsersBible (CC BY 2.0).**

"legal high." They have been employed by young people as a recreational hallucinogen with significant resulting illness. Overdoses can cause psychotic episodes. Moreover, commercially sold seeds are often treated with poisonous fungicides and other chemicals to promote their growth.

## 71.7 PROBLEMS

The worst weed called "morning glory" (better termed *field bindweed*) is *Convolvulus arvensis*. Other *Ipomoea* species are also serious invasive weeds in various tropical and subtropical counties. *Ipomea aquatica* is quite invasive in freshwater areas, albeit usefully harvested for food. *Ipomoea cairica* is especially invasive in terrestrial areas, albeit a useful ornamental vine (it is known as mile-a-minute vine, but the expression "mile a minute," indicative of fast spread, is employed for other invasive weeds, such as *Persicaria perfoliata* and *Mikania micrantha*). Species of *Ipomoea* that are invasive in some regions are sometimes prohibited from sale and/or planting under legislation controlling "noxious" weeds, and this is the case for several of the popular species discussed in this chapter.

Morning glories can be aggressive plants that climb over or wind around and suffocate other plants in a garden and so require control. Often, cultivars known to spread excessively are controlled by growing them in containers and by deadheading the flowers to prevent seeds from maturing and escaping.

Morning glory seeds are toxic, especially in large quantities. The entire plants have some potential to poison children, pets, and livestock. However, deer, groundhogs, rabbits, and other pests may consume the flowers.

## 71.8 CULTIVATION

### 71.8.1 Climate and Site

*Ipomoea* plants that are naturally annuals or are perennials grown as annuals can be planted from zones 2 to 11, while perennials can be grown outdoors overwinter from

FIGURE 71.12 Mile-a-minute vine (*Ipomoea cairica*) spreading as a weed in Thailand.

*Source:* **Photo by Shutterstock/WindOfHope.**

zones 9 to 11. Dwarf morning glory (*Convolvulus tricolor*) can also be grown as a perennial in zones 9 to 11, and as an annual in warmer climate zones. In zone 9 and near regions, the aboveground part (which is normally evergreen) may be killed by frost, while the underground part survives and regrows the plant. Morning glories often self-seed, so although they are often grown outdoors as annuals, gardeners need to be aware that the plants may persist in sites where they have been grown.

### 71.8.2 Propagation and/or Purchasing

Nurseries and garden centers stock both seeds and starter plants, which can also be obtained from online sources. Most cultivars need to be propagated vegetatively to breed true. This is done using the divided tuberous roots and by stem cuttings. In Northern areas, the tuberous roots of the perennials, which won't survive winters outdoors, can be excavated and stored indoors overwinter.

### 71.8.3 Planting

*Ipomoea* species grow well from seeds sown directly or started indoors 6 weeks or so in advance of the seasonal start. Some cultivars need to be established from cloned material to breed true. Unless the intent is to allow vine forms to trail over the ground, the plants need to be near a trellis, chain-link fence, pergola, arch, or other supporting structure. The smaller species grow well in containers, including hanging baskets.

### 71.8.4 Soil and Moisture Conditions

Morning glory species often tolerate poor, dry soils but do best in moist, well-drained ground. The appropriate soil pH range is 5.5 to 6.8. Limited fertilization is usually sufficient. Low-nitrogen/high-phosphorus fertilizers will promote flower rather than stem and foliage production.

### 71.8.5 Maintaining Plants

Pruning is not considered necessary, except to remove broken stems and to keep the plants in check. Deadheading (removing spent flowers) is also not really needed, except to keep seeds from developing and spreading.

## 71.9 FASCINATING FACTS, FABLES, FOLKLORE, AND FRIVOLITY

- Nectar-seeking hummingbirds are attracted to the flowers of some species of *Ipomoea*, especially blooms that are red or scarlet, such as those of *I. coccinea* (red morning glory).
- In Gabon, *I. involucrata* was thought to promote fecundity, and *I. cairica* to bring success to fishermen.
- In the 19th century, noted African American scientist George Washington Carver (1846–1943), much better known for developing new uses for the peanut, devised over a hundred ways of using the sweet potato and indeed is said to have had 500 ways to prepare them.
- Gerry Thomas, an executive of C. A. Swanson and Sons of Omaha, Nebraska, invented the frozen "TV dinner" in the early 1950s. The first version consisted of turkey, cornbread dressing and gravy, buttered peas, and sweet potatoes. Five million were sold in 1954, the first year of marketing. The first Swanson package was placed in an exhibit at the Smithsonian Museum in Washington, DC.
- Outhouses, in one form or another, are an ancient, practical invention. By the Victorian era, in much of Western culture, the "privy" came to be regarded

FIGURE 71.13 Hummingbirds visiting morning glory flowers: (a) Stripe-breasted starthroat (*Heliomaster squamosus*) on *Ipomoea purpurea*. (b) Rainbow starfrontlet (*Coeligena iris = Diphogena iris*) on *Ipomoea purpurea*. (c) St. Domingo hummingbird (*Sporadinus elegans*) on *Ipomoea indica*.

*Source:* **Illustrations in public domain. Gould, J., and H. C. Richter. 1861. *A monograph of the Trochilidæ, or family of humming-birds*, vol. 4.**

FIGURE 71.14 Morning glory vines growing on and "beautifying" an outhouse.

*Source:* Photo by Uff-da (CC BY 2.0).

as unsightly, in great need of camouflage. Morning glories have been regarded as ideal for the purpose, quickly covering at least some of the outbuilding.

## KEY PUBLICATIONS


Austin, D.F. 1998. Convolvulaceae, morning glory family. *Journal of the Arizona-Nevada Academy of Science* 30: 61–83.

Austin, D.F. 2013. Moon – flower (*Ipomoea alba*, Convolvulaceae) – medicine, rubber enabler, and ornamental: A review. *Economic Botany* 67: 244–262.

Fang, Z., A.M. Gonzales, M.L. Durbin, K.K.T. Meyer, B.H. Miller, K.M. Volz, et al. 2013. Tracing the geographic origins of weedy *Ipomoea purpurea* in the southeastern United States. *Journal of Heredity* 104: 666–677.

Gangopadhyay, M., A.K. Das, S. Bandyopadhyay, and S. Das. 2021. Water spinach (*Ipomoea aquatica* Forsk.) breeding. In: *Advances in Plant Breeding Strategies: Vegetable Crops*, Vol. 10, edited by J.M. Al-Khayri, S. Mohan Jain, and D.V. Johnson, 183–215. Springer, The Netherlands.

Meira, M., E. Pereira da Silva, J.M. David, and J.P. David. 2012. Review of the genus *Ipomoea*: Traditional uses, chemistry and biological activities. *Revista Brasileira de Farmacognosia* 22: 682–713.

Muñoz-Rodríguez, P., T. Carruthers, J.R.I. Wood, B.R.M. Williams, K. Weitemier, B. Kronmiller, et al. 2019. A taxonomic monograph of *Ipomoea* integrated across phylogenetic scales. *Nature Plants* 5: 1136–1144.

Muñoz-Rodríguez, P., J.R.I. Wood, T. Wells, T. Carruthers, A. Sumadijaya, and R.W. Scotland. 2023. The challenges of classifying big genera such as *Ipomoea*. *Taxon* 72: 1201–1215. https://doi.org/10.1002/tax.12887.

Wood, J.R.I., P. Muñoz-Rodríguez, B.R.M. Williams, and R.W. Scotland. 2020. A foundation monograph of *Ipomoea* (Convolvulaceae) in the New World. *Phytokeys* 143: 1–823. https://phytokeys.pensoft.net/article/32821/

Yadav, S., A. Hemke, and M. Umekar. 2018. Convolvulaceae: A morning glory plant review. *International Journal of Pharmaceutical Sciences Review and Research* 51: 103–117.


## Key Cultivation Websites

Convolvulaceae Unlimited. https://convolvulaceae.myspecies.info/.

Dave's Garden. *Growing Japanese Morning Glories*. https://davesgarden.com/guides/articles/view/4241.

E-how. *How to Transplant a Morning Glory Plant*. www.ehow.com/how_7546845_transplant-morning-glory-plant.html.

Happy Sprout. *How to Keep Your Beautiful Morning Glories from Growing Out of Control*. www.happysprout.com/gardening/control-morning-glories/.

Horticulture Magazine. *Morning Glory*. https://horticulture.co.uk/morning-glory/.

The Spruce. *How to Grow and Care for Sweet Potato Vine*. www.thespruce.com/sweet-potato-vines-4120149.

University of Missouri, Integrated Pest Management. *Morning Glory: Misunderstood and under Appreciated*. https://ipm.missouri.edu/MEG/2023/6/morningGlory-DT/.

WikiHow. *How to Grow a Sweet Potato Vine Indoors*. www.wikihow.com/Grow-Sweet-Potato-Vine-Houseplant.

WikiHow. *How to Grow Morning Glories*. www.wikihow.com/Grow-Morning-Glories.

WikiHow. *How to Make Morning Glories Bloom*. www.wikihow.com/Make-Morning-Glories-Bloom.

# 72 Nasturtium (*Tropaeolum* Species and Hybrids)

**Scientific Names:** The genus name *Tropaeolum* is based on the Latin *tropaeum*, meaning "trophy," which signified victory in battle. In classical times, this was often indicated by setting up a tree trunk on the battlefield and hanging captured helmets and shields upon it. The relationship with nasturtium is that gardeners used to grow the plants up poles, and Carl Linnaeus (1707–1778), who coined the genus name, thought the rounded leaves were like shields, and the flowers like spear-pierced, bloodstained helmets. (Consistent with this symbolism, nasturtiums came to represent conquest and victory in battle, after the plants were brought to Europe in the 16th century.)

**Common Names:** The word *nasturtium* is based on Latin *naris*, "nose," + *torquere*, "to twist," a reference to the unpleasant ("nose-twisting") odor of the plant. "Nasturtium" as a common name refers to species of the South American genus *Tropaeolum*, not to species designated by the scientific name *Nasturtium*, most notably the popular vegetable watercress (*N. officinale*). The curious transfer of the scientific name *Nasturtium* to the popular name nasturtium for an unrelated plant came about through the similar pungent aroma (due to isothiocyanates) of the leaves, flowers, and fruits of *Nasturtium* and *Tropaeolum*. Other, less-used names for nasturtium include canary creeper, climbing nasturtium, creeping canary, cress of India, cress of Peru, garden nasturtium, and Indian cress. In old England, garden cress (*Lepidium sativum*) was commonly used in salads, when leaves of nasturtium started to be used for the same purpose. To distinguish the two, nasturtium was called Indian cress (the Americas were often called the Indies) and cress of Peru. The rare name Capucine cress was based on resemblance of the flowers to the shape of the robes worn by Capucine monks.

FIGURE 72.1 Collections of colorful cultivars of nasturtium.

*Source:* (a) Photo by Osnat Amir (CC BY SA 4.0). (b) Photo by Shutterstock/AJstudio photography. (c) Photo by Ji-Elle (CC BY SA 3.0). (d) Photo by Shutterstock/Yuriy Chertok.

## 72.1 INTRODUCTION

Garden nasturtiums include a range of herbaceous plants ranging from dwarf shrubs to vines climbing to 3 m. They are well-known for producing large, beautiful flowers that

FIGURE 72.2 Paintings of colorful nasturtium cultivars.

*Source:* Public domain illustrations. (a) Vintage image by Dorothe from Pixabay. (b) *Paxton's Magazine of Botany and Register of Flowering Plants*, vol. 2, 1836. (c) Flame nasturtium (*T. speciosum*). Van Houtte, L. 1847. *Flore des serres et des jardin de l'Europe*, vol. 3, 1847. (d) *Deutsches Magazin für Garten-und Blumenkunde*, G. Weise, Stuttgart, 1854. (e) *Flore des serres et des jardins de l'Europe*, 1857. (f) Cultivar Hermine Grashoff, a variety with doubled red flowers. *The garden: An illustrated weekly journal of horticulture in all its branches*, vol. 20, 1881.


DOI: 10.1201/9781003586227-72

bloom for long periods. They have the added reputations of being edible and needing little care.

## 72.2 OVERVIEW

The genus *Tropaeolum* has about 90 species of annual or perennial herbs distributed in cool highlands from Mexico to Central Argentina and Chile. The species are usually frost-tender perennials often cultivated as annuals in regions with cold winters. The leaves are typically round, the leafstalk attaching at the center of the bottom of the leaf (reminiscent of water lily leaves). In many species, the petioles (leafstalks) twine, so the plants are climbing vines. Other species are erect or sprawling shrubs. *Tropaeolum majus* (usually just called nasturtium) and *T. minus* (dwarf nasturtium, small nasturtium) are the species most frequently grown as ornamentals. Most ornamental cultivars are hybrids of these two species and often others. Additional species are also sometimes grown.

*Tropaeolum majus* is a usually climbing, somewhat succulent annual herb with yellow, pink, mahogany, orange, red, or creamy white funnel-shaped flowers up to 7 cm wide. The spreading stems extend to 3 m, curling around any object they contact. The odor of the plant is mustard-like. Several of the ornamental forms available have double flowers and colors other than the common ones mentioned here. The plants produce three-segmented fruit, each portion with a single large seed. The species is apparently native to Andean South America, including Ecuador, Peru, and Columbia. However, it is difficult to determine its indigenous range, as it occurs in a feral state in many tropical and subtropical areas. Nasturtium is grown in many countries as an ornamental and sometimes escapes to the wild.

FIGURE 72.3 Paintings of colorful nasturtium cultivars shown in old garden catalogues.

*Source:* **Public domain illustrations. (a) Vick's 1898 Garden and Floral Guide, Rochester, NY. (b) Cultivar Chameleon Dwarf. W. W. Rawson & Co, 1900, Boston, MA. (c) John Lewis Childs Inc., 1897, Flora Park, NY. (d) A. W. Livingston's Sons Annual of True Blue Seeds, 1904, Columbus, OH. (e) D. M. Ferry & Co. 1889 catalogue, Detroit, MI. Photo by Boston Public Library (CC BY 2.0). (f) Perry Seed Store catalogue, 1896, Syracuse, NY.**

FIGURE 72.4 The floral spur of nasturtium (*Tropaeolum majus*).

*Source:* **(a) Sectioned flower. Photo by Frank Vincentz (CC BY SA 3.0). (b) Photo by Rüdiger (CC BY SA 3.0).**

Dwarf nasturtium (*Tropaeolum minus*), native to northwestern South America, is similar to *T. majus* (sometimes called tall nasturtium to distinguish it from dwarf nasturtium), differing in being short, non-climbing, having smaller flowers, the lower petals of which are spotted, and smaller fruits. Bush cultivars are likely to have substantial parentage of dwarf nasturtium.

Nasturtium flowers have a "spur," a hollow, tube-like extension at the back of the flower which produces nectar to attract pollinators. Spurs on flowers are generally intended to restrict entry of pollinators to those with size or proportions suitable to transfer pollen from flower to flower.

## 72.3 HISTORY

Conquistadors discovered nasturtiums in Peru and Mexico and brought plants (apparently *N. minus*) back to Spain in the 16th century. *Nasturtium majus* was brought to Europe by Dutch botanists in the late 17th century.

## 72.4 ECONOMIC IMPORTANCE

Nasturtium is one of many garden plants sold in the garden trade. It is a popular offering at garden centers, and although considered old-fashioned, the demand is dependable.

## 72.5 ORNAMENTAL USES

The tall climbing forms grow well on fences, trellises, and arbors and can be used to cover slopes. Trailing types look good in hanging baskets, window boxes, or cascading down walls. The bushy dwarf types (often no taller than 30 cm) are appropriate for beds, borders, edging, rock gardens, mass plantings, groundcover, and containers.

FIGURE 72.5 Ornamental uses of nasturtium: (a) Cut flowers in a vase. (b) Vines climbing on a fence. (c) Dwarf plants grown as a border beside a hedge. (d) Tall vines growing against a house.

*Source:* **Photos from Shutterstock: contributed by (a) Jaz Studio, (b, c, d) Coulanges.**

FIGURE 72.6 Nasturtium salads.

*Source:* **(a) Photo by Kaji Yukimi (public domain). (b) Photo by Karen Loxton from Pixabay (public domain).**

FIGURE 72.7 Mashua (*Tropaeolum tuberosum*), native to South America, where it is grown for its edible tuberous root: (a) A flowering vine. (b) Tuberous roots of different cultivars.

*Source:* **(a) *Paxton's Magazine of Botany and Register of Flowering Plants*, 1838, W. S. Orr and Co., London. (b) Photo by Michael Hermann (CC BY SA 4.0).**

## 72.6 OTHER USES

Nasturtium leaves and flowers have been employed for centuries as garnishes, condiments, and constituents of salads. Garden nasturtiums are also employed in pickling. The leaves, petals, and seeds have a hot, peppery taste. Flower buds and unripened fruit can be placed in vinegar and used like capers. Crushed flowers can be steeped in spiced vinegar, after which the mixture is strained and employed as a sauce. The leaves or shoot tips are sometimes used as a condiment, or in salads and sandwiches like cress, which they resemble in taste. Ripe, dried seeds can be ground up and used as seasoning. Nasturtiums are considered non-toxic to dogs, cats, and horses.

Although nasturtium is frequently consumed, certain precautions should be observed. Plants grown primarily as ornamentals have often been sprayed with insecticides or fungicides, and unless one is confident that they are safe, they should not be eaten. Nasturtiums attract aphids, tempting gardeners to employ chemical sprays. It should also be noted that herbs such as nasturtium that are generally used as ornamentals but are also used occasionally as food have generally not been well studied from a health and dietary viewpoint, so it is wise to limit use on any given occasion, as well as on a long-term basis. Those who take prescription medicines need to be extremely cautious about consuming plants that are not well-researched. The glucosinolates of *Tropaeolum* (that are responsible for the flavor) can be irritating to the digestive system of sensitive people. Nasturtium flower buds may provide excessive oxalic acid. The seeds are considered to represent the most potentially toxic part of the plant.

Mashua (*Tropaeolum tuberosum*) is grown in the High Andes for its tuberous root, used as a cooked vegetable. The species is attractive in flower and is sometimes grown as an ornamental.

## 72.7 PROBLEMS

In warm climates, nasturtium can become invasive. It is considered a weed in Hawaii, California, and Malta.

## 72.8 CULTIVATION

### 72.8.1 Climate and Site

Young plants are sensitive to frost and should not be transplanted to the outdoors until late spring. In climate zone

9 and southward, nasturtium can be grown as a perennial. The plants grow best in warm, sunny locations but will tolerate some shade. Size should be considered when planting nasturtiums: Bush types are best for smaller gardens or places where space is limited, while tall vines need space to grow.

### 72.8.2 Propagation and/or Purchasing

Most gardeners start nasturtiums from seed or purchase plugs with starter plants from garden centers. Nasturtium grows readily from seeds, which in temperate climates are preferably started indoors in pots or boxes for transplanting after danger of frost. Tip cuttings can be taken and rooted in late autumn, before a killing frost, and used to establish windowsill houseplants that can provide edible foliage and blossoms during the winter. Indoor plants require at least 4 hr of direct sunlight daily and grow best at temperatures ranging from 4 to 8°C at night to no more than 20°C during the day.

### 72.8.3 Planting

Because of the short season in temperate regions, nasturtium is usually started indoors. The plants should be in individual pots, usually fiber pots (plugs) that can be transplanted, since the roots are very sensitive to disturbance, which will be avoided if the entire plug is planted. Small cultivars can be planted 15 to 30 cm apart. High-climbing vines should be positioned near fences or other supports. To produce vines with mixed flower colors (which is very attractive), put young plants known to produce different flower hues near each other (60 to 90 cm apart is suggested for tall varieties). In warm climates, groundcover can be established from seeds directly put in the ground, and often they will grow as perennials and also self-sow new seedlings.

### 72.8.4 Soil and Moisture Conditions

Nasturtium is known to grow well in poor soils, but an ordinary garden loam, preferably somewhat sandy, is ideal. A pH range of 6.0 to 8.0 is acceptable. Very rich soil or excessive nitrogen encourages lush foliage but fewer blossoms. Authorities recommend not fertilizing the plants or at least avoiding fertilization. The soil should be fairly moist but well-drained. Once established, the plants are moderately drought-tolerant.

### 72.8.5 Maintaining Plants

Nasturtiums are considered to be low-maintenance plants. Trailing cultivars, especially in containers, tend to become leggy, and cutting them back is recommended to stimulate growth and improve appearance. Climbing vines and bush types do not require pruning.

## 72.9 FASCINATING FACTS, FABLES, FOLKLORE, AND FRIVOLITY

- The South American vegetable mashua (*Tropaeolum tuberosum*) has a reputation for being an anaphrodisiac (the reverse of an aphrodisiac). Sixteenth-century Spanish priests recorded that the Inca gave enormous amounts of mashua to their troops in the belief that this would make them forget their wives and concentrate on fighting. Studies of male rats fed on mashua tubers found that testosterone levels were halved.
- In Peru, the nasturtium was called "gold nugget" by the Incas, because of an Andean legend. It was said that an Incan man found a bag of gold nuggets and was attacked by a band of Spaniards. To hide the gold, he threw it into the woods, whereupon the Spaniards searched for it, only to find that the mountain gods had converted the gold into gold-colored nasturtium flowers. Later, the man recovered the gold, which he had fashioned into a tribute to the gods to be placed in the local temple.
- The "Victorian language of flowers" was a secret coded language in Victorian times, with flowers and plants symbolic of certain messages, so when the flower or plant was mentioned in a letter, those who knew the code could understand the hidden information. "Nasturtium" meant "patriotism."
- The "Elizabeth Linnaeus phenomenon" of apparently flashing flowers is an optical illusion whereby at dusk orange flowers surrounded by green leaves seem to emit small light flashes. The phenomenon was first described in an article based on nasturtium published in 1762 by Elisabeth Christina von Linné, a daughter of Carl Linnaeus, who devised the modern system for naming species.
- Two US presidents had particular regard for the taste of nasturtium. Thomas Jefferson (1743–1826), the third president of the United States, grew many culinary herbs at Monticello, his well-known estate in Virginia, but his favorite was nasturtium. Dwight D. Eisenhower (1890–1969), the 34th president of the United States, included nasturtium in his highly prized recipe for vegetable soup, a combination of vegetables, beef, lamb, and nasturtium.
- The spur of nasturtium flowers, which produces nectar to attract pollinators, may hide insects, so people collecting the fresh flowers to consume sometimes inadvertently eat the insects. The cultivar Whirlybird lacks this spur, making it easier to clean, so people who detest the thought of eating bugs sometimes grow Whirlybird to reduce the likelihood of bugs in their food.
- "Trap crops" are planted near crops that one wishes to protect against pests so that they will be diverted away and not damage the main crop. The

FIGURE 72.8 Flower fairies wearing nasturtium flower bonnets.

*Source:* **Public domain illustrations. (a) Gordon, E., and M. T. Rose. 1910. *Flower Children: The Little Cousins of the Field and Garden*. P. F. Volland & Company, Chicago. (b) Grandville, J. J. 1867. *Les fleurs animées (The Flowers Personified)*, vol. 1. Garnier frères, Paris. (c) Extracted from a 1900 French commercial trade card. (d) Extracted from an 1898 German commercial trade card.**

situation is like paying "protection money" to the pests but is often preferable to the use of harmful pesticides. Nasturtiums are very attractive to aphids and are sometimes used to attract these tiny sucking insects away from other plants.

- "Companion plants" are plant species that are planted near other species in the belief that at least one of the partners, but preferably both, will benefit from the association. Nasturtiums are sometimes planted in gardens near other plants in the belief that they repel whiteflies and cabbage beetles. The situation is like hiring private guards (nasturtiums in this case) for the garden.
- Learned writers in the past sometimes employed the phrase "to cast nasturtiums." This has nothing to do with nasturtiums but is a deliberate misinterpretation (a "malapropism"), punning the phrase "to cast aspersions."
- The conical shape of nasturtium flowers is so similar to old-fashioned bonnets that it has repeatedly inspired artists to show the blooms as hats in imaginative drawings of fairies.

## KEY PUBLICATIONS

Andersson, L. and S. Andersson. 2000. A molecular phylogeny of Tropaeolaceae and its systematic implications. *Taxon* 49: 721–736.

Christenhusz, M.J.M. 2012. *Tropaeolum minus*. *Curtis's Botanical Magazine* 29: 324–330.

Christenhusz, M.J.M. 2012. *Tropaeolum majus*. *Curtis's Botanical Magazine* 29: 331–340.

da Silva, T.P., J. da S. Lima, R.P.Q. Cavatte, F.L. Cuquel, and F.L. Finger. 2013. Physiology of flower development in *Tropaeolum majus* L. *Acta Horticulturae* 1002: 193–197.

Diamond, S.P., S.G. Wiener, and J.G. Marks, Jr. 1990. Allergic contact dermatitis to nasturtium. *Dermatologic Clinics* 8:77–80.

Duenas-Lopez, M.A. 2022. *Tropaeolum majus (Nasturtium)*. CABI Compendium. www.cabidigitallibrary.org/doi/pdf/10.1079/cabicompendium.54181.

Grau, A., R.N. Dueñas, C.N. Cabrera, and M. Hermann. 2003. *Mashua,* Tropaeolum tuberosum *Ruíz & Pav*. International Potato Center, Lima, Peru. 55 pp.

Hind, N. 2010. *Tropaeolum tuberosum*. *Curtis's Botanical Magazine* 27: 301–313.

Jakubczyk, K., K. Janda, K. Watychowicz, J. Łukasiak, and J. Wolska. Garden nasturtium (*Tropaeolum majus* L.) – a source of mineral elements and bioactive compounds. *Roczniki Państwowego Zakładu Higieny* 69: 119–126.

Johns, T., W.D. Kitts, F. Newsome, and G.H. Neil Towers. 1982. Anti-reproductive and other medicinal effects of *Tropaeolum tuberosum*. *Journal of Ethnopharmacology* 5: 149–161.

Lim, T.K. 2014. *Tropaeolum majus*. In: *Edible Medicinal and Non Medicinal Plants*, edited by K. Lim, 777–787. Springer, Dordrecht.

Mircea, D.M., R. Calone, R. Shakya, J. Zuzunaga-Rosas, R.E. Sestras, M. Boscaiu, et al. 2023. Evaluation of drought responses in two *Tropaeolum* species used in landscaping through morphological and biochemical markers. *Life* 13(4): 960. https://doi.org/10.3390/life13040960.

Prestes, D.K.P., F.L. Cuquel, and R.R.B. Negrelle. 2017. Ornamental potential of *Tropaeolum pentaphyllum* Lam. *Journal of Ornamental Horticulture* 23: 185–191.

Sparre, B. and L. Andersson. 1991. A taxonomic revision of the Tropaeolaceae. *Opera Botanica* 108: 1–139.

Traesel, G.K., C.D. Machado, C.A.S. Tirloni, J.V. Menetrier, F.A. dos Reis Lívero, E.L.B. Lourenço, et al. 2017. Safety assessment and botanical standardization of an edible species [*Tropaeolum majus*] from South America. *Journal of Medicinal Food* 20: 519–525.

## Key Cultivation Websites

Balcony Garden Web. *How to Grow Nasturtium in a Container*. https://balconygardenweb.com/how-to-grow-nasturtium-in-a-container/.

Curbwise. *Garden Nasturtium: A Guide on How to Grow & Care for Nasturtiums*. https://curbwise.ca/garden-nasturtium-a-guide-on-how-to-grow-care-for-nasturtiums/.

GardenersWorld.com. *How to Grow and Care for Nasturtiums*. www.gardenersworld.com/how-to/grow-plants/how-to-grow-nasturtiums/.

Old Farmer's Almanac. *How to Plant, Grow, and Care for Nasturtiums*. www.almanac.com/plant/nasturtiums.

Pacific Bulb Society. *Tropaeolum*. www.pacificbulbsociety.org/pbswiki/index.php/Tropaeolum.

The Spruce. *How to Grow and Care for Nasturtium*. www.thespruce.com/nasturtiums-cool-season-flowers-1402910.

University of Wisconsin, Horticulture, Division of Extension. *Nasturtium, Tropaeolum majus*. https://hort.extension.wisc.edu/articles/nasturtium-tropaeolum-majus/.

WikiHow. *How to Grow Nasturtiums*. www.wikihow.com/Grow-Nasturtiums.

WikiHow. *How to Grow Nasturtium Indoors*. www.wikihow.com/Grow-Nasturtium-Indoors.

# 73 Oleander (*Nerium oleander*)

**Scientific Names:** Several different explanations of the origin of the genus name *Nerium* have been advanced. The name may trace to the ancient Greek name for the plant, *nerion*, thought to derive from the Greek *neros*, "water," an allusion to frequent presence near rivers and streams.

**Common Names:** The most frequent vernacular name is "oleander," less frequently "nerium." "Oleander" comes from the Italian *oleandro*, a reference to the olive-like leaves. "Laurier rose" (the French name) is an occasional name (bay laurel, *Laurus nobilis*, is associated with oleander in several languages, perhaps reflecting ancient confusion of the two species for each other). "Rose bay" (rose-bay) usually refers to rhododendron species (also fireweed – *Chamaenerion angustifolium*) but is sometimes applied to oleander (once again, there appears to have been ancient confusion of these unrelated plants). "Yellow oleander" is *Cascabela thevetia*, a related, somewhat-similar plant also employed as an ornamental, and also extremely toxic.

## 73.1 INTRODUCTION

Oleander is a large, beautiful flowering shrub or small tree that is very resistant to heat, drought, wind, poor soil, salt spray, air pollution, and grazing pests, and it can be grown with very little care. It is hardly surprising that it is extremely popular in much of the world. However, it requires at least a subtropical climate to thrive, and it is extremely poisonous.

FIGURE 73.1 Oleander (*Nerium oleander*) flowering displays: (a) An oleander garden in the Mediterranean region. (b) An oleander hedge in Spain. (c) Plant in Italy trained into a tree form. (d) Plant in Turkey allowed to grow naturally as a shrub.

*Source:* **(a) Photo by MW from Pixabay (public domain). (b) Photo by Txemari (public domain). (c) Photo by Shutterstock/Shutterschock. (d) Photo by Zeynel Cebeci (CC BY SA 4.0).**

FIGURE 73.2 Uncommon, yellow-flowered forms of oleander (*Nerium oleander*).

*Source:* **(a) Photo by Shutterstock/Botond Horvath. (b) Cultivar Maria Gambetta. Photo by Peter Forster (CC BY SA 3.0).**

## 73.2 OVERVIEW

*Nerium oleander* is the only species in the genus *Oleander* ("*Nerium indicum*" is occasionally employed in the pharmacological literature). The species is widely planted in subtropical and tropical areas of the world. It grows naturally as an upright shrub but is often trained to grow as a small tree (i.e., with a single trunk) ranging from 2 to 6 m in height. Some dwarf cultivars are 60 to 120 cm tall. The evergreen leaves are long, narrow, smooth but leathery, and variegated in a few cultivars. The flowers are cream, pink, purple, red, or white (less commonly peach, salmon, or yellow), doubled and/or multicolored in some cultivars, clustered at the ends of branches. The blooms are fragrant in many cultivars. Depending on climate, blooming occurs from late spring to autumn, often with the heaviest flowering period occurring earlier in the season.

## 73.3 HISTORY

Oleander is thought to have originated in the Mediterranean Basin, but this is uncertain because the plant has been widely cultivated in much of the world, obscuring its natural range. Its cultivation extends back millennia to classical civilizations, but it is often not possible to be certain what species were grown in ancient times. Hundreds of cultivars are now available.

## 73.4 ECONOMIC IMPORTANCE

Oleander is one of the world's most commonly planted flowering woody ornamentals and is an important subject of horticultural industries in subtropical regions.

DOI: 10.1201/9781003586227-73

FIGURE 73.3 Oleander (*Nerium oleander*) beside several of the world's most famous historical monuments: (a) The Parthenon in Athens, Greece (constructed in 5th century BC). (b) The Colosseum in Rome (constructed AD 1st century). (c) The Taj Mahal in India (constructed mid-17th century).

*Source:* **(a) Photo by Claudia Schillinger (CC BY ND 2.0). (b) Photo by Shutterstock/ChiccoDodiFC. (c) Photo by Lienyuan Lee (CC BY 3.0).**

## 73.5 ORNAMENTAL USES

Oleander is widely employed as a residential shrub and hedge, for foundation plantings, borders, and gardens. In public areas, it is often grown as a street tree or shrub, frequently in parks. Because it is drought-resistant and hardy, oleander is widely planted along highways in warm climates, not just for beautification (often to screen off less-attractive scenery) but to stabilize the soil, provide some shade and protection from wind, and act as a noise and pollution barrier. Massed plantings on slopes and hillsides can provide beautiful landscape displays that also serve to stem erosion. Oleander is also often grown in planters and tubs, occasionally for indoor use as houseplants.

## 73.6 OTHER USES

Although toxic, oleander preparations have a long history of folk medicinal uses to treat a variety of conditions, including asthma, cancer, epilepsy, heart problems, indigestion, leprosy, malaria, painful menstrual periods, ringworm, and venereal disease, and to induce abortions. None of these treatments is considered appropriate today, but oleander extracts are under intensive study in the hope of discovering modern medicinal applications.

## 73.7 PROBLEMS

Oleander is very toxic if consumed, although its bitterness discourages eating it. Even a small amount ingested can lead to serious illness, even death, in humans, livestock, and pets. Symptoms may include heart rhythm disorders, nausea, vomiting, cramping, bloody diarrhea, confusion, dizziness, weakness, drowsiness, and visual disturbances. Contact with the sap can cause dermatitis and eye inflammation. Cut stems exude a sticky, irritating latex, so gloves should be worn when handling the plant. The smoke from burning oleander prunings can cause severe eye irritation. Using the branches as skewers for food is inadvisable. Pets and children should not have access to oleander.

FIGURE 73.4 Oleander (*Nerium oleander*) hedges beside highways: (a) In Italy. (b) In France.

*Source:* **(a) Photo by Dave Collier (CC BY 2.0). (b) Photo by Maarten Sepp (CC BY SA 3.0).**

FIGURE 73.5 Oleander (*Nerium oleander*) decorating and stabilizing strongly sloping hillsides in Montenegro.

*Source:* **Photo by Shutterstock/Iurii.b.**

FIGURE 73.6 Potted oleander (*Nerium oleander*): (a) Indoors. (b) Outdoors.

*Source:* **(a) Photo by Pressebereich Dehner Garten-Center (CC BY ND 2.0). (b) Photo by Libraryrachel (CC BY 2.0).**

Oleander is subject to leaf scorch, a bacterial disease that can seriously degrade and eventually kill the plant.

Oleander can be invasive in some hot climates. The cultivars with doubled flowers tend to be sterile and are much less likely to establish wild populations.

## 73.8 CULTIVATION

### 73.8.1 Climate and Site

Oleander grows well full-time outdoors in climate zones 8 to 11. It is tolerant of heat and drought. Oleander survives brief exposure to freezing temperatures (as often occurs in zone 8) and is capable of regrowth after significant frost damage to the foliage. The roots of some cultivars will regrow the plants killed aboveground by prolonged exposure to a temperature down to −9°C. A harder freeze usually kills the entire plant. In cold climates, oleander is sometimes grown in containers, stored in a dormant condition overwinter indoors in a cool place. The plant is also grown as a permanent houseplant (recommended minimum container sizes, depending on cultivar, are between 40 and 60 cm, and they should have adequate drainage holes). Full sun is strongly recommended, and while semi-shade is tolerated, growth suffers. Oleander withstands wind and salt spray and so is popular in coastal areas.

### 73.8.2 Propagation and/or Purchasing

Oleander cultivars are propagated primarily by cuttings, but seeds are also available. Most gardeners simply purchase potted plants from nurseries or other commercial sources.

### 73.8.3 Planting

Planting during cool weather is preferable – spring (when regrowth starts) or fall (after the plants go dormant). Oleanders are typically very large, fast-growing, multi-stemmed, mounded shrubs requiring considerable space in the landscape. If not regularly pruned and/or trained, a given cultivar in full sun can be expected to spread as widely as it is high, which, depending on cultivar, may be several meters. The root system is extensive, and unless given ample space, oleander will compete with nearby plants. Recommended spacing, depending on cultivar, varies from 1.2 to 4 m apart.

### 73.8.4 Soil and Moisture Conditions

Oleander is very tolerant of soil quality and drought but will grow more vigorously and produce better flowering displays if provided with good soil and moderate irrigation (especially if grown in a container). Nevertheless, the plants are known to do well in poor soil and are rather tolerant of different soil textures, but not prolonged soaking, and the soil must have good drainage. A neutral soil pH between 6.5 and 7.5 is recommended (5.0 to 8.3 is tolerated).

### 73.8.5 Maintaining Plants

Oleanders have a well-earned reputation as tough, independent plants. Once established, they can thrive with very little care, often lasting 20 years or more. However, they tend to eventually become unkempt, so pruning (at the end of the growth cycle) is recommended to keep them under control, whether in the ground or in containers. Many gardeners deadhead the spent flowers to prevent formation of seed heads, which are not particularly ornamental.

## 73.9 FASCINATING FACTS, FABLES, FOLKLORE, AND FRIVOLITY

- An old European superstition warned that oleander should never be kept in a bedroom or sickness would prevail. In Italy, it was believed that the plant should not even be kept in a house or it would bring disgrace and misfortune. Likely, the toxicity of oleander motivated these negative views of the plant.
- Cultivars with sweet-scented flowers are widely grown in India, and the blooms are used in certain religious Hindu ceremonies.
- "Mad honey," a poisonous honey produced by honeybees from the toxic nectar and pollen of *Rhododendron ponticum* and *R. luteum*, is used in Asia in Indigenous medicine. Widely circulated allegations that honey produced from oleander is also similarly poisonous appear to be unsupported. (Oleander flowers are nectarless, but they do produce pollen. Although it is widely stated that all parts of the plant are toxic, there does not seem to be any research on the toxicity of the pollen, and unless bees are provided with just one floral source, their pollen is a mixture of many species, so a few toxic pollen grains are very unlikely to be significant.) The belief that oleander produces toxic honey may trace to confusion of *Rhododendron* and *Nerium oleander* by ancient Greeks, such as Aristotle and Pliny the Elder, who wrote early accounts about mad honey.
- Galveston, Texas, is known as the "Oleander City" because the plant is so widely planted there. The city holds an annual oleander festival and is a center of oleander breeding.
- The flowers of oleander trap some insects (Broce and Ideker 1978). When collecting pollen, the mouthparts of the insects (especially flies) become lodged between the anthers, preventing withdrawal. In one case, an oleander shrub killed thousands of blowflies (Coleman 1935).
- Before the age of photography, skilled technical artists were widely employed in Europe from the late 18th to the early 20th centuries to illustrate scientific and horticultural publications and excellent works showing oleander were produced. In

FIGURE 73.7 Old European art based on flowering oleander. *Top row*: Technically accurate illustrations prepared for publications. *Bottom row*: Traditional impressionistic paintings prepared by world-famous artists.

***Source:*** **Public domain images. (a) Nineteenth-century advertising poster published by L. Prang & Co. Photo from Boston Public Library (CC BY 2.0). (b) *Curtis's Botanical Magazine*, vol. 43, 1815–1816. (c) *The Floral Magazine*, vol. 9, 1880. (d) Esser, P. 1910. *Die Giftpflanzen Deutschlands*. F. Vieweg, Braunschweig, Germany. (e) Painted by Ukrainian-born artist Jakob Koganowsky (1874–1926), active in Austria. (f) Painted by Swiss artist Johann Heinrich Müller (1825–1894). (g) Painted by Dutch artist Vincent van Gogh (1853–1890).**

parallel, European painters produced art that was often impressionistic and today often is too expensive for all but the wealthy. As shown here, both kinds of art have their charms.

## KEY PUBLICATIONS

Ananth, V.A. and R.S. Kumar. 2012. Effect of growth substances on growth and flower yield of nerium (*Nerium oleander* L.). *Indian Journal Plant Science* 1: 187–191.

Annasamy, K., S. Muthu Lakshmi, K.M. Sellamuthu, T. Thangaselvabai, J. Kannan, T.L. Preethi, et al. 2020. Effect of mulching, drip irrigation and fertigation on flowering, physiological and biochemical parameters of nerium (*Nerium oleander* L.). *International Journal of Chemical Studies* 8: 1058–1063.

Broce, A. and J. Ideker. 1978. Oleander flowers as insect traps. *Annals of the Entomological Society of America* 71: 628–629.

Coleman, E. 1935. More insect tragedies. Pollination of *Nerium oleander*. *Victorian Naturalist* 52(2): 20–22.

Edmunds, L.J. 2010. Good will and best practice – controlling oleander (*Nerium oleander*) in Parachilna Gorge, South Australia. In: *17th Australasian Weeds Conference Christchurch, New Zealand*, edited by S.M. Zydenbos, 26–30. New Zealand Plant Protection Society, Auckland. http://caws.org.nz/old-site/awc/2010/awc201010911.pdf.

Eggenberger, R. and M.E. Eggenberger. 1996. *The Handbook of Oleanders*. Tropical Plant Specialists, Cleveland, GA. 156 pp.

Farkhondeh, T., M. Kianmehr, T. Kazemi, S. Samarghandian, and M.R. Khazdair. 2020. Toxicity effects of *Nerium oleander*, basic and clinical evidence: A comprehensive review. *Human & Experimental Toxicology* 39: 773–784.

Henderson, L. 1992. Oleander: An invasive riverside shrub from the Mediterranean. *Veld & Flora* 78: 84–86.

Herrera, J. 1991. The reproductive biology of a riparian Mediterranean shrub, *Nerium oleander* L. (Apocynaceae). *Botanical Journal of the Linnean Society* 106: 147–172.

Lazzaro, L., E. Sarracco, R. Benesperi, and A. Coppi. 2018. A probable anthropic origin of *Nerium oleander* L. (Apocynaceae) population in Montecristo Island (Italy, Tuscany): Evidence from loci polymorphism and ISSR analysis. *Caryologia* 71: 50–57.

Pagen, F.J.J. 1987. *Oleanders; Nerium* L. *and the Oleander Cultivars*. Series of revisions of Apocynaceae XX. No. 87–2. Agricultural University, Wageningen, Netherlands. 113 pp. https://library.wur.nl/WebQuery/wurpubs/fulltext/290321.

Rajiv, G., M. Jawaharlal, S. Subramanian, D. Sudhakar, and D. Uma. 2018. Effect of plant growth retardants on the growth and flowering of nerium (*Nerium oleander* L.) cv. Red. *Chemical Science Review and Letters* 7: 875–879.

Sabatino, L., F. D'Anna, and G. Iapichino. 2019. Improved propagation and growing techniques for oleander nursery production. *Horticulturae* 5(3): 55. https://doi.org/10.3390/horticulturae5030055.

### Key Cultivation Websites

Gardener's Path. *How to Grow and Care for Oleander Shrubs*. https://gardenerspath.com/plants/ornamentals/grow-oleander/.

GardenersWorld.com. *How to Grow Oleander*. www.gardenersworld.com/how-to/grow-plants/how-to-grow-oleander/.

Horticulture Magazine. *How (and When) to Prune Oleander*. https://horticulture.co.uk/oleander/pruning/.

International Oleander Society. https://web.archive.org/web/20170705170538/http://oleander.org/.

The Spruce. *How to Grow and Care for Oleander*. www.thespruce.com/oleander-growing-guide-5203845.

University of Hawaii Extension. *Oleander*. https://scholarspace.manoa.hawaii.edu/server/api/core/bitstreams/f0154f99-ad13-47e9-a10b-d32e120d980f/content.

WikiHow. *How to Grow Oleander*. www.wikihow.com/Grow-Oleander.

WikiHow. *How to Grow Oleander from Cuttings*. www.wikihow.com/Grow-Oleanders-from-Cuttings.

WikiHow. *How to Prune Oleander*. www.wikihow.com/Prune-Oleander.

# 74 Osteospermum (*Osteospermum* Species and Hybrids)

**Scientific Names:** The genus name *Osteospermum* is based on the Greek *osteon*, "bone," and Latin *spermum*, "seed," descriptive of the achenes (single seeds along with an enclosing hard casing derived from the fruit wall; achenes are often interpreted as seeds, but botanically they are fruits).

**Common Names:** The name osteospermum (often mispelled osteosperumum) unambiguously identifies these plants. They are frequently called African daisy, blue-eyed daisy, Cape daisy, star of the veldt, and South African daisy, but these names also apply to other plants. Ostespermum is also called osteo. Often, *Osteospermum* species in the horticultural trade are labelled as dimorphotheca.

FIGURE 74.2 Variation of osteospermum flowers.

*Source:* **(a) Cultivar Dixie. Photo by Andy Mabbett (CC BY SA 4.0). (b) Photo by Christie Koehler (CC BY SA 2.0). (c) Photo by Callisto Schmidt from Pixabay (public domain). (d) *Osteospermum ecklonis*. Photo by Babij (CC BY SA 2.0). (e) Photo by PDPhotos from Pixabay (public domain). (f) Photo by Shutterstock/AngieC333.**

## 74.1 INTRODUCTION

Several genera of African plants with daisylike flowers are popular garden ornamentals. These are all often called "African daisies," and unless the genus is identified, it is frequently difficult to know which species is referred to when it is simply called African daisy. The principal genera of African daisies include *Arctotis*, *Gerbera* (treated in this book as a separate chapter), *Dimorphotheca* (most species of which have been transferred to *Osteospermum*), and *Gazania* (also treated in this book as a separate chapter).

FIGURE 74.1 Garden displays of osteospermum cultivars.

*Source:* **(a) Photo by Ruth Hartnup (CC BY 2.0). (b) Photo by Shutterstock/Julia Gardener. (c) Photo by Shutterstock/Bear-Fotos. (d) Photo by Seán A. O'Hara (CC BY 2.0). (e) Photo by Shutterstock/Buniak Andrii.**

The daisy family (Asteraceae, also called the Compositae and the sunflower family), with over 32,000 species, may be the largest of the plant families, although the orchid family (Orchidaceae) is thought to be about the same size. There are numerous ornamentals from the Asteraceae with daisylike flower heads (tiny florets on the outside with large petals, tiny florets on the inside apparently lacking petals, the entire heads usually interpreted by non-botanists as individual flowers). Most species have yellow flowers (and because the species are often difficult to identify, botanists often simply call them DYCs: "damn yellow composites"). Gardeners, of course, are attracted to colorful daisies, and osteospermums are among the most colorful.

## 74.2 OVERVIEW

There are about 70 *Osteospermum* species. These are native to southern and eastern Africa and the Arabian Peninsula. The garden forms are mostly cultivars or hybrids of *O. barberae*, *O. caulescens*, *O. ecklonis*, *O. fruticosum*,

DOI: 10.1201/9781003586227-74

O. *grandiflorum*, and *O. jucundum*. Osteospermums are usually grown as annuals in temperate climates, but many can be grown as tender perennial herbs or subshrubs in climates that are frost-free or mostly so. The plants are typically 30 to 60 cm in height, and some simply sprawl on the ground. Depending on cultivar, blooming is in spring to summer, and sometimes, reblooming occurs in the autumn. The "petals" of the daisylike flower heads may be white, peach, pink, purple, rose, or yellow, and bicolored varieties are also popular. Cultivars with doubled (extra) flowers are also available. Some varieties have strange, very slim petals ending in a spatula, sometimes described as spoon-shaped.

## 74.3 HISTORY

Osteospermum is a recently developed ornamental, first brought to Europe and the New World in the 19th century. In North America, a range of cultivars has been available only since the 1990s. Extensive hybridization has produced numerous cultivars in the last several decades. The flowers of the original strains closed up at night, and also during prolonged periods of cloudy weather. This was considered undesirable since it obscured the beauty of the flowers, and modern cultivars frequently remain open.

## 74.4 ECONOMIC IMPORTANCE

Although osteospermum has a short history of market development, it is rapidly increasing in popularity and appears poised to become an addition to the commonly planted garden ornamentals.

FIGURE 74.3 Closed osteospermum flowers. The flowers of wild plants and old cultivars often close in response to darkness and drought.

**Source: (a) Photo by Eric Chan (CC BY 2.0). (b) Photo by Paul C. Lee from Pixabay (public domain).**

FIGURE 74.4 Commercial production of osteospermum: (a) Cuttings established in small pots for sale in the spring. (b) Mature potted flowering plants raised in a greenhouse.

**Source: (a) Photo by Annette Taratuta (CC BY 2.0). (b) Photo by Shutterstock/BearFotos.**

## 74.5 ORNAMENTAL USES

Osteospermums are extremely popular plants for beds and containers. These usually low-growing plants are normally best positioned at the front of borders and intermixed with other ornamentals of similar size. They also look good in uniform beds. They are very popular cut flowers for flower arrangements.

## 74.6 OTHER USES

Before the age of modern medicine, virtually everywhere in the world people employed plants for medicinal purposes, and osteospermum is no exception. In Saudi Arabia, *O. vaillantii* plants were soaked in water, and the water

FIGURE 74.5 Examples of ornamental uses of osteospermum: (a) *Osteospermum fruticosum* growing in a wall-mounted container. (b) A boy holding a potted plant. (c) Growing along a median separating two highways.

**Source: (a) Photo by Fuentedelateja (CC BY SA 2.0). (b) Photo by Shutterstock/Paul B. Moore. (c) Photo by Shutterstock/Alybaba.**

FIGURE 74.6 Plantings of osteospermum in uniformly colored flowers.

**Source: (a) Photo by Bernard Spragg (public domain). (b) Photo by Emma Forsberg (CC BY 2.0). (c) Photo by Davie Bicker from Pixabay (public domain). (d) *Osteospermum fruticosum*. Photo by Shutterstock/Leobrave.**

consumed to treat fever, stomach ailments, and liver disorders. In South Africa, *O. herbaceum* poultices and infusions were applied to wounds and sore mouths and eyes. Sesquiterpene glycosides and other constituents from *Osteospermum* species are under investigation for possible medicinal applications.

## 74.7 PROBLEMS

Osteospermum has escaped from cultivation in some areas and has established, occasionally becoming weedy. Although numerous websites (including the ASPCA) state that it is not toxic to cats, dogs, and horses, it has caused the death of sheep and goats (see literature cited). Some websites also claim that osteospermum can be consumed by humans, but under the circumstances, this should not be attempted.

## 74.8 CULTIVATION

### 74.8.1 Climate and Site

Although native to hot countries, osteospermum flowering displays do not last long in strong heat. The plants can be grown as perennials in climate zones 9 and warmer. They can usually tolerate light frosts but will succumb to continuing frost. The plants are adapted to full sun but tolerate light shade.

### 74.8.2 Propagation and/or Purchasing

The best flowering varieties are propagated as cuttings and sold by nurseries and garden centers as rooted seedlings (in "plugs," "jiffy pots," or trays) in the spring. A few garden forms can be grown from seed, but most cultivars will not breed true. In frost-free climates, some varieties can self-seed.

### 74.8.3 Planting

A 30 cm spacing is frequently recommended.

FIGURE 74.7 Goat consuming osteospermum. As noted in the text, this is inadvisable.

*Source:* **Photo by Shutterstock/So vv.**

FIGURE 74.8 African daisy cartoons (osteosmerums are often called African daisies): (a) Giraffe. (b) Lion cub. (c) Elephant.

*Source:* **Photos from Shutterstock, contributed by (a) Po-melkomy, (b) Christina Li, (c) Teguh Mujiono.**

### 74.8.4 Soil and Moisture Conditions

Osteospermums are best provided with well-drained, slightly acidic (ideally 6.0–6.2), rich soil, preferably with added compost mixed in. Regular fertilizing will produce significantly better flowering displays. The plants are tolerant of dry soils, but this may deteriorate flowering. Osteospermums dislike high humidity. In containers, a commercial potting mix is recommended.

### 74.8.5 Maintaining Plants

Deadheading can significantly extend the flowering period.

## 74.9 FASCINATING FACTS, FABLES, FOLKLORE, AND FRIVOLITY

- Concrete sidewalks and driveways can be sources of lime that leach into adjacent soils, producing a higher pH level. They can also retain and radiate heat. The local microclimate near concrete can be sufficiently different to cause nearby plants to grow differently, especially plants like osteospermum that are quite sensitive to heat and do not like alkaline soils.

## KEY PUBLICATIONS


Allavena, A., A. Giovannini, T. Berio, A. Spena, M. Zottini, G.P. Accotto, et al. 2000. Genetic engineering of *Osteospermum* spp: A case story. *Acta Horticulturae* 508: 129–134.

Faccioli, P., P. Faccioli, V. Terzi, N. Pecchioni, T. Berio, A. Giovannini, et al. 2000. Genetic diversity in cultivated *Osteospermum* as revealed by random amplified polymorphic DNA analysis. *Plant Breeding* 119: 351–355.

Gawenda, I. and T. Debener. 2009. Genetic diversity of *Osteospermum* genotypes analysed by AFLP and chloroplast SSR markers. *European Journal of Horticultural Science* 74: 86–94.

Larsen, B.N. 1997. Large scale production of osteospermum: Ten years of development effort. *Proceedings of the International Plant Propagator's Society* 47: 317–318.

Norlindh, T. 1943. *Studies in the Calenduleae I. Monograph of the Genera* Dimorphotheca, Castalis, Osteospermum, Gibbaria and Chrysanthemoides. Published dissertation. Lund University, Sweden. 432 pp.

Sadler, R., T. Parker, G.A. Verboom, A.G. Ellis, M. Jackson, J. van Zyl, et al. 2022. A molecular phylogeny of Calenduleae (Asteraceae) supports the transfer of *Dimorphotheca polyptera* DC. to *Osteospermum* L. *South African Journal of Botany* 151: 234–245.

Tounekti, T., M. Mahdhi, and H. Khemira. 2019. Ethnobotanical study of indigenous medicinal plants of Jazan region, Saudi Arabia. *Evidence Based Complementary and Alternative Medicine* 2019: 3190670. https://doi.org/10.1155/2019/3190670.

Van Wyk, B.-E., H. de Wet, and F.R. Van Heerden. 2008. An ethnobotanical survey of medicinal plants in the southeastern Karoo, South Africa. *South African Journal of Botany* 74: 696–704.

Watt, B. and E. Bunker. 2019. *Case Notes. The Garden Plant, African Daisy, (*Dimorphotheca/Osteospermum*) Causing Cyanide Poisoning in Merino Wethers*. Flock & Herd (online). www.flockandherd.net.au/sheep/ireader/african-daisy.html.

Willingham, B.C. and G.A. White. 1973. Agronomic evaluation of prospective new crop species IV. *Dimorphotheca* and *Osteospermum* – sources of dimorphecolic acid. *Economic Botany* 27: 323–328.

Yue, J., Y. Dong, and S. Gong. 2022. The complete chloroplast genome of *Osteospermum ecklonis* (DC.) Norl. (Asteraceae: Asterodae: Calenduleae), an ornamental plant. *Mitochondrial DNA Part B* 7: 1377–1379.

## Key Cultivation Websites

Gardener's Path. *How to Grow Cape Daisies (Osteospermum)*. https://gardenerspath.com/plants/flowers/grow-osteospermum/.

Gardeners' World.com (Gardeners' World Magazine). *How to Grow Osteospermum*. www.gardenersworld.com/how-to/grow-plants/how-to-grow-osteospermum/.

Horticulture Magazine. *How to Propagate Osteospermum from Cuttings*. https://horticulture.co.uk/osteospermum/cuttings/.

Nature & Garden. *Osteospermum – Astounding Daisies Galore in the Garden*. www.nature-and-garden.com/gardening/osteospermum.html.

Osteospermum.com. Osteospermum website. www.osteospermum.com/about.htm.

The Spruce. *How to Grow and Care for African Daisy*. www.thespruce.com/growing-and-caring-for-african-daisies-1402912.

University of Massachusetts Extension Greenhouse Crops and Floriculture Program. *Production Guidelines for four Crops – Osteospermum, Angelonia, Calibrachoa & Ornamental Sweet Potato (*Ipomoea batatas*)*. https://ag.umass.edu/greenhouse-floriculture/fact-sheets/production-guidelines-for-four-crops-osteospermum-angelonia.

# 75 Pansy and Other Violets (*Viola* Species and Hybrids)

**Scientific Names:** The genus name *Viola* traces to the old Latin word for the plant, *viola*, which is based on the Greek word for violet, *ion* or *vion*. Io or Ione was a nymph in ancient Greek mythology, a mistress of Zeus, king of all other gods and men, who, to protect her from the jealousy of his wife, turned Io into a white cow. To console her and sweeten her diet, he turned her tears into violets for her to graze on.

**Common Names:** *Viola* species are called violets, and sometimes also violas. Several species unrelated to violets have "violet" in their names, such as African violets (*Saintpaulia*) and dog's tooth violet (*Erythronium dens-canis*). "Pansies" are popular cultivars of species in *Viola* section *Melanium*. "Heart's-ease" (possibly tracing to former use to treat heart ailments) is an occasional alternative name for pansy. The English word *pansy* evolved through the French *pensée* (meaning either "pansy" or "thought") from the Latin *pensare*, "to consider." It has been claimed that the French *pensée* is based on the resemblance of the flower to a face in deep thought. The word *pansy* is also applied to *Junonia*, a genus of butterflies, also known as buckeyes and commodores. Some of the species are quite colorful, reminiscent of pansy flowers.

FIGURE 75.1 Pansy butterflies (*Junonia* species). Many of the species have eyespots designed to confuse predators, reminiscent of the floral blotches of pansies: (a) Blue pansy (*J. orithya*). (b) Yellow pansy (*J. hierta*). (c) Gray buckeye (*J. grisea*). (d) Peacock pansy (*J. almanac*).

***Source:*** **(a) Photo by Anitava Roy (CC BY SA 4.0). (b) Photo by Charles J. Sharp (CC BY SA 4.0). (c) Photo by Robb Hannawacker (public domain). (d) Photo by Shutterstock/Boonchuay Promjiam.**

## 75.1 INTRODUCTION

Pansies and other violets (*Viola* species and hybrids) rank among the most popular herbaceous bedding and container plants in the world. Modern cultivars are easy to grow and produce spectacularly colorful flowers in the spring in cold climates. The common phrase "shrinking violet," applied to shy people, draws attention to the inconspicuous nature of violets, which are among the smallest of garden ornamentals. Nevertheless, the beauty of their flowers is scarcely rivalled.

## 75.2 OVERVIEW

*Viola* is a genus of worldwide distribution, centered on the north temperate regions, but also occurring in mountainous areas in the tropics. It has an estimated 600 species of perennial, rarely annual plants. It has been claimed that over 100 species are cultivated. About 80 species are native to North America. With so many species, many of which hybridize, identification can be difficult. *Viola* species are mostly herbaceous but are occasionally subshrubs or vines. They tend to be acaulescent (i.e., they lack an obvious stem, the leaves appearing to grow out of the ground in a basal rosette), but some have short stems. The leaves are usually simple, heart-shaped, scalloped, and arranged alternately.

FIGURE 75.2 Mixed collections of pansies (*Viola* × *wittrockiana*).

***Source:*** **(a) Photo by Mike Sutton (CC BY 2.0). (b) Photo by Mark F. Levisay (CC BY 2.0). (c) Photo by Shutterstock/April70.**

DOI: 10.1201/9781003586227-75

Most violets bloom in the spring. *Viola* flowers have five sepals and five petals. The five petals include two symmetrical pairs and one inferior scoop-shaped petal. In the *Melanium* group, the single inferior petal is enlarged and is thought to act as a landing platform for insect pollinators. The anthers of the stamens are clustered around the base of the solitary pistil, which has a style and stigma shaped like a club. Many species of *Viola* produce normal, showy, cross-pollinated (*chasmogamous*) flowers in the early spring and also develop closed, self-pollinating (*cleistogamous*) flowers that lack petals throughout the summer. This phenomenon occurs in about 60 families of plants, but *Viola* has more cleistogamous species than any other genus. In most such species, open flowers are large and showy, attracting insects for cross-pollination, whereas the closed flowers are much smaller and produce seeds by self-pollination and fertilization within the flower buds. Cleistogamy is insurance against the possibility that insects will not be available for pollination, with the result that seeds will not be produced. The lower pairs of petals in the cross-pollinated flowers are joined and form a spur (a hollow, tubular extension). Nectar is secreted into the bottom of the spur from the bases of the lowest two stamens. In trying to reach the nectar, insects brush against the stigma, often transferring pollen to it that has been acquired from previous visits to other flowers. Many species of violet turn their flowers toward the ground at night or when it is cloudy ("photonasty" is movement determined by light). The flowers of some *Viola* species are heliotropic (turn toward the sun), others not.

Violets bloom early in the season, furnishing nectar for bees and butterflies. They are often grown in butterfly gardens, not just for the nectar, but because the foliage is consumed by some caterpillars, notably fritillaries, which are large, attractive orange-and-brown summer butterflies sometimes confused with monarchs and viceroys.

The seeds are produced in capsules that break open to reveal three boat-shaped valves. Many species of *Viola* have two special methods for distributing their seeds. First, there is a short-distance dispersal method, by which seeds are explosively ejected from the capsules. As the valves of the capsule dry, an elastic tension is produced, and they suddenly open lengthwise, catapulting the seeds up to a few meters. Also, more than half of *Viola* species have a long-distance dispersal method, based on a mutually beneficial relationship with ants. *Myrmecochory* (sometimes spelled *myrmechory*) is seed dispersal by ants. The violet seeds have nutritious, pale-colored, oily attachments (elaiosomes) that attract the ants, which carry the seeds back to their underground nests. Seeds that the ants lose along the way or that become buried in the ants' tunnels often germinate far away from the parent plants. Violet seeds may also be distributed over great distances by birds.

The most important horticultural selections of *Viola* belong to *V.* section *Melanium*, which has as many as 100 species. Most cultivars are hybrids. Cultivated species of section *Melanium* are sometimes divided into horticultural (i.e., convenient but artificial gardening) groups. The term "pansy" is applied in different ways but is often restricted to the large-flowered domesticated forms of section *Melanium*. "Viola" is also applied in different ways – sometimes to all *Viola* species, but often to the small-flowered and more petite domesticated forms of section *Melanium*. These are often multi-flowered and grow in clumps (hence the alternate name "tufted pansies"). The most miniature violas are sometimes termed violettas.

#### 75.2.1 Garden Pansy (*Viola × wittrockiana*)

The most familiar cultivated violet is the garden pansy, *V. × wittrockiana* (commemorating the Swedish botanist V. B. Wittrock, 1839–1914, an early student of the *Melanium* violets). It is presumed to have arisen from hybridization between *V. lutea*, *V. tricolor*, and possibly *V. altaica* and others. Most of, perhaps all, the parent species are native

FIGURE 75.3 Fruit and seeds of *Viola*: (a) Opened capsule of garden pansy (*V. × wittrockiana*), showing three valves containing seeds. (b) Diagram of a capsule of *V. tricolor* from a 19th-century wall chart by Dodel-Port Atlas. (c) A seed of *V. elatior* tipped by a yellowish elaiosome, a fat body attracting ants to carry the seed to their nest.

**Source:** (a) Photo by Ruth and Dave (CC BY 2.0). (b) Photo by Paul K (CC BY 2.0). (c) Photo by Hans Stuessi (CC BY 4.0).

FIGURE 75.4 Uniform (single cultivar) beds of garden pansy (*Viola × wittrockiana*).

**Source:** (a) Photo by Shutterstock/Junrong. (b) Photo by Pasja1000 from Pixabay (public domain). (c) Photo by Liz West (CC BY 2.0). (d) Photo by Natasha de Vere (CC BY 2.0).

FIGURE 75.5 Illustrations showing the "pansy face" of numerous cultivars of *Viola*: (a) *Viola* × *wittrockiana*. (b) *Viola* × *wittrockiana*. (c) *Viola tricolor*.

*Source:* **(a) Photo by Rašo (CC BY SA 3.0). (b) Photo by Andrew Fogg (CC BY 2.0). (c) *L'Illustration horticole* (Gand, Belgium), vol. 1, 1854.**

to Europe and Asia Minor. Plants typically grow to about 20 cm in height. The flowers may be up to 10 cm in diameter and are single-colored, streaked, or patterned. Almost every flower color is available, including white and black (actually very dark purple). Most cultivars possess a well-defined blotch in the throat of the flower, often giving the bloom a remarkably face-like appearance, which British breeders in the early 19th century deliberately accentuated. Some cultivars have wavy or crinkled petals. This short-lived perennial or biennial is usually grown as an annual bedding plant in cold climates. It grows very well in containers.

### 75.2.2 Additional Pansy Hybrids

Pansy (*V.* × *wittrockiana*), which is thought to be a hybrid of *V. tricolor*, *V. lutea*, and *V. altaica*, has been hybridized with additional species to create a range of hybrid cultivars. Particularly popular is *Viola* × *williamsii*, a name referring to hybrids between *V. cornuta* and *V.* × *wittrockiana*.

### 75.2.3 European Wild Pansy (Field Pansy, *Viola tricolor*)

*Viola tricolor*, known as European wild pansy, miniature pansy, field pansy, and Johnny-jump-up, is perhaps the main ancestor of the garden pansy and has been grown since ancient Greek times. Until about 1800, it was the chief ornamental violet but has largely been replaced by the garden pansy. The species is naturalized in North America, where it is known as the wild or field pansy. This small pansy is well-known for its three-colored flowers that often combine different colors, such as yellow, blue, purple, red, and white. Cultivars of *V. tricolor* intergrade with cultivars of *V.* × *wittrockiana*, so distinguishing the two species is difficult.

### 75.2.4 Sweet Violet (Florist's Violet, *Viola odorata*)

*Viola odorata* is a member of *V.* section *Viola* and so is not closely related to the garden pansy previously discussed.

FIGURE 75.6 Cultivar Sorbet Sunny Royale of *Viola* × *williamsii* (hybrid horned pansy, tufted pansy, horned pansy), a cross between *V. cornuta* and *V.* × *wittrockiana*.

*Source:* **Photo by David J. Stang (CC BY SA 4.0).**

FIGURE 75.7 Illustrations (public domain) of the European wild pansy (*Viola tricolor*).

*Source:* **(a) Twamley, L. A. 1836. *The Romance of Nature*. Charles Tilt, London. (b) Redouté, P. J. 1833. Choix des plus belles fleurs et des plus beaux fruits. (c) Lindman, C. A. M. 1901–1905. Bilder ur Nordens Flora.**

FIGURE 75.8 Sweet violet (*Viola odorata*).

*Source:* **Public domain illustrations. (a) Sturm, J. 1900–1907. *J. Sturm's Flora von Deutschland*. Verlag von K. G. Lutz, Stuttgart. (b) Die Gartenwelt. 1897. G. Schmidt, Berlin. (c) Guillaumin, A. 1929. *Les Fleurs de Jardins, tome I: Les Fleurs de Printemps*. Paul Lechevalier, Paris.**

Sweet violet grows wild in Europe, Africa, and Asia and occurs as an occasional garden escape to roadsides and waste places across Canada and in much of the United States. Cultivars differ in stature (generally up to 15 cm tall). This low-growing perennial, apparently stemless herb arises from a creeping rhizome and spreads by long, rooting stolons (runners or creeping rhizomes). The leaves are heart-shaped. The flowers range considerably in size and color, with some forms having double flowers. The flowers are usually dark violet (varying to paler shades), rose, or white. Ornamental selections generally have much larger flowers than wild plants of the species. Unlike most species of *Viola*, the seeds of *V. odorata* are not forcibly ejected from the fruit at maturity. The sweet violet is employed in the florist trade and in gardens. There are many selections that have been made for these purposes.

## 75.3 HISTORY

Violets were among the first flowers cultivated, tracing back at least to Greece of 400 BC. Breeding began enthusiastically in Britain about 1800 and continued later in Europe and the United States. It is thought that prior to 1800, *Viola tricolor* was the principal cultivated species, but with the production of the hybrid *V.* × *wittrockiana*, the latter became dominant. Modern cultivars have mostly been bred to produce flowers quickly, although some of their wild progenitors often normally do not flower in their first year of growth. For decades, private companies have been breeding violets for larger flower size, color diversity, and tolerance to stresses (especially heat, which notably depresses growth). Like the rose, the violet has been adopted as a color because it is so well-known. Of course, roses come in other colors than red, and violets come in almost all other colors, not just blue.

FIGURE 75.9 Old paintings (public domain) of garden pansy (*Viola* × *wittrockiana*). Note that the flowers in each painting represent different varieties.

*Source:* **Public domain. (a) The Dutch book by Witte, H. 1868. *Flora: afbeeldingen en beschrijvingen van boomen, heesters, éénjarige planten, enz. voorkomende in de Nederlandsche tuinen.* Wolters, Groningen. Illustrated by Abraham Jacobus Wendel. (b) *The Floral Magazine* (London), vol. 6, 1867. (c) Published by L. Prang & Co., 1874. Photo by Boston Public Library.**

FIGURE 75.10 Old garden catalogues (public domain) featuring early cultivars of garden pansy (*Viola* × *wittrockiana*).

*Source:* **(a) John Lewis Childs (Floral Park, NY) 1910 catalogue. (b) James Vick's Sons (Rochester, NY) 1903 Garden & Floral Guide. (c) D. M. Ferry & Co. (Detroit, MI) 1889 catalogue. (d) W. Atlee Burpee Company (Philadelphia, PA) 1886 farm annual. (e) J. M. Thorburn (New York) 1904 catalogue. (f) John Lewis Childs (Floral Park, NY) 1900 catalogue.**

FIGURE 75.11 Illustration of the range of colors of modern pansy cultivars.

*Source:* **Photo by Shutterstock/Kella Carlton.**

An amazing range of multicolored flowers has been created by breeders.

Violets have been symbolically associated with the finest of human sentiments, most notably love. In general, violets have been considered a symbol of humility because they grew close to the ground. By the Victorian era, violets represented the qualities of an ideal wife: humility,

FIGURE 75.12 The romantic symbolism of violets: (a) A nineteenth-century print showing an affectionate couple, the young man dressed in the style of Napoleon, giving a bouquet of violets to his companion. (b) A vintage (ca. 1890) Valentine illustration of a boy greeting a young girl with violets (pansies), suggestive of love.

***Source:* (a) Public domain. (b) Photo by Shutterstock/Victorian Traditions.**

faithfulness, and modesty. The sweet violet (*V. odorata*), in particular, has been symbolic of modesty and humility. In more recent times, blue violets have been equated with faithfulness, fidelity, and love, and white violets with purity, candor, and innocence. Violets have long been used in love potions. The classical Greeks chose violet as their flower of fertility. In 17th-century Europe, candied violets were eaten as an aphrodisiac.

## 75.4 ECONOMIC IMPORTANCE

On a world basis, the value of flats of pansies produced annually for the bedding and container flower markets amounts to hundreds of millions of dollars.

## 75.5 ORNAMENTAL USES

Pansies and, to a much lesser extent, other violets are grown as edging or border plants and in containers, but they are also cultivated in massed display beds, commonly in mixed colors or in alternating color zones. Because of their low stature, they are not suitable for growing in mixtures with taller plants but can be associated with other low-growing plants. Some wild species of *Viola* are grown in rock gardens, but their economic significance is limited.

## 75.6 OTHER USES

### 75.6.1 Culinary Use

The use of violets as food is ancient, although very limited, extending back to early Greek and Roman times, when violets were used to flavor butter, oil, vinegar, and wine. Violet-flavored sherbet has been a favorite concoction of Asia Minor and other parts of the Middle East since very early

FIGURE 75.13 Commercial production and sale of pansies: (a) Flats of pansies for sale for the bedding and container markets. (b) A greenhouse full of pansies.

***Source:* (a) Photo by Eileenmak (CC BY 2.0). (b) Photo by Shutterstock/TammiMild.**

FIGURE 75.14 Landscape designs employing pansies: (a) An ornamental bed at the Jardin des Plantes in Paris, France. (b) Elaborate beds on the grounds of a castle in The Netherlands. (c) A complex bed in London, UK. (d) A teddy bear floral bed in Göteborg, Sweden.

***Source:* (a) Photo by David Monniaux (CC BY SA 3.0). (b) Photo by Shutterstock/Dirk M. de Boer. (c) Photo by Acabashi (CC BY SA 4.0). (d) Photo by Robande (CC BY SA 3.0).**

times, some considering this the national drink of Syria and Turkey. In the 15th century, violets were often used in sauces and soups and to make fritters, and some members of the royalty in Great Britain during the Victorian era were extremely fond of violet preparations.

Sweet violet (*V. odorata*) is one of the few violets with petals sufficiently strong that they hold their shape when candied. Candied violet flowers are used to decorate cakes and other confections, puddings, and ice cream, and fresh flowers are sometimes put in salads. The flowers can also be processed into syrup, jellies, and marmalade and added to gelatins, ices, vinegar, honey, wines, and salad dressings.

The young leaves of violets are rich in vitamins A and C and can be incorporated into salads. The flowers are also edible, raw or cooked. The leaves can be added to soup as a thickener, much like okra (the leaves of wild violets were used as a substitute for okra in the South during the American Civil War). The flowers and leaves of common pansies have a stronger flavor than other violets and are

FIGURE 75.15 *Viola* cultivars growing in containers. All are *V.* × *wittrockiana*, except (d), which is *V. tricolor.*

*Source:* **(a) Photo by Rept0n1x (CC BY SA 3.0). (b) Photo by 16999311 from Pixabay (public domain). (c) Photo by Shutterstock/Jana Milin. (d) Photo by Bob Collowan (CC BY SA 4.0).**

FIGURE 75.16 Culinary preparations with violets: (a) A violet-embellished saucer and teacup, with floating violets and pansies to flavor the beverage. (b) A cake decorated with fresh, edible pansy flowers. (c) Chocolate cake with candied violets. (d) Violet flowers garnishing a fruit smoothie served in a bowl.

*Source:* **(a) Photo by Shutterstock/Denise Kappa. (b) Photo by Shutterstock/ValdisOsins. (c) Photo by Lsalvay (CC BY SA 3.0). (d) Pxhere (public domain).**

eaten in salads or used as a garnish. The flowers are also used to make tea. For decorative effect, violet leaves and the fresh flowers can be used as an edible garnish.

"Violet syrup" allegedly from *Viola* is offered commercially (largely on the internet) for culinary usage. So-called violet syrup was sometimes used to flavor ice creams, candy, baked goods, and liqueurs, but these uses seem to be obsolete. Violet syrup is sometimes prepared by home cooks as a water extract of the flowers. It can be quite laxative.

### 75.6.2 Perfume Use

Some selections of sweet violet (*V. odorata*) have an exceptionally delightful odor and have been used as a source of perfume. Also, sweet-scented violets known as "Parma violets" (possibly *V. alba*) are cultivated in southern Europe as a source of essential oil for perfumery. These are hardy only in warm areas. Parma violets are sometimes thought to have been derived from *V. odorata* (which they very closely resemble), although there is evidence to suggest that they originated from other fragrant *Viola* species. "Violet absolute" is a solvent extract from the foliage of violets (*V. odorata* or Parma violets), employed in perfumery. Much so-called "violet" scent actually comes from orris "root" (rhizomes of *Iris* ×*germanica* var. *florentina* and *I. pallida*).

### 75.6.3 Medicinal Use

Although very rarely used in modern Western medicine, violets have been thought to have curative properties since antiquity. In classical Roman times, violets were prescribed for gout and spleen disorders. As well, violets were used to treat headaches and memory loss. Violets contain a chemical that is related to salicylic acid, the chief ingredient in aspirin, so that use for headache had some justification.

FIGURE 75.17 Perfume use of violets: (a) Parma violet, a traditional source of stock for perfumery. (b) Advertisement for 19th-century violet-based perfume.

*Source:* **Public domain images. (a) Watson, W. 1896. *Favourite Flowers of Garden and Greenhouse*, vol. 1. Frederick Warne & Co., New York. (b) Old advertising poster.**

(The phytohormone "salicylic acid, 2-hydroxybenzoate" functions as a defensive agent in many plants.) Violets were once commonly grown in gardens as a drug plant, primarily in Europe, but also in India. The medieval herbalists used violets as an antiseptic, to treat pain, and even for malignant tumors. Violets continue to be employed medicinally in herbal or folk medicine, especially in the Third World. The flowers are sometimes made into a jam reputed to soothe indigestion, and a syrup said to suppress a cough.

### 75.6.4 Objets d'Art

Violet flowers are exquisite, beautiful, and widely admired, making them particularly suitable for decorating valued items, particularly collectibles termed *objets d'art*. The French term *objet d'art* describes an ornamental work of art, especially small works that are three-dimensional, in contrast to paintings and large sculptures.

## 75.7 PROBLEMS

The seedpods, rhizomes, and roots of *Viola* species can be poisonous in large doses, causing severe gastroenteritis, nervousness, and respiratory and circulatory depression. These parts of the plant are strong laxatives and can induce nausea and vomiting. However, cultivated violets are not considered to be potentially dangerous for cats and dogs. Violets should not be used for culinary purposes, unless one is certain that they have not been sprayed with insecticides, herbicides, or fungicides. Harvesting wild plants for food use without being certain of their identity is dangerous.

## 75.8 CULTIVATION

### 75.8.1 Climate and Site

Pansies are suited to grow as winter-dormant perennials in climate zones 4 to 8 and can even flower during the winter in zones 9 to 11. However, they are often grown as annuals. *Viola* species are usually adapted to cool temperatures and grow well outdoors in the spring. Young plants tolerate some freezing, and if grown in containers, these can be temporarily protected inside. Excessive heat can cause some cultivars to become dormant, and in Southern climates, the cultivation of violets in the heat of summer is discouraged. In warm climates with limited frost, violets often rebloom in the fall.

Violets or pansies are not commonly grown as interior flowering potted houseplants. African violets and other small ornamentals that are more tolerant of shade are preferred. The domesticated *Viola* species usually grow well in full sun but often prefer part shade in hotter or sunnier climates. Many of the wild species that are suited for rock and wild gardens do best in part shade.

### 75.8.2 Propagation and/or Purchasing

Violets are easily started indoors from seed, beginning 8 to 12 weeks before transplanting outdoors. They may also self-seed in the garden, and some varieties will survive overwinter in warmer climates. Some cultivars are offered as hybrid seeds (i.e., produced by crosses between quite different parents), and for those wishing to minimize their labor, these can often be purchased as seedlings. Many of the most attractive cultivars are grown from cuttings (sometimes as offsets) and are best purchased as flats of young plants from nurseries. Pansies are so cheap that for most people, establishing plants from seeds isn't worth the effort.

### 75.8.3 Planting

In beds, the cultivated varieties can be spaced 15 to 20 cm apart. Trailing or spreading cultivars can be planted 25 to 30 cm apart. Trailing varieties display well in hanging baskets or tumbling over the edge of window boxes. Planting violets in the same soil consecutively for more than 3 years is inadvisable because of the possible buildup of fungi.

FIGURE 75.18 Objets d'art featuring violets: (a) Pansy-shaped gold and diamond brooch (ca. 1900). (b) A Tiffany table lamp, the lampshade decorated with pansy figures. Original Tiffany lamps such as this sell for millions of dollars. (c) Hand-painted Art Nouveau coffeepot decorated with a violet. This was manufactured in Germany around 1900.

*Source:* (a) Photo by Smithsonian Design Museum. (b) Photo by Tim Evanson (CC BY SA 2.0). (c) Photo by Presse03 (CC BY SA 3.0).

FIGURE 75.19 Girls planting pansies.

*Source:* (a) Photo by Dandelion Tea from Pixabay (public domain). (b) Photo by Shutterstock/Claudia Paulussen.

### 75.8.4 Soil and Moisture Conditions

Rich, well-draining, but moist soil with pH 5.4 to 5.8 is recommended. Humus-rich soil, such as a peat-based potting mix or garden soil with considerable added organic material, is best. A slow-release fertilizer may be added to the soil after planting. Some authorities recommend weekly or biweekly fertilization of violets, especially when grown in containers. Some drought may be tolerated, but best growth results from regular watering, the soil allowed to dry at least partly between waterings.

### 75.8.5 Maintaining Plants

To promote blooming and extend flowering, faded flowers should be removed ("deadheaded"), and leggy or overgrown plants may be cut back to about 10 cm in height.

## 75.9 FASCINATING FACTS, FABLES, FOLKLORE, AND FRIVOLITY

- The Greek dramatist Aristophanes (about 450–388 BC) referred to Athens in one of his plays as the violet-crowned city because the king was named Ion, which meant "violet." According to legend, Ion, the founder of Athens, led his people to Attica, where they were welcomed by naiads, water nymphs who could inspire men and who gave them violets as signs of their good wishes. The violet has been emblematic of Athens for many years. In the ancient Athenian games, the first prize was an award in the shape of a golden violet. Ancient Athenian houses, altars, statues, and brides were decorated with violets.
- As a derogatory slang term, "pansy" refers to a weak or effeminate male, or a male homosexual, meanings which appear to have become prominent about a century ago. Delicate, extremely attractive flowers naturally represent feminine characteristics, which may explain why certain ornamental plants have been associated with male homosexuality. "Violets" (both the flowers and the color) also have a long tradition of association with lesbianism (tracing to the ancient Greek poet Sappho from the island of Lesbos), but significantly less disrespectful, and violets have been employed as symbols of organizations and movements supportive of increased protection and rights for women in general.
- Ancient Greeks and Romans acquired the odd idea that violet wreaths relieved hangovers. Banquet tables were sometimes decorated with thousands of violets in the mistaken belief that the flowers could prevent drunkenness. Ironically, the Romans often drank a wine made from violet blossoms.
- In southern Germany during the Middle Ages, the violet was a symbol of spring. The discovery of the first violet to emerge triggered rejoicing. It was tied to a stake, and a festival was held in its honor.
- Before litmus (which is obtained from lichens) became widely employed to indicate pH, syrup of violets was used (it turns red with acids, green with alkalies).
- Reflecting their beauty, violets are employed as symbols of several political areas. The purple violet (marsh blue violet, *V. cucullata*) is the official provincial flower of New Brunswick. The violet (no particular species) is the national symbol of Poland and the regional symbol of North Portugal. The state flower of Wisconsin is the wood violet (*V. papilionacea*; "wood violet" is more commonly applied to an Old World species of violet). The state flower of New Jersey is the common meadow violet (*V. sororia*). The native American violet, *V. palmata*, was adopted in 1968 as the state flower of Rhode Island. Illinois adopted the "blue violet" as its state flower in 1908.
- Ionones, major chemicals (ketones) found in violet fragrance, have the ability to dull the sense of smell within a very short time, and this is why sweet-smelling violets seem to lose their fragrance quickly. The fragrance returns when the nose is rested. Shakespeare was aware of this when he wrote in *Hamlet*: "A violet in the youth of primy nature, Forward, not permanent; sweet, not lasting. The perfume and suppliance of a minute. No more."
- Inspired by the human face-like appearance of many pansy flowers, in the past, cartoonists created figures that expanded the human appearance of pansies.
- Women commonly wear beautiful flowers to enhance their own beauty, and violets are one of the leading choices. The famous romance of Queen Victoria (1819–1901) and Prince Albert (1819–1861) was associated with violets. She wore a posy

**FIGURE 75.20** Pansy characters shown by cartoonists: (a) Chromolithograph (late 19th century) showing three dancing female pansies. Published by L. Prang & Co. (b) A party of guests with pansy faces. (c) A lady with pansy decoration.

***Source:*** **Public domain. (a) Boston Public Library. (b) Pocci, F. 1876. In Bildern und Versen, Stroefer & Kirchner, München/New York. (c) Rose, M. T. In Gordon, E. 1910. *Flower children: The little cousins of the field and garden.***

of violets on her dresses. He never walked through the gardens at Windsor Castle without picking a bunch of violets. An even more celebrated love story that was intimately associated with violets concerns Napoleon Bonaparte (1769–1821) and Josephine (1763–1814). Napoleon's favorite flower was the violet, perhaps because it reminded him of his childhood in the woods of Corsica. He met Josephine at a ball, where she wore a coronet of violets and carried a bouquet of violets, which she threw to him from her carriage while departing. In memory of this, she wore a wedding gown embroidered with violets, and he gave her violets on each anniversary of their wedding. When he was sentenced to live out his days on the island of Elba, he told his followers that he would return in the spring, with violets. Napoleon's supporters sometimes determined people's political views by asking, "Do you like violets?" In 1815, Napoleon returned to the royal palace in Paris, showered by violets. Accordingly, he was called by his followers *Caporal Violette* ("Corporal Violet") and *Papa-Père la Violette* ("Daddy Violet"), and the violet was adopted as the emblem of the Imperial Napoleonic Party. After Josephine's death, Napoleon had her grave covered with violets. He kept some of them along with a lock of Josephine's hair in a locket near his breast until he died. Napoleon's lifelong grief over the loss of Josephine, however, did not prevent his remarriage to Eugenie, who he first saw at a ball, wearing a violet gown and violets in her hair. Like Josephine, she carried violets at her wedding and received bouquets of them at her anniversaries. When the French monarchy was restored, the wearing of violets was banned, and until 1874, French governments forbade any reproduction showing a violet, the symbol of Bonaparte supporters. Nevertheless, the violet flourished in the early days of the French Republic. France was flooded with postcards picturing innocent-looking violets, with Napoleon's portrait cleverly hidden among the flowers.

FIGURE 75.21 Examples of human silhouettes associated with drawings of violets. It requires some study before the faces become apparent: (a) *Violettes du 20 mars 1815*, by Jean-Dominique-Étienne Canu. In this widely circulated 19th-century drawing, there are hidden profiles of Napoleon (2, under the green leaf on the right, which resembles Napoleon's bicorne hat); his second wife, Marie Louise (1, facing Napoleon, on the far left, under the second violet); and their son, Napoléon François Joseph Charles Bonaparte (3, on the right of the central stems, next to the lower violet). (b) An early 19th-century drawing of a pansy with a hidden silhouette of Louis XVIII, King of France, from 1814 to 1824 (1 in the illustration), and silhouettes of other French nobility (labelled 2 to 5).

***Source:* Public domain illustrations. (b) is based on a Bequest of Mary Martin, 1938, to the Metropolitan Museum of Art.**

## KEY PUBLICATIONS


Baird, V.B. and F.S. Mathews. 1942. *Wild Violets of North America*. University of California Press, Los Angeles. 225 pp.

Ballard, H.E., Jr., K.J. Sytsma, and R.R. Kowal. 1998. Shrinking the violets: Phylogenetic relationships of infrageneric groups in *Viola* (Violaceae) based on internal transcribed spacer DNA sequences. *Systematic Botany* 23: 439–458.

Beattie, A.J. and N. Lyons. 1975. Seed dispersal in *Viola* (Violaceae): Adaptations and strategies. *American Journal of Botany* 62: 714–722.

Bernhardt, P. 1978. The wily violets: Schemes for survival. *Garden Journal (New York)* 2(3): 18–21, 32.

Brizicky, G.K. 1961. The genera of the Violaceae in the Southeastern United States. *Journal of the Arnold Arboretum* 42: 321–333.

Brown, K., M. Hoffman, and N. Ringer. 1999. *An Herbalist's Guide to Growing & Using Violets*. Storey Communications, Pownal, VT. 32 pp.

Chervin, J., T. Talou, M. Audonnet, B. Dumas, L. Camborde, M.-T. Esquerré-Tugayé, et al. 2019. Deciphering the phylogeny of violets based on multiplexed genetic and metabolomic approaches. *Phytochemistry* 163: 99–110.

Coombs, R.E. 2003. *Violets: The History and Cultivation of Scented Violets*, New edition. Batsford, London. 176 pp.

Coon, N. 1977. *The Complete Book of Violets*. Thomas Yoseloff, London. 147 pp.

Culver, D.C. and A.J. Beattie. 1980. The fate of *Viola* seeds dispersed by ants. *American Journal of Botany* 67: 710–714.

Farrar, E. 1989. *Pansies, Violas and Sweet Violets*. Hurst Publishing, Hurst, UK. 80 pp.

Fazeena, A.H.A. and M.A. Quamri. 2020. Banafsha (*Viola odorata* Linn.) – a review. *World Journal of Pharmaceutical Research* 9: 514–537.

Fuller, R. 1990. *Pansies, Violas & Violettas: The Complete Guide*. Crowood, Ramsbury, UK. 233 pp.

Gil-Ad, N.L. 1997. Systematics of *Viola* subsection *Boreali-Americanae*. *Boissiera* 53: 1–130.

Klaber, D. 1976. *Violets of the United States*. A.S. Barnes, South Brunswick, NJ. 208 pp.

Lim, T.K. 2014. *Viola × wittrockiana*. In: *Edible Medicinal and Non-Medicinal Plants*, Vol. *8: Flowers*, edited by T.K. Lim, 818–821. Springer, Dordrecht.

Little, R.J. and L.E. McKinney. 2003. *Viola*. In: *Flora of North America*, Vol. 6, online. www.efloras.org/florataxon.aspx?flora_id=1&taxon_id=134607.

Malécot, V., T. Marcussen, J. Munzinger, R. Yockteng, and M. Henry. 2007. On the origin of the sweet-smelling Parma violet cultivars (Violaceae): Wide intraspecific hybridization, sterility, and sexual reproduction. *American Journal of Botany* 94: 29–41.

Marcussen, T., L. Heier, A.K. Brysting, B. Oxelman, and K.S. Jakobsen. 2015. From gene trees to a dated allopolyploid network: Insights from the angiosperm genus *Viola* (Violaceae). *Systematic Biology* 64: 84–101.

McKinney, L.E. 1992. *A Taxonomic Revision of the Acaulescent Blue Biolets (Viola) of North America.* SIDA, Botanical Miscellany No. 7. Botanical Research Institute of Texas, Fort Worth, TX. 51 pp. + plates.

Mendonça de Carvalho, L., F.M. Fernandes, M. de Fátima Nunes, J. Brigola, N. Casbas, and C. Groves. 2013. History and cultivation of Parma violets (*Viola*, Violaceae) in the United Kingdom and France in the nineteenth century. *Harvard Papers in Botany* 18: 137–145.

Prager, S. 2020. *Four Flowering Plants That Have Been Decidedly Queered. The Queer History of the Pansy and Other Flowers.* https://daily.jstor.org/four-flowering-plants-decidedly-queered/.

Raskin, I. 2003. Role of salicylic acid in plants. *Annual Review of Plant Physiology* 43: 439–463.

Russell, N.H. 1965. Violets (*Viola*) of central and eastern United States: An introductory survey. *SIDA* 2: 1–113.

USDA. 1977. *Growing Pansies*, Revised edition. Agricultural Research Service, United States. Agricultural Research Service, U.S. Dept. of Agriculture, Washington, DC. 8 pp.

Valentine, D.H. 1962. Variation and evolution in the genus *Viola. Preslia* 34: 190–206.

Van der Kooi, C.J., P.G. Kevan, and M.H. Koski. 2019. The thermal ecology of flowers. *Annals of Botany* 124: 343–353.

Vidalie, H. 2009. Les pensées: Un marché de 500 cultivars (Pansies: A market of 500 cultivars). *PHM Revue horticole (Paris)* 515: 38–44.

Wachsmuth, B. 2012. Double scented violets. *Plantsman* 11: 232–235.

Wetherbee, A. 1986. History of violets. *The Herbarist* 52: 14–19.

Yockten, R. 2003. Phylogenetic relationships among pansies (*Viola* section *Melanium*) investigated using ITS and ISSR markers. *Plant Systematics and Evolution* 241: 153–170.

Zambra, G.L. 1950. *Violets for Garden & Market*, Revised edition. Collingridge, London. 79 pp.

## Key Cultivation Websites

The American Violet Society. https://avs.getcloseto.us/ (The site has been mostly dormant in recent times.)

Clemson University, Extension. *Pansy Diseases & Insect Pests.* https://hgic.clemson.edu/factsheet/pansy-diseases-insect-pests/.

GardenersWorld.com. *How to Grow Violets.* www.gardenersworld.com/how-to/grow-plants/how-to-grow-violets/#caring.

Horticulture Magazine. *Pansy.* https://horticulture.co.uk/pansies/.

*National Viola and Pansy Society of Great Britain.* https://floristpansyviola.wordpress.com/ (The site has been mostly dormant in recent times.)

The Spruce. *How to Grow and Care for Pansies.* www.thespruce.com/pansies-growing-a-cool-weather-favorite-1402913.

The Spruce. *How to Grow and Care for Violas.* www.thespruce.com/growing-violas-1402895.

*Success with Pansies in the Winter Landscape: A Guide for Landscape Professionals.* University of Georgia Extension Bulletin 1359. https://extension.uga.edu/publications/detail.html?number=B1359&title=Success%20with%20Pansies%20in%20the%20Winter%20Landscape:%20A%20Guide%20for%20Landscape%20Professionals.

Utah State University, Extension. *Growing Pansies.* https://extension.usu.edu/yardandgarden/research/growing-pansies.

Wikifarmer. *How to Care for Pansies in the Garden and Pots.* https://wikifarmer.com/how-to-care-for-pansies-in-the-garden-and-pots/.

WikiHow. *How to Cut from a Pansy Plant.* www.wikihow-fun.com/Cut-from-a-Pansy-Plant.

WikiHow. *How to Grow Pansies.* www.wikihow.life/Grow-Pansies#:~:text=Figure%20out%20when%20is%20the%20best%20time%20to%20plant%20your%20pansies.&text=So%20in%20most%20climates%2C%20the,fall%20or%20early%20spring%20bloom.

WikiHow. *How to Grow Violets.* www.wikihow.life/Grow-Violets.

WikiHow. *How to Make Pansies Bloom Again.* www.wikihow.com/Make-Pansies-Bloom-Again.

# 76 Passionflower (*Passiflora* Species and Hybrids)

**Scientific Names:** The name *Passiflora* is from the Latin *passio*, "passion," + *flos*, "flower." Detailed information is presented in the following on the basis of this association.

**Common Names:** The passionflower (passion flower, passion vine, flower of passion) acquired this name because Spanish missionaries to South America, when they became aware of it in the jungles of Brazil in the 16th century, thought the flower represented some of the objects associated with the Crucifixion or suffering of Christ. The "passion" in the name is based on the Passion of Christ, the period between the Last Supper and his death. For the Jesuit missionaries, it was the bloom of the mystic St. Francis of Assisi (1181 or 1182 to 1226, founder of the Franciscan orders), who saw a flower in a vision. The ten apostles who remained faithful to Jesus (leaving out Peter, who denied him, and Judas, who betrayed him) throughout the Passion are symbolized by the five petals and the five petal-like sepals, while a fringe of colored, hairlike filaments above the petals (botanically known as the corona) was thought to represent the crown of thorns that Jesus wore (or, by some, considered to symbolize a halo). The three spreading styles atop the stigma represented the three nails by which Christ was attached to the cross. The five hammer-like anthers on top of the stamens were thought by some to represent the hammer used to drive the nails; others interpreted the five stamens as the wounds of Christ (hence the name "flower of the five wounds" sometimes used in Peru, the West Indies, and elsewhere) and interpret the ovary (unfertilized seed case) as the hammer. Some have even interpreted the lobed leaves and long green vines as the hands and whips (scourges) of Christ's prosecutors, the column of the ovary as the pillar of the cross, the white tint of the flower as purity, and the blue tint as heaven. Supposedly, 72 corona filaments are present, reflecting the traditional number of thorns in the crown of thorns, and some even saw on the underside of the leaves dark spots representing the 30 pieces of silver, and the trinity in the three-lobed leaves. The Jesuit Fathers interpreted the eagerness of the Indians to consume the fruits as a heavenly sign that they were hungering for conversion from heathenism to Christianity and, with overbearing religious zeal, quickly succeeded in wiping out thousands of years of New World culture.

## 76.1 INTRODUCTION

Passionflowers are among the most beautiful flowers in the world, but they are not appropriate for cultivation in much of the world's cooler climates, nor are they low-maintenance. These vines are mostly subtropical and tropical and require a supporting latticework or alternative structure, along with regular pruning.

## 76.2 OVERVIEW

Of the more than 500 species of *Passiflora*, most are native to South America, but about 20 are found in Asia and Australia. The species occur primarily in Central and South America. Most are tendril-bearing herbaceous vines (sometimes becoming woody); a few are shrubs or trees.

*Passiflora caerulea* (blue passionflower, blue crown passionflower, or simply passionflower) is the species most

FIGURE 76.1 Examples of colorful flowers of passionflower (*Passiflora* species and hybrids). (a) *Passiflora incarnata*. (b) Unidentified species. (c) *Passiflora* 'Pierre Pomié'.

*Source:* **(a) Photo by Oliver P. Quillia (CC BY 3.0). (b) Photo by Lisa Ann Yount (public domain). (c) Photo by Igeco (CC BY SA 4.0).**

FIGURE 76.2 The two *Passiflora* species most commonly used as flowering ornamentals: (a) Blue passionflower (*P. caerulea*). (b) Maypop (*P. incarnata*).

*Source:* **(a) Photo by Shutterstock/Manfred Ruckszio. (b) Photo by Shutterstock/Gerry Bishop.**

DOI: 10.1201/9781003586227-76

FIGURE 76.3 Blue passionflower (*Passiflora caerulea*) in fruit. The species is grown primarily for its ornamental flowers, and the bland fruit is of limited quality for consumption, although also rather attractive.

***Source:*** **(a) Photo by Shutterstock/Irenestev. (b) Photo by Shutterstock/F. Neidl.**

FIGURE 76.4 Vintage technical botanical illustrations of *Passiflora* species and garden hybrids: (a) *Passiflora vitifolia.* (b) *Passiflora incarnata.* (c) *Passiflora* × *decaisneana.* (d) *Passiflora* 'Impératrice Eugénie'. (e) *Passiflora foetida.* (f) *Passiflora watsoniana.* (g) *Passiflora insignis.* (h) *Passiflora murucuja.*

***Source:*** **Public domain. (a) Karsten, K. W. G. H. 1869. *Florae Columbiae*, vol. 1. (b) Gottorfer Codex, 1649–1659. Photo by Hans-Simon Holtzbecker. (c) Weise, G. 1854. *Deutsches Magazin für Garten-und Blumenkunde*, Stuggart. (d) Witte, H., and A. J. Wendel. 1868. *Flora: Afbeeldingen en beschrijvingen van boomen, heesters, éénjarige planten, enz. voorkomende in de Nederlandsche tuinen* (Dutch book). (e) Denisse, E. 1843–1846. *Flore d'Amérique*, vol. 15. (f) L' *Illustration horticole*, vol. 36, 1889. (g) Walter Hood Fitch, in *The Florist and Pomologist* (1873). (h) Denisse, E. 1843–1846. Flore d'Amérique, vol. 139.**

widely planted as a flowering ornamental. It is an evergreen in warm and hot climates, and semi-evergreen or deciduous in cooler regions. *Passiflora incarnata* (Maypop) is hardier, growing outside of cultivation in much of the United States. It is also commonly grown as an ornamental. Individual flowers on plants of *Passiflora* often wilt quickly after they are pollinated.

## 76.3 HISTORY

Sixteenth-century Spanish explorers in South America brought passionflower species to Europe.

During the Victorian era (Queen Victoria's reign, 1837–1901), ornamental flowering *Passiflora* vines became very popular in England, and botanical illustrators of that time produced many masterpieces of technical botanical figures for illustrated garden journals, which were extraordinarily popular then (when photography was in its infancy). Hybrids were created using *P. caerulea*, *P. alata*, and other species to meet the demand for ornamental novelties.

## 76.4 ECONOMIC IMPORTANCE

*Passiflora* is most commercially important as a fruit crop, which is significant. As an ornamental, it is a specialty plant with a restricted market, largely in tropical and subtropical climates. Once established, it can be quite long-lived, so unlike annuals and short-lived perennials, its sales potential is limited. It also has limited sales as a cut flower.

## 76.5 ORNAMENTAL USES

Passionflowers are mostly very vigorous vines. They are wonderful for covering up unattractive fences but are best

FIGURE 76.5 Passionflower gardens in California. (a) Unidentified cultivar. (b) Grape-leaved passionflower (*Passiflora vitifolia*).

***Source:*** **(a) Photo by CasparGirl (CC BY 2.0). (b) Photo by Cultivar413 (CC BY 2.0).**

FIGURE 76.6 A woman with a passionflower garland.

*Source:* **Photo by Shutterstock/Freya-photographer.**

planted on walls and large garden gazebos. Most need substantial space, and they are difficult to confine, requiring regular pruning. Some of the species can be grown primarily for fruit, but the more ornamental cultivars tend to have relatively bland taste. Passionflower is an ideal plant for pollinator gardens. Bees, butterflies, and hummingbirds are attracted to the flowers, and butterfly caterpillars may eat the foliage. The beautiful flowers are not easy to obtain unless grown locally, but when available, they make very attractive garlands.

## 76.6 OTHER USES

About 60 species of *Passiflora* produce edible fruit, although only 10 to 12 have significant commercial importance. Passionfruit is grown widely throughout the tropics and subtropics. Commercial production occurs in South America, the Caribbean, Mexico, Australia, New Zealand, Southeast Asia, Taiwan, India, Africa, the Mediterranean area, Hawaii, and South Africa. There are small crops in California and Florida. The most important species is *P. edulis*, which is known simply as passionfruit (or passion

FIGURE 76.7 Passionfruit orchards: (a) In Colombia. (b) In India.

*Source:* **(a) Photo by Omar Uran (CC BY 2.0). (b) Photo by Jaseem Hamza (CC BY 3.0).**

FIGURE 76.8 Fruit of passionfruit species harvested for consumption: (a) Fruits of the principal species of *Passiflora* grown for consumption in Colombia. (b) Giant granadilla (*P. quadrangularis*) on sale in Trinidad and Tobago.

*Source:* **(a) Photo by John Ocampo (CC BY SA 3.0). (b) Photo by Kalamazadkhan (CC BY SA 4.0).**

fruit). It is native to South America, from southern Brazil through Paraguay to northern Argentina, and is widely grown, including plantations in Australia, Hawaii, and Mexico, with some cultivation in Southern California.

*Passiflora* species, like most plants, have been employed in Indigenous medicine in its native area, mostly in South America. In herbal medicine today, "passionflower" is promoted as a dietary supplement for anxiety and sleep problems, pain, heart rhythm problems, menopausal symptoms, attention-deficit hyperactivity disorder, skin burns, and hemorrhoids. The efficacy and safety of such uses have not been adequately verified.

## 76.7 PROBLEMS

Passionflowers are aggressive plants, requiring regular pruning. They can cover bushes and become very difficult to eradicate. Some species are invasive. For example, banana passionflower (*Passiflora tripartita* var. *mollissima*) was introduced from the Andes to the Hawaiian Islands as an ornamental about 1900 and is now smothering large areas of native forest on the islands of Hawaii and Kauai.

Despite the medicinal uses of many parts of the plants and the widespread consumption of the fruits, *Passiflora* species may produce significantly toxic cyanide compounds, including in unripe fruit. It has been recommended that immature fruits of all species should not be consumed because they are probably poisonous.

FIGURE 76.9 Weedy, invasive nature of passionflower: (a) A vine growing over the door and window of a house in Virginia, reflecting lack of pruning to keep the spreading plant in check. (b) A volunteer passionflower vine that has taken over a tree in Los Angeles. The fruit is reportedly insipid, reinforcing its unwelcomeness.

*Source:* **(a) Photo by Maggie McCain (CC BY SA 2.0). (b) Photo by RBerteig (CC BY 2.0).**

## 76.8 CULTIVATION

### 76.8.1 Climate and Site

*Passiflora* is basically tropical, most of the species adapted to hot climates, tolerating high heat but limited freezing. However, a few species reach northward and are relatively cold-tolerant.

*Passiflora caerulea*, one of the hardiest species, is considered appropriate for climate zones 6 to 10 and is claimed to be able to survive freezing winter temperatures down to –20°C. *Passiflora incarnata* (Maypop), which is even more cold-hardy (suited to zones 5 to 9), is also a popular choice, and it is resistant to dry conditions. Passionflowers are generally tolerant of high humidity but can develop fungal diseases if the air is not well circulated. They grow in full sun and part shade. *Passiflora caerulea* can grow 10 m in a season, and like most vines in the genus, it needs to be grown on a support system. The plants are often grown on arbors, trellises, walls, and fences, the roots established either in the ground or in containers. The tendrils cling to brickwork but are not considered to be damaging.

### 76.8.2 Propagation and/or Purchasing

If grown from seeds (in warm climates), sowing is usually done outdoors in the fall, while in the winter, cultivation can be in a cold frame or unheated greenhouse. Ornamental clones are reproduced from cuttings. Starter plants in containers are often available from nurseries.

### 76.8.3 Planting

When transplanting a potted plant purchased from a nursery, the crown of the plant should be positioned at soil level. Passionflower is best established at the base of a fence, trellis, or arbor, in situations where the vines have room to roam, keeping in mind that heavy pruning may be necessary to keep the vines in check.

FIGURE 76.10 Passionflower saplings ready to be planted in Nairobi, Kenya.

*Source:* **Photo by Kristinabudiati (CC BY SA 4.0).**

### 76.8.4 Soil and Moisture Conditions

Passionflowers grow best in moist but well-drained soils. They generally tolerate soils of different texture (sand to clay) and a pH of 6.0 to 8.0.

### 76.8.5 Maintaining Plants

Plants can be pruned after blooming to maintain a desired size. Deadheading of flowers is recommended. Caterpillar feeding on the foliage slows growth, but the plants are adapted to survive such herbivory.

## 76.9 FASCINATING FACTS, FABLES, FOLKLORE, AND FRIVOLITY

- As noted earlier, in England during the Victorian era, a craze for passionflowers developed. The enthusiasm for them spread to other countries, resulting in artists preparing masterpieces of art for the upper classes.
- In 1865, Queen Victoria (1819–1901) had a wreath of passionflowers sent to the funeral of

FIGURE 76.11 Artistic masterpieces of passionflowers prepared during the craze for them in the 19th century. (a) 1899 plant catalogue, Dingee & Conard Co., Pennsylvania. (b) Painting of an Indigenous female passionfruit seller by Peruvian artist Acuarela de Pancho Fierro (1807–1879). (c) Painting showing a passionfruit orchard in Indonesia. (d) Painting showing two snowcap hummingbirds and *Passiflora racemosa* in Panama by American artist Martin Johnson Heade (1819–1904). (e) Another painting by M. J. Heade of hummingbirds and passionflower.

*Source:* **(a, b) Public domain. (c) Tropenmuseum, Netherlands (CC BY SA 3.0). (d) Photo from Museum of Fine Arts, Boston (public domain). (e) Photo from Metropolitan Museum of Art (public domain).**

Abraham Lincoln (1809–1865, 16th president of the United States). The Queen also declared that the passionflower was the ideal flower for funerals. Gravestones with elaborately carved passionflower vines became very popular for the wealthy in England.

- The caterpillars (larvae) of several species of zebra butterflies can only eat passionflower leaves. Several passionflower species have adaptations to prevent damage by the caterpillars. Some plants disguise themselves by producing leaves that do not look like typical passionflower leaves. Some species have nectar-secreting glands to attract wasp and ant parasites of the larvae. Some species produce chemicals that poison the larvae. And still other species produce tiny egg-like particles on the leaves that send the message to the female butterfly that the leaves are already occupied so she should find another plant on which to lay her eggs.
- Some tropical passionflowers have a symbiotic relationship with ants, providing nectar via structures called extrafloral nectaries (nectar sources outside the flower), located at the base of each leaf, and in return receiving protection from predators by the ants.
- The symmetrical flowers of *Passiflora* are reminiscent of the face of a clock. In Israel, plants are known as "clock-flowers," and in Japan and Greece as "clock plants."
- *Passiflora caerulea* is the national flower of Paraguay.
- Passionfruit is a popular addition to cocktails. A "hurricane" is made with dark rum, passion fruit flavoring, and citrus juices, served in a hurricane glass (a footed glass that is bulbous at the bottom, tapering to a flaring cylinder at the top, used for blended or frozen tropical and specialty drinks). The origin of the beverage is obscure, but it has long been associated with Pat O'Brien's French Quarter Bar in New Orleans, Louisiana. A "Grand Passion" is a cocktail prepared with passion fruit liqueur, gin, and Angostura bitters.

## KEY PUBLICATIONS

Abreu, P.P., M.M. Souza, E.A. Santos, M.V. Pires, M.M. Pires, and A.F. Almeida. 2009. Passion flower hybrids and their use in the ornamental plant market: Perspectives for sustainable development with emphasis on Brazil. *Euphytica* 166: 307–315.

Bisht, T., V. Rana, and H. Bajaj. 2020. A review on genus *Passiflora*: An endangered species. *IOSR Journal of Pharmacy and Biological Sciences* 15: 17–21.

Boender, R. 2019. *Passionflowers: A Pictorial Guide*. Self-published. 297 pp.

Bugallo, V., S. Cardone, M.J. Pannunzio, and G. Facciuto. 2011. Breeding advances in *Passiflora* spp. (passionflower) native to Argentina. *Floriculture and Ornamental Biotechnology* 5: 23–34.

Cerqueira-Silva, C.B.M., O.N. Jesus, E.S.L. Santos, R.X. Corrêa, and A.P. Souza. 2014. Genetic breeding and diversity of the genus *Passiflora*: Progress and perspectives in molecular and genetic studies. *International Journal of Molecular Sciences* 15: 14122–14152.

Engels, G. and J. Brinckmann. 2016. Passionflower, *Passiflora incarnata* L., family Passifloraceae. *HerbalGram* 112: 8–17.

Janda, K., K. Wojtkowska, K. Jakubczyk, J. Antoniewicz, and K. Skonieczna-Żydecka. 2020. *Passiflora incarnata* in neuropsychiatric disorders – a systematic review. *Nutrients* 12(12): 3894. https://doi.org/10.3390/nu12123894.

Martin, F.W. and Y. Nakasone. 1970. The edible species of *Passiflora*. *Economic Botany* 24: 333–343.

Miroddi, M., G. Calapai, M. Navarra, P.L. Minciullo, and S. Gangemi. 2013. *Passiflora incarnata* L: Ethnopharmacology, clinical application, safety and evaluation of clinical trials. *Journal of Ethnopharmacology* 150: 791–804.

Popay, I. 2014. *Passiflora caerulea* (Blue Passionflower). CABI Compendium. https://doi.org/10.1079/cabicompendium.116172.

Ulmer, T. and J.M. MacDougal. (eds.). 2004. Passiflora*: Passionflowers of the World*. Timber Press, Portland, OR. 430 pp.

Vanderplank, J. 2000. *Passion Flowers*, 3rd edition. MIT Press, Cambridge, MA. 224 pp.

Vanderplank, J., E.G. Blanco, C. Feuillet, A. Frank, L. King, E. Kugler, et al. 2003. *The International* Passiflora *Register*. Passiflora Society International. 36 pp. www.researchgate.net/publication/268388602_The_International_Passiflora_Register_2003.

## Key Cultivation Websites

GardenersWorld.com. *How to Grow Passion Flowers*. www.gardenersworld.com/how-to/grow-plants/grow-guide-passion-flowers/.

Horticulture Magazine. Passiflora – *Passion Flower*. https://horticulture.co.uk/passion-flower/.

*The Journal & Newsletter of Passiflora Society International*. 2022. www.passiflorasociety.org/wp-content/uploads/Register-of-Passiflora-Cultivars-2018-2019_243-258.pdf.

Passiflora Online. www.passionflow.co.uk/.

Passiflora Society International. https://passiflorasociety.org/.

Royal Horticultural Society. *How to Grow Passion Flowers*. www.rhs.org.uk/plants/passion-flower/growing-guide.

The Spruce. *How to Grow and Care for Passionflower*. www.thespruce.com/passion-flowers-1403114.

WikiHow. *How to Grow and Harvest Passion Fruit*. www.wikihow.com/Grow-Passion-Fruit.

# 77 Peony (*Paeonia* Species and Hybrids)

**Scientific Names:** The genus name *Paeonia* is based on the Greek *paionia*, apparently derived from Paion, the physician of the classical gods, who employed peonies medicinally. Greek legend holds that Zeus transformed Paion into a peony flower to save him from his jealous teacher Asclepius.

**Common Names:** Peony is the universal English name. Names for particular species and classes of peonies are presented later.

## 77.1 INTRODUCTION

Peonies in bloom are stunning, so much so that many consider them the most attractive of all garden ornamentals. Reflecting their astonishing beauty, they have been called the "king of flowers" in China, the "queen of flowers" in ancient Greece, and the "prime minister of flowers" in Japan. Numerous renowned European artists, including Frédéric Bazille, Eugène Delacroix, Henri Fantin-Latour, Conrad Gesner, Paul Gauguin, Édouard Manet, and Pierre Auguste Renoir, produced peony paintings, and dating back many centuries, mostly unknown artists in China and Japan did the same.

## 77.2 OVERVIEW

*Paeonia*, the only genus in the family Paeoniaceae, has about three dozen species. These occur in four separate north temperate areas of Eurasia (eastern Asia, Central Asia, the western Himalayas, and the Mediterranean region), with two very localized (endemic) species, *P. brownii* in northern California, adjacent states, and British Columbia, and *P. californica* in California and adjacent Mexico. These two Western North American species constitute section *Onaepia*. Section *Moutan*, notable for *P. suffruticosa* (tree peony), discussed in detail later, includes six to nine species of shrubs of Central and western China. Several of these are rare and are considered endangered due to overcollection for ornamental and medicinal purposes. The remaining species, all herbaceous, are in section *Paeonia*, occurring in Europe, North Africa, and Asia – from Japan to Portugal and Morocco – but mainly in the Mediterranean region and eastern Asia. Three species of this section, *P. lactiflora*, *P. officinalis*, and *P. tenuifolia*, are discussed in detail in this chapter.

FIGURE 77.1 Bouquets of spectacular flowers of tree peonies (*Paeonia suffruticosa*). The flowers are notably doubled (i.e., with many extra petals), reminiscent of the most attractive roses.

*Source:* **(a) Photo by Shutterstock/Fusionstudio. (b) Photo by Shutterstock/MaskaRad.**

Most wild peonies grow in mild temperate climates, although the overall range extends from subtropical to subarctic. Peony species are adapted to overwintering in a dormant state, during which they are exposed to a period of cold temperatures. Flowering occurs in the spring and early summer. Peonies are usually quite drought-resistant and, in nature, often survive dry spells by going dormant. Both wild peonies and the cultivars are sun-loving plants, although many of the wild forms occur in open woodland or the edges of forests (no species is found in deep shade). Most of the species grow in moderately fertile, circumneutral, well-drained soils that are not compacted. Many are long-lived, capable of persisting for decades. They are naturally resistant to deer and rabbits, presumably because of bitter-tasting phenolic compounds in the foliage and flowers.

*Paeonia* species are stout branching perennial herbs and diffuse shrubs. Most of the herbaceous species are rhizomatous or have tuberous roots, which serve as the overwintering phase. The leaves are large and compound, occasionally finely dissected. The flowers are large, in the cultivated species often 15–30 cm across. Doubled blooms are common in the cultivars. The range of colors includes cream, peach, pink, purple, red, yellow, white, bicolor (two-colored), and multicolored.

Most peony cultivars grown in temperate regions belong to *P. lactiflora* and its hybrids. Peony specialists artificially classify these into various groupings on the basis of flower form: (1) Single – with one or two rows of petals and a center of functional stamens. (2) Semi-double – with three or more rows of petals (rarely less), and with a center of functional stamens. (3) Japanese and anemone – with a center of non-functional stamens: staminodes (stamens partially transformed into petals) in the Japanese category, petalodes (stamens more dramatically transformed into petals) in the anemone category. (4) Double – fully double, with many petals and often showing no stamens, or with hidden ones. Cultivars with doubled flowers are sometimes called French peonies. In a subtype of the double category, known as the "bomb" form, the central petals appear as a sphere, while the outer petals look like a plate.

In horticulture, less commonly cultivated herbaceous peony species are often called "species peonies." The two most widely cultivated kinds of species peonies are

DOI: 10.1201/9781003586227-77

FIGURE 77.2 Representative cultivars of the common Chinese peony, *Paeonia lactiflora*: (a) 'Coral Charm'. (b) Two-year-old girl beside 'L'Etincelante'. (c) 'Illiini Belle'. (d) 'Neon'.

*Source:* **(a) Photo by Poupou L'quourouce (CC BY SA 2.5). (b) Photo by Hinton Ampner (CC BY 2.0). (c) Photo by Daderot (public domain). (d) Photo by Andrey Korzun (CC BY SA 3.0).**

*P. officinalis* (European peony, native to Europe) and *P. tenuifolia* (cut-leaved peony, also native to Europe). Several dozen cultivars of both species have been bred.

### 77.2.1 *Paeonia lactiflora* (Chinese Peony, Common Garden Peony)

This species is the "standard" peony most often encountered. The wild *P. lactiflora* is a native from Tibet to China and Siberia and has many relatively small flowers on given stems. Thousands of cultivars have been produced by hybridization and selection. These range from 45 to 150 cm in both height and spread. Cultivars usually have only a few flowers per stem, and there has been selection for stems bearing just one large flower.

### 77.2.2 *Paeonia suffruticosa* (Tree Peony, Moutan)

The name "tree peony" is somewhat of a misnomer. The plants can take on the appearance of miniature trees (i.e., by developing a single main stem analogous to a tree trunk) but usually are multi-stemmed and grow into small to medium shrubs. Although woody ("woody peonies" would have been a better name than "tree peonies"), the stems are relatively brittle, especially when young. Tree peonies can grow to a height of 3 m but some cultivars are dwarfs, as short as 90 cm. Most mature to 1.2–2 m in height (and about the same width). Unlike herbaceous peonies, the aboveground stems are perennial, although the leaves are deciduous. There are hundreds of cultivars, but tree peonies are much less popular in North America compared to herbaceous peonies. Tree peonies are relatively tender and are not suitable for quite cold climates.

FIGURE 77.3 Representative illustrations of the tree peony, *Paeonia suffruticosa*.

*Source:* **(a) Photo by Leonisha (CC BY 2.0). (b) Painting from the Yokohama Nursery, Japan, 1900 catalogue (public domain image). The Yokohama Nursery exported peonies and other ornamentals to the United States from the early 1900s to the start of the Second World War. Its beautifully illustrated catalogues are collector's items, selling for thousands of dollars. (c) Painting from H. C. Andrews, 1806–1807, *The botanist's repository*, vol. 7, plate 448 (public domain image). (d) Photo by Jesse W900 (CC BY SA 4.0).**

FIGURE 77.4 The Itoh hybrid 'Bartzella'. As noted in the text, a single propagating division of this once sold for over $1,000.00.

*Source:* **Photos by F. D. Richards (CC BY 2.0).**

### 77.2.3 Itoh (Intersectional) Hybrids

Itoh or intersectional peonies are intermediate between herbaceous and tree species. They have larger, more numerous flowers, a longer flowering period, and stems that do not droop by comparison with most herbaceous peonies. Itoh cultivars produce overwintering buds aboveground (like tree peonies) but require relatively warm climates to survive aboveground overwinter. However, Itoh hybrids also produce buds underground (like herbaceous peonies), and these serve for overwintering in colder climates. Thus, Itoh cultivars are hardier than tree peonies, which are not well adapted to cold climates. During the 20th century, Itoh cultivars were in short supply but high demand. In the

1990s, 'Bartzella', an improved yellow form, sold for over \$1000.00 for a single division. Subsequent propagation by tissue culture has made Itoh hybrids much more affordable.

### 77.2.4 *Paeonia officinalis* (European Peony)

Aside from the principal ornamental peonies, *P. officinalis* is the most popular of the (so-called) "species peonies" grown in cold climates. The European peony, indeed, a native of Europe, has been cultivated there for 500 years, initially for medical purposes. Quite hardy cultivars are available, most with red or crimson flowers, several with moderately large doubled flowers up to 10 cm in diameter.

FIGURE 77.5 *Paeonia officinalis. Left*: Plants. *Right*: Dehiscing seedpods.

*Source:* **Photos by H. Zell (CC BY SA 3.0).**

FIGURE 77.6 *Paeonia tenuifolia.*

*Source:* **(a) Photo by Drew Avery (CC BY 2.0). (b) Photo by Clivid (CC BY 2.0). (c) Wild population in a protected area of Volgograd Oblast, Russia. Photo by Cryptobasis (CC BY SA 4.0).**

### 77.2.5 *Paeonia tenuifolia* (Cut-Leaved Peony, Fern-Leaved Peony)

This is the second-most popular "species peony" grown in cold climates. It is an early-flowering plant that grows only to about 50 cm in height and dies back to the ground a few weeks after flowering (depending on location, this may be in later summer or early fall). Despite its limited size and display, the cut-leaved peony is an old-time favorite. It has very attractive, lacy, finely cut, fernlike foliage and fragrant bloodred flowers. A few cultivars with doubled flowers are available. The species is ideal as a rock garden specimen or in gardens where the surrounding areas can be supplemented with flowering annuals.

## 77.3 HISTORY

Most but by no means all peony cultivars are of hybrid derivation, tracing the majority of their ancestry to China, where the plants have been grown for over 4,000 years. Since most cultivars are of interspecific hybrid origin, they are best recognized by hybrid names (*P.* × *lactiflora*, *P.* × *suffruticosa*, *P.* × *officinalis*, etc.), but this convention is rarely followed.

FIGURE 77.7 Old paintings showing the beauty of peony flowers. Peonies have been an extremely popular subject for artists in the past centuries: (a) *Peonies*, by American painter Charles Courtney Curran (1861–1942), famous for depicting women in various settings. (b) Peony flowers, painted by Jane Louden in 1841. (c) A three-panel mural, *A Modern Genji in the Peony Garden* (Hikaru Genji was the main character of an 11th-century Japanese tale), by Japanese print designer Utagawa Kunisada II (1823–1880).

*Source:* **Public domain.**

FIGURE 77.8 Old ceramic objects painted with peonies: (a) Double-sided teapot produced in China during the late 17th to the early 18th centuries. (b) Snuff bottle produced in China during the late 18th century. (c) Teapot produced in Vienna, Austria, about 1735. The floral drawings are stylized, but presumably, the Chinese plants are *Paeonia suffruticosa*, and the Austrian plant is *P. officinalis*.

*Source:* **Metropolitan Museum of Art, New York (public domain photos).**

*Paeonia lactiflora* (Chinese peony, common garden peony) is the dominant herbaceous species that has been domesticated. More than 5,000 cultivars have been recognized. This native of China, Tibet, and Siberia is the main parent of most cultivated peonies. It was not introduced into Europe from China until the late 18th century.

The parental species of cultivars of "tree peonies" have been difficult to ascertain, and it seems that several species are ancestral, and that there has been repeated domestication of them. In China, there appear to be cultivated groups of tree peonies that are taxonomically traceable to different wild species. In particular, "flare tree peonies" (so-named because there is an obvious flare at the petal base) are often recognized as representing hybrid cultivars of *P. suffruticosa* but actually trace much of their parentage to *P. rockii*. However, in Western nations, available cultivated tree peonies are usually recognized as "*P. suffruticosa*" and called "tree peony" or sometimes "moutan," a name of Chinese origin. This species, a native of China and neighboring regions of Asia, is a main parent of cultivated peonies that grow as woody shrubs, and about a thousand cultivars have been bred. Buddhist monks introduced tree peonies into Japan between the 5th and 8th centuries, where it became second in popularity only to the chrysanthemum. The Japanese bred varieties of exceptional beauty, concentrating on flowers that are not as doubled as are most Chinese varieties (and hence less prone to flower drooping). Tree peonies were introduced into Europe in the late 18th century, shortly after the importation of herbaceous peonies. Most tree peony cultivars were bred in China, a considerable number also in Japan, and a few in Western countries.

Itoh or intersectional peonies are intermediate between herbaceous species (mostly *P. lactiflora*) and tree species (often identified as *P. suffruticosa* or *P. lutea*). Toichi Itoh, a Japanese hybridizer, produced the first intersectional hybrid in 1948.

New, gorgeous semi-woody varieties have recently been created that are significantly more beautiful than traditional cultivars.

FIGURE 77.9 Commercial production and sales of peonies: (a) A peony farm in China. (b) Peonies in an outdoor flower market in Nice, France. (c, d) Peony flower farm in Salem, Oregon.

*Source:* **(a) Photo by Yobito kayanuma (CC BY SA 3.0). (b) Photo by Shutterstock/Veniamin Kraskov. (c, d) Photos by Travel Salem (CC BY ND 2.0).**

## 77.4 ECONOMIC IMPORTANCE

Peonies are extremely important in the international cut flower trade and of the florist trade. Leading commercial producers of cut peonies include Chile, Israel, The Netherlands, the United States, New Zealand, France, and Italy. In The Netherlands, it is not uncommon for a million peony stems to be sold in a single day. Christmas and the wedding season months of June and July are major sales periods. In the United States, peonies have become a traditional flower of Memorial Day (observed on the last Monday of May in honor of those who died while serving in the military).

## 77.5 ORNAMENTAL USES

Ornamental plants are predominantly grown for attractive flowers, and in this respect, peonies are virtually unsurpassed. Cultivars are available that produce huge flowers in a wide range of colors, and the flowers of many varieties have very attractive odors. The foliage of some cultivars is quite attractive in texture and color. Peonies are small to medium-sized shrubs that are adaptable to a wide range of landscaping applications. Most cultivars are relatively resistant to diseases, pests, drought, and Northern climates. They are very long-lived and, once established, require only a moderate degree of maintenance.

Peony blooms are a favorite vase flower of home gardeners. As cut flowers, peonies are very popular wedding, corsage, and special event flowers because of their spectacular size, colors, and attractive smell. Although all peonies can be employed as cut flowers, traditionally, *P. lactiflora* flowers that are large, doubled, and on long stems are preferred in the retail trade.

FIGURE 77.10 Large peony landscape plantings: (a) A peony border outside Kazan Cathedral in St. Petersburg, Russia. (b) A peony garden in Mukden Palace, China. Built in 1625, the site was occupied by the first three Quing emperors. (c) Tree peonies in Joruri-ji, a Buddhist temple in Japan. (d) Tree peonies in Higashimatsuyama City, Japan.

*Source:* **(a) Photo by Shutterstock/Baturina Yuliya. (b) Photo by Gary Lee Todd (public domain). (c) Photo by C.Stadler/Bwag (CC BY SA 4.0). (d) Photo by Keihin (CC BY SA 3.0).**

FIGURE 77.11 Peonies in containers: (a) Potted peonies decorating a temple in Japan. (b) Starter plants for transplanting. Such potted peonies are not intended for long-term cultivation in their pots.

*Source:* **(a) Photo by Sorrysorry (public domain). (b) Photo by Peggy Marco from Pixabay (public domain).**

For those wishing to harvest their own garden flowers, the stems should be cut just before the buds open (the buds become noticeably softer at this stage). Flowers can be floated in water or harvested with a long stem and placed in water in a vase (after removing the leaves below the waterline).

## 77.6 OTHER USES

The bark of the root of tree peonies (especially *P. suffruticosa*) has been employed in traditional Chinese medicine as a tonic, wound healer, preventer of fungal infections, analgesic, liver cleanser, and treatment for spasmodic conditions. The root of the common peony (*P. lactiflora*) has also been used frequently in herbal medicine of China, Japan, and Korea. During the Middle Ages in Europe, the seeds and roots of peonies, especially of the European peony (*P. officinalis*), were employed to cure many conditions, including bladder stones, diarrhea, epilepsy, jaundice, labor pains, lunacy, nightmares, and stomach ache. Today, peony bark continues to be widely used in traditional medicine of eastern Asia. The major constituents paeonin and paeoniflorin have been found to be useful in relieving inflammation and itching, and these and other glycosides of *Paeonia* roots are under investigation for use in new pharmaceuticals.

FIGURE 77.12 Cut flower usage of peonies: (a) A little girl with a big bouquet of peonies. (b) An elegant vase of white peonies. (c) A lady in a couture fashion dress with pink peonies, carrying a bouquet of peonies, near the walls of an ancient castle.

*Source:* **(a) Photo by Shutterstock/Myroslava Malovana. (b) Photo by Shutterstock/ArtMari. (c) Photo by Shutterstock/Anna Evlanova.**

## 77.7 PROBLEMS

There are a few gardening problems associated with peonies. Compared to annuals, they are slow to establish (especially if grown from seed). Compared to some perennials, peonies (especially the herbaceous kinds) have a relatively short flowering period. The flowers of most cultivars of the herbaceous kind (which dominate the peony world) often have weak stems, requiring support to prevent bending over to the ground. Peony breeders regularly produce exceptionally attractive cultivars that gardeners want, but these are often alarmingly expensive and often in short supply.

The literature and the internet often claim that oral consumption of peonies is toxic to humans, pets, and livestock, but research documenting this does not seem to be available, and authoritative guides to toxic plants uncommonly list peonies. Very rarely, dermatitis from handling the plants has been reported. In ancient China, peony roots and flowers were consumed both for medicinal and culinary purposes. In the Middle Ages of Europe, peony (*P. officinalis*) seeds were crushed and used as a seasoning, and peony roots were occasionally baked as a vegetable for the very rich. Today, the petals of *P. lactiflora* are sometimes consumed in salads or beverages. In China, they are parboiled and served as a sweet tea. The Chinese also employ the flowers of tree peonies in casseroles, cakes, teas, and drinks. Given the uncertainty about peony toxicity and the fact that the plants have medicinal constituents, they are best not consumed in significant amounts, either by humans or animals (both pets and livestock).

## 77.8 CULTIVATION

### 77.8.1 Climate and Site

Most herbaceous cultivars are well adapted to the colder regions of temperate climates, including hardiness zones 3 to 8, whereas tree peonies are rated to grow in zones 4 to 8, depending on cultivar. However, growing tree peonies in cold climates is a gamble and represents an investment in dollars and effort that primarily attracts peony enthusiasts. The hardier Itoh hybrids are a better choice for most cold areas. In zone 8 and southward, peonies may not receive sufficient chilling during the winter to flower in the subsequent spring.

Because they can live for many years, planting sites should be carefully chosen and prepared well. To avoid disease buildup, avoid planting where peonies have recently grown. Some peony specialists recommend preparing a "$100 hole" – one that will nurture the plants for decades (such a hole could be 45 cm deep and 60 cm wide). Guidebooks often recommend that tree peonies be planted in somewhat shady areas (to prevent bleaching of the flower colors by strong sunlight, to lower the risk of frost damage in early spring because the warmth prematurely stimulates growth, and to prevent excess drying) and in locations protected from wind (to prevent the petals from being blown away). In hot climates, peonies indeed benefit from some shade, but in most temperate region locations, sites should be sunny (at least 6 hr of full sun daily), although some shade is tolerated (however, shade inhibits flowering).

Some peonies grow very well in large containers and, indeed, in Asia potted tree peonies are popular. However, most cultivars develop roots that are too large for standard-sized pots. Peonies acquired in pots from stores are meant to be planted soon.

### 77.8.2 Propagation and/or Purchasing

The best option for buying peonies from commercial sources is to purchase potted plants sold and guaranteed by a local reputable nursery. They can be planted at any time of the season. It is also very advisable to have researched information on cultivars offered, since some can outlive the average human, and buying a peony can represent a lifelong commitment. Often, the most attractive cultivars will only be available from specialist peony suppliers. They normally ship bare-root planting stock in the fall, after the material has been divided. The best time to plant these is before freeze-up in the fall, allowing time to establish feeder roots.

Supermarket chains commonly (and inappropriately) supply bare-root material in the spring to address consumer demand for early-season planting. Offerings are often accompanied by photos of spectacular flowers, which frequently turn out to be misleading. Peony specialists have a very unfavorable opinion of the quality of much commercial bare-root peony stock (including herbaceous, tree, and Itoh classes), because there is a high rate of failure. Although often sold at "bargain" prices, the stock is generally too small, is often dried out, or the shoot buds have started developing foliage (all signs that the material should not be purchased).

*Paeonia lactiflora* and its hybrids are readily available from nurseries, garden outlets, and specialty mail-order sources, although some rare or new cultivars may require special order. In cold climates, Itoh and tree cultivars are considerably more difficult to obtain and are more expensive (sometimes hundreds of dollars for Itoh cultivars).

In cold climates, almost all tree peonies are supplied as bare-root, imported material, but potted stock is sometimes available. It is possible to import plants from specialty horticultural suppliers. Guaranteed potted plants are the best option, but availability is sometimes limited.

### 77.8.3 Planting

The several dozen species of peonies that are grown mostly by enthusiasts, specialists, and researchers are often

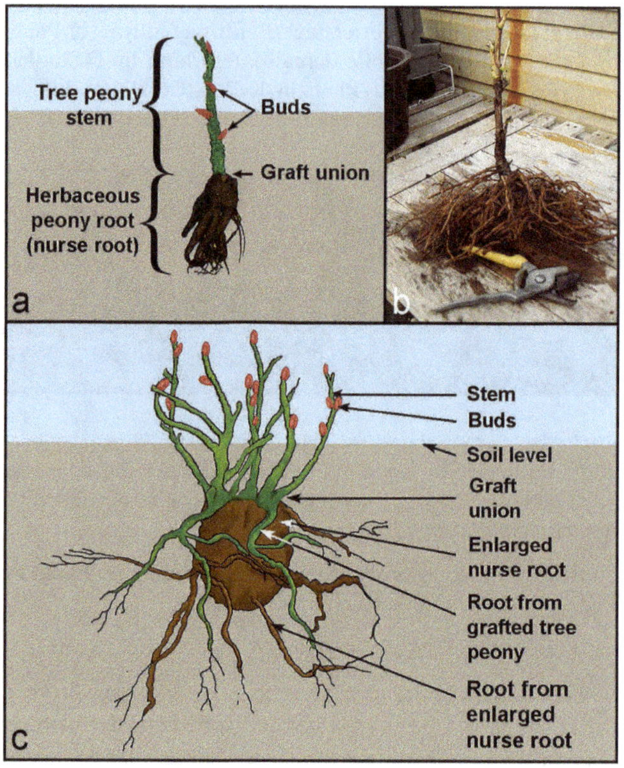


FIGURE 77.13 Development of the underground portion of a tree peony (*P. suffruticosa*), from planting to maturity: (a) Diagram showing how planting stock should be positioned in the soil. A young stem of a tree peony (in green; buds shown in red) has been grafted to a "nurse root" (in brown) of an herbaceous peony (*P. lactiflora*). As discussed in the text, the graft union should be 10–15 cm below soil level. (b) Excavated root system and a portion of an aerial stem of a well-established tree peony. (c) After several years, the grafted stem has developed its own roots and has produced new stems (all parts of the tree peony are shown in green). The old nurse root (shown in brown) has persisted but is no longer needed.

*Source:* **(b) Photo by F. D. Richards (CC BY 2.0). (a, c) Diagrams by B. Brookes.**

reproduced by seeds. Seeds are occasionally employed to establish the principal horticultural peonies, but of course, seeds will rarely reproduce the beautiful flowers of cultivars that have been maintained as clones. Peony seeds may have complex dormancy requirements, and if purchased, they should be from a reliable source that provides germination instructions. Collecting, conditioning, and storing peony seeds can be a demanding exercise, for which information is available in the references provided later. Herbaceous peonies usually produce flowers in their third year following seeding. When established from seeds, tree peonies generally require 5 to 7 years to produce flowers.

Minimum planting distance between peonies should be at least 80 cm, preferably 1 m, and 1.2 m in the case of tree peonies expected to grow well. Peonies are best planted at least 3 m from competing trees and shrubs, the roots of which are likely to compete significantly.

The horticultural industry and, indeed, most amateur gardeners usually reproduce herbaceous peony cultivars by division of the underground portion. This is best done when the plants have gone dormant in the autumn. The entire plant may be dug up. Herbaceous peonies should be divided into portions, each of which has several "eyes" (i.e., buds or meristems) as well as several storage roots. The same is true for tree peonies, but each division should also have a vigorous stem or two of the current season's growth. Herbaceous peony divisions are very widely available commercially as bare-root stock and as young potted plants (which can be established at any time of the season).

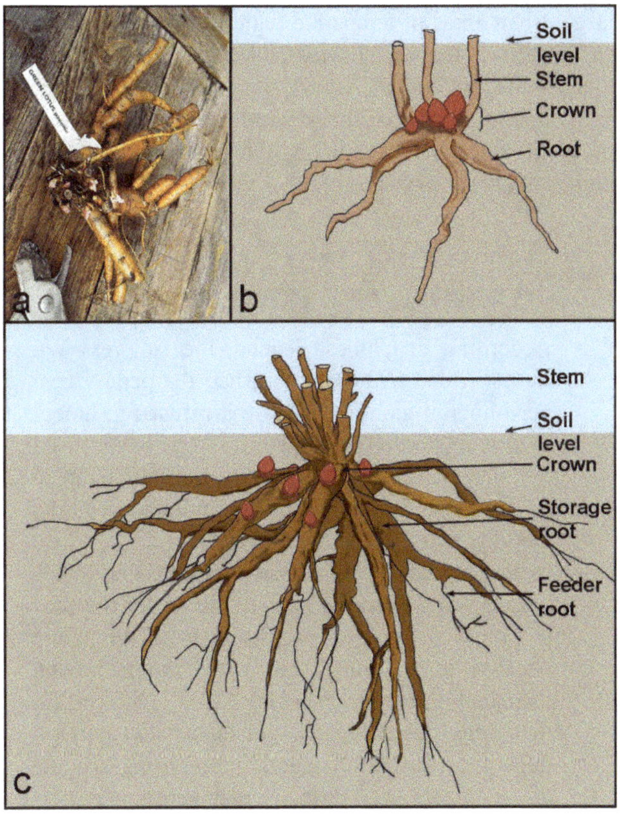


FIGURE 77.14 Planting a division of an herbaceous peony (*Paeonia lactiflora*): (a) A propagation division. (b) Diagram showing an appropriately planted division (buds are shown in red). The top of the highest bud should be 2.5–5 cm below the soil surface. (c) The underground portion of a plant that has been established for several years.

*Source:* **(a) Photo by F. D. Richards (CC BY 2.0). (b, c) Diagrams by B. Brookes.**

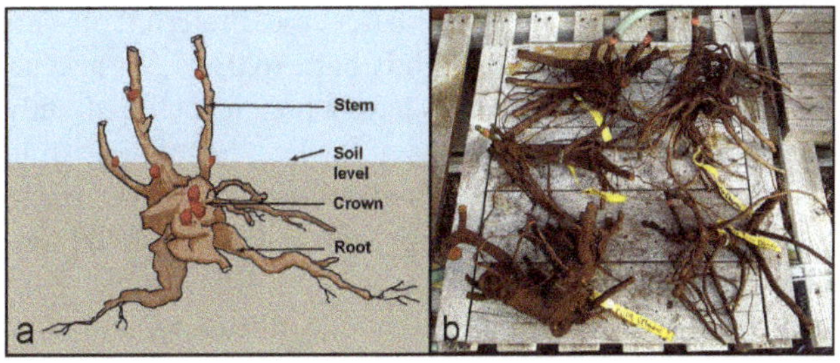


FIGURE 77.15 Divisions of Itoh hybrids employed as planting stock.

*Source:* **(a) Diagram (by B. Brookes) showing an appropriately planted division (buds are shown in red). (b) Photo by F. D. Richards (CC BY 2.0).**

The horticultural industry reproduces tree and Itoh peonies almost entirely by grafting cuttings onto the portions of underground organs of herbaceous peonies, a process that is slower than reproduction by divisions and requires skill. Micropropagation is currently becoming established and is expected to lower the price of these classes of peonies.

Herbaceous peonies often do not respond well to transplanting, unless dormant, and in any event, the deep, brittle roots are easily damaged, and up to 3 years may be required for an excavated plant to recover from the trauma.

Divisions should be planted shallowly – the highest buds not more than 5 cm below the soil surface (both shallower and deeper planting can retard or prevent production of flowers). Potted starter plants should be removed from the pot and planted so that the top of the soil from the pot is level with the soil in the planting site.

Planting stock of tree peonies is typically grafted onto a base of *P. lactiflora*. When planted, the graft union should be 10–15 cm below soil level, enabling the tree peony stem to develop its own roots. The stem/root graft union usually appears as a root bulge just below the graft union (caused by the root growing much faster than the grafted stem). To form a hedge, tree peonies can be planted 0.9 m apart.

Itoh hybrid (between herbaceous and tree peonies) planting stock will have erect basal stems with several buds attached to a crown which will usually have more buds than on the stems. The crown buds should be 5–10 cm below soil level, and (if possible) the lowest bud on the erect stems should be at soil level.

### 77.8.4 Soil and Moisture Conditions

Peonies can grow in infertile soils, but fertile loams are best. Both clay and sandy soils can be used, but supplementary fertilization is desirable in the latter case. Preferred pH varies from slightly acidic to slightly basic (6.0 to 7.5 is acceptable), although many peonies have a preference for slightly alkaline soil. Peonies are heavy feeders, benefitting from soil that has been fertilized with well-rotted (never fresh) manure worked well into the ground and from annual top-dressing with well-rotted manure (avoiding contact with the plant). Some growers recommend adding inorganic fertilizers, but never high-nitrogen fertilizers; the level of nitrogen should never be higher than the levels of potassium or phosphorus (e.g., 10–15–10 would be suitable). The fertilizer should be applied away from the main crown and distributed on the soil around the drip line of the foliage. Spring application as shoots emerge is desirable, although some gardeners also apply fertilizer in the autumn after the plants have gone dormant. Peonies need excellent drainage. They are intolerant of waterlogging, and if available sites are overly moist, planting in raised beds should be considered.

### 77.8.5 Maintaining Plants

*Peony hoops*. The flowers of most cultivars of the common peony (*P. lactiflora*) are notorious for flopping over and lying on the ground (tree peonies are quite resistant to this problem). This is because the flowers are poorly supported by their relatively weak stems. Wind and rain may cause the flowering stems to bend over, but often, large double-flowered blooms, especially of old-fashioned cultivars, droop to the ground simply because of their own weight. Bamboo stakes set around the plant and tied together by garden twine woven around the plants can be employed to support the plants. However, rust-resistant wire "cages," "hoops," "rings," or "supports" are widely employed to maintain the flowers in an erect position, installed in the spring at least before the plants have reached half their height. Peony hoops designed for the purpose are usually employed. Tomato ring supports should not be substituted because their diameter is too small.

FIGURE 77.16 Peony hoops employed to keep flowers from flopping over to the ground: (a) Hoops supporting peonies in a Japanese botanical garden. (b) A hoop supporting a pre-flowering peony in a garden bed in the UK.

*Source:* **(a) Photo by Tatebayashi Karino Botanen (CC BY SA 3.0). (b) Photo by Shutterstock/Paul Maguire.**

*Pruning and removing foliage in the autumn*. Some gardeners deadhead peonies after the flowers senesce, but this isn't necessary, and some cultivars develop attractive seed heads. Peonies (both herbaceous and woody) don't require pruning, except for damaged stems or branches. Foliage or branches that become senescent during the growing season should be removed to lessen the chances of fungal infection. The most serious disease is botrytis (gray mold), which thrives in cool, humid conditions; diseased portions should be cut away (using pruners dipped in a 10% bleach solution) and discarded. All senescent aboveground material can be removed after significant frost damage in late fall. Herbaceous peonies should be cut about 2.5 cm aboveground. Itoh hybrid peonies should be cut 5–10 cm aboveground. To avoid accumulation of disease organisms, the harvested material should be discarded.

*Winter survival and protection*. Where winter temperatures fall below −20°C, especially where snow cover is limited, an insulating mulch layer (hay and evergreen boughs are suitable) might be necessary to prevent loss of herbaceous peonies. Mulching is advisable in colder areas during the first year of establishment, removing the mulch when the shoots emerge. However, *P. lactiflora* peonies usually survive well with minimal or no winter care in most of the large urban areas of temperate regions. Some tree peonies can survive in the warmest parts of cold countries, but Itoh hybrids are preferable. A burlap wrap can be helpful. In areas that are marginal for growth, the shrubs can simply be cut off at ground level (this will lower the viability of the plants, but some may survive).

## 77.9 FASCINATING FACTS, FABLES, FOLKLORE, AND FRIVOLITY

- According to Chinese legend, the Empress Wu Zetian (ruled 690–705) banished the peony from her imperial gardens because it refused to bloom in winter. The eccentric empress was so incensed that she ordered all peonies of the capital, Chang'an (present-day Xi'an), the most populous city in the world at that time, to be destroyed.
- In the mythology of classical Rome, the woodpecker was viewed as sacred and under the protection of the god Mars, who guarded the peony because it had cured him when he had been wounded ("Woodpecker of Mars" has certain meanings, including the poisonous and hallucinogenic mushroom fly agaric.) According to some ancient Greeks and Romans, peonies could be collected easily at night because they gave off light. This was desirable because it was thought that peonies should only be dug up during darkness, for fear that if the pickers were seen by a woodpecker, the bird might peck their eyes out. This belief persisted into the Middle Ages in Europe, when people collecting peony roots for medicinal purposes were terrified that this might occur.

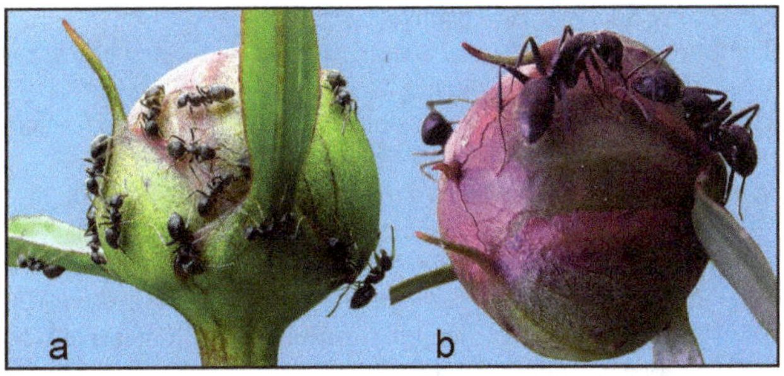


FIGURE 77.17 Ants on peony flower buds.

*Source:* (a) Photo by Shutterstock/Olekry. (b) Photo by Lee Bonnifield (CC BY 2.0).

- Demons were once thought to fly to the spot where bloodred peony flowers bloomed, but a bit of the root worn around the neck was believed to avert enchantment.
- The fourth month of the traditional Chinese calendar was called "Moon of the Peony."
- "Paeonia" is not only the genus name of peonies but also an ancient kingdom located mainly in the modern Republic of Macedonia, as well as part of northern Greece and western Bulgaria. In 168 BC, Paeonia was taken over and incorporated into the classical Roman Empire. The Paeonian language is extinct, and today the old country of Paeonia is known mainly for ancient coins, which are frequently offered by the collector's trade.
- The peony (no particular species) was adopted as the state flower of Indiana in 1957. In 1903, the tree peony (*P. suffruticosa*) was recognized as the official flower of China, but the modern People's Republic of China has not recognized an official national plant (the peony and the plum blossom, traditional emblems of China, are considered to be leading candidates for this honor).
- Ants (and, indeed, other insects) are commonly observed on peony flower buds. Tiny extrafloral nectaries produce a sweet liquid along the outside edges of the bracts covering the developing buds. Apparently, the ants and peonies have established a symbiotic relationship, the peonies furnishing a sweet carbohydrate to the ants, which, in turn, often drive off other insects. There is a common misconception that ants are necessary for peony flowers to open. To get rid of ants on peony buds to be used indoors as cut flowers, simply leave the cut stems in a vase outdoors for a period for the ants to leave.

## KEY PUBLICATIONS


Cheng, F. and X. Yu. 2008. Flare tree peonies (*Paeonia rockii* hybrids) and the origin of the cultivar group. *Acta Horticulturae* 766: 375–382.

Eason, J., T. Pinkney, J. Heyes, D. Brash, and B. Bycroft. 2002. Effect of storage temperature and harvest bud maturity on bud opening and vase life of *Paeonia lactiflora* cultivars. *New Zealand Journal of Crop and Horticultural Science* 30: 61–67.

Fearnley-Whittingstall, J. 1999. *Peonies: The Imperial Flower*. H.N. Abrams, New York. 384 pp.

Giguère, R. 2006. *Les Pivoines*. Editions de l'Homme, Montréal, QC. 320 pp. (In French.)

Halda, J.F. and J.W. Waddick. 2004. *The Genus* Paeonia. Timber Press, Portland, OR. 227 pp.

He, C.N., Y. Peng, Y.C. Zhang, L.J. Xu, J. Gu, and P.G. Xiao. 2010. Phytochemical and biological studies of Paeoniaceae. *Chemistry & Biodiversity* 7: 805–838.

Hong, D.-Y. 2010. *Peonies of the World: Taxonomy and Phytogeography*. Royal Botanic Gardens, Kew, UK. 302 pp.

Hong, D.-Y. 2011. *Peonies of the World: Polymorphism and Diversity*. Missouri Botanical Garden, St. Louis, MO. 94 pp.

Jakubowski, R., D. Hollingsworth, J. Nordick, H. Buchite, and C. Schroer. 2007. *Peonies 1997–2007*. American Peony Society, Gladstone, MO. 216 pp.

Ji, L., Q. Wang, J.A. Teixeira da Silva, and X.N. Yu. 2012. The genetic diversity of *Paeonia* L. *Scientia Horticulturae* 143: 62–74.

Kamenetsky-Goldstein, R. and X. Yu. 2022. Cut peony industry: The first 30 years of research and new horizons. *Horticulture Research* 9: uhac079. https://doi.org/10.1093/hr/uhac079.

Lee, B., Y.W. Shin, E.-A. Bae, S.-J. Han, J.-S. Kim, S.-K. Kang, et al. 2008. Antiallergic effect of the root of *Paeonia lactiflora* and its constituents paeoniflorin and paeonol. *Archives of Pharmacal Research* 31: 445–450.

McGeoge, P. and R. McGeorge. 2006. *Peonies*. David Bateman, Albany, New Zealand. 144 pp.

Nehring, A. and I. Nehring. 1960. *Peonies, Outdoors and In*. Hearthside Press, NY. 288 pp.

Page, M. 2005. *The Gardener's Peony: Herbaceous and Tree Peonies*. Timber Press, Portland, OR. 267 pp.

Qin, L., F.-Y. Cheng, and Y. Zhong. 2013. Advances in the in vitro culture and micropropagation of tree peonies during the past half century. *Acta Horticulturae* 977: 39–51.

Rogers, A. 1995. *Peonies*. Timber Press, Portland, OR. 296 pp.

Sang, T., D.J. Crawford, and T.F. Stuessy. 1997. Chloroplast DNA phylogeny, reticulate evolution, and biogeography of *Paeonia* (Paeoniaceae). *American Journal of Botany* 84: 1120–1136.

Shen, M., Q. Wang, X.N. Yu, and J.A. Teixeira da Silva. 2012. Micropropagation of herbaceous peony (*Paeonia lactiflora* Pall.). *Scientia Horticulturae* 148: 30–38.

Yang, Y., M. Sun, S. Li, Q. Chen, J.A. Teixeira da Silva, A. Wang, et al. 2020. Germplasm resources and genetic breeding of *Paeonia*: A systematic review. *Horticulture Research* 7: 107. https://doi.org/10.1038/s41438-020-0332-2.

Yu, X.-N., P.-P. Guo, G.-P. Lu, and Q.-X. Zhang. 2011. Optimum harvesting time of herbaceous peony buds for cutting flowers. *Journal of Forestry Research* 22: 137–140.

Yuan, J.H., A. Cornille, T. Giraud, F.Y. Cheng, and Y.H. Hu. 2014. Independent domestications of cultivated tree peonies from different wild peony species. *Molecular Ecology* 23: 82–95.

Zhou, S.L., X.H. Zou, Z.Q. Zhou, J. Liu, C. Xu, J. Yu, et al. 2014. Multiple species of wild tree peonies gave rise to the 'king of flowers,' *Paeonia suffruticosa* Andrews. *Proceedings Biological Sciences* 281. https://doi.org/10.1098/rspb.2014.1687.

Zhou, Z.-Q. 2006. Taxonomy, geographic distribution and ecological habitats of tree peonies. *Genetic Resources and Crop Evolution* 53: 11–22.

## Key Cultivation Websites

Note: There are dozens of peony societies. Only some examples are listed.

American Peony Society. www.americanpeonysociety.org/.

Canadian Peony Society. www.peony.ca/.

Garfinkel, A.R. and G.A. Chastagner. 2020. *A Grower's Guide to the Most Common Diseases of Peony in the United States*. Washington State University, Pullman, WA. 16 pp. file:///C:/Users/ottaw/Downloads/FS338E.pdf.

Heartland Peony Society. www.peonies.org/index.shtml.

The Peony Society. www.peonysociety.eu/.

The Spruce. *How to Grow Peonies*. www.thespruce.com/how-to-plant-peonies-1402864.

University of Wisconsin, Extension. *Peony*. https://hort.extension.wisc.edu/articles/peony/.

WikiHow. *How to Divide and Transplant Peonies*. www.wikihow.com/Divide-and-Transplant-Peonies.

WikiHow. *How to Grow and Care for Peonies*. www.wikihow.life/Grow-and-Care-for-Peonies#:~:text=Plant%20them%20in%20your%20garden,blossom%20into%20massive%2C%20beautiful%20flowers.

WikiHow. *How to Get Peonies to Bloom*. www.wikihow.com/Get-Peonies-to-Bloom.

WikiHow. *How to Grow a Peony in a Pot*. www.wikihow.com/Grow-a-Peony-in-a-Pot.

WikiHow. *How to Plant Peonies*. www.wikihow.com/Plant-Peonies.

# 78 Periwinkle (*Catharanthus* and *Vinca* Species)

**Scientific Names:** The genus name *Catharanthus* originates from the Greek *katharos*, "pure, unblemished," + *anthos*, "flower." The genus name *Vinca* is based on the Latin *pervincire* or *vincire*, "to bind or wind around," presumably referring to the way the creeping stems entwine anything in their path.

**Common Names:** The common name periwinkle is derived from the Old English word *perwince*, which, in turn, traces to the same root as mentioned earlier for *Vinca*. The periwinkles are sometimes called "myrtle," due to the resemblance of their leaves to the foliage of the true myrtle (*Myrtus communis*). Names like "blue periwinkle" are ambiguous, because the three principal cultivated species have been bred for flowers of various colors. There are many gardening articles on the web written by authors who seem unaware that the term *periwinkle* refers to different species of two different genera, which have different horticultural needs.

*Catharanthus roseus* is usually called Madagascar periwinkle, but other names include bright eyes, Cape periwinkle, graveyard plant, old maid, pink periwinkle, rose periwinkle, running periwinkle, and running myrtle. The species is often labelled as "vinca" (which can lead to confusion with *Vinca* species) and is often termed "annual vinca" because it is commonly grown as an annual in cool climates.

*Vinca minor* is called creeping periwinkle, dwarf periwinkle, trailing periwinkle, creeping myrtle, and running myrtle. *Vinca major* is called blue buttons, bigleaf periwinkle, greater periwinkle, and myrtle.

## 78.1 INTRODUCTION

*Catharanthus* and *Vinca* are closely related genera in the dogbane family (Apocynaceae), and indeed, in the past, both were classified within *Vinca*. They are perennial herbs (sometimes developing woody bases) known as periwinkles, which serve both as quite beautiful, low-growing flowering ornamentals and as very important sources of medicinal compounds.

## 78.2 OVERVIEW

Seven of the eight species of *Catharanthus* are native to Madagascar, one to India and Sri Lanka. The principal cultivated *Catharanthus*, *C. roseus* (previously known as *Vinca rosea*), has become widely naturalized around the world. It is an evergreen, frost-tender herb or subshrub growing as tall as 1 m, but frequently, it scrambles along the ground. In frost-free climates, the plant often develops a woody stem base. The flowers of the cultivars are lilac, pink, red, or white, often with contrasting throats. Flowering occurs from midsummer to frost. Some cultivars have doubled flowers.

FIGURE 78.1 *Catharanthus roseus.*

*Source:* (a) Mixture of cultivars. Photo by Cbaile19 (public domain). (b) Photo by Forest and Kim Starr (CC BY 2.0). (c) Photo by Carl Lewis (CC BY 2.0). (d) Photo by Jthapelo060 (CC BY SA 4.0). (e) Photo by Manuel M. V. (CC BY 2.0). (f) Photo by Faruk Nafiz Ertürk (CC BY SA 2.0).

FIGURE 78.2 *Vinca minor.*

*Source:* (a) Photo by H. Zell (CC BY SA 3.0). (b) Cultivar Josefine. Photo by Krzysztof Ziarnek, Kenraiz (CC BY SA 4.0). (c) Cultivar White Power. Photo by Krzysztof Ziarnek, Kenraiz (CC BY SA. 40). (d) Cultivar Atropurpurea kz02. Photo by Krzysztof Ziarnek, Kenraiz (CC BY-SA 4.0).

DOI: 10.1201/9781003586227-78

FIGURE 78.3 *Vinca major.*

**Source:** **(a) Photo by Forest and Kim Starr (CC BY 2.0). (b) Photo by Katya (CC BY SA 2.0). (c) Photo by PiPi (public domain).**

*Vinca* species are native to Europe, Northwest Africa, and Southwest Asia. Like *Catharanthus*, the plants are herbs or subshrubs but are usually shorter (no more than 70 cm aboveground), and their slender stems trail along the ground for up to 2 m, often rooting where they contact the soil. The foliage is evergreen in four of the approximately six species, including the two principal cultivated species, *V. major* and *V. minor.* The range of flower colors rivals that found in *Catharanthus. Vinca minor* is a low-growing (usually no higher than 15 cm) perennial producing a thick mat of interwoven trailing stems. It flowers from spring to summer. *Vinca major* closely resembles *V. minor* but is distinguished by (1) larger flowers (2.5–4.0 cm in diameter vs. 1.5–3 cm in *V. minor*), (2) more robust growth habit with more ascending flowering branches, and (3) larger leaves (4 to 9 cm long vs. 2 to 4.5 cm in *V. minor*) which have ciliate (hairy) margins, are not as leathery, and tend to drop or be damaged by very cold weather ("semi-evergreen" as opposed to the evergreen *V. minor*). A form with variegated leaves (i.e., with whitish areas, with little chlorophyll) has often been cultivated. Like *V. minor, V. major* usually flowers from spring to summer.

## 78.3 HISTORY

*Catharanthus roseus* was imported to Europe from Madagascar and grown as an exotic ornamental, often in hothouses, for over 200 years. In the 1950s, vincristine and vinblastine from *Catharanthus roseus* to treat childhood leukemia and Hodgkin's disease enormously increased the importance of the species. In recent decades, considerable breeding has occurred, with the result that a large range of diverse ornamental cultivars of *C. roseus* is now available. In Eurasia, *Vinca minor* and *V. major* were employed as medicinal and sometimes spiritual plants dating back at least to classical Roman times, and as garden ornamentals for many centuries. Both were introduced as ornamentals to North America in the 1700s.

## 78.4 ECONOMIC IMPORTANCE

The periwinkles are popular ornamentals in many countries, especially employed as groundcover. They represent a small part of the horticultural trade of the world, but a very important part of the production of natural medicinal compounds.

FIGURE 78.4 Old paintings (public domain) of cultivated periwinkles by European botanical illustrators: (a) *Catharanthus roseus.* (b) *Vinca minor.* (c) *Vinca major.* (d) *Vinca difformis* (medium periwinkle).

**Source:** **(a) Redouté, P. J. 1833. *Choix des plus belles fleurs et des plus beaux fruits.* Ernest Panckoucke, Paris. (b) Kops, J., and H. C. van Hall. 1832. *Flora Batava*, vol. 6. (c) Gleadall, E. E. 1834. *The beauties of flora.* Gleadall, UK. (d) *Curtis's Botanical Magazine* (London), vol. 139, 1913.**

FIGURE 78.5 Commercial aspects of *Catharanthus roseus*: (a) A field grown for harvest in Vietnam. (b) Production of potted plants for sale.

**Source:** **(a) Photo by Quyen Tran (CC BY SA 2.0). (b) Photo by Shutterstock/Joloei.**

## 78.5 ORNAMENTAL USES

Periwinkles are employed as groundcover, bedding, and edging. They are also grown in containers of different kinds (hanging baskets, window boxes, raised planters). Cut branches can be employed in vase arrangements. Certain cultivars can be grown as houseplants in brightly lit locations, but often, potted plants are grown short-term and discarded after their often long flowering period. Periwinkles have been traditionally grown in cemeteries as an evergreen maintenance-free groundcover. The plants grow well on slopes, sometimes serving to stabilize land and control erosion.

## 78.6 OTHER USES

For millennia, *Catharanthus* and *Vinca* periwinkles were employed in traditional herbal medicines. *Catharanthus roseus* is now grown commercially for the production of "vinca alkaloids" (sometimes called catharanthus alkaloids),

FIGURE 78.6 Landscape uses of *Catharanthus roseus*: (a) A garden of mixed cultivars in the Philippines. (b) A huge bed in a park in South Korea. (c) Contrasting garden beds beside a path. (d) A planter with mixed cultivars.

***Source:*** **(a) Photo by Judgefloro (CC BY SA 3.0). (b) Photo by UniCon (CC BY SA 3.0). (c) Photo by Shutterstock/Foto76. (d) Photo by Shutterstock/Yoshi0511.**

FIGURE 78.7 *Vinca major* in garden borders: (a) In Columbia. (b) In France.

***Source:*** **(a) Photo by Juan Carlos Caicedo (CC BY 4.0). (b) Photo by Michel Chauvet (CC BY SA 4.0).**

FIGURE 78.8 *Vinca major* stabilizing steep slopes against erosion. (a) In Hawaii. (b) In New Zealand.

***Source:*** **(a) Photo by Forest and Kim Starr (CC BY 2.0). (b) Photo by Tim Park (CC BY SA 4.0).**

especially vinblastine and vincristine, employed to treat cancer since the 1950s. Other alkaloids in *Catharanthus* and *Vinca* are also being employed medicinally.

## 78.7 PROBLEMS

The three periwinkle species that are widely cultivated are invasive in some regions. *Catharanthus roseus* is invasive in subtropical and tropical areas of the world (especially Africa) and is considered legally to be a noxious weed in Australia. The plant is extremely toxic and has caused human fatalities. Livestock and pets have also been poisoned. *Vinca major* and *V. minor* have proven to similarly be invasive in Australia, but also in cooler climates, including Canada, New Zealand, and the United States. Both species are considered to be mildly toxic by comparison with the much more toxic *C. roseus*.

## 78.8 CULTIVATION

### 78.8.1 Climate and Site

*Catharanthus roseus* originated in Madagascar, where it is adapted to long, hot, dry, sunny summers. It can be grown as an outdoor perennial in climate zones 10 and 11 but is usually cultivated as an annual in cooler areas. It grows well in full sun but tolerates part shade. Although it is very tolerant of hot, dry summers, cultivars have been selected that

FIGURE 78.9 Invasive *Vinca*: (a) *Vinca minor* in New York State. (b) *Vinca major* in Germany.

***Source:*** **(a) Photo by R. A. Nonenmacher (CC BY SA 4.0). (b) Photo by Eric Zeppenfeld (CC BY SA 3.0).**

FIGURE 78.10 *Vinca minor*, the most cold-tolerant of the periwinkles, covered by snow.

***Source:*** **Photo by Shutterstock/Daily Creativity.**

grow well in the cooler growing conditions of temperate regions. *Vinca minor* is suited to grow as a perennial in climate zones 3 to 8, *V. major* in climate zones 4 to 9, although some cultivars are less tolerant of Northern regions. Both *Vinca* species can tolerate full sun, grow best in part shade, and tolerate dense shade. *Vinca major* is significantly less cold-hardy and slightly less shade-tolerant than *V. minor.*

### 78.8.2 Propagation and/or Purchasing

Periwinkles are often available from nurseries as starter plants in plugs or sometimes as bare-rooted material. The plants can be grown from seeds and reproduced by stem tip cuttings. *Catharanthus roseus* sometimes self-seeds. *Vinca minor* and *V. major* regenerate from small pieces of the underground parts remaining in the ground and, once established, are difficult to eradicate.

### 78.8.3 Planting

Recommended planting distances for the periwinkles vary from 15 to 30 cm between plants. Dwarf cultivars can be planted relatively closely (advice is often provided for specific cultivars).

### 78.8.4 Soil and Moisture Conditions

*Catharanthus roseus* is adapted to dry, well-draining, low-fertility loams and is well adapted to the droughty and nutritionally deficient soils often found in subtropical gardens. Overfertilization produces luxuriant foliage at the cost of flowering. Recommended soil pH is 5.4 to 6.5. Very humid conditions, overwatering, and even overhead irrigation can be detrimental. Although quite drought-tolerant, better flowering results from preventing the soil from becoming very dry.

*Vinca minor* and *V. major* grow best in moist to slightly dry soil. They tolerate various soils, including loam, clay-loam, and rocky ground, but grow best with abundant organic material. The two *Vinca* species, like *C. roseus*, prefer slightly acidic soil (pH 6.0 to 6.5) but grow well with pH levels between 6.0 and 8.0.

### 78.8.5 Maintaining Plants

The periwinkles are considered to be low-maintenance plants. A fungus called phytophthora has caused serious infections of *Catharanathus roseus* in recent decades, and where it is present, it needs to be considered. Recommendations to keep the disease in check include ensuring good drainage, keeping plants at least 30 cm apart, mulching to reduce splashing from rainfall and irrigation water, removal of infected plants, and not replanting in the same bed year after year.

## 78.9 FASCINATING FACTS, FABLES, FOLKLORE, AND FRIVOLITY

- In olden times in England, consuming periwinkle foliage (which is toxic) was recommended to increase the love between a man and his wife (for this and other mythology, see Jun, R. F. 1884. Plant Lore, Legends, and Lyrics. www.gutenberg.org/files/44638/44638-h/44638-h.htm).
- The Italian name for *V. minor* is fiore di morte (flower of death), tracing to a Middle Ages European practice of placing stems of the plant on the heads of those to be executed. Also in Italy, another practice was to include the flowers in funeral wreaths for children, and as noted earlier, periwinkles are highly invasive in cemeteries, so they have several associations with death.
- Scottish clans commonly adopted sprigs of particular plant species as "clan badges" (also called plant badges). They were usually affixed to bonnets or pinned at the shoulder of a lady's tartan sash. *Vinca minor* is the apparent clan badge of the Brodie and Maclachlan clans, and perhaps others.
- The color "periwinkle" was based on the typical flowers of *Vinca minor.* The color was adopted for Crayola crayons in 1958.
- Periwinkle awareness ribbons are associated with advocacy for treatment of numerous medical problems, including eating disorders, pulmonary hypertension, and many cancers.

FIGURE 78.11 Periwinkle awareness ribbon.

*Source:* Niki K, MesserWoland (CC BY SA 3.0).

FIGURE 78.12 Periwinkle snails: (a) Old (public domain) illustration of a coastal tidal pool with marine animals, including periwinkle snails in the center. (b) Common periwinkle (*Littorina littorea*).

*Source:* (a) Wellcome Images. (b) Photo by Kristof Zyskowski (CC BY 4.0).

- As well as labelling plants, the term periwinkle is applied to many seashore and shallow freshwater mollusks, especially of the genus *Littorina*. These are edible and are commonly gathered and sold by "periwinklers."

## KEY PUBLICATIONS


Bean, C. and M.J. Russo. 1988. *Vinca major*. Bugwood Wiki. https://wiki.bugwood.org/Vinca_major.

Bergen, M.A. van and W. Snoeijer. 1996. *Catharanthus roseus*, the Madagascar periwinkle, a review of its cultivars. In: *Wageningen Agricultural University Papers* 96(3). 120 pp.

Chaturvedi, V., S. Goyal, M. Mukim, M. Meghani, F. Patwekar, M. Patwekar, et al. 2022. A comprehensive review on *Catharanthus roseus* L. (G.) Don: Clinical pharmacology, ethnopharmacology and phytochemistry. *Journal of Pharmacological Research and Developments* 4: 17–36.

Curry, H.A. 2012. Naturalising the exotic and exoticising the naturalised: Horticulture, natural history and the rosy periwinkle. *Environment and History* 18: 343–365.

Jhang, T., S. Dwivedi, and S. Sharma. 2022. Classical breeding and trait genetics in *Catharanthus*. In: *The Catharanthus Genome, Compendium of Plant Genomes*, edited by C. Kole, 35–83. Springer, Cham.

Kulkarni, R.N., K. Baskaran, and T. Jhang. 2016. Breeding medicinal plant, periwinkle [*Catharanthus roseus* (L) G. Don]: A review. *Plant Genetic Resources* 14: 283–302.

Landon, A.L. and T.J. Banko. 2005. Propagation of *Vinca minor* by single-node cuttings. *Journal of Environmental Horticulture* 23: 1–3.

Nejat, N., A. Valdiani, D. Cahill, Y.-H. Tan, M. Maziah, and R. Abiri. 2015. Ornamental exterior versus therapeutic interior of Madagascar periwinkle (*Catharanthus roseus*): The two faces of a versatile herb. *The Scientific World Journal* 2015: 982412. https://doi.org/10.1155/2015/982412.

Rani, N. 2022. Botanical features of periwinkle, *Catharanthus roseus*. In: *The Catharanthus Genome, Compendium of Plant Genomes*, edited by C. Kole, 1–14. Springer, Cham.

Rojas-Sandoval, J. 2015. *Catharanthus roseus (*Madagascar periwinkle*)*. CABI Compendium. www.cabidigitallibrary.org/doi/10.1079/cabicompendium.16884.

Shoba, V. and K. Pugazhendy. 2017. Ethnobotanical and phytopharmacological activities of Madagascar periwinkle (*Catharanthus roseus*) – a detailed review. *Indo American Journal of Pharmaceutical Research* 4: 4979–4986.

Stone, K.R. 2009. *Vinca major*, *V. minor*. In: *Fire Effects Information System* [Online]. U.S. Department of Agriculture, Forest Service, Rocky Mountain Research Station, Fire Sciences Laboratory. www.fs.usda.gov.

Van Der Heijden, R., D.I. Jacobs, W. Snoeijer, D. Hallard, and R. Verpoorte. 2004. The *Catharanthus* alkaloids: Pharmacognosy and biotechnology. *Current Medicinal Chemistry* 11: 607–628.


### Key Cultivation Websites


Greenhouse Production News. *Tips for Producing Vinca Bedding Plants* [Catharantus roseus]. https://gpnmag.com/article/tips-producing-vinca-bedding-plants/.

ProMix. *Growing Tough Crops:* Vinca (Catharanthus roseus). www.pthorticulture.com/en/training-center/growing-tough-crops-vinca-catharanthus-roseus/.

The Spruce. *How to Grow and Care for* Vinca minor *(Periwinkle)*. www.thespruce.com/vinca-minor-vines-2132217#:~:text=Tough%2C%20low%2Dmaintenance%2C%20and,known%20for%20its%20creeping%20habit.

The Spruce. *How to Grow and Care for Bigleaf Periwinkle*. www.thespruce.com/bigleaf-periwinkle-vinca-major-profile-5181703.

University of Georgia Extension. *A Guide for Commercial Production of Vinca* [Catharantus roseus]. https://extension.uga.edu/publications/detail.html?number=B1219&title=a-guide-for-commercial-production-of-vinca.

WikiHow. *How to Care for Madagascar periwinkle*. www.wikihow.life/Care-for-Madagascar-Periwinkle#Steps.

WikiHow. *How to Plant* Vinca minor. www.wikihow.life/Plant-Vinca-Minor.

# 79 Petunia (*Petunia*) and Relatives (*Calibrachoa, Petchoa*)

**Scientific Names:** The genus name *Petunia* (and the corresponding vernacular name petunia) is derived from a South American word, perhaps the Paraguayan Guaraní *pety*, applied to another member of the Solanaceae family, tobacco (the Old French word *pétun* means "tobacco"). Petunia is known as petunia or by a very similar word in most Roman alphabet languages. Almost all cultivated petunias (but usually not the calibrachoas) are assignable to the scientific name *P.* × *atkinsiana* (= *P.* × *hybrida*, an incorrect but widely used name).

**Common Names:** Most petunia cultivars are appropriately called "garden petunia" or simply petunia. Some plants with "petunia" in their name do not belong to the genus *Petunia* (notably, "Mexican petunia," species of *Ruellia*, sometimes grown as ornamentals). *Calibrachoa* (and sometimes the hybrid genus × *Petchoa*), discussed in this chapter, are often marketed as petunias and, indeed, are so closely allied and similar that they merit being considered to be petunias (and are so treated by some botanists). Information on the name calibrachoa is provided in the following text.

## 79.1 INTRODUCTION

By the latter half of the 20th century, petunias were so common and cheap that they suffered from widespread lack of respect. However, by the 21st century, the petunia had risen far above its former humble reputation, with the development of an astonishing range of floral and branching forms suitable for diverse applications. Moreover, once considered delicate, breeding has turned the petunia into a rugged, low-maintenance plant that is quite resistant to biotic and climatic stresses. Perhaps more than any other ornamental, the petunia has become the premier outdoor container plant for enhancing all forms of architecture from buildings to statues.

*Calibrachoa* is a closely allied genus of plants, not easily distinguishable from *Petunia*, and often not recognized as a separate genus in older guides to cultivated plants. For several decades, dozens of recently produced cultivars of *Calibrachoa* have been marketed (usually under the names petunia, calibrachoa, or million bells), usually for growth in containers. Most of these have smaller but more numerous and often brilliantly colored flowers compared to "true petunias." For practical purposes (because only experts can reliably distinguish *Petunia* from *Calibrachoa*), calibrachoa is appropriately considered to fall under the horticultural category of petunias. The most beautiful photos of "petunias" (as often shown in this chapter) are calibrachoas.

## 79.2 OVERVIEW

The genus *Petunia* consists of 14 or 15 species of South American herbs and shrubs growing up to 1 m tall. These occur in temperate and subtropical parts of Argentina, Brazil, Bolivia, Paraguay, and Uruguay, with a center of diversity in southern Brazil. Destruction of their native grasslands is currently endangering the survival of some of the narrowly distributed endemic species.

Garden petunias are tender perennials, usually grown annually, either from seed or cuttings. Reports in the literature regarding whether *Petunia* species are annual or perennial can be quite misleading. It is commonly stated that they are perennial but grown as annuals. In fact, in their native situations, most are naturally annual, but when protected in greenhouses, many can be grown as perennials. Some species with a distribution in both areas with freezing winter temperatures and areas with milder temperatures are likely flexible, annual in the former, biennial or perennial in the latter. In the related genus *Calibrachoa*, both annual and perennial species are known.

The chief ornamental feature of petunias is their funnel-shaped (trumpet-like) flowers, which in modern cultivars vary spectacularly in size (up to 15 cm in diameter) and number, whether the corolla is single or double, and in some

FIGURE 79.1 Collections of cultivated petunias.

*Source:* **(a) Photo by Shutterstock/Nina B. (b) Photo by Chris Rycroft (CC BY 2.0). (c) Photo by Dagoberta from Pixabay (public domain). (d) Photo by Shutterstock/Zigzag Mountain Art.**



DOI: 10.1201/9781003586227-79

cases, ruffled or fringed. Coloration may be solid or two-colored ("bicolor"), and there are various variegation patterns, such as spotting, speckling, longitudinal stripes, contrastingly colored venation, and differently colored edges (termed "picotee" and "frost") or centers ("star" patterns). Although petunias are available in virtually all colors (brown is absent, blue and black rare), flowers that are purple (the traditional royal hue) are especially popular (indeed, "petunia" is sometimes used to designate a particular shade of purple). Many cultivars, especially with white or violet flowers, have a pleasant fragrance. The foliage is hairy and sticky.

The wild species are mostly insect-pollinated, except for *P. exserta*, a red-flowered, hummingbird-pollinated species. Not surprisingly, garden petunias are attractive to bees and butterflies. Hummingbirds will visit garden petunias (which are often on lists of recommended hummingbird flowers) but usually find limited nectar available.

The appearance of garden petunias varies considerably. Some plants produce upright or mounding shapes; others cascade attractively from hanging baskets or produce spreading or trailing branches suitable as groundcover.

FIGURE 79.2 Examples of garden petunia flowers (cultivar names, when available, are in single quotes) illustrating the range of solid and mixed colors and single and doubled flowers.

*Source:* (a) 'African Sunset'. Photo by Les Serres Fortier (CC BY 2.0). (b) 'Cascadia Indian Summer'. Photo by Les Serres Fortier (CC BY 2.0). (c) 'Black Velvet'. Photo by Les Serres Fortier (CC BY 2.0). (d) Photo by Serge Melki (CC BY 2.0). (e) Photo by Rosendahl (public domain). (f) Photo by Joe deSousa (CC BY 2.0). (g) Photo by Aftabbanoori, CC BY 3.0. (h) 'Cascadia Rim Magenta'. Photo by Les Serres Fortier (CC BY 2.0). (i) Photo by HitroMilanese (CC BY 3.0). (j) Photo by Steve Slater (CC BY 2.0). (k) Photo by the Greenery Nursery (CC BY 2.0). (l) Photo by Mark Baylor (CC BY 2.0). (m) Photo by Aftabbanoori (CC BY 3.0). (n) 'Surfina Summer Double Rose'. Photo by Les Serres Fortier (CC By 2.0). (o) 'Sweet Sunshine'. Photo by 3268zauber (CC By 3.0). (p) Photo by Bradley West (CC BY 2.0).

Size or vigor also varies greatly among cultivars, ranging from short and compact ("dwarf") to tall. Petunias have been assigned to several standard horticultural categories, described in the following, although hybridization has obscured the differences.

### 79.2.1 Multiflora Petunias

This is the oldest-recognized class of petunias and includes somewhat-compact plants with relatively small flowers (5–7.5 cm wide) that are quite numerous. The plants are comparatively resistant to adverse weather conditions, including very hot, rainy, and windy situations. Multiflora petunias are particularly suitable for massed garden bed plantings in solid or mixed colors. Some cultivars have been bred with larger flowers approaching those of the size of the next category, grandiflora petunias, and it has become difficult to distinguish the two classes.

### 79.2.2 Grandiflora Petunias

This category arose in the 1880s as large-flowered mutants of the multiflora class and has become the most popular group of petunias. Grandifloras now include plants with flowers so large (9–15 cm in diameter) that they are relatively susceptible to wind and rain damage and are therefore more suitable for planters than for growth in garden

FIGURE 79.3 Multiflora petunias, notable for abundant, medium-sized flowers. (a) Sprawling against a brick wall. (b) Petunia 'Madness Royal'.

*Source:* (a) Photo by InAweofGod'sCreations (CC BY 2.0). (b) Photo by David J. Stang (CC BY SA 4.0).

FIGURE 79.4 Grandiflora petunias, notable for very large flowers. (a) Mixed cultivars in a planter. (b) Petunia 'Daddy Orchid'.

*Source:* (a) Photo by P. Kratochvil (public domain). (b) Photo by David J. Stang (CC BY SA 4.0).

beds. Most are upright, typically 30–40 cm in height, with trailing stems that tend to spread.

### 79.2.3 Floribunda Petunias

This group was produced in the 1950s by crossing grandifloras and multifloras, combining the larger flowers of the former and the weather resistance of the latter. Floribundas tend to be intermediate between grandifloras and multifloras, but there has been selection for relatively large flowers (7.5–9 cm).

### 79.2.4 Milliflora Petunias

These are compact, usually miniature plants, with petite blossoms (3.5–5 cm) that are extremely numerous. This group makes good edging and container plants and is suitable for small gardens.

FIGURE 79.5 Floribunda petunias. Petunia 'Stimoryne Sonya' $F_1$ 'Floribunda Picobella'.

*Source:* **Photo by Shutterstock/Volcko Mar.**

FIGURE 79.6 Milliflora petunias. Petunia cultivar Midnight.

*Source:* **Photo by Cephas (CC BY SA 4.0).**

### 79.2.5 Hedgiflora (Spreading, Groundcover) Petunias

These are short (10–15 cm high) but can spread up to 1.5 m. The plants are fast-growing, low-maintenance, and noted for heat and drought tolerance. They are commonly used as groundcover and look spectacular scrambling down sloping hillsides and over retaining walls. They also look attractive tumbling out of containers. Spreading petunias, such as "The Wave," "Suertunia," "Cascadia," and "Surfinia" (each of these includes a series of cultivars), are commonly used as bedding plants, groundcovers, and elevated container plants from which the stems trail downward. "Hedgiflora" is a subcategory of spreading plants which are grown close together to form a hedge 40–60 cm tall.

### 79.2.6 Petunia Relatives: Calibrachoa and Petchoa

The genus *Calibrachoa* (commemorating 19th-century Mexican botanist Antonio de la **Cal y Bracho**), with about two dozen (perhaps as many as 30) species, is very closely

FIGURE 79.7 Hedgiflora (spreading, groundcover) petunias: (a) Petunia 'Tidal Wave Silver'. (b) Supertunia petunia.

*Source:* **(a) Photo by David J Stang (CC BY SA 4.0). (b) Photo by Forrest and Kim Starr (CC BY 2.0).**

FIGURE 79.8 *Calibrachoa.* The phrase "Million Bells" is used as a general name for *Calibrachoa* cultivars as well as a particular series of cultivars: (a) Mixed cultivars. (b) 'Cabaret Hot Pink', a particular cultivar of *Calibrachoa.*

*Source:* **(a) Photo by Weissertier (CC BY 2.0). (b) Photo by Manuel m.v. (public domain).**

FIGURE 79.9 Garden beds made up of particular *Calibrachoa* cultivars. (a) *Calibrachoa* 'Million Bells'. (b) *Calibrachoa* 'Superbells Apricot Punch'. (c) *Calibrachoa* 'Mini Famous Light Pink Eye'. (d) *Calibrachoa* 'Mini Famous White'.

***Source:*** **(a) Photo by Carl Lewis (CC BY 2.0). (b) Photo by Serres Fortier (CC BY 2.0). (c) Photo by Serres Fortier (CC BY 2.0). (d) Photo by Serres Fortier (CC BY 2.0).**

FIGURE 79.10 The two major wild parental species of modern hybrid garden petunias: (a) *Petunia integrifolia*. (b) *Petunia axillaris*.

***Source:*** **(a) *Curtis's Botanical Magazine* (London), 1918, vol. 144 (= series 4, vol. 140), plate 8749. (b) Photo by Magnus Manske (CC BY SA 2.5).**

FIGURE 79.11 Doubled petunia flowers. (a) Unidentified cultivar. (b) Petunia 'Double Pirouette Red'. (c) Petunia 'Trailing Double Julia'.

***Source:*** **(a) Müller, I. (ed.). 1860. *Illustrierte Garten-Zeitung. E. Schweizerbart*, vol. 4, plate 12. Stuttgart, Germany. (b) Photo by HQ (CC BY SA 2.0). (c) Photo by HQ (CC BY SA 2.0).**

related to the genus *Petunia*, occurs substantially in the same regions of South America, and was included in the latter in the past. However, the two genera differ in chromosome number (14 in *Petunia*, 18 or 36 in *Calibrachoa*) and in a variety of anatomical features. Calibrachoas first became popular with the public in the 1990s, and since then, breeding has produced many kinds mimicking the classes of commercial petunias. "Million Bells" refers to a class of *Calibrachoa* hybrids, or to particular cultivars, the name alluding to numerous blooms resembling miniature petunia flowers. (Million Bells® and Super Bells® are trademarked names.) Other names include seaside (or seashore) petunia (a reference to presence in rocky coastal areas) and callies (a name also honoring Antonio de la **Cal y** Bracho).

Intergeneric hybrids between *Petunia* and *Calibrachoa* have been made and are assigned to the genus × *Petchoa* (a name combining elements of ***Pet**unia* and *Calibra**choa***). The foliage of these frequently lacks the objectionable sticky foliage of petunias.

## 79.3 HISTORY

Garden petunias are of hybrid origin, the first hybrid apparently having been generated in 1834 by a British nurseryman. Various species have been suggested to have been used historically to generate modern garden petunias, especially the white-flowered, night-fragrant *P. axillaris* (pollinated nocturnally by hawk moths) and the purple-flowered ("violet-flowered") *P. integrifolia* (*P. violacea*) (pollinated diurnally by bees). These plants were discovered in the late 1700s in South America and soon became cultivated in Europe.

Up until 1930, most petunias cultivars were developed sexually by mass selection (using large numbers of superior plants as breeding parents) and by vegetative propagation (maintaining superior clones). Subsequently, intensive selection for desired characteristics occurred, and the marketing of $F_1$ hybrids became important (first-generation hybrids can be distributed as seeds, which will establish superior plants but will not breed true, or as cuttings, which will remain true to type). Today, most commercially produced plants are generated by seeds, although vegetatively propagated cultivars have become quite popular.

Flowering in petunia is affected by photoperiod, but also by other factors, particularly temperature. Many of the species are "quantitative long-day plants," tending to flower with more than 13 hr of daylight. Flowering of many garden petunias has been selected to be resistant to length of daylight (i.e., although short days retard flowering, the delay has become less significant than in ancestral plants). True, day-neutral cultivars (completely unaffected by photoperiod) do not seem to be available. Because of its commercial importance, very extensive studies have been published on the genetic control of flowering.

As with numerous other ornamental plants, cultivars of petunia have been selected in which the flowers are "doubled" (i.e., extra petals have been added, usually by the conversion of stamens into petals), making the corolla more attractive. Cultivars with consistently doubled flowers first became available in the 1930s.

Although *Calibrachoa* has been popular for only a few decades, the plants have been cultivated for well over a century. Modern cultivated forms are mostly complex hybrids (sometimes labelled as "*Calibrachoa* × *hybrida*" by analogy with the obsolete name *Petunia* × *hybrida*) derived from crossing various species. Numerous varieties are now available.

## 79.4 ECONOMIC IMPORTANCE

The petunia is one of the most popular annual bedding plants in the world. In the United States, where it is the leading bedding plant, more than 140 million potted plants are sold annually, with a wholesale value exceeding $100 million. Petunias are very attractive to consumers because (1) they are usually among the cheapest annual planting stock available from nurseries; (2) they are available in a wide range of attractive growth patterns and flower characteristics and often also are pleasantly scented; (3) they are very easy to grow; (4) they flower continuously and prolifically from spring to autumn, often until killed by a hard frost; (5) they are disease- and pest-resistant; and (6) they are extremely versatile and suitable for diverse circumstances, including large floral beds to produce masses of color, borders for accents, and eye-catching hanging baskets and window boxes.

## 79.5 ORNAMENTAL USES

Petunias and calibrachoas are established, popular cultivated ornamentals employed primarily in beds, borders, and pots (especially hanging baskets). Petunia flowers are available cheaply in large numbers for much of the year and so are widely employed in decorating "floral sculptures" in large gardens and for parade floats.

FIGURE 79.12 Commercial production and sale of petunias and calibrachoas: (a) A greenhouse full of plants. (b) Hanging pots on sale in a nursery.

*Source:* **(a) Photo by Shutterstock/Exclusive Dn. (b) Photo by Shutterstock/Klever Ok.**

FIGURE 79.13 Landscape uses of petunias: (a) The garden of the Château de Chenonceau, a castle in France. (b) Fumio Kishida (Prime Minister of Japan) and Narendra Modi (Prime Minister of India) visiting a garden in Delhi, India, in March 2023. (c) A park in Beijing, China. (d) An enormous field of petunias at a commercial tourist site ("Mother Farm of Pink Sigh") in Futtsu City, Chiba Prefecture, Japan. (e) A border featuring different cultivars.

*Source:* **(a) Photo by Tim Sackton (CC BY SA 2.0). (b) Photo by Government of Japan (CC BY SA 4.0). (c) Photo by Lin Gaozhi (CC BY SA 4.0). (d) Photo by NakaoSodanshitsu (CC BY 2.0). (e) Photo by Shutterstock/Alybaba.**

FIGURE 79.14 Petunias in hanging pots: (a) Lancelot Restaurant in Vail, Colorado, sumptuously decorated with hanging pots of petunias. (b) Large hanging baskets on a light pole on a bridge over the River Thames, Reading, England. (c) A basket on a light pole in the city of St. Croix Falls, Wisconsin.

*Source:* **(a) Photo by rjones0856 (CC BY 2.0). (b) Photo by Shutterstock/Hans Wrang. (c) Photo by Shutterstock/Linda McKusick.**

## 79.6 OTHER USES

Beginning in the latter half of the 20th century, Petunia has become a widely employed model system for studies of gene regulation and genome structure, particularly with respect to reproductive characteristics, such as development

FIGURE 79.15 Petunias in pots.

*Source:* **(a) Photo by Shutterstock/Walter Erhardt. (b) Photo by Shutterstock/Ulianiculaie. (c) Photo by 4028mdk09 (CC BY SA 3.0).**

FIGURE 79.16 Petunias as decorative coverings of "floral sculptures": (a) A female mannequin dressed with petunias in a town square. (b) "Elon Muskox," a muskox sculpture at the front of the City Hall of Yellowknife (the capital city of Canada's Northwest Territories).

*Source:* **(a) Photo by Shutterstock/Zuzha. (b) Photo by Kazuha1029 (CC BY SA 4.0).**

of floral structure, floral pigments, and scent chemicals. Hundreds of research papers in these fields have been published. Among petunia's advantages for such studies are (1) simplicity of growth; (2) short life cycle (4 months from seed to seed); (3) large flowers, facilitating chemical sampling; (4) abundant seed production; (5) capability of both self- and cross-fertilization; (6) possession of an endogenous, very active, transposable element system, facilitating genetic screening; and (7) ease of genetic transformation and regeneration from leaf disks and protoplasts.

## 79.7 PROBLEMS

Although it is one of the world's most popular ornamentals, the petunia has image problems. Because it was the first to

FIGURE 79.17 Petunia in an onion patch.

*Source:* **Drawn by B. Brookes.**

be widely planted as a bedding plant and is one of the easiest to grow, it has become so common that many gardeners prefer less frequently encountered flowers. Indeed, seeds and planting stock are so cheap that many consider the petunia to be low-class, associated with the less-affluent. Even more vulgar are several pejorative sexual meanings associated with the petunia. Finally, the petunia is height-challenged, its low, ground-hugging stature not afforded the respect acquired by taller flowering plants. Despite its universal occurrence, there are almost no societies and festivals dedicated to the petunia, while roses, peonies, lilacs, and others are much more celebrated. The limited esteem in which the petunia has been held for many years is epitomized by the song "I'm a Lonely Little Petunia (In an Onion Patch)" (by J. Kamano, B. Faber, and M. Hartmann, published in 1946), that became wildly popular in the mid-20th century. An extract follows:

> I'm a lonely little petunia in an onion patch
> and all I do is cry all day
> Boo hoo, boo hoo
> The air ís so strong it takes my breath away
> feee – you
> Who put me in this bed?
> I'll bet his face is red
> I call him down with every teardrop that I shed.

## 79.8 CULTIVATION

### 79.8.1 Climate and Site

Petunias can be grown as perennials in climate zone 11 (but note the limitation on heat tolerance, described here) and even as far south as zone 8 as perennials if provided with protection. However, they are cold-sensitive. Petunias should not be planted out until there is limited danger of frost. They grow best at moderate temperatures, preferring cool conditions. Petunias react poorly to extreme heat. The plants are adapted to full sunlight. Partial shade is tolerated, but at least 5 hr of sunlight is recommended, although some shade may be preferable in very hot circumstances.

### 79.8.2 Propagation and/or Purchasing

The simplest way to obtain petunias for planting is to buy young material (plugs or young potted plants). As for all planting stock, look for young, vigorous, disease-free, pest-free plants in soil that is neither excessively wet nor dry. Short, compact plants will establish more readily than leggy, thin plants. Resist the tendency to buy the biggest plants with the most flowers, as these are slower to transplant. Read labels to learn of the plant's characteristics and maintenance requirements; if suitable information is not provided (often the case in supermarkets), purchase from a garden center where expert advice is offered. Many of the most attractive cultivars are sterile or do not produce many seeds and so are necessarily propagated from cuttings. Many cultivars are available that can be raised from seeds (10–12 weeks are required to establish seedlings before planting out; the tiny, dust-like seeds need light to germinate, so they should not be covered; warm temperatures (21–27°C) are best until germination, cooler thereafter (day: 18°C; night: 13–18°C). It should be understood that petunia seeds collected from plants in your garden may be sterile or may not come true to type since they are of hybrid origin (seeds from hybrids often produce plants that revert to the white or purple colors of ancestral plants).

### 79.8.3 Planting

In beds and borders, petunias should be separated by about 30 cm for most cultivars, 15 cm for miniature types, or 45–60 cm for the vigorous spreading kinds. If they are planted in pots, the containers should be large enough to provide room for root development (e.g., depending on variety, a 10 in (25 cm) pot can handle four plants; seven can be placed in a 20 in (50 cm) pot).

### 79.8.4 Soil and Moisture Conditions

Petunias grow best in moderately fertile soil, with a pH of 5.5 to 7.0. Good drainage is necessary (adding organic matter may be required to improve drainage), with soil maintained evenly moist but not waterlogged. Moderate fertilization should be provided (e.g., diluted, balanced fertilizer, such as 10–10–10, every 2 weeks). Deep watering once weekly is recommended for flower beds and borders during dry weather. Pots should be watered when the soil surface becomes dry (petunias of the spreading class, planted in the ground, tend to need frequent watering).

### 79.8.5 Maintaining Plants

Older cultivars should be deadheaded (removal of faded or wilted flowers to prevent the plant's energy going into seed formation) or pruned back to stimulate setting of flower buds, unpleasant tasks, since the foliage is sticky and spent flowers are mushy. Petunias produce new blooms at the ends of stems, so plants tend to become leggy and less productive of flowers as the season progresses, unless stems are cut back. Although newer cultivars may claim to be free of requirements for deadheading and pruning (which is mostly true for spreading petunias), some pinching back or shearing may be desirable in mid-season. Aphids are often present and can be reduced by blasting with water.

### 79.8.6 Calibrachoas

Calibrachoa cultivation is basically like petunias cultivation. Ornamental calibrachoas are hybrids that produce few seeds. Most of the popular cultivars are trademarked and cannot be legally reproduced by cuttings (the same is true for numerous other ornamentals, which are nevertheless often reproduced vegetatively without authorization). High pH can be detrimental to calibrachoa, and it has been recommended that pH be kept between 5.5 and 6.0.

## 79.9 Fascinating Facts, Fables, Folklore, and Frivolity

- Luther Burbank (1849–1926) was a genuine but erratic horticultural genius (often called "the plant wizard") when it came to creating amazing new hybrid varieties of plants. One of his alleged creations was a hybrid between a tobacco species (*Nicotiana wigandioides*) and garden petunia. He called the new plant *Nicotunia* and claimed that it combined the large, attractive flowers of petunia and the large leaves of tobacco. The plant was sold in the 1890s but subsequently disappeared. Whether *Nicotiana* and *Petunia*, in fact, can be hybridized by conventional sexual means is debatable. Some genetic transfer between the two genera is possible by modern technologies (see Glimelius and Bonnett 1986; Hinnisdaels et al. 1991), but apparently not by conventional sexual means.
- "Shanin," a floral preparation of *Petunia integrifolia* (*P. violacea*; one of the ancestors of garden petunias), has been claimed to have been used as a hallucinogen by Ecuadorian Indians. The report has not been validated and probably represents confusion with some of the truly hallucinogenic species in the Solanaceae family, such as those of *Datura* and *Iochroma*. Nevertheless, the rumor that garden petunias are hallucinogenic is widespread.
- Petunia corollas, even those that are white, have been demonstrated to develop functional chloroplasts that provide small but significant photosynthetic reserves to the plant.
- In 1990, geneticist Richard Jorgensen attempted to intensify the purple of petunia flowers by providing

FIGURE 79.18 (a) American horticulturist Luther A. Burbank and his claim to have hybridized petunia and a tobacco (*Nicotiana*) species. (b) Garden petunia, one of Burbank's alleged parents of a questionable hybrid. (c) Wild tobacco (*Nicotiana wigandioides*), the other alleged parent of Burbank's hybrid.

*Source:* **(a) Photo, taken in 1915, from the US Library of Congress Prints and Photographs division. (b, c) Photos from Whitson, J., R. John, and H. S. Williams. (editors). 1914. *Luther Burbank: His Methods and Discoveries and Their Practical Application*. Luther Burbank Press, New York. Vol. 2.**

FIGURE 79.19 (a) Petunia Pig costume. (b) Porky Pig and Petunia Pig at the 108th annual Golden Dragon Parade in Los Angeles in 2007. Sponsored by the Chinese Chamber of Commerce and celebrating Asian culture, the Warner Bros. Looney Tunes Merrie Melodies characters were featured because 2007 was the Year of the Pig.

*Source:* **(a) Photo by Starshyne09 (CC BY SA 2.0). (b) Photo by Jim Winstead (CC BY 2.0).**

the plants with extra copies of a purple pigment gene. Counterintuitively, some of the altered petunias were white or had white patches. The phenomenon, an example of what is now termed "gene silencing," remained a mystery (sometimes dubbed the "petunia mystery") until 1998, when Andrew Fire and Craig Mello discovered that gene silencing often occurs by "RNA interference" (RNAi). In 2006, the pair received the Nobel Prize for Physiology or Medicine for their discovery. *Gene silencing* refers to forms of gene regulation, particularly the reduction of expression of certain genes, and is becoming extremely important for therapeutic purposes. RNAi is an important natural way that cells regulate gene expression and may have evolved as a defense mechanism against cell invaders, such as RNA viruses. RNAi works by destroying messenger RNAs, which transport genetic information for protein synthesis. (For more information, see https://mcb.berkeley.edu/courses/mcb41/41-8.pdf.)

- Among the most popular floral names for females are Daisy, Daphne, Holly, Iris, Lily, and Rose. Although "Petunia" is occasionally found, it has been considered to be rather eccentric, best known for the cartoon character Petunia Pig, the girlfriend of Porky Pig.

## KEY PUBLICATIONS

Alaria, A., J.H. Chau, R.G. Olmstead, and I.E. Peralta. 2022. Relationships among *Calibrachoa*, *Fabiana* and *Petunia* (Petunieae tribe, Solanaceae) and a new generic placement of Argentinean endemic *Petunia patagonica*. *PhytoKeys* 194: 75–93.

Ando, T., H. Kokubun, H. Watanabe, N. Tanaka, T. Yukawa, G. Hashimoto, et al. 2005. Phylogenetic analysis of *Petunia* sensu Jussieu (Solanaceae) using chloroplast DNA RFLP. *Annals of Botany* 96: 289–297.

Butler, E.G., T. Robinson, and R.E. Schultes. 1981. *Petunia violacea*: Hallucinogen or not? *Journal of Ethnopharmacology* 4: 111–114.

Chen, S., K. Matsubara, T. Omori, H. Kokubun, H. Kodama, H. Watanabe, et al. 2007. Phylogenetic analysis of the genus *Petunia* (Solanacaeae) based on the sequence of the *HF1* gene. *Journal of Plant Research* 120: 385–397.

Dole, J.M., B.E. Whipker, and P.V. Nelson. 2002. Producing vegetative petunias and calibrachoa. *Greenhouse Production News*. www.plantgrower.org/uploads/6/5/5/4/65545169/producing_vegetative_petunias_and_calibrachoa_%E2%80%93_greenhouse_product_news.pdf.

Franceschinelli, E.V., J.R. Stehmann, C. Melo-Silva, and N.M. Neto. 2013. Preliminary study on genetic diversity of endemic and threatened species of *Petunia* (Solanaceae). *Darwiniana* 1: 220–226.

Fregonezi, J.N., L. Brandão de Freitas, S.L. Bonatto, J. Semir, and J.R. Stehmann. 2012. Infrageneric classification of *Calibrachoa* (Solanaceae) based on morphological and molecular evidence. *Taxon* 61: 120–130.

Ganga, M., S. Jayalakshmi, V. Jegadeeswari, K. Padmadevi, and M. Jawaharlal. 2011. *Petunia*. In: *Wild Crop Relatives: Genomic and Breeding Resources, Plantation and Ornamental Crops*, edited by C. Kole, 209–212. Springer-Verlag, Berlin.

Gerats, T. and J. Strommer. (eds.). 2009. *Petunia: Evolutionary, Developmental and Physiological Genetics*, 2nd edition. Springer, New York. 445 pp.

Gerats, T. and M. Vandenbussche. 2005. A model system for comparative research: *Petunia*. *Trends in Plant Science* 10: 251–256.

Glimelius, K. and H.T. Bonnett. 1986. *Nicotiana* cybrids with *Petunia* chloroplasts. *Theoretical and Applied Genetics* 72: 794–798.

Griesbach, R.J. 2007. *Petunia*. In: *Flower Breeding and Genetics*, edited by N.O. Anderson, 301–336. Springer, Dordrecht, The Netherlands.

Hinnisdaels, S., L. Bariller, A. Mouras, V. Sidorov, J. Del-Favero, J. Veuskens, et al. 1991. Highly asymmetric intergeneric nuclear hybrids between *Nicotiana* and *Petunia*: Evidence for recombinogenic and translocation events in somatic hybrid plants after "gamma"-fusion. *Theoretical and Applied Genetics* 82: 609–614.

Kanaya, T., H. Watanabe, H. Kokubun, K. Matsubara, G. Hashimoto, E. Marchesi, et al. 2010. Current status of commercial *Calibrachoa* cultivars as assessed by morphology and other traits. *Scientia Horticulturae* 123: 488–495.

Kulcheski, F.R., V.C. Muschner, A.P. Lorenz-Lemke, J.R. Stehmann, S.L. Bonatto, F.M. Salzano, et al. 2006. Molecular phylogenetic analysis of *Petunia* Juss. (Solanaceae). *Genetica* 126: 3–14.

Mäder, G. and L.B. Freitas. 2019. Biogeographical, ecological, and phylogenetic analyses clarifying the evolutionary history of *Calibrachoa* in South American grasslands. *Molecular Phylogenetics and Evolution* 141: 106614. https://doi.org/10.1016/j.ympev.2019.106614.

Shaw, J.M.H. 2007. A new hybrid genus for *Calibrachoa* × *Petunia* (Solanaceae). *Hanburyana* 2: 50–51.

Sink, K.C. (ed.). 1984. *Petunia*. Springer-Verlag, Berlin. 256 pp.

Strazzer, P., B. Verbree, M. Bliek, R. Koes, and F.M. Quattrocchio. 2023. The Amsterdam petunia germplasm collection: A tool in plant science. *Frontiers in Plant Science* 14: 1129724. https://doi.org/10.3389/fpls.2023.1129724.

Vandenbussche, M., P. Chambrier, S.R. Bento, and P. Morel. 2016. Petunia, your next supermodel? *Frontiers in Plant Science* 7: 72. https://doi.org/10.3389/fpls.2016.00072.

Zhang, X., I. Nakamura, and M. Mii. 2008. Molecular evidence for progenitorial species of garden petunias using polymerase chain reaction – restriction fragment length polymorphism analysis of the *CHS-j* gene. *HortScience* 43: 300–303.

## Key Cultivation Websites

GardenersWorld.com. *How to Grow Petunias*. www.gardenersworld.com/how-to/grow-plants/how-to-grow-petunias/.

Laidback Gardener. *2018: The Year of the Calibrachoa*. https://laidbackgardener.blog/2018/01/09/2018-the-year-of-the-calibrachoa/.

North Carolina State University Extension. *Calibrachoa × hybrida*. https://plants.ces.ncsu.edu/plants/calibrachoa-x-hybrida/.

The Spruce. *How to Grow & Care for Calibrachoa (Million Bells®)*. www.thespruce.com/calibrachoa-great-container-plants-4125238.

The Spruce. *How to Grow and Care for Petunias*. www.thespruce.com/how-to-grow-petunias-1402893.

Wikifarmer. *How to Grow Petunias at Home – Care of Petunias*. https://wikifarmer.com/how-to-grow-petunias-at-home-care-of-petunias/.

WikiHow. *Easy Ways to Keep Petunias Blooming All Summer*. www.wikihow.com/Keep-Petunias-Blooming-All-Summer#:~:text=Pruning%20and%20Maintenance,-Download%20Article&text=Pinch%20or%20cut%20each%20dying,the%20rest%20of%20the%20plant.&text=Deadheading%20prevents%20flowers%20from%20going,keep%20blooming%20in%20the%20summer.

WikiHow. *How to Grow Petunias*. www.wikihow.life/Grow-Petunias.

WikiHow. *How to Deadhead Petunias*. www.wikihow.com/Deadhead-Petunias.

WikiHow. *How to Prune Petunias*. www.wikihow.com/Prune-Petunias.

# 80 Phalaenopsis (*Phalaenopsis* Species and Hybrids)

**Scientific Names:** The genus name *Phalaenopsis* is based on the Greek *phalaina*, "moth," and *opsis*, "like," meaning "resembling a moth," reflecting similarity of the white and pink flowers (at least their round, flat floral segments) of many species to moths flying at twilight.

**Common Names:** Phalaenopsis (pronunciations: fayl-eh-NOP-siss; fal-en-OP-sis), phal, moth orchid.

## 80.1 SPECIAL SCIENTIFIC NAMING SYSTEM FOR ORCHIDS

While most well-known cultivated plants have unique common (vernacular) names, because there are so many species of orchids, their common names are frequently based in part or completely on their scientific names (which is why "*Phalaenopsis*" in this chapter is virtually equivalent to "phalaenopsis"). Hybrids are common in orchids, not just between two (or more) species, but sometimes between two (or more) genera, or between grexes ("greges"), which are discussed in the next paragraph.

Because so many (well over 100,000) hybrids have been generated among the species by humans, cultivar names are not systematically registered for orchids, unlike other ornamental plants. To bring order to the nomenclature of hybrid orchids, a special "grex nomenclature" has been adopted for orchid horticultural hybrids (most cultivated phalaenopsis are hybrids). The term "grex" comes from Greek meaning "flock" or "group" and refers to all the offspring as well as subsequent offspring of a hybrid cross. Grex names are chosen and registered by the breeders and are never Latinized. A complete name for a hybrid orchid cultivar usually consists of Genus name + Grex name + Cultivar name, for example, *Phalaenopsis* Hilo Lip grex 'Lovely'. In this case, "Hilo Lip" is a collective name for many cultivars (such names can consist of up to four terms), and 'Lovely' (in single quotation marks) is the particular cultivar. The word "grex" in the name is often omitted in informal usage and, indeed, is not often encountered in most literature or on the web. The Royal Horticultural Society (the URL is provided later) is the international registration authority for grex names.

## 80.2 INTRODUCTION

Many consider orchids to be the most beautiful and charismatic of all ornamental plants. A century ago, these "living jewels" were grown primarily as opulent symbols of wealth and status of the elite classes. Today, many orchids also grace the humblest of residences but still retain their status as the aristocrats of flowering plants. Modern phalaenopsis hybrid orchids are among the most graceful and spectacular, with blooms that last for months, superb adaptation to the low interior light and temperature conditions of homes, and tolerance of considerable maltreatment by those with very limited horticultural skills. Phalaenopsis now accounts for three-quarters of all orchids sold and represents a mass marketing industry with a global value of hundreds of millions of dollars annually. Aside from

FIGURE 80.1 Orchids from the late 19th century, when only the wealthy could afford such fabulous plants. None of these gorgeous flowers is a phalaenopsis, which in those early times was too delicate to survive transportation from wild collection, and too difficult to propagate in large numbers. The 16 orchid species shown in this lithograph are identified at: https://commons.wikimedia.org/wiki/File:Haeckel_Orchidae.jpg.

***Source:*** **Haeckel, E. 1904. *Kunstformen der Natur.* Bibliographische Institut, Leipzig, Plate 74.**

DOI: 10.1201/9781003586227-80

FIGURE 80.2 Botanical illustrations (public domain) of *Cattleya* flowers: (a) *Cattleya labiata*. (b) *Cattleya massaiana*. (c) *Cattleya praestans*. (d) *Laelia* × *Cattleya truffauttiana*. (e) *Cattleya* × *hardyana*. (f) *Laelia* × *Cattleya hippolyta*.

***Source:*** **(a) *Lindenia: Iconographie des orchidées* (Gand, Belgium), vol. 16, 1900. (b) Warner, R., B. S. Williams, T. Moore, and B. S. Williams. 1889. *The Orchid Album*, vol. 8. (c–f) *Dictionnaire Iconographique des Orchidees* (F. Havermans, Brussels), 1896–1907.**

FIGURE 80.3 *Cymbidium* flowers (a, b, c are public domain): (a) *Cymbidium parishii*. (b) *Cymbidium insigne*. (c) *Cymbidium dayanum*. (d) Mixed flowers.

***Source:*** **(a) Warner, R., B. S. Williams, T. Moore, and B. S. Williams. 1882. *The Orchid Album* (London), vol. 1. (b) *Curtis's Botanical Magazine* (London), vol. 136 (ser. 4 no. 6), tab. 8312, 1910. (c) *Curtis's Botanical Magazine* (London), vol. 128 (ser. 3 no. 58), tab. 7863.1902. (d) Photo by Shutterstock/Nnattalli.**

its economic importance, *Phalaenopis* is of considerable botanical interest because of the remarkable adaptations of orchids, and the genus has become the subject of very extensive multidisciplinary research.

In addition to *Phalaenopsis*, commercial orchids include species and hybrids of other genera, particularly *Brassavola*, *Calanthe*, *Cattleya*, *Coelogyne*, *Cymbidium*, *Dendrobium*, *Epidendrum*, *Laelia*, *Lycaste*, *Miltonia*, *Odontoglossum*, *Oncidium*, *Paphiopedilum*, and *Vanda*. *Cattleya* (the corsage orchid, illustrated here) is widely grown as a houseplant. *Cymbidium* (also illustrated here) is a top seller in international trade.

## 80.3 OVERVIEW

The genus *Phalaenopsis* includes about 60 species, mostly epiphytes, like most orchids. Also included are some "terrestrial" (rooted in the ground) species sometimes placed in *Doritis* (some botanists recognize *Doritis* as a separate genus and call the hybrid between it and *Phalaenopsis* "*Doritaenopsis*"). Most *Phalaenopsis* species are tropical or subtropical, but some species (sometimes placed in

FIGURE 80.4 Collections of phalaenopsis cultivars: (a) A private, residential collection. (b) Cultivars at a flower show.

***Source:*** **(a) Photo by Shutterstock/AdaCo. (b) Photo by Shutterstock/InnaFelker.**

FIGURE 80.5 Examples of floral displays of hybrid *Phalaenopsis* cultivars: (a) 'Elegant Deborah'. (b) 'Pandora'. (c) Unidentified cultivar. (d) 'Mino Lightning'.

***Source:*** **(a) Photo by Pressebereich Dehner Garten Center (CC BY ND 2.0). (b) Photo by Roger Fenton (CC BY 2.0). (c) Photo by B. Brookes. (d) Photo by Pressebereich Dehner Garten Center (CC BY ND 2.0).**

*Kingidium*) are montane. *Phalaenopsis* species are native from the Himalayan Mountains of India eastward to the Philippines and northern Australia. They are commonest in the hot, moist lowlands of Asia, from Assam through Burma, the Philippines, Malaysia, and Indonesia.

The leaves of *Phalaenopsis* species are located mostly at the base of the plants. They are typically dark green, leathery, and glossy, but some cultivars develop marbled or gray-green foliage. A single growing stem arises (once per season), the flower clusters located on alternate sides of the stem. Modern *Phalaenopsis* hybrids produce 12 to 20 or more flowers, and the flowering clusters of larger, mature plants tend to develop on branches. The flowers are available in virtually every color except blue but are usually pale pastel shades. White, pink, and striped flowers are widely marketed. The flowers of "harlequins" are usually spotted or blotched. "Multifloras" have abundant but small flowers. "Miniatures" are small plants with small flowers.

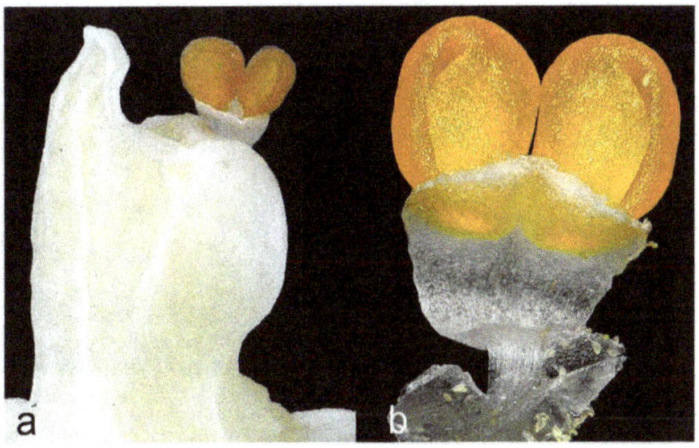


FIGURE 80.7 Sexual parts of a phalaenopsis flower: (a) Column. The white anther cap has been pushed back to reveal the pollinarium of the anther. The paired pollen masses of the pollinarium are yellow. (b) A pollinarium (a pair of pollen collections).

*Source:* **(a) Photo by Rassbak (CC BY SA 3.0). (b) Photo by Florapic (CCC BY SA 4.0).**

### 80.3.1 Floral Structure and Pollination

Orchids have very elaborate flowers adapted for pollination, usually by insects. Phalaenopsis flowers have three outer tepals (sepals), which are more or less identical and evenly spread out. There are two lateral, inner tepals (petals) and a highly modified inner median tepal (petal), the labellum or lip. The lateral petals often differ in coloration, patterning, and size from the labellum, which is three-lipped, the two side lobes raised at an angle to the front lobe. The lateral lobes in most species are usually tinged with yellow nectar guides and red splotches to attract pollinators. A raised area termed the callus occurs at the junction of the three lobes of the labellum. The labellum is fused to the foot of the column (a fusion of the male and female parts). Tropical epiphytic orchids like *Phalaenopis* have a single fertile anther, located at the tip of the column. A cap of tissue covers the anther. To promote cross-fertilization, a thin flap of tissue called the rostellum separates the anther from the stigma beneath. The orchid family (Orchidaceae) and the milkweeds (subfamily Asclepiadoideae of family Apocynaceae) are the only groups of plants that produce pollinia – masses of pollen grains, usually in pairs (a single pair in orchids is called a pollinarium), joined to a sticky pad (called a viscidium), which adheres to pollinators. *Phalaenopsis* has either two or four pollinia, which are transferred as pairs among flowers. A pollinator visiting the throat of a phalaenopsis flower positions itself on the callus, pushes back the anther cap, contacts and picks up a pollinarium while perhaps depositing a pollinarium it may have picked up previously. A single pollination results in hundreds of thousands of tiny seeds being formed, which are distributed by wind. A few phalaenopsis species are scented, but nectar is not produced and insects generally do not consume the pollen, so the flowers are rather deceitfully luring pollinators without rewarding them. Pollination is rather scarce in *Phalaenopsis* and many other orchids, and as a result, the blooms are naturally very long-lived – a characteristic that makes the plants very desirable as flowering ornamentals.

FIGURE 80.6 Parts of a phalaenopsis flower.

*Source:* **(a) Photo by André Karwath (CC BY SA 2.5). (b) Photo by Diegfo Delso (CC BY SA 3.0).**

### 80.3.2 Ecology

Orchids mostly originate from tropical and subtropical areas, with relatively limited seasonal fluctuations. Reflecting their warm, usually continually moist conditions, most species have evergreen leaves. Local temperatures usually range from 28°C to 35°C during the day, and from 20°C to 24°C at night. Most species grow on the vertical trunks of trees in shady areas, in diffused light under a forest canopy. A few species ("lithophytes") grow on rocks, and some occur in seasonally dry or cool locations. Most of the species have a single flowering time, but a few bloom twice a year or irregularly.

About 70% of all epiphytes are orchids, and this ecological niche results in many adaptive features, particularly with respect to water relations. At least in some species of *Phalaenopsis*, early growth stages exhibit conventional (C3)

FIGURE 80.8 *Phalaenopsis* species growing as epiphytes on trees: (a) *Phalaenopsis sumatrana.* (b) *Phalaenopsis amabilis.* (c) *Phalaenopsis speciosa.*

**Source:** **(a)** ***Lindenia: Iconographie des Orchidees*** **(Gand, Belgium), vol. 2, 1888–1891. (b) Blume, C. L. 1835–1849. Rumphia. C. G. Sulpke, Amsterdam. 4 vols. (c)** ***Lindenia: Iconography of Orchids*** **(Gand, Belgium), vol. 1, 1885–1888.**

photosynthesis. However, some (perhaps all) phalaenopsis species have developed Crassulacean acid metabolism (CAM) photosynthesis, which allows plants to keep their stomates closed to prevent excessive water loss during the hotter daylight hours but to absorb $CO_2$ through the opened stomates at night. This is a clever survival tactic for plants in arid environments, and although phalaenopsis generally grows in humid places, its roots have very limited access to water. CAM occurs in about 7% of vascular plants, mostly epiphytes and succulent xerophytes. Orchids with thick leaves usually undergo CAM photosynthesis, while those with thin leaves (less than 1 mm thick) typically perform C3 metabolism. Thick leaves are an adaptation for reducing water loss, and it is not surprising that this is correlated with CAM, which is also an adaptation for reducing water loss.

The majority of *Phalaenopsis* species (indeed, the majority of all orchids) grow as epiphytes on the trunks and limbs of trees, anchoring to the bark by aerial roots (hence the suitability of bark as a growth medium). As noted elsewhere in this chapter, phalaenopsis roots often grow out of the substrate contained in their pots, and these are often termed "air roots." About 300 species of orchids rely exclusively on photosynthesis that occurs in their roots, not their foliage. Air roots of phalaenopsis are somewhat photosynthetic, as revealed by the development of green coloration, and it has been shown that they develop chemicals that protect the exposed surfaces from ultraviolet radiation (Chomicki et al. 2015). The roots of epiphytic orchids absorb moisture, and a spongy, absorbent, hydrophilic coating of hollow thick-walled cells, termed *velamen*, aids in this role as well as in reducing water loss. Velamen cells of old roots are dead at maturity, resulting in a whitish or silvery appearance when dry (reflecting their high content of air; when wet, the velamen can be almost transparent). Young roots appear white when dry, green when moist. Photosynthetic cells are often present underneath the dead velamen and in the root tips, which are not covered with velamen.

## 80.4 HISTORY

During the Victorian era, orchids were "parlor plants," particularly desired by the wealthy of England. However, central heating was not well developed, and the wide fluctuation in temperature and humidity led to species of *Cattleya* and *Dendrobium*, which are resistant to such changes, becoming quite popular. Until the early 20th century, growing orchids from seeds was very difficult, most seedlings succumbing to fungal diseases – ironic, because orchids in nature require infection by a fungal symbiont to survive. Accordingly, most orchids were collected from the wild, and few survived the journey back to civilization. *Phalaenopsis* species proved too delicate, either to survive wild collection or growth in most homes, but became prized specimens that only the very rich could grow in special conservatories. After it was discovered in the early 20th century that orchid seeds germinated only in the presence of a symbiotic fungus, "symbiotic germination technology," involving inoculation of a sterilized growth medium with the fungus, made it easier to reproduce orchids sexually and to produce hybrids. However, large-scale production was

FIGURE 80.9 Some of the phalaenopsis species from subgenus *Phalaenopsis*, which includes the parents of most of the commercial hybrids. *Phalaenopsis amabilis* and *P. aphrodite* are the most significant parents of modern phalaenopsis hybrids. All illustrations are public domain. (a) *Phalaenopsis amabilis.* (b) *Phalaenopsis stuartiana.* (c) *Phalaenopsis schilleriana.* (d) *Phalaenopsis aphrodite.* (e) *Phalaenopsis pulcherrima.*

**Source:** **(a) Warner, R., and B. S. Williams. 1882.** ***The orchid album.*** **Williams, London, UK. (b) Sander, F. 1888–1894.** ***Reichenbachia.*** **H. Sotheran & Co., London, UK. (c) Lemaire, C. A. 1860.** ***L'Illustration horticole*****, vol. 10. (d) Lindley, J. ed. 1838.** ***Edward's Botanical Register*****, vol. 24. (e) Warner, R., and B. S. Williams. 1888.** ***The Orchid Album.*** **B. S. Williams, London, UK.**

still not possible because of continuing fungal infections. In 1922, it was discovered that, even in the absence of the fungal symbiont, orchid seeds could produce viable seedlings if sown on agar with appropriate salts and sugars (so-called "asymbiotic germination technology"). This technique allowed thousands of plants to be established from a single fruit, greatly stimulated the orchid industry, and led to the present domination or the orchid world by phalaenopsis.

Today, most phalaenopsis orchids in the marketplace are hybrids, selected for beauty, rapid growth, prolific blooming, and ease of culture. The parents of most of the hybrids are in subgenus *Phalaenopsis*, including *P. amabilis*, *P. aphrodite*, *P. equestris*, *P. lindenii*, P. *philippinensis*, *P. pulcherrima*, *P. sanderiana*, *P. schilleriana*, and *P. stuartiana*. These are native to warm and humid habitats and are the easiest to grow.

Subgenus *Polychilos* includes *P. amboinensis*, *P. bellina*, *P. gigantea*, *P. lueddemanniana*, *P. venosa*, and *P. violacea*. These also originate from warm, humid habitats and are the parents of some commercial hybrids.

FIGURE 80.10 Some of the phalaenopsis species from subgenus *Polychilos*, which includes the parents of several commercial hybrids. (a) *Phalaenopsis lueddemanniana*. (b) *Phalaenopsis violacea*.

***Source:*** **Public domain illustrations. (a) Linden, L., and Rodigas, E. 1887. *Lindenia: Iconographie des Orchidées*. F. Meyer-van Loo, Gand, Belgium. (b) Linden, L., and Rodigas, E. 1892. *Lindenia: Iconographie des Orchidées*. F. Meyer-van Loo, Gand, Belgium.**

FIGURE 80.11 Miniature *Phalaenopsis* species.

***Source:*** **Photo by Shutterstock/ZnmysteryPhoto.**

Miniature *Phalaenopsis* species from the subgenera *Parishinae* and *Aphyllae*, such as *P. gibbosa*, *P. lobbii*, *P. parishii*, and *P. wilsonii*, originate from seasonally cool and dry areas or relatively high elevations. They require cooler, dryer conditions than most phalaenopsis and are best suited to cultivation by specialists.

## 80.5 ECONOMIC IMPORTANCE

The global trade in orchids is estimated to be about US$500 million annually, most of this in cut rather than potted orchids. The flowering stems of phalaenopsis often are long-lasting and so are suitable for the cut flower industry. The cut orchid trade is dominated by eastern Asia, with Japan being the largest importer. The flowers of phalaenopsis orchids are so attractive they are often used in flower arrangements and corsages, although cattleyas are considered to be the most desirable corsage orchid, and *Cymbidium* and *Paphiopedium* have also been popular. Orchids are estimated to make up about 10% of the international fresh cut flower trade, with more than 40 countries exporting orchids and 60 importing them. Although the Orchidaceae is one of the largest plant families, with at least 25,000 species (representing about 10% of all flowering plant species), it has relatively few species that are commercially significant as ornamentals. *Phalaenopsis* represents the majority of orchid sales, although other important ornamental orchids include *Cymbidium*, *Dendrobium*, *Oncidium*, and *Vanda*.

In most of the Western world, orchids are grown primarily as indoor potted ornamentals, and in the horticultural trade, they are often classified as "tropical flowers" (most, but not all, in fact, originate from the tropics). This horticultural category is also dominated by species of *Anthurium* (anthuriums), *Alpinia* (gingers), *Heliconia* (heliconias),

FIGURE 80.12 Commercial production and sale of phalaenopsis: (a) A greenhouse full of phalaenopsis. (b) Orchids on sale. (c) A florist holding a white phalaenopsis.

***Source:*** **(a) Photo by Shutterstock/Intreegue Photography. (b) Photo by Andrea_44 (CC BY 2.0). (c) Photo by Shutterstock/Ground Picture.**

FIGURE 80.13 Phalaenopsis as a cut flower: (a) Cut stems hanging upside down in a greenhouse, awaiting packaging for transport to markets. (b) A bouquet in a fancy vase.

*Source:* **(a) Photo by Shutterstock/Dirk M. de Boer. (b) Photo by Shutterstock/Sichkarenko.com.**

*Strelitzia* (birds-of-paradise), and members of the Musaceae (banana), Bromeliaceae (pineapple), and Proteaceae (protea) families. Tropical flowers are grown in greenhouses in temperate areas (particularly The Netherlands) and in the open in tropical and subtropical countries (particularly in Africa, Hawaii, Central and South America, Southeast Asia, New Zealand, and Australia). Europe, Japan, and the United States are major importers. Importation of tropical flowers to cold-climate markets is usually more economical than growing them in the country, but it is often possible to ship young plants into Northern countries for "finishing" in local greenhouses prior to marketing.

In the United States, orchids have become the best-selling potted flowering plant, exceeding the poinsettia in annual sales. Taiwan is the world's largest producer of commercial orchids (although The Netherlands makes more revenue from its orchid crop). Most imports of orchids to the United States are from Taiwan, followed by The Netherlands and Thailand.

## 80.6 ORNAMENTAL USES

As a houseplant, many species and hybrids of *Phalaenopsis* offer many advantages: They are relatively easy to grow, are adapted to household temperatures, thrive in low light, bloom for long periods (often for months), and produce elegant sprays of beautiful flowers in a dazzling array of colors. Some cultivars are attractively scented (although most are not fragrant), and many produce interesting glossy green or marbled foliage. Plants are usually inexpensive and are very widely available – from grocery and big-box stores, garden centers, and mail-order suppliers. Although they may be so cheap that they can simply be discarded after flowering, phalaenopsis plants can live for many years. Unlike many of the most attractive plants that demand greenhouse culture, these beauties thrive on windowsills.

## 80.7 OTHER USES

*Phalaenopsis* belongs to the Orchidaceae, which has more than 25,000 species and is often considered to be the second

FIGURE 80.14 Potted phalaenopsis houseplants.

*Source:* **(a) Photo by Shutterstock/All for you friend. (b) Photo by ProFlowers (CC BY 2.0). (c) Photo by Shutterstock/Paul Atkinson.**

FIGURE 80.15 Vanilla plants. These vines are cultivated in tropical regions for their pods ("beans"), from which vanilla is extracted. *Vanilla planifolia* is the most widely grown species.

*Source:* **Public domain illustrations. (a) Vanilla vine, in flower, climbing on a tree. Photo from *Flore des serres et des jardins de l'Europe* (Ghent, Belgium), vol. 17, 1867–1868. (b) Chromolithograph (modified) published in the late 19th century by Bugbee & Brownell, Providence, Rhode Island. (c) Painting by MaX Fulcher (CC BY SA 2.0).**

largest plant family, after the daisy family, the Asteraceae, with perhaps 32,000 species. Aside from the ornamental value of orchids, the family has surprisingly limited economic value. The most important crop orchid is the genus *Vanilla*, several species of which are the source of genuine vanilla spice (artificial vanilla is employed in 95% of vanilla-flavored products).

## 80.8 PROBLEMS

As detailed by Chen (2020), phalaenopsis is a victim of its own commercial success. As the leading orchid in international trade (some claim that it is the world's most successfully marketed houseplant), there is enormous competition, resulting in fluctuating supplies, local overproduction, unpredictable markets, and unprofitability.

## 80.9 CULTIVATION

Orchids have a reputation for being difficult to grow, which is true for some, but not for phalaenopsis. It is relatively easy to cultivate, and there is a vast amount of information about how to do so and, indeed, how to grow other orchids. The information provided here is necessarily condensed (see the literature and websites mentioned later).

### 80.9.1 Climate and Site

*Phalaenopsis* has a relatively long juvenile phase to allow it to grow to a sufficient size to reproduce. Temperature is the primary environmental factor initiating flowering. It is widely and incorrectly believed that low nighttime temperatures induce flowering, but it has been shown that low daytime temperatures are the main stimulus. Commercial greenhouse producers commonly grow phalaenopsis for 1 to 2 years in a day/night temperature fluctuation of 28/26°C to promote vegetative growth and stimulate flowering with a chilling treatment of 30 to 60 days at 21/19°C. In the home, night temperature should not be lower than 15.5°C. The recommended day range is 24–29.5°C. The high temperatures employed by commercial growers produce faster growth but require closer control of humidity and air movement.

FIGURE 80.16 Offshoot plantlets ("keikis") produced by phalaenopsis: (a) Produced by a young vegetative plant of *P. pulchra*. (b) Produced on a flowering stem.

***Source:*** **(a) Photo by Smithsonian Institution (public domain). (b) Photo by Tangopaso (public domain).**

Orchids are basically houseplants, and containers are key to their success. The thick, succulent roots that arise from the base of the stem or lower nodes of phalaenopsis function mainly to adhere epiphytic plants to trees, although they also absorb some water and nutrients and carry out some photosynthesis. However, roots that contact a substrate (in contrast to aerial roots) are much more significant in absorbing water and nutrients and conducting photosynthesis, and so commercial growers prefer plants to dedicate their roots to growth in substrates. Some of the roots of phalaenopsis tend to grow out of their pots, especially evident with old plants. However, it has been shown that when the plants are grown in pots with a suitable medium, they tend to develop more roots in the substrate rather than out of it (i.e., aerially). Plants grown in bark develop more roots outside of the container than plants grown in sphagnum.

Because phalaenopsis mostly exhibits Crassulacean acid metabolism (CAM) photosynthesis, there can be complicated interactions between photoperiodic regime (number of hours of light per day), light intensity, relative humidity, and stage of maturity. These interactions can be important to commercial growers, who need to budget for control of climate and light intensity, but the hobbyist grower need not be concerned.

Growing the plants in greenhouses typically requires 70 to 85% shading. For the hobbyist, a light intensity of between 1,000 and 1,500 fc (foot-candles) (technically, photosynthetic photon flux between 200 and 300 $\mu mol/m^2/s$) usually suffices. Depending on location and time, this can represent between 10% and 15% of full sunlight. In homes, phalaenopsis grows well on a windowsill provided with bright light, but not direct sunlight, which can cause leaf scorch. However, in winter in Northern climates, a full Southern exposure may be best. Alternatively, broad-spectrum fluorescent lights can be employed, placed 15–30 cm above the leaves.

*Phalaenopsis* is considered to be a non-photoperiodic plant, although some varieties seem to require short days to induce flowering. For the orchid hobbyist, ambient seasonal light regimes usually suffice, although supplemental lighting during winter can be beneficial.

### 80.9.2 Propagation and/or Purchasing

Acquiring plants from orchid specialists, either at shows or by mail-order, is a relatively reliable way of ensuring quality. However, most phalaenopsis are cheapest at flower shops, at garden centers, and especially from supermarkets.

Although some orchids are still propagated from seeds (an ambitious undertaking best left to specialists), commercial orchids are mostly grown from cultured meristems (as noted in the following). The cost of phalaenopsis orchids is frequently as low as that of cut flowers, so some consumers simply discard them after blooming. Because it

FIGURE 80.17 Phalaenopsis in transparent containers.

*Source:* **(a) *Phalaaenopsis schilleriana* × *P. stuartiana*. Photo by Maja Dumat (CC BY 2.0). (b) Photo by Javier Mediavilla Ezquibela (CC BY SA 3.0).**

is single-stemmed, phalaenopsis cannot be divided like clump-forming orchids. However, a mature vertical stem that has grown to a reasonable length can be divided to produce new plants (so-called "stem propagation"). The old, remaining original plant is then usually discarded. Many species and cultivars produce plantlets ("keikis;" *keiki* is the Hawaiian word for "baby") with their own roots on the flowering stem, and these can be cut off and potted.

Commercially, there has been extensive research into the technology of orchid production, with emphasis on phalaenopsis. Orchids were the first plants to be propagated in large numbers by tissue culture, and micropropagation has become a standard method for mass production. "Mericloning" of orchids, referring to tissue culture of apical meristems (usually stem apices but sometimes also the growing points of roots), began in the 1960s and has become a very widely employed technique. The meristem is sliced into small segments which are cultivated in sterile nutrient culture in flasks until plantlets are developed to an appropriate size, and these are transferred like seedlings to pots.

### 80.9.3 Planting

In tropical climates, some orchids are planted outdoors in beds or on trees, but for the most part, orchids are grown in pots. Most people obtain their orchids from stores, and the chief planting issue is when and how to repot. Also, cuttings and viviparous plantlets (keikis, as explained in the preceding) may require planting.

Store-bought plants are usually provided in plastic pots. These are less preferable than terra-cotta pots, which dry faster and promote aeration. However, mass-produced clay pots are often relatively impervious, despite their reputation for porosity and not holding water excessively, and specialists often drill extra drainage holes. Pots designed for orchids have large holes at the bottom and slits (elongated holes) on the sides and are often referred to as "azalea pots" because azaleas, like phalaenopsis, prefer well-draining containers. Orchid specialists also employ other alternatives to promote root aeration, including the use of containers made with wickerwork or cloth. Shallower pots and hanging pots are other ways of promoting aeration. Repotting may be required every 1–3 years, when the plant becomes root-bound or the medium needs to be replenished.

When exposed to light, the roots are often greenish, reflecting the occurrence of some photosynthesis, although often they are whitish with age. The roots in transparent containers can also become quite green, although it is unclear how significant this is to the vigor of their growth. Commercial growers of phalaenopsis frequently employ translucent or clear containers, because their use has been shown to promote development of roots into, rather than out of, pots. The roots of retail potted plants are also more resistant to damage if they are inside, not outside, of the pots, so growers try to restrict growth of roots outside of the pots.

### 80.9.4 Soil and Moisture Conditions

Genuine "soil" is rarely employed in orchid compost. The media used to root phalaenopsis are very porous and coarse compared to the soil-like mixes used for conventional houseplants. Until the Second World War, peat was the commonest growth medium. For a period after the war, osmunda fiber (the root of *Osmunda*, especially the royal fern, *O. regalis*) was adopted, and later, bark-based media. Peat-based, soilless media have regained popularity today, although many specialists remain loyal to bark. While a very wide variety of potting media have been employed for phalaenopsis, store-bought plants are usually in peat moss, which requires much less watering than bark, a preferable medium. Commercial orchid potting media, formulated for good drainage, are usually ideal. These may be based on bark chips, hardwood charcoal, coarse peat moss, perlite, or other materials.

Bark-based media can be fertilized with a high-nitrogen fertilizer (because decomposition of the bark by bacteria is thought to use up nitrogen rapidly), while a balanced fertilizer is recommended for other substrates. A high-phosphorus fertilizer may be applied to promote flowering, but the merit of this has been disputed. A weaker-strength fertilization is preferable for cooler, more overcast seasons, when the plants grow less vigorously (although phalaenopsis tends to grow well year-round). Water-soluble commercial orchid formulations are available. To avoid burning roots, the substrate should be somewhat moist before applying fertilizers. Because of the possibility of harming the rather exposed roots, orchid specialists strongly recommend the application of dilute levels of fertilizer, generally quarter-strength. Slow-release fertilizers are also sometimes used. Potting media often show signs of decomposition within

1 year, and if this appears to be the case, the old medium should be removed from the roots and the new medium carefully placed among the roots.

Orchids like phalaenopsis have pseudobulbs (enlarged stems) which store water and so can tolerate periods of drought. However, phalaenopsis species are intolerant of substantial drying out, as they have limited means of storing water (just the succulent leaves) and are mostly adapted to areas that are rainy and misty. Larger pots require less watering, so oversize pots can be helpful (provided they are not overwatered). A bark-based medium dries out rapidly and should be watered every few days. By contrast, sphagnum moss is very water-retentive and does not require watering for at least a week. However, sphagnum tends to produce overly wet conditions, leading to fungal infection and decay. Orchid specialists often add extra drainage holes or expand the drainage holes of standard pots to promote drying. Salt-softened and distilled water should be avoided. Tepid water is recommended, although a frequent practice (sometimes criticized by specialists) is to simply put some ice cubes in the pot (to provide for meltwater over an extended time). Misting the plants can be somewhat helpful, but doing so with hard water can result in deposition of a mineral crust on the foliage. Watering in the morning is preferable, and the foliage should not be moist overnight, to prevent encouraging rot. Potting mixes should be kept evenly moist but allowed to dry out slightly between irrigations.

A humidity of 50–70% is suitable for most hybrid phalaenopsis. Humidifiers or misters are beneficial in dry situations, especially during the winter season, when it is very difficult to maintain suitably high humidity indoors. Nevertheless, the plants are usually just kept on a wet pebble tray with 2.5–5 cm of water, taking care to keep the water level in the tray above the pebbles (and so below the bottom of the pot) to prevent wicking of water to the roots. A disinfectant added to the water will discourage buildup of algae. Air movement is also beneficial. At higher humidities (such as in moist climates), air movement is strongly recommended. Some hobbyists keep plants in a transparent plastic tent in order to maintain humidity, and in this situation, it is especially important to provide air circulation.

### 80.9.5 Maintaining Plants

Phalaenopsis tends to produce flowers in the late winter and early spring but may develop some flowers year-round. After the flowers senesce, the plant can be encouraged to rebloom by cutting the flowering stem 1.3 cm above the first or second node with disinfected pruners. Only large and healthy plants should be stimulated to rebloom repeatedly.

## 80.10 FASCINATING FACTS, FABLES, FOLKLORE, AND FRIVOLITY

FIGURE 80.18 Beautiful women associating with beautiful orchids.

*Source:* **Illustrations from Shutterstock by artists (a) Forewer, (b) Liliya Butenko, (c) Weera S. Design, and (d) Julia Korobeynikova.**

- Orchids have a very long erotic reputation. The root word of "orchid" is the Greek *orkhis*, "testis," a reference to tubers (tuberous roots are present in a few species). The ancient Greeks viewed the plants as the incarnation of Orchis, the lustful son of a satyr and a nymph, who had attempted to rape a priestess and was turned into a flower after his death. Ancient Greek parents believed a man would father sons if he ate large, new orchid tubers, while a woman would give birth to daughters if she ate small tubers. The beauty and exotic nature of orchids led to their being associated with sensuality in the literature and mythology of various cultures. Orchid tubers have long been used as an aphrodisiac to boost a man's libido, and in parallel, the flowers have been employed as an enhancement to the formal attire of women to make them more attractive.
- Orchids vary tremendously in size. The smallest is likely *Platystele jungermanniodes*, which is less than 7 mm long. Several *Vanilla* species are longer than 30 m. *Grammatophyllum speciosum*, thought to be the largest orchid, produces clumps weighing as much as 2 t.
- During the Victorian era, orchids were ostentatious symbols of wealth flagrantly displayed by the rich. The astronomical prices paid to orchid poachers sometimes led to orchid collectors killing their competitors (Orleans 1998). More seriously, immense damage was done to orchid habitats. Some early collectors picked every plant in sight and burned the land to corner the market for a given species. To this day, illegal orchid poaching and smuggling represent dire threats to the survival of numerous species.

- Some species of *Phalaenopsis*, such as *P. violacea*, are unique with respect to petal development following fertilization of the flowers. In almost all plant species, petals senesce after pollination, but in certain phalaenopsis, they rejuvenate, becoming fleshy and green, and apparently producing enough chloroplasts to contribute to photosynthetic production.
- Orchid Island (Lanyu) is a tiny Taiwanese island with a population of about 4,000, named for its local species of *Phalaenopsis* on the southeastern coast of Taiwan. Unfortunately, most of the native flora has been decimated, and the location has become a dump for nuclear waste despite the protests of the mostly Indigenous Tao people. Lesser Orchid Island, a nearby uninhabited volcanic islet, is the refuge of a critically endangered endemic form of *P. equestris*. Military aircraft have used the site for target practice.
- Most orchids require regular attention, and orchid owners who have built up large and valuable collections often find themselves tied down to their "babies." In recent times, the profession of "orchid-sitting" has appeared to allow orchid enthusiasts to be away from home. Similarly, some nurseries offer a "boarding service," reminiscent of parents who send their children to camp while they vacation.
- *Phalaenopsis amabilis* is one of the three national flowers of Indonesia, officially recognized as the "flower of charm" in 1993.

## KEY PUBLICATIONS


Blanchard, M., R. Lopez, E. Runkle, and Y.-T. Wang. 2007. Growing the best phalaenopsis. Part 4: A complete production schedule. *Orchids* 2007(April): 266–271.

Blanchard, M.G. and E.S. Runkle. 2006. Temperature during the day, but not during the night, controls flowering of *Phalaenopsis* orchids. *Journal of Experimental Botany* 57: 4043–4049.

Blanchard, M.G. and Runkle, E.S. 2008. Container opacity and media components influence rooting of potted *Phalaenopsis* and *Doritaenopsis* orchids. *Acta Horticulturae* 788: 115–120.

Chen, C. 2020. The fundamental issue in the *Phalaenopsis* industry in the Netherlands. *Journal of Agriculture and Forestry* 67: 89–100.

Chomicki, G., L.P.R. Bidel, F. Ming, M. Coiro, X. Zhang, X. Wang, et al. 2015. The velamen protects photosynthetic orchid roots against UV-B damage, and a large dated phylogeny implies multiple gains and losses of this function during the Cenozoic. *New Phytologist* 205: 1330–1341.

Christenson, E.A. 2001. *Phalaenopsis: A Monograph.* Timber Press, Portland, OR. 330 pp.

Frowine, S.A. 2008. *Moth Orchids: The Complete Guide to Phalaenopsis.* Timber Press, Portland, OR. 204 pp.

Gordon, B. 1990. *Culture of the Phalaenopsis Orchid*, Revised edition. Laid-Back Publications, Rialto, CA. 188 pp.

Griesbach, R.J. 2002. Development of phalaenopsis orchids for the mass-market. In: *Trends in New Crops and New Uses*, edited by J. Janick and A. Whipkey, 458–465. ASHS Press, Alexandria, VA.

Griesbach, R.J. 2003. Orchids emerge as major world floral crop. *Chronica Horticulturae* 43: 7–9.

Guo, W.-J., Y.-Z. Lin, and N. Lee. 2012. Photosynthetic light requirements and effects of low irradiance and daylength on *Phalaenopsis amabilis. Journal of the American Society of Horticultural Science* 137: 465–472.

Harper, T. 2017. It's all in the genes: The 13 most important *Phalaenopsis* species used in hybridizing. *Orchids* 86(Suppl.): 24–37.

Hessayon, D.G. 2008. *The Orchid Expert.* Transworld, London. 128 pp.

Hsu, C.-C., H.-H. Chen, and W.-H. Chen. 2018. *Phalaenopsis.* In: *Ornamental Crops. Handbook of Plant Breeding*, Vol. 11, edited by J. Van Huylenbroeck, 567–625. Springer, Cham.

Ichihashi, S., T. Higuchi, H. Shibayama, Y. Tesima, Y. Nishiwaki, and K. Ota. 2008. Aspects of $CO_2$ uptake in the crassulacean acid metabolism orchid *Phalaenopsis. Acta Horticulturae* 766: 245–250.

Khosro, B., H. Zakizadeh, J.-A. Olfati, M. Vahedi, A. Kumar, and M. Indracanti. 2019. Recent advances in phalaenopsis orchid improvement using omics approaches. *Plant Tissue Culture & Biotechnology* 29: 133–149.

Lopez, R., E. Runkle, Y.-T. Wang, M. Blanchard, and T. Hsu. 2007.Growing the best phalaenopsis. Part 3: Temperature and light requirements height, insect and disease control. *Orchids* 2007(March): 182–187.

Noble, M. 1994. *You Can Grow Phalaenopsis Orchids*, Revised edition. McQuerry, Jacksonville, FL. 102 pp.

Orleans, S. 1998. *The Orchid Thief: A True Story of Beauty and Obsession.* Ballantine, New York. 284 pp.

Runkle, E.S. 2010. Environmental and hormonal regulation of flowering in *Phalaenopsis* orchids: A mini review. *Acta Horticulturae* 878: 263–267.

Runkle, E., Y.-T. Wang, M. Blanchard, and R. Lopez. 2007. Growing the best phalaenopsis. Part 1: An introduction to potted phalaenopsis orchids. *Orchids* 2007(January): 24–29.

Sweet, H.R. 1980. *The Genus Phalaenopsis.* Orchid Digest, Pomona, CA. 128 pp.

Trépanier, M., M.-P. Lamy, and B. Dansereau. 2009. Phalaenopsis can absorb urea directly through their roots. *Plant and Soil* 319: 95–100.

Tsai, C.C., Y.C. Chiang, S.C. Huang, C.H. Chen, and C.H. Chou. 2010. Molecular phylogeny of *Phalaenopsis* Blume (Orchidaceae) on the basis of plastid and nuclear DNA. *Plant Systematics and Evolution* 288: 77–98.

Wang, Y.-T. 2010. *Phalaenopsis* mineral nutrition. *Acta Horticulturae* 878: 321–333.

Wang, Y.-T., M. Blanchard, R. Lopez, and E. Runkle. 2007. Growing the best phalaenopsis. Part 2: Media, transplanting, water and nutrient requirements. *Orchids* 2007(February): 106–111.

Yeh, Y.-C., C.-Y. Hung, C.-J. Chang, and L.-H. Du. 221. Breeding of fragrant phalaenopsis. In: *Orchid Biotechnology*, Vol. 4, edited by W.-H. Chen and H.-H. Chen, 43–61. World Scientific, Singapore.

Yuan, S.C., P. Bolaños-Villegas, C.Y. Tsao, and F.C. Chen. 2021. The breeding of phalaenopsis hybrids. In: *The Orchid Genome*, edited by F.C. Chen and S.W. Chin, 29–40. Springer, Cham.


### Key Cultivation Websites

Note: There are hundreds of orchid societies and clubs worldwide, and most of these are interested in phalaenopsis orchids since these are the principal commercial forms. There are also thousands of firms supplying orchids. Accordingly, there are numerous websites providing information on phalaenopsis. The following sites are representative.

American Orchid Society. www.aos.org/default.aspx?id=1.

Canadian Orchid Congress. http://canadianorchidcongress.ca/.

De, L.C. 2022. Good agricultural practices of phalaenopsis orchids. *Biotica Research Today* 4(3): 173–177. www.researchgate.net/profile/L-De-2/publication/359337725_Good_Agricultural_Practices_of_Phalaenopsis_Orchids_Open/links/623598395b303e5c5aa86df1/Good-Agricultural-Practices-of-Phalaenopsis-Orchids-Open.pdf.

Orchid Mall. www.orchidmall.com/society.htm (Provides links to orchid societies around the world.)

Orchid Society of Great Britain. www.osgb.org.uk/.

Royal Horticultural Society. www.rhs.org.uk/ (Has numerous documents dealing with orchids.)

Royal Horticultural Society. *How to Grow Phalaenopsis (Moth Orchids).* www.rhs.org.uk/plants/phalaenopsis/growing-guide.

The Spruce. *How to Grow and Care for Phalaenopsis Orchids In.doors.* www.thespruce.com/phalaenopsis-orchids-definition-1902866.

University of Maryland, Extension. *Care of Phalaenopsis Orchids (Moth Orchids).* https://extension.umd.edu/resource/care-phalaenopsis-orchids-moth-orchids/.

WikiHow. *How to Care for Phalaenopsis Orchids (Moth Orchids).* www.wikihow.com/Care-for-Phalenopsis-Orchids-(Moth-Orchids).

# 81 Phlox (*Phlox* Species and Hybrids)

**Scientific Names:** The genus name *Phlox* is based on the Greek *phlox*, "flame," descriptive of the intense color of some of the plants.

**Common Names:** The common name phlox, like numerous economically important genera, is based simply on the genus name.

## 81.1 INTRODUCTION

Phlox species are very attractive flowering ornamentals, with long-lasting blooms available in an extensive range of colors and many cultivars suitable for numerous garden and landscape usages.

## 81.2 OVERVIEW

The 70 or so species of *Phlox* include both perennials and annuals, herbs and semi-woody shrubs, almost all native to North America (the only exception is *P. sibirica* of Russia). They occur in many habitats, including alpine tundra, open woodlands, and prairies. Depending on species, flowering occurs in spring, summer, or fall. Flower colors include blue, pink, red, violet, and white. The flowers are frequently fragrant. Many of the species and hybrids are grown as ornamentals, particularly the three mentioned in the following.

FIGURE 81.1 Garden phlox (*Phlox paniculata*).

*Source:* (a) Photo by Shutterstock/Pelevina Ksinia. (b) Photo by Bff (CC BY SA 3.0). (c) Cultivar Red Riding Hood. Photo by Forssman (CC BY SA 2.0).

Garden phlox (*P. paniculata*, also called fall phlox, perennial phlox, summer phlox, and panicled phlox) is the most commonly grown tall phlox species. It is an herbaceous, clump-forming perennial growing erect, ranging from 0.5 to 1.8 m in height, depending on cultivar. There are numerous cultivated varieties (mostly hybrids with other species). Garden phlox produces large, showy, fragrant clusters of flowers in dense, pyramidal, terminal clusters. The plants are available in a wide range of pastel to brilliant colors, including lavender, pink, orange, purple, red, rose, and white, and some are bicolored. Garden phlox usually flowers from midsummer to autumn, sometimes blooming for over 6 weeks. This traditional, sweetly perfumed garden favorite is sometimes found persisting untended at old homesites.

Moss phlox (*P. subulata*) is also called creeping phlox, moss pink, and mountain phlox. A similar but less-cultivated species, *P. stolonifera*, is also called creeping phlox, although most of the time the name creeping phlox refers to moss phlox. Moss phlox is low-growing (mat-forming) and is usually planted as a perennial groundcover. It grows no taller than 15 cm, producing slim, needle-like, evergreen or semi-evergreen leaves. The stems become woody as they age. Flowering occurs from late spring to summer. The blooms range widely in color and are sometimes bicolored.

Annual phlox (P. *drummondii*) is the only commonly planted annual phlox (some varieties can persist as short-term

FIGURE 81.2 Moss phlox (*Phlox subulata*).

*Source:* (a) Photo by Homelka (CC BY SA 3.0). (b) Photo by Famartin (CC BY SA 4.0). (c) Cultivar Kimono Pink-White. Photo by Salicyna (CC BY SA 4.0). (d) Cultivar Coral Eye. Photo by Andrey Zharkikh (CC BY 2.0).



DOI: 10.1201/9781003586227-81

FIGURE 81.3 Annual phlox (*Phlox drummondii*).

*Source:* **(a) Photo by Shutterstock/Bhupinder Bagga. (b) Photo by Raffi Kojian (CC BY SA 3.0). (c) Photo by Joanna Boisse (CC BY SA 4.0).**

perennials). It grows as tall as 50 cm. Its flowers have a wide color range, including blackish, cream, lilac, pink, purple, red, rose, and white. Cultivars with doubled flowers are available.

*Hesperis matronalis* is an herbaceous biennial or short-lived perennial known as dame's rocket and by other names. This native of Eurasia has been widely grown as a garden ornamental and has escaped to become an invasive weed in many locations, especially in North America. In North America, it is often confused with phlox species (dame's rocket has alternately arranged leaves and flowers with four petals, phloxes have opposite leaves and flowers with five petals).

FIGURE 81.4 Illustrations from covers of old garden catalogues featuring phlox cultivars. (a) John Lewis Childs 1891 seed catalogue, Floral Park, NY. The cultivars shown are of annual phlox (*Phlox drummondii*). (b) 1914 catalogue of Miss Ella V. Baines, the woman florist, Springfield, OH. The cultivars shown are of garden phlox (*P. paniculata*). (c) Dreer's 1902 garden calendar, Philadelphia, PA. The cultivars shown are of garden phlox (*P. paniculata*).

*Source:* **Public domain.**

FIGURE 81.5 Landscape uses of phlox: (a) A border with garden phlox (*Phlox paniculata*). (b) A rock garden with garden phlox (*P. paniculata*). (c) A large pot with moss phlox (*P. subulata*). (d) An elevated planter with moss phlox (*P. subulata*).

*Source:* **(a) Photo by Babij (CC BY SA 2.0). (b) Photo by Virginia Arboretum (CC BY ND 2.0). (c) Photo by Abirdv (CC BY 2.0). (d) Photo by Wendy Harman (CC BY 2.0).**

## 81.3 HISTORY

Garden phlox and creeping phlox were brought back to Europe by early explorers and have been in cultivation there since the late 1700s. Many of the early cultivars originated in England and Germany. Annual phlox wasn't described until 1835, but afterward it also became popular.

## 81.4 ECONOMIC IMPORTANCE

Phlox is a well-established, popular ornamental of moderate value to the horticultural industries of temperate climate countries.

## 81.5 ORNAMENTAL USES

Garden phlox is the most widely grown phlox species. It is planted around homes and employed in borders, massed plantings, and mixed ornamental gardens. Annual phlox is versatile in its uses, adaptable to flower beds, borders, planters, hanging baskets, and rock gardens. Rarely, it is grown as a temporary houseplant. Moss phlox is often employed in rock gardens, in containers, and as edging at the front of gardens. In spring and early summer, moss phlox festivals are held in Japan. Countless blooming plantings of *Phlox subulata* are arranged in intricate patterns and swirls. They are also superb for cut flowers. Phloxes are remarkably attractive to birds, butterflies, and bees and are commonly employed in gardens intended for wildlife.

**FIGURE 81.6** Moss phlox (*Phlox subulata*) spring festivals in Japan.

*Source:* **(a, b, c) Photos by Keihin Nikke (CC BY SA 3.0). (d) Photo by Ajari (CC BY 2.0).**

**FIGURE 81.7** Cut phlox flowering stems for use in vases: (a) A border collie with a basket of flowers. (b) A vase of flowers.

*Source:* **(a) Photo by Shutterstock/Eoghan. (b) Photo by Casablanca1911 (CC BY SA 3.0).**

## 81.6 OTHER USES

In areas of the United States where phloxes are native, they are occasionally employed in roadside beautification.

## 81.7 PROBLEMS

Powdery mildew (caused by several fungal diseases) is a significant hazard for garden phlox (much more so than the other main cultivated species), although recent cultivars have usually been bred for resistance.

## 81.8 CULTIVATION

The most commonly planted species are discussed here, but it should be understood that many of the other *Phlox* species and some hybrids are available that are especially adapted to cooler, shadier, or drier situations.

### 81.8.1 Climate and Site

Most phlox species are hardy quite far north. The commonly planted groundcover moss phlox (*P. subulata*) can

**FIGURE 81.8** Pollinators visiting phlox flowers: (a) A female ruby-throated hummingbird (*Archilochus colubris*) visiting a flower of garden phlox (*Phlox paniculata*) in Illinois. (b) A bumblebee visiting *P. paniculata*. (c) Monarch butterfly visiting phlox. (d) A clearwing hummingbird moth visiting phlox. (e) A green sweat bee on garden phlox in Missouri. All the other insects have a long proboscis that can reach the nectar, and the hummingbird's bill can do the same. Small bees such as sweat bees have a proboscis that is too short to reach the nectar, but they can pick up pollen. (f) An eastern tiger swallowtail butterfly visiting garden phlox.

*Source:* **(a) Photo by Shutterstock/Danita Delimont. (b) Photo by Karen Hine (public domain). (c) Photo by Paul VanDerWerf (CC BY 2.0). (d) Photo by ksblack99 (public domain). (e) Photo by Debbie Koenigs/US Fish and Wildlife Service (public domain). (f) Photo by Michele Dorsey Walfred (CC BY 2.0).**

**FIGURE 81.9** Annual phlox (*Phlox drummondii*), a native of Texas, purposely sown along Texas highways for roadside beautification.

*Source:* **Photo by Malcolm Manners (CC BY 2.0).**

be grown in hardiness zones 3 to 9 (sometimes also zone 2). Garden phlox (*P. paniculata*) is rated for zones 4 to 8. The annual *P. drummondii* can be grown throughout this range, since it need not survive the winter (it has been rated as growable as an annual in zones 2 through 11).

Garden phlox is best placed in a sheltered spot with full sun or partial shade. If planted near trees and shrubs, it will suffer from competition for moisture and nutrients. If planted next to solid fences or walls, where air circulation is poor, it may suffer from disease problems. Annual phlox

FIGURE 81.10 Phlox plants infected with powdery mildew fungus. The insert shows a mildewed leaf.

*Source:* **Photo by Shutterstock/SusaZoom.**

is usually a smaller plant ranging up to a height of 50 cm and can be treated like most sun-loving annuals of its size class. Moss phlox should be planted in full sun but is often placed under trees. Since it is very intolerant of wet soils, a slope is ideal.

### 81.8.2 Propagation and/or Purchasing

Many phlox cultivars will not reproduce true from seeds and are increased by division. Creeping phlox is easily propagated by division but can also be reproduced by stem cuttings and rooted stems. Often established phlox plantings self-seed in their location. Garden phlox and creeping phlox are typically purchased as potted plants from nurseries and garden centers and are also often available as bare-rooted plants for spring planting. Annual phlox is grown from seed.

### 81.8.3 Planting

The perennial phloxes should be spaced 45 to 90 cm apart. Tall varieties may need staking. Garden phlox is subject to fungal diseases, especially powdery mildew, and benefits from air circulation among the plants. Recommended spacing for annual phlox is 30 cm. Creeping phlox grows quickly to fill in interplant areas, so initial spacing is not important.

### 81.8.4 Soil and Moisture Conditions

Phloxes are adapted to well-drained, circumneutral (pH 6.0 to 7.0), organically rich soils maintained moist. Moss phlox can grow in sandy or rocky soil but grow better in loam.

### 81.8.5 Maintaining Plants

The perennial tall summer phloxes are best divided every 3 or 4 years. After the plants die down in the autumn, garden phlox should be cut to about 5 cm from the ground, and the materials discarded, to avoid mildew the next year at the site.

## 81.9 FASCINATING FACTS, FABLES, FOLKLORE, AND FRIVOLITY

- The odor of moss phlox (*P. subulata*) has been mistaken for that of marijuana, resulting in unwarranted police raids.
- The fictional humanoid Dr. Phlox of the 2001–2005 television series *Star Trek: Enterprise* (set in the 22nd century) played the role of chief medical officer. Originating from the Denobulan race (home planet Denobula Triaxa), he practiced a combination of futuristic medicine with primitive techniques from his homeland.
- "Nativars" are cultivars (particular domesticated varieties of a species or hybrid) intended to be grown in the area where their parent(s) are native. They are alleged to be better (than cultivars of foreign origin) for the local ecology because they interact more sustainably with local species (such as pollinators and herbivores) and are better adapted to the local climate (so they need less water, energy, and fertilizer to maintain). The concept is controversial but has sales value for phlox in North America, where it is native and where ecological sustainability has become a major concern. More useful is the practice of simply cultivating local wild species (especially rare ones) as ornamentals, which is much more likely to be ecologically beneficial than growing domesticated species, which almost always require appreciable energy to maintain and often become invasive weeds.

## KEY PUBLICATIONS


Bendtsen, B.H. 2019. Thinking outside the phlox. *Plant Review* 1: 48–49.

Farinas, C. 2020. *Understanding the Powdery Mildew Disease of the Ornamental Plant Phlox: Combining Applied and Basic Research*. Ph.D. dissertation. The Ohio State University, Columbus, OH. 120 pp.

Grant, V. 1959. *Natural History of the Phlox Family: Systematic Botany*. Springer, Dordrecht. 280 pp.

Harrison, C.S. 1910. *A Manual on the Phlox: For the Nurseryman and Amateur, Showing How to Grow Them, Also How to Originate New and Choice Varieties*. Webb Publishing Company, ST Paul, MN. 40 pp. https://books.google.ca/books?hl=en&lr=lang_en&id=2lpCAQAAMAAJ&oi=fnd&pg=PA8&dq=+Phlox&ots=ey06r3d0wJ&sig=KJBwjNchMTtWKQau1MXYl9DduBw#v=onepage&q=Phlox&f=false.

Hübner, I. and E. Murashko. 2017. *Summer-Flowering Phlox. Cultivars: Complete Checklist*. Deutschen Gartenbau-Gesellschaft, Berlin. 128 pp.

Locklear, J.H. 2011. *Phlox: A Natural History and Gardener's Guide*. Timber Press, Portland, OR. 315 pp.

Nevison, K. 2016. *The Role of Native Cultivars in the Ecological Landscape: Evaluating Insect Preferences and Nectar Quality in Phlox and Its Cultivars.* www.ecolandscaping.org/01/designing-ecological-landscapes/native-plants/the-role-of-native-cultivars-in-the-ecological-landscape-evaluating-insect-preferences-and-nectar-quality-in-phlox-and-its-cultivars/.

Rose, J.P. 2018. *Perspectives on* Phlox*: Evolutionary Trends in Polemoniaceae (Ericales) at Four Taxonomic Scales.* Doctoral thesis. The University of Wisconsin-Madison. 222 pp.

Wherry, E.T. 1955. *The* Genus Phlox. Morris Arboretum Monographs, Philadelphia, PA. 174 pp.

Zale, P.J. 2018. Developments in phlox in the U.S. *Plantsman* 17: 44–51.

## Key Websites

Clemson University Home & Garden Centre. *Phlox.* https://hgic.clemson.edu/factsheet/phlox/

GardenersWorld.com. *How to Grow Phlox.* www.gardenersworld.com/how-to/grow-plants/phlox-to-grow/

Gardening.org. *Phlox Full Growing Guide (Everything You Need to Know).* https://gardening.org/phlox-full-growing-guide/

Hawke, R.G. 2011. A comparative study of *Phlox paniculata* cultivars. *Plant Evaluation Notes* 35: 1–10. www.chicagobotanic.org/downloads/planteval_notes/no35_phloxpaniculata.pdf

Iowa State University, Extension. *Growing Phlox in Iowa.* https://hortnews.extension.iastate.edu/growing-phlox-iowa

New York Botanical Garden. *What Are the Best Phlox Plants Available, and the Most Mildew-Resistant Varieties?* https://libanswers.nybg.org/faq/222934

The Spruce. *How to Grow and Care for Garden Phlox (Tall Phlox).* www.thespruce.com/tall-garden-phlox-4117536

University of Minnesota, Extension. *Tall Garden Phlox.* https://extension.umn.edu/flowers/tall-garden-phlox

WikiHow. *How to Divide a Phlox Plant.* www.wikihow.com/Divide-a-Phlox-Plant

WikiHow. *How to Grow Phlox.* www.wikihow.com/Grow-Phlox

# 82 Plumeria (*Plumeria* Species and Hybrids)

**Scientific Names:** The genus name *Plumeria* commemorates Charles Plumier (1646–1704), an influential French botanist and Catholic monk of the Franciscan order who explored the New World.

**Common Names:** Plumeria is the most frequent English name encountered. The name "frangipani" is also often employed, widely interpreted as having originated from a 16th-century marquis of the Frangipani family in Italy who prepared a synthetic plumeria-like perfume for gloves. (The account is likely fictitious – note the 2019 publication by Tiernan.) The name "champak" is occasionally encountered, especially in eastern India and Bangladesh, reflecting identity of plumeria as one of the varieties of sacred champak flower recognized in Hinduism. Less-common names include nosegay, West Indian jasmine, and pagoda tree. "Native frangipani" is *Hymenosporum flavum*, a rainforest tree native to New Guinea, Queensland, and New South Wales. It is widely planted as an ornamental in Australia and can be confused with *Plumeria*. *Plumeria* (especially *P. obtusa*) is often grown as bonsai, leading to the following confusion. Another member of the dogbane family (Apocynaceae), *Adenium obesum* (best called desert rose), is also grown as bonsai, looks similar, is sometimes known as Japanese frangipani, and is often misidentified as plumeria. Climbing frangipani (*Chonemorpha penangensis*) is an unrelated vine.

FIGURE 82.1 *Plumeria rubra*. The name is applied to a wide variety of plants of hybrid origin created by plant breeding and represents the most common kind of cultivated plumeria. The "species" is suspected to exist only in cultivation and as escapes from cultivated plants. Although white flowers are most common, other floral colors are also found: (a) In Hawaii. (b) In Israel.

*Source:* **(a) Photo by Forest and Kim Starr (CC BY 2.0). (b) Photo by Zachi Evenor (CC BY SA 4.0).**

FIGURE 82.2 *Plumeria alba*: (a) In Columbia. (b) In Pakistan.

*Source:* **(a) Photo by Burkhard Mücke (CC BY SA 4.0). (b) Photo by Anbarasu Thiraviyam (CC BY SA 4.0).**

FIGURE 82.3 *Plumeria obtusa*, sometimes called "Singapore graveyard flower."

*Source:* **Photos taken in Hawaii by Forest and Kim Starr (CC BY 3.0).**

## 82.1 INTRODUCTION

Plumerias are easily grown shrubs and trees which produce numerous, long-lasting, gorgeous, colorful, extremely sweet-smelling flowers. With effort, they can be grown as portable plants that can be overwintered indoors, but these true tropical species will mostly be encountered in the hotter regions of the world.

## 82.2 OVERVIEW

*Plumeria* species are native to the Americas from Florida to Brazil. The number of species is uncertain, most specialists recognizing about a dozen. Identification of the species is challenging, and it appears that the classification of the genus is in need of study. There is some (controversial) evidence that the plants travelled from their native area in pre-Columbian times to other areas in the tropical world, where they seem to have been cultivated for many centuries prior to European colonization of the New World. Plumerias have become well-naturalized outside of their American homelands, particularly in Asia, Australia, and the Pacific Islands. Some species are evergreen or semi-evergreen, others deciduous (moist conditions promote retention of

DOI: 10.1201/9781003586227-82

FIGURE 82.4 Flower colors in plumeria.

*Source:* **All photo by John (CC BY SA 2.0).**

FIGURE 82.5 Eighteenth- and nineteenth-century paintings (public domain) of cultivated plumerias by European botanical illustrators. The addition of colorful butterflies and birds was intended to make the plates useful, not just for botanical magazines, but also as wall decorations: (a) Garden hybrid. (b) *Plumeria rubra*. (c) *Plumeria alba*.

*Source:* **(a) Merian, M. S. 1705. *Veranderingen der Surinaamsche Insecten*. Photo from University of Amsterdam. (b) Painting by German-born artist Georg Dionysius Ehret (1708–1770), who did much of his work in England and was considered by many to be the best botanical artist of his time. (c) Forbes, J. 1813. *Oriental memoires*.**

foliage). The leaves are leathery or fleshy, borne in clusters near the branch tips. The plants tend to have widely spaced, thick, succulent, brittle branches. The waxy flowers have tubular corollas, usually 5 to 10 cm long, with overlapping petals that give the flower a swirling or pinwheel appearance. Flower colors include orange, pink, red, white, yellow, and orange, and many blooms are bicolored. The flowers are most often white and very fragrant, especially at night, to attract moths for pollination. Their floral scent is reminiscent of citrus, gardenia, and jasmine.

## 82.3 HISTORY

Spanish explorers are presumed to have been the first to bring back plumeria to the Old World. Catholic missionaries are thought to have been important in transporting it to various parts of the world. Plumerias are extremely embedded in the culture of much of Asia, and it has been suggested (on the basis of old sculptures and art interpreted as plumeria) that somehow they arrived in Asia prior to Spanish colonization of the Americas. Plumerias are described in early British gardening books of the 1800s. Plumeria was introduced into Hawaii in 1860 and has become so entrenched that it is commonly thought to be native there. Similarly, the plants were introduced into Australia (at least by the late 1800s), where they also are widely established.

## 82.4 ECONOMIC IMPORTANCE

Plumeria is a popular flowering shrub in tropical areas, where it contributes significantly to horticultural industries. The flowers are produced on a substantial scale to produce leis and other floral decorations, and for extracts (mostly essential oils).

## 82.5 ORNAMENTAL USES

Plumeria are grown as ornamentals in tropical and subtropical areas worldwide for their attractive, fragrant flowers. In

FIGURE 82.6 Plumeria grown in containers. (a) *Plumaria obtusa*. (b) Unidentified species.

*Source:* **(a) *Plumaria obtusa*. Photo by Shijan Kaakkara (CC BY SA 4.0). (b) Photo by Shutterstock/You Touch Pix of EuToch.**

hot climates, they are grown as landscape plants but can be cultivated in large tubs. In cold areas, they need to be in containers that can be brought indoors when temperatures approach freezing. As an indoor plant, plumeria is much better suited to greenhouses than as houseplants, because they keep outgrowing their containers. There are dwarf forms, but they do not flower as well. The related desert rose (*Adenium obesum*) is similar but more suited to small containers.

*Plumeria* species have acquired considerable religious significance in Asia. They are often planted around both Hindu and Buddhist temples (the plants are sometimes called temple trees), and flowers are harvested and employed to decorate sacred statues.

## 82.6 OTHER USES

Plumeria flowers are employed to make leis and other forms of floral garlands in several Pacific islands, including the Cook Islands, Fiji, Hawaii, New Zealand, Samoa, Tahiti,

FIGURE 82.7 Plumeria planted at Asian religious sites: (a) A white-blooming plumeria tree in early spring near a statue of Buddha at the ruins of the ancient city Ayutthaya, the former capital of Thailand. (b) Plumeria beside a traditional Taoist Chinese monastery temple.

*Source:* **(a) Photo by Shutterstock/Matrix Reloaded. (b) Photo by Shutterstock/Dogora Sun.**

FIGURE 82.8 (a) A lei held in hands. (b) Hula girls waring leis around their necks.

*Source:* **(a) Photo by Shutterstock/AbElena. (b) Photo by Shutterstock/Deborah Kolb.**

and Tonga. The flowers are often used as hair adornments, floral necklaces, and floral headbands. The blooms are also widely used simply for their fragrance.

## 82.7 PROBLEMS

*Plumeria* species, like other members of the Apocynaceae (dogbane family), contain a milky latex sap with toxic compounds that can irritate skin and eyes in susceptible individuals. Cut flowers, leaves, or stems exude the sap when damaged. Some people who have consumed the (quite bitter) plant have experienced mild stomach pain and cramping. Pets and children are not considered to be in significant danger from plumeria, compared to many more toxic ornamentals.

FIGURE 82.9 A plumeria orchard employed to produce flowers for leis. The trees have been pollarded (the main trunk cut low so the plant produces many low branches) so the flowers are easily harvested at ground level.

*Source:* **Photo by Shutterstock/Reimar.**

FIGURE 82.10 Plumeria flowers worn as head ornaments: (a) Cambodian dancer. In traditional story-based dances in the country, the species of flower and where it is worn are associated with particular dramatic roles. (b) A beagle puppy. (c) Plumeria headdress.

*Source:* **(a) Photo by Michael Stout (CC BY SA 2.0). (b) Photo by Shutterstock/Kobkik. (c) Photo by Shutterstock/Zolotarevs.**

## 82.8 CULTIVATION

### 82.8.1 Climate and Site

Plumerias are suited to tropical climates, including the warmer parts of climate zone 10 and zone 11. They are often planted in zone 9 (which includes most of Florida), where they may suffer cold damage but often regrow if not killed to the ground. They grow best in full sun. Plumerias are noted for their tolerance of windy and salty coastal conditions.

### 82.8.2 Propagation and/or Purchasing

Plumeria can be propagated from seed, although many plants are sterile, and seeds may not be commercially available. The plants are usually reproduced by stem cuttings, which are necessary to maintain the appearance of most cultivars. Commercial growers often graft cuttings onto hardier rootstocks.

### 82.8.3 Planting

When plumeria is grown for commercial flower production, it is planted 3 m apart within rows, 4 to 5 m apart in rows. Homeowners are advised to separate their plantings 3 to 6 m to allow adequate root growth. The plants should be a minimum of 1.2 m from a wall or house (or they will grow top-heavy on the remote side). Plumeria grows well in tubs, which can be too heavy for many to move indoors to shelter overwinter.

### 82.8.4 Soil and Moisture Conditions

Plumeria grow best in well-drained, slightly acidic soil. The plants naturally flower during dry conditions, and concern has been expressed that irrigation before and during flowering could inhibit flowering. Nevertheless, most authorities recommend watering during dry spells (otherwise, dormancy may be induced). In a similar vein, timing of fertilization is controversial, some recommending fertilizing only after flowering, others during flowering. A phosphorus-rich fertilizer (which promotes floral development) is recommended. Plants in pots brought indoors can be stored semi-dormant with minimum moisture.

### 82.8.5 Maintaining Plants

Plumeria growing outdoors in hot climates are considered to be low-maintenance. Some plants are known to have reached hundreds of years of age. Keeping the plants growing outside of tropical regions is more demanding, since killed portions need to be pruned and specimens grown in containers in temperate areas need special care during winter.

## 82.9 FASCINATING FACTS, FABLES, FOLKLORE, AND FRIVOLITY

- White plumeria flowers are commonly associated with death in Bengali culture and are often exhibited during funerals and in cemeteries. In Malaysia, Indonesia, and the Philippines, they are similarly associated with cemeteries, but also the ghosts of passed people.
- Plumeria flowers have been adopted as emblems in several countries. *Plumeria alba* is the national flower of Laos. *Plumeria rubra* is the national flower of Nicaragua. Plumeria is the official flower of the city of Palermo in Sicily, Italy.
- Sailors shipping overseas from Hawaii during World War II would toss a plumeria lei overboard as they passed Diamond Head. If the lei floated ashore, it meant he would return; otherwise, not.
- In Polynesian culture (especially in Hawaii), a plumeria flower worn in the hair over the right ear is often an indication that the lady is single and looking, while worn on the left side means she is taken.

FIGURE 82.11 *Plumeria rubra* in the graveyard of Saint-Leu, Réunion island.

*Source:* **Photo by B. Navez (CC BY SA 3.0).**

FIGURE 82.12 Plumeria hair flowers worn over the right ear to indicate romantic availability, and over the left ear to advertise the reverse.

*Source:* **Photos from Shutterstock, contributed by (*left*) Arammart and (*right*) Ilsi Ruz. Adapted by B. Brookes.**

## KEY PUBLICATIONS


Criley, R.A. 2009. *Plumeria rubra*: An old ornamental, a new crop. *Acta Horticulturae* 813: 183–190.

Eggenberger, R. and M.H. Eggenberger. 2000. *Handbook of Plumeria*. Tropical Plant Specialists, Louisville, KY. 115 pp.

Haber, W.A. 1984. Pollination by deceit in a mass-flowering tropical tree *Plumeria rubra* L. (Apocynaceae). *Biotropica* 269–275.

Koeser, A.K., G. Hasing, and D. McLean. 2013. *Plumeria: Propagation from Cuttings*. University of Florida, Extension, Gainseville. 5 pp.

Mohan, J.S.S. and J.A. Inamdar. 1986. Ultrastructure and secretion of extrafloral nectaries of *Plumeria rubra* L. *Annals of Botany* 57: 389–401.

Perez, B.K. 2019. *Morphological and Molecular Approaches to Disentangling the Taxonomy of* Plumeria *Species (Apocynaceae)*. Doctoral dissertation. University of Hawai'i at Mānoa. 137 pp. https://scholarspace.manoa.hawaii.edu/server/api/core/bitstreams/f047aef7-450e-47de-880d-de6a8ce425ff/content.

Sharma, G., M.K. Chahar, S. Dobhal, N. Sharma, T.C. Sharma, M.C. Sharma, et al. 2011. Phytochemical constituents, traditional uses, and pharmacological properties of the genus *Plumeria*. *Chemistry* 8: 1357–1369.

Thornton, E.H. and S.H. Thornton. 1985. *The Exotic Plumeria (Frangipani)*. Plumeria Specialties, Houston, TX. 55 pp.

Tiernan, N.M. 2019a. A history of the plant name plumeria, commonly known as frangipani. *Personal Blog*. www.nicholetiernan.com/blog/2019/7/17/a-history.

Tiernan, N.M. 2019b. Using *Plumeria* pollination to engage student scientists. *The Tropical Garden* 74(2): 27–29.

Tiernan, N.M. 2021. *Taxonomy and Systematics of* Plumeria L. *(Apocynaceae) in the Caribbean Islands*. Doctoral thesis. Florida International University, Miami. 168 pp. https://digitalcommons.fiu.edu/cgi/viewcontent.cgi?article=6155&context=etd.

Tiernan, N., K. Nakamura, C. Burns, B. Jestrow, R.O. Prieto, and J. Francisco-Ortega. 2023. Phylogenetic relationships of Cuban and Caribbean *Plumeria* (Apocynaceae) based on the plastid genome. *Biological Journal of the Linnean Society* 140: 397–412.


### KEY WEBSITES


Blooming Anomaly. *The Ultimate Guide to Plumeria Care*. www.bloominganomaly.com/bring-hawaii-to-you-plumeria-care/

Criley, R.A. 2005. *Plumeria in Hawai'i*. College of Tropical Agriculture and Human Resources, University of Hawai'i at Mänoa. 57 pp. www.ctahr.hawaii.edu/oc/freepubs/pdf/of-31.pdf

The Plumeria Society of America. https://theplumeriasociety.org/

The Spruce. *How to Grow White Frangipani (Plumeria)*. www.thespruce.com/white-frangipani-plumeria-care-and-growing-guide-5197390

University of Wisconsin, Extension. *Plumeria*. https://hort.extension.wisc.edu/articles/plumeria/

WikiHow. *How to Care for Frangipanis*. www.wikihow.com/Care-for-Frangipanis

WikiHow. *How to Grow a Plumeria*. www.wikihow.com/Grow-a-Plumeria

WikiHow. *How to Grow Plumeria from Cuttings*. www.wikihow.com/Grow-Plumeria-from-Cuttings

WikiHow. *How to Prune Plumeria*. www.wikihow.com/Prune-Plumeria

WikiHow. *How to Plant Plumeria Seeds*. www.wikihow.com/Plant-Plumeria-Seeds

# 83 Poinsettia (*Euphorbia pulcherrima*)

**Scientific Names:** *Euphorbia pulcherrima* (*E. poinsettia* is an old synonym). *Euphorbia* is named for Euphorbus, a 1st-century Greek physician of an ancient African kingdom that was located in what is now modern Algeria. The epithet *pulcherrima* is Latin for "very beautiful."

**Common Names:** Poinsettia, common poinsettia. "Poinsetta" is a widespread misspelling, associated with the mispronunciation poin-SET-uh). Less-common names include Ataturk's flower (in Turkey; poinsettia was the favorite flower of Atakurk, the founder of modern Turkey), Christmas flower, Christmas plant, Christmas star, eastern flower, lobster flower, lobster plant (a name usually used for heliconia), (Mexican) flame leaf, painted leaf, points (a nickname encountered in the greenhouse trade), winter rose. The English name poinsettia commemorates Joel Roberts Poinsett (1799–1851), who was the first US minister to Mexico (ambassadors were not appointed until 1896) and an amateur botanist. He was also a co-founder of the National Institute for the Promotion of Science and the Useful Arts, one of the early efforts to initiate a US national history museum that led to the establishment of the Smithsonian Institution (based in Washington, DC), the world's largest complex of museums. "Wild poinsettia" is *Euphorbia heterophylla* var. *cyathophora* (or *E. cyathophora*), an annual or short-lived perennial native to North and South America, which is cultivated but is invasive and often viewed as a weed.

FIGURE 83.1 Joel Roberts Poinsett, prominent US politician and the namesake of poinsettia. He introduced the plant to the United States in 1828, initiating its worldwide popularization: (a) Photo of an oil painting in the possession of the US Department of Defense. (b) Statue of Poinsett in Greenville, South Carolina. Poinsett was very active in South Carolina, including Greenville.

*Source:* **(a) Public domain. (b) Photo by John Foxe (CC BY SA 4.0).**

## 83.1 INTRODUCTION

Except for Christmas trees, poinsettias have become the most popular winter season plant. Like Christmas trees, poinsettias are associated both with the religious and the secular aspects of Christmas. Almost unknown 150 years

FIGURE 83.2 Poinsettias decorating churches during Christmas: (a) Saint Martin of Tours Parish Church of Bocaue, Quezon City, Philippines. (b) St. Andrew Church, Sarasota, Florida.

*Source:* **(a) Photo by Ramon FVelasquez (CC BY SA 3.0). (b) Photo by Roger W (CC BY SA 2.0).**

FIGURE 83.3 Giant poinsettia "Christmas trees" fashioned by attaching potted plants to a supporting framework. These artistic creations are popular at Christmas worldwide: (a) An 8 m tall exhibit in Singapore. (b) In Japan. (c) In Germany.

*Source:* **(a) Photo by Choo Yut Shing (CC BY 2.0). (b) Photo by Scooped Tea (CC BY SA 4.0). (c) Photo by Brűhl (public domain).**


DOI: 10.1201/9781003586227-83

ago, they are now ubiquitous in homes and offices in much of the Americas, Europe, and other areas. There is something very special about the way poinsettias produce their vivid red bracts in midwinter, just in time to celebrate Christmas. Although sales are virtually confined to the Christmas period, the poinsettia is nevertheless the leading potted plant in many areas of the world.

Poinsettia is not grown for its attractive "flowers" but rather for its "bracts." In botany, a "bract" usually denotes a kind of leaf (typically smaller than the normal leaves) associated with flowers (the rare term "hysophyll" has the same meaning) or fruits. Brightly colored bracts that compensate for a lack of attractive true flowers occur in several ornamental plants other than *Euphorbia* – for example, *Bougainvillea*, *Cornus*, plants with spathes (a single large colored leaf associated with a collection of many flowers, especially in members of the Araceae, such as anthurium and peace lily), and bromeliads. Although poinsettia doesn't literally qualify as one of the world's leading ornamental "flowers," for practical purposes, this is an academic distinction.

## 83.2 OVERVIEW

*Euphorbia* is a huge genus of 2,000 or so species, several of which are grown as ornamentals. *Euphorbia pulcherrima*, the most widely grown, is indigenous to tropical and subtropical regions from western Mexico to Guatemala (the range now extending south to Costa Rica), occurring in deciduous forests at moderate elevations as well as in hot, seasonally dry forests. The species is commonly cultivated

FIGURE 83.5 Tree form of poinsettia: (a) In Africa. (b) In Egypt. (c) In Los Angeles, California.

*Source:* **(a) Photo by Pisecbesedil (CC BY SA 3.0). (b) Photo by Juliegomoll (CC BY 2.0). (c) Photo by Downtowngal (CC BY SA 4.0).**

FIGURE 83.4 Colorful collections of poinsettias.

*Source:* **(a) Photo by Shutterstock/Lisa F. Young. (b) Photo by Shutterstock/LiliGraphie. (c) Photo by Shutterstock/Amipi.**

in subtropical and tropical regions and has become naturalized in many areas. It is sometimes classified as an invasive alien. In North America, it occasionally has been found growing outside of cultivation in some of the southern US states.

This shrub or small tree develops several erect stems 0.6–4 m high. Depending on climate, the foliage is evergreen or deciduous. The normal leaves are dark green and dentate, but the upper ones are modified (as "bracts"), becoming red at maturity. The flowers are tiny, unisexual, lack petals and sepals, and occur in a cyathium – a greenish, cup-shaped cluster of flowers surrounded by very tiny bracts. Cyathia occur on the ends of branches, surrounded by 10 to 20 leaflike large colored bracts. Male flowers consist of a single red stamen, and several of these surround the solitary pistil in each cyathium. The female flower is exserted from the cyathium on a pedicel. Fused to one side of the cyathium is a nectar gland, appearing like yellow lips, this serving to reward pollinators (reportedly birds, especially hummingbirds, as well as various insects). The fertilized ovaries develop into fruits which explosively scatter the seeds. Poinsettia flowering and associated bract coloration are photoperiodic, occurring in response

FIGURE 83.6 Close-up of young poinsettia cyathia (flower clusters): *A* is unopened, *B* is opening, and *C* is recently opened. The single nectar gland associated with each cyathium has the appearance of yellow lips of an open mouth. The female flowers have not yet emerged (see Figure 83.7).

*Source:* **Photo by Philstone (CC BY SA 3.0).**

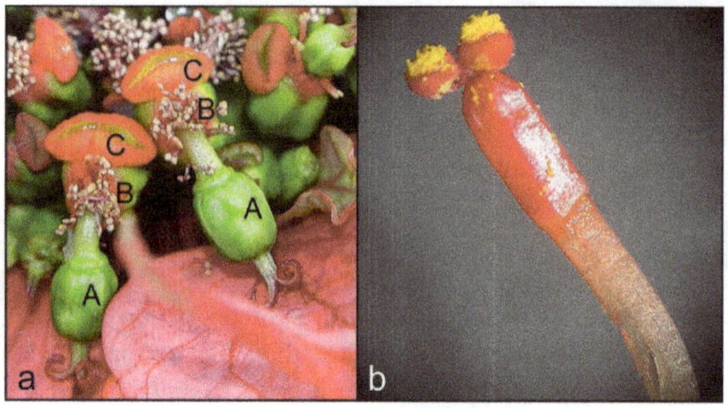


FIGURE 83.7 Mature flowers of poinsettia: (a) Poinsettia cyathia (flower clusters). *A* = ovary of the single female flower of each cyathium. *B* = male flowers (stamens) surrounding the single female flower. *C* = nectar gland. (b) Male flower (reduced to a single stamen with a bi-lobed anther) shedding pollen. As discussed in the text, plants shedding pollen are overmature and should not be purchased.

*Source:* **(a) Photo (public domain) by "Hardyplants at the Wikipedia project." (b) Photo by Aelwyn (CC BY SA 3.0).**

to shortening day length. In its native area of Central America, the plant flowers in winter (hence the association with Christmas), when days are short (i.e., nights are long, triggering flowering). The inconspicuous flowers fade after fertilization, but the bracts can remain colorful for weeks. In additional to the traditional red bract color, cultivars are available with bracts that are burgundy, pink, coral, orange, white, cream, yellow, pale green; marbled, mottled, or variegated; and even with wavy margins. Red remains the most popular poinsettia color, accounting for about 75% of North American sales; white (about 8%) and pink (about 6%) are next in popularity. The substantial display of the bracts compensates for the inconspicuous flowers, and indeed, non-botanists frequently refer to the bracts as petals.

In the poinsettia trade, tall, top-heavy plants are awkward to handle and are therefore very undesirable. Compared to the original wild ancestor, store-bought plants purchased for Christmas are compact and dwarfed, usually forced into bloom at a height of 0.3–0.6 m, and the stems have been pinched to promote branching. (In the past, it was common practice to place several cuttings in a pot to give the impression that one well-branched plant was present.) Commercial growers have probably selected low-growing forms, but height is also often limited by applying growth retardants. Some store-bought plants are capable of growing into tall shrubs if transferred to an appropriate environment.

FIGURE 83.8 Poinsettia cultivars differing in bract color (decreasing popularity: red, white, pink, yellow).

*Source:* **(a) Photo by Oregon State University (CC BY SA 2.0). (b) Photo by David Wagner (public domain). (c) Photo by Susan Reimer (CC BY 2.0). (d) Photo by Núria (public domain).**

### 83.2.1 The Remarkable Use of a Disease Organism to Produce Attractive Poinsettias

It may seem counterintuitive, but humans sometimes find diseased plants to be very attractive. Several dwarfed trees and shrubs in commerce are simply virus-infected plants maintained by cuttings. Foliage chimeras (i.e., with leaves variegated with white) of various species are also sometimes similarly diseased plants. Many of the early tulip cultivars with variegated petal colors were simply the result of viral infections; eventually, these were banned because they weakened the tulips and indeed threatened the tulip trade. However, as noted in the following, most modern

FIGURE 83.9 Poinsettia cultivars with variegated bracts.

*Source:* **(a) Photo by Erik Drost (CC BY 2.0). (b) Photo by SusanReimer (CC BY 2.0). (c) Photo by OwenITGuy (CC BY 2.0). (d) Photo by Shutterstock/Moolkum.**

poinsettias are, in fact, diseased plants, but the disease is highly beneficial for the industry.

In the 1980s, poinsettia breeder Gregor Gutbier grafted a poinsettia plant that was naturally poorly branched on top of a poinsettia plant that was highly branched. Very curiously, when he took cuttings from the poorly branched plant, they grew into highly branched plants. Apparently, others in the poinsettia trade have also utilized plants with the natural ability to produce highly desirable branching dwarfs. Later, Lee et al. (1997; see the bibliography) discovered that the highly branched dwarf poinsettias which have come to dominate the poinsettia trade were infected with a phytoplasma (a bacterium, notable for lack of cell walls), and that this disease is transmissible and *causes* the dwarfing and branching. (More precisely, it causes much-increased branching from a single pinching off of a terminal stem.) The infectious agent has been shown to be related to some other diseases (including Peach X and Spiraea Stunt) but is relatively benign in poinsettia. Today, the phytoplasma is deliberately introduced to poinsettia cultivars by grafting in order to stimulate the desired highly branched, compact habit.

## 83.3 HISTORY

*Euphorbia pulcherrima* was cultivated and, indeed, likely domesticated to some extent by the Aztecs in Mexico in pre-Columbian times. With Christianization of Indigenous people in Mexico, the Aztec tradition of using the plant in religious ceremonies was taken up in churches. In the early 1800s, the species was brought to both Europe and the United States, where its ornamental value was realized. By the middle of the 19th century, poinsettia was widely cultivated in both the New and Old Worlds, but primarily as a landscape shrub in subtropical climates. Northern Guerrero, Mexico, has been identified as the principal area from which modern poinsettia cultivars have been derived. In 1923, a poinsettia seedling grown by a Mrs. Enteman in Jersey City, New Jersey, became the basis of the cultivar 'Oak Leaf'. This plant proved to have excellent bract retention and to be exceptionally suitable for pot culture. It formed the basis of the potted poinsettia trade for 40 years. Selection of over 100 cultivars occurred during the 20th century, with emphasis on bract display. Bracts in cultivars are larger and more numerous than in most wild plants, and color variants have been selected.

Until the middle of the 20th century, the poinsettia was not well-known. Paul Ecke Jr. (1925–2002) is widely credited with transforming the species into the top-selling potted plant in the United States and an indispensable associate of Christmas. His father, Paul Ecke Sr., developed the first poinsettia that could be successfully grown as an indoor potted plant and was instrumental in founding the Paul Ecke Ranch, which was the world's largest producer of poinsettias in the early 21st century. Paul Ecke Jr. sent thousands of free plants to the major television stations to use

FIGURE 83.10 Poinsettias in art: (a) *Girl with Poinsettia*, painted in 1908. (b) John Lewis Childs (Floral Park, NY) 1923 fall catalogue. (c) Tiffany poinsettia-pattern hanging lamp. Genuine Tiffany lamps have sold for millions of dollars. (d) Poinsettia sculpture in Brisbane, Australia.

*Source:* **(a) US Library of Congress (public domain). (b) Public domain. (c) Photo by Daderot (public domain). (d) Photo by Kgbo (CC BY SA 4.0).**

as scenic backdrops for Christmas. During the Christmas season, television viewers were treated to backgrounds of hundreds of poinsettias on their favorite shows. Ecke also recognized that women are the chief purchasers of flowers, and he provided free plants to women's magazines. This inspired product placement hooked the public, and by 1986, poinsettias displaced chrysanthemums as the leading potted plant in the United States. The poinsettia is an outstanding example of commercial interests establishing a "tradition" of purchasing a commodity for a seasonal event. Today, poinsettias are well-known, admired ornamental plants, as reflected by their frequent appearance in the world of art.

## 83.4 ECONOMIC IMPORTANCE

"Stock plants" are simply rooted cuttings that can be used by garden center retailers for direct sales and for growing potted plants of many floricultural and some vegetable crops. Because of economic considerations (especially costs for labor, heating, and construction of greenhouses) and climate advantages, the production of stock plants has migrated "offshore" away from Northern regions to a very large degree. Central America has become the center of growing and breeding of new poinsettia cultivars, and of shipping small, rooted plants ("plugs") to commercial greenhouses all over North America and Europe. The greenhouses use these plugs to grow the plants in pots, pinching most cultivars to promote branching and using huge shade curtains to block off light, to provide the necessary induction of bract coloration.

Today, world production of poinsettias is controlled by a handful of firms. Growing poinsettias is a precise science, and numerous problems can arise (these are detailed in the references provided).

FIGURE 83.11 Commercial production and sales of poinsettias: (a) Poinsettia greenhouse. (b) Florist with poinsettia. (c) Sales display of poinsettias on shelves.

*Source:* (a) Photo by Shutterstock/Giancarlo Polacchini. (b) Photo by Shutterstock/BearFotos. (c) Photo by Liz West (CC BY 2.0).

Marketing poinsettias has become a very competitive business. Big-box stores and supermarkets have come to dominate poinsettia sales, in some cases selling the plants at retail prices that are actually less than the wholesale prices that many producers can afford. Sometimes plants are spray-painted (blue bracts are invariably painted) or cleverly interwoven with colorful foil, both rather deplorable trends that detract from the intrinsic beauty of the plants. Selling "designer" plants (those with unusual bract colors or color combinations), placing plants in unusual containers, and combining poinsettias with other plants are ways that some specialty flower shops are trying to remain viable. However, the mass marketing of poinsettia and, indeed, other ornamental plants is a very serious challenge for many independents today, some of whom are considering switching to alternative holiday plants.

Europe accounts for almost two-thirds of poinsettia world sales, and the United States for about one-third. In the United States, poinsettia is a leading produced potted plant (competing with orchids collectively and chrysanthemums collectively), accounting for about a third of sales of all flowering potted plants. California is the leading US grower, accounting for about a quarter of US production, but some plants are grown commercially in all American states.

## 83.5 ORNAMENTAL USES

Poinsettia's value is primarily as a seasonal short-term interior potted plant, which is very popular as a gift houseplant

FIGURE 83.12 (a) A well-grown specimen. (b) An elegant house in Spain. (c) Girl with potted plant. (d) A commercial recreational park in Thailand, the numerous pots slightly immersed in the ground.

*Source:* (a) Photo by Shutterstock/Leena Robinson. (b) Photo by Shutterstock/Brenda Kean. (c) Photo by Shutterstock/PicMy. (d) Photo by Shutterstock/BaYaoPix.

FIGURE 83.13 Poinsettia grown outdoors as a perennial in subtropical and tropical regions: (a) In Mexico. (b) In Myanmar. (c) In Canary Islands. (d) In Australia.

*Source:* **(a) Photo by Thelmadatter (CC BY 3.0). (b) Photo by Shutterstock/Matyas Rehak. (c) Photo by Frank Vincentz (CC BY SA 3.0). (d) Photo by Shutterstock/Rose Marinelli.**

and decoration for churches and large buildings. In warm countries, it is also employed as a flowering shrub and small tree, as well as an outdoor container plant.

## 83.6 OTHER USES

Poinsettia and other species of *Euphorbia* have been employed in Indigenous herbal medicine to treat a variety of conditions, and such applications are often touted in (unreliable) web-based articles. However, no significant usage occurs in modern Western medicine.

## 83.7 PROBLEMS

Depending on the cultivar, periods of uninterrupted darkness of 12 to 14 hr are needed daily for 7 to 10 weeks for bract coloration to develop. This requires considerable effort from commercial growers to supply the marketplace for potted plants that will flower during the Christmas period and is such a demanding exercise that very few people in the general public are willing to undertake the exercise.

For most of the last century, poinsettia was regarded as very toxic if consumed by people or pets, but in recent decades, this viewpoint has almost completely been abandoned. Nevertheless, it is recommended that the plant not be ingested, as substantial quantities of leaves could result in gastrointestinal upset. (In any event, there is no good reason to eat poinsettia.) Many species of *Euphorbia* contain a toxic, milky latex, and it has caused rashes and asthma-like symptoms in isolated cases. It should be kept in mind that several chemicals are employed in the commercial production of poinsettia, and these are potentially toxic.

## 83.8 CULTIVATION

Most people purchase potted plants for short-term flowering displays during the Christmas season. Purchased poinsettias remain attractive in the home for 2 to 4 weeks, but the beautiful display can be extended with excellent care to 8 weeks, or even longer with some varieties.

### 83.8.1 Climate and Site

Poinsettia is able to grow as a perennial outdoors in climate zones 10 to 15. However, in most of the world, the plant is grown as a houseplant. Potted poinsettia is very intolerant of cold and drafty locations (such as near heat vents, radiators, exterior doors, and open windows). Do not allow plants to touch cold windows. Minimum required temperature is 13–15°C. Ideal daytime temperatures are 15.5–21°C. Preferably, the plant should be moved to a cooler room (around 13–18°C at night. Indoors, bright but not direct sunlight is best for potted plants, at least 6 hr. Low light stimulates the loss of leaves. If a shady location is desired for best display while visitors are present, return the plant to its holding area afterward.

### 83.8.2 Propagation and/or Purchasing

Buy potted plants from a reputable source. If prior to purchase a plant has been stored in cold conditions or with inadequate attention to moisture, it will probably deteriorate rapidly. Look for aesthetic quality: dark-green foliage down to the soil line, bracts that are completely colored, an appearance that is balanced and symmetrical. Check for maturity: Ideally, the flower buds should be closed (i.e., unopened); the true flowers in the cyathia should be green or red-tipped and fresh-looking, and pollen should not be covering the bracts (indicative of overmaturity). Check for obvious health problems: fallen or yellowed leaves, drooping or wilted habit, and insects; ties or stakes should not be necessary. Check for poor display and storage conditions: Plants kept in sleeves tend to deteriorate; plants crowded together tend to drop their bracts; wet soil promotes fungal problems. Transporting plants from the shop to your home in the cold of winter (less than 10°C) can be very damaging; insist on completely wrapping the plant, preferably in plastic.

### 83.8.3 Planting

Advise is provided in the following for adventurous people willing to try to rejuvenate potted plants acquired for Christmas flowering.

### 83.8.4 Soil and Moisture Conditions

Poinsettias purchased for Christmas do not normally require fertilization. However, if kept for more than a month, to maintain the green foliage and promote growth, apply a low-nitrogen, high-potassium, soluble fertilizer monthly. Potted poinsettia is very intolerant of overly moist soil but requires continuous moderate moisture. Check soil moisture daily, water sparingly (but enough so that water flows from the bottom of the pot), and only when the surface of the compost begins to dry out (and is dry to the touch). This might require watering two or three times weekly. If the usual foil wrapping extends over the base of the pot, punch a hole through the bottom for drainage; discard water draining into the supporting dish (do not allow the pot to sit in water). Setting the container in a gravel-filled pan with water, or near other plants, can help maintain the humidity of the surrounding air. Alternatively, mist the foliage.

### 83.8.5 Maintaining Plants

Very few who receive a gift of poinsettia have the knowledge or patience to maintain the plant and stimulate it to reproduce its attractive coloration for another season, so the plants are regularly discarded. Accordingly, poinsettias are ideal for retailers, as consumers are quite willing to buy new ones annually. For most people, buying a new plant for Christmas is much easier than trying to resurrect the attractive bract display for next year. It is very unlikely that the attempt will result in a poinsettia of the original flower shop quality. Perhaps epitomizing the difficulties is the defunct rock band *Dead Poinsettia* of St. Paul, Minnesota. Do you really want to resurrect that sad-looking poinsettia, probably still in its colorful aluminum sleeve, after Christmas? Commercial poinsettias were grown under optimal light, humidity, and fertilization and sprayed multiple times with growth regulators. They have been carefully protected from numerous diseases and insects (poinsettias are whitefly magnets). And are you really going to be able to rotate the plants from the dark to the light at 7:00 a.m. and back to the dark at 5:00 p.m. every day for many weeks? Nevertheless, some readers will be attracted by the challenge. The following steps are recommended:

- After most of the leaves have fallen off, reduce watering (to keep plants on the dry side), and store the pot(s) in a cool (10–15°C), well-ventilated area (possibly a garage).
- In April, prune plants back to 10–15 cm.
- Repot in a larger-sized container (2.5–5 cm larger in diameter than the previous pot), into quality potting soil, preferably pasteurized (a recommended mixture: two parts peat moss, one part potting soil, and one part sand or perlite).
- Grow in a semi-shady, cool place over the summer, ideally at 15–18°C. The plant can be moved outdoors once the temperature is consistently above 15°C.
- Some people sink their pots into the soil outdoors to maintain them during the summer. This makes bringing the plants indoors in the fall more difficult, as they will extend their roots outside the pots (unless the pots are turned regularly).
- Fertilize regularly when new growth resumes, following the rate recommended on the brand of fertilizer being used. Reduce fertilization in December by half.
- To prevent the plant from growing too high, and to make it bushy, "pinch" growing tips every 4 to 6 weeks during the summer (much better: use a sharp knife to remove about 2.5 cm; wash away the secreted milky sap, and minimize exposure to skin). Do not prune after mid-August.
- If grown outside, bring plants indoors before the threat of cool weather (ensure insects are removed).
- For bracts to develop and color up, the plants should be kept in complete darkness (uninterrupted by occasional light) for a period of 12 to 14 hr daily for 8–10 weeks. For bracts to color up well, provide a constant temperature of about 13°C. A light-tight closet, box, or bag is normally used for the purpose by amateur horticulturists. Suggested schedule: Begin October 1, providing the dark period from 5:00 p.m. to 7:00 a.m. Once bract color develops in the top leaves, this strict schedule is no longer necessary; however, many horticulturalists recommend maintaining the procedure until the bracts are almost fully expanded and colored. Of course, light must be provided for growth for part of each day; uninformed people have killed their plants by leaving them in a dark closet for a month or more. In theory, in cold, Northern areas, the plant could be left in a window in a room that is not illuminated, but probably stray illumination at night coming through the window would interrupt the development of bract coloration.
- Instead of discarding the stems removed when the poinsettia is cut back in spring, the attempt may be made to root these in moist perlite or sand. However, better than using these relatively exhausted branches, take vigorously growing cuttings in early summer, as follows. Using new shoots 20–25 cm long, cut off the terminal 10–15 cm end with a sharp knife. Remove the lower leaves, and apply a rooting hormone to the bases. Place these in a moist rooting medium (clean, coarse sand, or a mix of sand and peat moss). Cover with glass or plastic to conserve moisture, and place in shade. Cuttings should produce roots in 3 to 4 weeks; transfer these to small pots (note that the roots are brittle).

## 83.9 FASCINATING FACTS, FABLES, FOLKLORE, AND FRIVOLITY

- The ancient Aztecs were the first to ascribe religious significance to the poinsettia. The red color of the bracts served as a reminder of the need for blood sacrifices to the gods. The Aztecs also considered the poinsettia to symbolize purity (its name, *cuetlaxochitl*, signified "flower that withers, mortal flower that perishes like all that is pure"). Considered a gift from the gods, poinsettias adorned the palace of Montezuma (1466–1520).

- In Christianity, poinsettia symbolizes the two main colors of Christmas: red, representing the blood of Jesus, and green, indicating the rebirth of life after winter. The star-like pattern of the bracts viewed from the tip of a branch is said to symbolize the Star of Bethlehem (hence the name "Christmas star").
- Moralistic stories are often associated with plants that have acquired religious significance. The most famous poinsettia story concerns a peasant girl in Mexico, too poor to contribute a significant gift to the Christ Child ceremony at the church. Trying her best, she picked a bouquet of weeds as her offering. Miraculously, her sincere effort was rewarded when the weeds were transformed into attractive poinsettias.
- While he was the US minister to Mexico (1825–1830), Joel Roberts Poinsett acquired a reputation for meddling in Mexico's internal affairs. The word *poinsettisimo* has persisted to the present in Mexico to express arrogance and high-handedness.
- December 12 is National Poinsettia Day in the United States, a holiday that was established in the mid-1800s to commemorate the death of Joel Roberts Poinsett.
- The city of Ventura in Southern California is nicknamed the Poinsettia City, reflecting the past importance of the crop in the area.
- The Poinsettia Bowl was a post-season football bowl game sanctioned by the NCCA (National Collegiate Athletic Association) played in San Diego (in Southern California) from 2005 to 2016 and originally played from 1952 to 1955 between military services teams.
- A poinsettia cocktail is a mixture of dry sparkling white wine, vodka (alternatively Cointreau or triple sec) and cranberry juice, the latter providing the red color for which the plant is famous.

FIGURE 83.14 (a) Poinsettia-based female costumes. (b) A lady in a poinsettia costume. (c) A vintage (1910) postcard.

**Source: (a) Gordon, E. 1910. *The Little Cousins of the Field and Garden*. Volland, Chicago (public domain). (b) Photo by Dave Walker (CC BY 2.0). (c) Public domain.**

FIGURE 83.15 William Jolly Bridge in Brisbane, Australia, painted in a poinsettia motif, in honor of the city's official flower.

***Source:* Photo by Brisbane City Council (CC BY 2.0).**

- Eighty percent of poinsettias are purchased by women.
- Latex from the poinsettia has been used as a depilatory (to remove hair).
- The red poinsettia was adopted in 1930 as the official flower of Brisbane, Australia. The William Jolly Bridge (completed in 1932 and named for a previous Lord Mayor), across the Brisbane River, is regularly decorated with poinsettias.

## KEY PUBLICATIONS

Bai, H.-N., D. Niu, and L.-J. Wang. 2009. RAPD analysis of different leaf colour in *Euphorbia pulcherrima*. *Journal of Northeast Agricultural University (Harbin, China)* 16(2): 32–36.

Buscher, F.K., J.L. Caldwell, P. Ecke, Sr., P. Ecke, Jr., and L.J. Herr. 1963. *Poinsettias: Culture and Use as Potted Plants*. State University Agricultural Experiment Station, Columbus, OH. 110 pp.

Ecke, P., III, J.E. Faust, J. Williams, and A. Higgins. 2004. *The Ecke Poinsettia Manual*. Ball Pub., Batavia, IL. 287 pp.

Grueber, K.L. 1985. *Euphorbia pulcherrima*. In: *Handbook of Flowering*, Vol. 2, edited by A.H. Halevy, 488–496. CRC Press, Boca Raton, FL.

Huang, P.-Y. 2007. Beneficial plant pathogens. *Chronica Horticulturae* 47(3): 9–14.

Ibáñez, M.D., M. Fernández-Nieto, M. Martínez, G.A. Cardona, J. Guisantes, S. Quirce, et al. 2004. Asthma induced by latex from 'Christmas flower' (*Euphorbia pulcherrima*). *Allergy* 59: 1127–1128.

Khan, R.F., R.P. Yadav, and A. Singh. 2021. Screening of pharmacological and biological properties of a euphorbious plant, *Euphorbia pulcherima*: A review. *International Journal of Pharmacognosy and Pharmaceutical Sciences* 3(2): 12–15.

Klug, S., G. Saleem, L. Hochuruk, and S. Marcus. 1990. Toxicity potential of poinsettia, is the plant really toxic? *Veterinary and Human Toxicology* 32: 368.

Lack, H.W. 2011. The discovery, naming and typification of *Euphorbia pulcherrima* (Euphorbiacaeae). *Willdenowia* 41: 301–309.

Lee, I.M., M. Klopmeyer, I.M. Bartoszyk, D.E. Gundersen-Rindal, T.-S. Chou, K.L. Thomson, et al. 1997. Phytoplasma induced free-branching in commercial poinsettia cultivars. *Nature Biotechnology* 15: 158–182.

Park, K.-R. and R.K. Jansen. 2007. A phylogeny of Euphorbieae subtribe Euphorbiinae (Euphorbiaceae) based on molecular data. *Journal of Plant Biology* 50: 644–649.

Runyon, R. 1980. Toxicity of fresh poinsettia (*Euphorbia pulcherrima*) to Sprague-Dawley rats. *Clinical Toxicology* 16: 167–173.

Santucci, B., M. Picardo, and A. Cristaudo. 1985. Contact dermatitis from *Euphorbia pulcherrima*. *Contact Dermatitis* 12: 285–286.

Stone, R.P. and W.J. Collins. 1971. *Euphorbia pulcherrima*: Toxicity to rats. *Toxicon* 9: 301–302.

Strømme, E. 1994. *The Scientific Basis of Poinsettia Production*. Agricultural University of Norway, Aas, Norway. 121 pp.

Taylor, J.M., R.G. Lopez, C.J. Currey, and J. Janick. 2011. The poinsettia: History and transformation. *Chronica Horticulturae* 51: 23–28.

Trejo, L, T.P. Feria Arroyo, K.M. Olsen, L.E. Eguiarte, B. Arroyo, J.A. Gruhn, et al. 2012. Poinsettia's wild ancestor in the Mexican dry tropics: Historical, genetic, and environmental evidence. *American Journal of Botany* 99: 1146–1157.

Whipker, B. 1999. *Commercial Poinsettia Production Guide*. North Carolina Commercial Flower Growers' Association, North Carolina State University, Raleigh, NC. 50 pp.

Winek, C.L., J. Butala, S.P. Shanor, and F.W. Fochtman. 1978. Toxicology of poinsettia. *Clinical Toxicology* 13: 27–45.

Zimmermann, N.F.A., C.M. Ritz, and F.H. Hellwig. 2010. Further support for the phylogenetic relationships within *Euphorbia* L. (Euphorbiaceae) from nrITS and *trn*L – *trn*F IGS sequence data. *Plant Systematics and Evolution* 286: 39–58.

## Key Cultivation Websites

*Avoiding Top Problems of Poinsettias*. www.gpnmag.com/avoiding-top-problems-poinsettias.

GardenersWorld.com. *Poinsettia Care: How to Keep a Poinsettia Alive*. www.gardenersworld.com/how-to/grow-plants/how-to-grow-poinsettias/.

Iowa State University, Extension. *Guide to Selection and Care of Poinsettias*. https://hortnews.extension.iastate.edu/guide-selection-and-care-poinsettias.

*The Poinsettia Pages*. http://urbanext.illinois.edu/poinsettia/index.cfm.

Royal Horticultural Society. *How to Grow Poinsettias*. www.rhs.org.uk/plants/euphorbia/poinsettia/growing-guide.

Sensorex. *How to Grow Poinsettias in a Greenhouse*. https://sensorex.com/poinsettias-greenhouse/.

The Spruce. *Poinsettia: Indoor Plant Care & Growing Guide*. www.thespruce.com/poinsettia-care-guide-1403587.

*The Texas Poinsettia Producers Guide*. http://aggie-horticulture.tamu.edu/ornamental/the-texas-poinsettia-producers-guide/.

University of Massachusetts, Extension. *Diseases of Poinsettias*. https://ag.umass.edu/greenhouse-floriculture/fact-sheets/diseases-of-poinsettias.

WikiHow. *How to Care for Poinsettia*. www.wikihow.com/Care-for-Poinsettias.

WikiHow. *How to Grow Poinsettia*. www.wikihow.com/Grow-Poinsettia.

WikiHow. *How to Keep Poinsettias Growing to Next Christmas*. www.wikihow.com/Keep-Poinsettias-Growing-To-Next-Christmas.

WikiHow. *How to Prune a Poinsettia*. www.wikihow.com/Prune-a-Poinsettia.

WikiHow. *How to Revive a Dying Poinsettia*. www.wikihow.com/Revive-a-Dying-Poinsettia.

# 84 Poppies (*Papaver* Species and Hybrids)

**Scientific Names:** The genus name *Papaver* is based on the classical Latin name for the plant. The word is related to the Latin *buro*, "to burn," and the Hebrew *bera*, "fire" or "burning," presumably because of the blazing red flowers of many of the species.

**Common Names:** The English word *poppy* is derived from the classical Latin name for the plant, discussed earlier. Many species in the poppy family (Papaveraceae) have the word *poppy* in at least one of their English names, such as California poppy (*Eschscholtzia californica*), celandine poppy (*Stylophorum diphyllum*), and the blue Himalayan poppies (*Meconopsis* species). Most ornamental plants called poppy are in *Papaver*, and sometimes species of *Papaver* are termed the "true poppies." The tropical water poppy (*Hydrocleys nymphodes* of the Alismataceae family) is often grown in ponds.

FIGURE 84.1 An early garden catalogue – Vick's 1894 Garden and Floral Guide – showing popular *Papaver* poppy species.

*Source:* **Public domain, modified by B. Brookes.**

## 84.1 INTRODUCTION

The genus *Papaver* has about 100 species of annual, biennial, and perennial herbs and some subshrubs, native mostly to temperate regions of the Northern Hemisphere. Many are grown as ornamentals. The four most widely cultivated groups are mentioned in this chapter. Poppies of the genus *Papaver* are well-established garden favorites, valued for the size and range of their very colorful flowers and the ease with which they can be cultivated.

## 84.2 OVERVIEW

Poppies are best known for the opium poppy, *P. somniferum*, which is employed legally for its therapeutic alkaloids and illegally as a recreational drug but is also the source of edible poppy seeds. There are also ornamental varieties often grown in gardens. Plants are occasionally found as weeds in many areas of the world, escaped from cultivation, which appears to have been initiated thousands of years ago in the Middle East. Drug varieties, seed varieties, and ornamental varieties have been selected to develop the features of interest, but even seed and ornamental cultivars can produce limited amounts of opium alkaloids. All three kinds can be termed "opium poppy," and indeed, varieties grown for seeds are often salvaged commercially for their content of opiates. In most Western nations, growing ornamental opium poppies is not subject to prosecution, but attempting to use these as a narcotic is (a very rare occurrence). The plants are usually 60 to 90 cm in height and have blue-gray leaves with toothed margins. The flowers typically are 10 to 12 cm in diameter, in shades of white, pink, purple, and red. The flowers of one variety particularly resemble doubled peony flowers, and such plants are termed "peony poppies." Opium poppy is grown as an annual in the cold North and a perennial in the warm South.

The Oriental poppy (*P. orientale*), a perennial from southwestern Asia (including Caucasus, northeastern Turkey, and northern Iran), is the most attractive of the ornamental *Papaver* species. Most garden cultivars are the result of hybridization with two other domesticated species, *P. pseudo-orientale* and *P. bracteatum*. The hybrids are taprooted, herbaceous perennials up to 1 m tall, which can live for decades in relatively cold climates to which they are adapted. They produce single flowers, on the end of stalks. The flowers are large (typically about 15 cm in diameter), with very thin (paper-like, almost-translucent, and very soft to the touch) petals in shades of orange, red, pink, lavender, or white, often with a large dark blotch in the center of the throat of the flower. Individual flowers may last only a day or two. The much-dissected foliage senesces shortly after flowering in early summer (an unattractive feature often

DOI: 10.1201/9781003586227-84

**FIGURE 84.2** Opium poppy (*Papaver somniferum*): (a) A commercial opium poppy field for drug production in England. (b) Portion of a stem showing leaves. (c–f) Flowers are arranged in sequence of increasing number of petals. (e) and (f) belong to the "peony-flowered" group.

*Source:* **(a) Photo by Marilyn Peddle (CC BY 2.0). (b) Photo by Magnus Manske (CC BY SA 3.0). (c) Photo by David Perez (CC BY 3.0). (d) Photo by Rahulkumarnimat (CC BY SA 4.0). (e) Photo by Shutterstock/Walter Erhardt. (f) Photo by Yewchan (CC BY SA 2.0).**

**FIGURE 84.3** Oriental poppy (*Papaver orientale*). (a) A field in Italy. (b) In a walled British garden. (c-f) Individual flowers.

*Source:* **(a) Photo by Valerio Clementi (CC BY SA 4.0). (b) Photo by Mark (CC BY 2.0). (c) Photo by Andy Morffew (CC BY 2.0) (d) Photo by Dguendel (CC BY 3.0). (e) Photo by Nino Barbieri (CC BY SA 2.5). (f) Photo by Jonathan Billinger (CC BY SA 2.0).**

concealed in gardens by smaller ornamentals), and the plants become dormant.

The corn poppy (*P. rhoeas*, also called red, Flanders, common, and field poppy) is an annual herb native to temperate areas of the Old World. The species has been hybridized to create garden cultivars called the Shirley poppy. The flowers come in white, pink, red, yellow, and other shades; are 5 to 10 cm in diameter; and may be single or doubled. The flowers sometimes have white centers versus the typical black of Oriental poppies.

The Iceland poppy (*P. nudicaule*, also called Icelandic poppy and Arctic poppy) is from the northern (boreal)

**FIGURE 84.4** Corn poppy (*Papaver rhoeas*). (a) In a garden. (b) A field in Japan. (c–f) Individual flowers.

*Source:* **(a) Photo by Okseok Choi (CC BY SA 4.0). (b) Photo by Keihin Nike (CC BY SA 3.0). (c) Photo by Jacinta lluch Valero (CC BY SA 2.0). (d) Photo by Anaka Juuyoh (CC BY 2.0). (e) Photo by Tim Green (CC BY 2.0). (f) Photo by David Perez (CC BY 3.0).**

**FIGURE 84.5** Iceland poppy (*Papaver nudicaule*): (a) In a public garden in New Zealand. (b) In a park in Japan. (c) A highly doubled flower.

*Source:* **(a) Photo by Denisbin (CC BY ND 2.0). (b) Photo by Keihin Nikke (CC BY SA 3.0). (c) Photo by Nicolas Gent (CC BY ND 2.0).**

parts of Eurasia and North America. (Some taxonomists have assigned the cultivated Icelandic poppy to *P. croceum*. Note: https://accs.uaa.alaska.edu/wp-content/uploads/Papaver_croceum_BIO_PACR80.pdf.) Most garden forms are derived from hybridization with other species. The plants are technically short-lived perennials but are usually grown as annuals. The garden forms have flowers that may be white, orange, yellow, pink, or multicolored.

## 84.3 HISTORY

Opium poppy (*P. somniferum*) was domesticated in Eurasia at least 5,000 years ago, if not much earlier, for its medicinal,

FIGURE 84.6 (a) An artificial poppy worn as a commemorative lapel flower. (b) A pug lover's tribute to Remembrance Day.

*Source:* **(a) Photo by Hobvias Sudoneighm (CC BY 2.0). (b) Photo by DaPuglet (CC BY SA 2.0).**

euphoric alkaloids and for its edible seeds. Poppies were among the many plants grown in ornamental gardens in Asia long ago. However, little cultivation of *Papaver* as an ornamental occurred in Western countries until *P. orientale* and the related species that contributed to its modern cultivars were brought to Europe in the 18th century. As with the majority of very popular garden ornamental varieties, the plants have been selected and hybridized to produce relatively larger, more colorful flowers that have been doubled (i.e., with extra petals), and with petals that may be fringed or ruffled.

An allusion to spilled blood, the red poppy flower has been the symbol of fallen warriors in many countries since ancient times. Following World War I, the red poppies of Flanders Fields (*P. rhoeas*) were widely adopted as a commemorative emblem. For example, it is the emblem of the American Legion. The first Poppy Day was held in Britain on November 11, 1921.

## 84.4 ECONOMIC IMPORTANCE

By a considerable margin, the chief economic value of poppies rests with the opium poppy, grown particularly for medicinal drugs, but also for edible seeds. However, the ornamental value of the garden forms is also substantial, and poppies are one of the world's favorite garden plants in temperate areas of the world.

## 84.5 ORNAMENTAL USES

Ornamental poppies are mostly planted in gardens and borders near private residences, and to decorate public parks and the land around commercial buildings. They are used to a lesser degree in containers (mostly outdoors). Dried flowering poppies and seedpods are often used in flower arrangements. Poppies are employed as freshly cut flowers but pose difficulties. When flowering stems of *Papaver* species are cut, they often ooze a milky latex, which can limit water uptake and so shorten vase life. The latex can be reduced by soaking in hot water for a minute or searing with a flame. Poppies are often used as personal floral decorations by women, and the latex from fresh flowers could stain clothing if care is not exercised.

FIGURE 84.7 Paintings by European artists showing the use of poppies for personal and home beautification: (a) *Straw Hat with Poppies*, by Italian painter Federico Andreotti (1847–1930). (b) *Bouquet with red corn poppies*, by German artist Elsbeth Müller-Kaempff (1869–1940). (c) *Girl with a Poppy*, by French painter Émile Vernon (1872–1920).

*Source:* **Public domain.**

FIGURE 84.8 The corn poppy (*Papaver rhoeas*) as a weed and a wildflower: (a) Growing as a weed in a grain field in Provence, France. (b) Growing as a result of roadside beautification planting in Alabama.

*Source:* **(a) Photo by Willyman (CC BY SA 2.0). (b) Photo by Brandi Sims (CC BY 2.0).**

The corn poppy, which has been said to be the weediest of all *Papaver* species, is used to beautify so-called natural areas (usually fields). In the south of the United States, it is extensively used on roadsides, as it reseeds easily. Some poppy species grow well in disturbed areas, which is what most weedy habitats are. However, they are often so beautiful that instead of being labelled by the pejorative "weeds," they are termed "wildflowers."

## 84.6 OTHER USES

Alkaloids are plant-produced compounds that contain nitrogen, are generally bitter, and often are very toxic. There are thousands of alkaloids, some of which are very important in foods. For example, piperine is responsible for the burning taste of black pepper, and caffeine is a stimulant in coffee and tea. The first alkaloid ever isolated chemically was morphine, the basic alkaloid from opium. The alkaloids from poppy are widely used in pharmacology, as narcotic analgesics, sedatives, sleep-inducing agents, and to counteract diarrhea, spasms, and cough. Indeed, opium poppy has been used as a tranquillizer and analgesic for millennia. Centuries of abuse of the drug as a recreational inebriant, and the well-known misery that has resulted, suggest that

FIGURE 84.9 Traditional opium production: (a) White latex exuding from a freshly cut opium poppy seedpod that has been slit to cause release of the milky substance. (b) An opium poppy field in Afghanistan. Notice that the seed capsules have been slit longitudinally and the exuded latex has turned brown.

*Source:* **(a) Photo by KGM007 (public domain). (b) Photo by Davric (public domain).**

FIGURE 84.10 Artists' conceptions (public domain) of Chinese opium dens: (a) In San Francisco. (b) In France.

*Source:* **(a) Detroit Publishing Company postcard series number 7339, 1903–1904. (b) Cover of *Le Petit Journal*, July 5, 1903.**

FIGURE 84.11 Examples of bakery uses of opium poppy seeds: (a) Poppyseed bagel. (b) The kolache, a traditional Czech sweet bun made with an egg and yeast dough and filled with preserves and poppy seeds.

*Source:* **(a) Photo by Stas Ovsky (public domain). (b) Photo by Petr Brož (CC BY SA 3.0).**

the opium poppy has been more of a curse on humanity than a blessing. However, the constituents of the plant are invaluable in medicine, and opium poppy is one of the most important medicinal plants.

The ripe seeds of *P. somniferum* (sometimes called maw seed) are eaten or pressed for poppy oil. The seeds have a mildly spicy, oily, agreeably nutty flavor and are used as a condiment on baked goods, rolls, and pastry. About half of the seed is a pale-yellow, fixed, tasteless oil useful as a salad oil, as it is less liable to become rancid than olive oil. The oil is also used to make margarine, as salad dressing, and as a cooking oil. Artists also use poppy seed oil as a drying oil, useful in paints and varnishes. The oil is also a component of some soaps. As an animal feed, poppy seed is commonly used as birdfeed, and after oil has been expressed from seeds, the oil cake has been fed to cattle.

## 84.7 PROBLEMS

Members of the Papaveraceae almost all contain alkaloids, and consequently, many are toxic. With the exception of the seeds of some species, all parts of the plants are toxic if consumed, and there are occasional reports of people and pets being poisoned.

The corn poppy of wartime remembrance (*P. rhoeas*), which came to prominence as a weed in Flanders Fields, is a common weed in Europe. Indeed, poppies often grow as weeds among cultivated crops, and in parts of England, poppy species are sometimes known as "corn rose" (note that "corn" in England generally means the principal cereal of a region, while "maize" designates the "corn" of North America).

## 84.8 CULTIVATION

### 84.8.1 Climate and Site

Opium poppies are best adapted to climate zones 6 to 7 but can grow from zones 3 to 8. Oriental poppies have also been rated as hardy in climate zones 3 to 8 but survive cold winters far better than opium poppies. Corn poppies are adapted to warmer zones (4 to 9). Iceland poppies are adapted to cooler zones (2 to 8). Most poppies are inappropriate for areas with very hot summers. In cool climates, poppies are best placed in full sun, with semi-shade recommended for more Southern locations. Oriental poppies are the species most likely to be grown as a perennial, and because the clump that they produce will expand, at least for the first few seasons, it is wise to allow 60 cm between plants. The annuals can be planted quite closely, depending on recommendations on seed packets. The taller poppies (opium poppy, Oriental poppy) are excellent for beds, borders, and cutting gardens. Corn poppy has proven to be ideal for beautifying fields (but has the potential to spread). The small Iceland poppy is an excellent choice for rock gardens and containers.

### 84.8.2 Propagation and/or Purchasing

Most poppies (including all the annuals) are started from seeds. Poppies are sensitive to being transplanted, and if started in small containers (preferably not in flats), care should be taken to minimize handling of the root balls.

Once established, the underground portion of Oriental poppies can be divided to produce new plants.

### 84.8.3 Planting

Poppy seeds are tiny and difficult to sow evenly. They can be diluted with sand to facilitate spreading well. They need to be planted shallowly so they can simply be scattered on a well-tilled soil surface and then barely covered with a thin layer of fine soil. To prevent washing away the tiny seeds, moisten with a watering can set to deliver a fine spray. When seedlings have established, thin to at least 15 cm apart.

### 84.8.4 Soil and Moisture Conditions

All poppies do best in moist but well-drained, slightly acidic, preferably organically rich soils. Opium poppies are notably drought-tolerant.

### 84.8.5 Maintaining Plants

Oriental poppies may need staking to keep from flopping over when in full bloom. Some authorities recommend removing spent flowers and seedpods to prolong flowering; others suggest it isn't worth the bother.

## 84.9 FASCINATING FACTS, FABLES, FOLKLORE, AND FRIVOLITY

- According to an ancient Greek myth, poppies sprung from the tears of the goddess Aphrodite when she mourned for her lover, Adonis, killed by a wild boar.
- An Asian legend contends that to keep from falling asleep, Buddha cut off his eyelids, and where they fell upon the earth, opium poppies sprouted.
- According to old superstitions, staring into the center of a poppy was believed to cause temporary blindness, while bringing a poppy flower into a house would cause illness.
- In the Middle Ages of Europe, several methods of predicting one's romantic future were based on poppies. The direction from which one's true love would appear could be determined by tossing a piece of poppy cake out of the door and sending a dog out to fetch it; the dog's path in returning would indicate where the person would come from. On St. Andrew's night (November 30), maidens would scatter poppy seeds behind them to stimulate dreams about their future husbands. Or a poppy capsule could be emptied, and a question of the heart written on a piece of paper put into the seedpod, which would be placed underneath one's pillow. The answer to the question would be revealed in a dream.
- Another Middle Ages belief was that poppy seeds hidden in the shoes of a bride would make her infertile.
- Still in the Middle Ages, an invisibility potion could be prepared by steeping poppy seeds in wine for 15 days. Drinking a glassful of this brew for 5 consecutive days while fasting was believed to make the person invisible at will.
- Throwing a handful of poppy seeds in the path of pursuing demons was believed to distract them, allowing one to escape. Apparently, the demons were compelled to stop and count the seeds. This method of protection was said to work as well for vampires, and poppy seeds were put into coffins to protect the dead person's body against preying vampires.
- Floral names were very popular during the late 19th century in America and England. Favorite names were Rose, Blossom, Daisy, Iris, Pansy, Fern, Viola, Violet, Zinnia, and Poppy.
- The "Victorian language of flowers" was a secret coded language in Victorian times in England, with flowers and plants symbolic of certain messages, so when the flower or plant was mentioned in a letter, those who knew the code could understand the hidden information. "Red poppy" meant consolation, "scarlet poppy" stood for fantastic extravagance, and "white poppy" translated as sleep or my antidote.
- "Capnomancy" is the observation of smoke to tell the future. The smoke is often from the burning of poppy seeds.
- The word "poppycock," meaning *nonsense*, is not related to poppies. The word is derived from the Dutch *papekak*, which, in turn, traces to the Indo-European words *pape*, "food," + *kak*, "feces," a combination that is obviously nonsensical.
- Etymologists have established that "to have a yen for" is an expression that arose in the United States and means to have an intense desire or longing for. "Yen" in this expression is based on the Chinese *yen*, opium smoke.
- The expression "tall poppies" refers to successful individuals bullied by those who are less successful. The "tall poppy syndrome" is the behavior of people attempting to cut people occupying high-status positions down to size. However, in Italian, *alto papavero* (high poppy or tall poppy) is a phrase applied to powerful people in politics, finance, or public administration.
- In 2004, the Royal Canadian Mint produced the world's first colored circulation coin. The silver-colored 25-cent piece featuring the red image of a poppy – Canada's flower of remembrance – inlaid in the center over a maple leaf. Thirty million copies of the two-color quarter were distributed to commemorate Canada's 117,000 war dead. In 2007, sensational accounts in the media reported that the unusual quarter led to an American Defense Department false espionage warning that the coin

contained nanotechnology-based radio transmitters being used to track people and spy on the United States. The supposed nanotechnology was actually a conventional protective coating applied to prevent the poppy's red color from rubbing off.

## KEY PUBLICATIONS


Baser, K.H.C. and N. Arslan. 2014. Opium poppy (*Papaver somniferum*). In: *Medicinal and Aromatic Plants of the Middle-East*, Vol. 2, edited by Z. Yaniv and N. Dudai, 305–332. Dordrecht, Springer.

Bernath, J. (ed.). 1999. *Poppy: The Genus* Papaver. CRC Press, Boca Raton, FL. 366 pp.

Butnariu, M., C. Quispe, J. Herrera-Bravo, M. Pentea, I. Sarac, A.S. Küşümler, et al. 2022. *Papaver* plants: Current insights on phytochemical and nutritional composition along with biotechnological applications. *Oxidative Medicine and Cellular Longevity* 2022: 2041769. https://doi.org/10.1155/2022/2041769.

Forsyth, A. 1986. Poppy culture. *Harrowsmith* 65 (January/February): 61–69.

Grey-Wilson, C. 1993. *Poppies: A Guide to the Poppy Family in the Wild and in Cultivation*. Timber Press, Portland, OR. 208 pp.

Grey-Wilson, C. 2017. Oriental poppies of gardens. *Plantsman* 16: 26–27.

Jan, Y. and L.A. Peer. 2021. *Papaver somniferum*: Phytochemistry, biological activity and toxicology; a review. *International Journal of Botany Studies* 6: 753–757.

Laura, G., M. Riccardo, and L. Virginia. 2021. Corn poppy, *Papaver rhoeas* L.: A critical review of its botany, phytochemistry and pharmacology. *Phytochemistry Reviews* 20: 227–248.

Mitich L.W. 2000. Corn poppy (*Papaver rhoeas* L.). *Weed Technology* 14: 826–829.

Salavert, A., A. Zazzo, L. Martin, F. Antolín, C. Gauthier, F. Thil, et al. 2020. Direct dating reveals the early history of opium poppy in western Europe. *Scientific Reports* 10: 20263. https://doi.org/10.1038/s41598-020-76924-3.

Saunders, N.J. 2013. *The Poppy. A Cultural History from Ancient Egypt to Flanders Fields to Afghanistan*. Oneworld Publications, London. 357 pp.

Small, E. 2010. Blossoming treasures of biodiversity 33. Non-narcotic drug poppies – benefits for people and biodiversity. *Biodiversity* 11(3 & 4): 73–80.

Solstad, H. 2011. Toward an understanding of arctic and alpine poppies. In: *Papers from the CAFF Flora Group (CFG) and Circumboreal Vegetation Map (CBVM) Workshops: 2009–2010*. https://oaarchive.arctic-council.org/bitstream/handle/11374/246/CFG_CBVM_Uppsala_Sweden_2009.pdf?sequence=1#page=42.

Tétényi, P. 1996. Opium poppy (*Papaver somniferum*): Botany and horticulture. In: *Horticultural Reviews*, Vol. 19, edited by J. Janick, 373–408. John Wiley & Sons, Hoboken, NJ.


## KEY WEBSITES


Dutch Grown. *Growing Guide: How to Grow* Papaver *(Oriental Poppy)*. www.dutchgrown.com/blogs/the-dutchgrown-blog/growing-guide-how-to-grow-papaver-oriental-poppy

GardenersWorld.com. *How to Grow Poppies*. www.gardenersworld.com/how-to/grow-plants/how-to-grow-poppies/#caring

Horticulture Magazine. *Poppy*. https://horticulture.co.uk/poppy/

Kokoro Garden. *How to Grow Icelandic Poppies – Stress Free!* https://thekokorogarden.com/blog/2019/3/7/how-to-grow-icelandic-poppies-stress-free

The Spruce. *How to Grow and Care for the Common Poppy*. www.thespruce.com/growing-common-poppy-papaver-rhoeas-5113751

The Spruce. *How to Grow and Care for Oriental Poppy Plants*. www.thespruce.com/oriental-poppy-flowers-2132558

The Spruce. *10 Types of Poppy Flowers & Growing Tips*. www.thespruce.com/poppy-plant-varieties-2132570

Utah State University, Extension. *Seed Poppy in the Garden [*P. somniferum *Grown for Edible Seeds]*. https://extension.usu.edu/yardandgarden/research/seed-poppy-in-the-garden

WikiHow. *How to Grow Poppies*. www.wikihow.com/Grow-Poppies

WikiHow. *How to Grow Oriental Poppies*. www.wikihow.life/Grow-Oriental-Poppies

WikiHow. *How to Plant Poppy Seeds*. www.wikihow.com/Plant-Poppy-Seeds#:~:text=Sow%20the%20seeds%20in%20spring,a%20good%20chance%20of%20frost.

# 85 Primrose (*Primula* Species and Hybrids)

**Scientific Names:** The genus *Primula* is based on the Latin *primus*, "first," reflecting the early spring flowering.

**Common Names:** Primrose, primula. The English word *primrose* traces to the late Middle English, the Old French *primerose*, and the medieval Latin *prima rosa*, literally "first rose," reflecting (as for the genus name *Primula*) early flowering. One of the meanings of "prim" is *demure*, which is descriptive of most primroses, but this is coincidental. The "evening primroses" are species of *Oenothera*. "Water primroses" are *Ludwigia* species. "Cape primroses" are *Streptocarpus* species. As noted later, some groups of primroses are known by their own names (such as the polyanthus and the auriculas), and there are some classes of primroses (such as drumsticks and candelabras).

## 85.1 INTRODUCTION

Primroses are among the most gorgeous of flowering plants, exhibiting an astonishing range of floral colors and shapes. They are easy to grow, with numerous very hardy cultivars. Their small size makes primroses particularly suited to limited spaces, both indoors and outside. They are excellent houseplants, widely grown as short-term disposables, but they can be maintained indefinitely indoors if provided with appropriate care.

## 85.2 OVERVIEW

Primulas are mostly herbs, usually perennial, but sometimes short-lived or rarely annual. They are native to the temperate Northern Hemisphere, extending southward to tropical Asian mountains and temperate southern South America. The Himalayas and western China are major centers of distribution, where hundreds of species occur in a wide variety of habitats, including damp alpine meadows, forests, valleys, mountainsides, and streamsides. *Dodecatheon* species have been transferred to *Primula*, which is now considered to include over 500 species.

Most primroses are small stemless herbs less than 30 cm tall and wide. The leaves are stalked (petiolate), oval or roundish, and arranged in a basal rosette. The foliage in the rosette is sometimes deciduous but is typically evergreen or persistent, although often deteriorated in cold, snowy environments. A stout peduncle (stem that will bear the flowers) usually arises from the rosette (but is obscure in many of the cultivars) and develops a solitary flower, or

FIGURE 85.1 A large bed of primroses.

*Source:* **Photo by Shutterstock/Tim Large.**

FIGURE 85.2 Paintings (public domain) of *Primula* species previously placed in *Dodecatheon*: (a) *Primula frigida* (previously *Dodecatheon frigidum*). (b) *Primula jeffreyi* (previously *Dodecatheon jeffreyi*). (c) *Primula meadia* (previously *Dodecatheon meadia*).

*Source:* **(a) Curtis, W. 1870. *Curtis's Botanical Magazine* (London), vol. 96, t. 5871. (b) Van Houtte, L. B. 1865. *Flore des serres et des jardin de l'Europe*, vol. 16, t. 1662. (c) *The Garden*. 1876, vol. 10, t. 41, p. 360.**

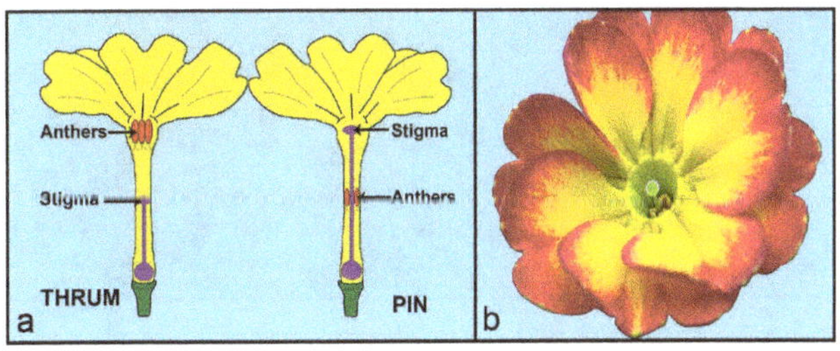


FIGURE 85.3 Heterostyly in *Primula*. (a) Diagrammatic contrast of Pin (long-styled) and Thrum (short-styled) flowers of *P. veris*. The calyx is not shown. (b) A Pin flower.

*Source:* **(a) Prepared by B. Brookes. (b) Photo by Nicolas Gent (CC BY ND 2.0).**

DOI: 10.1201/9781003586227-85

more usually a spherical cluster of flowers. The corolla tube opens into five spreading petals, which may be red, pink, purple, blue, white, or yellow. The five stamens are attached to the wall of the corolla tube. The pistil has a single slender style topped by a broad stigma. The flowers are typically 2–5 cm in diameter and often have an eye in the center that is differently colored than the petals. In the Primulaceae family, only species of *Cyclamen* and a few *Primula* species (including those formerly in *Dodecatheon*) share the characteristic of the corolla reflexing backward, making it appear as if the flower has been turned inside out. Many of the advanced cultivars have doubled flowers (i.e., with extra petals), sometimes resembling doubled rose flowers. The fruit capsule, if produced, has numerous small seeds. Many of the alpine species are small plants which are ideal for rock gardens and containers.

Heterostyly is the presence in a species of floral classes ("morphs") differing in relative position of stamens and styles. There are two classes of heterostyly: distyly, with two floral morphs, and tristyly, in which there are three morphs. *Primula* is known for distyly, producing two types of flowers in approximately a 50:50 ratio: long-styled flowers with anthers attached midway along the floral tube ("L-morph," or "pin"), and short-styled flowers with anthers attached at the top of the floral tube ("S-morph," or "thrum"). Crossing between but not within the two floral types produces viable seeds. Heterostyly has evolved in more than two dozen plant families. Heterostylous species frequently are programmed to reject pollen not just from the same plant but also from the same floral morph. About 90% of *Primula* species are distylous, whereas the remaining 10% produce self-compatible flowers of just one type. Charles Darwin, experimenting with primroses, was the first to explain that heterostyly is a mechanism promoting outbreeding. In 1876, he wrote: "I do not think anything in my scientific life has given me so much satisfaction as making out the meaning of the structure of heterostyled flowers." Darwin was responsible for the terms *pin* and *thrum*. He named pin flowers for the resemblance of the prominent style and stigma in the mouth of the flower to a dressmaker's pin, and thrum flowers for the resemblance of the anthers at the mouth of the flower to a "thrum" – an old weaving term for a tuft of thread.

FIGURE 85.4 Primrose cultivars with highly doubled flowers: (a) An 1890 chromolithograph showing seven different cultivars. (b) *Primula vulgaris* 'Scarlet Parade'. (c) *Primula sinensis*.

*Source:* **(a) Wellcome Images (CC BY 4.0). (b) Photo by HQ (CC BY SA 2.0). (c) *The Floral Magazine*, 1874, new ser. vol. 3. London, L. Reeve & Co. (public domain).**

FIGURE 85.5 Old botanical paintings (public domain) of English primrose (*Primula vulgaris*).

*Source:* **(a) *Revue horticole* (Paris), 1880. (b) *Revue horticole* (Paris), 1898. (c) Step, E., D. Bois, B. Herincq, and W. Watson. 1896. *Favourite Flowers of Garden and Greenhouse*, vol. 3. Frederick Warne, New York.**

FIGURE 85.6 Modern cultivars of English primrose (*Primula vulgaris*): (a) A display of English primroses in a park in London, England. (b) Cultivar Sirocco. (c) Cultivar Belarina Nectarine.

*Source:* **(a) Photo by Acabashi (CC BY SA 4.0). (b) Photo by Shutterstock/John R Martin. (c) Photo by Daderot (public domain).**

English primrose (*P. vulgaris*) is one of the most frequently cultivated primulas. It is also a main parent in the polyantha hybrids, discussed next, and the name English primrose is often employed as a generic name for many of the polyantha hybrids. Houseplant primroses are often the English primrose.

"Polyanthus" primroses (horticulturally termed *P.* × *polyantha*) are hybrid cultivars of *P. veris* and *P. vulgaris*. The word *polyanthus* is based on the Greek *polyanthos*, meaning "many-flowered." A large variety of colorful varieties is available, and these frequently dominate offerings at garden centers. Indeed, most primroses cultivated outdoors are these hybrids. The group has especially diverse flowers, some bicolored, some striped, and some doubled. The polyanthas are perennial, although often short-lived (grown as annuals or biennials), and are often established from seed or from young plants. The polyanthus primroses and its parental species are frequently grown as pot plants. In Japan, *P.* × *polyantha* was crossed with *P. juliae* to produce unique dwarf cultivars called Julian hybrids.

The orchid primrose (*P. vialii*), also known as Chinese pagoda primrose and poker primrose, is a particularly attractive short-lived garden perennial growing to 30–45 cm in height. At first sight, it does not look like a typical primrose. It produces spikes of tiny pinkish-purple flowers, but the uppermost flowers, which are unopened, are reddish. The species is relatively frost-tender but can be grown as an annual in cold climates.

Several other species are particularly well-known, as exemplified by the following. "Candelabra" primroses (such as *P. japonica*, *P. bulleyana*, and *P. florindae*) carry their flowers in whorls spaced on a main peduncle. Many

FIGURE 85.7 Polyanthus primroses (*Primula* × *polyantha*).

*Source:* (a) Photo by Foshie (CC BY 2.0). (b) Photo by Garry Knight (CC BY 2.0). (c) Cultivar Black Silver Lace. Photo by Shutterstock/Alex Manders. (d) Photo by Shutterstock/Kobus Peche.

FIGURE 85.8 Julian hybrid primroses (hybrids with *Primula juliae*).

*Source:* (a) Photo by Kanonn (CC BY ND 2.0). (b) Photo by Shutterstock/Kuma-mon.

FIGURE 85.9 Orchid primrose (*Primula vialii*).

*Source:* (a) Photo by Shutterstock/LapaiIrKrapai. (b) Photo by Garten-gg from Pixabay (public domain).

FIGURE 85.10 Candelabra primrose (*Primula japonica*).

*Source:* (a) Photo by Shutterstock/AngieC333. (b) Step, E., D. Bois, B. Herincq, and W. Watson. 1896. *Favourite flowers of garden and greenhouse*, vol. 3. Frederick Warne, New York. (c) Photo by Alh1 (CC BY ND 2.0). (d) *Revue horticole* (Paris), sér. 4, vol. 42, 1870.

FIGURE 85.11 Drumstick primrose (*Primula denticulata*).

*Source:* (a) Photo by Andreas (CC BY SA 2.0). (b) Photo by Agnieszka Kwiecień, Nova (CC BY SA 4.0). (c) Photo by Afanasovich (CC BY SA 4.0).

FIGURE 85.12 Auricula primrose (*Primula auricula*).

*Source:* (a) Witte, H. 1868. ***Flora: afbeeldingen en beschrijvingen van boomen, heesters, éénjarige planten, enz. voorkomende in de Nederlandsche Tuinen.*** **Wolteres, Groningen. (b) Redouté, P. J. 1827.** ***Choix des belles fleures.*** **Panckoucke, Paris. Notice the farina (whitish covering) on the foliage. (c)** ***The Floral Magazine*** **(London), vol. 6, 1867.**

of the cultivars have doubled flowers. Drumstick primrose (*P. denticulata*) is a widely planted garden species, with globular flower heads on 30 cm long peduncles (looking like drumsticks).

Some of the species develop a mealy white powder ("farina") on their surfaces. This is most pronounced in some of the natives of high-altitude habitats, where there is high irradiation. The farina appears to be protective, like sunblock lotion. The character has been particularly selected for its ornamental value in auricula primroses (*P. auricula*), which are often grown protected in glasshouses to maximize the white bloom for competitive flower shows. Auricula primroses have small clusters of flat-topped flowers and grows 15–20 cm in height. Numerous floral variations have been selected. Aside from the powdery coating, the plants are also notable for their distinctive thick leaves. Unusual for primroses, they are adapted to alkaline soil.

## 85.3 HISTORY

Primroses have been cultivated for centuries. As with other major ornamental plants, most cultivars grown today are hybrids, often with parentage tracing to several wild species.

FIGURE 85.13 Nineteenth-century paintings of Chinese primrose (*Primula sinensis*), which was much more popular in the past than today. Note: *Flora of China* (www.efloras.org/florataxon.aspx?flora_id=2&taxon_id=200017411) indicates that the correct name of *P. sinensis* is *P. praenitens*, a name change which is unfamiliar to most primrose specialists.)

**Source: Public domain illustrations. (a) Cover of a 19th-century catalogue. (b)** ***Deutsches Magazin fur Garten-und Blumenkunde*****, 1858, G. Weise, Stuggart. (c) Lindley, J. 1821–1826,** ***Collectanea botanica or, figures and botanical illustrations of rare and curious exotic plants.*** **Richard and Arthur Taylor, London. (d) Gartenflora (Berlin), Jahrg.41.1892.**

Although most varieties have Asian parentage, several European species were also domesticated over the last several hundred years, such as English primrose (*P. vulgaris*), also known as acaulis primrose), oxlip (*P. elatior*), cowslip (*P. veris*), and fairy primrose or baby primrose (*P. malacoides*).

## 85.4 ECONOMIC IMPORTANCE

Primroses represent one of the major, standard flowering commercial ornamentals, grown in numerous greenhouses for sale as houseplants and bedding and container plants. Many commercial interests internationally are involved in breeding and selling of primroses.

## 85.5 ORNAMENTAL USES

Primroses are widely employed as bedding and border plants, typically as low-growing herbs and as potted plants, especially as houseplants.

## 85.6 OTHER USES

The flowers and foliage of *Primula* primroses are considered to be edible, but most reports are based on *P. vulgaris*

FIGURE 85.14 Primroses being grown in a commercial greenhouse.

*Source:* **Photos by Vicky Brock (CC BY SA 2.0).**

FIGURE 85.15 Potted primroses.

*Source:* **(a) Photo by Shutterstock/New Africa. (b) Photo by Shutterstock/IfiStudio. (c) Photo by Shutterstock/New Africa. (d) Photo by Ursula55 from Pixabay (public domain).**

FIGURE 85.16 Landscapes employing primroses: (a) A bed in Scotland. (b) A bed in England. (c) A bed in The Netherlands. (d) A bed in England.

*Source:* **(a) Photo by Andrew Hendo (CC BY SA 3.0). (b) Photo by Leonora (Ellie) Enking (CC BY SA 2.0). (c) Photo by Pixelia from Pixabay (public domain). (d) Photo by Alh1 (CC BY ND 2.0).**

FIGURE 85.17 Edible flowers and foliage of *Primula* primroses: (a) A salad of leaves and flowers. (b) A garnish of flowers on a yeast loaf.

*Source:* **(a) Photo by Shutterstock/Marina Lohrbach. (b) Photo by Shutterstock/Natalia Greeske.**

and *P.* × *polyantha*. The leaves have been cooked in soup and employed for tea but vary in flavor, sometimes mild, but also bitter at times. Younger foliage is likely to taste better. The young flowers have been made into primrose wine. The flowers have been crystallized (traditionally with an egg white wash and a coat of sugar) and employed as a garnish. Primulas have a long history of folk medicinal usage and are known to possess a variety of chemicals which can affect metabolism, so caution is advisable, as some people might react negatively to consumption. The flowers are typically consumed in such small amounts that harm is unlikely.

## 85.7 PROBLEMS

The very attractive poison primrose (also known as German primrose; *P. obconica*) has leaves that can cause skin irritation, and indeed, this was a danger in the past, especially for florists. In modern times, the cultivar Touch Me is often marketed, since it lacks primine, the principal sensitizer causing itching.

## 85.8 CULTIVATION

A considerable proportion of gardeners consider primroses to be "disposable annuals," grown primarily as dainty container plants, especially as houseplants or as patio or window box plants. Nevertheless, there are numerous cultivars that are viewed as superlative garden plants.

### 85.8.1 Climate and Site

Most primulas are intolerant of summer heat and drought, and indeed, in such climates, the plants are often grown as annuals. By contrast, numerous primrose cultivars are extremely hardy and overwinter well in cold climates. Indeed, many require winter chilling to survive and produce flowers. Nevertheless, winter mulching is often recommended to minimize freezing damage. Depending on species and cultivar, primulas are hardy in parts of the range from zones 2 to 9, but most are hardy in zones 5 to 7. Zones suitable for the popular species are as follows:

FIGURE 85.18 Poison primula (*Primula obconica*).

*Source:* **(a) Chromolith, ca. 1897. Wellcome collection (CC BY 4.0). (b) Photo by Yewchan (CC BY SA 2.0). (c) Die Gartenwelt, 1897, G. Schmidt, Berlin. (d) Cultivar Libre Salmon. Photo by David Stang (CC BY SA 4.0).**

FIGURE 85.19 Primulas surviving early spring snow: (a) Cowslip (*Primula veris*). (b) English primrose (*P. vulgaris*).

*Source:* **(a) Photo by Shutterstock/Karen Faljyan. (b) Photo by Shutterstock/Gertjan Hooijer.**

*P. acaulis*, zones 4 to 8; *P. auricula*, zones 3 to 8; *P. denticulata*, zones 2 to 8; *P. japonica*, zones 4 to 8; *P. polyanthus*, zones 4 to 7; *P. veris*, zones 3 to 8; *P. vialii*, zones 5 to 8; *P. vulgaris*, zones 4 to 8.

Outdoors, primroses usually prefer partial shade and are often planted in masses near deciduous trees or in locations with north-facing exposure to the sun. However, planting under trees could be disadvantageous, as the tree roots may outcompete the primulas for moisture and soil nutrients. In cold Northern climates, the plants may need more exposure to the sun, although open locations may subject the plants to winter desiccation.

## 85.8.2 Propagation and/or Purchasing

The principal cultivated species and classes are sold in mass marketing outlets. Purchasing potted primroses when they are in flower guarantees the plant will reproduce the desired kinds of flowers (seeds sometimes are of hybrid origin, and the resulting plants segregate for flower color). The uncommon species are often available from rock garden societies and specialist online nurseries. The American Primrose Society (see "Key Cultivation Websites" guide, presented later) sells seeds to members.

Quite cool conditions (10°C or lower) are often required to germinate the tiny seeds (often carried out in a refrigerator) and establish seedlings. Many cultivars will self-sow outdoors. After flowering in the garden, primrose can easily be lifted and divided, which may be the only way of maintaining the genetic purity of some hybrids.

## 85.8.3 Planting

Primroses usually transplant well. The crowns of transplants should be positioned at soil level and at least 15 cm apart.

## 85.8.4 Soil and Moisture Conditions

Most cultivated primroses of Asian and/or woodland origin prefer slightly acidic (pH ca. 6.5), strongly organic soils, and some of these do best in constantly moist soils. Species of upper alpine origin tend to thrive in well-draining gritty soils resembling their high-mountain scree habitats.

## 85.8.5 Maintaining Plants

The doubled flower cultivars are especially attractive but tend to be relatively fragile. To maintain vigor, it has been recommended that the spent flowers be deadheaded (removed), and the plants divided every 2 or 3 years.

## 85.8.6 Care of Indoor Plants

The majority of potted houseplant primroses are purchased in supermarkets (hence the nickname "grocery store primroses"), often as gift plants. They are extremely attractive, relatively cheap, and compact and so appeal strongly to impulse buyers. Unfortunately, many are disappointed by the difficulties in maintaining the plants long-term. Primroses need moderate light and cool temperatures to thrive and do not respond well to relatively dark, overheated homes, especially in winter. They also require consistent moist but not soaked soil. To prolong good growth, the following procedures are recommended. Decorative wrapping or foil on the outside of the pot should be removed, as it tends to retard free drainage. The plants should be located in a cool (10 to 20°C) location, and the soil maintained neither too dry nor too wet. If a very bright location is unavailable, the plants may be placed under a fluorescent light, about 30 cm above the plants, for at least 12 hr a day. Fertilizer is usually not necessary for several months. Note that gift primrose houseplants (frequently *P. vulgaris*) are often forced to bloom for out-of-season sales, and when subsequently planted outdoors, they may lack the energy reserves needed to survive. In cold climates, the exercise of attempting to salvage houseplants for transplanting outdoors is likely to be unsuccessful.

## 85.9 FASCINATING FACTS, FABLES, FOLKLORE, AND FRIVOLITY

- One of the oddest primula variations is called “hose-in-hose.” In Elizabethan times in England (i.e., during the reign of Queen Elizabeth I, 1558–1603), it was fashionable for gentlemen to wear two pairs of “hose,” the upper hose often in the form of puffy knee-length trousers which were met by the lower hose (essentially a long stocking) on the lower leg. Hose-in-hose primulas seem to develop a flower from inside another flower, hence the name hose-in-hose. In fact, a mutation often observed in wild *Primula* causes the calyx of one flower to develop into another flower. Curiously, the floral mutation is linked to the S locus, which determines whether the flowers are pin or thrum.
- Reflecting the tradition of accentuating the attractiveness of women with beautiful objects, females are often shown decorated with flowers, including primroses, in allegoric paintings (i.e., art with figures reflecting a story or message).
- Many ships are named for flowers, and the unfortunate demise of one of these is notable. In 1981, a Hong Kong–registered cargo ship, MV *Primrose*, ran aground on a coral reef off North Sentinel Island in the Bay of Bengal, considered to be “the last island on Earth solely inhibited by noncontacted Native people.” The island is out of bounds under Indian law because of the hostility of the Natives. For a week, the crew was threatened by members of the Sentinelese tribe on the beach, wielding spears, shooting arrows, and attempting to board the freighter from canoes. Finally, the Indian navy rescued the crew. Less fortunate was a Christian missionary attempting to convert the Natives. In 2018, he was shot to death with arrows tipped with metal that the islanders had obtained from the Primrose wreck.
- Primroses are recognized culturally in several ways. Cowslip (*Primula veris*) is well-known in the UK and is the “county flower” of Northamptonshire, Surrey, and Worcestershire. The primrose (likely *P. vulgaris*, the English primrose) was the favorite flower of Benjamin Disraeli (1804–1881), Tory (Conservative) prime minister of England. This led to the founding in 1883 of the Primrose League, an organisation for spreading Conservative principles in Great Britain, which was influential until its disbandment in 2004. “Primrose Day” (April 19) in the UK marks the anniversary of Disraeli’s death. The much-repeated phrase “primrose path,” referring (often pejoratively) to a hedonistic route through life, was coined by Shakespeare, in *Hamlet* (in the very early 17th century).

FIGURE 85.21 Allegoric paintings (related to fictional characters) showing women decorated with primroses (public domain figures).

**Source: (a) Gordon, A. E. 1912. *Flower children; the little cousins of the field and garden.* (b) Painting titled *La Primevere/Polyanthus*, by Alphonse Mucha, 1899. (c) Cassell & Co., Ltd., London. 1909. Flowers from Shakespeare’s garden.**

FIGURE 85.20 “Hose-in-hose” appearance in fashion and in primula: (a) Hose-in-hose fashion in Elizabethan times. Painting (National Portrait Gallery, London) of Henry Wriothesley, 3rd Earl of Southampton (1573–1624), one of William Shakespeare’s patrons. (b) Hose-in-hose mutation in *Primula*.

**Source: (a) Photo by Ann Longmore-Etheridge (public domain). (b) Photo by Jacki-dee (CC BY ND 2.0).**

## KEY PUBLICATIONS

American Primrose Society (corporate author). 1967. *Pictorial Dictionary of the Cultivated Species of the Genus* Primula, 2nd edition. American Primrose Society, Tacoma, WA. 108 pp.

Barrett, S.C.H. 2019. 'A most complex marriage arrangement': Recent advances on heterostyly and unresolved questions. *New Phytologist* 224: 1051–1067.

Blasdale, W.C. 1948. *The Cultivated Species of* Primula. University of California Press, Berkeley. 284 pp.

Connolly, M., J. McCune, E. Dauncey, and C.R. Lovell. 2004. *Primula obconica* – is contact allergy on the decline? *Contact Dermatitis* 51: 167–171.

Fenderson, G.K. 1986. *A Synoptic Guide to the Genus* Primula. G.K. Fenderson, U.S. 213 pp.

Ganders, F.R. 1979. The biology of heterostyly. *New Zealand Journal of Botany* 17: 607–635.

Genders, R. 1964. *The Polyanthus*. Faber & Faber, Brantford, MA. 231 pp.

Gilmartin, P.M. 2015. On the origins of observations of heterostyly in *Primula*. *New Phytologist* 208: 39–51.

Halda, J.J. 1992. *The Genus* Primula *in Cultivation and the Wild*. Tethys, Denver, CO. 364 pp.

Jacquemyn, H., P. Endels, R. Brys, M. Hermy, and S.R.J. Woodell. 2009. Biological Flora of the British Isles: *Primula vulgaris* Huds. (*P. acaulis* (L.) Hill). *Journal of Ecology* 97: 812–833.

Karlsson, M.G. 2001. Primula culture and production. *HortTechnology* 11: 627–635.

Kato, J., M. Inari-Ikeda, M. Hayashi, J. Amano, H. Ohashi, and M. Mii. 2018. *Primula*. In: *Ornamental Crops. Handbook of Plant Breeding*, Vol. 11, edited by J. Van Huylenbroeck, 627–647. Springer, Cham.

Kelso, S. 2009. *Primula* Linnaeus. In: *Flora of North America North of Mexico*, Vol. *8: Magnoliophyta: Paeoniaceae to Ericaceae*, edited by Flora of North America Editorial Committee, 286–301. Oxford University Press, New York.

Lawson, E. 2019. *Primrose*. Reaktion Books, London. 256 pp.

Li, J., B. Dudas, M.A. Webster, H.E. Cook, B.H. Davies, and P.M. Gilmartin. 2010. Hose in Hose, an S locus – linked mutant of *Primula vulgaris*, is caused by an unstable mutation at the Globosa locus. *PNAS* 107: 5664–5668.

Lyall, H.G. 1959. *Hardy Primulas*. Transatlantic Arts, London. 88 pp.

Mast, A.R. and J.L. Reveal. 2007. Transfer of *Dodecatheon* to *Primula* (Primulaceae). *Brittonia* 59: 79–82.

Nowak, M.D., G. Russo, R. Schlapbach, C.N. Huu, M. Lenhard, and E. Conti. 2015. The draft genome of *Primula veris* yields insights into the molecular basis of heterostyly. *Genome Biology* 16: 12. https://doi.org/10.1186/s13059-014-0567-z.

Reveal, J.L. 2009. *Dodecatheon*. In: *Flora of North America North of Mexico*, Vol. *8: Magnoliophyta: Paeoniaceae to Ericaceae*, edited by Flora of North America Editorial Committee, 268–286. Oxford University Press, New York.

Richards, J. 2002. *Primula*, Revised edition. Batsford/Pavilion, London. 346 pp.

Smith, W.R., G. Forrest, and H.R. Fletcher. 1977. *The Genus* Primula. Cramer, Vaduz, Liechtenstein. 835 pp.

Ward, P. 1997. *Primroses and Polyanthus: A Guide to the Species and Hybrids*. Batsford, London. 160 pp.

## Key Cultivation Websites

American Primrose Society. https://americanprimrosesociety.org/.

GardenersWorld.com. *How to Grow Primroses and Primulas*. www.gardenersworld.com/how-to/grow-plants/how-to-grow-primulas/.

Horticulture Magazine. Primula *'Primrose'*. https://horticulture.co.uk/primula/.

National Auricula and Primula Society. www.auriculaandprimula.org.uk/.

Penn State, Extension. Primula *Diseases*. https://extension.psu.edu/primula-diseases.

*Primula World: A Visual Reference for the Genus* Primula. https://primulaworld.blogspot.com/.

Royal Horticulture Society. *Primroses and Polyanthus*. www.rhs.org.uk/plants/primula/primroses-and-polyanthus.

The Spruce. *How to Grow and Care for Primroses*. www.thespruce.com/primula-the-primrose-is-no-shy-wallflower-1402858.

U.K. House Plants. Primula *(All Indoor Species)*. www.ukhouseplants.com/plants/primula.

# 86 Ranunculus (*Ranunculus asiaticus*)

**Scientific Names:** The genus name *Ranunculus* is based on Latin for "little frog," the diminutive of *rana*, "frog," probably inspired by the waterside presence of many of the *Ranunculus* species.

**Common Names:** The vernacular name ranunculus, of course, is based on the scientific name and is very frequently employed specifically for *R. asiaticus*. This species is also called Persian ranunculus and Persian buttercup (*R. asiaticus* is indeed native to "Persia" (modern Iran), but more generally to the eastern Mediterranean area of southwestern Asia, southeastern Europe, and northeastern Africa). Most *Ranunculus* species are known as buttercups. The common name buttercup is speculated to have arisen from the idea that butter is yellow because cows consumed the usually yellow-flowered plants. Many species of *Ranunculus* (including *R. asiaticus*) are called crowfoot (the dissected leaf of such species is reminiscent of a crow's foot). Turban buttercup is an occasional name.

## 86.1 INTRODUCTION

The flowers of Persian ranunculus are incredibly attractive, as indicated by their frequent comparison to the blooms of camellias, peonies, and roses. While ranunculus is employed as garden and container ornamental plants, it is principally used in bouquets and flower arrangements and is one of the dominant staples of the cut flower industries.

FIGURE 86.1 Ranunculus (*Ranunculus asiaticus*) hybrids in flower.

*Source:* **(a) Photo by Shutterstock/Nora Connors Photography. (b) Photo by Shutterstock/Nnattalli. (c) Photo by 4028mdk09 (CC BY SA 3.0).**

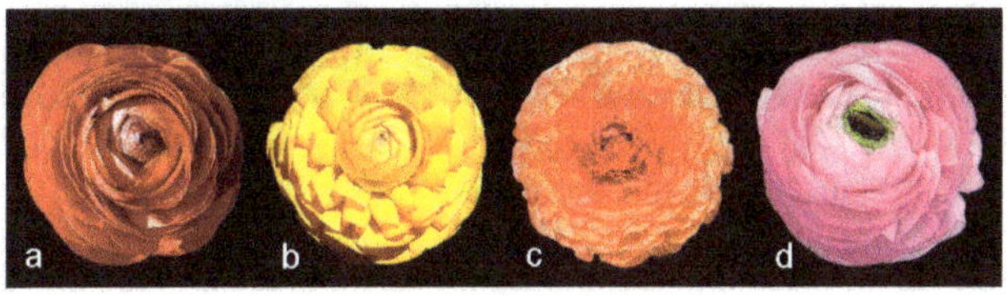

FIGURE 86.2 Examples of color variation in doubled flowers of Persian ranunculus (*Ranunculus asiaticus*) hybrids.

*Source:* **(a, b, c) Photos by Nino Barbieri (CC BY SA 2.5). (d) Photo by HQ (CC BY SA 2.0).**

## 86.2 OVERVIEW

*Ranunculus* is a genus of about 1,800 species of the Old World, or much fewer according to some authorities who classify it differently. It includes the buttercups and is especially known for the weedy creeping buttercup (*R. acris*) and meadow buttercup (*R. repens*). Cultivars have been selected of many of the buttercups, especially of the weedy species, but the Persian ranunculus highlighted in this chapter is the dominant ornamental.

Most garden varieties of Persian ranunculus grow to a height of 30 to 45 cm. The foliage is dissected (segmented, like parsley), especially the upper leaves. Unlike the wild plants which produce five petals, the cultivars usually have doubled flowers with multiple layers of petals packed together, closely resembling modern doubled roses. The doubled flowers frequently have over 100 petals. They are often about 5 cm in diameter but in some cultivars are sometimes over 15 cm. The flowers come in a rainbow of spectacular colors, including pink, purple, red, yellow, white, and any of numerous intermediate shades, and they are sometimes bicolored. Flowering occurs in the early season (late spring and early summer in temperate climates). The tuberous roots have an odd, claw-like appearance, often compared to miniature octopuses. A single large tuberous root can produce 5 to 10 stems and a total of over 50 flowers, but typically only about 4 are mature at a given time.

## 86.3 HISTORY

The Persian ranunculus appears to have been present in the gardens of the rulers of the Ottoman Empire as far back as the 13th century. However, its cultivation in England began in the 16th century. Subsequently, the popularity of the plant has varied over the centuries, but in recent times, thanks to advances in rapid transportation of flowers, ranunculus has become a star of the florist industries.

FIGURE 86.3 Old botanical illustrations (public domain) of Persian ranunculus (*Ranunculus asiaticus*). Notice the octopus-like appearance of the tuberous roots.

***Source:*** **(a) Step, E. 1896. *Favourite Flowers of Garden and Greenhouse.* (b) Guillaumin, A. 1929. *Les Fleurs de Jardins, tome I: Les Fleurs de Printemps*, Paul Lechevalier, 1929. (c) The Gottorfer Codex, a collection of paintings on vellum depicting flowers of the garden of Schloss Gottorf, created between 1649 and 1659. (d) Bessa, P. 1817–1827. *Herbier général de l'amateur*, vol. 8.**

FIGURE 86.4 Ranunculus flower farms: (a) In southern Israel. (b) In Southern California.

***Source:*** **(a) Photo by Shutterstock/Kavram. (b) Photo by Geraldine Knappenberger from Pixabay (public domain).**

## 86.4 ECONOMIC IMPORTANCE

While *Ranunculus asiaticus* is employed mostly by the cut flower industry, there are also substantial sales of potted plants and tuberous roots for establishing garden plants.

## 86.5 ORNAMENTAL USES

Flowering stems (best harvested when the flower buds are just starting to open) may last up to 2 weeks in vases. Of course, the plants can be grown at home, but most people purchase them from florists or other outlets, and the flowers are widely employed in uniform or mixed bouquets, as gifts, and for use in weddings and other celebrations. In addition to use as cut flowers, the plants are also suitable for beds, borders, and containers.

## 86.6 OTHER USES

*Ranunculus* species are under investigation for the possible pharmacological applications of their chemical constituents.

FIGURE 86.5 Cut flowers of Persian ranunculus.

***Source:*** **(a) Photo by Shutterstock/Evrymmnt. (b) Photo by Shutterstock/Agnes Kantaruk. (c) Photo by Shutterstock/Alona Gryadovaya. (d) Photo by Stefan Schweihofer from Pixabay (public domain). (e) Photo by Mansaku Ogikubo from Pixabay (public domain). (f) Photo by Alexia Schu from Pixabay (public domain).**

## 86.7 PROBLEMS

Ranunculus is toxic to livestock, pets, and people when consumed and can also cause dermatitis. A particular danger if taken orally is blisters in the mouth. Gloves are advisedly worn when handling the plants. After purchasing the tuberous roots, or storing them, care should be taken to prevent access by children and pets. The plants are bitter and have the advantage of not being attractive to deer.

## 86.8 CULTIVATION

### 86.8.1 Climate and Site

Persian ranunculus is winter-hardy in climate zones 8 to 11 but is grown as an annual in more Northern areas. Hot summer temperatures promote dormancy, which, of course, stops flowering. The plants should be situated in full sun or slight shade (some shade is preferable to avoid very hot afternoons). A site that is hard to access by animals and children is preferable since the plants are toxic. While

FIGURE 86.6 Persian ranunculus in containers.

*Source:* **(a) Photo by Shutterstock/Victoria Kurylo. (b) Photo by Shutterstock/Siete Vidas.**

FIGURE 86.7 Livestock consuming toxic weedy buttercups: (a) Stallions. (b) Calves.

*Source:* **(a) Photo by C Watts (CC BY 2.0). (b) Photo by Shutterstock/Anton Havelaar.**

not particularly suitable for use as houseplants, like most plants grown from "bulbs," the tuberous roots can simply be forced to produce an indoor display.

### 86.8.2 Propagation and/or Purchasing

Persian ranunculus can be grown from seeds, preferably started indoors, but this is unlikely to produce plants that will flower well the first season. Most home gardeners start with purchased tuberous roots. After the plants go dormant, the tuberous roots (usually labelled as tubers or bulbs) can be salvaged and stored for replanting the following spring. However, they do tend to lose vigor and are difficult to store successfully, and because they are relatively inexpensive, most gardeners choose to purchase new planting stock.

Nurseries frequently sell ranunculus "tubers" in four sizes. "Jumbo" tubers are from 7.0 to 7.9 cm in circumference. "Number One" tubers are 5.7 to 7.0 cm in circumference. "Number Two" tubers are 5.1 to 5.7 cm in circumference. "Number Three" tubers, infrequently marketed, are 2.9 to 5.7 cm in circumference. Jumbo roots can produce 30 to 40 flowers and are recommended. Smaller tuberous roots may produce only around 10 flowers. A plant grown from a Number Two tuber develops about 20 flowers, and a Number Three tuber yields about 12 flowers. The largest tubers should be employed in containers, but for mass plantings, the smaller tubers can produce a satisfactory display.

### 86.8.3 Planting

The tuberous roots can be planted 5 cm deep and 15 to 30 cm apart (larger cultivars should be separated by about 30 cm). The roots ("claws") naturally point downward, and the opposite (stem) end where they are joined should be pointing upward when planted in the ground. Many recommend soaking the tuberous roots for 24 hr (or at least several hours) before planting (some flower guides claim this is unnecessary). During the soaking period, the roots absorb considerable water and swell (if they are still alive).

### 86.8.4 Soil and Moisture Conditions

Persian ranunculus should be provided with moderately moist, well-drained, fertile soils, with a pH range of 6.1 to 7.6. Unamended clay is unsuitable; loamy soil is best.

### 86.8.5 Maintaining Plants

Deadheading away spent flowers will prolong the blooming period.

## 86.9 FASCINATING FACTS, FABLES, FOLKLORE, AND FRIVOLITY

- Respected Roman author, naturalist, and philosopher Pliny the Elder (23/24–79) apparently recommended employing buttercup species to treat

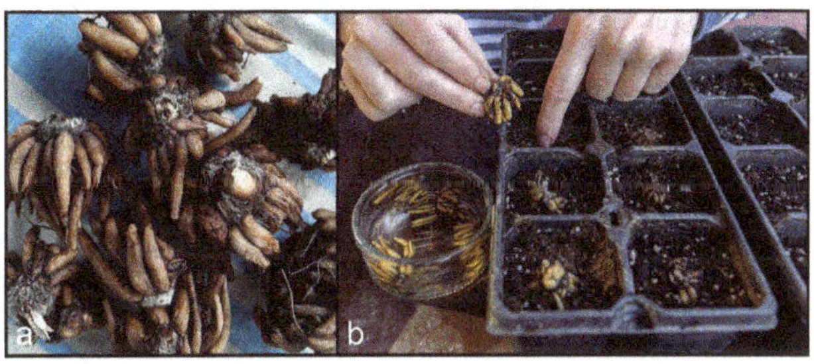

FIGURE 86.8 (a) Tuberous roots of Persian ranunculus. (b) Presoaked tuberous roots being planted into a plug tray.

*Source:* **(a) Photo by Carlfbagge (CC BY 2.0). (b) Photo by Shutterstock/Abo photography.**

leprosy, and down to medieval times, buttercup was used to treat blisters and other skin conditions. Ironically, buttercup sap produces skin irritation. Indeed, buttercup juice was one of the agents used by medieval beggars to blister their skin purposefully to arouse the sympathy of passersby. In herbal medicine, there is an old, widespread idea termed "doctrine of signatures" that holds that nature provides signals to people about the possible medical values of plants, so the fact that buttercup produces skin irritation may have been interpreted as a sign that it should be used to reverse the condition.

- Nineteenth-century fishermen sometimes poured buttercup tea on the ground to bring worms for bait to the surface.
- The petals of many buttercup species are shiny because of a reflective double layer of air just beneath the surface. The flowers are designed to focus the reflected light toward the center of the flower, apparently to create a warm environment attractive to pollinators.
- In English-speaking countries, a well-known tradition was to hold a buttercup under a person's chin, and if it glared yellow (which it would if well-lighted, as noted earlier), it meant he liked butter.
- *HMS* Pinafore *(or, The Lass That Loved a Sailor)* is a famous comic opera by Arthur Sullivan and W. S. Gilbert, with the first appearance in London, England, in 1878. One of the prominent characters was Little Buttercup, a rather stout, elderly "bumboat woman" whose occupation was to supply the officers and crew of the ship with little luxuries. The crew nicknamed her "Little Buttercup" because she was not at all like a pleasant pretty little flower. Her well-known aria includes the stanza:

  I'm called Little Buttercup – dear Little Buttercup,
  Though I could never tell why,
  But still I'm called Buttercup – poor Little Buttercup,
  Sweet Little Buttercup I!

FIGURE 86.9 Scenes from the comic opera *HMS Pinafore* by Sullivan and Gilbert, with special reference to "Little Buttercup": (a) An 1879 advertisement for an American production of the opera, housed at the US Library of Congress. (b) An 1878 advertising poster for the original theater presentation in London. (c) A 1926 cigarette card showing Little Buttercup on the HMS *Pinafore*. (d) An 1870s trade card by a Kentucky ham company, showing Little Buttercup with hams. (e) An illustration of Little Buttercup from Gilbert, W.S. 1913. *The story of HMS Pinafore*. G. Bell and Sons, London.

***Source:*** **Public domain illustrations.**

## KEY PUBLICATIONS


Aslam, M.S., B.S. Choudhari, M. Uzair, and A.S. Ijaz. 2012. The genus *Ranunculus*: A phytochemical and ethnopharmacological review. *International Journal of Pharmacy and Pharmaceutical Sciences* 4: 15–22.

Beruto, M., P. Martini, and S. Viglione. 2019. *Ranunculus asiaticus*: From research to production. *Acta Horticulturae* 1237: 117–128.

Beruto, M., M. Rabaglio, S. Viglione, M.-C. Van Labeke, and E. Dhooghe. 2018. *Ranunculus*. In: *Ornamental Crops*, edited by J. van Huylenbroeck, 649–671. Springer International, Cham, Switzerland. https://doi.org/10.1007/978-3-319-90698-0_25.

Dole, J. and H. Wilkins. 1999. *Ranunculus*. In: *Floriculture: Principles and Species*, edited by J. Dole and H. Wilkins, 792–797. Prentice Hall, Upper Saddle River, NJ.

Emadzade, K., B. Gehrke, H.P. Linder, and E. Hörandl. 2011. The biogeographical history of the cosmopolitan genus *Ranunculus* L. (Ranunculaceae) in the temperate to meridional zones. *Molecular Phylogenetics and Evolution* 58: 4–21.

Emadzade, K., C. Lehnebach, P. Lockhart, and E. Hörandl. 2010. A molecular phylogeny, morphology and classification of genera of Ranunculeae (Ranunculaceae). *Taxon* 59: 809–882.

Fusco, G.M., P. Carillo, R. Nicastro, L. Pagliaro, G.C. Modarelli, S. De Pascale, et al. 2023. The vernalization procedure influences growth and metabolism of tuberous roots of *Ranunculus asiaticus* L. *Preprints* 2023: 2023071297. https://doi.org/10.20944/preprints202307.1297.v1.

Horovitz, A. 1985. *Ranunculus*. In: *Handbook of Flowering*, Vol. 4, edited by H.A. Halevy, 155–160. CRC Press, Boca Raton, FL.

Karlsson, M. 2003, January. Producing ravishing ranunculus. *Greenhouse Product News*, 44–48.

Meynet, J. 1974. Research on ranunculus hybrid varieties for growing under protection. *Acta Horticulturae* 43: 191–196.

Ohkawa, K. 1986. Growth and flowering of *Ranunculus asiaticus*. *Acta Horticulturae* 177: 165–177.

Rauter, S. and M. Stock. 2023. *Ranunculus Cut Flower Production in Utah*. Utah State University Extension Factsheet. 6 pp. https://digitalcommons.usu.edu/cgi/viewcontent.cgi?article=3354&context=extension_curall.

Shahri, W. and I. Tahir. 2011. Flower development and senescence in *Ranunculus asiaticus* L. *Journal of Fruit and Ornamental Plant Research* 19: 123–131.

Van der Kooi, C., T. Elzenga, J. Dijksterhuis, and D. Stavenga. 2017. Functional optics of glossy buttercup flowers. *Journal of the Royal Society Interface* 14(127): 20160933. https://doi.org/10.1098/rsif.2016.0933.

## Key Cultivation Websites

Alaska Master Gardener Blog. *Growing Ranunculus in Southcentral Alaska.* https://alaskamastergardener.community.uaf.edu/2022/04/06/growing-ranunculus-in-southcentral-alaska/.

Clemson University Cooperative Extension. *Add Ranunculus to Your Fall Planting to-do List.* https://hgic.clemson.edu/add-ranunculus-to-your-fall-planting-to-do-list/.

Dutch Grown. *Growing Ranunculus – All You Need to Know!* www.dutchgrown.com/blogs/the-dutchgrown-blog/growing-ranunculus-all-you-need-to-know.

GardenersWorld.com. *How to Grow Ranunculus (Persian buttercup).* www.gardenersworld.com/how-to/grow-plants/ranunculus-bulbs/.

Royal Horticultural Society. *How to Grow Ranunculus.* www.rhs.org.uk/plants/ranunculus/growing-guide.

Shifting Roots. *How to Grow Ranunculus in Zone 3 (So They Don't Die!).* https://shiftingroots.com/how-to-grow-ranunculus-in-zone-3-so-it-doesnt-die/.

The Spruce. *How to Grow and Care for Ranunculus Flowers.* www.thespruce.com/ranunculus-7104375.

WikiHow. *How to Plant and Care for a Ranunculus.* www.wikihow.com/Plant-and-Care-for-a-Ranunculus.

# 87 Rhododendron (Azalea, *Rhododendron* Species, and Hybrids)

**Scientific Names:** The genus name *Rhododendron* is based on the Greek *rhodon*, "rose," and *dendron*, "tree."

**Common Names:** Rhododendron, rhody, azalea. Information on the use of the confusing name "azalea" is provided later in the text (the scientific name *Azalea* is obsolete). "Mock azalea" refers to *Adenium obesum* and sometimes other species unrelated to *Rhododendron*. "False azalea" is *Menziesia ferruginea* (which is also called mock azalea).

FIGURE 87.1 Huge public gardens featuring rhododendrons in Japan, where outstanding displays are found. (a) Kasama Tsutsuji Park. (b) Rhododendron festival in Tokyo. (c) Tutujigaoka Park. (d) Togakuin, the Azalea Temple.

*Source:* **(a) Photo by Keihin Kake (CC BY SA 3.0). (b) Photo by Peachbird (CC BY SA 3.0). (c) Photo by T. Kunikuni (CC BY 2.0). (d) Photo by Big Ben (CC BY SA 2.0).**

## 87.1 INTRODUCTION

Rhododendron (including "azalea") has been called "the king of shrubs," a reflection of the spectacular beauty of the large, colorful flowers which cover the plants when they are in full display. With over a thousand species representing a very wide range of habitats and tens of thousands of hybrid cultivars, rhododendrons have become indispensable ornamentals for landscaping large public areas as well as private residences, and they are even important as houseplants. In testimony to their attractiveness, picturesque gardens dedicated to rhododendrons have been established in many countries, and there are numerous annual festivals in their honor.

## 87.2 OVERVIEW

The genus *Rhododendron* is estimated to have over 1,000 species, indigenous mainly to the Northern Hemisphere. Rhododendrons are mostly native to Eurasia, North America, and northeastern Australia. The majority are in Asia, especially in the Himalayas and Malaysia. Over 400 species are endemic to China. Twenty-five species are established in North America north of Mexico. Rhododendrons occur from the arctic to the tropics, and from lowlands down to sea level, to alpine sites up to 5,800 m. They occupy quite diverse habitats, including woodlands, temperate rainforests, and tropical jungles. The plants are usually in markedly organic soils, in regions with high rainfall and humidity, and in temperate climates. They are often on sloping hills, so the soils tend to be well-drained. Unfortunately, more than a quarter of the 1,000 or so species of *Rhododendron* are under threat of extinction in their native habitats.

FIGURE 87.2 *Rhododendron arboreum*, native to Asia, but widely cultivated elsewhere. Sometimes the trees live for more than a century and reach over 30 m in height.

*Source:* **(a) Photo by Shutterstock/Emagnetic. (b) Photo by Jacinta lluch valero (CC BY SA 2.0) (c) Gigantic multi-trunked tree in Cornwall, UK. Photo by Carcharoth (CC BY SA 3.0).**

*Rhododendron* species are shrubs, small trees, or rarely, large trees. Some are tiny prostrate alpine shrubs no taller than 5 cm, while others are giant rainforest trees over 30 m in height. The foliage may be evergreen or deciduous, and



DOI: 10.1201/9781003586227-87

FIGURE 87.3 Rhododendron hybrid cultivars with doubled flowers (i.e., with extra petals).

*Source:* **(a) Photo by Shutterstock/K. Kemruji. (b) Photo by Raita Futo (CC BY 2.0).**

the leaves may be large (sometimes over 1 m in length) or small (less than 1 mm). About 600 species (subgenus *Rhododendon*) have tiny epidermal "scales," usually visible on the leaf undersurfaces (best seen with a magnifying glass). Species with scales are termed "lepidote"; those without are called "elepidote." Most species flower in the spring and/or early summer. The flowers are available in most colors, and there are cultivars with variegated flowers and/or doubled flowers. Most cultivars are of hybrid origin, often with parentage tracing to several species.

### 87.2.1 Azaleas

Species once placed in the genus *Azalea* are now assigned to *Rhododendron*. In common language, the name "azalea" is used in several ways to refer to certain species. Two subgenera of *Rhododendron*, usually distinguished by possession of flowers with only five stamens, are often called azaleas, while the remaining species with ten stamens are called rhododendrons. Sometimes, rhododendrons are said to usually be evergreen, and azaleas are claimed to be deciduous. However, evergreen "azaleas" constitute *Rhododendron* subgenus *Tsutsusi*, while deciduous "azaleas" make up subgenus *Penanthera*. There is no consistent character to distinguish an "azalea," and the word has become a source of confusion. Nevertheless, in the horticultural and retail worlds, there is often a preference for "azalea." Horticulturally, most cultivars are arbitrarily divided into evergreens and deciduous forms. "Azaleodendrons" are semi-evergreen hybrids (usually sterile) between deciduous "azaleas" and rhododendrons.

### 87.2.2 Vireya Species

Rhododendrons of section *Vireya*, often simply called vireyas and sometimes termed Malesians, include about 300 species of Malesia (the Malay Archipelago, Borneo, New Guinea, and adjacent regions). The group is named for the French pharmacist and natural historian Julien Joseph Virey (1775–1846). These evergreen species are too tender to grow outdoors in Northern regions, where they are raised as glasshouse specimens. Although most are "tropical," the majority are montane plants adapted to fairly cool conditions.

FIGURE 87.4 Illustrations (public domain) of selected species of vireya rhododendrons: (a) *Rhododendron retusum*. (b) *Rhododenron javanicum*. (c) *Rhododendron jasminiflorum*.

*Source:* **(a) *L'illustration horticole* (Gand, Belgium), vol. 2, plate 70, 1855. (b) *Curtis Botanical Magazine* (London), vol. 73: t. 4336, 1947. (c) L' Illustration horticole (Gand, Belgium), vol. 6, plate 203, 1859.**

FIGURE 87.5 Rhododendrons grown as houseplants.

*Source:* **(a) Photo by Kārlis Dambrāns (CC BY 2.0). (b) Photo by Manuel m.v. (CC BY 2.0).**

### 87.2.3 Houseplant Species

Some of the evergreen rhododendrons are among the most popular of flowering houseplants. "Florist's pot azaleas" are generated by specialty greenhouses. They are slow-growing and are forced to flower for seasonal sales, especially for Christmas. These are frequently hybrids with *R. simsii*, a species that is not cold-hardy but grows very well in containers and is useful as a houseplant. Belgium is the world's leading producer, in some years generating over $100 million in revenue. Belgian pot azaleas are primarily hybrids among *R. simsii*, *R. indicum*, *R. scabrum*, and *R. mucronatum* (all evergreens). Some of these (especially *R. indicum*) are also grown as bonsai.

## 87.3 HISTORY

Rhododendrons have been known since the times of the classical Romans but appear to have first been cultivated in Europe in the 17th century. One of the first domesticated species in Europe was *R. luteum*, a shrub sometimes more than 3 m tall. This is the only deciduous-leaved

FIGURE 87.6 *Rhododendron indicum* grown as bonsai.

*Source:* **(a) Photo by Shutterstock/Bernd Schmidt. (b) Photo by Sarah Stierch (CC BY 4.0). (c) Photo by Shutterstock/Bernd Schmidt.**

FIGURE 87.7 *Rhododendron luteum*: (a) Old cultivars. (b) A modern cultivar.

*Source:* **(a) Witte, H. 1868. Flora. Afbeeldingen en beschrijvingen van boomen, heesters, éénjarige planten, enz., voorkomende in de Nederlandsche tuinen. A. J. Wendel, Groningen, Netherlands (public domain). (b) Photo by Sgerbic (CC BY SA 4.0).**

*Rhododendron* native to Europe, and it is widely grown in western Europe.

Many species of *Rhododendron* can be easily hybridized. China and Japan have long been centers of breeding based on Asian species, and beginning about 1825, gardeners in Europe began extensive programs of interbreeding. In the United Kingdom, species from the eastern United States were the first subjects of breeding, followed by introductions from the Himalayas, Caucasus, and Turkey. Most of the early European hybrids are moderately cold-resistant and are often called "hardy hybrids." In the early 20th century, species from China and the eastern Himalayas were added to breeding programs in Western nations. Over 30,000 hybrids have been recognized, and in many cases, the pedigree of cultivars is unknown.

## 87.4 ECONOMIC IMPORTANCE

Rhododendrons are among the world's most popular flowering shrubs and, accordingly, are of considerable value in horticulture. Potted "azaleas" are popular gift houseplants and therefore valued by the florist industry.

## 87.5 ORNAMENTAL USES

*Rhododendron* species and cultivars are highly valued in landscaping, particularly as flowering shrubs and as ornamental evergreens. Tree forms and the larger shrubs make excellent isolated specimens. Rhododendrons are occasionally employed as hedges. As noted earlier, there is also considerable use of dwarf species as houseplants and as bonsai. Low-growing species are also often employed in rock gardens.

FIGURE 87.8 Commercial production and marketing of rhododendrons: (a) A field of plants flowering in the spring. (b) A greenhouse with plants.

*Source:* **(a) Photo by Shutterstock/Midfield Man. (b) Photo by Shutterstock/LightField Studios.**

FIGURE 87.9 Rhododendrons serving to landscape buildings. (a) "Spanish Steps," a landmark of Rome, Italy, famous as the longest and widest steps in Europe. (b) The historic town center of Westerstede, Germany. (c) A house with a straw roof in Dreibergen, Germany. (d) A well-landscaped house in Texas.

*Source:* **(a) Photo by Shutterstock/Gennaro Leonardi. (b) Photo by JoachimKohler-HB (CC BY SA 4.0). (c) Photo by Shutterstock/Josep M. Penalver Rufas. (d) Photo by Ms. Mannix (CC BY 2.0).**

## 87.6 OTHER USES

In Asia, *Rhododendron* flowers are occasionally encountered in food, mostly as a sour flavoring herb. Particularly in Korea, flowers have been employed to prepare wine, honeyed flower juice, and griddle cakes topped with blossoms. In the state of Himachal Pradesh, India, the flowers are used to flavor wines. However, such culinary uses are inadvisable because of toxicity, as noted in the next section.

"Labrador tea" refers to dried leaves employed as a beverage obtained from natural stands of *R. groenlandicum*, less often from *R. tomentosum* or *R. neoglandulosum* (the

FIGURE 87.10 Rhododendrons in large public parks: (a) In Kawasaki, Kanagawa Prefecture, Japan. (b) In the UK. (c) In China. (d) In Broenderslev, Jütland, Denmark.

***Source:* (a) Photo by Tak1701d (CC BY SA 3.0). (b) Photo by Shutterstock/Chrislofotos. (c) Photo by Mahlervv (CC BY SA 3.0). (d) Photo by Uwe1959 (CC BY SA 3.0).**

FIGURE 87.11 Rhododendrons in containers.

***Source:* (a) Photo by Shutterstock/Phovoir. (b) Photo by Shutterstock/Cristina Ionescu.**

FIGURE 87.12 Griddle cakes topped with rhododendron blossoms: (a) "Azalea pancakes" made with flowers of *Rhododendron mucronulatum* and prepared by pan-frying kneaded dough (a combination of glutinous rice flour and sugar) on a hot frying pan with vegetable oil. (b) Korean "flower cakes." In Korea, these are traditionally eaten on the traditional festival called Samjinnal, celebrated on March 3 in the lunar calendar.

***Source:* (a) Photo by Greenful (CC BY SA 2.0). (b) Photo by Shutterstock/Let Geo Create.**

three species formerly placed in the genus *Ledum*). This traditional beverage of Native inhabitants of North America is a minor commercial herbal product. Consumption should

FIGURE 87.13 Labrador tea (*Rhododendron groenlandicum*).

***Source:* (a) Illustration from Voight, F. 1854. *Sämmtliche Giftgewächse Deutschlands.* Leipzig, Germany (public domain). (b) Photo by Krzysztof Ziarnek, Kenraiz (CC BY SA 4.0). (c) "Labrador dog tea" by Shutterstock/Kitoumi.**

be limited because of the presence of ledol, a toxic sesquiterpene present in the essential oil. The degree to which grayanotoxins (discussed in the following) are also present is unclear, and indeed, it is probable that populations are available with low amounts of ledol and grayanotoxins.

Despite toxicity, *Rhododendron* species have been widely used in folk medicine systems in Asia, Europe, and North America, notably against gastrointestinal disorders, inflammation, pain, skin problems, and the common cold. In North America, *R. groenlandicum* has been employed by Indigenous people to treat diabetes, a usage which has been shown to have some validity.

## 87.7 PROBLEMS

*Rhododendron* and other genera of the Ericaceae family contain a class of toxins called grayanotoxins (named after the Japanese species *Leucothoe grayana*, commemorating the famous 19th-century American botanist Asa Gray). Fatal livestock (especially sheep and goat) kills occur, and very rarely, dogs and cats are affected. Honeybees are also occasionally poisoned (note Tiedeken et al. 2015). Caution is advisable to prevent access of domesticated animals to wild and cultivated plants, including houseplants. The relative degree of toxicity of different *Rhododendron* species is unclear, some authors stating that only certain species are significantly toxic, others implying that most, if not all, species are dangerous. North American evergreen *Rhododendron* species appear to be somewhat more toxic to livestock than deciduous species.

*Adenium obesum* (sometimes called *mock azalea* but better termed *dessert rose*), often grown as an ornamental and sometimes confused with "azalea," is extremely poisonous.

Humans occasionally inadvertently eat honey in which grayanotoxin-containing pollen and/or nectar from Ericaceae species occurs. According to legends, poisoning of soldiers by wild-collected contaminated honey determined the outcome of several wars among the classical Greeks, Romans, and Turks. So-called "mad honey" is deliberately produced in some regions of Asia (especially the Caucasus region of Turkey) from plants of the

FIGURE 87.14 Invasive *Rhododendron ponticum* in England: (a) In Lundy Island. (b) In Norfolk.

*Source:* **(a) Photo by Mary Gillham (CC BY 2.0). (b) Photo by Kolforn (CC BY SA 4.0).**

heath family, particularly from *Rhododendron* species, and employed for its alleged medicinal, aphrodisiac, or hallucinogenic properties. (It may be noted that all kinds of honey generally tend to have some medicinal value.) Symptoms of "mad honey disease" in people include low blood pressure, blurred vision, hallucinations, and seizures. In the 1700s, mad honey from the Black Sea region was exported to Europe and added to alcoholic beverages to increase intoxication. However, in recent times, sale of mad honey has been restricted in many jurisdictions.

*Phytophthora ramorum*, a fungus-like plant pathogen popularly called "sudden oak death" and "ramorum blight," has become a devastating disease of woodlands, commercial plantations, and nurseries in the last two decades. Probably from Asia (at least in the European population), it is now spreading in parts of Europe (especially the UK) and North America (especially California). This disease kills or weakens numerous species of trees and shrubs. Rhododendrons are susceptible but can live for many years in an affected state, spreading spores to other plants. Ramorum blight was discovered in Europe in the early 1990s on rhododendrons, and in the UK, the highly invasive *R. ponticum* is a very serious vector of the disease. Millions of trees in the Old and New Worlds have been destroyed, and ecosystems are being devastated. Control measures are difficult, and it appears that like several other epidemics harming plants, the world's vegetation may be permanently compromised.

Some rhododendron species are invasive, threatening native biodiversity. *Rhododendron ponticum* is a threat in New Zealand and western Europe and is considered to be the most serious plant pest in the UK.

## 87.8 CULTIVATION

It may be noted that there is not always unanimity of opinion on best practices for growing rhododendrons. Hundreds of articles are available online at the website of the American Rhododendron Society, experts offering opinions ranging from benign neglect to sophisticated analyses of fertilization, pruning, landscaping, and other topics.

### 87.8.1 Climate and Site

Rhododendrons are very popular shrubs in climates that are moist and are not exposed to extreme temperatures. Often, microclimates, such as frost pockets, determine survival. Rhododendrons are often planted near foundations, and the associated warmth frequently promotes survival in cold areas. For protection from severe winters, evergreens can be wrapped in burlap (with minimal contact to foliage). Rhododendrons tend to be shallow-rooted and are best planted away from trees that could provide competition. Most species thrive in partial shade (or full sun during part of the day, preferably in the morning). Although rhododendrons are shade-tolerant, they should not be placed in deep shade. Evergreens are best situated in locations sheltered from wind that might desiccate the foliage. Rhododendrons bloom for only a few weeks, but cultivars can be chosen so that flowering plants will be visible from spring to midsummer. Plants should be located where they are not objectionable during the non-flowering period.

### 87.8.2 Propagation and/or Purchasing

Commercial propagation of rhododendrons is carried out using stem cuttings, seeds, grafting, and micropropagation. Most species of *Rhododendron* that are not named hybrids or cultivars can be propagated from seeds by gardeners. Rhododendrons usually produce hundreds of tiny seeds in their capsules, which are naturally dispersed by wind after splitting of the fruits in the autumn. Seeds lose viability rapidly but will last if stored dry and cold.

In practice, nearly all rhododendron cultivars are propagated by cuttings, which are established as young potted plants for commercial sale. Ideally, plants sold locally should be the most appropriate cultivars for the locality, but today large marketers often distribute mass-produced biotypes, to be sold over large areas, for which they are often not completely suitable. Since rhododendrons can persist for many years, when considering new plantings, it is advisable to search for recently bred, particularly attractive cultivars bred for the conditions of the local climate, and to contact specialty suppliers.

### 87.8.3 Planting

To plant container-grown rhododendrons, remove the root ball from the container. If dry, soak until thoroughly moist. If pot-bound, score and loosen the fibrous roots. Place the top of the root ball slightly higher than the top of the hole (higher than surrounding ground surface). Firm-in soil around the root ball, and water-in well. Medium and large rhododendrons can be planted as isolated specimen plants or associated in beds with other shrubs. Low-growing plants are suitable near foundations or in front of taller plants.

### 87.8.4 Soil and Moisture Conditions

Like most species in the heath family (Ericaceae), rhododendrons are adapted to acidic soil, preferring a pH of 4.5–5.5

(although good growth often results up to a pH of 6.5, and species have been recorded growing at a pH >8.1). Drainage of water toward the plants from nearby alkaline soils is undesirable (sometimes, plants placed near houses suffer from alkaline seepage from the concrete foundations). Porous loam soils rich in organic matter are recommended. Good drainage is essential. If a site has clay and/or quite alkaline soil, raised beds should be considered. Fertile soils rarely require fertilizer, but in infertile soils, a complete fertilizer designed for acid-loving plants can be applied in limited amounts early in the season. Applying 5 cm of organic mulch (peat moss and wood chips are not recommended by some specialists) under the shrub annually tends to retain soil moisture and suppress weeds (avoid placing mulch near the trunk, as this can promote rot). Watering should be done only when the soil is moderately dry, but plants should not be subjected to drought.

### 87.8.5 Maintaining Plants

Deadheading after flowering is sometimes practiced. So is pruning, but this is not essential (although diseased and damaged branches should be removed).

### 87.8.6 Indoor Plants

Most indoor "azaleas" purchased from stores as seasonal houseplants are discarded after flowering, as they are difficult to rejuvenate. To prolong the flowering display, the plants should be maintained at cool temperatures (16–18°C), in well-lit areas out of direct sun, and in soil that is maintained damp but not soggy.

## 87.9 FASCINATING FACTS, FABLES, FOLKLORE, AND FRIVOLITY

- The Berserkers were ancient Viking warriors, reputedly invulnerable, enormously strong, and wildly frenzied in battle. In the Scandinavian sagas, Berserkers are often described as being fantastically ugly and guilty of monstrously depraved behavior, and their reputation terrified their opponents. The Danish historian Saxo Grammaticus (about 1150–1220) recorded the activities of a band of Berserkers in his *Gesta Danorum*:

  > The young warriors would harry and pillage the neighbourhood, and frequently spilled great quantities of blood. They considered it manly and proper to devastate homes, cut down cattle, rifle everything and take away vast hauls of booty, burn to the ground houses they had sacked, and butcher men and women indiscriminately. . . . [T]hey ravished other men's wives and daughters. . . . [S]carcely any place in the land was free from the imprints of their lust.

  It has been suggested that intoxication from Labrador tea (*Rhododendron tomentosum*) contributed to the deranged but effective wildness of the Berserkers.

FIGURE 87.15 Cartoons of Berserker Vikings, whose savage behavior has been attributed to drinking toxic rhododendron beverages.

***Source:* (a) Photo by Shutterstock/HitToon. (b) Photo by Shutterstock/Subarashii21. (c) Photo by Shutterstock/Macrovector.**

- The strong smell of Labrador tea was employed to discourage pests in past times. In Scandinavia, branches were placed among harvested cereals stored in barns to drive away mice. In Germany, branches were kept in bedrooms to discourage fleas from entering.
- Reflecting the toxicity of rhododendrons, sending a bouquet of flowering stems in a black vase was once considered to be a death threat.
- Because of their attractiveness, rhododendron species have frequently been chosen as official emblems of political regions. These include *R. aboreum* (national flower of Nepal), *R. ponticum* (state flower of Kashmir), *R. niveum* (state tree of Sikkim, India), *R. campanulatum* (state flower of Himachal Pradesh, India), *R. maximum* (state flower of West Virginia), and *R. macrophyllum* (state flower of Washington).
- The foliage of large-leaved evergreen rhododendrons that must survive winters often becomes tightly enrolled during freezing weather – a response that has been called a "natural thermometer." This phenomenon, termed "thermotropism" or "thermonasty," is an adaptation lowering water loss and desiccation when the plants cannot replenish the supply of moisture from their roots.
- During the Second World War, on August 6, 1945, the first atomic bomb used in war was dropped on Hiroshima, Japan, effectively obliterating the city and killing over 140,000 people. The skeletal ruins of a building near the epicenter have been maintained as a reminder of the destructive effects of nuclear warfare. Generally known as the Genbaku

FIGURE 87.16 Thermotropic foliage response of *Rhododendron maximum* to freezing temperatures by curling and folding down: (a) Normal, expanded leaves. (b) Rolled-up leaves.

***Source:*** **(a) Photo by James St. John (CC BY 2.0). (b) Photo by Nicholas A. Tonelli (CC BY 2.0).**

FIGURE 87.17 Rhododendron bordering the Peace Memorial ("Genbaku Dome") in Hiroshima, Japan, site of the first atomic bomb employed in war.

***Source:*** **Photo taken in May 2018 by Sky Ear (CC BY SA 4.0).**

Dome, the remains of the building are part of the Hiroshima Peace Memorial Park, designated as a UNESCO World Heritage Site in 1996. The Genbaku Dome is bordered by rhododendrons.

## KEY PUBLICATIONS

Antonelli, A.L., R.S. Byther, R.R. Maleike, S.J. Collman, and A.D. Davison. 1984. *How to Identify Rhododendron and Azalea Problems*. Washington State University Extension Bulletin 1229. 29 pp.

Blazich, F.A. and D.B. Rowe. 2008. *Rhododendron* L., rhododendron and azalea. In: *The Woody Plant Seed Manual. Agricultural Handbook* 727, edited by F.T. Bonner and R.P. Karrfalt, 943–951. U.S. Department of Agriculture Forest Service, Washington, DC.

Bowers, C.G. 1960. *Rhododendrons and Azaleas: Their Origins, Cultivation and Development*. Macmillan, New York. 525 pp.

Casati, M., T. Kichey, and G. Decocq. 2022. Monographs on invasive plants in Europe N° 7: *Rhododendron ponticum* L. *Botany Letters* 169: 213–236.

Chamberlain, D. 1996. *The Genus* Rhododendron*: Its Classification and Synonymy*. Royal Botanic Garden, Edinburgh, UK. 181 pp.

Cox, K. 2005. *Rhododendron & Azaleas, a Colour Guide*. Crowood Press, Ramsbury, UK. 240 pp.

Criley, R.A. 1985. Rhododendrons and azaleas. In: *CRC Handbook of Flowering*, Vol. 4, edited by A.H. Halevy, 180–197. CRC Press, Boca Raton, FL.

Cullen, J. 2005. *Hardy* Rhododendron *Species: A Guide to Identification*. Timber Press, Portland, OR. 496 pp.

Dampc, A. and M. Luczkiewicz. 2015. Labrador tea – the aromatic beverage and spice: A review of origin, processing and safety. *Journal of the Science of Food and Agriculture* 95: 1577–1583.

Davidian, H.H. 1982–1995. *The* Rhododendron *Species*, 4 Vols. Timber Press, Portland, OR.

Demircan, A., A. Keleş, F. Bildik, G. Aygencel, N.O. Doğan, and H.F. Gómez. 2009. Mad honey sex: Therapeutic misadventures from an ancient biological weapon. *Annals of Emergency Medicine* 54: 824–929.

Douglas, G.W. and J. Desrosiers. 2006. Conservation evaluation of Pacific rhododendron, *Rhododendron macrophyllum*, in Canada. *Canadian Field-Naturalist* 120: 169–174.

Eşen, D., E.T. Nilsen, and O. Yildiz. 2006. Ecology, competitive advantages, and integrated control of rhododendron: An old ornamental yet emerging invasive weed around the globe. In: *Floriculture, Ornamental and Plant Biotechnology*, Vol. 3, edited by J.A. Teixeira da Silva, 407–421. Global Science Books, East Sussex, UK.

Gibbs, D., D. Chamberlain, and G. Argent. 2011. *The Red List of Rhododendrons*. Botanic Gardens Conservation International, Richmond, UK. 129 pp. www.bgci.org/files/Worldwide/Publications/PDFs/RhododendronsRedList-lowRes.pdf.

Jansen, S.A., I. Kleerekooper, Z.L. Hofman, I.F. Kappen, A. Stary-Weinzinger, and M.A. van der Heyden. 2012. Grayanotoxin poisoning: 'Mad honey disease' and beyond. *Cardiovascular Toxicology* 12: 208–215.

Judd, W.S. and K.A. Kron. 2009. *Rhododendron*. In: *Flora of North America North of Mexico*, Vol. 8, edited by Flora of North American Editorial Committee, 456–473. Oxford University Press, New York.

Krebs, S.L. 2018. *Rhododendron*. In: *Ornamental Crops. Handbook of Plant Breeding*, Vol. 11, edited by J. Van Huylenbroeck, 673–718. Springer, Cham.

Kurashige, Y., J.-I. Etoh, T. Handa, K. Takayanagi, and T. Yukawa. 2001. Sectional relationships in the genus *Rhododendron* (Ericaceae): Evidence from matK and trnK intron sequences. *Plant Systematics and Evolution* 228: 1–14.

Li, S., A. Brault, M. Sanchez Villavicencio, and P.S. Haddad. 2016. *Rhododendron groenlandicum* (Labrador tea), an antidiabetic plant from the traditional pharmacopoeia of the Canadian Eastern James Bay Cree, improves renal integrity in the diet-induced obese mouse model. *Pharmaceutical Biology* 54: 1998–2006.

Linderman, R.G. and D.M. Benson. (eds.). 2014. *Compendium of Rhododendron and Azalea Diseases*, 2nd edition. American Phytopathological Society, St. Paul, MI. 144 pp.

McQuire, J.F.J., M.L.A. Robinson, and H.H. Davidian. 2009. *Pocket Guide to* Rhododendron *Species*. Kew Publishing, Royal Botanic Gardens, Kew, UK. 692 pp.

Nilsen, E.T. 1992. Thermonastic leaf movements: A synthesis of research with *Rhododendron*. *Botanical Journal of the Linnean Society* 110: 205–233.

Popescu, R. and B. Kopp. 2013. The genus *Rhododendron*: An ethnopharmacological and toxicological review. *Journal of Ethnopharmacology* 147: 42–62.

Postan, C. (ed.). 1996. *The Rhododendron Story: 200 Years of Plant Hunting and Garden Cultivation*. Royal Horticultural Society, London, UK. 224 pp.

Salley, H.E. and H.E. Greer. 1986. Rhododendron *Hybrids: A Guide to Their Origins: Includes Selected, Named Forms of* Rhododendron *Species*. Timber Press, Portland, OR. 391 pp.

Shi, Y., M. Zhou, Y. Zhang, Y. Fu, J. Li, and X. Yang. 2021. Poisonous delicacy: Market-oriented surveys of the consumption of *Rhododendron* flowers in Yunnan, China. *Journal of Ethnopharmacology* 265: 113320. www.sciencedirect.com/science/article/abs/pii/S0378874120332025.

Smith, A.W. n.d. *Rhododendrons and Azaleas for Ontario*. Ministry of Agriculture and Food, Toronto. 20 pp. www.rhodoniagara.org/rhodos/rhodos_omag45.pdf.

Tiedeken, E.J., P.A. Egan, P.C. Stevenson, G.A. Wright, J.F. Brown, E.F. Power, et al. 2015. Nectar chemistry modulates the impact of an invasive plant on native pollinators. *Functional Ecology* 30: 885–893.

Veen, T. Van. 1980. *Rhododendrons in America*, 2nd edition. Binford & Mort, Portland, OR. 180 pp.

Xia, X.-M., M.-Q. Yang, C.-L. Li, S.-X. Huang, W.-T. Jin, T.-T. Shen, et al. 2022. Spatiotemporal evolution of the global species diversity of *Rhododendron*. *Molecular Biology and Evolution* 39(1): msab314. https://academic.oup.com/mbe/article/39/1/msab314/6413646?login=true.

## Key Cultivation Websites

American Rhododendron Society. www.rhododendron.org/.

Azalea Society of America. https://azaleas.org/.

Gardenia. *Hardiest Azaleas and Rhododendrons*. www.gardenia.net/guide/Hardiest-Azaleas-and-Rhododendrons.

GardenersWorld.com. *How to Grow Azaleas*. www.gardenersworld.com/how-to/grow-plants/how-to-grow-azaleas/.

Grant, B.L. *Rhododendrons for Zone 4 Gardens – Types of Cold Hardy Rhododendrons*. www.gardeningknowhow.com/garden-how-to/gardening-by-zone/zone-4/rhododendrons-in-zone-4-gardens.htm.

Royal Horticultural Society. *How to Grow Rhododendrons*. www.rhs.org.uk/plants/rhododendron/growing-guide.

The Spruce. *How to Grow and Care for Rhododendron*. www.thespruce.com/rhododendron-growing-guide-7109537.

WikiHow. *How to Grow Rhododendron*. www.wikihow.com/Grow-Rhododendron.

WikiHow. *How to Plant Azaleas*. www.wikihow.com/Plant-Azaleas.

WikiHow. *How to Propagate Azaleas*. www.wikihow.com/Propagate-Azaleas.

WikiHow. *How to Prune Azaleas*. www.wikihow.com/Prune-Azaleas.

WikiHow. *How to Prune Rhododendrons*. www.wikihow.com/Prune-Rhododendrons.

# 88 Rose (*Rosa* Species and Hybrids)

**Scientific Names:** The genus name *Rosa* is based on the ancient Latin name for roses, *rosa*, which, in turn, originated from the Akkadian (the ancient language of Mesopotamia) word *russu*, "red."

**Common Names:** The English word "rose," like the genus name *Rosa*, traces to Latin, which is also the root of the name in other languages: *rose* in Danish, French, German, and Norwegian, and *rosa* in Italian, Portuguese, Russian, and Spanish. Not all plants with "rose" in their name are true roses. For example, "rose-apple" is *Syzgium jambos* of the Myrtaceae family. This tree is widely cultivated in tropical regions for its dry, crisp fruit that is insipid raw but prized for confections and jellies. The "Christmas rose" is *Helleborus niger*, a perennial herb of the Ranunculaceae family. Several ornamental cultivars of this are available, and the plant has also been used medicinally. However, all parts of the plant are extremely poisonous, and it should not be consumed. The "rose of Jericho" is any of several desert plants: *Anastatica hierochuntica* (mustard family) of northern Africa and Southwest Asia; *Selaginella lepidophylla* (club moss family) distributed from Texas and Arizona south to El Salvador; *Odontospermum pygmaeum* (sunflower family) of Asia Minor. These plants form tight balls when dry and unfold and grow under moist conditions. In the dry condition, they are tumbleweeds, scattering seeds as they roll about in the wind. Some plants called rose of Jericho, especially *Selaginella lepidophylla*, are also called "resurrection plants," because unlike most tumbleweeds, which are dead in the dry condition, the plants remain alive but dormant while they are tumbling about, and when provided with water, they begin to grow again. "Convertible rose" is lantana, discussed in this book.

The word "rose" has some special culinary meanings. "Stinking rose" is a euphemism for *garlic*. A "Russian rose" is a cocktail prepared with vodka, grenadine, and orange bitters. "China rose tea" is a blend of freshly cut rose petals and Chinese black tea, the resulting brew having a strong rose scent. Mei Kwei (known as rose tea) is a similar blend, combining Chinese black tea and rosebuds. In old lunch counter slang, "Pin a rose on it" meant add onion to an order. "Burn one, take it through the garden, and pin a rose on it" meant hamburger with lettuce, tomato, and onion.

## 88.1 INTRODUCTION

Over 2,500 years ago, the Greek poetess Sappho referred to the rose as "the queen of flowers." Today, rose species collectively are the most popular flowering ornamental plants in the world. From an economic perspective, roses rival the most valuable crops. From aesthetic and sentimental perspectives, however, roses are unrivalled – no other plants are so universally grown for their beauty, and none has equally inspired writers, artists, and lovers. "Rosarians" are devoted rose hobbyists.

The symbolism of the rose surpasses all other flowers. The perfect blossom represents love, beauty, youth, perfection, and immortality; withering of the flower reflects the ephemeral nature of beauty and youth; the thorns epitomize the pain of love and guilt; the white rose portrays purity and innocence and is often associated with the Virgin Mary; the red rose is the embodiment, on the one hand, of sensual desire, passion, and shame, and blood and sacrifice on the other; the yellow rose reflects perfect achievement, or sometimes jealousy, and can also mean joy and friendship (it meant infidelity and jealousy in Roman times but today often means friendship); a dark crimson rose signifies mourning; a dark pink rose stands for thankfulness; a

FIGURE 88.1 Colorful collections of rose flowers.

*Source:* (a) Photo by Carsten Sprotte (CC BY 2.0). (b) Photo by Shutterstock/LiliGraphie. (c) Photo by Shutterstock/Luz Rodas Arbelaez.



DOI: 10.1201/9781003586227-88

FIGURE 88.2 Roses – symbols of love and romance: (a) Roses dominating a bride's bouquet and a groom's boutonnière. (b) Young romance. (c) A just-married couple holding their wedding bouquet of roses. (d) Cupid surrounded by roses. (e) A gentleman with a huge bouquet of red roses for his lady friend. (f) A honeymoon hotel room bed decorated with rose petals.

*Source:* **(a) Photo by Shutterstock/Tnata 2009. (b) Photo by N-region from Pixabay (public domain). (c) Photo by Shutterstock/Aleksandra Zaitseva. (d) Photo by Katesupers from Pixabay (public domain). (e) Photo from Small, E., P. M. Catling, and B. Brookes. 2012. *Official Plant Emblems of Canada: A Biodiversity Treasure.* Public Works and Government Services Canada, Ottawa. (f) Photo by Mike Carter (CC BY ND 2.0).**

lavender rose means enchantment; an orange rose suggests fascination; a pale pink rose stands for grace and joy; a white and red rose together stand for unity; red and yellow roses mixed together say "Congratulations!"; yellow and orange roses together imply passionate thoughts; two roses of any color taped or wired together imply a commitment or forthcoming marriage; a rosebud signifies beauty and youth and a heart innocent of love; a thornless rose means love at first sight; and a single bloom of red rose can mean "I have just fallen in love" or "I still love you."

## 88.2 OVERVIEW

The genus *Rosa* consists of about 190 species of mostly deciduous (rarely evergreen) shrubs (occasionally with scrambling or trailing stems) distributed throughout the temperate and subtropical areas of the Northern Hemisphere. About 95 species occur in China. In North America, there are 15–20 native species, and perhaps 12–15 that are intro duced. The plants grow in a variety of habitats, usually in open situations. Some species are fairly widespread; others have narrow distributions and, indeed, are considered endangered. Depending on species, a range of soil types and tolerance of dry and wet soils is found.

Although commonly called "thorny" or "spiny," anatomically, roses have prickles or bristles. Thorns are modified branches or stems, developing from buds; spines are modified leaves or parts of leaves, developing at nodes. Both spines and thorns have vascular (transport) tissue. Prickles and bristles merely develop from the epidermis and perhaps also the underlying tissues. *Bristles* are flexible, weaker forms of prickles.

Natural rose colors include apricot, mauve, orange, pink, red, russet, white, and yellow. These vary in intensity, and there are cultivars with flowers that are variegated (segments of the petals differing in color). Naturally-blue roses do not exist in nature but are sometimes produced by dying white roses. There have been partially successful attempts to produce blue-flowered roses by genetic engineering. Naturally black roses also do not exist, but some so-called black roses are dark shades of red, purple, or maroon. The color of harvested "black roses" is sometimes accentuated with dye or by adding ink to the water of harvested flowers. A "rainbow rose" is created by splitting the stem base of a harvested rose and dipping each segment in differently dyed water; the dyes are drawn into the petals, which become differently colored. Rose flowers still attached to the plants have also sometimes been differentially colored by using hypodermic needles to inject dyes into the stems during flower development.

Most wild roses produce flowers only or predominantly at a particular time in the season. Domesticated roses have

FIGURE 88.3 Rose prickles: (a) An unidentified species. (b) *Rosa acicularis.* (c) *Rosa rugosa.*

*Source:* **(a) Photo by J. J. Harrison (CC BY SA 3.0). (b) Photo by Walter Siegmund (CC BY SA 3.0). (c) Photo by Thierry Pernot (CC BY SA 2.0).**

FIGURE 88.4 Rainbow roses, produced by injecting dyes into sectors of the floral stem.

*Source:* **(a) Photo by Shutterstock/Melslocum. (b) Photo by Lucy Rober (CC BY 2.0).**

usually been selected to produce flowers for longer periods. The term "remontant" (French for "blooming again") refers to roses that bloom repeatedly during the season. Many old rose cultivars bloomed at different, distinctive times of the season, but recent rose cultivars have been selected to be "ever-blooming," producing large floral displays more or less continuously. Most "old garden roses" bloom once; most modern roses produce a continuous flowering display for much of the season.

Some species have flowers with 200 or more stamens. Selection of mutations in which stamens have been converted to petals is the basis of so-called "doubling" of flowers, producing globes of tightly interleaved petals, which humans find much more attractive than the usual whorl of five petals in wild species.

The fragrance of rose flowers is one of their most endearing features, as reflected by Shakespeare's adage in *Romeo and Juliet*: "A rose by any other name would smell as sweet." The odor varies considerably, with an aroma that may be describe as floral, spicy, fruity, or musky, usually but not always pleasant.

FIGURE 88.5 Comparison of the five-petalled flower of a wild rose and the multi-petalled flower of a hybrid tea rose: (a) Dog rose (*Rosa canina*). (b) *Rosa* 'Johann Wolfgang von Goethe'.

*Source:* **(a) Photo by Aiwok (CC BY SA 3.0). (b) Photo by Huhu Uet (CC BY SA 3.0).**

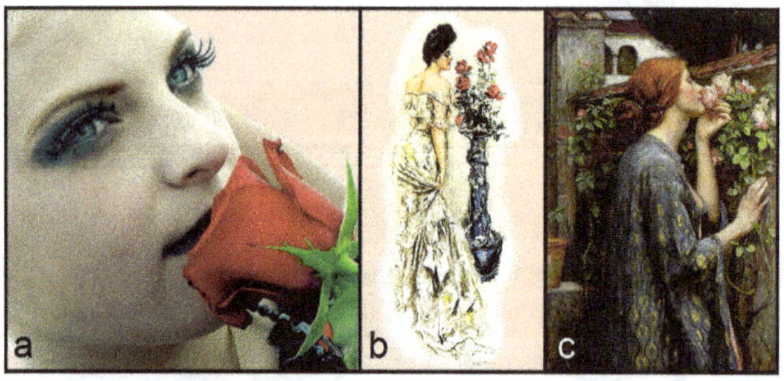


FIGURE 88.6 Appreciating the fragrance of roses.

*Source:* **(a) Photo by D. Sharon Pruitt (CC BY 2.0). (b) Vintage (1906) illustration (public domain). (c) Painting *The Soul of the Rose* by English artist John William Waterhouse (1849–1916) (public domain).**

FIGURE 88.7 Rose hips (fruits): (a) Hands full of rose hips. (b) Jars of rose hip jam.

*Source:* **(a) Photo by Shutterstock/Eremeev. (b) Photo by Shutterstock/Africa Studio.**

Rose fruits are commonly called hips. Botanically, these are false fruits, not true fruits. The fleshy, outer portion that surrounds the "seeds" is a "hypanthium," developed from the top of the flower stalk; in contrast, the flesh of true fleshy fruits is developed from one or more ovary walls. The "seeds" of a rose are, in fact, the true fruits; these are "achenes," defined as single-seeded, dry, indehiscent fruits in which the seed coat is separate from the fruit wall (sunflower "seeds" are another example of achenes). Rose fruits are often not produced in cultivars with flowers composed of densely packed petals, because the sexual organs are simply inaccessible for pollination.

### 88.2.1 Classification of Rose Cultivars

There is no universal system of classification of rose cultivars. There are various horticultural classifications, in some cases somewhat based on ancestry, but centuries of undocumented hybridization have made determination of genetic relationships very difficult. Moreover, hybridization has also obscured or eliminated differences among some categories that were once formally recognized, and new cultivars can be impossible to assign to categories based only on their appearance. In addition to ancestry, rose horticultural classifications are usually based on growth habit (size, whether a shrub or vine-like) and number, form, and disposition (display structure) of flowers. The two most influential rose organizations adopted different classifications. In 1979, the World Federation of Rose Societies (WFRS) adopted a new system, but this was not accepted by the American Rose Society (ARS), which has its own system. The British Association of Rose Breeders has recognized 30 categories. However, the categories are often problematic. For example, climbing roses or miniature roses can often fit into several categories. The ARS system is the most widely adopted. It categorizes garden roses into three groups: "species roses," "old garden roses," and "modern roses," discussed in the following. Species roses are the ancestors of old garden roses, which are, in turn, the ancestors of modern roses.

### 88.2.2 Species Roses

So-called "species roses" (often referred to as "wild roses") are wild roses (possibly slightly changed by domestication) which are cultivated. Typically, their flowers only have five petals, but frequently, there has been some selection (especially for color and doubling of flowers). It has been said that species roses, unlike the majority of garden roses, will reproduce true from seed, but this is not entirely accurate, since many cultivars referred to as species roses are in fact hybrids, and these (and indeed some non-hybrid forms) are propagated vegetatively as clones. Although species roses are usually somewhat domesticated and frequently exist as named cultivars, they are much closer to natural wild species than are most cultivated roses, and accordingly are hardier, often tolerating poor soil, some shade, diseases, and insects, and accordingly, they require relatively limited care. Most originated from northern Europe and are fairly tolerant of cold climates. Many have interesting or large fruits (hips) that often persist into winter, attracting wildlife. They also tend to have pleasant floral odors. However, their flowers are smaller and not as attractive as most garden roses, and they usually bloom only once during the season. Commonly grown species roses include Austrian yellow rose, Banksian rose, dog rose, Gallic rose, musk rose, Scotch rose, and sweetbriar (scientific names are given in the section on history).

### 88.2.3 Old Garden Roses

In 1867, J. B. Guillot of France hybridized a "hybrid perpetual" and a "tea" rose to produce what is considered to be the first "hybrid tea" cultivar, 'Madame Bravy'. The ARS and, indeed, many rose horticulturists now widely categorize garden roses produced prior to 1867 (as well as roses produced after 1867 but using only old cultivars as parents) as "old garden roses" (sometimes called "antique roses"), and new hybrids produced after 1867 as "modern roses." Old garden roses usually bloom just once. The flowers are typically white, mauve, pink, or red-purple (in the past, crimson red and yellow were not available) and are often sweetly scented. Like species roses, many are tolerant of stresses, including cold, and are relatively undemanding and easy to care for (compared to modern roses).

Common groupings of old garden roses include:

1. *Old European roses (cold-hardy, usually blooming only once):* (a) Alba roses (likely the progeny of dog and Damask roses); (b) Centifolia roses (cabbage roses; based on *R.* × *centifolia*); (c) Gallica roses (French roses, Provence roses; the oldest garden roses, tracing to classical Greek and Roman times; extremely cold-hardy; derived from *R. gallica*); (d) moss roses (with moss-like glandular hairs on stems and buds; these are mostly mutations of Centifolia and Damask); (e) Damask (summer Damask roses flower once, like groups (a)–(d); autumn Damasks have a limited repeat blooming in the fall; derived from *R.* × *damascena*).
2. *Chinese roses (imported into Europe about the middle of the 18th century; repeat bloomers, not cold-hardy):* (f) China rose (*R. chinensis*); (g) tea rose (*R.* × *odorata*).
3. *Hybrids of European and Chinese roses:*
   - (h) Portland (possibly a China × Gallica hybrid).
   - (i) Noisette (an old class of largely climbing roses with long slender canes that thrive in warm to hot climates; these originated around 1800 in Charleston, South Carolina, derived from *R. moschata* and *R. chinensis* 'Old Blush').
   - (j) Bourbon (a cross of China rose and Damask autumn rose).
   - (k) Hybrid perpetual (an old class dominant during the late Victorian and Edwardian eras, with over 1,000 cultivars; derived from recrossing Bourbons and Portlands with China rose).

### 88.2.4 Modern Roses

Modern roses include the most attractive of roses, producing richly colored flowers. They are hybrids of European and Chinese roses, like the last-mentioned group, but produced after 1867. Common groupings are noted in the following (but it needs to be emphasized that contemporary cultivars are frequently not clearly assignable to just one class):

1. *Hybrid tea.* By far, the most popular of roses, these upright shrubs are characterized by large solitary flowers on long stems and high-centered flower shape (long inner petals forming a central cone); used mostly as garden roses and cut flowers. The French hybridizer Joseph Pernet-Ducher introduced brilliant yellow and copper tones by crossing with *R. foetida* in the late 19th and early 20th centuries. This type is predominant at florist shops.

FIGURE 88.8 Close-up photos of flowers of modern rose cultivars: (a) Cultivar Heat Wave. (b) *Rosa fedtschenkoana*. (c) Cultivar Harison's Yellow. (d) Cultivar Morden Centennial.

***Source:*** **(a) Photo by T. Kiya (CC BY SA 2.0). (b) Photo by Salicyna (CC BY SA 4.0). (c) Photo by Barra (CC BY SA 3.0). (d) Photo by T. Kiya (CC BY SA 2.0).**

2. *Polyantha.* A class of roses created by the French breeder Jean Baptiste Guillot in the early 1970s from crosses between climbing varieties of *R. multiflora* (an Asian rose, then identified as *R. polyantha*) and *R. chinensis*. These present delicate flowers in sprays well above their foliage.
3. *Floribunda (hybrid polyanthas derived from crossing polyanthas and hybrid teas).* (*Floribunda* literally means "abundant flowers.") These produce clusters of 3 to 15 smaller flowers on a stem (compared to single flowers of hybrid teas) on relatively low-growing, compact plants.
4. *Grandiflora.* Similar to hybrid teas and floribundas (and derived as a cross between them), but producing more flowers per stem than hybrid teas and fewer and larger flowers than floribundas. Mostly tall-growing (up to 2 m), these are currently uncommon.
5. *Shrub.* (All roses are shrubs, and this inaccurately named category has been applied in different ways, including any modern bushy roses that do not fit into other categories, or more specifically, David Austin roses, also known by the inaccurate name English roses.) Roses originated by the contemporary British hybridizer David Austin, in which the repeat blooming of modern roses has been combined with the form and fragrance of old garden roses. Shrub roses are available in various shapes and sizes for various landscape applications. Many are low-growing and bear flowers over a long season and are tough, are disease-resistant, and require less attention than other roses.

### 88.2.5 Rose Classes Based on Habit

In addition to the aforementioned, the following classes based substantially or only on "habit" (in biology, this term refers to a characteristic growth form) are also commonly recognized:

*Climbers (ramblers, trailers).* Forms with long canes that are appropriately trained on fences, trellises, pillars, wall, or other supports. They have long, arching branches with numerous flowers and are ideal for covering structures.

*"Cut rose" varieties.* These have been bred for long stems for use in bouquets and vases. This makes them susceptible to wind, so they are unsuitable as garden plants and are often raised in greenhouses and protected locations.

*Miniatures.* Typically less than 30 cm and rarely higher than 45 cm in height, these (genetically) dwarf plants are ideal for containers. (About 1920, the dwarf cultivar *R. chinensis* 'Pompon de Paris' became the basis of breeding many additional dwarf roses, facilitated by dwarfness being determined by one dominant gene.) A very wide range of cultivars is now available. These flower continuously and are ideal for containers, borders, and small spaces.

*Patio roses (dwarf, cluster-flowered roses).* Small rosebushes that are larger than miniature roses.

FIGURE 88.9 Climbing roses.

**Source: (a) Photo by Jassiaa from Pixabay (public domain). (b) Photo by Georges Seguin/Okki (CC BY SA 3.0). (c) Photo by HeungSoon from Pixabay (public domain).**

FIGURE 88.10 Miniature roses: (a) A collection of potted miniature roses. (b) A bed of *Rosa* cultivar Little Sunset. (c) *Rosa* cultivar Coco.

**Source: (a) Photo by Supernan (CC BY 2.0). (b) Photo by Shutterstock/Gary Matuschka. (c) Photo by Shutterstock/Gary Matuschka.**

*Tree roses (standards).* A graft of one or more (usually hybrid tea roses) on top of an elongated, bare stem ("trunk," generally 90–120 cm tall) of a vigorous rose (which is often grafted onto a basal hardy rootstock). These are visually stunning but require care to survive through the winter (they may need to be excavated and buried in cold climates).

Many rose cultivars grow considerably more vigorously when the stems are grafted onto the base ("rootstock") of other species, and this practice is very common. Species roses are not grafted onto hardy rootstocks, nor are most old garden roses, climbers, and miniature roses. Most modern roses (including most of the popular hybrid teas) require hardy rootstocks, but there is a movement to grow some on their own roots.

FIGURE 88.11 Tree roses.

*Source:* (a) Photo by Shutterstock/Vladimir Konstantinov. (b) Photo by Karsten Ratzke (public domain).

FIGURE 88.12 Covers of old rose catalogues.

*Source:* Public domain. (a) 1896. John Lewis Childs (Floral Park, NY) 1896. (b) John Lewis Childs (Floral Park, NY) 1902. (c) Dingee & Conard (West Grove, PA) 1899. (d) Dingee & Conard (West Grove, PA) 1884. (e) Peter Henderson (New York) 1915. (f) Ella V. Baines (Springfield, OH) 1915.

## 88.3 HISTORY

Rose cultivation traces back thousands of years in temperate regions of Eurasia, notably China, Mesopotamia, Persia, Egypt, Greece, and Rome. The plants were often used in ancient times predominantly for medicine and fragrance, but also for ornament. Today, ornamental roses are sometimes planted for their decorative fruit or as hedging and groundcover but are grown mostly for floral display.

Until about 2 centuries ago, European and Chinese domesticated roses were selected independently. In the 18th century, roses from China were imported in substantial numbers to Europe. The Chinese roses have a much longer blooming period and other desirable characteristics lacking in roses that were cultivated in Europe, but require a warmer climate. The introduction of Chinese roses strongly stimulating the cultivation of roses as ornamental plants and the creation of new cultivars. France's Empress Josephine (1763–1814), wife of Napoleon Bonaparte, adored roses and financed the collecting and planting of over 250 varieties in her garden at Malmaison (a manor house 12 km west of central Paris, where Josephine also imported and maintained hundreds of exotic plants and animals). Although "rosariums" (rose gardens) have been known for millennia, Josephine is considered to have developed the first modern display rose garden. Pierre Joseph Redouté (1759–1840), the great Belgian flower portraitist, was commissioned by Josephine to paint her roses, and many believe that his book of engravings, *Les Roses*, is the best-known rose book in the world. Rose cultivation and the development of new cultivars subsequently expanded greatly. It is claimed that 30,000 to as many as 100,000 cultivars have been recorded, and perhaps as many as 18,000 are commercially available. Most cultivars have been bred during the last 2 centuries, and hundreds of new varieties are being introduced annually.

Almost all roses grown today originated as cuttings and represent clones of identical plants, each clone tracing back

FIGURE 88.13 Old paintings of roses: (a) Cultivar Decoration de Geschwindt. (b) Cultivar Duke of Edinburgh. (c) Cultivar Maréchal Niel. (d) Cultivar Gloire de Dijon. (e) Cultivar Baronne Prévost. (f) Cultivar Reynolds Hole (*top*) and cultivar Francois Michelon (*bottom*). (g) Cultivar Antoine Ducher. (h) Cultivar Professeur Ganiviat.

*Source:* Public domain. (a) *Journal des Roses* (Paris) 1896. (b) *The Floral Magazine* (London), vol. 7, 1868. (c) Charles Antoine Lemaire (1866): https://artvee.com/dl/rose-the-marechal-niel/. (d) *Flore des serres et des jardins de l'Europe* (Gand, Belgium), vol. 9, 1853–1854. (e) *Flore des serres et des jardin de l'Europe* (Gand, Belgium), vol. 8, 1853. (f) *The Garden* (London), vol. 11, 1877. (g) *The Floral Magazine* (London), vol. 6, 1867. (h) *Journal des roses* (Paris), May 1896.

FIGURE 88.14 Pierre Joseph Redouté (1759–1840), who was born in Belgium and spent much of his career in France and is considered by many to be the greatest rose artist. One of his paintings of *Rosa gallica* is shown on the left, his painting of *R. sulfurea* on the right.

***Source:* All illustrations are public domain.**

FIGURE 88.15 Marie Antoinette, Empress of France, posing with a rose. As noted in the text, she greatly stimulated the development of roses. Both paintings (which are public domain) are by French artist Elisabeth Louise Vigée Le Brun (1755–1842).

***Source:* (*Left*) Photo by the French Ministry of Culture. (*Right*) Photo of painting by Y. Caradec (CC BY SA 2.0).**

to a single seed which grew into a plant with a superior floral display. In recent times, that foundational seed is usually the hybrid of two cultivars that similarly once produced superior roses. This process of continual improvement of roses traces back thousands of years. When you look at a garden rose, you see the cumulative artistry of the finest gardeners in history and the judgment of millions. Aside from selection for floral features (petal doubling, size, color, longevity, etc.), hybridization by gardeners has been the most important consideration in the evolution of modern roses. Most garden roses are hybrids, and in many cases, their ancestry is uncertain. In addition to professional plant breeders, many amateurs conduct rose hybridization as a hobby. Hybridization is the key to understanding why roses turned into the world's leading floral ornamental in recent centuries. The development of transportation and commerce provided access to the world's wild and cultivated species and hence facilitated the extensive hybridization which is the basis of the numerous cultivars that have been bred. The huge natural gene pool of the genus *Rosa* (many species are very variable, with forms adapted to various ecological stresses) and the natural ability of the species to hybridize (barriers to interbreeding are frequently weak) have allowed mankind to generate almost endless variations that meet the needs of different climates and, indeed, the changing tastes of fashion.

The following six wild species, substantially of Asian origin, are the most important forerunners of modern cultivated roses:

- *R. chinensis* (China rose, Bengal rose; native to China)
- *R. foetida* (Austrian yellow rose, Persian yellow rose; Asia)
- *R. gallica* (Gallic rose, French rose, apothecary rose; Eurasia)
- *R. moschata* (musk rose; known only in cultivation)
- *R. multiflora* (Japanese rose, multiflora rose; Asia)
- *R. rugosa* (Japanese rose, Turkestan rose; Asia)

There are numerous other rose species that are cultivated or have contributed to the breeding of domesticated varieties. Significant wild ancestors of some rose cultivars include:

- *R. arkansana* (prairie rose; North America)
- *R. arvensis* (field rose; Eurasia)
- *R. banksiae* (Banksian rose; Asia)
- *R. bracteata* (Macartney rose; Asia)
- *R. canina* (dog rose; Eurasia)
- *R. fedtschenkoana* (no English name; Asia)
- *R. luciae* (*R. wichuraiana* Crép.) (memorial rose, Wichura's rose; Asia)
- *R. moyesii* (Mandarin rose; China)
- *R. pendulina* (Alpine rose; Europe)
- *R. phoenicia* (fenicierros, Middle East)
- *R. rubiginosa* (eglantine, sweetbriar; Eurasia)
- *R. sempervirens.* (evergreen rose; Eurasia, Africa)
- *R. setigera* (prairie rose; North America)
- *R. spinosissima* (*R. pimpinellifolia*) (Scotch rose, burnet rose; Eurasia)

Many hybrids of garden origin have been used to generate additional hybrids. The following have been particularly important:

- *R.* × *alba* (white rose, white rose of York; parentage unclear, perhaps *R. gallica* and other species)
- *R.* × *centifolia* (cabbage rose; a complex hybrid of *R. gallica*, *R. moschata*, *R. canina*, and *R.* × *damascena*)
- *R.* × *damascena*. (Damask rose; *R. gallica* × *R. moschata*; Turkish origin)
- *R.* × *gigantea* (*R* × *odorata* nothovar. *gigantea* [a *nothovariety* is an intervarietal hybrid]; Asian origin)
- *R.* × *odorata* (tea rose; *R. chinensis* × *R.* × *gigantea*; Chinese origin)

## 88.4 ECONOMIC IMPORTANCE

Roses grown for ornament and cut flowers are the basis of a huge, worldwide industry. Roses are favorite gift flowers on holidays, especially Valentine's Day. Roses grown for extracts employed for perfumes, flavoring, and medicinal purposes are of secondary but significant value.

## 88.5 ORNAMENTAL USES

In both residential and public gardens, shrub roses are the chief choices for landscape beds and foundation plantings beside buildings. Hardy roses (such as the dog rose and the Turkestan rose) make excellent, relatively carefree hedges. Rambling roses are often employed as groundcover, and climbing roses (the same as rambling roses) are frequently grown on fences, trellises, and against walls, sometimes to enhance appearance or just to conceal less-attractive features. Container-grown roses are ideal for patios, balconies, and window boxes and are also sometimes employed as houseplants. Cut flowers for use in bouquets and floral arrangements are a staple of the florist industry.

FIGURE 88.16 Commercial production and sale of roses: (a) A rose nursery in Arizona. (b) A rose greenhouse. (c) Pots of colorful cut roses at a flower market in Saigon, Vietnam. (d) A florist working with roses.

*Source:* **(a) Photo by Shutterstock/Paul R. Jones. (b) Photo by Shutterstock/Zoran Zeremski. (c) Photo by Marco Verch (CC BY 2.0). (d) Photo by Shutterstock/Agnes Kantaruk.**

FIGURE 88.17 Vintage Valentine's Day postcards featuring roses.

*Source:* **Public domain.**

FIGURE 88.18 Large rose gardens: (a) In Buenos Aires, Brazil. (b) In Lower Austria. (c) In Landkreis Osnabrück, Germany. (d) In Gera/Ronneburg Germany.

*Source:* **(a) Photo by Roberto Fiadone (CC BY SA 2.5). (b) Photo by Henry Kellner (CC BY SA 4.0). (c) Photo by Corradox (CC BY SA 4.0). (d) Photo by Anja Leidel (CC BY SA 3.0).**

FIGURE 88.19 Rose beds enhancing large buildings: (a) A mansion in The Netherlands. (b) Mansion in Bamberg, Germany. (c) Saint Isaac Cathedral, Saint Petersburg, Russia.

*Source:* **(a) Photo by Angelegea from Pixabay (public domain). (b) Photo by Tilman 2007 (CC BY SA 3.0). (c) Photo by Shutterstock/Popova Valeriya.**

## 88.6 OTHER USES

### 88.6.1 Perfume

A few roses yield important essential oils used in perfumery and extracts employed as spices. To make 1 kg of costly rose essence for use in fragrances and perfumes, thousands of kilograms of rose flowers are required. The perfumery

FIGURE 88.20 Cut rose flowers.

*Source:* **(a) Photo by Shutterstock/MaraZe. (b) Photo by Shutterstock/Pixel-Shot. (c) Photos by ProFlowers (CC BY 2.0).**

FIGURE 88.21 Rose houseplants.

*Source:* **(a) Photo by Shutterstock/Ortis. (b) Photo by Shutterstock/Pixel-Shot. (c) Photo by Shutterstock/Laddawan.**

FIGURE 88.22 Collecting flowers from Damask rose (*Rosa* × *damascena*) in Bulgaria for preparation of extracts and perfumes.

*Source:* **(a) Photo by Jdblack from Pixabay (public domain). (b) Photo by Shutterstock/TSViPhoto.**

### 88.6.2 Medicine

Roses were important medicinal plants during the Middle Ages, used to reduce fever, inflammation, pain, and to stop any kind of excessive flow, be it tears, diarrhea, or hemorrhage. The popularity of the rose eventually resulted in its elevation to the status of a wonder drug, employed even for epilepsy, tuberculosis, goiter, and gout. During plagues, rose fragrance was used to purify the air in public places. Apothecaries even used rose galls, employing them to induce urination. Rose hips still have significant market value as sources of medicinal preparations. Rose hip oil, which is pressed from rose hips of some species, is employed medicinally. It is high in essential fatty acids, which contribute to skin health, and so is incorporated into skin care products. Extracts from rose hips have been shown to alleviate rheumatoid arthritis. Rose hips also have health values for pets and livestock. They are fed to pet guinea pigs – rodents which (like humans) cannot synthesize their own vitamin C (they are also fed to chinchillas in the mistaken belief that they also must have supplemental vitamin C). Rose hips are fed to horses in the form of a dried powder, with the intention of improving coat condition and stimulating hoof growth.

### 88.6.3 Food

The use of rose fruits as food is minor but nevertheless has a long history. Neolithic Old World archaeological sites dating back to 5,000 BP show evidence of humans intentionally collecting rose fruits for food, and they were also consumed by Indigenous people of North America. Rose hips are often covered with hooked, hairy outgrowths which should be removed before the fruits are cooked. In most species, the hairs on the outside are less troublesome than the fuzz on the inside. De-seeding the hips and removing the fuzz and hairs are tedious tasks. The principal species used for rose hips are dog rose and Turkestan rose. Rose flowers have limited but interesting culinary uses. Drinks have been perfumed with roses, and rose petal wines have been made since classical times. Rose water, used for flavoring sweet dishes, is the result of distilling rose flowers with a agent attar of roses is an oil which rises to the surface of rose water left to stand and kept cool. Damask rose is the main source. It is grown commercially mainly in Bulgaria and Turkey (hence the alternative names Bulgarian rose and Turkish rose). Farmers commonly provide dried petals or extract (termed "rose water"). Cabbage rose is grown for rose oil in North Africa and France.

FIGURE 88.23 Rose petal culinary preparations: (a) Tea. Dried rose buds are in the saucer. (b) Jam flavored with rose petals. (c) Almond cake with rosewater glaze and sugared rose petals.

*Source:* **(a) Photo by Marco Verch (CC BY 2.0). (b) Photo by Spasimir Pilev (CC BY SA 4.0). (c) Photo by Sarah Stierch (CC BY 2.0).**

little water. The Damask rose, mentioned in the preceding text, is the main source. It is also used to produce rose petal jam. The French rose provides flower petals that are often crystallized, preserved in syrup, or dried and used to flavor teas, beverages, cakes, honey, and liqueurs. Rose water, rose petal jam, rose petal–flavored honey, rose-flavored candy, and rose-flavored vinegar are still marketed, albeit on a very limited scale. To a minor extent, the food industry uses rose extracts to flavor beverages, candy, ice cream, bakery goods, gelatin desserts, and jelly.

## 88.7 PROBLEMS

Some roses have escaped from cultivation and become quite invasive. Worldwide, the most troublesome species are *R. canina*, *R. multiflora*, *R. rubiginosa*, and *R. rugosa*. In North America, the most invasive species include *R. bracteata*, *R. laevigata*, R. *multiflora*, and *R. rugosa*.

The geraniol in the oil of roses may cause contact dermatitis in people handling rosebushes. For most people, the "thorns" (prickles, as explained earlier) are the chief source of concern, and gloves and protective clothing are advisable when handling roses.

The most beautiful roses, like the highest-yielding crops, have been highly selected, meaning, their genetic base is often relatively narrow. Moreover, the most attractive cultivars are clones, which are genetically uniform. Both considerations tend to make the finest roses susceptible to attack by microorganisms (especially fungi and bacteria), as well as by herbivores (especially insects). Chemical pesticides are widely employed as control measures, which is harmful to biodiversity and the environment.

## 88.8 CULTIVATION

### 88.8.1 Climate and Site

Roses can be grown in most climates, except the tropics. Unfortunately, most modern roses are not well-adapted to cold environments (many roses can be grown as annuals in very cold areas, a practice that doesn't allow most cultivars to grow large enough to develop good displays). In cold climates, the dominant challenge to roses is winter hardiness, and native species have often been employed as breeding sources of cold-tolerant genes. The Explorer rose series, bred in Canada, is widely considered to be the world's most successful breeding program for winter-hardy roses. To survive or thrive in given locations, particular cultivars need to be planted at certain depths and provided with winter protection, and this information should be consulted when purchasing the plants. Miniature roses are useful as potted houseplants, but in contrast to the universal long-term cultivation of roses in gardens, potted miniature roses are usually purchased as gifts to be grown as short-term, throwaway plants.

FIGURE 88.24 Invasive *Rosa rugosa* (sometimes called "beach rose") on the beaches of the East Coast of the United States: (a) Cape Neddick Light Station, Maine. (b) Lewis Bay, Mid Cape Cod Dunes, Massachusetts.

***Source:*** **(a) Photo by Lupinemomma (CC BY SA 4.0). (b) Photo by EgorovaSvetlana (CC BY SA 4.0).**

Most roses prefer full sunlight, although they tolerate slight shade, and afternoon shade is recommended in hot climates. In very windy areas, roses benefit from shelter, such as provided by a fence or wall. A warm wall also is helpful in cold climates. Planting roses near trees or shrubs should be avoided, because the latter can compete with them for water, nutrients, and light.

### 88.8.2 Propagation and/or Purchasing

Roses can be established by seeds, cuttings, grafting, or budding; by layering; and by division. Except for the use of seeds, the other methods require skill and are usually best left to professional horticulturalists. Growing roses from seeds can be frustrating. Seeds need to be stratified (cold-hardened, by exposure to temperatures around freezing, often for 3–4 months), and germination may be irregular. Seeds obtained from dealers will usually have had this requirement met. Most people establish roses from purchased bare-rooted plants or potted starter plants.

### 88.8.3 Planting

Cultivars grow to different sizes, and planting distances vary accordingly. Rosebushes should not be crowded, which restricts airflow around the plants and encourages fungal diseases. Roses are best transplanted in early spring, when there is limited risk of freezing, but before regrowth begins. Bare-rooted roses should be placed in the center of the hole, with the knobby graft union slightly below the soil level. Container-grown roses should be placed so that the graft union will be 2 to 5 cm below the soil line after the soil settles.

### 88.8.4 Soil and Moisture Conditions

The ideal rose soil is a compost-rich, deep loam. Although the soil should retain moisture, it must also be well-drained. Roses generally prefer a slightly acid soil (pH 6.0–7.0). Fertilizers formulated specifically for roses are available, and the instructions should be followed. A multipurpose fertilizer may be applied once in the early spring and at least once during the summer, when blooming begins. In cool and cold climates the final application of fertilizer should not be later than midsummer, to allow the plants to harden for the winter. A mulch around the base of roses is helpful to control weed growth. When watering, inasmuch as possible, wetting the foliage should be avoided.

### 88.8.5 Maintaining Plants

Pruning is important for roses to promote healthy, strong growth. The best time to prune roses is late winter or early spring, before growing begins. Pruning is also often done in the autumn after the first frost. Suckers that arise from the underground hardy rootstock must be removed. Dead or diseased branches should also be eliminated. One or both canes that rub or cross each other should be trimmed. Canes that are thinner than a pencil should be considered for removal, as well as those that do not fit the desired shape. Some varieties benefit from being cut down to the base, but others are less-tolerant. Ideally, instructions are available for the cultivars being grown. Experienced gardeners often prune at a 45° angle, cutting above an outward-facing node to encourage the roses to grow outwardly.

In cold climates, it may be desirable to cut back shrubs to about 60 cm to reduce damage from winter wind and snow. Hay or straw can be added to the base for insulation and removed in the spring.

## 88.9 FASCINATING FACTS, FABLES, FOLKLORE, AND FRIVOLITY

- Roses have been a heraldic emblem on the shields of soldiers since the time of the Romans. The War of the Roses was a series of battles (1455–1487) between two branches of the royal family Plantagenet in England. The House of Lancaster, represented by the red rose, defeated the House of York, represented by the white rose. The white rose is reputed to have been a form of *R. canina* (also said to be *R.* × *alba*, a hybrid involving *R. canina*). The two houses combined with the marriage of Henry VII of Lancaster to Elizabeth of York, Henry becoming the first Tudor king. He adopted a red-and-white Tudor rose as his symbol, and this is now the national flower of England. A rose cultivar called 'York and Lancaster' which develops both red and white flowers is currently marketed. This has been identified as a cultivar (named in 1551) of the briar rose, *R. rubiginosa*.
- The Roman emperor Nero (37–68) reputedly used up the equivalent in modern currency of $100,000.00 in cut roses to decorate a single night's feast. He is also believed to have had pipes installed under banquet plates to allow his guests to be spritzed with rose scent between dinner courses.

FIGURE 88.25 *The Roses of Heliogabalus*, by Dutch painter (who settled in the UK) Lawrence Alma-Tadema (1836–1912).

***Source:*** **Public domain. Photo by Pérez Simón Collection.**

- The debauched Roman emperor Elagabalus (Heliogabalus, 204–222) is said to have accidentally killed several of his supper guests by showering them with so many rose petals that they suffocated.
- Some privileged Romans slept on beds of rose petals, giving rise to the saying "It's not a bed of roses."
- Roses are the most popular of gravestone-inscribed flowers. In Victorian cemeteries where a husband and wife were buried next to each other, the gravestone of the wife often showed a hand holding a lily (symbolic of innocence and purity), while that of the husband showed a hand with a rose. Children's gravestones often showed rosebuds as symbols of innocence, and sometimes the stem of the rosebud was carved as broken, suggesting their short lives.
- The Cherokee rose (*R. laevigata*) is a native of China which was imported into the United States and now grows wild in the South. According to legend, it acquired its name when a Cherokee maiden carried one of these roses to her lover, a soldier who was wounded and convalescing in a tent. The white-flowered Cherokee rose has long been used as a peace symbol following fighting. During the American Civil War, it was often planted as a memorial on the graves of the fallen.
- Roses have been very popular choices as floral symbols of political regions, although often a particular species is not identified. Alberta and Iowa chose a "wild rose" as their emblem. The rose is the New York state flower. Georgia adopted the white Cherokee rose (which was once also the state flower of North Carolina). The District of Columbia is associated with the American beauty rose. North Dakota chose the prairie rose. Roses have national significance in Bulgaria, England, Iran, Luxembourg, and the Maldive Islands in the Indian Ocean. The rose served as the national

flower of Honduras from 1946 to 1969, but an orchid was adopted as the national flower in 1946. In 1986, the rose became the "national floral emblem" of the United States, when President Reagan signed the proclamation in the White House Rose Garden. According to one newspaper report, the president "rose to the occasion."

- Since the time of the ancient Egyptians, roses have been associated with secrecy. According to Greek myth, Cupid, shocked by the conduct of his mother, Venus, took her flower, the rose, to Harpocrates, god of silence, to keep her scandal confidential. This story is said to have brought about the expression "sub rosa," referring to activities conducted confidentially. Roses were customarily suspended over the dinner table as a sign that confidences revealed under the influence of wine were to be kept secret. For many years, the ceilings of European dining rooms featured formed plaster roses, with the same symbolism. The plaster ornament in the center of a ceiling is still known as "the rose" or "rosette," even if it is no longer in the form of a rose. Beginning in the Middle Ages, a wood carving of a rose was placed in the space over confessional doors in Catholic churches, once again symbolizing confidentiality.
- The rose is extremely significant in Islam. According to legend, the rose arose from perspiration that fell from the brow of the prophet Mohammed. When he later saw a rose, he kissed it, pressed it to his eyes, and declared that it was part of God's glory. In the Koran, the king of Babylon, Nimrod, cast Abraham into a burning pyre. Abraham declined help from the angel Gabriel, relying on his faith, and as a result, Allah turned the fire into a bed of roses. Rose water was used to purify mosques profaned by infidels.
- The rose is the principal flower associated with the Virgin Mary, a transfer from its association with the Roman goddess Venus. The Virgin Mary was called "the rose without thorns." Rosaries are used to count prayers to Our Lady. The first "rosaries" were made out of pounded rose petals moulded into beads and threaded (the word *rosary* is derived from the Latin *rosarium*, "rose garden"). Rosaries were later made out of rose hips, and eventually out of various materials. The scent of the beads was supposed to remind one of the roses in Mary's garden. However, early Christian Europe suppressed the rose as a symbol associated with Christianity, because of previous associations with sensuality. In those early times, prostitutes were forced to wear roses as a sign of disgrace.
- "Rose windows" are decorated circular windows, often glazed with stained glass, popular in churches, tracing to the 10th century in Europe, and to earlier use in classical Roman architecture.

FIGURE 88.26 Interior view of rose windows: (a) South rose window of Notre-Dame de Paris Cathedral, construction initiated in 1163. (b) Chapelle des Pénitents d'Espalion, Aveyron, France, constructed in 1700. (c) Himmelfahrtskirche, Dresden-Leuben, Germany, erected in 1901.

*Source:* **(a) Original public domain photo by Albertus Teolog, re-imaged by Eusebius (CC BY SA 3.0). (b) Photo by Père Igor (CC BY SA 4.0). (c) Photo by Schwalbe (CC BY SA 3.0).**

Several outstanding examples occur in Notre Dame Cathedral in Paris, constructed in the mid-13th century. The phrase "rose window" is thought to reflect an association with the Virgin Mary and her symbol, the rose.

- Western North American Indians thought wild roses repelled ghosts and evil spirits. Rose branches were scattered around the house and yard of recently deceased persons to prevent the return of their ghost, and relatives and friends drank a tea made from rose branches to protect them from it. A rose branch was used to sweep out the grave before burial to prevent the living from being drawn down.
- The following children's nursery rhyme traces to the bubonic plague, known as the Black Death, that was carried by fleas that lived on the black rat. The plague devastated Europe in 1347–1351 and remained endemic in England until 1666. The death rate was high – between 20% and 25% of the English population died in the period 1348–1349. "Ring a ring o'roses" (or often, "Ring around a rosey") refers to a pinkish circle that would form on a victim's body prior to turning black. Flowers or posies were thought to purify the air. "Achoo, Achoo" refers to sneezing, said to occur just prior to death. "We all fall down" is death.

  Ring a ring o' roses,
  A pocket full of posies,
  A-tishoo! A-tishoo!
  We all fall down.

- During World War II, rose hips were eaten for their content of vitamin C while supplies of foreign fruit were limited. The British Ministry of Health distributed "National Hip Syrup." Similarly, German submarines and ships were supplied with rose hip syrup.
- In England, it was found that the vitamin C content of rose hips increases toward the north, with

four times as much in plants grown in Scotland compared to those in southern England.

- In 1939, Frances Meilland, a rose specialist in France, found a rose with magnificent pale gold blossoms growing from one seed he had nurtured. He sent cuttings to a Pennsylvania rose grower, which were transported on the last American plane that got out of France in November 1940, a step ahead of the invading Nazis. The cuttings were used to propagate the Peace rose, which many experts consider the best ever developed. It blooms on more than 30 million bushes throughout the world.
- The rose is associated with numerous old beliefs concerning brides and marriage. In old England, girls picked a moss rose (various species with sepals of "mossy" appearance, particularly the moss rose, also known as the cabbage rose, *R.* × *centifolia*), wrapped it in white paper, and hid it. If it was still fresh in a month, she would wear it to church with the expectation that her future husband would appear and take it from her. Another Middle Ages European belief was that if a young unmarried woman dreamed of red roses, she would marry a handsome man. In Normandy (France), the headdress of white roses worn by a bride was laid out on the honeymoon bed following the wedding; the fading of the roses was symbolic of the fading of the life of the bride as a maiden. An old belief in France was that a maiden would develop rosy cheeks if she pricked her finger with a thorn from a red rosebush and buried a drop of her blood beneath it. In Switzerland, the bridal wreath, generally made with roses, was set on fire by the mistress of ceremonies at the wedding feast. If the wreath burned briskly, it was taken as a sign of good fortune, and vice versa.
- The origin of the often-shown image of a red rose clenched by its stem between the teeth is obscure. Since the rose has been associated with love and passion since ancient times, performers playing amorous figures probably adopted the practice to symbolize their emotion. The opera *Carmen* (premiered in 1875) sometimes illustrated this practice. By the early 20th century, a stereotype was established often showing tango dancers with a rose between their teeth. Famously, in 1921, *The Four Horsemen of the Apocalypse*, the bestselling silent film of all time, featured the movie star and sex symbol Rudolph Valentino (1895–1926) dancing the tango with a rose in his mouth.
- It has been claimed that the world's largest rosebush is in Tombstone, Arizona. The plant has been identified as *Rosa banksiae* and nicknamed Lady Banksia. It was imported from Scotland in 1885. The plant produces more than 200,000 blossoms in full bloom and spreads over an arbor under which more than 150 people can be seated comfortably. The treelike plant has a trunk circumference of 4.1 m, stands 2.7 m high, and covers an area of 743 m$^2$.

FIGURE 88.27 Peace rose flowers.

*Source:* **Photo by John (CC BY SA 2.0).**

FIGURE 88.28 Images of the widespread practice (indicating passion) of holding a red rose in the mouth by its stem: (a–c) Tango dancers. (d–f) Animals.

*Source:* **Images from Shutterstock, contributed by (a) Francesco Abrignani, (b) Kit8.net, (c) Attia Ster, (d) Tanya Rybakova, (e) Hannamariah, and (f) Topdigipro.**

FIGURE 88.29 The world's largest rose, the "Lady Banksia," in Tombstone, Arizona.

*Source:* **(a) Photo by Stephanie (CC BY ND 2.0). (b) Photo by A. Rancid Amoeba (CC BY SA 2.0).**

- The Latin that is used by botanists to describe and name species of plants has sometimes been referred to by the uncomplementary phrase "dogrose Latin."
- The naming of roses has become a commercial venture. Some companies offer clients the possibility of having new rose varieties named for them, for a price sometimes ranging up to $75,000.00. New rose varieties are being named for corporations ('Barbie' for the doll; 'Weight Watcher's Success' rose), causes ('Arthritis' rose, 'Veterans' Honor' rose), and of course, famous personalities (including 'Billie Graham', 'Céline Dion', 'Chris Evert', 'Vidal Sassoon', 'Rosie O'Donnell', 'Barbra Streisand', and 'George Burns').
- 'Dr. Huey' is a rose cultivar bred in 1914 that is the most widely used rootstock for grafted roses. It is a cold-hardy hybrid tea with rambling stems and red doubled flowers. 'Dr. Huey' has become the most widely observed rose in North America, because roses grafted on it frequently die, and the rootstock sends up shoots that replace the more attractive rose that once grew on it.
- The phrase "flower sculpture" may refer to a variety of artistic creations, including topiary (shearing of shrubs into shapes, especially of animals). Flower sculptures made of harvested flowers pinned to and covering structures are especially popular. These may be stationary or mobile (such as the floats discussed next).
- The Rose Parade or Tournament of Roses Parade is an annual parade held in Pasadena, California, on New Year's Day. The parade includes elaborate, flower-covered floats and is followed by the Rose Bowl, a major college football game.

FIGURE 88.30 The conquest and elimination of cultivars grafted onto hardy rootstocks by the rootstocks: (a) The 'Dr. Huey' rose, the most widely employed rootstock to which less-hardy rose cultivars are grafted. Frequently, the latter die off, and the Dr. Huey rose vigorously develops stems and flowers. (b) A beautiful rose cultivar grafted to a vigorous rootstock (note the swelling, indicating the location of the graft) being suppressed by a sucker from the rootstock.

*Source:* **(a) Photo by Malcolm Manners (CC BY 2.0). (b) Prepared by B. Brookes.**

FIGURE 88.31 Artistic rose-covered creations: (a) A rose-covered display in India, in the form of a Saraswati veena. (This is an ancient Indian plucked veena, named after the Hindu goddess Saraswati. A *veena* is a stringed instrument reminiscent of a sitar.) (b) A heart-shaped floral arrangement composed of roses.

*Source:* **(a) Photo by Goodwin Solly (CC BY SA 4.0). (b) Photo by 4028mdk09 (CC BY SA 3.0).**

FIGURE 88.32 Some of the floral floats exhibited in the Tournament of Roses Parade: (a) A float titled "Doing Good in the World," sponsored by the Rotary Club. (b) A float titled "No Call Too Small," sponsored by the City of Alhambra Fire and Emergency Department. (c) A float titled "Books Bring Us Together," sponsored by the UPS Store.

*Source:* **(a) Photo by Prayitno/Thank You (CC BY 2.0). (b) Photo by Thank You (CC BY 2.0). (c) Photo by Prayitno/Thank You (CC BY 2.0).**

## KEY PUBLICATIONS

American Rose Society. 2000. *Ultimate Rose*. Dorling Kindersley, New York. 160 pp.

Austin, A. (contributor). 1999. *Botanica's Roses: The Encyclopedia of Roses*. Welcome Rain, New York. 704 pp.

Beales, P. 1985. *Classic Roses: An Illustrated Encyclopaedia and Grower's Manual of Old Roses, Shrub Roses, and Climbers*. Holt, Rinehart, and Winston, New York. 432 pp.

Bruneau, A., J.R. Starr, and S. Joly. 2007. Phylogenetic relationship in the genus *Rosa*: New evidence from chloroplast DNA sequences and an appraisal of current knowledge. *Systematic Botany* 32: 366–378.

Cairns, T. 2000. *Modern Roses XI. The World Encyclopedia of Roses*. Academic Press, San Diego, CA. 638 pp. + CD-ROM.

Conart, C., N. Saclier, F. Foucher, C. Goubert, A. Rius-Bony, S.N. Paramita, et al. 2023. Evolution of scent genes in roses. *Acta Horticulturae* 1368: 361–370.

Crockett, J.U. 1978. *Roses*. Time-Life Books, Alexandria, VA. 160 pp.

Dobson, B. and P. Schneider. 2002. *Combined Rose List, 2002: Roses in Commerce, Rose Registrations Update, Hard--to---Find Roses & Where to Find Them-*. Peter Schneider, Mantua, OH. 240 pp.

Hall, T. 2021. *The Kew Gardener's Guide to Growing Roses: The Art and Science to Grow with Confidence*. Royal Botanic Gardens, Kew, UK. 144 pp.

Harmon, D.D., H. Chen, D. Byrne, W. Liu, and T.G. Ranney. 2023. Cytogenetics, ploidy, and genome sizes of rose (*Rosa* spp.) cultivars and breeding lines. *Ornamental Plant Research* 3(1). www.maxapress.com/article/id/6434ac39fa6c58182a2a7135.

Hessayon, D.G. 1996. *The New Rose Expert*, Revised edition. Transworld Publishers, London. 144 pp.

Krüssman, G. 1981. *The Complete Book of Roses*. Timber Press, Portland, OR. 436 pp.

MacPhail, V.J. and P.G. Kevan. 2009. Review of the breeding systems of wild roses (*Rosa* spp.). *Floriculture and Ornamental Biotechnology* 3(Special issue 1): 1–13.

Macoboy, S. 1993. *The Ultimate Rose Book*. H.N. Abrams, New York. 472 pp.

Malécot, V. and K. Debray. 2023. Towards a reclassification of the genus *Rosa*. *Acta Horticulturae* 1368: 347–350.

McCann, S. 1993. *The Rose: An Encyclopedia of American Roses, Rosarians, and Rose Lore*. Stackpole Books, Harrisburg, PA. 198 pp.

McKeon, J.C. 1995. *The Encyclopedia of Roses: An Organic Guide to Growing and Enjoying America's Favorite Flower*. Rodale Press, Emmaus, PA. 192 pp.

Moody, M. and P. Harkness. 1992. *The Illustrated Encyclopedia of Roses*. Timber Press, Portland, OR. 304 pp.

Phillips, R. and M. Rix. 1988. *Roses*. Random House, New York. 224 pp.

Quest-Ritson, C. and B. Quest--Ritson--. -2003-. *American Rose Society Encyclopedia of Roses*. DK, New York. 448 pp.

Richer, C. and C.G. Davidson. 2004. *Winter-Hardy Roses: Explorer, Parkland and Prairie Series*, Revised edition. Agriculture and Agri-Food Canada, Ottawa, ON. 56 pp.

Roberts, A.V., T. Debener, and S. Gudin. 2003. *Encyclopedia of Rose Science*, 3 Vols. Elsevier Academic Press, Amsterdam, The Netherlands.

Scanniello, S. and T. Bayard. 1990. *Roses of America: The Brooklyn Botanic Garden's Guide to Our National Flower*. H. Holt, New York. 224 pp.

Schneider, P. 1995. *Taylor's Guide to Roses*, Revised edition. Houghton Mifflin CO., Boston, MA. 451 pp.

Shanley, P. and P. Kukielski. 2010. *The Sustainable Rose Garden: A Reader in Rose Culture*. Newbury Books, Havertown, PA. 271 pp.

Svejda, F. 1984. *New Winter-Hardy Roses and Other Flowering Shrubs*, Revised edition. Agriculture Canada, Ottawa, ON. 36 pp.

Svejda, F. 2008. *The Canadian Explorer Roses*. National Roses Canada, London, ON. 58 pp.

Wylie, A.P. 1954. The history of garden roses. Part 1. *Journal of the Royal Horticultural Society* 79: 555–571.

Wylie, A.P. 1955a. The history of garden roses. Part 2. *Journal of the Royal Horticultural Society* 80: 8–24.

Wylie, A.P. 1955b. The history of garden roses. Part 3. *Journal of the Royal Horticultural Society* 80: 77–87.

## Key Cultivation Websites

American Rose Society. www.rose.org/.

Canadian Rose Society. www.canadianrosesociety.org/.

GardenersWorld.com. *How to Grow Roses*. www.gardenersworld.com/how-to/grow-plants/how-to-grow-roses/.

*How to Prune a Mini Rose Bush*. www.wikihow.com/Prune-a-Mini-Rose-Bush.

Iowa State University, Extension. *Rose Types and Cultivars*. https://hortnews.extension.iastate.edu/rose-types-and-cultivars.

Pikes Peak Wholesale Florist. *How to Handle Fresh-Cut Roses as a Florist*. https://pikespeakfloral.com/care-handling/roses/.

Purdue University, Extension. *Roses*. www.purdue.edu/hla/sites/yardandgarden/wp-content/uploads/sites/2/2016/10/HO-128.pdf.

Santa Clarita Valley Rose Society. http://scvrs.homestead.com/.

The Spruce. *Roses*. www.thespruce.com/roses-4127748 (Provides links to numerous sites.)

WikiHow. *How to Care for Roses*. www.wikihow.com/Care-for-Roses.

WikiHow. *How to Cut Roses*. www.wikihow.com/Cut-Roses.

WikiHow. *How to Deadhead Roses*. www.wikihow.com/Deadhead-Roses.

WikiHow. *How to Grow Roses*. www.wikihow.com/Grow-Roses.

WikiHow. *How to Grow a Rose Bush from a Single Stem or Cutting*. www.wikihow.com/Grow-a-Rose-Bush-from-a-Single-Stem-or-Cutting.

WikiHow. *How to Grow Roses from Seed*. www.wikihow.com/Grow-Roses-from-Seed.

WikiHow. *How to Keep [Cut] Roses Fresh*. www.wikihow.com/Keep-Roses-Fresh.

WikiHow. *How to Make [Cut] Roses Last Longer*. www.wikihow.com/Make-Roses-Last-Longer.

WikiHow. *How to Plant Roses*. www.wikihow.com/Plant-Roses.

WikiHow. *How to Prepare Soil for Roses*. www.wikihow.com/Prepare-Soil-for-Roses.

WikiHow. *How to Propagate Roses*. www.wikihow.com/Propagate-Roses.

WikiHow. *How to Properly Prune Roses in the Summer*. www.wikihow.com/Properly-Prune-Roses-in-the-Summer.

WikiHow. *How to Properly Water Roses*. www.wikihow.com/Properly-Water-Roses.

WikiHow. *How to Prune Knockout Roses*. www.wikihow.com/Prune-Knockout-Roses.

WikiHow. *How to Prune Roses on a Trellis*. www.wikihow.com/Prune-Roses-on-a-Trellis.

WikiHow. *How to Prune Roses in Winter*. www.wikihow.com/Prune-Roses-in-Winter.

WikiHow. *How to Prune Shrub Roses*. www.wikihow.com/Prune-Shrub-Roses.

WikiHow. *How to Prune Tea Roses*. www.wikihow.life/Prune-Tea-Roses.

WikiHow. *How to Prune Tree Roses*. www.wikihow.com/Prune-Tree-Roses.

WikiHow. *How to Replant a Rose*. www.wikihow.com/Replant-a-Rose.

WikiHow. *How to Save a Dying Rose Bush*. www.wikihow.com/Save-a-Dying-Rose-Bush.

WikiHow. *How to Take Rose Cuttings*. www.wikihow.com/Take-Rose-Cuttings.

WikiHow. *How to Trim Rose Bushes in Fall*. www.wikihow.com/Trim-Rose-Bushes-in-Fall.

World Federation of Rose Societies. www.worldrose.org/.

# 89 Royal Poinciana (*Delonix regia*)

**Scientific Names:** The genus name *Delonix* is based on the Greek *delos*, "conspicuous," and *onyx*, "claw," a reference to the clawed petals. (In zoology, a "claw" is typically a sharp hook on the end of a toe; however, in technical botany, a clawed petal is not a petal with such a claw – it is a petal that is abruptly constricted toward its base. Virtually every presentation on the internet that discusses this for royal poinciana confuses the zoological and botanical meanings.)

**Common Names:** The most frequent English name is royal poinciana (the species was formerly placed in another genus, *Poinciana*, named for Phillippe de Longvilliers de Poincy, a 17th-century governor of Saint Kitts, that is, Saint Christopher Island). *Flamboyant* (tree) is also frequently applied. Less-common names include *peacock tree and flower* (also used for *Caesalpinia pulcherrima*), *royal peacock*, *gold mohar*, and *royal gulmohur*. *Flame tree* (and the similar names *fire tree* and *flame of the forests*) is also encountered but also used for several other ornamental trees. "Poinciana" can refer to different plants – for example, "dwarf poincinia" is *Caesalpinia pulcherrima*, and "yellow poinciana" is *Peltophorum pterocarpum*. The name *royal poinciana* reliably denotes *Delonix regia*, but most other English names do not.

## 89.1 INTRODUCTION

Even in tropical regions, where impressively flowering trees are common, royal poinciana often produces the most eye-catching display in the neighborhood. Many consider it to be one of the most beautiful trees in the world. Regrettably, its cultivation is largely restricted to the hottest countries.

FIGURE 89.1 Royal poinciana (*Delonix regia*) on street sides, the most common ornamental usage of the trees: (a) In Indonesia. (b) In Brazil. (c, d) In Australia.

*Source:* **(a) Photo by Farhan Perdana (CC BY 2.0). (b) Photo by Luis Carlos Martinelli (CC BY SA 4.0). (c, d) Photos by John (CC BY SA 2.0).**

## 89.2 OVERVIEW

*Delonix regia* is native to Madagascar, where its survival in the wild is seriously threatened because of forest overexploitation. This umbrella-shaped tree grows as tall as 20 m in height, although usually it is closer to about 12 m. Impressively, it can have a spread of up to 23 m. Under favorable conditions, the trunk can have a circumference of 2 m (some trees in Australia have a girth of almost 5 m). Royal poinciana grows fast, often 1.5 m in height annually, and it often lives over 50 years. It produces a spectacular display of flowers that is so attractive the species is cultivated worldwide in tropical and subtropical countries and has become naturalized in some regions (i.e., it grows outside of cultivation). The semi-evergreen leaves (dropped in winter when the soil is dry but virtually evergreen if maintained year-round in moist soil) are up to 60 cm long and quite dissected (fernlike) – divided into leaflets which, in turn, are also divided into smaller leaflets. The flowers are large, with five reddish-orange or scarlet (occasionally yellow) spoon-shaped petals. One of the petals is larger than the others and has markings that may be yellow or white. The four smaller petals are 5 to 8 cm long. Individual trees are in flower for about a month during the summer, the time

FIGURE 89.2 Plant parts of royal poinciana (*Delonix regia*): (a) Different perspectives of a flower. (b) Foliage. (c) Seedpods.

*Source:* **(a) Photo by D. L. Schuiling (published in 2018 book *TropCrop: An Introduction to Crops Grown in the Tropics and Subtropics*; CC BY SA 4.0). (b) Photo by Tony Hisgett (CC BY 2.0). (c) Photo by Chris Hunkeler (CC BY SA 2.0).**

DOI: 10.1201/9781003586227-89

FIGURE 89.3 Paintings (public domain) of royal poincinia (*Delonix regia*) in flower.

*Source:* **(a) Illustration by Krumao on FreePik. (b) Norton, E. H. 1893. Brazilian flowers drawn from nature in the years 1880–1882 in the neighborhood of Rio de Janeiro. (c) Hoola van Nooten, B. 1880. *Fleurs, fruits et feuillages choisis de l'ille de Java: peints d'apres nature.***

FIGURE 89.4 Uncommon, yellow-flowered royal poinciana (*Delonix regia*): (a) In Hawaii. (b) In Australia. (c) In Florida.

*Source:* **(a) Photo by Ben Gantz (CC BY 2.0). (b) Photo by Jan Smith (CC BY 2.0). (c) Photo by Marc Averette (CC BY 3.0).**

dependent on local conditions. The flat seedpods are as long as 80 cm, becoming dark brown and woody at maturity and containing several dozen seeds.

## 89.3 HISTORY

*Delonix regia* was discovered in Madagascar by botanist Wensel Bojer in the early 19th century (the previous name *Poinciana regia* was assigned to Bojer's collection). Subsequently, the species became widely cultivated, and probably today, royal poinciana is cultivated in every tropical country of the world.

## 89.4 ECONOMIC IMPORTANCE

Royal poinciana is mostly employed as one of the most important ornamental trees of the tropics. It is so widely grown that it has limited potential for expansion, but maintaining and replacing the aging plants are chief activities of local horticulturalists.

## 89.5 ORNAMENTAL USES

Royal poinciana is mostly planted in parks and avenues and along streets as an ornamental flowering tree and as a shade

FIGURE 89.5 Royal poinciana (*Delonix regia*) as a park tree: (a) In Australia. (b) In Mexico. (c) In Australia.

*Source:* **(a) Photo by Siggy Nowak from Pixabay (public domain). (b) Photo by Nacasma (CC BY 3.0). (c) Photo by John (CC BY SA 2.0).**

FIGURE 89.6 Royal poinciana (*Delonix regia*) as a residential tree: (a) In Israel. (b) In Australia.

*Source:* **(a) Photo by Avishai Teicher (CC BY SA 2.5). (b) Photo by John (CC BY SA 2.0).**

tree. It is also employed as a residential tree, although it is too large for many properties. In tropical climates, shade trees can be essential for livestock and, in particular settings, also for people, such as in school compounds and outdoor transportation hubs, and the trees are often deliberately planted for this purpose.

## 89.6 OTHER USES

The wood from the trees is employed to a minor extent in light construction and for fence posts, firewood, and crafts. The immature pods are edible and have potential for use as a dietary protein source for humans and livestock (although the pods are considered mildly toxic and the seeds are said to be safer when cooked). The leaves are also rich in protein and are used to some extent to feed livestock. The flowers are consumed by a variety of wild animals.

FIGURE 89.7 Animals employing royal poinciana (*Delonix regia*) as a food: (a) A female olive-backed sunbird (*Nectarinia jugularis plateni*) in Indonesia. (b) A bonnet macaque (*Macaca radiata*) in India, where it is native. (c) A honeybee.

***Source:*** **(a) Photo by Unox/Batin E. Tuwuilu (CC BY SA 3.0). (b) Photo by Praveenp (CC BY SA 4.0). (c) Photo by Jimmy Chuang (CC BY SA 2.0).**

FIGURE 89.8 Examples of impressive spreading of royal poinciana (*Delonix regia*): (a) Canopy of a street tree spreading over the roadway in Puerto Rico. (b) A tree in Hawaii. Notice the formidable, shallow, spreading, trunk-buttressing roots, which can break up pavement.

***Source:*** **(a) Photo by José Astacio (CC BY SA 2.5). (b) Photo by Forest and Kim Starr (CC BY 2.0).**

## 89.7 PROBLEMS

Although royal poinciana is widely employed as a shade tree and a residential lawn tree, grass and other plants grow very poorly under the wide-spreading canopy. The flowers can produce substantial litter when they senesce and fall off the plants. The large bean-like pods persist through winter but drop off in the spring, also producing litter. Royal poinciana is considered invasive in Australia, Cuba, Hawaii, several African countries, some of the Pacific islands, and the Galapagos Islands.

## 89.8 CULTIVATION

### 89.8.1 Climate and Site

Royal poinciana is adapted to the frost-free parts of subtropical areas and tropical regions, which include climate zones 10 to 12 (and, debatably, the warmer areas of zone 9). Prolonged exposure to temperatures below 7°C is not survived. The plants flower best in climates with a pronounced dry winter season and do not do well where the winters are rainy. The trees thrive in full sun but tolerate some shade. Choosing a site is important, because the branches and very robust roots can spread 12 m from the trunk, potentially damaging walls and pavement and extending into private and public property. A protected site is helpful, as the branches are subject to breakage in high winds, and the shallow root systems facilitate toppling the tree in storms. This tree is much too large for smaller properties and can even overwhelm medium-sized yards.

### 89.8.2 Propagation and/or Purchasing

Both germination of seeds and cuttings are used to propagate royal poinciana, the latter method much preferable to maintaining cultivars. As with most ornamental trees, buying a sapling or young tree greatly speeds up establishing royal poinciana. Although it is fast-growing, it may not start blooming for 5 to 10 years.

### 89.8.3 Planting

The tree is basically unsuitable for containers, except when grown for bonsai, in which case repotting every 2 years is recommended. As noted earlier, when it is planted near buildings, streets, and other paved areas, substantial distances are required. It is often recommended that the tree should not be located closer than 10 m from concrete and asphalt roads, sidewalks, or building foundations. When the trees are planted in a row, a spacing of 4 to 5 m (or more) is advisable. As with all trees, standard recommendations for size of excavation, soil provided, fertilization, and watering during early growth should be followed.

### 89.8.4 Soil and Moisture Conditions

Suitable soils are well-drained, sandy or loamy, with a circumneutral pH, although the trees have been found in soils with pH between 4.9 and 10.6. Although drought-tolerant, royal poinciana grows better when watered during dry spells. However, in the winter season, avoiding irrigation encourages the tree to drop its leaves and consequently redevelop vigorous, new foliage along with much more abundant flowering in the following spring.

### 89.8.5 Maintaining Plants

Pruning (in spring, before new growth begins) to develop fewer but stronger branches is recommended. Also, low-hanging limbs should be removed to allow movement under the tree.

## 89.9 FASCINATING FACTS, FABLES, FOLKLORE, AND FRIVOLITY

- An old legend holds that when Jesus was crucified, there was a small royal poinciana tree nearby with

FIGURE 89.9 Vintage scenic postcards (public domain, dated 1930–1945) from Florida featuring royal poinciana. These are printed on linen, popular in early times, although the clarity is compromised.

*Source:* **Photo from Boston Public Library.**

white flowers that became bathed in his blood, transforming the species permanently to produce red flowers.

- The seeds within the pods of royal poinciana are relatively free to move, and in the Caribbean and African regions, the pods are employed as maracas or rattles by Indigenous musicians.
- The royal poinciana is the official tree of the Commonwealth of the Northern Mariana Islands, that is, the 14 Northern Mariana Islands, an unincorporated territory of the United States in the northwestern Pacific Ocean.
- Flowering in royal poinciana can be diminished or absent if illuminated by streetlights at night. Sometimes, flowers develop only on the sides of the canopy, away from the light source.
- Royal poinciana is a member of the bean family (Fabaceae). Most of the 18,000 species develop root swellings called nodules, within which bacteria of the genus *Rhizobium* are engaged in the "fixation" of nitrogen. In this process, inert nitrogen is taken out of the air and converted to ammonia, a form that plants can employ in metabolism. Comparatively little natural fixation of nitrogen is associated with all other plant families combined. This symbiotic association is responsible for the enormous improvement of the fertility of soils, indeed more than all fertilization carried out by the world's agricultural activities. However, royal poinciana and many of its relatives are not associated with nitrogen fixation (reports to the contrary appear to be in error).
- Since 1937, the city of Miami, Florida, has held an annual royal poinciana festival during the first week of June. It has been estimated that there are more royal poinciana trees growing in the city than in Madagascar, the home country of the species.
- Scenic attractive postcards have long been a staple of the tourist industry, enticing potential visitors to come to a location and contribute to the local economy. In tropical countries and locations, royal poinciana has long been a key subject of postcards.

## KEY PUBLICATIONS


Babineau, M. and A. Bruneau. 2017. Phylogenetic and biogeographical history of the Afro-Madagascan genera *Delonix*, *Colvillea* and *Lemuropisum* (Fabaceae: Caesalpinioideae). *Botanical Journal of the Linnean Society* 184: 59–78.

CABI Compendium. 2017. *Delonix regia (Flamboyant)*. https://doi.org/10.1079/cabicompendium.1852.

Cooke, G.R. 2017. The poinciana: Icon of a Queensland summer. *Australian Garden History* 29(2): 4–7.

Du Puy, D.J., P.B. Phillipson, and R. Rabevohitra. 1995. The genus *Delonix* (Leguminosae: Caesalpinioideae: Caesalpinieae) in Madagascar. *Kew Bulletin* 50: 445–475.

Gilman, E.F. and D.G. Watson. 1993. *Delonix regia,* Royal poinciana. University of Florida Cooperative Extension Service, Fact Sheet ST-228. https://hort.ifas.ufl.edu/database/documents/pdf/tree_fact_sheets/delrega.pdf.

Jain, S., A. Vaidya, N. Jain, V. Kumar, and A. Modi. 2021. Bioactive compounds of Royal poinciana (*Delonix regia* (Hook.) Raf.). In: *Bioactive Compounds in Underutilized Vegetables and Legumes*, edited by H.N. Murthy and K.Y. Paek, 1–20. Springer, Cham.

Lewis, G. 2020. *Delonix regia*: Leguminosae. *Curtis's Botanical Magazine* 37: 324–331.

Lim, T.K. 2012. *Delonix regia*. In: *Edible Medicinal and Non-Medicinal Plants*, edited by T.K. Lim, 617–623. Springer, Dordrecht.

Rivers, M.C., N.A. Brummitt, E.N. Lughadha, and T.R. Meagher. 2011. Genetic variation in *Delonix* sl (Leguminosae) in Madagascar revealed by AFLPs: Fragmentation, conservation status and taxonomy. *Conservation Genetics* 12: 1333–1344.

Schroder, E.C. 1984. *Delonix regia*: Confirmation of its non-nodulating character. In: *Advances in Nitrogen Fixation Research*, edited by C. Veeger and W.E. Newton, 359. Springer, Dordrecht.

Singh, S. and S.N. Kumar. 2014. A review: Introduction to genus *Delonix*. *World Journal of Pharmacy and Pharmaceutical Sciences* 3: 2042–2055.

Vozzo, J.A. 2003. *Delonix regia* (Bojer ex. Hook) Raf. Species Descriptions, Reforestation, Nurseries and Genetic Resources Net, USDA, USA. https://citeseerx.ist.psu.edu/document?repid=rep1&type=pdf&doi=2cfe7a27df202a60d01cd46f176a09d0a226b2ea.


### Key Cultivation Websites

Agricultural Magazine. *How to Grow and Care for Royal poinciana: Propagation and Planting Instructions*. https://agriculturalmagazine.com/how-to-grow-and-care-for-royal-poinciana-propagation-and-planting-instructions/.

Brisbane Trees and Gardens. *Delonix regia*. https://brisbanetrees.com.au/poinciana/.

Green Life. *How To Grow and Care for Flame Tree (Royal poinciana)*. https://green-life.blog/how-to-grow-and-care-for-flame-tree-royal-poinciana/.

The Spruce. *How to grow and care for flame tree (Royal poinciana)*. www.thespruce.com/royal-poinciana-growing-profile-3269287.

University of Florida. *Royal poinciana*. https://gardeningsolutions.ifas.ufl.edu/plants/trees-and-shrubs/trees/royal_poinciana.html.

Urban Garden Gal. *How to Keep Poinciana Trees Small*. www.urbangardengal.com/keep-poinciana-tree-small/.

WA Scene. *How to Grow Poinciana Trees*. www.wascene.com/home-garden/flowers-and-plants/growing-poinciana-trees/.

# 90 Sacred Lotus (*Nelumbo nucifera* and Hybrids)

**Scientific Names:** The genus name *Nelumbo* traces to the Sinhalese (an Indic language of Ceylon) name for the plant: *nelumbo* or *nelun*. The term *nucifera* in the scientific name is Latin for *nut-bearing*, referring to the nutlike seeds produced by the plant. *Lutea* in the scientific name *N. lutea* (American lotus) means "yellow," for the flower color, the best character for distinguishing the two species of *Nelumbo*.

**Common Names:** The common name lotus is derived from the Latin *lotus* and Greek *lotos*, a name used for several plant species in ancient times. Today, the name is applied not just to sacred lotus and its relative American lotus but also to some species of *Nymphaea* (which are among the plants called water lilies). "Snow lotus" refers to species of the genus *Saussurea*. Especially confusing, the genus name *Lotus* refers to quite different plants, including species employed as livestock forage. Sacred lotus is often simply called lotus but is known by many other names, including Chinese arrowroot, East Indian lotus, Hindu lotus, Indian lotus, Oriental lotus, and Egyptian bean, and it is often referred to as a lily or water lily.

FIGURE 90.1 Sacred lotus stands: (a) In an arboretum in Germany. (b) In Australia. (c) A lady harvesting flowers in Vietnam.

*Source:* **(a) Photo by Hans-Dieter Warda (CC BY SA 3.0). (b) Photo by Bernard Dupont (CC BY SA 2.0). (c) Photo by Tuan Anh Nguyen from Pixabay (public domain).**

FIGURE 90.2 Nineteenth-century illustrations of sacred lotus.

*Source:* **Public domain. (a) Blanco, F. M. 1880–1883. *Flora de Filipinas*. (b) *Seikei Zusetsu*, vol. 28, 1804. Photo by Leiden University Library. (c) Setsuan, Y. 1876. *Album Depicting Birds, Flowers, Landscapes, and Flower Pots*. Photo by Walters Art Museum.**

## 90.1 INTRODUCTION

The sacred lotus is the most extensively employed ornamental of the so-called "water lilies." This class of plants is characterized by rooting in shallow water and producing large, colorful flowers that float on or extend above the water surface. Most water lily species are in the "water lily family," the Nymphaeaceae, and lotus has been placed there in the past but is now segregated into its own family, the Nelumbonaceae (the "lotus-lily family").

*Nymphaea* water lilies (treated in the chapter "Water Lilies") are available in a remarkable range of colors, and these compete for the attention of those interested in water gardens. Frequently on the internet, photos of *Nymphaea* water lilies are misleadingly labelled as "lotus."

## 90.2 OVERVIEW

The sacred lotus is an herbaceous perennial plant native to the warmer parts of Asia (Iran to Japan) and Australia but introduced and established from escapes from cultivation in much of the warm-temperate world. It is an aquatic plant, growing from a tuberous rhizome (underground storage stem, swollen with starch like potato) that is rooted in the soil of shallow water (of ponds, lakes, or slow-moving rivers) or sometimes in the adjacent shoreline mud. In water, the leafstalks (petioles) extend upward from the rhizomes, the roundish leaf blades (laminas) either floating on the water or extending above the water. The long petioles attach at the middle of the leaf blade underside. The tops of the leaf blades are concave (a feature that makes them appropriate for use as dinner plates). The petioles can exceed 2 m in length (allowing them to reach the surface of water this

DOI: 10.1201/9781003586227-90

deep), and the blades can approach 1 m in width but usually are not larger than 60 cm. The leaf blades may float on deep water but rise high above shallow water. The leaf surfaces are extremely water-repellent, and one can often see glittering water droplets that have beaded up on the leaves. The flowers are white or pink; red and other colors occur in some cultivars, especially in hybrids with American lotus. The flowers are solitary, fragrant, large (up to 30 cm in diameter), and displayed above the water on long floral stalks that extend downward to the sausage-like rhizomes. Very large flowers can have thousands of petals. Blooming occurs in midsummer. In most plants, individual flowers last only a few (2 to 5) days, but a patch of plants can be in flower for 3 months. After the petals fall off, a conical seed head that looks like the rose (nozzle) of a watering can, or an ice cream cone, is left and becomes large and woody. At the top (the widest part), there are openings to small chambers, each containing a seed.

FIGURE 90.3 Flower colors of sacred lotus.

*Source:* **(a) Photo by RebexArt (CC BY SA 3.0). (b) Photo by Yoshikazu Takada (CC BY 2.0) (c) Photo by Smithsonian Institution, NMNH, Botany, WVU Herbarium (public domain).**

FIGURE 90.4 American lotus (*Nelumbo lutea*): (a) Painting (public domain) showing American lotus (*Nelumbo lutea*) by American artist Charles Courtney Curran (1861–1942). (b) In a pond in Parc Floral de Paris, in Vincennes, France. Most of the flowers have matured into young fruits.

*Source:* **(a) Photo from Daniel J. Terra Collection. (b) Photo by Moonik (CC BY SA 3.0).**

The so-called "roots" (actually rhizomes, that is, underground stems) are thick and tuberous and creep in the substrate at the bottoms of ponds or slow-moving streams. They are 60–120 cm long and 6–9 cm in diameter. They are cylindrical but pinched in at regular intervals (at the nodes, that is, where the leaves arise) and so look like a string of sausages. They have long air chambers, which provide for oxygen transportation to and carbon dioxide from the underground, underwater parts of the plant. The air chambers show up as an ornamental pattern of large and small holes in cross sections of the rhizomes, which are commonly served in restaurants.

American or yellow lotus (*N. lutea*) is native to Ontario in Canada, most of the eastern United States, Mexico, Central America, Colombia, and the West Indies. True to its name, its flowers are yellow. Some hybrid cultivars with sacred lotus have been bred. There are far fewer cultivars of American lotus compared to sacred lotus. It is an extremely attractive species, but not nearly as widely cultivated as sacred lotus. It can also be an aggressive weed and has been introduced outside of its native range.

## 90.3 HISTORY

Sacred lotus has been cultivated in Asia for thousands of years, for its edible seeds and rhizomes, and as an ornamental. More than a thousand cultivars have been named, many recognized in China within the last half-century. Cultivars have been selected for three basic uses: ornament, edible rhizomes,

FIGURE 90.5 The religious importance of sacred lotus in Hinduism. The three greatest deities, Brahma (four-headed), Vishnu, and Shiva, are shown seated on lotuses with their consorts, Saraswati, Lakshmi, and Paravati, respectively.

*Source:* **Illustration (public domain) by unknown artist in northwest India, about 1770. Victoria and Albert Museum.**

FIGURE 90.6 Skillfully crafted artificial lotus flowers prepared as religious tributes, reflecting the sacred nature of the plant.

*Source:* **(a) Photo by Jennifer (CC BY 2.0). (b) Photo by Shutterstock/Cortyn.**

FIGURE 90.7 Economic importance of sacred lotus edible seed heads in Asia: (a) Top of seed head showing seeds each in its own cavity. (b) A woman in Vietnam harvesting seed heads from a lotus pond. (c) A girl selling lotus seed heads in Cambodia.

*Source:* **(a) Photo by Rumpleteaser (CC BY 2.0). (b) Photo by Lucas Jans (CC BY SA 2.0). (c) Photo by Lynhdan (CC BY ND 2.0).**

and harvest of seeds. The flowers of wild varieties usually have fewer than 25 petals, while those of ornamental forms usually have been "doubled" and have more than 50 petals (over 16,000 petals have been counted in some flowers).

The sacred lotus is one of the most venerated of plants and is a cultural icon. It is considered divine in Asian and Oriental cultures, symbolizing perfection, immortality, and enlightenment. Often called the sacred lotus of India, it is revered by Buddhists and Hindus, who often carve lotus flower ornaments into their religious buildings. The sacred lotus is the national flower of India and is part of the country's national emblem (a scene which also has lions, a bull, an elephant, and a horse). It is also the national flower of Vietnam and has been considered to be one of several plants considered to have national significance in Iran. Sacred lotus is also the holy flower of Buddhism, symbolizing purity and divine birth. The rise of the flower from its watery environment is said to be symbolic of the evolution of human consciousness. In China, the sacred lotus is highly respected in Confucianism, and lotus roots and seeds are used as good luck charms. The American lotus also has a long spiritual tradition. It is esteemed by several Native American cultures and is believed to have mystic powers.

## 90.4 ECONOMIC IMPORTANCE

Water gardens are considerably more difficult to establish and maintain compared to terrestrial gardens, so that cultivation of water lily plants is a niche market. Sacred lotus is cultivated almost exclusively as an ornamental in North America and Europe. However, in Asia, it is primarily cultivated as a food plant, mostly for its edible, starchy rhizomes (underground stems), but also for its edible seeds, and sometimes for the spicy stamens from the flowers. Sacred lotus has been cultivated in China for at least 3,000 years. It is grown in paddies similar to those for rice. The majority of edible sacred lotus rhizomes are grown in China, where hundreds of thousands of hectares are cultivated. As a vegetable and spice, sacred rhizome is a very important economic plant in Asia.

FIGURE 90.8 Dried sacred lotus stamens collected from the flowers and employed to make herbal tea.

*Source:* **Photo by Shutterstock/Sasimoto.**

FIGURE 90.9 Mughal lotus water gardens. The Mughal Empire was a Muslim domain centered in northern India that controlled much of South Asia between the 16th and 19th centuries.

*Source:* **(a) Illustration by Shutterstock/NS Vector. (b) Illustration by Shutterstock/Artistsayeed.**

## 90.5 ORNAMENTAL USES

Lotus is widely cultivated in water gardens. In Asia, the plant is often allowed to flourish in rivers and lakes in urban areas because it is so ornamental and respected. The flowers are extensively harvested and employed as decorations, especially for religious celebrations. The curious seed heads are widely employed in floral arrangements.

FIGURE 90.10 Sacred lotus allowed to clog a waterway in Zhaoqing (a city of 4 million in southeastern China) because it is so attractive in flower.

*Source:* **Photo by Ilee Wu (CC BY ND 2.0).**

FIGURE 90.11 Rhizome of sacred lotus. Note air channels. (a) Whole and sliced segments. (b) Cooked preparation.

*Source:* **(a) Photo by FotoosRobin (CC BY SA 2.0). (b) Photo by Kent Wang (CC BY SA 2.0).**

## 90.6 OTHER USES

All parts of sacred lotus are edible, especially the starchy rhizomes, which represent the principal economic product in Asia. In Asian cooking, the roots are often employed in the same range of foods as potatoes in Western cuisine but are also used in pickled dishes. The seeds, flowers, stems, and foliage are also consumed in Asia. American lotus rhizomes were also eaten by Indigenous people in North America. Lotus is not sold in most stores of Western countries but is often available in stores specializing in Asian foods. A type of flour is prepared from the roots in China (and is sometimes available in Western markets). This is employed much like cornstarch as a thickener for soups and sauces, to make sweet dishes and soft-textured cakes, and to give a crispy coating to deep-fried foods. An odd feature of this flour is that it turns purple when combined with liquid.

Lotus seeds (often called "lotus nuts") are eaten raw or boiled, roasted, candied, or made into jam. The seeds are traditionally served in tea at weddings in China because they are a symbol of fertility. The seeds are also used to flavor a variety of dishes, including sweet soups and desserts. Young, fresh lotus seeds look like peanuts with green skins, while dried lotus seeds have dark-brown skins like filberts. Dried lotus seeds need to be blanched and peeled before eating, and a toothpick or needle is employed to push out the bitter green heart. Canned seeds require no preparation. In India, lotus seeds are popped like popcorn. Lotus paste, made from sweetened cooked lotus seeds, is often available in cans in Asian markets and is used in cooking, especially as a pastry for mooncakes.

Lotus leaves are often used by Asians to add flavor and fragrance to food and are dried and employed as a tamale-like wrap for dishes like glutinous rice. "Lotus packets" are small bundles of food, typically sticky rice, meat, chicken, and vegetables wrapped in lotus or similar leaves and steamed. These are popular in Asia, especially China. The leaves are frequently dried and used as plates.

As for most widespread plants, the *Nelumbo* species have been extensively employed in folk medicine in their native distribution range. In traditional Asian medicine, lotus seeds and rhizomes are used medicinally, in combination with other herbs, to treat diarrhea, dizziness, dysentery, fever, sunstroke, liver disease, and stomach problems. The whole plant is used to relieve food poisoning; the seeds and rhizomes to treat hyperlipidemia (elevated cholesterol in the bloodstream); the seeds alone for cholera, diarrhea, fever, indigestion, insomnia, nausea, and nervous disorders; and the flower stalks in combination with other herbs to treat bleeding of the uterus. In China, lotus roots and seeds are consumed as a tonic. The chemical components of sacred lotus are under very extensive examination for medicinal applications.

## 90.7 PROBLEMS

In warm climates, sacred lotus can spread aggressively and is best confined to containers and ponds. In some jurisdictions, it is regulated as an invasive alien. Like several other aquatic weeds, it can form dense mats of floating vegetation which inhibit native aquatic species and negatively affect boating, angling, and swimming. In some parts of the United States, it is recommended that lotus species not be grown outdoors – even in a container – because it escapes containment either by dispersal of seeds or extension of rhizomes. The rhizomes are reported to sometimes travel more than 9 m in a single season.

## 90.8 CULTIVATION

### 90.8.1 Climate and Site

The species is considered hardy north to climate zone 4, but this is a rather arbitrary evaluation since it assumes that the rhizomes, protected at the bottom of ponds, will not experience freezing temperatures. Sacred lotus is adapted to full sun. The plants grow normally in calm waters, as is usually provided by water gardens and small ponds.

### 90.8.2 Propagation and/or Purchasing

Rhizomes (commonly termed *bulbs*, *tubers*, and *roots* in the trade) are preferable starting material, breeding true to type. Modern nurseries divide rhizomes so as to contain at least one bud, but they usually grow such small segments to provide larger starting stock for the public. If grown from seeds, their extremely hard coats prevent germination, unless one end is carefully removed (e.g., with a hacksaw) to expose a slight amount of white endosperm (nutritive tissue). The chipped seeds should be placed in water at 20–25°C for two or three days, then placed about 2 cm deep in a pot of damp compost standing in a shallow bowl of water at 20–30°C. The seedlings need to be grown under light as high as possible. They can be transferred outdoors to shallow water and hopefully will survive the winter. Much more conveniently, rhizomes can be purchased from suppliers.

### 90.8.3 Planting

Rhizomes or young plants can be placed in the muddy margin of a pond, where they can spread into deeper water. Alternatively, the rhizomes can be planted in boxes of soil sunken in water gardens or ponds.

### 90.8.4 Soil and Moisture Conditions

The plant can grow well in out-of-water containers of organically rich loams that are kept moist but won't look as attractive as they do when displayed in small bodies of water. Sacred lotus can grow in water up to 2.5 m deep. The preferred pH range is 6.0–6.5. Specialized aquatic fertilizers for water plants are available. Alternatively, some aquatic gardeners package fertilizer (especially nitrogen-rich formulations) in a permeable material and push this down into the root zone.

FIGURE 90.12 Sacred lotus growing in a container.

*Source:* **Photo by Shutterstock/Asharkyu.**

### 90.8.5 Maintaining Plants

In areas with freezing temperatures in the winter, survival of the rhizomes may require assistance. Ponds can be heated so they won't freeze. If the rhizomes are growing in containers, in the fall they can be moved into deeper water, which won't freeze to the bottom. Rhizomes can be excavated and brought indoors for overwintering. For those unwilling to provide such care, the plants are sometimes grown as expendable (albeit expensive) annuals.

## 90.9 FASCINATING FACTS, FABLES, FOLKLORE, AND FRIVOLITY

- According to Hindu religious tradition, the goddess of love, Lakshmi, was born as Padma (one of her incarnations) as a lotus which arose from the primordial sea. From the lotus flower, which was pure gold and had a thousand petals, there was born the creator god, Brahma. Padma had two sons, representing mud and moisture, the ingredients of fertile soil. This association of the lotus

FIGURE 90.13 The Hindu goddess Lakshmi holding and standing on lotus flowers.

*Source:* **Painting by Raja Ravi Varma (1848–1906). Public domain.**

with plant fertility was paralleled by an association of the plant with human fertility in India.

- In China, lotus seeds are also a symbol of fertility. According to traditional belief, women who wish to give birth to baby boys should consume foods rich in lotus seeds.
- In Greek mythology, lotus-eaters were a people, described in the ancient Greek tales of the Odyssey, who ate the fruit of the "lotus tree" (possibly jujube, date palm, or elm, but not the sacred lotus) and then forgot their friends and homes, lost all desire to return to their native land, and lived in a drugged, lazy state. Hence, the term "lotus-eater" is applied to a lazy person devoted to pleasure and luxury, and "lotus land" is a pejorative name for Southern California.
- The seeds of scared lotus are extraordinarily long-lived. Seeds about 1,300 years of age have been germinated.
- Sacred lotus is one of the very few plants known to generate heat. The flower buds open slightly on the first day of flowering, and pollinating insects, especially beetles, are trapped inside overnight. Heat inside the flowers can reach 36°C at night, despite much lower outside temperatures. It has been speculated that the heat keeps the beetles warm overnight, facilitating pollination the next day when they are released. Alternatively, the floral heat could help release volatile compounds to attract the flying insects.
- The large leaves have been employed as dinner plates.
- The upper surfaces of sacred lotus foliage are water-repellent. When water droplets fall on the leaves, numerous small wax-covered hairs spaced more closely than the diameter of a small water drop keep most of the water from contacting the leaf surface, and the water droplets roll away. Such surfaces have been shown to be "self-cleaning," because water droplets travelling across the surface pick up dirt particles and remove them. The phenomenon, termed "the lotus effect," and ultrahydrophobicity inspired Wilhelm Barthlott, director of the Botanical Gardens at the University of Bonn, Germany, to create a self-cleaning, hydrophobic paint.
- The lotus (Padmasana) position is a power yoga exercise employed in Zen and other meditative disciplines. The feet are placed on the opposite upper thighs, the soles of the feet turned up, and the spine straight. The position (very difficult for most beginners) is said to be based on resemblance to the lotus flower. This classical cross-legged sitting meditation pose traces to ancient India and is often assumed by Buddha in statues and paintings (often shown seated on a giant lotus flower).
- The sacred lotus is especially respected in Vietnam, where it is symbolic of strength, capable of growing in a stressful environment – swampy, muddy water often subjected to stormy weather. According to legend, the more opaque and muddy the water, the more beautiful the lotus flower.

FIGURE 90.14 Figure of the Buddha (ca. 450 BCE), whose teachings form the basis of Buddhism. He is shown in the conventional lotus position, seated on and surrounded by lotus blossoms.

***Source:* Image by Peace, love, happiness from Pixabay (public domain).**

FIGURE 90.15 Sacred lotus (*Nelumbo nucifera*) in Vietnam, admired by a young girl.

***Source:* Photo by Yves Picq (CC BY SA 3.0).**

## KEY PUBLICATIONS

Bangar, S.P., K. Dunno, M. Kumar, H. Mostafa, and S. Maqsood. 2022. A comprehensive review on lotus seeds (*Nelumbo nucifera* Gaertn.): Nutritional composition, health-related bioactive properties, and industrial applications. *Journal of Functional Foods* 89: 104937. https://doi.org/10.1016/j.jff.2022.104937.

Borsch, T. and W. Barthlott. 1993. Classification and distribution of the genus *Nelumbo* Adans. (Nelumbonaceae). *Beitraege Biologie Pflanzen* 68: 421–450.

Griffiths, M. 2009. *The Lotus Quest. In Search of the Sacred Flower*. St. Martin's Press, New York. 338 pp.

Guo, H.B. 2009. Cultivation of lotus (*Nelumbo nucifera* Gaertn. ssp. *nucifera*) and its utilization in China. *Genetic Resources and Crop Evolution* 56: 323–330.

Hall, T.F. and W.T. Penfound. 1944. The biology of the American lotus *Nelumbo lutea* (Willd.) Pers. *American Midland Naturalist* 31: 744–758.

Li, R., J. Li, S. Wang, and R. Wang. 2022. Novel insights into floral thermogenesis: In vivo analyses of mitochondrial dynamics in *Nelumbo nucifera* flowers. *International Journal of Molecular Sciences* 23(19): 11950. https://doi.org/10.3390/ijms231911950.

Lim, T.K. 2016. *Nelumbo nucifera*. In: *Edible Medicinal and Non-Medicinal Plants*, edited by T.K. Lim, 55–109. Springer, Cham.

Lin, Z., C. Zhang, D. Cao, R.N. Damaris, and P. Yang. 2019. The latest studies on lotus (*Nelumbo nucifera*) – an emerging horticultural model plant. *International Journal of Molecular Sciences* 20(15): 3680. https://doi.org/10.3390/ijms20153680.

Liu, L., Y. Li, J. Min, Y. Xiang, and D. Tian. 2019. Analysis of the cultivar names and characteristics of global lotus (*Nelumbo*). *Hans Journal of Agricultural Science* 9: 163–181. (In Mandarin)

Matthews, P.G. and R.S. Seymour. 2006. Anatomy of the gas canal system of *Nelumbo nucifera*. *Aquatic Botany* 85: 147–154.

Obando, W.O., P. Biles, L.A. Gettys, K. Tilt, K. Moore, F.M. Woods, et al. 2012. American lotus: More than just a pretty face. *Aquatics* 34(3): 7–9.

Seymour, R.S. 1997. Plants that warm themselves. *Scientific American* 276(3): 104–109.

Seymour, R.S. and P. Schultze-Motel. 1998. Physiological temperature regulation by flowers of the sacred lotus. *Philosophical Transactions: Biological Sciences* 353: 935–943.

Sharma, S.C., and A.K. Goel. 2000. Environmental degradation and ex-situ conservation of *Nelumbo nucifera*. In: *Environmental Stress: Indication, Mitigation and Eco-Conservation*, edited by M. Yunus, N. Singh, and L.J. de Kok, 406–410. Kluwer Academic Publishers, Dordrecht, The Netherlands.

Shen-Miller, J., J.W. Schopf, G. Harbottle, R.J. Cao, S. Ouyang, K.S. Zhou, et al. 2002. Long-living lotus: Germination and soil gamma-irradiation of centuries-old fruits, and cultivation, growth, and phenotypic abnormalities of offspring. *American Journal of Botany* 89: 236–247.

Small, E. and P.M. Catling. 2011. Blossoming treasures of biodiversity 15. Asian lotus (*Nelumbo nucifera*) and American lotus (*N. lutea*), spiritual symbols, valuable foods, and successful invaders. *Biodiversity* 6(1): 25–30.

Wang, Z., Y. Cheng, M. Zeng, Z. Wang, F. Qin, Y. Wang, et al. 2021. Lotus (*Nelumbo nucifera* Gaertn.) leaf: A narrative review of its phytoconstituents, health benefits and food industry applications. *Trends in Food Science & Technology* 112: 631–650.

## Key Cultivation Websites

GardenersHQ.com. *Nelumbo Plant Growing & Care Guide for Gardeners*. www.gardenershq.com/Nelumbo-Waterlily-Lotus.php.

Home Gardeners. *Sacred Lotus – How to Grow and Care for Lotus Plants*. https://homegardeners.in/lotus-plants/.

Loch Ness Water Gardens. *How to Plant a Lotus*. https://lochnesswatergardens.com/blogs/pondblog/how-to-plant-a-lotus.

PondLotus.com. *Growing Lotus Outside the Pond*. https://pondlotus.com/pages/lotus-outside-the-pond.

PondLotus.com. *How to Grow Sacred Lotus from Seed*. https://pondlotus.com/pages/buy-sacred-lotus-seeds.

The Spruce. *How to Grow Beautiful Lotus Flowers in Your Water Garden*. www.thespruce.com/lotus-plant-7964244.

University of Florida, Extension. *Nelumbo nucifera, Lotus, Sacred Lotus*. https://edis.ifas.ufl.edu/publication/FP424.

WikiHow. 2020. *How to Make a Lotus Pond*. www.wikihow.com/Make-a-Lotus-Pond.

WikiHow. 2022. *How to Grow Lotus Flower*. www.wikihow.com/Grow-Lotus-Flower.

# 91 Salvia (*Salvia* Species and Hybrids)

**Scientific Names:** The genus name *Salvia* originates from the Latin *salvia*, probably tracing to the ancient Roman name for the culinary and medicinal herb *Salvia officinalis*, which is called sage, garden sage, and common sage. This species is often considered to be the world's most popular flavoring and seasoning herb. The word *sage* is based on the Latin adjective *salvus*, "safe and healthy," and the verb *salvere*, "to heal" or "to save." These Latin terms are also the root of the words *salve* and *salvation*.

**Common Names:** The ornamental plants of *Salvia* are usually termed "salvia," although many of the 1,000 or so species are also called "sage" (so are some unrelated species). In the past, the word *sage* employed alone has normally referred to *S. officinalis*, but in recent years, it has often been employed for the hallucinogenic *S. divinorum*, employed as a recreational drug. "Sagebrush" refers to shrubby species of *Artemisia* native to semi-arid plains and mountain slopes of western North America. "Wood sage" refers to species of *Teucrium*. Similarly, several other plants quite unrelated to *Salvia* have "sage" in their names.

FIGURE 91.1 Mixed color flowering displays of scarlet sage (*Salvia splendens*), the most popular ornamental salvia: (a) At a flower park in Japan. (b) In a border beside a stone wall.

***Source:*** **(a) Photo by Harum Koh (CC BY SA 2.0). (b) Photo by Shutterstock/Irina N.**

## 91.1 INTRODUCTION

Salvias include a huge range of ornamental herbs and shrubs which are greatly admired for their extremely attractive floral displays, as well as comparatively limited need for maintenance. Several of the species are widely employed to enhance public landscapes as well as private gardens.

## 91.2 OVERVIEW

Most species of *Salvia* are small herbs or shrubs. The classification of the genus is debated, often considered to be made up of about a thousand species, mostly from South America, with some in Eurasia. However, future botanists may choose to split *Salvia* as currently understood into several genera. Like other members of the mint family (Lamiaceae, Labiatae), salvias have square stems. Most cultivars range in height from 30 to 120 cm, although some of the wild species are much taller. The tubular flowers are two-lipped, borne on vertical spikes. Some cultivars have bicolored flowers.

*Salvia splendens*, called bedding sage and scarlet sage (both names also applied to other species), is the most widely grown ornamental salvia. In its native Brazil, it is a perennial adapted to subtropical conditions and growing as tall as 8 m, but the cultivated forms are dwarfs varying in height from 20 to 75 cm and raised as annuals in cold climates. Cultivars with brilliant scarlet-red flowers are most popular, but the blooms are also available in lavender, orange, purple, salmon, and white. The plants flower from summer to autumn.

## 91.3 HISTORY

Natural hybridization (which takes place spontaneously in nature) is widespread among *Salvia* species, but so is artificial hybridization (conducted by people, outside of nature), which has been responsible for the generation of numerous ornamental cultivars. Often, the parents of garden cultivars are not known (in the past, gardeners frequently did not prepare good records or wanted to keep the information secret). Scarlet sage (*S. splendens*), the most important ornamental species, was collected from Brazil in the early 19th century, and numerous horticultural forms and hybrids have been created.

## 91.4 ECONOMIC IMPORTANCE

Salvias are popular garden plants of substantial importance to the horticultural industry. The ornamental cultivars are grown particularly from cuttings, so there is a dependable annual demand for supplying starter plants. In addition,



DOI: 10.1201/9781003586227-91

FIGURE 91.2 Some of the commonly cultivated ornamental species of *Salvia*: (a, b) *Salvia* cultivar Hot Lips. These “Mexican salvias” (hybrids of *Salvia microphylla*, the baby sage or Graham’s sage, and other species, including the hybrid *S. microphylla* × *S. greggii* = *S.* × *jamensis*) are evergreen shrubs of the Southern United States and Mexico – climates where they can be grown as perennials. (c) *Salvia* × *superba* cultivar Adrian. This hybrid species of domesticated origin is of uncertain ancestry. In garden centers, it is frequently labelled as *S. nemorosa* (a name that belongs to another species). It is a sterile hybrid, maintained by cuttings. The plant is hardy in climate zones 3 to 8 and usually is grown as a perennial (and so is sometimes labelled “perennial salvia”). (d, e) Blood sage (*Salvia coccinea*, also called tropical sage), a native of southern North America and South America. Unfortunately, it is also called scarlet sage – the same name applied to *S. splendens*; it is also called Texas sage, a name also applied to ornamental *Leucophyllum frutescens*. It is grown as an annual in cold climates but can be grown overwinter in climate zones 9 and southward. (f) Mealycup sage (*Salvia farinacea*, also called mealy sage and mealy blue sage), a native of the southernmost United States and Mexico. It is named for the mealy-white (sometimes purple) floral sepals covered with felted hairs. In climate zones 7 and 8, it grows as an herbaceous perennial, dying back to the ground in winter. In zones 9 and southwards, it can be grown as an evergreen perennial.

**Source:** **(a) Photo by Manuel m.v. (CC BY 2.0). (b) Photo by Shiva Shenoy (CC BY 2.0). (c) Photo by Salicyna (CC BY SA 4.0). (d) Photo by Forest and Kim Starr (CC BY 2.0). (e) Photo by Carl E. Lewis (CC BY 2.0). (f) Photo by Suzannedoig (CC BY SA 4.0).**

sage and marjoram (which are *Salvia* species) are major flavoring herbs of importance to agriculture.

## 91.5 ORNAMENTAL USES

Salvias are employed for mass plantings, borders, containers, and (to a minor extent) cut flowers.

## 91.6 OTHER USES

Numerous salvias are grown in gardens as culinary herbs. The principal salvias cultivated as flavoring herbs include garden sage (*S. officinalis*), rosemary (*S. rosmarinus*), Greek

FIGURE 91.3 Paintings (public domain) by botanical illustrators of old garden cultivars of *Salvia*: (a) *Salvia splendens*. (b) *Salvia farinacea*. (c) *Salvia coccinea*. (d) *Salvia officinalis*. (e) *Salvia elegans*. (f) *Salvia uliginosa*.

**Source:** **(a) Step, E., D. Bois, B. Herincq, and W. Watson. 1896. *Favourite Flowers of Garden and Greenhouse*, vol. 3. Frederick Warne, New York. (b) *Revue horticole*, Librairie agricole de la maison rustique, Paris, 1873. (c) Lindley, J. 1853. Paxton’s flower garden, vol. 2. (d) Köhler, E. 1883–1914. Köhler’s Medizinal-Pflanzen in naturgetreuen Abbildungen mit kurz erläuterndem Texte. Fr. Eugen Köhler, Gera-Untermhaus. (e) *Revue horticole* (Paris), sér. 4, vol. 45, 1873. (f) Fitch, M. S., and J. N. Fitch. 1914. *Curtis’s Botanical Magazine*, vol. 140, London.**

FIGURE 91.4 *Salvia splendens* on a commercial flower farm in Taipei, Taiwan.

**Source:** **Photo by Shutterstock/Kit Leong.**

FIGURE 91.5 Ornamental beds of scarlet salvia (*Salvia splendens*). (a) At an entertainment park in Japan. (b) In a residential garden.

*Source:* **(a) Photo by Foooomio (CC BY 2.0). (b) Photo by OakleyOriginals (CC BY 2.0).**

FIGURE 91.6 Species of *Salvia* employed as flavoring herbs: (a) Common sage (*Salvia officinalis*), a shrub native to the Mediterranean area, can be grown as a perennial as far north as climate zone 5 (also zone 4 with protection). In colder areas, it survives overwinter by its roots, the aboveground parts killed by frost (often, it is grown as an annual in cold climates). Mostly grown as a flavoring herb, the plant is nevertheless quite ornamental. A single plant usually provides enough sage for the average family. (b) Pineapple sage (*Salvia elegans*), a perennial shrub native to Mexico which is usually grown as an annual in cold climates but can be grown overwinter in zone 7 and southward. It is both a culinary herb (the scent reminiscent of pineapple) and a garden ornamental. (c) Rosemary (*Salvia rosmarinus*, known by the scientific name *Rosmarinus officinalis* until 2017). This shrub with white, pink, purple, or blue flowers is a native of the Mediterranean region. It is widely employed as a flavoring herb, but also for extracts employed in perfumery. Most cultivars can be grown as perennials in climate zone 8 and southwards, and some cultivars will survive overwinter as far north as zone 6.

*Source:* **(a) Photo by Javier Martin (public domain). (b) Photo by Beverly Buckley from Pixabay (public domain). (c) Photo by JackieLou DL from Pixabay (public domain).**

sage (*S. fruticosa*; also called Turkish sage), blue sage (*S. clevelandii*; also called Jim sage and Cleveland sage), pineapple sage (*S. elegans*), bluebeard sage (*S. viridis*; also known as Joseph sage), clary sage (*S. sclarea*; also called muscatel sage), and apple sage (*S. pomifera*). The seeds of some species, notably *S. columbariae* (California chia, golden chia), are edible and have been employed for food. Chia has become a popular trendy specialty grain. The essential oil of *S. lavandula* (Spanish sage) is employed to flavor commercial food preparations.

Salvias are frequently visited by hummingbirds, butterflies, and bees and are often grown in pollinator or wildlife gardens.

FIGURE 91.7 Pollinators visiting salvia flowers: (a) Honeybees. (b) Male broad-tailed hummingbird (*Selasphorus platycercus*) in Colorado. (c) Common rose butterfly (*Pachliopta aristolochiae*) of southern Asia.

*Source:* **(a) Photo by Beverly Buckley from Pixabay (public domain). (b) Photo by Shutterstock/Danita Delimont. (c) Photo by Shutterstock/Small1.**

## 91.7 PROBLEMS

Some salvias have been alleged to have minor toxicity, and some have tendencies to become invasive weeds. Compared to the damage caused by some of the other species discussed in this book, these are relatively minor considerations.

## 91.8 CULTIVATION

### 91.8.1 Climate and Site

Most salvias are adapted to hot summer conditions but differ considerably in ability to overwinter in cold climates. Many are grown as annuals in cold climates but can be grown as perennials in warmer climate zones. Depending on species and cultivar, salvias can be grown as perennials in zones 5 to 11. The popular *S. splendens* can be grown as a perennial in zones 10 and southward. Most salvias do best in full sun; some tolerate partial shade.

### 91.8.2 Propagation and/or Purchasing

Named varieties (especially of annual bedding-type salvias) often will not breed true from seeds and are reproduced by cuttings (usually obtained as starter plants from nurseries and garden centers). Some cultivars are sterile hybrids and do not produce seeds. Salvias often self-propagate by the seeds they develop at a particular site.

### 91.8.3 Planting

Planting distances vary considerably, depending on cultivar, and whether grown as an annual or perennial. The popular *S. splendens*, mostly grown as an annual, can be planted at a spacing of 25 cm. When grown as perennials, the larger cultivars of *Salvia* need a minimum spacing of 60 to 90 cm.

### 91.8.4 Soil and Moisture Conditions

Most salvias are adapted to dry conditions and well-drained, slightly acidic soils. Many grow well with minimal or no fertilization, but plants in soil that has had compost added or receives regular but limited fertilization usually produce better flowering displays.

### 91.8.5 Maintaining Plants

Deadheading away faded flower spikes encourages continuing flowering and is strongly encouraged by many gardening experts. Perennial salvias are best divided and/or pruned in early spring.

## 91.9 FASCINATING FACTS, FABLES, FOLKLORE, AND FRIVOLITY

- *Salvia divinorum*, usually called salvia, although known by many colloquial names, is a native plant of Mexico. Whether it is genuinely wild there or simply the product of escapes from ancient domestication is unclear. It has been long employed by Indigenous people as an entheogen (a psychoactive preparation employed for spiritual or sacred purposes) and as an herbal medicine. Its foliage, or extracts from the leaves, are smoked, chewed, or otherwise consumed to produce a hallucinogenic state. Beginning in the 1990s, salvia progressively became a popular recreational inebriant like marijuana, employed by millions. Possession and/or such usage is illegal, or at least strongly discouraged, in many countries because of concern about potentially harmful effects. However, there are efforts to find legitimate medical applications, and exceptions are often made for legitimate research. The plant, which may be over 1 m in height, has flowers with white petals and blue sepals. It rarely produces seeds and is mostly reproduced by cuttings. Curiously, it is the only one of the 1,000 or so species of *Salvia* documented to have significant psychoactive effects.
- Burning aromatic plants to produce smoke for ritualistic purification of a location (smudging) is widespread in history. Independently, in ancient Greece and North America, ceremonial traditions of burning local *Salvia* species developed. Among North American First Nations, the most common smudging materials are tobacco, sage, cedar, and sweetgrass.
- *Salvia hispanica* is one of several species called "chia." It is native from Mexico to Peru and has long been harvested for its edible seeds. Its mucilaginous, sticky seeds are widely sprouted on clay or other porous materials in the shape of animals and sold as "chia pets."

FIGURE 91.8 *Salvia divinorum*, a hallucinogenic plant traditionally employed as a spiritual aid in Mexico, used in recent decades as a recreational drug: (a) Growing in a forest in San Jose Tenango, Sierra Mazateca, Oaxaca, Mexico. (b) In a private facility. Note the small plants being established as cuttings from the large mother plants.

***Source:* (a) Photo by Alan Rockefeller (CC BY 4.0). (b) Photo by Artofactivism (CC BY SA 3.0).**

FIGURE 91.9 Chia pets, prepared by growing seedlings of *Salvia hispanica* on animal figurines: (a) Alligator. (b) Dog. (c) Elephant.

***Source:* (a) Photo by Jeremy Noble (CC BY 2.0). (b) Photo by Andrew Ratto (CC BY 2.0). (c) Photo by Richie Diesterheft (CC BY 2.0).**

## KEY PUBLICATIONS

Cambron, M. 2016. *A Comparison of Historical and Current Use of Salvia divinorum in the United States and Mexico.* www.lakeforest.edu/news/a-comparison-of-historical-and-current-use-of-salvia-divinorum-in-the-united-states-and-mexico.

Claßen-Bockhoff, R., P. Wester, and E. Tweraser. 2003. The staminal lever arm mechanism in *Salvia* – a review. *Plant Biology* 5: 33–41.

Clebsch, B., and C.D. Barner. 2008. *The New Book of Salvias.* Timber Press, Portland, OR. 344 pp.

Drew, B.T., J.G. González-Gallegos, C.-L. Xiang, R. Kriebel, C.P. Drummond, J.B. Walked, et al. 2017. *Salvia* united: The greatest good for the greatest number. *Taxon* 66: 133–145.

Kintzios, S.E. (ed.). 2000. *Sage: The Genus* Salvia. CRC Press, Boca Raton, FL. 328 pp.

Lara-Cabrera, S.I., M. de la Luz Perez-Garcia, C.A. Maya-Lastra, J.C. Montero-Castro, G.T. Godden, A. Cibrian-Jaramillo, et al. 2021. Phylogenomics of *Salvia* L. subgenus

*Calosphace* (Lamiaceae). *Frontiers in Plant Science* 12:725900. https://doi.org/10.3389/fpls.2021.725900.

Ortiz-Mendoza, N., E. Aguirre-Hernández, I. Fragoso-Martínez, M.E. González-Trujano, F.A. Basurto-Peña, and M.J. Martínez-Gordillo. 2022. A review on the ethnopharmacology and phytochemistry of the neotropical sages (*Salvia* subgenus *Calosphace*; Lamiaceae) emphasizing Mexican species. *Frontiers in Pharmacology* 13. https://doi.org/10.3389/fphar.2022.867892.

Rivera, D., C. Obón, and F. Cano. 1994. The botany, history and traditional uses of three-lobed sage (*Salvia fruticosa* Miller) (Labiatae). *Economic Botany* 48: 190–195.

Sharma, Y., J. Fagan, and J. Schaefer. 2019. Ethnobotany, phytochemistry, cultivation and medicinal properties of garden sage (*Salvia officinalis* L.). *Journal of Pharmacognosy and Phytochemistry* 8: 3139–3148.

Sutton, J. 1999. *The Gardener's Guide to Growing Salvias*. Timber Press, Portland, OR. 160 pp.

Tychonievich, J. and R.M. Warner. 2011. Interspecific crossability of selected *Salvia* species and potential use for crop improvement. *Journal of the American Society for Horticultural Science* 136: 41–47.

Whitcomb, S. *Salvia divinorum, Herb of Mary, the Shepherdess*. Southern Illinois University Carbondale, Ethnobotanical Leaflets. https://opensiuc.lib.siu.edu/cgi/viewcontent.cgi?referer=&httpsredir=1&article=1368&context=ebl.

Whittlesey, J. 2014. *The Plant Lover's Guide to Salvias*. Timber Press, Portland, OR. 220 pp.

Zimmer, K. and K. Junker. 1985. *Salvia splendens*. In: *Handbook of Flowering*, Vol. 4, edited by A.H. Halevy, 265–272. CRC Press, Boca Raton, FL.

## Key Cultivation Websites

Clemson University Home & Garden Information Center. *Salvia*. https://hgic.clemson.edu/factsheet/salvia/.

Garden Chronicle. *How to Grow Perennial Salvias – Sage*. https://gardenchronicle.com/how-to-grow-salvias-sage/.

GardenersWorld.com. *How to Grow Salvia*. www.gardenersworld.com/how-to/grow-plants/how-to-grow-salvia/.

Horticulture Magazine. *Salvia*. https://horticulture.co.uk/salvia/.

Royal Horticultural Society. *Salvia Cuttings*. www.rhs.org.uk/plants/salvia/cuttings.

The Spruce. *How to Grow Red Salvia*. www.thespruce.com/red-salvia-scarlet-sage-salvia-splendens-2132144.

The Spruce. *10 Popular Species of Salvia Plants*. www.thespruce.com/common-types-of-salvia-flowers-annual-and-perennial-4767399.

University of California Master Gardeners of Butte County. *Sage Advice: Welcome Salvias into Your Garden*. https://ucanr.edu/blogs/blogcore/postdetail.cfm?postnum=28226.

University of Minnesota, Extension. *Salvia*. https://extension.umn.edu/flowers/salvia.

WikiHow. *How to Grow Salvias*. www.wikihow.life/Grow-Salvias.

WikiHow. *How to Prune Salvias*. www.wikihow.com/Prune-Salvias.

# 92 Snapdragon (*Antirrhinum majus*)

**Scientific Names:** The genus name *Antirrhinum* is derived from the Greek *antirrhinon*, which has been interpreted as "calf's snout," usually understood as an allusion to the odd shape of the seed capsule.

**Common Names:** "Dragon" in the common name *snapdragon* (also spelled *snap-dragon* and *snap dragon*) is based on a fancied resemblance of the flower to a dragon's face. "Snap" is based on the way that the two lips of the flower open and close when the flower is laterally squeezed and released. Snapdragon is also called *dog flower* and has been termed *lion's-mouth*, *rabbit's mouth*, *toad's mouth*, and similar names reflecting the perception of the flower as an animal with a prominent muzzle armed with threatening jaws. Summer snapdragon is *Angelonia angustifolia*, which is very reminiscent of common snapdragon.

## 92.1 INTRODUCTION

Snapdragons have been popular garden and cut flower plants for centuries. Improved cultivars are now available in multiple flower colors and shapes, producing extremely attractive, long-lasting floral displays. The plants come in different sizes, suitable for a variety of landscape uses, and are easy to grow.

FIGURE 92.2 Variation in flower color of garden snapdragon cultivars.

*Source:* (a) Photo by Michael Apel (CC BY SA 2.5). (b) Photo by SAplants (CC BY SA 4.0). (c) Photo by Ruth Hartnup (CC BY 2.0). (d) Photo by Shadowmeld Photography (CC BY SA 4.0). (e) Photo by HQ (CC BY SA 2.0). (f) Photo by Randy Robertson (CC BY 2.0).

## 92.2 OVERVIEW

Just what species should be included in *Antirrhinum* is debated. The species are native to rocky areas of Europe, North Africa, and also (depending on how the genus is defined) North America. There is a developing consensus that only the 20 or so Old World species should be considered to belong to *Antirrhinum*. Of these, the common snapdragon (*A. majus*) represents the vast majority of cultivated snapdragons. This herbaceous perennial (usually grown as an annual) ranges in height from rock garden forms only 15 cm high to tall forms a meter or more in height. The species is a native of the Mediterranean region but frequently escapes from cultivation, and it has become naturalized in much of the temperate world. Snapdragon is best known for its beautiful flowering display, consisting of a tall spike with numerous flowers that are prominently two-lipped, the two parts of the tubular corolla pressed together. Strong bumblebees land on the lower lobe, force open the two lips, and crawl into the throat to reach the nectar (honeybees are too small to gain access). Cultivars have flowers ranging through solid colors and combinations of lavender, orange, peach, pink, red, yellow, and white – almost everything except blue. Flowers are available in both pastel shades and bright colors. Some cultivars have fringed or ruffled petals, some have doubled flowers (with extra petals), and some lack the normal two-lipped structure (see "peloria," discussed later). The flowers open progressively from the bottom toward the top, and the flowering period can last

FIGURE 92.1 Beds of flowering snapdragons.

*Source:* (a) Photo by Shutterstock/Toeizuza Thailand. (b) Photo by Patrickgerardlynch (CC BY SA 4.0). (c) Photo by Lazaregagnidze (CC BY SA 3.0).

DOI: 10.1201/9781003586227-92

all summer in cooler climates. The blooms are pleasantly fragrant.

## 92.3 HISTORY

Snapdragons (presumably wild) have been claimed to have been grown by the classical Romans, but there is more substantially documented evidence of cultivation of domesticated varieties in Europe at least since the 15th century. Snapdragons are a traditional cottage garden flower. They are, nevertheless, not considered old-fashioned, because breeding has generated numerous quite spectacular cultivars. The Old World species of *Antirrhinum* can be crossed to produce fertile hybrids, and many cultivars are hybrids with *A. majus*.

## 92.4 ECONOMIC IMPORTANCE

The snapdragon is one of the important floral crops in both domestic and international trade of many temperate zone countries. It is a very popular garden plant, with considerable demand in the seed, nursery, and cut flower trades.

## 92.5 ORNAMENTAL USES

Snapdragon is extensively cultivated as an ornamental plant, as massed plantings in private and public gardens,

FIGURE 92.3 Botanical paintings (mostly from the 19th century) of snapdragon cultivars.

***Source:*** **Public domain. (a) John Lewis Childs's (Floral Park, New York) 1924 nursery catalogue. (b) *The Floral Magazine*, 1880, London, Reeve & Co. (c) *Revue horticole*. 1865. Librairie Agricole de la Maison Rustique, Paris. (d) *The Garden, an Illustrated Weekly Journal of Horticulture*, vol. 35, 1889. (e) Hibberd, S., and F. E. Hume. 1879. *Familiar Garden Flowers*. Cassell, Peter, Galpin, & Co., London. (f) Thomé, O. W. 1885. *Flora von Deutschland, Österreich und der Schweiz*. Gera, Germany. (g) *Deutsches Magazin fur Garten- und Blumenkunde*; G. Weise, Stuttgart, Germany. 1849.**

FIGURE 92.4 Snapdragons growing on a commercial flower farm.

***Source:*** **Photo by Shutterstock/Mike Flippo.**

FIGURE 92.5 Major horticultural uses of snapdragons: (a) A designed lawn plot. (b) A large massed garden planting. (c) Containers. (d) Cut flowers.

***Source:*** **(a) Photo by Ghislain Mary (CC BY SA 2.0). (b) Photo by Viaouest (CC BY SA 4.0). (c) Photo by Inga Munsinger Cotton (CC BY 2.0). (d) Photo by Shutterstock/Ramona Georgescu.**

in borders, against foundations, and in containers. It is also employed as a cut flower and, in some countries, is among the top ten commercial cut flowers.

## 92.6 OTHER USES

Snapdragon seeds have been employed to a minor extent as a source of edible oil, notably in Russia.

## 92.7 PROBLEMS

Snapdragons are subject to a variety of diseases and insects. They are intolerant of heat and, in hot weather, will grow poorly and may drop flowers.

## 92.8 CULTIVATION

### 92.8.1 Climate and Site

Snapdragons require full sun or at least light shade (semi-shade is appropriate for hot conditions). They can be grown as tender perennials in climate zones 7 to 10, and with protection, they may be overwintered as far north as zone 5. They are grown as annuals in both cold and semi-tropical regions. However, snapdragons are significantly intolerant of heat and do poorly during summer heat spells. Height and size need to be considered. Dwarf and trailing forms are appropriate for containers and rock gardens, intermediate-sized varieties for beds, and tall cultivars as background for shorter plants.

FIGURE 92.6 "Snapdragons" in Norwich England, reflecting the centuries-long tradition of honoring "Snap, the dragon."

*Source:* **Photos by Martin Pettitt (CC BY 2.0).**

### 92.8.2 Propagation and/or Purchasing

Snapdragons are usually propagated by seeds but can be reproduced by cuttings. Young plants in flats and fiber pots for transplanting are marketed at garden centers.

### 92.8.3 Planting

Plants should be spaced 15 to 25 cm apart, depending on size.

### 92.8.4 Soil and Moisture Conditions

Snapdragons are suited to slightly acidic, well-drained, rich loam soils, but not unamended clay. Fertilization with a balanced fertilizer is usually recommended to promote flowering. The soil should be kept moist.

### 92.8.5 Maintaining Plants

Tall plants may need to be staked. Deadheading will prolong flowering.

## 92.9 FASCINATING FACTS, FABLES, FOLKLORE, AND FRIVOLITY

- Since the early 15th century through to the 1950s, Norwich, England, held a procession featuring a dragon nicknamed "Snap." Old masks, effigies, and paintings of the "snapdragon" have become collector's items. Today, the city continues to hold dragon-themed events and is extensively decorated with imaginative artistic Norwich snapdragon sculptures and statues.
- Snapdragon was the name of an 18th-century game, now obsolete, that required players to eat raisins after snatching them out of burning brandy.
- Dating to ancient Greece and Rome, snapdragons were thought to be magical. According to old European mythology, snapdragons are protective against curses, sorcery, and witchcraft and can counter illness and poisoning. They were planted near the gates of European castles to ward off evil and enemies. During the Renaissance, snapdragons were thought to lead to glory, honor, and social status and were worn on one's sleeve in the belief that this would favor personal advancement.
- Peloric flowers are aberrant symmetrical forms of flowers in species in which the flowers are normally asymmetrical. This is often because of a gene mutation, and in some species, including snapdragon, horticulturalists have selected these unusual forms as new varieties. Without much understanding of what genes were and how they affected characteristics, Charles Darwin, the father of evolution, explored the inheritance of peloria in snapdragon.
- Snapdragon Stadium is a 35,000-seat outdoor stadium in San Diego, California, completed in 2022, hosting college football and other sporting events. The name is unrelated to the flower. Snapdragon (part of Qualcomm Technologies Inc.), the commercial sponsor which purchased naming rights to the stadium ($45 million for 15 years), is a company in San Diego with mobile technology platforms branded as Snapdragon.
- Dried seedpods of snapdragon are eerily reminiscent of miniature shrunken human skulls.
- Gregor Mendel (1822–1884, born in Brünn, now Brno, Czech Republic), an Augustinian prelate, conducted pioneering genetic studies of inheritance with peas, published in 1862. His work stimulated additional studies employing plants, notably with garden snapdragon.
- Test question in a college genetics course: When red-flowered snapdragon plants are crossed with white-flowered snapdragon plants, all the offspring have

FIGURE 92.7 Seedpods of snapdragon look like tiny faces when green and resemble miniature human skulls when dried.

**Source: (a) Photo by Barakat 2011 (CC BY SA 4.0). (b) Photo by K. Chaikovska (CC BY SA 4.0). (c) Photo by David (CC BY 2.0). (d) Scanning electron photomicrograph by Karin Kones (CC BY SA 2.0).**

FIGURE 92.8 Amazing floral variation observed in inheritance studies of snapdragon.

**Source: Baur, E. 1911. *Einfuhrung In Die Experimentelle Vererbungslehre [Introduction to Experimental Heredity]*. Gebrüder Borntraeger, Berlin.**

pink flowers. Based on these results, which of the following inheritance patterns determined flower color? (a) incomplete dominance, (b) multiple allele, (c) polygenic, (d) sex-linked. (A is correct.)

## KEY PUBLICATIONS

Albach, D.C., H.M. Meudt, and B. Oxelman. 2005. Piecing together the "new" Plantaginaceae. *American Journal of Botany* 92: 297–315.

Farooq, S., A. ul Haq, M.L. Lone, S. Parveen, F. Altaf, and I. Tahir. 2022. Salicylic acid enhances the vase life and improves the postharvest attributes – a case study of *Antirrhinum majus* L. *East African Scholars Journal of Agriculture and Life Sciences* 5: 87–96.

Harte, C. 1974. *Antirrhinum majus* L. In: *Handbook of Genetics*, edited by R.C. King, 315–331. Springer, Boston, MA.

Hudson, A., J. Critchley, and Y. Erasmus. 2008a. Cultivating *Antirrhinum*. *Cold Spring Harbor Protocols*. https://doi.org/10.1101/pdb.prot5051.

Hudson, A., J. Critchley, and Y. Erasmus. 2008b. The genus *Antirrhinum* (snapdragon): A flowering plant model for evolution and development. *Cold Spring Harbor Protocols*. https://doi.org/10.1101/pdb.emo100; http://cshprotocols.cshlp.org/content/2008/10/pdb.emo100.

Hudson, A., J. Critchley, and Y. Erasmus. 2008c. Propagating *Antirrhinum*. *Cold Spring Harbor Protocols*. https://doi.org/10.1101/pdb.prot5052.

Lim, T.K. 2014. *Antirrhinum majus*. In: *Edible Medicinal and Non Medicinal Plants*, edited by T.K. Lim, 633–639. Springer, Dordrecht.

Moor, U. and S. Vabrit. 2002. Snapdragon (*Antirrhinum majus*) cut flowers – past and future in horticulture. In: *Environment and Crop Production*, edited by R. Dris, S.M. Jain, and I.A. Khan, 299–305. CRC Press, Boca Raton, FL.

Oyama, R.K. and D.A. Baum. 2004. Phylogenetic relationships of North American *Antirrhinum* (Veronicaceae). *American Journal of Botany* 91: 918–925.

Ramadan, M.F. and H. El-Shamy. 2013. Snapdragon (*Antirrhinum majus*) seed oil: Characterization of fatty acids, bioactive lipids and radical scavenging potential. *Industrial Crops and Products* 42: 373–379.

Tolety, J. and A. Sane. 2011. *Antirrhinum*. In: *Wild Crop Relatives: Genomic and Breeding Resources*, edited by C. Kole, 1–14. Springer, Berlin.

Vinod K. and S.K. Bhattacarjee. 2006. *Antirrhinum*. In: *Advances in Ornamental Horticulture*, Vol. 1, edited by S.K. Bhattacharjee, 234–246. Pointer, Jaipur, Rajasthan, India.

## KEY CULTIVATION WEBSITES

Clemson University, Extension. *Snapdragons for Fall Blooms*. https://hgic.clemson.edu/snapdragons-for-fall-blooms/.

Coker, C. 2018. *Snapdragon for the Farmer Florist*. Mississippi State University Extension 3301. https://extension.msstate.edu/sites/default/files/publications/publications/p3301_web.pdf.

GardenersWorld.com. *How to Grow Snapdragons* (Antirrhinum). www.gardenersworld.com/how-to/grow-plants/how-to-grow-snapdragons-antirrhinum/.

Gilman, E.F., R.W. Klein, and G. Hansen. 2018. Antirrhinum majus *Snapdragon*. University of Florida (Gainesville), Extension. https://edis.ifas.ufl.edu/publication/FP044.

Horticulture Magazine. Antirrhinum – *Snapdragons*. https://horticulture.co.uk/antirrhinum/.

Huo, H. and J. Chen. 2021. *Planting and Propagation of Snapdragons in Florida*. University of Florida Extension. https://edis.ifas.ufl.edu/publication/EP549.

Lewis, M., M. Stock, T. Maughan, B. Black, and D. Drost. 2021. *Snapdragon Cut Flower Production in Utah*. Utah State University Extension Paper 2187. https://digitalcommons.usu.edu/extension_curall/2187.

National Cut Flower Centre. *Snapdragons (Antirrhinum majus) as a Cut Flower Grown in Polythene Tunnels*. Snapdragons (Antirrhinum majus) as a cut flower crop grown in polythene tunnels.pdf (projectblue.blob.core.windows.net)

The Spruce. *Snapdragons: Plant Care & Growing Guide*. www.thespruce.com/growing-and-caring-for-snapdragons-1402915#:~:text=Snapdragons%20do%20best%20in%20rich,late%20fall%20or%20early%20winter.

WikiHow. *How to Grow Snapdragons*. www.wikihow.com/Grow-Snapdragons.

# 93 Stock (*Matthiola incana* and Relatives)

**Scientific Names:** The genus name *Matthiola* is named for the very influential Italian botanist-physician Pietro Andrea Mattioli (1501–1577), personal doctor of Ferdinand II, Archduke of Austria, and Emperor Maximilian II, Holy Roman Emperor of Austria. Mattioli believed that the scent of stock promoted love and lust.

**Common Names:** The common name stock (sometimes *stocks*) usually refers to cultivars of *M. incana* but is sometimes applied to all 50 or so species in the genus. The odd name "stock" has been explained in different ways – perhaps a reference to the "stocky," that is, woody, stems when grown as a perennial (an old English meaning of "stock" is *tree trunk*; the word is also found as "stoc" in German and Dutch). A more speculative explanation is that *stock* originates from use as a temporary stock of flowers that are produced in the early season. Other names for *M. incana* include Brompton stock (*Brompton* is an old, popular, late-blooming variety with doubled flowers), common stock, garden stock, hoary stock (*incana* in the scientific name means "hoary," a reference to the hairy, hence grayish, surface of the leaves), ten-week (tenweeks) stock (for varieties that flower rapidly), and gillyflower (or gilliflower, a name also employed for other aromatic ornamentals, tracing to an ancient Greek name for the spice known as clove; the spicy scent of stock is reminiscent of clove). *Stock* is sometimes confused with *wallflower*, also discussed in this book.

## 93.1 INTRODUCTION

Several of the species of *Matthiola*, including annuals and perennials, herbs and subshrubs, are grown as ornamentals, valued particularly for their scent. Evening stock or night-scented stock (*M. longipetala*), an annual, is extremely fragrant, but of limited floral beauty ("unsightly" according to one author). However, the dominant cultivated species is *M. incana*. It is a well-established garden and florist's staple, valued for its wide range of colorful, fragrant flowers displayed in impressive spires.

## 93.2 OVERVIEW

*Matthiola incana* is native to southern Europe but has become naturalized in many areas of the world. The species is mostly grown as an annual, but the wild ancestor is usually a biennial or perennial, and cultivated varieties have been selected that are employed as biennials (especially in short seasons that limit growth in the first year) or as perennials (in warmer locations). The plants produce a taproot, and the lower stems can become quite woody when grown as a perennial. In flower, the plants are usually 25 to 60 cm in height, sometimes reaching 90 cm. Some dwarf cultivars are quite low (less than 30 cm). The flowers of cultivars are fragrant, apricot, blue, cream, lavender, pink, purple, red, yellow, or white, occasionally bicolored, and many varieties have semi-doubled or doubled blooms. Some varieties produce a single stem bearing a dense cluster of flowers; others are multi-stemmed with more scattered blooms. The flowers open sequentially from the bottom of the spike upward. *Matthiola incana* is a cool-weather plant, often blooming from early spring into summer.

## 93.3 HISTORY

*Matthiola incana* was introduced into England in the 16th century, and very soon double-flowered selections became popular. Like other early garden selections that became staples in European gardens centuries ago, it was exported to many temperate countries throughout the world.

## 93.4 ECONOMIC IMPORTANCE

Stock is highly valued by the international cut flower industry and florists throughout the world. It is also of considerable value to the horticultural industry supplying starter plants for bedding.

FIGURE 93.1 Flowering plants of common stock (*Matthiola incana*).

*Source:* **(a) Photo by Shutterstock/Anjo Kan. (b, c) Photos by Yewchan (CC BY SA 2.0).**

DOI: 10.1201/9781003586227-93

FIGURE 93.2 Paintings (public domain) by botanical illustrators of (a, b, c) common stock (*Matthiola incana*) and (d) night-scented stock (*M. longipetala*).

*Source:* **(a) Step, E., D. Bois, B. Herincq, and W. Watson. 1896. *Favourite Flowers of Garden and Greenhouse*, vol. 3. Frederick Warne, New York. (b) Sowerby, J. E. 1863. *English botany, or coloured figures of British plants*, ed. 3, vol. 1. (c) Guillaumin, A. 1929. *Les Fleurs de Jardins, tome I: Les Fleurs de Printemps*, Paul Lechevalier. (d) Lindley, J. and J. Ridgway. 1839. *Edwards's botanical register*, vol. 25.**

FIGURE 93.3 Commercial production of stock (*Matthiola incana*): (a) In a greenhouse. (b, c, d) On flower farms in California, south of Los Angeles.

*Source:* **(a) Photo by Shutterstock/Lina Ermolaeva fotoart. (b, c, d) Photos by Harold Litwiler (CC BY 2.0).**

FIGURE 93.4 Cut flowers of stock (*Matthiola incana*).

*Source:* **(a) Fresh-cut bouquets. Photo by Tracy Ducasse (CC BY 2.0). (b) Photo by Shutterstock/Anjo Kan. (c) Photo by Shutterstock/Life Morning.**

FIGURE 93.5 Garden beds of stock (*Matthiola incana*).

*Source:* **(a) Photo by Shutterstock/Hiyaphoto. (b) Photo by Shutterstock/Alybaba.**

FIGURE 93.6 Garden stock (*Matthiola incana*) grown in containers.

*Source:* **(a) Photo by Shutterstock/Klever Ok. (b) Photo by Shutterstock/Victoria Kurylo.**

## 93.5 ORNAMENTAL USES

Stock is extensively grown as an ornamental garden plant, employed in beds, borders, and containers, and especially valued as a cut flower.

## 93.6 OTHER USES

Stock flowers are considered to be edible, with a taste compared to cinnamon, cloves, and radish. Like most edible

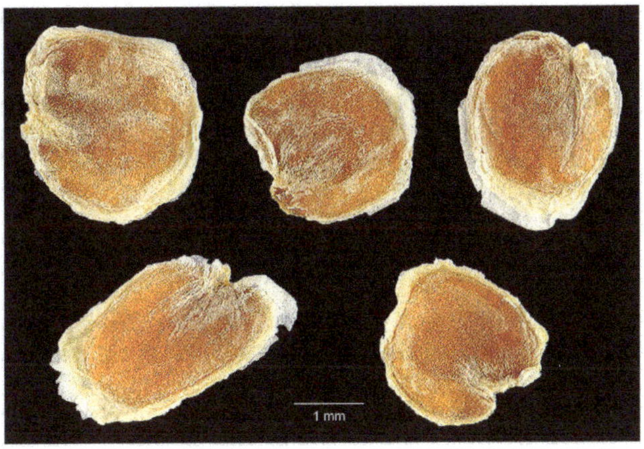


FIGURE 93.7 Seeds of *Matthiola incana*, shown to have potential dietary value for reducing cholesterol.

*Source:* **Photo by Steve Hurst, Agricultural Research Service Laboratory, US Department of Agriculture (public domain).**

flowers, they are mostly used as a garnish to decorate salads and other dishes, and the genuinely tasty parts of the flowers are the delicate petals. The dried flowers are marketed as an herbal tea, claimed to have medicinal value. More than half of the weight of the seeds of *Matthiola incana* is made up of linolenic acid oil, which has potential for reducing cholesterol levels.

## 93.7 PROBLEMS

The garden *Matthiola* species have become minor invasive weeds in some locations. Because of its strong scent, people with sensitivities to perfumes may be affected by the fragrance of stock.

## 93.8 CULTIVATION

### 93.8.1 Climate and Site

As an annual, stock can be grown in virtually all climate zones. Some cultivars are suitable for cultivation as a perennial in zones 7 to 10. However, stock declines rapidly in hot weather. Full sun or light shade is appropriate. *Matthiola incana* is naturally adapted to maritime conditions and usually grows well in coastal regions.

### 93.8.2 Propagation and/or Purchasing

Plants can be grown from seed. Cultivars with mostly doubled flowers are more or less sterile and are often grown from cuttings. For many gardeners, it is convenient to purchase starter plants from nurseries and garden centers.

### 93.8.3 Planting

Stocks are often planted at a spacing of 30 to 45 cm, but sometimes closer, to accentuate the collective floral display. Generally, larger cultivars require more space.

### 93.8.4 Soil and Moisture Conditions

Stock grows best in moist, well-drained, circumneutral or slightly alkaline soil (pH 6.8 to 7.5) of medium fertility.

### 93.8.5 Maintaining Plants

Some gardeners sow seed at intervals (e.g., every 2 weeks) to produce flowering displays over an extended interval.

## 93.9 FASCINATING FACTS, FABLES, FOLKLORE, AND FRIVOLITY

- Stock with doubled flowers (with extra petals) is much preferred by gardeners, and the genetics that control the production of such flowers is now understood. However, in the 17th century in England, it was recommended that just after a full moon, a stock plant grown in poor soil should be transplanted to rich soil in order to produce doubled flowers (see US Department of Agriculture 1917 Yearbook of Agriculture). A more recent belief held that the most doubled blooms would be produced on plants seeded in April under a full moon (the early planting would be when the weather was cool, which would favor good growth).
- In Sicily, St. Joseph Day (in Western Christianity, held on March 19, in honor of Saint Joseph, husband of the Virgin Mary and nominal father of Jesus Christ) is accompanied by celebrations with decorated horses. To suppress the associated unpleasant smells, fragrant stock flowers are also attached to the horses (Romano, D., A. Tribulato, S. Toscano, and D. Scuderi. 2013. Ethnobotanical uses of Brassicaceae in Sicily. *Acta Horticulturae* 1005: 197–204.).

FIGURE 93.8 Folkloristic traditionally decorated Sicilian horse, as employed on St. Joseph Day and, as noted in the text, de-odorized with stock flowers.

*Source:* **Photo by Shutterstock/Dario Lo Presti.**

## KEY PUBLICATIONS


Çelikel, F.G. and M.S. Reid. 2002. Postharvest handling of stock (*Matthiola incana*). *HortScience* 37: 144–147.

Cockshull, K.E. 1985. *Matthiola incana*. In: *Handbook of Flowering*, Vol. 3, edited by A.H. Halevy, 363–367. CRC Press, Boca Raton, FL.

Dole, J.M. and H.F. Wilkins. 2005. *Matthiola*. In: *Floriculture: Principles and Species*, 2nd edition, edited by J.M. Dole and H.F. Wilkins, 682–687. Prentice Hall, Upper Saddle River, NJ.

Gowler, Z. 1998. *A Taxonomic Revision of the Genus* Matthiola *R. Br. (Cruciferae) and Related Genera*. Ph.D. thesis. University of Edinburgh. Edinburgh, Scotland. 338 pp. https://era.ed.ac.uk/handle/1842/12049?show=full.

Irani, S.F., M. Arab, M. Norouzi, and M. Lotfi. 2016. Genetic diversity of stock (*Matthiola incana* L.) cultivars based on cytogenetic characteristics. *Asian Journal of Advanced Basic Sciences* 4(2): 65–73.

Jaén-Molina, R., J. Caujapé-Castells, J.A. Reyes-Betancort, H. Akhani, O. Fernández-Palacios, J. Pérez de Paz, et al. 2009. The molecular phylogeny of *Matthiola* R. Br. (Brassicaceae) inferred from ITS sequences, with special emphasis on the Macaronesian endemics. *Molecular Phylogenetics and Evolution* 53: 972–981.

Lim, T.K. 2014. *Matthiola incana*. In: *Edible Medicinal and Non-Medicinal Plants*, Vol. 7, edited by T.K. Lim, 633–637. Springer, Dordrecht.

Pui, D.A., M. Ardelean, M. Cordea, V. Mitre, and A. Nicoară. 2007. Heritability of several quantitative characters of interest in gillyflower (*Matthiola incana*) breeding. Bulletin of University of Agricultural Sciences and Veterinary Medicine Cluj-Napoca. *Horticulture* 64(1/2): 54–57.

Yaniv, Z., D. Schafferman, I. Shamir, and Z. Madar. 1999. Cholesterol and triglyceride reduction in rats fed *Matthiola incana* seed oil rich in (n-3) fatty acids. *Journal of Agricultural and Food Chemistry* 47: 637–642.


### Key Cultivation Websites


Gardener's Path. *How to Grow Stock: A Cottage Garden Staple*. https://gardenerspath.com/plants/flowers/stock-cottage-garden-staple/.

GardenersWorld.com. *How to Grow and Care for Stock Flowers*. www.gardenersworld.com/how-to/grow-plants/stock-flowers/.

Mississippi State University Extension. *Stock* (Matthiola incana) *for the Farmer Florist*. http://extension.msstate.edu/publications/stock-matthiola-incana-for-the-farmer-florist.

The Spruce. *How to Grow Stock Flowers*. www.thespruce.com/stock-flowers-growing-guide-5189057.

Utah State University Extension. *Stock Cut-Flower Production in Utah*. https://digitalcommons.usu.edu/cgi/viewcontent.cgi?article=3275&context=extension_curall.

WikiHow. *How to Grow Night Scented Stock*. www.wikihow.com/Grow-Night-Scented-Stock.

WikiHow. *How to Grow Stock Flowers*. www.wikihow.com/Grow-Stock-Flowers.

# 94 Sunflower (*Helianthus* Species and Hybrids)

**Scientific Names:** The genus *Helianthus* is based on the Greek words *helios*, "sun," and *anthos*, "flower," referring to the flowering heads. *Helianthus annuus* is the principal economic species and the main subject of this article. *Annuus* in the scientific name is Latin for "annual."

**Common Names:** Sunflower (occasionally *sun flower* or *sun-flower*), common sunflower. Usually (as in this article), the unqualified word "sunflower" refers to *H. annuus* or hybrid cultivars mainly with *H. annuus* parentage. The name sunflower is an allusion to the similarity of the flower to the sun's shape, having a circular head around which there are emanating rays (petals; as noted later, "ray florets" in *Helianthus* are tiny flowers each with one large petal located on the outer circumference of the flower head). "Mexican sunflower" is *Tithonia rotundifolia*, a perennial that is often cultivated as an annual ornamental. False sunflower is *Heliopsis helianthoides*, grown as a perennial ornamental. Other unrelated plants with "sunflower" in their name include Oregon sunflower (*Balsamorhiza sagittata*) and desert sunflower (*Geraea canescens*). "Wild sunflower" can refer to species in several genera.

In some languages, the common name for sunflower translates as "turn with the sun" (e.g., French: *tournesol*; Spanish: *girasol*). As explained later, while many green plants, particularly the sunflower, grow toward light sources ("phototropism") and, during early growth, bend toward the east in the morning and toward the west in the evening, once the sunflower opens its flowers, floral phototropism no longer occurs. (This should be obvious: in the Northern Hemisphere, the sun sets in the west, but flowering sunflower heads permanently face east, not west.) Nevertheless, the belief is extremely widespread that the sunflower head actually follows the sun.

## 94.1 INTRODUCTION

As an ornamental, the sunflower is valued for its spectacular, gigantic flower heads. The species also happens to be one of the world's leading food plants. Its importance in society is reflected by widespread historical respect for the plant. The sunflower was worshipped by the Incas, who viewed it as a symbol of the sun. It is the national flower of Ukraine, the state flower of Kansas, and one of the city flowers of Kitakyushu, Japan. The sunflower has also been employed as a symbol of the Vegan Society, founded in the UK in 1944. In Victorian times, the sunflower became a symbol for the Aesthetic Movement, a reaction against the Industrial Age. The flower's image was carved into chair backs, glazed onto vases, and emblazoned on iron railings. Other historical activities represented by sunflowers include the Spiritualist Movement (a church initiative which began in mid-19th-century United States), and the Sunflower Movement (a 2014 mass protest in Taiwan).

FIGURE 94.1 Variation in sunflower (*Helianthus annuus*) flowers (flower heads): (a) Sunflower cultivar Harlequin. (b) Sunflower cultivar Chianti. (c) Sunflower cultivar Vanilla Ice. (d) Sunflower cultivar Pro Cut Yellow Lite. (e) A bouquet of mixed cultivars.

***Source:*** **(a) Photo by S. G. S. (CC BY SA 4.0). (b) Photo by Rob Duval (CC BY SA 3.0). (c) Photo by Rob Duval (CC BY SA 3.0). (d) Photo by Dwight Siple (CC BY 2.0). (e) Photo by Swallowtail Garden Seeds (CC BY 2.0).**

## 94.2 OVERVIEW

*Helianthus* is a genus of approximately 80 species native to North America and Central America, except for three South American species. About half are annual, half perennial. Most ornamental sunflower cultivars are annuals. Perennial ornamental sunflowers are usually hybrids identified as *Helianthus* × *multiflorus*, but a few other species are available. Today, over 2,000 cultivars of sunflowers have been recognized. Many of the species hybridize readily, making identification difficult. *Helianthus annuus* is the main species of interest. It is a course, hairy, fast-growing annual herb, typically 1–3 m in height, with stiff upright stalks. The common sunflower is native to western North America and Mexico, but weedy forms and escapes from cultivated plants have been found in every state of the United States, every province of Canada (except Newfoundland and Labrador), and much of the world. The species is grown in many areas of the world and has escaped and become

DOI: 10.1201/9781003586227-94

established in many countries. The wild plants grow in prairies and other grasslands, old fields, pastures, roadsides and railroads, savannas, and along forest edges and riverbanks. The wild sunflower is adapted to prairies and open, dry habitats. The taproot sometimes penetrates to a depth of 3 m, which facilitates growth during drought. Wild sunflowers grow best in sunny, moist locations, especially in disturbed ground. Soils occupied vary in texture, range in pH from somewhat acidic to slightly basic, and are usually well-drained.

Domesticated oilseed sunflowers typically develop a single large flower head and are much less branched than their wild relatives. Ornamental forms have been available for centuries, grown mostly as garden plants. In the early 1990s, the sunflower became an important cut flower and an occasional ornamental potted plant in the florist industry. "Dwarf" oilseed types are no higher than 1.4 m tall and have been grown as garden ornamentals. However, they are considered less suitable for use as garden ornamentals and for cut flowers than biotypes recently selected specifically for the latter uses.

Ornamental sunflower cultivars are sometimes divided into three habit (appearance) categories: non-branching, producing one single stem with a comparatively large terminal head; branching, with a terminal head as well as many axillary shoots and heads; and spray, with many small flower heads. Some garden ornamental sunflowers have been selected for their very large, tall stems (sometimes over 5 m), others (especially pot plants) for their short stems. "Mammoth" (Russian oilseed) varieties can have flower heads over 30 cm in diameter. Plants grown for cut flowers are generally short. Plants grown for sale in pots are also short, but this is sometimes the result of growth retardants. About 1990, the Japanese introduced male-sterile hybrids which do not shed pollen and hence became popular because they were not as messy (shedding seeds) as conventional sunflowers. Ornamental cultivars often have red rather than yellow ray florets, occasionally orange, bronze, or cream, or combinations of yellow, orange, and red. White ray flowers have also been observed as mutants. More recently, ornamental forms with disc flowers transformed ("doubled") to ray flowers have been created, the flower heads resembling doubled chrysanthemums. It may be noted that in addition to chrysanthemums, flowers of several other species of the Asteraceae can look quite similar to those of sunflowers, and photos on the web are frequently misidentified.

FIGURE 94.3 Sunflower 'Teddy Bear', a cultivar with doubled inflorescences (heads with mostly or entirely ray flowers). (a) A potted plant. (b) A doubled flower head. (c) 'Teddy Bear' flower heads shaped into a teddy bear.

*Source:* **(a) Photo by Shutterstock/Marina Kuchmenko. (b) Photo by Prenn (CC BY SA 3.0). (c) Photo by Shutterstock/Peter Etchellstock.**

FIGURE 94.2 Large sunflowers: (a) Tall plants. (b) A large flower head. (c) Another large flower head.

*Source:* **(a) Photo by Shutterstock/Melinda Nagy. (b) Photo by Felicíssimo e Ramires (CC BY SA 3.0). (c) Photo by Shutterstock/Olivkairishka.**

### 94.2.1 The Sunflower Head (Capitulum): A Marvel of Evolutionary Engineering

The apparent "flower" of species in the Asteraceae (Compositae) family is actually a head (compact cluster) of miniature flowers ("florets"). In many of the "radiate" species, including the sunflower, there are two kinds of flowers, the inner (central) ones, with quite inconspicuous petals, and the outer (peripheral) ones, with a large strap-shaped petal facing outward. Reminiscent of the rays of the sun, the large petals facing outward are called "ray petals." *Helianthus* flower heads are usually conspicuous, typically with yellow ray florets on the outside and yellow or maroon disc florets in the center. In *Helianthus*, the ray flowers are always sterile, while the inner flowers produce the seeds. The sunflower head has more flowers than any other genus in the daisy family. There are often between 1,000 and 8,000 individual flowers present in the flowering head.

Living things often grow in mathematical patterns, best known of which are "Fibonacci series." Fibonacci numbers

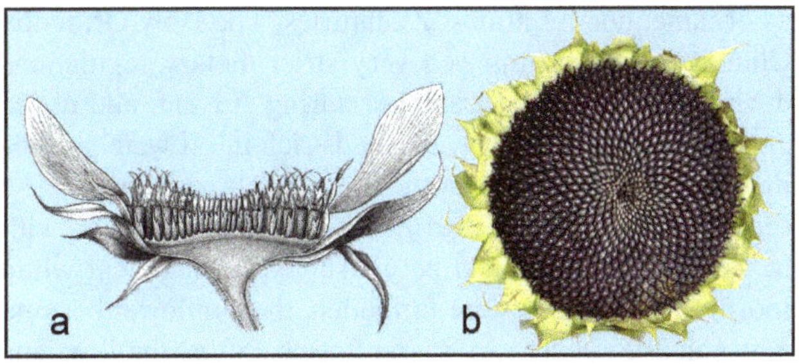


FIGURE 94.4 Flower/fruit head of sunflower (*Helianthus annuus*): (a) Sectioned flower head. (b) Fruiting head with mature seeds. Note intersecting spirals.

*Source:* **(a) Public domain illustration from Engler, H. G. A., and K. A. E. Prantl (eds). 1889. *Die natürlichen Pflanzenfamilien*, ed. 1. Wilhelm Engelmann, Leipzig, Germany. Vol. 4(5): 88. (b) Photo by Amada44 (CC BY 3.0).**

are a sequence of numbers with several fascinating properties. The most evident property is that each term is the sum of the two previous terms: 0 +1 = 1, 1 + 1 = 2, 1 + 2 = 3, 2 + 3 = 5, 3 + 5 = 8, etc., leading to: 0, 1, 1, 2, 3, 5, 8, 13, 21, 34, 55, 89, 144, 233, 377, 610, 987, 1597 . . . Another property has to do with the "golden ratio" or "divine ratio," which is approximately 0.62 (for example, a picture is said to look best if its height is 0.62 times its width). The sequence of ratios between every consecutive pair of Fibonacci numbers is alternatively greater than, or less than, the golden ratio and gets closer and closer to it (for example, 1/1 = 1.00, 1/2 = 0.500, 2/3 = 0.667, 3/5= 0.6000, 5/8 = 0.625, 8/13 = 0.615). Fibonacci numbers were first discussed by an Italian mathematician, Leonardo of Pisa (1175–1250), whose nickname was Fibonacci. Fibonacci numbers are most apparent in plants in *phyllotaxis*, the spiral patterned arrangement of organs on plants, such as the leaves on stems. Demonstrate this for yourself with the stem of any herbaceous plant. Start at the bottom with a given leaf, move up the stalk, counting the leaves until you reach a leaf that is directly above the first one (do not count the first one), and you will have a Fibonacci number. In addition, the number of times you have circled the stalk will be another Fibonacci number. Leaf arrangements in most plant species demonstrate a ratio of Fibonacci numbers, the one representing the number of units (leaves in this case) before one gets to the same position, and the other the number of spirals to get to the same position. The arrangements of seeds in sunflower heads spectacularly demonstrate Fibonacci ratios, but these are more difficult to count than leaves on a stem. In a good specimen of a sunflower, observe two sets of spirals superimposed or intertwined, one spiral turning clockwise, and the other counterclockwise, with each seed filling a dual role by belonging to both spirals. The number of clockwise spirals and the number of counterclockwise spirals are successive Fibonacci numbers. A sunflower head may have Fibonacci ratios of 21/34, 34/55, 55/89, or 89/144. These patterns are believed to arrange the seeds in the most efficient ways possible to minimize supporting tissues.

### 94.2.2 Heliotropism

*Heliotropism* is aerial movement resulting in "sun-tracking." During early growth, sunflower stems tilt toward the sun, but mature flower heads do not. It is a widespread misconception that sunflower heads track the sun's diurnal movement across the sky. At maturity, sunflower plants usually face east. It has been speculated that the flower heads face east because this prevents overheating of the seeds by excessive exposure to the hot sun. Studies have shown that the movements of sunflowers are more complex than previously appreciated and follow a circadian (daily) rhythm. Young flower stems face east at dawn and slowly turn west as the sun moves across the sky, and during the night, they slowly return to their eastern positions to resume the cycle. The movements are due to growth rates on the west side being low during the day and higher at night, resulting in increased photosynthesis and local heat attractive to pollinators (for a video, see www.npr.org/sections/thetwo-way/2016/08/05/488891151/the-mystery-of-why-sunflowers-turn-to-follow-the-sun-solved). Regardless of the flowers, some phototropism may be exhibited by the leaves.

*Photoperiodism* is an independent phenomenon related to sunlight. Photoperiodism with respect to flowering time refers to how the relative length of daylight and night daily

FIGURE 94.5 Mature sunflowers in commercial oilseed fields. The flowering heads are all permanently fixed facing east.

*Source:* **(a) Photo by WikiImages from Pixabay (public domain). (b) Photo by HUNGQUACH679PNG from Pixabay (public domain).**

stimulates plants to come into flower. Most sunflower cultivars are day-neutral (flowering is not induced by day length), but flowering tends to be faster, and stems shorter, with shorter days.

## 94.3 HISTORY

Although there have been claims that the common sunflower (*H. annuus*) was domesticated as early as 5,000 BC in Mexico (note Pohl 2008), there is well-supported evidence that there was a single domestication center, located in the eastern United States, from where the domesticated sunflower was taken to other areas of North America. The sunflower was grown for millennia as a "seed" (technically an achene, a kind of dry fruit) crop by Indigenous peoples from Mexico to southern Canada. Spanish explorers brought sunflowers to Europe in 1510, where they were grown mostly as ornamentals for almost 2 centuries. The Holy Orthodox Church of Russia imposed very strict dietary regulations during the 40 days of Lent and during Advent, and in the early 19th century, nearly all foods rich in oil were prohibited. But the sunflower was not on the list, perhaps because the church authorities had not yet become familiar with it. Because the oil could be consumed during Lent while most other oil foods were forbidden, the sunflower became a principal food crop in Russia. In the 1800s, the Russians selected varieties with oil contents of nearly 50% – twice as high as then current American varieties. Modern sunflowers are mainly the annual herb *H. annuus* but have often been hybridized with other species of *Helianthus*.

FIGURE 94.6 Sunflower (*Helianthus annuus*) cultivation by Aztecs in ancient Mexico.

***Source:*** **Photo by Zina Deretsky, US National Science Foundation.**

FIGURE 94.7 Very early *Helianthus* ornamental hybrid cultivars.

***Source:*** **Public domain illustrations. (a) Cover of 1915 Peter Henderson & Co., New York, garden catalog, featuring ornamental sunflowers. (b) *Revue horticole* (Paris), sér. 4, vol. 89, 1917 (doubtfully identified as *H. debilis* and *H. cucumerifolius*). (c) Van Houtte, L. B., *Flore des serres et des jardin de l'Europe* (Gand, Belgium), vol. 15, 1845 (doubtfully identified as *H. californicus*).**

## 94.4 ECONOMIC IMPORTANCE

The chief concern in this chapter is the ornamental importance of sunflower. *Helianthus annuus* has traditionally been widely grown as a garden and border ornamental (tall varieties can be used as a screen), but in recent times, cultivars for use as cut flowers and in containers have been bred. These ornamental uses now constitute significant market niches. Sunflowers for commercial harvest of cut flowers are produced both outdoors (as a seasonal crop) and also as a year-round glasshouse crop in some countries.

FIGURE 94.8 Commercial sale and production of ornamental sunflowers: (a) A florist selling sunflower cut flowers. (b) A worker caring for potted sunflowers in a greenhouse.

***Source:*** **(a) Photo by Shutterstock/Svechkova Olena. (b) Photo by Shutterstock/BearFotos.**

FIGURE 94.9 Sunflower ornamental cut flowers: (a) A vase with cut flowers. (b) A lady with cut flowering stems. (c) A lady with a bouquet.

***Source:*** **(a) Photo by Katzenfee50 from Pixabay (public domain). (b) Photo by Shutterstock/Olena Antonenko. (c) Photo by Shutterstock/Fusionstudio.**

FIGURE 94.10 Jerusalem artichoke (*Helianthus tuberosus*): (a) A field of plants in Russia. (b) Flowers. (c) Freshly dug tubers. (d) Tubers.

*Source:* **(a) Photo by Shutterstock/Sergey Bezgodov. (b) Photo by Miguel Vieira (CC BY 2.0). (c) Photo by Shutterstock/IMarzi. (d) Photo by Allispossible.org.uk (CC BY 2.0).**

Sunflower is the world's fourth most important oil crop (after soybean, palm, and Canola), producing about 13% of all vegetable oil. *Helianthus tuberosus*, Jerusalem artichoke or sunchoke, is the best-known cultivated *Helianthus* after *H. annuus* but is much less important economically. Its tubers are employed for food and occasionally as a source of carbohydrates for fermenting into alcohol. It is also sometimes grown as an ornamental.

## 94.5 ORNAMENTAL USES

Ornamental sunflowers are usually grown as annuals in garden beds and borders. Small cultivars such as *H. annuus* 'Teddy Bear' are suitable for containers.

## 94.6 OTHER USES

As noted earlier, sunflower seeds are one of the world's leading food crops. The main production areas are Russia, Ukraine, Eastern Europe, the European Union, China, Argentina, and the United States (mostly North and South Dakota and Minnesota). Annual world production usually exceeds 20 million t. Most of the harvested "seeds" (technically "cypselas," the achene-like fruit of the Asteraceae family) are crushed for oil consumed by humans, with the remaining protein-rich "oil cake" or "press cake" constituting excellent livestock feed. Sunflower oil is notably unsaturated, hence considered nutritionally superior to many other vegetable oils. About 5% of the world sunflower harvest is eaten as "nuts" or "kernels" in snacks. Confectionery-type seeds are relatively large, with kernels (embryos) that can be separated much more easily from the relatively thick hulls (pericarps; "shells") compared to oilseed-type seeds, which have kernels firmly attached to their thinner hulls. Confectionery-type seeds also usually have white-striped hulls (sometimes all-white or all-black, especially in the Old World), in contrast to seeds of oilseed cultivars, which are typically solid black or gray. Sunflower seeds (usually

FIGURE 94.11 Landscape uses of sunflowers (*Helianthus annuus*): (a) A temple in Nakhon Ratchasima, Thailand, fronted by sunflowers. (b) A large border of sunflower cultivar Buttercream at the Chicago Botanic Garden. (c) A garden plot of sunflowers. (d) Sunflowers beside a house.

*Source:* **(a) Photo by Shutterstock/Nicepix. (b) Photo by Cultivar413 (CC BY 2.0). (c) Photo by Rob and Stephanie Levy (CC BY 2.0). (d) Photo by Emma Forsberg (CCC BY 2.0).**

FIGURE 94.12 Ornamental sunflowers in containers: (a) Potted plants. (b) A window box of sunflowers.

*Source:* **(a) Photo by Maja Dumat (CC BY 2.0). (b) Photo by Shutterstock/Elena Kirey.**

FIGURE 94.13 "Seeds" (achenes) of sunflower (*Helianthus annuus*): (a) A handful of confectionery (white-striped) seeds. (b) Hulled (confectionery-type) seed (the pericarp or "shell" has been removed, leaving the embryo or "kernel"). (c) Confectionery-type striped seed, showing white stripes. (d) Oilseed-type black seed.

*Source:* **(a) Photo by Wetwebwork (CC BY 2.0). (b, c) Photos by Kaldari (public domain). (d) Photo by Amada44 (CC BY 3.0).**

FIGURE 94.14 Wild birds consuming sunflower seeds from the fruiting heads: (a) American goldfinch (*Spinus tristis*). (b) Red-winged blackbird (*Agelaius phoeniceus*). (c) Great tit (*Parus major*).

*Source:* **(a) Photo by Watts (CC BY 2.0). (b) Photo by Shutterstock/Jim Schwabel. (c) Photo by Shutterstock/Karel Gallas.**

FIGURE 94.15 Sunflowers cultivated beside a nuclear power plant in Temelin, Czech Republic.

*Source:* **(a) Photo by Shutterstock/Vaclav Volrab. (b) Photo by Shutterstock/Jaroslava V.**

the oilseed type) are also widely employed as birdfeed (both wild and domesticated birds love sunflower seeds). In addition to its culinary uses, sunflower oil is employed for producing lubricants, soaps, paints, and varnishes. Sunflower hulls represent about 15% of the weight of the seeds, and they are used for livestock bedding and occasionally as a source of fiber for cattle feed. The plants themselves are sometimes fed as forage or silage. Sunflowers have been used in phytoremediation, most notably to remove radioactive elements from the Chernobyl nuclear disaster. Sunflowers are often planted beside nuclear powerplants, as a precaution.

## 94.7 PROBLEMS

In some regions (especially in the United States), some sunflower species are considered to be noxious weeds. If garden plants are allowed to produce seeds, there may subsequently be an undesired accumulation of seeds and seedlings in adjacent areas. Sunflower seeds are allelopathic, leaching toxins that suppress germination of nearby seeds and reduce growth of some other species, including grasses. Some individuals are susceptible to allergic reactions from contacting sunflowers. The seed

FIGURE 94.16 Pest herbivores consuming sunflower seed heads: (a) Chipmunk. (b) Squirrel.

*Source:* **(a) Photo by NatureFriend from Pixabay (public domain). (b) Photo by Garten-gg from Pixabay (public domain).**

heads of sunflowers are very attractive to wildlife, especially birds, and this is often considered to be a benefit. However, squirrels and chipmunks can become pests, damaging the appearance of the plants and rendering them useless as a source of food.

## 94.8 CULTIVATION

### 94.8.1 Climate and Site

When sunflowers are grown as annuals, winter conditions do not need to be considered. Plants grown as perennials are adapted to climate zones 4 to 8, depending on cultivar. Sunflowers grow well in hot, long summers. In cooler locations, starting seeds indoors 2 weeks in advance of the local traditional planting date may be considered.

Sunflowers, true to their name, grow best in full sun. Giant cultivars may need staking in windy areas, and of course, when mature, they may produce excessive shade for good growth of adjacent plants. Accordingly, tall sunflowers are often grown in the back of garden beds. Short cultivars are suitable for borders, and dwarf cultivars are ideal for pots. Because of disease (notably powdery mildew) and insect problems, it is generally not recommended to grow ornamental sunflowers in the same area for more than two consecutive years.

### 94.8.2 Propagation and/or Purchasing

Packages of seeds of ornamental sunflowers are available from supermarkets and nursery supply outlets, and most garden seed catalogues list at least a dozen, often several dozen, cultivars. Seeds of especially interesting recently bred cultivars often need to be obtained from specialty online catalogues. Young potted plants or trays of plugs are frequently available from garden centers.

Most ornamental sunflower varieties available to home gardeners today are hybrids. Those who find that they have grown a particularly huge plant or one with a huge seed head often collect the seeds to grow next year. However, as is frequent with hybrid plants, the seeds may not produce plants with as admirable characteristics as the parental plant.

### 94.8.3 Planting

Sunflowers are easily grown annual garden plants. The seeds may be started indoors or sown directly outdoors (about 2.5 cm deep) after danger of frost, preferably in well-prepared beds. Giant cultivars may be planted as much as 1 m apart. Large cultivars should be 45 cm or more apart, and medium-sized varieties at least 30 cm between plants.

Several perennial species are cultivated as ornamentals, but some are suitable mostly for relatively warm climates. Like the annuals, they are also usually established from seed, but new plants can be established by dividing old plants (in any event, to maintain vigor, perennials should be divided every 2 to 4 years).

### 94.8.4 Soil and Moisture Conditions

Although sunflowers are drought-tolerant and capable of growing in a wide range of soil conditions, fertile, moist, well-drained soil is best. Addition of well-cured compost is beneficial. The ideal pH preference range is 6.5–7.5. Fertilizers rich in potassium are recommended for stem strength, especially for cut flower production. Excessive fertilization, especially with nitrogen, can stimulate vegetative growth at the expense of flower production.

### 94.8.5 Maintaining Plants

Gardeners who want a continuous display of flowers from midsummer to autumn frost sometimes plant seeds every few weeks. Although sunflowers are quite drought-tolerant, they benefit from watering during dry spells.

### 94.8.6 Harvesting Flowers

In north temperate climates, outdoor-grown sunflowers are in flower from midsummer to frost. Home growers, of course, can harvest when flowers are fully open, at their peak display. Professional growers harvest earlier, cutting stems when petal color is visible, but well before all the petals are unfolded. Stems are best harvested during cooler periods of the day and when flowers are free from dew. The cut stems should be stripped of all but a few leaves, recut, and placed in clean water (ideally acidified or with a floral preservative, employed for its disinfectant, not its nutritional, value, as for many other cut flower species). Simply using a clean vase and regularly changing the water will prolong flower life. Older cut flower cultivars have a vase life of 7–10 days; the newer, pollen-free types can last up to 15 days.

## 94.9 FASCINATING FACTS, FABLES, FOLKLORE, AND FRIVOLITY

- American Indians used sunflower as a hunting calendar. When the plants were tall and in bloom, the buffalo were fat, and the meat good.
- A number of Native American tribes of the southwestern United States, including the Hopi, extracted yellow and purple dyes from wild sunflowers, which were used to color basketry and decorate bodies.
- Stems of wild sunflowers were used by Native Americans to construct hoods over baking stoves.
- During the reign of the Czars in Russia, sunflower seeds were a staple food for many. Soldiers in the field were sometimes issued 1 kg of sunflower seeds as "iron rations" and at times had little else to eat.
- Sunflower stems were used to fill lifejackets before the advent of modern flotation materials.
- "Sunflowers instead of missiles in the soil would ensure peace for future generations." This was a statement by US Secretary of Defense William J. Perry on June 4, 1996, while celebrating Ukraine officially giving up its nuclear weapons. Perry joined Russian and Ukrainian defense ministers in a ceremony, during which sunflowers were planted on a former missile base in the Ukraine. The symbolism of using sunflowers arose from their use to soak up irradiation-contaminated soil resulting from the Chernobyl nuclear plant explosion. Ironically, Ukraine and Russia, the world's leading sunflower producers, are at war, and sunflowers have become symbols of Ukrainian resistance against the Russian invasion.
- Sunflower plants taller than 9 m have been recorded. By contrast, the shortest recorded mature sunflower was just over 5 cm tall and was grown in Oregon using a bonsai technique.
- Species of the sunflower family are popular with butterflies because the wide flower head makes a good landing platform and the numerous individual flowers make for a high probability of finding nectar. Monarch butterflies are commonly seen sucking nectar from sunflowers during their fall migration.
- Some botanists contend that the largest family of flowering plants is the sunflower family, with estimated number of species ranging from about

FIGURE 94.17 Ukrainian children in traditional costumes demonstrating their patriotism with sunflowers.

*Source:* **(a) Photo by Shutterstock/Dobra Kobra. (b) Photo by Shutterstock/Pandorabox.**

24,000 to 32,000 species. The sunflower family is rivalled in size only by the orchid family (Orchidaceae), some specialists considering it as the largest family, while others think the legume family (Leguminosae or Fabaceae) has as many species.

- During their peak growing period, sunflowers can grow as much as 30 cm per day.
- Breeders have created sunflower plants with heads that face downward so that birds find it harder to reach the seeds.
- Dutch painter Vincent van Gogh (1853–1890) is famous for his 12 paintings of sunflowers, which are known for their vibrant yellow hues. It has been speculated that the artist misperceived yellow because of overindulgence in absinthe and digitalis. More likely, he was one of the first to employ the superior yellow oil paints first made available in the 19th century.

FIGURE 94.18 Sunflowers and Vincent van Gogh: (a) One of van Gogh's many famous paintings titled *Sunflowers*, housed in the National Gallery of London. (b) A painting (1888) by Paul Gauguin (1848–1903) titled *The Painter of Sunflowers: Portrait of Vincent van Gogh*. Housed in the van Gogh Museum, Amsterdam.

*Source:* **(a) Public domain photo. (b) Public domain photo.**

FIGURE 94.19 Imaginative art (public domain) featuring sunflowers.

*Source:* **(a) Crane, W. 1889. *Flora's Feast, A Masque of Flowers*. Cassel & Company, London. (b) Grandville, J. J. 1867. *Les fleurs animées*, vol. 1. Garnier, Paris. (c) Tyra Kleen (1874–1951), Swedish artist, author, and women's rights activist. (d) A soap powder trade card prepared by Charlotte Perkins Gilman (1860–1935) from the collections of the US National Library of Medicine.**

FIGURE 94.20 Sunflower sculptures: (a) Edwin Watts Southwind Park, Springfield, Ill. (b) University of Costa Rica, San José de Costa Rica. (c) Musée des beaux-arts, Montréal.

*Source:* **(a) Photo by Mike Linksvayer (public domain; a sundial is in the foreground). (b) Photo by Carlos Reusser (CC BY 2.0). (c) Photo by Jeangagnon (CC BY SA 3.0).**

- Aside from highly valued paintings of sunflowers by the great European artists such as van Gogh, sunflowers have been featured in children's books and in sculptures.

## KEY PUBLICATIONS

Atamian, H.S., N.M. Creux, E.A. Brown, A.G. Garner, B.K. Blackman, and S.L. Harmer. 2016. Circadian regulation of sunflower heliotropism, floral orientation, and pollinator visits. *Science* 353(6299): 587–590.

Azania, A.A.P.M., C.A.M. Azania, P.L.C.A. Alves, R. Palaniraj, H.S. Kadian, S.C. Sati, et al. 2003. Allelopathic plants. 7. Sunflower (*Helianthus annuus* L.). *Allelopathy Journal* 11: 1–20.

Blackman, B.K., M. Scascitelli, N.C. Kane, H.H. Luton, D.A. Rasmussen, R.A. Bye, et al. 2011. Sunflower domestication alleles support single domestication center in eastern North America. *PNAS* 108(34): 14360–14365.

Blacquière, T., N. Straver, and D. van den Berg. 2002. Possibilities for using photoperiodism to program flowering of sunflowers (*Helianthus annuus*) in the greenhouse and in the open. *Acta Horticulturae* 580: 101–109.

Ciarka, D., H. Gawronska, U. Szawlowska, and S.W. Gawronski. 2009. Allelopathic potential of sunflower. I. Effects of genotypes, organs and biomass partitioning. *Allelopathy Journal* 23: 95–109.

Cvejić, S., S. Jocić, and E. Mladenović. 2016. Inheritance of floral colour and type in four new inbred lines of ornamental sunflower (*Helianthus annuus* L.). *Journal of Horticultural Science and Biotechnology* 91: 30–35.

Garfinkel, A.R. and K.L. Panter. 2014. Year-round greenhouse production of cut sunflowers in the Rocky Mountain West. *HortTechnology* 24: 743–748.

Giuffrida, F., C. Cassaniti, and C. Leonardi. 2009. Effects of cultivation practices on sunflower production as cut flower. *Acta Horticulturae* 807: 699–703.

Grams, S. 2023, September. Radioactive sunflowers? *Natural News Blogs*. https://naturalnewsblogs.com/radioactive-sunflowers/.

Haba, P.D.L., L.D.L. Mata, E. Molina, and E. Agüera. 2014. High temperature promotes early senescence in primary leaves of sunflower (*Helianthus annuus* L.) plants. *Canadian Journal of Plant Science* 94: 659–669.

Hayata Y. and Y. Imaizumi. 2000. Effect of photoperiod on flower bud development of ornamental sunflower (*Helianthus annuus* L.).

*Journal of the Japanese Society of Horticultural Science* 69: 708–710.

Heiser, C.B. 1976. *The Sunflower*. University of Oklahoma Press, Norman, OK. 198 pp.

Hladni, N, and Miladinović, D. 2019. Confectionery sunflower breeding and supply chain in Eastern Europe. *OCL [Oilseeds & Fats, Crops & Lipids]* 26: 29. www.ocl-journal.org/articles/ocl/full_html/2019/01/ocl190002/ocl190002.html.

Lentz, D.L., M.D. Pohl, J.L. Alvarado, S. Tarighat, and R. Bye. 2008. Sunflower (*Helianthus annuus* L.) as a pre-Columbian domesticate in Mexico. *PNAS* 105: 6232–6237.

Martínez-Force, E., N.T. Dunford, and J.J. Salas. 2015. *Sunflower: Chemistry, Production, Processing, and Utilization*. AOCS Press, Urbana. 728 pp.

Mladenović, E., S. Cvejić, J. Čukanović, G. Žeravica, and S. Jocić. 2016. Evaluation of sunflower genotypes for ornamental use. *Contemporary Agriculture* 65: 39–43.

Mladenović, E., S. Cvejić, S. Jocić, J. Čukanović, M. Jocković, and G. Malidža. 2017. Variability of morphological characters among ornamental sunflower collection. *Genetika* 49: 573–582.

Organisation for Economic Co-operation and Development. 2004. *Consensus Document on the Biology of* Helianthus annuus L. *(Sunflower)*. OECD, Paris. www.oecd.org/biotrack.

Pallez, L.C., J.M. Dole, and B.E. Whipker. 2002. Production and postproduction studies with potted sunflowers. *HortTechnology* 12: 206–210.

Peel, L. 1997. *The Ultimate Sunflower Book: Gardening, Cookery, Craft*. William Morrow/HarperCollins, New York. 112 pp.

Plant Biosafety Office, Canadian Food Inspection Agency. *The Biology of* Helianthus annuus *L.* www.inspection.gc.ca/plants/plants-with-novel-traits/applicants/directive-94-08/biology-documents/helianthus-annuus-l-/eng/1330977236841/1330977318934.

Pohl, M.D. 2008, April 29. Sunflower debate ends in Mexico, researchers say. *Florida State University News*. https://news.fsu.edu/news/science-technology/2008/04/29/sunflower-debate-ends-mexico-researchers-say/.

Poncavage, J. 1996. *Totally Sunflowers*. Celestial Arts, Berkeley. 94 pp.

Radanović, A., D. Miladinović, S. Cvejić, M. Jocković, and S. Jocić. 2018. Sunflower genetics from ancestors to modern hybrids – a review. *Genes* 9: 528. https://doi.org/10.3390/genes9110528.

Schilling, E.E. 2006. *Helianthus*. In: *Flora of North America North of Mexico*, Vol. 21, edited by Flora of North America Editorial Committee, 141–169. Oxford University Press, New York.

Schneiter, A.A., G.J. Seiler, J.F. Miller, L.D. Charlet, and J.M. Bartels. (eds.). 1997. *Sunflower Technology and Production*. Agronomy Series No. 35. American Society of Agronomy, Madison. 834 pp.

Seiler, G.J. 1992. Utilization of wild sunflower species for the improvement of cultivated sunflower. *Field Crops Research* 30: 195–230.

Seiler, G., T. Gulya, and L.F. Marek. 2023. Fifty years of collecting wild *Helianthus* species for cultivated sunflower improvement. *Helia* 46(78): 1–51.

Short, K., C.L. Etheredge, and T.M. Waliczek. 2017. Studying the market potential for specialty cultivars of sunflower cut flowers. *HortTechnology* 27(5): 1–7.

Smith, B.D. 2014. The domestication of *Helianthus annuus* L. (sunflower). *Vegetation History and Archaeobotany* 23: 57–74.

Steinitz, B. and Y. Shchori. 1981. Sunflowers as cut flowers. *Gartenbauwissenschaft* 46(1): S. 9–12.

Wien, H.C. 2016. Pinching ornamental sunflowers increases cut stem yield and reduces flower size. *HortTechnology* 26: 762–766.

## Key Cultivation Websites

Canadian Wildlife Federation. *Native Sunflowers – Joy in a Flower*. http://cwf-fcf.org/en/resources/encyclopedias/flora/native-sunflowers.html.

GardenersWorld.com. *How to Grow Sunflowers*. www.gardenersworld.com/how-to/grow-plants/how-to-grow-sunflowers/#propagate.

The Spruce. *How to Grow and Care for Sunflowers*. www.thespruce.com/growing-sunflowers-1402916.

SunflowerSelections.com. www.sunflowerselections.com/.

*Tips on Growing a Giant Sunflower*. https://kewgardener.com/tips-on-growing-a-giant-sunflower/.

WikiHow. *How to Care for Sunflowers*. www.wikihow.com/Care-for-Sunflowers.

WikiHow. *How to Get Rid of Sunflowers*. www.wikihow.com/Get-Rid-of-Sunflowers.

WikiHow. *How to Grow Sunflowers*. www.wikihow.com/Grow-Sunflowers.

WikiHow. *How to Harvest Sunflower Seeds*. www.wikihow.com/Harvest-Sunflower-Seeds.

WikiHow. *How to Plant & Care for Sunflowers in Containers*. www.wikihow.com/Grow-a-Sunflower-in-a-Pot.

WikiHow. *How to Plant Sunflower Seeds*. www.wikihow.com/Plant-Sunflower-Seeds.

WikiHow. *How to Protect Sunflowers from Squirrels*. www.wikihow.com/Protect-Sunflowers-from-Squirrels.

WikiHow. *How to Prune Sunflowers*. www.wikihow.com/Prune-Sunflowers.

# 95 Sweet Pea (*Lathyrus odoratus* and Hybrids)

**Scientific Names:** The genus name *Lathyrus* originates from the Greek *lathyros*, the prefix *la-* meaning "very," and the suffix *-thyros* meaning "passionate." *Odoratus* in the name *L. odoratus* is Latin for *fragrant*.

**Common Names:** Sweet pea should not be confused with sweet pea shrub (*Polygala* × *dalmaisiana*) or the common edible pea (*Pisum sativum*). The common name sweet pea (sometimes spelled sweet-pea) has been claimed to have been first used by the English poet Keats:

> Here are sweet peas, on tip-toe for a flight:
> With wings of gentle flush o'er delicate white,
> And taper fingers catching at all things,
> To bind them all about with tiny rings.
>
> – John Keats, in "I Stood Tiptoe on a Hill," 1817

FIGURE 95.1 Modern sweet pea (*Lathyrus odoratus*) cultivars grown closely together.

*Source:* **(a) Photo by Shutterstock/Andrew Fletcher. (b) Photo by Liz West (CC BY 2.0). (c) Photo by Tony Hisgett (CC BY 2.0). (d) Photo by Mike Peel (CC BY SA 4.0).**

## 95.1 INTRODUCTION

Sweet pea, a member of the pea family (Fabaceae), is a low-growing annual ornamental vine that produces extremely colorful, intensely scented flowers. In England, where the greatest enthusiasm for the plant developed, sweet pea has been called "the queen of annuals."

## 95.2 OVERVIEW

The sweet pea is a fast-growing herbaceous vine 0.5 to 2 m (sometimes to 3 m) in length, climbing by twining tendrils, or simply sprawling on the ground if supports are unavailable. A narrow, continuous wing appears on either side of the stem. The species is indigenous to southern Italy, the Aegean Islands, and Sicily. However, it is grown worldwide and has become introduced (i.e., naturalized) in numerous countries. Hundreds of cultivars are marketed. The flowers are about 3 cm in diameter and tend to cluster at the ends of the stems. Flowers are available in shades of blue, mauve, pink, purple, red, and white, as well as bicolored combinations (yellow flowers are notably absent). Cultivars with doubled flowers (with extra petals) have also been bred. The flowers appear in midsummer. Pealike pods develop in fall and are hairy, up to 6 cm in length, ripen to brown, and contain several globular gray-brown to black seeds.

The genus *Lathyrus* has more than 150 species, native to temperate areas of the world. In addition to *L. odoratus*, several other *Lathyrus* species are employed as ornamentals, especially everlasting pea (*L. latifolius*, sometimes called perennial sweet pea), a perennial, unscented relative of the sweet pea, hardy to climate zone 5. *Lathyrus* species are potentially quite poisonous, and the "sweet" in the name *sweet pea* is highly misleading. Unfortunately, the phrase "sweet pea" is frequently applied to the edible garden pea (*Pisum sativum*).

FIGURE 95.2 Anatomical features of sweet pea (*Lathyrus odoratus*): (a) Two-winged stem. (b) Leaves and tendrils. (c) Pod.

*Source:* **(a) Photo by Frank Vincentz (CC BY SA 3.0). (b) Photo by Shutterstock/Anneka. (c) Photo by Tanya Impeartrice (CC BY ND 2.0).**

## 95.3 HISTORY

Sweet peas have been cultivated at least since the 17th century in Europe. Scottish nurseryman Henry Eckford


DOI: 10.1201/9781003586227-95

FIGURE 95.3 Botanical paintings of old garden cultivars of sweet pea.

*Source:* Public domain illustrations. (a) A postcard published by Edward H. Mitchell, San Francisco, California, in 1910. (b) Same as (a), but published in 1911. (c) Painting by Pierre Joseph Redoute (1759–1840) in his 1883 publication *Choix des plus belles fleurs* (*The Most Beautiful Flowers*). (d) Curtis, W. (ed.) 1788. *The Botanical Magazine.* (e) Biodiversity Heritage Library. (f) Moninckx, J. 1682–1709. *Moninckx Atlas*, vol. 4. (g) *Revue horticole* (Paris), sér. 4, vol. 86 (1914). (h) An 1878 chromolithograph, after H. Moon. From Wellcome Collection.

FIGURE 95.4 Everlasting pea (*Lathyrus latifolius*): (a) Plants spreading over the ground. (b) A vine bearing pods. (c, d, e) Public domain illustrations of garden cultivars.

*Source:* (a) Photo by Katrin Schneider (CC BY SA 4.0). (b) Photo by Trystan Harpold (CC BY 4.0). (c) Miller (Müller), J. S., and M. B. Borckhausen, 1804. *Illustratio systematis sexualis Linnaei* (German edition). (d) *Garden. An illustrated weekly journal of horticulture*, vol. 13, 1878. (e) Redouté, P. J. 1833. *Choix des plus belles fleurs et des plus beaux fruits.* Ernest Panckoucke, Paris.

FIGURE 95.5 Illustrations from old seed catalogues featuring sweet pea cultivars.

*Source:* (a) *Annual of True Blue Seeds*, published in 1905 by A. W. Livingston's Sons, Columbus, OH. (b) Childs 1922 catalogue, published by John Lewis Childs, Floral Park, NY. (c) "Everything for the Garden" 1902 catalogue of Peter Henderson & Co. (d) Vick's Garden and Floral Guide, 1898 catalogue of James Vick's Sons, Rochester, NY. (e) Dreer's Garden 1902 calendar published by Henry A. Dreer, Philadelphia, PA. (f) Wm. Elliott & Sons, Seedsmen, 1897 catalogue, published by Wm. Elliott & Sons, New York.

(1823–1905) is credited with greatly improving the sweet pea in the 19th century, producing more than 100 cultivars. Late 19th-century and early 20th-century seed catalogues featured spectacular illustrations of sweet pea cultivars, reflecting the great enthusiasm for these ornamentals. Modern hybrids have bigger, more colorful flowers and stronger vines but sometimes lack the fragrance of older cultivars.

## 95.4 ECONOMIC IMPORTANCE

Sweet pea is a well-established, popular garden ornamental throughout the temperate world. There is a strong demand for seeds, supplied by the horticultural industry.

## 95.5 ORNAMENTAL USES

Sweet peas are superb as groundcover. They are also appropriate as vines for fences but are too small to compete with perennial vines for covering tall structures and walls. Dwarf varieties are best for containers, including hanging baskets. Sweet peas provide excellent, fragrant cut flowers.

## 95.6 OTHER USES

The only significant economic value of sweet pea is as an ornamental. *Lathyrus* species are employed as food for

FIGURE 95.6 A field of flowering red sweet peas on a seed farm in Lompoc, California.

*Source:* Photo by Shutterstock/Joojoob27.

FIGURE 95.8 Sweet peas as cut flowers. Harvesting sweet pea flowers from garden plants not only provides beautiful bouquets for use in homes but also stimulates additional flowering.

*Source:* (a) Photo by Cody and Maureen (CC BY 2.0). (b) Photo by Angelbattle Bros (CC BY ND 2.0). (c) Photo by Downtown-gal (CC BY SA 4.0).

FIGURE 95.7 Common landscape uses of sweet peas: (a) Groundcover. (b) Grown as a border against a wall. (c) Grown on supports.

*Source:* (a) Photo by Kenpei (CC BY SA 3.0). (b) Photo by Acabashi (CC BY SA 4.0). (c) Photo by Aftabbanoori (CC BY SA 3.0).

humans and feed for livestock, despite the toxicity, and it has been speculated that in the future, breeding could eliminate the poisonous chemicals, allowing the species to be employed for various purposes.

## 95.7 PROBLEMS

All parts of the plants are toxic to humans, pets, and livestock. In particular, consuming large amounts of seed for weeks to months can result in paralysis, slow and weak pulse, breathing difficulties, and convulsions. Because the seedpods resemble edible sweet peas, and because the name suggests edibility, people unfamiliar with the plant are sometimes tempted to eat the seeds. A few seeds consumed just once are unlikely to be harmful and should not cause panic.

Sweet pea is considered invasive in New Zealand and is thought to have the potential to become a weedy pest in other countries.

## 95.8 CULTIVATION

### 95.8.1 Climate and Site

As an annual, sweet pea is not particularly limited by climate zone, although in colder latitudes seeds need to be started indoors well in advance of planting outdoors, and in hotter areas, the plants are limited by their inappropriate adaptation to high heat. Sweet pea is adapted to full sun and light shade. In very hot climates, semi-shade in the afternoon is recommended. To avoid soil depletion, changing locations or fortifying the soil each year is sometimes practiced. Bush types (which produce short vines) do not require supports, and dwarf types grow well without supports in containers and in borders. While most long-vine types can be allowed to sprawl over the ground, better displays result when stakes, trellis, or netting systems are provided. Chain-link fences are ideal, providing slim wire that tendrils can easily wrap around. Bamboo poles, canes fashioned from tree branches, and chicken wire are often employed. Even a few cables stretched through the planting area can improve the appearance. Protection from wind is advisable for the taller varieties.

FIGURE 95.9 Weedy sweet pea invading a wetlands nature reserve near Giethoorn, Netherlands.

*Source:* **Photo by Gouwenaar (public domain).**

FIGURE 95.10 Support systems for sweet pea vines: (a) Growth on a mesh (chicken wire) fence. (b) On a garden arch. (c) On a pergola. (d) On a network of bamboo poles.

*Source:* **(a) Photo by Shutterstock/Intreegue Photography. (b) Photo by Husqvarnaimagery (CC BY 2.0). (c) Photo by Shutterstock/Roger de Montfort. (d) Photo by Karen Bryan (CC BY ND 2.0).**

### 95.8.2 Propagation and/or Purchasing

Sweet pea is normally grown from seeds, although in warm (i.e., long-season) climates, cuttings are also sometimes employed.

### 95.8.3 Planting

Seeds or seedlings can be planted at a spacing of 15 to 20 cm.

### 95.8.4 Soil and Moisture Conditions

Sweet pea has a reputation as a strong feeder and is best grown in humus-rich, fertile, well-drained soil, preferably with alkaline pH (greater than 7.0, ideally 7.5). Fertilizer can be added during the growth period (but incorporating well-rotted compost in the soil prior to planting is strongly recommended). Like other plants of the pea family, sweet pea roots are associated with bacteria that transform inert atmospheric nitrogen into usable compounds, so their need for nitrogen fertilization is reduced.

### 95.8.5 Maintaining Plants

To promote branching, the tops of the vines can be pinched off once they are at least 10 cm high. Deadheading away spent flowers is recommended to prolong flowering. Of course, just harvesting the flowering branches to display in vases achieves this purpose.

## 95.9 FASCINATING FACTS, FABLES, FOLKLORE, AND FRIVOLITY

- Old European superstitions hold that for best growth, sweet pea should be planted on Good Friday (the Friday preceding Easter Sunday, the date determined by local Christian authorities). In Ireland, it was thought that sweet pea would become larger and more fragrant if planted on St. Patrick's Day (March 17).
- The word *sweet* is often employed to indicate sentiment instead of taste sensation. "Sweet pea" is a colloquial term of endearment for something or someone who is lovely or sweet in character, presumed to be based on the beauty of the sweet pea plant. The phrase is most familiarly applied to the baby Little Swee'Pea in the comic based on the character Popeye.
- Grass pea or blue sweet pea is *Lathyrus sativus*, which is often grown in Asia and Africa for

FIGURE 95.11 Competitive sweet pea flower show at Chelsea, UK.

*Source:* **Photo by La Citta Vita (CC BY SA 2.0).**

human and livestock consumption because it is very resistant to drought and is high in protein. Unfortunately, the seeds contain a neurotoxin (the same as in sweet pea) that causes a degenerative disease called lathyrism if consumed in large amounts over a long period. Extremely large numbers of people have been affected, especially when famine conditions left no alternative foods.

- Throughout the world, competitive "flower shows" provide opportunities for both professional and amateur gardeners to exhibit their finest plants. In the case of vines such as sweet pea, the plants are much too awkward to transport as potted plants, so freshly harvested flowering branches are shown in identical containers of water. Strict rules need to be followed, so that the size of the bouquets is comparable.

## KEY PUBLICATIONS

Brakie, M. 2020. *Plant Guide for Perennial Pea* (Lathyrus latifolius). USDA-Natural Resources Conservation Service, East Texas Plant Materials Center, Nacogdoches, TX. 5 pp. https://plants.usda.gov/DocumentLibrary/plantguide/pdf/pg_lala4.pdf.

Crane, D.B. 1910. *The Book of the Sweet Pea*. John Lane, New York. 136 pp. https://books.google.ca/books?hl=en&lr=lang_en&id=7sdEAAAAIAAJ&oi=fnd&pg=PA1&dq=+%22sweet+pea%22&ots=sHjIytShGI&sig=pVjcdqqpOkQBcRTD7hlFmD3EkA4#v=onepage&q=%22sweet%20pea%22&f=false.

Jansen, A., A. Vermeulen, A.W. van Toorenenbergen, and P.H. Dieges. 1995. Occupational asthma in horticulture caused by *Lathyrus odoratus*. *Allergy and Asthma Proceedings* 16(3): 135–139.

Murfet, I.C. and J.J. Ross. 2019. *Lathyrus odoratus*. In: *Handbook of Flowering*, Vol. 3, edited by A.H. Halevy, 246–250. CRC Press, Boca Raton, FL.

Murray, B.G. and K.R.W. Hammett. 1998. New sweet pea (*Lathyrus odoratus*) cultivars via interspecific hybridization with wild *Lathyrus* species. *Acta Horticulturae* 454: 115–118.

Parsons, R. 2019. *Sweet Peas: An Essential Guide*, 2nd edition. Crowood Press, Ramsbury, UK. 192 pp.

Rice, G. 2002. *The Sweet Pea Book*. Timber Press, Portland, OR. 144 pp.

Rojas-Sandoval, J. 2017. *Lathyrus odoratus* (Sweet Pea). CABI Compendium. https://doi.org/10.1079/cabicompendium.31584.

Small, E. and P.M. Catling. 2004. Blossoming treasures of biodiversity 14. Grass pea (*Lathyrus sativus*). Can a last resort food become a first choice? *Biodiversity* 5(4): 29–32.

Vaz Patto, M.C. and D. Rubiales. 2014. *Lathyrus* diversity: Available resources with relevance to crop improvement – *L. sativus* and *L. cicera* as case studies. *Annals of Botany* 113: 895–908.

## KEY CULTIVATION WEBSITES

Almanac. 2023. *How to Plant, Grow, and Care for Sweet Pea Flowers*. www.almanac.com/plant/sweet-peas.

Garden Design. *Sweet Pea Flowers: A Growing Guide*. www.gardendesign.com/flowers/sweet-pea.html.

Horticulture Magazine. Lathyrus *'Sweet Peas'*. https://horticulture.co.uk/lathyrus/.

Johnny's Selected Seeds. 2016. *Sweet Pea Production*. Growing Guide 8171. www.johnnyseeds.com/on/demandware.static/-/Library-Sites-JSSSharedLibrary/default/dwfb8c600c/assets/information/sweet-pea-production.pdf.

Lewis, M., M. Stock, T. Maughan, D. Drost, and B. Black. 2019. *Sweet Pea Cut Flower Production in Utah*. Utah State University Extension. https://digitalcommons.usu.edu/cgi/viewcontent.cgi?article=3079&context=extension_curall.

Royal Horticultural Society. *How to Grow* Lathyrus. www.rhs.org.uk/plants/lathyrus/growing-guide.

The Spruce. *Sweet Pea Flowers: Planting, Care & Growing Guide*. www.thespruce.com/sweet-peas-the-queen-of-the-annuals-1402917.

WikiHow. *How to Grow Sweet Peas*. www.wikihow.com/Grow-Sweet-Peas.

WikiHow. *How to Harvest Sweet Pea Seed Pods*. www.wikihow.com/Harvest-Sweet-Pea-Seed-Pods.

# 96 Trumpet Creeper (*Campsis* Species and Hybrids)

**Scientific Names:** The genus name *Campsis* is based on the Greek *kampsis*, "bending," a reference to the bending stamens of the flowers.

**Common Names:** The most frequent common name for the plants is *trumpet creeper*, the first word descriptive of the shape of the flowers, the second referring to the viny growth pattern. The plant is also called trumpet vine and trumpet flower. (Some botanists reserve the word "vine" for herbaceous climbing or trailing plants, in contrast to "lianas," which are the same, but woody.) Numerous other species of ornamental vines with trumpet-shaped flowers are sometimes similarly termed, and misidentification is frequent. (For example, see the chapter on angel trumpet.) Infrequent names for *C. radicans* include hummingbird vine and trumpet honeysuckle. The names Devil's shoelaces (Devil's shoestrings) and hellvine have been applied to weedy *Campsis* plants. The name "cow-itch vine" reflects the redness and itching of some people after contacting the plant. Trumpet bush is *Tecoma stans*, a member of the same family (Bignoniaceae) as trumpet vine (*Tecoma* species are sometimes called trumpet flowers).

## 96.1 INTRODUCTION

Trumpet creepers are large impressive vines with spectacular, colorful, exotic-looking flowers. They are cultivated from north temperate areas to the subtropics, and often the chief complaint against them is that they simply grow too well. Moreover, they require so little attention that it has been said that anyone incapable of growing them should not attempt to grow any plant.

FIGURE 96.1 Chinese trumpet creeper (*Campsis grandiflora*) in South Korea: (a) Climbing up the front of a multi-story building. (b) In front of a one-story building.

*Source:* **(a) Photo by Dalgial (CC BY SA 3.0). (b) Photo by Dae Jeung Kim from Pixabay (public domain).**

FIGURE 96.2 American trumpet creeper (*Campsis radicans*) supported by a gazebo in Philadelphia.

*Source:* **Photo by Laura Blanchard (CC BY SA 2.0).**

FIGURE 96.3 Hybrid trumpet creeper (*Campsis* × *tagliabuana*) climbing up a light standard in Central California.

*Source:* **Photos by David Prasad (CC BY SA 2.0).**

## 96.2 OVERVIEW

Two species of *Campsis* are recognized. *Campsis grandiflora* (Chinese trumpet creeper) is native to China and Japan in eastern Asia, and *C. radicans* (American trumpet creeper) is native to eastern North America. Their hybrid, *C.* × *tagliabuana*, is also frequently cultivated (the name *tagliabuana* commemorates brothers Alberto Linneo Tagliabue and Carlo Tagliabue, who were Italian nurserymen). The plants are perennial, deciduous (sometimes semi-evergreen in warm climates), woody vines which climb up to 13 m (rarely up to 20 m) by aerial roots (threadlike

DOI: 10.1201/9781003586227-96

FIGURE 96.4 Aerial (adventitious) roots at a stem node (where the leafstalks originate) of Chinese trumpet creeper (*Campsis grandiflora*). These specialized threads serve to anchor the vine to objects on which it climbs.

***Source:*** **(a) Photo by Rosser1954 (CC BY SA 4.0). (b) Photo by Vikipediy Velykyy (CC BY SA 4.0).**

FIGURE 96.5 Multiple trunk development in *Campsis radicans*. The "trunks" are not like those of trees but are relatively slim, comparable to mature woody grapevines: (a) Several separated trunks that arose by suckers at the base of the plant. (b) A clump of just a few trunks, most of the basal suckers having been pruned away from the initial planting.

FIGURE 96.6 Fruit pods and winged seeds of Chinese trumpet creeper (*Campsis grandiflora*).

***Source:*** **Photo by Shutterstock/Jiang Tianmu.**

FIGURE 96.7 Variation of flower color among cultivars of American trumpet creeper (*Campsis radicans*).

***Source:*** **(a) Photo by Shutterstock/Martin Dlugo. (b) Photo by Shutterstock/RukiMedia. (c) Photo by Ettore Balocchi (CC BY 2.0). (d) Photo by Manuel M.V. (CC BY 2.0).**

extensions growing from the stems) which penetrate cracks in surfaces, tying the vine to them. The plants (if not pruned) naturally produce "suckers" that develop into woody trunks. The flowers are large (the flared part is up to 10 cm or more in width and 8 cm in length), trumpet-shaped or funnel-shaped, and in cultivars are mostly orange or red, but also pink, yellow, and bicolored. The plants bloom from summer to autumn. The fruits are plump, bean-like pods 7–20 cm long, initially green but maturing to brown, and containing numerous wind-dispersed winged seeds. The Asian species has somewhat larger or more flaring flowers produced in larger clusters than the North American species, but the differences between the species are imperceptible to most people. *Campsis radicans* is suited to growth in more Northern regions, whereas *C. grandiflora* grows better in more Southern areas, and the flowers of the Chinese species are larger, more open, and more attractive.

## 96.3 HISTORY

American trumpet creeper (*C. radicans*) was brought to England in the early 17th century. Chinese trumpet creeper

FIGURE 96.8 Nineteenth-century paintings (public domain) of cultivated trumpet creepers by European botanical illustrators: (a) *Campsis radicans.* (b) *Campsis radicans.* (c) *Campsis radicans.* (d) *Campsis grandiflora.* (e) *Campsis grandiflora.* (f) *Campsis* × *tagliabuana.*

*Source:* **(a, d) Witte, H., and A. J. Wendel. 1868. *Flora: afbeeldingen en beschrijvingen van boomen, heesters, éénjarige planten, enz. voorkomende in de Nederlandsche Tuinen.* Wolters, Groningen. (b) Duhamel du Monceau, H.-L., and Redouté, P. J. 1801–1819. *Traité des arbres et arbustes que l'on cultive en France en pleine terre.* Paris. (c) Kraus, J. C. 1802. *Afbeeldingen der fraaiste, meest uitheemsche boomen en heesters.* J. Allart, Amsterdam. (e) *Curtis's Botanical Magazine* (London), vol. 34, 1811. (f) *Revue horticole* (Paris), sér. 4 (1852–1974).**

(*C. grandiflora*) was not recognized in Europe until the late 18th century, but the species was cultivated in China and Japan for centuries. A large-flowered hybrid of the two species, labelled Mme Galen, was introduced about 1889 by the Tagliabue nursery of Italy.

## 96.4 ECONOMIC IMPORTANCE

The demand for trumpet creepers is relatively limited, because there are numerous other similar flowering vines that are cultivated as ornamentals. Also, the plants can be extremely long-lived (note the record cited later of 153 years of age), and a single plant can spread so extensively that there is little need to purchase more plants.

## 96.5 ORNAMENTAL USES

Trumpet creepers are very impressive climbing vines, employed mostly to decorate walls, fences, pergolas, and other strong supporting structures. Plants can be used in containers, the stems cascading over the sides. Occasionally, the plants are used as groundcover.

FIGURE 96.9 Examples of landscape uses of Chinese trumpet creeper (*Campsis grandiflora*). All photos are from Pixabay, show scenes in South Korea, and are public domain: (a) Flowering vines cascading over a wall painted with murals. (b) Growing over a stone wall. (c) Growing flush against a stone wall.

*Source:* **(a) Photo by Chunseong River. (b) Photo by Jaesung. (c) Photo by Jeung Kim.**

FIGURE 96.10 Examples of landscape uses of American trumpet creeper (*Campsis radicans*): (a) Vines artistically arranged on either side of the entrance to a building. (b) Cultivar Flamenco cascading over a brick wall at the Chicago Botanical Garden. (c) Situated beside an artificial waterfall at the Chicago Botanical Garden.

*Source:* **(a) Photo by Shutterstock/Raulbaldean. (b) Photo by Cultivar413 (CC BY 2.0). (c) Photo by Cultivar413 (CC BY 2.0).**

## 96.6 OTHER USES

The nectar-rich flowers of trumpet creepers are very attractive to birds, especially hummingbirds, as well as butterflies and long-tongued bees, and the plants are often employed

FIGURE 96.11 Paintings of the attractiveness of trumpet creeper flowers to birds. (a) Illustration by Elzza. (b) Painted in the Florida Keys by John James Audubon (1785–1851), who identified the bird as the mangrove hummingbird (it has been identified as the black-throated mango, *Anthracothorax nigricollis*).

***Source:* (a) Photo by Shutterstock/Elzza. (b) Audubon, J. J. 1827–1838. *Birds of America*, Plate 184.**

FIGURE 96.12 Ruby-throated hummingbirds (*Archilochus colubris*) visiting flowers of American trumpet creeper (*Campsis radicans*).

***Source:* (a) Photo by Shutterstock/Mike Truchon. (b) Photo by Gerry Buckel (CC BY ND 2.0). (c) Photo by Kelly Colgan Azar (CC BY ND 2.0). (d) Photo by Abbeyprivate (public domain).**

in pollinator gardens. Some birds simply like to nest in the dense foliage. The natural geographical distributions in the summer of the ruby-throated hummingbird and the American trumpet creeper almost match, and the June-to-September blooming period of the plant coincides with the hummingbirds' migration south. The bird is the only native eastern North American hummingbird, and the plant is the only native trumpet creeper, so it seems the two have co-evolved (although bees also pollinate the flowers).

## 96.7 PROBLEMS

*Campsis radicans* is an aggressive weed in parts of the United States, spreading to roadsides, fields, and woods. *Campsis* × *tagliabuana* is less invasive, and *C grandiflora*

FIGURE 96.13 American trumpet vine (*Campsis radicans*) behaving as an invasive weed: (a) Conquering a tall pine tree. (b) Overwhelming a utility pole.

***Source:* (a) Photo by Shutterstock/FlorinRO. (b) Photo by John D. Byrd (CC BY 3.0).**

is not considered significantly so. Aside from the damage the American trumpet creeper causes as an invasive weed, there are several minor issues of concern, primarily to homeowners. (1) Once established, cultivated plants can be difficult to eradicate. (2) Ants often occur on trumpet vines and are objectionable to some gardeners. (Nectar is produced not only in the flowers but also by special glands called extrafloral nectaries on the leafstalks and other plant parts, which attract the ants.) (3) The clinging aerial roots of the plants are designed to penetrate cracks and crevices of the bark of host trees to hold the vine in place. When placed against building surfaces, they can cause significant damage. (4) The thick, woody stems can grow between, and damage, house siding panels and roof shingles. (5) The sap can cause skin irritation on contact, and indeed, the foliage is slightly toxic (which is advantageous in repelling consumption by deer and rabbits).

## 96.8 CULTIVATION

The possibility that planting *Campsis radicans* will contribute to the local invasive weed problem needs to be considered. In regions where there is concern, the danger of vegetative spread can be limited by growing the plants in containers and preventing the seedpods from maturing (although they are often considered to be attractive). The hybrid trumpet creeper (*C.* × *tagliabuana*) is less invasive, and cultivars have been bred with very low seed production.

### 96.8.1 Climate and Site

*Campsis radicans*, the hardiest species, is rated for growth in climate zones 4 to 10. *Campsis grandiflora* is suited to zones 7 to 11, and the hybrid between the two species (*C.* × *tagliabuana*) can grow from zones 6 to 10. Trumpet creepers grow well in full sun, but also in part shade (they grow naturally in woodlands). Toward the northern limit of their tolerance, the site should be as warm as possible (e.g., against a wall that is heated by the sun) and fully exposed to the sun. *Campsis radicans* is grown quite far north for a species that is basically native to warm climates, and like other southern species that are cultivated well to the north of their natural areas of distribution, it leafs out in very late spring and drops its leaves with the first frosts.

### 96.8.2 Propagation and/or Purchasing

The plants can be grown from seeds, although cultivars (especially the hybrid *C.* × *tagliabuana*) will not reproduce the parental form accurately. Professional growers usually use cuttings to generate new plants. Trumpet creepers are somewhat of a specialty item at nurseries and garden centers and are not always easily available. *Campsis* × *tagliabuana*, not the parental species, is most likely to be available in trade. Because early growth can be slow and flower production may not occur for several years, plants are usually cultivated from starter stock. Once established, the plants spread by suckers produced from the rhizomes (underground stems) and also may self-seed (*C. radicans* generally requires cross-pollination to set seed). If allowed to grow laterally on the ground, the stems root at intervals (this is known as "layering").

### 96.8.3 Planting

Trumpet creepers need considerable space. A single plant can spread remarkably, but when planting more than one, a spacing of 1 to 2.5 m is recommended. For groundcover, a spacing of 2.5 m is appropriate. Trumpet creepers grow into massive vines that must be provided with strong supports, which could be poles, pillars, fences, pergolas, arbors, or heavy trellises. The vines are often grown on trees (their natural habitat), but they can be detrimental to them, easily toppling small trees. In early growth, it may be necessary to train the vines to grow in desired directions and to tie them to their support.

### 96.8.4 Soil and Moisture Conditions

Wild plants of *C. radicans* in North America have been recorded growing in diverse soils, sand, loam, or clay, wet to dry, with a pH range of 3.7 to 6.8 (in cultivation, they also grow well with pH up to 8.0). Although trumpet creepers can grow in low-quality soil, for better growth, they should be provided with loamy soil that is moderately fertile and moist but well-drained. However, cultivating the plants in very rich soil can produce excessive growth.

### 96.8.5 Maintaining Plants

Trumpet creepers are sometimes grown as bushes or hedges by heavy pruning but are usually cultivated as climbers. The vines can become quite rampant and are best controlled by limiting the growth of the main branches and cutting back the number of main stems that arise from the ground. Toward their northern climatic limit, the plants have a short growing season, and so they do not need much trimming. Pruning is best done while the plants are dormant. Desirable cultivars are often grafted to hardy rootstocks, and it is important to remove shoots ("suckers") from the latter, or they may outcompete the cultivar.

## 96.9 Fascinating Facts, Fables, Folklore, and Frivolity

- American and Chinese trumpet creepers exhibit a well-known classical plant geographical distribution pattern, with one species in eastern Asia and a closely related species in eastern North America. There are about 70 counterparts among other plant species. The most widely accepted explanation is that a temperate forest was widely distributed in the Northern Hemisphere tens of millions of years ago, but several million years ago, climatic cooling exterminated most of the plants, except those that survived in the two regions.
- A worldwide record of 2.05 m for trunk girth (circumference, measured at a height between 1.30 m and 1.50 m, excluding species with multiple trunks) for American trumpet creeper was recorded for a vine estimated to have been 154 years of age, in France, in 2016 (www.monumentaltrees.com/en/trees/campsisradicans/records/).
- In Korea, there are "birth flowers" for every day of the year (not just the months). The trumpet creeper is the birth flower for August 6 and symbolizes honor. (For a guide to all 366 birth flowers, including for February 29, see www.creatrip.com/en/blog/8543).
- Computerized spellcheckers are often unkind to scientific names. *Campsis radicans* is sometimes altered to "campus radicals."

## Key Publications

Bertin, R.I. 1982. Paternity and fruit production in trumpet creeper (*Campsis radicans*). *American Naturalist* 119: 694–709.

Bertin, R.I. 1982. Floral biology, hummingbird pollination and fruit production of trumpet creeper (*Campsis radicans*, Bignoniaceae). *American Journal of Botany* 69: 122–134.

Dohse, L. and D. Elston. 2009. Botanical briefs: Trumpet vine (*Campsis radicans*). Close encounters with the environment. *Cutis* 83: 177–178.

Edwards, J.T. and L.R. Oliver. 2004. Emergence and growth of trumpetcreeper (*Campsis radicans*) as affected by rootstock size and planting depth. *Weed Technology* 18: 816–819.

Elias, T.S. and H. Gelband. 1975. Nectar: Its production and functions in trumpet creeper. *Science* 189(4199): 289–291.

Elias, T.S. and H. Gelband. 1976. Morphology and anatomy of floral and extrafloral nectaries in *Campsis* (Bignoniaceae). *American Journal of Botany* 63: 1349–1353.

Oates, K.M., T.G. Ranney, D.H. Touchell, and Z. Viloria. 2014. *Campsis* × *tagliabuana* 'Chastity': A highly infertile triploid trumpet vine. *HortScience* 49: 343–345.

Van Nest, B.N., A.A. Edge, M.V. Feathers, A.C. Worley, and D. Moore. 2021. Bees provide pollination service to *Campsis radicans* (Bignoniaceae), a primarily ornithophilous trumpet flowering vine. *Ecological Entomology* 46: 117–127.

Wen, J. and R.K. Jansen. 1995. Morphological and molecular comparisons of *Campsis grandiflora* and *C. radicans* (Bignoniaceae), an eastern Asian and eastern North American vicariad species pair. *Plant Systematics and Evolution* 196: 173–183.

Wennerberg, S. 2004. *United States Department of Agriculture Plant Guide:* Campsis radicans. https://plants.sc.egov.usda.gov/DocumentLibrary/plantguide/pdf/pg_cara2.pdf.

## Key Cultivation Websites

GardenersWorld.com. *How to Grow* Campsis *(Trumpet Vine)*. www.gardenersworld.com/how-to/grow-plants/campsis-trumpet-vine/.

Horticulture Magazine. Campsis *'Trumpet Vines' Plant Care & Growing Tips*. https://horticulture.co.uk/campsis/.

Housing.com. *Trumpet Vine: Health Benefits, Uses, and Care Tips*. https://housing.com/news/trumpet-vine/.

Royal Horticultural Society. *How to Grow* Campsis. www.rhs.org.uk/plants/campsis/growing-guide.

Southern Living. *How to Grow Trumpet Vine Without it Taking Over Your Garden*. www.southernliving.com/garden/vines/invasive-trumpet-vine.

The Spruce. *How to Grow and Care for Trumpet Vine*. www.thespruce.com/stop-spreading-of-trumpet-vines-2132896.

WikiHow. *How to Kill a Trumpet Vine*. www.wikihow.com/Kill-Trumpet-Vine.

WikiHow. *How to Prune Your Trumpet Vine*. www.wikihow.com/Prune-a-Trumpet-Vine.

# 97 Tulip (*Tulipa* Species and Hybrids)

**Scientific Names:** The collective species *Tulipa gesneriana* includes most cultivated ("garden") tulips. These are sometimes identified as *T. hybrida* hort. or *T.* × *hybrida*, which are pseudo-scientific horticultural names not employed in formal taxonomic nomenclature. The genus name *Tulipa* is based on the Ottoman Turkish *tülbend*, ultimately tracing to the Persian *dulband*, meaning "turban," almost universally interpreted as a reference to the tulip flower. The folds of a turban, reminiscent of tulip petals (especially when they overlap, as in young, unopened flowers), may be the basis of the name.

**Common Names:** Tulip, garden tulip, Dutch tulip. Tulip tree is *Liriodendron tulipifera* of North America. Indian tulip tree is *Thespesia populnea*, found on tropical coasts. Cape tulips are two toxic weedy species of *Moraea* native to South Africa and introduced elsewhere by escapes from cultivation.

## 97.1 INTRODUCTION

Tulips are unique. No other ornamental plant provides a comparably dazzling, exuberant display of eye candy when planted en masse. Tulip beds can paint stretches of landscapes in astonishingly vivid and glorious colors. But individual tulips also have remarkable charm. Despite its simple structure, the tulip is so elegant and endearing in form that it is the plant most often drawn by young children. And of course, the tulip is the quintessential star of the spring garden. Plump, gorgeous flowers swaying in the breeze are a welcome harbinger of a new season of growth.

## 97.2 OVERVIEW

There may be over 100 species of tulips (*Tulipa* species), although the exact number is uncertain. Tulips are perennial herbaceous plants of the lily family (Liliaceae) with large, showy flowers. Wild tulip species range from southern Europe and North Africa to northwestern China. The center of distribution is Central Asia, particularly Kazakhstan. The plants occur in hilly or mountainous areas in temperate climates with long, cool springs followed by dry summers. Tulip species overwinter by bulbs, which in the spring produce a small number of basal, strap-shaped leaves with prominent parallel venation. The plants also develop flowers in the spring, on erect stalks that are 10–70 cm tall, depending on species. A tulip grown from seed usually requires 5 to 7 years to produce its first flower. Vernalization (treatment needed to induce flowering, typically 15 to 24 weeks of cold exposure) is required to stimulate the ability to flower in tulip bulbs. Most tulip species and cultivars have only one floral stalk ("peduncle"), and this has a solitary flower (a few tulips have more than one flower on the peduncle). The flowers are usually cup- or star-shaped, with three petals and three sepals, these six segments so similar that they are usually collectively referred to as tepals (collectively "perianth") or petals. The outer three tepals are generally broader at the base and more rounded, while the inner three are narrower at the base and often wedge-shaped. Most garden tulips have little or no scent, although a few fragrant cultivars are offered by the nursery trade. Nectaries are absent. The fruit capsules contain many (often 200 to 300) small seeds, in two rows in each of the three segments.

FIGURE 97.1 Tulip beds with a wide range of floral colors.

*Source:* **(a) Photo by Susan-lu4esm from Pixabay (public domain). (b) Photo by Hans from Pixabay (public domain).**

In the horticultural trade, a variety of underground organs composed of stem tissue modified for food storage are often termed bulbs or characterized as bulbous. True bulbs are condensed stems made up mostly of scaly, fleshy storage leaves, which are attached below to a "basal plate," beneath which roots emerge (the flattened base of a tulip bulb is usually easily observed). True bulbs are, for the most part, simply buds surrounded by swollen leaves. Tulip produces true bulbs, and so do onion, lily, and narcissus (nearly all true bulbs are in the monocot group of flowering plants). Other so-called bulbous organs include corms, tubers, and rhizomes. These are incorrectly called bulbs, and in all cases, they are mostly made up of undifferentiated tissues, so that a cross section does not show layers as in true bulbs. Although tulips are perennial, the bulbs are annual, all the stored energy being depleted to develop new plants in the spring.

Many garden tulips are self-fertile, but the pollen from the six stamens starts to shed before the stigma matures, promoting outcrossing. However, most modern tulip cultivars are sterile, and when seeds are available from cultivars, they usually produce plants with inferior floral displays. Accordingly, almost all garden tulips are grown from daughter bulbs (sometimes called "bulblets," or more colorfully, "maidens") that develop as offsets from the axils

DOI: 10.1201/9781003586227-97

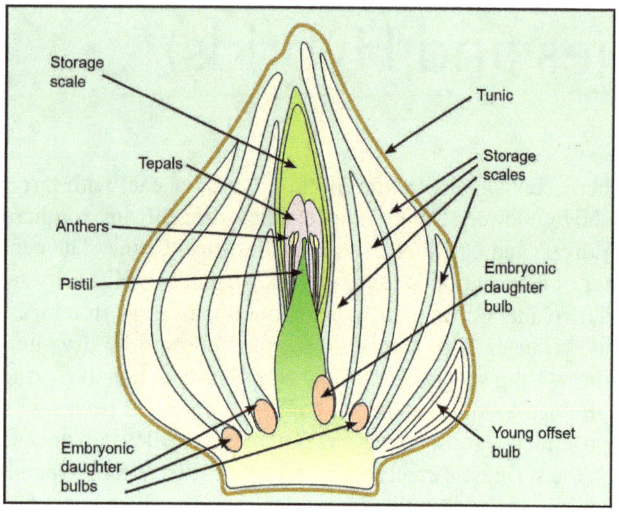


FIGURE 97.2 Anatomy of a sectioned tulip bulb. Most of the bulb is made up of storage scales (specialized leaves). Note the anthers (male parts) and pistil (female part) of the embryonic flower in the center of the bulb. Compare Figure 97.17.

*Source:* **Illustration by Retired Electrician (public domain).**

of the fleshy leaves of the maternal bulb. Typically, two to three offset bulbs are produced annually. Commercial tulip farms excavate and grade all their tulip bulbs at the end of the season. Smaller bulbs usually result in poorer tulips and so are not marketed. Instead, the small bulbs are replanted (for several years, if required) until they result in bulbs large enough to be sold. Tulip bulbs are standardly measured by circumference (in centimeters) in a horizontal plane at the middle height of the bulb. For most cultivars, the recommended ideal top size for tulips is 12 cm or more (some cultivars have naturally smaller bulbs). Tulip bulbs are usually slightly flattened on one side, and on that side there may be a small scar where the bulb was attached to the previous year's flowering stem. A smooth, brownish layer, the tunic, covers the bulb, and tulip breeders favor cultivars with tough tunics to protect them during handling. A long section of a tulip bulb reveals four or five layers (each representing a fleshy leaf) which surround the embryonic shoot that will produce the flowering stem.

Like many other low-growing plants that flower in the early spring and become dormant in early summer, most tulip species take advantage of the limited competition from other plants in the early season. The bulbs supply energy, which allows the leaves to develop quickly before taller plants produce their foliage and shade out the tulips. The presence of both seeds and bulbs provides reproductive insurance for tulips, the bulbs representing a way of staying in an area that has proven to be conducive to tulip growth, and the many seeds representing a means of colonizing distant areas.

The garden tulip is a spring-flowering ornamental, a given plant maintaining its flower in good condition for only a week or so (about the same limit for a bouquet in a vase). An early-season heat wave can cause tulips to dramatically abbreviate their flowering period. However, different cultivars are available, which can provide a range of flowering times extending over 2 months. Planting bulbs in semi-shady areas can lengthen the flowering period.

Although tulips are perennials, in cultivation, most garden tulips are either (1) treated as annuals (the offset daughter bulbs are effectively treated like seeds, to be replanted for the ensuing season) or (2) regarded as short-lived perennials because they deteriorate rapidly and should be discarded after a few years. Tulips are always grown as annuals in the coolest season of otherwise very hot climates, as bulbs become degraded rapidly, and there isn't the necessary cold period for vernalization and induction of flowering. Tulips are also often grown only for one season in cold climates, simply to avoid the work involved in harvesting and storing the offset bulbs.

Almost all colors can be found in tulip flowers, with the most notable exception of blue, the bluest ones showing tinges of violet. Flowers with solid (uniform) color rather than mixed coloration remain the most popular kinds. In a survey of tulip bouquet color preferences by the American Floral Marketing Council, men purchased (in decreasing order) red, yellow, pale pink, hot pink, white, and purple tulips, while women bought pale pink, pale purple, hot pink, peach, red, and yellow. Overall, red is the top-selling color for all tulip sales.

Although the color black is associated with evil and danger (the blackness of night has always seemed threatening to humans), black flowers have been especially desired for centuries. However, black flowers are very rare in nature and, indeed, in cultivation. Black tulips have been elusive. 'Queen of Night' is regarded as the cultivar that is closest to producing black flowers, but the hue is a dark maroon. *The Black Tulip* is a celebrated novel by the famous French writer Alexandre Dumas (1802–1870). The plot concerned a competition in 17th-century Holland to grow the blackest tulip, and how greed for a prize of 100,000 guilders led to disaster for the competitors. At that time, no available tulip was really blackish, and Dumas provided a rather silly explanation for how to produce a black tulip: Simply spray the bulb with black dye, and the flowers will develop black coloration.

FIGURE 97.3 Black tulips. In reality, black tulips are deep maroon or purple.

*Source:* **(a) Photo by Atilin (CC BY SA 3.0). (b) Cultivar Paul Scherer. Photo by Homelka (CC BY SA 4.0). (c) Photo by Virginie-l from Pixabay (public domain).**

### 97.2.1 Classification of Garden Tulips

Most modern tulip cultivars belong to the collective species *T. gesneriana*. Over 3,000 cultivars have been recognized historically, over 2,000 still available. Almost all tulip cultivars are clones, reproduced year after year by genetically identical offset bulbs. Most cultivars are complex hybrids, the result of crossing of older cultivars, some of which may have been interspecific crosses. Informal (i.e., non-Latinized) horticultural classifications of these into groupings trace back centuries. The most commonly used current system of grouping tulip cultivars was devised by England's Royal Horticultural Society and the Royal General Bulb Growers Society in 1967. Fifteen classes were recognized; 11 of the groups deal basically with *T. gesneriana* and are loosely based on height, flower shape, and blooming time, and the remaining four groups are concerned with other species or hybrids. Numerous websites and books reproduce the traditional 15-class tulip classification. However, groups have become increasingly difficult to delimit because of extensive breeding, selection, and hybridization, and many cultivars have ancestry tracing to more than one of the traditional groupings.

FIGURE 97.4 Darwin tulips. These beautiful flowers are mostly obsolete, replaced by "Darwin hybrid" tulips.

*Source:* **(a) Photo by Bob Jenkin (CC BY 2.0). (b) Photo by William Copeland McCalla family fonds/Provincial Archives of Alberta (public domain). (c) Front cover of 1916 Dingee & Conard Company (West Grove, PA) flower catalog (public domain). (d) Front cover of 1928 John Lewis Childs (Floral Park, NY) flower catalog (public domain).**

"Darwin tulips," named after the father of evolution, Charles Darwin, were first marketed at the beginning of

FIGURE 97.5 Tulips with semi-doubled flowers. In such cultivars, some central stamens are present, while all the stamens have been transformed into petals in fully doubled flowers.

*Source:* **(a) Photo by 35731065 from Pixabay (public domain). (b) Photo by Victoria Vasilieva (CC BY SA 2.0).**

FIGURE 97.6 Tulips with doubled flowers (i.e., with numerous petals, the result of transformation of all the stamens).

*Source:* **(a) Cultivar Icoon. Photo by Shutterstock/Ritvars. (b) Cultivar Yellow Pomponette. Photo by Shutterstock/Veronique Stone. (c) Photo by Nursil from Pixabay (public domain). (d) Photo by Ellen26 from Pixabay (public domain).**

the 20th century. These vigorous, tall, well-colored tulips became very popular subsequently, with hundreds of varieties recognized, but widespread hybridization substantially obliterated the distinctiveness of the Darwin tulip class. "Darwin Hybrid Tulips" are hybrids (mostly sterile triploids, that is, with three sets of chromosomes) of Darwin cultivars and *T. fosteriana*. These very popular tulips are characterized by exceptional height, large beautiful flowers, and vigor (the plants sometimes last for several years without annual replanting).

Most cultivated tulips have flowers of traditional appearance, that is, with six unaltered tepals. However, several kinds of tulip flower with modified corolla features have been selected. Each of the following is often recognized as a distinctive tulip class: (1) Doubled flowers (i.e., with extra petals) have been bred, some cultivars having flowers that are remarkably similar to the doubled flowers of peonies and roses. (2) "Parrot" cultivars have laciniate or scalloped petals (also often ruffled, curled, or twisted) that are reminiscent of feathers. (3) Fringed cultivars have petals with fringes (especially on the distal parts). (4) Lily-flowered cultivars have flowers that are quite reminiscent of garden lily flowers, with pointed petals, and they tend to flower fairly late in the season. (5) Most cultivars bear only one flower per stem; "multi-flowering tulips," however, develop a branching stem that bears several flowers (typically three to seven), each on its own stalk.

FIGURE 97.7 Parrot-class tulips (note feather-like petals).

*Source:* **(a) Cultivar Estella Rijnveld. Photo by Rick Ligthelm (CC BY 2.0). (b) Photo by Merre57 from Pixabay (public domain). (c) A painting of the 'King of France' tulip in Hortus nitidissimis (1750–1786, by C. J. Grew and G. D. Ehret). Photo by Shanta Rohse (CC BY 2.0). (d) Photo by MJO from Pixabay (public domain). (e) Photo by Shutterstock/Piaffe Photography. (f) Cultivar Flaming Parrot. Photo by Shutterstock/Alex Manders.**

FIGURE 97.8 Cultivars with fringed petals.

*Source:* **(a) Photo by Samara34 from Pixabay (public domain). (b) Photo by Yeşilelma from Pixabay (public domain). (c) Photo by Plukje from Pixabay (public domain).**

FIGURE 97.9 Lily-flowered tulips (note the long pointed, reflexed tepals and the urn shape of the flowers).

*Source:* **(a) Photo by Michela Simoncini (CC BY 2.0). (b) Photo by Laura Nolte (CC BY 2.0). (c) Photo by Andreas160578 from Pixabay (public domain). (d) Photo by Elstef from Pixabay (public domain).**

## 97.3 HISTORY

Classical Greek, Roman, and Byzantine civilizations were apparently unacquainted with tulips. The first evidence of interest in them traces back at least a thousand years in Central Asia, primarily in Anatolia and Persia. Tiles with pictures of tulips appeared in the 11th century in Anatolia, and garden forms were probably selected by the 12th century. However, few records of tulip cultivation are available prior to the 16th century. From Turkey, tulips were first imported into Europe in the mid-16th century, at the conclusion of which they were particularly grown in Holland. It has been suggested that tulips were introduced into Spain about the end of the 11th century, 500 years before the time that is commonly accepted for the first European introductions.

Tulip cultivars are derived from at least 15 wild species, some more or less directly from one species, many with ancestry tracing to more than one. Central Asia is considered to be the cradle that gave birth to the majority of domesticated tulips. *Tulipa gesneriana* is the principal species from which most garden varieties have been derived. There are some cultivars of *T. fosteriana* (emperor tulip), and also many cultivars which are hybrids of *T. gesneriana* and *T. fosteriana*. In addition, cultivars (and hybrids) of several other wild tulip species are grown as garden ornamentals and are popularly known as "species tulips," less commonly as "botanical tulips." These include *T. bakeri*, *T. clusiana*, *T. greigii*, *T. kaufmanniana* (water lily tulip), *T. tarda*, and others.

### 97.3.1 Tulip Breaking Virus

In 17th-century Holland, tulip plants with unusually colored flowers became very popular. The flowers were characterized by very attractive mosaic color patterns on the petals, often resembling flames, bars, stripes, streaks, or featherings. The plants were referred to as "broken tulips," because their floral patterns were "broken" by comparison with the solid floral colors of most cultivars that were available. Not until the 1920s was it discovered that a virus, tulip breaking virus, is responsible for this condition (the nature of viruses was not appreciably clarified until the 1890s). The symptoms of this virus were first described in 1576 by Carolus Clusius, a Flemish professor of botany at Leiden. Long considered to be the earliest recorded plant virus, it is now thought that the symptoms of another virus, tobacco leaf curl virus, were recorded as early as the 9th century. Tulip breaking virus was being spread among tulips by the peach potato aphid (*Myzus persicae*), which was widespread because of the popularity of peach trees. Tulip color is determined by anthocyanins immersed in a base of yellow-white pigments. The virus irregularly suppresses the anthocyanins, allowing the yellow-white pigments to show through. After it was realized in the 1920s that the virus infection weakened tulips, which consequently produced fewer and smaller offset bulbs, broken tulips were systematically eliminated, along with most of the virus. Plants with multicolored flowers resembling broken tulips are available today, the result of breeding and not the presence of the virus. Some of the classical broken tulip varieties are still extant, but trade in virus-infected plants is forbidden in most countries where tulips are grown (the virus is lethal to lilies, so its presence in gardens is very undesirable). The virus-infected, broken tulips were often referred to as "Rembrandt tulips," presumably a complement to the 17th-century Dutch master, reflecting the apparently artfully painted flowers. (Although Rembrandt tulips were illustrated by many of the famous Dutch painters, Rembrandt himself never painted them.) The phrase "Rembrandt tulips" is also applied to modern, virus-free cultivars with colorful flowers reminiscent of the old broken tulips.

FIGURE 97.10 Plants infected by tulip breaking virus: (a) Painting of a 17th-century "broken" tulip (infected by tulip breaking virus). This particular cultivar (Semper Augustus) was the most celebrated and expensive tulip ever sold – 10,000 guilders for one bulb, debatably equal to as much as $500,000 in modern currency (see text). The cultivar is extinct and only existed in a dozen or so specimens in the 1620s, mostly in the garden of Dr. Adriaen Pauw, a tulip fancier and the director of the new East India Company. To magnify the attractiveness of his plants, he built an elaborate gazebo around them, with mirrors that multiplied the beauty of the flowers.

*Source:* **(a) Photo from Norton Simon Museum. (b) Photo by Shutterstock/Kazakov Maksim.**

FIGURE 97.11 Nineteenth-century illustrations (public domain) of popular old tulip cultivars with floral patterns probably reflecting infection by tulip breaking virus.

*Source:* **(a) Loudon, J. 1841. The ladies' flower-garden of ornamental bulbous plants. (b) Eeden, A. C. van. 1872–1881. *Album van Eeden, Haarlem's flora, afbeeldingen in kleurendruk van verschillende bol: en knolgewassen.* (c) *Flore des serres et des jardins de l'Europe* (Gand, Belgium), vol. 16, 1865–1867.**

FIGURE 97.12 Rembrandt variegated tulips, reminiscent of "broken" tulips (infected by tulip breaking virus), but the floral patterns are determined by genetics.

*Source:* **(a) Photo by Anncapictures from Pixabay (public domain). (b) Photo by Stux from Pixabay (public domain). (c) Cultivar Helmar. Photo by Shutterstock/Alex Manders. (d) Photo by Laila from Pixabay (public domain).**

FIGURE 97.13 Illustrations of the tulip mania period: (a) A 17th-century painting showing the sale of extremely valuable tulip bulbs (artist unknown, painting housed in Musée des beaux-arts de Rennes). (b) An 1882 painting by J.-L. Gérôme illustrating an incident during the 17th-century tulipomania craze in The Netherlands. Soldiers are trampling tulip flowerbeds in a governmental attempt to stabilize the tulip market by limiting the supply.

*Source:* **(a) Photo by Ptyx (CC BY SA 4.0). (b) Photo by Walters Art Museum (CC BY 3.0).**

### 97.3.2 Tulipomania

Tulip bulbs were first imported into Holland from Constantinople in 1593. Amsterdam merchants displayed their success by surrounding their estates with flower gardens and became enchanted with the tulips, which quickly became a status symbol for the aristocratic class. With competition for novelty colors in the early 17th century, especially as exhibited by the broken tulips described in the preceding text, prices for elite specimens climbed to exaggerated values. Tulips often sold for more money than commoners earned in a year. Lured by tales of sudden wealth, people traded their land, livestock, farms, homes, and life savings, sometimes just for a single tulip. By the fall of 1635, rather than purchasing actual tulips, investors turned to offerings in market exchanges, which were much like modern stock markets (the practice came to be known as "wind trade"). The so-called "colleges" in which trading transactions occurred were generally taverns, and it was customary at the conclusion of every trade for a fee (termed "wine money") to be used to buy wine and beer for everyone present. Speculation rose to alarming levels in 1636, and it soon became evident that tulips had been greatly overvalued. In 1637, the tulip market crashed, prices plummeted, speculators went bankrupt, and the frenzy for tulips ended.

The tulip financial folly has become known as "tulip mania," or tulipomania, and represents one of the first "economic bubbles." During the early 17th century, the most expensive tulip bulbs sold for as much as 750 Dutch guilders, alleged by some to be equivalent to about $35,000.00 in modern currency. (The guilder was replaced by the euro in 2002.) A top-selling tulip could purchase the equivalent of 15 fat pigs, 4 metric tons of butter, the work of a skilled laborer for 3 years, or a country estate. Tulipomania is a most interesting story and has become widely cited by economists as representative of the dangers of financial bubbles eventually bursting. However, several recent analyses have suggested that both the purchasing power of fancy tulips and the enthusiasm for unwise investment in them during tulipomania have been significantly exaggerated. In any event, some economists have concluded that foolish investment can be beneficial for investors, so long as there are further investments by even foolisher speculators ("the greater fool theory;" there are analogies with pyramid and Ponzi schemes). Despite the rather shameful aspects, tulipomania may have contributed to the ultimate establishment of The Netherlands as the dominant commercial tulip-growing nation.

A second, albeit smaller, rage for tulips took place in Turkey. The "Tulip Period" or "Tulip Era" refers to a peaceful, prosperous time from 1718 to 1739 in Ottoman history, when a craze for tulips developed among Ottoman court society. Once again, the elite competed for the best tulips. Prices rose dramatically in the late 17th century, peaking in 1726–1727, at which time the state intervened and set limits on tulip prices.

## 97.4 ECONOMIC IMPORTANCE

The tulip is the world's most important ornamental flower cash crop, with more than 10,000 ha in production worldwide. The Netherlands is the most significant producer of tulips, and clarification of the country's name may be helpful for some readers. The Netherlands (strictly the Kingdom of The Netherlands) is frequently referred to as "Holland" in the wide sense, a usage that is often considered undesirable

FIGURE 97.14 Commercial tulip farms (chiefly producing bulbs): (a) In The Netherlands. (b) In The Netherlands. (c) In Skagit Valley, Washington, USA. (d) In Tasmania, Australia.

*Source:* **(a) Photo by Enoch111 from Pixabay (public domain). (b) Photo by Alessandro Vecchi (CC BY 3.0). (c) Photo by Jimdrinkwine from Pixabay (public domain). (d) Photo by Neridabates from Pixabay (public domain).**

FIGURE 97.15 Commercial greenhouse production of potted tulips.

*Source:* (a) Photo by Shutterstock/MakSan. (b) Photo by Shutterstock/Scharfsinn.

FIGURE 97.16 Commercial marketing of tulip bouquets.

*Source:* (a) Photo by Shutterstock/Pressmaster. (b) Photo by Jernej Furman (CC BY 2.0).

FIGURE 97.17 Tulip bulbs: (a) Several bulbs, some sectioned. (b) Bulbs for sale in a shop in The Netherlands.

*Source:* (a) Photo by Mariluna (CC BY 3.0). (b) Photo by Congerdesign from Pixabay (public domain).

except in a historical context. Holland, in the narrow sense, is located in the midwestern, coastal part of The Netherlands and existed as a separate country until 1581. It now constitutes North Holland and West Holland, two of the 12 provinces of The Netherlands. "The Netherlands" is often rendered "the Netherlands," authorities differing on capitalization of the "T." Just "Netherlands" is mostly encountered in lists, although some have argued that "The" should always be omitted.

Almost all tulips are raised from bulbs, which are produced on tulip farms. Tulip bulbs are produced commercially in many lands. The Netherlands has the largest area devoted to tulip farms (over 6,000 ha), followed by the UK, Denmark, and Japan, with a dozen or so other countries producing appreciable acreages. The Netherlands also leads the world in tulip exports, and indeed almost half of the flower bulb area in the country is devoted to tulips. Three billion bulbs are produced every year, two-thirds for export. Why did Holland (western Netherlands) become the predominant tulip grower and trader? It has been suggested that the monotone gray dress that was popular in the 17th century (because of Calvinism's religious intolerance of showy personal displays) predisposed the population to being attracted to the gaudy colors of tulips and led to the Dutch love affair with the flower. The sandy soils of old Holland, when appropriately fertilized, are ideal for growing tulips (offset bulbs develop very well and are easily harvested from sand). During the 1600s, The Netherlands was a major sea power, responsible for about half of Europe's shipping trade, so it was natural to seize on the export of tulip bulbs as a marketing opportunity. And the Dutch were astute traders (most notably in 1626, when Dutch colonists purchased Manhattan Island from Indians for goods estimated to be worth only $24).

There is a huge domestic and international trade in fresh-cut flowering tulip stems sold as bouquets for exhibition in homes. (There is also a much smaller market for potted tulips for both indoor and outdoor display.) Forcing of bulbs to produce cut flowers out-of-season is the basis of this industry and occurs primarily from early December to May in the Northern Hemisphere. Valentine's Day (February 14), Easter (varies between March 22 and April 25), and Mother's Day (different dates; second Sunday in May in Canada and the United States) are key sales periods for cut tulips. Most forcing of tulip bulbs occurs in cool greenhouses, often employing soilless production systems (such as hydroponics). However, in some places, the climate permits tulips to be grown outdoors in raised beds to produce flowers naturally for spring sales. Recently, Southern Hemisphere countries such as Australia, New Zealand, and Chile have taken up cut tulip production to provide an out-of-season supply to the Northern Hemisphere.

## 97.5 ORNAMENTAL USES

Tulips are widely grown as ornamentals in large outdoor flower beds by government agencies and private companies, and by homeowners in private gardens. To a lesser but still significant extent, tulips are marketed as cut flowers for bouquets, and as potted houseplants (usually as gift plants intended to be discarded after flowering).

To encourage offset bulb production, tulip farms cut off the flowers, which are accordingly available to decorate "floral sculptures" and parade floats.

## 97.6 OTHER USES

Up until the 17th century in Europe, gardens were primarily devoted to medicinal and edible plants. When tulips were first introduced to European countries, people sometimes used them medicinally (for example, as a remedy for

FIGURE 97.18 Impressive landscapes featuring tulips: (a) The Eiffel Tower in Paris, viewed from tulip beds in Trocadéro (situated on the opposite side of the Seine River from the Eiffel Tower, Trocadéro is home to spectacular gardens). (b) Old buildings bordered by a huge bed of tulips in Amsterdam, Netherlands. (c) An elegantly designed tulip bed in Kashiwa City, Japan.

*Source:* **(a) Photo by Shutterstock/Neirfy. (b) Photo by Shutterstock/Olena Znak. (c) Photo by Kurofune (CC BY SA 4.0).**

FIGURE 97.19 Spectacular tulip landscape designs at the Keukenhof, a 32 ha ornamental garden in the Netherlands featuring millions of tulips.

*Source:* **(a) Photo by Krysi@ (CC BY 3.0). (b) Photo by Shutterstock/Olena Znak. (c) Photo by Shutterstock/Akshay-PhotOvation.**

flatulence). Tulip flowers have been (and still are) occasionally used in culinary dishes and to flavor wine.

## 97.7 PROBLEMS

During the Second World War, famine compelled some people in The Netherlands to eat tulip bulbs, although they are not safe to consume. Tulip bulbs are considered to be toxic to dogs, cats, and horses, but many mammals appear to be immune. Rabbits and rodents (notably squirrels, mice, and rats) are very fond of the bulbs and are capable of causing extensive damage to tulip beds. Skunks, raccoons, voles, and birds sometimes also consume the bulbs.

FIGURE 97.20 Cut flowers of tulip: (a) A woman with a tulip bouquet. (b) A vase of tulips. (c) A vase of tulips. (d) A woman harvesting tulips from a flower garden. (e) A woman harvesting tulips from a commercial nursery.

*Source:* **(a) Photo by Shutterstock/SvetlanaFedoseyeva. (b) Photo by Buntysmum from Pixabay (public domain). (c) Photo by Shaferlens/Dave Shafer (CC BY 2.0). (d) Photo by JillWellington from Pixabay (public domain). (e) Photo by AndyBellm from Pixabay (public domain).**

FIGURE 97.21 Potted tulip houseplants.

*Source:* **(a) Photo by Tdfugere from Pixabay (public domain). (b) Photo by Koczot from Pixabay (public domain). (c) Photo by Alexas Fotos from Pixabay (public domain).**

Some people are allergic to tulips, suffering dermatitis and breathing problems. Tuliposides and tulipalins are the compounds responsible for the allergies. Tulipalin A is particularly known for inducing occupational dermatitis in tulip bulb sorters and florists; it is present in all parts of the plant, but the highest concentration is in the bulbs.

## 97.8 CULTIVATION

### 97.8.1 Climate and Site

Most tulips are grown from dormant bulbs, usually planted in later autumn, in climates that provide a cold period

FIGURE 97.22 Tulip bulbs forced to produce flowers indoors in soil-less containers.

*Source:* **(a) Photo by Steinchen from Pixabay (public domain). (b) Photo by Shutterstock/New Africa.**

FIGURE 97.23 Living floral sculptures (created by affixing cut flowers to a supporting form): (a, b, c) Whimsical creations made of tulip flowers in May 2012, from the Springfields Festival Gardens, Spalding, Lincolnshire, the site of the largest annual tulip festival in the UK – (a) aviator, (b) snail, (c) teddy bear. (d) A parade float in Haarlem, Netherlands.

*Source:* **(a) Photo by Jim Linwood (CC BY 2.0). (b) Photo by Paul Stainthorp (CC BY SA 2.0). (c) Photo by Jim Linwood (CC BY 2.0). (d) Photo by Shutterstock/Tatiana Popova.**

("cold stratification"), required to break dormancy. Recommendations for the cold requirement vary from 12 to 14 weeks at 0 to 13°C to 8 to 10 weeks at 4 to 7°C. In warm

FIGURE 97.24 Animals consuming tulips: (a) A squirrel eating a tulip bulb. (b) A goat snacking on a tulip.

*Source:* **(a) Photo by Dave Lundy (CC BY ND 2.0). (b) Photo by Shutterstock/Zanna Pesnina.**

climates incapable of providing a sufficient period of cold or for developing interior houseplants, the bulbs need to be chilled (e.g., in a refrigerator) before planting. In climate zones 9 and 10, tulips are usually grown as annuals. Some suppliers furnish pre-chilled bulbs. Too-early planting sometimes results in premature sprouting and death by frost. Although tulips are naturally perennial, so-called "perennial tulips" (also labeled perennializing, reblooming, or naturalizing) are relatively well-suited to developing larger, vigorous new bulbs. Tulips are suited to grow best as perennials in climates with moist, cool, or cold winters and warm, dry summers. Depending on cultivar and climate, tulips can be grown well as perennials in zones 3 to 8. Zones 4 to 6 are best for growing tulips as perennials. Nevertheless, all tulip beds eventually require rejuvenation with newly acquired bulbs, and often, tulips are grown as annuals regardless of climate.

It is helpful to mulch tulip beds for the winter using material that does not pack down (which could inhibit emergence in the spring). Remove or reduce mulch in the spring; thick covers can suppress growth. Keep in mind that mulches can be an invitation to small mammals to nest in the location and devour the bulbs.

Plant tulips in full sun, preferably where they will receive a minimum of 4 to 6 hr of bright sunlight when the foliage is green. Light shade is tolerated, and some midday shade is often recommended to avoid drought stress at the hottest time of the day. Tulips can be planted near deciduous trees and shrubs (which won't shade the flowers in the early spring). Shady locations prevent vigorous development of offset bulbs, which are needed for reproduction in the following season. Plant in sheltered locations, but not near heated basements. Most garden tulips are ideal for beds and borders, and many of the smaller species tulips are ideal in rock gardens. Dwarf forms need to be located in containers or in locations that facilitate viewing.

For outdoor display, tulips are almost always laid out in beds or borders, except for dwarf *Tulipa* species grown in rock gardens. Single tulips are not particularly attractive, and groupings of not less than a dozen or so are

FIGURE 97.25 Tulips in late-spring snow.

*Source:* (a) Photo by Ellen26 from Pixabay (public domain). (b) Photo by u_poq98d8i from Pixabay (public domain).

recommended. Planting tulips in grass is strongly discouraged. Planting in mixed beds with other ornamental species is often done to avoid the unkempt and unoccupied appearance of the bed after the tulips die down.

Ideally, to prevent buildup of disease organisms, tulips should be planted in a new location each year, rotating with other plants, waiting 5–7 years before planting again. Some gardeners will remove the top layer of soil before replanting tulips in the same area. For most people, finding a new location for tulips after they have been grown in the same site for several years is a good idea. Excavate and discard diseased plants as soon as possible.

### 97.8.2 Propagation and/or Purchasing

Almost all tulips are grown today from bulbs, not from seeds. Tulips require vernalization to produce offset bulbs. This is achieved in nature by cold-season dormancy. When tulip bulbs are grown in warm climates that lack sufficiently cold winters, offset bulbs are fewer and smaller and are of limited suitability for production of tulips in the following season. In such warm climates, tulips are often grown as annuals, the plants discarded after they flower. Most tulip plants tend to produce poorer plants the second season, even if the climate is ideal, unless the offset bulbs are excavated and replanted. Offset tulip bulbs can be employed to grow new plants in the next season. The offsets can be lifted out of the soil after the leaves have died, and stored in a cool (preferably 2–10°C), well-ventilated place until replanted in the late fall (store in paper bags, never in plastic). However, most people prefer to simply buy new bulbs.

Bulbs should be purchased that are "top size" (with a circumference of at least 11 cm, although the bulbs of species tulips are often only 7–9 cm). They should be firm to the touch and free of dents, blemishes, and mold (especially on the basal plate). Small nicks and loose skin are not significant.

Robust cultivars capable of persisting for several seasons are best, although degeneration of the bed is inevitable (in particular, small offset bulbs tend to produce inferior, crowded plants). Tulips with considerable parentage of *T. fosteriana* (such as the Darwin hybrid tulips) and species tulips (such as Greigii and Kaufmanniana tulips) are known for their ability to perennialize, that is, to regrow and rebloom for several years after planting.

Flowering time of cultivars should be considered. Generally, the labels of packaged tulips will indicate whether the cultivars are intended to flower in early spring, early to mid-spring, mid-spring, or late spring. This information should also be available on the pictured example that accompanies bulbs purchased in bulk from bins.

### 97.8.3 Planting

Plant bulbs in the fall, before deep frost hardens the ground, but when the weather will remain sufficiently cool, to prevent initiation of shoot growth. Usually, in climate zones 3 to 5, September is appropriate; in zones 6 and 7, October is suitable; and in zones 8 to 9, November or December is best.

Tulip bulbs should be planted pointed end up. Purchased bulbs should be accompanied by recommendations for planting depth. Generally, plant at least 10–15 cm deep, but larger bulbs are best planted deeper (rule of thumb: plant at a depth that is three times the length of the bulb). Deeper planting can discourage rodents from digging up bulbs. After planting, thoroughly water the soil. Bulbs are usually laid out 10–16 cm apart, depending on the size of the cultivar (not the size of the bulb). Closer planting can produce more impressive displays but smaller plants. To produce uniform height, bulbs should be about the same size. Beds showing flowers of one solid color are very attractive, as are beds of contrasting color blocks, but randomly or systematically mixing colors can also be attractive.

### 97.8.4 Soil and Moisture Conditions

The soil should be deep, fertile, humus-rich, and well-drained, with a pH 6 to 7. Sandy soil is best (but compost should be added), and heavy clay soils are unsuitable (unless amended with organic material or vermiculite). If available soil is quite inappropriate, consider bringing in soil and creating raised beds. The soil should be thoroughly tilled to a depth of 30–45 cm before planting. Tulips should be watered well right after fall planting to encourage root formation before freeze-up (but very late planting will merely delay emergence in the spring). Tulips need regular moisture from first appearance in the spring until the foliage yellows in the fall. Once the plants are dormant, discontinue watering. Avoid overwatering. Keeping the foliage and flowers damp for prolonged periods can encourage fungal disease, such as botrytis. Soaker hoses are often preferable to overhead watering.

### 97.8.5 Maintaining Plants

To encourage development of offset bulbs, work some quality compost and a slow-release nitrogen fertilizer into the bed before planting (do not place fertilizer in direct contact with the bulbs). To encourage reblooming of established plants, apply a complete fertilizer with fast-release nitrogen in the spring, just as shoots are emerging. Or just use a "bulb food" intended for tulips. Highly nitrogenous fertilizers and fresh manure should not be applied, as they encourage disease and vegetative growth rather than flowering.

When maintaining the tulips as perennials, "deadhead" after flowering, cutting away fading flowers and developing seed heads, which deplete food reserves from the developing offset bulbs (a third of the bulb's energy can be diverted into seed formation). Also, remove the leaves after flowering, waiting until they are quite yellow or brown, and dry or limp (and easily pulled free), as their photosynthesis provides storage carbohydrates for the developing offset bulbs. Tulip may need about 6 weeks to die back before leaves may be removed.

## 97.9 FASCINATING FACTS, FABLES, FOLKLORE, AND FRIVOLITY

- In the 19th and early 20th centuries, characters (usually females) in illustrated children's books were often shown in flower costumes. As one of the most common of ornamental plants, tulip costumes were often shown.
- The tulip is recognized as the official flower of Turkey (and is accepted as the emblem of Istanbul). It is also considered to be the unofficial national flower of The Netherlands. Several cities have adopted tulips as their emblem (either officially or by common consent). The tulip was declared to be the official flower of Ottawa, Canada, in 2001. The cultivar 'Orange Wonder' became the emblem of Albany, New York, in 1948. Japanese cities recognizing tulips as their emblem include Nigata City and Yubetsu Town.
- In 1980, J. W. S. Van der Wereld, a Dutch horticulturalist who had Parkinson's disease, produced a red-and-white tulip, which in 1981 he named the 'Dr. James Parkinson' tulip, in honor of the man who described the medical condition. Subsequently, in 2005, the "Red Tulip" was recognized as the worldwide symbol of the disease at the 9th World Parkinson's Disease Day Conference in Luxembourg.
- Calvinism is a major branch of Western Christianity that follows practices of John Calvin (1509–1564) and other Protestant Reformation theologians who separated from the Roman Catholic Church. Five chief doctrinal points of Calvinist theology are sometimes referred to by the acronym TULIP. This stands for "total depravity" (everyone is born into sin), "unconditional election" (God elects to give mercy to those receiving salvation through Christ), "limited atonement" (atonement is limited to some), "irresistible grace" (the irresistible call to grace by God), and "perseverance of the saints" (all who are truly saintly will be saved).
- Sultan Ahmed III (1673–1736), who came to rule the Ottoman Empire in 1703, was a tyrant who lived in a luxurious lifestyle while his people suffered. He was quite obsessive about tulips, insisting that his guests harmonize their clothing with the flowers at his tulip festivals. Showing up in the wrong costume was a very bad idea, since his gardeners also served as state executioners.
- Sheik Mohammed Lalizari (*lalizari* means "tulip lover") was another 18th-century Turkish ruler with a fanatical enthusiasm for tulips. A manuscript he authored mentioned 1,323 kinds of tulips, and he had more than 500,000 plants in his gardens.
- In The Netherlands, presenting a bouquet of less than ten tulips as a gift is considered to be a gross social blunder.
- It has been estimated that the 3 billion tulip bulbs produced annually in The Netherlands, if planted 10 cm apart, would circle the globe at the equator seven times.
- It is possible to grow a tulip from bulb to flowering plant in total darkness.
- Cut tulip stems turn toward the light, so fresh flowers in a vase may become awkwardly displayed. The stems often grow 2.5 cm in length while in the vase, causing drooping over the edge of the vase. To maintain an attractive arrangement, rotate the vase a quarter turn each day and straighten any drooping stems.
- "Tiptoe through the Tulips," also known as "Tiptoe Thru' the Tulips with Me," is a popular song published in 1929. American singer Tiny Tim resurrected the popularity of the song in 1968. The expression "Tiptoe through the tulips" indicates the fragility of the flowers, and indeed, public tulip displays sometimes have signs discouraging walking on the plants.

FIGURE 97.26 Illustrations (public domain) of characters wearing tulip costumes.

*Source:* (a) Grandville, J. J. 1867. *The Flowers Personified*. (b) Gordon, E., and M. T. Rose. 1910. *Flower Children*. P. F. Volland, Chicago. (c) A 1910 vintage postcard.

FIGURE 97.27 A sign in a public tulip garden requesting visitors to avoid "tiptoeing through the tulips."

*Source:* **Photo by Valerie Fenwick (CC BY SA 2.0).**

FIGURE 97.28 Scenes from the Ottawa (Canada) Annual Tulip Festival: (a) Roadside tulip bed. (b) Red tulips beneath the Peace Tower on Parliament Hill. (c) Some of the 150 5 ft fiberglass tulips created for the 2008 Canadian Tulip Festival on the occasion of the 150th anniversary of Ottawa being chosen as Canada's capital in 1858.

*Source:* **(a) Photo by Paul Shannon (CC BY SA 2.5). (b) Photo by Kasia/Catherine Bulinski (CC BY ND 2.0). (c) Photo by ElPadawan (CC BY SA 2.0).**

- During the Second World War, when The Netherlands was occupied by the Nazis, the princesses of the Dutch Royal family, Princess Juliana (queen between 1948 and 1980), and her daughters, Beatrix and Irene, took refuge in Ottawa, Canada's capital, from 1940 to 1945. A third princess, Margriet, was born in Ottawa's Civic Hospital in 1943 (Margriet was named for the "marguerite," species of daisylike flowers worn as symbols by the Dutch resistance). The Canadian government proclaimed that the maternity ward was "extraterritorial," as sole Dutch citizenship was necessary for the purpose of royal succession. To thank Canada for providing a safe haven and to acknowledge the important role of Canadian troops in liberating the Netherlands, in the fall of 1945, Princess Juliana presented Ottawa with 100,000 tulip bulbs and promised an annual gift of tulips for the rest of her life. Upon her death in 1994, an agreement between Canada and The Netherlands provided that the annual gift of 20,000 bulbs would continue in perpetuity. Several thousand of the bulbs are dedicated to the Civic Hospital where Margriet was born (so was the first author of this book). A huge tulip festival is held in Ottawa annually.

## KEY PUBLICATIONS

Ágoston, J., A. Almási, K. Salánki, and L. Palkovics. 2020. Genetic diversity of potyviruses associated with tulip breaking syndrome. *Plants* 9(12): 1807. https://doi.org/10.3390/plants9121807.

Baker, C., W. Lemmers, E. Sweeney, and M. Pollan. 1999. Tulipa*: A Photographer's Botanical*. Artisan, New York. 296 pp.

Bermejo, J.E.H., and E.G. Sánchez. 2009. Tulips: An ornamental crop in the Andalusian middle ages. *Economic Botany* 63: 60–69.

Botschantzeva, Z.P. 1982. *Tulips: Taxonomy Morphology, Cytology, Phytogeography and Physiology*. A.A. Bolkema, Rotterdam, The Netherlands. 230 pp. [Translated from the 1962 Russian edition.]

Christenhusz, M.J., R. Govaerts, J.C. David, T. Hall, K. Borland, P.S. Roberts, et al. 2013. Tiptoe through the tulips – cultural history, molecular phylogenetics and classification of *Tulipa* (Liliaceae). *Botanical Journal of the Linnean Society* 172: 280–328.

De Hertogh, A.A. 1996. *Holland Bulb Forcer's Guide*, 5th edition. The International Flower-Bulb Center, Hillegom, The Netherlands. 587 pp.

De Hertogh, A.A., L.H. Aung, and M. Benschop. 1983. The tulip: Botany, usage, growth, and development. *Horticultural Review* 5: 45–125.

Goldgar, A. 2007. *Tulipmania: Money, Honor, and Knowledge in the Dutch Golden Age*. University of Chicago Press, Chicago. 425 pp.

Hall, A.D. 1929. *The Book of the Tulip*. Martin Hopkinson, London. 224 pp.

Hall, A.D. 1940. *The Genus* Tulipa. The Royal Horticultural Society, London. 171 pp.

Heath, B. and B. Heath. 2001. *Tulips for North American Gardens*. Bright Sky Press, New York. 144 pp.

Hirschey, M. 1998. How much is a tulip worth? *Financial Analysts Journal* 54(4): 11–17.

King, M. 2005. *Gardening with Tulips*. Timber Press, Portland, OR. 192 pp.

Liddle, S. and M. Malak. 1999. *Canada's Capital Region – A Capital of Tulips*. National Capital Commission, Ottawa, ON. 92 pp.

Lodewijk, T. 1979. *The Book of Tulips*. Vendome Press, New York. 128 pp. [Translated from the 1978 Dutch edition.]

Marasek-Ciolakowska, A., M.S. Ramanna, and J.M. van Tuyl. 2011. Introgression of chromosome segments of *Tulipa fosteriana* into *T. gesneriana* detected though GISH and its implications for breeding virus resistant tulips. *Acta Horticulturae* 886: 175–182.

Miller, W.B. 2002. A primer on hydroponic cut tulips. *Greenhouse Product News* 12(8): 8–12.

Pavord, A. 1999. *The Tulip*. Bloomsbury, New York. 388 pp.

Pollan, M. 2002. *The Botany of Desire*. Random House, New York. 271 pp.

Scheepen, J. van. 1995. Cultivar groups in the genus *Tulipa* L. (Liliaceae). *Acta Horticulurae* 413: 137–143.

Scheepen, J. van. 1996. *Classified List and International Register of Tulip Names*. Royal General Bulbgrowers' Association, Hillegom, The Netherlands. 623 pp.

Stefanaki, A., T. Walter, and T. van Andel. 2022. Tracing the introduction history of the tulip that went wild (*Tulipa sylvestris*) in sixteenth-century Europe. *Scientific Reports* 12(1): 9786. www.nature.com/articles/s41598-022-13378-9.

Van Eijk, J.P., L.W.D. van Raamsdonk, W. Eikelboom, and R.J. Bino. 1991. Interspecific crosses between *Tulipa gesneriana* cultivars and wild *Tulipa* species: A survey. *Sexual Plant Reproduction* 4: 1–5.

Van Raamsdonk, L.W.D. and T. de Vries, T. 1996. Cultivar classification in *Tulipa* L. (Liliaceae). *Acta Botanica Neerlandica* 45: 183–198.

Van Raamsdonk, L.W.D., W. Eikelboom, T. De Vries, and T.P. Straathof. 1997. The systematics of the genus *Tulipa* L. *Acta Horticulturae* 430: 821–828.

Veldkamp, J.F. and B.J.M. Zonneveld. 2012. The infrageneric nomenclature of *Tulipa* (Liliaceae). *Plant Systematics and Evolution* 298: 87–92.

Wilford, R. 2006. *Tulips, Species and Hybrids for the Gardener*. Timber Press, Portland, OR. 211 pp.

Wilson, B. 2022. *Tulipa: The Taxonomy and Evolutionary History of the Genus and Its Impact on Conservation Priorities in Central Asia*. Ph.D. thesis, University of Cambridge Repository. 258 pp. https://doi.org/10.17863/CAM.94432.

Zonneveld, B.J.M. 2009. The systematic value of nuclear genome size for "all" species of *Tulipa* L. (Liliaeae). *Plant Systematic and Evolution* 281: 217–245.

## Key Cultivation Websites

Amsterdam Tulip Museum. https://amsterdamtulipmuseum.com/.

Dutch Grown. *How to Grow Tulips*? www.dutchgrown.com/blogs/the-dutchgrown-blog/how-to-grow-tulips.

GardenersWorld.com. *How to Grow Tulips*. www.gardenersworld.com/how-to/grow-plants/how-to-grow-tulips/.

Iowa State University, Extension. *All About Tulips*. https://hortnews.extension.iastate.edu/all-about-tulips.

The Spruce. *How to Plant, Grow, and Care for Tulip Flowers*. www.thespruce.com/tulips-planting-and-growing-tulips-1402137.

The Spruce. *When to Plant Tulip Bulbs to Get the Most Beautiful Spring Blooms*. www.thespruce.com/when-to-plant-tulip-bulb-for-spring-blooming-8349062.

University of Maryland – Extension. *Production of Tulips as Cut Flowers*. https://extension.umd.edu/resource/production-tulips-cut-flowers.

WikiHow. *Everything You Need to Know About Growing Tulips for Beautiful Blooms*. www.wikihow.com/Care-for-Tulips.

WikiHow. *How to Force Tulip Bulbs in Water*. www.wikihow.com/Force-Tulip-Bulbs-in-Water.

WikiHow. *How to Grow Tulips*. www.wikihow.life/Grow-Tulips.

WikiHow. *How to Grow Tulips in Pots*. www.wikihow.com/Grow-Tulips-in-Pots.

WikiHow. *How to Plant Tulip Bulbs*. www.wikihow.com/Plant-Tulip-Bulbs.

WikiHow. *How to Take Care of Fresh Tulips in a Vase*. www.wikihow.com/Care-for-Fresh-Cut-Tulips.

# 98 Wallflower (*Erysimum cheiri*)

**Scientific Names:** The genus name *Erysimum* originates from the classical Greek *eryo*, "to draw out," in allusion to the ability of some of the species to produce blisters (an occasional name for wallflower is blister cress). Alternatively, the root Greek word has been interpreted as *eryso*, "to ward off or cure," a supposed reference to medicinal properties. Previously, *E. cheiri* was known as *Cheiranthus cheiri*, from the Greek *cheir*, "hand," a reference to an old custom dating to the Middle Ages whereby wallflowers were carried in small bouquets by hand to festivals and other events. Sometimes the scientific name is presented as *Erysimum* × *hybrida* to emphasize that hybridization with other species has contributed to modern cultivars.

**Common Names:** The common name wallflower is based on the habit of the plant of growing against walls, an adaptation exhibited in the wild by growing against steep hills and cliffs. True to its name, it is sometimes found growing in small pockets of soil on stone walls. The name gilliflower or gilyflower is sometimes applied to wallflowers, but also to several other ornamental plants.

FIGURE 98.2 Displays of flowering wallflowers (*Erysimum cheiri*).

*Source:* **(a) Photo by Shutterstock/Eduard Valentinov. (b) Cultivar Red Jep. Photo by Acabashi (CC BY SA 4.0). (c) Photo by Shutterstock/Vipul 1989. (d) Cultivar Hilden. Photo by Notafly (CC BY SA 4.0). (e) Cultivar Orange Bedder. Photo by David Monniaux (CC BY SA 3.0). (f) Photo by Florapic (CC BY SA 4.0).**

## 98.1 INTRODUCTION

The wallflower is an old-fashioned but still popular spring-flowering ornamental widely planted in gardens for its fragrant flowers, available in a wide range of colors. The plant is tough, requiring significantly less care than most other low-growing plants valued for their blooms.

## 98.2 OVERVIEW

*Erysimum* is a Northern Hemisphere genus of about 150 species, most of which are in Eurasia. Several are cultivated, but *E. cheiri* is the predominant species that is grown. The wallflower is commonly termed an herb, although it can develop into a woody subshrub. It is perennial, frequently grown as a biennial, and most often as an annual. The flowering stems are 15–80 cm tall. The plant is a native of southern Europe (often, Greece and the Aegean area are identified as the original homeland) but grows widely elsewhere as an escape from cultivation. The leaves are narrow, up to 15 cm long. The foliage is evergreen in areas where the plants can grow overwinter but is killed back by frost at the transition of climate zones 7 and 8 (the plants may overwinter as biennials in zones 6 and 7). The flowers are fragrant; brown, burgundy, cream, maroon, orange, purple, red, yellow, or white; bicolored and/or doubled in some of the many cultivars. Depending on variety, flowering occurs in the early spring to early summer.

FIGURE 98.1 Wallflowers (*Erysimum cheiri*) growing on walls.

*Source:* **Photos from Pixabay (public domain) by (a) 445693 and (b) Markus Baumeler.**

## 98.3 HISTORY

Wallflowers became popular in Great Britain at least by the 16th century, and in early times in much of Europe. Of the more than 150 species in the genus *Erysimum*, *E. cheiri* is the principal garden plant, but *E. linifolium* is also grown. Cultivar Bowles Mauve, with pale purple flowers, is the most popular variety of *E. linifolium*.

## 98.4 ECONOMIC IMPORTANCE

Wallflowers are garden bedding staples, valued by the horticultural industry in temperate regions of the world.



DOI: 10.1201/9781003586227-98

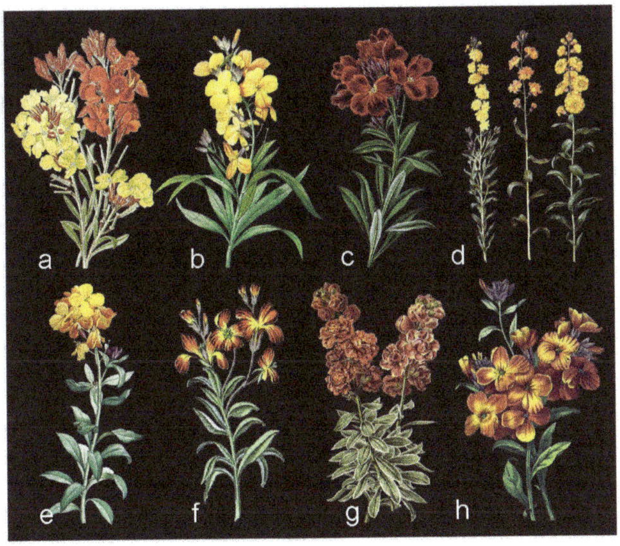

FIGURE 98.3 Paintings of wallflower (*Erysimum cheiri*) from old garden catalogues and botanical publications (public domain).

*Source:* (a) Step, E., D. Bois, B. Herincq, and W. Watson. 1896. *Favourite Flowers of Garden and Greenhouse*, vol. 1. Frederick Warne, New York. (b) Redouté, P. J. 1833. *Choix des plus belles fleurs et des plus beaux fruits*. Ernest Panckoucke, Paris. (c) Bessa, P. (artist). In *Herbier général de l'amateur*, vol. 8. Paris. (1817–1827). (d) Holtzbecker, H.-S. Between 1649 and 1659. *Gottorfer Codex* (a four-volume collection of paintings, housed in Denmark). (e) Redouté, P. J. 1805. *La Botanique de J. J. Rousseau*. Paris. (f) Anonymous. 1836. *Flora and Thalia; or, Gems of flowers and poetry*. Carey, Lea, and Blanchard, Philadelphia. (g) *Revue horticole, Librairie agricole de la maison rustique*, Paris, 1869. (h) Drapiez, P. A. J., and M. Bory de Saint-Vincent. 1837. *Dictionnaire classique d'histoire naturelle*. Meline, Cans, Brussels.

## 98.5 ORNAMENTAL USES

Wallflowers are primarily garden plants grown in beds, in borders, in rock gardens, and along paths. They are also employed as a cut flower. Some sprawling cultivars can be employed as groundcover in warm climates. Wallflowers are frequently planted among spring-flowering bulbs, enhancing the overall display.

## 98.6 OTHER USES

Wallflowers are extensively employed in herbal medicine, especially in Asia.

## 98.7 PROBLEMS

Wallflowers do not overwinter in cold areas (where they are usually grown as annuals), and when grown as perennials, they tend to expire in a few years at most. *Erysimum* species are reputedly sensitive to fungal and bacterial infection and are often discarded after flowering to prevent disease from developing.

FIGURE 98.4 *Erysimum linifolium*: (a) Cultivar Bowles Mauve. (b) A variegated cultivar. (c) A female cuckoo bumblebee (*Bombus vestalis*) visiting flowers of *E. linifolium* in England.

*Source:* (a) Photo by Shutterstock/John B Hewitt. (b) Photo by Raffi Kojian (CC BY SA 3.0). (c) Photo by Martin Cooper (CC BY 2.0).

FIGURE 98.5 Garden displays of wallflower (*Erysimum cheiri*).

*Source:* (a) Photo by Shutterstock/Mabeline72. (b) Photo by Jean Jones (CC BY 2.0). (c) Photo by Shutterstock/Peter Turner Photography.

## 98.8 CULTIVATION

### 98.8.1 Climate and Site

In warm climates (zones 8 to 10), wallflowers can bloom nearly year-round and may be grown as short-term perennials. In zones 5 and warmer, they can be grown as a biennial, but in zones 6 and cooler, they are mostly grown as annuals. Wallflower is adapted to full sun and part shade. Because of the possibility of diseases persisting in the soil, sites that have recently grown other species in the cabbage family (Brassicaceae) are best not planted with wallflower.

### 98.8.2 Propagation and/or Purchasing

Wallflower is reproduced both by seed and stem cuttings.

### 98.8.3 Planting

Seedlings established indoors in one's residence or available from garden centers can be planted after danger of frost. In warm areas (zones 8 to 10), seeds can be planted in the fall for spring germination. Spacing (or thinning out of seedlings produced from seeds planted in the garden) of 15 to 45 cm is recommended depending on cultivar.

### 98.8.4 Soil and Moisture Conditions

In nature, the wild plants are adapted to survive in dry, shallow, rocky soils. In gardens, the plants are not particular with regard to the texture of the soil (sand, loam, or clay), but good drainage is very important. The recommended soil pH is 6.1 to 7.3, but wallflowers are noted for their tolerance of basic soils (up to pH 9.0). Some authorities recommend not fertilizing at all; others suggest adding compost to the soil before planting and fertilizing lightly thereafter.

### 98.8.5 Maintaining Plants

Deadheading is recommended to prolong flowering.

## 98.9 FASCINATING FACTS, FABLES, FOLKLORE, AND FRIVOLITY

- As noted earlier, wallflower is sometimes called gilliflower or gillyflower, but so are other plants with fragrant flowers. The name is based on the French *giroflée*, tracing to the Greek *karyophyllon*, applied to clove (dried flower buds of *Syzygium aromaticum* used as a spice) because of its scent. It appears that in medieval times in England, tenant farmers often paid their rent with spices sometimes termed gilliflowers.
- It is believed that wallflowers were planted near windows on the walls of castles and manor houses so that the fragrance would enter the buildings and hide the less-desirable odors present.
- Applied to people, a *wallflower* is a shy introvert who avoids interacting with people at social events. The application of the flower's name to this bashful behavior has been explained as reflecting the appearance of the plant growing against a wall, paralleling a shy person's retreating from others until (s)he is against a wall.

FIGURE 98.6 A human wallflower.

***Source:* Photo by Shutterstock/GoodStudio.**

## KEY PUBLICATIONS


Al-Shehbaz, I.A. *Erysimum*. Flora of North America online. www.efloras.org/florataxon.aspx?flora_id=1&taxon_id=112142.

Erum, A., A. Mohd, M.A. Jafri, M.A. Ahmed, and A.W. Yousuf. 2017. Phytochemical and ethnopharmacological review of "Tudri surkh" (*Cheiranthus cheiri*). *World Journal of Pharmaceutical Research* 6: 352–359.

Fakhrzad, F., A. Jowkar, A. Shekafandeh, and M.J. Kermani. 2023. Tetraploidy induction enhances morphological, physiological and biochemical characteristics of wallflower (*Erysimum cheiri* (L.) Crantz). *Scientia Horticulturae* 308: 111596. https://doi.org/10.1016/j.scienta.2022.111596.

Gómez, J.M. and F. Perfectti. 2010. Evolution of complex traits: The case of *Erysimum* corolla shape. *International Journal of Plant Sciences* 171: 987–998.

Keeler, K. 2022. *The Attractive Wallflowers, Genus Erysimum*. https://khkeeler.blogspot.com/2022/06/the-attractive-wallflowers-genus.html.

Moazzeni, H., S. Zarre, B.E. Pfeil, Y.J.K. Bertrand, D.A. German, I.A. Al-Shehbaz, et al. 2014. Phylogenetic perspectives on diversification and character evolution in the species-rich genus *Erysimum* (Erysimeae; Brassicaceae) based on a densely sampled ITS approach. *Botanical Journal of the Linnean Society* 175: 497–522.

Mosleh, G., B. Parmis, A. Azadi, Z. Abolhassanzadeh, S.V. Hosseini, and A. Mohagheghzadeh. 2019. Wallflower (*Erysimum cheiri* (L.) Crantz) from past to future. *Research Journal of Pharmacognosy* 6: 85–95.

Osuna-Mascaró, C., R. Rubio de Casas, J.M. Gómez, J. Loureiro, S. Castro, J.B. Landis, et al. 2023. Hybridization and introgression are prevalent in Southern European *Erysimum* (Brassicaceae) species. *Annals of Botany* 131: 171–184.

Ouarmim, S., C. Dubset, and E. Vela. 2013. Morphological and ecological evidence for a new infraspecific taxon of the wallflower *Erysimum cheiri* (Brassicaceae) as an indigenous endemism of the southwestern Mediterranean. *Turkish Journal of Botany* 37(6): 8. https://doi.org/10.3906/bot-1212-4.

Weeks, S. 2010. Perennial Erysimum. *Plantsman* 9(1): 32–37.


### Key Cultivation Websites


Better Homes & Gardens. *How to Plant and Grow Wallflower*. www.bhg.com/gardening/plant-dictionary/annual/wallflower/.

Garden Chronicle. *How to Grow* Erysimum – *Wallflower*. https://gardenchronicle.com/how-to-grow-erysimum-wallflower/.

Gardeners HQ. Erysimum *Plant Growing & Care Guide for Gardeners*. www.gardenershq.com/Erysimum-Alpine-Wallflower.php.

GardenersWorld.com. *How to Grow Wallflowers*. www.gardenersworld.com/how-to/grow-plants/how-to-grow-wallflowers/.

Horticulture Magazine. Erysimum *Wallflowers*. https://horticulture.co.uk/erysimum/.

Plantura Magazine. *Wallflowers: Planting, Pruning & Overwintering*. https://plantura.garden/uk/flowers-perennials/wallflowers/wallflowers-overview.

The Spruce. *How to Grow and Care for Wallflower Plants*. www.thespruce.com/wallflower-plant-profile-4799246.

# 99 Water Lily (*Nymphaea* Species and Hybrids; *Victoria amazonica*)

**Scientific Names:** The genus name *Nymphaea* is derived from the Greek *nymphaia* and the Latin *nymphaea*, the ancient names for water lily, inspired by the nymphs of Greek and Roman mythology. Nymphs were usually maidens associated with nature. For example, the Naiads were freshwater nymphs. The genus name *Victoria* commemorates Queen Victoria (1819–1901; the genus was named for her in 1837, the year she became queen of the UK and Ireland).

**Common Names:** Water lily is also spelled *water-lily* and *waterlily*. Some unrelated plants have adopted the lifestyle of growing rooted in the underwater mud of freshwater and producing emergent or floating leaves and flowers. Species of *Nymphoides* (Menyanthaceae family), for example, are sometimes cultivated as "water lilies." However, the best-known water lilies are in the water lily family (Nymphaeaceae). This contains six genera (*Barclaya, Euryale, Nuphar, Nymphaea, Ondinea*, and *Victoria*), containing about 70 species. The genus *Nelumbo* (with just two species) was once placed in the water lily family (which it very closely resembles) but is now placed in its own family, the Nelumbonaceae. Sacred lotus (*Nelumbo nucifera*) is still often called a water lily, and because it is the most extensively grown species of all plants called water lilies, it is treated in a separate chapter in this book. The name "lotus" is applied not just to sacred lotus and its relative American lotus but also to some species of *Nymphaea* (the principal "water lilies" discussed in this chapter). For example, the "blue lotus of the Nile" and the "blue lotus of India" are *Nymphaea caerulea* and *Nymphaea nouchali*, respectively. As clarified in the chapter on sacred lotus, the name lotus is also applied to quite unrelated plants.

FIGURE 99.1 *Nymphaea* water lilies and the water nymphs that inspired their name: (a) *Hylas and the Water Nymphs*, a painting (public domain) by British artist Henrietta Rae (1859–1928). According to a Greek and Roman legend, the youth Hylas was abducted by water nymphs. (b) Illustration by Iaroslava Daragan.

*Source:* **(a) Public domain. (b) Photo by Shutterstock/Iaroslava Daragan.**

## 99.1 INTRODUCTION

As noted earlier, "water lilies" is a general phrase applied to a variety of plants, but particularly to three cultivated genera, including the sacred lotus (*Nelumbo nucifera*), treated in its own chapter, and to *Victoria* and *Nymphaea* of the Nymphaeaceae, treated in this chapter. Although many gardeners will not grow water plants because they seem too demanding, water gardens are nevertheless popular, and the water lilies are justly termed the queens of ornamental water plants. Water lilies are the stars of large, expensive constructed artificial ornamental pools as well as natural ponds that have been adapted to grow plants. However, as discussed later, the average person can easily grow water lilies in small containers, and cultivars are available for most climates.

## 99.2 OVERVIEW

Plants in the water lily family are aquatic herbs that are rooted in soil which is usually covered by water (they sometimes grow in mud), and they produce leaves and flowers which float on or emerge above the surface (the leaf blades sometimes develop underwater). The rooted base produces rhizomes (usually defined as belowground horizontal stems), stolons (which are similar, but aboveground), or tubers (swollen stem tissue for storage, typically of starch reserves). All of these serve to produce plants (with foliage and flowers) and to overwinter the plants, normally protected underwater. The species are almost all perennials, and they occur very widely, both in temperate and tropical regions. The flowers are on long stalks and are frequently big and seem to have many petals (technically, "tepals," a term applied collectively to sepals and petals) and many stamens. Two of the genera in the Nymphaeaceae, *Victoria* and *Nymphaea*, are widely cultivated as aquatic ornamentals.

The genus *Victoria* has three species. The giant water lily (*Victoria amazonica*), the principal cultivated species, has "lily pads" (floating leaves) that are up to 3 m across and are often exhibited in large greenhouses (they are too large and difficult for most gardeners). The Bolivian water lily (*V. boliviana*) has the world's largest water lily leaf.



DOI: 10.1201/9781003586227-99

FIGURE 99.2 Giant Amazon water lily (*Victoria amazonica*): (a) Huge floating leaves of Amazon water lily dominating the surface of a large water garden in a public greenhouse. (b) A heron perching on a giant Amazon water lily leaf.

*Source:* **(a) Photo by Alex Lomas (CC BY 2.0). (b) Photo by Ludwig Bickel from Pixabay (public domain).**

FIGURE 99.3 (a) Illustrations of giant Amazon water lily (*Victoria amazonica*) growing in tropical lakes in South America. View of shoreline growth. (b) Figure bordered by two Guianan tribesmen. A lithographic reproduction of an engraving by M. Gauci, ca. 1814. (c) A lithographic reproduction of an engraving by M. Gauci, ca. 1841. (d) Hummingbirds visiting a flower. (e) View of mid-lake growth.

*Source:* **(a) *Curtis's Botanical Magazine* (London), Plate 4275, vol. 73, 1847 (public domain). (b) Wellcome Images (CC BY 4.0). (c) Wellcome Images (CC BY 4.0). (d) Gould, J. 1861. *A monograph of the Trochilidæ, or family of humming-birds*, vol. 5. Taylor and Francis, London (public domain). (e) Johnston, H. 1911–1912. *Marvels of the Universe, a Popular Work on the Marvels of the Heavens, the Earth, Plant Life, Animal Life, the Mighty Deep*, vol. 1. Hutchinson, London (public domain).**

Most cultivated "water lilies" are any of the numerous species and hybrids of the genus *Nymphaea* (although sacred lotus is more widely cultivated than any single species of plants called water lily). About 40 species (as many as 80 species by some authors) of *Nymphaea* are recognized. Flower colors include white, yellow, orange, pink, red, and (usually mostly in tropical species) blue, produced in late spring and/or summer. Species adapted to temperate regions ("hardy water lilies") usually open their flowers during the day and close in the evening. The more tender tropical species often open their flowers, instead, at night and close in the morning. The tropical species often have larger flowers in a greater range of colors than the hardy species. The flowers usually last 3 or 4 days and subsequently may sink below the water surface and develop fruits. The leaves are usually round, with a notch where the leafstalk attaches to the leaf blade and are referred to as "lily pads" when they float on water. Fertilized flowers produce berry-like fruits.

*Nymphaea odorata* and *N. alba* are the most popular cultivated species of *Nymphaea*. The wild species are usually white-flowered, but cultivars (and hybrids) with flowers of other colors are available. *Nymphaea alba* is a native of North Africa and Eurasia, capable of growing in a wide range of climates and producing large (20 cm or more wide), slightly fragrant flowers. *Nymphaea odorata* (variously named, including American white water lily) is

FIGURE 99.4 Flower color variation in *Nymphaea*.

*Source:* **(a) Photo by Brenda Bonebrake from Pixabay (public domain). (b) Photo by Tsiegretlop (CC BY 2.0). (c) Photo by Anastasia Limareva (CC BY 2.0). (d) Photo by Andy Mondragon (CC BY 3.0). (e) Photo by David Short (CC BY 2.0). (f) Photo by Trisorn Triboon (CC BY SA 4.0). (g) Photo by Mauro Halpern (CC BY 2.0). (h) Photo by Roland Zh (CC BY SA 3.0).**

FIGURE 99.5 Popularly cultivated species of *Nymphaea*: (a) *Nymphaea odorata*. (b) *Nymphaea alba*. (c) *Nymphaea lotus*. (d) *Nymphaea nouchali*.

*Source:* **(a) Photo by Roland Zh (CC BY SA 3.0). (b) Photo by H. Zell (CC BY SA 3.0). (c) Photo by Ilee Wu (CC BY ND 2.0). (d) Photo by Bernard Dupont (CC BY SA 2.0).**

a native of the eastern United States, with big (sometimes 13 cm or more wide) fragrant flowers, widely grown in public and private ponds in warm temperate regions. Some dwarf (often called "pygmy" and "miniature") species (with small flowers and small overall size) are marketed for cultivation in pots.

## 99.3 HISTORY

Water lilies of the Nymphaeaceae have been cultivated in water gardens since ancient times. Hundreds of ornamental cultivars have been bred in recent decades.

*Victoria amazonica* was first recognized by Europeans in the early 19th century and became the subject of enormous public interest in specially built greenhouses housing interior ponds featuring the giant water plant. The craze started when Joseph Paxton, the head gardener for the Duke of Devonshire, built a special glasshouse for the "queen of aquatics" (as it was called) at Chatsworth, the Duke's country estate about 300 km from London. Paxton dressed his 7-year-old daughter, Annie, in a fairy costume and set her on the largest of the floating leaves. An image of the scene was widely circulated, and in 1849, almost 50,000 people visited the Chatsworth greenhouse.

FIGURE 99.7 Annie Paxton standing on a leaf of the giant water lily (*Victoria amazonica*). This sketch of the 7-year-old daughter of Joseph Paxton was based on a demonstration in the special greenhouse he built at Chatsworth, England, for the plant.

*Source:* **Figure published in *The Illustrated London News* in 1849. Colorized by B. Brookes.**

## 99.4 ECONOMIC IMPORTANCE

Water lilies are specialty garden plants, not usually marketed in most garden centers and plant outlets. There is a niche market for those who cultivate the plants in home gardens, but a more substantial industry associated with large pools of water lilies displayed in public and private institutions. Maintenance of larger pools can be a source of considerable employment.

FIGURE 99.6 Nineteenth-century paintings (public domain) of cultivated *Nymphaea* water lilies by European botanical illustrators: (a) Garden hybrids. (b) *Nymphaea caerulea*. (c) *Nymphaea andreana*. (d) *Nymphaea gigantea*. (e) *Nymphaea lotus*. (f) *Nymphaea alba*.

*Source:* **(a) *Revue horticole* (Paris), vol. 67, 1895. (b) Thornton, R. J. 1804. *The Temple of Flora, or Garden of Nature*. T. Bensley, London. Courtesy of the Metropolitan Museum of Art. (c) *Revue horticole* (Paris), vol. 68, 1896. (d) *Flore des serres et des jardins de l'Europe* (Belgium), vol. 7, 1851–52. (e) Step, E., D. Bois, B. Herincq, and W. Watson. 1896. *Favourite Flowers of Garden and Greenhouse*, vol. 1. Frederick Warne, New York. (f) Paxton's flower garden (London), vol. 1, 1882.**

FIGURE 99.8 Harvesting wild stands of *Nymphaea pubescens* for sale in markets: (a) In Vietnam. (b) In Bangladesh.

*Source:* **(a) Photo by Quang Nguyen Vinh from Pixabay. (b) Photo by Shaafimahin (CC BY SA 4.0).**

FIGURE 99.9 Commercial nursery production of *Nymphaea* water lilies: (a) Greenhouse with water lilies in large concrete tanks. (b) Plants in water-filled pots, grown outdoors for sale.

***Source:*** **(a) Photo by Shutterstock/Peeradontax. (b) Photo by Binh Dang Nam from Pixabay (public domain).**

FIGURE 99.10 Lavishly constructed water lily pools attached to opulent estates: (a) A castle in Poland. (b) Bodnant Gardens, North Wales, UK. (c) Holcombe Court, Devon, UK. (d) Villa Balbiano, Lake Como, Italy.

***Source:*** **(a) Photo by Twisted Realm (CC BY 2.0). (b) Photo by Joanna Rycerz from Pixabay (public domain). (c) Photo by Alison Day (CC BY ND 2.0). (d) Photo by Sergio Cerrato from Pixabay (public domain).**

## 99.5 ORNAMENTAL USES

Water lilies are necessarily grown in natural or constructed ponds. Extremely impressive and artistic water lily ponds, usually made of concrete, are associated with expensive estates. Natural or adapted water lily ponds are often found in public parks, where they are often associated with water birds. Homeowners can repurpose children's wading pools and tubs for water lilies or purchase relatively small structures intended for the purpose.

## 99.6 OTHER USES

With the growing importance of sustainability, "water gardens" in recent years have combined aquatic plants and fish, the plants utilizing the waste produced by the fish as natural fertilizer, and the fish employing the plants as protective

FIGURE 99.11 Simple in-ground water lily containers.

***Source:*** **(a) Photo by Art Anderson (CC BY SA 3.0). (b) Photo by Matthew T. Rader (CC BY SA 4.0).**

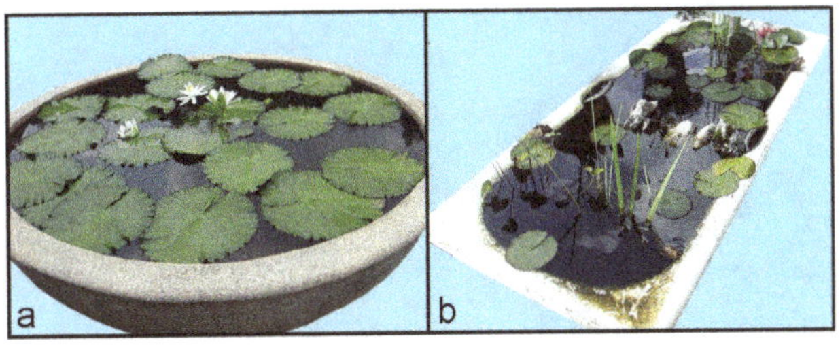

FIGURE 99.12 Simple aboveground water lily containers: (a) A large waterproof planter. (b) An old bathtub.

***Source:*** **(a) Photo by Raffi Kojian (CC BY SA 3.0). (b) Photo by Spider Dog from United Kingdom (CC BY 2.0).**

FIGURE 99.13 Sustainable ornamental fish and water lily pools.

***Source:*** **(a) Photo by Yehudit Garinkol Pikiwiki Israel (CC BY 2.5). (b) Photo by Inspired Images from Pixabay (public domain).**

cover from the sun. Koi and goldfish are the most commonly maintained ornamental fish in ponds, the former more suited to larger ponds and with the ability to hibernate outdoors over cold winters. Adding fish to water gardens is recommended to eliminate mosquito larvae, although predators sometimes learn to go fishing at such sites.

## 99.7 PROBLEMS

Depending on species and portion of the plant, water lilies can be toxic due to the presence of alkaloids. Nevertheless, in many countries, various parts of the plants (rhizomes, foliage, flowers, seeds) were consumed, notably as a famine

FIGURE 99.14 Invasive water lily infestations: (a) *Nymphaea pubescens* in Satla District, Bangladesh. (b) Wular Lake, Kashmir Valley, India.

*Source:* **(a) Photo by Lonely Explorer (CC BY SA 4.0). (b) Photo by Imran Rasool Dar (CC BY SA 4.0).**

FIGURE 99.15 The Trevi Fountain in Rome, built in the style of ancient Roman nymphaea fountains.

*Source:* **Photo by Diliff (CC BY 3.0).**

food, and frequently after boiling, to reduce toxicity. Extracts from blue lotus (*N. caerulea*) are sometimes employed at high doses to produce euphoria and hallucinations.

Water lilies can be invasive, although several other plants grown as aquatic ornamental have proven to be much more serious.

## 99.8 CULTIVATION

### 99.8.1 Climate and Site

The so-called "hardy" species that grow naturally in temperate regions are cold-tolerant as far north as climate zone 3 (when protected from freezing underwater) and are heat-tolerant as far south as zone 11. They are able to tolerate water temperatures between 10 and 27°C. The tropical, tender species are usually adapted to growing well in water temperatures of at least 18°C (preferably 21°C), and they can tolerate water as hot as 35°C. They are cultivated as permanent outdoor plants in zones 9 to 11 and can frequently bloom year-round in warmer climates. In colder areas, tender species can be grown year-round in protected pools, with minimum overwinter water temperatures of 10°C (generally, this will be provided indoors, but a pond heater can be employed outdoors). Alternatively, the tubers or rhizomes can be excavated in late fall and overwintered in a cool (ca. 10°C), damp condition. All water lilies should be grown in full sun, in still water. Depending on climate and resources, water lilies can be grown in large natural bodies of water, artificially constructed ponds, or containers which can be quite small for some species.

### 99.8.2 Propagation and/or Purchasing

Seeds can be planted (in containers of water with a soil base), but growth is slow and will not reproduce the desired flowers of the majority of ornamental forms. Most cultivars need to be propagated from rhizomes, purchased from larger garden centers, or by order from specialty sources. Established plants can be divided and replanted. Often, offsets are produced which can be planted. *Nymphaea micrantha* and some of its hybrid progeny are "viviparous," producing plantlets at the junction of the leafstalk and leaf blade, and these can be employed to grow new plants.

### 99.8.3 Planting

Rhizomes can be planted just under the soil surface of ponds. Alternatively, they can be placed in containers, maintained in the water. Aggressive plants can be kept from taking over a pool by keeping them confined in containers

### 99.8.4 Soil and Moisture Conditions

Rhizomes can be placed in underwater loamy garden soil, which is best covered with fine ("pea") gravel or sand. Potting soils with loose humus, peat, vermiculite, or perlite are inappropriate, because these can simply float away. When maintained in containers, the soil can be kept in place by covering the base of the pot with loose burlap that retains the soil but allows leaves and flower stalks to grow through. It may be necessary to weight down the containers with rocks. Freshly planted rhizomes in containers are best maintained in early growth by adjusting the level of water above the plants (by controlling water level in the pond or by placing the containers on adjustable supports). Small rhizomes should be about 10 cm below the water surface, large rhizome about 50 cm below the surface. Special aquatic fertilizers can be applied to ponds, or pellets put in containers, according to manufacturer's instructions.

### 99.8.5 Maintaining Plants

Deadheading of spent flowers will prolong flowering. Senescent foliage should be removed to minimize decay in the pond. Mosquito reproduction in still water is a significant issue, which may be addressed by adding fish, aeration, or circulating the water.

## 99.9 FASCINATING FACTS, FABLES, FOLKLORE, AND FRIVOLITY

- In ancient times, "nymphaea" were originally natural grottoes with running water, allegedly inhabited by nymphs. Throughout their empire, the Romans built monumental fountains in imitation of these, especially in the eastern provinces. Water was supplied by aqueducts emptying into pools, which citizens could employ to draw water. The structures were typically lavishly decorated with sculptures and statuary. There were 20 or more nymphaea in Rome, but none has survived. The Trevi Fountain, completed in 1762, closely resembles a Roman-style nymphaeum.

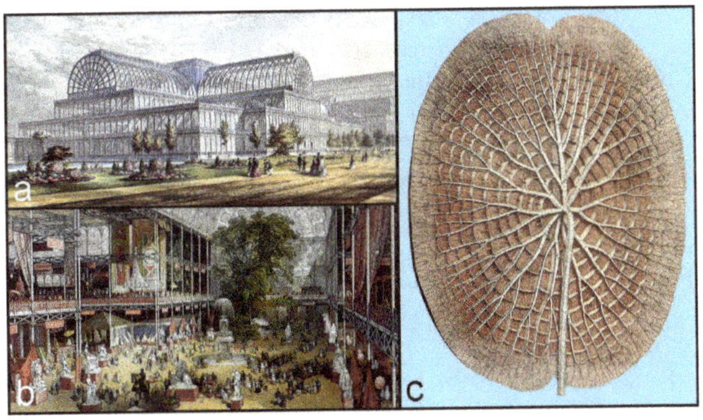

FIGURE 99.16 Parallel support framework of the giant Crystal Palace greenhouse in London and the leaf venation of the giant Amazon water lily (*Victoria amazonica*). (a) Exterior view of the Crystal Palace. (b) Interior view of the Crystal Palace. (c) Underside of leaf of the giant Amazon water lily (*Victoria amazonica*).

**Source: Public domain illustrations. (a) Nash, J., L. Haghe, and D. Roberts. 1852. *Dickinson's Comprehensive Pictures of the Great Exhibition of 1851.* Dickinson Brothers, London, UK. (b) McNeven, J. 1851. The transept from the Grand Entrance, Souvenir of the Great Exhibition, William Simpson (lithographer), Ackermann & Co. (publisher). (c) A chromolithograph prepared by William Sharp (1803–1875) and published in 1854 by Boston, Dutton & Wentworth in a booklet written by John Fisk Allen.**

- The ancient Romans thought that boys could be turned into eunuchs by forcing them to drink crushed *Nymphaea* in vinegar for 10 consecutive days.
- The Crystal Palace was a giant glass and cast-iron structure erected in Hyde Park, London, England, to house the Great Exhibition of 1851. At that time, it was the largest and most expensive "greenhouse" ever built. The building was 41 m high and occupied 72,000 m$^2$ on the ground floors. The iron skeleton was inspired by the venation pattern of the leaf of the giant Amazon water lily, *Victoria amazonica*. The architect, Joseph Paxton, stated, "Nature was the engineer. Nature had provided the leaf with longitudinal and transverse girders and supports that I, borrowing from it, have adapted in this building" (www.lookandlearn.com/blog/25162/a-giant-water-lily-inspired-the-structure-of-paxtons-crystal-palace/).
- Water lilies (*Nymphaea* species) are the national flowers of Bangladesh, French Guiana, and Sri Lanka (disputedly, *N. nouchali*, under the synonym *N. stellata*). The giant water lily (*Victoria amazonica*) is the national flower of Guyana. (Websites often confuse *Nymphaea* and *Nelumbo* ["lotus"] as "water lily national flowers.")
- "Nymphaeamaniacs" are extremely enthusiastic water lily enthusiasts.
- Oscar-Claude Monet (1840–1926) was a French painter and founder of impressionist painting. He was one of the world's most famous painters during the second half of the 20th century. Monet created a large water lily garden on his property, which he employed as a model for many of the more than 250 paintings of water lilies that he created.

FIGURE 99.17 Claude Monet and water lilies (public domain photos): (a) *Bridge over a Pond of Water Lilies*, painted in 1899 by Claude Monet. (b) Portrait of Claude Monet (at about 26 years of age) painted in 1867 by French artist Carolus-Duran (1837–1917).

**Source: (a) Photo by Daniel Schwen, taken at the Metropolitan Museum of Art, New York. (b) Photo by Monet Collector, Musée Marmottan Monet.**

## KEY PUBLICATIONS


Bertol, E., V. Fineschi, S.B. Karch, F. Mari, and I. Riezzo. 2004. *Nymphaea* cults in ancient Egypt and the New World: A lesson in empirical pharmacology. *Journal of the Royal Society of Medicine* 97: 84–85.

CABI Compendium. 2016. Nymphaea lotus *(White Egyptian lotus).* www.cabidigitallibrary.org/doi/10.1079/cabicompendium.115821.

Conard, H.S. 1905. *The Waterlilies: A Monograph of the Genus* Nymphaea. Carnegie Institution of Washington, Washington, DC. 279 pp.

Emboden, W.A. 1978. The sacred narcotic lily of the Nile: *Nymphaea caerulea. Economic Botany* 32: 395–407.

Emboden, W.A. 1982. The mushroom and the water lily: Literary and pictorial evidence for *Nymphaea* as a ritual psychotogen in Mesoamerica. *Journal of Ethnopharmacology* 5: 139–148.

The Gardens Trust. 2018. *Joseph, Victoria and the Amazing Leaf [History of* Victoria amazonica*].* https://thegardenstrust.blog/2018/07/14/joseph-victoria-and-the-amazing-leaf/.

Kameni, P.M., D.P.D. Dzeufiet, D.C. Bilanda, M.F. Mballa, N.Y.S. Mengue, T.H. Tchoupou, et al. 2019. *Nymphaea lotus* Linn. (Nymphaeaceae) alleviates sexual disability in L-NAME hypertensive male rats. *Evidence-Based Complementary and Alternative Medicine* 2019: 8619283. https://doi.org/10.1155/2019/8619283.

Kiranmai, B., M. Sandhyarani, and A.K. Tiwari. 2023. Water lily (*Nymphaea nouchali* Burm. f): An ancient treasure of food and medicine. *Pharmacognosy Research* 15: 226–234.

Nakkuntod, M., S. Srinarang, and K.W. Hilu. 2019. Systematics of water lilies (genus *Nymphaea* L.) using 18S rDNA sequences. *International Journal of Bioengineering and Life Sciences* 13: 85–90.

Nash, H. and S. Stroupe. 2003. *Complete Guide to Water Garden Plants.* Sterling, New York. 224 pp.

Schimpf, M., T. Ulmer, H. Hiller, and A.F. Barbuto. 2021. Toxicity from blue lotus (*Nymphaea caerulea*) after ingestion or inhalation: A case series. *Military Medicine* usab328. https://doi.org/10.1093/milmed/usab328.

Selvakumari, E., A. Shantha, C.S. Kumar, and T.P. Prabhu. 2016. Phytochemistry and pharmacology of the genus *Nymphaea. Journal of Academia and Industrial Research* 5(7): 98–108.

Slocum, P.D. 2005. *Waterlilies and Lotuses.* Timber Press, Portland, OR. 260 pp.

Smith, L.T., C. Magdalena, N.A.S. Przelomska, O.A. Pérez-Escobar, D.G. Melgar-Gómez, S. Beck, et al. 2022. Revised species delimitation in the giant water lily genus *Victoria* (Nymphaeaceae) confirms a new species and has implications for its conservation. *Frontiers in Plant Science.* www.frontiersin.org/articles/10.3389/fpls.2022.883151/full?utm_source=mp-fotoscapes.

Wiersema, J.H. 1988. Reproductive biology of *Nymphaea* (Nymphaeaceae). *Annals of the Missouri Botanical Garden* 75: 795–804.

Zhang, L., F. Chen, X. Zhang, Z. Li, Y. Zhao, R. Lohaus, et al. 2020. The water lily genome and the early evolution of flowering plants. *Nature* 577: 79–84.


### Key Cultivation Websites


BBC Gardeners' World Magazine. *How to Plant a Water Lily.* www.gardenersworld.com/how-to/grow-plants/how-to-plant-a-water-lily/.

Green Blog: University of California Division of Agriculture and Natural Resources. *Mosquito Management for Ponds, Fountains and Water Gardens.* https://ucanr.edu/blogs/blogcore/postdetail.cfm?postnum=10847#:~:text=Larvicides%20containing%20spores%20or%20metabolites,larvae%20within%20a%20few%20days.

The International Waterlily & Water Gardening Society. https://iwgs.org/nymphaea-and-nelumbo-registration/.

The Pond Guy. *How to Plant Water Lilies.* www.thepondguy.com/learning-center/how-to-plant-water-lily/.

Royal Horticultural Society. *How to Grow Waterlilies.* www.rhs.org.uk/plants/waterlilies/growing-guide.

The Spruce. *How to Grow and Care for Water Lilies and Lotus.* www.thespruce.com/the-incomparable-waterlily-and-lotus-1403525#:~:text=Water%20lilies%20(Nymphaea)%20are%20jewels,centered%20around%20a%20radial%20notch.

University of Missouri, Integrated Pest Management. *Water Lilies: Easier Than You Think.* https://ipm.missouri.edu/MEG/2017/5/Water_Lilies/.

WikiHow. *How to Grow Water Lilies.* www.wikihow.com/Grow-Water-Lilies.

# 100 Wisteria (*Wisteria* Species and Hybrids)

**Scientific Names:** The genus name *Wisteria* was named for Caspar Wistar (1761–1818), a professor of anatomy at the University of Pennsylvania.

**Common Names:** The vernacular name wisteria is also often spelled wistaria (but wisteria is much more frequent). Occasionally (especially in Europe), the name glycine is applied (although this risks confusion with soybean, which is in the genus *Glycine*). Several unrelated plants have wisteria (or wistaria) in their names. Water wisteria is *Hygrophila difformis*. Rhodesian wisteria and wild wisteria are *Bolusanthus speciosus*. Scarlet wisteria refers to species of *Sesbania*. Evergreen wisteria is *Millettia reticulata*.

## 100.1 INTRODUCTION

Wisterias are woody twining vines producing extraordinarily attractive, massive flower displays. There are about ten species, natives of temperate eastern Asia and the eastern United States. Japanese wisteria (*W. floribunda*) and Chinese wisteria (*W. chinensis*) are the most commonly planted species.

FIGURE 100.1 Mature gigantic Japanese wisteria (*Wisteria floribunda*) vines in Ashikaga Flower Park, Ashikaga, Tochigi, Japan. The park is famous for its massive wisterias, in bloom from late April to mid-May. The plants are supported by extensive pergola/pillar/trellis systems: (a) The largest wisteria in Japan, planted about 1870, covering approximately 2,000 m$^2$. (b) A white-flowered form. (c) A huge wisteria viewed under night illumination.

*Source:* **(a) Photo by Hiroaki Kaneko (CC BY SA 3.0). (b) Photo by Slimhannya (CC BY SA 4.0). (c) Photo by Manish Prasbhune (CC BY 2.0).**

## 100.2 OVERVIEW

Wisterias climb by means of the young growing stems rotating around structures that they encounter. (These stems are often mislabeled as "tendrils"; true tendrils are slender, wire-like, twining, special branches.) Chinese and Japanese wisterias can climb to 30 m in height (especially in mild climates), and much longer laterally. Some old specimens (plants sometimes live for over a century) can develop trunks exceeding 40 cm in diameter (often, main stems near the ground coalesce very intimately to form what appear to be large trunks). The foliage is annual (i.e., deciduous), lasting one season. The leaves are compound, with up to 13 leaflets on Chinese wisteria and up to 19 on the Japanese plant. These species flower in the spring, as the foliage is expanding. The flowers occur in very conspicuous hanging clusters. The blooms of cultivars are usually shades of lavender or lilac, less often white, and occasionally pink. The flowers are pleasantly fragrant. The Chinese and Japanese species have produced fertile hybrids (called *W.* ×*formosa*)

FIGURE 100.2 Professional 19th-century botanical illustrations (public domain) of flowering branches of the two principal ornamental wisteria species: (a) Japanese wisteria (*Wisteria floribunda*). (b) Chinese wisteria (*W. sinensis*).

*Source:* **(a) Van Houtte, L. B. 1873. *Flore des serres et des jardin de l'Europe*, vol. 19. (b) The 1868 Dutch-language book *Flora afbeeldingen en beschrijvingen van boomen, heesters, éénjarige planten, enz. voorkomende in de Nederlandsche tuinen* by H. Witte, illustrated by A. J. Wendel, published by Wolters, Groningen.**

DOI: 10.1201/9781003586227-100

FIGURE 100.3 Compound "trunk" of an ancient plant of Japanese wisteria (*Wisteria floribunda*). The trunk has been formed by coalescence of several woody vine stems that twisted around each other.

*Source:* Photo by Sakaori (CC BY SA 4.0).

in much of the southeastern United States, and distinguishing the hybrids and the two parental species can be difficult.

## 100.3 HISTORY

Chinese and Japanese wisterias have been grown for well over a thousand years in their Asian homelands, but both countries were largely inaccessible to Westerners until the 18th century. Subsequently, the two species were introduced to the West. American wisteria (*W. frutescens*) was imported to England in 1724, but its popularity diminished once the more attractive Asian species became established.

## 100.4 ECONOMIC IMPORTANCE

Wisterias are significant commercial horticultural subjects in warm-temperate countries.

## 100.5 ORNAMENTAL USES

Wisterias are employed extensively in warm climates to landscape public parks, large buildings, and residences. They are often trained to grow on trellises attached to walls, on pergolas and arches, as specimens trimmed to grow in the form of shrubs and trees, as large plants in containers, and as bonsai.

## 100.6 OTHER USES

Constituents of wisterias are under investigation for their potential medicinal applications. In Asia, various parts of native wisterias were employed as herbal medicines, and the foliage, flowers, and seeds were even consumed as food (but note the information regarding toxicity presented next).

FIGURE 100.4 Chinese wisteria (*Wisteria sinensis*) trained to grow on the walls of large buildings: (a) Magdalene College, Cambridge, England. (b) Beacon Hill, Boston, Massachusetts.

*Source:* (a) Photo by Andrew Dunn (CC BY SA 2.0). (b) Photo by M2545 (CC BY SA 3.0).

FIGURE 100.5 Wisteria vines trained to grow on arches: (a) Diagram of an arch supporting wisteria. (b) In Jackson, Mississippi.

*Source:* (a) Husqvarna UK (CC BY SA 2.0). (b) Photo by Natalie Maynor (CC BY 2.0).

FIGURE 100.6 Chinese wisteria (*Wisteria sinensis*) trained to grow as a small tree (with a single trunk). In these examples, supports are provided to keep the trunks erect. (a) Planted specimen in England. (b) Plant in Venice, Italy.

*Source:* (a) Photo by Charles Miller (CC BY 2.0). (b) Photo by Vvlasenko (CC BY SA 3.0).

## 100.7 PROBLEMS

Both Japanese and Chinese wisterias are serious invasive weeds in various regions, including the Southern United States. Wisterias are comparable to pythons: They are capable of strangling the life out of trees, albeit much more slowly. In some regions, efforts are made to dissuade the planting of the Asian ornamental wisterias. An additional

FIGURE 100.7 Wisteria grown as bonsai.

*Source:* **(a) Photo by Macassar (CC BY SA 4.0). (b) Photo by Precue (CC BY 3.0).**

FIGURE 100.8 Invasive Chinese wisteria (*Wisteria sinensis*) in South Carolina: (a) Growing as a roadside weed. (b) A tree trunk strangled by a wisteria vine (notice how growth is suppressed just below the encircling vine, while the tree trunk expanded just above it).

*Source:* **(a) Photo by Nadiatalent (public domain). (b) Photo by Carlfbagge (CC BY 2.0).**

problem of wisterias is their toxicity. The seeds are poisonous and have been a cause of gastroenteritis in children and pets. Other parts of the plant also contain toxins, but to a lesser extent.

## 100.8 CULTIVATION

### 100.8.1 Climate and Site

Chinese wisteria is cold-hardier than the Japanese species (it tolerates climate zone 4) and can be grown well as far north as climate zone 5. It can also be grown as far south as zone 10 but might not receive sufficient cold in the winter to induce dormancy, such "stratification" required for a good spring flower display. Some recent cultivars of American wisterias (in contrast to the Asian species) are claimed to be suitable as far north as zone 3. Wisterias prefer to be in full sun (the Chinese species is more shade-tolerant than Japanese wisteria). Wisterias growing on trees can become so heavy that they cause tree limbs to break. They can even kill smaller trees by shading and strangling them. Wisterias should not be planted where they will spread over shrubs and trees.

### 100.8.2 Propagation and/or Purchasing

Although plants can be grown from seeds, the attractive features of particular cultivars are unlikely to be reproduced well, and the plants may take decades to flower. Most gardeners purchase cuttings that have been rooted or grafted to hardy stock.

### 100.8.3 Planting

Wisterias should be provided with a sturdy structure, such as a pergola, espalier, arch, or even just a pole or fence, on which to grow. A frequent mistake is to underestimate the weight of mature vines, with the result that flimsy structures collapse. The vines can be grown on trees, but this can damage the trees and make it difficult to prune the vines. They are frequently grown against house walls and on porch pillars, but there is potential of their extending into and damaging rain gutters, roof aerials, and the like.

### 100.8.4 Soil and Moisture Conditions

Fertile, moist, well-drained soil is preferred, although Chinese wisteria is relatively tolerant of poor soils. Wisterias can withstand a soil pH range of 5.0 to 8.0. Like most other members of the pea family (Fabaceae or Leguminosae), wisteria roots are symbiotically associated with rhizobia, a group of "good" soil bacteria that infect the roots and inhabit consequent root swellings called nodules. The bacteria take inert nitrogen out of the atmosphere and "fix" it (transform it to a usable chemical form). This is an invaluable service, since nitrogen is the key plant nutrient limiting plant growth. When excess nitrogen is provided in fertilizer, it can disturb nodule development and suppress flowering. Some horticulturalists recommend not fertilizing wisterias.

FIGURE 100.9 Wisteria roots with nodules housing rhizobial bacteria, which fix inert atmospheric nitrogen (convert it to chemical compounds that the plant can metabolize), reducing or eliminating the need for nitrogen fertilizer.

*Source:* **Photo by Rowan Adams (CC BY SA 4.0).**

### 100.8.5 Maintaining Plants

Anyone tempted to plant a wisteria should recognized that it is accompanied by a serious need to prune it at least yearly. Training wisterias to form small trees or conform to a trellis, espalier, or gazebo requires regular cutting away and directing the future path of branches. Detailed information on these aspects is given in the section on key websites. In addition, plants may set seeds, and unless the seedpods are removed, a forest of seedlings may be produced around the vines.

## 100.9 FASCINATING FACTS, FABLES, FOLKLORE, AND FRIVOLITY

- Not all vines climb by spiraling of their stems around objects (some climb by tendrils, some by suckers, some by hooks), and the following comment pertains to vines in which the stems and branches twine. The spiraling pattern of such vines around objects is described as left-handed (anticlockwise viewed from below, as if one were viewing the face of an analog clock in the sky) or right-handed (clockwise viewed from below). Japanese wisteria twines counterclockwise (left-handed) when viewed from below, while the right-handed Chinese wisteria twines clockwise. Numerous publications mistakenly state exactly the opposite, because they do not follow the convention of describing spiral pattern from the perspective of earliest time (the start point, that is, the base of the plant, not the top of the plant). Most spiraling vine species wind clockwise (the same is true for tendrils of plants that climb by tendrils). Why most twining vines are right-handed is as unclear as why most people are right-handed. In some species (such as cucumber), some plants spiral one way; others do the opposite. Whether in the Northern or Southern Hemisphere is irrelevant, although numerous publications liken vine spiraling to the direction of water spiraling when a toilet is flushed.
- It is widely understood that in sexual reproduction, the male and female contribute equally to the genes of their offspring (although male Y chromosomes can have fewer genes than female X

FIGURE 100.10 (a) Defining the direction of vine spiraling. (b) Left-twining stem of Japanese wisteria (*Wisteria floribunda*). (c) Right-twining stems of Chinese wisteria (*W. sinensis*). (d) The main double winding staircase of the *Titanic* (which sank in 1912), illustrating how defining the direction of a spiral depends on the viewpoint of the observer, just as whether a wisteria vine spirals clockwise or counter-clockwise depends on which end of the vine is viewed.

***Source:*** **(a) Figure prepared by B. Brookes. (b) Photo by Andreas Rockstein (CC BY SA 2.0). (c) Photo by Dénes Emőke (public domain). (d) 1911 Olympic and Titanic White Star Line brochure (public domain).**

FIGURE 100.11 Examples of wisteria in art: (a) Painting titled *A Maiden under the Wisteria*, by French artist Émile Vernon (1872–1920). (b) Wisteria table lamp, ca. 1902, by Tiffany Studios, designed by Clara Driscoll, from the collection of the Virginia Museum of Fine Arts. (c) View of Mount Fujiyama, Japan, framed by wisteria. Painting by English artist Marianne North (1830–1890).

***Source:*** **(a) Public domain. (b) Photo by Fopseh (CC BY SA 4.0). (c) Public domain.**

chromosomes). However, this is true for the cell's nuclear genes, not necessarily for other gene-containing cellular constituents, such as mitochondria and (in plants) chloroplasts. Most flowering plants get their mitochondria and chloroplasts (and their genes) from the maternal parent, but in wisteria, the seeds inherit chloroplasts only from the male (pollen) parent.

- The world's largest known wisteria, a Chinese variety, was planted in 1894 in Sierra Madre, California. The plant was recognized by Guinness World Records as the world's largest blossoming plant.

## KEY PUBLICATIONS

Compton, J. 2015. The genus *Wisteria* Nuttall. *Curtis's Botanical Magazine* 32: 183–192.

Compton, J.A. 2015. *Wisteria sinensis* on the slow boat from China: The journey of *Wisteria* to England: Leguminosae (Fabaceae). *Curtis's Botanical Magazine* 32: 248–293.

Compton, J.A. and H.W. Lack. 2012. The discovery, naming and typification of *Wisteria floribunda* and *W. brachybotrys* (Fabaceae) with notes on associated names. *Willdenowia* 42: 219–240.

Compton, J. and C. Lane. 2019. *RHS Wisteria: The Complete Guide*. Royal Horticultural Society, London. 304 pp.

Hokanson, S.C., S. McNamara, K. Zuzek, M. Zins, and N. Rose. 2015. *Wisteria frutescens*, American wisteria, 'Betty Matthews' First Editions® Summer Cascade™. *HortScience* 50: 317–319.

Kaufman, S. 2016. *Wisteria sinensis* (Chinese wisteria). Invasive Species Compendium, CABI. www.cabi.org/isc/datasheet/56852.

Smyth, D.R. 2016. Helical growth in plant organs: Mechanisms and significance. *Development* 143: 3272–3282.

Stone, K.R. 2009. *Wisteria floribunda*, *W. sinensis*. In: *Fire Effects Information System*, U.S. Department of Agriculture, Forest Service, Rocky Mountain Research Station, Fire Sciences Laboratory (Producer). www.fs.usda.gov/database/feis/plants/vine/wisspp/all.html.

Trusty, J.L., B.G. Lockaby, W.C. Zipperer, and L.R. Goertzen. 2007. Identity of naturalised exotic *Wisteria* (Fabaceae) in the south-eastern United States. *Weed Research* 47: 479–487.

Trusty, J.L., B.G. Lockaby, W.C. Zipperer, and L.R. Goertzen. 2008. Horticulture, hybrid cultivars and exotic plant invasion: A case study of *Wisteria* (Fabaceae). *Botanical Journal of the Linnean Society* 158: 593–601.

Valder, P. 1995. *Wisterias: A Comprehensive Guide*. Timber Press, Portland, OR. 160 pp.

Xia, M.Q., R.Y. Liao, J.T. Zhou, H.Y. Lin, J.H. Li, P. Li, et al. 2019. Phylogenomics and biogeography of *Wisteria* (Fabaceae), with implication on plastome evolution of inverted repeat-lacking clade. *Research Square (Preprint)*. https://assets.researchsquare.com/files/rs-5542/v1/50318c54-6d51-417c-908b-58632aedae25.pdf?c=1631827326.

## KEY CULTIVATION WEBSITES

Horticulture Magazine. *Wisteria*. https://horticulture.co.uk/wisteria/.

Royal Horticultural Society. *How to Grow Wisteria*. www.rhs.org.uk/plants/wisteria/growing-guide.

Royal Horticultural Society. *Wisteria: Pruning*. www.rhs.org.uk/plants/wisteria/pruning-guide.

The Spruce. *How to Grow American Wisteria*. www.thespruce.com/american-wisteria-plant-profile-5181640.

The Spruce. *How to Grow and Care for Wisteria*. www.thespruce.com/wisteria-growing-guide-5207953.

Utah State University, Extension. *Wisteria in the Garden*. https://extension.usu.edu/yardandgarden/research/wisteria-in-the-garden.

WikiHow. 2021. *How to Prune Wisteria*. www.wikihow.com/Prune-Wisteria.

WikiHow. 2022a. *How to Build a Trellis for Wisteria*. www.wikihow.com/Build-a-Trellis-for-Wisteria.

WikiHow. 2022b. *How to Get Wisteria to Bloom*. www.wikihow.com/Get-Wisteria-to-Bloom.

WikiHow. 2022c. *How to Grow Wisteria*. www.wikihow.com/Grow-Wisteria.

# 101 Zinnia (*Zinnia* Species and Hybrids)

**Scientific Names:** The genus name *Zinnia* is based on German botany professor Johann Gottfried Zinn (1727–1759). He was also an ophthalmologist, one of the first scientists to accurately describe the structure of the eyeball. (For information on this adventuresome scientist, see www.friendsofthedd collinshouse.org/flowers).

**Common Names:** Ornamental zinnias, as will be noted, are the result of hybridization, primarily among three species, but the parentage of most garden varieties traces primarily to *Zinnia elegans*, sometimes termed common zinnia.

## 101.1 INTRODUCTION

Zinnias are one of the major garden ornamentals, familiar to most gardeners of the world. It has been said that only dahlias, peonies, and roses can compete for size of bloom, intensity of color, and showiness. However, none of the latter (which are perennials) is as easy to grow or as long-blooming.

## 101.2 OVERVIEW

The 20 or so species of *Zinnia* are native from the southwestern United States to Argentina, occurring particularly in Mexico. They are annual or perennial herbs or small shrubs. The garden varieties range in size from dwarf to over 1 m in height. Zinnias are members of the daisy family (Compositae or Asteraceae), characterized by flower heads usually made up of tiny flowers ("florets") of two types: (1) with what appears to be a single relatively large petal, these florets in a ring on the periphery of the head, and (2) with florets with very inconspicuous petals, located within the outer ring. The flower heads of garden zinnias, usually interpreted as single flowers by non-botanists, are available in an enormous variety. They are categorized in size and form as mini, button, giant, dahlia-type, cactus-type, pompom, and cut flower–type. Colors are often vibrant and virtually unlimited, except for blue and brown. The flowers (i.e., flower heads) are borne singly atop their stalk. As with other extremely popular flowering ornamentals of the daisy family, the basic wild-type "single" flower head resembles a conventional daisy (i.e., with the appearance of a ring of petals surrounding a circle). However, advanced cultivars have been selected to become "semi-double" (additional prominent petals added within the outer circle of prominent petals) or entirely "doubled" (having the appearance of a disk or sphere entirely covered with prominent petals crammed together).

FIGURE 101.1 Flower variation in garden zinnias.

*Source:* (a) Photo by Media Mike Hazard (CC BY 2.0). (b) Photo from Pixabay (public domain). (c) Photo by Drilnoth (public domain). (d) Photo by Swallowtail Garden Seeds (CC BY 2.0).

FIGURE 101.2 Nineteenth-century botanical illustrations (public domain) of garden zinnias. Note the mix of cultivars with single, semi-doubled, or doubled flowers, available long ago.

*Source:* (a) Step, E. 1896–97. *Favourite Flowers of Garden and Greenhouse*, vol. 2. Frederick Warne & Co, London. (b) A chromolithograph published by L. Prang & Co. in 1861–1897, painted by E. T. Fisher. Photo by Boston Public Library. (c) Flore des Serres et des Jardins de l'Europe, vol. 13, 1858.

## 101.3 HISTORY

The relatively unattractive *Z. peruviana* was brought to Europe by the Spaniards in the 1500s, but not until the more desirable *Z. elegans* was imported in the late 18th century did European development of zinnia proceed enthusiastically. By the 1920s, zinnias had become a very popular garden flower. Most cultivars are derived from *Z. elegans*, but *Z. angustifolia* and *Z. haageana* have also contributed to hybrids and cultivars.



DOI: 10.1201/9781003586227-101

FIGURE 101.3 Paintings of zinnia varieties offered in early 20th-century garden catalogues (public domain). Note the popularity of large, doubled flower heads.

*Source:* **(a) Ferry's 1920 Seed Annual. (b) John Lewis Childs 1921 catalogue. (c) Currie's 1937 garden annual.**

FIGURE 101.4 Commercial zinnia flower farms: (a) A zinnia farm in Andong, South Korea. (b, c) Hafa (Flower) Farm near Saint Paul, Minnesota.

*Source:* **(a) Photo by Sieon Nam from Pixabay (public domain). (b, c) Photos by Media Mike Hazard (CC BY 2.0).**

## 101.4 ECONOMIC IMPORTANCE

Zinnias are well-established in the international horticultural trade and are one of the regular offerings in the nursery trade and the cut flower industry.

## 101.5 ORNAMENTAL USES

Zinnias are widely employed as annual bedding and border plants, especially in massed plantings of mixed varieties. The small or dwarf cultivars are especially suitable for containers and rock gardens, and the long-stemmed forms are ideal as cut flowers (with vase life up to 2 weeks), either from residential "cutting gardens" or commercial fields. The plants are considered non-toxic for both people and pets.

FIGURE 101.5 Dedicated zinnia gardens: (a) In Spain. (b) In North Carolina. (c) In Poland.

*Source:* **(a) Photo by Jebulon (public domain). (b) Photo by Selena N. B. H. (CC BY 2.0). (c) Photo by Mirosław Kiwior (CC BY 3.0).**

FIGURE 101.6 Zinnia cut flowers, a major component of the floral industry.

*Source:* **Photo by Shutterstock/Troy D. Duncan.**

FIGURE 101.7 Zinnias as pollinator magnets, ideal for wildlife gardens: (a) A monarch butterfly (*Danaus plexippus*) on zinnia, in Virginia. (b) A bumblebee on zinnia, in British Columbia. (c) A female ruby-throated hummingbird on a zinnia, in Minnesota.

*Source:* **(a) Photo by Judy Gallagher (CC BY 2.0). (b) Photo by Gldosa (CC BY SA 4.0). (c) Photo by Mike Kestell (CC BY 2.0).**

Zinnias are extremely attractive to butterflies, bees, and hummingbirds and so are often planted in "butterfly gardens" or "wildlife gardens." Doubled flowers do not supply much, if any, pollen and may also not produce the nectar that single flowers produce, so they are not ideal for attracting pollinators. Accordingly, single-flowered zinnias are preferable for wildlife gardens.

Some regions employ plants along roadways for beautification, for soil stabilization, and as habitats and food for wildlife. Zinnia is ideal for this purpose because, once planted, it can survive virtually without care because of its resistance to drought, heat, and intense sunlight.

## 101.6 OTHER USES

In the native American range of wild zinnia species, the plants have been employed in folk medicine. Today, there is extensive research in progress attempting to find potential modern medical applications.

## 101.7 PROBLEMS

Zinnias are "garden classics," so well-established that some people consider them to be old-fashioned, despite the regular release of novel varieties.

## 101.8 CULTIVATION

### 101.8.1 Climate and Site

Zinnias are frost-sensitive, which is why they are grown in temperate regions as annuals. Minimum daytime temperature for good performance is 16°C; ideal temperatures are 23–28°C. Most zinnias are adapted to full sun. In semi-tropical areas (hardiness zones 9 and warmer), some cultivars will persist through the winter, and in such locations, they may benefit from protection from hot afternoon sunlight. Zinnias are subject to fungal foliar diseases, and places that provide good air circulation are recommended to help prevent this.

FIGURE 101.8 Remarkable roadside beautification plantings of zinnia by the North Carolina Department of Transportation.

*Source:* **Photo by NCDOTcommunications (CC BY 2.0).**

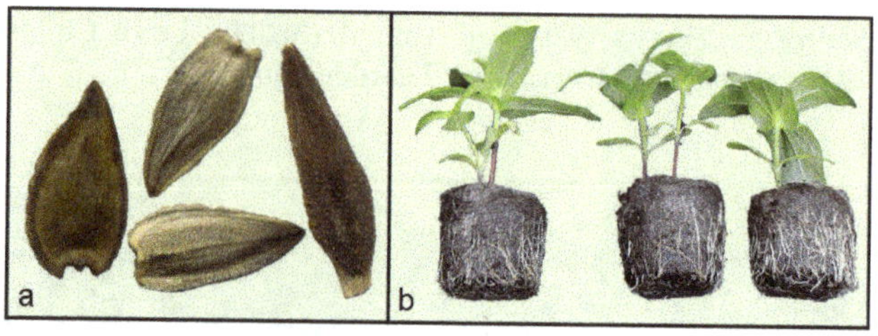

FIGURE 101.9 Zinnia seeds and seedlings: (a) Seeds. (b) Seedlings in fiber transplant pots.

*Source:* **(a) Photo by John Alan Elson (CC BY SA 4.0). (b) Photo by SGS (CC BY SA 4.0).**

### 101.8.2 Propagation and/or Purchasing

Zinnias are grown from seeds, offered by most garden supply sources. Buyers need to decide whether they wish a particular, uniform variety or a package of seeds that will produce a range of different flower types and a rainbow of color combinations. Seeds are often started indoors (16–18°C is recommended) in advance of the planting season, as they should not be placed in the soil until it is warm. If planted too early or in limited light, the seedlings become spindly and growth is retarded. Young plants are also widely purchased in flats and small fiber pots for transplanting.

### 101.8.3 Planting

Zinnia seedlings started in small fiber pots should be transplanted before they become large (at which stage they transplant significantly less well, as the roots respond poorly to disturbance – a phenomenon termed "transplant shock").

### 101.8.4 Soil and Moisture Conditions

Like the majority of ornamental plants, well-drained, rich, loamy, circumneutral (pH 5.5 to 7.5 is acceptable) soil. Unlike most cultivated plants, once the young plants have been established, they should receive water only when conditions become quite dry, as zinnia is adapted to long, hot, dry conditions. Waterlogged soil is very detrimental. Occasional fertilization with a balanced fertilizer is usually all that is required (or simply employing well-composted soil).

### 101.8.5 Maintaining Plants

Before planting and when the seedlings are small, weeds should be removed (mass planting of seeds will suppress weeds once the plants grow thickly). Zinnias require little care after they are planted. As for numerous flowering garden plants, deadheading (removing) the spent flowers prolongs the flowering period. The side branches of flowering stems of plants grown specifically for cut flowers are sometimes removed to encourage long stems that will fit in vases. Collecting seeds from advanced cultivars of most species is inadvisable because they won't breed true, but since zinnias look spectacular in mixed colors, it is a good idea to harvest seeds in the autumn to plant next season.

## 101.9 FASCINATING FACTS, FABLES, FOLKLORE, AND FRIVOLITY

- Southwestern US Indigenous people considered wild zinnias to be sacred and medicinal. The Navajo tribes viewed the zinnia as a goddess, and according to legend, she sent the flower to plant among their crops to protect them against evil. The Pueblo people reputedly fed the blossoms to their children to make them more intelligent.
- Curanderos or traditional healers of Brazil put zinnia leaves on top of a patient's head to cure madness.
- The Savoy Theatre in London is famous as the home of W. S. Gilbert and Arthur Sullivan's Savoy operas, first opened in 1881. Almost half a century later, the interior was remodeled, and according to *The Observer* newspaper, the color scheme was based on five colors seen in the zinnia beds of Hyde Park.
- Zinnia was the state flower of Indiana from 1931 to 1957. The carnation had been the state flower dating to 1913, replaced by the flower of the tulip tree in 1923. Representative Laurence Baker of Kendallville, a commercial peony grower, is credited with persuading the government to change the state flower to the peony in 1957.
- In 2016, a zinnia bloomed on the International Space Station. Some reports mistakenly alleged that it was the first flower produced extraterrestrially, but as early as 1982, Russians had flowered rockcress in space.
- A zinnia named "My Lucky Lady" measured 3.86 m in height – a record when reported in 2008 (www.guinnessworldrecords.com/world-records/tallest-zinnia).
- Because they are so easy to grow, zinnias are recommended not just for experienced gardeners but also for beginners and children (explaining the occasional name "youth-and-old-age").

FIGURE 101.10 Planting "Youth-and-Old-Age" (alternate name of zinnia).

*Source:* **(a) Illustration by Shutterstock/Gbmotiongfx. (b) Illustration by Shutterstock/Tanasab.**

## KEY PUBLICATIONS


Armitage, A.M. 1985. *Zinnia elegans* and *Z. angustifolia*. In: *Handbook of Flowering*, Vol. 4, edited by A.H. Halevy, 548–552. CRC Press, Boca Raton, FL.

Bag, M.K., P. Ranjan, and J.K. Ranjan. 2017. *Zinnia*. In: *Commercial Ornamental Crops: Cut Flowers*, edited by R.L. Mishra and S. Misra, 535–544. Kruger Brentt, UK.

Carleo, J., N. Polanin, and M. Melendez. 2011. *Cut Flowers for the Market and Home Garden – Zinnia*. Rutgers, The State University of New Jersey, Agricultural Station Fact Sheet FS1161. 3 pp. https://njaes.rutgers.edu/pubs/publication.php?pid=FS1161.

Gomaa, A.A.R., M.N. Samy, S.Y. Desoukey, and M. Kamel. 2019. A comprehensive review of phytoconstituents and biological activities of genus *Zinnia*. *Journal of Advanced Biomedical and Pharmaceutical Sciences* 2: 29–37.

Grissell, E. 2020. *A History of Zinnias: Flower for the Ages*. Purdue University Press, Lafayette, IN. 286 pp. (e-publication)

Jones, T. 2022. *The Zinnia Breeder's Handbook: A Comprehensive Guide for Flower Farmers, Gardeners and Enthusiasts*. Self-published. 148 pp.

Kalinowski, J., E.P. Moody, and J.M. Dole. 2022. Improving hydration and vase life of cut *Zinnia*. *Scientia Horticulturae* 293: 110661.

Maughan, T., M. Stock, and M. Lewis. 2020. *Zinnia Cut Flower Production in Utah*. Paper 2124. https://digitalcommons.usu.edu/extension_curall/2124.

Metcalf, H.N. and J.N. Sharma. 1971. Germ plasm resources of the genus *Zinnia* L. *Economic Botany* 25: 169–181.

Stimart, D. and T. Boyle. 2007. Zinnia. *Zinnia elegans*, *Z. angustifolia*. In: *Flower Breeding and Genetics: Issues, Challenges and Opportunities for the 21st Century*, edited by N.O. Anderson, 337–360. Springer, Dordrecht.

Stimart, D.P., T.H. Boyle, and V.M. Terry-Lewandowski. 1987. Genetic and physiological studies of *Zinnia elegans*, *Z. angustifolia*, and their interspecific hybrids. *HortScience* 22: 526–690.

Torres, A.M. 1963. Taxonomy of *Zinnia*. *Brittonia* 15: 1–25.

Weddle, C. 1945. The elegant zinnia. *National Horticultural Magazine* 24(4): 83–89.


### Key Cultivation Websites


GardenersWorld.com. www.gardenersworld.com/how-to/grow-plants/how-to-grow-zinnias/.

Goldwyn, G. 2022. *How to Harvest Zinnia Seeds*. www.bybrittanygoldwyn.com/harvest-zinnia-seeds/.

Minneopa Orchards. 2022. *How to Harvest Zinnia Seeds: Everything You Need to Know*. https://minnetonkaorchards.com/how-to-harvest-zinnia-seeds/.

Mississippi State University, Extension. *Zinnias*(Zinnia elegans) *for the Farmer Florist*. https://extension.msstate.edu/publications/zinnias-zinnia-elegans-for-the-farmer-florist.

Royal Horticultural Society. *How to Grow Zinnia*. www.rhs.org.uk/plants/zinnia/growing-guide.

The Spruce. *How to Plant and Care for Zinnia Flowers*. www.thespruce.com/zinnias-a-burst-of-hot-flower-colors-from-zinnias-1402919.

The Spruce. *How to Plant Zinnia Seeds*. www.thespruce.com/how-to-plant-zinnia-seeds-3986368.

University of Minnesota, Extension. *Zinnia*. https://extension.umn.edu/flowers/zinnia.

University of Missouri, Integrated Pest Management. *Zinnia: Not the Flower Grandma Grew*. https://ipm.missouri.edu/MEG/2017/7/zinnia/.

Urban Leaf. 2021. *How to Grow Zinnia Indoors*. www.geturbanleaf.com/blogs/edible-flowers/growing-zinna-indoors.

WikiHow. 2022. *How to Plant Zinnias*. www.wikihow.com/Plant-Zinnias.

# Vernacular Index

Note: Page numbers in *italics* indicate a figure on the corresponding page.

## T



## V



## W



## X



## Y



## Z

# Scientific Index

Note: Page numbers in *italics* indicate a figure on the corresponding page.

## D



## E



## F



## G

## H



## I



## J



## K

L



M



N

## O



## P



## Q

## R



## S



## T



## U



## V

# General Index

Note: Page numbers in *italics* indicate a figure and page numbers in **bold** indicate a table on the corresponding page.

## G



## H



## I



## J



## K



## L



## M



## N



## O



## P



## R



## S

For Product Safety Concerns and Information please contact our EU
representative GPSR@taylorandfrancis.com
Taylor & Francis Verlag GmbH, Kaufingerstraße 24, 80331 München, Germany

www.ingramcontent.com/pod-product-compliance
Lightning Source LLC
LaVergne TN
LVHW080110011026
849160LV00013B/1063

* 9 7 8 1 0 4 1 0 0 2 8 4 0 *